The Rise of Metallurgy in Eurasia

The Rise of Metallurgy in Eurasia

Evolution, Organisation and Consumption of Early Metal in the Balkans

Edited by

Miljana Radivojević, Benjamin W. Roberts, Miroslav Marić, Julka Kuzmanović Cvetković and Thilo Rehren

Archaeopress Archaeology

ARCHAEOPRESS PUBLISHING LTD
Summertown Pavilion
18-24 Middle Way
Summertown
Oxford OX2 7LG

www.archaeopress.com

ISBN 978-1-80327-042-5
ISBN 978-1-80327-043-2 (e-Pdf)

DOI: 10.32028/9781803270425

Cover: Ljiljana Dinić; Copper hammer-axe, type Pločnik, c. 4600 BC
(from Pločnik, Serbia) - Julka Kuzmanović Cvetković.
Inner back cover: Reconstruction of the world's earliest copper smelting. Green flames come from the extraction of metal from malachite. Experiments at Pločnik, Serbia (2013) - Marko Djurica

Copyeditor: Kate Sharpe

To the memory of Borislav Jovanović, our colleague, friend and inspiration

(1930 – 2015)

Contents

List of Authors

Antonović, Dragana (PhD) is a Principal Research Fellow at the Institute of Archaeology, Belgrade (Serbia) and a rare specialist for ground stone industries working on prehistoric assemblages from south-eastern Europe. She has rich experience in archaeological research and was member of numerous projects for excavations and analyses of archaeological material, including the sites such as Vinča, Starčevo, Grivac, Pločnik, Belovode and many more. Her research is related to the study of the production of stone tools and weapons from the territory of Serbia with the aim of reconstructing the processes of supplying raw materials for their production and technology of production of those objects, as well as their use, ritual and symbolic significance in prehistoric societies. She has authored over 50 peer reviewed articles published in international and Serbian journals and edited volumes and four books about Neolithic stone tools and prehistoric copper objects from Serbia. The last few years she has dedicated to exploring prehistoric mining in Serbia as the director of an archaeological project of investigation at a mine on the Prljuša-Mali Šturac site on Rudnik mountain (2011-ongoing). Within *The Rise of Metallurgy in Eurasia* project, she led the specialist team for stone tool technology.

Amicone, Silvia (PhD) is a Research Scientist at the University of Tübingen, Germany, within the Competence Centre Archaeometry Baden-Wuerttemberg (CCA-BW), and an Honorary Research Associate at University College London (UCL) Institute of Archaeology. She completed her AHRC-funded doctoral research at the UCL Institute of Archaeology, as a member of the international project *The Rise of Metallurgy in Eurasia*. As a pottery analyst specialising in pottery technology in contexts of intense socio-cultural innovation, Silvia has contributed to several projects in the Balkans and the Mediterranean area and is an active member of the Ceramic Technology Research Network at the UCL Institute of Archaeology. In 2016 she was the holder of the Fitch Bursary award at the British School at Athens, where she carried out interdisciplinary research on cooking pottery production and consumption in the Greek and Roman site of Priene (Turkey). Since 2014, Silvia has promoted the creation of the first research network for pottery technology in the Balkans, in cooperation with the Institute of Balkan Studies and the Serbian Academy of Arts and Sciences.

Bonato, Enrica (PhD) is a Junior Researcher at the Institute of Planetary Research, German Aerospace Center (DLR) in Berlin (Germany). She is currently setting up the Planetary Sample Analysis Laboratory (SAL) for the geochemical and mineralogical analyses of extra-terrestrial materials, with the main aim of analysing material from sample return missions from the Moon and asteroids. Enrica's research interests cover different fields, from planetary science to archaeometry with particular focus on ceramic and lithic analyses. She is an archaeological scientist by training, and she gained her BSc at the University of Padova and her MSc at the UCL Institute of Archaeology during which she specialised in ceramic petrography, mineralogy and geochemistry. For over 6 years she was based at the Natural History Museum (NHM) in London where she worked as microanalyst and mineralogical sample preparator, as well as researcher. During her time at the NHM she also gained a PhD in Planetary Science in collaboration with the University of Glasgow. For *The Rise of Metallurgy in Eurasia* project Enrica was leading the geochemical analysis of chipped stone industry.

Bulatović, Jelena (PhD) is a researcher at the Department of Historical Studies, University of Gothenburg (Sweden). She was formerly a research associate at the Laboratory for Bioarchaeology, Faculty of Philosophy, University of Belgrade (Serbia). Her research interest in the broadest sense is zooarchaeology. She is especially interested in the nature of human-animal interrelationships in the Central and Western Balkans, mostly during the later prehistory – from the Early Neolithic to the Late Iron Age. Her work is focused on detecting strategies of exploitation of domestic animals. The main questions she is trying to answer in her research include how the strategies of animal exploitation changed over time at the local and regional level. She gained her BA, MA and PhD degrees in archaeology from the University of Belgrade. Within her doctoral research, Jelena used animal remains from three archaeological type-sites: Vinča-Belo Brdo, Pločnik and Bubanj, to investigate zooarchaeological aspects of the social and cultural changes in the Central Balkans during the Neolithic-Eneolithic transition in the mid-5th millennium BC. She has been collaborator on several Serbian and international (bio)archaeological projects and has authored and co-authored over 30 peer-reviewed publications. For the *Rise of Metallurgy in Eurasia* project, Jelena conducted zooarchaeological analysis on the assemblage from the site of Pločnik.

Dimić, Vidan (PhD) is a Research Associate at the Archaeological Institute in Belgrade. He completed his undergraduate, masters and PhD studies at the Department of Archaeology, Faculty of Philosophy in Belgrade (Serbia). The main focus of his research is polished and abrasive stone tools, such as the exploitation, processing, and use of stone raw materials and tools in prehistory, especially in the Neolithic of Central Balkan. In recent years, he incorporated the archaeological experiment as a complementary approach in his study of the production process and use of polished stone tools. An important part of that research is the analysis of traces of use. For the past eight years, Vidan Dimić has been an associate on a project that investigates prehistoric copper mining and metallurgy at the Prljuša-Mali Šturac site on the Rudnik Mountain in Serbia. Dr Dimić has been the author or co-author of scientific papers related to Neolithic polished and abrasive stone tools manufacture and use, archaeological experiments that include flint and polished stone tools, hammer-stones used for copper mining, as well as Eneolithic copper mining technology at the site of Prljuša-Mali Šturac. Dr Dimić was one of two specialists for polished and abrasive stone tools on *The Rise of Metallurgy in Eurasia* project.

Dimitrijević, Ivana (PhD) is a Research Assistant at the Institute of Archaeology in Belgrade, Serbia. She acquired her bachelor, masters, and PhD degrees in archaeology from the University of Belgrade. Her research is focused on the Neolithic of southeast Europe, with specific interest in human-animal relations – economy, food practices and social role of animals in the Neolithic settlements. Ivana has been an associate on the project investigating Middle Morava Valley in Neolithisation of Southeast Europe for more than a decade. Dr Dimitrijević was one of three specialists for analysing animal remains on *The Rise of Metallurgy in Eurasia* project. She conducted analyses of the faunal assemblage from the site of Belovode.

Filipović, Dragana (PhD) is a Postdoctoral Researcher in the Collaborative Research Centre 1266 of the Institute for Pre- and Protohistory, Kiel University (Germany). She coordinates archaeobotany-related activities, including the analysis of plant remains from Neolithic and Bronze Age sites in central and northern Europe. She possesses bachelor and master's degrees in archaeology from the University of Belgrade (Serbia), MPhil from the University of Nottingham (UK) and PhD in Archaeobotany from the University of Oxford (UK). She was an independent researcher and a research associate at the Institute for Balkan Studies of the Serbian Academy of Sciences and Arts, where she oversaw the implementation of archaeobotanical investigations in Serbia, publication of the results, teaching, field and laboratory training, designing and participation in public outreach. Her main field of interest is the archaeology of plant production and consumption, for which she uses archaeobotany, ethnoarchaeology, plant stable isotope chemistry and plant ecology. She has been involved in the Çatalhöyük Research Project, NEOTECH, several ERC-funded initiatives (TOTL, AGRICURB, EUROFARM, Hidden Foods, BIRTH, The Fall, PLANTCULT) and excavations in Turkey, Serbia, Portugal, Slovakia and Germany. For *The Rise of Metallurgy in Eurasia* project, she developed field sampling and processing routine and conducted archaeobotanical analysis.

Furholt, Martin (PhD) is Professor at the Institute of Archaeology, Conservation and History at the University of Oslo, Norway. Before he was working as Research Fellow and Lecturer at Kiel University. His main research interests are the social and political organisation, mobility and community composition, local and regional social networks of Neolithic and Bronze Age communities in Southeast Europe, Central Europe, and Northern Europe. He conducted his PhD research on Baden Complex materials in Poland and Czech Republic, and his Habilitation thesis on the Neolithic and Chalcolithic of the Aegean Region. He is currently conducting fieldwork on 6th millennium BC Neolithic settlement in Slovakia and Serbia, and publishes papers related to the ongoing 3rd millennium BC migration debate in Europe. For *The Rise of Metallurgy in Eurasia* project, Dr Furholt was a member of the geophysics team.

Ibragimova, Elmira (BA) worked at the Department of Archaeology at the State Historical Museum (Moscow, Russia) from 2008 to 2014, first as a Curator, then as a Research Associate. At that time her main research interests encompassed chipped stone industries of the Neolithic and post-Neolithic periods at the Near East, Caucasus, with a focus on lithic technology. For *The Rise of Metallurgy in Eurasia* project Elmira conducted research on the chipped stone collections from Belovode and Pločnik.

Jablanović, Aleksandar (MA) works as a researcher on different archaeological projects in Serbia and abroad. He acquired his masters at the Department of Archaeology at the Faculty of Philosophy in Belgrade. His main interests are related to geophysical survey of prehistoric sites of the Vinča culture as well as their spatial organisation. In 2017 he published the monograph 'Kremenite njive' (Flint Fields, in Serbian) which provided synthesis of all excavation and survey results related to this archaeological site of the Vinča culture. Within *The Rise of Metallurgy in Eurasia* project, he worked on geophysical survey in both sites and accompanying data processing as well as flotation of samples obtained from the cultural layer during archeological excavations. In recent years, he has also been presenting the cultural heritage (cultural tourism) of Serbia, primarily archaeological sites and cultural and historical units.

Kuzmanović Cvetković, Julka (Magistar) was a Principal Curator at the National Museum of Toplica in Prokuplje for 40 years. She has graduated from the Faculty of Philosophy, Department of Archaeology.

The emphasis of her work at the Homeland Museum of Toplica was research of the sites and creation of archaeological inventory of sites in the Toplica district. She was a co-director of Pločnik excavation projects for almost two decades, as well as a co-author of numerous publications that stemmed from these excavation campaigns. With the goal of presenting the Neolithic to the global public, she created and carried out the experimental reconstruction of the life on Pločnik, and hence established a unique archaeo-park, active to the present day. The reconstruction includes building four houses for specialists and one communal building from natural materials and using ancient techniques, which are currently a tourist attraction in the region, and beyond. Julka also carried out numerous copper smelting experiments with the aim to popularise archaeometallurgical research on this site. Within *The Rise of Metallurgy in Eurasia* project, she was a project partner, Co-Director of fieldwork research in Pločnik, co-editor of the monograph and a specialist for figurines from both Belovode and Pločnik.

Marić, Miroslav (PhD) is a Research Associate at the Institute of Balkan Studies of the Serbian Academy of Science and Arts (Belgrade) with primary research interests in Neolithic and Chalcolithic of southeast Europe, mainly the Vinča period. A major area of his research is settlement organisation and patterning, communication and movement and past landscapes in the Danube-Tisza interflow. He possesses a Bachelor, Masters and PhD in Archaeology from the University of Belgrade. Over the years he has been involved with research activities within the Vinča Project, The Times of their Lives project, CONPRA project and other national and international research projects. For *The Rise of Metallurgy in Eurasia* project, he developed the strategy and methodology for field excavations in Pločnik and Belovode, conducted C14 sampling and Bayesian modelling for both sites and statistical analyses for pottery and jointly edited the monograph.

Mertl, Patrick (MA) is a Researcher at the University of the Saarland (Saarbrücken), Germany, teaching digital methods in archaeology. He studied Pre- and Protohistory, Near Eastern Archaeology and Christian Archaeology/Byzantine Art History at the Johannes Gutenberg-University at Mainz, Germany. His research interests lie on landscape archaeology across multiple epochs and the use of geophysical applications and GIS in archaeological prospection. He was involved in the large-scale geophysical prospections at the Serbian sites within *The Rise of Metallurgy in Eurasia* project, while still being a student-research assistant at the Römisch-Germanischen Kommission (RGK). After his graduation he worked as a site director in cultural heritage management and as a researcher at the Johannes Gutenberg University (JGU) Mainz. Since 2014 he is providing geomagnetic surveys as a private contractor.

Milić, Marina (PhD) obtained her BA and MA degrees at the University of Belgrade (Serbia) and PhD at the Institute of Archaeology, University College London (UK). She is currently an Irish Research Council Postdoctoral Fellow at the School of Archaeology, University College Dublin (Ireland). Her work focuses on the questions of interaction and mobility of Neolithic communities throughout South-eastern Europe. These themes have been explored through analyses of obsidian artefacts and other material culture used extensively throughout the Neolithic. She worked on a number of projects in Turkey, Greece, Serbia and Egypt. For *The Rise of Metallurgy in Eurasia* project, Marina conducted obsidian analysis.

Mirković-Marić, Neda (PhD) is an Archaeologist at the Intermunicipal Institute for the Protection of Cultural Monuments Subotica, Serbia. Her research interests include cultural heritage studies, pottery studies, experimental archaeology, and social boundaries. She has participated in numerous archaeological projects and has collaborated with different institutions in Serbia for more than two decades, as well as on international projects such as *The Rise of Metallurgy in Eurasia* as a regular member of the excavation team, and lead specialist on pottery analysis and typology for both Belovode and Pločnik. Since 2014 she has been one of the directors of the 'Systematic archaeological research of Gradište-Iđoš project (ARISE project - Iđoš site and its environment)'. Since 2017, she has been the director of the ongoing project 'Archaeological Map of Northern Bačka Region (Serbia)' and the director of many rescue archaeological excavations, as well as director of the rescue excavations of nine endangered sites on the magistral pipeline 'Turkish stream' in Kanjiža, Senta and Ada municipalities in Serbia in 2019.

Müller, Johannes (PhD) is a Professor and Director of the Institute for Prehistoric and Protohistoric Archaeology at Kiel University, Germany. He is the founding director of the Johanna Mestorf Academy, Speaker of the Collaborative Research Centre 'Scales of Transformation: Human-environmental Interaction in Prehistoric and Archaic Societies' and of the Excellence Cluster 'ROOTS – Social, Environmental, and Cultural Connectivity in Past Societies'. He conducts research on Neolithic and Bronze Age Europe, including the challenge of interlinking natural, social, life sciences, and the humanities within an anthropological approach of archaeology. Intensive fieldwork was and is carried out in international teams, e.g., on Tripolye mega-sites in Eastern Europe, the Late Neolithic tell site of Okolište in Bosnia-Hercegovina, different Neolithic domestic and burial sites in Northern Germany, and Early Bronze Age sites in Greater Poland. Ethnoarchaeological fieldwork has been conducted, e.g., in India. Within the Kiel Graduate School 'Human Development in Landscapes', now the Young Academy of ROOTS, and the Scandinavian Graduate School 'Dialogues of the Past', Johannes Müller promotes international PhD projects. He is the PI of the synERC xSCAPES on material minds and materiality. For *The Rise of Metallurgy in Eurasia* project, Professor Müller was a member of the geophysics team.

Müller-Scheeßel, Nils (PhD) is a Scientific Editor at Kiel University, Germany. He obtained his PhD in 2006 at the University of Tübingen. His particular interests include the means and meaning of the disposal of the dead, landscape archaeology, the possibilities of remote sensing, and quantitative methods. For *The Rise of Metallurgy in Eurasia* project, Dr Müller-Scheeßel was a member of the geophysics team.

Nikolić, Mladen (PhD) is an Assistant Professor at the Department of Computer Science, Faculty of Mathematics, University of Belgrade (Serbia). He received his MSc and PhD degrees from the same faculty while working on problems of automating deductive reasoning in artificial intelligence. Currently his research interests focus on various areas of machine learning, primarily computer vision, fairness in artificial intelligence, graph neural networks, reinforcement learning, and applications in geosciences. He is also interested in archaeology and anthropology, focusing on modelling population dynamics and migrations during Neolithic demographic transition. He has contributed to several projects funded by Defence Advanced Research Projects Agency (DARPA), Swiss National Science Foundation, European Research Council, and Serbian Ministry of Science on various topics in machine learning and archaeology. For *The Rise of Metallurgy in Eurasia* project, Dr Nikolić was a member of the paleodemography team.

Orton, David (PhD) is a Senior Lecturer in Zooarchaeology in the Department of Archaeology, University of York, UK. His doctoral research focused on the links between herding practices and the gradual development of large, settled communities in the central Balkan Neolithic. He subsequently worked as a researcher on a number of projects, ranging from Halaf-period zooarchaeology in south-east Turkey to the (post)medieval fish trade in north-west Europe, and returning to Neolithic research in the former Yugoslavia with a role on the ERC-funded EUROFARM project and as lead archaeozoologist on *The Rise of Metallurgy in Eurasia* project. Throughout all this he has maintained a variety of side-projects, most notably a long-standing role as zooarchaeologist for the West Mound Project at Çatalhöyük. At present his research focuses on medieval fishing and urban fauna, particularly rats. He currently coordinates the SeaChanges ITN, a major international PhD training network bridging archaeology and marine biology.

Pendić, Jugoslav (MA) is a Junior Researcher at the BioSense Institute of Novi Sad University. He obtained his MA degree at the University of Belgrade, where he is currently enrolled at the PhD programme. His work has been focused on providing input on Cultural Heritage Management stratiegies during a development driven project life-cycle: feasibility studies, management of design edits, compliance auditing, environmental statement and assessment report production. His research so far included application of large area geophysics and remotely sensed data to analysis of archaeological landscapes of Neolithic, Chalcolithic and Bronze Age of central and south Balkans. Additionally, he specializes in material science studies in archaeology (he is a trained operator of instruments with ionising radiation sources – pXRF and uCT). He is highly proficient in 3D scanning techniques, production pipelines and product delivery, primarily for Cultural Heritage documentation, analysis and visualization. Between 2017-2020 he was a project associate of ERC funded project BIRTH, where he contributed to project goals through spatial analysis of intra-project data, material analysis, digitalization and reconstruction of spatial data from grey literature. During *The Rise of Metallurgy in Eurasia* project he co-managed excavations at Pločnik, assisted geophysical data acquisition and processing for the sites of Pločnik and Belovode, as well as provided technical support during the monograph writing process.

Pernicka, Ernst (PhD) was a Professor for Archaeometry at the Institute of Geosciences at Heidelberg University and is now a Senior Professor for Archaeometry at the University of Tübingen. He studied chemistry and physics at the University of Vienna, Austria, and then spent twenty years as senior researcher at the Max-Planck-Institute for Nuclear Physics at Heidelberg. In 1997 he moved to the University of Technology Bergakademie Freiberg as professor for archaeometallurgy and from 2004 to 2013 he was Professor for Archaeometry at the University of Tübingen. Generally, he has developed and applied scientific methods to objects, concepts and questions of cultural history. Among those are physical, chemical and geochemical methods for age determination and provenance analysis of archaeological finds. The focus of his present research is the origin and development of metallurgy in the Old World. Ernst Pernicka has also established the Curt-Engelhorn-Zentrum Archäometrie in Mannheim and is presently the senior and managing director. Professor Pernicka was a Project Partner on *The Rise of Metallurgy in Eurasia* project.

Porčić, Marko (PhD) is an Associate Professor at the Department of Archaeology, Faculty of Philosophy, University of Belgrade Serbia. He completed his PhD at the University of Belgrade in 2010. The subject of his thesis was the inference of household and settlement population size, as well as social structure, from the remains and inventories of the Late Neolithic Vinča culture houses. His main research interests include paleodemography, evolutionary archaeology (cultural transmission theory) and the social and demographic aspects of the Balkan Neolithic. Dr Porčić was also a Senior Researcher on the ERC funded BIRTH project (directed by Prof. Dr S. Stefanović) where he was in charge of the paleodemographic work package, with an aim to reconstruct the population dynamics and demographic aspects of the Neolithic transition in the Central Balkans based on the radiocarbon evidence and computer simulation. He published a book about the theory and methods of archaeological demography. He is currently working on a research which involves the implementation of models of cultural transmission through computer simulation in order to understand the patterns of material culture variation in space and time. For *The Rise of Metallurgy in Eurasia* project, Dr Porčić led a paleodemography team.

Quinn, Patrick (PhD) is a Principal Research Fellow at the UCL Institute of Archaeology, UK, and a leading expert in the field of ceramic analysis. His book on ceramic petrography is essential literature for any serious studies of ceramics. He has over 20 years of experience developing and undertaking research projects on ancient pottery and other ceramics from different archaeological periods and geographical regions. These include studies of ceramic production and distribution patterns in prehistoric Greece, as well as migration and cultural interaction in pre-contact North America. Patrick was one of the first western scholars to analyse scientifically the ceramic sculptures of the Terracotta Army in China. Patrick coordinates the Ceramic Technology Research Network at the Institute of Archaeology. He has an exceptional track record in high quality publications, including three books and over 40 peer review journal articles. He has taught specialist courses on ceramic analysis since 2005, including a highly successful intensive course on ceramic petrography, which draws participants from all continents. Patrick conducts commercial analysis of ceramics from various periods and parts of the globe. Dr Quinn was one of the ceramic specialists and PhD supervisors involved in *The Rise of Metallurgy in Eurasia* project.

Radivojević, Miljana (PhD) holds a Lectureship in Archaeomaterials at the UCL Institute of Archaeology, UK, where she acquired her PhD in Archaeometallurgy. During her previous studies and research posts at the Universities of Belgrade, Cambridge and UCL she has developed a strong research profile in both fieldwork excavations and laboratory analysis of material culture, specifically technology of early metal making. She specialised in the emergence of early copper making in the Balkans before expanding research collaborations across the Chalcolithic and Bronze Age Europe and northern Eurasia. Dr Radivojević has published on the origins of metallurgy in the Balkans and southwest Asia, relationship of metallurgy and pottery technologies, invention, innovation and transmission of copper and tin bronze metallurgy across southeast Europe, aesthetics of ancient metal objects, as well as co-developed a novel method of re-assessing archaeological phenomena using complex networks analysis of metal supply systems in the Balkans. Dr Radivojević was one of the authors of *The Rise of Metallurgy in Eurasia* project, then Research Associate, Project Manager, and the lead editor of this volume.

Radloff, Kai (MA) is a member of the Berlin Graduate School of Ancient Studies, where he is enrolled in the programme 'Landscape Archaeology and Architecture'. He studied Archaeology and History of the Roman Provinces and Pre- and Protohistory at the Goethe-University in Frankfurt, Germany and at the University of Vienna, Austria. His main fields of interest are Archaeological prospection and the analysis of past landscapes with regard to socio-political aspects. In his studies, he is focussing on the spatial organisation of different communities and their regional networks. After graduating, he worked in the technical department of the Römisch-Germanischen Kommission (RGK) for several years, where he contributed to *The Rise of Metallurgy in Eurasia* project as a member of the team conducting geophysical surveys at the site of Belovode. Currently, he is doing research as a PhD candidate on the border landscape along the Lower Rhine in the Roman era.

Rajičić, Milica (BA) is an archaeologist who acquired her degreed from the University of Belgrade. In 2020 she started the ARCHMAT-Erasmus Mundus Joint Master in Archaeological Material Science at the University of Évora. During and after her undergraduate studies she participated in many archaeological projects and excavations in Serbia, developing her interest in various archaeological materials. During the fieldwork research for the project *Rise of Metallurgy in Eurasia*, Milica was part of the team for on-site analyses of pottery on the site of Pločnik.

Rassmann, Knut (PhD) is Head of the Department of Survey and Excavation Methodology at the German Archaeological Institute, before he was Scientific Editor for the Bericht der Römisch-Germanischen Kommission (RGK). He obtained his PhD in 1991 at the Freie Universität Berlin. His main interests are Landscape Archaeology, survey and excavation methodology, Neolithic, Copper and Bronze Age in Europe. For *The Rise of Metallurgy in Eurasia* project, Dr Rassmann led the geophysics team.

Rehren, Thilo (PhD) is the A.G. Leventis Professor for Archaeological Sciences at the Cyprus Institute in Nicosia, Cyprus, and Director of its Science and Technology in Archaeology and Culture Research Center (STARC). Following a Masters in Mineralogy and a PhD in Petrology (Volcanology) he worked for nine years at the Deutsches Bergbau-Museum in Bochum, Germany, where he helped to establish the Institut für Archäometallurgie. In 1999 he was appointed Professor for Archaeological Materials and Technologies at the UCL Institute of Archaeology in London, from where he was seconded from 2011 to 2016 to establish UCL Qatar as an academic department of UCL based in Doha, Qatar, focusing on Archaeology, Museum and Gallery Practice, and Conservation Studies. His main interest lies in the study of production processes for metals and glass, and the related technical ceramics involved in this. Among other topics, he has worked on early gold and silver production in Ecuador and Bolivia, respectively, on African iron smelting and glass making, on crucible steel production in Central Asia and India, on faience, silver and copper production in China, on Egyptian LBA bronze casting and glass making, on Roman to early modern brass making in western Germany, on 19th century platinum production in Russia, and more recently on the distribution and recycling of Roman to Byzantine glass and copper and silver metallurgy in Morocco. The majority of his more than 30 PhD students have succeeded in careers in archaeology, holding positions in the Americas, Africa, the Middle East, Asia and across Europe. Professor Rehren formally held the AHRC grant underpinning *The Rise of Metallurgy in Eurasia* project (PI) and edited the monograph.

Rittner, Martin (PhD) is currently working as an Application Scientist for TOFWERK AG, developing Laser Ablation applications for the icpTOF Time-of-Flight mass spectrometer in their Demo Lab in Thun, Switzerland. Martin earned his Mag. rer. nat. (MSc) degree in Geology at the University of Innsbruck, Austria. During his PhD at Royal Holloway, University of London, he started to work more in isotope geochemistry and laser ablation (LA) analyses. He was running routine LA analyses at the London Geochronology Centre, UCL, London, a facility that is also occasionally used for research by members of other faculties and universities. This is where opportunity to analyse the samples described in this study arose. After two postdoc positions at UCL, Martin moved from academia into industry, but continues research in microanalytical methods for mainly geoscientific applications. For *The Rise of Metallurgy in Eurasia* project Martin carried out the geochemical analysis of chipped stone industry.

Roberts, Benjamin W. (PhD) an Associate Professor in the Department of Archaeology at Durham University. Prior to this, he was the Curator for the European Bronze Age collections in the British Museum. He completed his PhD on the origins of metallurgy in western Europe at Cambridge University. He has spent over twenty years researching and publishing on early metallurgy and metal objects across Europe and beyond including the major edited volume *Archaeometallurgy in Global Perspective: Methods and Syntheses*. He is currently the leading the Leverhulme funded *Project Ancient Tin: Did British tin sources and trade make Bronze Age Europe?* He is also lead archaeologist on numerous smaller projects in Britain ranging from Bronze Age funerary sites to shipwrecks to chronologies. Dr Roberts was one of the authors of *The Rise of Metallurgy in Eurasia* project, Co-PI and the Co-Director of fieldwork research in Serbia.

Savić, Marija (MA) is a Research Associate at the Institute for Serbian Culture in Priština/Leposavić (Serbia). She is currently a PhD student at the Faculty of Philosophy at Belgrade University. At the Institute, her main projects are in the Serbian province of Kosovo and Metohija. Her interests are in Ancient History, Roman archaeology, especially Roman military archaeology and history. Over the course of *The Rise of Metallurgy in Eurasia* project, Marija was engaged as a member of the pottery processing team on Belovode and Pločnik.

Scholz, Roman (MSc) is an employee of the Römisch-Germanischen Kommission (RGK) and responsible for field research. He has been working for the German Archaeological Institute (DAI) since 2011 and specializes in prospecting and methodology development. He studied excavation technology and then landscape archaeology at the University of Technology and Economics and the University of Potsdam. Since 2013 he has been involved in research on the Neolithic tell settlements in eastern Croatia and also accompanies the investigations on the large Copper Age settlements in the Republic of Moldova through experimental archaeological work. He is currently doing his PhD on the settlement in Bapska in Croatia at the University of Kiel. For *The Rise of Metallurgy in Eurasia* project, he was a member of the geophysics team.

Vitezović, Selena (PhD) is a Senior Research Associate at Institute of Archaeology, Belgrade. She obtained her bachelor, MPhil and PhD at Faculty of Philosophy, University of Belgrade. Her research is focused on Mesolithic, Neolithic and Eneolithic periods and on topics related to technology, economy and human-animal relations, in particular osseous technologies. She participated at numerous excavations projects as archaeologist and bone tool specialist in Serbia, Bulgaria and Croatia, including the sites of Vlakno (Croatia), Vinča-Belo Brdo, Bubanj, Pavlovac (Serbia), Nova Nadezhda, Slatina (Bulgaria) and many more. She published over 50 articles in journals and chapters in monographs and edited volumes. For *The Rise of Metallurgy in Eurasia project* she worked as a bone tool specialist for assemblages from Pločnik and Belovode.

Foreword by Evgeniy N. Chernykh

I have been waiting for this book for more than four decades! When I published 'Mining and Metallurgy in Ancient Bulgaria' (*Gornoe Delo i Metallurgiya v Dreivneishei Bolgarii, in Russian)* in 1978, I was hopeful that the impetus given to Balkan archaeometallurgy with it and the excavation projects in Rudna Glava, Durankulak or Varna would have resulted in more in-depth analysis of metallurgical materials that could then reveal the much sought-after origins of Balkan metallurgy and settle the debate on independent vs diffusionist perspectives. It took another 30 years for the first secure evidence for copper smelting to show up in the now much cited publication by Radivojević *et al.* (2010a), which ultimately gave way to this monograph and set out the vision for the *Rise of Metallurgy in Eurasia* project. While reading it, I have been personally humbled to learn that my 1978 book was the main inspiration for pursing the origins of metallurgy research by Radivojević, Roberts, Rehren and their team, and I therefore remain grateful for the honour to being amongst the first readers of this outstanding scholarly achievement.

This is a major publication both on the evolution of early metallurgy in the 'Old World' and on the Central Balkan archaeology of the 6th and 5th millennia BC. It provides a highly detailed investigation into the dating, technology and organisation of metallurgy in the Vinča culture, which covers the mentioned chronological spread. The excavation and recording of the sequences at key sites of Belovode and Pločnik, with accompanying analysis of the ore and metal analysis from the earlier projects throughout the Balkans presents a clear archaeological and archaeometallurgical framework. The radiocarbon dating programme that follows meticulous excavation and recording, as well as Bayesian modelling, yields the much welcome confirmation of c. 4900 BC as extremely secure evidence for one of the earliest known copper smelting in the world. The monograph, however, goes beyond the two key sites, and far beyond the last major Vinča site publication (e.g. Selevac in 1990 by Ruth Tringham). The scale and diversity of evidence presented and reviewed offers a holistic perspective on early metallurgy and society in the wider Balkan area.

The emphasis on incorporating the broader archaeological / societal context is clear from the beginning, and leads to the excavation reports, surveys and post-excavation analyses at Belovode and Pločnik with an exceptional level of detail, specialist reports and encyclopaedic knowledge on the variety of topics, including information that has been missing in the few publications over the previous decades of exploration at both sites. The highlights of these sections are most certainly the confirmations of the secure age of the metal smelting activities very close to the beginning of the 5th millennium BC, as already argued in Radivojević *et al.* (2010a), together with the importance of black and green ores for metal extraction. Another highlight comes from meticulous research on the connection of pottery and metallurgical technology, successfully targeting yet another several decades old scholarly dilemma on the connectedness of these two pyrotechnologies. The richness of accompanying research on plants and animals, various stone technologies and landscape features offers an unprecedented contextualisation of the development of metallurgical knowledge that takes the readership beyond the focus on workshops, to people – their everyday lives and interaction with the environment and other communities at the time. While admittedly the geophysical surveys on both sites reveal a largely complete perspective on their spatial organisation, the excavation campaigns at both sites were naturally far smaller in scale owing to the depth and complexity of the archaeological sequences at both sites.

The heavy focus on the excavation methodology, findings and analytical procedures on materials from Belovode and Pločnik in the first three sections of the monograph stands as a kind of a final excavation report on its own; yet the following section provides an invaluable and very extensive collection of synthetic chapters on the Neolithic-Chalcolithic in the Central Balkans by emerging and leading specialists. This section almost stands as a second volume of this monograph, whose readership will likely be far wider than the previous sections focused on two key sites. As the most precious point that I take with me is the conclusion on the importance of community cooperation in performing the metallurgical activities, and the value of shared household spaces and practices across the Chalcolithic villages. It also stands in contrast to much repeated ideas on the emergence of elites and social inequality in the Balkans, which does not have much support in the evidence from across the region (see also Porčić 2019).

Finally, the global perspective section engages in a stimulating discussion on the place of early Balkan metallurgy in the interpretation of global early metallurgies and argues convincingly for the need to shift paradigms from unfruitful pursuits into the origins of social inequality to the narratives tailored to each context individually and with a strong focus on detailed technological analysis that go beyond reporting tables with compositional data. The cutting-edge approaches such as complex networks analysis reveal a completely novel world of research opportunities to explore patterns of cooperation, as means to re-evaluate the concept of archaeological culture, as well as to probe established ideas, such as metallurgical provinces established in my previous work. I am pleasantly surprised by how much these novel approaches underline and expand the potential of defining metallurgical provinces on the basis of shared technological knowledge and I will be looking forward to future explorations on this topic.

There is certainly no shortage of material for future research and debate. If anything, the lead authors of this section set a bold challenge for prospective research by indicating points that they wish they could have done within this project (e.g. environmental analysis), and invite new generations to benefit from their fully accessible excavation and analytical database in exploring novel avenues for research. The final points that come closer to my work target the crucial role that Balkan metallurgy played in starting the wave of technological changes across the Eurasian Steppe in the Bronze Age, which authors mention as yet another challenge to address with greater analytical detail and focus on transmission of the metal production knowledge.

Finally, what this monograph has largely shown goes beyond the Balkan archaeology and metallurgy, and that is the power of archaeomaterials research in exploring the topics of technological invention, innovation, and its transmission. From the micro detail of identifying manganese-rich copper ores as the primary source of the world's first metal to the macro detail of revealing patterns of cooperation amongst metal using communities, this project sets a very high bar for any similar research in the future and highlights the necessity for multidisciplinary research and hence cuts to the core of what archaeology is: a multi-faceted endeavour that keeps growing and benefiting from cross-disciplinary achievements. This monograph fully owns its title, as much as it will own the shelves of a global readership for many decades to come.

Moscow, 18.5.2021

Foreword by Barbara S. Ottaway

Most studies of the beginning of the earliest European/Eurasian metallurgy have centered on the Balkans. Anyone interested or working in this field had to master the complex local prehistory of Bulgaria, Romania, Hungary, Serbia and Croatia of the late Neolithic, Eneolithic or Chalcolithic period. Much of this literature was not available in English.

This has now been elegantly resolved by the publication of the 'The Rise of Metallurgy in Eurasia', edited by Miljana Radivojević, Benjamin W. Roberts, Miroslav Marić, Julka Kuzmanović Cvetković and Thilo Rehren. In this monograph the reader will find succinct summaries of the archaeology and periodisation of these cultures in the Balkans in clear and very readable English.

This forms the background to new excavations at two sites: Belovode and Pločnik, which have been executed in exemplary fashion. The digitised methodological excavation techniques, supported by numerous radiocarbon-dated events and layers, has produced a great amount of new information. This not only allows the authors to date the start of new developments and activities, but it also provides a precise absolute chronology for the occupation of both sites.

The wealth of cultural heritage of the Balkans has been enriched by the many specialists presenting their post-excavation results of Belovode and Pločnik. This provides much needed information on subsistence strategies through archaeobotanical and zooarchaeological results. A study of settlement patterns, using geomagnetic data, has given us the estimates of overall size of the settlements, number of houses as well as estimated population size of the later Vinča period.

However, the most exciting results are that both Belovode and Pločnik have provided clear evidence of the earliest, experimental stages of metal production using green-black copper ores in the early levels. This phase was followed by more successful, sustainable smelting of the same copper ore. Furthermore, it has been shown that the smelting of copper ore at both sites was a natural progression of working with green malachite, used for the production of beads, in the earliest levels of the settlements.

This, together with the Vinča settlers experience in pyrotechnological knowledge, gained in the production of black-burnished pottery, convincingly argues for a local, independent development of metal production at Belovode and Pločnik. Furthermore, the progression of metallurgical experience gained by the settlers reflects their ability and competence in performing the entire *chaîne opératoire* of smelting, melting and working copper.

The authors are to be congratulated in successfully carrying through and publishing this ambitious project. The entire data has also been made available online, encouraging further research using their results.

Exeter, 24.5.2021

Foreword by Stephen J. Shennan

Since at least the 1960s the Balkan Copper Age has come to be seen as a salient topic in European prehistory, one that has to be covered in introductory Archaeology courses. The reason for this is the extraordinary richness of its archaeological record. Its settlement record is outstanding, its figurines striking, its painted pottery of extraordinary elaboration and its large numbers of copper axes remarkable. The initial impression of its exceptional nature that emerged from large-scale excavation programmes after WWII was only confirmed when the gold inventories of the burials in the Varna cemetery began to be published in the 1970s. However, there was also another aspect of the Balkan Copper Age that attracted attention: its date. Using his knowledge of the stratigraphic sequences from recent Balkan excavations together with the gradually increasing number of radiocarbon dates and his awareness of the need for their recalibration, Colin Renfrew, in his famous 1969 paper in *Proceedings of the Prehistoric Society*, showed that the long assumed chronological correlation between the Vinča culture and the Troy Early Bronze Age was fallacious. The developments of the Balkan Copper Age, most importantly the metallurgy itself, were at least a millennium earlier than previously believed, and much earlier than the sources from which they had been thought to derive: they were autonomous.

In *The Rise of Metallurgy in Eurasia* Radivojević and Roberts and their colleagues offer a compelling new vision of Balkan Copper Age metallurgy and its significance, while confirming Renfrew's claim of its autonomy. The work is a model of 21st century interdisciplinary research. It combines question-oriented fieldwork, deep technical knowledge of metal-making materials and processes and their archaeological residues, and of the analytical methods used to obtain information about them. These go together with a mastery of the archaeological record of the Copper Age Balkans. The volume integrates the results of the fieldwork programme and of the analyses of the material it produced with those of many specialist analytical publications that various combinations of the volume's contributors have produced in recent years, to produce a synthesis that justifies its title. I cannot begin to do justice to all its different elements and their importance in this foreword.

The six questions that provided the agenda for the project, addressing the 'how' and the 'why' of early metallurgy as well as the 'what' and the 'when' as the project leaders put it, are systematically addressed and answered. Understanding the 'how' of early Balkan copper smelting is especially difficult because it has left so few traces in the form of slag, while the furnaces seem to have been no more than sherd-lined holes in the ground, whose remains are evanescent and therefore difficult to find. The project has overcome these difficulties and made major advances in the use of archaeological evidence for identifying the processes of metallurgical innovation. It has confirmed the independent beginning of copper smelting in the Balkans at ~5000 BCE based on the use of distinctive black-and-green coloured copper minerals, visible traits that provide a phenomenological basis for understanding the choices made by the first metal producers. The innovative smelting of these minerals, confirmed by evidence from slag analysis, resulted in the development of a process that continued in use for hundreds of years, using ore sources in eastern Serbia. The recognition of the importance of this black-and-green colour choice also neatly makes sense of the appearance of early tin bronze as a result of the exploitation of similarly-coloured copper-tin ore, as well as the non-use of less colourful copper minerals.

The consistency of the methods used is one of the many interesting features of the organisation of copper production that has emerged from the project, as is the evidence of the use of multiple ore sources, even if the famous excavated early mine at Rudna Glava remains an enigma. Another is the results of the copper smelting experiments that were carried out, showing that the process would have needed a team of people to bring it to a successful conclusion. Even if one of these team members had more expertise (including magical knowledge) than others, it is hard to see how the process could have been kept a secret, as many have argued in the past. In fact, the authors suggest, the evidence that pieces of copper ore were found very widely across the two sites excavated indicates that every household was probably producing metal. Given the small scale of any given production event in a hole-in-the-ground furnace, less than needed to produce a single axe, they must have been frequent. Knowledge would have been community wide. Together with the fact that different communities used ore from the same sources, this helps to explain the widespread uniformity in production techniques. Even if details passed on to other communities were inexact or incorrect, the possibilities would ultimately have been limited by what worked. Importantly, knowledge of the technology was sufficiently widespread that metal production did not cease with the end of the Vinča culture but continued elsewhere in the Balkans till the end of the 5th millennium: it was resilient rather than fragile in Valentine Roux's terms (2010).

In her studies of stone-bead making and of the use of rotative kinetic agenda to make wheel-turned pots Roux (e.g. Roux et al. 1995) emphasises the length of apprenticeship involved in becoming an expert – time taken that effectively excludes them to at least some degree from other activities. The widespread distribution of the evidence from Belovode and Pločnik would seem to suggest that copper smelting was not a specialist activity in the same way, perhaps because it did not require the acquisition of highly controlled embodied know-how taking years to acquire, for example the skill to produce exceptionally long flint blades. Once the key discoveries had been made and a successful recipe developed, just about anyone could do it if they had access to the relevant ores, and there were no barriers to entry in the form of expensive or difficult to build installations.

As Radivojević and Roberts argue in their final chapter, the recognition that metal smelting was not an exclusive activity goes a long way towards undermining the long-assumed association of metallurgy with the growth of social elites and the established narratives that follow from this assumption. This is one of the major outcomes of their project. It is also in keeping with other recent arguments (Porčić 2019a) against the existence of high levels of social inequality in the Balkan Copper Age. In a similar vein, they see metallurgy as only one element in what I would call the economic growth that characterised the Vinča period and the Balkan Copper Age more generally, visible in increased population, increased production and circulation of a range of material items, and the scale of cattle-keeping. Finally, they suggest, their work demonstrates not *the* autonomous development of copper production but *an* autonomous development, characterised by the specific technology they describe.

In bringing together the massive amount of research that leads to these novel conclusions, this volume forms the baseline for all future studies not just of Balkan copper metallurgy but of any study of early metallurgical innovation.

London, 30.3.2021

Acknowledgements

The individuals and organisations involved in the project and subsequent writing of this book go far beyond the list of contributing authors and their respective institutions. First and foremost, we wish to acknowledge our gratitude for the generous support received from the UK AHRC major project grant that underpinned the entire project - *The Rise of Metallurgy in Eurasia: Evolution, organisation and consumption of early metal in the Balkans* (AH/J001406/1), and the reviewers both for that grant, and the current publication for their encouragement and helpful comments.

We remain indebted to all the institutions that kindly provided access to the sites, archives and materials that made the research process smooth and enjoyable: The National Museum in Belgrade, Serbian Ministry of Culture, Homeland Museum in Petrovac on Mlava, Homeland Museum in Prokuplje, Homeland Museum in Priboj on Lim, The Municipality of Prokuplje, Serbia Railways, Faculty of Technology and Metallurgy Belgrade, Fire Brigade in Prokuplje, Planinka Kuršumlija, and the Institute for Archaeo-Metallurgical Studies. We are immensely grateful to Jasmina Živković, Bojana Borić-Brešković, Gordana Grabež, Aca Djordjević, Aleksandar Kapuran, Jelena Živković, Nina Drakulović, Saša Živanović, Igor Jovanović, Moma Cerović, Jovan Mitrović, Marija Svilar, Radmila Gavrilović, Loïc Boscher, Bastian Asmus, Simon Timberlake, Kunlong Chen, Siran Liu, Pira Venunan, Andrijana Pravidur, Ana Franjić, Dimitrije Cvetković, Marko Djurica, Peter Northover and Željko Kamberović for their kind support for the duration of the research project in Serbia and the UK. Special thanks go to Duško Šljivar and Dušan Borić – our debates gave inspiration and strength to the entire team to persevere in delivering continuing research excellence.

The *Rise of Metallurgy in Eurasia* team were not only scholars, but also the local workers at both Belovode and Pločnik, whose long experience in supporting the archaeological research of these sites prior to 2012 was tremendously helpful in identifying 'the little green ones' [malachite and copper slag] under the scorching Serbian sun during the excavation seasons of 2012 and 2013. Hence, we proudly present here our local team from Belovode: Milan Radosavljević, Milan Bogdanović, Zoran Ilić, Dragiša Panić, Radiša Panić, Dane Nikolić, Milorad Tanasković, Danijel Ljubomirović, Saša Bogdanović, Miladin Milutinović, Stefan Miloradović, Milan Popović, Milan Blagojević, Dragan Stojiljković, Filip Jovanović, Mlađo Jovanović, Milija Jovanović, Toma Mijatović, and Žika Mijatović. The local team from Pločnik were: Dejan Miladinović, Jelena Miladinović, Zoran Miladinović, Slobodan Lukić, Milivoje Lukić, Dejan Vesović, Milan Rakić, Zoran Ristić, Zoran Lukić and Jovan Vidojević. A special mention goes to Slavoljub and Snežana Radivojević, for their technical support in the field and accommodation for the team members in Prokuplje.

A comprehensive list of all the acts of kindness, good will, patience and understanding sent our way, especially while finally finishing this monograph during several lockdowns over the course of the global pandemic in 2020-2021 would span many pages. We thank everyone involved. We acknowledge the financial support for the monograph production process from the Department of Archaeology, Durham University and financial support for Open Access from the University Libraries of Durham and University College London. We are especially grateful to Ljiljana Dinić (née Radivojević) for preparing illustrations and designing the cover and to Kate Sharpe for encouraging us throughout and for meticulously copy-editing the entire monograph.

Lastly, but most importantly we remain humbled by each other's colleagiality and friendship, and grateful to the good fortune that brought us together to set a milestone in the research of the origins of metallurgy. We are aware, now more than ever, of the growing complexity of the topic and of our gaps in knowledge. This monograph is far from being a final statement. We hope instead that it serves as an invitation for scholars to join us in engaging with the many questions, approaches, data sets and opportunities for discovery that early metallurgy offers for decades to come.

Photo by Marko Djurica

Part 1
Introduction

Chapter 1

The birth of archaeometallurgy in Serbia: a reflection

Julka Kuzmanović Cvetković

The history of the study of archaeometallurgy in Serbia begins in Pločnik at the moment when massive copper tools were discovered in the Neolithic layer, in 1928. These tools were finished products, with no clues as to where or how they were made, or to where the ore originated. The beginnings of archaeometallurgy in this region were a major research topic for the academic Borislav Jovanović, whose doctoral thesis was entitled 'The Metallurgy of the Eneolithic Period of Yugoslavia' (Jovanović 1971). While exploring Rudna Glava, a Neolithic copper mine, he discovered the foundations of the craft (Jovanović 1978, 1980); in Pločnik, where metal-cast objects were discovered, he found its conclusion.

Jovanović's research led him to suggest what was to be confirmed much later by exact analyses - that copper metallurgy originated in the Balkans (Radivojević *et al.* 2010a). He received affirmation for this claim from the scientific world, initially by Colin Renfrew in the UK, who had also considered the metallurgy of copper indigenous to the Balkans (Renfrew 1969). Following in Jovanović's footsteps in Serbia, Savo Derikonjić discovered Neolithic mines at Jarmovac near Priboj (Derikonjić *et al.* 2011). In 1995, at the site of Belovode, Dušan Šljivar found materials later to be identified as the earliest copper slags (Radivojević *et al.* 2010a), then Dragana Antonović excavated the Eneolithic mine of Mali Šturac (Antonović and Vukadinović 2012), and several other copper items were found at Pločnik (Šljivar 2006). Finally, Miljana Radivojević (then an MSc student) confirmed what Jovanović had hypothesised back in 1971 by performing in depth technological analysis on archaeometallurgical materials from Belovode, Pločnik, Vinča, Gomolava and Gornja Tuzla (Radivojević 2012). Perfecting her knowledge of archaeometallurgy through Master's and doctoral studies at University College London (UCL) Institute of Archaeology with the help of her mentor Thilo Rehren, Radivojević expanded and developed the story of Pločnik with the major project that has resulted in this monograph. This chapter provides a brief history of archaeometallurgy in Serbia and explains the context within which the recent project was undertaken between 2012 and 2015.

In 1928, when copper tools were discovered accidentally during the building of the Niš-Priština railway in the village of Pločnik, 42 km from Niš, they were sent, together with pieces of ceramic pots, to the National Museum of Belgrade. The following year, a rescue excavation was undertaken, which recovered another cluster of copper tools. At the time, it was believed that although the copper artefacts were found in the Neolithic layer of the settlement, they had been stored at a later date, in the Eneolithic (or Chalcolithic) era. Indeed, the entire site was considered to be the Copper Age by the Head of Research, Miodrag Grbić. The work was published as 'Pločnik-Aeneolithische Ansieldung' (Grbić 1929). Many years later, two further concentrations of copper objects were discovered: one in 1964, during the digging of foundations for a wool processing facility and a second in 1968, during engineering work on the Prokuplje-Kuršumlija road (Stalio 1964; Stalio 1973).

Further research at Pločnik, led by Blaženka Stalio, was carried out intermittently between 1960 and 1978. This resulted in the discovery of an abundance of baked clay material, and another metal tool—a chisel—was found on the floor of a house feature. The results of the survey were only partially interpreted; Stalio did not deal more thoroughly with the copper finds although she published an article on the third group of tools (Stalio 1964).

In 1973, when a fourth cluster of tools was discovered at Pločnik (Stalio 1973), Borislav Jovanović had his research running at Rudna Glava, a group of prehistoric mines where copper ore was extracted during the Vinča culture period. Jovanović laid the foundations for the study of archaeometallurgy in Serbia, based on his research of the oldest mining shafts in Serbia, where Vinča pottery amphorae and other characteristic material was discovered (Jovanović 1982).

As there were several sites with traces of ancient metallurgy, Jovanović worked with The Institute for the Protection and Scientific Study of Cultural Heritage of the Republic of Serbia to establish the Commission for Cultural Goods of Archaeology and Industrial Archaeology which, in turn, initiated the publication of the 'ARHEOMETALURGIJA' (Archaeometallurgy) magazine in 1995. The sixth issue of this magazine, in 1998, was entitled 'The Oldest Metallurgy of Copper on the site of Pločnik near Prokuplje, a settlement of Vinča culture' (Šljivar and Kuzmanović Cvetković 1998a). It was dedicated to the discoveries at Pločnik. A poster was designed representing Rudna Glava, ore mining and the final products, a copper chisel and an axe from Pločnik.

Our own co-operation with Borislav Jovanović began in Aleksandrovac in 1993, at the Annual Conference of the Serbian Archaeological Society. We discussed the grooved stone bats / hammerstones found in the Toplica region and in the village of Mačina near Prokuplje, together with Vinča ceramics. These were a clue to prehistoric mining. With my colleague Dušan Šljivar, who succeeded Blaženka Stalio at the National Museum in Belgrade, I started making plans for the renewal of research at Pločnik (representing the National Museum of Toplica). It seemed to us that the site had not been fully explored and interpreted. But it was not at all simple. One of the aggravating circumstances was the opinion of Milutin Garašanin, who had created the periodisation of the Vinča culture and placed Pločnik in the earlier phase of Vinča culture, and who considered the Neolithic to be completely explained. In a book written by Miodrag Tomović, based on a conversation with Garašanin, I read his (Garašanin's) opinion that Neolithic sites were being excavated by curators of provincial museums in order to fill the display cases with material! I must admit that this statement angered me at the time. I respected the Professor's opinion but could not agree with it.

When the 600th anniversary of Prokuplje was commemorated in 1995, a scientific gathering of 'Prokuplje in Prehistory, Antiquity and the Middle Ages' was held at the National Museum of Toplica in Prokuplje. This gathering was attended by archaeologists working on the sites within the Municipality of Prokuplje, including Petar Petrović (then Director of the Archaeological Institute), Borislav Jovanović, and my colleague Jovan Glišić with the Department of Archaeology from Faculty of Philosophy in Belgrade. When visiting Pločnik, Borislav and Jovan, our colleagues who had explored Vinča culture sites for years but had not previously had a chance to see Pločnik, were delighted with the almost 4 m high profile - layers revealed by the Toplica river. In short, one of the conclusions of the gathering was that research at Pločnik should certainly be continued, especially with respect to archaeometallurgy, and that this research should be organised by the National Museum of Belgrade and the National Museum of Toplica. The following year, we cleaned up a 30 m profile to clarify the vertical stratigraphy. By hiking around the site, we identified the wider area it had occupied.

The next year, 1997, was important because we found an eyewitness, Radovan Zdravković, to the excavations carried out in 1927, and were able to clarify the conditions of the discovery of the third tool store (Šljivar and Kuzmanović Cvetković 1998b). Radovan explained that he had found copper tools on the wider surface of the trench, in a deep layer of ash. When we opened a trench immediately beside the foundations of his house in the village of Pločnik, we discovered a workshop and two magnesite axes. Radovan gave us a chisel that he had kept as a memento, bringing the total number of known copper artefacts from Pločnik at that time to 38.

We had previously learned that the findings from the railway building were not grouped together, so it was clear to us that this was not another storeroom. The copper chisel from the trench found in 1978 was the first indication that the copper objects were from the cultural layer, from the Gradac Phase of the Vinča culture (Šljivar 1996). By charting the finds, we saw that all but one of the copper items were found in the western part of the site, and concluded that this was the metallurgical, craft-focussed area of the settlement (Šljivar and Kuzmanović Cvetković 1997a).

At the end of 1997, a meeting of the prehistoric section of the Serbian Archaeological Society was held at the National Museum in Belgrade, and reports from the excavation of prehistoric sites, primarily Belovode and Pločnik, were on the agenda. The meeting was chaired by Milutin Garašanin and Borislav Jovanović. Dušan Šljivar, the Head of the Research, spoke about Belovode, and I presented the latest results from Pločnik. There was a coffee break and then Borislav arrived at the meeting. I still remember how he told us that Milutin was furious, since what we were talking about ruined his chronology.

We conducted no excavations during 1999; this was the year of the NATO bombing of Serbia, and the country was slowly recovering after four months of air strikes. At the annual assembly of the Serbian Archaeological Society, in Belgrade, in his introductory speech, Garašanin spoke about the metallurgical aspect of Vinča culture.

In 2000, the proceedings of the scientific gathering in Prokuplje were finally published, and a launch was organised at the National Museum in Belgrade. One of the speakers at the event was Milutin Garašanin. Others included Miloje Vasić, Director of the Archaeological Institute, Dragoslav Marinković, Director of the National Museum of Toplica, and academic researchers Borislav Jovanović and Ivana Popović. Most of the meeting was yet again dedicated to Pločnik which was discussed by both Garašanin and Jovanović. At the end of the official part of the launch, we showed the professor the monumental head of a figurine, accidentally discovered at Pločnik and a pitcher, both finds from the Gradac

Phase (Kuzmanović Cvetković and Šljivar 1998). He was thrilled when he saw and held the finds for the first time. He praised our work and wished us luck in our further work. We were very relieved, being aware that we were introducing changes into already established schemes, to know that Professor Garašanin agreed with our findings.

At the same time, my colleague Dušan Šljivar was conducting research at Belovode, near Petrovac on the Mlava, expecting to find traces of copper processing, since significant quantities of copper ore, malachite, had been found on the site.

Also in 2000, during excavations at Pločnik in November, we found a piece of copper chisel in a trench in the western part of the site. The chisel was a sensation. That was also the year in which the research team was joined by Miljana Radivojević, an archaeology student from Prokuplje with local origins in the village of Mačina nearby Prokuplje, where Vinča culture grooved hammerstones were found, and who had decided to engage in archaeometallurgy.

We found a further chisel in 2004 and then another in 2007, on the floor of the workshop in Trench 20 (Šljivar and Kuzmanović Cvetković 2009a). During those years, a piece of slag—a product of copper smelting—was also found at Belovode (Radivojević *et al.* 2010a; Radivojević and Kuzmanović Cvetković 2014). All the pieces were coming together. We were hoping for more extensive excavations, and to benefit from international cooperation and recognition. Our colleague Miljana Radivojević travelled to London during that period, to continue studying archaeometallurgy at the UCL Institute of Archaeology, where we were hoping that she would be able to undertake scientific analyses.

During the spring of 2010, Radivojević and her mentor from the UCL Institute of Archaeology, Thilo Rehren, arrived at the National Museum of Belgrade. They took more samples for analyses from the site of Belovode, and Professor Rehren gave a lecture at the Faculty of Philosophy. In the autumn of the same year, at the time of excavations in Pločnik, we held the Second Balkan Metallurgical Workshop in Prolom Banja. Radivojević gathered a team of experts from Britain, Germany, Bulgaria and, of course, Serbia. The organisers were the National Museum of Toplica and CRAM (Center for Research of Archaeological Material, Faculty of Philosophy, Belgrade). JSC 'Planinka', a tourism organisation from Kuršumlija, with whom we had already established links at Pločnik, enabled us to work smoothly at their business premises. Some important researchers of European archaeometallurgy were there with us, at 'the end of the world', as Ernst Pernicka wittily remarked. It was a validation of previous efforts and the promise of new opportunities for studying the beginnings of metallurgy in Serbia.

At the end of 2011, news arrived that a large project *The Rise of Metallurgy in Eurasia: Evolution, organisation and use of early metal in the Balkans* had been approved in the UK, supported by the Arts and Humanities Research Council. On the UK side the holder was UCL Institute of Archaeology; the partners from Serbia were the National Museum of Belgrade, the National Museum of Toplica and the Homeland Museum from Priboj. The project would last three years, from 2012 to 2015, with excavations conducted in the first two years at Belovode, Pločnik and Jarmovac near Priboj. The results would be summarised, and a monograph would be prepared in the third year. The agreement was signed at the National Museum in Belgrade, attended by the British Ambassador to Serbia, the Director of the UCL Institute of Archaeology, researchers Thilo Rehren, Ernst Pernicka and Ben Roberts, and from the Serbian side by the Minister of Culture, myself as then-Director of the National Museum of Belgrade, and Savo Derikonjić, the Director of the Homeland Museum from Priboj. As a student from Serbia undertaking doctoral studies in London, Miljana Radivojević provided the link between the two teams: she had created the concept and shaped the entire project with the help of colleagues who believed in this project and wanted it to succeed.

It seemed as though our dreams were becoming reality. Finally, we had the opportunity to excavate for another two years and then to carry out numerous analyses, the facilities for which we simply didn't have in Serbia. It was a chance for an international team of experts to study the entire course of copper ore processing from the mine in Priboj, through Belovode where copper slag had already been found, to Pločnik which had finished products – i.e. the entire production chain.

One trench per site was investigated at both Belovode and Pločnik. The excavations produced expected results: at Pločnik, a part of the house, an abundance of ceramic material, tools made of bone and stone, and two pieces of copper jewellery – a link and a ring. Almost the same team worked at both sites, including manager Dušan Šljivar, Miljana Radivojević, Benjamin Roberts, Silvia Amicone, Miroslav Marić, Neda Mirković, Milica Rajičić, Aleksandar Jablanović, Jugoslav Pendić, Marija Savić, Nina Drakulović, Saša Živanović, Marija Svilar, Jovan Mitrović and Jasmina Živković (from the Petrovac on Mlava Museum).

Reconnaissance and recording of the mine at Jarmovac were also undertaken by the German Mining Museum in Bochum team led by Dr Peter Thomas. Ground penetrating radar surveys were carried out at Belovode and Pločnik by a team from the Roman-Germanic commission led by Knut Rassmann, in order to determine the spatial plan and extent of both Belovode and Pločnik.

The final part of the research was an experiment in ore smelting at the Pločnik site, within the archaeological park. The experiment was designed and organised by Miljana Radivojević. The team that performed the experiment included experts from the UK, Germany, Canada, China, Thailand, Italy and Serbia. Employees of the National Museum of Toplica also took part, as well as local residents from the villages of Pločnik and Blace, who had also worked on the excavations. Over a period of two days and one night, 12 experiments were performed, in the presence of numerous colleagues from Serbia, Bosnia and Bulgaria. The experiment was a great attraction: an event for all of us and a spectacle for the audience since they were given the opportunity to participate and contribute.

We smelted copper in the same manner as in prehistoric times, surrounded by teams of people blowing air through the pipes made by our colleague Milija Rakić out of elderberry wood, and also using bellows made of goat skin as well as a ceramic cylinder (aka Serbian sulundar) made of clay. We smelted the copper and then poured it into a clay mould to produce a small tool and a ring. After the excavations were completed, we began processing the material and conducting additional analyses, the results of which are now presented here.

This monograph is far from a final say on the lives of metallurgists of Belovode and Pločnik. Rather, what is hoped is that it shifts the story of these sites into well-researched and thoroughly analysed settlements that will offer inspiration for many more generations to come. We are hopeful that they will pick up our leads and do more and better than we did, particularly empowered with more new technologies and analytical opportunities. This is for them.

The bibliographic reference for this chapter is:

Kuzmanović Cvetković, J. 2021. The birth of archaeometallurgy in Serbia: a reflection, in Radivojević, M., Roberts, B. W., Marić, M., Kuzmanović Cvetković, J., and Rehren, Th. (eds) *The Rise of Metallurgy in Eurasia*: 3–6. Oxford: Archaeopress.

Chapter 2

The Rise of Metallurgy in Eurasia: Evolution, organisation and consumption of early metal in the Balkans: an introduction to the project

Thilo Rehren, Miljana Radivojević and Benjamin W. Roberts

The study of early metallurgy has many aspects and has, accordingly, taken many forms and foci (Rehren and Pernicka 2008 and literature therein). Some scholars have documented the morpho-typological evolution of artefact types and some have explored the role of metals in creating social hierarchies, in storing and displaying wealth, or the more transcendent role of metals in a variety of rituals. Other researchers are fascinated by the skills and technical achievements of the metalworkers and their intangible heritage as expressed in intricate castings, ingenious manufacturing methods and elaborate surface decorations. Yet others study the transformation of rocks and ores to metal as documented in the slags and furnace fragments or try to trace the geological origins of metal objects, as a proxy for the movement of people, materials, and ideas. The investigation of ancient mining extends well beyond the field of archaeometallurgy, with mines for flint, pigments, precious stones and salt all pre-dating metal smelting, and quarrying for building stone exceeding metal mining both in scale and value generation (e.g. Schauer *et al.* 2020). This range of interests inevitably implies the application of a multitude of methods, borrowed from a host of mother disciplines, adjusted and refined to form the interdisciplinary field of archaeometallurgy. It also makes any holistic project both a daunting prospect and an exercise in interdisciplinary diplomacy.

The Rise of Metallurgy in Eurasia project did not, of course, spring into existence in a vacuum. The subject of early metallurgy in the Balkans has attracted scholarly attention for almost a century and was closely associated with early 20th century investigations of Vinča-Belo Brdo, the eponymous settlement of the Vinča culture (*c.* 5400 – 4600 BC) (Vasić 1932-1936), the discovery of metal artefacts at the tell settlement of Pločnik (south Serbia) (Grbić 1929), and the excavation of Vinča-style pottery in copper mining shafts at Jarmovac in southwestern Serbia (Davies 1937). The Balkan Peninsula, and specifically its northern part, subsequently became a major focus for scholarship concentrating on early mining and metallurgy, as manifest, for instance, in:

- The excavation of the copper mining sites of Rudna Glava in Serbia and Ai Bunar in Bulgaria (Chernykh 1978a; Jovanović 1971, 1980, 1982), which were the subject of pioneering provenance studies (Pernicka *et al.* 1993; 1997). These two sites were identified as the central nuclei of the Carpatho-Balkan Metallurgical Province (CBMP), which has served as a highly influential model in understanding community inter-connections across the Balkans and the Eurasian Steppes (Chernykh 1978b, 1992, 2013; Chernykh and Kuzminykh 1989; Chernykh *et al.* 2004; Kohl 2007; Koryakova and Epimakhov 2007; Kuzmina 2008; Yang *et al.* 2020). The abundance of copper deposits and the general richness of polymetallic veins across the Balkans has been discussed at length as crucial for early access to minerals and experimentation (e.g. Bogdanov 1982; Janković 1967, 1977, 1982; Jelenković 1999; Monthel *et al.* 2002; Neubauer and Heinrich 2003; Pernicka *et al.* 1993, 1997; Sillitoe 1983) and it is worth noting that this rich metallogenic profile still supports a key industry in the modern era in the region.
- The application of radiocarbon dating and, subsequently, archaeometallurgical research, which together revealed both the earliest known dates and the characteristics of copper metallurgy, and accompanying evidence for the independent invention of this technology in the Balkans (Glumac 1991; Jovanović 1980; Jovanović and Ottaway 1976; Pernicka *et al.* 1997; Renfrew 1969; Ryndina and Ravich 2000, 2001; Todorova 1978). The recent analysis of copper slag at the eastern Serbian Vinča culture site of Belovode, dating to *c.* 5000 BC (Radivojević 2013; Radivojević and Kuzmanović Cvetković 2014; Radivojević and Rehren 2016; Radivojević *et al.* 2010a), subsequently reignited the debate around the multiple inventions of metallurgy across Eurasia (see Montero-Ruiz *et al.* 2021; Pearce 2015; Pernicka 2020; Radivojević 2015; Roberts and Radivojević 2015; Roberts *et al.* 2009; Rosenstock *et al.* 2016). Accordingly, the

Balkans now has the earliest known evidence for metallurgy with respect to:

1. *lead,* smelted probably from the end of the 6th millennium BC (Radivojević and Kuzmanović Cvetković 2014) but more regularly used from the mid-5th millennium BC in the central Balkans (Glumac and Todd 1987) and later in the eastern Balkans (Hansen *et al.* 2019);
2. *copper,* smelted from *c.* 5000 BC onwards in eastern Serbia (Radivojević 2013; Radivojević and Kuzmanović Cvetković 2014; Radivojević and Rehren 2016; Radivojević *et al.* 2010a);
3. *gold,* used from *c.* 4650 BC onwards in eastern Bulgaria (Higham *et al.* 2007, 2018; Krauss *et al.* 2014, 2017; Leusch *et al.* 2014, 2015);
4. *bronze,* smelted as a natural alloy from *c.* 4650 BC in southern Serbia and found across Bulgaria (Chernykh 1978b; Radivojević *et al.* 2014a, 2014b; Radivojević *et al.* 2013);
5. and, probably, *silver,* produced by cupellation rather than occurring naturally, by the end of the 5th/early 4th millennium BC in Greece (Maran 2000; Muhly 2002).

- The scholarly tradition—best exemplified by the *Prähistorische Bronzefunde* series—of constructing detailed typo-chronologies of the many early metal objects, primarily (*c.* 4300) copper implements, which were then placed at the core of archaeological narratives in the Balkans and the surrounding regions (e.g. Antonović 2014a; Chernykh 1992; Diaconescu 2014; Driehaus 1952-55; Govedarica 2001; Heeb 2014; Kuna 1981; Patay 1984; Ryndina 2009; Schubert 1965; Taylor 1999; Todorova 1981; Vulpe 1975; Žeravica 1993).
- The discovery and excavation of the spectacular cemetery at Varna in Bulgaria, still unparalleled in metal volume, upon which major debates relating to the existence (or not) of elites and the dynamics of inequalities in the 5th and 4th millennium BC have since been played out (e.g. Biehl and Marciniak 2000; Chapman 1991, 2013; Chapman *et al.* 2006; Crnobrnja 2011; Fol and Lichardus 1988; Hansen 2013a; Higham *et al.* 2018; Ivanov 1978a; Klimscha 2014; 2020; Krauss *et al.* 2017; Leusch *et al.* 2017; Müller 2012; Porčić 2012a, 2019a; Reingruber 2014; Renfrew 1978a, 1986; Slavchev 2008).

It is still the case in Balkan prehistory that metallurgy is understood mostly through the lens of copper mining and the typology and distribution of metal (mainly copper and gold) artefacts, although this reflects only two ends of the metal production process. Production debris such as slags or crucibles, despite their rarity (and infrequent recovery in the field and subsequent analysis) in the archaeological record during the Chalcolithic, provides far more information about the metal-making recipes, and the transmission of metallurgical knowledge or ore provenance than the morphology of the final products or their origins (cf. Hauptmann 2014; Killick 2014; Martinón-Torres and Rehren 2008; Martinón-Torres and Rehren 2014; Ottaway 1994; 2001; Rehren 2003, 2008; Rehren *et al.* 2007). Slag, a by-product of metal extraction, is a vitreous, usually amorphous and often magnetic material that typically contains traces of all components contributing to its formation, while remaining largely resistant to post-depositional processes and dislocation (Bachmann 1982). Slags can be found as free pieces but also attached to the walls of crucibles, furnaces, or ceramic fragments known as 'slagged sherds', as is the case for early metal production in the Balkans (Radivojević and Rehren 2016; Rehren *et al.* 2016).

Since the 1990s, the deteriorating political situation in the Balkans hugely disrupted many early metal-orientated archaeological and archaeometallurgical research projects in the region. The negative impact on fieldwork, publications and collaborations has only recently been reversed, as evidenced, for instance, by the success and growth of the *Balkan Early Metallurgy Symposia (BEMS)* meetings in London, UK (2007); Prokuplje, Serbia (2010); Sozopol, Bulgaria (2013); and Targu Jiu, Romania (2015). This upsurge can also be seen in the continued prominence of metallurgical research within the festschrifts of major Neolithic-Copper Age Balkan archaeologists whose students and colleagues now occupy prominent positions in archaeological museums, university departments and resesearch institutes (e.g. Forţiu and Cîntar 2014; Stefanovich and Angelova 2007; Ţerna and Govedarica 2016). Metal-orientated scholarship is also very evident, not only in the classic and still influential conference proceedings published as *Die Kupferzeit als Historische Epoche* (Lichardus 1991a), but also in more recent proceedings from three major international conferences published on the region: *The Neolithic and Eneolithic in Southeast Europe* (Schier and Draşovean 2014); *Neolithic and Copper Age between the Carpathians and the Aegean Sea: Chronologies and Technologies from the 6th to the 4th Millennium BC* (Hansen *et al.* 2015); and *Der Schwarzmeeraum vom Neolithikum bis in die Früheisenzeit (6000-600 V. Chr)* (Schier and Nikolov 2016). It is however less prevalent in the most recent conference *Formation and Transformation of Early Neolithic Lifestyles in Europe* (Krauss et al. 2020). All reflect the persistent depth and influence of German scholarship – and the increasing use of English in publications.

Narratives on the emergence and evolution of Balkan metallurgy have always been modelled against developments in the Near East (or more precisely Southwest Asia) following a much embraced trend in scholarship from the late 19th century onwards that proclaimed *'Ex Oriente Lux'* or 'light from the east' to explain the emergence of 'European civilisation', as argued influentially by Montelius (1899) and Childe (1930). The transmission of farming technologies, products and practices as well as new pyrotechnologies such as ceramics from the Near East to Anatolia and onwards to the Balkans in the mid-7th millennium BC is, indeed, very well evidenced and clearly established in both past and more recent scholarship (see de Groot 2019; Shennan 2018; Whittle 2018). This earlier confirmation of the *'Ex Oriente Lux'* model has consistently created a strong intellectual paradigm for a Southwest Asian metallurgical origin that only a few individuals such as Renfrew (1969), Jovanović (1971), Ottaway (Jovanović and Ottaway 1976) and Todorova (1978) have challenged, arguing instead for the independent origins of Balkan metallurgy. Against this backdrop we appreciate immensely the resumption of the Belovode and Pločnik excavations with the clear agenda of Šljivar and Kuzmanović Cvetković (1998) to demystify the Vinča metallurgy debate despite scholarly resistance at the time. The decade-long, painstaking archaeological work of D. Šljivar and J. Kuzmanović Cvetković at Pločnik, and of D. Šljivar at Belovode with their *Archaeometallurgy in the Vinča Culture Project* provided the foundations for our project. Pločnik, in particular, had been known for almost a century as a major metal-yielding site (Grbić 1929), even though the majority of copper finds from the site stem from hoards rather than potential production contexts. However, it was D. Šljivar's work at Belovode and his joint work with J. Kuzmanović Cvetković in Pločnik that eventually enabled the identification of the first real copper smelting slag from a Vinča culture site (Radivojević *et al.* 2010a). It was their joint early mentoring of one of us (MR) which ultimately led to the formation of the current research team. While J. Kuzmanović Cvetković fully participated in bringing this monograph to fruition, we deeply regret that for reasons beyond our control D. Šljivar felt unable to join us for writing up this project. His absence from the team was not for lack of trying from our side to integrate him, and we put on record here our debt of gratitude for his important early work at Belovode and beyond, his mentoring and full support of our early studies, and the essential role of the National Museum in Belgrade and its staff for the study of the early metallurgy in Serbia and the wider Balkans.

Our interdisciplinary research project, *The Rise of Metallurgy in Eurasia: Evolution, organisation and consumption of early metal in the Balkans*, funded by the UK's Arts and Humanities Research Council (AHRC) connected Serbian, UK and German institutions in what was, at the time, the largest archaeometallurgical undertaking in the Balkans and beyond. It drew together UCL Institute of Archaeology, Durham University (UK), National Museum in Belgrade, Homeland Museum of Toplica in Prokuplje, Homeland Museum of Priboj on Lim in Priboj (Serbia), Curt-Engelhorn Centre for Archaeometry in Mannheim, Roman-Germanic Commission in Frankfurt, and the German Mining Museum in Bochum. This was also an international team, intellectually and practically led by a Serbian scholar (MR). Three key debates were shaping our questions. Firstly, is there a single origin of metallurgy in Eurasia, or several? Then, how did pre-existing technical know-how influence and inspire the emergence of metallurgy? And lastly, how was this early metallurgy organised across the *chaîne opératoire* of metal production and use, and integrated across a range of metals and alloys? From these debates or research themes we developed six specific questions, aiming to understand how metallurgy was transmitted and adopted by different communities across the Balkans and beyond:

1. How did the mineralogical and technological basis for early metal production in the Balkans emerge and evolve during the 6th-5th millennia BC?
2. To what extent was metallurgy related to pottery technology and production, and how did pre-existing technological knowledge influence the emergence of metallurgy?
3. How were ore sources, smelting, and casting connected and organised?
4. Where did the smelted metals circulate?
5. What metal types were being made and how did these evolve?
6. Was there a close relationship between ore sources, metallurgical technology, and artefact types?

These research questions were prompted by new insights gained during the early work at Belovode. Having identified a smelting site within a settlement meant that we now knew what to look for, at least in broad terms. We also knew to look for complex networks of ores and metals moving across the wider Balkans, and not to limit ourselves to a single region. And we also understood how little evidence survived from a much larger picture, and the need to add more pieces to the jigsaw (Pernicka 1990; Rehren 2014; Taylor 1999). We aimed to integrate the smelting in its wider *chaîne opératoire*, by adding two precursor technologies: the mining for minerals, and the firing of black-burnished pottery, the latter long seen as having laid the foundation for metal smelting. We were constrained, as always, by the usual limitations of archaeological research: the nature of suitable, accessible archaeological sites; the financial and time limitations imposed by our (most generous!) funding body; the assembled expertise of the team; and the unpredictable nature of luck (or lack thereof) that dominates so much fieldwork.

The resulting project fieldwork spanned two seasons of surveys and excavations at Belovode and Pločnik in 2012 and 2013 by the core project team, as well as extensive geophysical surveys at both sites by project partners from the Roman-Germanic Commission (RGK). There at the copper mining site of Jarmovac in 2013 by the core project team in collaboration with project partners at the Priboj on Lim Homeland-Museum (Savo Derikonjić) and the German Mining Museum Bochum, which will be published separately (Thomas *et al.* in preparation). Beyond the fieldwork, the processes involved in copper smelting were explored and evaluated in a series of experimental reconstructions at Pločnik conducted in 2013. Also, metal production debris from the earlier excavations of the site of Gornja Tuzla (Bosnia and Herzegovina) were included in the program of laboratory analysis and will be addressed in detail separately. Rather than simply focussing only on the metallurgical remains, the extensive post-excavation programme deliberately encompassed the analysis of all material and environmental remains by a wide range of specialists. At both Belovode and Pločnik, the entire stratified sequence of excavated activities was subject to radiocarbon dating. For the first time ever, we have directly dated metallurgical activities in a Vinča culture settlement.

The Rise of Metallurgy in Eurasia project also included a funded PhD to undertake the petrographic and compositional analysis of pottery from Belovode and Pločnik as an integral part of research into pottery and metal technology. This doctoral thesis examined the selection of clays, tempers, and firing conditions with an emphasis on black burnished pottery predating the emergence of metallurgy. The results of the PhD (Amicone 2017) are presented within this volume (Chapters 14, 29, and 43) and have also been published elsewhere (Amicone *et al.* 2020). The research also led to a workshop and edited volume (Amicone *et al.* 2019).

The long gestation period of this volume was partly the result of the changing professional circumstances and rapidly increasing responsibilities of the lead editors, but also due to the conscious expansion of its scope. We realised that the traditional format of the excavation monograph with only a limited discussion of the newly discovered evidence in the broader regional context was neither sufficient nor satisfactory. This was resolved by the addition of thematic surveys of specific phenomena from across the Balkans that also offered an international platform to a new generation of mainly Serbian specialist scholars.

This volume reports the outcomes of the project, the ways in which we arrived at them, and the further work we hope will be done, either by ourselves or by others. The outcomes differ from those we expected when we started and further questions arise as we reach a close. As such, this volume is a waymark on the winding road that is 'research'. We place it at the mercy of the reader and hope it will contribute to a better understanding of the archaeology of early metallurgy and society in the Balkans.

The bibliographic reference for this chapter is:

Rehren, Th., Radivojević, M. and Roberts, B. W. 2021. The Rise of Metallurgy in Eurasia: Evolution, organisation and consumption of early metal in the Balkans: an introduction to the project, in Radivojević, M., Roberts, B. W., Marić, M., Kuzmanović Cvetković, J., and Rehren, Th. (eds) *The Rise of Metallurgy in Eurasia*: 7–10. Oxford: Archaeopress.

Chapter 3

Balkan metallurgy and society, 6200–3700 BC

Miljana Radivojević and Benjamin W. Roberts

This chapter reviews the pre-existing evidence and interpretations for early mineral use and metallurgy in the Balkans from the earliest use of copper minerals at *c.* 6200 BC (Late Mesolithic-Early Neolithic) to *c.* 3700 BC (end of the Chalcolithic). It presents the empirical and intellectual foundations upon which the data, analyses and interpretations of *The Rise of Metallurgy in Eurasia* project builds. The early metallurgy in this region encompasses the production, distribution and consumption of copper, gold, bronze, lead and silver, all being either pure metals or a natural alloy (tin bronze)[1]. The chapter initially defines the geographical and temporal scope under consideration before evaluating the archaeological and metallurgical evidence in relation to: mineral exploitation; mining; smelting, metals and metal artefacts; and metal circulation. Following each of these sub-sections is a summary of how *The Rise of Metallurgy in Eurasia* project oughtto contribute to this aspect of metallurgical activity, setting this in relation to the project's six research questions as presented in Chapter 2. The chapter concludes by highlighting the dominant interpretative narratives relating to early metallurgy, metallurgists and societies in the Balkans that *The Rise of Metallurgy in Eurasia* project will evaluate, against all the available and relevant archaeological and metallurgical data.

Geographies and chronologies

The cultural, historical and geographical complexities within the widely used term 'Balkans' (Todorova 1997) and the influence of these upon archaeological research (cf. Gori and Ivanova 2017) is acknowledged. For the purposes of this chapter, although we use the geo-political term 'Balkans', defined by the Adriatic Sea to the west, the Ionian and Aegean seas to the south (including southeast and southwest) and the Black Sea to the east, we focus only on those sites that have evidence of mining and metal production and/or use during the indicated time frame. According to the current political divisions of this space, we recognise these sites as located in the states of Serbia, Bulgaria, Romania, Hungary, Bosnia & Herzegovina, Northern Macedonia and Greece, with evidence of the heaviest concentration of metal production and consumption present in the first four of these countries (see Figure 1, Table 1). However, due to the nature of the evidence and the current debates, archaeological and archaeometallurgical research from surrounding geographical regions will also be drawn upon throughout the chapter.

The relative chronological frameworks spanning the Balkans during the absolute date range of this chapter (*c.* 6200–3700 BC) are notoriously complex, largely due to the accumulation of over a century of scholarly traditions that have varied significantly. For instance, in order to avoid confusion, the period related to the emergence of metallurgy throughout southeast Europe is referred to here as the Chalcolithic, replacing the Eneolithic (as used in the former Yugoslavia and Romania) or the Copper Age (as used in Hungary). The potential confusion is especially pertinent with regard to the use of the term Eneolithic by former Yugoslav archaeologists, defined as starting with the beginning of the use of metals from the mid-late 5th millennium BC which, for Bulgarian archaeologists, correlates with the Middle/Late Chalcolithic period, when metals had been widespread for centuries (e.g. Todorova 1995). To facilitate navigation through the various labels used by Balkan archaeologists for the same phenomenon, we will adopt the term 'Chalcolithic' throughout this text, and also the (relative) Chalcolithic periodisation in the Balkans (Early, Middle, Late and Final), as elaborated by Bulgarian scholars.

The application of radiocarbon dating in the past few decades and, more recently, Bayesian statistics, has significantly influenced and strengthened the independent and relative temporal frameworks for Balkan prehistory between *c.* 6200 and 3700 BC (e.g. Bojadžiev 2002; Forenbaher 1993; Georgieva 2012; Görsdorf and Bojadžiev 1996; Higham *et al.* 2007, 2018; Krauss 2008; Krauss *et al.* 2014, 2017; Lazarovici 2006; Luca 1999; Müller 2012; Patay 1974; Pernicka *et al.* 1993, 1997; Orton 2017; Radivojević *et al.* 2010a; Schier 1996, Todorova 1981, 1995; 2014a; Vander Linden *et al.* 2014; Weninger *et al.* 2009; Whittle *et al.* 2016). This is especially true of recent intensive radiocarbon dating and Bayesian modelling of entire stratigraphic sequences at selected, well-excavated sites. Major radiocarbon dating projects

[1] Natural alloy refers to metal alloys produced from smelting complex ores— in this case, copper-tin bearing ores— as opposed to those produced by exposing two or more metallic elements to high temperature treatment through co-smelting, cementation or alloying of metals, ores, or metallic mixtures (i.e. speiss).

across Neolithic Europe led by Alastair Whittle (Whittle *et al.* 2002, 2016; Whittle 2018) have encompassed the sequences of the Balkan Neolithic-Chalcolithic Age sites of Uivar in Romania and Vinča Belo-Brdo in Serbia (Draşovean *et al.* 2017; Draşovean and Schier 2021; Tasić *et al.* 2015, 2016a). A further radiocarbon dating project across Late Neolithic-Early Bronze Age Greece and Bulgaria led by Zoï Tsirtsoni has also recently been completed (Tsirtsoni 2016b). These two major projects are further complemented by a range of smaller radiocarbon dating projects at specific sites such as Okolište in Bosnia (Müller et al. 2013a) and including earlier dating programmes at Belovode and Pločnik in Serbia (Radivojević *et al.* 2010a; Radivojević and Kuzmanović Cvetković 2014). The addition of an extensive number of radiocarbon dates at both Belovode and Pločnik by *The Rise of Metallurgy in Eurasia* project on the newly excavated sequences, enables not only a refinement of the respective site chronologies but also of the broader chronologies of metal and ceramic pyrotechnology and Vinča settlement activity (see Chapters 11, 14, 26, 29).

However, there frequently remains an absence of extensive radiocarbon dating at the majority of late 7th to early 4th millennium BC sites and, invariably, at potential copper mining sites and depositions of metal objects across the Balkans. It is still, therefore, the relative chronological frameworks based on ceramic types and archaeological cultures, frequently identified a century ago, whose absolute date ranges are constantly being refined, as occurred recently with Vinča culture ceramics (cf. Whittle *et al.* 2016). Furthermore, the emergence of rival national traditions of archaeological scholarship in the 20th century across the Balkans has frequently meant that virtually identical archaeological phenomena whose distribution crosses modern national borders have been assigned different nomenclatures, an example being the Starčevo–Körös–Criş cultural complex. Körös and Criş are the names of the same river after which an Early Neolithic cultural phenomenon was named in Hungarian and Romanian respectively whilst Starčevo is the type site in northern Serbia. This results in regional scholarship being

Table 1. Relative and absolute chronology for copper mineral (malachite) and metal-using cultures / archaeological complexes in the 'core' metallurgical zone (Serbia, Bulgaria, parts of Romania) between 6200 BC and 3700 BC. Chronological framework largely based on Schier (1996; 2014), Boyadžiev (1995; 2002) and Whittle et al. (2016). Green font = use of copper minerals (i.e. malachite beads); red = metallurgical materials (i.e. metal artefacts, slags).

Period	C14 dates	Vojvodina	Central Balkans	West Bulgaria	South Bulgaria	Muntenia	North-east Bulgaria	Black Sea Coast (west)
Proto Bronze Age	**3200**	Boleráz	Cernavoda III	**Galatin**	**Yagodina**	Cernavoda III	**Usatovo**	
Final Chalcolithic	**3700**	**Salcuţa IV** **Bodrogkeresztúr** **KSBh**				Cernavoda I	**Cernavoda I/Pevets**	Cernavoda I
Late Chalcolithic	**4100**	**Tiszapolgár / KSBh**				**Kodžadermen-Gumelniţa-Karanovo VI**	**Kodžadermen-Gumelniţa-Karanovo VI**	**Varna III**
Middle Chalcolithic	**4450**	**Vinča D**		**Krivodol-Salcuţa-Bubanj hum (KSBh)**	**Karanovo VI** Marica IV	**Boian-Spanţov**		**Varna I** **Varna II** **Hamangia IV**
Early Chalcolithic	**4600**	**Vinča D** **Vinča C**		**Gradešnica** **Dikilitash-Slatino**	**Marica III-Karanovo V**	Boian-Vidra	Poljanica	**Sava / Hamangia III**
Late Neolithic	**5000**	**Vinča B** **Vinča A**		Kurilo/Akropotamos Topolnica	Karanovo IV Karanovo III	Boian III	Hotnica	**Hamangia II** **Hamangia I**
Early Neolithic	**5500** **6200**	**Starčevo** **Lepenski Vir III**						

subsequently tasked with identifying the connections between these culture-historical sequences and then proposing new nomenclatures that integrate the pre-existing terms.

It is therefore not uncommon to see debates on the connections between the emergence of metallurgy and the Gradac Phase of Vinča culture ceramic sequence, or the relationship between the development of metallurgy and the widespread graphite painted decoration on the ceramics of the Kodžadermen-Gumelniţa-Karanovo IV (KGK IV) cultural complex (e.g. Amicone *et al.* 2019, 2020b; Garašanin 1994/1995; Jovanović 1971, 1994, 2006; Radivojević *et al.* 2010a; Radivojević and Kuzmanović Cvetković 2014; Renfrew 1969; Spataro *et al.* 2019; Spataro and Furholt 2020; Todorova 1995; Todorova and Vajsov 1993). As is now widely acknowledged in Balkan and world prehistory, the creation of spatial and temporal frameworks through the identification of similarities and differences in materials and practices continues to evade researchers; straightforward explanations are unlikely (cf. Gori and Ivanova 2017; Roberts and Vander Linden 2011; Shennan 2013). It would seem inevitable that, despite well-argued proposals for abandoning relative typologies and cultures in the Balkans due to improved and increased independent scientific dating techniques (Tsirtsoni 2016a), they will very likely endure into future generations of archaeological scholarship.

For the purpose of *The Rise of Metallurgy in Eurasia* project, we use the available relative and absolute dating spanning *c.* 6200–3700 BC throughout the Balkans. We identify six periods reflecting the changing characteristics in the metallurgical evidence that enable questions surrounding metallurgical origins, development and societal inter-relationships to be addressed. These are: *Early Neolithic* (*c.* 6200–5500 BC), *Late Neolithic* (*c.* 5500–5000 BC), *Early Chalcolithic* (*c.* 5000–4600 BC), *Middle Chalcolithic* (c. 4600–4450 BC), *Late Chalcolithic* (*c.* 4450–4100 BC) and *Final Chalcolithic* (*c.* 4100–3700 BC). It should be stressed, however, that in certain areas there are insufficient modern, published excavations, archaeometallurgical analyses and/or resolution of radiocarbon dating for the framework to be evaluated. Our strongest focus remains, therefore, on the modern-day territories of Serbia, Bulgaria, Romania and Hungary (see Figures 1–3), as the core area of activities related to mineral use and metallurgy. The majority of the periods used in the chronological scheme for this article have been employed in earlier frameworks. The identification of a new 150 year long period, spanning the mid-5th millennium BC (*c.* 4600–4450 BC, Middle Chalcolithic), reflects recent dating and current interpretations centred on the iconic—and still currently unparalleled—site of Varna, Bulgaria and the possibility that the site is a reflection of a relatively short regional phenomenon encompassing distinctive metal production and consumption evidenced by a growth in wealth, amongst other observations (e.g. Biehl and Marciniak 2000; Chapman 2013; Schier 2014a).

Mineral exploitation

Copper *minerals* in the archaeological record potentially represent only copper *ores*. Ore is a culturally defined term referring to agglomerations of minerals from which the extraction of one or more metals is seen as a profitable action, largely in pre-industrial times (e.g. Radivojević and Rehren 2016; Rapp 2009; Rehren 1997). In other words, what the modern mining industry considers the economically feasible exploitation of mineral resources today differs from what prehistoric miners saw as an acceptable investment of labour. The significance of this distinction in the context of metallurgical activities has been raised by Muhly (1989), who compared the relationship of malachite (copper carbonate) and copper metallurgy at an archaeological site to that of haematite (iron oxide) in a cave painting with iron metallurgy.

Evidence suggests that use of copper mineral and native copper in neighbouring Anatolia and the Near East occurred much earlier than in the Balkans. The earliest example dates back to the 11th millennium BC Epipalaeolithic burial site of Shanidar Cave, where a malachite bead was deposited as a grave offering (Bar-Yosef and Deborah 2008; Solecki *et al.* 2004: 96). By the 9th millennium BC, native copper and copper minerals were increasingly worked, as at the settlement of Çayönü Tepesi in eastern Turkey, a site which also yielded evidence for the annealing of native copper (Maddin *et al.* 1999; Özdoğan and Özdoğan 1999). This settlement was conveniently located near the rich copper mineralisation Ergani Maden but the exploitation of this source has not yet been demonstrated. By 6000 BC, the use of copper minerals had spread beyond its 'core' zone in Anatolia and northern Mesopotamia to the Levant (Golden 2010), Transcaucasia (Kavtaradze 1999; Courcier 2014), the Balkans (Glumac and Tringham 1990; Thornton 2001; see below), Iran (Pigott 1999; Thornton 2009; Helwing 2013) and Pakistan (Kenoyer and Miller 1999; Hoffmann and Miller 2014). The strong association of intensive copper mineral use and agriculture is apparent and has been advocated as inherently related to the strong symbolism of its green colour in relation to crop fertility (Bar-Yosef Mayer and Porat 2008). The study by Bar-Yosef Mayer and Porat (2008: 8549, Table 1) also showed that copper minerals were not the only 'green' option for the Near Eastern (Pre)Neolithic communities, since ornaments made of apatite, turquoise, amazonite or serpentinite were also sought for their visual properties.

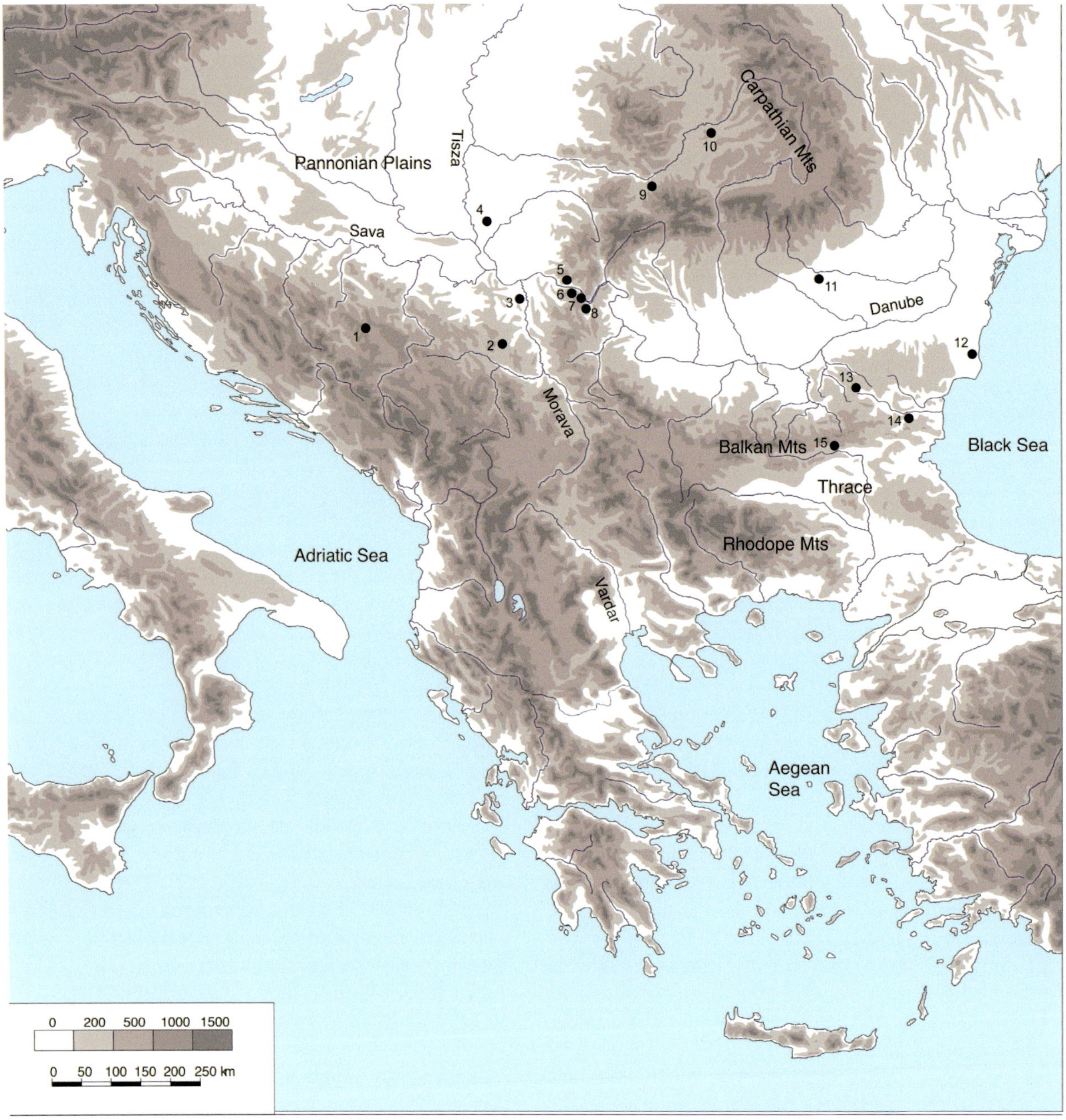

Figure 1. Map of the Early Neolithic sites (*c.* 6200–5500 BCE) mentioned here. 1- Obre I; 2- Divostin; 3- Zmajevac; 4- Szarvas 23; 5- Gornea; 6- Lepenski Vir; 7- Vlasac; 8- Rudna Glava; 9- Balomir; 10- Iernut; 11- Cernica; 12- Durankulak; 13- Ovcharovo I; 14- Usoe I; 15- Karanovo; 16- Kolubara-Jaričište. (map CC BY-NC-ND 4.0 J. Pendić and M. Radivojević)

The earliest evidence for the use of copper minerals in the Balkans occurs during the transition to the Early Neolithic (or the emergence of Starčevo-Criş-Körös culture groups) in *c.* 6200–5500 BC, with evidence spanning the Carpathian Basin, Moldavia, western Ukraine and northern Balkans (Bognár-Kutzián 1976: 70–73). The earliest exploitation of copper minerals was possibly by hunter-fisher-gatherer communities (likely mixed with the early farming population migrating from Anatolia, see for instance Mathieson *et al.* 2018), as indicated by samples discovered at Lepenski Vir, Vlasac and Kolubara-Jaričište (Figure 1), and dated to *c.* 6200 BC (Radivojević 2015: 325; Srejović and Letica 1978: 11–14). The processing of copper minerals and native copper developed within the subsequent Neolithic Starčevo-Criş-Körös culture groups, which mostly produced beads from malachite $[Cu_2CO_3(OH)_2]$ or azurite $[Cu_3(CO_3)_2(OH)_2]$. In addition to malachite and azurite beads from Lepenski Vir and Divostin I (Glumac1988: 460; Radivojević 2012; Srejović 1969: 173; Srejović 1972: 146), similar items were found in the cemeteries of Cernica in southern Romania and Durankulak in northeastern Bulgaria, and settlements of Obre I in Bosnia, and Ovcharovo I and Usoe I in Bulgaria (Figure 1) (Cantacuzino and Morintz 1963: 72–75, Figure 28, 18, 19; Pernicka *et al.* 1997: 44; Sterud and

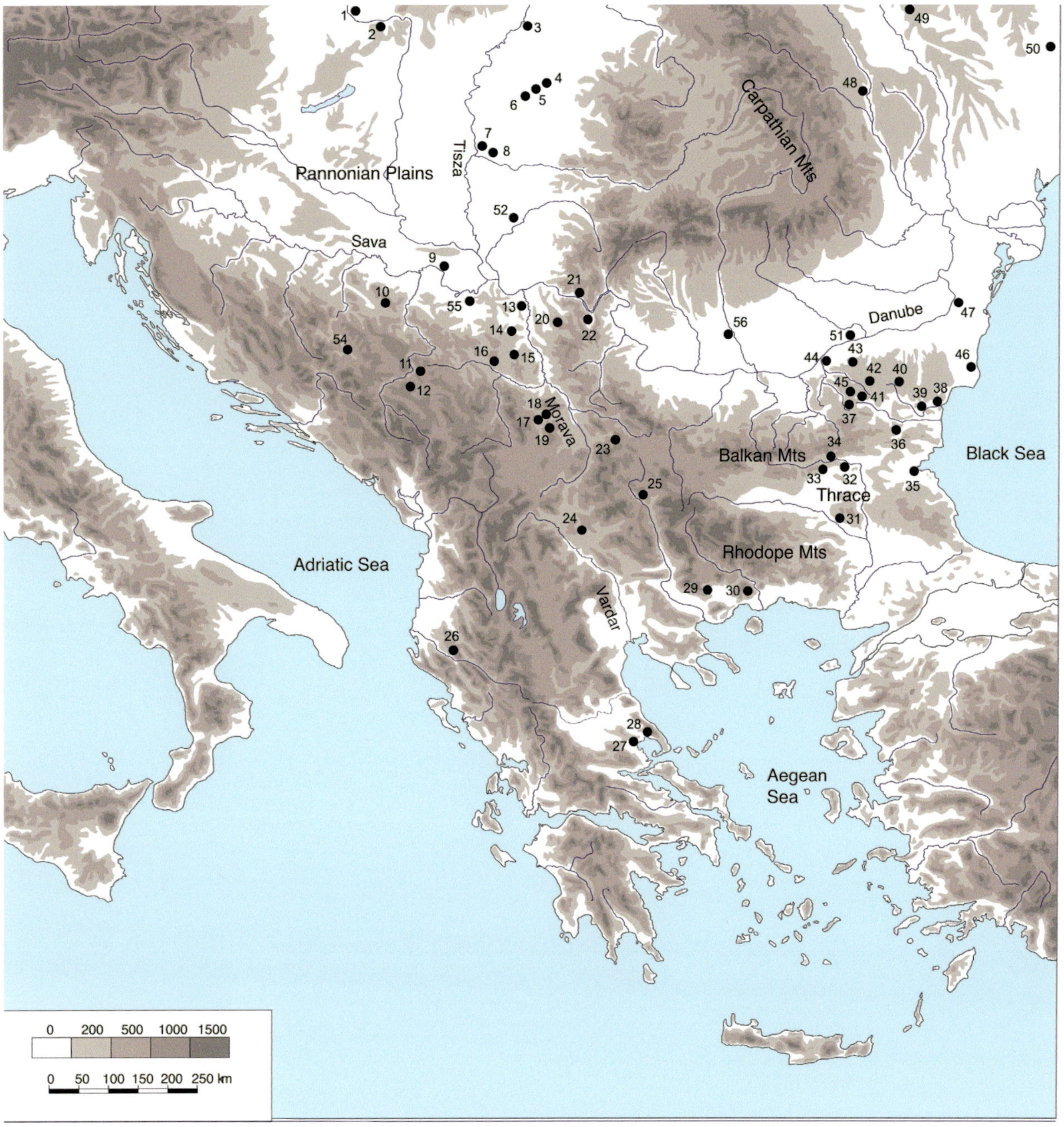

Figure 2. Map of the Late Neolithic / Early Chalcolithic sites (*c.* 5500–4600 BCE) mentioned here. 1- Mlynárce; 2- Neszmély; 3- Csöszhalom; 4- Hérpály; 5- Berettyószentmárton; 6- Zsáka- Markó; 7- Hódmezővársahely-Kopáncs-Kökénydomb; 8- Gorsza; 9- Gomolava; 10- Gornja Tuzla; 11- Stapari; 12- Jarmovac; 13- Selevac; 14- Mali Šturac; 15- Divostin; 16- Ratina; 17- Pločnik; 18- Merovac; 19- Mačina; 20- Belovode; 21- Gornea; 22- Rudna Glava; 23- Hisarluka; 24- Anzabegovo; 25- Slatino; 26- Kamnik; 27- Dimini; 28- Sesklo; 29- Sitagroi; 30- Dikili Tash; 31- Maritsa; 32- Azmashka Mogila; 33- Ai Bunar; 34- Karanovo; 35- Medni Rid; 36- Golyamo Delchevo; 37- Targovište; 38- Varna; 39- Devnja; 40- Vinitsa; 41- Ovcharovo; 42- Radingrad; 43- Kubrat; 44- Ruse; 45- Polyanica; 46- Durankulak; 47- Cernavodă; 48- Izvoare I; 49- Lukavrublevetskaya; 50- Karbuna; 51-Pietrele; 52-Foeni; 53-Ždrelo; 54 – Okolište; 55- Stubline; 56- Reşca (map CC BY-NC-ND 4.0 J. Pendić and M. Radivojević).

Sterud 1974: 258; Stratton et al. 2019; Todorova 1981: 4; Vlassa 1967). Lumps and flakes of copper minerals were also identified in the settlements of Zmajevac in eastern Serbia and Szarvas 23 in Hungary (Chapman 1981: 131; Chapman and Tylecote 1983: 375; Comşa 1991: 51; cf. Bailey 2000: 210) (Figure 1). Malachite beads and copper minerals are also commonly found in early Vinča culture settlements (pre-5000 BC) at the sites of Belovode, Pločnik, Vinča-Belo Brdo, Selevac and Medvednjak (Figure 2), and occur continually until the abandonment of the settlements, and throughout other, later Vinča culture manifestations, such as Gomolava (Glumac and Tringham 1990; Radivojević 2012; Radivojević and Kuzmanović Cvetković 2014; Šljivar 1993-2009; Šljivar and Kuzmanović Cvetković 1996-2009). Significantly, at Divostin II (Vinča D phase), malachite beads and a

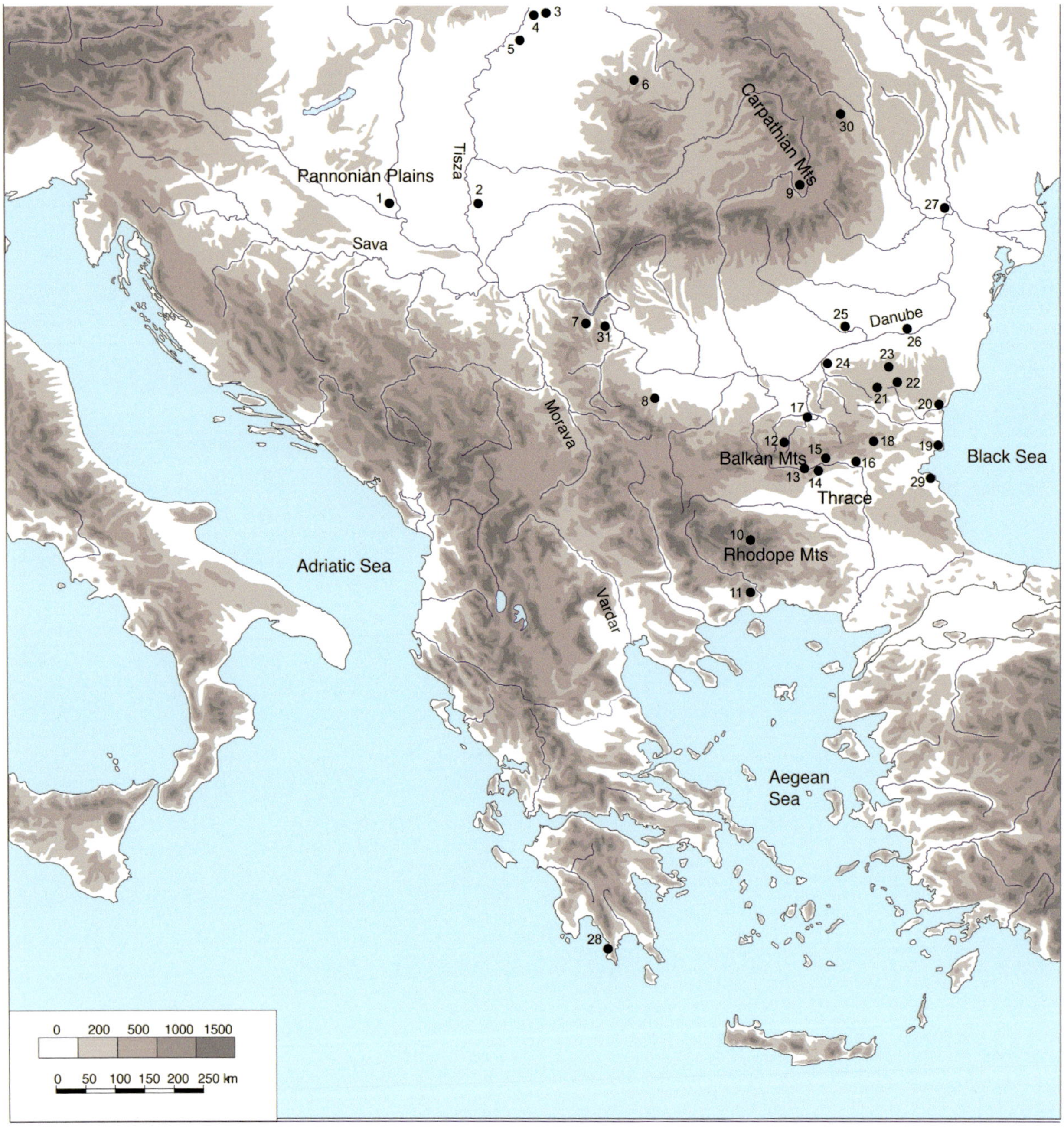

Figure 3. Map of the Middle, Late and Final Chalcolithic sites (4600–3700 BCE) mentioned here. 1- Zengővárkony; 2- Tiszapolgár-Basatanya; 3- Tibava; 4- Lucska; 5- Tiszapolgár-Hajdúnánás Road; 6- Moigrad; 7- Lazareva cave; 8- Gradeshnitsa; 9- Ariuşd; 10- Dolnoslav; 11- Dikili Tash; 12- Hotnica; 13- Bereketska Mogila; 14- Ai Bunar; 15- Karanovo; 16- Chatalka; 17- Kačica; 18- Smjadovo; 19- Kasla- Dere; 20- Varna; 21- Kodžadermen; 22- Bubanj; 23- Mečkjur; 24- Ruse; 25- Vidra; 26- Gumelniţa; 27- Traian; 28- Alepotrypa Cave; 29-Akladi Cheiri; 30-Poduri; 31-Kmpije; 32- Bubanj (map CC BY-NC-ND 4.0 J. Pendić and M. Radivojević).

metal bracelet were predominantly found in a group of large houses (McPherron and Srejović 1988) and were interpreted as a possible indication of the higher status of the occupants on the basis that larger households would have a larger labour force available to create a surplus and therefore an economic advantage (Porčić 2012a).

The provenance analyses of most of the Lepenski Vir, Vlasac and Vinča culture sites minerals indicated the use of local sources, predominantly Majdanpek in eastern Serbia, then Ždrelo (near Belovode, Figure 2) and an as yet unidentified copper source consistent with most of the Pločnik minerals and metal artefacts and copper slags from Belovode (Pernicka *et al.* 1993: 6, 1997: 93 ff.; Radivojević *et al.* 2010a: 2784, Figure 10; Radivojević 2012: 393 ff.). The only securely dated source where there is evidence for copper mineral exploitation within this period is the site of Rudna Glava, in Serbia (Figure 2) where copper mining

Figure 4. Copper mineral and metallurgical evidence from mid-6th – mid-5th millennium BC. a) typical malachite bead on the left and a black-and-green mineral on the right (Belovode); b) copper slag from Belovode; c) ceramic sherd with copper slag overflowing its top, most likely used to line a hole-in-the-ground smelting installation (Belovode); d) ceramic sherd with copper slag (Foeni); e & f) ceramic nozzle, potentially used for (s)melting (Bubanj, see also in Bulatović 2015) (adapted after Radivojević 2012; CC BY-NC-ND 4.0 M. Radivojević & Lj. Radivojević).

activities intensify from *c.* 5500 BC in parallel with copper mineral use at nearby Belovode (Radivojević and Kuzmanović Cvetković 2014; O'Brien 2015). Importantly, the inhabitants of this site distinguished between pure green copper *minerals* (malachite) used for bead making and 'tainted' black-and-green copper *ores* for the smelting charge (Figure 4a). This argument was further strengthened by the identification of a distinctive lead isotope signature for the bead minerals and smelting ores at Belovode, indicating existing knowledge of the different properties of these materials (Radivojević *et al.* 2010a: 2784). Similar practices of distinguishing between pure green copper minerals and black-and-green manganese-rich copper ores have been detected throughout the entire Vinča occupation at the sites of Vinča-Belo Brdo and Pločnik (Figure 2, Figure 4a), indicating a similar awareness of the material properties of these distinctive copper occurrences for *c.* 800 years across all these settlements (Radivojević and Rehren 2016: 215).

In light of this existing evidence for the exploitation of copper minerals and potential copper ores and the debates surrounding their interpretation, *The Rise of Metallurgy in Eurasia* project retrieved, recorded and spatially mapped each and every fragment of copper mineral excavated at Belovode and Pločnik. This provided the most comprehensive possible range of stratified samples for analysing and evaluating patterns of exploitation in space and time relating to the provenance, selection, organisation and uses of the minerals/ores and contributing to Research Questions 1, 3 and 6 (see Chapters 11 and 26).

Mining

The only copper mining activities in the Balkans that have been securely dated to the 5th millennium BC occur at two sites: Ai Bunar in central Bulgaria (Chernykh 1978a; Pernicka *et al.* 1997) (Figure 2), and Rudna Glava in eastern Serbia (Jovanović 1971; O'Brien 2015). Rudna Glava (Jovanović 1980, 1982) is the earliest documented copper mine, not only in Europe but across Eurasia, and is radiocarbon dated to c. 5400-4650 BC (O'Brien 2015; Pernicka et al. 1993; Radivojević and Kuzmanović Cvetković 2014). It consisted of eight groups of mine shafts with access platforms, all following veins rich in magnetite, chalcopyrite and carbonate copper ores. Near the entrances and inside the shafts, hoards of distinctive Vinča culture-style ceramics were found, dating from the Gradac Phase, broadly between the early and mid-5th millennium BC (the later dates associated with a prolonged Gradac Phase in the southern Serbian sites), together with stone mallets and deer antler tools. Mining activities at Rudna Glava peaked around *c.* 5000 BC. Despite this extensive evidence for mining, no analyses prior to *The Rise of Metallurgy in Eurasia* project had confirmed that any analysed metal artefact from the Balkans was made from the Rudna Glava copper ores (Pernicka *et al.* 1993: 2; Pernicka *et al.* 1997: 143; for opposing view see Jovanović 1993; see Chapter 41 for the current state of research). Having visited and surveyed the area around Rudna Glava in various capacities from 2009, we contend that ancient mining evidence at this site survived only as a result of Rudna Glava not being particularly rich in malachite. Conversely, Majdanpek, the most abundant deposit of copper, which has lasted into modern times, is only 12 km away from Rudna Glava, and provenance analysis has long indicated Majdanpek copper to be one of the main sources for Vinča copper implements (Pernicka *et al.* 1993, 1997; Radivojević 2012).

Ancient mining activities are also known from several localities within Serbia, some potentially dating to the 5th millennium BC. These are: Ždrelo in eastern Serbia (near Belovode), Mali Šturac in central Serbia, Medvednik in western Serbia and Jarmovac in the southwest (Figure 2) (Antonović 2014a; Jovanović 1971; Pecikoza 2011; Radivojević *et al.* 2010a). In Mali Šturac on Mt. Rudnik in central Serbia, grooved stone mallets resembling those from Rudna Glava were recovered, leading scholars to believe that they were of the Vinča culture provenance (Bogosavljević 1995; Jovanović 1983). More recent and ongoing excavations at this site yielded additional material that roughly dates this mine to the mid to late 5th millennium BC (Antonović and Vukadinović 2012; Antonović *et al.* 2014). Furthermore, grooved stone tools, identical to those discovered at Rudna Glava and Mali Šturac, were found during field surveys of the Vinča culture settlements of Mačina and Merovac (Figure 2), both situated in the vicinity of the ore-rich deposits at Mt. Kopaonik and Radan in southern Serbia (Kuzmanović Cvetković 1998; Radivojević 1998); these are comparable with stone tools at mining sites throughout Europe (cf. De Pascale 2003; O'Brien 2015).

The Mid-Late Chalcolithic (*c.* 4600–(4450)–4100 BC) was dominated by exploitation of Bulgarian sources, predominantly at Ai Bunar in Bulgaria (Chernykh 1978b: 54–75). This source, near Stara Zagora in central Bulgaria (Figure 2), was much larger and had more productive efficiency than Rudna Glava, with shafts up to 500 m long. The material associated with the site belongs to the Kodžadermen-Gumelniţa-Karanovo (KGK) VI cultural complex and is therefore relatively dated to the mid to late 5th millennium BC (Chernykh 1978a, 1978b, 1992). Metal from Ai Bunar is known to have travelled long distances within the Balkans, as far south as Thessaly, and as far north as the northern Black Sea coast (Chernykh 1978b: 122, 263; Gimbutas 1977: 44; Pernicka *et al.* 1997; Radivojević and Grujić 2018; Renfrew 1972: 308, Figure 16/2). However, the results of provenance analysis suggested the exploitation of more than one copper deposit in this period, for example, that from Medni Rid, in southeastern Bulgaria (Pernicka *et al.* 1997: 143–146). The most recent excavations in this location revealed materials from Roman and later times, although some indicate exploitation activities by the communities of the KGK VI cultural complex (Leshtakov 2013), also supported by recently analysed metal production evidence from Akladi Cheiri (Figure 3), a settlement nearby. Metal production at this site is argued to date to the late KGK VI, or broadly to the middle of the 2nd half of the 5th millennium BC, based on the typology and ornamentation of pottery found in the same pit as metallurgical evidence (Rehren *et al.* 2016: 207; Rehren *et al.* 2020). The exploitation of the Medni Rid ores may have started earlier, given the finds of late 6th millennium BC malachite in nearby settlements or its use for making metal items in the Karbuna hoard which was found in a typical Tripolye A pot (*c.* 4700–4600 BC) (Figure 2) (cf. Pernicka *et al.* 1997; Rehren *et al.* 2016).

Rather than being mined, from *c.* 4650 BC Balkan gold was most likely collected from river streams as alluvial (washed) nuggets that had eroded from primary deposits (Avramova 2002; Boyadžiev 2002; Makkay

1995: 70). This suggestion has been elaborated recently in a study of Varna gold, where Leusch *et al.* (2014) presented a diversity of different gold, copper and silver ratios in the excavated gold artefacts, explaining them as originating from natural compositional variations in (alluvial) gold nuggets. This in turn demonstrates that various gold occurrences were exploited for the making of Varna gold, and possibly acquired through a well organised gold supply network (Leusch *et al.* 2015). Such supply networks also procured copper, *Spondylus*, carnelian, marble, serpentinite, long yellow flint blades (superblades) and other prestige commodities unearthed as paraphernalia in the Varna cemetery (Leusch *et al.* 2017).

In light of the existing evidence for mining, *The Rise of Metallurgy in Eurasia* project analysed the trace elements and lead isotopes of the newly excavated copper mineral and metal artefacts at Belovode and Pločnik in order to explore their provenance in relation to known and potentially unknown copper mining sites (Chapter 41). The project also excavated the potentially Chalcolithic mining site at Jarmovac (southwest Serbia) in conjunction with the Priboj on Lim Homeland Museum and the German Mining Museum in Bochum. This complex of ancient mines was first mentioned by Davies (1937) who identified Vinča culture sherds in one of the shafts. The site was previously excavated by the local museum authorities, who discovered an associated settlement with a Late Vinča culture phase (Vinča D) only 300 m away (Bunardžić *et al.* 2008: 86; Derikonjić 2010). The new excavations sought to obtain evidence of mining activity, stratified samples which would enable radiocarbon dates, and the analysis of the trace element and lead isotopes of the copper ores being mined for comparison to existing results in the region. The results will be published elsewhere; they detail the recovery of an antler pick fragment from a stratified sequence radiocarbon dated to the mid-4th millennium BC combined with trace element and lead isotope analyses demonstrating that the mine was most likely used to produce copper objects in the 5th millennium BC (Thomas *et al.* in preparation). All these results contribute to Research Questions 1, 3 and 6 of the project although it is highly likely that further mining sites remain to be discovered along the rich metallogenic belt running through the Balkans (cf. Janković 1977).

Smelting

Smelting evidence from the Early Chalcolithic is extremely scarce and mostly constrained to the Vinča culture phenomenon in the central Balkans (Table 1, Figure 4b-d). Prior to more analytically extensive studies of the early metal production debris from a selected number of Vinča culture sites (Glumac 1991; Radivojević 2007, 2012), there was mention of a copper slag lump from the settlement site of the Anzabegovo in the eastern part of Northern Macedonia (Figure 1), dated to *c.* 5200 BC (Gimbutas 1976a), although this has never been chronologically or analytically verified. Moreover, in its relative regional vicinity, though in the valley of Strymon River at the Greek-Bulgarian border, is the site of Promachon-Topolnica, which has yielded indicative field structures ('hollows' with traces of copper), of which the most important is a small clay crucible with a spout, dated broadly to the first half of the 5th millennium BC (Koukouli-Chryssanthaki et al. 2007: 51, Fig. 7.4). While the authors reported that the crucible contained traces of non-slagging copper processing with distinct traces of heavy burning, no analysis are made available, which makes its more accurate identification challenging. A similarly vague situation applies to the situation in the site of Stapari (Figure 2), where an alleged lump of slag was dated relatively to within the late Vinča culture phase (Jurišić 1959). Pieces of 'greenish slag resulting from intense fire' were reported at depths of ▼6.2, ▼6.4 and ▼7.0 m at Vinča-Belo Brdo (M. Vasić excavations), however no further analysis or details of these finds are available (Antonović 2002: 36, note 59). Microstructural, chemical and isotopic analysis of copper slag and other production evidence from the sites of Belovode, Vinča-Belo Brdo (N. N. Tasić excavations), Pločnik, Gornja Tuzla and Selevac are the first secure evidence for sustained metallurgical activities within the Vinča culture and highlight its role as the core archaeological phenomenon in the evolution of Balkan metallurgy (Glumac and Todd 1991; Govedarica 2016; Radivojević 2012; Radivojević and Rehren 2016; cf. Čović 1961).

The estimated chronological sequence of the finds studied by Radivojević (2015) starts with the Belovode slags at *c.* 5000 BC (until *c.* 4600 BC), which overlaps for around 200 years with the Vinča-Belo Brdo production evidence (dated in the range of *c.* 4800–4600 BC). Copper smelting took place at the settlement of Gornja Tuzla for up to 200 years after both Belovode and Vinča were abandoned (*c.* 4400 BC). Both macro- and micro-analytical approaches demonstrate that copper smelting was, in total, practiced throughout *c.* 600 years, with remarkable similarities in the level of expertise and technological choices, but with clear differences in the composition of the ores smelted. The striking detail that underlines the chemistry of ores smelted at the sites of Belovode, Vinča, and Gornja Tuzla is their dominant colour: whatever the exact minerals that were present in the ore charge, they most likely had strong colours in the range of green/blue (vivianite, arthurite, apatite, scorodite), and violet (strengite), in addition to black and green Mn-rich malachite

(Belovode and Vinča only). Such a conclusion has been corroborated by a detailed inspection of slag matrices and residual ores found in them. It is also important to underline that the indicated ores were not necessarily copper rich ones. Rather, they had striking green/blue/violet colours that attracted the Vinča prospectors (Radivojević and Rehren 2016: 225 ff.). Although it is not clear from the analyses whether black and green minerals were selected separately or as a mixed ore, the conclusion that emerges from the analytical discussion is the presence of a common knowledge regarding the suitability for smelting of distinctively coloured mixed minerals. Noteworthy is the ongoing analytical study of slagged sherds and a metal object from the site of Foeni in Romania (Figure 2), contemporary with the Vinča culture metal production which, in principle, confirms similar technological choices for early metal extraction across the Balkans (Draşovean 2006; Radivojević *et al.* in preparation).

The Vinča culture metal production practice fits well within the 'ephemeral model' of Chalcolithic metallurgy in western Eurasia (Bourgarit 2007); the individual slags weigh a little less than 10 g in total (see example in Figure 4b). This is commonly explained by the use of much cleaner ore at the early stages, resulting in a 'slagless' or nearly slagless metallurgy (cf. Craddock 1995). Depending on the relative proportions of (slag-forming) dark components in the ore and pure green mineral, a large amount of copper may have formed with only a small quantity of slag; this is the favoured scenario in the recent analytical studies (Radivojević and Rehren 2016: 227 ff). Of note though is the discovery of a lead-based slag cake in the undisturbed horizon dated to 5200 BC at Belovode, and weighing nearly 800 g. As this is a unique find currently unsupported by evidence for sustainable lead metal production, it will be addressed in detail in the 'Lead and silver' section below.

The structures in which smelting took place—so-called smelting 'furnaces'—are evidenced primarily by slagged sherds at both Belovode and Gornja Tuzla (example in Figure 4c–d), which suggest the presence of a hole-in-the-ground installation lined with broken pottery. Such installations were possibly operated by using blowpipes or bellows, where five to six blowpipes would normally suffice to bring the temperature to around 1100–1200 ºC (cf. Rehder 1994: 221). The only indication of how these blowpipes may have looked is found in the ceramic nozzles recovered from the sites of Bubanj (Figure 4e-f) and Kmpije in Bor (Figure 3). In the absence of any other evidence, the hole-in-the-ground installations appear to be the only technological possibility for primary metal production in the early to mid-5th millennium BC (see Figure 4c, 4d). In addition, although analysed crucibles are absent from the record, their presence must be assumed, given that they would have been needed for (re)melting and casting of the thousands of heavy metal objects known from this period. There is a possibility though, to identify as crucibles two oval, ladle-like vessels with a short, vertically pierced handle and secondary traces of firing, from the site of Reşca-Dâmbul Morii, in Vallachia (Romania, Figure 2), (Stefan 2018: 119, Table VII/1, 2). These have not yet been analysed, and their context is still under discussion, although argued to belong to the Vădastra culture horizon, which dates between 5200 and 5000 BC; nevertheless, this is the closest potential clue to how crucibles might have appeared during this period. Curiously, the casting moulds that are firmly contextualised for the vast number of metal implements produced in this period are also absent from the archaeological record (Kienlin 2010: 42 ff; Heeb 2014).

Copper production evidence is still scarce in the Mid-Late Chalcolithic (*c.* 4600–4100 BC), although it is documented in more settlements than for the previous period. In Bulgaria, copper smelting evidence comes from the sites of Dolnoslav, Chatalka and Akladi Cheiri (Figure 3), all dated to the mid to late 5th millennium BC (Ryndina *et al.* 1999; Rehren *et al.* 2016, 2020). All three sites yielded crucibles, amongst other finds, although only the Dolnoslav and Akladi Cheiri examples were preserved. The crucible from Dolnoslav was a vessel with an oval plan, round base and 10–25 mm thick walls (Ryndina *et al.* 1999); the well-preserved example from Akladi Cheiri had a similar flat oval base though the frontal part slightly profiled as a spout (Rehren *et al.* 2016: 207, Figure 2). Microstructural and compositional analyses of the Dolnoslav crucible indicated smelting of polymetallic ores (a mix of malachite with primary copper ores), which were rich in zinc and lead oxide (Ryndina *et al.* 1999: 1066, Table 2). The dominant presence of zinc and lead in the slag matrix, together with copper oxide in trace amounts, presents a copper smelting technology that is different and possibly more efficient than that encountered in the Vinča culture with slags rich in manganese oxide (Radivojević 2015: 332, Table 2). The Akladi Cheiri example on the other hand was for (re)melting: its inside was contaminated with copper and no other evidence for gangue elements such as iron, cobalt or sulfur (Rehren *et al.* 2020: 152). These are the earliest crucibles in the Balkans and become more common only from the mid-4th millennium BC Baden culture in the north-central Balkans (Ecsedy 1990: 224; Glumac and Todd 1991; Radivojević *et al.* 2010b).

More examples, although not analytically confirmed, come from the late 5th millennium BC Tiszapolgár culture cemetery of Tibava (Figure 3), in the form of a cylinder vessel with a crude inner surface, identified amongst pottery grave goods (Andel 1958: Plate I/7;

Šiška 1964: 317, Figure 12/5). It was thought to be a melting pot but was never analysed (Bognár-Kutzián 1972: 134). Another crucible described as a 'vessel covered with blue verdigris and with two small copper crumbs' was discovered among grave goods in the cemetery of Tiszapolgár-Hajdúnánás Road (Figure 3) and has been unfortunately lost (Bognár-Kutzián 1972: 98, 134). The ceramic vessels widely interpreted as crucibles at Cucuteni A2 and B1 levels at the site of Poduri-Dealul Ghindaru in Romania (Figure 3) (Mareş 2002: 85, 138–139, Table 64/8) have yet to be subjected to archaeometallurgical analyses so cannot be considered as confirmed metal production evidence despite the presence of two copper ingots from the same site (Monah *et al.* 2002). A similar situation is encountered in Sitagroi (phase III, roughly contemporary with KGK VI), where an assemblage of thirty-six slagged sherds, accompanied by copper metal artefacts, present a convincing evidence for local copper smelting as well as a distinctively similar slagging pattern to contemporaneous Akladi Cheiri for instance (Renfrew and Elster 2003: 306). These sherds are yet to be subjected to detailed technological analysis.

The hole-in-the-ground smelting installations identified earlier in the Vinča culture sites find parallels at the site of Akladi Cheiri, near Sozopol in Bulgaria (Figure 3), where an exceptional discovery of 300 ceramic sherds with traces of firing and slag adhered to them testifies to intensive metal production activities in the foothills of the Medni Rid copper deposits, dated to later phases of the KGK VI complex (Rehren *et al.* 2016). This mid to late 5th millennium BC metallurgical workshop contains fragmented slagged sherds (that possibly lined a hole in the ground), associated slags and the melting crucible mentioned above and similar to those hypothesised at the Vinča culture sites of Belovode and Gornja Tuzla (Radivojević and Rehren 2016). Analysis of slag from Akladi Cheiri revealed features already observed for early copper smelting, such as a high degree of variability in glassy to micro-crystalline slag matrix and the formation of inclusions, paired with equally varied redox conditions. The presence of fayalite, clusters of magnetite with matte and copper metal, copper sulphides, olivine crystals, delafossite and cuprite across the studied assemblage speaks of unstable firing conditions during the smelt and different levels of exposure of the slagged sherds during the smelting events (Rehren *et al.* 2020), all of which are known features of the earlier examples from the Vinča culture. More slagged sherds from the mid to late 5th millennium BC come from the site of Kmpije in Bor in eastern Serbia (Figure 3), where a slagged sherd was discovered in association with copper metal artefacts. Analytical work is underway in collaboration with one of the authors (MR) and I. Jovanović from the Mining Museum in Bor.

Interestingly, a piece of slag was deposited as a grave offering in the late 5th millennium BC Lengyel culture cemetery of Zengővárkony (Figure 3), Hungary. It was found in a well-contextualised grave of a middle-aged woman, together with numerous ceramic vessels and two pure copper spiral bracelets on each arm (Dombay 1939; Glumac and Todd 1991: 14). Slag analyses revealed mineral phases of cuprite and cassiterite, copper metal prills with a significant content of tin, ranging from 0.4 to 37 wt%, as well as relevant concentrations of tin in slag silicates (Glumac and Todd 1991: 14). Ottaway and Roberts (2008: 197) discuss this find as accidental co-smelting of copper and tin-bearing ores since, compositionally, it predates tin-alloys in the region. The recently discovered piece of tin bronze foil at the site of Pločnik in south Serbia, dated to *c.* 4650 BC (Radivojević *et al.* 2013) opens the possibility for the potential intentionality of the Zengővárkony copper-tin slag. In terms of cultural significance, Glumac and Todd (1991: 15) argue that copper smelting might have had a ritual role for the community buried at Zengővárkony. This view is supported by the discovery of more copper smelting debris in the early 4th millennium BC Lengyel culture burial site of Brzec Kujawski, central Poland.

In light of this existing evidence for smelting copper ores, *The Rise of Metallurgy in Eurasia* project retrieved, recorded, spatially mapped and analysed each fragment of copper slag excavated at Belovode and Pločnik as well as any associated finds and features (see Chapters 11 and 26). In addition, experimental reconstructions of copper smelting processes were performed to evaluate different interpretations of the existing evidence in relation to the creation of metal. Whilst these will be published elsewhere, the experimental reconstructions have provided strong evidence in relation to the extensive and co-ordinated labour and the material and pyrotechnological expertise required to reproduce early copper smelting. The results from the excavated slag analyses and experimental reconstructions enable the multiple debates surrounding the technology and organisation of copper smelting spanning Research Questions 1, 3 and 6 to be addressed.

Ceramic and metal pyrotechnologies

It is important to note that the smelting of copper ores was by no means the earliest application of pyrotechnology in either the Balkans or Anatolia. The transmission of ceramic forms and pyrotechnology from Anatolia to the Balkans occurred from *c.* 6600 BC, with ceramic production and consumption subsequently being extensively practiced and developed by early farming communities (Amicone *et al.* 2019; de Groot 2019; Spataro and Furholt 2020). Given that this process started around 1500 years before the earliest evidence for metallurgy in the Balkans or elsewhere, it leads us to the issue of the interdependence of pottery and metal pyrotechnologies.

The most common question with regards to this relationship is whether the ability to create and manage high temperatures (exceeding *c.* 1000 °C) could have led to the transformation of copper ore to copper metal. Earlier studies of Vinča pottery indicated that potters were not achieving temperatures beyond *c.* 900 °C (1083 °C is the melting temperature for copper) (Kingery and Frierman 1974: 204–205). Compositional analysis of the Vinča culture pottery revealed that all fine, medium and coarse fabrics were made of low calcareous clay (less than 6% CaO) and was normally fired under reducing conditions below 800 °C (Maniatis and Tite 1981: 73). In contrast, Renfrew's (1969) suggestion of a direct connection between the production of graphite painted ceramics and the invention of copper metallurgy provided a new explanatory framework. This pyrotechnological transfer model was subsequently also advocated by Gimbutas (1976a); however, this claim was not investigated from a pyrotechnological comparative perspective until *The Rise of Metallurgy in Eurasia* project nearly four decades later.

Frierman (1969) reported a two-step process of firing graphite-painted pottery, broadly similar to the two-step process of the earliest metal smelting reconstructed by Radivojević and colleagues (2010b: 2777). Specifically, experiments showed that graphite burns at 725 °C in an oxidising atmosphere, leading Frierman (1969: 43) to assume that pots coated with the graphite slip were fired in an oxidising atmosphere up to *c.* 500 °C or 600 °C, after which the atmosphere for the remainder of the firing had to be strongly reducing over a prolonged period to preserve the graphite. The use of a slow firing process under the reducing conditions is further corroborated by the evenness and the black colour of the resulting surfaces. This is important to the broader debate as the principle of two-step firing also applies to the reduction of copper ores to copper metal, however in reverse order: chemical reduction from ore to metal requires reducing conditions and relatively low temperatures from *c.* 700 °C upwards (Budd 1991), while the melting of the copper metal requires temperatures in excess of 1080 °C but has fewer constraints on the redox conditions. Graphite use and decoration principles in the Late Neolithic and Chalcolithic Balkans have been extensively documented (Chokhadzhiev 2000; Gaul 1948; Leshtakov 2005; Todorova 1986; Todorova and Vajsov 1993). Whilst cones of graphite were used to decorate pottery (cf. Gaul 1948: 98; Ryndina and Ravich 2000: 16–17), the possibility of graphite-rich moulds being used for metal casting is also speculated, arguing that craftspeople understood the protective role of graphite against oxidation of freshly cast metal. *The Rise of Metallurgy in Eurasia* project specifically sought to address the debates surrounding the interdependence of pottery and metal technology in Research Question 2, through a dedicated PhD thesis. This analysed the excavated ceramics from Belovode and Pločnik in order to evaluate the Vinča potters' pyrotechnological skills, especially the temperatures achieved and the control of the firing atmosphere conditions, and the pyrotechnological technologies involved in graphite decoration (Amicone 2017; Amicone *et al.* 2020b) (see Chapters 14, 29, 43).

Metals and metal objects

In contrast to the fragmentary and copper-orientated mining and smelting evidence across the Balkans from *c.* 5000–3700 BC, recent research has produced a far greater quantity and quality of data relating to the creation of different metals by smelting, melting and alloying and of different forms in those metals by casting, annealing and cold/hot working.

Copper

Artefacts made of native copper appear in the Balkans only at the start of the Late Neolithic (*c.* 5500–5000 BC), but most have been only relatively dated and their cultural provenience is debatable. One out of three such artefacts, a fragmented copper object from Iernut (Horedt 1976; Lazarovici 1979, 2014; Mareş 2002), a site located deep in the Carpathian Mountains in Romania (Figure 1), has been ascribed to the last phase of the Starčevo-Criş-Körös phenomenon (mid-6th millennium BC). A 14 cm-long double pointed awl, discovered at the site of Balomir (Figure 1), is the earliest identified implement made of native copper in the Balkans (Vlassa 1967: 407, 423, Figure 6). It is relatively dated to the mid-6th millennium BC, around the same time as a fishhook from the site of Gornea in the Danube Gorges (Lazarovici 1970: 477). While it is challenging to distinguish between the use of native copper and that made of smelted copper ores, Pernicka (1990) argues that increased concentrations of cobalt (Co) and nickel (Ni) are a useful indicator of copper artefacts made of smelted copper. He synthesised Balkan and Anatolian copper metal artefacts trace element data and compared these to the analyses of native copper from the mentioned regions. The Co and Ni concentrations in native copper (approximately <20ppm) are extremely low in comparison to much higher concentrations of these elements in both Balkan and Anatolian copper metal artefacts (Pernicka *et al.* 1997: 124, 159–160, Figure 23, Table A3a). Interestingly, a few copper implements from Pločnik show borderline concentrations of Co and Ni (Pernicka *et al.* 1997: 147–148, Table A1), which might indicate their origin to be native copper. There is no evidence for the exploitation of native gold or silver in this period, despite the geological potential throughout the Balkans (e.g. Jovanović 2001).

Copper metal jewellery and small tools appear alongside malachite beads and pendants in the early 5th millennium BC in the Balkans. These are usually found in settlements and cemeteries located in the lower Danube basin and further towards the northern Black Sea coast, such as Gomolava (Brukner 1977), Gornea (Lazarovici 1970: 477), Cernavodă (Berciu 1967: 53) and Izvoare I in Romania (Vulpe 1957: Figures 72/3; 85/5,6), or Lukavrublevetskaya on the Dniester (Bognár-Kutzián 1976: 71; cf. Bibikov 1953) (Figure 2). Also, a copper metal bead from the site of Dikili Tash I in northern Thessaly is speculated to be made either of native or of smelted copper (Séfériadès 1992a: 114). Noteworthy is the unique context of Gomolava metal found with some of the deceased in this male-only cemetery dated to 4700–4650 BC, including copper beads buried with an infant. Ancient DNA analysis has shown that the individuals in the cemetery are of the same lineage, prompting assumptions of copper metal related to an inherited status in the case of the infant (Brukner 1980; Stefanović 2008).

The earliest *smelted* copper metal implements originate from the sites of Pločnik in south Serbia, Slatino in western Bulgaria, Devebargan-Maritsa in northern Thrace and Durankulak on the Black Sea coast (Figure 2) (Pernicka *et al.* 1997: 48, 72, 131, Table A1; Radivojević 2012); of these, only Durankulak is a cemetery, while the others are settlements. The difference between the artefacts made from native copper or smelted copper ores lies in the trace element analyses; while objects made of the latter contain relevant readings of cobalt and nickel as discussed above, the concentrations of these elements in the former are close to non-detectable (Pernicka 1990; Pernicka *et al.* 1997). In burials, copper implements were usually accompanied by *Spondylus* and *Dentalium* beads, or bone, clay or marble figurines, as in the Devnja cemetery in the Bulgarian Black Sea coast (Todorova-Simeonova 1971: 23–25). One of the most impressive collections of massive copper implements comes from Pločnik, where 38 copper metal artefacts were discovered (Antonović 2014a; Grbić 1929; Radivojević and Kuzmanović Cvetković 2014; Šljivar *et al.* 2006; Stalio 1964). These include: four hammer-axes of Pločnik type, 25 chisels, a copper ingot bar and a pin, altogether weighing *c.* 16 kg. They are a unique and exceptional assemblage of early copper metal and, based on the most recent AMS dating of the context of a fragmented copper chisel to *c.* 5040–4840 BC (Radivojević and Kuzmanović Cvetković 2014: 23; Whittle *et al.* 2016), they are one of the earliest in this part of Eurasia. Seventeen copper metal tools from Pločnik were studied for their chemical composition and lead isotopes (Pernicka *et al.* 1993), revealing an unexpected complexity of ore/metal exchange networks. At least three different copper deposits from eastern Serbia, Macedonia and across Bulgaria provided metal for their production. The only closely comparable collection is that from the Rakilovci hoard in western Bulgaria, where a total of nine copper metal implements were recovered from a ceramic pot (Mihaylov 2008).

A further exceptional copper metal assemblage comes from the site of Karbuna, in Moldavia: among 852 precious objects discovered as a hoard in a typical Tripolye A pot (*c.* 4700–4600 BC); 444 were made of pure copper (Dergachev 2004; Sergeev 1963: 135; Videiko 2004). Significantly, the hoard included two massive copper implements, one being a hammer axe of Pločnik type (broadly dated to the early to mid 5th millennium BC) (see also Diaconescu 2014). The considerable volume of the find and distinctive typology of its contents suggest close associations with contemporary cultures in both Serbia and Bulgaria (Chernykh 1991: 581, 587). Chernykh (1966: 53–58, 86–88, 1978b: 122; 1991: 387, 581) argues that the Karbuna hoard metal could have come from Ai Bunar, while Pernicka speculates that it might have derived from Medni Rid in eastern Bulgaria, since artefacts from northeastern Bulgaria, southeastern Romania and further to the northeast fit well with the compositional pattern of Medni Rid (Pernicka *et al.* 1997: 141). Interestingly, based on its distinctive chemical signature, this metal was probably recycled and traded further north towards the Volga valley and into the steppes (Chernykh 1991: 587–588, Table 5).

In the northwestern Balkans, the early appearance of copper artefacts is more modest and accumulated mostly in the Great Hungarian Plain and Transdanubia, or the Tisza-Hérpály-Csöszhalom group, Železiowce and Lengyel cultures (Ecsedy 1990: 220; Scharl 2016). Copper jewellery, awls and chisels come from the sites of Mlynárce (Novotný 1958: 28), Neszmély, Csöszhalom or Hódmezővársahely-Kopáncs-Kökénydomb (Bognár-Kutzián 1963: 331–333), all located along or near the major rivers in this area. Metal artefacts are also recorded at sites located along the Tisza River and closer to the Carpathian Mountains, such as the settlements of Hérpály, Berettyószentmárton and Zsáka-Markó, or the Gorsza cemetery (Bognár-Kutzián 1963: 331–336, 487).

During the second half of the 5th millennium BC, the production of massive copper metal implements flourished in eastern Bulgaria, in contrast to the central Balkans, after the collapse of the Vinča culture. This can be followed archaeologically from the mid-5th millennium BC Hamangia IV phase (Table 1) (Boyadžiev 2002: 67), when mass metal consumption is reflected by the exceptionally rich grave goods recovered from burials in Varna and Durankulak, including both copper and gold objects (Ivanov 1978a, 1978b, 1988a, 1988b;

Ivanov and Avramova 2000; Todorova 2002a). Massive copper implements and gold decorations were also found in settlements, for example at Hotnica, Ruse, Kasla-Dere, Gumelniţa or Vidra, all set along or in the hinterlands of the lower Danube (Bognár-Kutzián 1976: 71; Gimbutas 1977: 44).

Comprehensive typological schemata have been developed to track the appearance of specific types of massive copper implements. Changes in the morphology of hammer-axes are particularly interesting as they appear to be related to specific regions across the Balkans. For instance, hammer axes of the Pločnik type are generally associated with the Vinča culture (these start to appear in the Early Chalcolithic), the Vidra type with the north central Bulgarian sites (associated with the KGK VI), while the Čoka-Varna type is characteristic of the northeastern Bulgarian sites (Varna culture) (Antonović 2014ac; Chernykh 1978b; Govedarica 2001; Kuna 1981; Novotna 1970; Radivojević 2006; Schubert 1965; Todorova 1981; Vulpe 1975; Žeravica 1993). Importantly, lead isotope analyses of the late 5th millennium BC Vidra and Čoka-Varna hammer-axes showed that they were made of the same metal (Pernicka *et al.* 1997: 94–98, 105–106, 142, Table 3), indicating that there was no relationship between a metal source and the tool type. The strong preference for a specific tool type regardless of source potentially suggests that particular technological choices reflect the identity of a producer or consumer group.

Scholarly debates regarding Final Chalcolithic period in the Balkans (*c.* 4100–3700 BC) have traditionally been dominated by narratives of a societal collapse in eastern and central Bulgaria as indicated by a substantial reduction in archaeologically visible (and dated) settlements (Kienlin 2010; Weninger *et al.* 2009). Metal provenance data also support this interpretation with a noticeable shift in copper supply networks from the eastern to the central Balkans, where it not only continues but also intensifies with the exploitation of novel sources in the Carpathian Basin (Pernicka *et al.* 1997; Schalk 1998; Siklósi et al. 2015; Siklósi and Szilágyi 2019). The presence of increasing numbers of copper objects in the northern Alpine region (cf. Bartelheim 2007; Cevey *et al.* 2006; Kienlin 2010, 2014; Klassen 2000; Scharl 2016; Turck 2010), as well as throughout the neighbouring central Mediterranean region (Dolfini 2013, 2014) provides evidence for the emergence of other copper industries outside the 'core' Balkan region but still, however, associated with the Balkan sources (Höppner *et al.* 2005).

Copper production rapidly changes in the late 5th / early 4th millennium BC with metal production re-emerging in the central and northwestern Balkans, shown in the increase of metal consumption in the Bodrogkeresztúr culture and intensified exploitation of eastern Serbian and western Bulgarian copper sources (Pernicka *et al.* 1997: 98–101, 105–106, Table 3). As a consequence, the earliest metal using cultures emerge north of the Alps, such as at Mondsee or Pfyn (Kienlin 2010; Krause 2003; Ottaway 1989). In contrast to the quantity of copper implements, production evidence is extremely rare and understudied. Cultures of the late 5th millennium BC Great Hungarian Plain produced the first known metal knives, and massive copper implements are found in both settlements (e.g. Lucska) and cemeteries (e.g. Tibava) (Bognár-Kutzián 1972: 140; Hansen 2013b; Šiška 1964: 7 ff.; Todorova 1995: 90). One of the most exceptional collections of metal artefacts in this region comes from the late 5th millennium BC

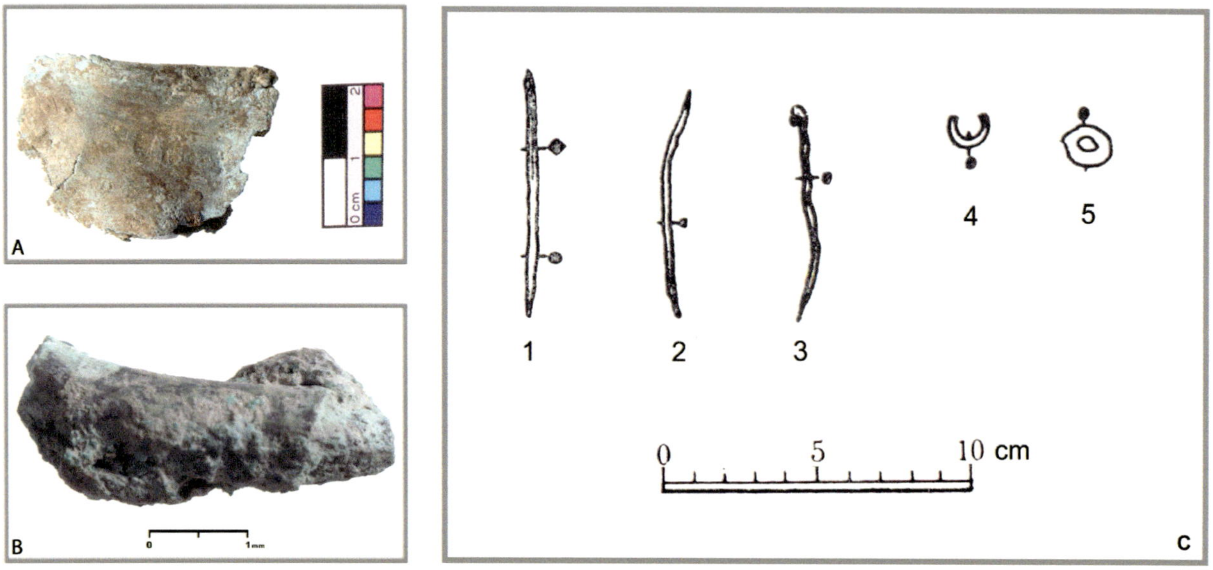

Figure 5: A selection of the 5th millennium BC tin bronze artefacts from the Balkans. a) Pločnik foil; b) Gomolava ring; c) 1- Bereketska Mogila, 2- Gradeshnica, 3, 5-Ruse, 4-Karanovo (adapted after Chernykh 1978b: Tables 15/24,42; 18/30; 19/4,7; Radivojević *et al.* 2013).

Lengyel culture cemetery of Zengővárkony (Figure 3), where a large number of spiral copper metal bracelets, rings and malachite beads were deposited as grave goods (Dombay 1939: 50–64; Dombay 1960: 75–144; Ecsedy 1990: 212–218). Of note are the cemeteries, like Rákóczifalva-Bagi-föld, where the uneven distribution of grave goods, including copper and gold objects, potentially indicates a degree of social inequality (Csányi *et al.* 2009) as evaluated across the Eastern Carpathian Basin by Siklósi (2013).

Tin bronze

The recent excavation and archaeometallurgical analysis of a tin bronze foil at Pločnik from a secure context radiocarbon dated to *c.* 4650 BC (Radivojević *et al.* 2013) revealed the emergence of tin bronze metallurgy at this time. The compositional analyses of the Pločnik tin bronze foil indicated that copper ore including stannite (Cu_2FeSnS_4), a copper-tin bearing mineral, or its secondary weathering products was the probable raw material used for making this natural alloy with *c.* 12wt% Sn and relevant traces of As, Fe, Co and Ni (see Radivojević *et al.* 2013: 1035, Table 1). This means that the earliest known tin bronze artefact was not made by alloying two elements (copper and tin) but rather by smelting a copper-tin bearing ore.

There are 14 additional tin bronze artefacts known from the mid-late 5th millennium BC, however these finds only occurred together in what appears to be a short-lived tin bronze horizon in the Balkans based on geochemistry that links them with the Pločnik foil. Twelve finds originate from the Bulgarian sites of Ruse, Karanovo, Gradeshnitsa, Smjadovo, Zaminec and Bereketska Mogila (Chernykh 1978b; Pernicka *et al.* 1997), and two from the Serbian sites of Gomolava and Lazareva Cave (Ottaway 1979; Tasić 1982; Glumac and Todd 1991: 15). The assemblage of awls, rings, needles, borers, and a rod from Bulgaria and Serbia (Figure 5) has tin concentrations ranging from 1–10 wt%, with consistently significant levels of lead, arsenic, nickel, cobalt, iron and gold (Chernykh 1978b: 112, 339, 342–343, 351–352, Tables 15/24, 42, 18/30, 19/4,7; Pernicka *et al.* 1993: 190, Table 3; Pernicka *et al.* 1997: 70, 121–126, 156, Table A1).

In terms of context, the majority of these finds remain under question with the exception of a borer from Ruse that originated from the secure, primary context of a child's burial (Glumac and Todd 1991: 15; Pernicka *et al.* 1997: 125–126). Despite having different chemical compositions, these tin bronzes typologically match contemporary regional counterparts in pure copper. Yet, their form is culturally and chronologically non-distinctive, thus offering little information about their exact provenance. The Pločnik tin bronze foil was therefore crucial in determining their chronology based on a common unique chemical signature (Radivojević *et al.* 2013); this, and the fact that no other tin bronze artefacts are known in the Balkans before the 3rd and 2nd millennia BC (Chernykh 1978b; Schickler 1981; Pernicka *et al.* 1997; Pare 2000), make it very unlikely that these early finds are intrusions from later layers.

Another artefact with relevant tin content (*c.* 1.5 wt%) and minute concentrations of silver and nickel in the copper base, originates from a mid to late 5th millennium BC phase in Dikili Tash II (Figure 3) (Sériadès 1992a: 114–115, Table 12). The trace element signature of this object is not, however, consistent with the tin bronze assemblage from Bulgaria and Serbia, although it was probably also made by co-smelting malachite and tin-rich ore. A tin-rich slag piece from the late 5th millennium BC cemetery of Zengővárkony, the only technological debris of its kind at the time, adds more chronological certainty for the production of tin bronzes mentioned above. Yet, the context of this particular artefact remains uncertain (Pernicka *et al.* 1997: 125).

Radivojević and colleagues (2013: 1040) further argued that the golden hue of fifteen 5th millennium BC Serbian and Bulgarian tin bronze artefacts, which contain between *c.* 1 wt% and *c.* 12 wt% Sn, must have been critical to their value, particularly as these artefacts were roughly contemporary with the emergence of the earliest known gold artefacts, unearthed in the cemeteries of Varna and Durankulak in Bulgaria (cf. Avramova 2002; Dimitrov 2002; Higham *et al.* 2007; 2018; Ivanov 1988b; Krauss *et al.* 2017; Leusch *et al.* 2014; Todorova and Vajsov 2001). The importance of the new colour palette at the time has been emphasised in detailed compositional analyses by Leusch *et al.* (2014), which showed that not all gold items in the Varna cemetery had the same shade of yellow. We may assume that the rarity of objects coloured in these new shades in the 5th millennium BC Balkans might have dictated their limited production, but also that demand for them was both social and technological, since tin bronzes in particular disappear with the collapse of the KGK VI and related cultural phenomena at the end of the 5th millennium BC in the Balkans. In contrast to the tin bronzes, there is currently no evidence for arsenical copper objects and their production in the Balkans prior to the early to mid-4th millennium BC (e.g. Antonović 2014a; Chernykh 1978b; McGeehan-Liritzis and Gale 1988; Nerantzis *et al.* 2016; Pernicka *et al.* 1997).

In order to investigate the golden hue argument in greater detail, Radivojević *et al.* (2018) designed a Cu-As-Sn colour ternary diagram based on an extensive set of experiments that yielded 64 binary and ternary metal pellets further exposed to colorimetric analysis.

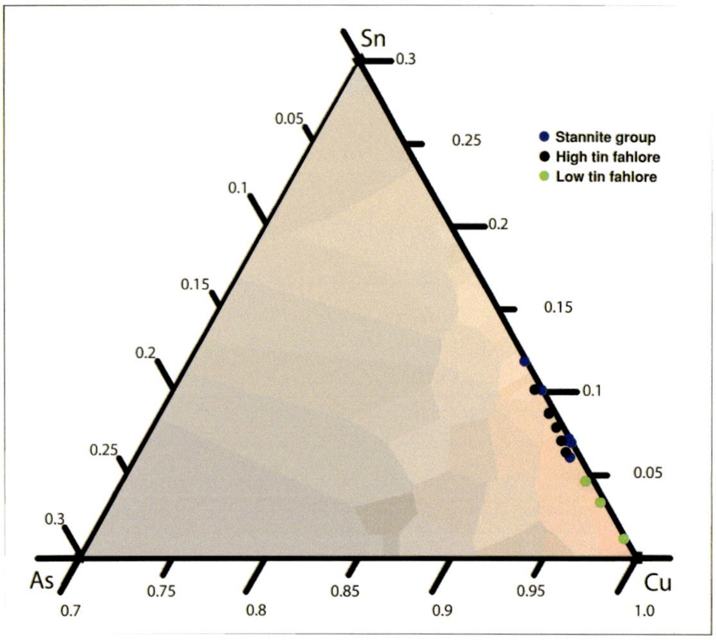

Figure 6: The mid to late 5th millennium BC Balkan bronzes plotted against the Cu-As-Sn ternary colour diagram (100 wt% Cu – 30 wt% As – 30 wt% Sn corner, see Figure 8). Fifteen artefacts split into three groups indicate a variety of colour changes, significantly visible after *c.* 5wt% Sn on the Cu-Sn axis. Data from Radivojević *et al.* (2013: 1035, Table 1); image from Radivojević *et al.* (2018: 118, Figure 12).

The analysis of 15 5th millennium BC Balkan tin bronzes were then plotted on this colour diagram in three distinctive groups, based on the likely mixture of ores other than malachite in the smelting charge: stannite, high-tin fahlore and low-tin fahlore (data from Radivojević *et al.* 2013: 1035, Table 1). Figure 6 indicates that the stannite and high-tin fahlore group (12 artefacts, Sn range between 6 wt% and 12 wt%) had a visibly emphasised golden hue when produced, as opposed to the low-tin fahlore group where, although colour change would have been noticeable, it was not as significant as for those above *c.* 5 wt% (Radivojević *et al.* 2018: 115–118, Figure 12).

It was therefore concluded that the group of 5th millennium BC Balkan tin bronze artefacts, in particular the assemblage of stannite and high-tin fahlore group, must have appeared significantly different, aesthetically, from pure copper artefacts, since the addition of tin increased the golden hue of their resultant colour. With such a different appearance, it is very likely that the production of the 5th millennium BC Balkan tin bronzes was dictated by the demand for the 'exotic' golden colour at the time, or by the pursuit of its closest imitation, which supports the claims in the original publication (Radivojević *et al.* 2013; see also Radivojević *et al.* 2014a). It is also very interesting that the shape of the Pločnik foil (Figure 5a) indicates that it was wrapped around an object, which could have been a pottery vessel or a stone, wood, bone or copper object.

If we seek inspiration for the use of foils at that time (*c.* 4650 BC), metal foils with the same, golden colour (see Figure 7a) are found in abundance in Varna burials. The most notable examples come from the rich burials 36 and 43, including the (infamous) golden 'penis sheath' (Leusch *et al.* 2017).

Ryndina and Ravich (2001: 4, Figure 3) maintain that the provenance of the Balkan early tin bronze artefacts may have been associated with local sources, since their chemical signatures correlate well with those of copper metal that circulated in Transylvania, Hungary and northern Yugoslavia, extending towards Moldavia and Ukraine. Conversely, Pernicka *et al.* (1997: 141) argue that the tin bronze artefacts they analysed did not fall within the isotopic range of the majority of artefacts from the 5th millennium BC. In her doctoral thesis, Radivojević (2012) showed that the provenance of the Pločnik tin bronze foil was highly consistent with the rest of the Pločnik copper implements.

While the provenance of these artefacts remains to be explored in future publications, it is important to emphasise that the information we have assembled thus far speaks in favour of the limited use of tin bronzes across the Balkans in the late 5th millennium BC. Furthermore, it is essential to remember that, although the excavation methodology used in their discovery varied, the excavators were not aware of the relevance of tin bronze objects based on their appearance (the green patina would be similar to that on pure copper objects), and hence they could not have been biased in their recording. If anything, these items were mislabelled as 'usual copper' until chemical analysis showed otherwise, which was initially the case with the Pločnik tin bronze foil. Although this early use of copper-tin ores to make natural alloys has only started to emerge in the literature, special caution is needed regarding claims that involve superficial or rapid analyses and insufficiently elaborated contextual evidence.

Gold

The appearance of thousands of small decorative objects made of gold dates from the mid-5th millennium BC in northeastern Bulgaria, southeastern Romania and northern Thessaly (Makkay 1991; Higham *et al.* 2007; Krauss *et al.* 2017). Although the gold from the cemetery of Varna I is claimed as the earliest known (dated most recently between 4690 and 4330 cal. BC) (Krauss *et al.* 2016), there are earlier uses of gold ornaments (although not as securely dated) in the Varna II cemetery (Todorova and Vajsov 2001: 54), as well as in the cemetery of Durankulak (Avramova 2002: 193, 202, Table 24; Dimitrov 2002: 147). The Durankulak finds are, for instance, dated to the Hamangia IV phase, between *c.* 4650/4600–4550/4500 BC (Bojadžiev 2002: 67). Gold also appears in more modest quantities in sites located in the lower Danube basin: Vidra (Dumitrescu 1961: 80), Hotnica (Jovanović 1971: 37), Traian, Gumelniţa (Dumitrescu 1961: 70–71, 80–81) or deep in the Carpathians, as in Ariuşd (Makkay 1995: 74) (Figure 3).

The most exceptional collection, including *c.* 3100 gold objects (and 160 copper implements), however, comes from the cemetery of Varna I, weighing *c.* 6.5 kg in total (Biehl and Marciniak 2000; Fol and Lichardus 1988; Ivanov and Avramova 2000; Leusch *et al.* 2017). The volume of the collection and the range of techniques applied in its production deserves special attention here. Around 70 of the 320 burials (inhumations, deposits, symbolic/ cenotaph graves) contained gold artefacts ranging from one item to 990 objects (totalling 1.5 kg of gold) in a single burial, no. 43. Of 61 graves with gold artefacts, 34 were symbolic / cenotaph, 10 male, 13 female (?) and 4 disturbed (Biehl and Marciniak 2000: 186). Leusch *et al.* (2017: 112, Table 2) indicate cenotaphs as the richest graves, followed by male and then female burials. It is notable that no comparable range of prestige items and status markers have been found in adjacent settlements, which do not exhibit evidence for structural hierarchies or inequalities. Hence, scholars agree that the Varna cemetery served several local communities of an unspecified scale, rather than just a single settlement (Biehl and Marciniak 2000; Chapman *et al.* 2006; Ivanov 1988b; Lichardus 1991b; Renfrew 1978a).

The Varna gold collection includes a range of decorative artefacts made of small beads, appliques and sheets. Although made of native gold, the varying naturally occurring concentrations of copper and silver in the golden nuggets exploited resulted in golden objects having many different shades of gold (Figure 7d) from white, via yellow, to light pink (Leusch *et al.* 2014:175, Figure 11b). Overall, silver concentrations range between 5 and 45%, and copper between 0.05 and 2.5% (Leusch *et al.* 2017). Leusch *et al.* (2016: 108, Figure 7.8) use the Pt/Pd ratio to discriminate between four different groups of gold in the Varna assemblage (300 objects analysed in total), which may be indicative of discrete geological resources, suppliers or workshops. While any of these scenarios need further research, the recent discovery of placer gold deposits near Varna (Yovchev 2014) points to potential regional resources being exploited at the time.

The artefacts buried in these graves include awls, chisels, cushion stones, stone adzes, flint scrapers, hammer axes and antler tools. A sound case has been made that these might have been the tools of artisans. The deposition of such items alludes to the significance of artisans and crafting for the community at Varna (Leusch *et al.* 2017: 118). Anthropological analysis of one of the richest burials, no. 43, has shown that the male individual, aged between 50 and 65 years, had pathological conditions related to squatting and hard work, particularly to great robusticity of the lower arm muscle attachments, which supports the interpretation of this individual as an artisan or craftsperson rather than as 'royalty' (Leusch *et al.* 2017).

To further contextualise the paraphernalia related to crafts at Varna I, 122 out of 226 burials have items identified as tools that have never been used. These tools are as common as any other object deposited in the burials. Two potential imitations of objects are also present in the collection, adding to the assemblage of artisan tools: a copper pick (imitation of an antler pick?) and a golden 'penis sheath' (a likely imitation of a tuyère?) (Leusch *et al.* 2017: 107, Table 1). The latter has been famously claimed as unearthed between the thighs of the individual in Burial No. 43, which led to its interpretation as a penis sheath. However, its original position was beside the right thigh (Biehl and Marciniak 2000: 186; Ivanov 1988b: 55, Figure 25; Leusch *et al.* 2014: 168, 177, Figure 4a). An alternative interpretation, that it was an imitation or gilding of a tuyère, has typologically close parallels with clay imitations from across sites in Bulgaria (Kubrat, Goljamo Delčevo), Romania (Pietrele, Radovanu) and Serbia (Bubanj, Kmpije) (Figure 4e-f) (Bulatović 2015: 12, Table II/13; Comşa 1990; Hansen 2009; Lichardus 1988; Lichardus 1991a: 174; Todorova 1982). The idea of gilding is equally interesting, given that this golden object had two perforations at the base, indicating that it was stitched to another item, hence potentially serving as an ornament. Leusch *et al.* (2017: 114) claim that the item could not have been a tuyère imitation since the output vent has a wider diameter than the clay models; nevertheless, imitations do not need to be exact copies. Finally, if the item was used as a foil decoration for clay tuyères, it would fit well with the practice of working with gold foil in the Varna cemetery (see Figure 7a) (Leusch *et al.* 2015).

Careful examination of a total of 300 golden objects analysed within the Varna-project (led by E. Pernicka) revealed different shaping techniques applied with hammers, punches and doming blocks, chisels used for chasing and parting, conical points for perforations, and sand, stones, ashes and siliceous plants used for finishing and polishing. Little is known, however, about the production debris of gold making. Similarly to native copper, native gold would not produce any slags. Casting equipment required a similar set of tools to those needed for copper processing:

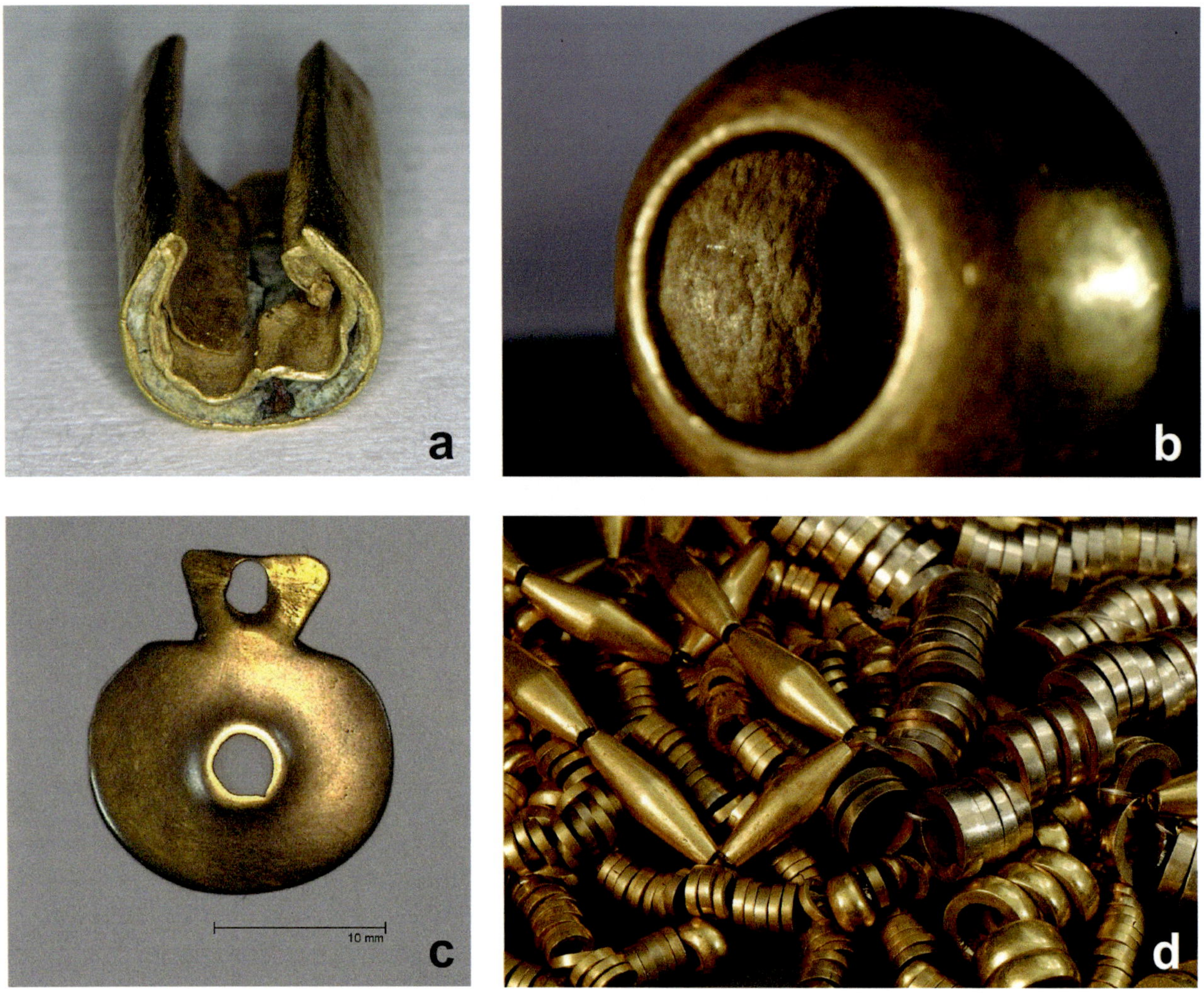

Figure 7. A selection of gold objects from Varna. a) Sheet-gilded copper bead from burial no. 41; b) Gold bead from burial no. 4 with a hollow body made with lost wax casting technique; c) The ring-idol from grave no. 271 is the earliest known gold-copper alloy (*c.* 50 wt% gold, 14 wt% silver, and 36 wt% copper); d) Gold beads with different shades of gold due to the variable silver content. The silvery beads (top right) from grave no. 43 contain on average 58 wt% gold, 40 wt% silver, and 2 wt% copper (adapted after Leusch *et al.* 2014: 167, 175, Figure 3a, 10b, 11a-b; c_CC BY-NC-ND 4.0 by B. Armbruster and V. Leusch).

crucibles, casting moulds, hearths and tuyères (Leusch *et al.* 2015). The exquisite craftsmanship required for making these objects is showcased using techniques borrowed from copper working, together with complex casting techniques, to produce three of the world's first examples of alloying, gilding and lost-wax casting (Figure 7a-c). A small group of gold-copper alloys was found to contain copper content exceeding *c.* 30 wt%, which is significantly higher than the naturally occurring concentrations within native gold (Hauptmann *et al.* 2010) and hence implies intentional alloying (ring idol example in Leusch *et al.* 2014: 175, Figure 11a) (Figure 7c). A copper bead from grave no. 41 was sheet-gilded (Figure 7a) (Leusch *et al.* 2014: 167, Figure 3a), probably to bring up the much sought-after golden colour, while a hollow and solid globular bead from another burial was produced using a lost-wax technique (Figure 7b) (Leusch *et al.* 2014: 175, Figure 10b). This bead is the earliest known record of a lost wax cast object and predates the spoked wheel shaped native copper amulet from the site of Mehrgarh (Pakistan) by as much as 500 years (Thoury *et al.* 2016). The amulet came from the Early Chalcolithic horizon on this site broadly dated between 4500 and 3600 BC, the authors settling on *c.* 4000 BC as the likely date for the emergence of lost wax casting in the far eastern end of the Iranian Plateau.

The mastery of gold production did not only include the production of gold objects, but also extended to the decoration of non-metal objects (like pottery) with gold. Éluère and Raub (1991: 13) investigated the technology of gold coating on a large plate recovered from one of the rich Varna graves, and showed that its gold layer consisted of natural Au-Ag alloy with *c.* 7% Ag, and some copper. After coating, no polishing tools were used, as this may have removed the gold. Éluère and Raub (1991: 19) speculated that washed (alluvial) gold dust was applied onto a plant glue which covered a ceramic surface in a process called sintering, which welded together particles without requiring a liquid stage (Raub 1995: 247–248). The tradition of decorating pottery with gold extends into the Krivodol-Salcuța-Bubanj Hum complex in southwestern Bulgaria / southeastern Serbia, continuing well into the first centuries of the 4th millennium BC (Bulatović *et al.* 2018; Gajić-Kvaščev *et al.* 2012).

During the late 5th to early 4th millennium BC, the production of gold artefacts shifted towards the west Carpathian Basin, where gold pendants and decorations appeared within the late Tiszapolgár, Lasinja and Bodrogkeresztúr cultures (Dumitrescu 1961: 92–93). Gold ornaments of varying size were deposited as grave offerings in the cemetery of Tibava, in Slovakia (Šiška 1964: 332) or in hoards, as at Moigrad, in Romania (Dumitrescu 1961: 71) (Figure 3). Gold metal from this period amounts to a total of *c.* 5–6 kg of extant objects (Makkay 1991: 119–120); of these the most impressive is the heaviest golden object currently recorded from the Balkan Chalcolithic, a 31 cm-diameter disc from the Moigrad hoard that weighs *c.* 800 g (Makkay 1989).

Silver and lead

Objects made of silver emerge in parallel to those fashioned from gold, although to a lesser extent. Only a few pieces, of unknown context, originate from the Carpathians (Makkay 1991), while in Greece, hundreds of small items of silver jewellery have been found (Maran 2000; Muhly 2002). The richest find is a hoard from the Alepotrypa Cave (Figure 3) on the Mani peninsula in Greece, dated roughly between the mid-5th and early 4th millennium BC (Muhly 2002: 78; Papathanasiou *et al.* 2018); other sites with silver ornaments were discovered in the islands of Crete and Lemnos. One of the large silver pendants from the Alepotrypa Cave has a distinctive shape: it is circular, with a central perforation and a pierced suspension tab; as such, it resembles a slightly earlier golden counterpart from the cemetery of Varna. There is no contemporary evidence for silver production, with the earliest evidence for litharge fragments coming from Limenaria, Thassos and northern Greece, dating to the early 4th millennium BC (Nerantzis *et al.* 2016; Papadopoulos 2008).

The earliest processing of lead ore in the Balkans is documented at the site of Belovode, where a large slag 'cake' (Figure 8) was recovered from an undisturbed and secure context associated with 5200 BC (Radivojević and Kuzmanović Cvetković 2014; Šljivar *et al.* 2012). Microstructural analysis conducted by the first author of this chapter reveals well formed—and once molten throughout—fayalitic slag with magnetite, matte inclusions and droplets of lead metal, which suggests the use of complex lead ore and would require temperatures in excess of 1100 °C. Most importantly, it could not have been made by chance (Radivojević and Rehren 2019). While there are no preserved lead objects known currently from this site, the only contemporary evidence in the broader 'Old World' sphere is the lead bracelet from layer 12 at Yarim Tepe I in Iran (Merpert and Muncaev 1987). The results of chemical (qualitative) analyses conducted by E.N. Chernykh (Merpert and Munchaev 1972) speak of pure lead metal as the base, some silver and traces of iron. This suggests the use of a lead ore of high purity, like galena, or native lead which is very rare (Patterson 1971). However, without an exact quantification of the silver content it is difficult to say which type of lead ore was used. Tylecote (1962: 76) reported that lead can be smelted easily from galena by a 'simple fire', which possibly refers to the melting point of lead at 328 °C. Interestingly, the wider Levant / Northern Syria region hosts some of the earliest known lead objects in the world, at least from the late 5th millennium BC onwards (cf. Yahalom-Mack *et al.* 2015).

The use of lead minerals (for beads) has also been documented at the Vinča culture sites of Autoput,

Figure 8. Slag 'cake' from the site of Belovode, eastern Serbia, discovered in a context dated to 5200 BC. Compositional analysis revealed lead metal to be the likely product of the smelt (photo CC BY-NC-ND 4.0 M. Radivojević).

Selevac and Opovo in Serbia and Donja Tuzla in Bosnia, in all cases in horizons that end in 4500/4400 BC at the latest (Glumac and Todd 1987; Vogel and Waterbolk 1963; Quitta and Kol 1969). As such, these artefacts, together with the lead slag, predate the use of lead ores at the site of Pietrele (set at *c.* 4400–4300 BC), erroneously claimed as the first and only evidence of lead ore processing in the Balkans (Hansen *et al.* 2019). The biconical crucibles in question are a very interesting find and are apparently present in at least two Chalcolithic Romanian sites besides Pietrele: they are small biconical objects (*c.* 6 cm in diameter on average) with a narrow opening at the top, yet with inconsistent traces of heating across the discovered assemblage. The purpose of these crucibles is yet to be resolved, as it remains unclear what the smelting of galena (PbS) in such a way produced. Hansen *et al.* (2019) dwell on the possibility of manufacturing a colouring agent, a yellow or red lead oxide, which would fit with the earlier practice of painting pottery in the Vinča culture (e.g. Gajić-Kvaščev *et al.* 2012; Mioč *et al.* 2004).

Miloje Vasić described two interesting situations that might have indicated the presence of smelting (lead) installations at the site of Vinča-Belo Brdo. The first refers to finds from 1913, when several ellipsoid-shaped shallow pits were discovered within a small area at ▼8.1 to ▼8.9 m (this translates into Vinča A phase in this settlement, *c.* 5300–5200 BC), the largest being 2.1 x 0.5 x 0.1 m in size (Antonović 2002: 35–36, note 60, Figure 3). Their walls were *c.* 8 cm thick, and they were intensely fired only in the centre and filled with soot and ash in the bottom. A galena bead was identified in the vicinity of one of these features. Similar shaped shallow installations were used for lead smelting in the village of Vinča (near Belgrade) in the early 20th century; this prompted Vasić to propose a similar function for these pits (cf. Antonović 2002).

Conclusion

In light of this existing evidence for metals and metal objects, *The Rise of Metallurgy in Eurasia* project retrieved, recorded and spatially mapped each metal object excavated at Belovode and Pločnik, as well as any associated finds and features. All metal objects were subsequently analysed to determine their metal composition, metal provenance and techniques of manufacture. Given the earlier discoveries of copper and tin bronze objects at Pločnik and copper smelting slag and lead ore at Belovode, it was important to explore, where possible, where the different metal production evidence and objects were placed on each site and how these may have been organised. This pertains especially to Research Questions 1, 5 and 6.

Metal circulation

Analyses of metal objects and the large scale of copper production during the 5th millennium BC prompted Chernykh (1978b, 1992, 1997, 2008b, 2008a) to define the Carpatho-Balkan Metallurgical Province (CBMP) as a distinctive (and the earliest) technological and cultural entity, from where metallurgical knowledge was carried eastward in staged migrations over the following *c.* 4000 years. 'Metallurgical Provinces' (MPs) represent large interconnected systems of shared metallurgical technology, trade and exchange, which encompassed areas of up to a few million km^2 across Eurasia, and which lasted for a few thousand years. On a practical level, they were linked through: i) shared utilisation of morphologically defined ornaments and implements; ii) common principles of metalmaking with the availability of or access to the same ore resources; and iii) comparable dating. Chernykh (1992: 7) goes further, making a fine distinction between metallurgical *province* and metallurgical *foci*, the latter of which refers to smaller-scale regions where similar metal artefacts were produced by a group of skilled craftsmen over a certain period. The current understanding of the extents of metallurgical provinces currently relies on the growing database of compositional analyses (nearly 120,000; see Chernykh 2008a) and associated datable materials from between the Adriatic and the northern forests of Mongolia. As such, the MPs are detached from the concept of culture, and may encompass an area of up to 8 million km^2 and endure over long periods of time.

The CBMP area included several cultural phenomena in the northern Balkans and the Carpathian Basin related to the emergence and spread of copper metallurgy during the 5th millennium BC, and most notably around the mid-5th millennium BC, which is defined as the 'metal boom' phase by Chernykh (1978b; 1991; 1992). This province spanned *c.* 1.3–1.4 million km^2 at the peak period of metal production. Chernykh (2008b: 76) distinguished three groups of the 5th and early 4th millennium BC metal-producing and consuming cultures. The first, 'core' group (Butmir, Vinča C/D, Lengyel, Karanovo V-Maritsa, KGK VI, Varna, Tiszapolgár, Bodrogkeresztúr cultures, see also Figure 10), broadly includes the central, eastern and northern Balkans and spans over *c.* 500 years across an area of 0.75–0.8 million km^2. A second group is represented by the Tripolye-Cucuteni culture, which extended from the Carpathian Mountains to the Dniester and Dnieper regions, with centres in modern-day Moldova and western Ukraine, and occupied *c.* 0.16-0.18 million km^2, while a third group consists of communities occupying the steppes to the north and northeast of the Black Sea coast.

Several technological features emerge as common to the cultures within reach of the entire CBMP area: a similar set of classes and types of products, similar technology of working and (pure copper) metal composition (Chernykh 1978b, 1992; Ryndina and Ravich 2000, 2001). The most recent technological and metallographic study showed that the massive copper implements from the Vinča culture sites of Pločnik were worked in the same way as those from KGK VI and Varna sites in northeastern Bulgaria (Radivojević 2012), confirming the existence of a shared technological principle (or recipe) for metal working across the Balkans, in place from the very beginnings of the 5th millennium BC. Radivojević (2012) further observed that the shared metallurgical tradition, mirrored in the specific technique for finishing the massive copper implements across the Balkans reveals that the network of metalsmiths was resistant to various cultural collapses (like Vinča or KGK VI), and that it probably existed outside the remits of archaeological cultures as defined by distinctive material traits. The study and subsequent publications relating to the invention, innovation and cultural transmission of metallurgical knowledge in the 5th millennium BC Balkans (Radivojević 2015; Radivojević *et al.* 2013; Radivojević and Kuzmanović Cvetković 2014; Radivojević and Rehren 2016) support the concept of the metallurgical province as an entity independent of particular cultural phenomena, and highlights shared technological knowledge as the key to understanding the social dynamics of this period. This concept needs further probing in relation to (extractive) production and all aspects of the metallurgical *chaîne opératoire* in order to interpret the nuanced detail of the knowledge transmission.

Extensive programs of compositional analyses indicate that the 5th millennium BC metal artefacts in the Balkans were made of almost pure copper (e.g. Chernykh 1978b; Junghans *et al.* 1968; Radivojević and Grujić 2017; Pernicka *et al.* 1993, 1997; Radivojević 2012; Radivojević and Grujić 2018: Table S1), which is why the trace element signature, along with the lead isotope analyses, have proved particularly useful for indicating plausible sources of metal. The Early Chalcolithic period was dominated by sources in eastern Serbia, probably Majdanpek, although other outcrops in this region, like Ždrelo, could also have been exploited (Radivojević *et al.* 2010a). Bulgarian sources, like Ai Bunar, become active only towards the mid-5th millennium BC, and are associated with the earliest copper implements from southern and northeastern Bulgaria (Chernykh 1978a; Pernicka *et al.* 1997: 93, Table 3;). An important point arising from the available provenance data is the existence of multi-producer and multi-consumer networks of copper from the early stages of metallurgical development, as is the case with the Vinča culture sites of Belovode and Pločnik. While provenance analyses of copper slags from Belovode indicate trace elements highly consistent with 16 Chalcolithic copper metal implements found mostly along the lower Danube, similar analyses of copper implements from Pločnik point to a minimum of three different copper deposits that provided metal for their production (Pernicka *et al.* 1993, 1997: 93–94, 105–106, Table 3; Radivojević *et al.* 2010a).

Provenance (lead isotope and trace element) analyses of several hundred copper artefacts from the mid to late 5th millennium BC indicate the use of local Balkan sources, amongst which the signature of Ai Bunar was predominant (Pernicka *et al.* 1997: 117, Figure 20, Table 3). All copper artefacts analysed by Pernicka *et al.* (1997) were assigned to ten distinctive lead isotope grouplets each relating to a particular deposit, a group of spatially tight deposits or to the same geochronological unit (not necessarily spatially close). These grouplets are therefore not sufficiently well characterised to allow predictions of the exact location of origin of the smelted metal. The information on grouplets is then paired with that for clusters (derived from clustering of trace element signatures) and used together with archaeological information to ensure the best estimate of metal provenance (Pernicka *et al.* 1997). Given the widespread presence of copper metal implements from various copper deposits across this region, we may assume that these local sources were shared among different cultural groups. There is, indeed, a prevalence of KGK VI material culture in the ancient mines of Ai Bunar, however the distinctive chemical signature of this source is found in nearly one quarter of Middle-Late Chalcolithic copper objects analysed thus far.

Another distinctive provenance signature is ascribed to Majdanpek in eastern Serbia, although this deposit was more intensively exploited in the Early and Final Chalcolithic. Pernicka *et al.* (1997) observed large shifts in copper supply throughout the Balkan Chalcolithic in the provenance data. For example, the copper in a significant number of analysed artefacts from the first half of the 5th millennium BC originates from the Majdanpek copper field. It is almost absent from the Middle and Late Chalcolithic artefacts but becomes a dominant source again in the Final Chalcolithic (Pernicka *et al.* 1997: 106). These changes go hand in hand with the known cultural dynamics at the time: the use of eastern Serbian sources decreases sharply with the end of the Vinča culture, while the exploitation of Ai Bunar and other Bulgarian deposits intensifies with the rise of the KGK VI, Varna and Krivodol-Salcuţa-Bubanj Hum cultural phenomena. As noted above, with the collapse of these cultures—commonly ascribed to an environmental catastrophe (Weninger *et al.* 2009) but remaining the subject of considerable debate (see

Tsirtsoni 2016a)—the eastern Serbian deposits became more actively used again, followed by the appearance of the Bodrogkeresztúr groups.

More recently, Radivojević and Grujić (2018) developed a unique approach to investigating the networks and dynamics of copper supply between *c.* 6200 and *c.* 3200 BC, based on the currently available datasets from Pernicka *et al.* (1993, 1997), Radivojević (2012) and the project presented in this monograph (see Chapter 41), including 410 copper-based objects from 79 sites (all made freely available in Table S1 in Radivojević and Grujić, 2018)[2]. The authors applied a complex networks approach, using a modularity maximisation method (Blondel *et al.* 2008) in order to explore the structure of the most densely connected sites through the strength of copper supply, trade or exchange links. They identified three highly interconnected systems—community structures or 'modules'—composed of supply networks that reflect organisation of the copper industry and, effectively, social and economic ties in the Balkans between *c.* 6200 and *c.* 3200 BC (Radivojević and Grujić 2018: 116, Figure 6). The intensity of algorithmically calculated social interaction revealed three main groups of communities that appeared spatiotemporally and statistically significant: the resulting structures held a strong resemblance to at least three dominant economic and social cores of copper industry in the Balkans across *c.* 3000 years, traditionally defined as Vinča, KGK VI and Varna, and Bodrogkeresztúr (Figures 9 and 10). Importantly, the complex wiring topologies of these three modules were quantified independently of cultural, chronological and geographical attributes.

Besides suggesting spatiotemporal patterning, this resemblance showed that algorithmically calculated community structures currently represent the most precise mathematical model available for identifying such archaeological phenomena. The dynamics of copper exploitation, production and consumption practices reflected closely those of recorded social interactions for the time and region studied. Although Radivojević and Grujić (2018) did not suggest that metallurgy-related practices were the sole factor in defining interactions such as collapses or rises of cultural complexes, their research indicates that these industries must have been sufficiently powerful to play a major role in their shaping.

Radivojević and Grujić (2018) also observed the selective formation of network ties amongst site populations in relation to both specific regional copper sources (e.g. eastern Serbian Majdanpek, central Bulgarian Ai Bunar) and communication routes (e.g. lower Danube), as well as their association with either seemingly 'monopolised' (e.g. Bodrogkeresztúr) or 'open-market' (e.g. KGK VI) organisation of copper supply networks across the periods analysed (Figure 11). These results are consistent with previous research on metal provenancing in the Balkans (Pernicka *et al.* 1993, 1997; Radivojević 2012). Importantly, this study also indicated an overall tendency for communities identified as archaeological cultures to maintain their own regional network of copper exploitation, production, exchange and consumption. In this light, metal recycling practices are plausible, although they may have happened within specific regional networks of copper supply (or 'modules'). In such cases, recycling would not be easily identified in provenance analyses, as this activity homogenises the metal pool – and if the metal were coming from a single source or deposit, the signature would stay the same regardless of the reuse or recycling process.

This is not to say that modules or archaeological phenomena identified in this way did not cooperate amongst themselves. Quite to the contrary, there were links (see Figure 9) *between* the modules although these were not as strong as those *within* them. Knowledge of metalmaking spread through these links across the Balkans from the Vinča culture 'core' centre. It expanded and collapsed along with the rise and fall of the cultural complexes, but it never ceased to be practiced. Although the dataset beyond 3200 BC was not targeted in the networks research, we are aware of continuing metalmaking practices at the fringes of the 'core' metallurgical area: the western Balkans, eastern Alps, Slovakian Alps, Carpathians (both Transylvania and Moldova) and well into the Caucasus Mountains, all arising during the early to mid-4th millennium BC with copper and arsenical copper production (e.g. Antonović 2014a; Bognár-Kutzián 1972; Courcier 2014; Dolfini 2014; Hansen 2013b; Höppner *et al.* 2005; Novotna 1970; Radivojević *et al.* 2010b; Roberts *et al.* 2009; Ryndina and Ravich 2012; Scharl 2016; Vulpe 1975).

Given this existing evidence for metal circulation, *The Rise of Metallurgy in Eurasia* project evaluates the metal composition and metal provenance of the new copper finds from Pločnik and Belovode in the light of networks modelling and related interpretations to metal circulation. This addresses Research Questions 4 and 6.

Metallurgy, metallurgists and societies

In order to re-investigate the interpretations of early metallurgy, metallurgists and societies, as per the aim of *The Rise of Metallurgy in Eurasia* project, it is not only necessary to build interpretations from all the available and relevant data which is gathered and analysed by new and innovative approaches, but also to acknowledge and critically evaluate the existing narratives that

[2] Data also available at: https://www.repository.cam.ac.uk/handle/1810/265760

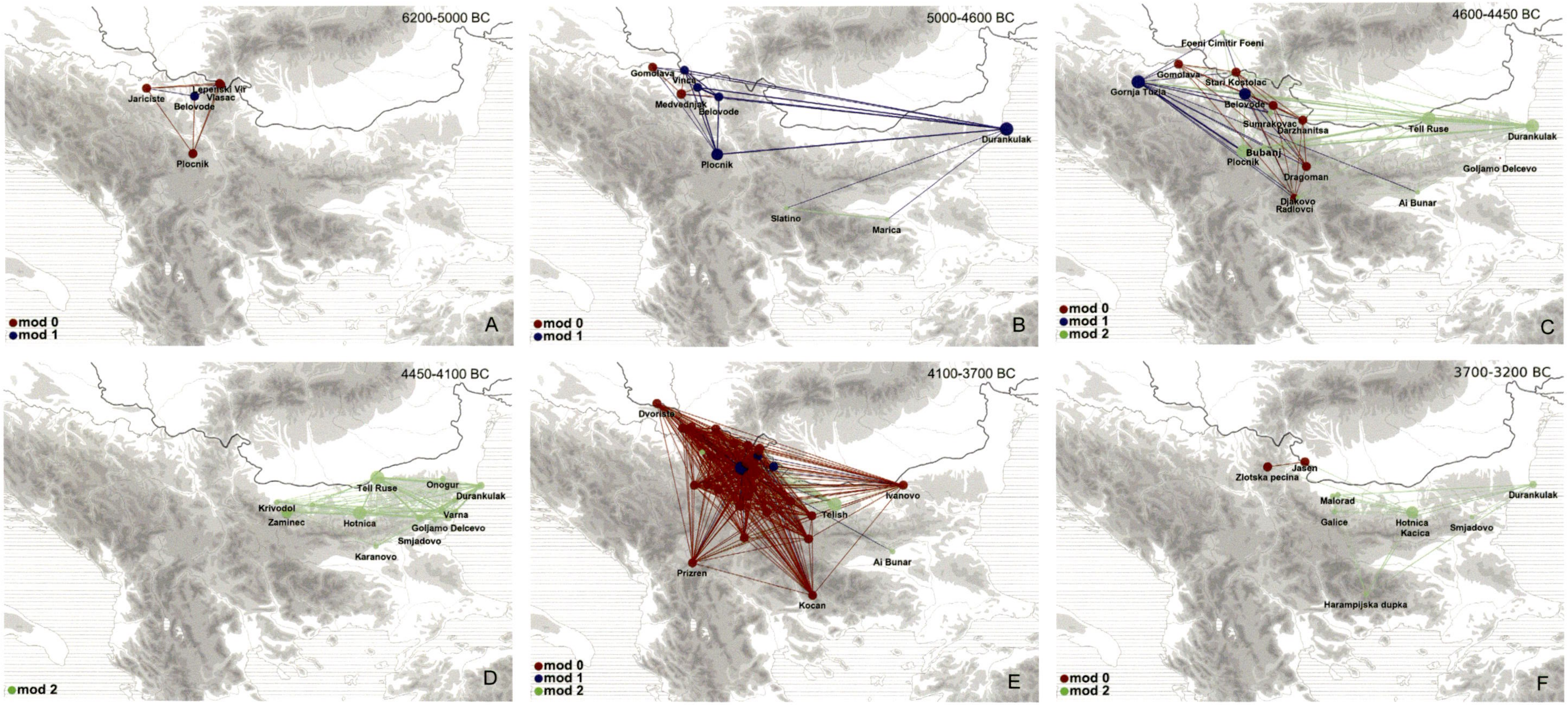

Figure 9. Networks of copper supply throughout *c.* 6200–3200 BC in the Balkans. (a) Period 6200–5000 BC illustrates the early core of supply networks for copper mineral-only artefacts; (b) Period 5000–4600 BC is dominated by the supply networks of Module 1 (proxy for Vinča culture, see Figure 9); (C) Period 4600–4450 BC is dominated by the developing Module 2, which emerged in parallel with the slow disappearing supply regional networks of Modules 0 and 1; (D) Period 4450–4100 BC demonstrates the supremacy of Module 2 in the east Balkans (proxy for Kodžadermen-Gumelniţa-Karanovo VI cultural complex, Figure 9); (E) Period 4100–3700 BC shows the rise of supply networks of Module 0 in central Balkans (eastern Serbia, proxy for Bodrogkeresztúr culture, Figure 9) following the collapse of the eastern Balkan networked systems by 4100 BC; (F) Period 3700–3200 BC presents a picture of nodes scattered in eastern Balkans, altogether reflecting the incoherent set of available data (after Radivojević and Grujić 2018: Figure 6).

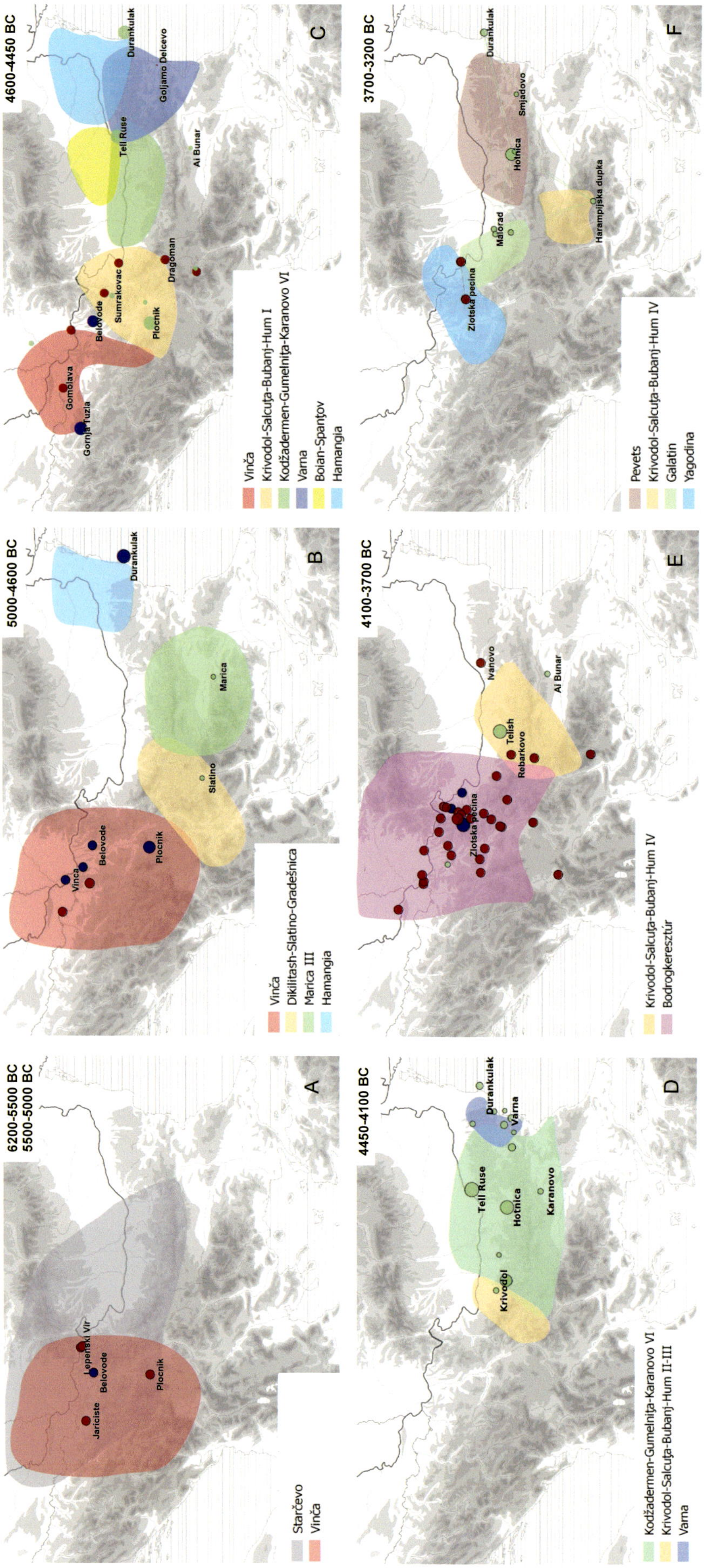

Figure 10. Distribution of archaeological cultures / copper-using societies in the Balkans between c. 6200 and c. 3200 BC, with the most relevant sites. Note colour-coding and size of nodes consistent with the module colour (Module 0 – red, Module 1 – blue, Module 2 – green) (after Radivojević and Grujić 2018: Figure 7).

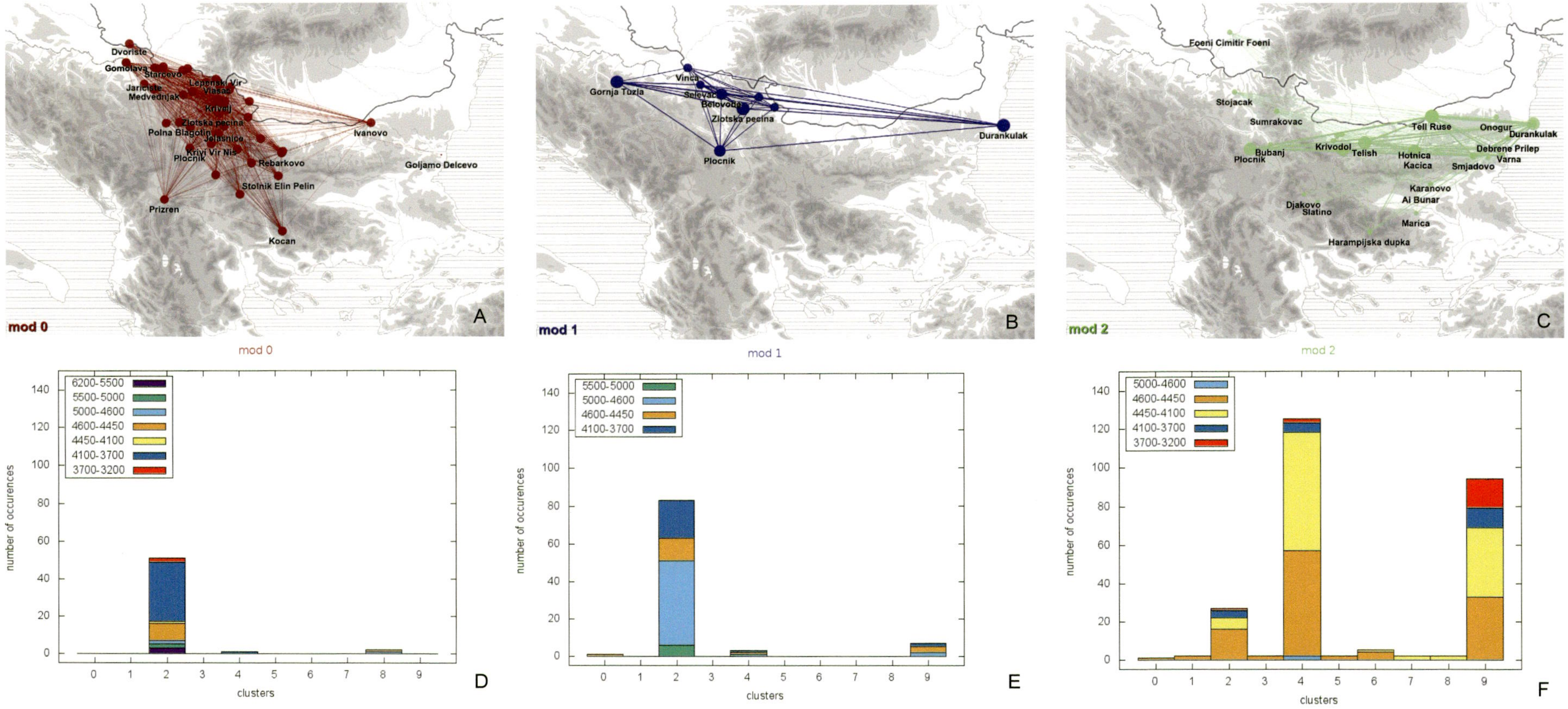

Figure 11. Three individually presented modularity structures of copper producing and exchanging communities in the Balkans (*c.* 6200 – *c.* 3200 BC) paired with diagrams illustrating the frequency of ten different chemical clusters within each of these modules, throughout different periods. (A, D) Module 0 is represented with 50.5 % of nodes in the total network and three chemical clusters only, of which No. 2 is predominant and covers *c.* 6200–3200 BC. (B, E) Module 1 is represented with 11.8 % of all nodes and four chemical clusters. Within the chronological span of *c.* 5500–4450 BC and *c.* 4100–3700 BC, chemical cluster No. 2 is the most dominant, while clusters 0, 4 and 9 have a minor presence. (C, F) Module 2 includes 37.6% of nodes in the total network and includes all ten chemical clusters (0–9). Chronologically it covers the period between *c.* 5000 and *c.* 3200 BC, with two divisions (*c.* 4600–4450 BC and *c.* 4450–4100 BC) representing together 85% of all artefacts in this module (after Radivojević and Grujić 2018, Figure 5).

dominate the discourse. The relationship between early metallurgy, metallurgists and societies in the Balkans has been the subject of extensive and wide-ranging scholarship (e.g. see review in Kienlin 2010). This has invariably concentrated upon the proposed significant impact of metallurgy on the societal themes of social complexity and craft specialisation, especially in relation to the emergence or (self-) identification of elites, as both producers and/or consumers, across the region.

The interpretative narratives in which early metallurgy, metallurgists and societies are deeply embedded can be defined accordingly:

1. The knowledge and expertise relating to production of metal represented a technological revolution.
2. The invention and/or innovation of metallurgy impacts significantly upon the social, political and ritual lives of the farming communities across the Balkans.
3. The knowledge and expertise relating to metallurgy was restricted to specialist individuals who practiced in relative secrecy and held a distinct and elevated status.
4. The properties of metal objects—whether hardness, lustre and/or colour—ensure that they are fundamentally and consistently desirable and valuable to the farming communities across the Balkans.
5. The production, circulation and consumption of metals was integral to the creation and maintenance of elite status and identity in farming communities across the Balkans.

Each of these five inter-related interpretative narratives builds on 19th century ideas that equated (pyro-) technological abilities with societal development within a social evolutionary scheme (Díaz-Andreu 2007; Pearce 2019; Roberts and Radivojević 2015; Rowley-Conwy 2007). The consequence for scholarship regarding the Balkans from *c.* 6200–3700 BC are ongoing debates as to whether a Copper, Eneolithic or Chalcolithic Age represents a distinct historical epoch (Lichardus 1991a; Schier 2014a) and the extent to which metals, elites and social complexity are inter-related (e.g. Hansen 2012, 2013b; Kienlin 2010; Kienlin and Zimmermann 2012). It is also inevitable that the contemporary and historical and contemporary perceptions of the metals involved are influential, with the gold at Varna leading to narratives of the emergence of wealth and social differentiation (e.g. Ivanov and Avramova 2000; Renfrew 1978a, 1986), with copper throughout the Balkans leading to narratives of technological and industrial production, distribution and scale (Chernykh 1992; Ryndina 2009), and with lead in the Balkans and elsewhere being largely ignored as a low value and technologically uninteresting material.

It is therefore not surprising that the interpretation of the life of the prehistoric Balkan communities (and especially of the 5th millennium BC) has frequently been influenced by the conventional, metal-orientated approaches in archaeological research in the area, even with the rapid growth of settlement, landscape and environmental research, and interpretational perspectives. Unsurprisingly, it derived from a seductive idea that the presence of craft specialisation indicated the presence of a complex social organisation (Childe 1950), and that the technology is tightly correlated with the increase in social complexity (e.g. Childe 1944; Morgan 1985 [1877]; White 1959). This notion led to the pursuit of centralised decision making in any society with metallurgical practice, making the Balkan case—with the earliest traces of metal making and the earliest large-scale production and circulation of metal ornaments and implements—a fertile ground to justify the advent of highly specialised knowledge with accumulation of individual wealth and emerging hierarchy (e.g. Renfrew 1986; Hansen 2013b).

This metal-construct is still frequently dominant in scholarship, in defining the (elite) socio-economic dynamics of prehistoric communities at the time. This is despite the fact that other materials such as ceramics, flint, polished stone, obsidian and spondylus (e.g. Amicone *et al.* 2020a; Ifantidis and Nikolaidou 2011; Klimscha 2016, 2020; Milić 2015; Spataro 2018; Whittle *et al.* 2016; Windler 2018; 2019) were also comparably, if not much more extensively, sourced, shaped, traded and/or deposited in settlements and graves prior to, and along with, metal objects. It is evident that, especially in the last decade, many major Balkan Neolithic-Chalcolithic projects have explicitly sought to go beyond traditional metal-orientated perspectives, especially given the infinitely larger scale and depth of the non-metallurgical archaeological and environmental record. This is reflected in recent syntheses, whether encompassing the Balkans (Chapman 2020) or the Black Sea region (Ivanova 2013). In particular, research engaging with population dynamics, subsistence strategies, settlement practices, and responses to local and regional environmental and climatic change is thriving (e.g. Benecke *et al.* 2013; Chapman and Souvatzi 2020; Filipović *et al.* 2017, 2018; Gaastra *et al.* 2018, 2019; Ivanova 2012; 2020; Ivanova *et al.* 2018; Marić 2013a, 2015, 2017; Müller 2012, 2017; Orton 2010; Orton *et al.* 2016; Porčić 2011, 2012a, 2020; Porčić *et al.* 2016; Silva and Vander Linden 2017). It is also worth noting that several major, modern excavation and survey projects of Neolithic-Chalcolithic sites in the Balkans such as Okolište (Bosnia), Uivar (Romania), and Vinča (Serbia) (Draşovean and Schier 2021; Müller *et al.* 2013a; Schier 2014b; Tasić *et al.* 2016a) have yet to reveal substantial metal objects or evidence for metal production. However, where metal objects and/or metallurgical remains are found as for instance at Pietrele (Romania) (Hansen and Toderaş 2012; Hansen *et al.* 2019), familiar interpretative narratives relating to metals and elites are proposed (Hansen 2012, 2013a; Klimscha 2020).

The primary challenge, at least at the broader interpretative scale, in investigating the origins, development and societal inter-relationships of early metal objects and metallurgy in the Balkans. What this means in practice is to analyse and interpret the metal-orientated evidence, not in technological or intellectual isolation, but in relation to the other practices and activities of communities living in the region in the Late Neolithic and throughout the Chalcolithic (*c.* 6200–3700 BC).

The bibliographic reference for this chapter is:

Radivojević, M. and Roberts, B. W. 2021. Balkan metallurgy and society, 6200–3700 BC, in Radivojević, M., Roberts, B. W., Marić, M., Kuzmanović Cvetković, J., and Rehren, Th. (eds) *The Rise of Metallurgy in Eurasia*: 11–37. Oxford: Archaeopress.

Chapter 4

The Vinča culture: an overview

Benjamin W. Roberts, Miljana Radivojević and Miroslav Marić

This chapter reviews the archaeological evidence for the Vinča culture, the broader archaeological context for the majority of the metal production and metal artefacts extensively explored in Chapter 3, as well as for the sites of Belovode and Pločnik, whose investigation forms the core of *The Rise of Metallurgy in Eurasia* project. The chapter will provide a lengthy introduction to the current data and interpretations of the Vinča culture that are subsequently developed in far greater detail in the thematic overviews by many of the leading specialists in later chapters (Chapters 39–52). This monograph seeks to address, at least in part, the absence of a dedicated synthesis of the Vinča culture since Chapman's (1981) monograph (see Chapman 2020b for a critical reflection).

The concept of archaeological cultures remains problematic in European prehistory in terms of definition and interpretation, yet extremely resilient in the absence of comparable empirically orientated alternatives (Roberts and Vander Linden 2011). Due to competing national traditions of scholarship, the culture history groupings and terminologies are strikingly complex in the later prehistoric Balkans (Gori and Ivanova 2017; Tsirtsoni 2016a). As such, the chapter explores the historiography and complex debates that surround the archaeological and temporal definitions of the Vinča culture. The importance of the Vinča culture lies not only in the evidence of early metallurgy but also in the evidence for the expansion of material culture production and circulation, the intensification of agriculture and increase in sedentism and settlement growth, which are all subsequently reviewed. The chapter concludes by examining past and present interpretations of the communities who lived and died within what we now term the Vinča culture.

Defining the Vinča culture

The Vinča culture is a Late Neolithic/Early Chalcolithic phenomenon, which lasted from *c.* 5350/5300 BC to *c.* 4500 BC across the northern and central Balkans and is fundamentally defined by ceramic types (Porčić 2020; Whittle *et al.* 2016). It occurs across a large area of the Balkans (Figure 1), which includes all of present-day Serbia, the Romanian Banat, Transylvania and parts of Oltenia, western Bulgaria, eastern Macedonia, eastern parts of Slavonia and Bosnia and the southern Hungarian region of Baranya. Much of the general information about this culture is known from research by several national and international teams, starting with the seminal work of Vasić (1932–1936), but also Childe (1929), and many others (e.g. Borojević 2006; Chapman 1981; Fewkes 1936; Garašanin 1951, 1979; Gimbutas 1976a; Holste 1939; Jovanović 1971; Markotić 1984; McPherron and Srejović 1988; Orton 2008; Porčić 2009b; Renfrew 1970; Srejović 1963, 1984a; Srejović and Tasić 1990; Tringham and Krstić 1990a). Regarding the cultural historical surroundings, this phenomenon shows strong links with the contemporaneous Karanovo (phases III to Kodžadermen-Gumelniţa-Karanovo VI) in Bulgaria, Precucuteni-Tripolye A in Moldavia and Ukraine, Dimini in Greece, and the late manifestations of the Starčevo culture and early Sopot culture in eastern Croatia.

The origins of the Vinča culture are still elusive; opinions on this issue have traditionally been divided between advocates of colonisation from Anatolia, based on typological parallels with the black burnished ware from this area (Childe 1929; Garašanin 1973; Jovanović 1962; Milojčić 1949; Schachermeyr 1955) and proponents of local development (Chapman 1981; Kaiser and Voytek 1983; Leković 1990; Makkay 1990; Markotić 1984; Renfrew 1969, 1970; Srejović 1988; Todorova 1978; Tringham 1971). While the diffusionists' argument favours an external influence in explaining the origins of the Vinča culture, the 'autochthons' rely on the fact that the Vinča culture territory was culturally preceded by the Starčevo-Körös-Criş complex, and that there is strong evidence for local development of settlements, ceramic typology and stratigraphy. Whilst a major review of the existing radiocarbon dates highlights the rapid spread of the indicative biconical and black burnished ceramics throughout the Central Balkans, it takes neither side, and indeed seeks to move beyond these traditional and binary debates (Whittle *et al.* 2016). Furthermore, neither the north nor the south of the Vinča culture 'potscape' exhibit notably earlier dates (Whittle *et al.* 2016). More recently, Porčić (2020) reviewed the 'origins' evidence across the Balkans, with a particular emphasis on radiocarbon dates. He highlighted the earlier Western Anatolian evidence for black-burnished ware (Çevik 2018; Özdoğan 2011) and its presence in the eastern Balkans several centuries prior to *c.* 5300 BC. He also stressed the

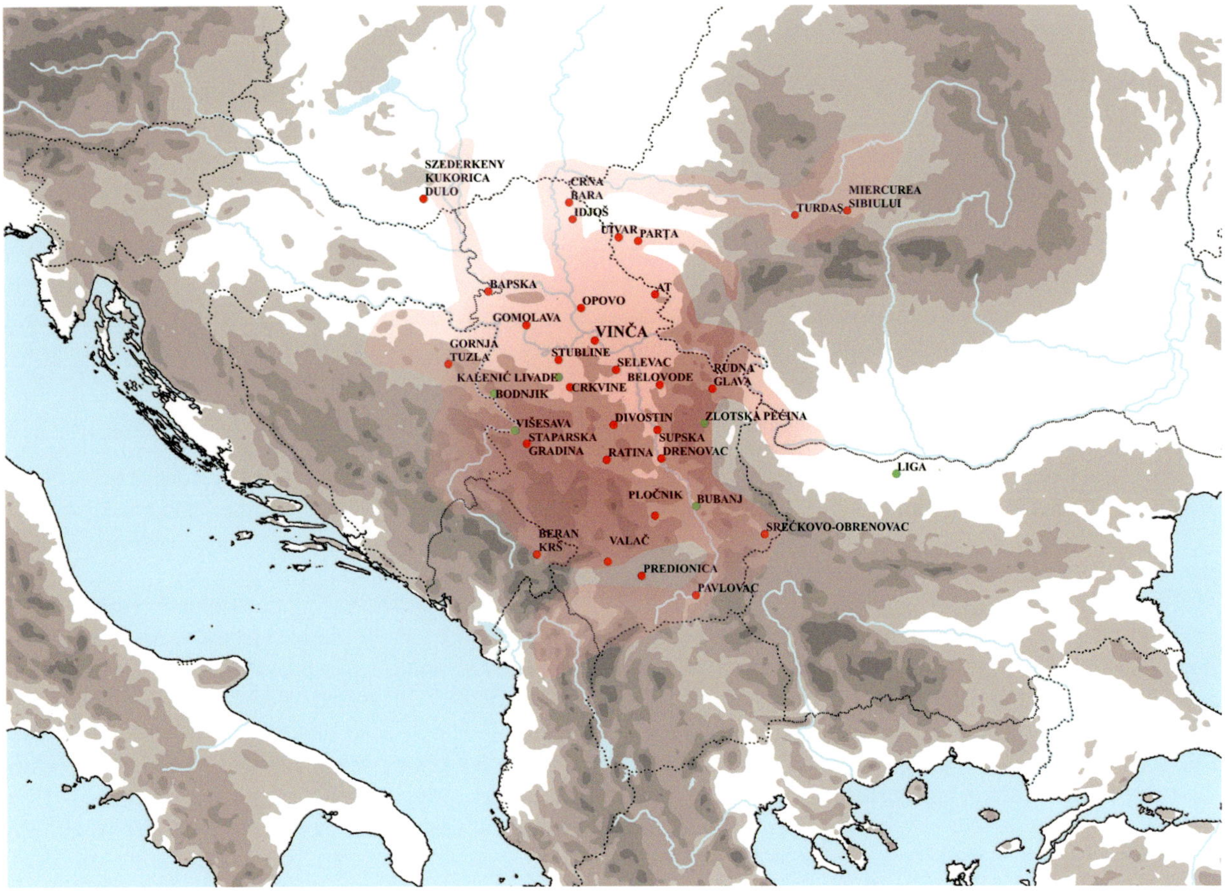

Figure 1. The distribution of the Vinča culture (shaded) with Vinča sites (red dots) and later settlements (green dots) (Middle Chalcolithic). Prepared by J. Pendić and M. Marić.

significant increase in the settlement type and size from the flat (pit-) settlements, typically across 1–2 ha at Starčevo culture sites (though larger sites exist) to the rectangular wattle and daub settlement structures typically across 5–10 ha at Vinča culture sites. There are exceptions to this broader trend, with several Starčevo culture sites such as Nosa (Garašanin 1961) and Ludaš Budžak (Sekereš 1967) near Subotica and Zlatara near Ruma (Leković 1988) in Serbia and Szentgyörgyvölgy-Pityerdomb (Bánffy and Sumegi 2011) in Hungary also yielding evidence of wattle and daub structures. However, the radiocarbon dating evidence at least partially undermines evidence of settlement or ceramic continuities whilst the modelled radiocarbon dates indicate a demographic decline in the centuries prior to the emergence of the Vinča culture (Porčić 2020), albeit one that is not evident until several centuries later in the eastern Balkans. Whether this can, indeed, be related to any incoming population—as aDNA studies on small samples could suggest (e.g. Hervella *et al.* 2015; Mathieson *et al.* 2018)—remains to be seen.

The terminology of the Vinča culture is yet another matter of dispute, closely related to the acknowledgement of the scope of metallurgical activities within this archaeological culture. The majority of ex-Yugoslav archaeologists refer to the Vinča culture as a Late Neolithic manifestation, while Bulgarian archaeologists acknowledge it as partly an Early Chalcolithic occurrence (e.g. Todorova 1978). However, Milojčić (1949: 108) argued that the Vinča phenomenon was Chalcolithic (or Eneolithic) starting from the Gradac Phase, a view supported by Čović (1961: 127–128), and later by Garašanin (1994/1995: 17). On the other hand, Jovanović and Ottaway (Jovanović 1971; Jovanović and Ottaway 1976) shared the opinion that the entire culture was already a Chalcolithic phenomenon. The term 'Late Neolithic' for the Vinča culture nevertheless remained firmly established in the ex-Yugoslav literature, despite the fact that knowledge of the Vinča culture metallurgy has advanced since the last century (cf. Radivojević, *et al.* 2010 and literature therein). Whilst the authors would advocate for the use of the term 'Early Chalcolithic' for the entire Vinča culture, it is used throughout this monograph in conjunction with the Late Neolithic where necessary, with an intention to facilitate interconnections across various regional publications and excavation reports.

Vinča culture chronologies

The chronology of the Vinča culture was initially established on the basis of ceramic typology (see detailed review and re-dating in Whittle *et al.* 2016) and the stratigraphic sequence of the type-site Vinča-Belo Brdo which has recently been re-dated (Tasić *et al.* 2015, 2016a, 2016b). Of the abundant typological schemata (Berciu 1961; Chapman 1981; Garašanin 1951, 1979; Holste 1939; Lazarovici 1979, 1981; Lenneis and Stadler 1995; Menghin 1931; Milojčić 1949; Parzinger 1993; Schier 1991, 1996, 2000), the most popular are those of Garašanin (1951) and Milojčić (1949). Garašanin (1951) divided the Vinča culture into an early phase, Vinča-Tordoš (I and II) and a late one, Vinča-Pločnik (I, IIa, IIb), with an intermediate phase called 'Gradac' (Table 1) added later, in the mid-1970s (Garašanin 1979). The other major periodisation, introduced by Holste (1939) and further developed by Milojčić (1949), was based on the use of alphabetical letters rather than sites (Vinča A–D with subdivisions). Milojčić's periodisation is the preferred choice in this monograph, however sparingly used in agreement with the Garašanin's proposal. The spatial division of the Vinča culture has also been the subject of debate with several regional and local groups defined across its extent. Of these, the most widely used conventional division is that of Garašanin (1951) who identified seven regions with different lifetimes (classical, South-Moravian, Kosovian, East-Bosnian, Transylvanian, Oltenian and Srem-Slavonian).

Early Vinča pottery reflects a combination of three distinct ceramic traditions: northern with incised band decoration (LBK and Moldavian-Ukrainian complex), southern (Anatolian- Balkan) with dark burnished ware, and a local background of impressed and barbotine of the Starčevo culture (Chapman 1981: 53; Garašanin 1951: 63; Milojčić 1949: 106; Schachermeyr 1955). This is also reflected in the appearance of ceramic types with no immediate predecessors from the previous period. However, some ceramic styles, such as coarse ware (with impressed, incised or barbotine decoration) and painted ware, are well grounded in the preceding Starčevo culture, and continue to be used throughout the early sequences of the Vinča culture (Chapman 1981; Schier 1996). The early Vinča culture (Vinča A–B1) also saw the emergence of the black burnished ware style, typical throughout its sequences in black-topped (Figure 2a), black-burnished and black-polished varieties (Figure 2b) (Chapman 1981; Garašanin 1951).

Vessel shapes in the early Vinča phases (Vinča A–B1) are characterised by biconical bowls, some of which are pedestal, and carinated bowls with longer detached rims, usually decorated with shallow channelling technique and incised decoration-ribbons often filled with round points made using a sharp tool (Figure 3a) (Schier 1996). The most dramatic change in vessel shape and decoration is observed at the beginning of the Gradac Phase (Vinča B2–C1): among the many new forms are tri- or four-partite vessels with cone-

Table 1. Overview of alternative typological schemes for Vinča ceramics (after Schier 1996; Whittle *et al.* 2016; Figure 2).

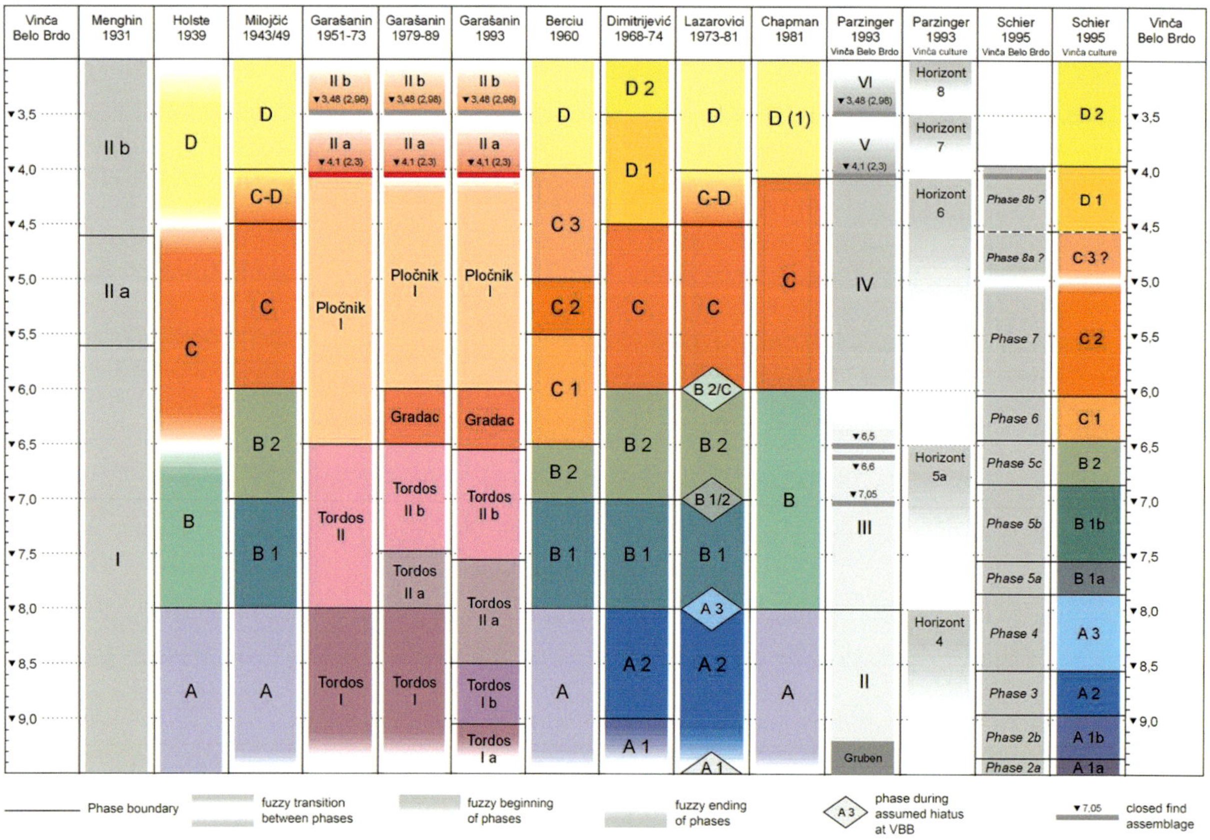

Figure 2. Vinča culture pottery: a) a black-topped 'fruit stand' (early Vinča); b) a black polished three-legged bowl (late Vinča (C–D); after Nikolić 2008: Cat. 166, 176).

Figure 3. Vinča culture pottery: a) an amphora decorated with incised ribbon; b) a shallow channelled amphora (after Nikolić 2008: 261, 264).

shaped necks and protruding shoulders, and the so-called Gradac dishes: conical bowls with thickened rims channelled on the interior (Schier 1996). These shapes also occur throughout the later Vinča phases (Vinča C–D). Beside vessel shapes, this change is also visible in the decoration methods and motifs, with a decline in incised ribbons filled in with points made with a sharp tool, sporadically replaced by stamped, rounded dents. Bi-chromatic, rainbow and black topped vessels disappear from assemblages, and the dominant decoration technique gradually becomes channelling, polishing and burnishing (Figure 3b).

From the late 1970s, more radiocarbon dates became available (Breunig 1987; Ehrich and Bankoff 1992; Lenneis and Stadler 1995; Obelić *et al.* 2004; Renfrew 1969; Schier 1996, 1997, 2000; Srdoč *et al.* 1975, 1977, 1987) which, in recent years, have culminated in a major critical review and modelling of existing radiocarbon dates for Vinča ceramics (Whittle *et al.* 2016) supported by the extensive sampling, radiocarbon dating and Bayesian modelling of the type site sequence at Vinča-Belo Brdo (Tasić *et al.* 2015, 2016a, 2016b) and Uivar, Romania (Draşovean *et al.* 2017). This significantly expanded database of radiocarbon dates has

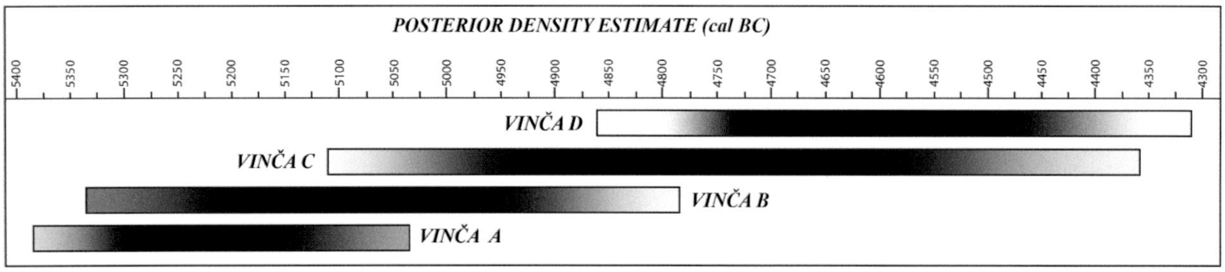

Figure 4. Schematic diagram showing the currency of the different phases of Vinča ceramics proposed by Milojčić (1949). The darker the shading the more probable that a ceramic phase was present in a particular 25-year period. (Derived from Model 1). (Whittle *et al.* 2016; Figure 37)

subsequently enabled demographic modelling (Porčić 2020). Whittle *et al.* (2016) place the start of the Vinča A phase (as exemplified with the type site of Vinča-Belo Brdo at *c.* 5400/5300 BC, while Vinča B starts around 5200 BC. The highest probability end date for Vinča B1 is *c.* 5000/4950 BC, which marks the beginning of the Gradac Phase. The Gradac Phase was an episode between Vinča B and C, probably lasting for 50–100 years, at least at the site of Vinča-Belo Brdo. Vinča C ended in *c.* 4850/4800 BC, while the end of the Vinča culture falls around 4500 BC (Figure 4). The re-dating of the entire sequence at Belovode and Pločnik—which, as sites, span the entire Vinča culture—by *The Rise of Metallurgy in Eurasia* project enables further refinements to the overall Vinča culture chronology as presented in Chapter 37.

Traditional theories on the disappearance of the Vinča culture relate to violent encounters with communities related to the Bubanj-Salcuţa-Krivodol (BSK) cultural complex from Oltenia, or the bearers of the Baden culture (Garašanin 1979: 204–205) as evidenced by the widespread burning and apparent abandonment of major settlement sites. This transition to Middle Chalcolithic (or mid-5th millennium BC), marked in culture-historical terms by the Bubanj-Sălcuţa-Krivodol (BSK) group in southern and central Serbia and Tiszapolgár and Bodrogkeresztúr group finds on the edge of the Pannonian plain and in the river valleys of Sava and Danube tributaries, appears to be a highly dynamic process. However, the period remains relatively poorly understood and the core debate over changes or continuities in settlement practices needs to be addressed in more detail.

The first point to consider is what settlement activity can be identified. As Ristić-Opačić (Ристић-Опачић 2005) demonstrated in her analysis of Vinča settlement topography and chronology, the Vinča D phase has the lowest number of newly formed settlements identified. Yet rather than representing a straightforward decline in past settlement activity, this trend is largely the product of archaeological excavations concentrating overwhelmingly upon earlier, established settlement sites that have a prolonged existence. Those settlements dating to the second half of the 5th millennium BC have been less extensively excavated and dated compared to earlier settlement sites. Given that, on current evidence, the mid-5th millennium BC onwards in the central Balkans frequently sees the construction of new, single phased settlements primarily being founded away from existing Vinča culture sites, as demonstrated in the area of Mačva in western Serbia (Tripković and Penezić 2017), it is easy to see how the shift towards a less archaeologically visible settlement activity could be incorrectly interpreted as abandonment or collapse.

The Late Vinča D site of Crkvine, near Mali Borak provides a potentially instructive case study. It consists of a series of spaced wattle and daub structures on a hillock with steep sides above a local stream, a tributary of Kolubara River in western Serbia (Marić 2011), representing a broader shift from river terraces to more elevated terrains which are far less visible in the landscape to archaeologists, creating the false impression of little or no settlement activity. Neither should it be assumed that this period comprised only small-scale settlements with sites such as Stubline near Obrenovac (Crnobrnja 2014) and Drenovac near Paraćin (Perić *et al.* 2016) showing clear evidence of the aggregation of inhabitants with several hundred wattle and daub structures visible in geomagnetic surveys, albeit in a rather different spatial organisation.

The second point to consider is evidence of settlement continuity. Given the underlying culture-historical framework of the settlement debate, an important starting point is the mid- 5th millennium BC settlement of Kalenić Livade in western Serbia which comprises a single wattle and daub rectangular structure with a mixed ceramic assemblage that ranges from BSK to Tiszapolgár and Bodrogkeresztúr and evidence of a settled farming economy (Blagojević 2005; Trbojević 2005). This highlights not only the intensive contacts between various communities of the region (Parkinson *et al.* 2004) but also the problematic issues with defining them by their pottery. In terms of settlement activity, Vinča continuities can be seen at sites such as Bubanj-Humska Čuka (Bulatović and Milanović 2020; Bulatović *et al.* 2018) and potential continuities at Bodnjik-

Družetić (Živanović 2013). Even at the site of Vinča-Belo Brdo, Bodrogkeresztúr graves have been recovered succeeding the Vinča settlement activity (Jevtić 1986). When considered from a broader Balkan perspective and beyond, settlement continuities throughout the second half of the 5th millennium BC at multi-layered sites in Bulgaria and the Pannonian Plain are well documented (Porčić 2019a) as are settlements in caves and other naturally and otherwise protected locations across eastern Serbia and western Bulgaria (e.g. Borić *et al.* 2018, Merkyte 2005; Nikolov 1975; Nikolov 1984). It is clear that a more detailed evaluation of the settlement evidence undermines the traditional explanations of Vinča collapse and/or conquest during the mid-5th millennium BC, thus reflecting the broader trends in Neolithic-Chalcolithic Balkan scholarship (Porčić 2019a; Tsirtsoni 2016b), but far more research is required to explain the transformations that occurred.

The Gradac Phase

The Gradac Phase (Vinča B2–C1) is of particular interest in defining the Vinča culture, as it marks the change in material culture, settlement activity and, most importantly here, pyrometallurgical activities. Garašanin (1990: 12–15, 1973: 103, 1979: 152) inserted this phase between Vinča- Tordoš IIb (B2) and Vinča- Pločnik I (C1) (▼6.5 m–▼6.1 m) in the type-site, admitting, however, that it did not separate as well in the classical Vinča culture as in its southern variants. The appearance of the Gradac Phase, although not immediate across the whole Vinča culture, is marked with the house destruction horizon in several sites, an increased number of settlements erected at more dominant positions, intensification of elaborated monumental figurine production as well as introduction of new pottery forms (Garašanin 1991). Jovanović's (1994) subsequent periodisation of the later Vinča culture phases is noteworthy as it builds on the significance of the Gradac Phase, dividing it into three sub-phases confined to the Morava valley, namely Gradac I–III. Gradac I is synchronised with Vinča B2/C1, as exemplified by Rudna Glava and associated pottery hoards (Vinča B2/C1). Gradac II relates to the disappearance of the Vinča culture further in the Danube basin and the final settlement horizon in Divostin (Vinča C2, D1–2). Gradac III is associated with the longer-surviving southernmost areas of the Vinča culture and its Kosovian variant in southern Serbia.

The most distinctive traits of the Gradac Phase material culture are ceramic plates with a thickened rim, single-handled carinated jugs, so-called Vidovdanka figurines (Figure 5), voluminously modelled and with a polygonal face, triangular ceramic altars with deer or ram heads (Figure 6) and the introduction of graphite-painted pottery (Garašanin 1979: 174). Todorova (1978: 30) related metallurgy in Thracian Bulgaria to the expansion of the graphite painting of black burnished pottery. This technique created a silver-like surface brilliance, thus potentially resembling the lustre of metal objects (Chapman 1981: 138; Chokhadzhiev 2000: 97; Schachermeyr 1955: 133). Graphite painting, as well as the crafting of the black burnished ware long thought to be connected to early metallurgy (Gimbutas 1976a; Renfrew 1969), required mastery in controlling firing conditions, in which Vinča potters were exceptionally skilful. The darker fabric pottery was fired in a reducing atmosphere between 700° and 900° C, while both reducing and oxidising atmospheres were combined to attain a multicoloured effect (such as black-topped pottery) (Goleanu *et al.* 2005: 945; Varvara *et al.* 2008: 10). The ceramic pyrotechnologies of the Vinča culture are addressed in detail in Chapters 14, 29 and 43 whilst the inter-dependence of metal and ceramic technologies is explored in Chapters 43 and 52. The discovery of the early mining site of Rudna Glava along with new forms of Vinča pottery (Jovanović 1982) convinced Garašanin to acknowledge the Gradac Phase as the beginning of mining activities and hence the Early Chalcolithic within the Vinča culture (Garašanin 1994/1995: 17) (see also Chapter 1).

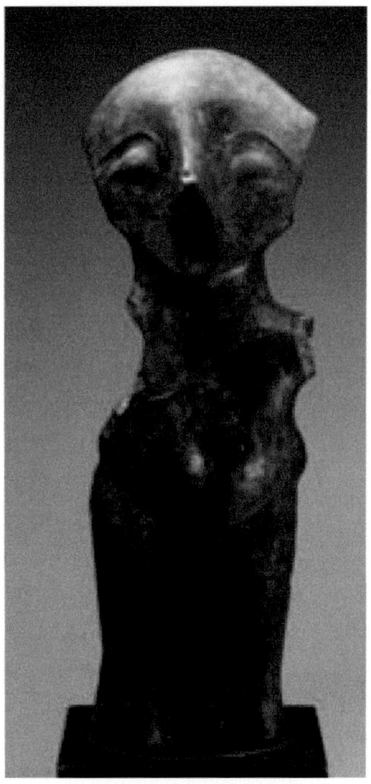

Figure 5. The Vidovdanka, discovered at Vinča-Belo Brdo (after Nikolić 2008: 222).

The nature and extent of the cultural change that marked the beginning of the Gradac Phase has not been studied thoroughly, let alone the reasons for its varying magnitude across the Vinča culture. The most intensive changes in material and settlement activity during this phase were exhibited in the southern variants of the Vinča culture, the south-Moravian and Kosovian (Jovanović 2006: 222; cf. Vasić 1911), which led some scholars to believe that the Gradac Phase was a short-lived phenomenon of a merely typological character. However, as noted above, Jovanović identified three distinctive stages of the Gradac Phase, based on the longer lifetime of Vinča culture sites south of the Danube and along the Morava Valley (Jovanović 1994). Gradac Phases I–III also follow the development of the Vinča culture metallurgy, starting with the mining activities in Rudna Glava and intensifying with the settlement production of massive copper implements, as seen in Divostin and Pločnik.

Figure 6. A triangular ceramic 'altar' with deer or ram head (after Nikolić 2008: 172, Figure 68).

In the wider region, in terms of relative chronology, the Gradac Phase is contemporary with the following cultures: Maliq Ia, Gradeshnitsa III A-B, Poljanica, Sava, Vidra (Boian III), Pre-Cucuteni I, Hamangia III, Maritsa V, Dikili Tash II, Paradimi IV, classic (late) Dimini, Sitagroi II, Szakalhat, Tisza (transition), Sopot B and Železiowce (Garašanin 1994/1995: 15–16). In terms of broader interconnections beyond metallurgy, the typical Gradac one-handled jugs are found as far as east Bulgaria (Durankulak) or the Turkish Thracian coast (Toptepe) (Jovanović 2006: 223–224; cf. Özdoğan and Dede 1989: 22–23) whilst the triangular altars (?) with ram heads also occur in central Bulgaria and the lower Danube (Gimbutas 1976b: 88–89). Jovanović (2006), unlike Garašanin (1994/1995), thought that the metallurgy was the driving force behind the Gradac Phase and accompanying cultural changes in the region; however, both agreed that its origins should be sought in the 5th millennium BC cultures of the north Balkans. This debate over the relationship between early metallurgy and the Gradac ceramic phase is addressed in Chapter 52.

Vinča settlement and subsistence

The evidence for settlement in the Vinča culture comprises the pit structures (Bogdanović 1988; Marić and Mirković-Marić 2011) and wooden framed, wattle and daub houses above the ground in larger settlement sites with longer durations of occupation. Whilst the shape and size of pit structure can vary significantly, at both the Vinča tells and flat settlements, the houses are rectangular or squarish in shape with visible internal organisation (Tripković 2009a). The structures were usually up to *c.* 10 m^2, with occasional ancillary structures (>10 m^2) (Chapman 1981: 60; cf. Cook 1972). Inside the settlements, a food preparation area is usually located around thermal structure and silos (Borojević *et al.* 2020), while spaces for practicing crafts are commonly found both inside houses and in the outer yards (Chapman 1981: 63–68). The evolution in size is noticeable over time, with houses evolving into 100 m^2 multi-roomed constructions at the sites of Gomolava and Divostin (cf. Brukner 1980; McPherron and Srejović 1988; Porčić 2009a, 2019b).

The usual settling location for the Vinča culture groups were river terraces or plateaux, hillocks in waterlogged landscapes, or hill slopes near streams; dominant hillfort settlements are rare as they are mostly associated with the later phase of the period and less traversable terrains (Garašanin 1979). The Vinča culture communities inhabited a wide variety of soil types, from highly arable locations to seasonally flooded ones (Chapman 1981: 84–116). The overlapping of buildings resulted in tell-type sites, as seen at Vinča-Belo Brdo (Tasić 2005; Vasić 1932–1936) or Gomolava (Brukner 1980; 1988). However, these are rare, as most multi-layered Vinča period sites do not show the typical traits of tells, i.e. a prominent mound-like central area surrounded by thinner archaeological layers. Rather, Vinča culture sites tend to be horizontally dispersed settlements, as is the case with the 'open' flat type of Selevac (Tringham and Krstić 1990a) or Pločnik (Grbić 1929; Šljivar and Kuzmanović Cvetković 1998a).

There has been considerable debate concerning the spatial scale of settlements with the seminal analysis by Chapman (1981) identifying three groupings: 1.0–1.9 ha; 4.0–4.9 ha and 20–29 ha with accompanying estimations of populations of 50–300, 200–500 and 1200–2500 people respectively. The conclusion was that Vinča culture settlement sites were agricultural villages whose expansion was limited by their food production capabilities. As highlighted in the magisterial survey across southeast Europe by Lichter (1993), excavations and surveys had revealed relatively little about the spatial organisation of settlements in the region which naturally limited the critical evaluation of Chapman's

(1981) conclusions. In the decades since, these debates have been re-invigorated by extensive geophysical surveys of the frequently burnt houses at Vinča culture settlement sites, notably at Crkvine-Stubline (Crnobrnja *et al.* 2009), Drenovac (Perić *et al.* 2016) and Bordjoš (Medović *et al.* 2014); by geophysical surveys at contemporary and comparable settlements sites across southeast Europe (see review in Chapter 38); and by more sophisticated methodological approaches (e.g. Porčić 2012a, 2019a, 2019b) (see Chapter 40) with the conclusion that larger settlements such as Divostin, Belovode, Pločnik, and probably Vinča-Belo Brdo, may have had populations of *c.* 1000 people but probably fewer than *c.* 2000 people (Porčić 2019a).

Scholarly understanding of the subsistence strategies of the communities of the Vinča culture have traditionally been limited by the rarity of excavation recovery strategies, such as sieving, that would enable archaeobotanical and zooarchaeological remains to be identified and interpreted (see Chapters 20–21, 34–35, and 50–51). Even with the more widespread adoption of improved fieldwork methods, there are still relatively few sites from which to extrapolate broader food production trends (Borojević 2006; Filipović and Tasić 2012). Vinča communities grew domesticated crops such as einkorn, emmer, barley, lentil, pea and flax/linseed (see Chapters 20, 34, 50) and reared domesticated animals such as cattle, pig, sheep and goats (see Chapters 21, 35, 51). The higher proportion of cattle in settlements, particularly males, suggests the importance of cattle possession in terms of wealth, while their symbolic role is indicated by the so-called 'bucrania shrines' (Orton 2008; Tripković 2007).

Wild plants such as edible fruits and nuts were gathered and wild animals such as red and roe deer were hunted. Crops were stored in ceramics, storage pits and potentially organic bags at settlement sites (Filipović *et al.* 2018). The main evidence for salt production comes from Gornja Tuzla in east Bosnia, located a few kilometres away from a rich rock-salt mine. Here, conical coarse ware with elongated feet appeared only in the Vinča culture sequence and were presumably related to the salt production (Benac 1961: 50 ff.; Čović 1961: 90, 115–116). Whilst the evidence for subsistence practices varies slightly in terms of proportions and occasionally species across different sites, the overwhelming trend in food production and consumption is one of continuity throughout the duration of the Vinča culture.

Vinča craft production

Craft production in the Vinča culture has frequently been debated in terms of an increase in standardisation and specialisation, an increased diversity in the forms and materials involved, and an increase in quantity (Earle 2018; Tringham and Krstić 1990a; Vuković 2011; Vuković 2020). Scholarship has traditionally concentrated on widely discovered inorganic materials such as ceramics (Amicone *et al.* 2019, 2020; Spataro 2018), polished stone, obsidian, flint (Antonović *et al.* 2005; Antonović 2003; Milić 2015; Šarić 2015), and metal (Radivojević *et al.* 2013 and literature therein; Radivojević *et al.* 2010a) (see Chapters 11, 14, 16, 18, 19, 26, 29, 31, 33, 41, 43, 45, 47–49). Recent research has also transformed understandings of organic materials such as bone (Vitezović 2013b, 2018; Vitezović and Antonović 2020) (see Chapters 17, 32, 46) and shell (Windler 2018). Yet despite the emphasis on craft specialists and specialisation, many scholars have concluded, after fairly exhaustive research, that the majority of the forms and technologies associated with their particular material were likely to have been widely known and practiced and were therefore not made by highly specialised craftspeople (Amicone *et al.* 2020; Kaiser and Voytek 1983). Experimental archaeological reconstructions have enabled a clearer understanding of the processes involved, with a firing experiment showing that the entire range of pottery found at Vinča sites could be produced without using a proper kiln (Svoboda *et al.* 2004/2005; Vuković 2018a). The identification of a specialised craft 'workshop' in any material is, perhaps unsurprisingly, a rarity.

The importance of colour in Vinča craft production and material culture has been consistently highlighted over the last two decades (Chapman 2011). Whether evidenced in the selection and extensive continued use of white coloured materials for tools and ornaments in stones such as quartz and magnesite or bones and shells (Antonović 2003; Vitezović *et al.* 2017) (see Chapters 16, 17, 31, 32, 45, 46), or the consistent selection of green and black copper-rich ores for both copper and tin bronze metallurgy (e.g. Radivojević and Rehren 2016; Radivojević *et al.* 2013) (see Chapters 11, 26, 41), or the pale yellow, grey or black shades of ceramics created through different firing conditions (Amicone *et al.* 2020; Chapman 2006) (see Chapters 14, 29, 43), the colourful aesthetics in craft materials were evidently important. This is certainly not unique to the Vinča culture. The pre-existing networks of long-distance circulation via rivers in the central Balkans of *Spondylus*/*Glycymeris* shells from the Aegean (Bajnóczi *et al.* 2013; Dimitrijević and Tripković 2002; Windler 2018, 2019) and obsidian from the Carpathian Basin (Milić 2015; Tripković and Milić 2008) (see Chapter 19, 49) are also strongly evident in the earlier phases of the Vinča culture. The subsequent identification of similar networks in copper across the central Balkans and beyond (Radivojević and Grujić 2018) (see Chapter 2) implies at least a partial continuity in these inter-connections.

Interpreting Vinča culture communities

The communities who comprised what archaeologists now term the Vinča culture were farmers living in settlements that, at certain sites, had a population of more than 500 but fewer than 2000 (Porčić 2019a) (see Chapter 40). There is no evidence for any proto-urbanism (Gaydarska *et al.* 2020) nor specialised military, religious or administrative centres, thus reflecting the broader Balkan area during the Neolithic-Chalcolithic Age (Chapman 2010; 2020a; Lichter 2014; Porčić 2019b; Reingruber 2014). The existence of larger buildings at settlements has generated considerable debate concerning their potential as the residence of elites (Chapman 2010; Lichter 2014).

It is difficult to evaluate any interpretations of social hierarchies against the evidence for the funerary record as there are only two known cemeteries for the entire Vinča culture, one at the early Vinča site of Botoš (Marinković 2010) and burials at the late Vinča settlement at Gomolava (Stefanović 2008a). Both exhibit indications of differentiation among buried individuals, in terms of grave goods and sex respectively (Grbić 1934; Milleker 1938; Stefanović 2008b). Single skeleton or cremation burials, including also partial findings of human skulls or a mandible, have been found at Vinča, Potporanj (Garašanin 1979), Parţa (Lazarovici *et al.* 2001), Pločnik (Bogosavljević *et al.* 2019) and Belovode (Šljivar *et al.* 2006). Recent analysis of the Gomolava individuals indicated the possibility of a kin relationship among buried individuals (Čuljković *et al.* 2002; Stefanović 2008b: 97). The social interpretation of these burials is potentially important with differences between individuals indicated by the grave goods. The placing of a malachite bead necklace as a grave good in a child's burial indicates a social position designated by birth rather than age. Similar patterns have also been identified, albeit with far more extensive funerary evidence, in contemporary cemeteries in northeast Bulgaria (Lichardus 1988: 93–100). This interpretation, although based on a limited number of analyses, could be significant for our understanding of the Vinča culture society.

Human representation is far more extensively evidenced than human burial in the Vinča culture. The evolution of figurines can be followed in relations to face shaping: from triangular in the early periods, polygonal in the Gradac Phase, to ornitomorphic in the later phases (Garašanin 1979; Gimbutas 1982; Hansen 2007). Figurine production reached its culmination, and hence the greatest variety, during the Gradac Phase, and then slowly decreased in quality in the later Vinča periods (Garašanin 1979; Tasić 2008a). In terms of function, figurines are described as votive offerings for deities (Gimbutas 1982) or fertility symbols (Letica 1964), although the concept of an 'Earth Mother' is disputed by some authors (Tasić 1973; Ucko 1969).

The evidence for weaponry and human conflict in the Vinča culture remains ambiguous (Chapman 1999a), partially as a result of the very limited evidence of funerary practices. The majority of excavated Vinča culture settlements do have evidence for extensive burning and many had large ditches constructed at their boundaries. However, this does not necessarily support interpretations of inter-group tensions (cf. Müller 2012). Experimental replications indicate that house burning could not have occurred without a significant investment in labour, which indicates a deliberate and potentially ritual act (Chapman 1999b, 2020b; Lichter 2016; Stevanović 1997).

The extensive scholarly and political debates over the existence of a Vinča culture script based on incised markings on ceramic sherds has been analysed in considerable depth by many authors such as Winn (1981), Starović (2004, 2005) and Merlini (2005). Yet, the evidence and logic of interpretations has recently been argued to be questionable (Porčić 2019a). This does not, however, diminish the importance of the repetitive appearance of a designated set of incisions in clay (on pots or figurines) throughout Vinča culture settlements, which remains a subject of continuing interest.

Over the last 40 years, the interpretation of Vinča culture communities has been re-evaluated from their culture-historical origins as a representation of a people by a wide range of scholars, primarily from the Balkans, Germany, America and Britain, who span many theoretical approaches (Chapman 2020b; Chapman and Souvatzi 2020; Porčić 2019a). There are recurring themes across the different approaches, such as identity, inequality, (subsistence) economics, (long distance) exchange and social complexity. The intellectual breadth of scholarship encompassing the Vinča culture communities means that the social agency of fragmentation and circulation patterns in artefacts, as innovatively and influentially proposed and evaluated by Chapman (2000; 2020a) and Chapman and Gaydarska (2007), co-exists with the evaluation of inequalities with the nuanced and highly stimulating application of the Gini co-efficient model (Porčić 2019a). The modelling of settlement location selection in the landscape by individual Vinča farming communities (Marić 2017) can be complemented by the modelling of potentially related house orientations across the Vinča and Linearbandkeramik culture areas, which together almost span the breadth of continental Europe (Hofmann and Müller-Scheeßel 2020). This diversity of approaches and perspectives means that there is no single vision for what the Vinča culture represents in terms of the communities who lived and died in the central Balkans for the duration of this phenomenon.

The bibliographic reference for this chapter is:

Roberts, B. W., Radivojević, M. and Marić, M. 2021. The Vinča culture: an overview, in Radivojević, M., Roberts, B. W., Marić, M., Kuzmanović Cvetković, J., and Rehren, Th. (eds) *The Rise of Metallurgy in Eurasia*: 38–46. Oxford: Archaeopress.

Chapter 5

Introduction to Belovode and results of archaeometallurgical research 1993–2012

Miljana Radivojević

The site of Belovode (44°18'42.34"N, 21°24'27.09"E) is located near the village of Veliko Laole, *c.* 140 km southeast of Belgrade (MAP) and lies on a windy plateau with the eponymous spring running through the settlement. The location is typical for a Vinča culture settlement: a large rolling plateau of ellipsoidal shape at an altitude of *c.* 200 m, suitable for agricultural activities as well as cattle breeding in the dense forests and pastures (Šljivar *et al.* 2006: 251–252). The nearby Mlava River runs deep into the volcanic mountain range called Homolje, which lies within a zone of primary copper mining and metallurgy (Krajnović and Janković 1995).

The site has been excavated since 1993 by the National Museum of Belgrade and the Museum in Požarevac (Šljivar and Jacanović 1996a, 1996b, 1997a; Jacanović and Šljivar 2003; Šljivar *et al.* 2006; Šljivar 2006). Given that the publication record for Belovode has been mainly limited to attempts to interpret and explain archaeometallurgical activities, a more detailed account of the history of research at the site will be the focus of this chapter.

In 2010, the site of Belovode received wide international recognition following a study of five copper slag pieces, identified as the earliest in the world (Radivojević *et al.* 2010a). Further analyses of archaeometallurgical materials excavated up to 2009 (Radivojević 2012, 2013; Radivojević and Kuzmanović Cvetković 2014) led, in 2012, to the establishment of one of the largest ever international collaborative projects focusing on Eurasian archaeometallurgy[1].

During almost two decades of excavations at Belovode, led by the National Museum in Belgrade, four building horizons were recognised within *c.* 3 m of cultural stratigraphy (Belovode A–D, Figure 1). These were found to correlate well with the entire Vinča culture sequence: Vinča Tordoš (A to B1) and the Gradac Phase (I–III) (Jovanović 1994; Šljivar 1993–2009; Šljivar and Jacanović 1996b) (see Chapter 4).

The internal phasing of Belovode was established on the basis of distinctive ceramic typology (Garašanin 1951, 1973), including locally recognised pottery variations (Arsenijević and Živković 1998; Šljivar and Jacanović 1996b). At its earliest stages, Belovode was most likely inhabited by Starčevo groups, as indicated by occasional finds of late Starčevo pottery. Several excavated dwellings yielded a great abundance of pottery sherds, covering all Vinča culture phases, along with stone tools, obsidian blades, and decorative items made of malachite (copper carbonate), bone and precious stones. Malachite, pottery and bones with green stains, and numerous 'fired surfaces' are also characteristic of this settlement and will be explored in detail below. The later occupation of this site by the Vinča culture ended by fire in *c.* 4650 BC, a practice that appears to be common not only for this culture, but across the Balkans at the time (Stevanović 1997). By the end of the 4th millennium BC, a section of the site was briefly re-occupied by the Late Chalcolithic Kostolac culture, recorded both in the excavated materials as well as by C14 dating (Jacanović and Šljivar 2003: 298; Radivojević and Kuzmanović Cvetković 2014). By 2011, *c.* 430 m^2 of the site had been excavated through 17 trenches, usually 25 m^2 in size (Figure 1).

Nine accelerator mass spectrometry (AMS) radiocarbon dates have recently been obtained from animal bones from Belovode, confirming the expected Vinča culture chronological span (cf. Radivojević and Kuzmanović Cvetković 2014; Whittle *et al.* 2016; also see Chapter 37, this volume). The probability distribution for the beginning of the Vinča occupation of Belovode indicates a date of *c.* 5350 BC, while the boundary for the end of the Vinča culture use of the site is set at *c.* 4650 BC. Of particular significance here is the dating of the earliest stratigraphic evidence for the extractive metallurgy in Belovode, which starts at around 5000 BC; this is currently the earliest secure date for copper metal production anywhere (Radivojević *et al.* 2010a). Importantly, it coincides with the intensive mining activities in Rudna Glava, which culminated in around *c.* 5000 BC in the vicinity of this site (*c.* 50 km).

The mining site at Rudna Glava does not, however, appear to have been exploited by the inhabitants of

[1] https://gtr.ukri.org/project/D208DC64-842F-4E99-9C9E-248D8185D75A

Figure 1. Plan of the Belovode settlement with locations of excavation trenches by 2011. Prepared by J. Pendić.

Belovode; another copper source, discovered at Ždrelo (Figure 2), *c.* 10 km away from Belovode, has been argued to be the most likely candidate according to lead isotope analysis (Radivojević *et al.* 2010a: 2781, Figure 10). It is notable that a total of nine Vinča culture settlements have been found in the wider catchment area of Belovode, prompting scholars to propose their association with the mining and metallurgical activities both at Belovode and in the wider area (Šljivar and Jacanović 1996b). This notion remains to be explored in future research.

Archaeometallurgical materials up to 2009

All materials considered below were discovered at the site of Belovode before 2009 (see Table 1). The majority of the studied collection consists of copper minerals and malachite beads, while the rest of the assemblage includes individual slag samples, slagged ceramic sherds and copper metal artefacts. The sample size appears small in comparison with the amount of technological debris (and slags in particular) in later prehistoric periods; however, it covers a crucial period in the evolution of metallurgy in Europe and, as a coherent assemblage, is unprecedented in size, quality and resolution. In order to address metal production in the Vinča culture, activities related to copper mineral use and pyrometallurgical activities are described here in three distinctive stages of the process: copper mineral processing, (s)melting debris, and the making and working of finished metal objects.

All copper minerals studied here are recognised as archaeological since they originated from archaeological sites (in contrast to geological minerals from the mines). Although bead minerals and ores in this study are both typically malachite, a rationale to distinguish between these was developed in a previous study of material from Belovode (Radivojević *et al.* 2010a) and relates to their differentiation in the technological treatment applied during processing. Thus, minerals most likely used for bead making at Belovode (i.e. 'cold' processing; Figure 3a) are referred to as 'copper minerals', while those most likely used for copper smelting (or 'hot' processed) will be termed 'copper ores'.

In this study, copper ores are assumed to include significant manganese content, as first indicated by previous chemical analyses of copper production evidence from Belovode (Radivojević *et al.* 2010a). Macroscopically, these ores appear green and black, where green comes from the colour of malachite and black from the manganese content (Figure 3b).

a

b

Figure 2a) Open-cast mine (?) at Ždrelo; b) Entrance at the shaft-hole in Ždrelo (photo by M. Radivojević).

Table 1. Study material from Belovode, arranged by analytical number. All samples starting with M (except for M3, M6, M10 and M35) have already been studied and presented in Radivojević 2007).

No.	Analytical No.	Year	Field label	Field context	Type of Material
1	Belovode 3	2007	Trench 13, spit 14	Household	Copper mineral
2	Belovode 9	1995	Trench 3, spit 12	Household	Malachite bead
3	Belovode 10	2001	Trench 8, spit 22	Household	Malachite bead
4	Belovode 12	2007	Trench 13, spit 10	Household- workshop?	Malachite bead
5	Belovode 13	2003	Trench 10, spit 27	Household	Malachite bead
6	Belovode 18	2007	Trench 13, spit 10	Household	Copper mineral
7	Belovode 23	2001	Trench 8, spit 23	Household	Malachite bead
8	Belovode 30a, 30c	1995	Trench 3, spit 5	(Building) waste pit	Slagged ceramic sherd
9	Belovode 31a, 31b	1995	Trench 3, spit 6	(Building) waste pit	Slagged ceramic sherd
10	Belovode 33b	2008	Trench 14, spit 15, surface 4	Household	Copper mineral
11	Belovode 34a	2008	Trench 14, spit 3	Household	Copper minerals from an amphora (3 pieces)
12	Belovode 131	1995	Trench 3, spit 6	(Building) waste pit	Copper slag
13	Belovode 134	1995	Trench 3, spit 7	(Building) waste pit	Copper slag
14	Belovode 136	1995	Trench 3, spit 5	(Building) waste pit	Copper slag
15	Belovode 154	1999	Trench 7, spit 6	Household - workshop?	Malachite bead
16	Belovode M3	1995	Trench 3, spit 8	(Building) waste pit	Copper mineral
17	Belovode M6	1995	Trench 3, spit 10	(Building) waste pit	Copper metal droplet
18	Belovode M10	1995	Trench 3, spit 19	Household	Copper mineral
19	Belovode M14	2002	Trench 9, spit 11	Household	Copper metal droplet
20	Belovode M32	1995	Trench 3, spit unknown	Household	Malachite bead
21	Belovode M35	1995	Trench 3, spit 17	Household	Copper mineral
22	Belovode M12	1998	Trench 6, spit 10	Household	Copper mineral
23	Belovode M13	2000	Trench 7, spit 18	Household	Copper mineral
24	Belovode M17	2004	Trench 10, spit 8	Household	Copper mineral
25	Belovode M20	1995	Trench 3, spit 2	(Building) waste pit	Copper slag
26	Belovode M21	1995	Trench 3, spit 4	(Building) waste pit	Copper slag
27	Belovode M22 (a,b)	1995	Trench 3, spit 5	(Building) waste pit	Copper slag
28	Belovode M23	1995	Trench 3, spit 7	(Building) waste pit	Copper slag

Figure 3. a) Typical bead malachite from Belovode; b) Typical black and green copper mineral from Belovode (Radivojević and Rehren 2015: Figure 2)

'Smelting' of copper ores refers to the primary extraction process, where the produced metal was usually further purified by refining or melting. Fundamentally, copper smelting can be separated into two discrete steps: the reduction of copper ore to copper metal, which requires reducing conditions and temperatures from *c.* 700 °C upwards (Budd and Ottaway 1991), and the melting of the copper metal, which requires temperatures around and in excess of 1083 °C. The melting or second step in this process has fewer constraints on the redox (reduction-oxidation) conditions, while smelting requires an oxygen partial pressure of less than $10^{-5.5}$ atm (cf. Elliott 1976). Thus, the balance between the two opposing tendencies, high temperature and reducing conditions, is the key to the successful smelt.

In the first step, the mineral (typically malachite) decomposes:

$$Cu_2CO_3(OH)_2 \rightarrow 2CuO + CO_2 + H_2O \ (< c.\ 500\ °C)$$

and the incomplete burning of charcoal provides the carbon monoxide which drives the reduction of copper oxide to copper metal:

$$2CuO + CO \rightarrow Cu_2O + CO_2$$

$$Cu_2O + CO \rightarrow 2Cu + CO_2$$

At temperatures < 1083 °C, copper precipitates in the solid state and forms a porous, spongy, powdery mass, still mixed with gangue minerals (Hauptmann 2007: 222; 2020). However, although relatively low temperatures are unlikely to allow for the separation of slag, there may well have been pockets of higher temperature directly in front of the blow pipes that led to localised fusion and partial melting of slag and metal. Thus, copper that formed during the first step was only transformed chemically and needed higher temperatures for a physical change from the solid powdery mass into liquid metal. The higher temperatures also facilitated full separation from gangue minerals, resulting in the formation of slag.

For the full melting of copper, temperatures of around *c.* 1100 °C needed to be maintained for a period sufficient to allow fully molten conditions to be reached throughout the charge. Such temperatures require an installation which retains and concentrates the heat generated from the fuel within the reaction container (Rehren 2003). This container can, for both stages, be a ceramic vessel such as a crucible or furnace, or a simple hole in the ground leaving little identifiable structure. Significantly, it is possible that both steps were performed in the same container or structure, facilitating the blending of one step into the other. They may also have been conducted separately, depending on various constraints originating from the physical, social or environmental spheres.

The common forms of smelting debris found at archaeological sites are installations, slagged ceramics and slags. Technical ceramics (crucibles, furnace remains, or *tuyères*) are particularly interesting for studying past metallurgical processes as they reflect technological choices. The most informative part of a crucible or an installation fragment is the slag attached to its walls, as is present on several 'slagged sherds' from Belovode (Figure 4b). One of the most valuable materials for studying metal production at the site are the small slag pieces (eight in total), discovered together with slagged sherds in a single trench (No. 3) (for a typical example see Figure 4a).

Slag is a waste product of high-temperature metallurgical processes, which 'collects' all unwanted substances or impurities from ores, the furnace lining and the charcoal ash within the smelting, casting or alloying systems. Chemically, it is a solution of molten oxides. Two major oxide components of metallurgical slags are silica (SiO_2) and iron oxide (typically the more reduced ferrous oxide, FeO), which are combined in wide varieties with elements such as manganese, magnesium or zinc. Other common constituents are lime (CaO) and alumina (Al_2O_3), followed by a number of minor compounds which contribute to the overall chemical composition. A slag of optimum composition has a low energy of formation, a low melting point and a high degree of fluidity (e.g. low viscosity) (Bachmann 1980: 118, 1982: 10). Slag can represent a single phase (glass), although ancient slag is most commonly constituted of a wide range of crystals. The newly formed compounds in the ancient slag are usually referred to as 'phases' or 'crystals' rather than minerals, even if they are structurally identical, since they are formed by anthropogenic processes.

Slag is ideal for studying past pyrometallurgical activities because it typically contains information about all components affecting its formation. The slag chemistry preserves details of the conditions of the smelt and composition of ores, indicates which metal was extracted, the contribution of fuel ash, potential fluxes, and even the design of installations (Rehren *et al.* 2007). Moreover, slags are ideal for study because they are highly resistant to weathering conditions and are usually accessible for invasive analysis by archaeological scientists. In the archaeological record, early slags are typically formed on top of the metal in a crucible or lined container (Tylecote *et al.* 1977: 307) and may thus be found as free pieces or attached to the walls of an installation. However, the free slag samples can be easily overlooked or mistaken for clay or minerals, as was the case with copper slags from Belovode (Radivojević *et al.* 2010a: 2779). Also, the slags from early periods were often crushed in the search for metallic prills for further refining, making them archaeologically less visible.

The artefact group at Belovode consists of malachite beads and copper metal items. Malachite beads were commonly processed in a series of 'cold' shaping techniques, applied

Figure 4. a) Belovode slag (sample 134); b) Belovode slagged sherd (sample 31b)

in several technological steps. The ornament (bead) making process starts with a raw nodule of mineral which, once roughly shaped (but not perforated), is recognised as a roughout. Subsequent fine chipping creates a preform or blank, and this is commonly followed by drilling to produce the final bead shape (Lankton *et al.* 2003: 16). Notably, the order of this sequence varies both culturally and in relation to the materials used. The variety present within the copper metal artefacts group is particularly informative for the metallurgical *chaîne opératoire* in the Vinča culture, allowing the investigation of different sequences of production and their interpretation within wider cultural, environmental and physical surroundings.

All technological analyses were carried out at the Wolfson Archaeological Science Laboratories at the UCL Institute of Archaeology, by the primary author of this paper, and under the supervision of Professor Thilo Rehren (then UCL Institute of Archaeology, currently at The Cyprus Institute in Nicosia, Cyprus). Below I present the main results, while the full dissertation paper is available in Appendix B_Ch5.

Results

Copper mineral use

Malachite beads, pendants and 'green' copper minerals are present from the earliest occupation of Belovode and continue throughout all building horizons. These most numerous finds at the site are usually uncovered and mixed with ash and pieces of charcoal. Other contexts include house floors, storage jars, ceramic sherds (with malachite adhered to them), or workshops within a household. Two such areas in Trenches 12 and 13 together yielded *c.* 2.5 kg of copper minerals throughout all building horizons, equivalent to almost one third of the total weight of malachite finds discovered at this site[2].

Ten samples from the copper minerals group were analysed, three of which came from the so-called 'metallurgical' Trench 3 (Radivojević *et al.* 2010a); the rest originated from various household contexts (Radivojević and Rehren 2016) (see Table 1). All samples, barring M3, M13, M17, 33b and 3, were confirmed to be malachite (copper carbonate) with significant levels of manganese in their composition (Radivojević 2013 18 ff., 2015). The remaining copper mineral samples were sourced from a vein containing cuprite with copper sulphides. The small size of all mineral samples (*c.* 1–3 cm) may imply that they were beneficiated, i.e. crushed to facilitate smelting.

Significantly, both oxidic and sulfidic minerals present a common feature: their colour is distinctively black (or dark) and green (see Figure 3), which is argued to reflect an intentional choice of a colour that appealed to the Belovode metallurgists (Radivojević and Rehren 2016). The only group of minerals from Belovode that were not black-and-green were those used for malachite bead making, implying that colour-coding was a very important aspect for copper mineral selection. Although the geological environment of eastern Serbia offers more than one kind of copper mineral (Krajnović and Janković 1995), the Belovode copper craftsmen seem to have selected only pure green or black/dark and green minerals.

Figure 5. Stone mallet from Trench 7 at Belovode (after Šljivar *et al.* 2006: plate I/4)

[2] The author weighed all malachite from Belovode for the purpose of her PhD research completed in 2012 at the UCL Institute of Archaeology.

Two stone mallets, each with a central groove, were discovered in the context of workshops in Trenches 1 and 7 (Figure 5) (Šljivar *et al.* 2012: 259, Plate I/4), and could offer clues about tools used during processing. Similar finds from Rudna Glava (Jovanović 1982) suggest such tools may also have been used for mining.

(S)melting debris and installations

Charred surfaces with malachite, copper mineral powder, adhering to fragmented ceramic sherds and grooved stone mallets are common in household contexts in Belovode. A few small pottery vessels of conical shape and coarse fabric from Trenches 7 and 8 (Vinča B1 horizon, Figure 6) also had green minerals attached to the outer surface however analyses have shown that these were not crucibles (Radivojević 2007). In addition, a fragment of ceramic mould discovered on the site surface, is thought to originate from the latest layer of occupation (Šljivar *et al.* 2006: 259, Plate I/3).

Two shallow pits rimmed with ceramic sherds and a burnt layer of clay from Trenches 10 and 13, a Vinča B1 building horizon, are identified as early furnaces by the excavator (Figure 7), based on comparison with similar small hearths discovered at Durankulak in Bulgaria from the late 5th millennium BC (Šljivar *et al.* 2006: 253, 260, Plate II/4; cf. Dimitrov 2002). Elongated cylindrical ceramic forms (so-called 'chimneys') with a diameter of about 20 cm, a reconstructed height of up to 80 cm, and open at both ends (Figure 8) have been tentatively linked to these rimmed pits in the ground and, thus with the smelting operation (Šljivar 2006; Šljivar *et al.* 2006: 253). Nevertheless, neither of these 'chimneys' showed convincing traces of use in the smelting process after compositional analyses (Radivojević *et al.* 2010a: 2779).

Metal production evidence

Pyrometallurgical activities at Belovode are represented by only eight individual copper slag samples, four slagged ceramic sherds, and a copper droplet (metal) sample from Trench 3, all of which demonstrate sustained smelting activities taking place at the outskirts of the site. Another pyrometallurgical situation has been recovered in Trench 9; both will be

Figure 6. a, b) Fragment of ceramic vessel with green powder adhering to its bottom (labelled as Belovode 30ch, Trench VIII, spit 12, Belovode 2001 campaign, reported in Radivojević 2007); c, d) Bel29 Fragment of ceramic vessel with light green and brown matrix adhered to the outer walls (labelled as Belovode 26, Trench VII, spit 12, Belovode 2000 campaign, reported in Radivojević 2007).

Figure 7. Shallow pit lined with ceramic sherds from Trench 10 (upper left corner) (Photo courtesy of Duško Šljivar, National Museum Belgrade, Serbia).

presented in more detail below. These materials were usually discovered in areas filled with ash, charcoal, charred wood or stone constructions and mainly represent an outdoor activity.

Trench 3

Trench 3 (dimensions 8 m x 2 m) yielded evidence for copper smelting activities throughout the final, D horizon of occupation i.e. across the entire Gradac Phase sequence starting in *c.* 5000 BC (Radivojević *et al.* 2010a). This phase of the Vinča culture is known to have lasted longer in the Morava Valley settlements than in those situated nearer to the Danube (Jovanović 1994) and, at this site, most likely covers the late Vinča culture sequence, dated to *c.* 5000–4650 BC.

The Belovode D horizon, represented in this trench by materials from a waste pit (Šljivar and Jacanović 1996a), includes all finds from spits 1–11, including various archaeometallurgical debris (Table 2). Thousands of ceramic finds were unearthed in this horizon alone, many of which are diagnostic of the Gradac Phase in general, for example, the conical bowls with a thickened rim channelled on the interior, or tri- and four-partite vessels with a cone-shaped neck and protruding shoulder, usually accompanied by ornaments with typical incised ribbon decoration (Arsenijević and Živković 1998; see also Schier 1996; Schier 2000).

The slag pieces collected from this trench are vitrified, strongly magnetic and green-stained droplets, not exceeding 1 cm in length (example in Figure 4a). They appear to have been highly viscous and very rich in copper metal, however with no signs of crushing in pursuit of copper metal prills. This may have been due to their small size and weight, since all eight samples weigh, in total, less than 10 g. Given that, in appearance, these slag pieces resemble (green) malachite as a result of the corrosion of the copper metal prills entrapped in them, it is possible that the green colour facilitated their recognition in the field excavations, leading to a biased recovery in favour of more copper-rich pieces, and that those without green staining were more likely overlooked, as mentioned above (cf. Radivojević *et al.* 2010a: 2779).

The green staining on fragmented slagged sherds comes from the contact of these samples with the metallurgical process. The slagged mass is surrounded by heavily vitrified areas, which appear along the edges of the studied samples (Figure 4b), but also extend across their cross-sections. The latter implies that these sherds were most likely used in a fragmented form during the metallurgical process. A copper metal droplet (M6), found in addition to this assemblage, provides a more complete account of the types of metallurgical activities conducted in the workshop in Trench 3.

All metallurgy-related materials were discovered sealed with building waste such as the remains of house daub, domestic pottery and animal bones. Notably, in spit 10, which belongs to this building horizon, two shallow, rock-lined constructions, indicated as fireplaces were identified as potentially linked with metallurgical debris in excavation records. The stratigraphic evidence related to the earliest slag piece is dated to *c.* 5000 BC; the smelting evidence, according to the excavation reports, continued until the abandonment of the site in *c.* 4650 BC.

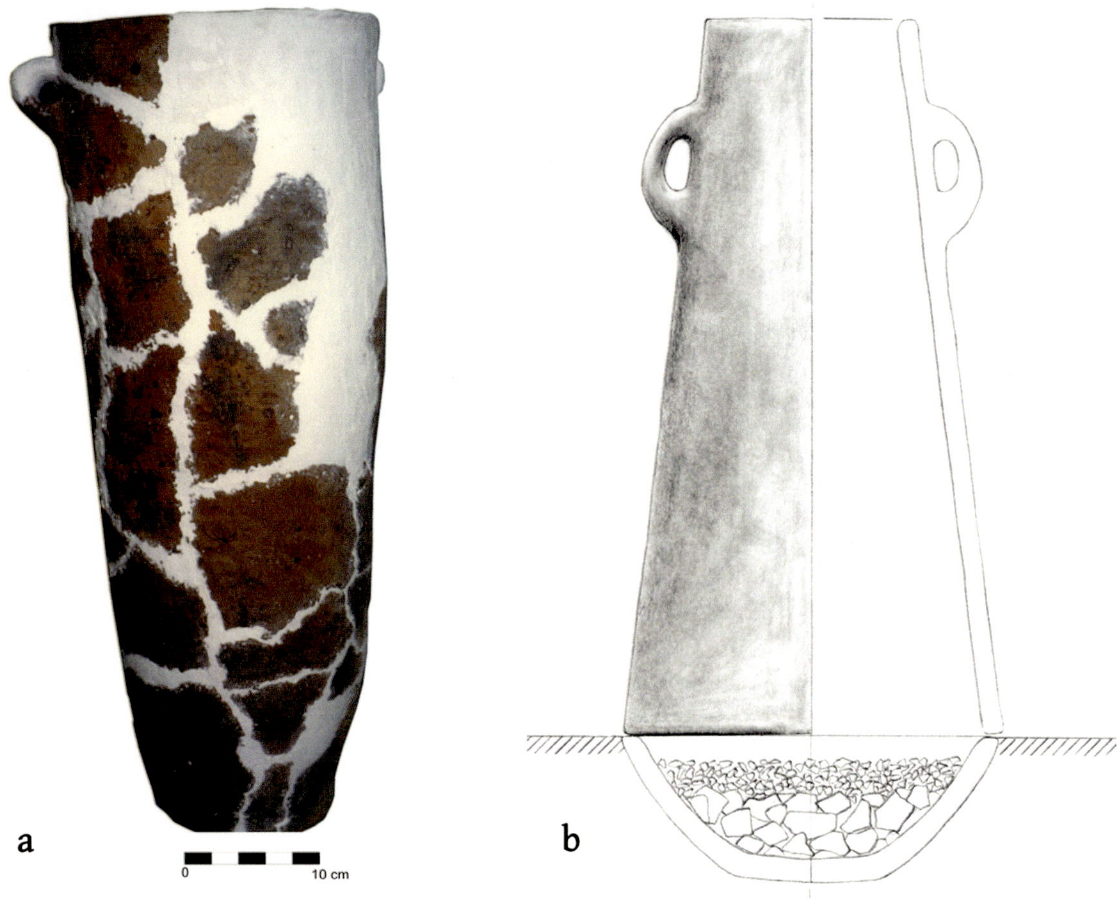

Figure 8. a) Reconstructed ceramic 'chimneys' from Belovode. Note the marks of secondary burning. (Photo courtesy of Duško Šljivar, National Museum Belgrade, Serbia); b) Tentative reconstruction of the smelting installation at Belovode (after Šljivar *et al.* 2006: plate II/5).

Table 2. Study materials from Belovode 'metallurgical' Trench 3, arranged by spits.

No.	Analytical No.	Year	Context	Type of Material	Chronology/building horizons
1	Belovode M20	1995	Trench 3, spit 2	Copper slag	Gradac Phase, Belovode D
2	Belovode M21	1995	Trench 3, spit 4	Copper slag	Gradac Phase, Belovode D
3	Belovode M22 (a, b)	1995	Trench 3, spit 5	Copper slag	Gradac Phase, Belovode D
4	Belovode 136	1995	Trench 3, spit 5	Copper slag	Gradac Phase, Belovode D
5	Belovode 30 (a,c)	1995	Trench 3, spit 5	Slagged ceramic sherd	Gradac Phase, Belovode D
6	Belovode 31 (a,b)	1995	Trench 3, spit 6	Slagged ceramic sherd	Gradac Phase, Belovode D
7	Belovode 131	1995	Trench 3, spit 6	Copper slag	Gradac Phase, Belovode D
8	Belovode 134	1995	Trench 3, spit 7	Copper slag	Gradac Phase, Belovode D
9	Belovode M23	1995	Trench 3, spit 7	Copper slag	Gradac Phase, Belovode D
10	Belovode M3	1995	Trench 3, spit 8	Copper mineral	Gradac Phase, Belovode D
11	Belovode M33b	1995	Trench 3, spit 8	Malachite bead	Gradac Phase, Belovode D
12	Belovode M6	1995	Trench 3, spit 10	Copper metal droplet	Gradac Phase, Belovode D
13	Belovode 9	1995	Trench 3, spit 12	Malachite bead	Vinča B1, Belovode C
14	Belovode M35	1995	Trench 3, spit 17	Copper mineral	Vinča B1, Belovode C
16	Belovode M10	1995	Trench 3, spit 19	Copper mineral	Vinča B1, Belovode B

Compositional and contextual analyses of archaeometallurgical assemblage from Trench 3 indicated that they were probably part of a minimum of three smelting episodes (Radivojević and Rehren 2015: 22 ff.). This assumption is further strengthened by the compositional analyses, which indicate strong correlations in the ore signature for production evidence, i.e. manganese and zinc. Manganese is specifically known for the advantageous chemical-physical properties of its oxides, which enable an easier reduction of copper ores to metal and slags (Huebner 1969: 462, Figure 3). The aesthetic aspect of copper and manganese rich ores (being black and green) has been argued to be particularly important, as this colour code appears to dominate the Vinča culture metallurgy of both copper and tin bronze making (Radivojević *et al.* 2013).

All slag-based samples appear very heterogeneous under the microscope with dross areas (copper-oxide-based phases) dominating the microstructure (Radivojević and Rehren 2015: 16). Other major phases are metal prills, spinels and delafossite, all of which are embedded in a glassy matrix (for comparison and detail, see Chapter 11). The co-occurrence of these phases suggests that the gas atmosphere had a partial oxygen pressure that was low enough to reduce cuprite to copper metal (Elliott 1976) and was therefore sufficient to smelt copper. The working temperatures of the smelting systems across all slag-based samples is estimated to have been just over 1083 ºC, according to the fully molten state of copper metal prills embedded in them.

Trench 9

Pyrometallurgical activities are also recorded in Trench 9 (dimensions 5 m x 5 m) (Jacanović and Šljivar 2003). In this trench, spits 21–7 yielded typical Vinča culture material that corresponds with all four building horizons at Belovode (A–D). Archaeological materials found in spits 6–1 belonged, most probably, to the latest manifestation of the Vinča D phase (cf. Jovanović 1994).

The use of copper minerals occurred regularly throughout the building horizons Belovode A–C, excluding Belovode D and the succeeding, not yet well-defined, horizon. Of particular note here is spit 11, which yielded a copper metal droplet (M14, see Table 1). This was discovered in the context of the regular appearance of ceramic pedestal bowls, typical of the Vinča A to Gradac Phase. Chronologically, and in relative stratigraphic terms, it can be correlated with the early Gradac Phase and, effectively, with the start of metallurgical activities in Trench 3, dated to *c.* 5000 BC (cf. Radivojević *et al.* 2010a). Interestingly, early Belovode horizons in this trench yielded Vinča culture figurines with modelled appliques: necklaces with perforated disc-pendants (Šljivar *et al.* 2012: 31, Plate III/1-4) resembling gold applications from the late 5th millennium BC burials and settlements in Bulgaria and Ukraine. Similar examples have also been discovered at the site of Vinča-Belo Brdo (Tasić 2008a: 151, Figure 58).

An unusually large, round slag cake was unearthed in spit 18 (Belovode 40) and argued as firmly contextualised within the Vinča B1 Phase (Šljivar *et al.* 2012: 33–34, Pl. VIII/1)(Radivojević and Kuzmanović Cvetković 2014) (see also Chapter 3, this volume). Preliminary analyses revealed that this artefact has significant concentrations of lead (Radivojević and Rehren: 2019), which may suggest the production of this metal even before copper; future analysis is expected to shed more light on this unusual archaeometallurgical evidence. In terms of absolute dating, the metal droplet (M14) could be dated to *c.* 5000 BC, while the slag cake (Belovode 40) may be ascribed the date of *c.* 5200 BC (Radivojević 2013: 17; also Chapter 3, this volume).

Copper mineral and copper metal artefacts

Malachite beads occur throughout all horizons at Belovode and vary in size from 4 mm to 1.5 cm in diameter. One exception is a deltoid pendant with perforation found in Trench 1, Vinča A1 horizon. Beads selected for this study were found related to various copper minerals, slag pieces, workshop activities or dwelling structures. They occur as whole artefacts or fragments, implying the potential presence of a bead workshop at the site. Some Belovode beads have undergone mineralogical study, which confirmed the presence of malachite, with traces of tenorite and kolwezite (Jović 1996). Compositional analyses further suggested the use of rather pure malachite for bead making, with main impurities being low contents of iron and zinc oxides, as well as manganese in one instance (Radivojević 2012: 306, Table 48).

Both copper metal droplets (M6 and M14) were initially thought to be copper minerals until their cross-sections revealed a dark red phase (copper metal) surrounded by a thick, light green layer of corrosion (Radivojević *et al.* 2010a; Radivojević 2013: 28, Figure 18). The amorphous shape of the metal phase in both samples indicates that it was once molten copper, however it is now heavily corroded due to post-depositional processes. Both droplets most likely cooled rapidly, as indicated by porosity holes and cracks throughout the investigated polished sections.

In addition to these metal droplets, two further copper artefacts have been identified as belonging to the Vinča culture occupation of Belovode: a copper chisel and a bun-shaped metal ingot (Šljivar *et al.* 2006: 252, 269, Plate I/1, 2). Since these were found in the vicinity of the site, they could belong to the Late Chalcolithic occupation, as already indicated by the late 4th millennium BC dates from the top horizon in a defined area of this settlement.

Discussion

Extensive microanalytical examination of copper minerals and production debris indicate that, together, they form a coherent assemblage, largely linked through significant manganese and zinc content, which came from the gangue minerals in an exploited copper deposit.

It is noteworthy that, regardless of how insignificant the small amount of slag from Belovode may appear, it fits the overall picture of rather ephemeral production evidence in pre-Bronze Age metallurgy (Craddock 2001: 152). The early slags of the 5th and 4th millennium BC mostly look the same: they weigh hardly more than a few grams each and reach nut size at most (cf. Hauptmann 2007: 158); it is not surprising that, size-wise, the earliest documented slags sit at the lower end of the range of slag 'heaps' identified thus far (Bourgarit 2007: 4, Table 1).

The Belovode smelting installation debris also fit well the 'ephemeral model' of Chalcolithic metallurgy. Slagged sherds from the site suggest the presence of a hole-in-the-ground installation lined with broken pottery. However, none were discovered in situations related to a hearth or a similar detectable feature in the field. Hence, it may be hypothesised that the copper smelting installation was too ephemeral to survive *c.* 6000–7000 years of post-depositional processes and took the form of a shallow indentation in the soil, lined with ceramic sherds. Such an installation was possibly operated by using blowpipes or *tuyères*, where five to six blowpipes would normally suffice to bring the temperature to around the 1100–1200 ºC needed to (s) melt copper (cf. Rehder 1994: 221).

The chemical fingerprint of the ore used for smelting copper at Belovode indicates a strong chemical association of Mn and Zn, with some Co, Ca and Fe (Radivojević 2013: 21 ff.). It is likely, therefore, that the ores used could have been a paragenesis of copper-zinc-manganese, with some other elements coming from the attached primary copper sulphide mineralisation (such as S, Fe) or gossan. Their inclusion in the ore charge may not have been intentional, but most likely evolved as a natural consequence of the stratigraphy of weathered copper sulphide ore bodies (cf. Rostoker *et al.* 1989: 85).

With regard to the copper droplets M6 and M14, the structural difference between them results from M6 containing a sulfur-rich phase. Thus, it may be assumed that the smelted ore in M6 was originally a mineral combination of copper sulphides (chalcocite) and oxides (or carbonates). This assumption is corroborated by the presence of a similar mineral combination at Belovode (3 and 33b), which suggests that it was these types of 'mixed' minerals that were included in smelting activities at the site (Radivojević 2013: 30).

Conclusion

The overview of activities related to copper mineral use and extractive metallurgy at the Vinča culture site of Belovode suggests that metal smiths at the settlement were covering several stages of the metallurgical *chaîne opératoire*. The analytical highlights underlying this narrative are the compositional connection of copper minerals and production debris from Trench 3 through manganese-rich copper ores, the potential presence of a minimum of three separate smelting events in Trench 3 and identification of both slagging and non-slagging events in the earliest context of metallurgical development at the site.

The black and green minerals (both oxidic and sulfur-rich) selected by the Belovode miners predate the earliest documented smelting event at this site, indicating that they were potentially experimented with during the first centuries of occupation. This experimentation, although probably unsuccessful, could be recognised in sample M6, which contains molten copper, but also some residual sulfur-rich phases. The distinctively coloured black and green copper minerals became copper ores only at the dawn of the 5th millennium BC, and their smelting is attested by the strong presence of manganese and zinc in the chemical signatures of glassy slag matrices as well as in other newly formed phases in the metal production samples.

The combination of analytical results and the available fieldwork data facilitated recognition of potentially three separate smelting events in Trench 3, one of which, represented by M6, was most likely unsuccessful. Firstly, the data demonstrates sustainable smelting activities during at least the latest building horizon in Belovode (*c.* 5000–4600 BC), while sample M6 was produced somewhat earlier. Secondly, the data shows a similar technological principle for the slagging events, but also highlights that the early beginnings of metallurgy were not producing slag, as previously assumed by Craddock (1995). Still, it was not long before the Belovode metal smiths optimised metal extraction by producing minute concentrations of slag, documented at the turn of the 5th millennium BC (Radivojević *et al.* 2010a).

It appears that the Belovode metal smiths were aware of the properties of black and green copper minerals, knowledge that possibly developed over the course of a few centuries. This understanding related not only to manganese-rich copper minerals, but also to those rich in sulfur, indicating that it was the distinctive colour of the minerals that prompted their selection and subsequent smelting. The colour appeal of copper ores has already been argued to be the most significant sensory aspect for the invention of early metallurgical activities (Radivojević 2015) and the Belovode example stands out as the earliest currently known in the line of evidence assembled for the Vinča culture.

Small-scale smelting reactors emerge as the principal technological choice in the copper metal production, not only for Belovode but also throughout the Vinča culture. The minute size of slags and ephemeral evidence of smelting installations has also been discussed in Radivojević and Rehren (2016), who pointed towards a similar technological principle of copper production being present across the Vinča culture settlements of Belovode, Vinča, and Gornja Tuzla. The smelting process at all three sites can be generally characterised by the use of mixed copper ores selected for their colour, which were smelted in moderately reducing/partially oxidising conditions in an ephemeral 'hole-in-the-ground' installation. The process was conducted at all three Vinča culture sites with remarkably similar mastery, resulting in copper metal probably being produced in globules together with small quantities of highly viscous slag; some of this slag formed in direct contact with already-broken pottery sherds.

Although the Vinča culture production evidence studied here represents only very few of these episodes, the replication of the production pattern across all three sites within different occupational sequences indicates that the level of expertise achieved remained relatively unchanged and potentially stagnant across an estimated period of six centuries. The process slowly evolved into smelting of more complex copper ores only towards the end of this culture, as attested by the use of colourful complex copper-tin bearing ores for making tin bronze objects at the site of Pločnik around the mid-5th millennium BC (Radivojević *et al.* 2013). The distinctive role of black-and-green ores in early Balkan metallurgy therefore emerges as a key factor to support arguments for its independent development, as well as supporting claims about its unique technological trajectory.

One of the most significant outcomes of the results discussed here is to demonstrate the potential of a materials science approach for incorporating insufficiently contextualised materials into an integrated technological and archaeological narrative. The lack of precise contextual information is a result of insufficiently detailed documentation for the Belovode excavation. Despite careful excavation, records are not always clear and, unfortunately, do not offer sufficient information regarding the formation of the site. Since the majority of field documentation between 1993 and 2012 remains unpublished, relevant contexts for this study are limited to relative spits within individual trenches. This has provided only limited information on the spatial distribution of metallurgical debris and its relation to the spatially closest settlement features. Importantly, most of these features originated from arbitrary units, where the relationships are obscured and not always straightforward. This chapter therefore focused on the material properties of a variety of excavated artefacts, highlighting important connections between them from this particular perspective, rather than relying on a distribution plan of small trenches scattered across the site.

A further problem in relation to the state of research at Belovode arises from the lack of AMS data for specific contexts related to diverse metallurgical activities. AMS data for this site, recently synthesised in Whittle *et al.* (2016), do not offer sufficiently high resolution to distinguish copper mineral use and pyrometallurgical processes at Belovode (and other Vinča culture settlements), nor do they provide direct dating evidence for the majority of metallurgy-related finds. Therefore, the temporal analysis of metallurgical activities presented here has been mainly dependant on the relative chronology— based on specific pottery forms—and the conventional periodisation of the Vinča culture across the entire region. Such an approach, although potentially not without errors, currently provides the only feasible contextual framework for most metallurgy-related samples considered here.

An understanding of the relationship between the Gradac Phase and the rise of metallurgy is crucial for our interpretation of this process within the Vinča culture: the earliest stratigraphic evidence for copper smelting is discovered at the beginning of this phase and is also contemporary with the intensified activities in Rudna Glava, as well as the earliest dated copper implement from Pločnik. The changes in material culture that follow this phase are a phenomenon common across the whole Balkan region (cf. Garašanin 1994/1995) and will require particular attention for the interpretation of the origins of metallurgy in this part of Eurasia in future discussions.

The introduction of metallurgy evidently influenced other aspects of material culture at this time. The most important association is, beyond doubt, with pottery production, such as the conjunction of the appearance of black burnished ware and graphite painted decoration with the emergence of metallurgy, which has been discussed at length (e.g. Renfrew 1969; Gimbutas 1976a), and studied in detail within this volume (see Amicone *et al.* Chapters 14, 29 and 43, this volume) and in Amicone *et al.* (2020). Another valuable observation comes from Belovode, where modelled applications on figurines resemble contemporaneous metal jewellery from sites located along the lower Danube and further towards the northern Black Sea coast (see examples in Dumitrescu 1961; Sergeev 1963). This highlights the importance of the Danubian communication route for the spread of metallurgy across the Balkans, also highlighted in the most recent application of complex networks methodology by Radivojević and Grujić (2018).

The following chapters will explore in greater detail both the technological and archaeological circumstances for the appearance of the world's earliest metallurgy, providing a more nuanced contextualisation and chronological framework for the invention widely argued as the trigger for broader material and social changes, both in the Vinča culture and beyond.

The bibliographic reference for this chapter is:

Radivojević, M. 2021. Introduction to Belovode and results of archaeometallurgical research 1993–2012, in Radivojević, M., Roberts, B. W., Marić, M., Kuzmanović Cvetković, J., and Rehren, Th. (eds) *The Rise of Metallurgy in Eurasia*: 47–59. Oxford: Archaeopress.

Appendix

Appendix avaliable online as part of Appendix B at https://doi.org/10.32028/9781803270425/AppendixB_Ch5

Chapter 6

Introduction to Pločnik and the results of archaeometallurgical research 1996–2011

Miljana Radivojević

The site of Pločnik (43°12'35.72"N, 21°21'50.42"E) is situated beneath the eponymous modern village, 19 km west of the town of Prokuplje in south Serbia and 300 km south of the capital, Belgrade. It is set at about 300 m above sea level on the left bank of the Toplica river, whose shifting course presently erodes away the estimated 3.60 m thick cultural layer of the site (Stalio 1960: 34; 1962: 21). The village is surrounded by good quality agricultural land (Chapter 23, this volume) and thermal springs, and has good communication routes along the river Toplica. This is the major watercourse in this part of Serbia, which springs from Kopaonik, a mountain approximately 50 km away from Pločnik, whose rich iron veins were exploited in Roman and medieval times (Bogosavljević *et al.* 1988, 1989; Bogosavljević-Petrović and Tomović 1993; Bogosavljević-Petrović 1995; Mrkobrad 1989; Mrkobrad *et al.* 1989). More than 50 sites with archaeometallurgical installations from both periods were recognised around toponymic places like Suvo Rudište (in Serbian: 'Dry Mine') or Bakarnjača (in Serbian: 'Copper-rich'), indicating intensive metallurgical activities in the past.

The archaeological settlement of Pločnik was initially discovered in 1928, when the first group of metal artefacts was found during the building of the railway line between the towns of Prokuplje and Kuršumlija. Excavation campaigns commenced later the same year and then, after a considerable pause, continued between 1960 and 1978, under the jurisdiction of M. Grbić and B. Stalio respectively, both of the National Museum Belgrade (Grbić 1929; Stalio 1960, 1962, 1964, 1973). Most recently, field excavations resumed in 1996 under the joint supervision of D. Šljivar (National Museum Belgrade) and J. Kuzmanović Cvetković (Museum of Toplica, Prokuplje) (Šljivar 1996, 1999, 2006; Šljivar and Kuzmanović Cvetković 1997a, 1998a, 1998b, Šljivar *et al.* 2006; Kuzmanović Cvetković 1998). Over the course of two years, starting in 2012, field excavations at Pločnik were carried out under the joint supervision of the National Museum in Belgrade, the Museum of Toplica in Prokuplje and the UCL Institute of Archaeology.

Grbić's and Stalio's campaigns uncovered an area of *c.* 1800 m^2, to which another *c.* 650 m^2 from an ongoing campaign adds up to a total of *c.* 0.2 ha explored thus far. The estimated size of the uppermost cultural layer of the Vinča village at Pločnik was estimated to be *c.* 100 ha (Šljivar and Kuzmanović Cvetković 1998b: 80). Most recent geophysical prospection established its more accurate size to be almost a third of the initial estimate, as discussed in more detail in (Chapter 24, this volume).

The unique and abundant ceramic finds from the site inspired Milutin Garašanin to name the late Vinča culture phase after Pločnik (I, IIa and IIb) (Garašanin 1951: 12), corresponding to Vinča C, D1 and D2 (Milojčić 1949). Twenty massive copper implements, discovered by chance and also during the excavations in the 1928 campaign, prompted Grbić (1929) to entitle the first site publication: 'Pločnik, eine Prähistorische Ansiedlung aus der Kupferzeit' ('Pločnik, a prehistoric settlement of the Copper Age'). He assumed that Pločnik was an important metallurgical centre that existed at the dawn of the Copper Age and maintained intensive exchange networks with other contemporary communities in the region, such as those occupying Vinča-Belo Brdo, Gradac or Butmir (Grbić 1929: 7, 18). Nevertheless, these copper implements, having been unique and isolated finds from a location remote from the Near East, did not appeal to Garašanin as being of genuinely Vinča culture origin (Garašanin 1951; 1973: 185). He referred to them as intrusive 'hoards' belonging to the Early Eneolithic/Chalcolithic Bubanj-Hum culture[1] which, in his opinion, succeeded the Vinča culture sequence at the site of Pločnik. Stalio, coming across two more groups of massive metal artefacts in Pločnik, agreed that they belonged to the Bubanj-Hum culture (Stalio 1960: 35, 1964: 39–40).

In 1978, a single, well-contextualised copper metal implement was discovered associated with a Gradac Phase feature (dwelling?) but was only published almost two decades later (Šljivar 1996; Šljivar and Kuzmanović Cvetković 1997a). The resumed excavation campaign (from 1996) brought more evidence for dating metallurgical activities at Pločnik firmly within the Vinča culture sequence. In total, three building horizons were identified on the site, all of which

[1] This would translate into Middle Chalcolithic given the periodisation of contemporary cultures in Bulgaria, see Chapter 3.

belonged solely to the Vinča culture, with no detected intrusions from later periods (e.g. Šljivar 1996: 94; Šljivar and Kuzmanović Cvetković 1998a). Horizon III (the Gradac Phase) yielded a well-contextualised copper chisel typologically resembling some of the previous chance metal finds from Pločnik (Šljivar *et al.* 2006: 255, Plate VIII/3). This, and the chisel from 1978 (Šljivar 1996), demonstrated the Vinča culture origins for metallurgical activities in Pločnik.

The site of Pločnik has recently received the first AMS dates (see details in Radivojević and Kuzmanović Cvetković 2014: 17–18). The probability distribution for the start of the Vinča culture occupation in Pločnik was 5290–5140 cal. BC, with the highest probability of around 5200 BC. The highest probability for the end of the settlement is *c.* 4650 BC, suggesting it was active for *c.* 600 years. Significantly, one of the AMS dates is closely related to the copper chisel (here sample labelled as Pločnik 216, Table 1) (Šljivar *et al.* 2006: 255, Plate VIII/3). The preceding context is dated between 5040–4860 cal. BC (95% probability), thus marking the *terminus ante quem* for this and other metal artefacts in Pločnik, along with the start of the Gradac Phase on this site. This is consistent with the beginning of this phase at other sites (e.g. Belovode, Rudna Glava), placing it securely in the first century of the 5th millennium BC.

Three building horizons at Pločnik correspond to Vinča A, B1 and B2 respectively (Šljivar 1996). Horizons I and II (*c.* 2.5 m thick) include massive remains of dwelling structures (up to 6.5 m in length) and related postholes, including wide pits filled with ash, charcoal and bone fragments in the top layers, and including ceramic fragments characteristic of Vinča A and B1 phases (Šljivar and Kuzmanović Cvetković 1997a: 106). The Gradac Phase is represented by a cultural layer of about 1 m thickness at Pločnik, which, at around 0.4 m, exhibited visible damage due to agricultural activities (Šljivar and Kuzmanović Cvetković 1997a: 104). Dwelling structures are intersected with wide pits filled with charcoal and ash in several successive layers, some with bone and ceramic fragments in the upper layers (Figures 1 and 2) (Šljivar and Kuzmanović Cvetković 1998a: 5–6).

As for Belovode, the publication record for the 14 years of most recent excavations at Pločnik has been limited to attempts to interpret and explain archaeometallurgical activities. This volume presents a more detailed account of metallurgical activities on the site. In addition, a set of samples related to metallurgy from the 1998–2009 campaigns, from seven trenches in total (see Table 1), was analysed in depth (Radivojević 2012). The most extensively sampled is Trench 20 which, in the Gradac Phase of occupation, yielded exceptional evidence for the Vinča culture metallurgy (Šljivar and Kuzmanović Cvetković 2009a).

The studied collection consisted of copper minerals, malachite beads and copper metal artefacts. A single droplet (sample Pločnik 52) is likely evidence of primary copper production (Radivojević and Rehren 2016:

Table 1. Study materials from the site of Pločnik.

No.	Analytical No.	Excavation year	Field label	Field context	Type of Material
1	**Pločnik 43**	2007	Trench 20, spit 10	Workshop	Malachite bead
2	**Pločnik 51**	2006	Trench 19, spit 23	Household	Copper mineral
3	**Pločnik 52**	2000	Trench 14, spit 10	Household	Copper metal droplet
4	**Pločnik 54 (b, m)**	2002	Trench 16, spit 19	Household	Copper mineral and a malachite bead (blank)
5	**Pločnik 57**	2006	Trench 19, spit 13	Household	Copper mineral
6	**Pločnik 63**	2008	Trench 21, spit 5	Dwelling structure	Metal sheath/cover
7	**Pločnik 65**	2008	Trench 21, spit 6	Dwelling structure	Malachite bead
8	**Pločnik 66**	2004	Trench 18, spit 7	Stone structure	Malachite bead
9	**Pločnik 67**	2007	Trench 20, spit 7	Workshop	Copper artefact
10	**Pločnik 69**	2007	Trench 20, spit 4	Workshop	Copper metal droplet
11	**Pločnik 71**	2007	Trench 20, spit 7	Workshop	Copper mineral
12	**Pločnik 72 (b, m)**	2007	Trench 20, spit 3	Workshop	Copper mineral and malachite bead (roughout)
13	**Pločnik 73**	2007	Trench 20, spit 7	Workshop	Copper artefact
14	**Pločnik 75**	2007	Trench 20, spit 7	Workshop	Copper artefact
15	**Pločnik 143**	2004	Trench 18, spit 7	Stone structure	Copper artefact
16	**Pločnik 145**	2007	Trench 20, spit 7	Workshop	Copper artefact
17	**Pločnik 207**	2009	Trench 22, spit 17	Household	Malachite bead
18	**Pločnik 209**	2009	Trench 22, spit 15	Household	Copper mineral
19	**Pločnik 216**	2000	Trench 14, spit 7	Stone structure	Copper chisel

Figure 1. Southeast section of Pločnik, facing the river. Note the massive cultural layer bearing remains of dwelling structures intersected with pits (Photo by Jugoslav Pendić)

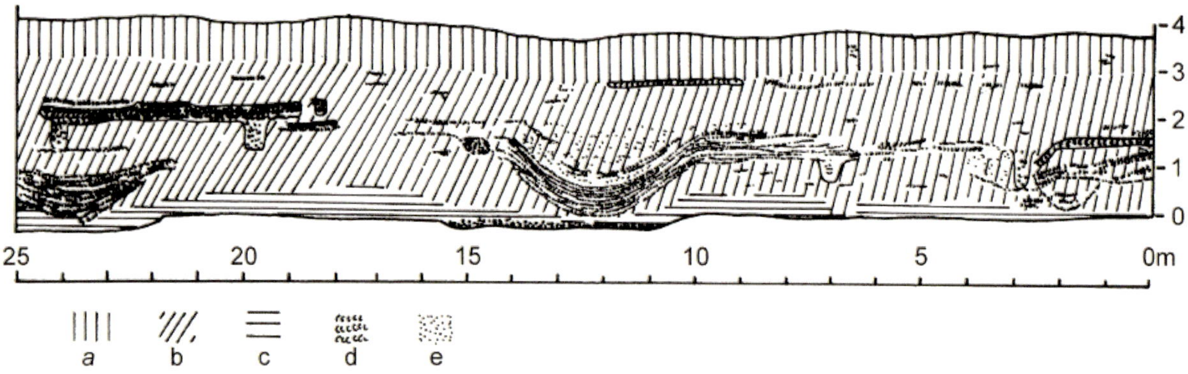

Figure 2. Southeast profile of Pločnik: a- dark brown soil; b- light brown soil; c- sterile soil; d- daub; e- ash (after Šljivar and Kuzmanović Cvetković 1998a: 4)

220). The same analytical methodology was applied as described in the introductory chapter on Belovode (see Chapter 5, this volume). Activities related to copper mineral use and metallurgy are also described here in three distinctive stages of the process: copper mineral processing, (s)melting debris, and copper mineral and metal artefacts. All technological analyses were carried out at the Wolfson Archaeological Science Laboratories at the UCL Institute of Archaeology.

Results

Copper mineral use

The use of copper minerals is evident from the early period of this site: green lumps of malachite are found scattered across the settlement as at Belovode, usually outside potential dwelling features, in the so-called 'workshop' areas. These workshops usually consisted of stone structures and the amorphous remains of floors for which insufficient evidence is present to ascribe them to either economic or dwelling structures. Šljivar and collaborators (2006: 256 ff.) reported finds of lumps of copper minerals of varying size, with altered structure, porous and mixed with charcoal, hence associating them with metallurgical activities. In addition to copper minerals, the use of cinnabar has also been recently confirmed (Gajić-Kvaščev *et al.* 2012), although used in the 'cold' process, that is without high temperature treatment like copper.

Six copper minerals were analysed in greater detail (Pločnik 51, Pločnik 54m, Pločnik 57, 7 Pločnik 1, Pločnik 72m, and Pločnik 209. See Table 1) Two of these (Pločnik 71 and Pločnik 72m), come from Trench 20 and belong to the Gradac Phase horizon (the trench was only excavated to this level). Other copper mineral finds were sampled in order to provide information on their use in earlier phases of Pločnik occupation. It is worth noting that all samples found in the field were of sizes varying between a few mm and a few cm, possibly indicating that they were already beneficiated and prepared for further processing.

The copper minerals form two distinctive groups: oxide and sulphide minerals (the latter only present in Pločnik 72). All oxide minerals are black and green and, excepting sample Pločnik 71, contain copper and manganese contents in ratios varying from 1:1 to 2:1 in favour of copper (Radivojević and Rehren 2016: 213 ff.). In addition, the ternary plot of MnO/ZnO/CuO (Radivojević and Rehren 2016: 15, Figure 5) clearly illustrates the striking compositional similarity of all studied minerals to those from Belovode. Sample Pločnik 72 contains significant sulfur readings contained in mineralisation of chalcocite (Cu_2S), pyrite (FeS_2) and sphalerite (Zn, Fe)S, besides copper oxide/carbonate content. Macroscopically, it appears more solid than the green-and-black oxide minerals although it, too, has distinctively coloured cross-sections in shades of green and grey with metallic lustre.

(S)melting debris

The only production evidence discovered at the site is sample Pločnik 52 (Figure 3a), which was uncovered on the floor of a collapsed burnt structure (Trench 14, spit 10), directly dated to *c.* 5040–4840 BC (Radivojević and Rehren 2016: 220, Figure 8). Sulfur-rich copper droplet, Pločnik (52), was initially thought to represent copper mineral, until the cross-section revealed a dark-red phase mixed with a thick, light-green layer. This amorphous piece does not exceed *c.* 1 cm in length. Compositional analysis revealed a complex structure of copper-based compounds, predominantly copper oxides (dross, tenorite), followed by a copper metal phase. The amorphous shape of the metal phase indicates that this sample was once molten copper, however it is now heavily corroded due to various post-depositional processes.

A copper metal droplet, Pločnik 69 (Figure 3b), was also initially assumed to be a copper mineral, due to its blocky shape and thick, light-green patina, until it exhibited a copper metal phase surrounded by a rich dark-red phase of copper oxides in the cross section (Figure 4). This, together with porosity holes in its polished section, indicates an intensive reaction with the atmosphere during cooling which, overall, points to this piece once being molten copper metal. The absence of any other phases but copper marks this sample as melting, rather than smelting debris, and it is the closest in nature to sample M14 from Belovode (Radivojević *et al.* 2010a; Radivojević 2012; Radivojević and Kuzmanović Cvetković 2014: 15).

Copper mineral and metal artefacts

Malachite beads of different sizes and shapes were recovered throughout all horizons of occupation in Pločnik, six of which were selected for the present study (see Table 1). A bead roughout (Pločnik 72) and a whole bead (Pločnik 43) were discovered in Trench 20, and together with two other whole beads (Pločnik 65,

Figure 3. a) Pločnik 52, droplet; b) Pločnik 69, droplet.

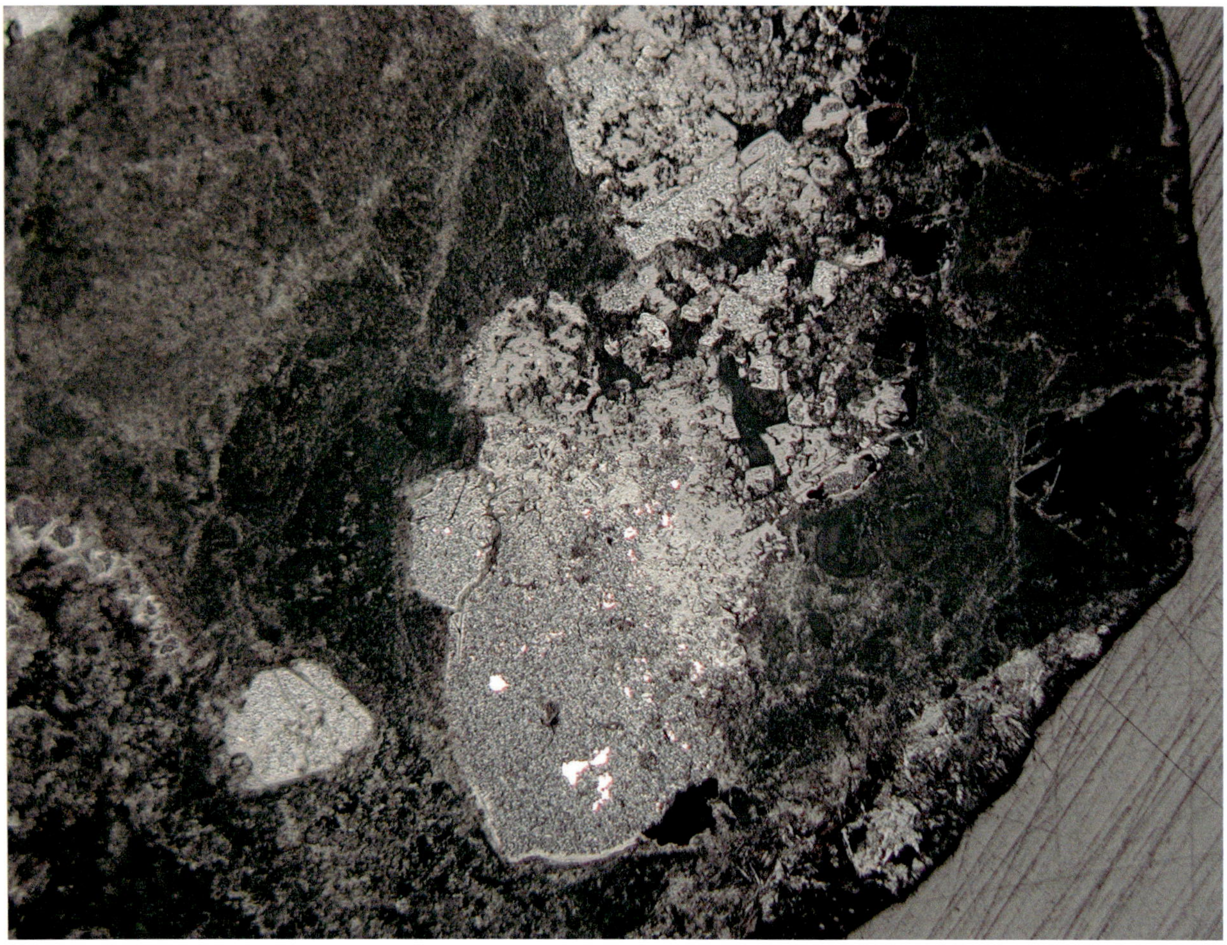

Figure 4. Photomicrograph of Pločnik 69 under plane polarised light. Note the copper metal phase (bright yellow islands) embedded in copper dross and surrounded by corrosion (green) (width 6.4 mm, magnification 25x).

Pločnik 66; see Figure 5) most likely date to the Gradac Phase (horizon III) in Pločnik (Šljivar and Kuzmanović Cvetković, 1996-2009). A bead blank (Pločnik 54b) and two whole beads (Pločnik 43, Pločnik 207) precede these finds (Table 1). The Pločnik beads cover only the last two building horizons (II and III) and appear spatially associated with copper minerals (Pločnik 54b, Pločnik 72b, Pločnik 207), a workshop (Pločnik 43, Pločnik 72b), or metal artefacts (Pločnik 65, Pločnik 66). Typologically, the beads can be roughly divided into two distinctive categories: (circular) cylindrical (Pločnik 65) and flat disc (Pločnik 43, Pločnik 66 and Pločnik 207) (cf. Wright and Garrard 2003; Wright *et al.* 2008).

Visually, all copper mineral ornaments exhibit a thick, light-green corrosion layer, with occasional macroscopically-visible oolithic formations on the surface, as in Pločnik 43 (Figure 5g). All beads are, judging by the bright green colour, made of malachite, and only four survived as finished objects (Pločnik 43, Pločnik 65, Pločnik 66 and Pločnik 207); the rest are found in various stages of processing (bead production stages after Lankton *et al.* 2003). A roughout (Pločnik 72b) belongs to the early stages of bead reduction. It is a subcircular artefact shaped into a bead form and has a shallow indentation in the middle (Figure 5e). Pločnik 54b represents a tabular preformed, but not finalised, bead blank: it is flaked and chipped around the edges and shows traces of drilling from both sides (Figure 5a/b). These samples contain important information for the Vinča culture bead-making technology: beads were most likely initially shaped into form (sawing, chipping, abrading), then drilled, and finally polished (Radivojević 2012).

The drilling process in finished beads is represented by fully preserved bead, Pločnik 66. Bead Pločnik 54b appears to have been drilled halfway through and then turned, so that the perforation could be completed from the opposite direction (Figure 5a/b), but the perforation was not finalised. Rotary drilling from both directions has been experimentally confirmed to produce regular perforations (cf. Gorelick and Gwinnett 1990). This was identified within this assemblage in Pločnik 66 (Figure 5d). The fine drill holes may have been the result of using a finer drilling tool (a bow?). Although no traces of polishing were observed on these beads, we can assume that the final products were finished in this way as it is a known practice in the Neolithic tradition

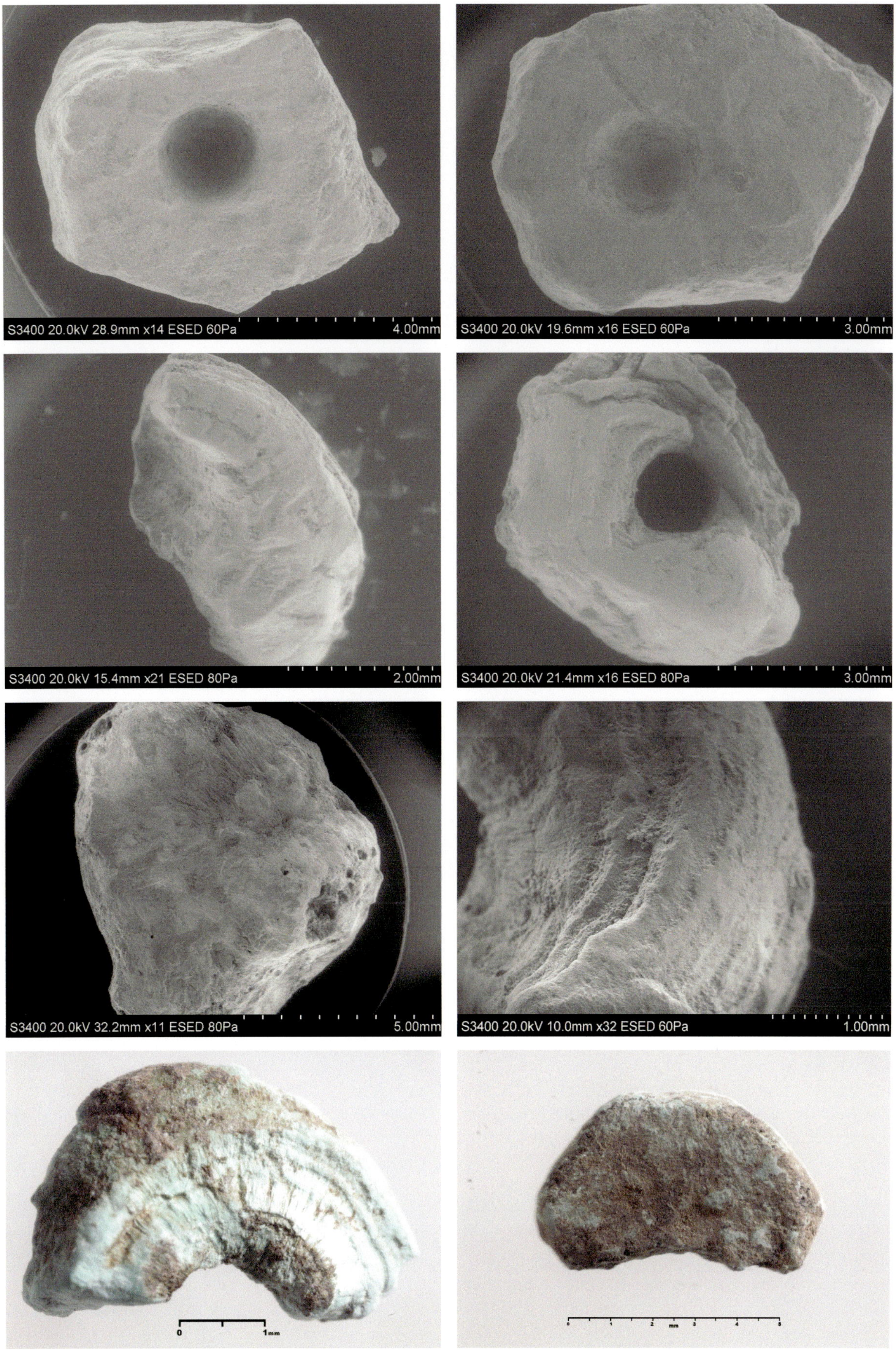

Figure 5. a) Pločnik 54b side 1; b) Pločnik 54b side 2; c) Pločnik 65; d) Pločnik 66; e) Pločnik 72m; f) Pločnik 43; g) Pločnik 43; h) Pločnik 207.

of bead making in the wider western Eurasia region (Lankton *et al.* 2003).

Chemical analyses reveal that the major constituents of all the artefacts are CO_2 and CuO, with the majority of readings compositionally closest to the malachite formula $[Cu_2(CO_3)(OH)_2]$, which contains, besides water, 71.9 wt% of CuO and 19.9 wt% of CO_2. Silica readings are particularly important with regard to different values detected on the inside surface of the bead perforation, as they may indicate the use of stone tools for drilling. An elevated silica content was observed on the inside surface of Pločnik 43 and Pločnik 54b, in amounts two to three times higher than on the outside surface (Radivojević 2012). Such readings may imply the use of stone tools for both drilling and polishing of beads and pendants, however no such tools have yet been recognised in the context of these finds.

The most significant discoveries at Pločnik thus far are four groups of exceptionally massive copper metal implements, numbering 34 in total (Grbić 1929; Stalio 1964; Šljivar 1999; Šljivar *et al.* 2006). To this collection, four more implements can be added, all well-contextualised within the Gradac Phase occupation of this settlement (Šljivar 1996; Šljivar and Kuzmanović Cvetković 1996–2009). The term 'hoard' was introduced by Grbić (1929) however, but to avoid confusion they will be referred to here as 'groups'.

Group I was amongst objects donated by the Directorate of Yugoslav Railways in 1928, discovered during the building works for the railway station in Pločnik, at a presumed depth of *c.* 0.8–1.0 m (Grbić 1929). Stalio (1964) terms this group 'Hoard 1'. Šljivar and collaborators (Šljivar *et al.* 2006: 254) indicate that it consisted of nine objects, seven of which were made of copper metal: two hammer axes (Pločnik type), two chisels, two complete and one fragmented bracelets, and two stone axes made of magnesite (Figure 6).

Group II consisted of 18 objects, 13 of which were copper metal tools: one hammer-axe (Pločnik type), 12 chisels and five stone axes made of magnesite. Grbić (1929: 8–9) reported that they were found in the vicinity of a destroyed 'furnace', scattered over an area 5 m wide, at a depth of *c.* 1 m (see also Šljivar *et al.* 2006: 255).

Similar conditions were recorded for Group III: copper and stone tools were uncovered, scattered in an area of 5.0 x 0.5 m, at 0.7 m depth (Stalio 1964: 35). The find consisted of 11 objects, nine of which were made of copper metal: one hammer-axe (Pločnik type), five chisels, a bracelet, a pin with a forked end, a copper ingot bar and two stone axes made of magnesite (Figure 6). The resumed excavation of 1996 took place in the area of the Group III discovery, unearthing dozens of ceramic materials belonging to the early Gradac Phase, a small rectangular stone structure and one stone axe made of magnesite, identical to that accompanying the metal objects in Group III (Šljivar and Kuzmanović Cvetković 1998b: 82). The owner of the excavated land also donated a copper chisel to the National Museum Belgrade. He found the chisel during the building of his family house, in the vicinity of the Group III location.

Group IV was found in 1968 during the building of the Nis-Priština railway, at a depth of *c.* 0.3 m. It includes 14 objects: five copper chisels, eight stone axes made of magnesite and one miniature pottery vessel (a casting pot?) (Figure 6) (Stalio 1973). The casting (?) pot, however, did not show assumed traces of metalworking upon inspection by the author at the National Museum in Belgrade.

Together, the four Pločnik groups of metal finds amount to: four Pločnik type hammer-axes, 25 chisels, four bracelets, a copper ingot bar and a pin, weighing in total *c.* 16 kg (Šljivar *et al.* 2006: 255). As such, they are a unique and exceptional find of copper artefacts and, according to the most recent AMS dating, one of the earliest in this part of Eurasia (Radivojević and Kuzmanović Cvetković 2014: 23). Seventeen of the copper metal tools from Pločnik have previously been studied for chemical composition and lead isotopes, the results of which revealed an unexpected complexity of ore/metal exchange networks: at least three different copper deposits from east Serbia, Macedonia and across Bulgaria provided metal for their making (Pernicka *et al.* 1997: 93–94, Table 3).

Significantly, all finds originated from a single area of Pločnik, which excavators also call the 'industrial' zone due to the thick ash-and-charcoal deposit frequently encountered in Horizon III. A superbly preserved figurine head from the 'industry zone' was discovered with half of the face painted lengthwise with a light grey slip prior to firing (Figure 7). The excavators believe that the colour effect on its face symbolised the light (ground) and dark (underground) environments of early miners' life in Pločnik (Kuzmanović Cvetković and Šljivar 1998; Šljivar and Kuzmanović Cvetković 1998a).

The recent excavation campaign (1996–2011) uncovered two distinctive situations that yielded evidence for a potential metallurgical workshop in Pločnik: one in Trench 20 and the other in Trench 21 (Šljivar and Kuzmanović Cvetković 2009a; Radivojević *et al.* 2013). In Trench 20, a structure whose contours were followed over a 6 x 6 m area, appeared at a relative depth of 0.8 m, with a surface filled with rubble, stones, numerous pottery fragments and several metal artefacts and metal casting debris. A rectangular furnace (?), measuring 1.4

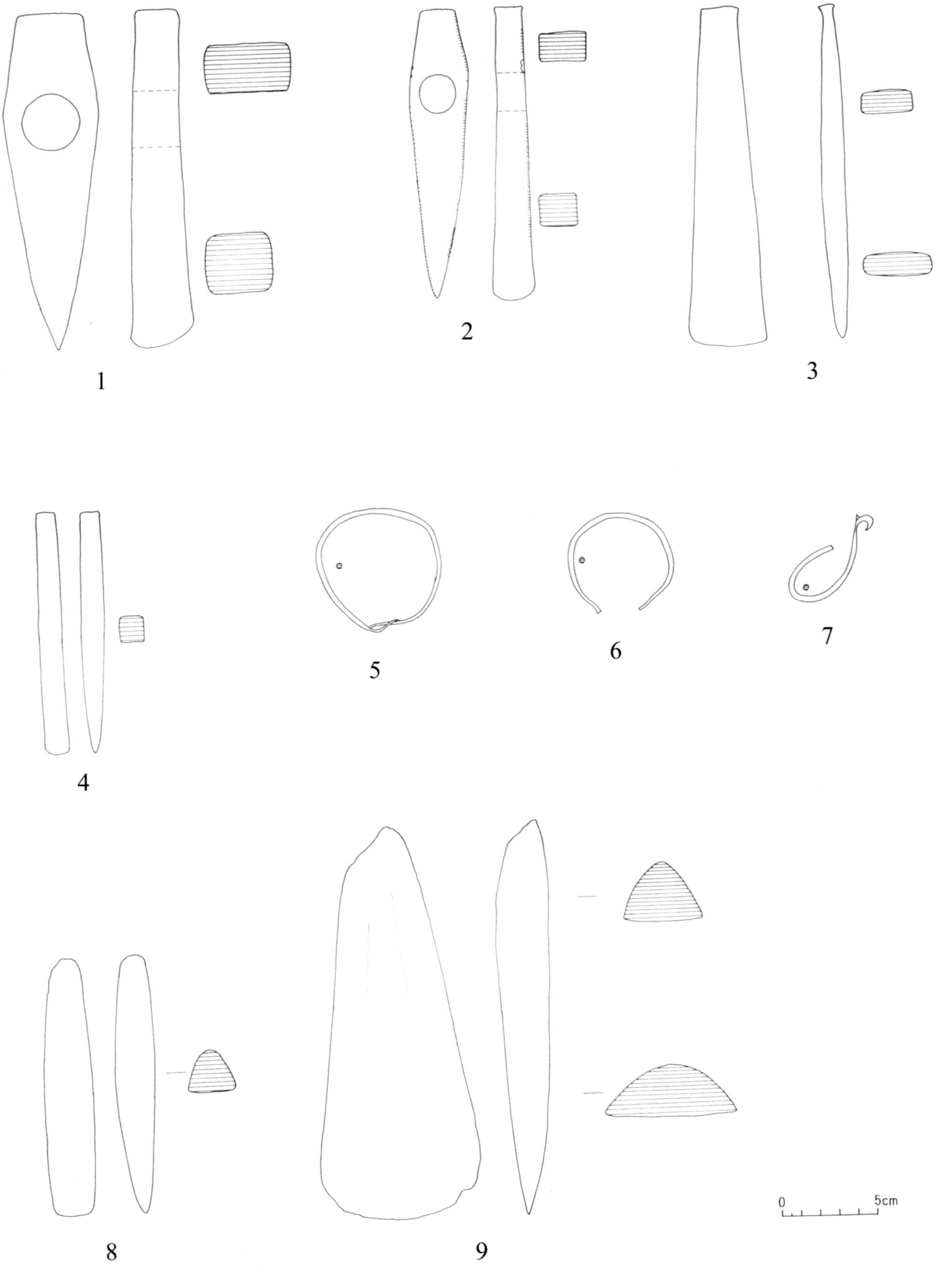

Figure 6. Four Pločnik groups: a. Group I (complete) (after Šljivar *et al.* 2006: 261–265).

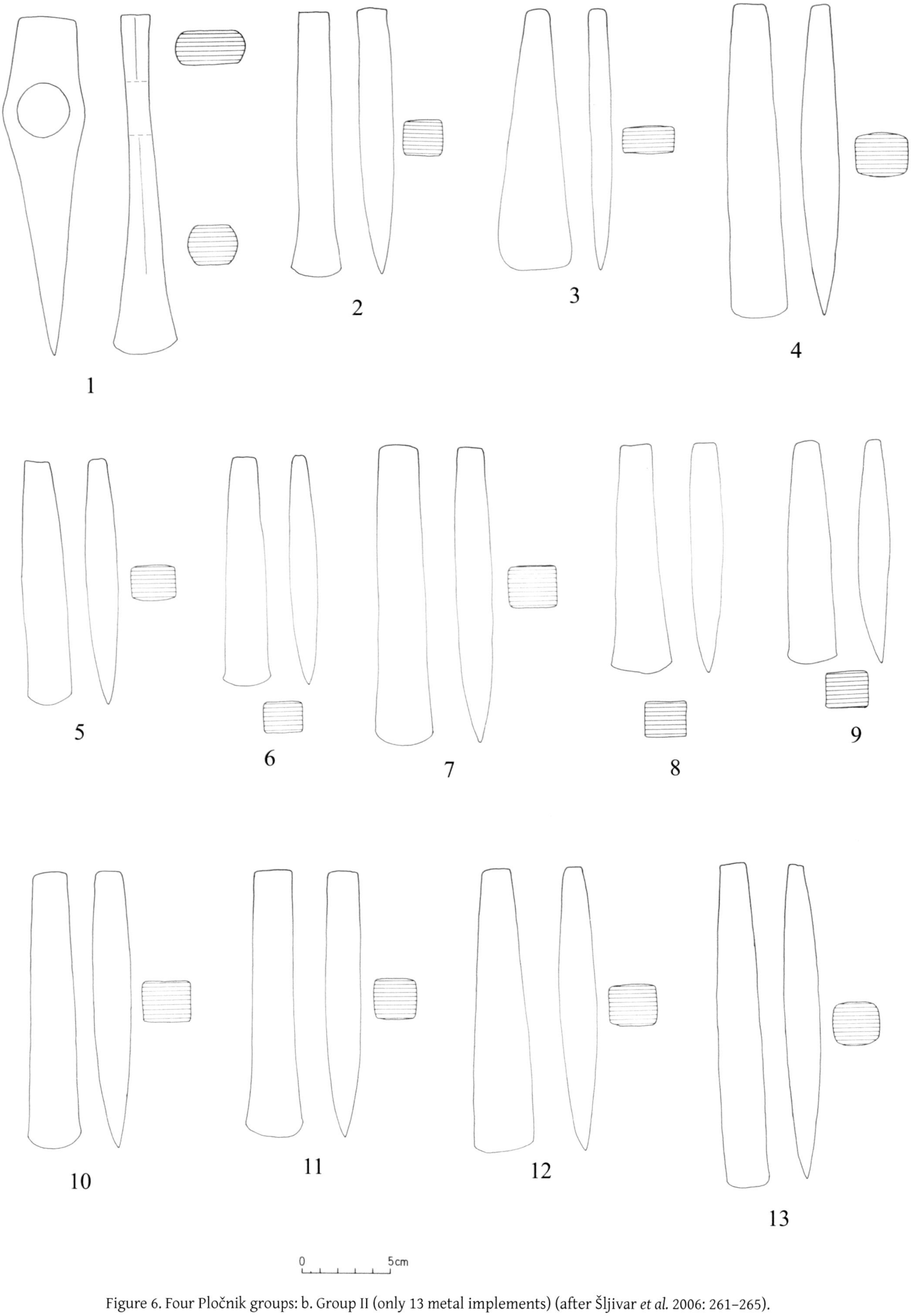

Figure 6. Four Pločnik groups: b. Group II (only 13 metal implements) (after Šljivar *et al.* 2006: 261–265).

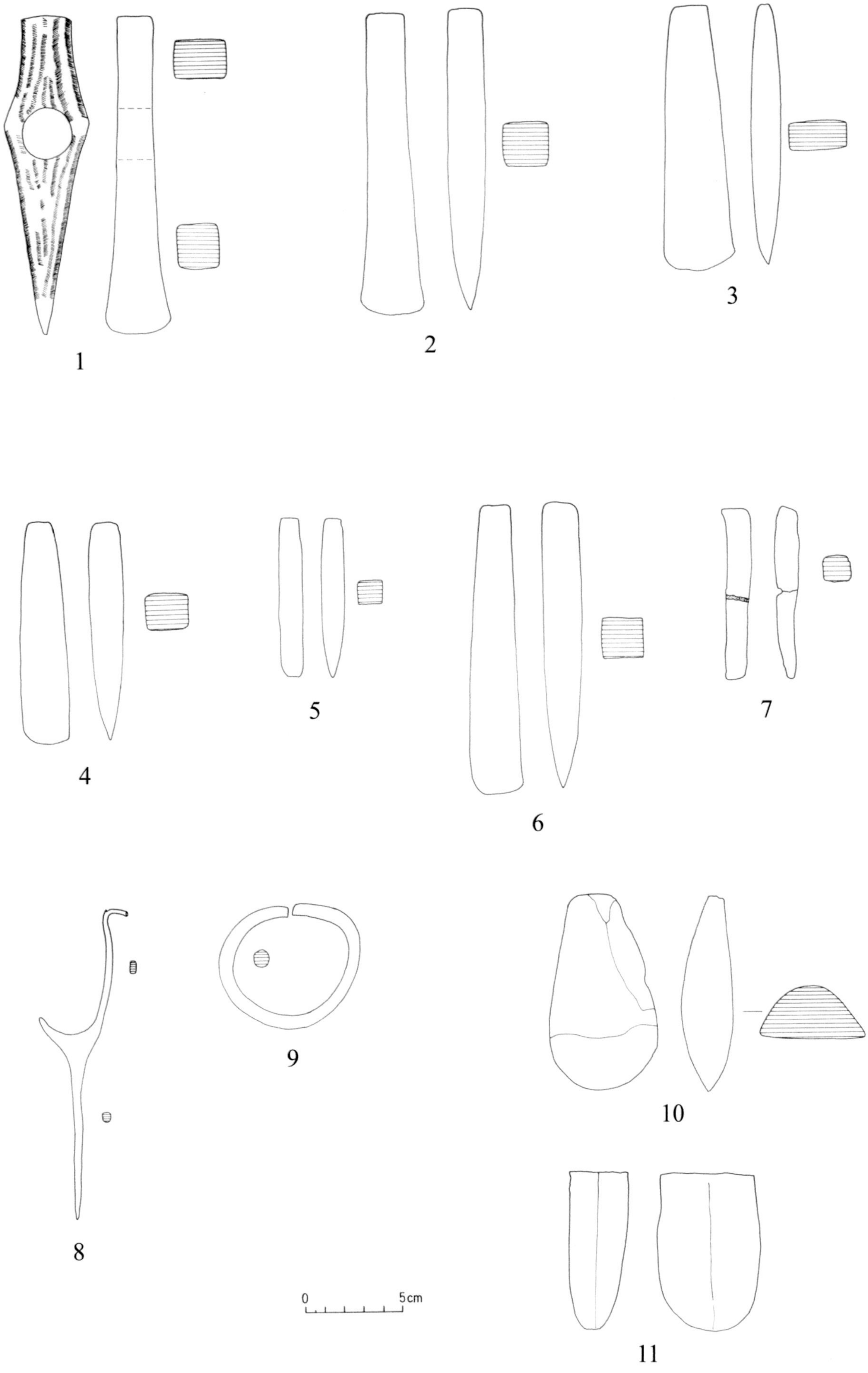

Figure 6. Four Pločnik groups: c. Group IV (complete) (after Šljivar *et al.* 2006: 261–265).

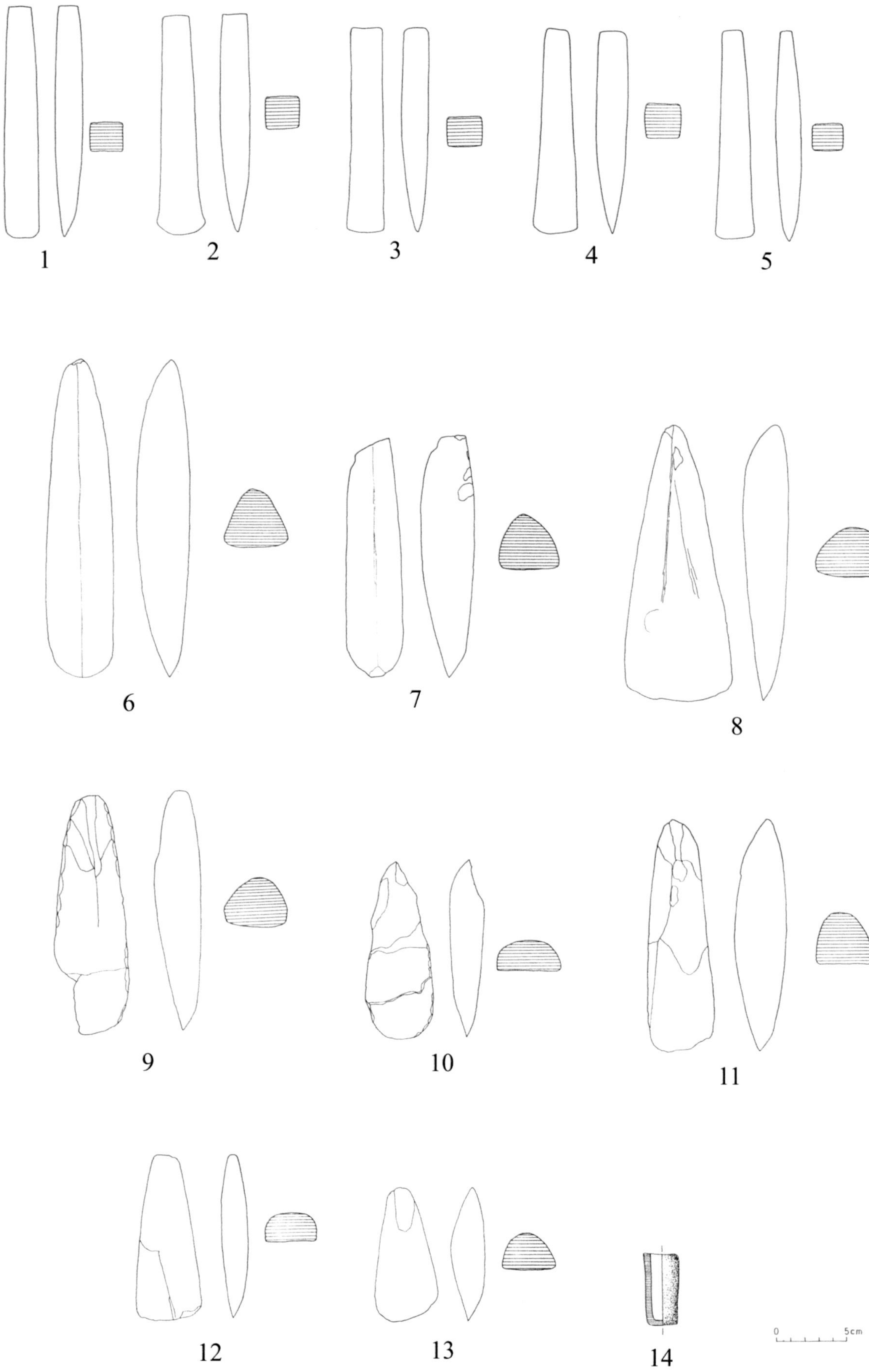

Figure 6. Four Pločnik groups: d. Group III (complete) (after Šljivar *et al.* 2006: 261–265).

Figure 7. A figurine head from Pločnik, painted lengthwise. Dimensions (in cm): height 20, face 10.7; face width 16.3; and neck width 11.6. (Courtesy of J. Kuzmanović Cvetković, Museum of Toplica, Prokuplje).

x 1.4 m dominated this structure, with massive walls preserved up to 0.5 m in height (Figure 8), visibly repaired several times (several clay layers), and with traces of intense firing. This, along with the discovery of a massive copper chisel (Pločnik 145; Figure 9), a fragment of a tool/ornament (?) (Pločnik 67), a fragmented bracelet (?) (Pločnik 73), a folded metal sheet (Pločnik 75) and copper minerals (one studied here, Pločnik 71) (see Table 2), led excavators to assume that the structure represented a metallurgical workshop (Šljivar and Kuzmanović Cvetković 2009a: 61).

Other finds in this structure included stone tools and large ceramic vessels, such as amphorae, jugs and similar water (?) containers. Of particular interest were ceramic 'tubes' (Figure 10), which did not contain indications of use in metallurgical processes but resembled the 'chimneys' from Belovode (Radivojević and Kuzmanović Cvetković 2014). The furnace (?) remains, although impressive in shape and size, did not offer sufficient evidence to confirm a metallurgical function. However, the coincidence of fragmented metal objects in the same structure could indirectly imply its use for casting or melting, for example. A comparable structure was discovered a year later in Trench 21. Its relationship with a securely contextualised tin

Figure 8. A workshop in Trench 20. Note the rectangular feature (furnace?) in the upper left corner. (Courtesy of Julka Kuzmanović Cvetković, Museum of Toplica, Prokuplje).

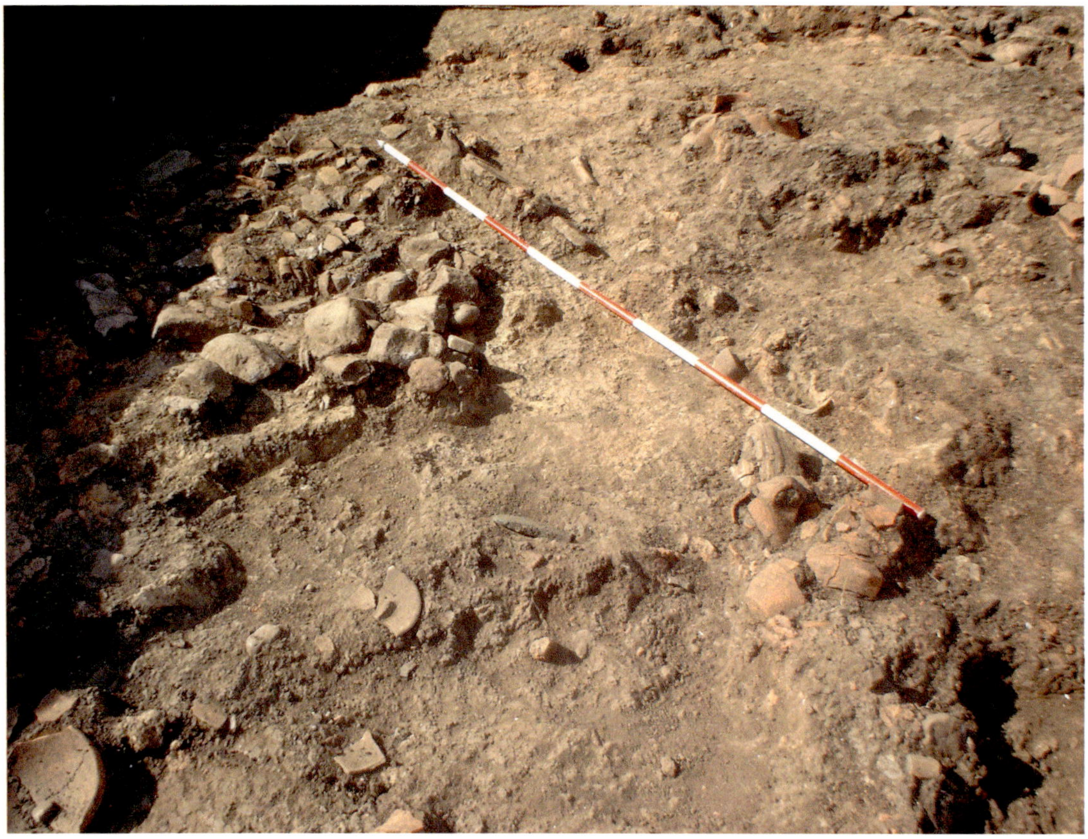

Figure 9. A copper chisel (Pločnik 145) *in situ* in Trench 20.
(Courtesy of J. Kuzmanović Cvetković, Museum of Toplica, Prokuplje).

Figure 10. A ceramic 'tube' (far left) from Trench 20.
(Courtesy of J. Kuzmanović Cvetković, Museum of Toplica, Prokuplje).

Table 2: Study materials from Trench 20 at Pločnik arranged by excavation levels and proposed chronology/building horizons.

No.	Analytical No.	Excavation year	Context	Type of Material	Chronology/building horizon
1	**Pločnik 72 (b, m)**	2007	Trench 20, spit 3	Copper mineral and malachite bead	Gradac Phase, horizon III
2	**Pločnik 69**	2007	Trench 20, spit 4	Copper metal droplet	Gradac Phase, horizon III
3	**Pločnik 67**	2007	Trench 20, spit 7	Copper artefact	Gradac Phase, horizon III
4	**Pločnik 71**	2007	Trench 20, spit 7	Copper mineral (flakes)	Gradac Phase, horizon III
5	**Pločnik 73**	2007	Trench 20, spit 7	Copper artefact	Gradac Phase, horizon III
6	**Pločnik 75**	2007	Trench 20, spit 7	Copper artefact	Gradac Phase, horizon III
7	**Pločnik 145**	2007	Trench 20, spit 7	Copper artefact	Gradac Phase, horizon III
8	**Pločnik 43**	2007	Trench 20, spit 10	Malachite bead	Vinča B1, horizon II

bronze foil and resemblance to a feature discovered in Trench 20 prompted researchers to argue for a similar metallurgical function (Šljivar and Kuzmanović Cvetković 2009b; Radivojević *et al.* 2013).

A detailed set of analyses was performed on seven copper and tin bronze artefacts (Table 1) (Radivojević 2012; Radivojević *et al.* 2013). Pločnik 143, the copper chisel (Figure 11), was discovered within a stone structure in Trench 18, together with a stone axe made of magnesite and a small pottery vessel, all of which correspond with the beginning of the Gradac Phase in Pločnik (Šljivar *et al.* 2006: 255–256). A massive copper chisel (Pločnik 145) was unearthed in a similar workshop setting in Trench 20, along with two fragmented pieces of a tool (Pločnik 67) and a bracelet (Pločnik 73) respectively, a folded metal sheet (Pločnik 75, see Figure 11), a copper mineral (Pločnik 71), and several stone tools. All of these were discovered scattered across the same surface, in spit 7 and in the vicinity of the 'furnace' (Šljivar and Kuzmanović Cvetković 2009a: 58–60). Pločnik 216 is a small, fragmented chisel discovered above a stone structure outside the potential dwelling feature in Trench 14, dated between 5040 and 4840 BC (95.4% probability). The emergence of copper metal artefacts in Pločnik thus coincides with the start of the Gradac Phase on this site.

All metal artefacts from this site studied by Pernicka and collaborators (1993, 1997) showed a copper composition of 99.9 wt% on average. A new set of the electron microprobe examination of another set of samples (see Table 1) confirmed these results, barring the tin bronze foil (Pločnik 63). The pure copper composition is followed by low trace element contents, most relevantly of iron, sulfur, gold and nickel, in varying ratios (Radivojević 2012).

The common microstructural feature of all samples is that they present a yellow/orange (copper) metal body, with green corrosion products developing on its edges. The main metal body shows a residual as-cast structure, preserved in the microstructure of the copper-copper oxide eutectic with α-grains of copper (see example in Figure 12a). Optically, the α-grains of copper are characterised by their highly reflective bright colour, embedded in the eutectic structure with grey particles within a bright matrix of metal grains.

The major differences among the sets of metal artefacts analysed stem from the varying combinations of working techniques, which appear carefully designed to respond to the desired function of the object in question. Pločnik 67 (a fragmented tool/ornament?), Pločnik 73 (a copper bracelet?), and Pločnik 75 (a folded metal sheet) display fully recrystallised structure and traces of several cycles of cold working and annealing (Radivojević 2012). The only two massive copper implements examined for microstructure in the most recent study are Pločnik 143 and Pločnik 145 (Figures 11a/e, 12b), both of which bear similar small traces of finishing work towards their tip and on the surface (Radivojević 2012). These massive copper chisels display only slight post-annealing working, which could be equally ascribed to either intentional hardening of the tip and along the surface area, or to hardening during the use of these objects. It could also be interpreted as the result of intense hot working with continuous re-heating during a forging process of some duration (cf. Kienlin 2010).

A tin bronze foil (Pločnik 63) was excavated from an undisturbed context, on the floor of a dwelling structure next to the likely copper metal workshop at the site, about 1 m from a fireplace, and was enclosed in several late Vinča culture pottery vessels (Radivojević *et al.* 2013: 1033, Figure 2). Compositional analysis demonstrated that this metal foil was made of a complex alloy of copper and tin, along with significant concentrations of elements including As, Sb, Co, Ni, Pb and Fe. This securely contextualised find comes from a single, undisturbed occupation horizon at Pločnik, dated to *c.* 4650 BC. This date is, according to the field evidence, the *terminus ante quem* for the Pločnik foil at present. The tin bronze foil from Pločnik is, therefore, the earliest known tin bronze artefact anywhere in the world (Radivojević *et al.* 2013: 1032).

Figure 11. a) Pločnik 143, a copper chisel; b) Pločnik 67, a fragmented tool/ornament? c) Pločnik 75, a folded metal sheet; d) Pločnik 73, a fragmented bracelet; and e) Pločnik 145, a copper chisel.

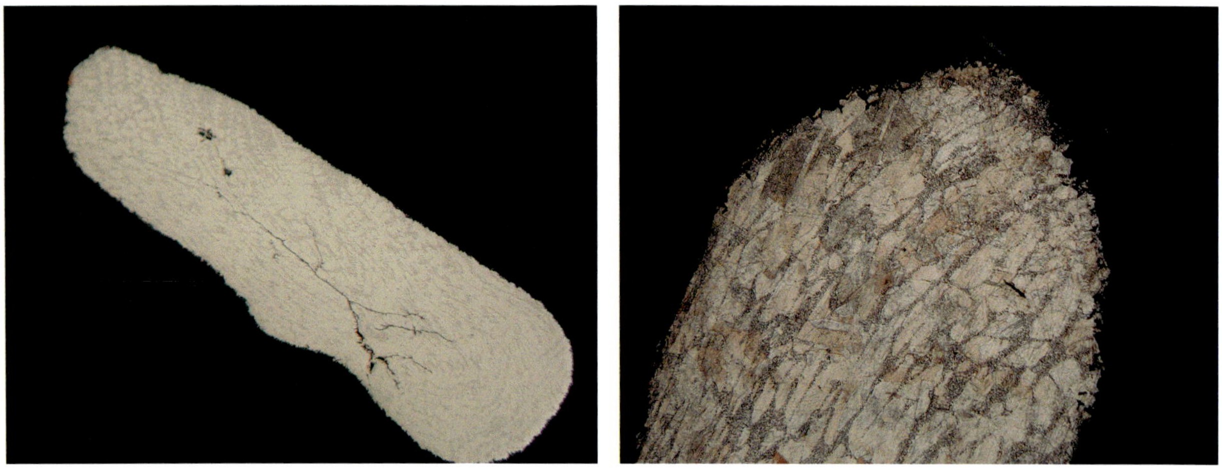

Figure 12. a) Photomicrograph of the unetched section of Pločnik 67 under plane polarised light. Note the intensive working on one side (left) exhibited by the elongated grains, as opposed to the oxide-inclusion abundant on the right side of the object (magnification 50x, width 3.2 mm); b) photomicrograph of the etched section of Pločnik 75 (a folded metal sheet) under plane polarised light (magnification 50x, width 3.2 mm).

Discussion

The archaeometallurgy of the site of Pločnik has been the most visible feature of this settlement in all previous studies, with very little attention paid to the everyday life of Pločnik communities beyond attempts to describe unique finds, like the figurine head mentioned above (Figure 7; Kuzmanović Cvetković and Šljivar 1998). However, metal producing and working activities in the site of Pločnik offer an insightful view into the life of a society with emerging pyrotechnology during a time of major cultural, economic and material changes in the 6th and 5th millennium BC in the Balkans (see Chapter 3, this volume).

Based on the recent analyses of archaeometallurgical materials from the site, the excavated items cover all major activities of metallurgical *chaîne opératoire.* The use of both oxide and sulphide copper minerals attests to the knowledge of Pločnik smiths of the properties of these materials, and this is further confirmed by the presence of both smelting and melting droplets discovered within a domestic context: Pločnik 69 was part of workshop activities in Trench 20, while Pločnik 52 indicates smelting of a sulfur-rich copper ore at another location within the settlement (Trench 14).

The presence of a sulfur-rich copper droplet (Pločnik 69) and sulphide copper mineral (Pločnik 72m) potentially implies that the latter was part of the ore batch used in smelting activities. The remarkable morphological similarities between these samples further strengthen this assumption. This, in effect, could imply that the craft workers of Pločnik were intentionally selecting copper sulphide minerals with an appealing green tint and that they were probably aware of their properties (Radivojević 2015; Radivojević and Rehren 2016).

The production techniques employed in malachite bead making are evidently consistent with the Neolithic stone bead industry (cf. Kenoyer *et al.* 1991; Lankton *et al.* 2003; Wright *et al.* 2008). This is best reflected in the various reduction sequences of bead manufacturing, identified across different domestic contexts: 1) the reduction of nodules into roughouts; 2) the shaping of roughouts into blanks by further flaking, sawing and rough grinding; 3) perforation; 4) the final shaping; and 5) the final polishing (Wright *et al.* 2008: 140). The Pločnik bead production material is particularly interesting since it could potentially represent workshop stocks or stored merchants' goods (Kenoyer *et al.* 1991: 57). The distribution pattern of finished and semi-finished beads in Pločnik nevertheless appears more scattered than concentrated at this site (see Table 1), hence implying limited— if any—administrative control over the bead production. However, bearing in mind the small quantity of these finds, as well as the lack of sufficiently integrated contextual evidence within the settlement, the presence of a specialised bead workshop remains a matter of speculation.

The striking feature of these ornamental artefacts is the strong preference for the green colour. It has already been emphasised in previous research that pure green beads were used for minerals, while black and green were more utilised for copper metal extraction (Radivojević 2015; Radivojević and Rehren 2016). The same pattern is apparent at Pločnik. and is corroborated by provenance studies from the site of Belovode, which also indicated two different mineral sources for bead making and copper smelting at this site as well (Radivojević *et al.* 2010a: 2784), confirming that these practices were part of a shared system of values among the Vinča culture communities.

A common feature of the metal artefacts from Pločnik is that, regardless of their composition, it was their function and shape that dictated the combination of techniques applied in their making and working, suggesting a good level of understanding of different material properties of copper, tin, and bronze. All seven copper metal artefacts studied here were made of high purity molten copper metal; a detailed comparison of trace element data from other Vinča culture metal artefacts indicates that they were very likely made of metal smelted from copper ores (Radivojević 2012). The ores typically had low (but diagnostic) levels of impurities, allowing their differentiation from native copper. The tin bronze foil is yet another reminder of remarkable skills and control over different material properties of the Pločnik smiths. This object was made from natural alloy, produced from complex copper-tin bearing ores, and worked at temperatures at least twice as high as those required to make copper (Radivojević *et al.* 2013). It testifies that the Pločnik smiths were aware of the different requirements of the newly acquired metal and developed skills in order to master these.

Conclusions

This overview of activities related to copper mineral use and extractive metallurgy at the site of Pločnik reveals the very close similarities in terms of technology with those established for the sites of both Belovode and Vinča-Belo Brdo (Radivojević and Kuzmanović Cvetković 2014; Radivojević 2015; Radivojević and Rehren 2016). Particularly striking is the common preference for the green and black copper minerals used for malachite bead making and the green copper minerals, used for copper smelting.

In terms of the field interpretation of these minerals, it is important to emphasise their domestic use. Only in the Vinča culture (from the Gradac Phase onwards) can we observe household-based pyrometallurgical activities. Significantly, some minerals studied here co-occur with bead-making activities, with no indication for their use as bead nodules. The distinction between the bead minerals and ores ('metallurgical minerals') has already been mentioned and will be further discussed in later chapters in the light of the production debris and malachite bead analyses.

Although the most convincing evidence for the distinctive roles of green and black and green copper minerals comes from the site of Belovode, there is a great likelihood that black and green minerals were used to make the (s)melting droplets Pločnik 52 and Pločnik 69. The excavations at Pločnik, however, have produced the greatest number of finished artefacts discovered in groups with others implements, which prompts us to assume a predominantly consumer role for the settlement within the broader organisation of metal production and distribution.

Regarding the organisation of metal production, there is little field evidence reported thus far that can shed light on any kind of specialisation or different lifestyles of occupants of dwellings, with or without evidence of metal working or consumption. Trenches 20 and 21 revealed unusual square features in dwelling objects that were termed 'furnaces', since there were metal objects in their vicinity (Šljivar and Kuzmanović Cvetković 2009a). Although no direct evidence was found to prove this assumption, the coincidence of these features with semi-worked objects, as in Trench 20, raises the likelihood of them being used for metalworking activity. Remains of a fragmented tool/ornament (Pločnik 67), a fragmented bracelet (Pločnik 73), and a folded metal sheet (Pločnik 75), together with a well-preserved massive copper implement (Pločnik 145) in a single structure in Trench 20 further suggest that this object could have been occupied by a metal smith. This workshop setting was, nevertheless, not used for primary metal production but, based on the present evidence, possibly only for the casting and/or repair of metal tools.

The rectangular structure from Trench 20 was also used for malachite bead making, as evidenced by the presence of a bead roughout (Pločnik 72b). The bead blank (Pločnik 54b) from Trench 16 a few metres from the metal workshop in Trench 20, may suggest the presence of yet another bead making workshop.

The quantity and type of pottery assemblage found in structures from Trenches 20 and 21 has not been previously reported as different to any other dwelling discovered at the site of Pločnik during the most recent excavation campaigns (e.g. Šljivar *et al.* 2006; Šljivar and Kuzmanović Cvetković 1998a, 1998b; Šljivar and Kuzmanović Cvetković 1996-2009). This theoretically implies that, for instance, metalworkers in Pločnik did not lead a different life to that of a farmer or, more accurately, that these two roles were not exclusive. The concept of a specialised metallurgical workshop appears later in prehistory (e.g. Bronze Age), and earlier suggestions of the presence of specialists need to be treated with caution until they are supported by a more detailed research on site formation (see Chapter 3, this volume).

Based on current evidence, the crafters from Belovode and Pločnik were each covering different ends of the same metal making process, which strengthens the likelihood of their potential collaboration. The lead isotope match of the Belovode slag samples with the Pločnik copper chisel (Pločnik 216) strengthens this assumption (Radivojević 2012); this will be explored in detail in Chapter 41, and in future publications. The quantity and quality of collected and sampled materials is currently unprecedented in academic work and provides an excellent resource for studying the emergence, evolution and transmission of metallurgical skills both within the Vinča culture and across the Balkans.

As with Belovode, there is a lack of AMS data on specific contexts related to diverse metallurgical activities at Pločnik (cf. Radivojević and Kuzmanović Cvetković 2014: 25). Also, as at Belovode, metallurgical activities are visible only from the start of the Gradac Phase, which highlights the need for Vinča archaeologists (and others) to investigate the circumstances and mechanisms through which this particular cultural phase emerges. It is important to emphasise that metallurgy was not the only novelty occurring within the Gradac Phase, and that over-arching social and economic changes are detected throughout the Balkans around the early 5th millennium BC (cf. Garašanin 1994/1995).

The following chapters will investigate in more detail the claims and interpretations stemming from commendable pioneering work at the site of Pločnik, conducted by Grbić, Stalio, Šljivar and Kuzmanović Cvetković. We will attempt to give, for the first time since the start of the Pločnik excavations, a general overview of all subsistence and economic activities at this site, and we will address the broader questions of the everyday life of the metal-producing and consuming communities of the Vinča culture and beyond.

The bibliographic reference for this chapter is:

Radivojević, M. 2021. Introduction to Pločnik and the results of archaeometallurgical research 1996–2011, in Radivojević, M., Roberts, B. W., Marić, M., Kuzmanović Cvetković, J., and Rehren, Th. (eds) *The Rise of Metallurgy in Eurasia*: 60–76. Oxford: Archaeopress.

Chapter 7

Excavation methodology for the sites of Belovode and Pločnik

Miroslav Marić, Benjamin W. Roberts and Jugoslav Pendić

The goals of the field excavations were dictated by the overall goal of the project: an understanding of the emergence of metallurgy in the Balkans at the turn of the 6th millennium BC. Excavations at the sites of Belovode and Pločnik were focused on settlement remains in order to identify the copper smelting installations used. To this end, pre-excavation research was conducted in cooperation with the National Museum in Belgrade, the Museum of Toplica, Prokuplje, and the Museum of Petrovac na Mlavi, to identify the optimum locations for the trenches. Both sites have been archaeologically researched in the recent past (for Belovode see Šljivar and Jacanović (1996a, 1997a) and for Pločnik see Šljivar (1996) and Šljivar and Kuzmanović Cvetković (1997a)). These campaigns yielded dozens of metallurgical finds which illustrated the entire smelting procedure, from raw ore to finished implements. For the new excavations, trench locations for each site were based on the distribution of metallic finds, primarily fragments of copper slag. At Belovode, Trench 18 was opened between previous Trenches 3 and 17 (Figure 1), both of which had yielded numerous samples of copper slag (Radivojević *et al.* 2010a). Similar reasoning lay behind the location of Trench 24 at Pločnik, which was positioned based on the occurrence of metallic finds from Trenches 20 and 21 (Figure 2) in previous campaigns (Šljivar *et al.* 2012). A geophysical (magnetic) survey was also consulted to better pinpoint the location of possible underground structures.

Project time and budgetary constraints led to a decision to excavate a 5 x 5 m trench at each site, but this proved to be only partially possible. Almost immediately, the need to expand trenches became apparent on both sites, but the subsequent expansions were strictly limited to include important structures that needed to be excavated in full. Ultimately, Trench 18 at Belovode had a surface area of 33 m^2, whilst at Pločnik, Trench 24 was almost doubled to 45 m^2. Both trenches were aligned with Magnetic North. As no previous square or sector grid had been employed on the site it was decided to use absolute coordinates expressed in the Serbian National Grid System (a Gauss Krueger Transverse Mercator 3° projection using a Hermannskogel datum). All coordinates recorded were expressed in metric values. Vertical values, also expressed on a metric scale, were obtained from a fixed network of

Figure 1. Aerial Image of Trench 18 location at Belovode (red) between Trenches 3 and 17 (yellow).

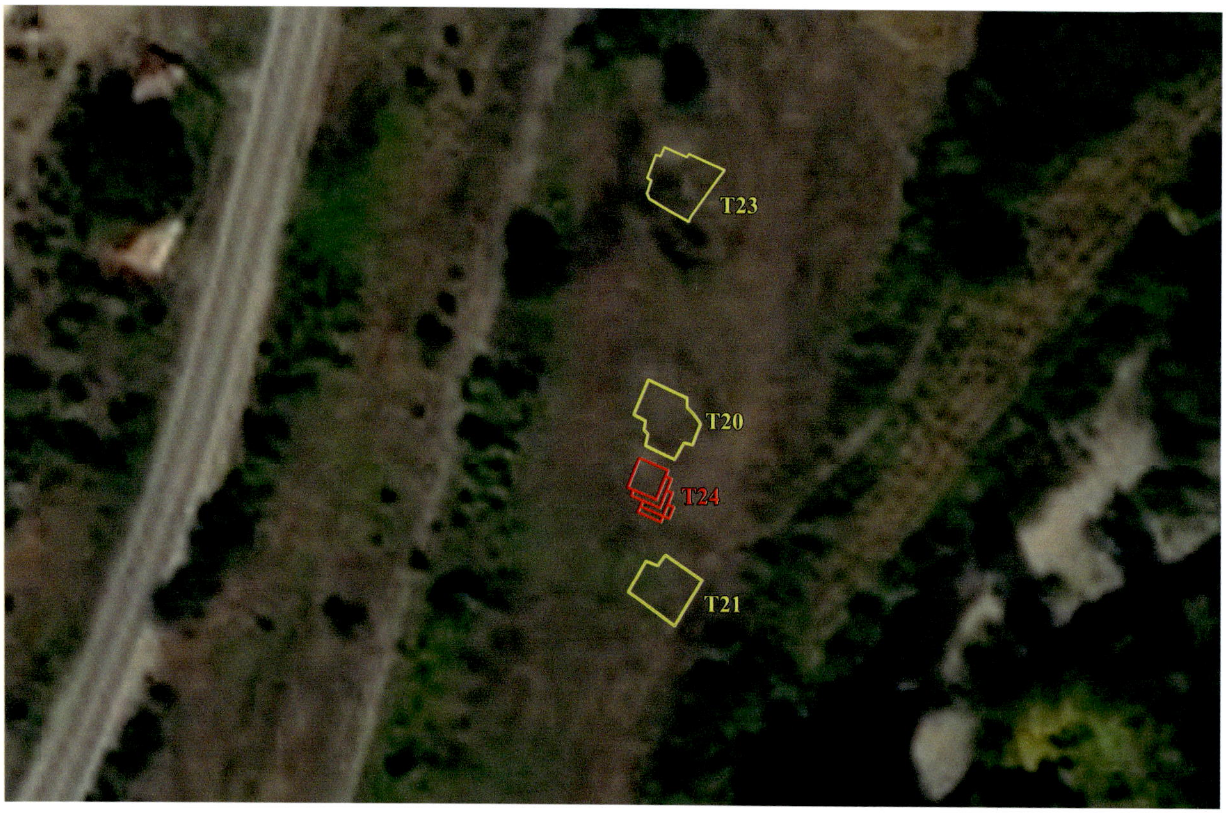

Figure 2. Aerial image of Trench 24 location at Pločnik (red) between Trenches 20 and 21 (yellow)

ground control points maintained by the Serbian Geodetic Authority. Measurement data was acquired using a total station Leica TCR705 instrument in IR-fine mode, with declared precision of 2 mm + 2 ppm.

Each trench was given a unique numeric marker, following the system of previous excavations; the trenches were not subdivided into smaller units. Previous excavation methodology on both sites included the use of an arbitrary spit system of 10 cm, until recognisable features were encountered. This approach was also adopted for the 2012–13 campaign. Each vertical 10 cm spit was numbered in sequence, beginning with the surface. The relative depths of the spits were maintained in order to allow direct comparison of movable finds on various parts on the site but, due to terrain contours, this did not necessarily mean that, for example, spit 5 in each individual trench was located at same absolute depth. When detected, individual features were named according to function and assigned a numerical value (e.g. house 1, pit 1). Prior to the excavations, it was decided to combine this system with single context recording and excavation using the natural stratification of the site. In practice, this usually meant that 10 cm spits were employed within the same subsoil class, but defined and distinguishable features were not excavated using the same spit. Rather, they were excavated by proper deconstruction of the stratification sequence, regardless of its depth.

To facilitate excavations, the team employed workers from the local population, who were trained by the previous Site Director of both sites, Dušan Šljivar, of the National Museum in Belgrade. Picks, spades and shovels were used for most of the excavations, with clods broken into smaller pieces within the trench before being examined for finds. Identified features were cleared and excavated using smaller spades, trowels, spatulas and, occasionally, dentistry tools. Handheld magnets were used regularly to search for metallic evidence in the excavated soil before it was removed from the trench.

Soil samples were taken from each spit and feature for macro-botanical analysis. These consisted of approximately 5 kg soil per spit or feature, or the total content where applicable. Additional soil samples of similar size were taken from areas where metallurgical finds were registered during the excavation process. These were wet sieved on site before being processed further. Dry sifting of spoil was not undertaken due to time and financial constraints, although soil from several features was roughly sifted in situ. Although the recovery rate of smaller finds may have been impaired by using this larger grain methodology when compared to some contemporary studies in the region and period, the thickness of the archaeological deposit and the limited excavation season did not allow for a slower, more detailed pace of excavations, which would certainly have resulted in improved recovery rates and decreased bias towards larger individual pieces and numerically dominant categories of finds.

The excavation methodology and recording procedures were adapted from those developed since 1998 at the Neolithic site of Belo Brdo in Vinča, where the approach focused primarily on the digital acquisition of data. As part of the Belo Brdo team for almost two decades the principal author of the chapter was involved in the development, testing and implementation of this excavation methodology and recording procedures, and successfully adapted them to facilitate both Belovode and Pločnik excavations.

Spatial data recording was performed with an electronic distance meter device and included recording point data for individual finds together with a series of point data representing the spatial extent and characteristics of detected features. Each spit level was recorded separately with the four corners of the trench marked A to D, clockwise from the northernmost point. The spatial extents of features were later transformed into shape files using specialised GIS software. This made post-excavation processing and analysis quick and widely available to the specialists working on the movable finds. Relational stratigraphy was enforced using a simplified Harris matrix (Harris 1989) which has proven sufficiently robust for sites with complex vertical stratigraphy (Harris *et al.* 1993). Digital imaging was also undertaken, using both orthogonal and oblique photography, in combination with photogrammetry, to document individual features and spits. Post excavation processing of the recorded orthogonal imagery enabled the production of scaled technical sketches and vertical section plans, whilst the use of commercial photogrammetric software allowed the creation of interactive 3D models, now available in the electronic repository associated with this book.

During field excavations, movable finds were separated by class and treated accordingly. Most finds, including vessel fragments, animal bone, and stone were not recorded individually, but a small percentage of 'study inventory' and 'C-finds' were singled out based on rarity and significance. Most such finds were given an additional field inventory ID or 'C number'. This class included, but was not limited to, figurine and altar fragments, metallurgical finds, polished stone and bone tools, fragmented and complete chipped stone tools, bone and/or stone jewellery, and amulets. These were stored separately from the main body of finds, were described according to type and function, and were photographed and drawn to scale. Analysis of these finds forms the core of the specialist reports found in the later chapters of this volume and were the most representative and typical material discovered on both sites.

Finally, it should be mentioned that cross-referencing of different classes of movable finds and their feature of origin was maintained during the entire process from field excavation to typological and statistical analysis in order to better understand and interpret individual contexts. It is our firm belief that the results presented in this volume represent the best possible interpretation of the excavations, based on the experience drawn from field research, various laboratory analyses, and processing performed in the post excavation period. All excavation documents are available in Appendix A.

Appendix A – Excavation data for Belovode and Pločnik (seasons 2012 and 2013)

Avaliable online at
https://doi.org/10.5522/04/14769990

The bibliographic reference for this chapter is:

Marić, M., Roberts, B. W. and Pendić, J. 2021. Excavation methodology for the sites of Belovode and Pločnik, in Radivojević, M., Roberts, B. W., Marić, M., Kuzmanović Cvetković, J., and Rehren, Th. (eds) *The Rise of Metallurgy in Eurasia*: 77–79. Oxford: Archaeopress.

Photo by Miljana Radivojević

Part 2
Belovode

Chapter 8

Belovode: landscape and settlement perspectives

Miroslav Marić

Belovode region

The modern region of Braničevo (Figure 1), in which the site is located, lies approximately 100 km east of Belgrade, the capital of Serbia and has its administrative centre in the city of Požarevac. Braničevo is bounded on the west by the Velika Morava river, to the north by the Danube, to the east by the River Pek and to the south by a mountain range including the Homolje, Beljanica and North Kučaj mountains. The area is *c.* 85 km across from north to south and east to west, with a total territory of *c.* 3855 square km. The difference in elevation between the lowest and the highest areas in the region is 1255 m, with the lowest land at 60 m above sea level, whilst the highest mountain peak (Beljanica) reaches 1336 m above sea level. The stark differences across the landscape of the region enables the distinction of several geographic units. The Požarevac Morava region centres on the flow of the Velika Morava river and the city of Požarevac and is predominantly flat terrain (Figure 1, left). The second largest geographic unit is the Stig and Danube, the area to the northeast of Požarevac, delimited by the Mlava river in the west and the Homolje mountains in the southeast. This area, although mostly flat, also encompasses undulating tracts of elevated terraces and plateaux. Another unit, the area of Zvižd, lies to the east of the Stig plain and is centred on the Kučevo valley with the gold-bearing Pek river flowing through the centre. With the exception of several river valleys (the Pek and Tumane valleys being the largest), this area is predominantly hilly with peaks reaching 400 m above sea level. The markedly mountainous region of Homolje forms the eastern and southern boundary of the Braničevo region, centred on the town of Žagubica (Figure 1, lower right). This unit includes the Homolje valley around the upper part of the Mlava river, but in a wider sense it also encompasses the Homolje mountains.

The modern climate of the area is either 'mountainous' or 'mildly continental', depending on the geographic setting of the landscape, and was most likely very similar in the prehistoric period. The winters can be very bitter and lengthy, whilst the summer is mostly hot and dry with occasional local showers. The average rainfall is around 50 l per m^2 (Stojić and Jacanović 2008: 21).

The soil in the area is very fertile, the most productive being that in the Stig region. Aside from alluvial plains and rivers, lower portions of the region consist of Chernozem soils with slopes and hills being covered with vertisols and cambisols (Stojić and Jacanović 2008: 14). Hills and elevated terrains are colonised by beach and oak forests providing good habitats for abundant wildlife. In the regions of Stig, and Braničevo in general, aeolian accumulative formations occur sporadically, with loess deposits and smaller areas of windblown sand. A significant proportion of water from the Pek river is absorbed by the sand, reducing its volume in its lower course compared to the middle course. These sediments form under the influence of the dominant wind in the region, the so called *Košava* wind.

Three major rivers of the Braničevo region have a broadly similar direction of flow—predominantly southeast to northwest—and join with the Danube (Figure 1), which flows from west to east, at the northern boundary of the region. The Morava and the Mlava, the two larger rivers, have distinctly meandering courses. The course of the Velika Morava was regulated in the late 19th and throughout the 20th century. The Velika Morava flood plain played a key role in the area from the earliest Neolithic period. It is wider than any other such floodplain in central Serbia and, due to its size and pronounced meandering, the annual deposition of fresh alluvium was effectively limited to the central parts of the plain (Chapman 1990: 27). The course of the Pek, the third river, also meanders, but its valley is significantly more constricted by nearby hills, making its paleo-course significantly narrower than the other rivers. The Danube is the northern delimiter which separates the Braničevo region from the edge of the South Pannonian plain, but the influence of the Pannonian climate and winds extends south into the Braničevo region. The predominant winds are eastern and the Braničevo region is known to have *c.* 170 windy days per year (Stojić and Jacanović 2008: 13). The hydrographic network of this regions' central Mlava river is unevenly developed. In the mountainous region it rises with springs which evolve into streams and smaller rivers that form a profoundly dense network of waterways. This is particularly true of the Homolje mountains area and the contact zone of Beljanica and Homolje where numerous springs and founts exist. The Mlava and Velika Morava are separated by the low-level Požarevac ridge.

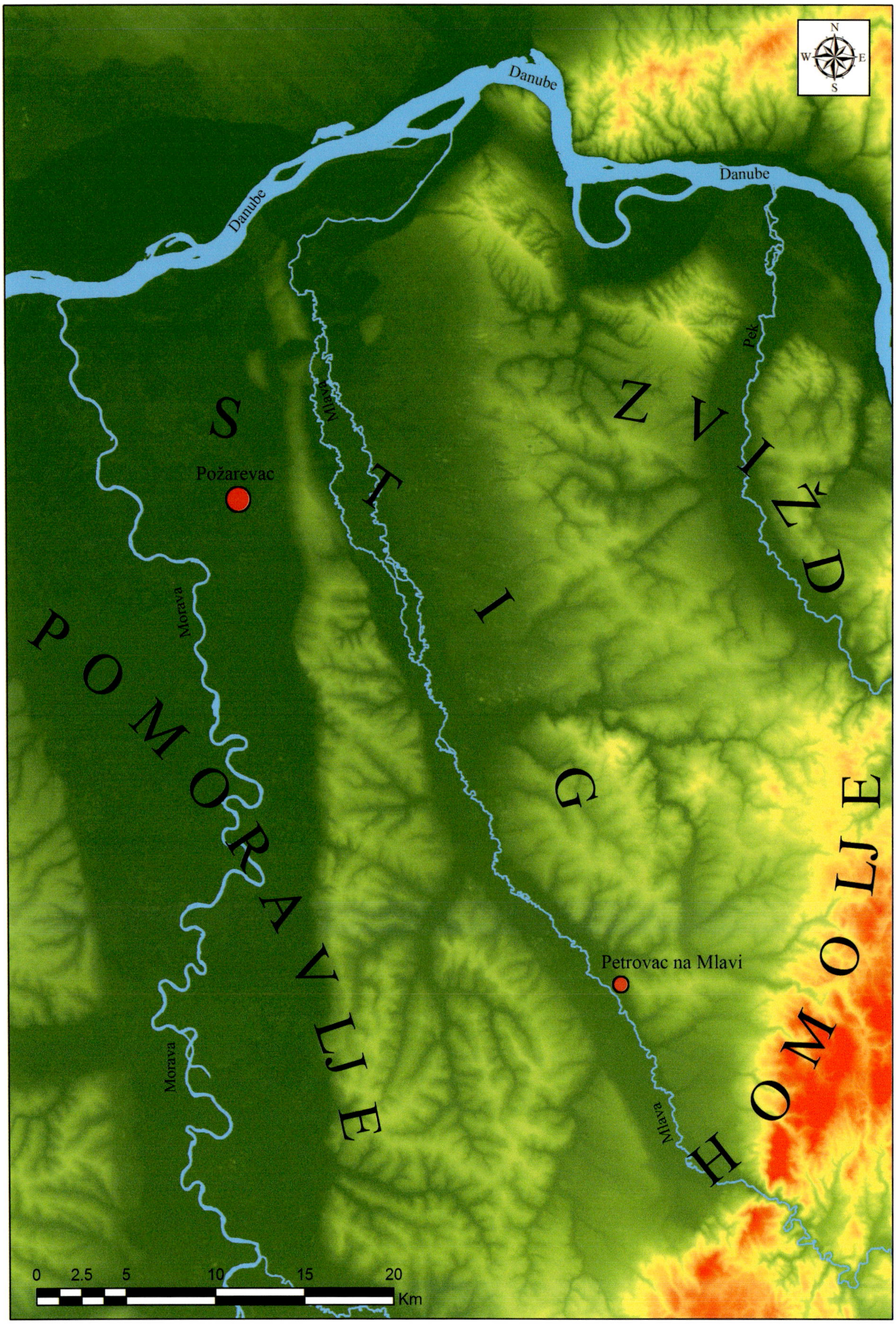

Figure 1. The Braničevo region.

The Neolithic in the Braničevo region

Human occupation of the region is attested archaeologically from the Mesolithic period (Lepenski Vir culture) and the mid-8th millennium BC (Srejović 1972; Radovanović 1996); it was probably settled during the Palaeolithic, although the archaeological evidence is currently lacking. For the early Neolithic period, the Starčevo communities are known from 11 sites in total (Garašanin and Garašanin 1951; Jacanović 1988), however the rich landscape most likely supported many more as yet undetected settlements. The almost millennium-long Early Neolithic period was replaced by the Late Neolithic manifestation known as the Vinča culture, with 26 known settlements, including Belovode, now recorded from various chronological stages (Figure 2; Stojić and Jacanović 2008: 45). Unfortunately, only a handful of sites have been researched in detail, Belovode being among them (Šljivar and Jacanović 1996c, 1997c), as well as Viteževo (Šljivar and Jacanović 1995) and Orašje (Marić 1951). Opencast mining operations at Kostolac in the north of the region have also revealed several Late Neolithic sites like such as Lugovi, Selište and Čair, but these are not published in detail (Jacanović 1988).

The state of research is sufficient, however, to allow us to extrapolate certain conclusions about the Late Neolithic settlements in the region. From the position of known sites (Figure 2) it can be determined that they were located close to one another, between 1.5 and 6 km apart, which translates to a walk taking between 0.5 and 1.5 hr. The shortest distance recorded is between the sites of Čair and Mali Grad (Figure 2, top), i.e. 1.5 km in a direct line. The settlements were located on slightly elevated plateaux, set close to a source of water (Table 1). Due to the meandering nature of the major waterways, it is possible that many sites have been either eroded or buried by the waters of the Velika Morava, the Mlava and the Pek in the six millennia since the Neolithic but some sites, like Orašje, Toplik (Stojić and Jacanović 2008) or Bresje (Jacanović 1985), located between 100 and 300 m from the meanders of the Velika Morava and the Mlava, may have remained undamaged owing to the vertical distance to the water surfaces, which ranges from 1.5 to more than 10 m.

The settlements were also placed on elevated river terraces that were either flat or mildly sloping (Figure 3) and, occasionally, were even set against the hilly background (Table 1), for example at Kod česme in Poljana and Selo in Simićevo (Spasić 1993). Several settlements are also known to have been placed in elevated positions, dominating the surrounding landscape, like the sites of Mali Grad in Stari Kostolac (Stojić and Jacanović 2008: 267), Zbegovište in Oreškovica (Jacanović 1988: 119), and Ladne Vode in Rečica (Jacanović 1988: 117). The latter two sites can be dated to the later phases of the Vinča culture and follow a trend similar to that existing in the central territory of the Vinča culture, attested at numerous other sites including Gradac near Zlokućane (Stalio 1972), Crkvine in Mali Borak (Arsić *et al.* 2010), Crkvine in Stubline (Crnobrnja 2009) and others. Another interesting feature of the Neolithic in the region is the utilisation of the Ostrvo, the large river island on the right bank of the Danube, located between Dubravica and Kličevac (Figure 2, top right). At least two Late Neolithic period settlements existed on this 20 km-long island: Selište, near Old Kostolac (Jacanović 1988: 114), a long-term settlement with at least three metres of cultural deposits, and Hrastova Humka, a tell-type

Table 1. Landscape characteristics of Late Neolithic sites in the Braničevo region.

Site no.	Degree of slope	Aspect	Soil type	Distance to water (m)
Belovode	2.988	West	Cambisols	124.935
1	4.196	East	Fluvisols	402.691
2	0	Flat	Fluvisols	>1000
3	5.031	North	Fluvisols	903.940
4	3.891	East	Fluvisols	519.246
5	6.497	Southwest	Fluvisols	462.940
6	0	Flat	Fluvisols	360.942
7	12.951	Northwest	Arenosols	194.042
8	0	Flat	Phaeozems	507.788
9	0	Flat	Gleysols	270.440
10	1.512	Southeast	Phaeozems	88.905
11	7.861	West	Phaeozems	568.170
12	0	Flat	Fluvisols	216.743
13	2.696	Northwest	Fluvisols	160.554
16	0	Flat	Phaeozems	268.056
18	5.188	Southwest	Cambisols	243.120
19	0	Flat	Fluvisols	615.667
20	4.554	Southeast	Gleysols	338.804
21	8.788	West	Cambisols	241.720
22	4.871	East	Cambisols	174.275
23	6.576	West	Fluvisols	100.529
24	5.904	West	Cambisols	199.769
25	0	Flat	Cambisols	186.547
26	14.369	South	Cambisols	301.690
28	7.657	South	Fluvisols	228.079
29	6.249	North	Cambisols	454.950
31	26.510	South	Cambisols	55.311
32	31.273	West	Cambisols	>1000

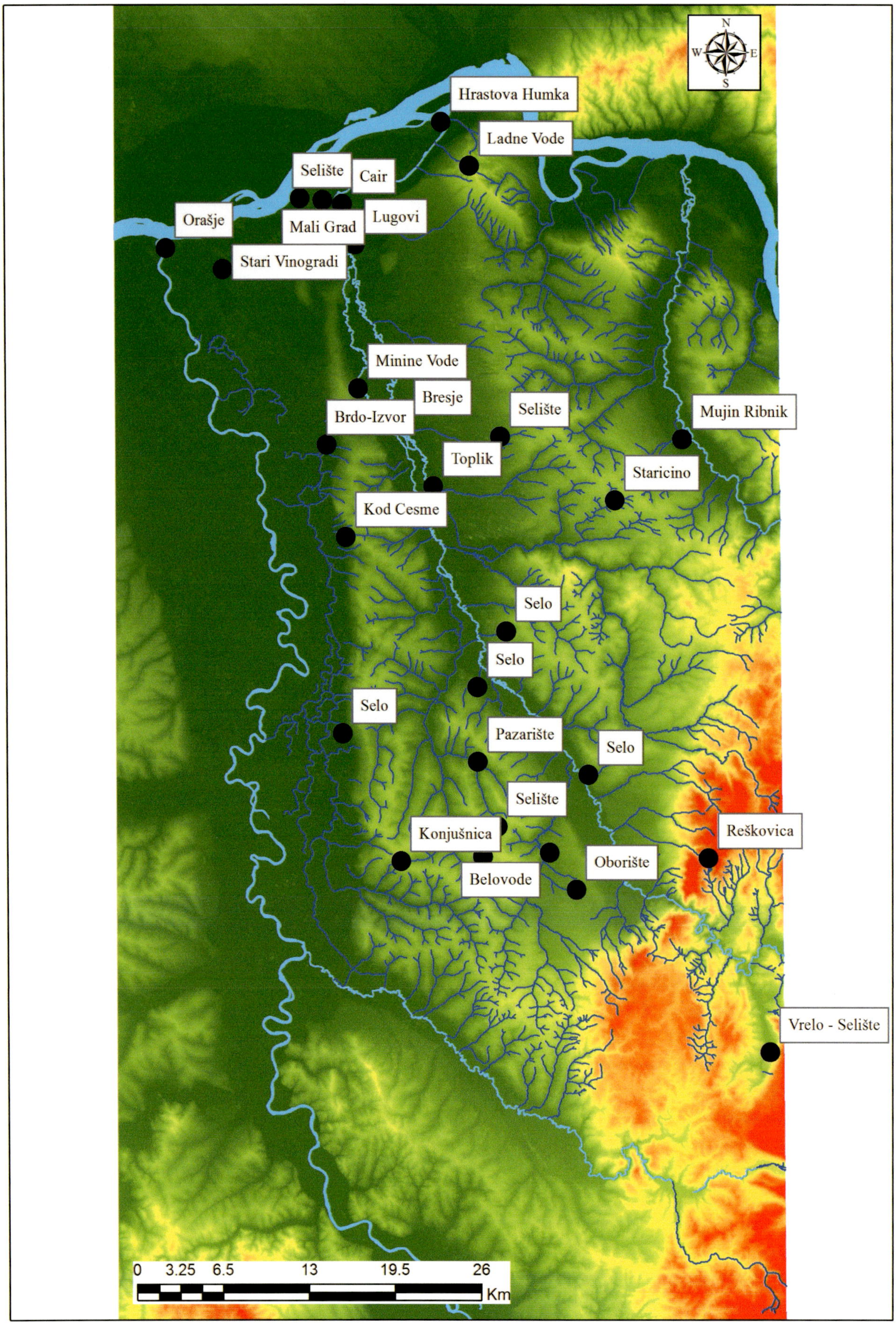

Figure 2. Late Neolithic sites in the Braničevo region.

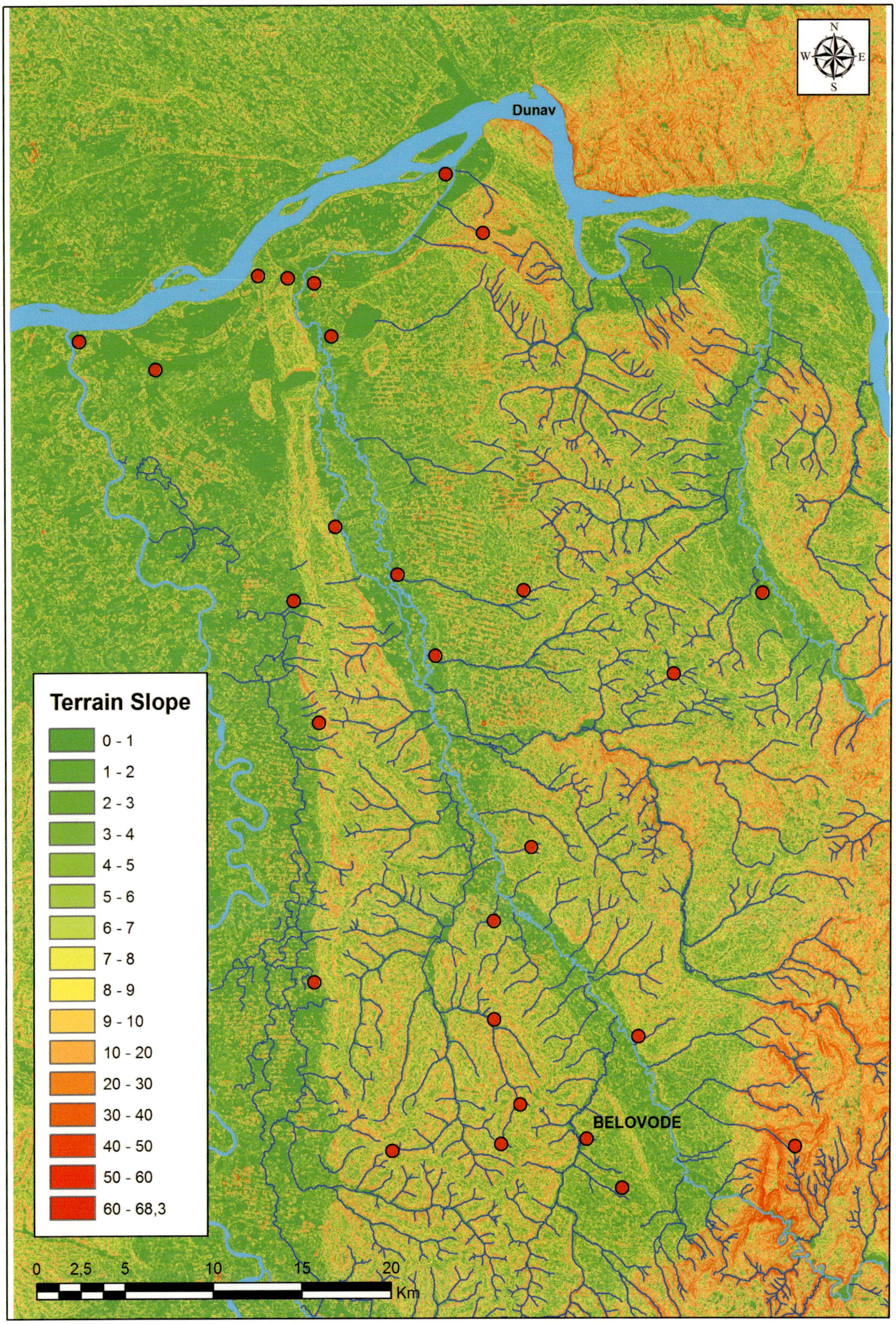

Figure 3. Terrain slope values in the Braničevo region.

settlement dated to the Vinča C–D phases (Jacanović 1988: 116). These sites indicate that, although perhaps not in its present-day shape, the island existed in the Neolithic, and was sufficiently large and dry enough to provide for long term settlements.

Table 1 illustrates that, of known Late Neolithic sites, almost one third (32.1% or nine sites) are found on cambisol, the principal soil type of the region (*c.* 55% of total soil coverage). It is surprising, however, that even more (11 sites; 39%) are found on fluvisols, which account for only 25.6% of the total soil available in the Braničevo region. Phaeozems, the most fertile soils in the region, and just 7% of the total area, were also settled with four known sites, and there is a likelihood that further Late Neolithic settlements existed in this fertile zone around the lower course of the Mlava (Figure 4). If we divide the number of known sites by the percentage of the coverage of the individual soil type on which they are found, an interesting relationship becomes apparent with respect to the site density (Table 2). It must be noted here that, as cambisols appear in less-accessible areas of Braničevo, the detection of Late Neolithic sites may be biased towards more easily accessible, flat terrain in the river valleys of the principal waterways in the region. The dominance of sites located on soils with above-average fertility (53% are located on the two most fertile soil classes) indicates a strong dependence on soil cultivation in the Late Neolithic, but also allows for the existence of specialised settlements, either due to a particular economy (e.g. transhumance or raw material procurement) or due to unstable conditions towards the end of the Vinča societies (the 'elevated retreat' sites of the late phases).

Table 2. Site density according to soil types in the Braničevo region.

Soil type	Soil coverage (%)	Number of sites	Site density
Fluvisols	25.6	11	0.429
Cambisols	55.0	9	0.16
Phaeozem	7.0	4	0.57

Belovode and its surroundings

The site of Belovode is located about 10 km to the southwest of Petrovac na Mlavi (Figure 5), on an elevated plateau above the Busur River, a tributary of the Mlava River. The settlement was placed between two springs that join with the Busur about 1 km from their origin (Šljivar and Jacanović 1996c; Rassmann *et al.*, Chapter 9, this volume). The central part of the settlement is nestled between the springs and is the most densely populated area. The plateau, approximately 600 x 200 m in size, is defined by three steep sides (Figure 6) that drop sharply towards the springs and the Busur. The site is located on podzolised cambisol, whilst the Busur valley consists of alluvial-diluvial soil. The high level of production that can be achieved on cambisols, in combination with the alluvial soils found to the west of the site, provides a powerful foundation for prolonged crop production with above-average yields, allowing long-term occupation of the area.

The settlement plan (Rassmann *et al.*, Chapter 9, this volume, Figure 5) shows the densely populated southern area of the site, centred between the two streams that flow at 90 degrees to the course of the Busur river. Several smaller, enclosure ditches occur to the east and the west of the central plateau and the settlement extends further east towards Bikova Bara, where it terminates with a large enclosure ditch. The settlement appears to have originated at the southern side of the plateau and, at some point, extended rapidly away from the Busur valley. The densest part of the settlement has confirmed stratigraphy of over three metres in places, but it cannot be identified as a tell settlement as it lies on an already-elevated plateau and shows no sign of a tell bulge. As is the case with Pločnik, the settlement was not restricted by the space available for expansion. This is clearly evidenced in its eastern portion, where structures are wider apart with larger spaces in between, most likely gardens or similar open areas.

The position of the settlement and its longevity may be somewhat puzzling if considered on a broader scale. The site is not located in the valley of the Mlava river, the arterial pathway through the region of Stig and Homolje, but rather on the opposite side of the ridge, facing away from the Mlava. When viewed on the topographic map of the region (Figure 2), however, it becomes apparent that the site is just 15 km away from the valley of the Resava river, one of the principal tributaries of the Velika Morava river, which is a key routeway during the entire Neolithic period, connecting the south of the Balkans with the Pannonian plain in the north. This side access to the Velika Morava is easily traversable via the valleys of either the Busur or the Busur and Đurinac rivers. From the valley of the Resava, at the point closest to the Busur and Đurinac valleys, the Velika Morava is only a further 16 km away. Another possible corridor towards the Velika Morava lies to the west, via the networks of the Čokordin river and its tributaries. Several sites of the Late Neolithic period, such as Zbegovište and Konjušnica (Figure 2), are known in this area, which could confirm this corridor as the preferred routeway to the Velika Morava.

Concluding remarks

Although interpretation of the late Neolithic period in the Braničevo region is limited by the number of known sites, there are clear signs of differentiation in terms

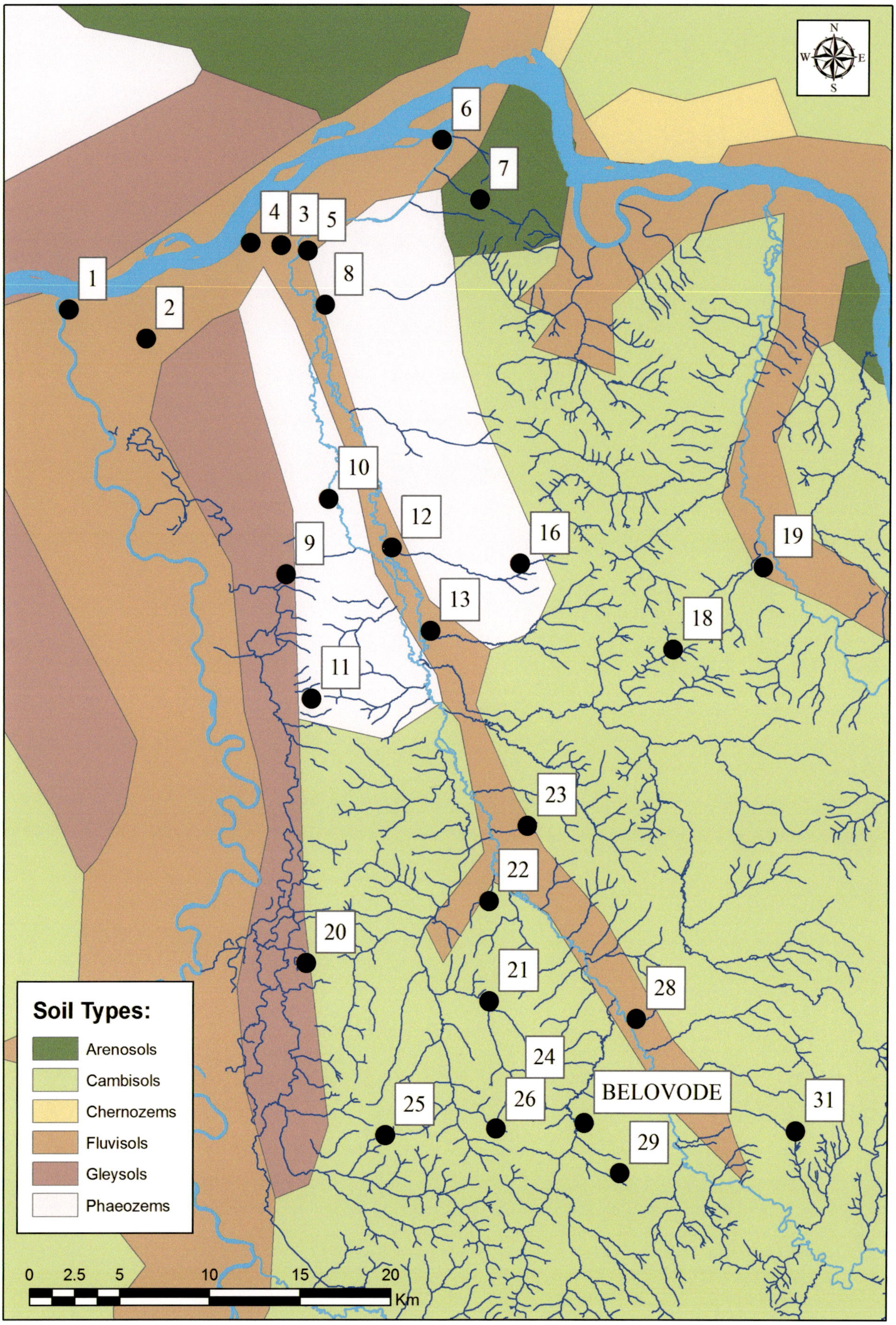

Figure 4. Soil types in the Braničevo region.

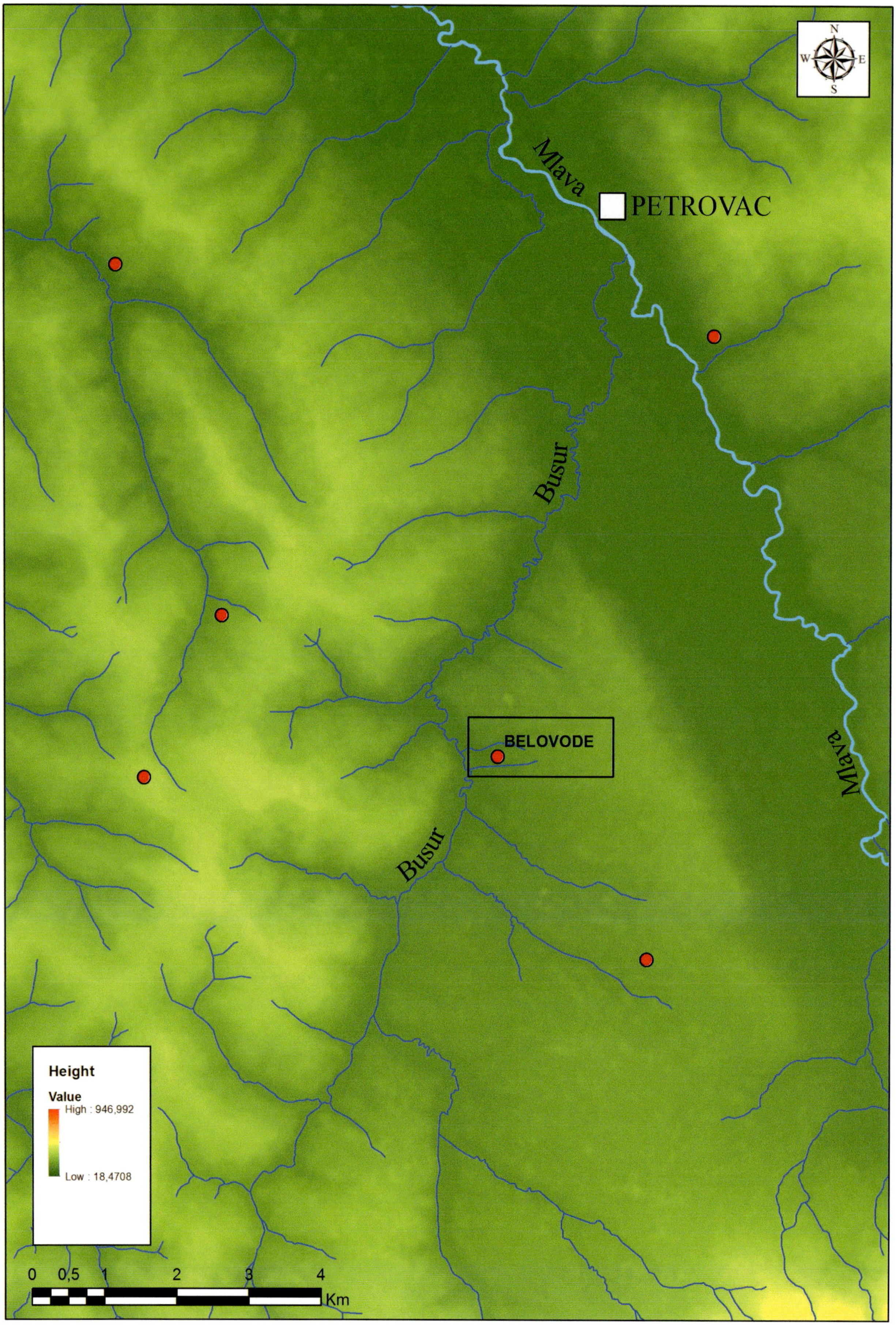

Figure 5. The location of Belovode and neighbouring Late Neolithic sites.

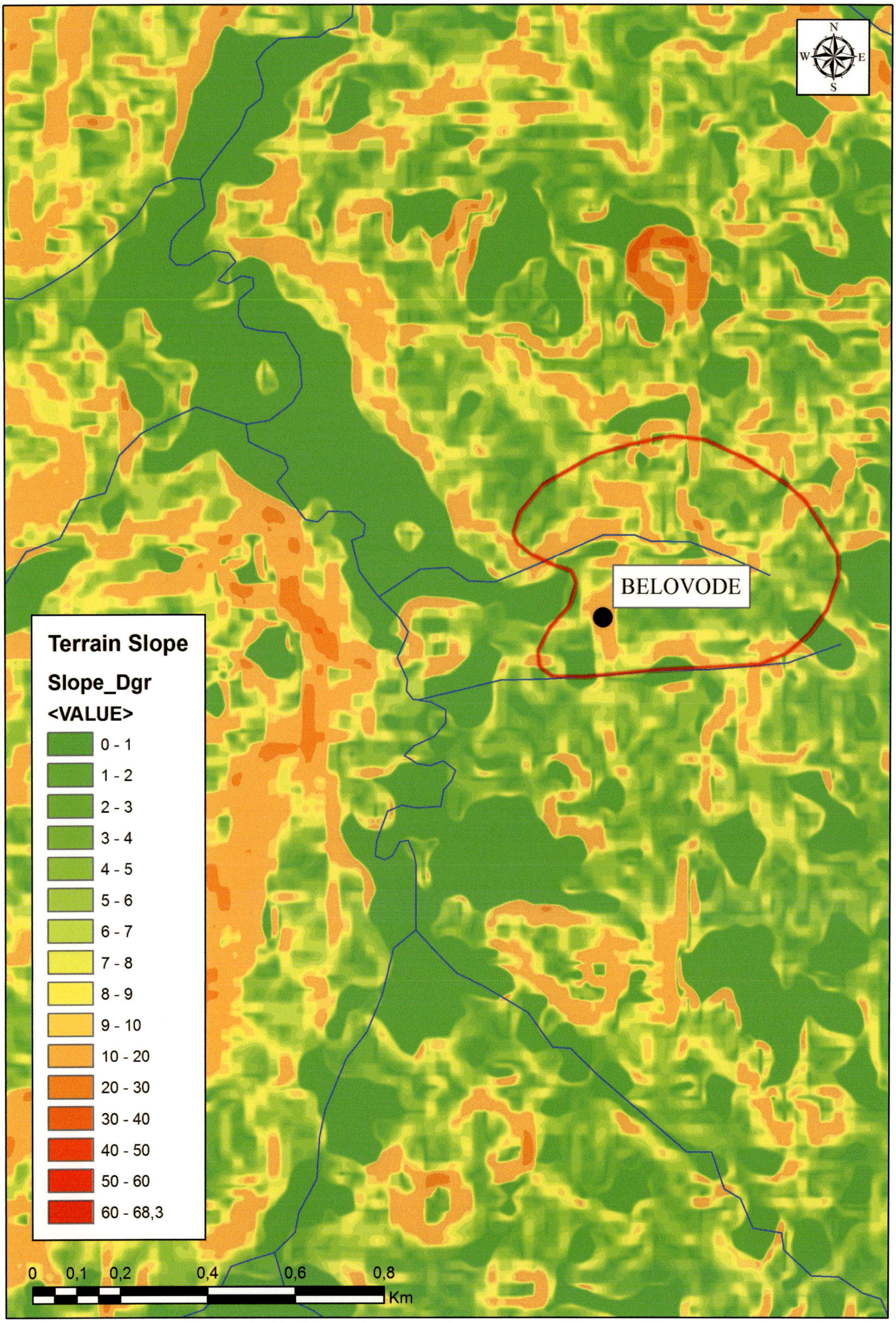

Figure 6. Terrain slope values around Belovode.

of settlement location. Bordering the flat plains of Vojvodina on the north, the region shows influences of the local manifestation of the Vinča culture, especially in its early phases, but later slowly incorporates the traditions of the Morava valley, transmitted from the central and southern variants of this culture. Both the extent and the thickness of the settlement layers indicate prolonged occupation of chosen locations and the tendency to nucleate and create hierarchy within the settled areas. Whether this reflects the establishment of an organised social, politic, or economic domination remains under consideration, as the intra-societal organisation and hierarchy of the Late Neolithic Vinča is still somewhat unknown for the central Balkans. The size of the settlements, ranging between 30 and 40 ha, is often double the size of the settlements in the Vojvodina plains, but these may not represent the largest sites in their respected regions. Recent research in other areas of central Serbia (Crnobrnja *et al.* 2009; Crnobrnja 2011; Tripković 2007; Tripković 2013) and also in Romania (Parța, Uivar) indicate that structures had notable differences in their functions based on inventory, size, organisation and construction methods. The proposed existence of shrines and economic structures speak to the level of specialisation present in the settlement which, by the end of the Late Neolithic, appears to have reached a high degree.

The positioning and the longevity of Belovode is, in part if not largely, due to its hinterland, with abundant sources of minerals and stone, usually within an easily traversed distance (15–20 km). The manner of procurement and the existence of networks can be sensed in the work of some authors writing on the origins of obsidian (Tripković and Milić 2008) and copper (Radivojević and Kuzmanović Cvetković 2014), but it should be made clear that the majority of the procurement remained on the level of individual settlements, especially with everyday commodities, albeit with certain exceptions (Amicone, Chapter 14, this volume). It was not, however, the rich hinterland alone that enabled the long-term existence of massive settlements like Belovode. The fertile brown cambisol that dominates most parts of central Serbia, together with the phaeozems in the Braničevo region, combined with the primitive wooden ards most likely available by the late 5th millennium BC (Bogaard 1981: 31), enabled annual cultivation of crops in the immediate vicinity. The crops become an increasingly important part of the diet, as evidenced in the numerous charred remains recovered across sites of this period (Filipović *et al.* 2018). With this new soil type becoming available for cultivation and the annual replanting of preserved seeds, crop yields would have increased. At the same time, the dependence of an annual deposition of fertile alluvium decreased, making settled life in a fixed location much easier. The introduction of bread wheat, the most productive type in the long line of domesticated wheat, and its dominance towards the end of the Late Neolithic, strengthens this presumption. Hunting and foraging traditions did not cease, however, and the woods and brushes remained an important source of food until the very end of the Neolithic, as evidenced by finds of wild fruits on other sites of the period (Filipović and Tasić 2012).

The existence of long-term settlements in fixed positions must also have changed local perceptions of the landscape surrounding them, leading to the establishment of semi-permanent or permanent networks of pathways for trade and exchange of both goods and people (e.g. inter-settlement marriages). Both Belovode and Pločnik are located in terrain that could be considered ideal for expedient travel. The Mlava and Toplica river valleys are arterial corridors that link a major channel of movement around the central Balkans: the Morava river system. Spreading from the far south towards the Danube, this truly pan-European corridor operated since the earliest time. These riverine routeways are dotted with numerous sites from the Late Neolithic period, many of which contain traces of long-term settlements (e.g. Pavlovac, Drenovac, Slatina). The Mlava valley also connected eastern Serbia with the lower course of the Morava river, enabling fast exchange of raw minerals from the Homolje mountains and beyond with other regions occupied by people of Vinča traditions. On the other hand, the Toplica river valley was the principle connection between the Kosovo plain and the core territory of the Vinča period in central Serbia.

The context surrounding the decline of these large settlements remains unresolved. It is well established that all the Late Neolithic sites in the Central Balkans area thus far excavated were destroyed in large fires together with their entire assemblages, but the circumstances that led to these conflagrations are not clearly understood. Some zooarchaeological analyses performed on a Late Neolithic assemblage from Belo Brdo in Vinča suggest that, in the late phase, there may have been an environmental crisis that led to increased exploitation of unusual meat resources like dogs or turtles (Dimitrijević 2006: 252–253, 255). It can be hypothesised that numerous large, permanent settlements, gradually increasing in size, may have become a strain on the surrounding environment, and on available technology and resources, leading to a decrease in the availability of food and commodities that may have fractured a society still highly dependent on the surrounding landscape leaving it unable to fulfil demands for resources without abandoning their long-established settlements.

Acknowledgements

I would like to express my gratitude to Julka Kuzmanović Cvetković, museum advisor and the director of the National Museum of Toplica in Prokuplje, Dragan Jacanović, archaeological custodian of the National Museum in Požarevac, and Jasmina Živković, archaeological custodian of the Zavičajni Muzej in Petrovac for their kind contribution regarding the locations of the late Neolithic sites in the Pločnik and Belovode regions and the phasing of individual sites.

The bibliographic reference for this chapter is:

Marić, M. 2021. Belovode: landscape and settlement perspectives, in Radivojević, M., Roberts, B. W., Marić, M., Kuzmanović Cvetković, J., and Rehren, Th. (eds) *The Rise of Metallurgy in Eurasia*: 83–93. Oxford: Archaeopress.

Chapter 9

Belovode: geomagnetic data as a proxy for the reconstruction of house numbers, population size and the internal spatial structure

Knut Rassmann, Roman Scholz, Patrick Mertl, Kai Radloff, Jugoslav Pendić and Aleksandar Jablanović

Introduction

The Vinča culture settlement Belovode was surveyed over the course of two campaigns, in 2012 and 2013. In 2012, we covered the area of 19 ha with a 16-channel magnetometer (SENSYS MAGNETO®-MX ARCH). In 2013, a 5-channel magnetometer (SENSYS MAGNETO®-MX ARCH) was used for smaller areas (covering a total of 6.6 ha). The survey revealed around 550 characteristic anomalies that can be classified as burnt houses, and several linear anomalies that presumably represent ditches. Their distribution demarcates an area that stretched over *c.* 33 ha. Within this is a core area comprising *c.* 21 ha which demonstrates a higher and (more variable?) density of house anomalies. The ridge of the hill in the southern part of the settlement exhibited the highest house density, a fact that perhaps necessitated the orientation of the buildings into regular rows. By contrast, the northern areas exhibited large house clusters between which were areas having no characteristic archaeological signature. The latter presumably represent areas which were without houses in prehistory.

To put the survey area (*c.* 26 ha) into perspective, our data are sufficiently representative to facilitate the reconstruction of the settlement's internal structure. A key benefit of the recent work was to make available precise data from the northern area for the first time. Former calculations of the settlement's size (at a 100-ha scale) overestimated the real dimensions (Radivojević *et al.* 2010a: 2278). Crucial for such calculations of the settlement size are several ditches in the northern and western area. These indicate several chronological phases at the site. In light of this, it is very likely that the burnt houses visible in the geomagnetic data correspond to different use phases at the settlement (i.e. they were not all contemporaneous). The geomagnetic data also indicate differences in the sizes of houses, as well as in their orientation and in the spatial organisation and structure of the settlement. Kernel Density Estimations of houses and daub remains reveal the presence and location of settlement areas with higher densities. The recognition of these patterns is crucial for the reconstruction of maximal and minimal house groups and acts as the basis for estimating the size of the local population.

Geomagnetic survey of Belovode was first undertaken in the early 2000s in the limited southern area of the site but was never published. We resumed this work in 2012 and 2013, with an aim to cover large areas of the settlement and its periphery. Our objectives were to deliver information on the size and the spatial structure of the settlement, a valuable approach that would allow the excavation results to be placed within a broader regional context. Former surveys at other Late Neolithic settlements like Uivar (Schier and Draşovean 2004), Okolište (Hofmann *et al.* 2007), Drenovac (Perić *et al.* 2016), Borđoš mound near Novi Bečej, Serbia (Medović *et al.* 2015), and Crkvine near Stubline (Crnobrnja 2011) demonstrated the potential of the Balkan Late Neolithic settlements for geomagnetic research.

Belovode provides a good case study as the entire site was used as a farmland until recently, allowing survey of the full settlement area. This is particularly valuable as Belovode can be used as a reference for the prehistoric settlement sites of Okolište and Pločnik, which are partly covered by modern villages. Nonetheless, the highly subdivided agrarian landscape at Belovode complicated the general course of geomagnetic survey. Some areas are gardens, others are small fields, with only a few larger fields. These circumstances required the application of two different systems. For larger fields, we worked with a 16-channel magnetometer and in smaller fields with a 5-channel magnetometer (Figure 1). Despite using two instruments and carrying out two campaigns, the survey of Belovode remains incomplete as some areas were not accessible during the two seasons, but the 26 ha area revealed representative data that enabled the reconstruction of the internal spatial structure, house numbers and population size. Although there was no direct evidence of metalworking, these data enable us to set the archaeometallurgical research within the broader context of the spatial structure at settlement level.

Figure 1. Overview of the surveyed area and the instruments used in the 2012 and 2013 campaigns.

Methodology and data processing

Both the 5-channel and 16-channel magnetometers (SENSYS MAGNETO®-MX ARCH) were manufactured by Sensys GmbH, Bad Saarow, Germany. They are made entirely from fibre-reinforced plastic. Both systems included FGM-650B tension band fluxgate vertical gradiometers with 650 mm sensor separation, a ±3000 nT measurement range and 0.1 nT sensitivity.

16-Channel magnetometer

The 16-channel magnetometer was mounted on a vehicle-drawn cart. The gradiometers were set at 0.25 m intervals on a 4 m-wide sensor frame. The vehicles housed both power supply and data processing hardware. MAGNETO®-MX compact 16-channel data acquisition electronics with 20 Hz sampling frequency were used for data acquisition with Trimble RTK-DGPS georeferencing (base/rover combination). With speeds of approximately 12–16 km per hour and a sample rate of 20 readings per second, the system provided xyz-data on a mesh of 0.25 m by approximately 0.3 m.

5-Channel magnetometer

The 5-channel magnetometer was mounted on a hand-propelled carriage. The gradiometers were set at 0.25 m intervals. A walking pace of *c.* 4–5 km per hour yielded a mesh of 0.25 m by approximately 0.06–0.08 m. The prospection areas were first marked out using a Leica DGPS (GX 1000). The rectangular areas ware prospected in a zig-zag pattern. In order to maintain the correct orientation, each 2.5 m was marked by a cord. The interpolation of the measurements along the cord was based on the distances recorded by an odometer.

Data processing

Use of this new generation of geomagnetic instruments enabled us to prospect large areas in relatively short time periods. One challenge, however, was to analyse the data produced within an appropriate time frame. A second difficulty was the production of reliable data in order to facilitate the straightforward comparison of different sites. Such cross-comparisons require consistency in data processing and software tools and could only be accomplished through close cooperation with project partners. The first level of analysis was based on commercial software; in the later stages of analysis, we used open source software to facilitate this cooperation.

The SENSYS MonMX, DLMGPS and MAGNETO®-ARCH software package was used for data acquisition, primary data processing, interpolation and export. Each track contained measurements produced by the five or sixteen channels and the DGPS date and was saved separately. For Belovode, this resulted in 460 tracks. The tracks recorded by SENSYS MonMX were then imported into the DLMGPS software. To check the data in the field, we used MAGNETO®- ARCH for initial geomagnetic map imaging. For more detailed interpolation, we used OASIS montaj. In order to export the data from DLMGPS to OASIS montaj, a simple text file was produced which contained the vertical gradient (z) as nanotesla values and the measured track and the number of the probes as shown in Table 1.

Post processing was completed with OASIS montaj 8. The results were exported as a Surfer 7 file (compatible with GIS software). The use of Surfer 7 files enables the user to modify threshold, colour scale and (in combination with the rich choice of available raster and vector tools) to produce spatial analysis of the date. This further processing was done in QGIS 2.6. Each digitised anomaly was assigned an ID number for further analytical steps. This is especially important when features are added to the specific ID, albeit on different layers (e.g. electromagnetic data and aerial photography).

Methodological remarks

The geomagnetic map of the settlement at Belovode (Figure 2) shows many settlement features (especially houses). While a spatial analysis of those features has the potential to reveal the size and spatial structure of the settlement, it should also consider uncertainties regarding the chronology of the features under study. Radiocarbon dates from Belovode trace the settlement's use over a period of approximately 700 years from 5350–4650 BC (Radivojević *et al.* 2010a: 2778; Chapter 37, this volume). Excavations on the promontory suggest that the geomagnetic anomalies of burnt houses are likely to belong to the latest settlement horizon. The chronology of the house cluster in the northern settlement is less clear. Despite the uncertainties outlined, our intention was to produce an approximate model based on the maximal extent of the settlement, and support or negate the inclusion of surface collection and target excavations on the forthcoming field research agenda.

The ambiguity of the geomagnetic picture can be minimised by a more advanced exploration of the geomagnetic data, in which each individual house anomaly is analysed in detail as a single unit. The analysis should integrate a perspective on the spatial context of the 'objects', (e.g. houses or storage pits). As Hillier and Hanson (1984) emphasised: '...in talking about buildings, we need not only to talk about objects, but also about systems of spatial relations.' If we apply this rationale to our analysis, we can classify the orientation of houses, observe how they are embedded within the greater spatial layout of the site, quantify the variation in size of each object within a house cluster and later, as a result of these analyses, compare the various house clusters to reveal the differences and similarities between them.

Some of these characteristics (e.g. size and orientation) may reflect differences in chronology, as has been indicated by the excavation results from Late Neolithic

Table 1. Exported txt-file with raw data.

x-coordinate	y-coordinate	nT	Track file	Probe number
34340748.010	5140263.576	-59.5	fa123.prm	1
34340748.092	5140263.340	-3.7	fa123.prm	2

Figure 2. Overview of geomagnetic data for the whole site.

settlements like Kundruci and Okolište (Furholt 2012: 14f.; Hofmann 2013: 366). A similar relationship was assumed by Medović *et al.* (2015) based on geomagnetic data from the Vinča settlement of Borđoš near Novi Bečej. Figure 3 outlines this analytical framework and the specific properties of the anomalies as well as their information content in order to illustrate the workflow through which geomagnetic date can be set into a broader analytical context.

Results: geomagnetic data

Promontory

The focus of the excavations since 1993 has been the settlement area that lies on the promontory. The topographical situation of the site is similar to that of a hillfort (Figure 4). In addition to the *c.* 180 houses, the geomagnetic survey revealed three linear anomalies.

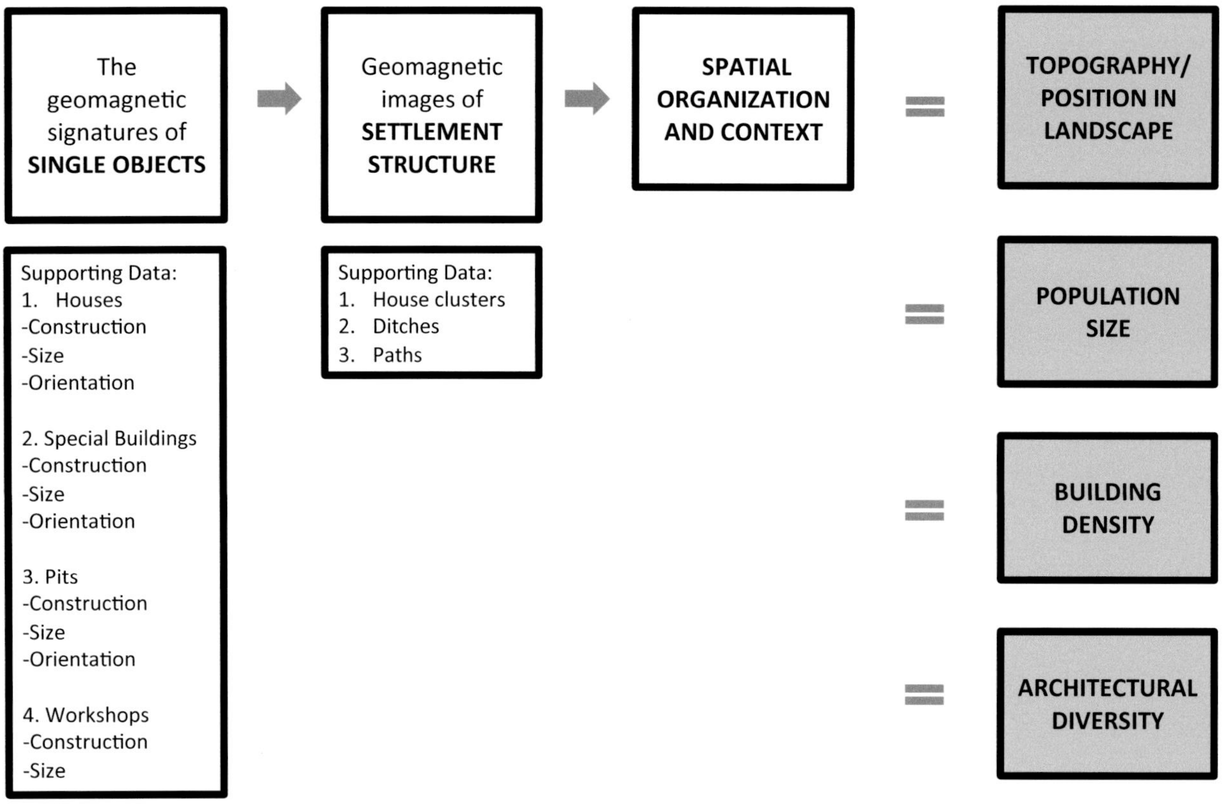

Figure 3. Diagnostic flowchart for geomagnetic data of architectural elements for the reconstruction of the spatial layout of ancient settlements.

Houses were oriented in rows, mainly orientated NE–SW, as was the case both at Okolište (Hofmann *et al.* 2007: 56, Figure 7) and at Drenovac (Perić *et al.* 2016). However, this is in contrast with the principle orientations noted at the sites of Pločnik (with its main NW-SE axis) and Crkvine, where orientations varied from NNW–SSE to NNE–SSW (Crnobrnja 2011: 133, Figure 2). The housing density at Belovode is comparable to that at Crkvine and is somewhat less than that recorded at the sites of Okolište or Pločnik. The preservation of the houses varies, as seen in Figure 5. The majority are marked by concentrations of large pieces of daub generating a dynamic range of >6 nT. By contrast, on the western edge of the promontory some houses were revealed, not by large accumulations of daub, but rather by clearly visible house walls. We assume that in these cases the houses were originally burnt like the others but that the daub had since completely eroded.

Two linear structures cross the western periphery of the survey area; a third protects the eastern access to the promontory. In the neighbouring area to the east, a further double linear anomaly is visible (Figure 4). These linear features are small, having a diameter varying between 0.5 and 2 m. Only those structures with widths of at least 2 m can be classified as ditches. Those that are smaller than this possibly represent palisade trenches. The selection of the promontory site in combination with the construction of a fortification indicates that the founders of the settlement needed to protect their community. The maximal house concentration and the existence of a fortification could reflect intergroup conflicts on a larger scale.

Eastern periphery

An additional eight linear anomalies were revealed in the eastern part of the settlement (Figure 6). Once again, their signatures varied. Only Feature 6 (2 m width) could reasonably be classified as a ditch. The density of buildings in the eastern part of the site is much lower than on the promontory. The thickness of the settlement layers in this area is 2 m (indicated by measurements made from a single auger hole (Figure 5). The data provide a preliminary point of reference for the settlement duration for this part of the site. Comparable data are available for Vinča-Belo Brdo and Okolište. For Vinča-Belo Brdo, Schier reconstructed a sedimentation rate of 1–2 cm per year; shortly after the founding of the tell-site the rate reached 7 cm, and then only in part (Schier 2001: 378, Figure 4). The sedimentation rate at the site of Okolište was considerably lower than 1 cm per year (Hofmann 2013). The 2 m sequence at Belovode indicates long-term settlement activities for more than 200 years, confirmed by absolute dates for samples taken on earlier excavations and again in 2012 and 2013 (Whittle *et al.* 2016: 22–23, Figure 9; Chapter 40, this volume).

Figure 4. Overview of the geomagnetic data on the promontory in the southern settlement area. Three ditches are labelled.

Northern plateau

Around 270 houses were revealed on the northern plateau. These cluster in three principal house groups (Figure 7) in which houses are organised in rows in the same layout as that on the promontory and, as in that area, they appear to have been preserved by fire (Figure 8). The magnetic anomalies include a wide range of shapes, sizes and mean nTs. Overall, the pattern is different from that of the regularly burnt houses of the Cucuteni-Tripolje culture (Rassmann *et al.* 2014: 209, Figure 14). These differences may indicate variations in the size, construction, and destruction process of the houses by fire (Figure 9). There is, however, some evidence for unburnt houses (Figure 8B). In contrast to the probable eroded houses on the western edge of the promontory, the house floors here are marked by a 'shadow' with slightly higher values from 2–5 nT. In the northwestern area, a small gap is the only indication of an entrance (Figure 8A). However, we can use this data to reconstruct a path that crossed between the different house rows. The main result of

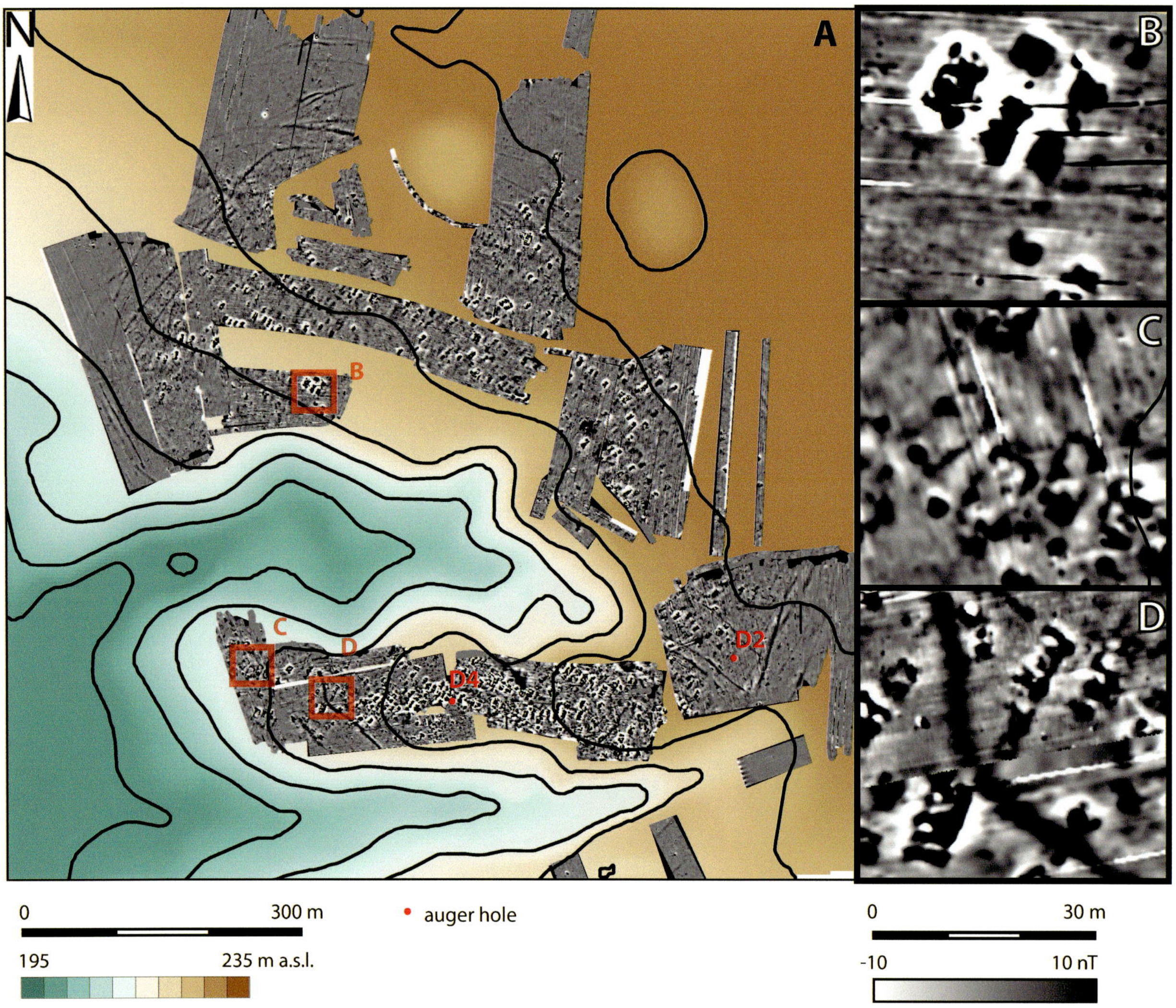

Figure 5. Geomagnetic results demonstrating the differences in anomalies for houses produced destroyed by fire or well-preserved. A) Overall plan showing locations of B, C and D (red squares); the location of the auger holes with chemical analysis (D2 and D4) are also shown; B) Heavily burnt and well-preserved houses in the northern settlement; C) Burnt and eroded houses on the western slope of the promontory; D) Burnt houses on the promontory in the southern area of the settlement.

our survey is the indication that the overall settlement area was considerably smaller than originally thought. Moreover, the houses located inside this area were divided into three main house groups characterised by uneven distribution densities.

Analysis and interpretation

In the first stage of our analysis, the characteristic anomalies were digitised by hand. To evaluate the results, the team worked independently in Belgrade (Pendić) and in Frankfurt (Rassmann). The results differed slightly; Pendić classified 578 anomalies as houses, whereas Rassmann found a total of 556 (see Figure 10). The digitised maps also revealed pit anomalies besides the identified houses. To minimise the subjective factor, we established a specific workflow for the automatic classification of daub (see Figure 11). With the GRASS 'r.contour.level' tool, the geomagnetic data were converted into isolines by 1 nT-steps (Figure 11B). The isolines were compared with the underlying geomagnetic raster data (Surfer 7.grd file). In both Pločnik and Belovode, house anomalies were closely correlated with the 6 nT line. The 6 nT-lines were therefore selected and then converted into polygons (Figure 11C). By spatial queries we generated characteristics (area size, mean of nT-values) for these polygons in the attribute table of the respective shp-file. The statistical analysis of the house size and the size of the daub area correlated with each other. The example shown in Figure 11C demonstrates the difficulties of defining the exact shape of a house (especially if that house was not completely burnt or it was heavily impacted by erosion). In this way, we measured the size of the daub below the reconstructed houses in order to obtain more reliable data.

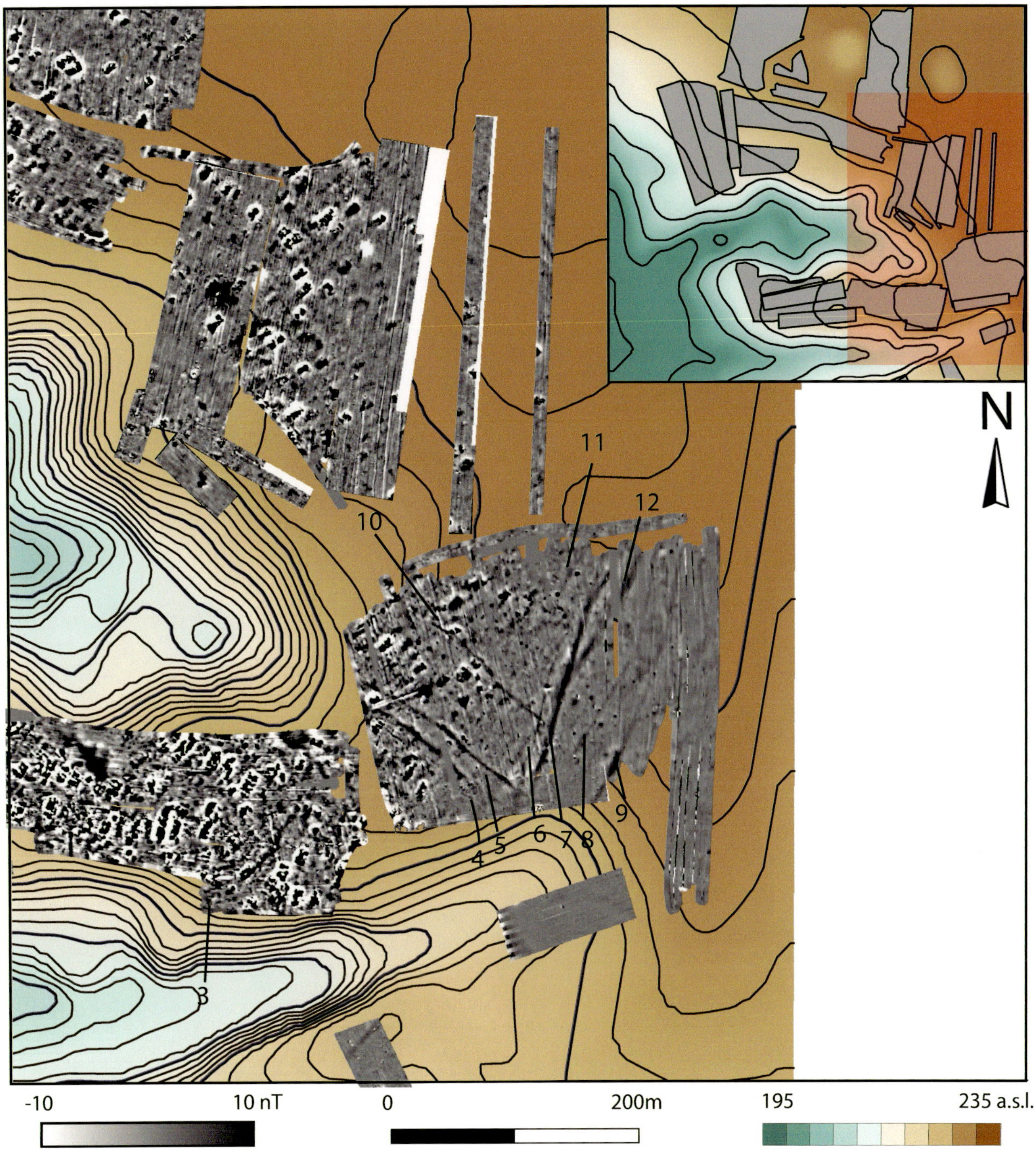

Figure 6. Overview of the geomagnetic data in the eastern settlement area.

The scatter plot (Figure 9C) visualises the correlation of both data sets. The histogram of the size of the reconstructed house areas (Figure 9B) shows the mean to be around 50 m² however at least two peaks are likely. More clearly defined are the multiple peaks in the histogram of daub area size (Figure 9A). As mentioned above, different house size could indicate a chronological difference. An increase in the size of earlier houses has been noted at several sites in the central Balkans, including Kundrici (Furholt 2012: 14f.), Okolište (Hofmann 2013: 366), Divostin (Tripković 2009b, 2013) and Stubline (Crnobrnja 2011). By contrast, other sites demonstrate decreasing size in the earlier phase, such as at Gomolava (Tripkovic 2013; Brukner 1988, 2002; Petrovic 1992) and Banjica (Tripković 2007, 2013). While small houses are present (ranging from 30–40 m²), other final Vinča sites tend to have larger houses (Porčić 2012b). Based on the multiple peaks in the daub areas, we mapped two classes of houses by the daub size: <20 m² and >20 m². The larger houses were

Figure 7. Overview of the geomagnetic data in the northern settlement area.

situated within the area of highest house density on the promontory. They were also concentrated towards the centre of the house clusters on the northern plateau.

The large-scale surveys enabled the analysis of similar spatial phenomena over the entire surface of the settlement, revealing patterns such as the four principal house groups mentioned above. To validate our model, we quantified the variation of house density by Kernel Density Estimation (KDE). The value and potential of this method has been well-proven in various other archaeological applications (Herzog 2012). The mapping of house density marks the first step towards a more general spatial analysis within a wider context. To evaluate the density map for houses we also calculated the daub areas. This procedure was based on the centroids of both features (reconstructed houses versus 6 nT polygons) and was completed by a bi-weighted KDE with a radius of 40 m. Both maps clearly show the four areas with higher densities

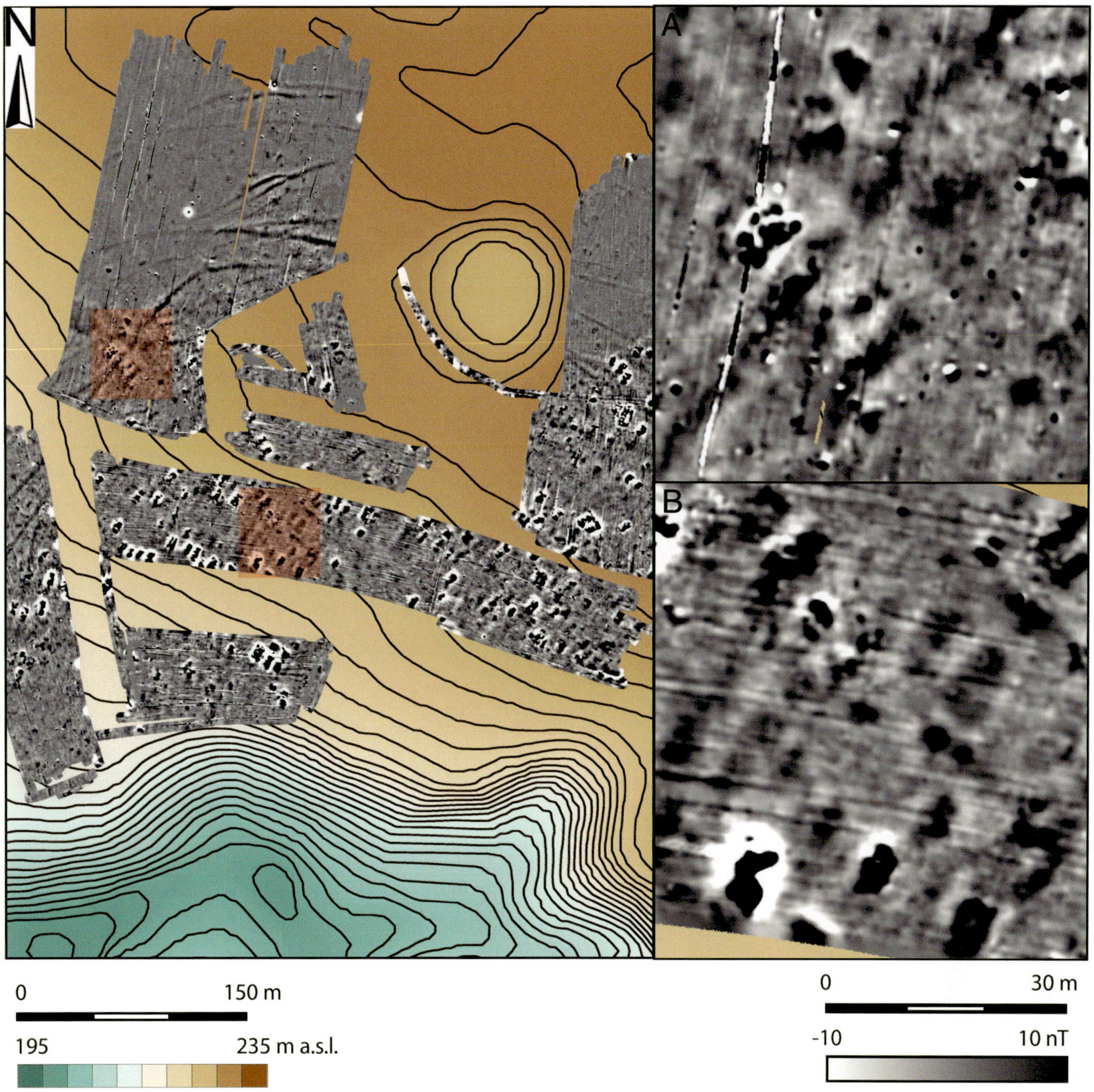

Figure 8. A) Entrance in the northwestern settlement; B) Area with burnt and probable unburnt houses in the northern part of the settlement.

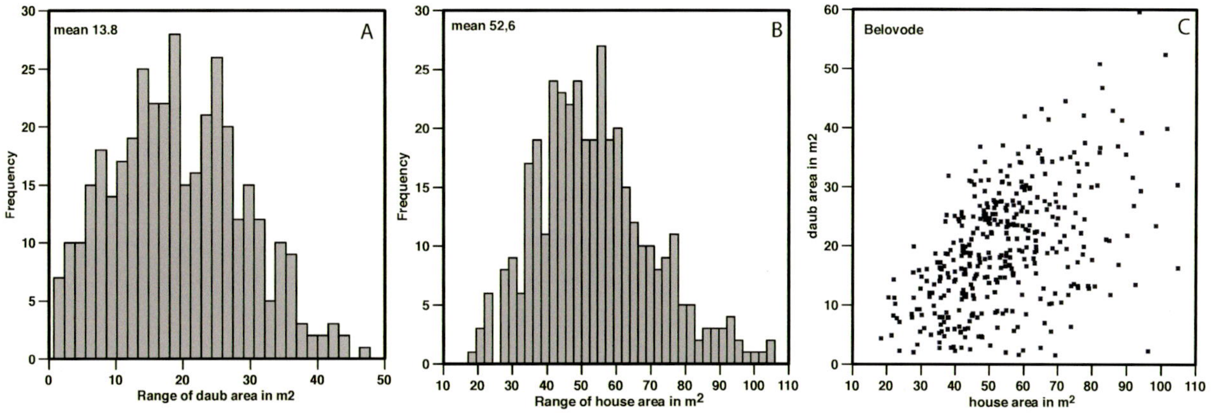

Figure 9. Analysis of the size of the daub area (6nT line) in relation to the size of reconstructed houses: A) Histogram of the daub area size; B) Histogram of the size of the reconstructed houses; C) Scatter plot of the size of houses and the underlying daub area.

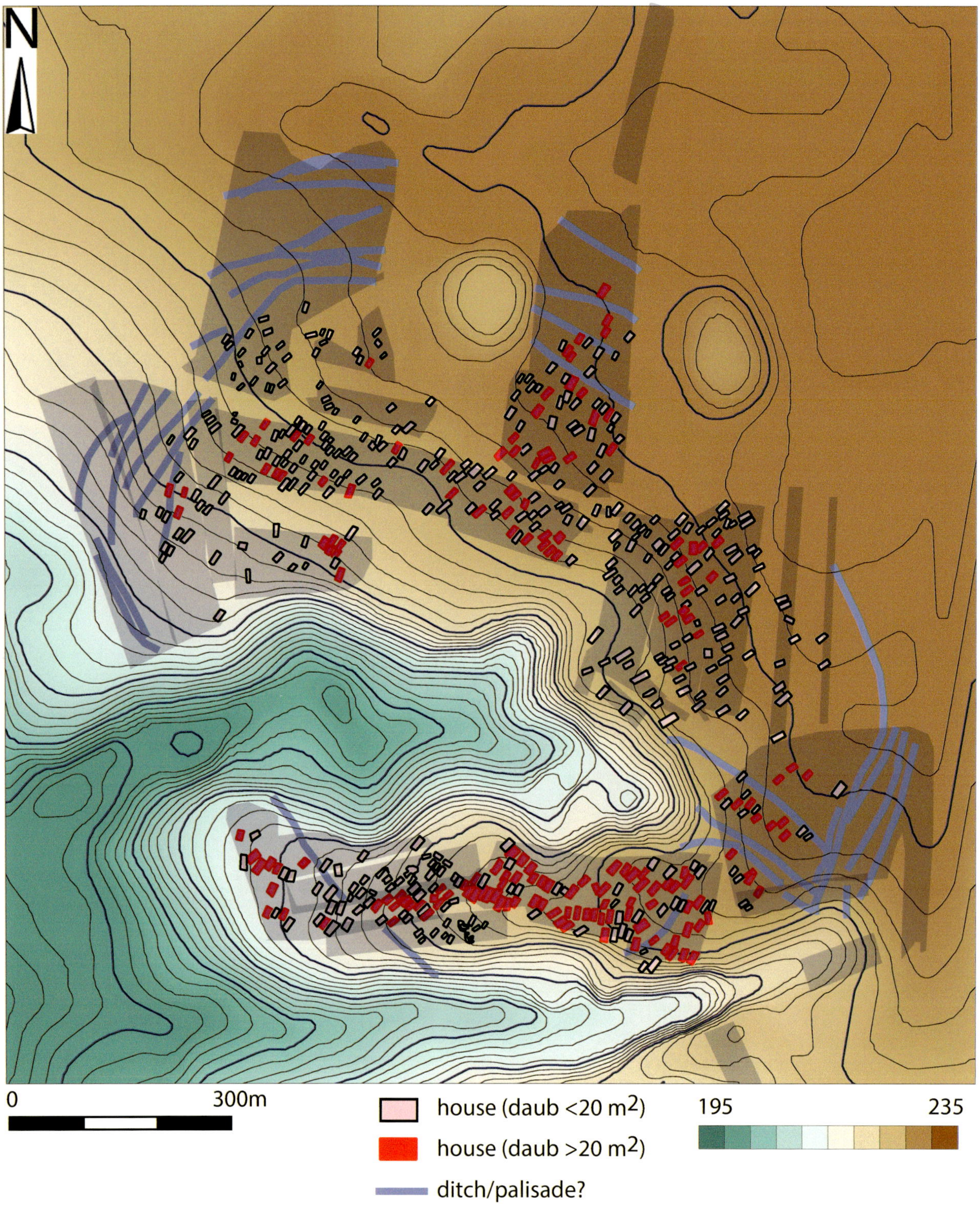

Figure 10. Interpretation of the geomagnetic data.

of architectural features (Figure 12). One maximal house group is located on the promontory with three principal groups on the northern plateau. The overall size of these house groups is between 4 and 5 ha. They are surrounded by zones characterised by a remarkably low density of building remains.

A crucial observation is the variation of the house density within these house groups. These structures vary in KDE both in terms of houses as well as daub areas. When combined with the reconstructed houses, we can reconstruct smaller minimal house groups (sized at approximately 2000–4000 m^2) inside these

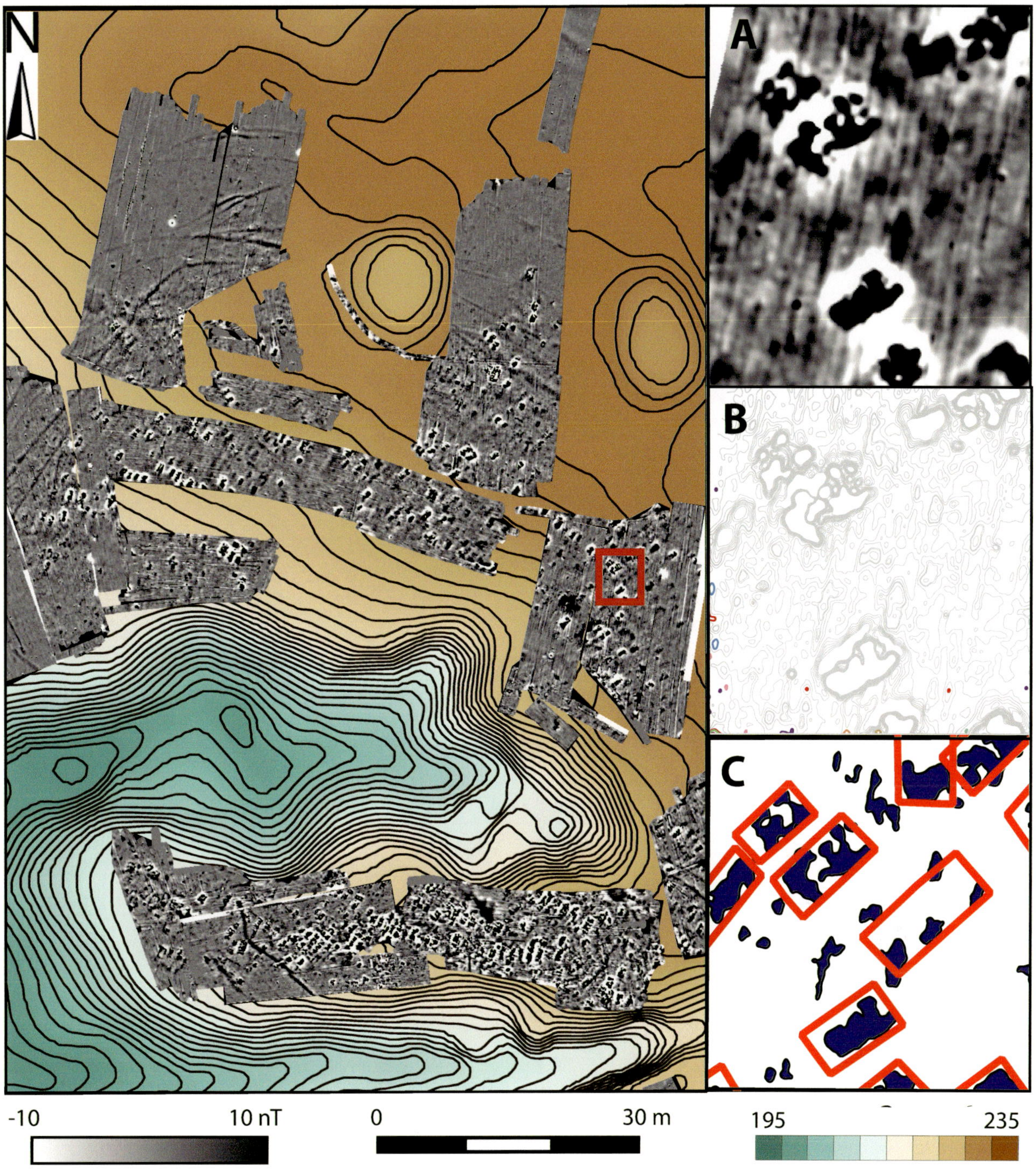

Figure 11. Analytical steps used to explore the data. A) Geomagnetic data; B) ± 20 nT Polylines; C) 6nT Polygons and reconstructed houses.

main house groups. The minimal house groups were constructed on a scale that is often observed on Copper Age and Bronze Age settlements (cf. Rassmann *et al.* 2014; Müller-Scheeßel *et al.* 2016). The crucial question is how many subgroups belonged to a larger unit? The first indication of the answer to this puzzle is provided by the variation in the building density inside these large groups and the spatial layout of the reconstructed house rows. A rough estimation is based on available space. For the large groups (4–5 ha in size) we calculate a maximum of ten subgroups (Figure 13). The number of large groups and the approximation of 36 subgroups can be used to calculate the upper limits of ancient population size. The crucial factors within these calculations are the number of subgroups and the number of residents per subgroup. To make the calculations, we must return once again to the available space. If one assumes an area of *c.* 140 m for one house, a single house group would have consisted of *c.* 12–20 houses. In the Copper Age sites of the Cucuteni-Tripolje culture, studies have found house groups comprising 8–12 houses

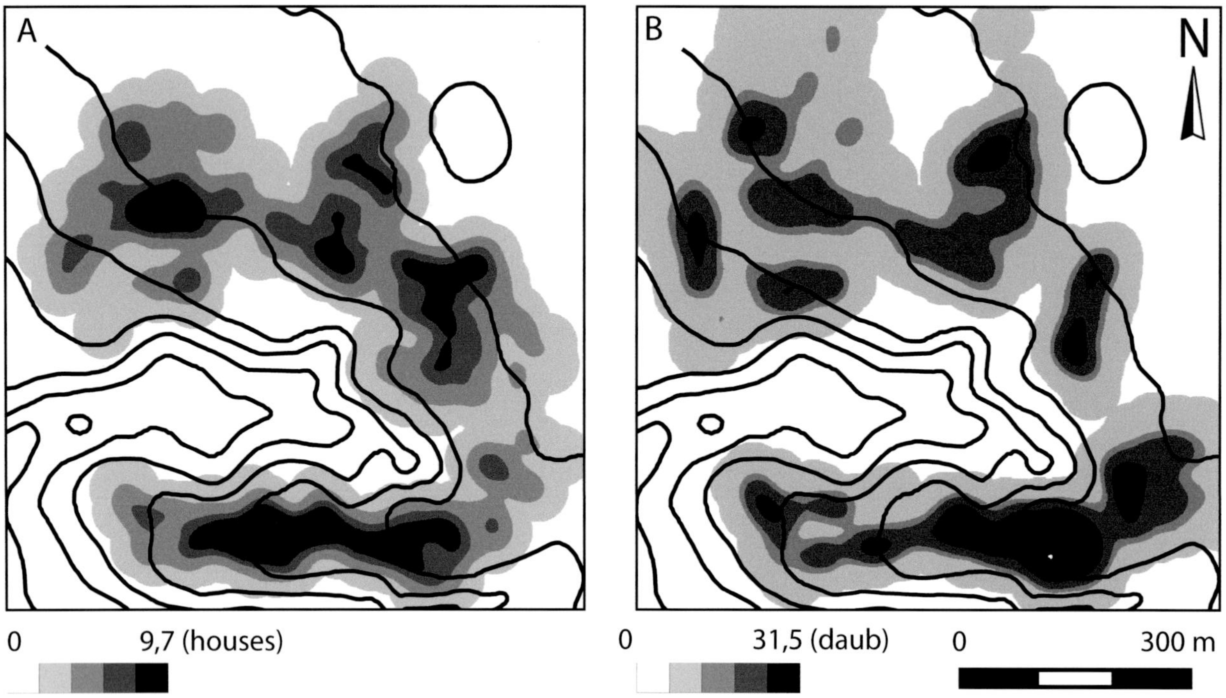

Figure 12. Visualisation of house clusters by Kernel Density Estimation (KDE, bi-weighted interpolation, radius 40 m). Interpolation based on the centroids of reconstructed houses. A) KDE of the house centroids; B) KDE of the centroids of the house daub.

(Rassmann *et al.* 2014). Similar dimensions were revealed by modelling the settlement dynamics in the Visoko Basin around Okolište (Müller *et al.* 2013a). This study applied a holistic approach to analyse the interaction between central settlements and those on the periphery. Valuable data for the reconstruction of house groups were also generated by the analysis of satellite settlements like Kundruci with its 7–10 houses (Furholt 2012: 216). When these house group data are combined with estimates of an average of five persons per household, the size of a house group can be calculated to be in the 40 to 60-person range, which is close to ethnographic data for minimal lineage (Hahn 2012: 35).

To extend these formulae a little further, a large group consisted of a maximum of 10 x 40 residents. Considering the four large groups, this gives a combined total of 1600 residents when the settlement was at its largest. If we only take into account the 36 house groups with their estimated 40 residents, we come to a number of up to 1440 residents (Figure 13). This number is in the same range as that produced for the Porčić and Nikolić mathematical model for the population size at Belovode (see Chapter 40, this volume). Such mathematical modelling allows an approximation of population change dynamics, which are crucial for understanding demography. In dealing with demographical data we must consider that populations were highly dynamic (cf. Shennan 2000: 812 f.). Therefore, our computation should also be qualified in the future.

Conclusions

The analysis of the geomagnetic data from the surveys can be used to reconstruct both the spatial structure and the upper limits of the settlement's population size. The first model forms an approximation based on spatial layout and structure during a stage at which the settlement reached its largest extent. In order to take probable dynamics and changes in the settlement history into account, we must qualify that model in the future.

Auger holes on the promontory and in the eastern settlement area revealed settlement layers of varying depth. The analysis of the settlement along more general lines should therefore be included on the immediate field research agenda in order to accomplish the important task of setting the indications for early metallurgy within a broader social-economic context.

The developments within the settlement are only understandable in relation to the changes that took place within the site's surrounding landscape. Our experience with the Okolište Project illustrates the great potential of combining site and landscape analysis. Future fieldwork should also, therefore, examine Vinča sites in relation to the landscape in which they were situated. The combination of geomagnetic survey and drilling programs for the investigation of different house groups, surface collection, and targeted excavation, has the potential to produce high resolution data for the reconstruction of settlement history and to expand our knowledge of the lives of past societies.

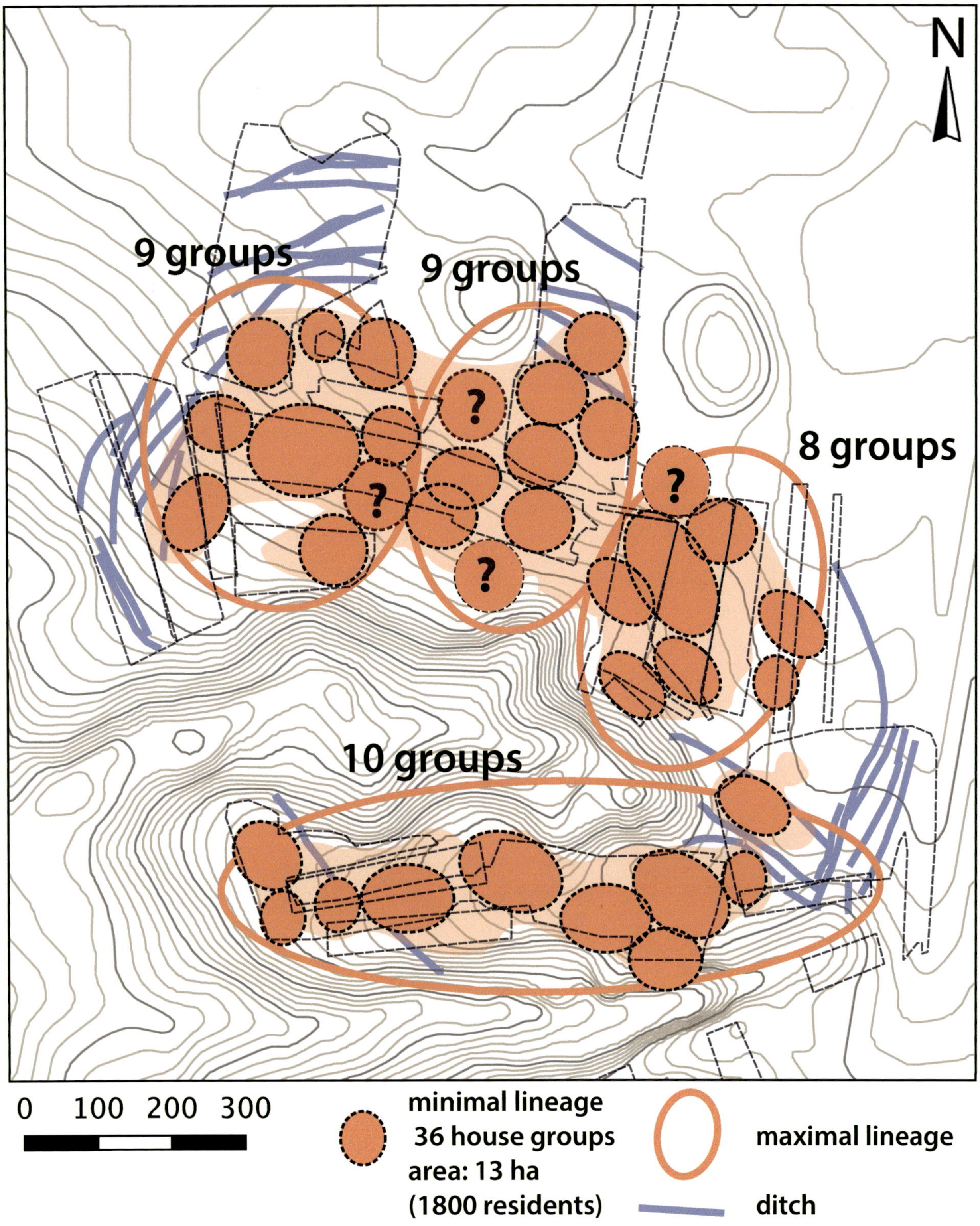

Figure 13. Hypothetical model of the settlement when at its largest extent.

The bibliographic reference for this chapter is:

Rassmann, R., Scholz, R., Mertl, R., Radloff, K., Pendić, J. and Jablanović, A. 2021. Belovode: geomagnetic data as a proxy for the reconstruction of house numbers, population size and the internal spatial structure, in Radivojević, M., Roberts, B. W., Marić, M., Kuzmanović Cvetković, J., and Rehren, Th. (eds) *The Rise of Metallurgy in Eurasia*: 94–107. Oxford: Archaeopress.

Chapter 10

Belovode: excavation results

Miroslav Marić, Benjamin W. Roberts and Miljana Radivojević

The Neolithic–Chalcolithic site of Belovode covers approximately 40 ha (Figure 1). In the two fieldwork campaigns of 2012 and 2013, only 31.5 m^2 was excavated due to the archaeometallurgical focus of the project. The trench was positioned on the eastern platform of the settlement, where previous excavations had uncovered significant metallurgical evidence in Trenches 3 (Šljivar and Jacanović 1997c, Radivojević *et al.* 2010a) and 17, which are located to the north and the south of Trench 18 respectively. A 5 x 5 m area was opened in the 2012 season and then, based on the preliminary spatial analysis of metallurgical finds, in 2013 the trench was slightly expanded with a 2 x 3 m extension on the eastern side.

The 2.3 m of archaeological layers in Trench 18 are not the thickest found at the site. In the central part of the plateau, the stratigraphy comprised 4 m (Šljivar and Jacanović 1996a: 55). Such variation mostly reflects the occurrence of pits in the earliest occupation horizon of the site. These tend to be dug well into the natural layer. It can also be assumed with a high degree of certainty that the outer extents of the site have significantly thinner occupation evidence than those towards the centre. This is due to the settlement gradually increasing in size throughout the duration of occupation at the site and shown by the geophysical evidence of several enclosure/defensive ditches that were, on several occasions, overlaid by wattle and

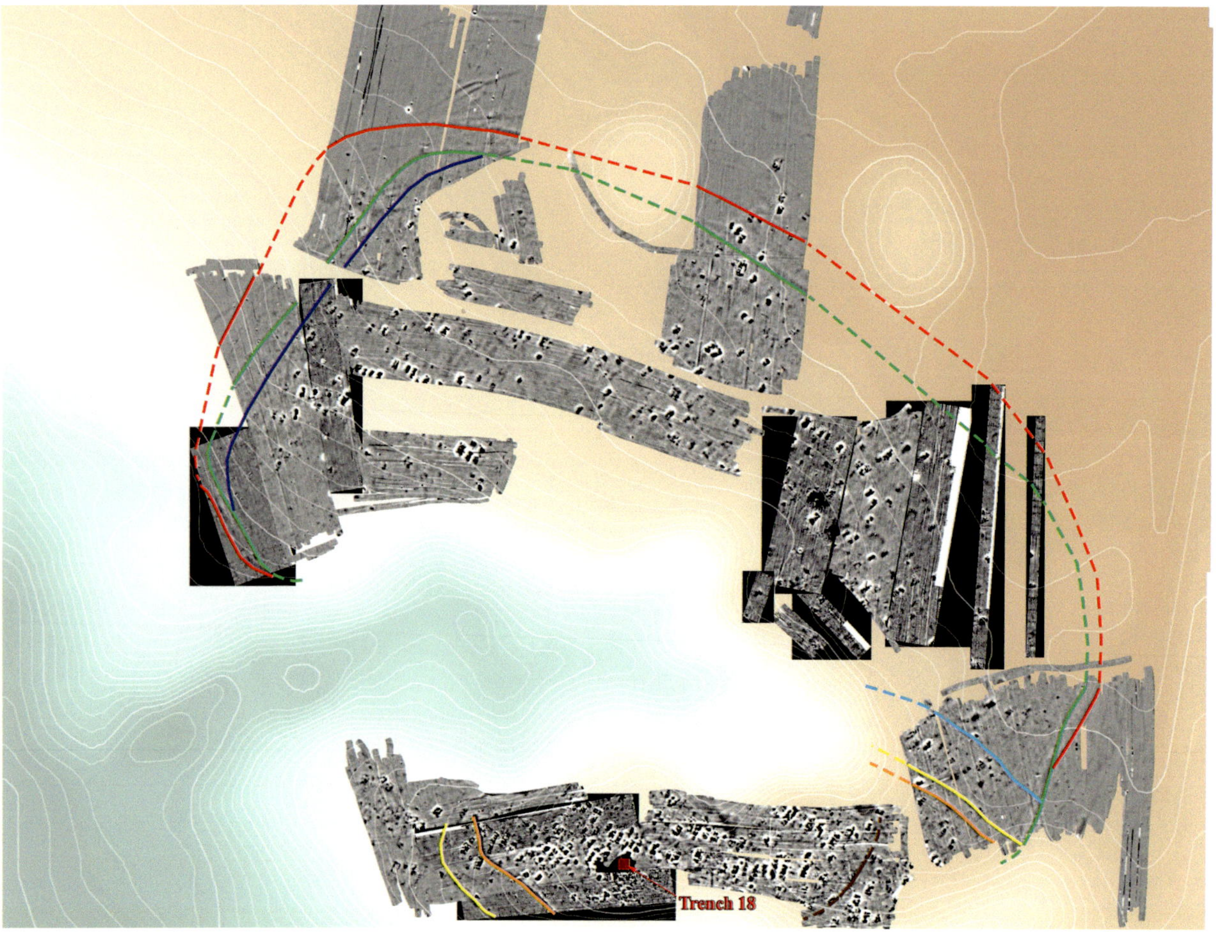

Figure 1. Geophysical survey of the settlement.

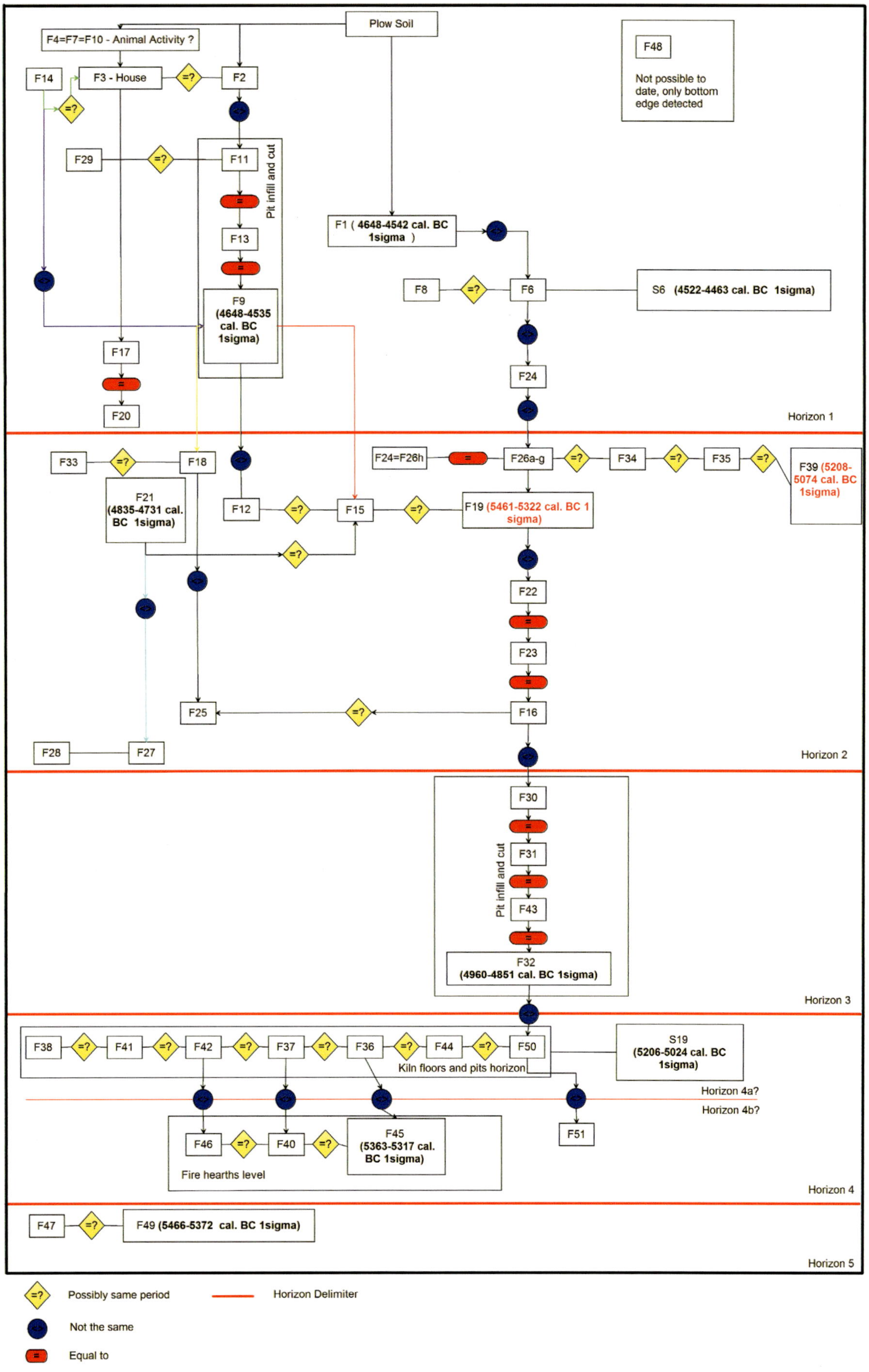

Figure 2. Relative stratigraphy of Trench 18.

daub structures. It should also be emphasised that the position of the settlement on a sloping plateau and the perennial agricultural activity may have led to significant soil erosion the extent of which could not be established in full by the small-scale excavations.

The single context excavation and recording method employed at the site led to the establishment of multiple phases within the trench. Based upon the vertical stratigraphy and superimposition of the contexts discovered we devised a relative stratigraphic sequence of five settlement horizons (Figure 2). These horizons are marked with numbers between 1 and 5, with Horizon 1 being chronologically the latest and Horizon 5 the earliest. It must be noted that these horizons do not imply specific chronological phasing within the framework of any of the existing relative chronology periodisations (e.g. Garašanin 1979, 1993; Schier 1995), but rather represent the discernible settlement construction phases in this particular part of Belovode. A more precise relative and absolute chronology of the site is explained in Chapter 39 of this volume.

This chapter presents the use of occupation space revealed by Trench 18 using a catalogue and description of the key features. As the only wattle and daub structure (Feature 3) is found at the very end of the occupation layers, we are quite certain that this portion of the site can be characterised as a predominantly non-dwelling area of the settlement. Here, in parallel to the description of features, we will illustrate the temporal and spatial contexts, starting from the latest horizon and moving towards the earliest. Further detail regarding the excavation results can be accessed in the Appendices (see Chapter 7)

Structural features in Trench 18

The archaeological remains found in Trench 18 are numerous. During the two excavation seasons, 51 features were discovered within the trench. These can be subdivided into several classes and include: a wattle and daub structure; several pits; multiple hearths and ash bins; several pottery concentrations; and a circular structure comprising six sub-oval post holes. Some of the structures show clear signs of superimposition as described below but, for the sake of brevity, only selected structures will be detailed individually in this chapter.

Horizon 1

Feature 1 was discovered at the border between the plough soil and the archaeological layers. It comprises a concentration of pottery fragments belonging to one larger vessel – an amphora with two (possibly four) looped handles on the middle section. Within the widest area of the amphora, the belly, several unworked stones were discovered, possibly the original content of the vessel. It cannot be said with certainty why these stones were kept in the vessel.

Feature 2 was found at the bottom of spit 4 and is another concentration of pottery shards belonging to several vessels although, unlike Feature 1, the vessel profiles could not be identified *in situ* with certainty. Several unworked stones were discovered in the southeast corner of the feature,

Feature 3 is a rectangular structure made of wattle and daub and was found at the base of spit 3, in the northwest corner of the trench and possibly extends under the west profile. It measured 3 x 3.2 m (Figure 3) but was damaged near the western profile of the trench and it may well have been larger. Based on the preserved section, the structure was oriented north-south with a declination towards the east of about 18°. Massive fragments of wall daub were fired bright red; beneath the structure lay the remains of vessels fired in the same fire that destroyed it. The vessels found on the structure floor were smashed *in situ*. Amongst the debris were two large grindstones as well as several smaller stone tools and 16 pieces of malachite. No heating installation or domed kilns were found in the remains. On removal of the orange fired floor level, the imprints of wooden planks that had been part of the floor construction became visible (Figure 4). The planks were placed parallel to each other and were oriented northwest-southeast, i.e. perpendicular to the longer wall of the structure.

No foundation trench or post holes could be detected, but least some post holes would have been needed to support the wattle construction of the wall. This kind of quadratic structure is not very common on Vinča culture sites, but at least one similar feature was detected during excavations at Belo Brdo (Tasić 2007). This had similar dimensions of 3 x 2.8 m, and also had several groups of *in situ* vessels on the floor but no evidence for cooking or heating installation. The lack of a heating or cooking installation could identify this structure as a storage feature, most likely related to a larger dwelling structure located nearby.

Feature 6 is a small, bowl-shaped (?) pit, 75 cm in diameter and 35 cm deep. The pit was detected as an oval area of orange daub mixed with pottery sherds, charcoal and ash. The infill was very similar to the surface material. This feature is unique due to its association with metallurgical debris (see Chapter 11, this volume). Slagged sherds, free slag pieces, metal droplets and even a copper metal artefact a bit further away, all demonstrate high temperature activities happening in this horizon. More closely, slagged sherds in the zone of burning confirm previous assumptions about the nature of the earliest copper smelting

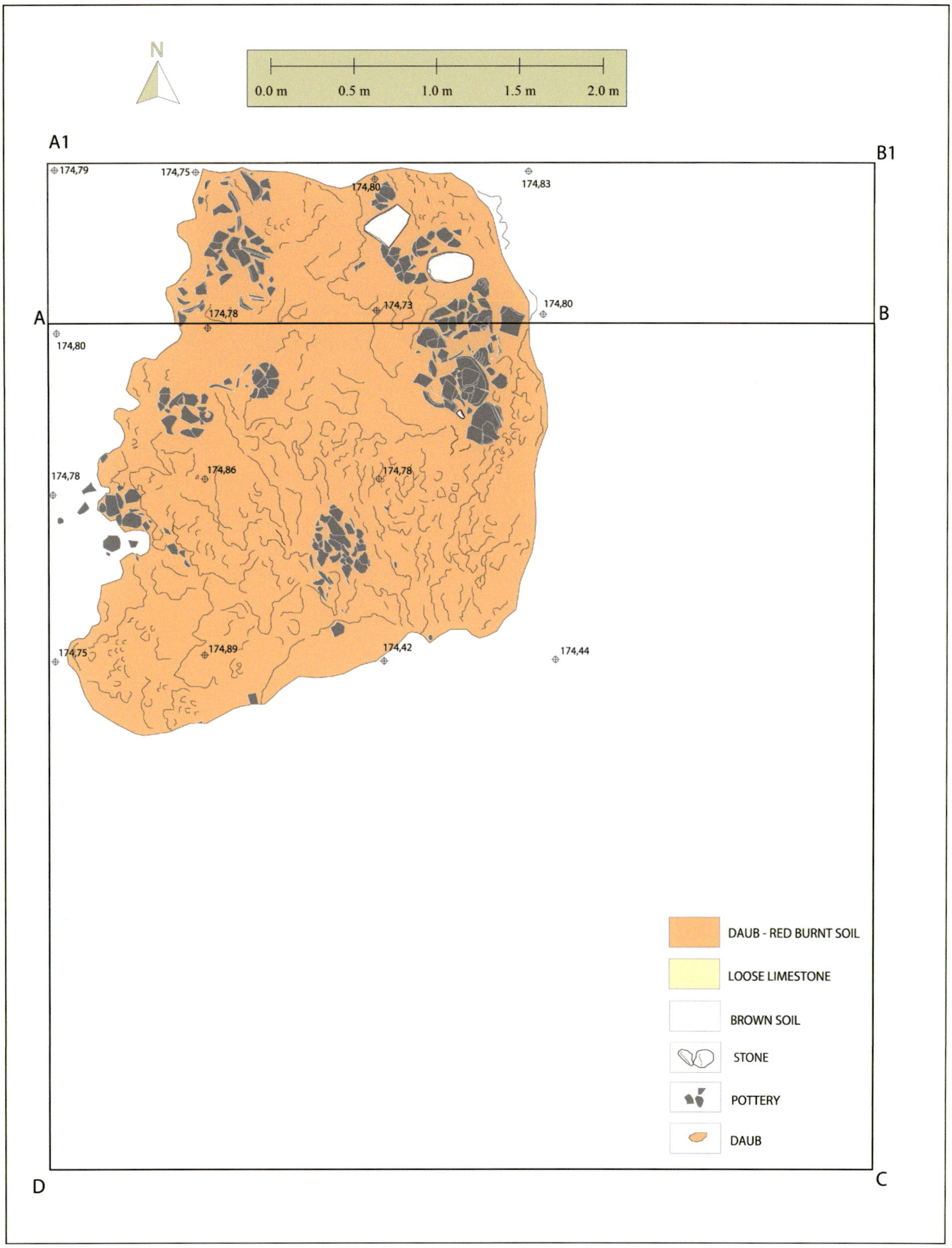

Figure 3. Feature 3: object of wattle and daub with *in situ* finds.

installations – that they were hole in the ground lined with fragmented ceramic sherds. The fact that none of these sherds can be identified as crucibles, and that in form and nature of contact with pyrometallurgical activities they resemble previously studied samples (Radivojević and Rehren 2016) show that there is a consistency in metal smelting technology across the site of Belovode. Although these hole-in-the-ground

Figure 4. Imprints of underfloor wooden planks beneath Feature 3.

installations were previously only assumed, this is the first time we are able to discern them in direct association with metal production on this site.

It could be hypothesised that the pit contained refuse from an unidentified burning that occurred nearby.

Feature 8 is a concentration of five pots which were found next to the southeast corner of the trench in spit 6 but extends further into the eastern profile of the trench. The vessels were found *in situ*, and a polished stone chisel was discovered inside Vessel B. The vessels include various forms and functions and can broadly be divided into those used for cooking and for storage. When the vessels were removed, a small, irregularly shaped pit was detected in spit 7.

Feature 9 is a large oval pit next to the south profile of the trench which was detected and excavated from spit 6. This steep-sided pit measured 2.6 x 1.9 m with a maximum depth of 1.2 m (Figure 5). The infill of the pit consisted predominantly of dark brown soil with daub, ash and charcoal fragments, and occasional yellow clay lumps. The pit contained a high quantity of pottery shards, animal bones and several fragments of malachite, figurines and a miniature cup. The nature of finds and the quality of infill indicates that this feature can be treated as a refuse pit, possibly connected to Feature 3 to the northwest.

Feature 20 is a sub-oval feature similar in shape to Feature 8, with evidence of burning, daub, animal bones and pottery fragments. The longer axis of the feature is oriented northwest-southeast and measures 1.45 m. The feature is only a few cm thick. No clear function can be attributed.

Horizon 2

Feature 16 is a concentration of daub, irregular in shape, and damaged on several sides by two pits (Features 9 and 19). It consists of the fragments of a destroyed kiln (or several of them) and is easily distinguishable by dome and floor fragments found in the debris (Figure 6). The remainder of the feature consists of white ash and charcoal. The layer was between 10 and 15 cm thick, and at its north-south axis measured 1.82 m, whilst at its west-east axis it measures just under 1 m in diameter. It is potentially part of Feature 32, a pit with various infills (see below).

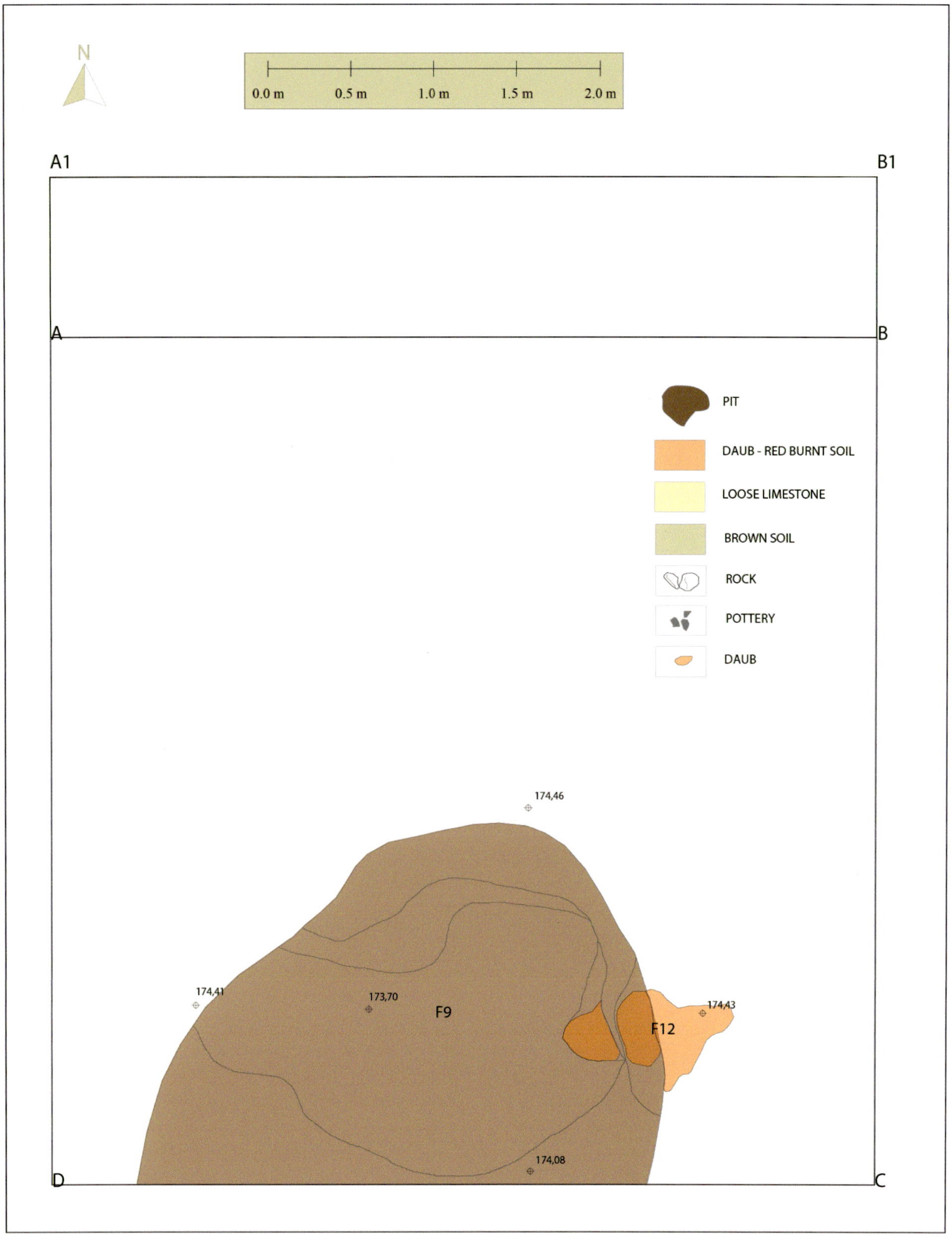

Figure 5. Feature 9 outline.

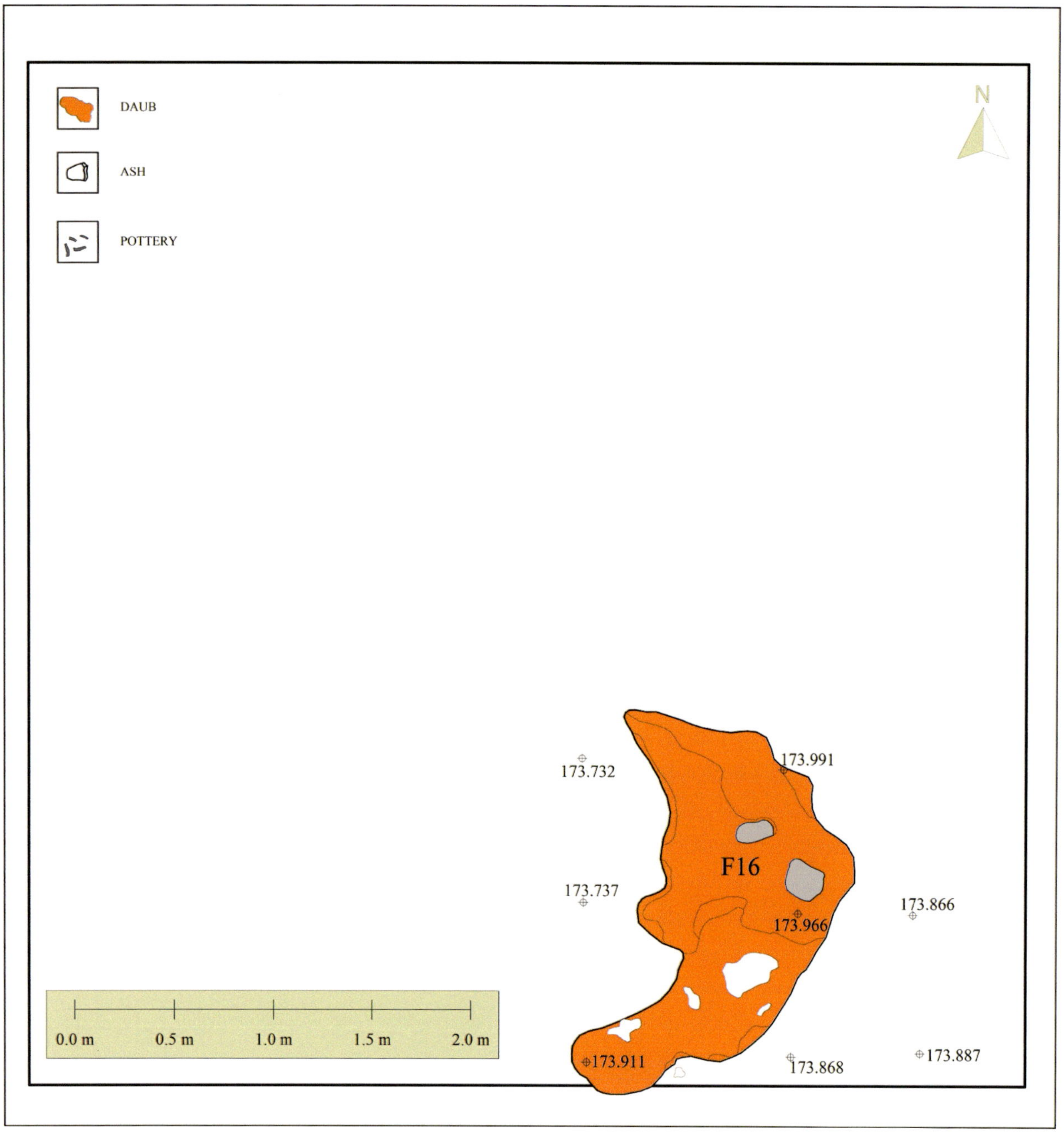

Figure 6. Feature 16: daub concentration of a destroyed kiln.

Feature 18 is a circular pit filled with the remains of orange, baked daub. It was cut by the later pit, Feature 9, in its southeast part. The diameter of the pit is between 1.11 and 1.12 m in all directions, whilst its depth is around 30 cm (Figure 7). The walls of the pit were burned to a black colour, so it can be assumed that the daub was still hot when deposited into the pit. The fragments most likely originate from the wall of a structure rather than a kiln. A large amount of charred chaff and seeds were retrieved from in between the fragments, as a macro botanical sample.

Feature 19 is a large elliptical pit located in the eastern part of the trench. The longer axis of the pit is oriented northeast to southwest. The top of the pit was detected in spit 8 as an area of mixed daub, pottery, stone, and charcoal. The pit dimensions are 3.1 x 1.8 m with a maximum depth of about 70 cm (Figure 8). Its sides are funnel shaped, narrowing towards the bottom. The finds consist mainly of pottery and animal bones, indicating that the feature was used as a refuse pit. Most of the finds were found concentrated at the very bottom. A significant discovery within this feature was a significant quantity of thermally altered malachite. It is unclear whether this represents a direct by-product of a nearby smelting operation or displaced smelting refuse deposited in this location upon the clearance of a smelting installation in a different portion of the site.

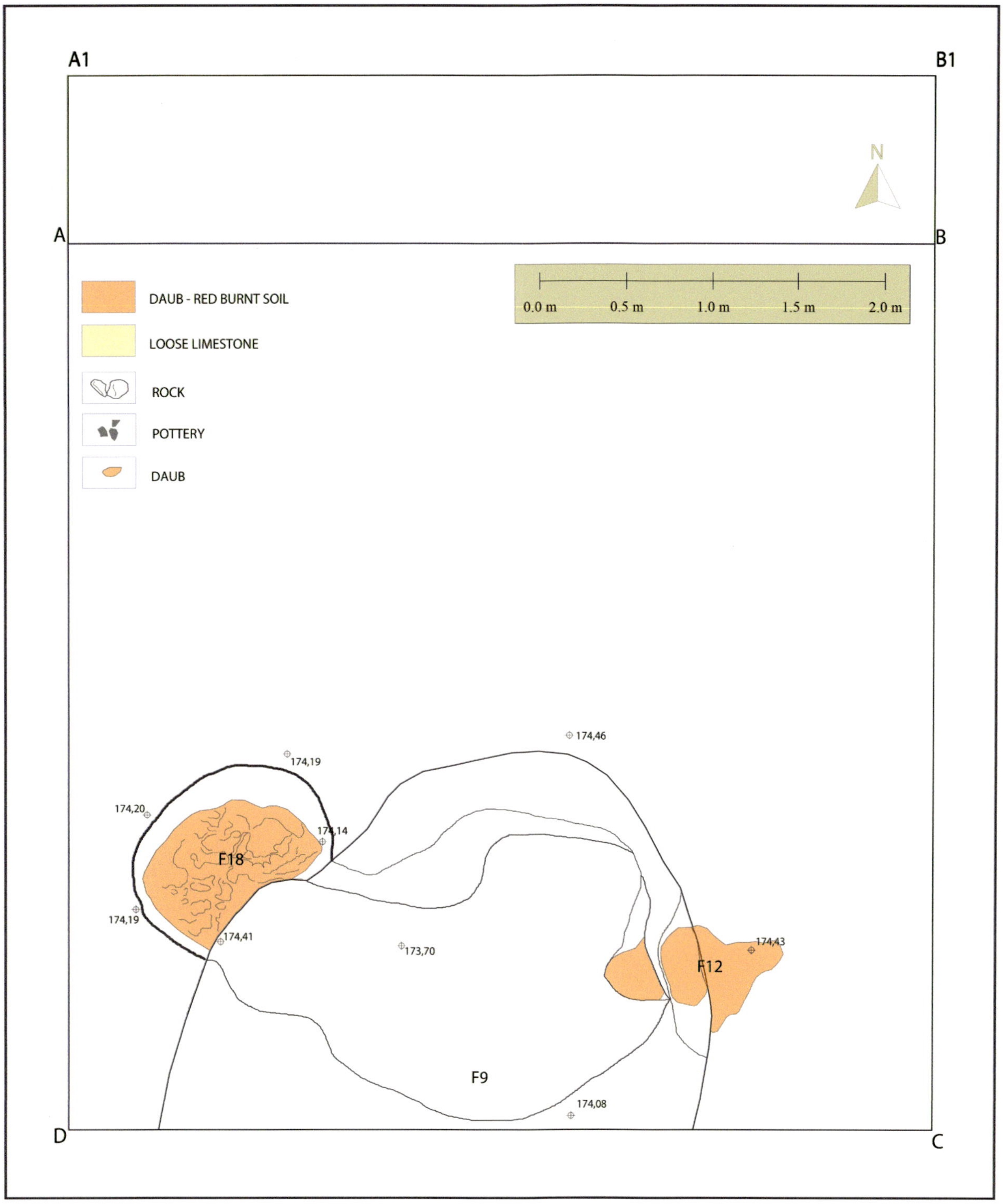

Figure 7. Feature 18 and its relationship with Feature 9.

Feature 21 is an oval pit next to the north profile of the trench. It was detected in spit 12 and is 1.6 m at the longer axis and 1.4 m at the shorter axis (Figure 9). Several malachite pieces and importantly, two copper metal droplets were recovered from the pit, as well as one fragment of obsidian. The feature could be interpreted as a refuse pit. It is striking that the content was very compact and difficult to excavate, most likely due to the pressure of debris from wattle and daub structure Feature 3 that lay on top of it. These metal droplets stand out as the earliest directly documented metal production debris during the excavation campaigns 2012/13. They also further confirm the dating of the original findings of copper slag (*c.* 5000

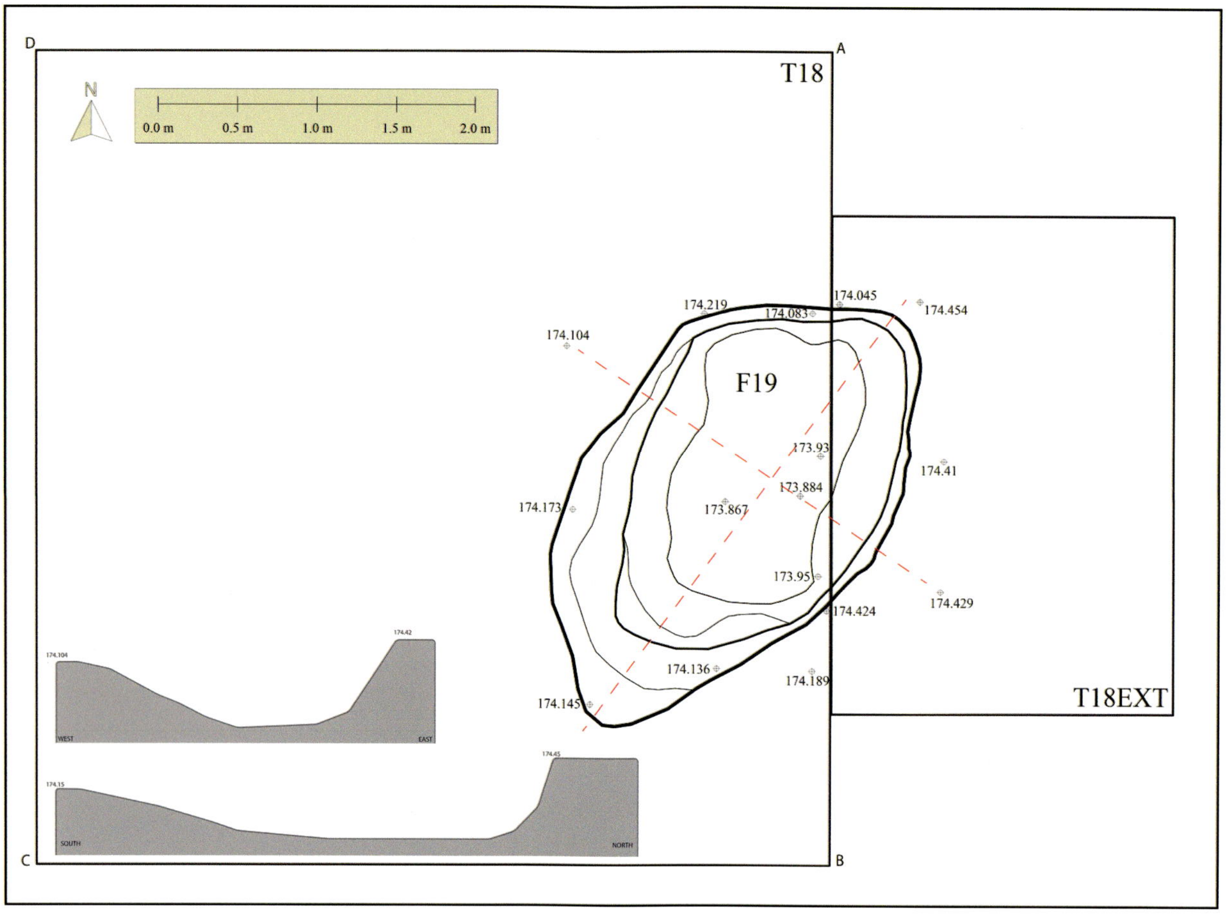

Figure 8. Feature 19 and cross-sections.

BC), published in Radivojević *et al.* (2010), hence leaving no doubt about the currently earliest documented metallurgy in the world.

Feature 25 is a large, irregular, rectangular ash bin, damaged by the later cut of Feature 9. It measures 0.9 m in length and is approximately 60 cm wide Located in spit 13, it consisted of white and grey compact ash, orange burnt daub, and pieces of charcoal.

Feature 26a-g was first detected during the excavation of Feature 19 which it cuts (Figure 10). It consists of six individual vertical post holes arranged in a circle. Nearby, another circular hole was detected in the same spit, but it most likely does not belong to this feature. The diameters of the holes vary between 25 and 45 cm. They clearly formed the base for a circular wooden structure but nothing more can be stated with confidence (Figure 11). One possible interpretation could be that the feature represents a lookout post, as it was located towards the edge of the settlement.

Feature 27 is a small elliptical ash bin filled with white-grey ash, charcoal and small daub fragments. It was partially damaged by later activity. It measures just under 50 x 30 cm and is 3–5 cm deep.

Feature 35 is a circular hearth with a diameter 26 cm and depth of 10 cm discovered in spit 13. It is located close to the northeast corner of the Trench 18 extension (Figure 12). It has a thin, orange, baked wall and is filled with white ash and charcoal. An interesting find from the bottom half of the feature is part of the wall of a ceramic vessel, which was found against the side and base of the feature.

Feature 39 was found in immediately southeast of Feature 35 and is an elliptical ash bin with excavated dimensions of 25 x 40 cm and depth of 5–10 cm (Figure 12). The bin was only partially excavated as it extended under the eastern profile of Trench 18 extension. It contained white and grey compact ash mixed with charcoal and occasional orange daub lump. It is without doubt related to the activity conducted in Feature 35.

Horizon 3

Features 30, 31, 32 and 43 comprise a large pit which was partially excavated in the eastern extension of the trench. Its edges were detected in Feature 19, making of later date. The exact shape of the pit could not be detected as it extended under both the eastern and

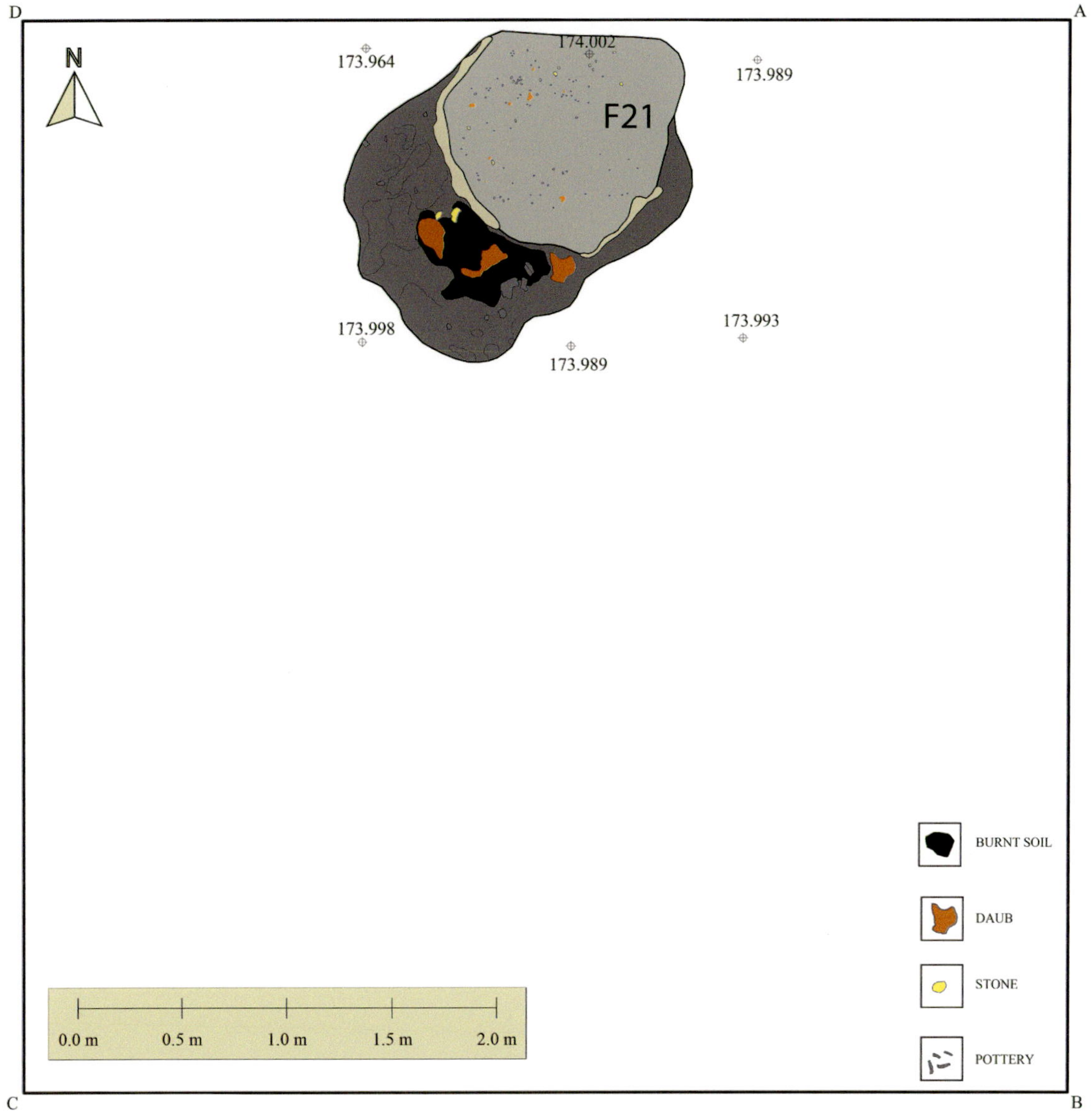

Figure 9. The appearance of Feature 21 in spit 12.

Figure 10. Feature 26 upon detection and when subsequently emptied.

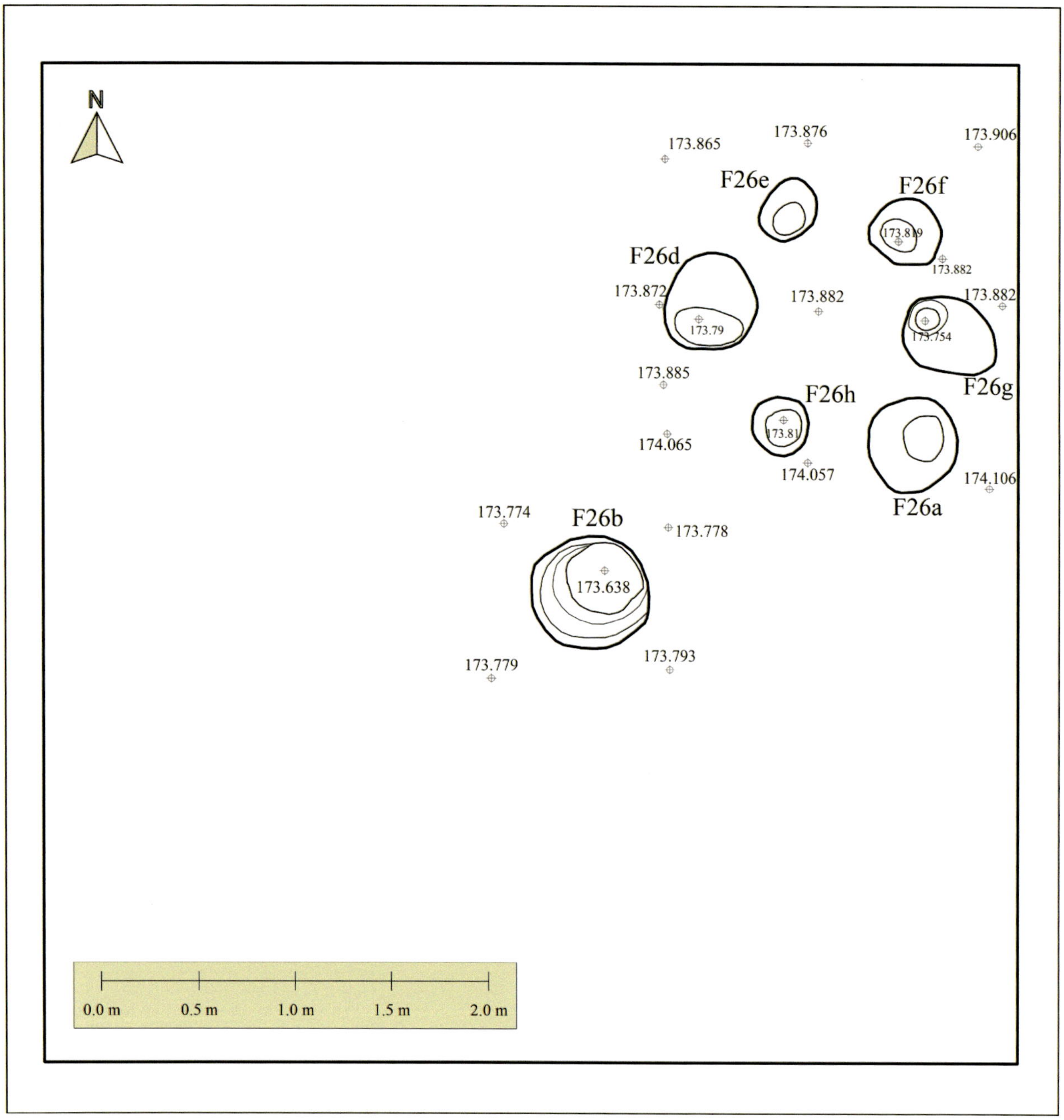

Figure 11. Feature 26 upon final excavation.

southern profiles of trench extension (Figure 13). The detected dimensions are 3.3 x 1.4 m, but it is clearly somewhat larger. It was first detected in spit 12 and extended as far as spit 17. The infill of the pit consisted of several different types of soil, including a compact, predominantly red and orange burnt layer of soil with lots of fragments of daub, charcoal and ash (Feature 43). This varied infill could indicate that the pit was in use for a longer period. In the deepest part, next to the southeast corner of the trench extension, it was 80 cm deep, which would indicate a rather large feature. The infill consists of various items, predominantly pottery and animal bones, but a significant quantity of malachite fragments was also recovered.

Horizon 4 (a and b?)

Horizon 4a

Features 36, 41, 44 and 50 comprise four kiln floors detected at the bottom of spit 17 and extending into spit 18 throughout the entire area of the trench (Figure 14). The kiln floors are not found *in situ* but are re-deposited; they are all disposed of in approximately the same horizon (between 173.30 and 173.40 m). They do not seem to originate from the same kiln and, whilst Features 36 and 41 are somewhat compact, Features 42 and 50 are not. The same character and the similar horizon of these features indicates that they belong to the same process.

Figure 12. Features 35 and 39 in the east extension of Trench 18.

Features 37 and 38 are two smaller pits. Feature 37 is elliptical whilst Feature 38 is circular. They were discovered in spit 17 in the southwest corner of the trench. The infill of the pits differed in content and colour from the surrounding area. The infills consisted of grey, compact soil with daub and charcoal fragments. The area around the features was predominantly brown in colour. Very few finds were discovered in the infill. It is possible that both pits are the remains of clay outcrops used for pottery production or another similar activity.

Horizon 4b

Features 40, 45 and 46 are three short lived hearths which were discovered *in situ*. Feature 45 and Feature 46 are elliptical in shape whilst Feature 40 is circular (Figure 15). Feature 45, which is the best preserved of the three, is 50 cm on the long axis and 38 cm on the short axis. Feature 46 could be only partially excavated as some of it remained under the north profile of the trench. All the hearths have thin, orange baked walls on the perimeter and are filled with the original content of the last firing episode, which consists of white ash and black charcoal mixed with occasional lumps of orange daub. Similar features are known from the early Vinča culture sites of Masinske Njive and Jaričište 1 in the Kolubara region, but both sites are currently unpublished or partially published (Marić 2013b).

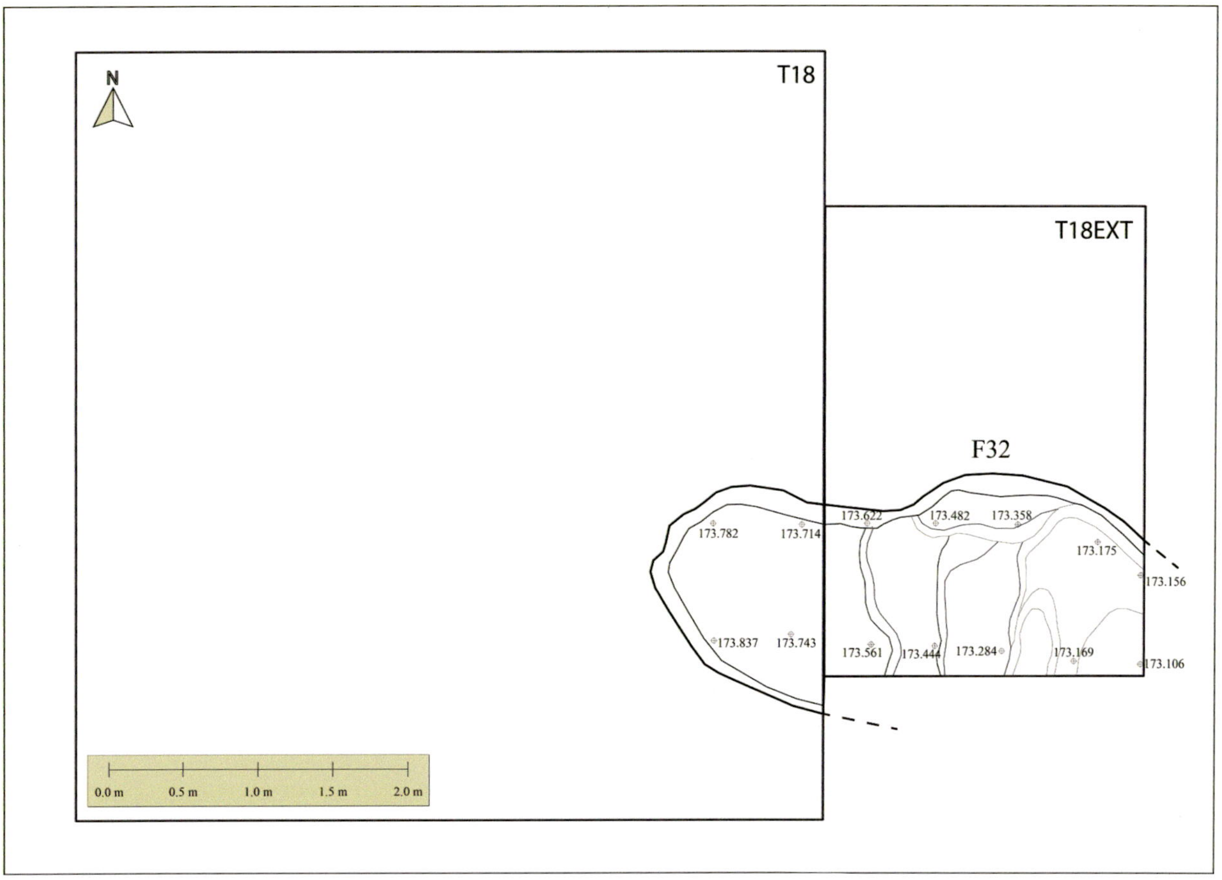

Figure 13. The outline of Feature 32.

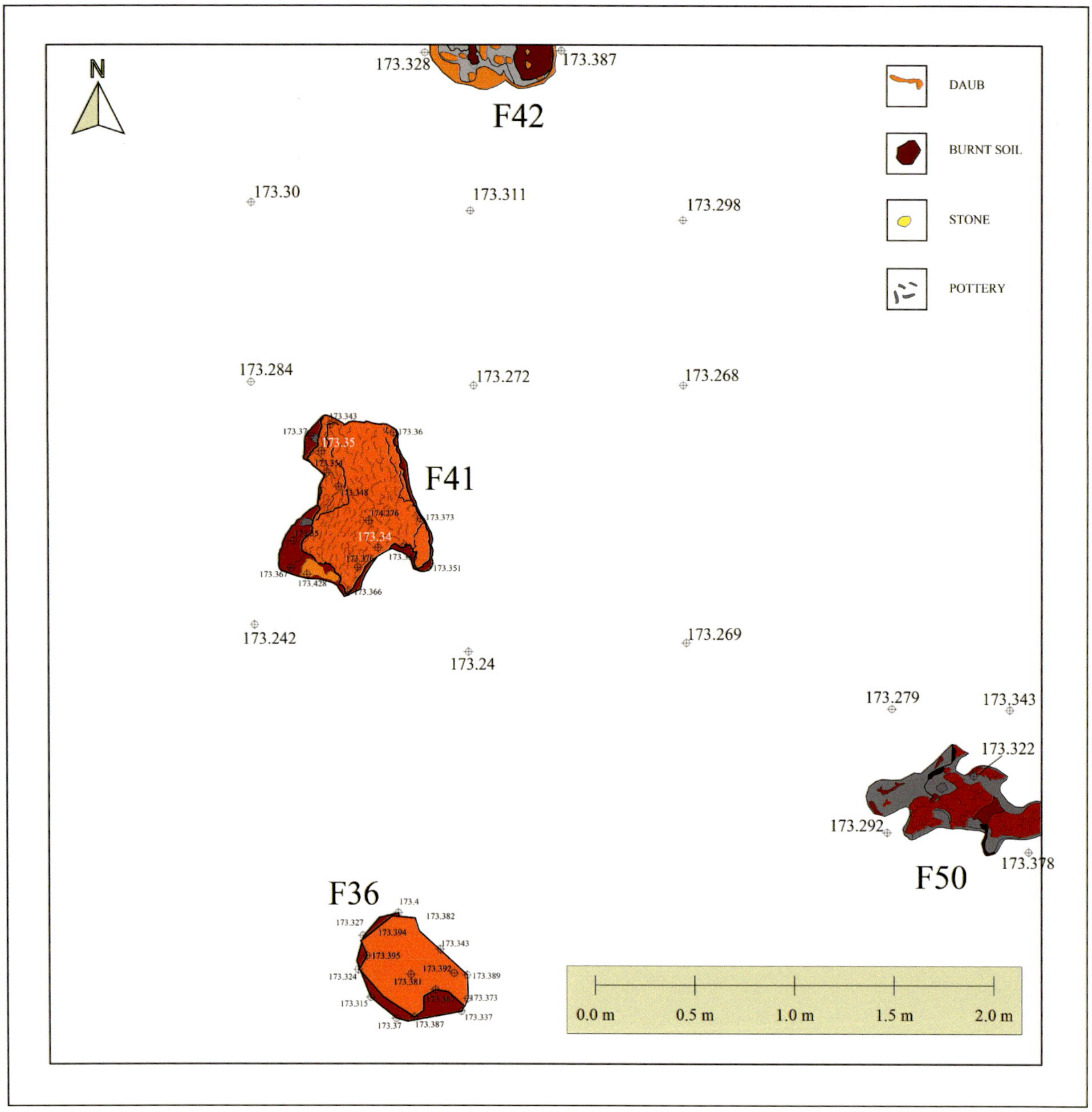

Figure 14. The kiln floors horizon. Features 36, 41, 42 and 50.

Horizon 5

Feature 47 is a large, irregularly shaped feature, with its long axis being oriented northwest-southeast (Figure 16). It is 3.1 m long and 1.12 m wide at its broadest extent. Its depth does not exceed 20 cm. The presence of only very few finds in the infill indicates that the feature was not backfilled but rather left to fill through natural causes. It is possible that the feature is a clay outcrop used for extracting raw material, potentially for either pottery or in construction. The presence of Feature 49 in its immediate vicinity can potentially corroborate this interpretation.

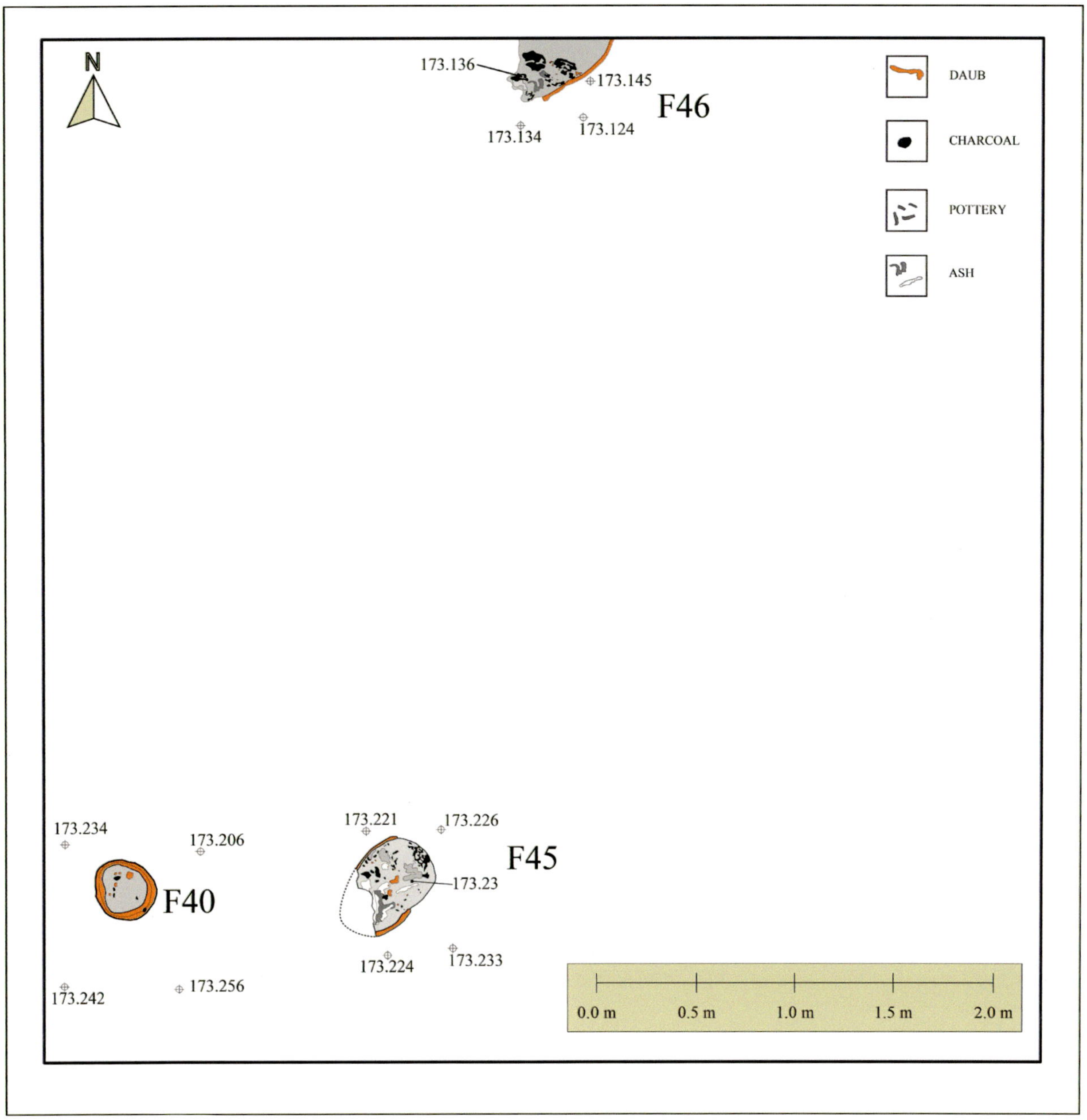

Figure 15. Hearth horizon. Features 40, 45 and 46.

Feature 49, a large hearth, was found to the northwest of Feature 47 (Figure 16). It was only partially excavated as it extended under the north profile of the trench. This feature consists of orange burnt soil along the walls encompassing black burnt soil in the centre. In the excavated part, it was roughly circular in shape. Aside from a large fragment of a stone there were no other finds in the feature. The absolute dating of the feature puts it at the very beginning of Vinča culture at the turn of the 5th millennium BC.

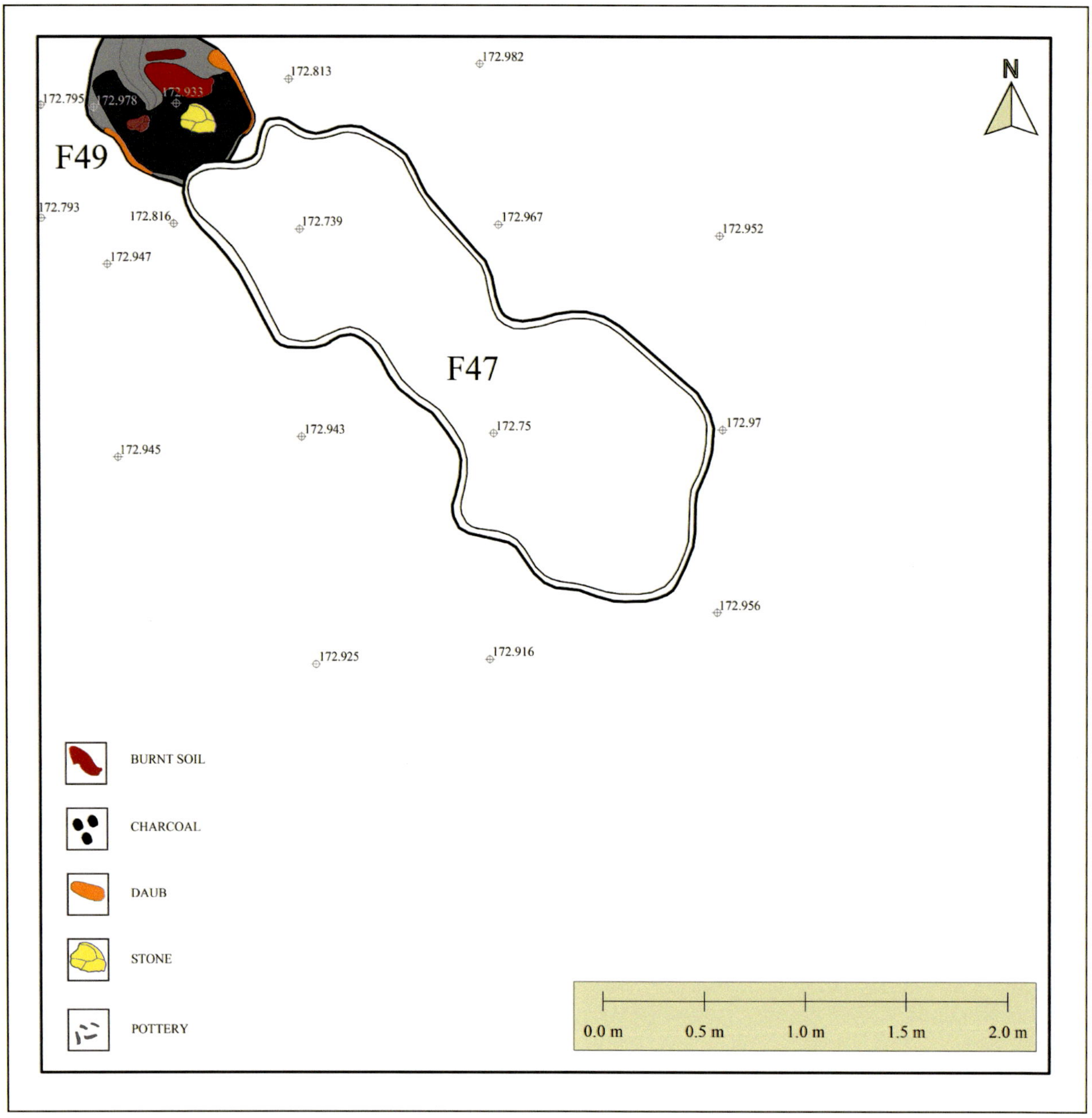

Figure 16. Features 47 and 49: the earliest structures in Trench 18.

The bibliographic reference for this chapter is:

Marić, M., Roberts, B. W. and Radivojević, M. 2021. Belovode: geomagnetic data as a proxy for the reconstruction of house numbers, population size and the internal spatial structure, in Radivojević, M., Roberts, B. W., Marić, M., Kuzmanović Cvetković, J., and Rehren, Th. (eds) *The Rise of Metallurgy in Eurasia*: 108–122. Oxford: Archaeopress.

Chapter 11

Belovode: technology of metal production

Miljana Radivojević and Thilo Rehren

Metal production evidence yielded during the excavation campaigns 2012 and 2013 in Belovode shows similar characteristic to the samples from the site studied and published previously (Radivojević 2012, 2013, 2015; Radivojević and Kuzmanović Cvetković 2014; Radivojević and Rehren 2016; Radivojević *et al.* 2010a). These are predominantly malachite mineral and ore samples, most likely roughly beneficiated (no samples larger than 2–3 cm in length, see Appendix B_Ch11), and very importantly, without any significant spatial pattern in the excavated area of Trench 18 or its extension (T18ext henceforth). These minerals were discovered in all areas, whether in living or economic spaces, inside the dwellings and other features, and across the excavation spits, which is why they have also been found by previous excavation campaigns (Šljivar 1993–2009).

During the 2012 and 2013 campaigns, Trench 18/ T18ext, yielded *c.* 1300 malachite and azurite minerals, including malachite beads but excluding sherds with traces of malachite. The uncovered copper mineral samples have macroscopically similar characteristics to samples from Belovode: a prevailing number of green (malachite) minerals with black/dark specs, others that are more purely green, with an occasional blue mineral (azurite) (see Figure 1).

The initial number of 14 samples related to metal production for this settlement (See Chapter 5, this volume), was expanded and enriched, with ten new fragments of production debris: slagged sherds (B23/12 and B47/12/1), individual slags (B24/12/2 and B47/12/2), metal droplets (Bf21/12, B29/12, B47/12/3, Bf43/13 and Bf56/13) and a fragment of a metal artefact (B71/12) (Tables 1 and 2). Of these, B23/12, B24/12/2, Bf21/12 and B47/12/(1-3) were all found in the eastern corner of Trench 18, where a surface covered with ashes, charred and burnt soil (Feature 6) emerged from the base of spit 5 (where B23/12, B24/12/2 and B29/12 were found) and continued throughout spit 6 (with Bf21/12 and all B47/12 in the vicinity) (Figure 2).

This cluster also included a copper mineral B46/12, the fragment of a metal artefact (B71/12) and a sherd stained with malachite (B62/13), coming from Trench 18ext (see Appendix B_Ch11 and Table 1). Feature 6 has a distinctive bowl-shaped appearance, with initially scattered signs of burnt and charred soil narrowing down to what looks like a pit (Figure 3), *c.* 75 x 35 cm in size in the upper part, and 66 cm in length at the bottom. Given the spatial association of metallurgical debris from spits 5 and 6 with the work area of Feature 6, and the indicative bowl-shaped feature it is suggested that these all comprise a single unit/context. Interestingly, the whole eastern area of Trench18, including its extension, appears to be an economic area in which various activities took place throughout all Belovode horizons (see Chapter 5). Direct ^{14}C dates are associated with Bf21/12 and all B47/12 samples (both through animal bones, see Chapter 37, Table 1; here in Table 2). In Chapter 37, Marić *et al.* model the site chronology using Bayesian statistical method, which combines both the radiocarbon dates and the relative stratigraphy recorded during the excavation. These modelled dates are presented in Table 2 and will be referenced when discussing the dating of metallurgical samples at Belovode. Of note in this context is that ^{14}C dates for spit 6 (Bf21/12 and B47/12/1-3) are argued to be less secure than direct dates for Feature 6 (see Table 2), as the latter is a well-defined feature in the Belovode stratigraphy, and the former comes from the spit layer, and could easily have been intrusive (See Chapter 37, Figure 2).

Hence, the modelled dates for samples Bf21/12 and B47/12/1-3 are associated with Horizon 1a, which starts at 4818–4693 cal. BC (95.4% prob.) and ends at 4702–4541 cal. BC (95.4% prob.), or possibly lasts from 4776–4709 to 4689–4550 cal. BC (68% prob.). Samples B23/12, B24/12/2, and potentially B29/12, can also be closely (and more directly) associated with modelled dates for F6, which starts at 4702–4541 and ends at 4600–4404 cal. BC (95.4% prob.), or possibly has a span of 4689–4550 to 4572–4481 cal. BC (68% prob.). Overall, a rough estimate places the activities related to this assemblage in the 47th century BC. The fragment of a metal artefact, B71/12, although part of the same cluster in T18ext, is associated with a dwelling structure (Feature 3) rather than the production assemblage, and hence could be dated slightly earlier, starting at 4818–4693 and ending at 4702–4541 cal. BC (95.4% prob.), or possibly between 4776–4709 to 4689–4550 cal. BC (68% prob.).

Figure 1. Typical copper minerals found in the site of Belovode: a) malachite; b) azurite and c) black and green manganese-rich copper mineral.

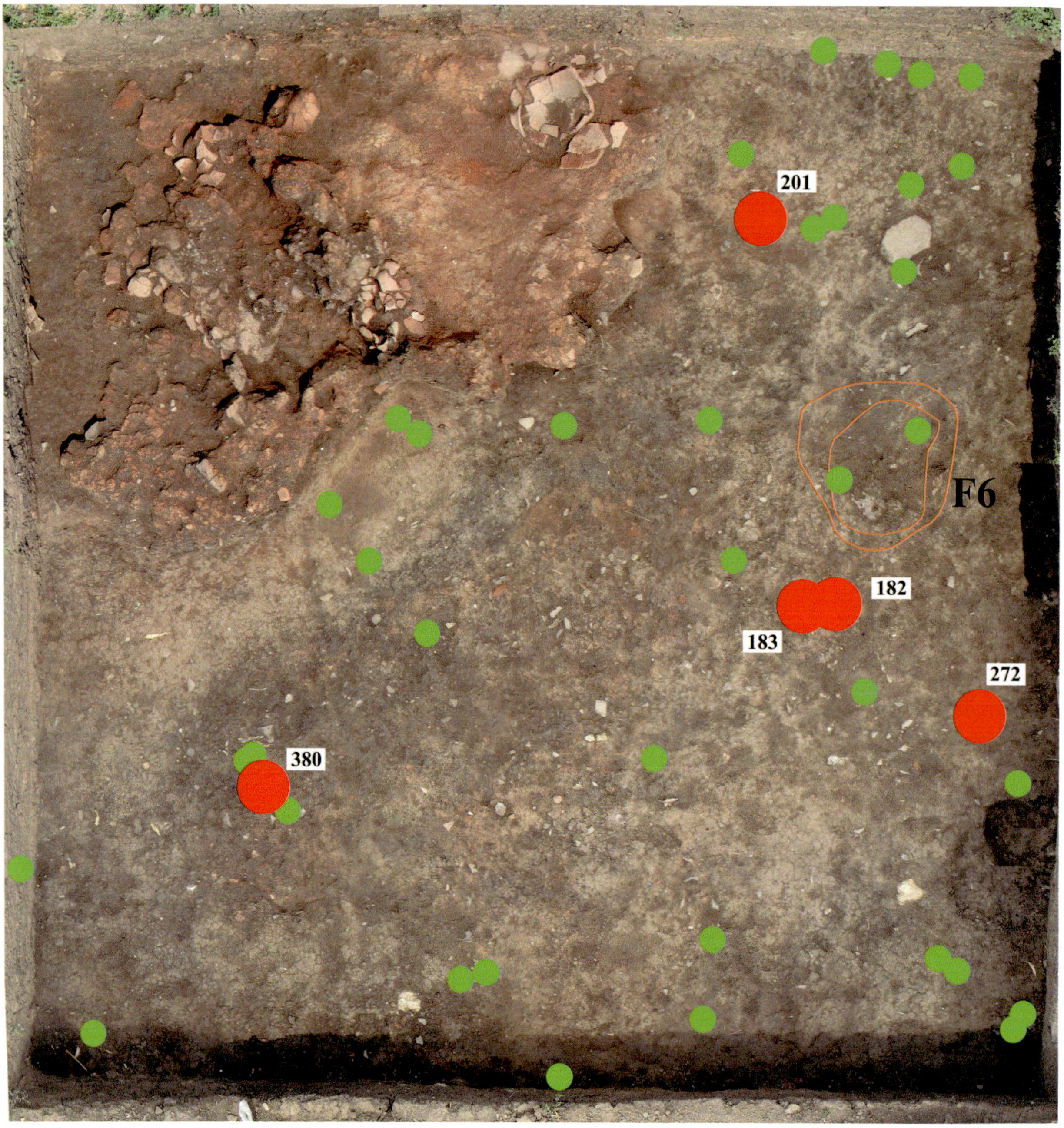

Figure 2. Trench 18 situation in spits 5 and 6, with indicated Feature 6, metallurgical finds as red spots (EDM numbers, see Table 1) and malachite finds as green. EDM 182: B23/12; 183: B24/12/2; 201: B29/12; 272: B47/12/1-3, Bf21/12; and 380: B71/12 (prepared by M. Marić).

Table 1. The list of minerals and metallurgical materials from excavation campaigns Belovode 2012 and 2013. Note an indicated subset analysed in depth with various analytical instruments.

No	trench	spit	find no.	EDM	type of material	metallography	OM	SEM-EDS	EPMA	LIA	NAA	LA-ICP-MS	weight (g)	reasoning for provenance analysis
B21/13	18ext	S4	1378	420	malachite (green)					X	X			top layer malachite
B23/12	18	S5	77	182	slagged sherd		X	X	X	X	X	X	11.04	slagged sherd
B24/12_2	18	S5	78	183	slag		X	X	X			X	0.42	slag
B29/12	18	S5	88	201	metal droplet		X	X					0.65	
B62/13	18ext	S5	1495	480	sherd with malachite									
Bf7/12	18	S5	70	166	malachite (black and green)		X	X					0.64	
Bf21/12	18	S6	112	272	metal droplet	X	X		X	X	X	X	3.78	metal droplet
Bf22/12/2	18	S6	116	276	mineral		X	X		X	X		5.07	workshop area mineral
Bf121/12	18	S6	111	271	haematite		X	X					1.73	
B41/12	18	S6	102	258	malachite (black and green)		X	X						
B46/12	18	S6	111	271	malachite (green)		X	X					0.66	
B47/12(1/2/3)	18	S6	112	272	slagged sherd (1) + slag (2) + metal droplet (3)		XXX	XXX	X/2	X/3	X/3	X	(1) 22.47; (3) 0.83; (2) 0.36	production evidence / metal
B52/12	18	S6	124	285	malachite bead									
B71/12	18/F3	S6	159	380	metal artefact	X	X	X	X			X		
B68/13	18ext	S6	1567	490	malachite (green)									
Bf25/12	18	S6	124	285	malachite bead									
B78/12	18/F4	S7	168	429	malachite		X	X					0.21	
B80/12	18	S7	173	434	malachite									
B171/12	18	S7	312	969	malachite bead									
B91/13	18ext	S8	1645	526	azurite									
B198/12	18	S9	1104	361	sphalerite									
Bf111/12	18	S9	1085	353	malachite bead									
B42/13	18	S11	1417	453	malachite bead									
Bf77/12	18/F9	S11	256	760	malachite bead									
Bf30/13	18	S11	1407	443	malachite bead									
Bf43/13	18/F21	S12	1492	477	metal droplet	X	X	X		X	X			metal droplet
B108/13	18	S13	1691	550	malachite (black and green)					X	X			associated with Bf43/13
C_B5/13	18ext	S13	2201	762	malachite bead									
B143/13	18	S13	1762	598	malachite									
B350/13	18ext/F39	S13	2429	881	malachite					X	X			Near ash feature 39

Table 1 continued. The list of minerals and metallurgical materials from excavation campaigns Belovode 2012 and 2013. Note an indicated subset analysed in depth with various analytical instruments.

No	trench	spit	find no.	EDM	type of material	metallography	OM	SEM-EDS	EPMA	LIA	NAA	LA-ICP-MS	weight (g)	reasoning for provenance analysis
B408/13	18ext	S13	2446	882	ash		X	X						
B409/13	18ext	S13		2266/71	pot F35									
Bf56/13	18/F21	S13	1608	506	metal droplet		X	X						
Bf111/13	18ext/13	S13	2282	803	malachite bead									
Bf125/13	18ext	S14	2350	863	silverish ore									
B95/13	18ext	S14	1532	669	malachite bead									
B155/13	18	S14	1785	612	copper mineral					X	X			a different type of a copper mineral
B361/13	18ext/F32	S15	2496	899	malachite bead									
C_B7/13	18	S17	2319	836	Malachite bead									
C-B10/13	18	S18	2510	905	green stone bead flot									
B385/13	18	S19	2596	937	malachite					X	X			last malachite occurrence
B387/13	18	S19	2598	939	malachite									

Table 2. Direct dating of metallurgical materials from Belovode

No	trench	spit	find no.	EDM	horizon	type of material	associated feature	modelled C14-date 1σ	modelled C14-date 2σ	directly dated feature/spit (2σ)
B23/12	18	S5	77	182	1a	slagged sherd	F6	4689-4550 to 4572-4481 cal. BC	4702-4541 to 4600-4404 cal. BC	MAMS23380 - 4724-4555 cal. BC
B24/12_2	18	S5	78	183	1a	slag	F6	4689-4550 to 4572-4481 cal. BC	4702-4541 to 4600-4404 cal. BC	MAMS23380 - 4724-4555 cal. BC
B29/12	18	S5	88	201	1a	metal droplet	F6	4689-4550 to 4572-4481 cal. BC	4702-4541 to 4600-4404 cal. BC	MAMS23380 - 4724-4555 cal. BC
Bf21/12	18	S6	112	272	1a	metal droplet	F6 (S6)	4776-4709 to 4689-4550 cal. BC	4818-4693 to 4702-4541 cal. BC	MAMS23380 - 4724-4555 cal. BC MAMS22069 - 4547-4455 cal. BC
B47/12(1/2/3)	18	S6	112	272	1a	slagged sherd (1) + slag (2) + metal droplet (3)	F6 (S6)	4776-4709 to 4689-4550 cal. BC	4818-4693 to 4702-4541 cal. BC	MAMS23380 - 4724-4555 cal. BC MAMS22069 - 4547-4455 cal. BC
B71/12	18/F3	S6	159	380	1b	metal artefact	F3 base	4776-4709 to 4689-4550 cal. BC	4818-4693 to 4702-4541 cal. BC	MAMS23376 - 4778-4611 cal. BC
Bf43/13	18/F21	S12	477	1492	2	metal droplet	F21	4899-4798 to 4776 -4709 cal. BC	4950-4756 to 4818-4693 cal. BC	MAMS23378 - 4872-4720 cal. BC MAMS22079 - 4847-4722 cal. BC
Bf56/13	18/F21	S13	506	1608	2	metal droplet	F21	4899-4798 to 4776 -4709 cal. BC	4950-4756 to 4818-4693 cal. BC	MAMS23378 - 4872-4720 cal. BC MAMS22079 - 4847-4722 cal. BC

Figure 3. Feature 6 in Trench 18, once emptied. Note charred and burnt soil with traces of ashes (white).

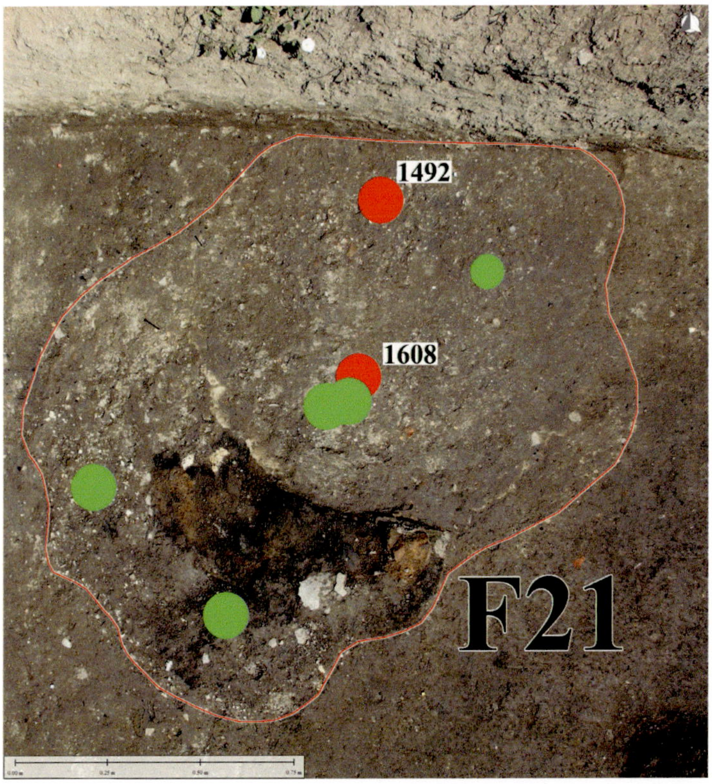

Figure 4. Feature 21. Note red dots for B43/13 (No. 1492) and B56/13 (No. 1608).

Significantly, the two copper metal droplets from Feature 21 (Bf43/13 and Bf56/13), spits 12 and 13 respectively (see Tables 1 and 2, Figure 4), are currently the earliest directly dated evidence for metal production in T18ext. They are directly associated with the dates for Feature 21, which starts at 4950–4756 and ends at 4818–4693 cal. BC (95.4% prob.), or possibly has a span of 4899–4798 to 4776–4709 cal. BC (68% prob.). In the discussion below we place these within the 49th century.

During the excavation we were not able to identify any solid structure that could have been used for metallurgical activities (such as a furnace or smelting installation). However, a vessel bottom found embedded in the soil, with traces of ashes and charred surfaces around it might have been related to a firing procedure involving copper smelting, especially if considered in the context of a nearby ash bin that partially extended under the eastern profile of the trench extension (see Figure 5). Figure 5c-d shows a fragmented vessel bottom, *c.* 20 cm

Figure 5. Feature 35, with traces of ash and charred soil (a, b) and the vessel bottom dug into the soil (c) and once excavated and cleaned (d).

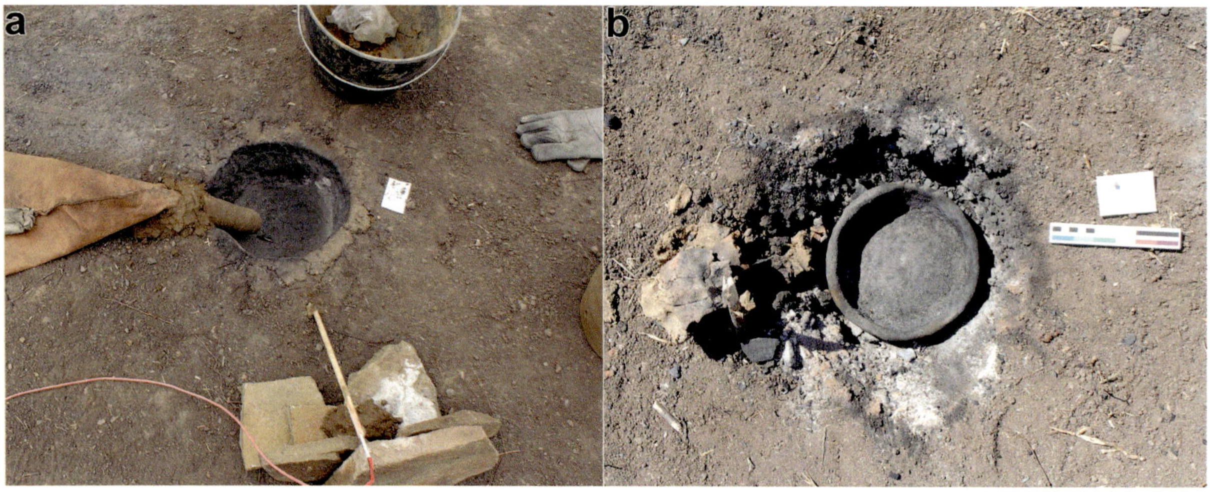

Figure 6. Experimental reconstruction of Feature 35 (see Figure 5) in 2013. a) A smelting installation set up with a tuyère (not an original find but hypothesised for this experiment); b) post-smelting situation. Note no adhering slags or other remains in the vessel, despite a successful smelt taking place in it.

in diameter, as part of Feature 35, surrounded by ashes, burnt soil and charcoal (Figure 5a-b). The surface was analysed with pXRF but there were no detectable traces of metal contamination, hence its connection with the smelting process is only assumed due to the unusual field situation. Yet, the same Belovode horizon (2) and excavation spit (13) produced the earliest find of metal production in the Trench 18 excavations: Bf56/13, a metal droplet (and associated Bf43/13 droplet). The direct date for Feature 35 indicates its start at 5003–4845 cal. BC (95.4% prob.), or possibly 4987–4857 cal. BC (68% prob.) (Chapter 37, Table 1), hence very close to the 49th century BC, the dating of the metal droplets in Feature 21 (Table 2).

In September 2013, the excavation team ran a series of copper smelting experiments and reconstructed this particular vessel bottom feature as part of the smelting installation (Figure 6a). Whilst we successfully extracted metal in this installation, there was no slag or any traces of production left attached to the vessel bottom (Figure 6b). This prompted us to conclude that such an installation could indeed have been used to smelt metal at Belovode, without exhibiting commonly encountered evidence, such as slag or green staining, either on the inside or adhering to its walls. This experiment will be reported elsewhere.

Methodology

The sampling strategy initially involved selecting materials on the basis of their visual appearance, and their response to a magnet (slag and slagged materials), in the field and laboratory. The research collection was catalogued, measured and photographed prior to sample preparation and analysis (Table 1). The following protocol for sample preparation has already been published (Radivojević and Rehren 2016) but is presented here for convenience in a shortened form. Samples selected for microstructural and compositional study were cut to size (where necessary) using a fine, diamond-coated circular saw. They were then washed with water, dried and mounted in epoxy resin. The mounted blocks were then ground using abrasive paper (1200 and 2400 grit) and polished using diamond pastes (down to 1 μm and 0.25 μm). The polished blocks were washed in an ultrasonic bath and rinsed with ethanol between each grinding and polishing stage. The initial analytical stage consisted of reflected light microscopy (OM), with photomicrographs taken on Leica and Olympus microscopes at 25x, 50x, 100x, 200x, 500x and 1000x (Table 3a). Polished blocks of artefacts (Bf21/12 and B71/12) were prepared for metallographic examination using ammonia hydrogen peroxide as an etchant, made from equal proportions of ammonia (NH_4OH), water and 3% H_2O_2.

For the following compositional analysis, samples were carbon-coated for examination under the Scanning Electron Microscope with Energy Dispersive Spectrometry (SEM-EDS, Oxford Instrument's INCA X-cite) and Electron Probe Micro Analysis (EPMA), respectively, both at an accelerating voltage of 20 kV. The EPMA was used only for the samples which contained a distinctive copper metal phase. All analyses were conducted by the first author, barring EPMA, which were conducted by Kevin Reeves, former technician at the Wolfson Archaeological Science Laboratories, UCL Institute of Archaeology, London. Table 1 includes information pertaining to samples analysed using OM and SEM-EDS analysis. The acquired data are corrected against certified reference materials (CRM), analysed under the same conditions as the metallurgical materials on Superprobe JEOL- JXA-8600. A correction factor was applied only in cases where divergence (relative error) was higher than 10% (Table 3b). The corrected values are reported below, and uncorrected 'raw' data provided in the Appendix B_Ch11.

Where possible, EPMA was used for analysing copper metal phases, enabling the detection of elements present at the 10 ppm (0.001%) level. Seventeen elements were searched for in all samples (Table 3a), with the analytical background of these adjusted during analysis with respect to the CRM for alloyed copper. Each sample had eight to eleven runs to obtain the most precise data. In order to assess the true presence of trace elements in copper metal phases in various samples, all gained values were assessed in relation to measurements acquired for copper alloy CRM; a correction factor was applied only in cases where divergence (relative error) was higher than 5% (Table 3c). During data analysis and interpretation, the threshold for trusted values for trace elements was established at ≥ 100 ppm. Although the data are reported as μg/g (microgram per gram), we shall use ppm (parts per million) to discuss them throughout this monograph.

A subset of metal production and finished metal artefacts samples were sent to Curt-Engelhorn-Centre for Archaeometry (Mannheim, Germany) for high resolution / low threshold analysis of metal phases within these samples with a Thermo iCAP Q inductively coupled plasma mass spectrometer coupled with a Resonetics laser ablation system (Table 3a). The following protocol was followed: samples were prepared on a specimen holder. The laser system was adjusted to a spot size of 73 μm for the pre-ablation cleaning step and 58 μm for ablation at 10 Hz and an energy density of 5 mJ. Helium was used as carrier gas (600ml/min). The plasma power was set to 1400 W. Cool, Auxiliary and Argon gas flow were adjusted to 13.0, 0.7 L/min and 0.9 L/min, respectively. The following isotopes were determined: ^{24}Mg, ^{27}Al, ^{29}Si, ^{31}P, ^{52}Cr, ^{55}Mn, ^{57}Fe, ^{59}Co, ^{60}Ni, ^{63}Cu, ^{68}Zn, ^{75}As, ^{82}Se, ^{107}Ag, ^{100}Mo (^{100}Ru), ^{111}Cd, ^{113}In (^{113}Cd), ^{115}In (^{115}Sn), ^{118}Sn, ^{121}Sb, ^{125}Te, ^{197}Au, ^{208}Pb, ^{209}Bi. Further isotopes were monitored but not quantified: ^{101}Ru, ^{103}Rh, ^{104}Pd (^{104}Ru), ^{105}Pd, ^{106}Pd (^{106}Cd), ^{189}Os, ^{191}Ir, ^{195}Pt. Line ablations with a length of 300-400μm were performed. Data were collected in time resolved mode, including around 30s integrated gas blank signal before each sample signal. The data acquisition sequence consists of blocks up to three samples each ablated five times which were enclosed by a blocks of solid reference materials. Quantification was carried out using ablation yield correcting factors with sum normalisation (Lin *et al.* 2016). A set of different external standards (BAM 211, BAM 227, BAM 376, BAM 375 and NIST 400 as well as NIST610 and NIST 612) were used to quantify major, minor and trace elements (Hawkins *et al.* 2016; Walaszek *et al.* 2013).

Table 3a: Analytical instruments used in this study, aim of analysis and relevant analytical parameters.

Instruments	Aim of Analysis	Analytical Parameters
Reflected Polarized Light Microscopy (Leica DMLM and Olympus BX60)	Phase identification and visual characterisation of microstructure	Plane polarized light and crossed polarized light were applied to examine phases in samples, their colour, homogeneity, porosity and inclusions (shape, size and uniformity). Cross-polarized light was also applied for internal reflection and identifying the composition of phases present. The microscope was equipped with a Nikon digital camera, with highest magnification of 1000x.
SEM-EDS Scanning Electron Microscopy with Energy Dispersive Spectrometry (Superprobe JEOL- JXA-8600)	1. Phase identification in samples using electron images and area/ point analyses 2. Quantitative compositional analyses of observed phases 3. Observation of the relationships between phases on the basis of their atomic number contrast	Backscattered electron (BSE) imaging used. All materials analysed on JXA-8600. The accelerating voltage was 20 kV, with average dead-time of 35-40 % and working distance of 10 mm. All data are presented as normalized with stoichiometrically added oxygen, if not otherwise stated. The iron content is presented as FeO, which here stands for total iron (both valencies).
EPMA Electron Probe Micro Analysis (Superprobe JEOL-JXA-8100)	Compositional analysis of copper metal phases in all samples (down to trace element level)	All samples analysed at an accelerating voltage of 20 kV, beam current 50 nA, with average dead-time of 35-40 % and working distance of 10 mm. The following elements were checked for: Se, Zn, Cu, Fe, As, Ag, Cl, Te, S, Au, Sn, Bi, Co, Sb, Ni, Mn, Pb. All data presented as wt% and ppm, with the trusted values for the latter established at ≥ 10 ppm.
LA- ICP-MS Thermo iCAP Q inductively-coupled plasma mass spectrometer coupled with a Resonetics laser ablation system (ArF, 193 nm). Located externally at the Curt-Engelhorn-Centre for Archaeometry, Mannheim, Germany	Trace element analysis (low detection threshold) of copper metal phases in production debris and selected metal artefacts	The parameters of the ICP-MS were optimized to ensure a stable signal with a maximum intensity over the full range of masses of the elements and to minimize the formation of oxides and double-ionized species. The following elements were checked for: Co, Ni, Cu, Zn, As, Ag, Sb, Te, Pb, Bi, Mg, Al, Si, Mn, Cd, Sn, Se, P, Fe. The data is presented as wt% and ppm. Mg, Al, Si, Mn, Cd and Sn were below 10 ppm; Se below 30 ppm, and P and Fe below 50 ppm; the rest of the readings are reported in the indicated table in the text.

Results: Technology of metal making at Belovode

A total of fifteen objects were selected for in-depth analysis: five minerals/ores and ten examples of production evidence (slags, slagged sherds, metal droplets and a metal artefact fragment). All presented evidence for copper-based metallurgy, which in its nature and quantity is consistent with previous studies on archaeometallurgical materials from this site (Radivojević 2007, 2012, 2013, 2015; Radivojević and Rehren 2016; Radivojević *et al.* 2010a).

Processing: archaeological minerals

Five mineral samples were analysed (see Table 1) for microstructure and composition: Bf7/12, Bf22/12/2, B41/12, B46/12 and B78/12. All are copper based, with Bf22/12/2 exhibiting pure copper oxide (Figure 7), and the rest presenting a mixture of pure green (malachite) occasionally mixed with oolithic structures that contain high MnO content, reaching just below 40 wt% (Table 4, Figure 7). These results are in line with the findings of previous studies on Belovode minerals,

Table 3b. SEM-EDS compositional data of certified reference materials (CRM) for basalt glasses: BIR1, BCR-2, BHVO-2, given in wt%. All measured values are presented against certified average values for CRMs, with correction value calculated only for relative errors above 10% divergence. The averages of all correction values and correction factor are given in the bottom two lines; the correction factor was applied for calculating real values for analysed samples.

	Na_2O	MgO	Al_2O_3	SiO_2	P_2O_5	K_2O	CaO	TiO_2	MnO	FeO
	wt%	wt%	wt%	wt%	wt%	wt%	wt%	wt%	wt%	wt%
BIR1	1.45	8.71	14.35	49.86			13.54	1.24	0.08	10.76
BIR1	1.35	8.7	14.59	49.71			13.58	0.99	0.11	10.99
BIR1	1.47	8.42	14.5	49.85			13.72	1.08	0.32	10.65
average measured (1)	*1.42*	*8.61*	*14.48*	*49.81*			*13.61*	*1.10*	*0.17*	*10.80*
certified value (2)	*1.82*	*9.70*	*15.50*	*47.96*			*13.30*	*0.96*	*0.18*	*11.30*
absolute error (3) = (1)-(2)	*-0.40*	*-1.09*	*-1.02*	*1.85*			*0.31*	*0.14*		*-0.50*
*relative error in % (4) = (3/1)*100*	***-27.87***	***-12.66***	*-7.04*	*3.71*			*2.30*	***12.99***		*-4.63*
correction value (5) = (4)/100	***-0.28***	***-0.13***	*-0.07*	*0.04*			*0.02*	***0.13***		*-0.05*
BCR-2	2.49	3.28	12.51	55.56	0.61	1.94	7.49	2.6	0.27	13.25
BCR-2	2.65	3.15	12.44	55.52	0.45	2.06	7.52	2.57	0.25	13.38
BCR-2	2.41	3.04	12.17	56.32	0.49	1.91	7.63	2.38	0.27	13.38
BCR-2	2.56	3.36	12.55	56.06	0.46	1.89	7.47	2.35	0.2	13.11
average measured (1)	*2.53*	*3.21*	*12.42*	*55.87*	*0.50*	*1.95*	*7.53*	*2.48*	*0.25*	*13.28*
certified value (2)	*3.16*	*3.59*	*13.50*	*54.10*	*0.35*	*1.79*	*7.12*	*2.26*	*0.24*	*13.80*
absolute error (3) = (1)-(2)	*-0.63*	*-0.38*	*-1.08*	*1.77*		*0.16*	*0.41*	*0.22*		*-0.52*
*relative error in % (4) = (3/1)*100*	***-25.02***	***-11.93***	*-8.72*	*3.16*		*8.21*	*5.41*	*8.69*		*-3.92*
correction value (5) = (4)/100	***-0.25***	***-0.12***	*-0.09*	*0.03*		*0.08*	*0.05*	*0.09*		*-0.04*
BHVO-2	1.77	6.54	12.41	51.51	0.22	0.54	11.65	3.01	0.22	12.12
BHVO-2	1.73	6.42	12.53	51.39	0.22	0.64	11.7	2.95	0.23	12.19
BHVO-2	1.73	6.26	12.59	51.63	0.02	0.54	12.22	2.94	0.2	11.86
average measured (1)	*1.74*	*6.41*	*12.51*	*51.51*	*0.15*	*0.57*	*11.86*	*2.97*	*0.22*	*12.06*
certified value (2)	*2.22*	*7.23*	*13.50*	*49.90*	*0.27*	*0.52*	*11.40*	*2.73*	*0.20*	*12.30*
absolute error (3) = (1)-(2)	*-0.48*	*-0.82*	*-0.99*	*1.61*		*0.05*	*0.46*	*0.24*		*-0.24*
*relative error in % (4) = (3/1)*100*	***-27.34***	***-12.85***	*-7.91*	*3.13*		*9.30*	*3.85*	*7.98*		*-2.02*
correction value (5) = (4)/100	***-0.27***	***-0.13***	*-0.08*	*0.03*		*0.09*	*0.04*	*0.08*		*-0.02*
calculating correction factor										
	Na_2O	MgO						TiO_2		
BIR1	-0.28	-0.13						0.13		
BCR-2	-0.25	-0.12								
BHVO-2	-0.27	-0.13								
average	-0.27	-0.12						0.13		
correction factor ALL	***1.27***	***1.12***						***0.87***		

Table 3c. EPMA compositional data of certified reference materials (CRM) for copper alloys (bronzes): 32XSN6 and 33XGM29, given in wt%. All measured values are presented against certified average values for CRMs, with correction values calculated for relative errors above 5%. The averages of all correction factors are given in the bottom two lines; the correction was applied to all EPMA readings of metal phases presented below.

	Ni	Cu	Se	Co	As	Zn	Mn	Fe	Ag	Sn	Bi	S	Sb	Pb	Cd	Au
	wt%	wt%	wt%	wt%	wt%	wt%	wt%	wt%	wt%	wt%	wt%	wt%	wt%	wt%	wt%	wt%
32X SN6 average observed (1)	0.218	87.436	0.001	0.770	0.489	1.069	0.003	0.113	1.071	7.026	0.080	0.048	0.297	1.202	0.077	0.010
certified value (2)	0.203	86.390		0.655	0.764	1.170	0.002	0.099	1.159	7.310	0.158	0.018	0.323	1.559	0.090	0.007
absolute error (3)= (1)-(2)	0.01	1.05		0.11	-0.27	-0.10	0.00	0.01	-0.09	-0.28	-0.08	0.03	-0.03	-0.36	-0.01	0.00
relative error in % (4) = (3/1)*100	**6.67**	1.20		**14.91**	**-56.21**	**-9.43**	**27.27**	**12.39**	**-8.27**	-4.04	**-97.50**	**62.50**	**-8.94**	**-29.71**	**-16.97**	**24.74**
correction value (5) = (4)/100	**0.07**			0.15	-0.56	-0.09	0.27	0.12	-0.08		-0.98	0.63	-0.09	-0.30	-0.17	0.25
33XGM29 average observed (1)	0.030	89.383	0.000	0.004	0.000	4.236	0.004	0.014	0.003	6.232	0.008	0.010	0.000	0.050	0.005	0.000
certified value (2)	0.029	89.360				4.230		0.010	0.003	6.120	0.002	0.002		0.050		
absolute error (3)= (1)-(2)	0.00	0.02				0.01		0.00	0.00	0.11	0.01	0.01		0.00		
relative error in % (4) = (3/1)*100	3.99	0.03				0.15		**28.17**	**18.75**	1.80	**75.32**	**76.92**		-0.60		
correction value (5) = (4)/100								0.28	0.19		0.75	0.77				
calculating correction factor	Ni	Cu	Se	Co	As	Zn	Mn	Fe	Ag	Sn	Bi	S	Sb	Pb	Cd	Au
average correction value	0.07			**0.15**	**-0.56**	**-0.09**	**0.27**	0.20	0.05		-0.11	0.70	**-0.09**	**-0.30**	**-0.17**	0.25
correction factor	**0.93**			0.85	1.56	1.09	0.73	0.80	0.95		1.11	0.30	1.09	1.3	1.17	0.75

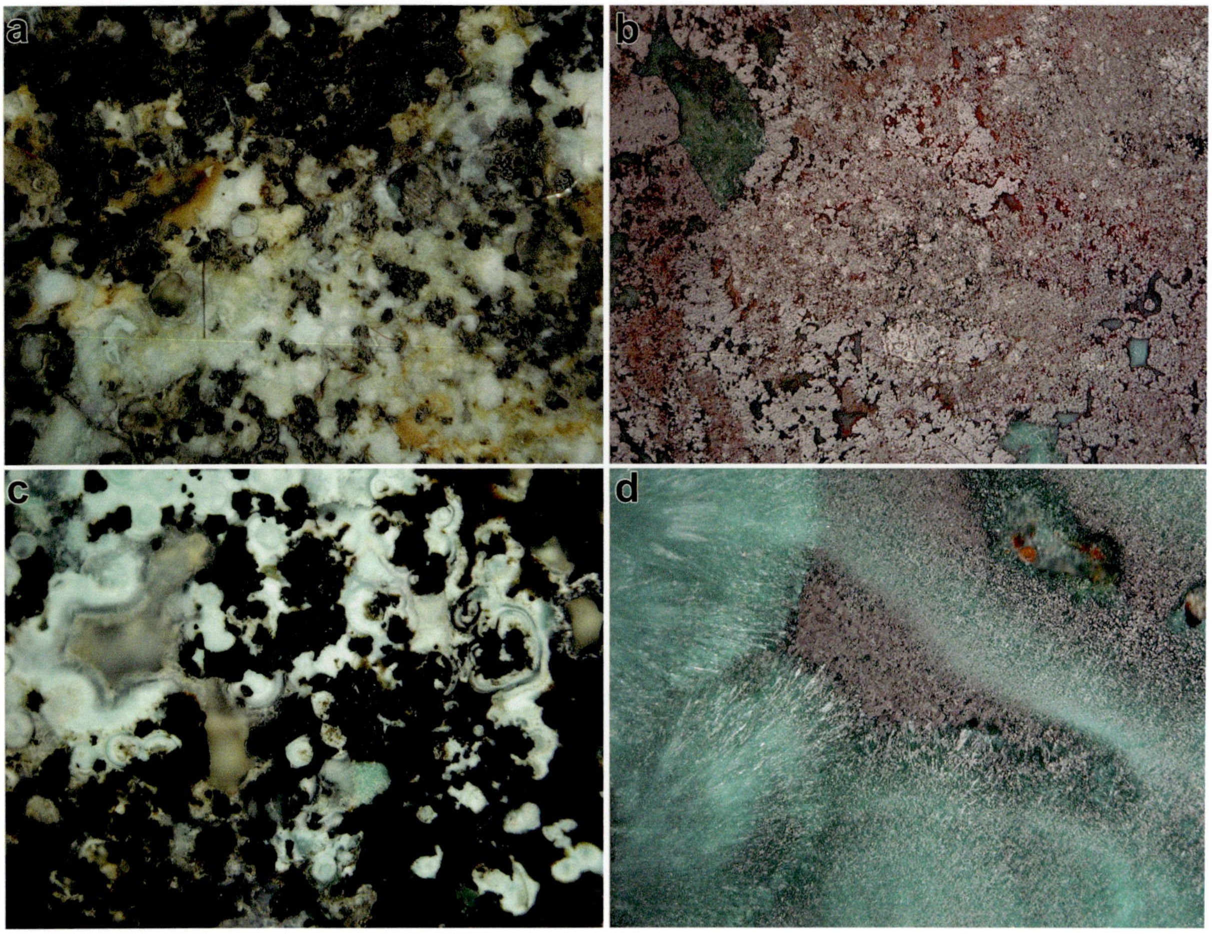

Figure 7. Photomicrographs of copper minerals from Belovode, cross polarised light. a) Bf7/12, magnification 50x, width 3.2mm. Note black (Mn-rich) and green (Cu-rich) phases; b) Bf22/12/2, magnification 50x, width 3.2mm. Note reddish/violet phase of tenorite and cuprite; c) B41/12, magnification 100x, width 1.6mm. Note black (Mn-rich) and green (Cu-rich) phases; B46/12, magnification 50x, width 3.2mm. Note pure green (copper carbonate – malachite) phase.

Table 4. SEM-EDS compositional data for oolithic (dark) phases in minerals Bf7/12 and B41/12. All values are averages of ten to fourteen analyses of each sample / phase and normalised to 100%.

	Al_2O_3	SiO_2	P_2O_5	K2O	CaO	MnO	FeO	CoO	NiO	CuO	ZnO
	wt%	wt%	wt%	wt%	wt%	wt%	wt%	wt%	wt%	wt%	wt%
Bf7/12	9.0	15.8	2.7	0.2	2.1	39.3	0.9	0.1	0.1	28.6	1.2
B41/12	6.9	14.0	2.6	0.0	2.0	38.0	0.2	0.0	0.0	36.1	0.1

where manganese-rich black and green copper minerals feature prominently in the collection of beneficiated ores used for copper smelting on the site (Radivojević 2007, 2012, 2013, 2015; Radivojević and Rehren 2016; Radivojević *et al.* 2010a) and beyond, in Vinča-Belo Brdo. Previous analyses have also shown that sulfidic minerals are also a common inclusion in secondary copper ores from Belovode.

Production: slagged sherds and slags

Installations/slagged sherds

The two slagged sherds (B23/12 and B47/12/1) show close technological and spatial association with the cluster of production debris around the Feature 6 area in Trench 18 (Figure 2, Table 2). Given these links, we are inclined to ascribe the entire assemblage to Horizon 1b (see Chapter 10), which is in line with the foundations of Feature 3, or the dwelling feature in Trench 18. Both finds are from the bottom of spit 5 and the beginning of spit 6; given the artificial nature of excavation spits, the association is without doubt. Also, the microstructural and compositional relationship presented below strengthens our assumption that these belong to a single or two closely associated smelting events.

Visual inspection of both objects (Figures 8 and 9) shows that the ceramic bodies were unaffected by heat treatment exceeding the initial firing temperature, apart from the single sections in both examples that exhibit a bloated surface stained with a mass of green and grey appearance. These stained surfaces are spread across the broken sections, implying that the sherds were fragmented prior to the contact with metallurgical activities. This is the same scenario already observed for four slagged sherds discovered in previous excavations (see Chapter 5, this volume) and extensively reported elsewhere (e.g. Radivojević 2013, 2015; Rehren *et al.* 2016). Hence, as with these previous examples, these sherds most likely formed a lining for a hole-in-the-ground smelting installation, leading to some localised burning and bloating of the ceramic and droplets of slag and metal adhering to them. This is further corroborated by the presence of ashes, burnt soil, clay and charred surfaces in the charred pit, Feature 6, associated with these finds; in Figure 3 we see that Feature 6 was deep enough to possibly represent such a hole-in-the-ground installation. Two slagged sherds found in its vicinity strengthen this hypothesis.

The ceramic fabric of Belovode 23/12 and 47/12/1 appears optically dark grey, well kneaded and tempered with abundant quartz grains, amounting to almost 50 vol% (Figures 10a and 10d). The dense paste looks 'dry', with little indication of potential collapsing. Most of the quartz grains closer to the area of intense vitrification have lost their angularity and decomposed (Figure 10b); this is followed by coarse bloating pores. Noteworthy is the formation of slag on top of these sherds; while in some areas it is clear that the slag had only a short liquid contact with the ceramic surface (Figure 10c), others exhibit a stronger interaction of the slag and ceramic body (Figures 10b and 10d). This is yet another indicator that these sherds were not used as a crucible, but for lining a hole in the ground; the surfaces that show close interaction were most likely closer to the 'hot spot' in the metallurgical installation.

Bulk chemical analyses of the ceramic bodies indicated the use of similar clay for both samples (Table 5), where the consistent ratio of silica to alumina (approximately 5:1) and similar readings of iron oxides, lime, potash and magnesia suggest a common origin for the clay. These were, however, parts of two different pottery vessels,

Figure 8. Slagged sherd B23/12. Note green stain mass adhering to the top and cross sections.

Figure 9. Slagged sherd B47/12/1. Note green staining on one section only.

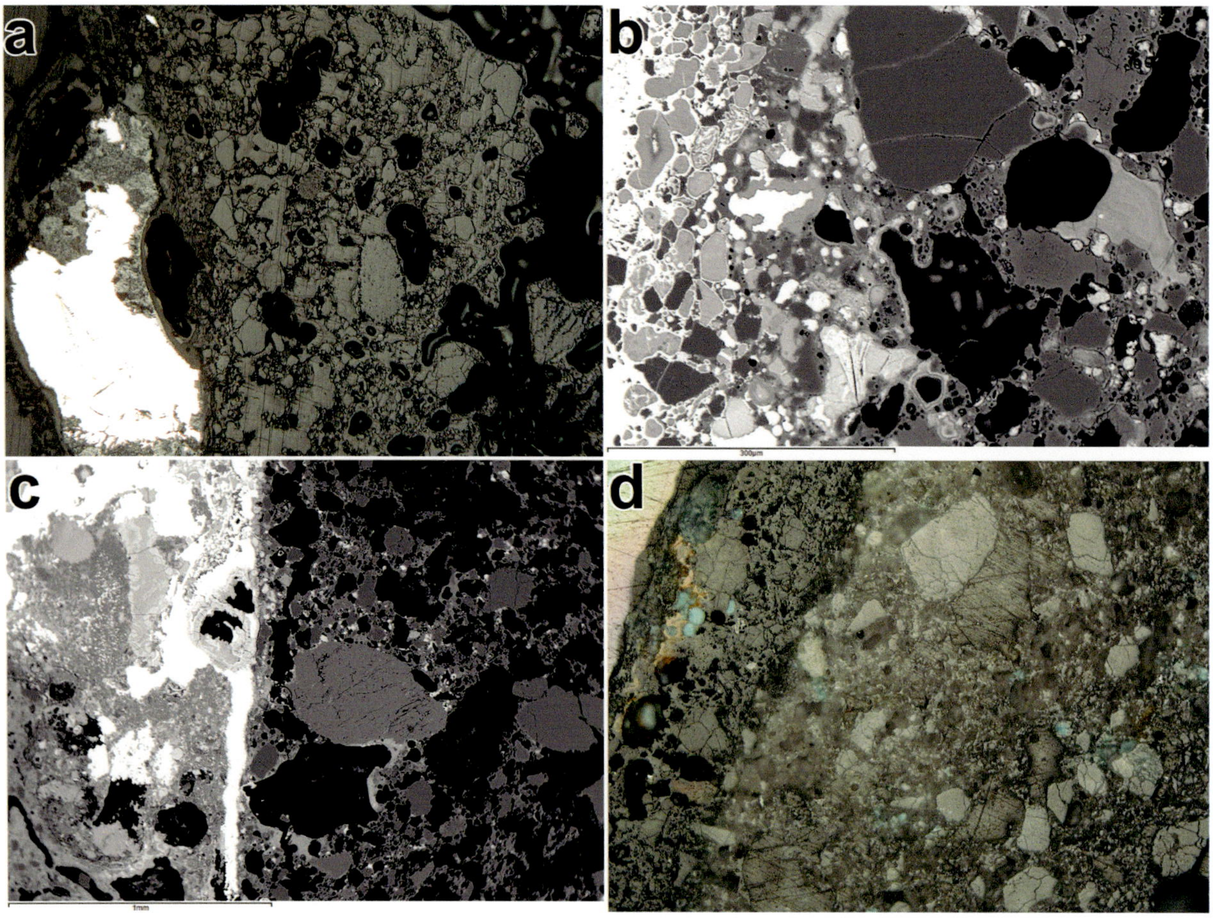

Figure 10. a) Photomicrograph slagged sherd B23/12 taken under plain polarised light, 50x magnification, 3.2 mm width. Note bright metal phase stuck on top of the 'cold' ceramic body; b) Backscattered electron image of a section of B23/12. Note highly reactive ceramic body (right, with decomposing grey quartz grain) and slag (left, with bright metal phases); c) backscattered electron image of a section of B23/12. Note poor reaction of metal phase with ceramic body; d) Photomicrograph of slagged sherd B47/12/1, taken under plain polarised light, 50x magnification, 3.2mm width. Note 'cold' ceramic body moderately fused with slag (left, with green corrosion products).

and were not found in association in the field. The 'hot' or bloated sections of the sherds (Table 5) exhibit a slightly different picture, with greater variability of the silica to alumina ratio (on average 4:1, though it varies from 3:1 to 7:1). Elevated readings of copper (up to 6.8 wt% CuO) and iron oxides (up to 12.4 wt% FeO^t) and sulfur (only in B47/12/1) represent the ores included in the smelt, while increased potash and lime concentrations would have been due to fuel ash contamination.

Optical microscopy and SEM-EDS analyses revealed a heterogeneous structure of slag and copper-rich materials in both samples (23/12 and 47/12/2). The slag contains newly formed phases and metal prills in a glassy matrix, with corrosion products developing on the edges and inside porosity holes (Figure 11). The main inclusions suspended in the slag matrix are copper 'dross', delafossite and iron-rich spinels (the latter two only for B23/12), suggesting at least the primary smelting nature of the process related to B23/12 (Figure 11a). Both samples are dominated by copper oxide and copper metal prills (Figure 11b and 11c). The chemical analyses of the bulk glassy slag matrix were conducted through area analysis in B23/12, while in B47/12/2 this was not possible due to the uneven distribution of slag and its small volume on this sherd's surface. Spot analyses were conducted to aid the understanding of slag chemistry in both samples, in areas relatively free of copper-based components (Table 6).

The chemistry of the green-stained and bloated outer surface of all slagged sherds revealed significant contamination with copper (up to 27 wt%, with increased levels of MnO and ZnO in B23/12) and fuel ash (elevated readings of CaO, P_2O_5, K_2O, MgO), clearly different from the ceramic composition (Table 5). These data correlate well with the already published slagged sherd and slag samples from Belovode, which include significant readings of MnO and ZnO, and increased values of FeO (Radivojević 2013, 2015; Radivojević and Rehren 2016; Radivojević *et al.* 2010a). Analyses

Table 5. SEM EDS compositional data for ceramic body of slagged sherds B23/12 and B47/12/1, sections unaffected (cold) and affected (bloated) by high temperatures. All values are averages of two to twenty-seven analyses of each sample / phase and corrected with factors based on CRM analysis; the uncorrected data is reported in the Appendix.

	Na_2O	MgO	Al_2O_3	SiO_2	P_2O_5	SO_3	K_2O	CaO	TiO_2	MnO	FeO	CoO	NiO	CuO
	wt%	wt%	wt%	wt%	wt%	wt%	wt%	wt%	wt%	wt%	wt%	wt%	wt%	wt%
B23/12 cold	2.2	1.5	14.8	71.6	0.1	0.0	3.0	1.4	0.7	0.0	5.3	0.0	0.0	0.0
stdev s	*0.6*	*0.2*	*1.3*	*2.5*	*0.3*	*0.0*	*0.4*	*0.4*	*0.4*	*0.0*	*0.7*	*0.0*	*0.0*	*0.0*
B47/12/1 cold	1.4	1.5	12.8	73.5	0.4	0.0	3.0	1.0	0.7	0.0	6.2	0.0	0.0	0.0
stdev s	*0.1*	*0.3*	*0.2*	*2.4*	*0.5*	0.0	0.0	0.3	0.3	0.0	1.5	0.0	0.0	0.0
	Na_2O	MgO	Al_2O_3	SiO_2	P_2O_5	SO_3	K_2O	CaO	TiO_2	MnO	FeO	CoO	NiO	CuO
	wt%	wt%	wt%	wt%	wt%	wt%	wt%	wt%	wt%	wt%	wt%	wt%	wt%	wt%
B23/12 bloated	4.1	1.3	19.9	63.9	0.0	0.0	4.8	2.5	0.4	0.0	3.8	0.0	0.0	0.2
stdev s	*1.7*	*1.1*	*3.1*	*3.5*	*0.0*	*0.0*	*1.2*	*1.1*	*1.0*	*0.0*	*3.2*	*0.0*	*0.0*	*0.4*
B47/12/1 bloated	1.8	1.7	14.7	66.5	1.3	0.4	2.6	1.6	0.7	0.0	6.1	0.0	0.0	3.1
stdev s	*0.5*	*0.3*	*1.7*	*5.0*	*1.3*	*1.0*	*1.2*	*0.4*	*0.4*	*0.0*	*1.9*	*0.0*	*0.0*	*1.8*

Table 6. SEM EDS compositional data for slag matrix in slagged sherds B23/12 and B47/12/1, both as spot analysis (relatively free of inclusions) and as area analysis (bulk slag analysis in areas of c. 100 x 100 microns in B23/12). All values are averages of three to twenty-nine analyses of each sample / phase and corrected with factors based on CRM analysis; the uncorrected data is reported in the Appendix.

	Na_2O	MgO	Al_2O_3	SiO_2	P_2O_5	SO_3	K_2O	CaO	TiO_2	MnO	FeO	CoO	CuO	ZnO
	wt%	wt%	wt%	wt%	wt%	wt%	wt%	wt%	wt%	wt%	wt%	wt%	wt%	wt%
B23/12 spot	0.7	1.9	10.3	47.2	1.7	0.0	4.8	12.0	0.3	3.4	4.2	0.2	13.3	0.4
stdev s	*0.7*	*0.8*	*3.8*	*10.1*	*2.4*	*0.0*	*2.0*	*7.0*	*0.3*	*4.1*	*2.2*	*0.5*	*7.0*	*0.7*
B23/12 area	1.4	1.4	13.8	70.6	0.2	0.0	5.1	1.0	0.8	0.0	3.8	0.0	2.3	0.0
stdev s	*0.4*	*0.3*	*3.4*	*8.0*	*0.3*	0.0	2.6	0.3	0.6	0.0	0.3	0.0	0.6	0.0
B47/12/1 spot	0.0	1.8	11.6	45.4	1.2	0.0	3.3	7.9	0.6	0.0	6.6	0.0	21.7	0.0
stdev s	*0.0*	*0.2*	*1.9*	*6.7*	*2.2*	*0.0*	*1.1*	*9.0*	*0.1*	*0.0*	3.8	*0.0*	*2.8*	*0.0*

presented here reinforce the hypothesis of a preference for manganese-rich black and green ores for metal extraction, both at Belovode and across the Vinča culture sites.

The difference in (bulk) chemical composition between the two sherds is also reflected in the formation of phases in the slag matrix. While slagged areas in B47/12/1 are commonly rich in copper 'dross' suspended in slag matrix (Figure 10c), B23/12 also presents manganese- and iron-rich spinels and delafossite (Table 7, Figure 11a). Iron-rich spinels commonly form grey cubic crystals embedded in a glassy matrix, corresponding to the general formula AB_2O_4, where A could be magnesium, zinc, or manganese; B may be aluminium, iron or chromium; and O is oxygen. Judging by its composition, the phase that formed in B23/12 is similar to the mineral franklinite, $(Zn, Mn^{2+})(Fe, Mn^{3+})_2O_4$. This is the same as seen in the slagged sherd Belovode 31b, recovered from Trench 3 and published previously (Radivojević 2013: 22).

Delafossite ($Cu^{1+}Fe^{3+}O_2$) is commonly recognised optically as straight grey lathes (Figure 11a). In nature, it is not usually found as a primary mineral, but rather near the base of the oxidised zone of copper deposits. The co-appearance of delafossite and cuprite is indicative of oxidising conditions during the smelt, at around the partial oxygen pressure required to reduce copper from cuprite (Bachmann 1982; Müller *et al.* 2004: 40). The debate on whether delafossite indicates melting or smelting conditions has bodies of evidence supporting both arguments but previously published metallurgical debris from Belovode strongly suggests

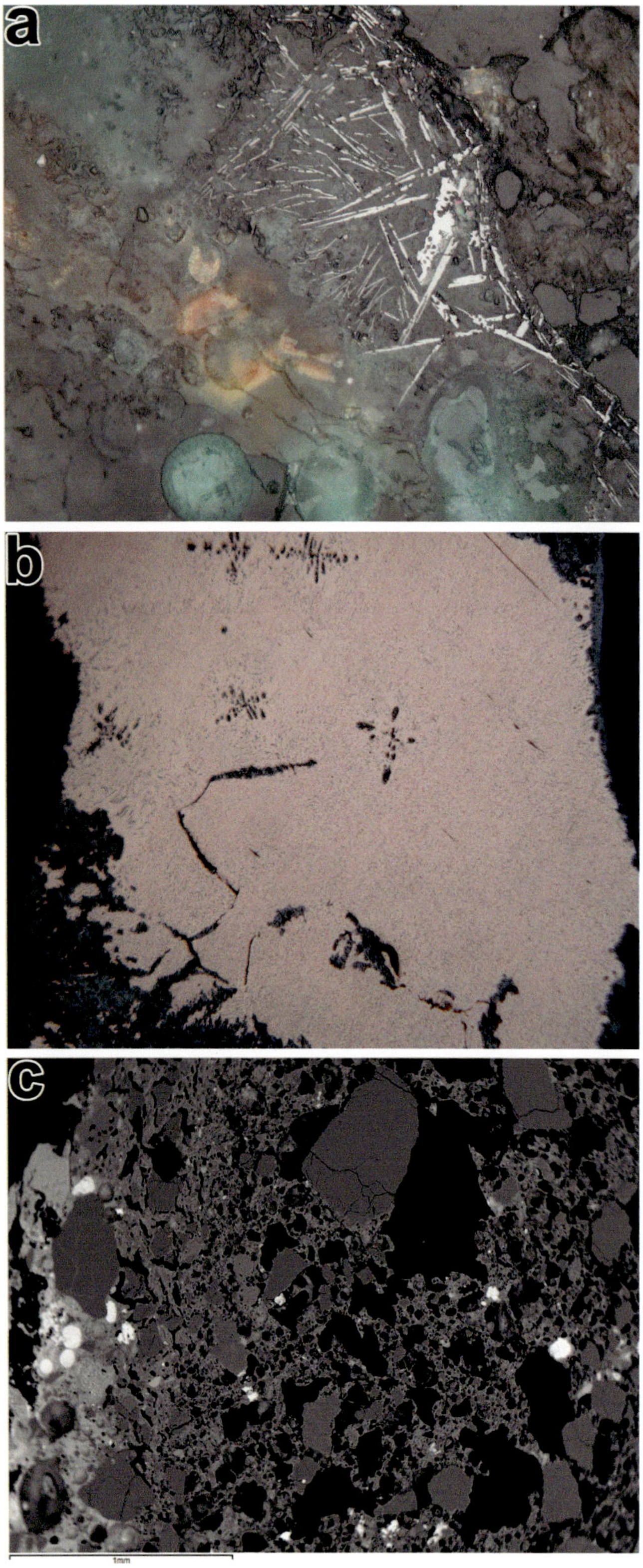

Figure 11. a) Photomicrograph of slagged sherd B23/12 taken under cross polarised light, 500x magnification, 0.34 mm width. Note light grey lathes of delafossite suspended in the glassy slag matrix; b) Photomicrograph of copper metal in slagged sherd B23/12 taken under cross polarised light, 200x magnification, 0.85 mm width. Note the copper-coper oxide eutectic texture and some dendritic cuprite; c) Backscattered electron image of slag section (left, with bright metal prills) in B47/12/1. Note slag forming from fused ceramic with decomposing quartz grains.

moderately reducing conditions indicative of copper smelting (Radivojević 2013; Radivojević and Rehren 2016). The presence of these newly formed phases further corroborates the argument that B23/12 had contact with slag forming in a process that involved manganese-rich copper ores, with relevant readings of zinc and nickel, as in previously published evidence (Radivojević 2013: 22, Table 6). Sample B47/12/1 does not offer the same indication and may have been either part of a different metallurgical process (but not a crucible) or, more likely, simply farther from the 'hot spot' in the lined hole-in-the-ground smelting installation in Feature 6 in Trench 18.

Finally, the copper metal phase in both samples demonstrates the nature of the metal produced (Figures 11b and 11c). In B23/12 we see a smaller size of dendritic structures in the metal phase with a pronounced copper–copper oxide eutectic, which could indicate a faster cooling rate (Figure 11b). The very presence of fully molten copper saturated with oxygen suggests exposure to temperatures of around 1070 °C or above for both slagged sherds. In addition, electron microprobe examination of pure copper metal phases in the slag matrix of B23/12 revealed a significant string of trace elements that might be helpful in discerning the type of copper ores used (Table 8). Notable readings indicate S, Ni, Mn, Sn, Bi and Pb and, less significantly, iron content in the copper metal phase, the significance of which to understanding of the slagging process and reducing conditions of the smelt has been discussed at length (Craddock 2001; Craddock and Meeks 1987; Tylecote *et al.* 1977) Concentrations of manganese of up to 150 ppm, nickel reaching 200 ppm and tin on average 70 ppm indicate a distinctive compositional pattern of the copper ore charge, the origins of which will be discussed in Chapter 41. The LA-ICP-MS analysis (Table 8a) are largely consistent with these readings, most notably the prominent Ni content.

In summary, the slagged sherds present further evidence for being used as the lining for a hole-in-the-ground smelting installation, reinforcing previous archaeometallurgical research on this site. Strengthening this argument is the firm contextual association with Feature 6, which includes burnt soil, charred surfaces and ashes, shaped in a bowl-like structure in the excavation layers (Figures 2 and 3). The conditions of the smelt were moderately reducing, as determined by the valency of newly formed copper and iron phases in the glassy matrices of slagged sherds:

Table 7. SEM EDS compositional data for MnFe spinels and delafossite in slagged sherd B23/12. All values are averages of two to nine analyses of each sample / phase and normalised to 100%.

	MgO	Al_2O_3	SiO_2	P_2O_5	SO_3	K_2O	CaO	TiO_2	MnO	FeO	CoO	NiO	CuO	ZnO
	wt%	wt%	wt%	wt%	wt%	wt%	wt%	wt%	wt%	wt%	wt%	wt%	wt%	wt%
B23/12 MnFe spinel	2.9	3.7	1.0	0.0	0.0	0.1	0.8	0.0	28.3	53.1	3.8	0.4	3.8	2.5
stdev s	*0.0*	*1.1*	*0.5*	*0.0*	*0.0*	*0.2*	*0.4*	0.0	*5.3*	*7.3*	*0.1*	*0.5*	*1.7*	*0.3*
B23/12 delafossite	0.1	1.3	4.6	0.7	0.0	0.1	0.7	2.7	4.9	32.0	0.0	0.0	52.6	0.0
stdev s	*0.3*	*0.6*	*5.9*	*1.1*	*0.0*	0.3	0.7	0.4	2.2	4.3	0.0	0.0	4.1	0.0

Table 8. EPMA compositional data of different metal prills in the slag matrix of B23/12 slagged sherd (selected significant trace element values), given in wt%. Values above c. 0.01 wt% (100 ppm) are considered reliable based on CRM measurements; values below this are indicative only. All data are corrected for values obtained from the reference material, using a procedure reported in the methodology section. Values sought but not found at levels above c. 0.01 wt% were indicated as not detected (n.d.).

	S	Mn	Fe	Co	Ni	As	Ag	Sn	Te	Pb	Bi	Analytical total
	wt%	wt%	wt%	wt%	wt%	wt%	wt%	wt%	wt%	wt%	wt%	
B23-12	0.009	0.015	0.002	0.001	0.008	n.d.	n.d.	0.033	n.d.	0.007	0.032	100.80
B23-12	0.007	n.d.	0.004	n.d.	0.010	0.084	n.d.	n.d.	n.d.	n.d.	n.d.	100.94
B23-12	0.014	n.d.	0.002	0.003	0.010	n.d.	n.d.	0.011	0.012	n.d.	n.d.	100.71
B23-12	0.000	n.d.	n.d.	n.d.	0.001	n.d.	n.d.	n.d.	n.d.	0.038	n.d.	100.97
B23-12	0.006	n.d.	0.004	n.d.	0.021	n.d.	n.d.	0.004	0.014	n.d.	n.d.	100.18
B23-12	0.015	n.d.	n.d.	0.001	0.001	n.d.	n.d.	0.001	0.024	n.d.	n.d.	100.74
B23-12	0.003	0.007	n.d.	0.003	0.012	n.d.	n.d.	n.d.	n.d.	n.d.	0.008	100.32
B23-12	0.005	0.007	n.d.	0.006	0.000	n.d.	n.d.	n.d.	n.d.	0.004	n.d.	100.59
B23-12	0.007	0.013	0.001	n.d.	0.013	0.031	0.004	0.011	n.d.	n.d.	0.027	100.44
B23-12	0.001	0.004	0.002	n.d.	0.007	n.d.	n.d.	0.010	n.d.	0.022	n.d.	99.44

Table 8a. LA-ICP-MS analysis of copper metal phases in production debris and artefacts. Values sought but not found above the indicated detection limit were treated as not detected (n.d.).

	Cu	Co	Ni	Zn	As	Ag	Sb	Te	Pb	Bi
	%	*µg/g*	*µg/g*	*µg/g*	*µg/g*	*µg/g*	*µg/g*	*µg/g*	*µg/g*	*µg/g*
detection limit		*1*	*20*	*5*	*37*	*4*	*5*	*2*	*2*	*1*
B23/12	100.0	10	67	n.d.	n.d.	6	n.d.	n.d.	3	n.d.
B24/12	100.0	16	38	69	n.d.	7	n.d.	n.d.	n.d.	n.d.
Bf21/12	100.0	14	89	15	n.d.	12	n.d.	n.d.	n.d.	n.d.
B47/12	100.0	15	97	12	n.d.	12	n.d.	n.d.	n.d.	n.d.
B71/12	100.0	n.d.	84	23	n.d.	7	n.d.	n.d.	n.d.	n.d.

dross including cuprite (Cu_2O), tenorite (CuO) and delafossite ($CuFeO_2$). The formation of spinels with a strong presence of manganese, and manganese, nickel and tin in trace element analysis of the copper metal all provide further arguments for primary extraction of copper ores, as well as being indicative of the composition of gangue minerals in the smelted ores.

Free slag pieces

Two free slag pieces (B24/12/2 and B47/12/2) were discovered associated with the two slagged sherds described above, B23/12 and B47/12/1 respectively (Figures 12 and 13). Both are therefore part of the production debris assemblage related to Feature 6. The

Figure 12. Free slag sample B24/12/2. Note green staining on top of (red) copper metal phase.

Figure 13. Free slag sample B47/12/2. Note green staining and porous outer surface.

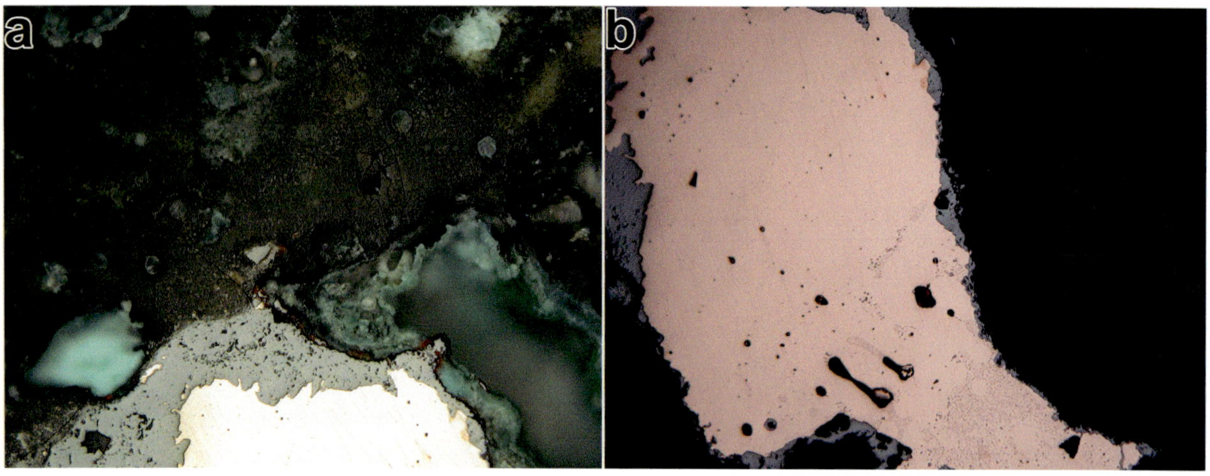

Figure 14. a) Photomicrographs of slag B24/12/2 taken under cross polarised light, magnification 100x, width 1.6mm. Note bright yellow copper metal phase (bottom) surrounded by copper oxide (grey), corrosion products (green), and glassy slag matrix (dark grey) with small grey particles dispersed in it (spinels); b) Photomicrographs of bright metal phase in slag B24/12/2 taken under plain polarised light, magnification 100x, width 1.6mm. Note the well-developed copper–copper oxide eutectic.

Table 9. SEM-EDS compositional data of slag matrices in B24/12 and B47/12/2, both as spot analysis (relatively free of inclusions) and as area analysis (bulk slag analysis in areas of c. 100 x 100 microns). All values are averages of two to twenty analyses of each sample / phase and corrected with factors based on CRM analysis; the uncorrected data is reported in the Appendix.

	Na_2O	MgO	Al_2O_3	SiO_2	P_2O_5	SO_3	K_2O	CaO	TiO_2	MnO	FeO	CoO	CuO
	wt%	wt%	wt%	wt%	wt%	wt%	wt%	wt%	wt%	wt%	wt%	wt%	wt%
B24/12/2 spot	0.0	0.8	5.2	38.6	7.8	0.0	0.0	5.6	0.0	3.4	11.9	0.0	0.0
stdev s	*0.0*	*0.1*	*0.7*	*1.9*	*0.0*	*0.0*	*0.0*	*1.5*	*0.0*	*1.3*	*2.0*	*0.0*	*2.3*
B24/12/2 area	0.0	1.3	7.2	33.8	14.3	0.0	0.2	8.1	0.1	3.3	10.7	0.4	0.0
stdev s	*0.0*	*0.4*	*1.3*	*5.3*	*2.4*	*0.0*	*0.4*	*1.8*	*0.2*	*2.3*	*3.9*	*0.5*	*3.0*
B47/12/2 spot	0.2	2.4	13.7	49.3	0.1	0.4	2.6	9.4	0.7	0.1	4.7	0.0	0.0
stdev s	*0.5*	*0.4*	*4.0*	*3.6*	*0.3*	*1.1*	*1.6*	*11.9*	*0.5*	*0.2*	*1.9*	*0.0*	*6.5*
B47/12/2 area	1.2	1.4	12.5	59.5	0.5	0.0	2.9	1.9	0.9	0.0	4.7	0.0	0.0
stdev s	*1.1*	*0.1*	*0.7*	*3.9*	*0.6*	*0.0*	*0.2*	*0.7*	*0.5*	*0.0*	*0.8*	*0.0*	*5.4*

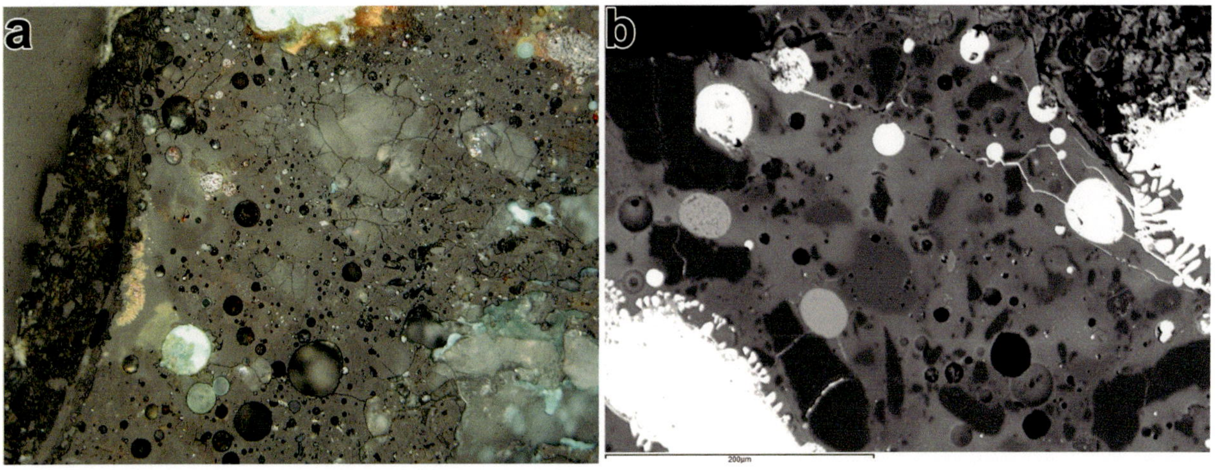

Figure 15. a) Photomicrograph of slag matrix in slagged sherd B47/12/1, taken under cross polarised light, 100x magnification, 1.6 mm width. Note green corrosion developing in place where rounded metal prills of copper formed; b) Backscattered electron image of glassy slag section in B47/12/1. Note bright metal phase and decomposing dark grey quartz grains.

slag specimens are vitrified, strongly magnetic and green-stained droplets, not exceeding 1 cm in length (Figures 12 and 13). They solidified from a highly viscous melt into amorphous grey samples with light green stains, gas holes, and corroded metal in places. Their combined weight is less than 1 gram (B24/12/2 is 0.46 g and B47/12/2 0.36 g), which fits the current knowledge of the appearance and volume of Belovode free slag pieces as published previously (Radivojević and Rehren 2016; Radivojević *et al.* 2010a).

Microstructural and compositional analyses revealed heterogeneous slag matrices and their newly developed phases (Figures 14 and 15). These samples consist primarily of copper 'dross' and copper metal phases/prills and manganese- and iron-rich spinels (only B24/12/2) suspended and unevenly distributed throughout a mostly glassy slag matrix (Figure 14a). The bulk slag glass composition is relatively even in both area and spot analysis mode (Table 9), with one difference being elevated lime averages in the spot analysis for B47/12/2. This slag is consistent overall with the slag in the slagged sherds, with a ratio of silica to alumina of 4–5:1. The major difference between these two slag pieces is the significantly enriched matrix with fuel ash components (particularly P_2O_5) and ore charge (manganese, copper and iron oxides). Both samples present high readings of fuel ash (magnesia, potash and lime) in comparison to the 'cold' unreacted ceramic bodies of slagged sherds (Table 5) and have copper contamination rising to 27 wt%; the evidence is clearly indicative of these samples being molten pieces of ceramic mixed with ore gangue material fluxed by fuel ash.

Both free slag samples contain areas rich in copper 'dross': copper oxide phases varying from tenorite and cuprite to corrosion products (Figures 13 and 14). Distinguished by its typical bright red / orange internal reflection, cuprite is mostly present as convoluted

Table 10. SEM-EDS compositional data of MnFe spinels slag B24/12/2. All values are averages of four analyses of this phase and corrected with factors based on CRM analysis; the uncorrected data is reported in the Appendix.

	MgO	Al_2O_3	SiO_2	P_2O_5	SO_3	CaO	TiO_2	MnO	FeO	CoO	NiO	CuO	ZnO
	wt%	wt%	wt%	wt%	wt%	wt%	wt%	wt%	wt%	wt%	wt%	wt%	wt%
B24/12/2 MnFe spinel	6.4	6.3	1.4	1.1	0.0	0.0	0.8	0.3	20.6	56.6	2.6	0.0	2.9
stdev s	*2.5*	*2.1*	*1.3*	*1.6*	*0.0*	*0.7*	*0.3*	*2.9*	*3.8*	*0.9*	*0.0*	*1.6*	*0.6*

Table 11. EPMA compositional data of different metal prills in slag matrices of B24/12/2 and B47/12/2 free slag samples (selected significant trace element values), given in wt%. Values above c. 0.01 wt% (100 ppm) are considered reliable based on CRM measurements; values below this are indicative only. All data are corrected for values obtained from the reference material, using a procedure reported in the methodology section. Values sought but not found at levels above c. 0.01 wt% were indicated as not detected (n.d.).

	S	Mn	Fe	Co	Ni	As	Ag	Sn	Te	Pb	Bi	Analytical total
	wt%	wt%	wt%	wt%	wt%	wt%	wt%	wt%	wt%	wt%	wt%	
B24/12/2	0.013	0.003	0.003	0.006	0.008	0.062	n.d.	n.d.	0.024	n.d.	n.d.	100.31
B24/12/2	0.020	n.d.	0.002	0.005	0.006	n.d.	0.007	n.d.	n.d.	0.014	0.017	100.23
B24/12/2	0.012	n.d.	0.015	0.009	n.d.	n.d.	0.004	0.023	n.d.	0.027	0.027	99.01
B24/12/2	0.011	0.001	0.001	n.d.	0.001	n.d.	0.007	n.d.	n.d.	n.d.	n.d.	98.88
B24/12/2	0.005	0.001	0.002	0.005	0.011	0.012	n.d.	n.d.	n.d.	0.004	n.d.	100.22
B24/12/2	0.005	n.d.	n.d.	0.009	0.009	n.d.	0.007	n.d.	0.005	n.d.	n.d.	98.64
B24/12/2	0.013	0.007	n.d.	0.003	n.d.	0.009	n.d.	n.d.	0.010	0.042	n.d.	99.29
B24/12/2	0.005	0.001	0.008	0.004	0.004	n.d.	n.d.	n.d.	0.008	n.d.	n.d.	99.93
B24/12/2	0.009	0.005	0.005	0.002	0.014	n.d.	0.004	0.014	n.d.	n.d.	0.004	100.76
B24/12/2	0.008	0.009	0.002	n.d.	0.004	n.d.	n.d.	0.024	0.031	n.d.	n.d.	100.63
B47/12/2	0.008	n.d.	0.062	0.009	0.015	n.d.	0.007	n.d.	0.011	n.d.	0.003	100.24
B47/12/2	0.009	n.d.	0.042	0.001	n.d.	n.d.	0.008	0.008	n.d.	0.035	n.d.	99.80
B47/12/2	0.013	0.001	0.018	0.004	0.004	n.d.	0.008	0.006	0.034	n.d.	0.006	99.79
B47/12/2	0.004	0.014	0.001	n.d.	0.007	n.d.	n.d.	n.d.	0.016	n.d.	0.019	100.04
B47/12/2	0.013	0.003	0.022	n.d.	0.008	n.d.	0.005	0.002	n.d.	n.d.	n.d.	100.32
B47/12/2	0.006	n.d.	0.012	0.001	0.007	n.d.	0.003	0.006	n.d.	0.025	n.d.	100.20
B47/12/2	0.018	n.d.	0.012	n.d.	0.018	n.d.	n.d.	n.d.	0.009	n.d.	n.d.	99.17
B47/12/2	n.d.	n.d.	0.004	n.d.	n.d.	n.d.	n.d.	n.d.	n.d.	n.d.	0.028	100.26

agglomerations or dendritic crystals; these appear yellow and red under cross polarised light (Figure 15a). Manganese- and iron-rich spinels (Table 10) appear only in sample B24/12/2 and are compositionally consistent with the same phase observed in slagged sherd B23/12/2 (Table 7), with relevant readings of copper, cobalt and zinc oxides. Finally, the copper-copper oxide eutectic (Figures 14b and 15b) in both free slag samples confirms a high temperature process in excess of 1070 °C. Electron microprobe examination of these pure copper metal phases in B24/12/2 and B47/12/2 (Table 11) reveals a fairly consistent intake of sulfur, nickel, manganese, cobalt, tin and iron. The presence of iron in copper metal, as mentioned above, indicates sufficiently reducing conditions to facilitate this process (cf. Craddock and Meeks 1987). These readings are also consistent with the trace element pattern seen in associated slagged sherds B23/12 and B47/12/1, overall indicating a specific signature of the ore charge, with the consistent selection of manganese-rich copper ores that contain nickel and cobalt, some remnants of primary copper minerals (iron, sulfur) and potentially associated with polymetallic deposits that contain arsenic, tin, lead and bismuth.

The closest geological formations that fit this description are located in eastern Serbia, in the wider area of the Bor mining region, with dominant types of copper ore including porphyry copper, and massive sulfide deposits with significant Pb, Zn, and Au mineralisation (Janković 1967; Jelenković 1999; Monthel *et al.* 2002; Neubauer and Heinrich 2003). The porphyry deposits at Bor also carry enargite (copper-arsenic-sulfide), a common mineral for this locality (Sillitoe 1983).

Making and working: Copper mineral and metal artefacts

The total assemblage of artefacts from Belovode includes 12 mineral ornaments, two metal droplets and, for the first time on this site, a fragment of a finished artefact. They are consistent with copper-based metallurgy and, very importantly, metal smelting, melting and working activities, thus completing the evidence for the full metallurgical *chaîne opératoire* at Belovode (*sensu* Ottaway 2001).

Copper mineral artefacts (ornaments)

A total of 12 mineral ornaments were found, of which all but one (C-B10/13, green stone ring bead) are made of malachite of various levels of purity (Figure 16). Typologically, the malachite beads can be roughly divided into three distinctive categories: circular or cylindrical (B52/12, B171/12), flat disc (Bf77/12, Bf30/13, C_B5/13, Bf111/12, Bf111/13, B361/13) and ring beads (B42/13, B95/13, C_B7/13) (cf. Wright *et al.* 2008; Wright and Garrard 2003). As such they are consistent with previously discovered and studied beads from the site of Belovode (Radivojević 2007, 2012). Visually, all beads exhibit a thick, light green corrosion layer, with occasionally macroscopically visible oolithic formations on the surface, as in Bf111/12, Bf111/13 and B95/13 (Figures 16f, 16g, and 16j). Only a few complete beads survived, such as C_B7/13 (Figure 16k). Although we have not conducted compositional analyses of these artefacts, macroscopic observation suggests that at least three out of 11 (Bf111/12, Bf111/13 and Bf30/13) might have relevant manganese oxide readings (dark phases), which is not uncommon, given the abundant use of black and green copper minerals and ores in Belovode.

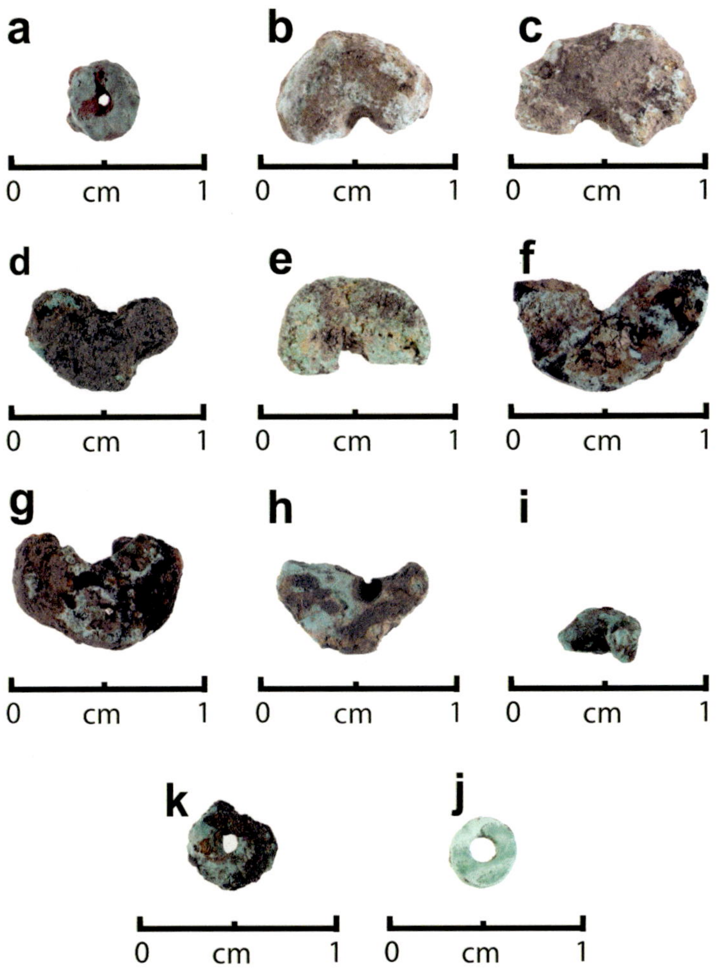

Figure 16. Malachite beads. Circular a) B52/12; b) B171/12; Flat disc c) Bf77/12; d) Bf30/13; e) C_B5/13; f) Bf111/12; g) Bf111/13; h) B361/13; Ring beads i) B42/13; j) B95/13; k) C_B7/13.

Copper metal droplets

The droplet category of metallurgical debris can include semi-molten and fully molten pieces of ore/metal, and technically covers a wider range of activities that could have produced them. Droplets can be the result of anything from an attempt to smelt metal to accidentally lost debris from smelting, melting or casting. We treat them here as a single category as they are unified by one aspect: they have not been worked and hence occur in nebulous forms with many noticeable porosity holes; in the field they occur covered in green corrosion products, just as any other copper metal artefact (Figure 17). Five copper droplets from excavation campaigns of Trench18/T18ext were found in two distinctive clusters: one directly associated with the assumed metallurgical pit, Feature 6, and the other with another distinctive and well defined feature, Feature 21, which is interpreted as a sealed refuse pit (see Chapter 10). The former includes Bf21/12, B29/12, B47/12/3 (Figure 17), of which Bf21/12 and B47/12/3 were recovered from the same (EDM) spot (see Tables 1 and 2), while the latter yielded Bf43/13 and Bf56/13, together with pieces of malachite and a piece of obsidian (Chapters 10 and 19).

In the first group of droplets (Bf21/12, B29/12 and B47/12/3), B29/12 is morphologically and structurally different (Figures 17c-d) from the other two, which both have a thoroughly developed green patina on their surface (Figures 17a-b and 17e-f). B29/12 contains predominantly dross and some remnants of what seems to have once been a fully developed metallic phase (Figure 18, Table 12). Optically, this dross is also confirmed with bright

Figure 17. Copper metal droplets associated with Feature 6. a-b) Bf21/12; c-d) B29/12; e-f) B47/12/3. All are classified as droplets following the criteria outlined above.

red internal reflections under cross polarised light, characteristic of cuprite (Cu_2O) (Figure 18b). B29/12 could thus be interpreted as resulting from either a melting event (as indicated by the lack of iron reading as seen in Table 12), or a smelting event of copper ore of high purity, which was followed by full oxidation leading to the dross formation. Experiments with copper smelting and refining have yielded evidence of this form of droplet, attached to the crucible wall due to insufficiently high temperatures (i.e. only just above the melting point), or the presence of impurities (S. Timberlake, personal communication). Artefacts of this shape are also known from previous analyses of metallurgical debris from Belovode (Radivojević 2013; Radivojević and Rehren 2016) (see Chapter 5).

Bf21/12 and B47/12/3 both present fully molten copper metal bodies, with no traces of working. Their

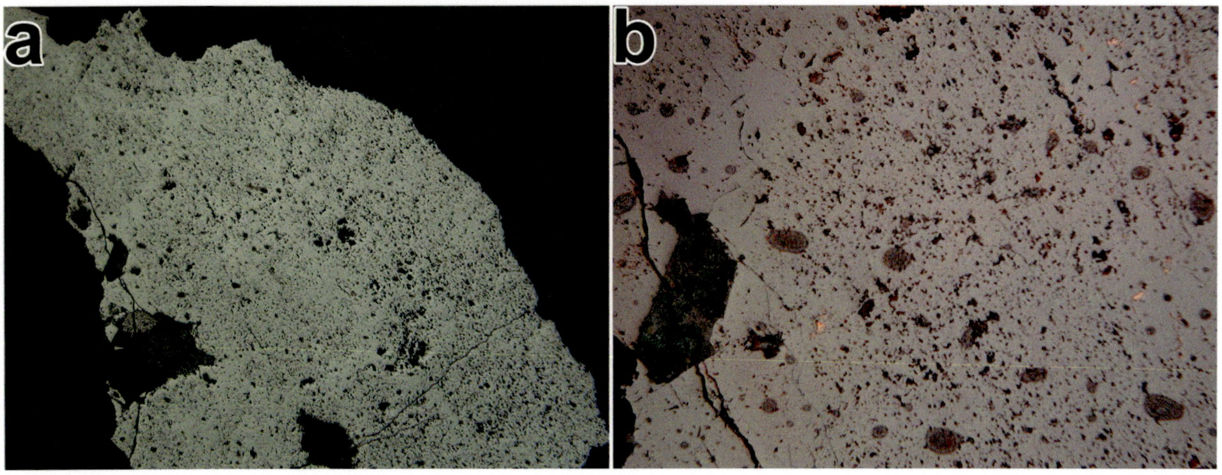

Figure 18. a) Photomicrograph of fully oxidised copper metal droplet B29/12 taken under plain polarised light, 25x magnification, 6.4 mm width. Note grey phase of predominantly copper oxide ('dross'); b) Photomicrograph of copper metal droplet B29/12 taken under cross polarised light, 100x magnification, 1.6 mm width. Note small bright specks of copper metal (pale pink) in the dross.

Table 12. SEM-EDS compositional data of copper rich phases in Belovode B29/12, B47/12/3 and Bf43/13. All values are averages in at% of two to thirteen analyses of each sample / phase.

	O	S	Fe	Cu
	at%	at%	at%	at%
B29/12 metal phase	2.5	0.0	0.0	97.5
B29/12 dross	32.3	0.0	0.0	67.7
B47/12/3 metal phase	3.9	0.0	0.0	96.1
B47/12/3 dross	31.8	0.0	0.0	68.2
Bf43/13 metal phase	2.2	0.0	0.0	97.8
Bf43/13 dross	30.6	0.0	0.0	69.4

micrographs (Figure 19) both exhibit a pale yellow/ orange copper (metal) body, with green corrosion products developing on its edges. The optically bright phase shows a residual as-cast structure, preserved in the microstructure of the copper–copper oxide eutectic with well-developed α grains of copper. These alpha grains are, in both cases, characterised by their highly reflective bright colour and dendritic shape in the as-cast eutectic ($Cu+Cu_2O$) structure with grey oxide particles within a bright matrix (example in Figure 19a). In Figures 19a-d we do not see any deformation, but rather the opposite; the α grains are present, unaffected by any mechanical action and preserving the fully developed dendrites. The porosity and gas holes in Figures 19a-d are consistent with the oxygen-rich nature of the metal, which could have been a result of discarding these pieces of molten metal in ambient air while still hot. Bf21/12 and B47/12/3 are therefore candidates for being interpretation as casting debris, and not only because of their association with other metallurgical finds. This assumption is, in the case of Bf21/12, further strengthened by the analyses indicating a very low presence of iron (15 ppm on average; Table 13), which for freshly-smelted metal would have been higher, as seen in the earlier analyses presented above (cf. Craddock and Meeks 1987). Their association with the smelting evidence of B47/12/1 and B47/12/2 is therefore interesting and either indicates the widely varying redox conditions typical of early metallurgical operations, or they represent debris from different metallurgical activities that were taking place on the same spot (smelting, melting and casting), in the pit, Feature 6.

The two metal droplets, Bf43/13 and Bf56/13, mentioned above, were associated, within a sealed refuse pit (Feature 21) and are securely dated to *c.* 49th century BC, which makes them the earliest metal artefacts discovered in the excavated section of Trench 18/T18ext. Similar to the case of copper metal droplet Belovode M14 (see Chapter 5) (Radivojević 2013), these were initially assumed to be copper minerals, due to their blocky shape and a thick, light-green patina covering their surfaces (Figure 20). Both samples are about 1 cm long and exhibit a rich dark-red phase in the cross section (see Bf43/13). The matrices of both samples share general characteristics: predominantly grey copper-rich dross with preserved speckles of bright yellow copper metal and porous corrosion products on the edges (Figure 21). The porosity and cracks throughout the polished sections in both cases suggest an intensive reaction with the atmosphere during cooling, which could indicate, as was likely the case with B29/12, that they were discarded while the metal was still hot. They could also have originated from a smelting event of rather pure copper ores, or refining (melting).

Compositionally, these two samples differ. While Bf43/13 consists only of copper metal and dross (copper oxides)

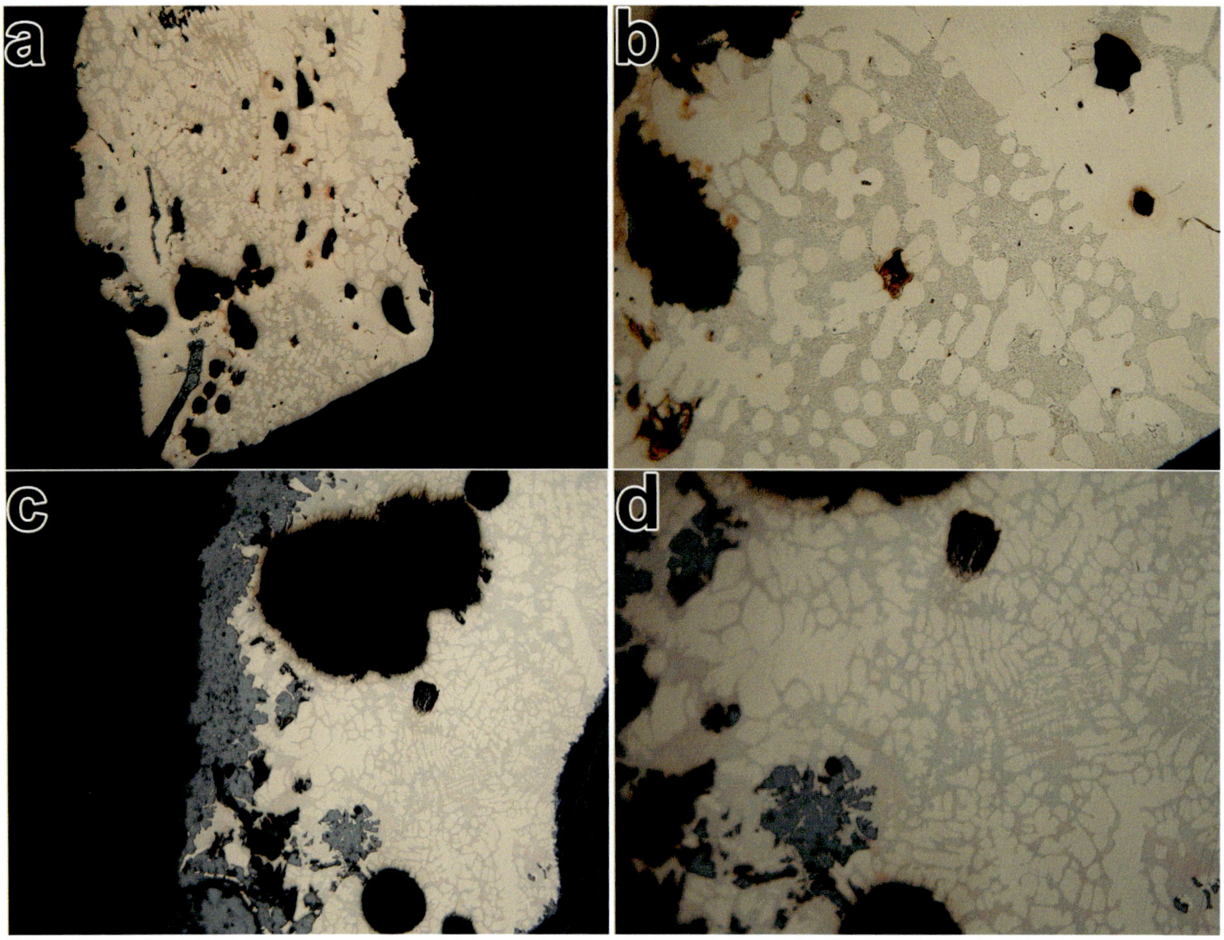

Figure 19. a) Photomicrograph of Bf21/12 taken under plain polarised light, magnification 25x, 6.4 mm width, etched with ammonia hydrogen peroxide. Note bright yellow alpha grains of copper surrounded by copper – copper oxide eutectic; b) Photomicrograph of Bf21/12 taken under plain polarised light, magnification 100x, 1.6 mm width, etched with ammonia hydrogen peroxide; c) Photomicrograph of B47/12/3 taken under plain polarised light, magnification 100x, 1.6 mm width; d) Photomicrograph of B47/12/3 taken under plain polarised light, magnification 200x, 0.85 mm width.

Figure 20. Copper metal droplets associated with Feature 21. a) Bf43/13; b) Bf56/13.

Table 13. EPMA compositional data of copper metal phase in Bf21/12 (selected significant trace element values), given in wt%. Values above c. 0.01 wt% (100 ppm) are considered reliable based on CRM measurements; values below this are indicative only. All data are corrected for values obtained from the reference material, using a procedure reported in the methodology section. Values sought but not found at levels above c. 0.01 wt% were indicated as not detected (n.d.).

	S	Mn	Fe	Co	Ni	As	Ag	Sn	Te	Pb	Bi	Analytical total
	wt%	wt%	wt%	wt%	wt%	wt%	wt%	wt%	wt%	wt%	wt%	
Bf21/12	0.005	0.006	0.003	0.003	0.012	n.d.	n.d.	n.d.	n.d.	n.d.	0.030	99.75
Bf21/12	0.005	n.d.	0.002	n.d.	0.016	n.d.	n.d.	0.001	n.d.	0.014	0.026	99.82
Bf21/12	0.004	0.005	0.002	0.002	0.013	0.129	n.d.	n.d.	0.005	0.021	n.d.	100.25
Bf21/12	0.003	0.001	0.002	0.003	0.025	0.095	n.d.	0.015	0.016	n.d.	n.d.	99.56
Bf21/12	0.008	0.002	n.d.	0.009	0.016	n.d.	n.d.	n.d.	0.015	n.d.	0.001	99.28
Bf21/12	0.009	0.008	0.002	0.001	0.017	n.d.	n.d.	0.053	n.d.	0.005	n.d.	99.08
Bf21/12	n.d.	0.012	n.d.	0.001	0.003	n.d.	n.d.	n.d.	n.d.	n.d.	n.d.	99.38
Bf21/12	0.001	n.d.	0.001	0.002	0.018	n.d.	0.003	0.017	n.d.	n.d.	n.d.	98.77
Bf21/12	n.d.	0.007	n.d.	0.005	0.006	n.d.	n.d.	0.014	0.008	n.d.	n.d.	99.39
Bf21/12	0.004	0.001	0.005	0.006	0.012	0.017	n.d.	n.d.	n.d.	0.022	0.038	99.36
Bf21/12	0.011	n.d.	n.d.	n.d.	0.011	n.d.	0.001	n.d.	n.d.	n.d.	n.d.	98.18

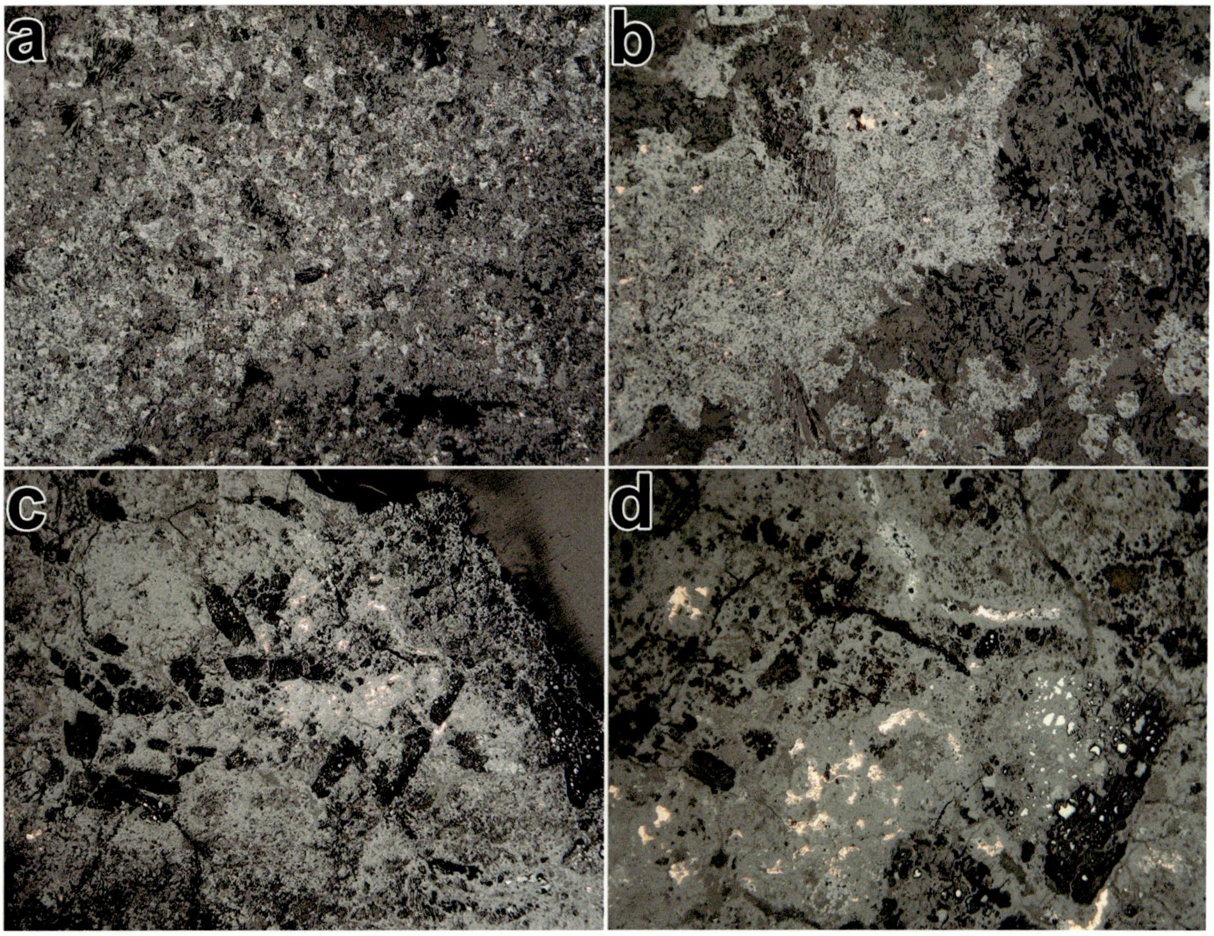

Figure 21. a) Photomicrograph of almost fully oxidised copper metal droplet Bf43/13 taken under plain polarised light, magnification 25x, 6.4 mm width. Note bright yellow speckles of copper metal in grey copper dross; b) Photomicrograph of copper metal droplet Bf43/13 taken under plain polarised light, magnification 200x, 0.85 mm width. Note bright yellow prills of copper metal in grey copper dross; c) Photomicrograph of the almost fully oxidised copper metal droplet Bf56/13 taken under plain polarised light, magnification 25x, 6.4 mm width. Note bright yellow speckles of copper metal in grey copper dross; d) Photomicrograph of copper metal droplet Bf56/13 taken under plain polarised light, magnification 200x, 0.85 mm width. Note bright yellow phase of copper metal in grey copper dross, and a cluster of white pale angular and partially rounded phases of copper sulfide.

Table 14. SEM-EDS compositional data of various copper-rich phases in Bf56/13. All values are averages in at% of one to seven analyses of each sample / phase.

	O	S	Fe	Cu
	at%	at%	at%	at%
Cu metal phase	8.1	0.0	2.7	89.2
Cu2O	33.3	0.0	2.5	64.2
CuO, tenorite	50.0	0.0	1.7	48.3
FeS2, pyrite	0.0	66.6	32.9	0.5
Cu2S, chalcocite	0.0	35.7	3.5	60.8
CuFeS2, chalcopyrite	0.0	52.2	16.5	31.4

(Table 12), Bf56/13 contains various sulfur-rich phases alongside molten copper metal (Figure 21d, Table 14). These diverse phases indicate that the copper ore was a mix of primary and secondary copper minerals (such as chalcopyrite, chalcocite). The dominant presence of chalcocite, for instance in rounded prills, rather than its natural angular form, and its embeddedness in copper dross favours of this interpretation. It is not clear if this sample was intended for any practical use; the presence of sulfur-rich inclusions would have rendered the copper metal brittle and unusable, so this artefact may have been discarded early in the process. It is notable is that similar sulfur-rich examples of early experimentation with smelting copper were identified at Belovode in previous studies (Radivojević 2013). They were then identified as demonstrating 'slagless' copper smelting, and this interpretation also can be applied to Bf43/13 and Bf56/13, as well as B29/12.

Copper metal fragment

Sample B71/12 is the only worked copper metal artefact ever found in the site of Belovode. Although it originates from the same spit (6) as the rest of the metal production debris, it is associated with Feature 3, which is a dwelling (Figure 2). The object itself is around 1 cm long and is identified as a fragment (Figure 22a). Metallographic examination (Figure 22b-d) shows a pale yellow/orange copper metal body, with green corrosion products developing on its edges. This bright metal phase shows as-cast structure preserved in the microstructure of the copper–copper oxide eutectic (bright with grey dots) with α grains of copper (bright). The elongation of the α grains of copper and the eutectic is due to mechanical deformation during working of the artefact. This elongation pattern persists all along the length of the sample (Figure 22c) and partially at right angles across the width, indicating locations where the heaviest force was applied. Based on the criteria given by Rostoker and Dvorak (1990: 16), the reduction of thickness could have been around 20%, suggesting at least several cycles of cold reduction working followed by annealing. The reduction in thickness indicated by the deformed grains and eutectic seems not to have caused intolerable brittleness of this object (Figure 22d).

The electron microprobe examination of this object (Table 15) shows broadly a similar trace element pattern for both this artefact and Bf21/12 (Table 13), but also includes other metal phases of production debris (Tables 8, 11). Bismuth, arsenic, tin and sulfur feature as predominant trace element (*c.* 100 ppm on average), followed by lead, nickel and iron, although in slightly lower quantities. This is probably because B71/12, as the finished artefact, must have gone through more cycles of re-melting than the droplets. The relevant readings of sulfur provide an indicator of the primary copper ore mix with the secondary minerals representing the choice of ore charge.

Discussion and conclusion

The metallurgical evidence from the site of Belovode presents all main elements of the *chaîne opératoire* of metal making, from ore selection, via experiments and successful smelts, to melting, re-melting and working an artefact. The highlight from the assemblage is the association of slagged sherds, slags and metal droplets within Feature 6, which may come close to reflecting how the Vinča culture smelting installations might have looked (Figures 2 and 3). The slagged sherds were not found lining the feature but the analysis indicates that they were exposed to high temperatures (reaching *c.* 1100 °C), and Feature 6 has the closest indication of such temperatures in association with metallurgical debris. Given the ephemeral nature of these early hole-in-the-ground installations, even in experimental settings (see Chapter 4), a burnt pile of soil and charcoal is the best evidence likely to be found from the onset of metallurgical practice at the site of Belovode, or elsewhere.

The sealed refuse pit, Feature 21, with two metal droplets (Bf43/13 and Bf56/13) and dated to the 49th century BC, is another important element for understanding metallurgical activities at Belovode, as it is closest to the earliest dated evidence for copper smelting at the site (Radivojević *et al.* 2010a). Radivojević *et al.* (2010a) made the most of the available absolute dating in their article by associating the pit in which the slags were found with the stratigraphically closest data point in an adjacent trench, hence the stated uncertainty. Our new dates (Chapter 37) are directly associated with metallurgical remains, and the fact that Feature 21 can be securely dated within the 49th century BC (see Table 2) is reassuring for the validity of the *c.* 5000 BC date previously proposed.

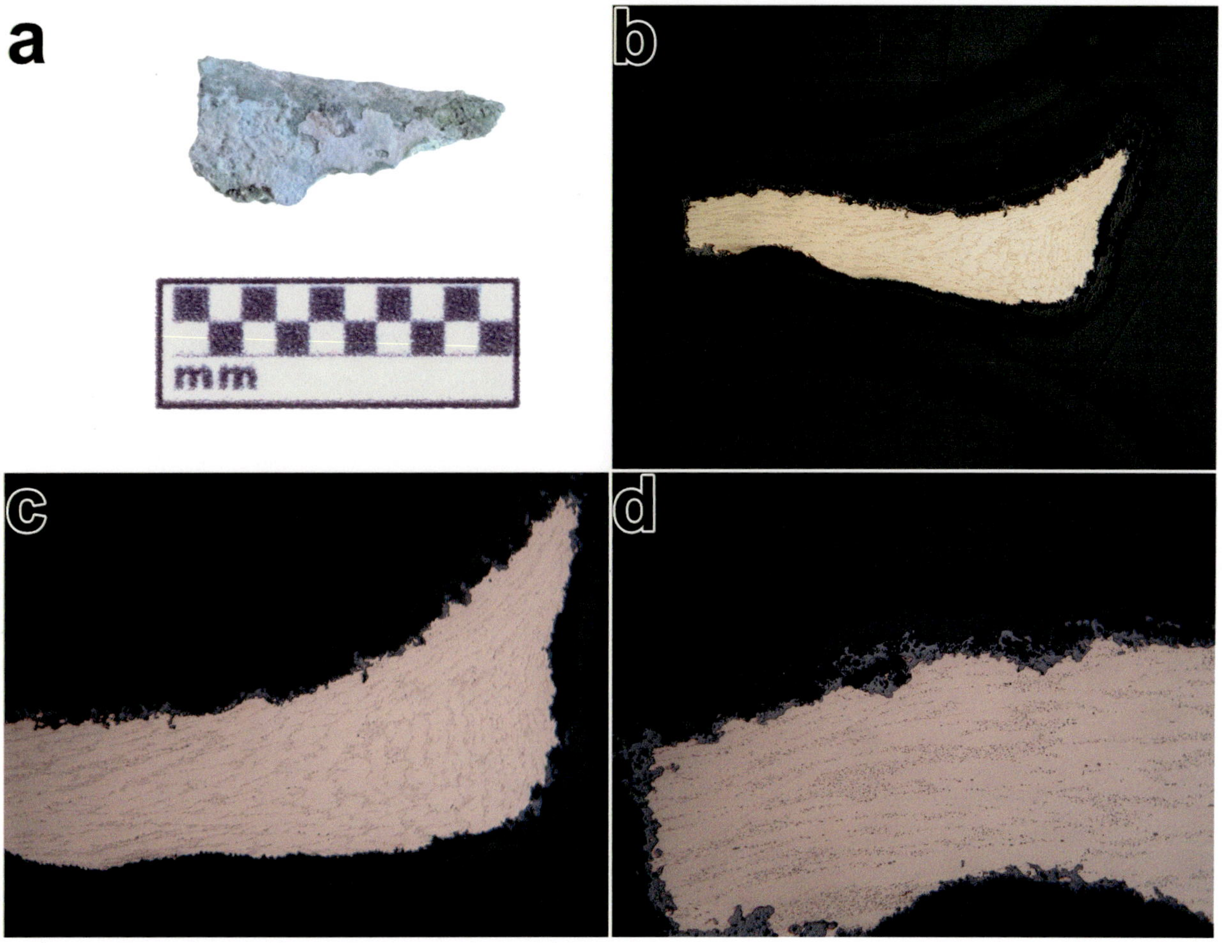

Figure 22. a) Fragment of a copper metal artefact, B71/12; b) Photomicrograph of B71/12 taken under plain polarised light, magnification 50x, 3.2 mm width, etched with ammonia hydrogen peroxide; c) Photomicrograph of B71/12 taken under plain polarised light, magnification 100x, 1.6 mm width, etched with ammonia hydrogen peroxide. Note elongated bright alpha grains of copper suspended in copper-copper oxide eutectic; d) Photomicrograph of B71/12 taken under plain polarised light, magnification 200x, 0.85 mm width, etched with ammonia hydrogen peroxide. Elongation of alpha copper grains and copper-copper oxide eutectic reveals mechanical deformation by working.

Table 15. EPMA compositional data of copper metal artefact B71/12 (selected significant trace element values), given in wt%. Values above c. 0.01 wt% (100 ppm) are considered reliable based on CRM measurements; values below this are indicative only. All data are corrected for values obtained from the reference material, using a procedure reported in the methodology section. Values sought but not found at levels above c. 0.01 wt% were indicated as not detected (n.d.).

	S	Mn	Fe	Co	Ni	As	Sn	Te	Au	Pb	Bi	Analyical total
	wt%	wt%	wt%	wt%	wt%	wt%	wt%	wt%	wt%	wt%	wt%	
B71/12	0.005	0.000	0.005	0.000	0.018	0.000	0.000	0.012	0.000	0.000	0.027	100.55
B71/12	0.012	0.000	0.000	0.003	0.001	0.000	0.000	0.019	0.000	0.000	0.010	100.43
B71/12	0.007	0.000	0.000	0.000	0.004	0.000	0.008	0.019	0.011	0.005	0.000	100.49
B71/12	0.008	0.012	0.010	0.003	0.000	0.000	0.000	0.000	0.000	0.018	0.026	100.70
B71/12	0.017	0.000	0.004	0.000	0.025	0.033	0.005	0.000	0.000	0.034	0.030	100.70
B71/12	0.002	0.000	0.000	0.002	0.000	0.075	0.015	0.000	0.000	0.000	0.000	100.58
B71/12	0.005	0.002	0.000	0.000	0.016	0.000	0.044	0.046	0.000	0.000	0.043	100.86
B71/12	0.025	0.004	0.000	0.006	0.005	0.000	0.000	0.000	0.020	0.000	0.000	100.63
B71/12	0.010	0.005	0.007	0.000	0.010	0.000	0.019	0.002	0.000	0.005	0.027	100.76
B71/12	0.000	0.000	0.002	0.002	0.004	0.005	0.000	0.000	0.009	0.000	0.000	98.84

Also reassuring is that the conclusions from previously conducted studies on Belovode metallurgy (Radivojević 2007, 2012, 2013, 2015; Radivojević and Rehren 2016; Radivojević *et al.* 2010a) are fully consistent with the microstructural and chemical examination of the assemblage as presented in Table 1. There are, however, variations regarding the composition of ores, most strongly indicated by the results of EPMA analysis of metals and metal phases in slags in Trench 18/18ext. This is visible in comparisons with data published by Radivojević and Rehren (2016: 223, Table 4) where sulfur appears either only sporadically in slags, or is not detected at all, while in B24/12/2 and B47/12/2 it has a consistent presence of up to 150 ppm (Table 11). A similar situation is shown for B23/12 (Table 8). A novel element not previously detected is tin (Sn), which is present in all EPMA-analysed samples (Tables 8, 11, 13, 15). It is therefore suggested that the batch of ore used in association with the excavated household in Trench 18/18ext was a 'standard' one, with black and green manganese-rich copper ores, which additionally contained primary copper minerals (hence the sulfur content), as well as some concentrations of tin, amongst other trace elements. Significantly, all but one of the samples examined with the electron microprobe come from the same horizon (1b), the exception being associated with Feature 6.

The known preference for black and green manganese-rich copper ores is demonstrated in the ternary plot of CaO/MnO/K_2O, together with previously published Belovode metal production data (Figure 23). There are two distinctive clusters: one (Ellipse 1) that associates lime readings with potash, and the other (Ellipse 2), characterised by the manganese-rich chemical signature in slag which comes from the black and green ores mentioned above. Ellipse 1 reflects the fuel contamination, with beech ash readings used to estimate what the fuel might have contributed to slag formation (data from Jackson and Smedley 2004: 39, Table 4). It is interesting to note that most readings for B47/12/1 and B47/12/2 fall into the fuel ash group, while free slag B24/12/2 shows the greatest relative contribution from manganese. A similar conclusion has been reached in previous studies, pointing to the depletion of ore signature in slagged sherds in comparison to free slag samples, due to a higher fuel ash and ceramic contribution to the slag (Radivojević 2012; Radivojević and Rehren 2016: 224, Fig. 11).

The ternary plot of components understood to represent typical pottery ($SiO_2/Al_2O_3/TiO_2$), fuel ash ($CaO/MgO/P_2O_5/K_2O$) and ore ($FeO/MnO/ZnO/NiO/CoO/As_2O_3/SnO_2/Sb_2O_3$) contamination in the glassy slag matrices (re-cast as Cu-free) in Figure 24 illustrates two points: the difference of slag composition from the ceramic composition, and the technological similarity of early metal making in Belovode, and across the Vinča culture. The data used cover all analyses of Vinča culture slag matrices conducted thus far (Radivojević 2007, 2012), and includes previous analyses of Belovode pottery ('cold' sections of slagged sherds) and here presented 'cold' sections of slagged sherds B23/12 and B47/12/1 (Table 5). Figure 24 shows that the new Belovode analyses are consistent with previously analysed Belovode samples, as well as with slag matrices from the sites of Vinča-Belo Brdo and Gornja Tuzla. The strong cluster of all glassy matrices in the silica+alumina+titania corner implies that they were predominantly formed by these acidic oxides. This in effect explains why the slags (and specifically the free slag samples) are highly viscous, despite the significant input of basic oxides such as lime, iron and copper oxides (Davenport *et al.* 2002: 63). It is also apparent that some readings in the newly analysed Belovode assemblage have much stronger intake from ceramic elements, something that has been observed already

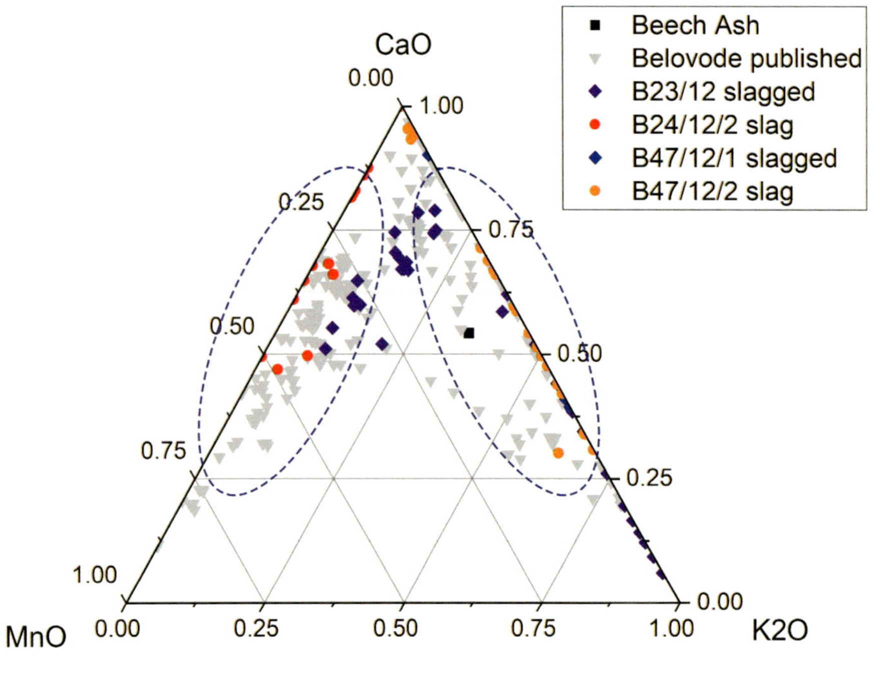

Figure 23. The K2O-CaO-MnO plot of slag matrices in Belovode 23/12, 24/12/2, 47/12/1, 47/12/2 and all previously conducted comparable Belovode data (Radivojević 2007, 2012). Beech ash data from Jackson and Smedley (2004: 39, Table 4) to simulate fuel ash.

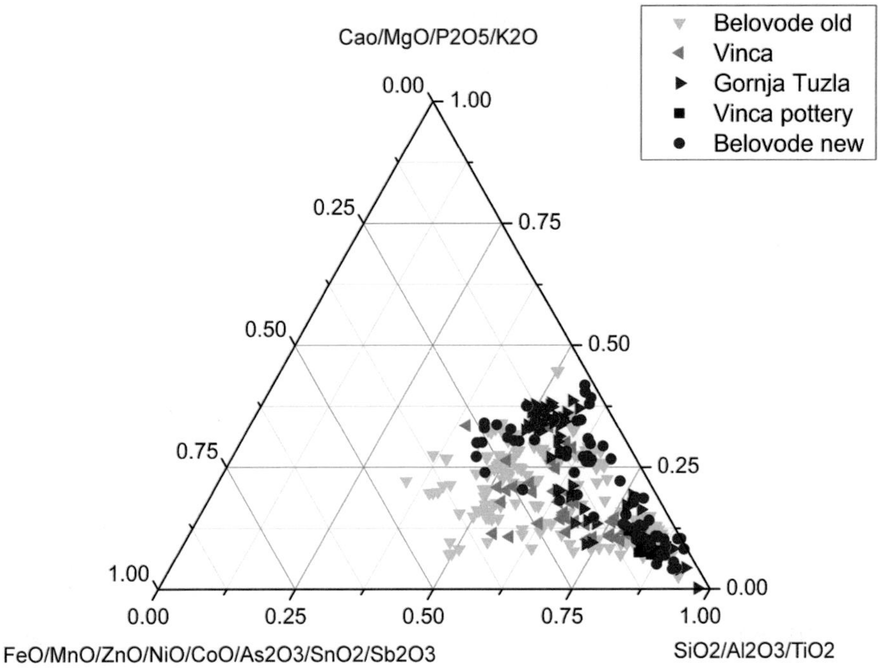

Figure 24. $SiO_2/Al_2O_3/TiO_2$-$CaO/MgO/P_2O_5/K_2O$-$FeO/MnO/ZnO/NiO/CoO/As_2O_3/SnO_2/Sb_2O_3$ ternary plot for all Belovode and Vinča culture metallurgical slag matrix data and ceramic values for previously studied Belovode pottery and 'cold' pottery studied here (data from Radivojević 2007, 2012).

in the analyses of slag on slagged sherds (see Tables 5 and 6). Equally, the stronger fuel ash/ore intake is predominantly related to free slag samples.

One of the most significant aspects revealed by this ternary plot is the similarity of technological parameters used in copper smelting across the studied Vinča culture sites. The slags are highly viscous, though different from ceramic composition, and products of the smelting process of this nature of persist throughout *c.* 600 years in the Vinča culture. The exceptions to the rule are, however, virtually slagless smelting events preserved in the structure of B29/12, Bf43/13 and Bf56/12. Interestingly, B29/12 and Bf43/13 did not contain iron or manganese (Table 12), which are crucial for the generation of slag (Bachmann 1982).

The newly analysed assemblage from Belovode fits well into the known picture of early copper smelting practices in this part of the world, using the same type of manganese-rich black and green copper ores and employing the same principles of smelting, with temperatures kept just above the threshold required for metal extraction (1100 °C). The latter is demonstrated by the evidence for molten copper metal giving a minimum temperature of around 1070 °C for the copper–copper oxide eutectic, and the high viscosity, heterogeneity and incompletely liquefied slag glassy matrices limiting the range towards higher temperatures, which has already emerged as the benchmark of the earliest copper smelting technology.

Appendix

Appendix avaliable online as part of Appendix B at https://doi.org/10.32028/9781803270425/AppendixB_Ch11

The bibliographic reference for this chapter is:

Radivojević, M. and Rehren, Th. 2021. Belovode: technology of metal production, in Radivojević, M., Roberts, B. W., Marić, M., Kuzmanović Cvetković, J., and Rehren, Th. (eds) *The Rise of Metallurgy in Eurasia*: 123–151. Oxford: Archaeopress.

Chapter 12

Pottery from Trench 18 at Belovode

Neda Mirković-Marić, Marija Savić and Milica Rajičić

Introduction

The place of Belovode within the Vinča culture, and especially within its metallurgical development, has been widely discussed (Radivojević *et al.* 2010a; Šljivar *et al.* 2006).

The 2012 and 2013 excavations of Trench 18 produced a total of *c.* 50,000 ceramic fragments encompassing *c.* 35,500 non-diagnostic and *c.* 14,500 stylistically and typologically indicative ceramic fragments (e.g. rims, handles, bottoms, and ornamented belly fragments). The analysis of the assemblage was conducted with the aim of establishing the relative ceramic sequence at the site which, together with the stratigraphic sequence and new radiocarbon dates, would enable the creation of a detailed typo-chronological framework for Belovode.

An analysis of the ceramics previously excavated at the site by D. Šljivar and associates was published only in summary and according to nine vessel types (Arsenijević and Živković 1998: 281–291) (Figure 1). The current analysis conducted on the pottery from Trench 18, excavated as part of the *Rise of Metallurgy in Eurasia* project, applied the existing classifications, but the attributes recorded (containing fabric, colour, surface treatment, and decoration) were modified to include metrical characteristics concerning rim percentages and diameters, fragment weights and total belly weights (cf. Orton and Hughes 2013). In addition, the new pottery database encompassed all excavated sherds, unlike earlier work on previous trenches where the ceramic assemblage was selective with some fragments discarded in the field.

Excavations of Trench 18 at Belovode revealed five building horizons (cf. Chapter 10, this volume). This chapter provides an overview of the key characteristics of the assemblage, including the types and varieties of vessels and handles, ornamentation techniques and motifs, and surface treatments. It is presented according to the five horizons and key features. Code numbers for vessel types are indicated throughout in parentheses.

Macroscopic analysis of pottery from Trench 18

Horizon 1

The most numerous ceramic fragments in this horizon are from bowls. These represent 67% of the entire horizon assemblage (Figures 2 and 3). Fragments from conical bowls with straight walls (100, Figure 12.2) are the most common, accounting for 15% of the bowl sherds. Those from bowls with rounded walls (101, Figure 12.3) are less numerous at 6%, while those from bowls with a thickened and profiled rim account for 5%. Sherds from rounded forms are also numerous: 16% are from hemispherical bowls (108), 12% are from spherical bowls (107), with 1% from spherical bowls with a short cylindrical neck (106).

The most frequently represented bowls in this horizon are those with turned-in rims (117), with 22% of the fragments (Figures 12.1). Other biconical bowls are also represented, but only by a small percentage (1–4%) of sherds. These are from types having: short (111, Figure 12.4) or long (113, Figure 12.5) vertical necks and angular shoulders; short (112, Figure 12.7) or long (114, Figure 12.8) vertical necks and rounded shoulders; concave

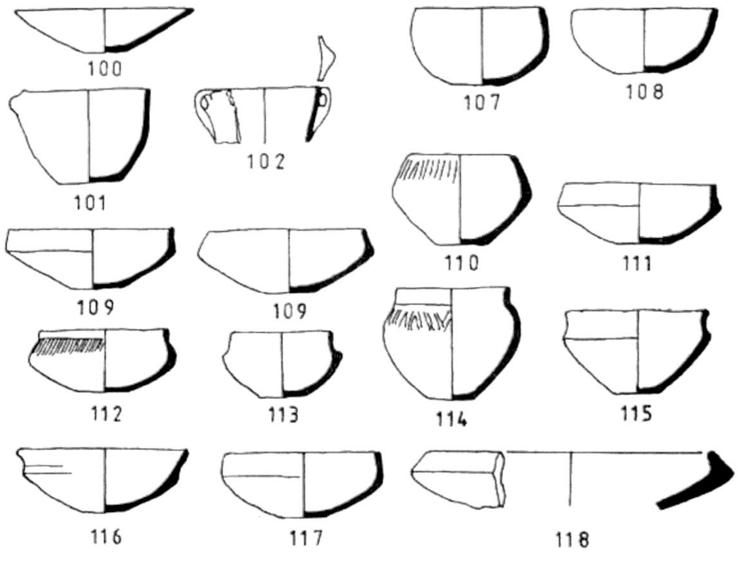

Figure 1. Bowl types according to Арсенијевић and Живковић 1995.

The Rise of Metallurgy in Eurasia (Archaeopress 2021): 152–169

(115) and funnel-shaped (116, Figure 12.6) upper cones; and with a massive angular shoulder (118).

Fragments from beakers with a conical neck and handles on the rim (221, Figure 13.1) occur occasionally comprising just 1% of the horizon assemblage. In addition, in the layers corresponding to this horizon, there are sherds from biconical and conical beakers with conical necks and ribbon handles (223).

Jugs (340) are not represented in the features of the horizon, but sherds are present in a few of the corresponding spits. Amongst these are fragments of distinctive jug types: with elongated biconical profiles, and long cylindrical necks with strap handles (341, Figure 13.3); with sharper biconical profiles, massive conical necks, and strap handles (343); and others with sharper biconical profiles, massive conical necks, and strap handles on the rims (344), with several fragments from Spits 05, 06, and 07 and from Spits 02, 03, 05, and 06 from the extension of the trench.

Fragments from plates and amphorettaes (320, Figure 13.4) are numerous, each forming 5% of the assemblage. Fragments from plate forms that are conical with a profiled rim (605, Figure 13.2) are especially numerous.

For storage vessels, fragments of amphorae are the most common, comprising 16% of the whole horizon assemblage. The most numerous (10%) are from biconical amphorae with conical necks (306, Figure 13.5-6), while those from funnel-shaped (307, Figure 13.7) and cylindrical necks (305) are present in smaller quantities. Pithoi sherds account for 11% of the assemblage; those from biconical types with conical necks (404) are the most numerous.

Of the fragments from cooking vessels, those from cooking pots (420) make up 7% of the horizon assemblage, the most numerous being from pots with rounded conical profiles (421); fragments of cooking pans (500) are less abundant (1%). Sherds from lids also occur, from a flat form with a handle (704).

Fragments from the so-called 'chimneys' (450, Figure 13.8) occur as single specimens in Spits 02 and 03.

The most common surface decorations are polished lines and slanted, arched, and horizontal channelling (Figures 12.8, 13.2, 4, 7). Incised ornamentation is rare. Tongue and wart-shaped (Figures 12.4) projections, buttons, lunate protuberances, and pinched and ellipsoid-shaped protrusions also occur as ornamentation.

The handles are frequently band-like (Figure 13.1, 3) or tongue-shaped (Figures 12.2, 3), but button-shaped and band-like handles with expanded ends are also common. Other variations are horizontal, elbow-, crescent-, and horn-shaped, or have a button-like protuberance, a 'pinch', or are double wart-shaped, tunnel-shaped, saddle-shaped, cylindrical, coil-shaped or are protome shaped handles.

Surface treatment is dominated by polishing on the bowls and on amphorae.

The most important features of this horizon are F1, F2, F3, F6, F8, F9, and F20 (Chapter 10, this volume). Features F1, F2, and F8 are identified as pottery concentrations. Diagnostically significant pottery from F1 includes rim fragments belonging to bowls (100, 117, 116, 107, 108, 112); amphora (300, 306); beakers (341); and cooking pots (421, 424). F2 yielded diagnostically significant rim fragments from bowls (100, 107, 108, 112, 114, 116, 117); beakers (221); amphorae (300, 303, 306, 307); cooking pots (421, 424); and pitoi (401). F8 is also identified as concentration of pots (Chapter 10, this volume). Diagnostically significant material from this feature includes fragments of pot bellies decorated with polishing and channelling, rim fragments from bowls (100, 101, 107, 116, 117); jugs (344); cooking pots (420, 422); cooking pans (500); and plates (604).

F3 is a rectangular structure made of wattle and daub. The lack of a heating or cooking installation leads to the conclusion that this could be a storage feature related to a larger dwelling structure located nearby (Chapter 10, this volume). Several pottery concentrations, labelled A–K, were identified on the floor of the structure, as follows: A- the remains of a biconical amphora with a conical neck (306) and a biconical pithos with a conical neck (404); B- pithos remains (400), a biconical pithos with a conical neck (404), a biconical amphora with a conical neck (306), and spherical and hemispherical bowls (107, 108); C- the remains of a biconical amphora with a conical neck (306); E- fragments of a biconical amphora with a conical neck (306) and a biconical amphora with a funnel-shaped neck (307); F- the remains of a biconical amphora with a conical neck (306), fragments of amphorettae (320), a conical bowl (100), spherical and hemispherical bowls (107, 108), and straight lids with a handle (704); H- a biconical pithos with a conical neck (404) and fragments of a conical bowl (100); J- fragments of amphorettae with a biconical profile, conical neck and funnel-shaped rim (324); and K- the remains of a biconical amphora with a conical neck (306) and a biconical bowl with a high cylindrical neck and carinated shoulder (113). The content of the house also includes fragments of different bowls (types 100, 101, 104a, 104b, 106, 107, 108, 111, 117, 118); amphorae (300, 306, 307); pithos (400, 403, 404); lids (700); and beakers (221). The pottery material on the floor of the smaller structure, consisting mainly of the storage and serving vessels—with an absence of the cooking ware—could confirm the hypothesised storage function of the feature.

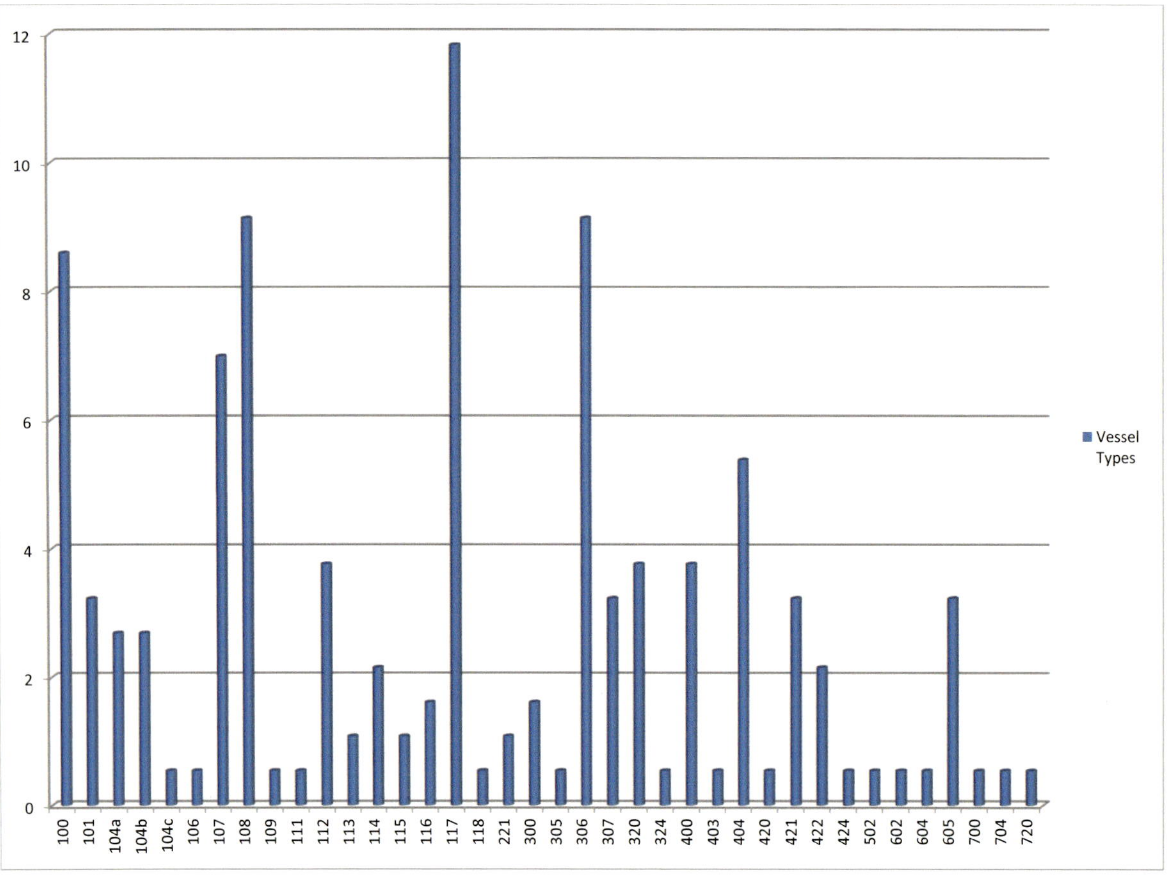

Figure 2. Vessel types represented in Horizon 1.

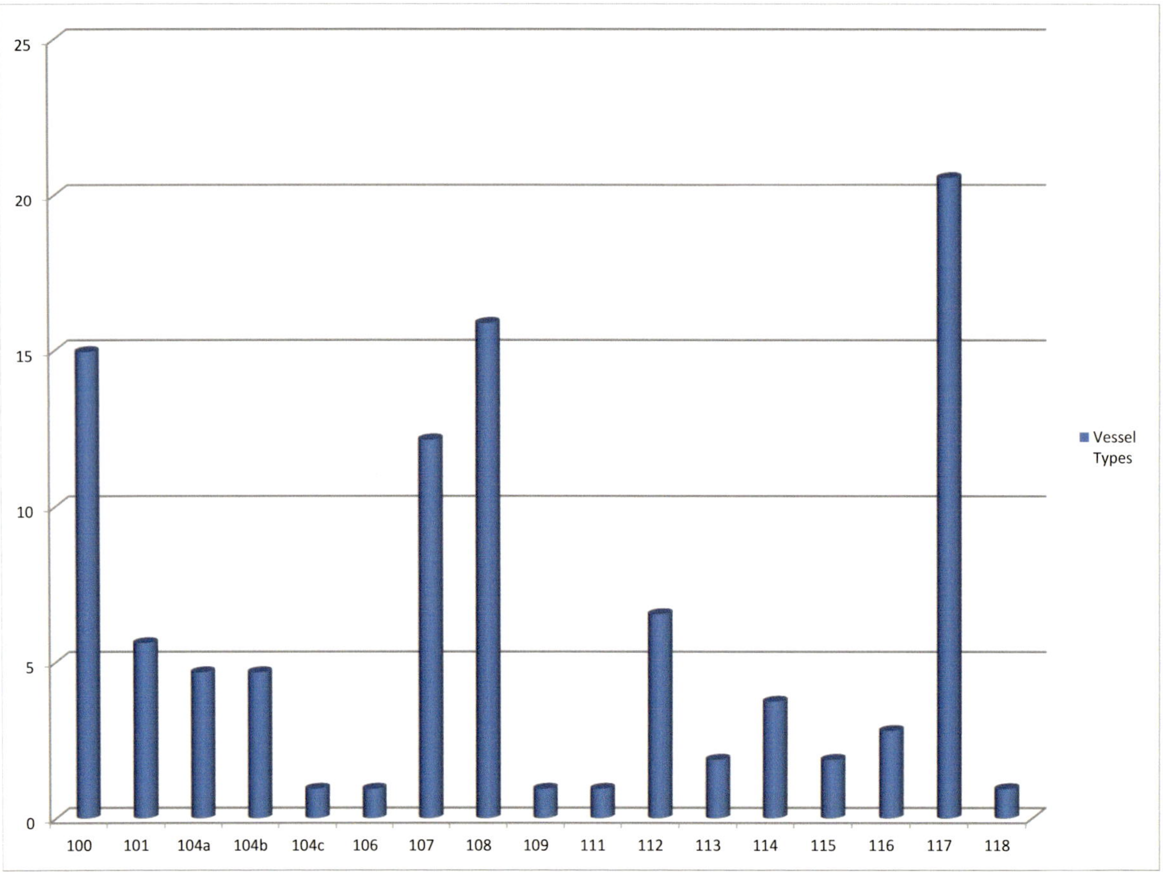

Figure 3. Bowl types represented in Horizon 1.

F9 is a large oval pit (Chapter 10, this volume). Diagnostically significant material from this feature includes fragments of bowls (100, 101, 102, 104a, 104b, 104d, 107, 108, 106, 109, 110, 111, 112, 113, 114, 115, 116, 117); spouted bowls (160); beakers (221); pedestal bowls (201); amphorae (300, 305, 306, 307); amphorettae (320, 324); jugs (341); pithoi (400, 401, 404); cooking pots (420, 421, 422, 423, 424, 420, 426); and plates (602, 605). The character of the pottery material indicates that this feature is a refuse pit.

F20 is a sub-oval feature (Chapter 10, this volume). The diagnostically significant pottery material from this feature includes fragments of pot bellies decorated with polishing and channelling; rim fragments belonging to bowls (101, 116, 106, 100, 104a); amphorae (307 and 306); amphorettae (320); cooking pots (420); and plates (604). This feature is also a refuse pit.

Horizon 2

This horizon is dominated by bowl fragments. These account for 54% of the entire horizon assemblage (Figures 4 and 5). The most numerous are from conical forms. Sherds from conical bowls with straight walls (100) make up 14% of all bowls in the horizon; 9% are from conical bowls with slightly curved walls (101); and 2% are from those with rim-handles (102). Just 1.7% are from conical bowls with profiled or thickened rims. Of the fragments from rounded bowls, those from hemispherical types (108) occur most frequently and comprise 10% of all bowl fragments in the horizon, while those from spherical bowls (106, Figure 14.8; 107) account for 7%. A large number (17%) of sherds are from bowls with a funnel-shaped profile (116, Figure 14.1-2).

The most frequently represented bowls in the horizon are those with a turned-in rim (117); fragments of this type make up 18% of the assemblage. Other biconical bowls are also represented including: biconical bowls with a short upper cone (109) with 5% of the assemblage; with a short (112, 4%, Figure 14.4), or long cylindrical neck (114, 3%) and rounded shoulders, and those with short (111, 1%) and long (113, 2%) cylindrical neck and angular shoulders. Those from bowls with a concave upper cone (115, Figure 14.3) are not common (2%).

Plates (600) are represented by 3% of the horizon assemblage, the most numerous fragments being from those with a conical profile and thickened rim (605). Amphorettae (320, Figure 14.5) are represented by 3% of the horizon assemblage.

Jugs of a form with elongated biconical profile, long cylindrical neck and a strap handle (341) are represented by 2% of the horizon assemblage; a form with a sharp biconical belly and a massive conical neck and a rim strap handle (344) is also represented.

Pedestal beakers (200, Figure 14.6) are scarce. Biconical beakers with a conical neck and rim handles (221) are represented by individual examples. Biconical beakers, both with a cylindrical neck with ribbon-shaped handles (222) and w with a conical neck and band-shaped handle (223) also occur in the spits of the horizon.

Of the storage vessels, amphorae (300) are represented by 14% of the horizon assemblage. The most numerous sherds are from amphorae that are biconical with a conical neck (306; 9% of the assemblage), and with funnel-shaped neck (307, Figure 14.7; 4%). Pithoi (400) fragments make up about 4% of the assemblage.

In contrast, sherds from containers for food preparation are numerous, especially those from cooking pots (13% of the assemblage), the most numerous being from pots with a rounded conical (421) or rounded biconical profiles (424). At less than 1% of the assemblage, sherds from casserole dishes (500) are rare, as are lids and plastic vessels.

The so-called 'chimneys' (450) are represented by 2% of the horizon assemblage.

Ornamentation in the form of polished lines is common, while sherds in certain features are dominated by channelling decoration which occurs in horizontal, slanting, vertical, and arcade forms (Figure 14.2). Engraved ornamentation rarely occurs (Figure 14, 5, 8). Black topped and barbotine techniques appear occasionally. In addition, wart-shaped, pinched, lunate, ellipsoidal, horn, tongue-shaped protuberances and ellipsoidal flattening are represented on shoulders.

Handles are frequently strap-shaped (Figure 14.7), but also numerous are forms that are roll-shaped (Figure 14.4) with button-shaped endings and ribbed handles. Handles also occur in forms such as prismatic, tongue-shaped (Figure 14.1), wart, fan-shaped, crescent, button-shaped (Figure 14.8), and striped with expanded ends.

Surface treatment is dominated by burnishing on the bowls, amphorae, amphorettae, with some examples of smoothing and polishing.

The most important features of this horizon are F18, F19 and F21. F18 is a circular pit filled with the remains of fired daub (Chapter 10, this volume). The diagnostically significant pottery material includes fragments of pot bellies decorated with polishing and channelling techniques; rim fragments belonging to bowls (100, 115, 101, 116, 106, 104a); amphorae (306, 307); amphorettae (320); cooking pots (420); and plates (604). The fragmentation and character of the pottery material indicates that this feature is a refuse pit.

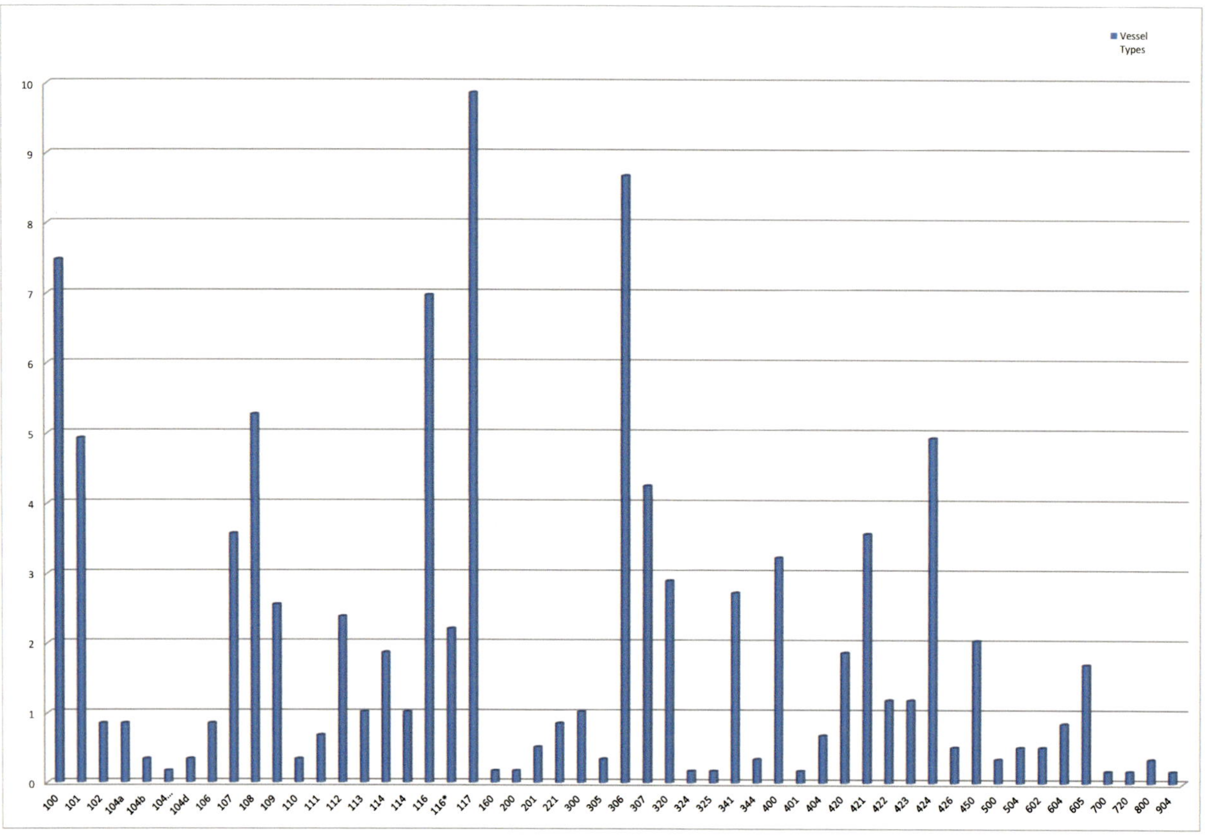

Figure 4. Vessel types represented in Horizon 2.

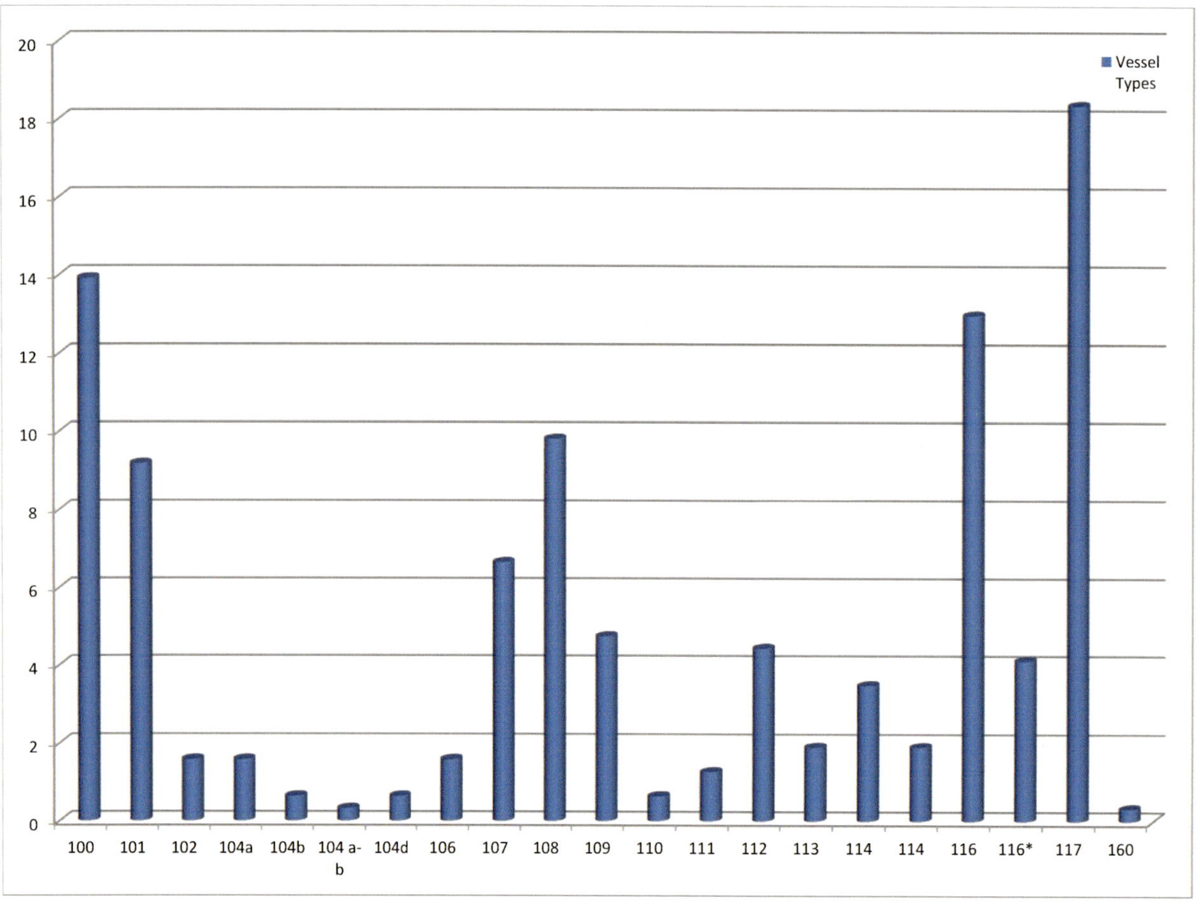

Figure 5. Bowl types represented in Horizon 2.

F19 is a large elliptical pit located in the eastern part of the trench (Chapter 10, this volume). The diagnostically significant pottery material includes rim fragments belonging to bowls (100, 101, 102, 111, 106, 117, 109, 113, 108, 107, 116, 109, 112, 114, 107, 101a, 104a,b,c, 112, 114); spouted bowls (160); pedestal bowls (200); amphorae (307, 306, 300, 305, 304); amphorettae (324, 321, 320); cooking pots (424, 421, 425, 420, 422); pithoi (400, 404); plates (602, 604, 605); and a miniature vessel. Most of the fragments are of fine ware, but the overall contents suggest that this was also a refuse pit.

F21 is an oval pit detected in spit 12 (Chapter 10, this volume). The diagnostically significant pottery material from F21 includes fragments of pot bellies decorated with polishing and channelling; rim fragments belonging to bowls (107, 104a, b, 117, 107, 108, 111, 116, 112, 100, 114, 101, 104b); pedestal bowls (201); amphorae (307, 306); amphorettae (320); cooking pots (421, 424, 420, 422); and plates 605).

Horizon 3

The most numerous fragments in Horizon 3 are from bowls, represented by 49% of the horizon assemblage (Figures 6 and 7). Sherds from conical bowls are numerous and include those from straight walled types (100), represented by 10% of the assemblage; slightly curved walled types (101; 12%); those with rim handles (102; 1%); those with a thickened rim (104a, Figure 15.1; 3%); and those with a profiled thickened rim (104b, Figure 15.2; 10%). Rounded forms are represented by a small number of fragments with those from spherical bowls with a short cylindrical neck (106), hemispherical bowls (107, Figure 15.6) and the globular bowls (108) bowl types, which are the most frequent, accounting for 9% of the total number of rounded bowl fragments in this horizon.

Of the fragments from biconical bowls, the most frequent are from types with a turned-in rim (117); these account for 15% of the assemblage bowl fragments. There are numerous other bowl types represented, such as those with a short (112, Figure 15.5) and long (114) cylindrical neck and rounded shoulders (14% of the assemblage) or funnel-shaped (116; 9%). Others are represented in smaller proportions, such as those with a short upper cone (109, Figure 15.3; 3%), with a short (111, Figure 15.4; 2%) or long (113; 3%) cylindrical neck and square shoulder, while those with a concave neck (115) contribute only 1% of the assemblage.

Pedestal beakers (200), including those with a hollow conical neck (201) are represented by only 2% of the assemblage. A small percentage of sherds are from biconical beakers with a conical neck and peripheral handles (221). Jugs are not represented, either in features or in horizon spits.

Fragments of amphorettae (320) are numerous, represented by 7% of the horizon assemblage and sherds from plates (600, Figure 15.7) account for 4%.

Of the storage vessels, amphorae (300, Figure 15.8) are represented by 16% of the fragments in the horizon assemblage, with the largest number (8%) being from biconical amphorae with a funnel-shaped neck (307). By contrast, pithoi (400) are rare, represented by just 3% of the horizon assemblage.

Vessels for food preparation such as cooking pots are numerous, represented by 17% of the assemblage. This includes those with a rounded conical profile (421; 6%) and biconical pots with a conical upper part (422; 8%).

Ornamentation is dominated by channelling, predominantly slanted, but also horizontal and vertical (Figure 15.2, 6, 7). Polishing is well represented in the form of lines, but only on the sherds from certain features. Black topped techniques (Figure 15.4) and organised barbotine are common; incised decoration is relatively rare.

Handle types vary and include band-shaped, button and tongue shaped, protomes (Figure 15.2), coiled, turban handles, striped handles with expanded ends, warts and ovals.

Burnishing is present on bowls, pedestal beakers and amphorettae, with polishing on bowls, amphorae, pedestal beakers and amphorettae.

In this horizon there is still a high prevalence of the later vessel types, although the representation of a wide range of different types of biconical bowls highlights the flourishing of the Vinča material culture. Conical bowls with profiled rims are common as are early types of bowls with a cylindrical neck and rounded shoulder. Amphorae are very numerous in this horizon, represented by 16% of the assemblage.

The significant features of this horizon are F30, F31, F32 and F43, which together comprised a large pit. The fragmentation and character of the pottery material indicate that this was a refuse pit. The amount of diagnostically significant pottery fragments from F30 is not large. and includes fragments of pot bellies decorated with channelling and polishing.

The material from F31 includes fragments of pot bellies decorated with polishing and incised marks; rim fragments belonging to bowls (107, 112, 106, 101, 104b, 100, 116, 107); pedestal bowls (201); amphorae (306, 307); amphorettae (324); cooking pots (422, 424, 421, 425); and lids (703).

The material from F32 includes fragments of pot bellies decorated with polishing and incised marks;

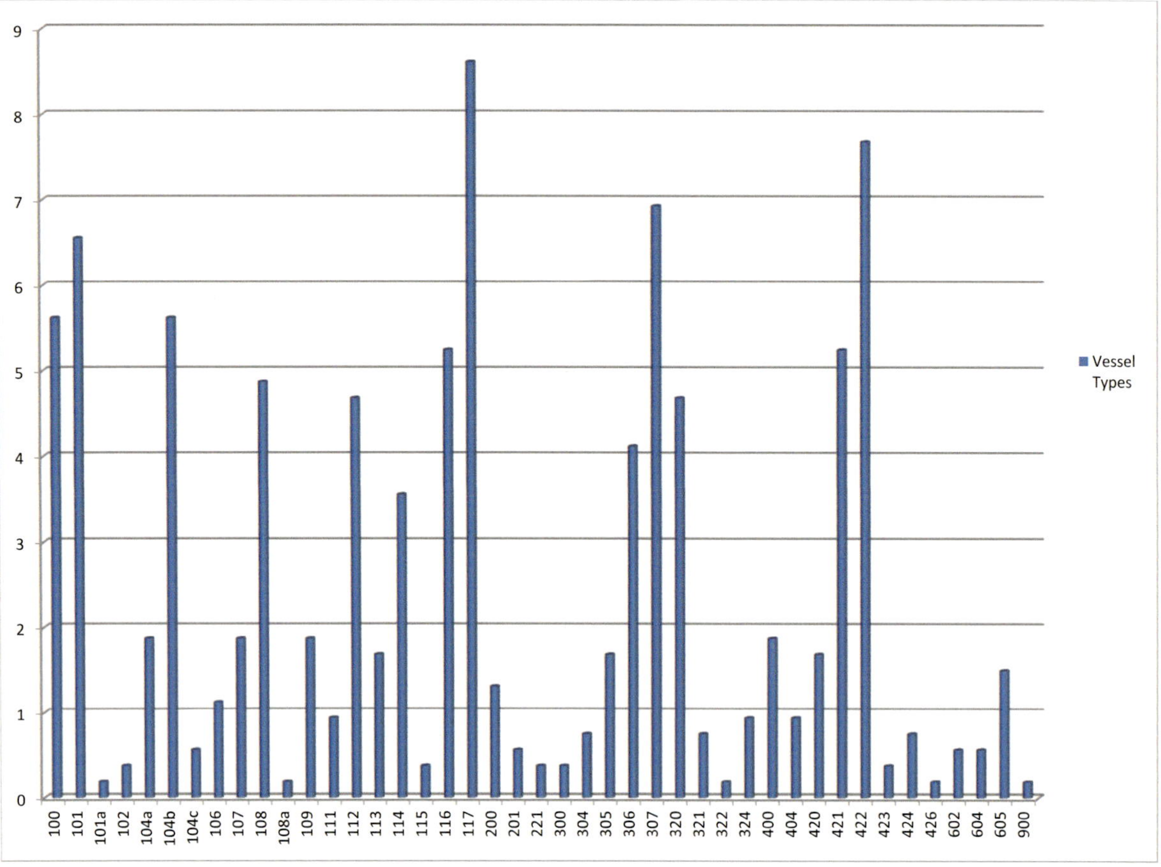

Figure 6. Vessel types represented in Horizon 3.

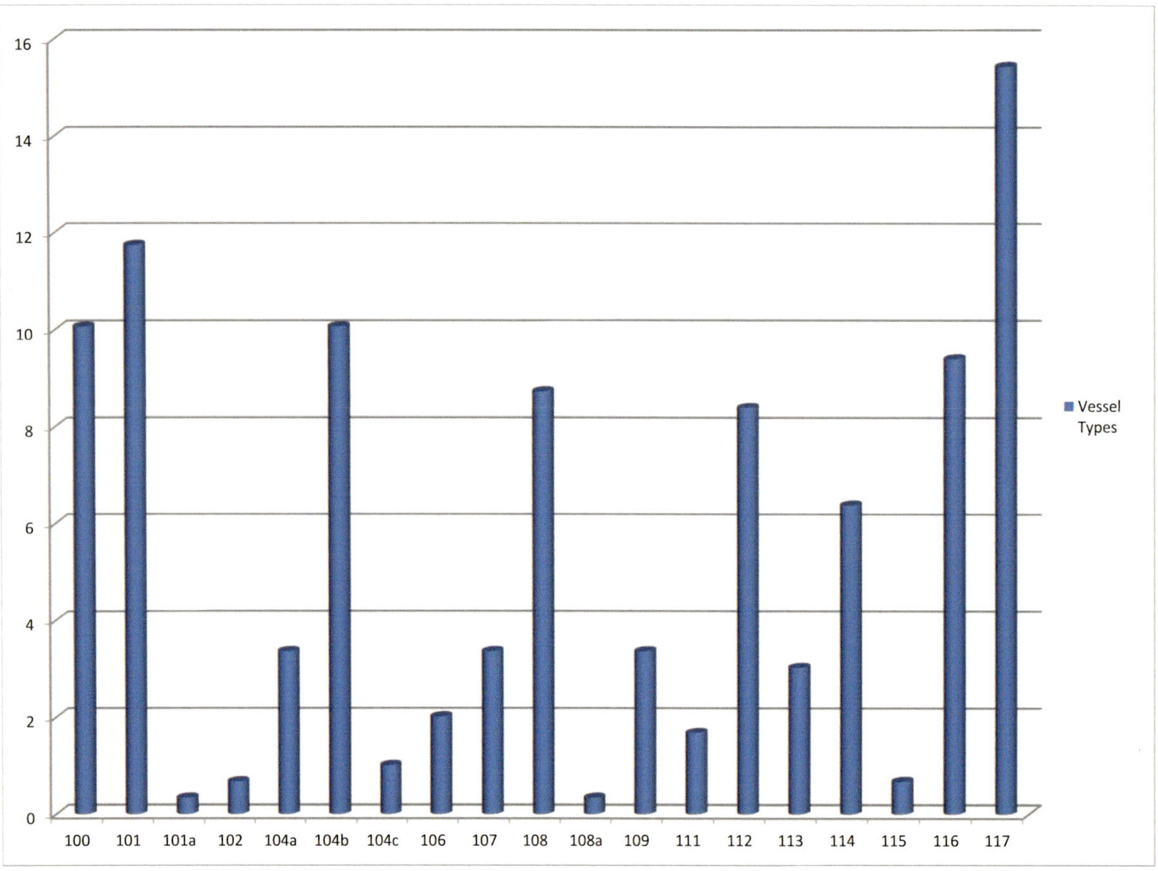

Figure 7. Bowl types represented in Horizon 3.

rim fragments belonging to bowls (104b, 101, 106, 113, 104b, 100, 107, 115, 112, 100, 116, 108, 104b, 100, 104a, 101, 108a); pedestal bowls (200); amphorae (304, 305, 306, 307); cooking pots (422, 421, 420); plates (605); and prosopomorphic lids (720).

The material from F43 includes fragments of pot bellies decorated with channelling and polishing; fragments of various handles and bottoms; and rim fragments belonging to bowls (100, 101, 104b, 109, 110, 111, 112, 116, 115); pedestal bowls (200); amphorae (304); amphorettae (320); cooking pots (420, 421, 422); and plates (605).

Horizon 4

The most common ceramic type in Horizon 4 is the bowl, represented by 47% of the entire horizon assemblage (Figures 8 and 9). Fragments from conical bowls are the most common, with those from bowls with straight conical walls (100) and those with slightly curved conical walls (101) each accounting for 18% of the assemblage; fragments from bowls with a thickened rim (104a) and profiled rim (104b) are less frequent, at 3% and 12% respectively. Sherds from rounded forms are numerous. Those from spherical forms with a short neck (106) account for 5% of the horizon assemblage, spherical (107, Figure 16.1) for 4% and semi-spherical bowls (108) for 5%.

Of the fragments from biconical bowls, those from bowls with a short biconical neck and rounded shoulder (112, Figure 16.2) are most numerous (10%). Although a large number of other types are represented in the horizon, fragments of these appear in smaller percentages: those from bowls with a short upper cone (109, Figure 16.4) make up 2% of the assemblage; those from bowls with upper and lower cones of the same height (110) make up 7% of the assemblage. Biconical bowls with a short (111, Figure 16.3) and a long (113) cylindrical neck and angular shoulder in the upper part are represented by 3% and 2% of the assemblage respectively; those with a concave (115) or funnel-shaped (116) upper part are presented by 4% and 6% respectively. Bowls with a spout (160) also appear in this horizon.

Conical plates with thickened rim (604) and with profiled rim (605) are represented by 3% of the assemblage. Amphorettas (320) are not numerous, represented by 3% of the assemblage, with two variants: those with a conical neck and funnel-shaped rim (324), and those with a concave upper and conical lower part (325).

Pedestal beakers (200) are very well represented by 6% of the assemblage; with fragments from pedestal beakers with a hollow conical foot (201) accounting for 1%.

Of the storage vessels, amphorae (300) are represented by 6% of the assemblage, and pithoi (400) by just 1%. The most common amphorae are biconical with a funnel-shaped neck (307).

Of the cooking vessels, pots are represented by 14% of the assemblage, the most numerous sherds are from cooking pots with a biconical profile (422, Figure 16.5), represented by 8% of the assemblage, while those from cooking pots with a biconical profile (424) and rounded conical profile (421) are equally represented but to a lesser degree.

Fragments from lids are fewer in number with those from a type of flat lid (703) being dominant.

The most common decorative technique is barbotine (Figure 16.6), but evidence for the black topped technique is also frequent. Channelling is common (Figure 16.1-2, but polishing is rare, as is imprinted or incised ornamentation. There are button-shaped, tongue, wart (Figure 16.1) and ellipsoidal handles, as well as examples with an ellipsoid flattening on the shoulder. Handles are frequently band-shaped and tongue-shaped, as well as button-shaped.

Surface treatment is usually burnishing on the bowls, pedestal beakers, amphorettae, with polishing the most common technique on amphorae, bowls, pedestal beakers; smoothing is also used.

Significant features of Horizon 4a are F37 and F38 and from Horizon 4b, F40, F45, and F46 (Chapter 10, this volume).

The diagnostically significant pottery material from F37 includes fragments of handles and pot bellies; several rim fragments belonging to bowls (113); and cooking pots (422). The quantity of pottery fragments is not substantial, and this feature is identified as a clay outcrop. The material from F38 includes fragments of handles and pot bellies; several rim fragments belonging to bowls (104b, 110, 111); and pithoi (400). The quantity of pottery fragments is not great, and this feature is also identified as clay outcrop. The amount of diagnostically significant pottery fragments from F40, F45 and F46 (all hearths) is negligible.

Horizon 5

There are few finds from Horizon 5. Fragments of bowls form almost the entire horizon pottery assemblage, accounting for 72% (Figures 10 and 11). Bowls with a short upper part (109), and those with equally high upper and lower parts (110) are represented by 25% of the bowl fragments in the horizon assemblage. The same proportion are from bowls with a short

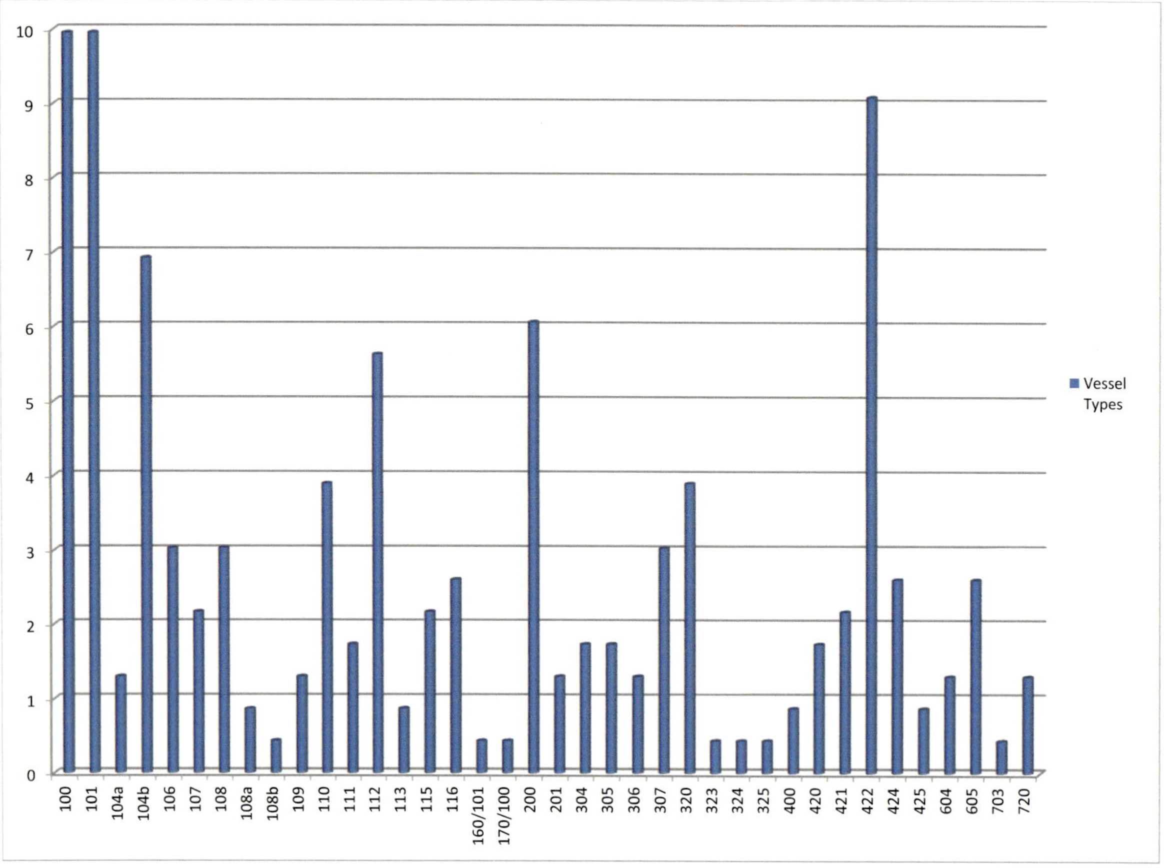

Figure 8. Vessel types represented in Horizon 4.

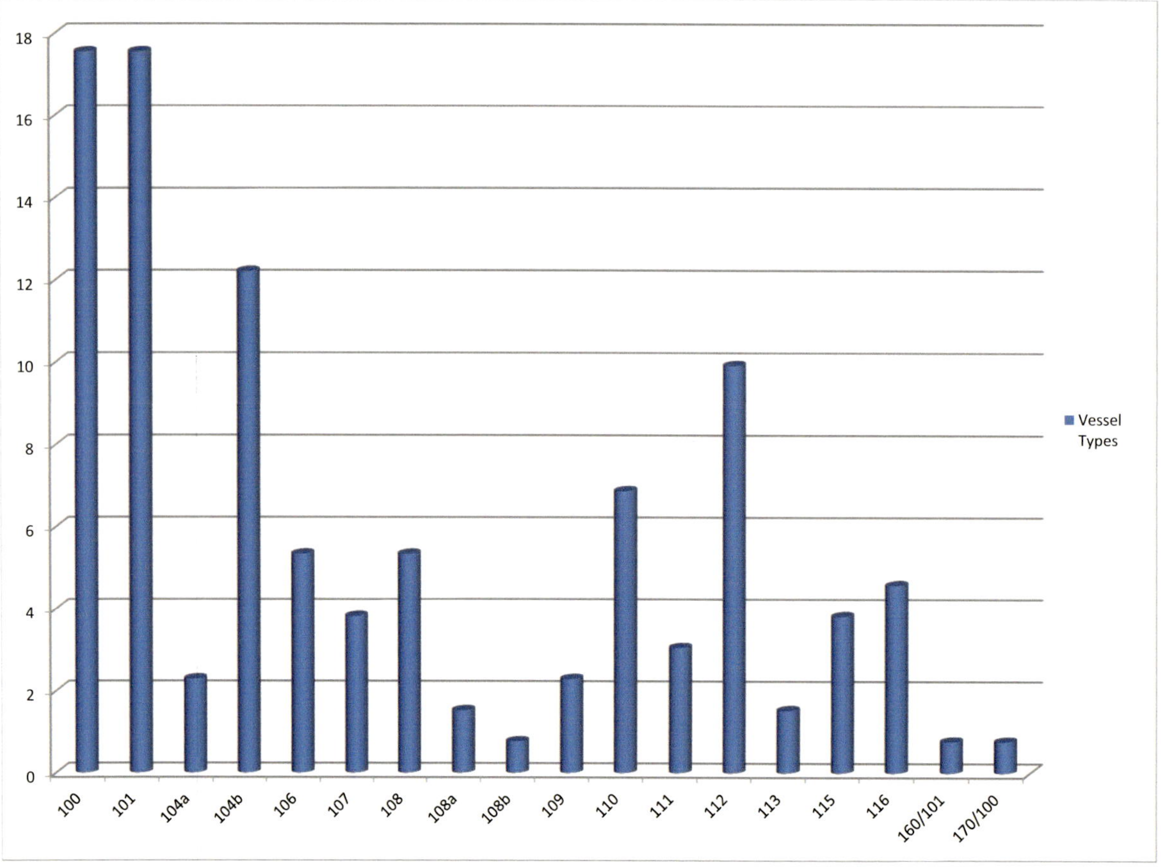

Figure 9. Bowl types represented in Horizon 4.

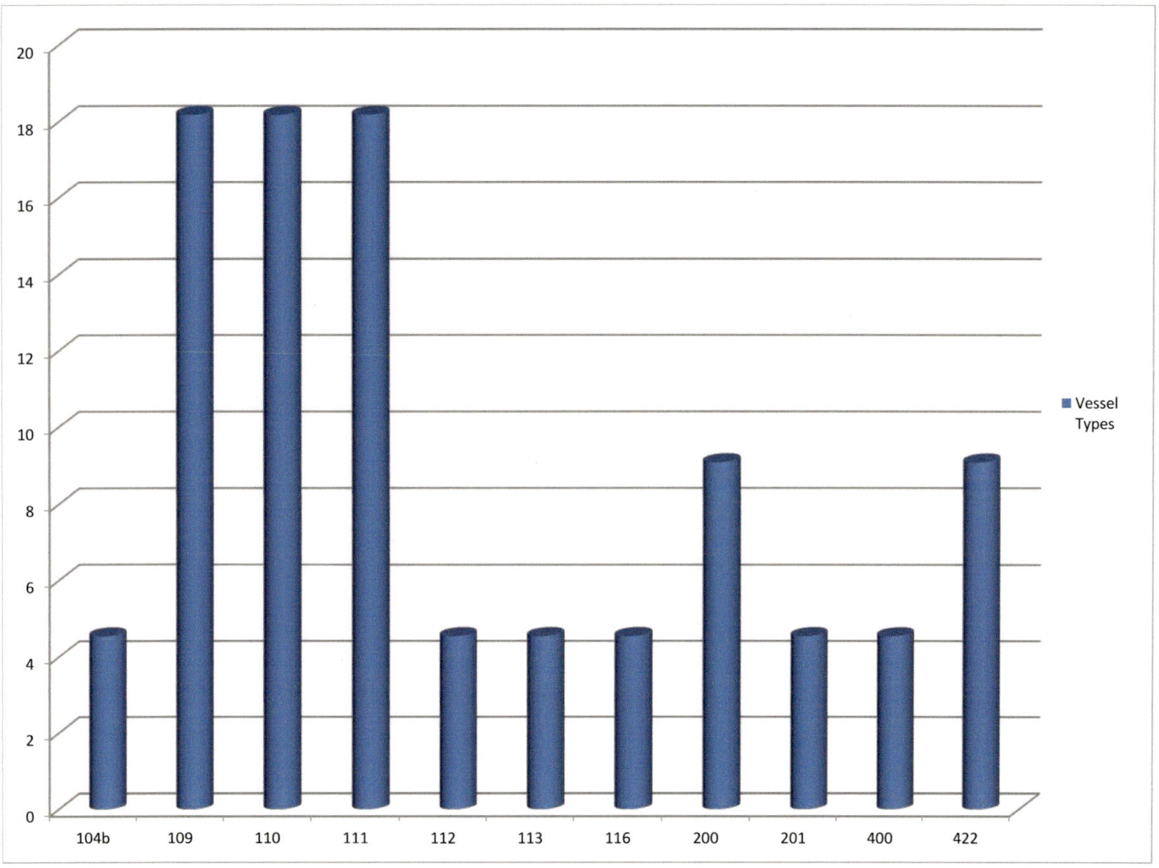

Figure 10. Vessel types represented in Horizon 5.

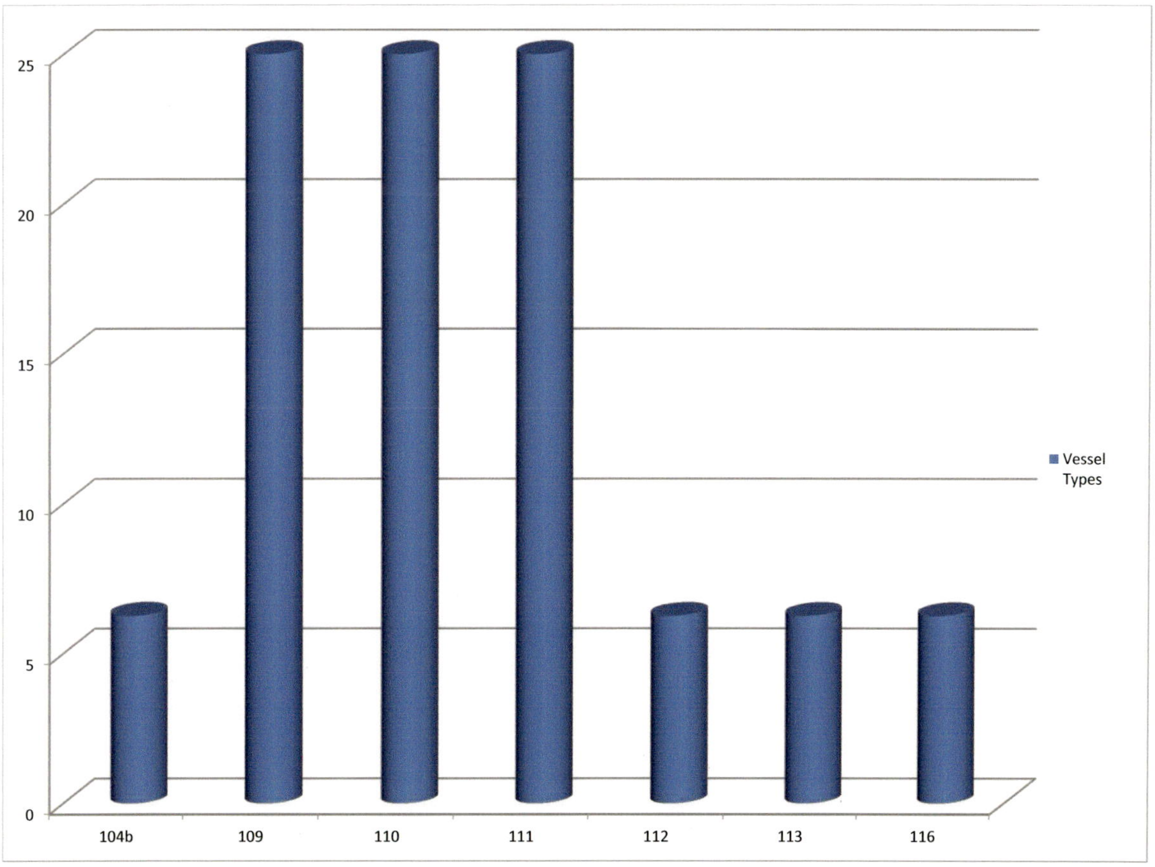

Figure 11. Bowl types represented in Horizon 5.

cylindrical neck and angular shoulder (111). Bowls with a long cylindrical neck and angular shoulder (113) are represented by 6%, as are those with funnel-shape profile (116) and a short cylindrical neck and rounded shoulder (112). Pedestal beakers are numerous, represented by 9% of the horizon assemblage, with the majority (5% of the assemblage) having a hollow conical foot (201). Storage vessels in the form of pithoi (400) are represented by 5% of the horizon assemblage and those for food preparation, biconically profiled pots (422), are represented by 9%.

The most frequent ornamentation is barbotine (Figure 16.7), both organised and unorganised, as well as incised marks (Figure 16.8). Band-shaped handles are the most frequent. The surface treatment on bowls and beakers is dominated by burnishing and polishing.

The amount of diagnostically significant pottery fragments from significant features in this horizon, i.e. from the pit, F47, and the hearth, F49, is negligible.

Relative chronology

Horizon 1

Pottery from Horizon 1 at Belovode corresponds to the last phases of the Vinča culture, as represented on many sites. The main characteristic of the assemblage is the presence of vast numbers of bowls with turned-in rims.

At the Belo Brdo site, the final phase of the Vinča culture is represented by material from the spits above 4.1 m (Vasić 1936a; Garašanin 1979) and corresponds to other sites of classical variants: Crkvine in Stubline, Crkvine in Mali Borak (Spasić 2011: 101–146), Jakovo-Kormadin (Glišić and Jovanović 1960: 113--39), Obrež-Beletinci (Brukner 1962: 89–122, Tables IV–VI), as well as Gomolava Ib in Vojvodina (Brukner 1981: 33–37, Table IV).

The material at Belovode has analogies with late Vinča culture settlements in the Grivac VI horizon (Nikolić 2004: 213–216), the Divostin IIb horizon (Madas 1988: 143–71, Figures 6.2–6.21; Bogdanović 1990: 99--07), and Spits 3, 2 and 1 at Supska (Garašanin and Garašanin 1979: 40–41, Tables I–VII). The Kosovo variant has analogies in later horizons at sites such as Predionica and Valač (Garašanin1998: 80–84).

Horizon 2

The large number of funnel-shaped bowls (116) represented in Horizon 2 are characteristic of the phases from Vinča Pločnik I to IIa (cf. Garašanin 1979: 175, Figure 13). There remain, however, a large percentage of sherds from bowls with a turned-in rim. At Vinča, this bowl type is present between the depths of 5.9-4.0 m and 3.9-2.0 m (Vasić 1936a) and is dominant after 4.1 m (Vasić 1936a: 78, Plates 108 and 119), from the house at 4.1 m (2.3 m) (Vasić 1936a: 73, 74, Plates 182, 190, and 191a–d), from 3.2 and 3.4 m (3.7 and 3.9 m), 3.1 m (3.6 m), 2.9 m (3.4 m), and 1.5 m (2.0 m). Some have curved walls, inside and out (depth 2.2 m, 3.3 m), whilst others have angular shoulders and therefore correlate more closely with a type with a short upper cone (depth 3.8 m, 2.8 m, 2.6 m, 2.0 m, 3.3 m). Bowls with a turned-in rim are present in the inventory of the house in Vinča from 2.98 m (3.48 m). This type is present to a depth of 5.7 m (5.2 m) (Vasić 1936a: Plate 126).

In the typology developed by W. Schier, bowls with funnel-shaped profiles are labelled types 251 and 252. These are present in Phases 5b (7.5-6.9 m at Vinča) and 5c (6.8-6.5 m at Vinča) (Schier 1996: 146, Figure 8,). They occur more frequently in Phase 6, which represents the Gradac Phase (6.4-6.0 m at Vinča) and continue to develop in Phase 7 (6.0-5.0 m at Vinča) (Schier 1996: 146). Bowls with funnel-shaped profile appear in Supska in Spits 3–5, from the Gradac Phase to the Vinča Pločnik IIa phase (Garašanin and Garašanin 1979: Tables VI/5; VII/ 2; X/2, 4, 5; XIV 4). At Grivac, this shape is characteristic of the Grivac VI Phase, represented by types 222 and 223: deep, almost funnel-shaped bowls on which the very short shoulder emphasises the biconical profile (Nikolić 2004: 214). Together with the open, deep bowls with a short or reduced upper cone (types 218, 219, 220) they represented 25% of the overall assemblage of fragments from biconical bowls (Nikolić 2004: 214).

Funnel-shaped bowls also occur at Selevac, in the horizon of the 1977–78 Trenches VII and IX. These include types 346, 358, 355, 369 which are numerous in the Gradac and Vinča Pločnik I phase (Vukmanović and Radojčić 1990: 296).

Horizon 3

This horizon can be dated to the Gradac Phase at the earliest and to Vinča Pločnik I or Vinča Pločnik I/II at the latest. Many bowl types—particularly biconical variations—emphasising the flourishing of the Vinča material culture.

The horizon between 7.5 and 4.5 m at Vinča represents the peak period of the Vinča culture (Tasić and Tomić 1969: 57). On the eponymous site, this phase is characterised by bowls with a cylindrical neck and reinforced rounded shoulder; bowls with a concave upper cone; deep bowls with a conical neck and accentuated shoulder (Schier 1996: 146, Figure 7: S242, S243, S52, Figure 8: S244, S54). Within the classical variants we find analogies in the Gomolava Ia, ab phase in Vojvodina (Brukner 1981: 33–37, Tables I– III).

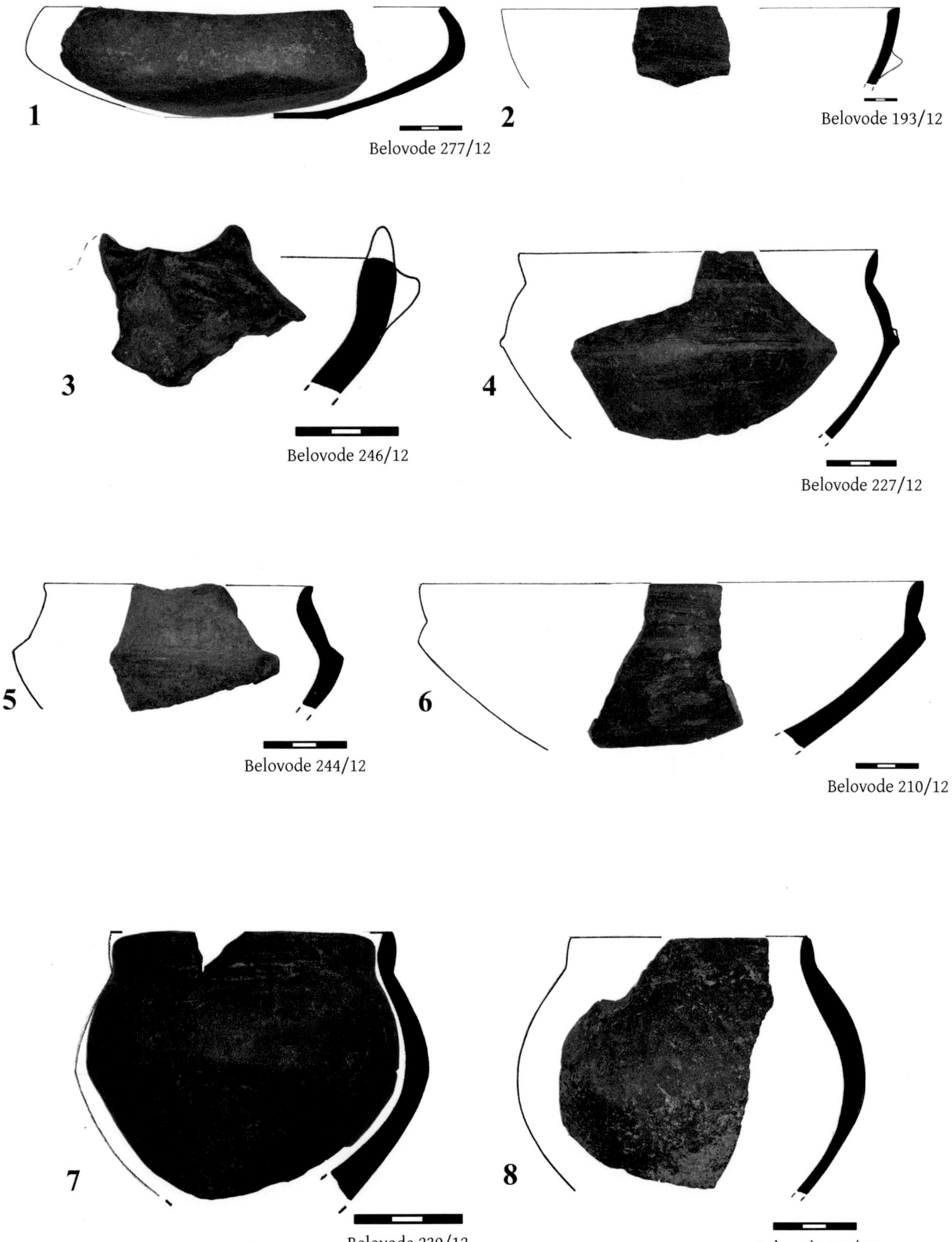

Figure 12. Most characteristic vessel types of Horizon 1.

Figure 13. Most characteristic vessel types of Horizon 1.

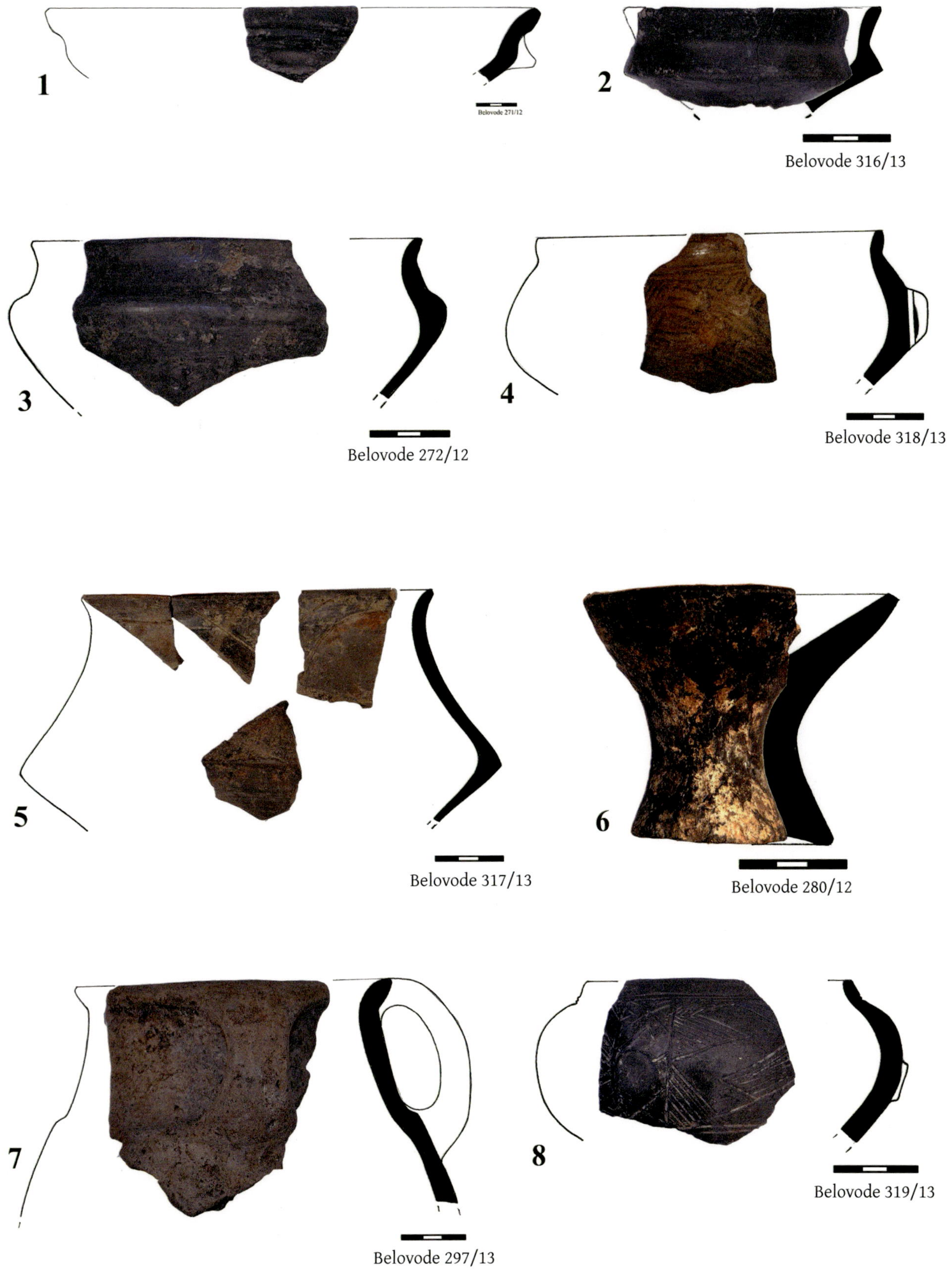

Figure 14. Most characteristic vessel types of Horizon 2.

There are also analogies within the Grivac V horizon (Nikolić 2004: 209–213, 220–226), the older Vinča horizon in Crnokalačka bara (Tasić and Tomić 1969: 60), Spits 4, 5 and 6 in Supska (Garašanin and Garašanin 1979: 39, Tables XIV–XX), Horizons V–VII at Selevac (Vukmanović and Radojčić 1990: 289–316), at Gradac (Stalio 1972: Tables I, II, IV, V), in material within the Spits 5–7 excavated in 1985/1986 at the Motel Slatina site, and in material from Spits 5 and 6 from Trenches I–III from 1962/1963 (Perić 2006: 243).

This stage also corresponds to the house horizon at Predionica, as well as the settlement of Fafos I, Mitrovica in the Kosovo variant (Garašanin 1998: 80–82).

Horizon 4

A substantial representation of bowls with rounded shoulders and a short cylindrical neck as well as pedestal beakers place this horizon in the Vinča Tordoš II phase. The presence of a large number of shards from conical bowls with thickened rims also occurs in the Gradac Phase of the classical variant (Garašanin 1979).

At Vinča, the Vinča Tordoš II phase occurs in spits between 8.5 and 6.5 m (Garašanin 1993: 7–20). According to Schier (1996), this includes Phases 4–5c in which rounded bowls with a short neck (Figure 5, S135, S136), biconical bowls with rounded shoulders and a vertical neck (Schier 1996: Figure 6, S153, S163) and those with curved or angular shoulders with a very long, sometimes concave neck (Schier 1996: Figure 7, S178, S169, S185) are frequently represented. Bowls with a rounded shoulder, including types 136, 143, 152, and 163, reach their peak during Phases 4 and 5a, which occurs between 8.5-8.0m and 7.8-7.6m at Vinča (Schier 1996: Figure 6, 145, 146). This represents Vinča Tordoš IIa according to Garašanin's periodisation (Garašanin 1993). The material has analogies at Supska Spit 7 (Garašanin and Garašanin 1979: Plates XXI-XXVIII) and in the construction horizons I-IV at Selevac (Vukmanović and Radojčić 1990: 295–296, Figure 9.3, Types 314–316).

Horizon 5

The vessel types represented in this horizon at Belovode are a major part of the ceramic production of the oldest phase of the Vinča culture, characterised by a small repertoire of vessel types and use of barbotine ornamentation, the main trait of the older Starčevo tradition (Nikolić 2004: 216). At Vinča, pottery from the oldest phase, Vinča Tordoš I (Garašanin 1979: 152) occurs in spits between 9.3 m and 8 m and is characterised by three types of bowls: shallow bowls with a very short upper cone (Vasić 1936a: Plates 30, 31, 62–68, Schier 1996: Figure 5, S21, S28), shallow open bowls with a short rounded shoulder (Vasić 1936a: Plates 14, 18, 87, 33c; Schier 1996: Figure 5, S124 and S126) and deep bowls of spherical shape with a short neck ring (Vasić 1936a: Figure 18, No. 93/98; Schier 1996: Figure 5, S141, S146, S135, Figure 6, 137, S132, S135 3-4). Schier classifies these bowls into 17 types (Schier 1996: Figures 5, 6). In the Grivac typology they are represented by Types 201, 202, 203, and 205 (Nikolić 2004: 217).

The settlement of Trnovača-Bajbuk (Jovanović 1965: 35, Tables I–VI) belongs to this phase. Gornea in Romania has phases Vinča A1, A2, and A3 (Lazarovici 1979: Table 9). In Hungary, sites with analogous phases include Versend-Gilencsa (Jakucs et al. 2016) and Szederkény-Kukorica-Dűlő (Jakucs and Voicsek 2015) while in Romania, Maroslele Pana (Paluch 2011).

Starčevo pottery tempered with chaff and ornamented with barbotine is occasionally present within the horizon at Belovode.

Discussion and conclusions

Based on the statistical analysis of material from Trench 18 and analogies from other Vinča culture sites, confirmed by new radiocarbon dates (Chapter 39, this volume), we can confirm that the entire range of Vinča culture is indeed present at Belovode. The pottery assemblage includes examples that match the main characteristics defining each of the individual phases of the Vinča culture (Whittle *et al.* 2016; Shier 1996).

Certain characteristics of the Belovode material, especially the jugs and beakers with handles (Jovanović 2006: 221–235), link the settlement with the South Morava regional variant of the Vinča culture (Garašanin 1979: 188) during the so-called Gradac Phase. Beakers do not appear in a high proportion of the assemblage and are limited to the middle horizons for Belovode.

The contexts from Trench 18 are mostly refuse pits. The only other feature detected in the trench, marked as Feature 3 is only partially excavated. Our knowledge of the usage of the pots within a household is therefore limited. The statistics available provide information about the pottery production (Chapter 14, this volume) and give some insight into typological preferences and the functional and metric characteristics of some vessel types.

The percentages given in the previous sections demonstrate that there is a disproportionate representation of vessels used for serving and consumption compared with those for cooking and storage. The representation of the bowls within the whole assemblage is very high. This is typical for many sites and has been the subject of research by several authors, using functional characteristics and ethnoarchaeology to explore why the remains of so

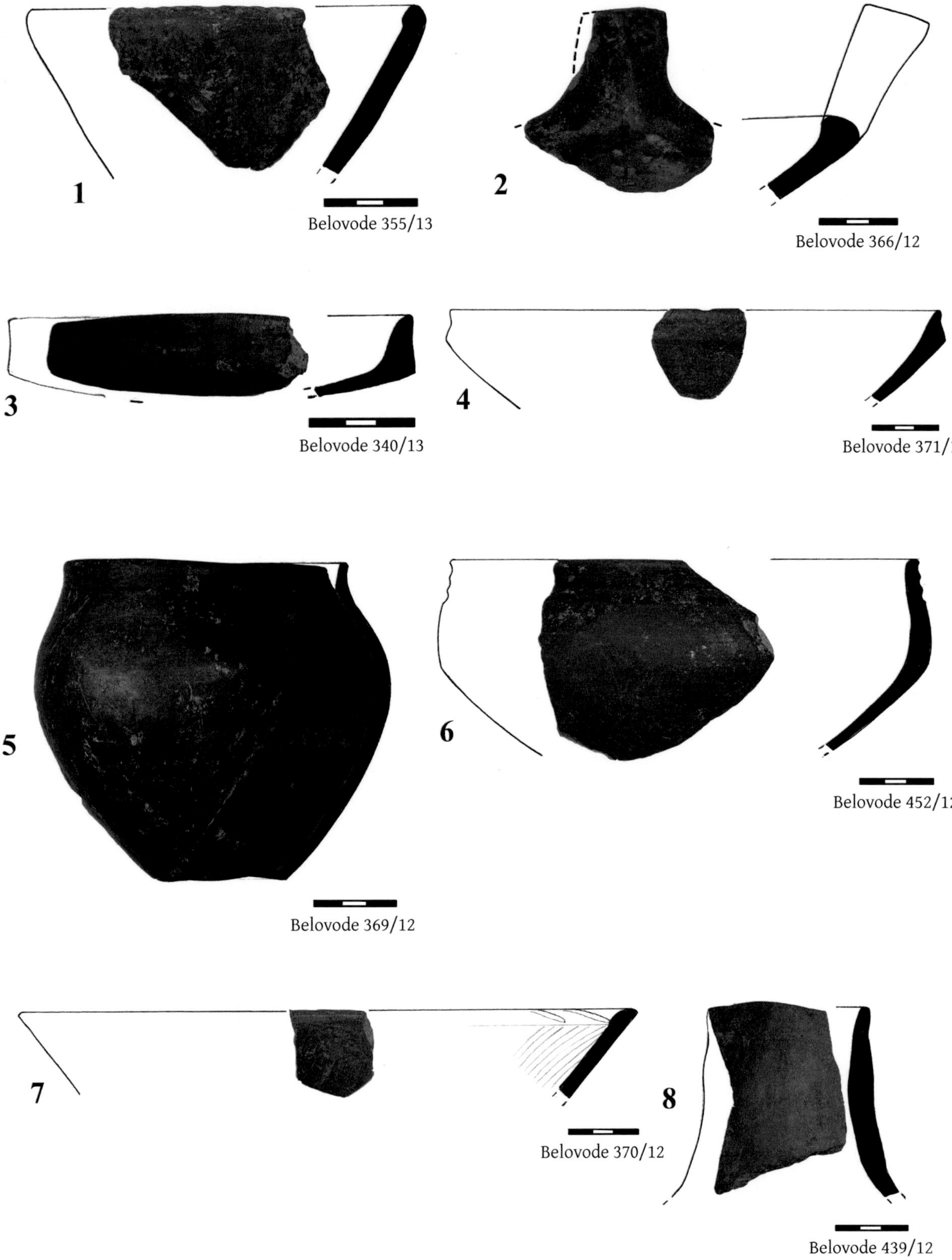

Figure 15. Most characteristic vessel types of Horizon 3.

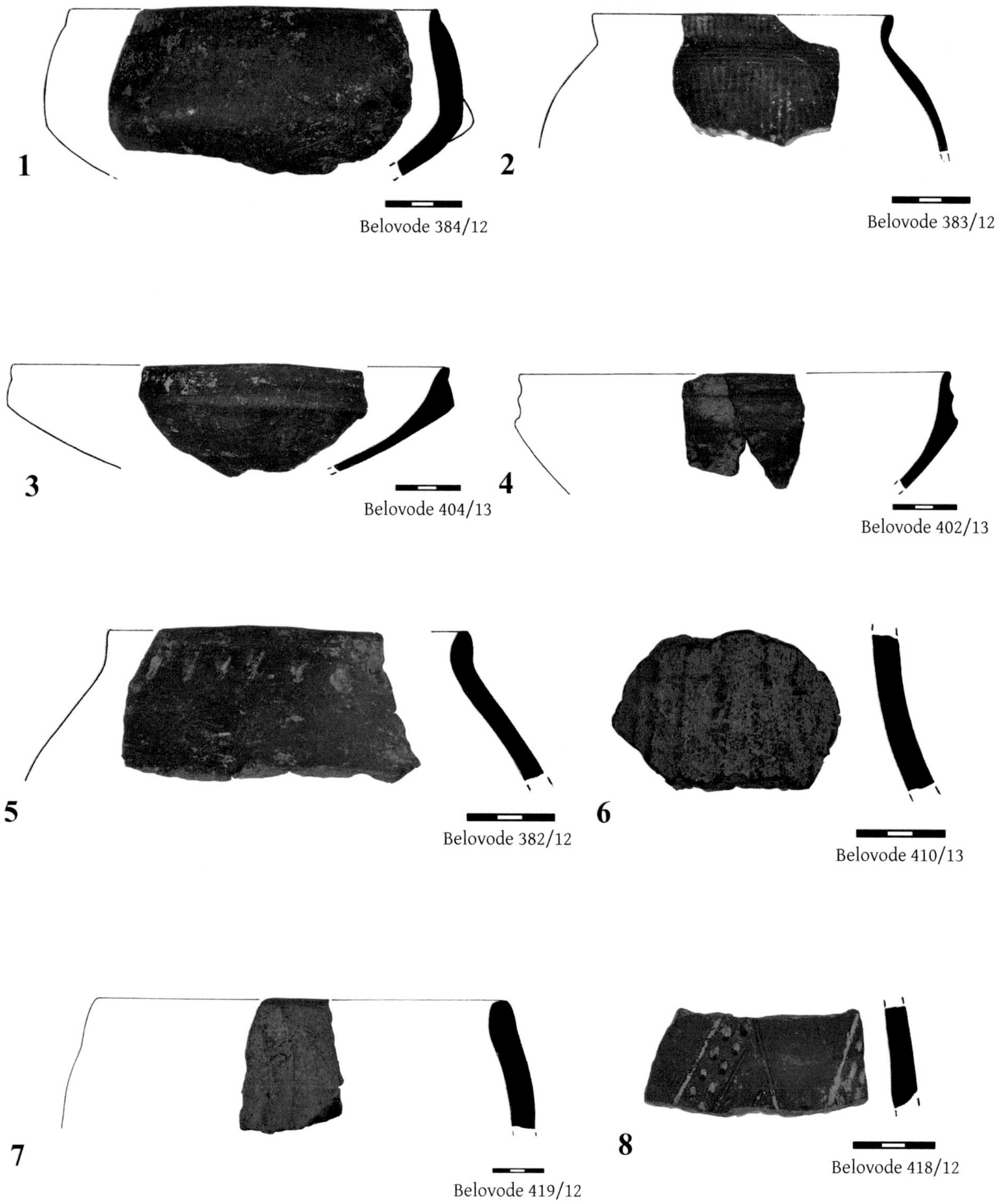

Figure 16. Most characteristic vessel types of Horizons 4 and 5.

many serving vessels occur in the archaeological record. The problem could be explained by the physical characteristic of the serving and consumption vessels, which have thinner walls that are more prone to damage. Bowls are also handled more often more and moved around, which may also lead to increased breakage. The dimensions of the vessels (height, weight and volume) also determine longevity (Shott 1996), with larger vessels being more static and therefore lasting longer (Vuković 2010: 13).

The properties of the fine ware include porosity, hardness, strength and thermal properties. Fine pottery shows a high degree of hardness and strength and low porosity. Low porosity causes low permeability, good thermal conductivity and low resistance to thermal shock (Sillar 2003: 175), high heating effectiveness and low cooling effectiveness. Based on these properties, fine pottery seems to be especially suitable for a storage function, especially storage of liquids (low porosity), transport (resistance to mechanical stresses), and mechanical processing of food (high hardness and resistance to abrasion); they are unsuitable for thermal food processing (low resistance to thermal stress). The high number of sherds belonging to bowls in comparison with other vessel classes may also imply that they had other functions, perhaps for the temporary storage of food or for mechanical preparation of food and beverages (Vuković 2010: 12).

The small number of cooking pans among the cooking vessels could indicate preferences concerning food preparation. Surface treatment of the Belovode vessels varies. Usually, the vessel surfaces are well preserved, but post depositional processes often affect them. The surface of the vessel could be smoothed, burnished or polished. While cooking vessels (pots and pans) as well as pithoi, i.e., vessels for storage of solid food, have smoothed surfaces, burnishing and polishing appear on vessel classes traditionally considered to be used for serving and consumption (bowls, beakers, jugs), but also on vessels for the storage of liquid (amphorae and amphorettae). Burnishing and polishing of the bowls and amphoras has a practical function, making the vessel less porous. This process compacts particles on the surface, which makes it harder and more resistant to abrasion. Handles also have primarily functional characteristics. Most of the storage vessels, and the amphoras and amphorettae, have larger functional handles (in different shapes) for easier handling or for hanging. Cooking pots and cooking pans also have handles, usually tongue-shaped, for easier handling, primarily during the cooking process. The handles on the bowls take many different forms. Often rudimentary, they are smaller than on other vessels and have a decorative character, but also make the slippery burnished or polished vessel easier to grip.

The bibliographic reference for this chapter is:

Mirković-Marić, N., Savić, M. and Rajičić, M. 2021. Pottery from Trench 18 at Belovode, in Radivojević, M., Roberts, B. W., Marić, M., Kuzmanović Cvetković, J., and Rehren, Th. (eds) *The Rise of Metallurgy in Eurasia*: 152–169. Oxford: Archaeopress.

Chapter 13

Chronological attribution of pottery from Trench 18 at Belovode based on correspondence analysis

Miroslav Marić and Neda Mirković-Marić

In this chapter, correspondence analysis (CA) will be used to illustrate a degree of correlation between bowl assemblages and individual features in Trench 18 at Belovode. The comparison will be made between individual chronological settlement horizons defined using a combination of the absolute dates and the relative stratigraphy of the trench. Correspondence analysis (CA) is a statistical method that illustrates a dependence of data rows and columns in databases, improving the possibility of interpretation and the detection of patterns that exist within data sets (Greenacre 1998). This method diminishes the total number of dimensions required to show individual data by breaking down total variability of the table and defining the least required number of dimensions needed to detect variability in assemblages. The primary goal is to determine the existence of a group of variables that are similar in their variations.

The starting premise is that if the gradual evolution of pottery types is to be assumed, then there ought to exist a high degree of correlation between horizons that are chronologically close (e.g. between Horizons 1 and 2, Horizons 2 and 3, and so on). From an analytical perspective, the same technology and CA script will be used as in the Pločnik analysis (Alberti 2013a). An increase in the number of articles (e.g. Kjeld Jensen and Høilund Nielsen 1997; Smith and Neiman 2007; Baxter and Cool 2010; Alberti 2013b) regarding the application of this statistical procedure in archaeology shows its great potential, especially for pottery specialists. For the sake of brevity, in this chapter a series of graphs generated by the R script will be presented and commented upon in order to show the similarity of features and possible identical chronological attribution.

Horizons 1 (4768–4542 cal. BC) and 2 (4835–4731 cal. BC)

The last two horizons at Belovode belong to the period of the late Vinča culture, known as the Vinča D (or Pločnik II) Phase (Whittle *et al.* 2016: 22–23, Figure 9), often divided into two sub-phases (Garašanin 1993). For the Morava Valley and its immediate surrounding, Jovanović (1994) proposes a different, tripartite relative chronology synchronised with the type-site sequence, according to which Horizons 1 and 2 would fall into the Gradac II Phase. Whilst Horizon 1 at Belovode has an abundance of 20 features, including the burned wattle and daub structure (F3), Horizon 2 is significantly less represented with only nine features. The situation is similar for the number of absolute dates obtained (5:1), but the relative stratigraphy recorded during the excavations provides a solid structure for the division between the two horizons.

The correlation coefficient calculated from a contingency table of features (rows) with associated pottery types (columns) is rather high at 0.93, indicating the existence of significant dependency in the data (Greenacre 2007: 28, 61; Healey 2013: 289–290). The number of dimensions that explain more that the average inertia is four, which indicates the number that should be retained in order to explain the data variability (Figure 1).

In order to interpret the CA of the first two horizons (i.e. the final horizons of the Neolithic settlement), the CA script provides symmetric plots for each of the dimensions considered relevant to the analysis. These plots illustrate the relative positions of row and column points with regard to the average values (denoted as the 0.0, 0.0 point), i.e. they illustrate the variability of individual rows and columns with respect to the average values. The closer the points are to the 0.0 point, the less variance they possess. In the case presented here, we interpret the relative position of row points (features) in the space defined by column points (pottery types), in order to understand the similarity of features based on the proportion of pottery types present in each of them. In order to make this possible it is necessary to first observe which pottery types contribute to which dimension. Looking at Figure 2, top left, it is clear that the major contributor to Dimension 1 is pottery type 108, with type 112 contributing just barely. The situation is somewhat different with Dimension 2 (Figure 2, top right), with types 104b, 107 and 117 being the major contributors, the latter contributing the most. In Dimension 3, the situation is again different, with types 100, 104b 115 and 117 being the major contributors (each also important in Dimension 2). Finally, Dimension 4 is defined by types 100, 101, 104b,

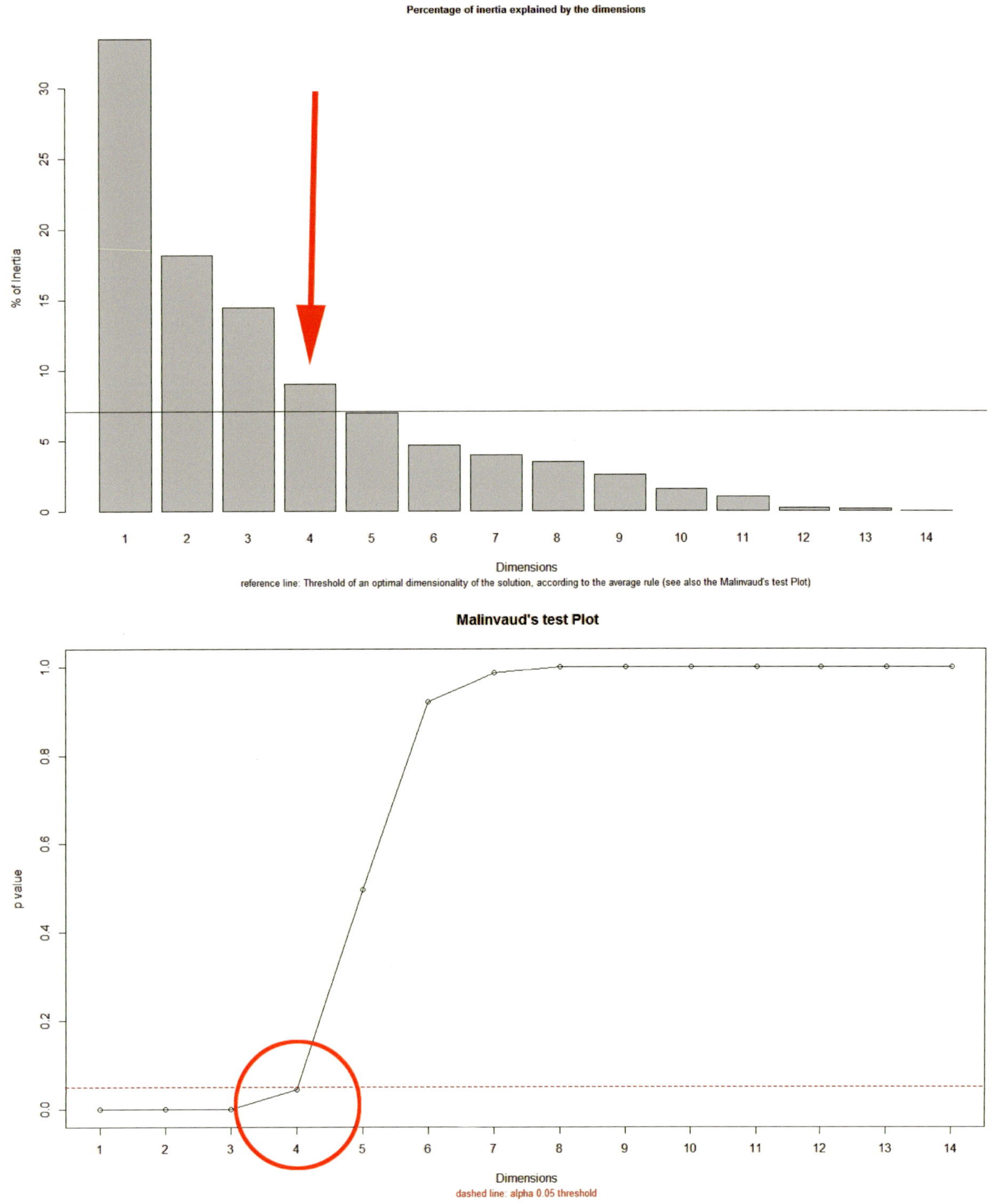

Figure 1. Percentage of inertia explained by dimensions in Horizons 1 and 2.

106, 109, 113, 115, 116 and 117 (types 100, 104b and 117 also being important in Dimensions 2 and 3).

With the important pottery types identified, the optimal 2D symmetric map can be examined (Figure 3). Since there were four dimensions retained, there are three maps, as the number of possible combinations of individual elements is always one less than the number of dimensions retained. The first optimal symmetric map (Figure 3, top) shows the first two principal axes, representing Dimensions 1 and 2. Unfortunately, these two dimensions explain only 51.67% of the total inertia (Dimension 1 – 33.49%; Dimension 2 – 18.18%). Looking at the graph, it can be seen that if all the feature points were vertically projected onto the horizontal axis they would fall very close together, except for features

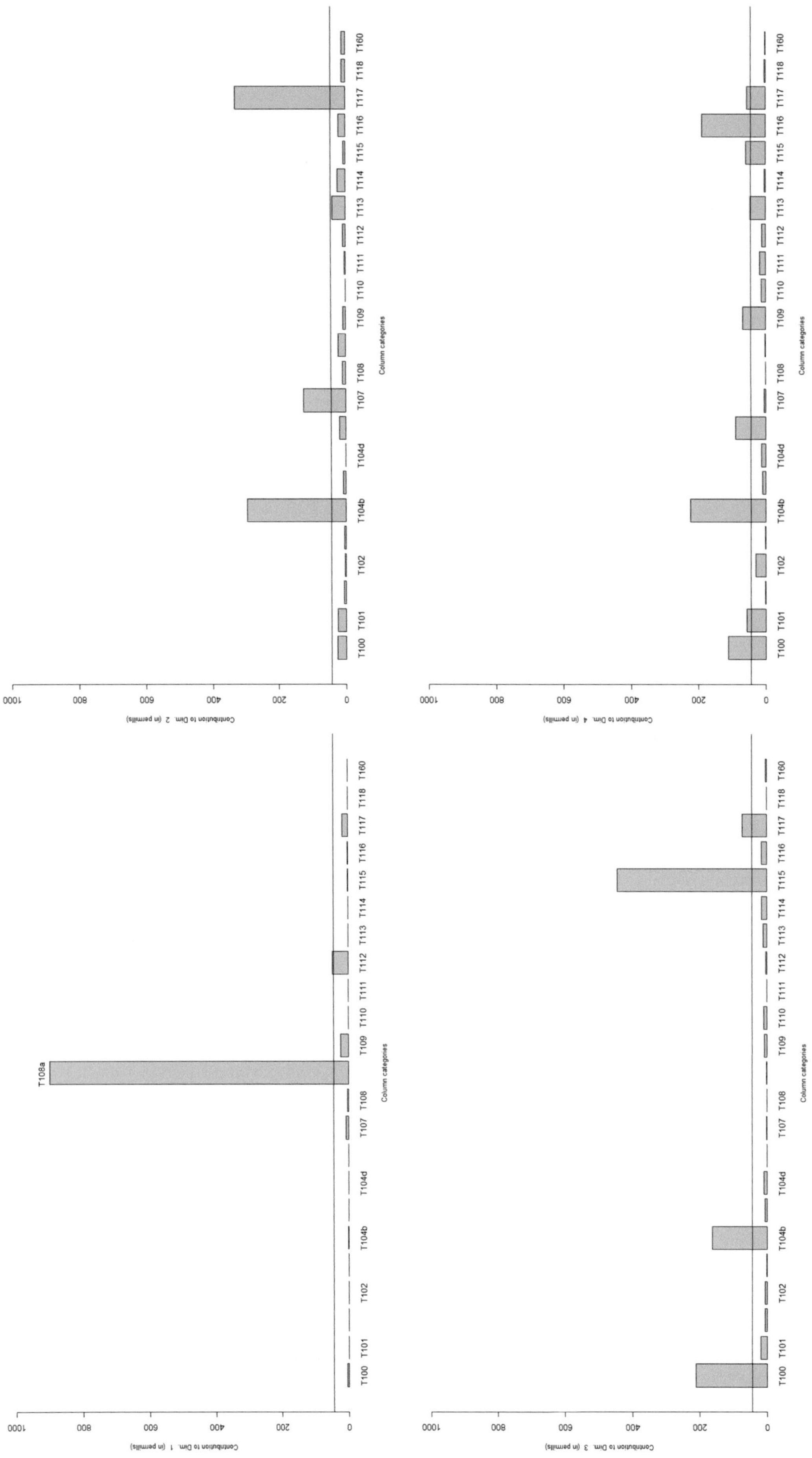

Figure 2. Major pottery type contributors to dimensions in Horizons 1 and 2.

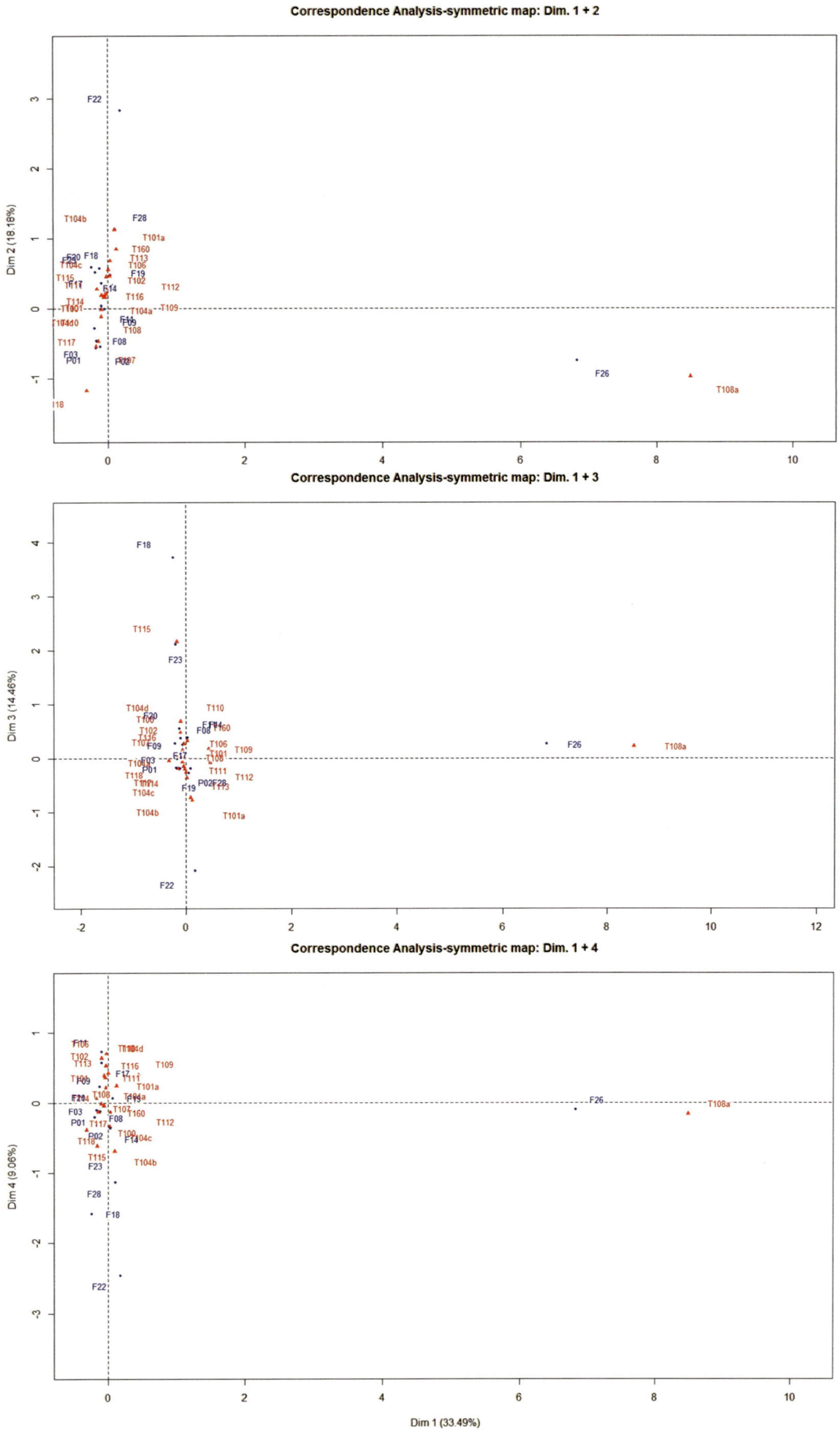

Figure 3. CA symmetric maps for all dimensions retained.

F22 and F26 which are found on the extreme top and extreme right of the graph, indicating them to be outliers. For feature F26, this is further accentuated by the occurrence of pottery type 108a to the right of it, explaining its extreme position perfectly. It is no surprise that this feature is not similar to the rest of the horizon features as it is a circular arrangement of several postholes, and the recovered pot sherds represent the infill of the postholes, i.e. disturbed soil. Similar can be said for feature F22, a small pit with burnt soil and pottery found in the infill, also disturbed soil. The rest of the features in the two horizons analysed show greater similarity with each other, clearly evident from their concentration in the immediate vicinity of the average profile. However, a clear division is visible on the vertical axis, representing Dimension 2. The difference is based on types 107 and 117 which define features F03, P01, P02 and F08, opposite to type 104b which appears more in features F18 and F19. Furthermore, the most similar profile of finds, i.e. the collection of finds with the least variability compared to the average of the horizons being analysed, can be found in features F9 and F11 (F11 being an integral part of F9).

If we look at the second optimal symmetric map (Figure 3, middle), which represents Dimensions 1 and 3 (explaining an additional 14.46% of inertia) a similar pattern occurs. Whilst most of the features can be plotted close together on the horizontal axis indicating little difference amongst them, several outliers exist. Aside from features F22 and F26, already identified as outliers in the previous symmetric map, features F18 and F23 stand out here, clearly associated with pottery type 115, one of the major contributors to Dimension 3. Again, a clear division is visible in the remaining data, with features P01, P02, F03 having greater than the average amount of types 104b and 117 and features F09, F20 having more than the average presence of type 100.

The optimal symmetric map of Dimensions 1 and 4 (explaining an additional 9.06% of inertia) indicates that features F18, F22, F26 and F28 are outliers, whilst the remaining feature points plot close to each other on the vertical axis, showing great similarity between themselves, and some variability with respect to the vertical axis, with types 104b and 117 being the major contributors to features P01, P02, F03 and types 100, 101, 106 and 116 appearing more often than average in features F09, F11 and F17.

Based on the similarity of pottery profiles in the features, cluster analysis identified six clusters (Figure 4) in the assemblage of Horizons 1 and 2. Except for the outliers grouped into separate clusters (Clusters 1, 5 and 6), the remaining clusters are all located much closer to the average profile value (0,0 point), however, the level of clustering differs. Cluster 1 (Figure 4, black), consists of two features, F18 and F23. It is very densely packed in Dimensions 1 and 2, but spaced somewhat apart in other analysed dimensions, indicating a degree of variance. The densest cluster in all retained dimensions is Cluster 2 (Figure 4, red) with three features (F03, P01 and P02) found immediately around the cluster mark indicating significant similarity between them. Cluster 3 (Figure 4, green) is somewhat different, as the included features are found almost opposite to each other around the cluster mark in Dimensions 1 and 2, indicating greater similarity between features F9 and F11 on one side and features F14, F17, F19 and F20 on the other, whilst in Dimension 4, features F14 and F17 are on the outer limits of the cluster, the first showing similarities with Cluster 2. Finally, Cluster 4, close to the average profile in Dimensions 1 and 3, is plotted as an outlier in the factor map of Dimensions 1 and 2, and 1 and 4.

Horizons 2 (4835–4731 cal. BC) and 3 (4987–4851 cal. BC)

The absolute dates of the second horizon place it at the beginning of the Vinča D (or Vinča Pločnik II) Phase, whilst the two available dates for Horizon 3 are synchronous with the Vinča C (Gradac I) Phase, usually placed between 5000/4950 and 4850 cal. BC (Whittle *et al.* 2016: 22–23, Figure 9), of particular interest as the period when copper was introduced into the daily lives of the Vinča population. The six features of this horizon (four actually being different infills of the same pit) represent the only detected activity of this horizon in Trench 18.

The correlation coefficient of the contingency table formed from features of Horizons 2 and 3 is high (0.957), similar to the previous analysis, but the number of dimensions that explain this inertia is smaller, at just two, according to the average rule test, or one, according to the Malinvaud's test (Figure 5). In order to make the analysis possible at all, two dimensions were retained.

The symmetric plots of Dimensions 1 and 2 (explaining 72.67% of the variation) show that the closest to the average profile of these two horizons can be found in feature F18 (Figure 6), with the largest variation found in features F22, F31 and F43, plotted at various extremes of the plot. When compared against specific bowl types it becomes apparent that each of the outliers is defined by a set of one of several bowl types that are distinct in appearance when compared to the rest of the assemblage found in these horizons. However, it must first be seen which of the pottery types contribute to individual dimensions. Figure 7 shows the contributions for both dimensions, and it can clearly be seen (Figure 7, top) that the major bowl types contributing to Dimension 1

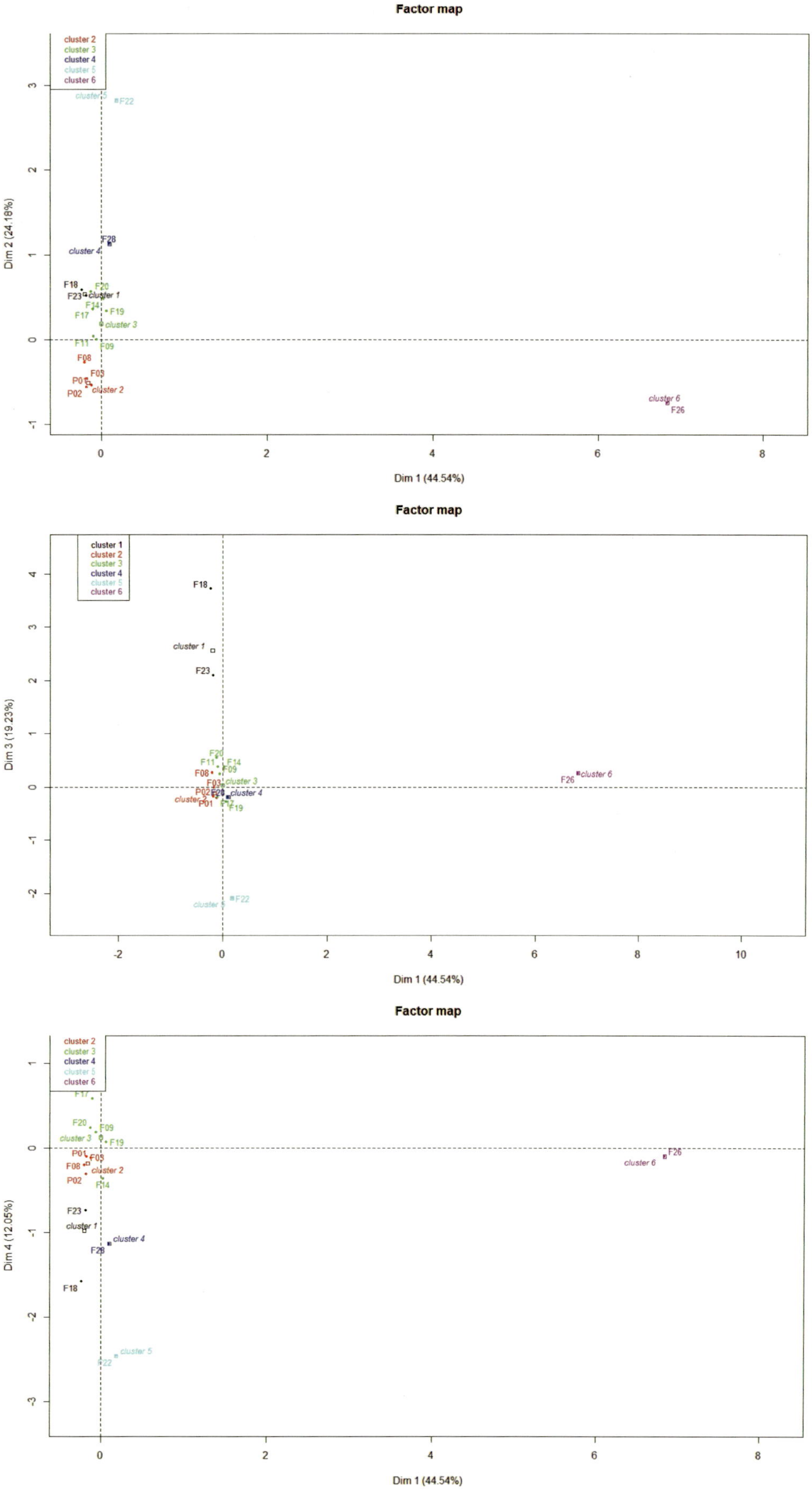

Figure 4. Cluster mapping based on similarity of pottery profiles in the features.

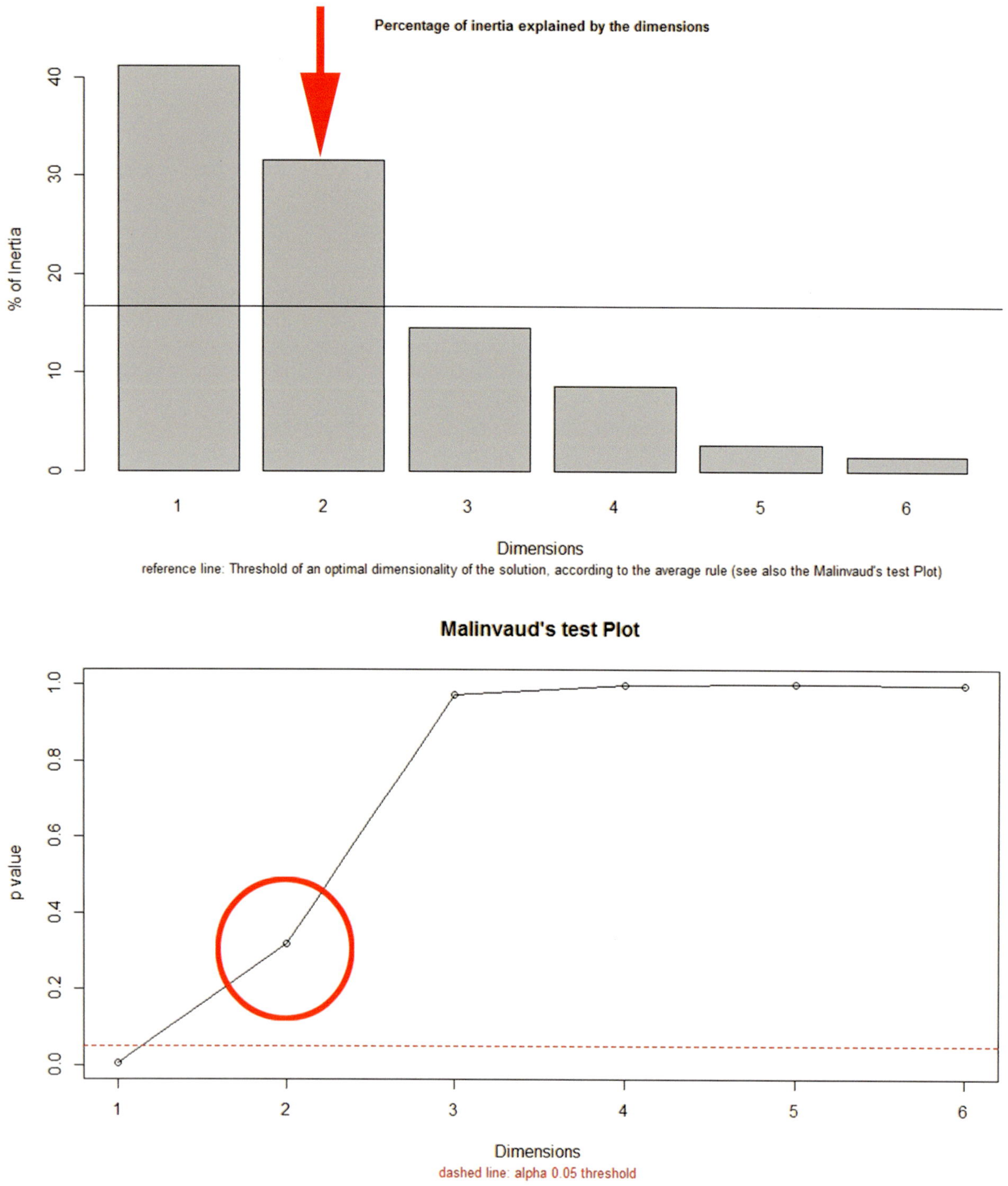

Figure 5. Percentage of inertia explained by dimensions in Horizons 2 and 3.

are 100, 107, 110 and 116, which surpass the threshold level in permills (Greenacre 2007: 82). On the other hand, major contributing bowl types in Dimension 2 are 104b, 108 and 108b, 109, 110 and 111. Such clear distinction shows that the only well-represented bowl type is 110, appearing in both dimensions.

Finally, if we interpret the clustering of the features in these two horizons based on their similarity (Figure 8), the result is rather interesting. The CA script used for the analysis identified three clusters, comprising of mixed features. Cluster 1 (features F23 and F31) consists of two features, the latter being the top infill of the pit

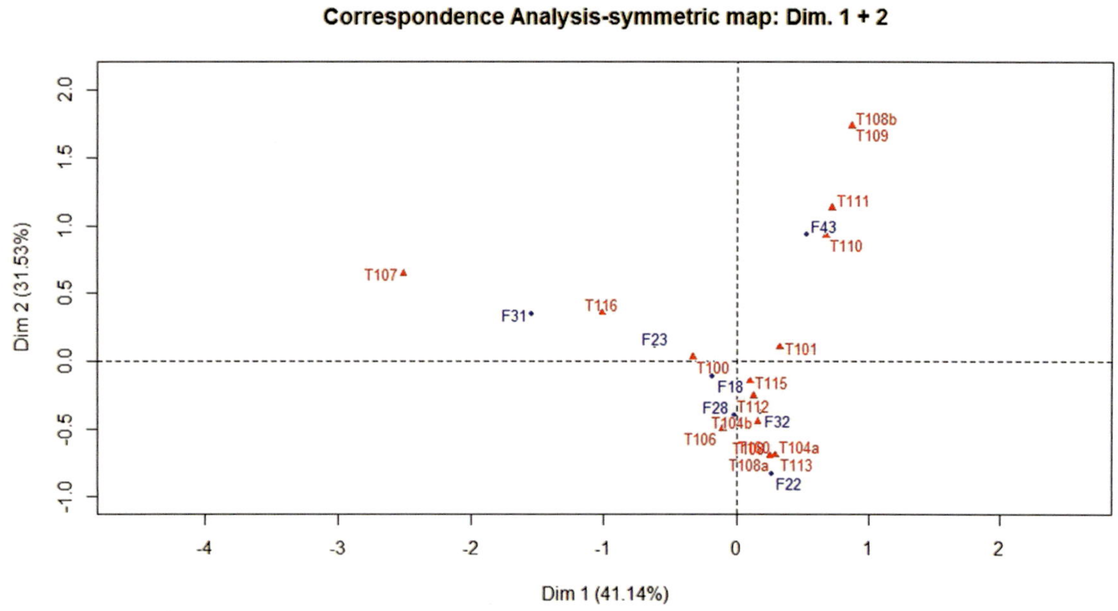

Figure 6. CA symmetric maps for retained dimensions in Horizons 2 and 3.

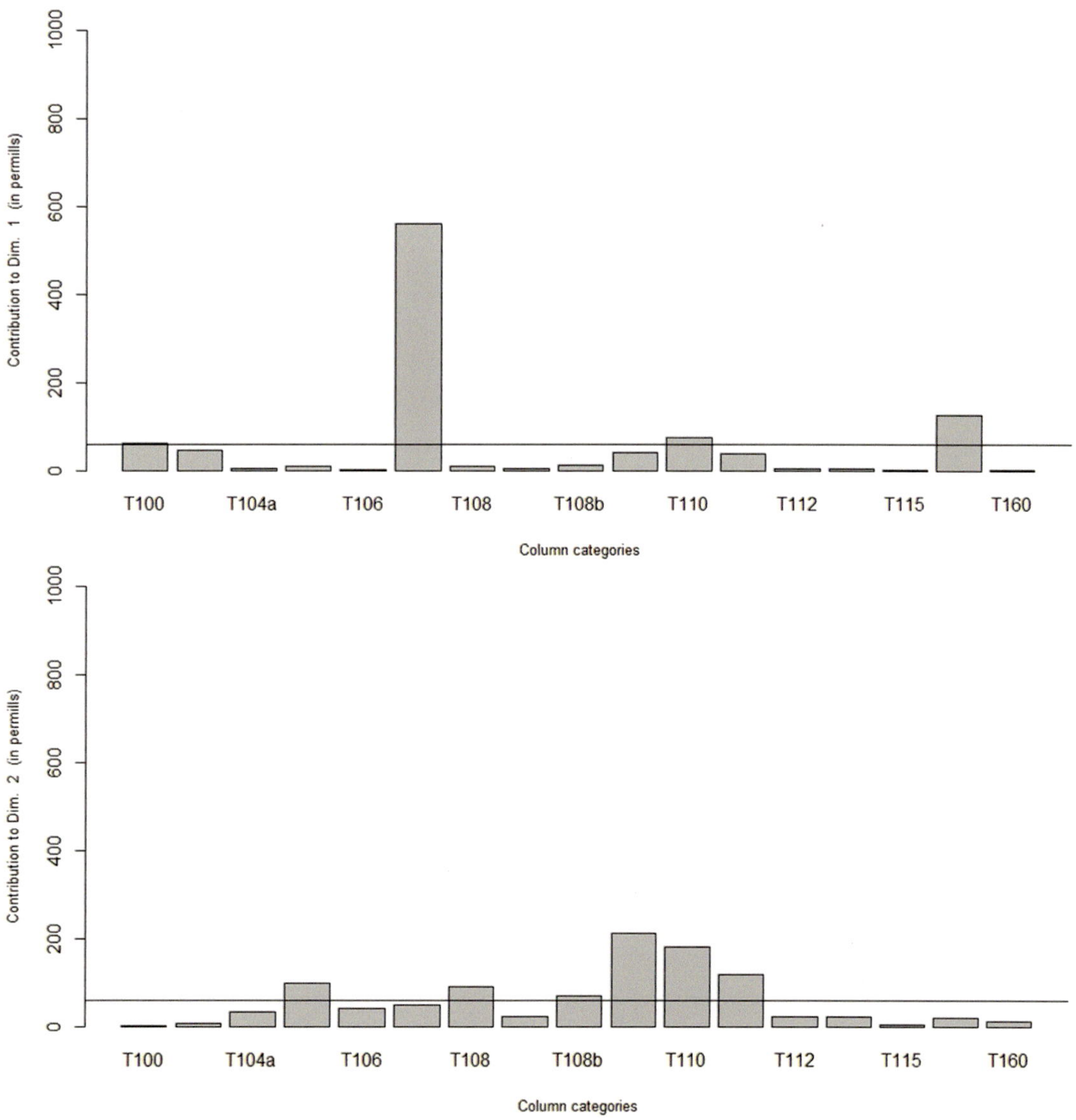

Figure 7. Major pottery type contributors to Dimensions 1 and 2 in Horizons 2 and 3.

marked as F32. It is no surprise that these two features show similarities as feature F31 represents the final infill phase of feature F32, which would indicate a later date. However, the similarity is not very pronounced, signified by the positioning of the cluster marker much closer to feature F31.

Cluster 2, which comprises four features, three (F18, F22 and F28) belonging to Horizon 2 and one (F32) to Horizon 3, is an interesting amalgam (Figure 8, red). Whilst the clustering of three features belonging to Horizon 2 is not unexpected (all of them being a part of the same period of activity) the appearance of feature F32, a major part of the infill of a pit, is unusual. Furthermore, the cluster marker closest to this feature point indicates that the cluster is only loosely defined, based on weak similarities in the finds. One possible explanation is that feature F28, a shallow pit, consists of an assemblage of pottery fragments that originate from an older horizon originally located below the pit cut, i.e. that it contains pottery fragments that do not represent the horizon, but rather are residual remains of older activities deposited in a younger pit.

Finally, Cluster 3 (Figure 8, green) consists of a single feature, F43, which is the earliest deposit in a pit located below feature F32. This feature, consisting of a mixture of yellow clayish soil, burnt daub, charcoal, ash and movable finds obviously belongs to an earlier activity, differing significantly from the rest of the features directly above it, and created by later depositions in the same pit cut. It is no surprise that the pottery assemblage from this feature is different from the rest of the pit, as it probably represents residual refuse deposited in a trash pit early in the phase.

Horizons 3 (4987–4851 cal. BC) and Horizon 4a (5206–5024 cal. BC)

The next horizon pair to be analysed is Horizon 3 (Vinča C/Gradac Phase) and Horizon 4 which, based on absolute dates, can attributed to the Vinča B2 Phase. The later horizon in Trench 18 consists mainly of secondary deposited kiln floors and several shallow, mostly empty pits, thus it was necessary to analyse the spits associated with the horizon instead of features, as fewer than 20 typologically identifiable pieces were recovered from the features in Horizon 4a (compared to over 150 individual bowl fragments identified in spits S18 and S19). The correlation coefficient is high (0.815). Both the average rule test and the Malinvaud's test suggest two dimensions should be retained (Figure 9); these explain 81.1% of inertia in the bowl assemblage.

Several bowl types contribute significantly to Dimension 1, including types 100, 101, 104b, 107, 109, 110, 111 and 113 (Figure 10, top), whilst the second dimension is contributed by types 101, 104b, 107 and 116 (Figure 10, bottom). It must also be noted that the only under-represented feature in these two horizons is feature F43, whilst the other features have over 60% representation in the percentage of inertia for Dimensions 1 and 2.

The symmetric plot of principal dimensions (Figure 11) shows that feature F43 has the most similar bowl assemblage with respect to the average bowl type profile of combined horizons, which explains its under-representation. The other features and spits in the analysis are clearly placed on opposite sides and show a

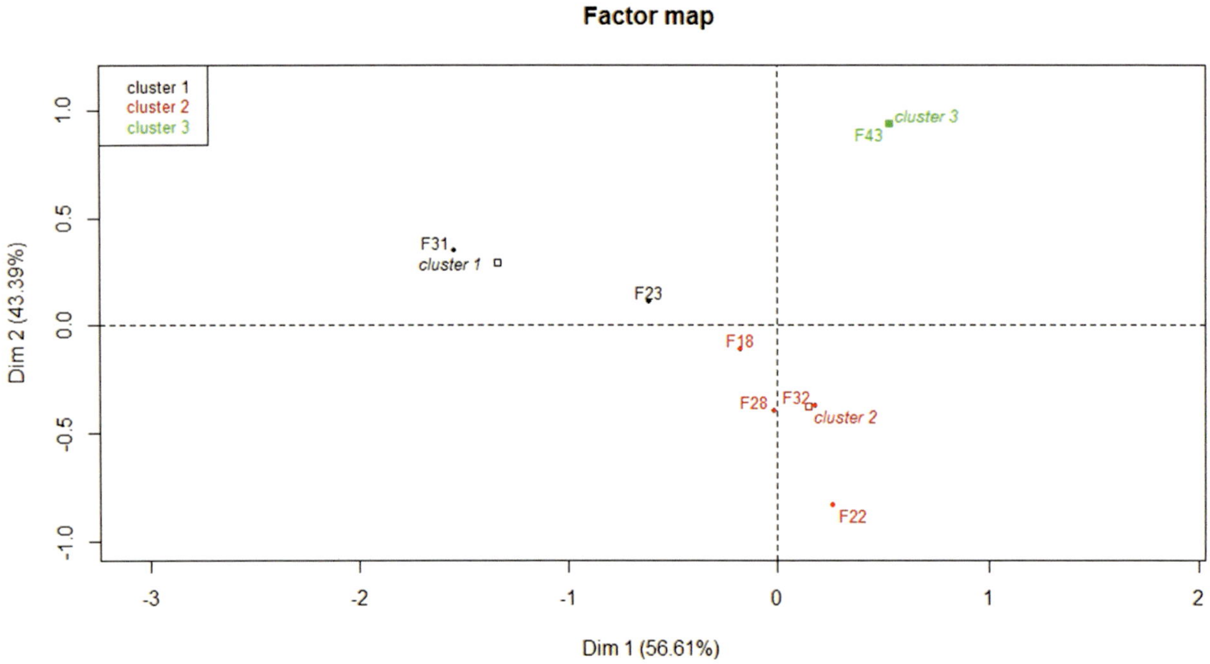

Figure 8. Cluster mapping based on similarity of pottery profiles in the features in Horizons 2 and 3

large variance. On the left side of the Dimension 1 axis lie spits S18 and S19, whilst on the right side are features F31 and F32, which show the existence of a greater than average presence of several bowl types (types 109, 110, 111, 113 and 114 for spits S18 and S19; types 107 and 116 for feature F31; and types 101, 104b, 108a and 160 for feature F32). This interesting division shows a clear distinction and a process of change between the Vinča B2 and Vinča C phases, which may reflect other changes in the Vinča communities in the period between the end of the Neolithic and the introduction of copper.

The results of the CA analysis have already shown the clear grouping of the features, which is also reflected in the cluster analysis as can be seen in the factor map (Figure 12). The clustering is identical to the results of the symmetric plot, with three distinct clusters appearing on opposite sides of the plot.

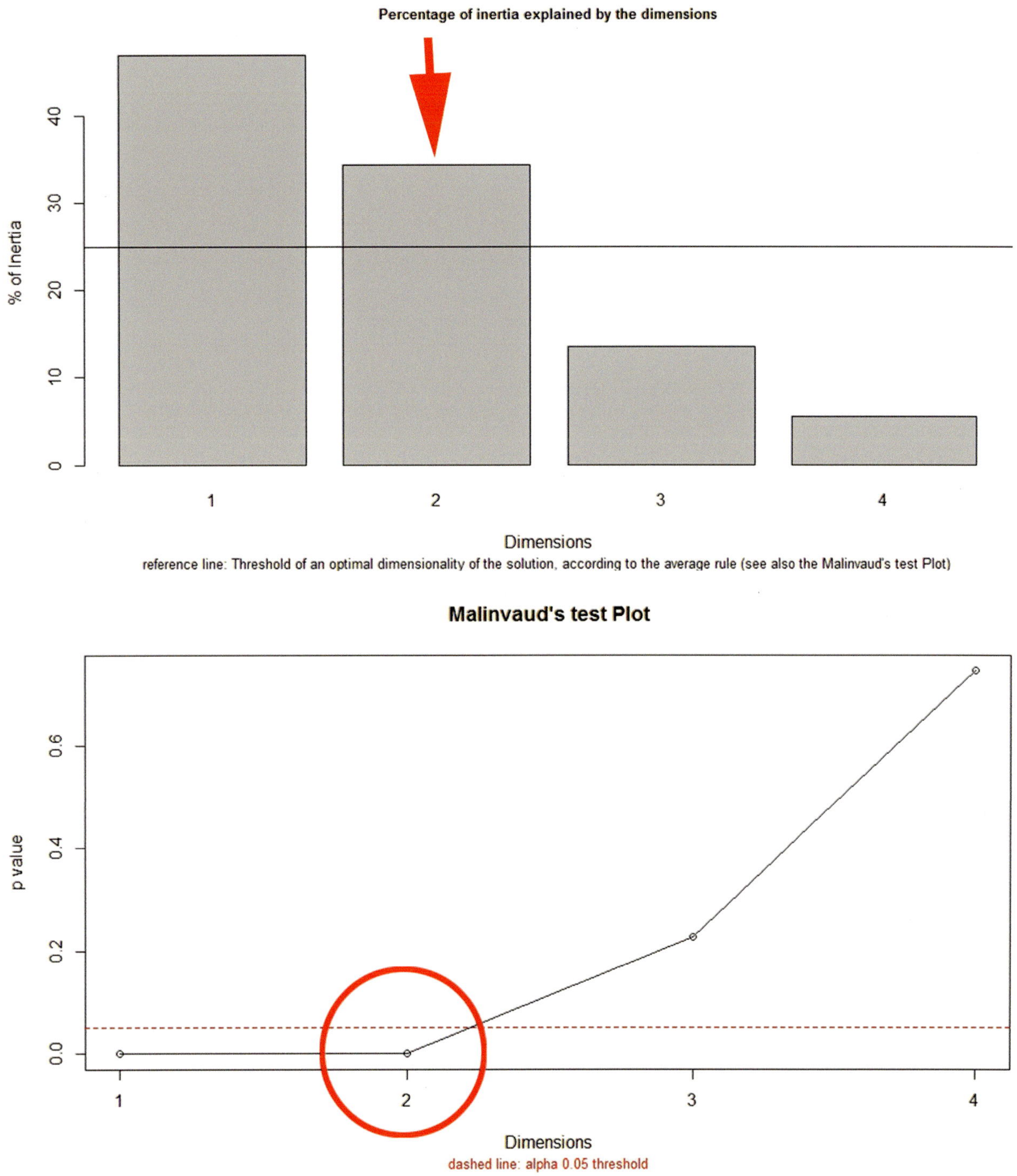

Figure 9. Percentage of inertia explained by dimensions in Horizons 3 and 4.

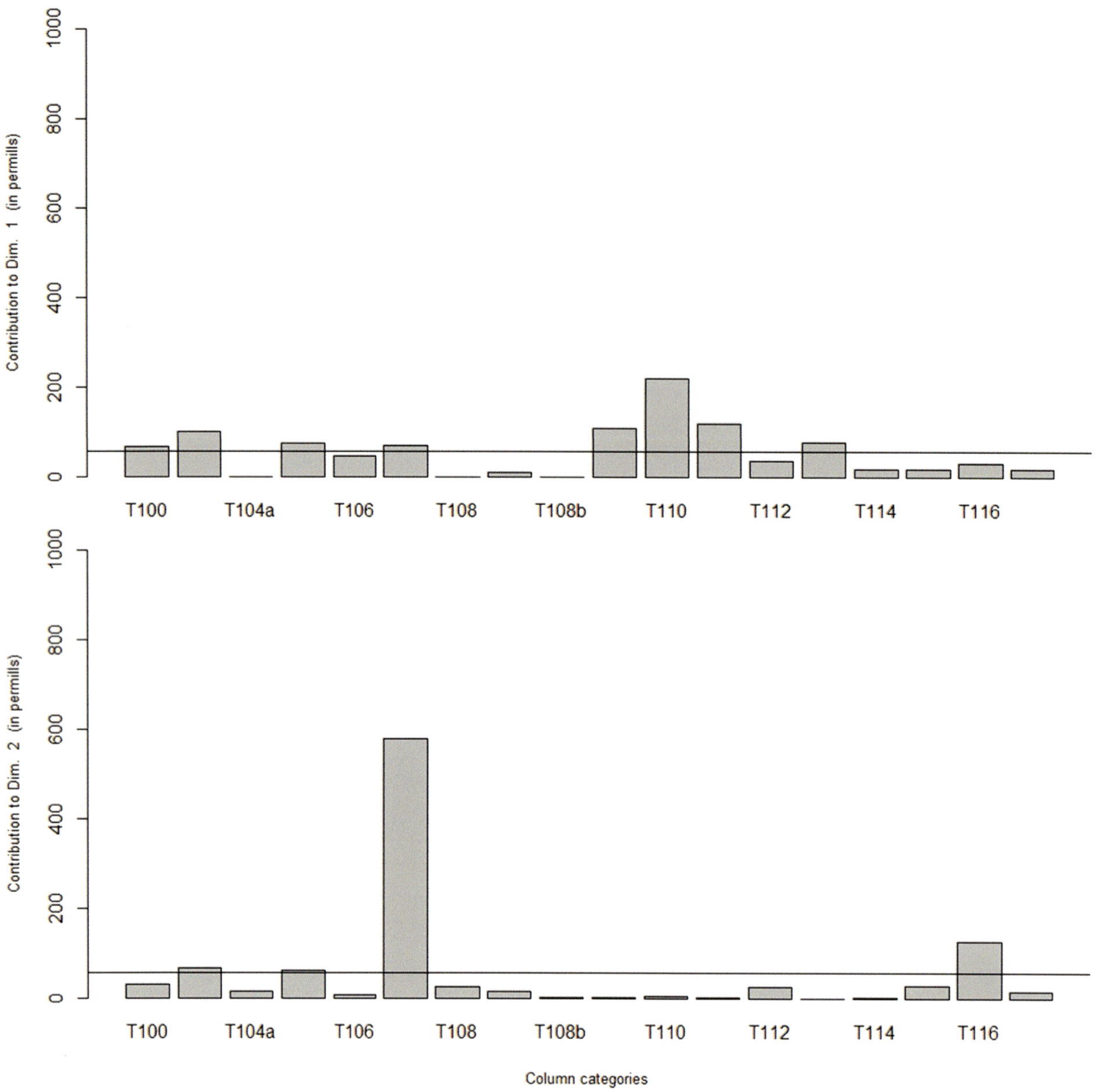

Figure 10. Major pottery type contributors to dimensions in Horizons 3 and 4.

Horizon 4a (5206–5024 cal. BC) and Horizon 4b (5360–5200 cal. BC)

In the traditional chronology of Vinča culture, Horizon 4b would belong to the Vinča B1 sub phase or even the end of the Vinča A Phase (Whittle *et al.* 2016: 22–23, Figure 9), depending on the range of dates available. As the relative stratigraphy of Trench 18 mimics that of Vinča culture, ideally with two sub horizons, it is our deep conviction that, in this case, Horizon 4b should be attributed to the period of Vinča B1. The lack of features (only three, all hearths) prompted an analysis of the spits associated with this horizon (namely spit S20). However, the number of identifiable bowl fragments in Horizon 4b remains small (25 in total), as this part of the site was not intensely used in the earliest period. The correlation coefficient is significantly lower than in previous analyses (0.448), but still statistically significant. The average rule test and the Malinvaud's test (Figure 13) however, indicate that just one (or none in the case of latter) dimension should be retained ($\lambda_1 = 0.13336784$, $\chi^2 = 35.99025$, $p = 0.208397$) demonstrating a significant difference amongst the values in the contingency table. This is best illustrated in a plot of row contribution to individual axis (Figure 14), showing a discrepancy between spit S18 being the sole contributor to Dimension 1 and spits S19 and S20 contributing solely to Dimension 2. Finally, the same can be seen in the symmetric map (Figure 15) with each spit being far from the average profile, having little in common.

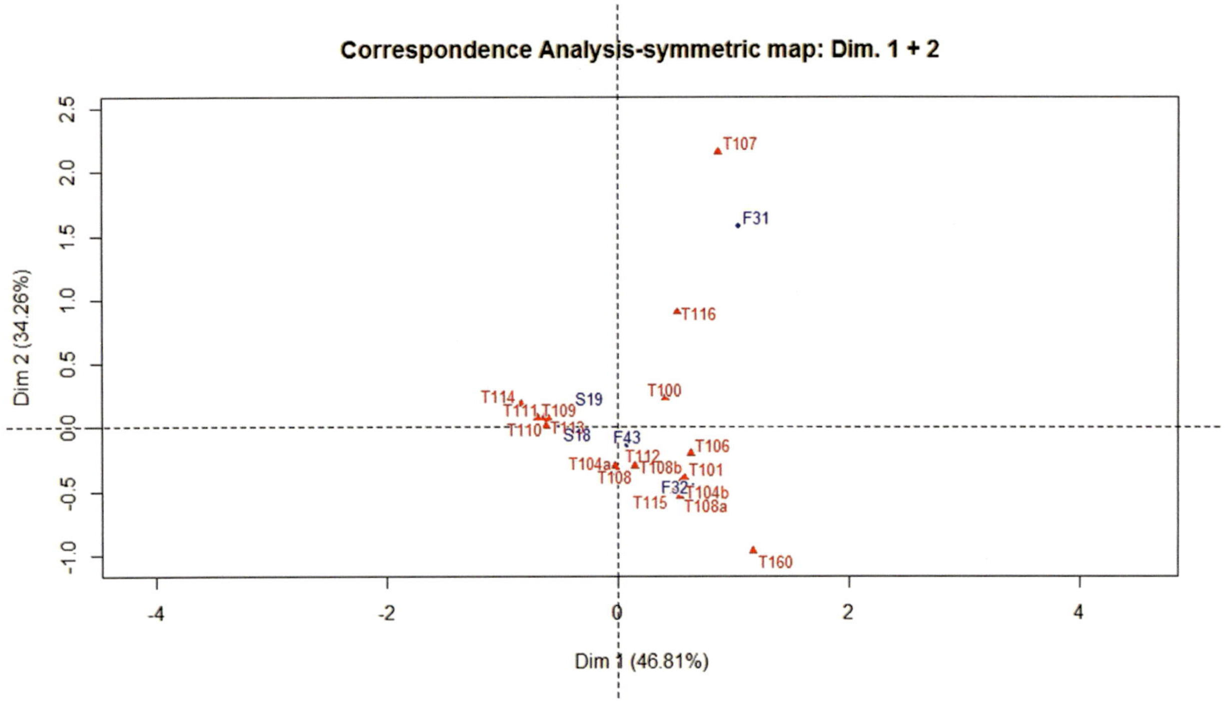

Figure 11. CA symmetric maps for Dimensions 1 and 2 in Horizon 3 and 4.

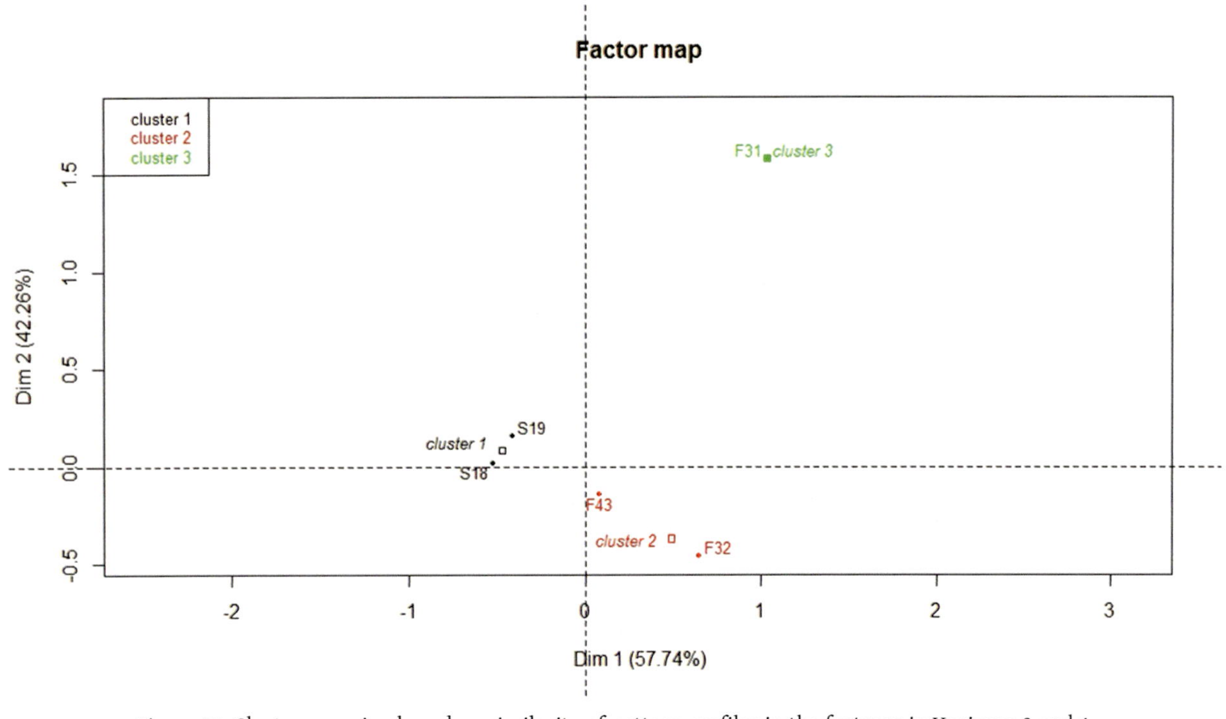

Figure 12. Cluster mapping based on similarity of pottery profiles in the features in Horizons 3 and 4.

However, the profile points for spits S19 and S20 plot to the left of the average profile (on the Dimension 1 axis), which would indicate a degree of similarity, albeit insufficient. The greater than average presence of bowl types 108a, 115, and 116 set spit S19 apart from spit S20, where there is a greater than average presence of types 101, 107 and 109. Opposite to these, spit S18 is characterised by a greater than average presence of types 100, 104a-b, 112 and 113.

Horizon 4b (5360–5200 cal. BC) and Horizon 5 (5466–5372 cal. BC)

The very first occupational horizon at Belovode, marked as Horizon 5, is dated to Vinča A Phase (Whittle *et al.* 2016: 22–23, Figure 9) based on absolute dates (see Chapter 37). The part of the site where Trench 18 is located was not extensively used in this period with only two features being detected (F47, an irregular

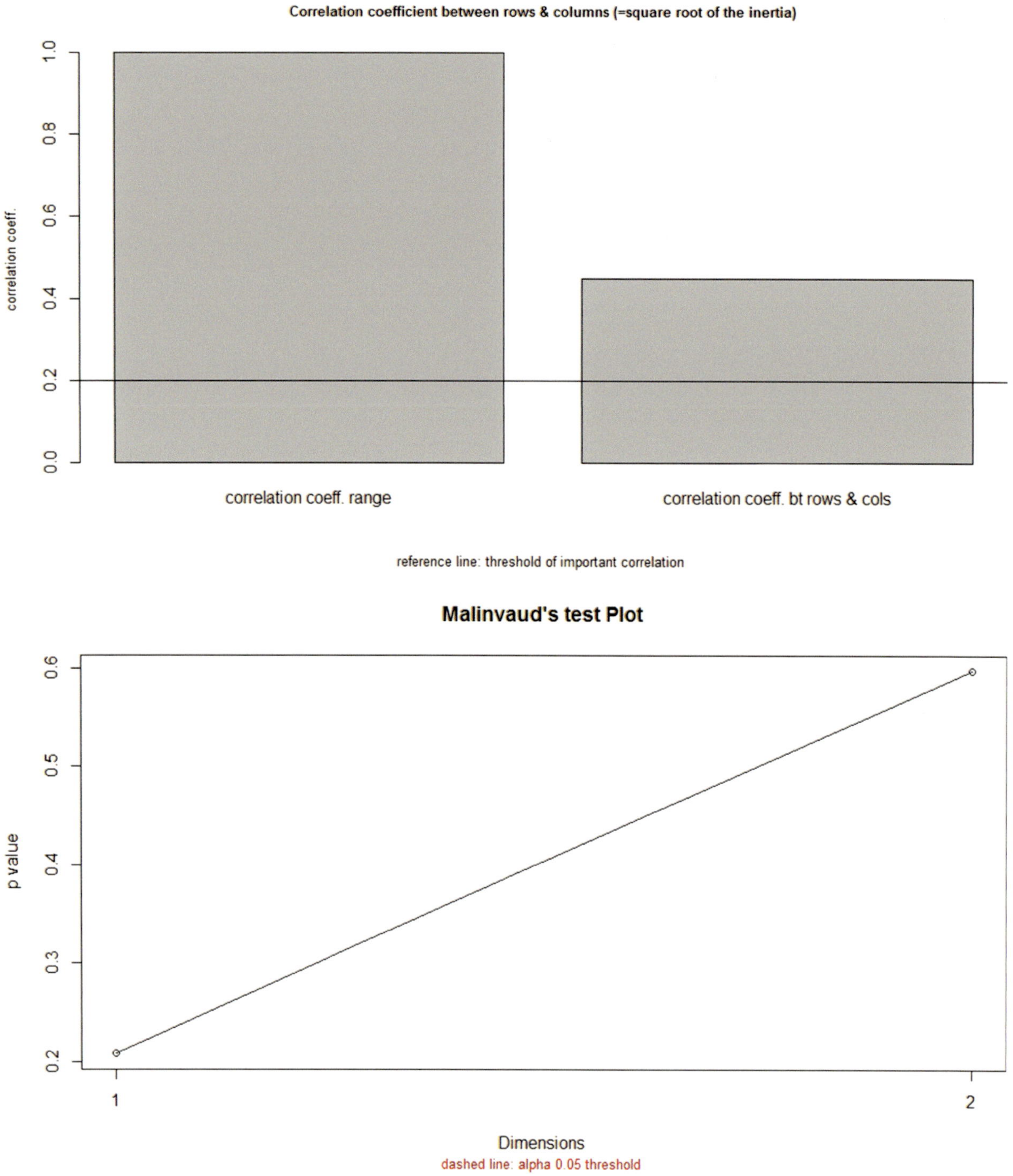

Figure 13. Correlation coefficient and average rule test for Horizons 4a and 4b.

shallow pit, possibly a clay outcrop and F49, a hearth/fire pit on the northwest edge of F47). This absence of activity resulted in an absence of pottery shards (only seven typologically distinct bowl fragments were discovered), and correspondence analysis for these two spits was not possible. The only possible way to include Horizon 5 in the analysis was through a bulk analysis of the entire assemblage of Trench 18, identical to that performed for the Pločnik material. The results were not surprising, identifying spit S21 as an outlier in all dimensions, on the far right of the symmetric map (Figure 16). It is also interesting to note that, in all five retained dimensions, the most similar bowl profiles to that of spit S21 are those of spits S18, S19 and S20 which precede it, as illustrated by their position with respect to principal axis 1 (horizontal axis).

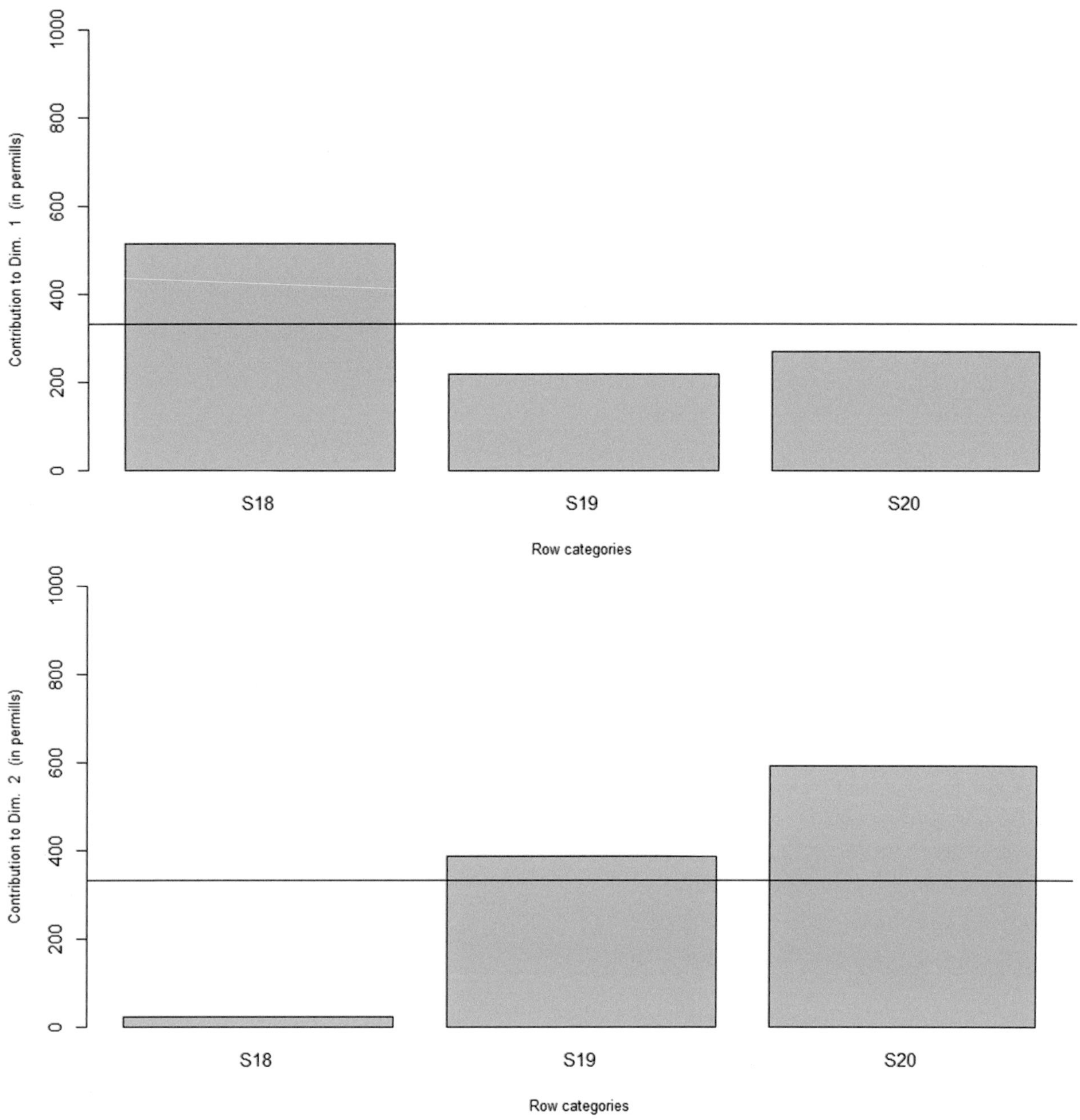

Figure 14. Feature contributions to retained dimensions.

Concluding remarks

The correspondence analysis of individual horizons at Belovode has shown that, in most cases, such a procedure is possible but certain limitations do exist. The dwindling numbers of identifiable bowl fragments towards the earlier phases of occupation in the analysed part of the settlement presented a problem for several aspects of the analysis, the first being the quality of the sample to be used. Although some statistical tests set this number at 5% of the total population, other authors (e.g. Drennan 2009: 127) suggest that this rule of thumb may not always be applied with the same certainty and without scrutiny. It can be argued that a larger sample is always better, but in the case of archaeological trenches this need not be the case, given that individual trenches may not actually adequately represent the site assemblage, especially when the trench does not contain the whole occupation sequence of the site.

The other potential problem in the application of correspondence analysis lies more in the realm of the principles guiding the establishment of certain chronological phases on the basis of specific pottery forms. This is best illustrated in the work of Jovanović (1994) on the relative chronology of late Vinča culture sites in the Morava basin. In his paper, he divides the late phase of the culture in this region into three sub phases. These are not based on the most common vessel type (the bowl), but rather on a comparison of jugs and amphorae,

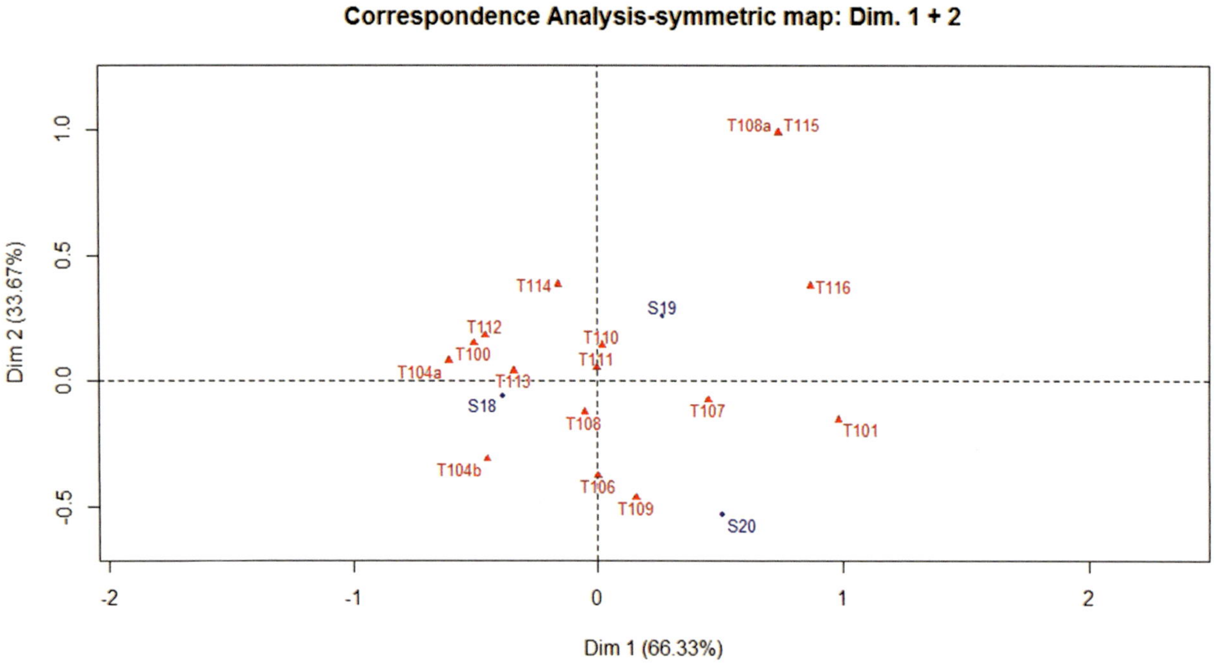

Figure 15. CA symmetric map for retained dimensions in Horizons 4a and 4b.

which are usually disregarded as chronologically insensitive in relative chronologies based on vessel types. It can be argued, perhaps, that more vessel types aside from bowls should be included in contingency tables for a better overview of the variation in the assemblage. Without adequate preselection of types, however, it may prove aggravating, as introducing more vessel types may lead to increased ambiguity and require the retention of more dimensions.

Another problem may lie in the frequency of certain vessel types in different occupational horizons, since the same space may not have had the same function through time, thus retaining different archaeological evidence.

Finally, if we compare the clustering of features in the whole assemblage over the five retained dimensions, five separate clusters appear. Whilst Cluster 1 in Dimensions 1 and 2 (explaining 67.4% or over two thirds of the inertia) reflects almost exactly the relative chronology established in the excavations, Cluster 2 contains certain discrepancies. Aside from correctly placed features like F18, F23, and F28, features F20, F31 and F32 are found out of position as established in the excavations. Whilst feature F20 may truly belong to this cluster and to Horizon 2 instead of Horizon 1 (being vertically situated below feature F3), features F31 and F32, two infills of the same pit, should be more similar to an older horizon. It may be possible that these feature bear greater similarities to the later features as they were deposited later in the time span of Horizon 2, but without additional absolute dates it cannot be said with certainty. Cluster 3, with a single member, feature F22, may represent an outlier containing an under-represented sample, similar to Cluster 4 with feature F26. The latter is certainly a case of disturbed context, being a feature consisting of six postholes arranged in a circle, a part of a construction of perishable (wood) material. In conclusion, Cluster 5 reflects the situation in the lower part of the trench, with spits S18, S19 and S20 tightly grouped around the cluster marker, indicating a similarity between these features. Of particular interest is the appearance of feature F43 in this cluster, which could support the notion of it representing older remains deposited in a later dug hole. On the opposite side of the factor map, the profile point position of spit S21 indicates a difference in bowl assemblage from other members of the cluster but is sufficiently similar to be included. An identical situation occurs in other retained dimension maps.

The application of correspondence analysis in order to identify similarities between undated features may prove a valuable additional tool for better understanding site sequences and interactions. It is clear, however, that larger surfaces with more features need to be explored using consistent methodology for a satisfactory sample to be obtained. Only such an approach can lead to better, more comprehensive results.

The bibliographic reference for this chapter is:

Marić, M. and Mirković-Marić, N. 2021. Chronological attribution of pottery from Trench 18 at Belovode based on correspondence analysis, in Radivojević, M., Roberts, B. W., Marić, M., Kuzmanović Cvetković, J., and Rehren, Th. (eds) *The Rise of Metallurgy in Eurasia*: 170–185. Oxford: Archaeopress.

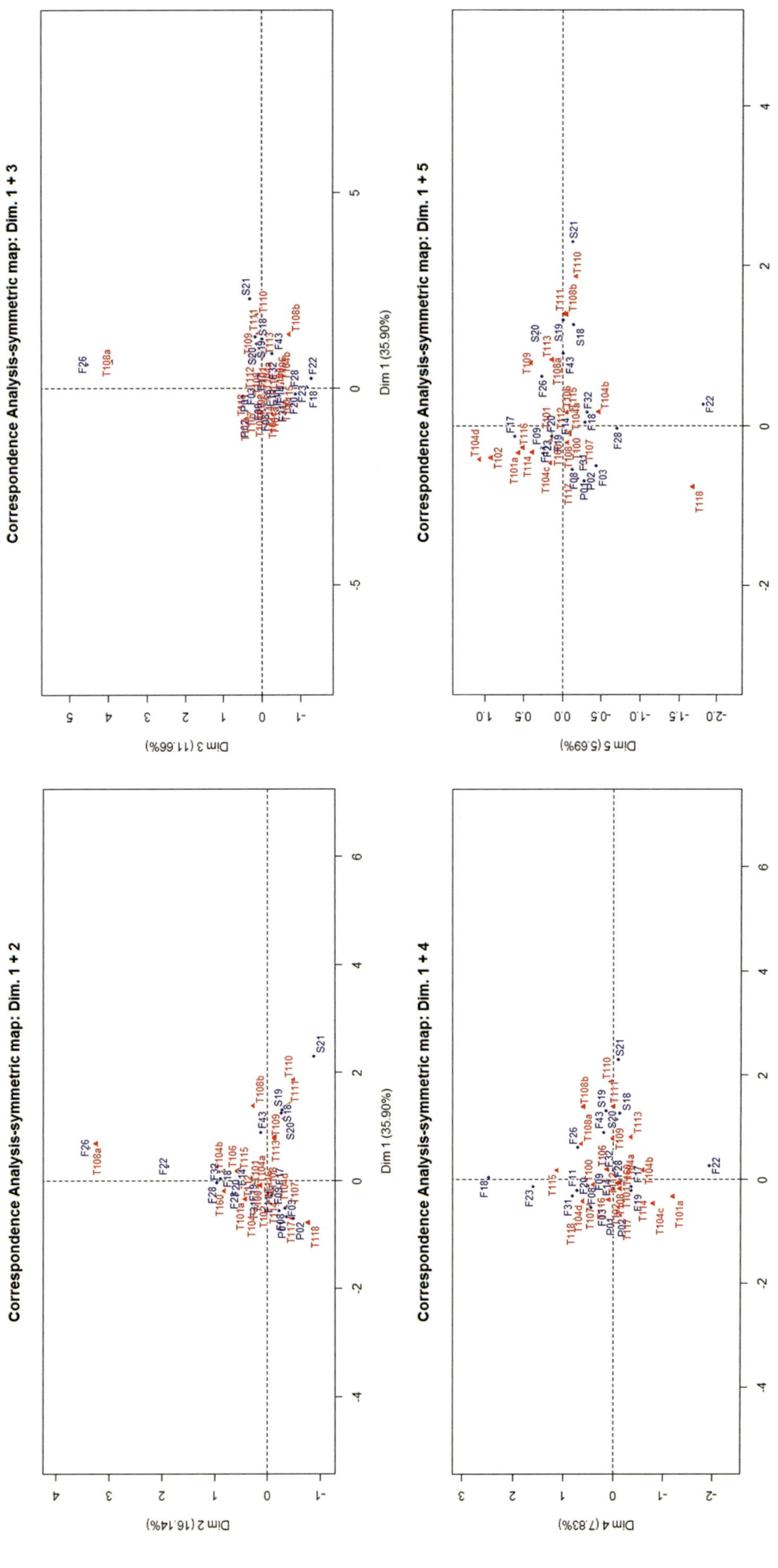

Figure 16. CA symmetric maps for all retained dimensions in Horizons 1 to 4.

Chapter 14

Belovode: technology of pottery production

Silvia Amicone

Introduction

The Serbian Neolithic/Chalcolithic site of Belovode has yielded hundreds of thousands of pottery sherds, along with some of the earliest known evidence of copper smelting. The rich material culture of this site holds significant potential for the study of Vinča pottery craft technology during the transition into the metal age. The site is also well known for dark-burnished pottery (also known as Black Burnished Ware). During the Late Neolithic, this pottery tradition has a widespread distribution across the Balkans (Bonga 2013: 133–178; Chapman 2006; 2007: 296; Holmberg 1964). According to Garašanin (1954), dark-burnished pottery may have originated in Anatolia, where there have been finds of aesthetically similar pottery, however a technological link to the Balkan examples has never been thoroughly investigated. Other scholars argue convincingly that the style could have evolved independently in the Balkans (Chapman 2006; Childe 1936/1937: 29). Dark-burnished pottery is a characteristic component of Vinča material culture and is found from the earliest phases of development; its pyrotechnology has been considered a precursor to metal smelting (Gimbutas 1976a; Kaiser *et al.* 1986).

This chapter focuses on the development of pottery production technology leading up to and following the emergence of copper metallurgy at this important site. Our primary aim is to identify possible technological changes concomitant with the introduction of metalworking to Vinča society. To achieve this, we sampled a selection of pottery sherds representing the different techno-stylistic groups present at Belovode across several chronological phases (cf. Marić and Roberts, Chapter 37, this volume). In particular, sampling included material from the Gradac Phase, which records the beginnings of metalworking at the site, as well as sherds from the pre-metal phases Vinča A–B. These specimens were studied using an integrated analytical approach that incorporates archaeological and archaeometric analysis, including thin section petrography, x-ray powder diffraction (XRPD), and scanning electron microscope (SEM). All conclusions given here are the preliminary results of this ongoing research.

Methods

Ceramic assemblages recovered from Trench 18 at Belovode were studied by the team of pottery specialist working on site (cf. Mirković *et al.*, Chapter 12, this volume). Ceramic artefacts were classified macroscopically according to their production techniques, surface treatment, decoration, fabric and typology. Macroscopic analysis was conducted on fresh fractures using a 10x magnifying glass. The presence or absence of calcareous inclusions was tested with 10% hydrochloric acid.

Following this initial, overall classification, 151 ceramic samples were chosen for more detailed, macroscopic technological characterisation (Table 1 and Appendix B_Table1/Ch14). A selection of these samples was also analysed using thin section petrography, XRPD, and SEM.

Table 1. Number of samples analysed with petrography and XRPD according to building horizons.

Petrography			XRPD		
Belovode 18			Belovode 18		
Number of samples	Horizon	Absolute Chronology	Number of samples	Horizon	Absolute Chronology
82	1	4817-4400 cal. BC	14	1	4817-4400 cal. BC
36	2	4951-4762 cal. BC	7	2	4951-4762 cal. BC
10	3	5140-4859 cal. BC	3	3	5140-4859 cal. BC
16	4	5452-5054 cal. BC	3	4	5452-5054 cal. BC
7	5	5648-5338 cal. BC	2	5	5648-5338 cal. BC

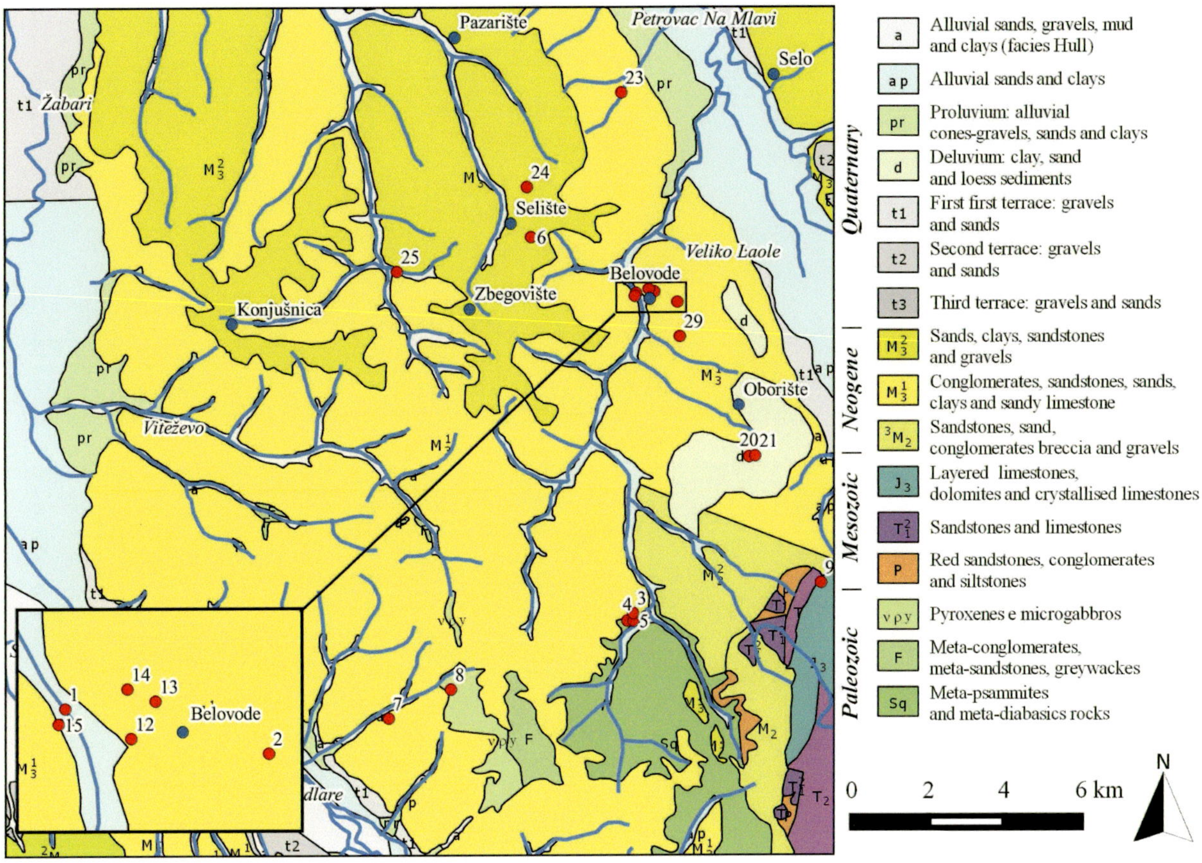

Figure 1. Geological map of the area surrounding Belovode. Red dots indicate the location of specimen samples. (Map by Enrico Croce)

Finally, a prospection of raw materials for pottery making was carried out in the area surrounding the site. Belovode is situated on the Neogene strata, consisting of conglomerate, sandstone, mudstone and sandy limestone (Figure 1). Superficial and recent deposits include alluvial sand and clay from the nearby river Bosur. From the samples collected during the survey, a selection was analysed by petrography, providing important information on the nature and variability of the available raw sources of material for the production of pottery at the site.

Results of the petrographic analysis

One hundred and fifty-one pottery ceramic samples from Belovode were analysed by petrography (see Appendix B_Table2/Ch14). These sherds represent different types of vessels and belong to the five building horizons defined during the excavation of Trench 18 (Marić and Roberts, Chapter 37, this volume). Based on the nature of the aplastic inclusions, matrix, and voids, it was possible to divide the samples into three fabric groups.

Group A: Metasedimentary Rock Fabric (Figures 2 and 3)

This heterogeneous group consists of coarse-grained (subgroup A1) to fine-grained (subgroup A2) samples. It is characterised by poorly sorted sub-angular to sub-rounded inclusions of quartz, polycrystalline quartz, plagioclase and k-feldspar that could be derived from metasedimentary rocks. In addition to quartz and feldspars, the principal mineral inclusions are muscovite, chert, and a minor quantity of epidote and amphibole and, occasionally, calcite and mudstone. In the coarser-grained specimens it was possible to observe the presence of fragments of metasedimentary rocks. The matrix was non-calcareous, and the colour varied from light yellow and bright orange to dark red in plane polarised light (PPL). In cross polarised light (XP), colours ranged from light brown and dark reddish to grey. These characteristics suggest that the raw material employed for producing these pots was a secondary, sedimentary clay.

Significant compositional and textural variation is included in this group, and it can be subdivided into several smaller subgroups. Notable variations include samples BEL 193, BEL 249, BEL 275, BEL 295, BEL 301, BEL 310, and BEL 330, which show a bimodal distribution of the inclusions that could be evidence of tempering (e.g. Figure 2c,d,e). In addition, BEL 10, and BEL 82 (Figure 2f) can be singled out for their coarse-grained inclusions (maximum size 1.5 mm) and for the higher proportion of metasedimentary rock fragments. Moreover, BEL

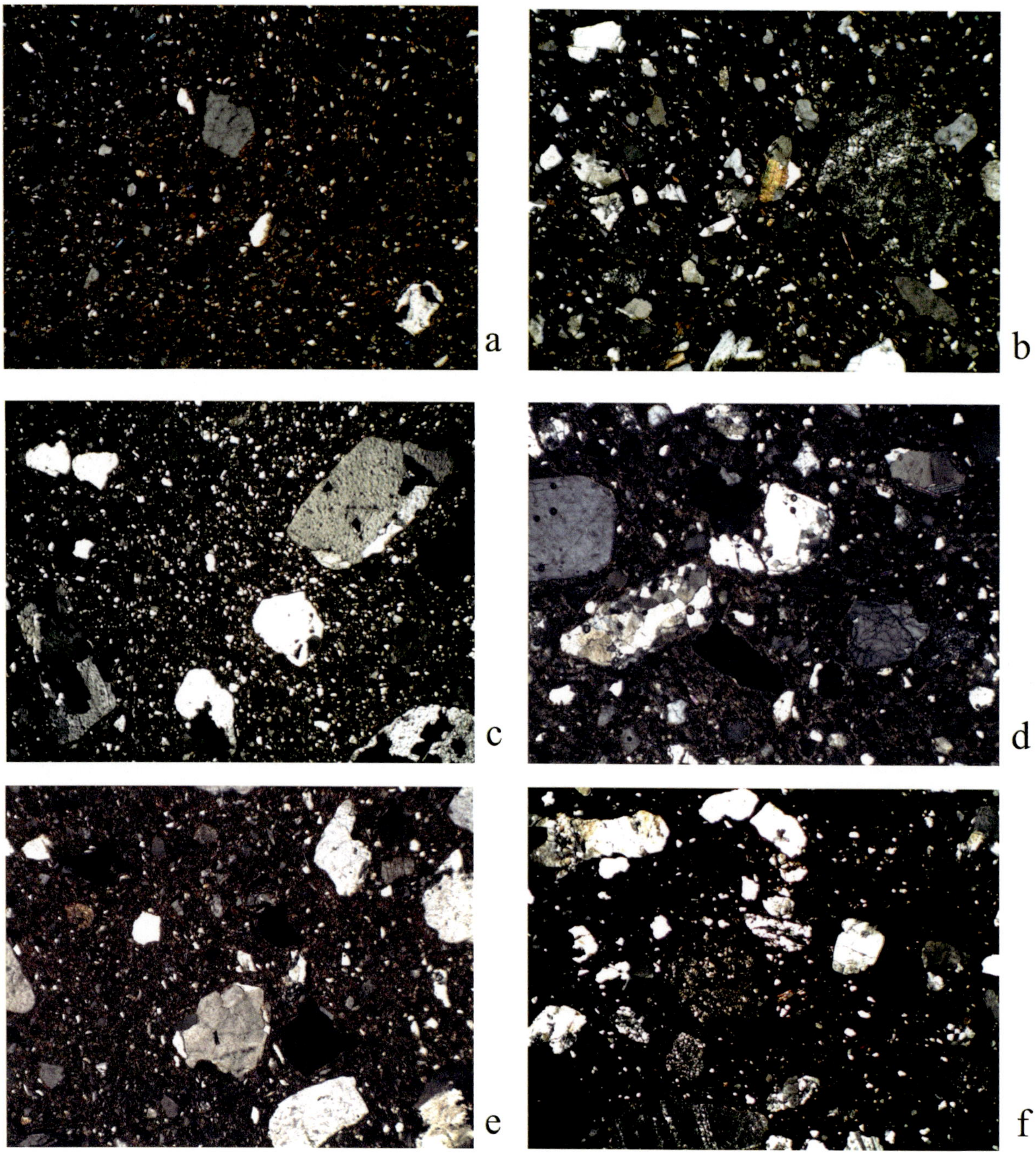

Figure 2. Thin section photomicrographs of selected ceramics from Belovode analysed in this study: a) Fabric Group A with abundant quartz, plagioclase and fragments of metamorphic rocks, fine XP; b) Fabric Group A XP, coarse; c) Fabric Group A, bimodal distribution (BEL 310) XP; d) Fabric Group A, bimodal distribution (BEL 193) XP; e) Fabric Group A, bimodal distribution (BEL 275) XP; f) Fabric A (BEL 10) XP. Image width = 3 mm (a, b, d, e) except c, f, = 6 mm.

4 and BEL 67 (Figure 3a) show possible evidence of mudstone tempering, while BEL 104 (Figure 3b) could have been produced with the addition of organic material. Finally, BEL 87, BEL 300, and BEL 324 (Figure 3c) show presence of limestone fragments which could have been added as temper and BEL 318 (Figure 3d) is characterised by abundant clay pellets. The optical activity of the clay matrix in the ceramics varies between samples. Several sherds, particularly those of dark-burnished ware, display a very low optical activity suggesting that these were fired to relatively high temperatures; other specimens are characterised by a medium to high optical activity that indicates lower firing temperatures.

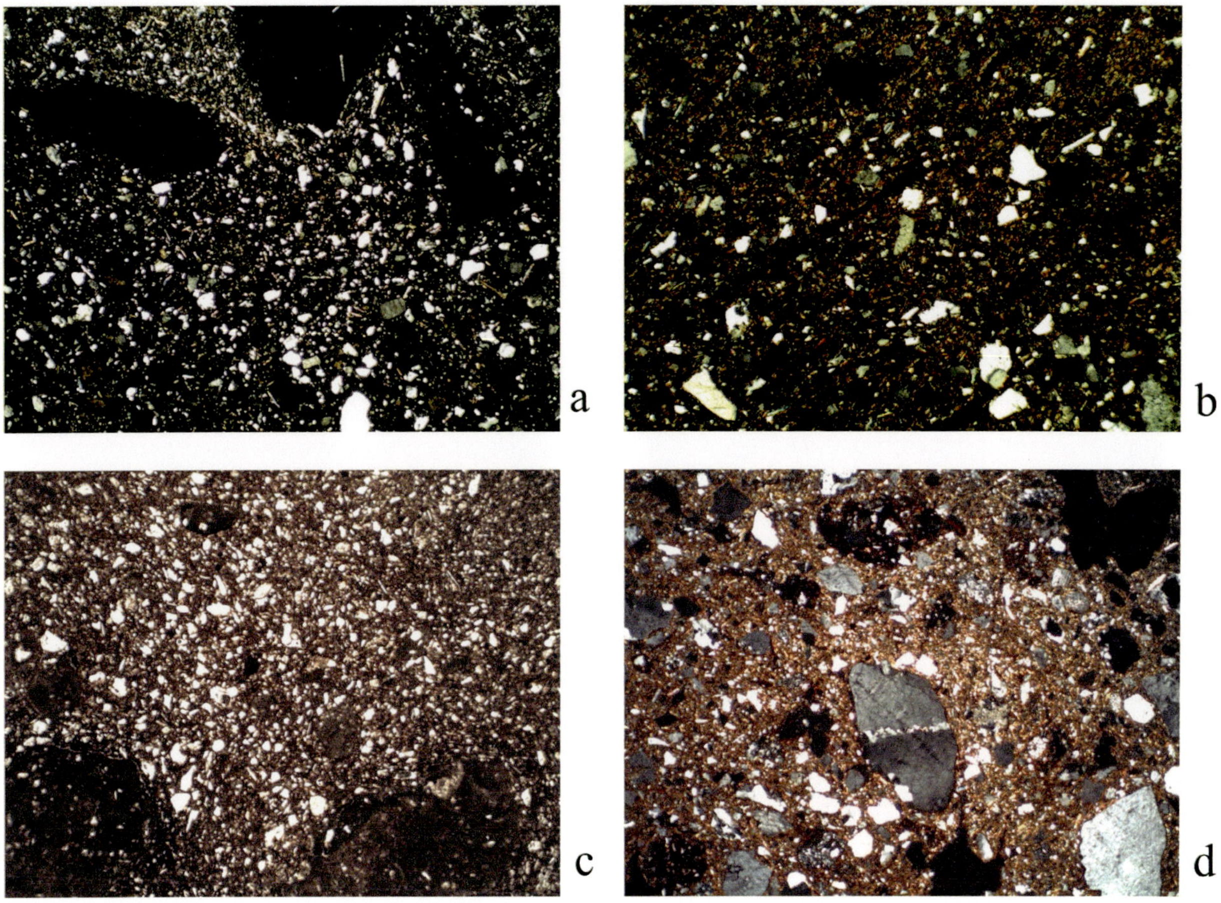

Figure 3. Thin section photomicrographs of selected ceramics from Belovode analysed in this study: a) Fabric Group A with mudstone (BEL 67) XP; b) Fabric Group A with organic material (BEL 104) XP; c) Fabric Group A, limestone tempering (BEL 300) PPL; d) Fabric Group A, abundant clay pellets (BEL 318) XP. Image width = 6 mm (a, c, d) except b = 3 mm.

Group B: Fossiliferous Fabric (Figure 4a,b)

This is a heterogeneous group of fine- to coarse-grained samples dominated by the presence of shell fragments and microfossils. Mineral inclusions include quartz, polycrystalline quartz, plagioclase, muscovite, calcite, chert and, in minor quantities, epidote and amphibole. In the coarser samples (BEL 97, BEL 98, BEL 108, BEL 110, and BEL 114) it was possible to observe the presence of metasedimentary rocks and mudstone fragments. In PPL they appeared light yellow in colour; in XP they were yellow-to-dark brown. The characteristics observed for this group suggest that the material employed for manufacturing these vessels was a mixed secondary clay material derived from the weathering of shelly limestone.

Group SR: Organic Tempered Fabric (Figure 4c).

This petrographic group includes three samples from the earliest horizons (BEL 303, BEL 334, and BEL 348) characterised by organic tempering and a black core. Predominant mineral inclusions include quartz, muscovite, and plagioclase. Common mineral inclusions comprise amphibole, chert and opaques. Less frequently, fragments of metasedimentary rocks also occur. Their colour in PPL is light yellow to grey, and yellow-to very dark grey in XP. Based upon these mineral inclusions and rock fragments, the raw material used to produce these ceramics appears to be a secondary clay.

Along with these three petrographic groups it was possible to single out several outliers. Samples BEL 15 and BEL 31 are fine-grained samples (Figure 4d), characterised by the presence of well-sorted and small (mean size 0.1 mm) sub-angular inclusions of quartz; BEL 43 and BEL 60 (Figure 4e) show a very fine clay matrix to which a coarse sand (maximum size 2.8 mm) composed of metamorphic rocks quartz and plagioclase has been added; BEL 53, BEL288 and BEL 321 (Figure 4f) are characterised by a very well-sorted coarse fraction (mean size 1mm), rich in metasedimentary rocks amphibole and epidote; BEL 155 is a medium-coarse sample (Figure 4g) characterised by the presence of serpentinite; and BEL 247 and BEL 325 (Figure 4h) are tempered with mica-schist.

Figure 4. Thin section photomicrographs of selected ceramics from Belovode analysed in this study: a) Fabric Group B with shells and microfossils XP; b) Fabric Group B with shells, microfossils and metamorphic rocks XP; c) Fabric Group Starčevo with organic tempering PPL; d) BEL 31, very fine matrix XP; e) BEL 43 XP; f) BEL 53 XP; g) BEL 155 with serpentinite XP; h) BEL 325 with mica-schist XP. Image width = 3 mm (a, d, e, f, g, h) except b, c = 6 mm.

Results of the XRPD and SEM analyses

Twenty-nine samples from Belovode were analysed by XRPD diffraction. These included dark-burnished pottery as well as other ceramic types. The results (Table 2) revealed that most of the samples contain illitic clay, which can be distinguished by a main diffraction peak at 10 Å d spacing and further peaks at increasing 2θ° with variable intensities. Some samples exhibit a weak diffraction peak corresponding to a d-value of approximately 14 Å which points to either chlorite or montmorillonite.

Overall, the other main mineralogical assemblages that were detected through XRPD analysis comprised quartz, felspars, and calcite. In addition, the presence of amphibole was verified its main characteristic peak at about d ~ 8 Å. Finally, three samples (BEL 46, BEL 123, and BEL 224) show weak peaks of haematite, cristobalite, and spinel.

The sequence of mineralogical changes within ceramics during firing can be used to determine firing temperatures reached during ceramic production (Figure 5). In order to identify the interval of temperatures at which ceramics were fired, particular minerals are considered as indicators for mineralogical changes that occur in the firing process (Maggetti 1982; Maritan 2004; Nodari *et al.* 2007). These minerals include, for example, haematite, magnetite, cristobalite,

Table 2. Summary of the XRPD results (DB=dark-burnished).

Sample	Chronological Horizon	DB	Optical Activity	Qtz	Fsp	Cc	Am	Ill	Msc	Hem	Cri	Spl	Temp
BEL 31	1 (C-D)	X	high	X	X			X					< 900 °C
BEL 46	1 (C-D)		absent	X	X					X	X	X	> 1000 °C
BEL 52	1 (C-D)		weak	X	X	X		X					< 900 °C
BEL 68	1 (C-D)	X	moderate	X	X			X					< 900 °C
BEL 94	1 (C-D)	X	moderate	X	X			X					< 900 °C
BEL 95	1 (C-D)	X	high	X	X			X					< 900 °C
BEL 101	1 (C-D)	X	high	X	X	X		X					< 900 °C
BEL 109	1 (C-D)	X	weak	X	X		X	X					< 900 °C
BEL 115	1 (C-D)		moderate	X	X	X		X					< 900 °C
BEL 116	1 (C-D)		high	X	X	X		X					< 900 °C
BEL 118	1 (C-D)	X	weak	X	X			X					< 900 °C
BEL 123	1 (C-D)		absent	X	X					X	X	X	> 1000 °C
BEL 132	1 (C-D)		weak	X	X	X		X					< 900 °C
BEL 162	2 (Gradac-C)	X	moderate	X	X		X	X					< 900 °C
BEL 163	2 (Gradac-C)	X	high	X	X			X					< 900 °C
BEL 169	1 (C-D)	X	weak	X	X			X					< 900 °C
BEL 176	2 (Gradac-C)	X	weak	X	X			X					< 900 °C
BEL 198	2 (Gradac-C)	X	weak	X	X			X					< 900 °C
BEL 219	2 (Gradac-C)	X	moderate	X	X		X	X					< 900 °C
BEL 221	2 (Gradac-C)		absent	X	X	X		X					< 900 °C
BEL 224	2 (Gradac-C)		absent	X	X					X	X	X	> 1000 °C
BEL 288	4 (A)		absent	X	X			X					< 900 °C
BEL 289	3 (B1-B2)	X	moderate	X	X			X					< 900 °C
BEL 290	4 (A)		absent	X	X			X					< 900 °C
BEL 295	4 (A)		weak	X	X			X					< 900 °C
BEL 299	3 (B1-B2)	X	moderate	X	X			X					< 900 °C
BEL 300	3 (B1-B2)	X	high	X	X	X		X					< 900 °C
BEL 303	5 (Starčevo/A)		weak	X	X			X					< 900 °C
BEL 334	5 (Starčevo/A)		weak	X	X			X					< 900 °C

mullite, calcite, montmorillonite, clay minerals and feldspars. The diffractograms of most samples (Figures 6 and 7) revealed the presence of illite, which indicates that the maximum firing temperature must have been below 850–900°C (Kulbicki 1958; Maggetti 1982). Two of the dark burnished samples (BEL 18-52 and BEL 18-101) and the cooking pot (BEL 18-116) revealed the presence of calcite in their diffraction pattern (Figures 8 and 9) which suggests that these vessels could have not been fired at temperatures above 850°C, at which point calcite decomposes (Maggetti 1982; Maritan 2004). The overall composition of the samples suggests that they were probably fired to a maximum temperature of between 750 and 850°C. Three samples, however

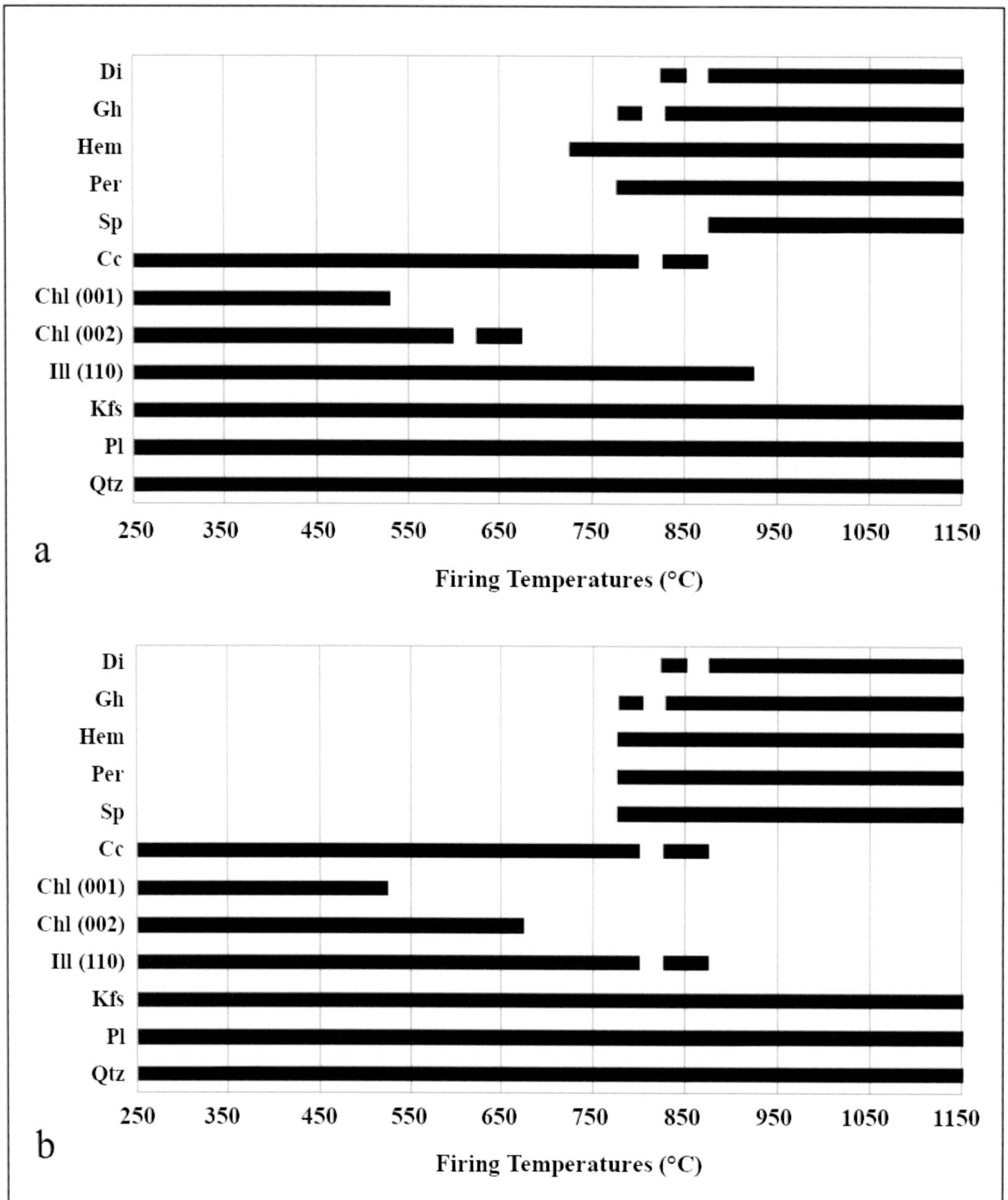

Figure 5. Bar charts showing mineralogical changes that take place during ceramic firing: a) oxidising atmosphere. Di: diopside; Gh: gehlenite; Hem: hematite; Per: periclase; Sp: spinel; Cc: calcite; Chl: chlorite; Ill: illite; Kfs: feldspar potassium; Pl: plagioclase; Qtz: quartz; b) reducing atmosphere. Di: diopside; Gh: gehlenite; Per: periclasio; Mag: magnetite; Sp: spinel; Cc: calcite; Chl: chlorite; Ill: illite; Kfs: feldspar potassium; Pl: plagioclase; Qtz: quartz. (Modified after Maritan 2004, p. 304, Fig. 7.)

(Figure 10, BEL 46, BEL 123, and BEL 224), showed the presence of mineral phases (cristobalite and spinel) normally associated with higher temperatures. They also show the presence of haematite which, along with magnetite, when present in non-calcareous clays begins to nucleate at about 550°C under oxidising and reducing conditions respectively. In calcareous clays, it begins to nucleate at about 725–750°C. None of the other samples show the presence of these two minerals, but they may occur in most specimens at levels below the limits of detection.

The observations based on the XRPD results were confirmed by the SEM analysis. With this technique, fresh fractures of fourteen samples were examined in order to assess their degree of vitrification (Maniatis and Tite 1981). All samples (Table 3 and Figure 11) showed an initial degree of vitrification that closely matches the temperature estimated by XRPD. Only samples BEL 46, BEL 123, and BEL 224 (Figures 11e,f,g) show an extensive or continuous degree of vitrification, confirming that these samples were fired to higher temperatures, of around 1000°C.

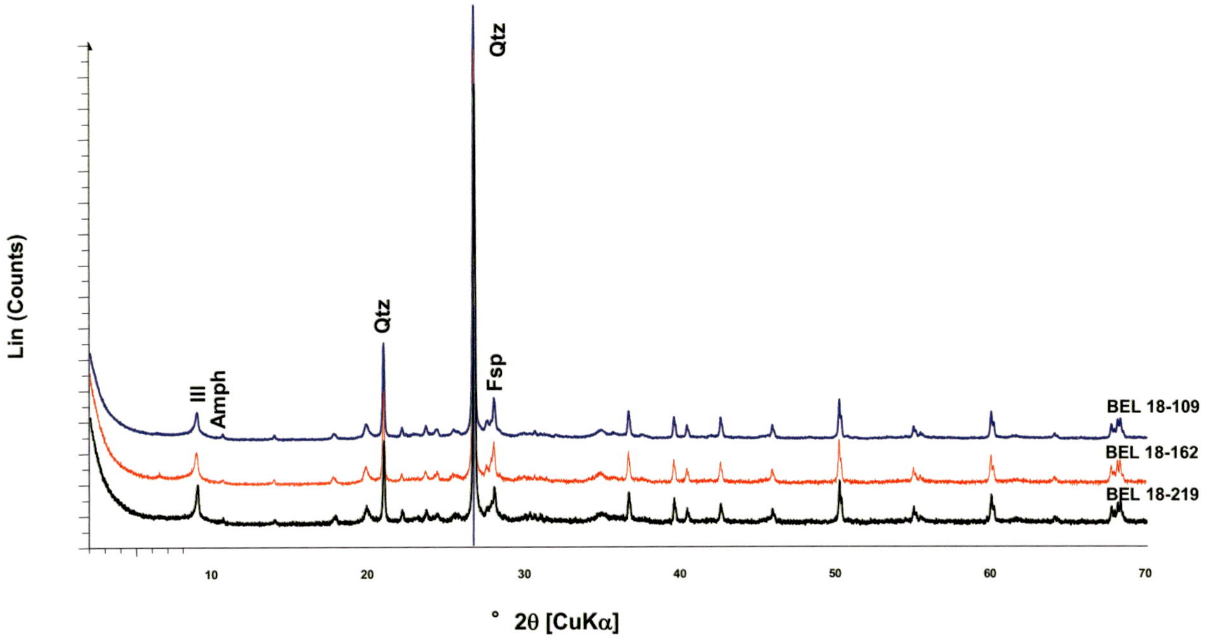

Figure 6. Diffractograms of samples BEL 18-219, BEL 18-162, and BEL 18-109 (dark-burnished ware). Ill: illite; Amph: amphibole; Qtz: quartz; Fsp: feldspar.

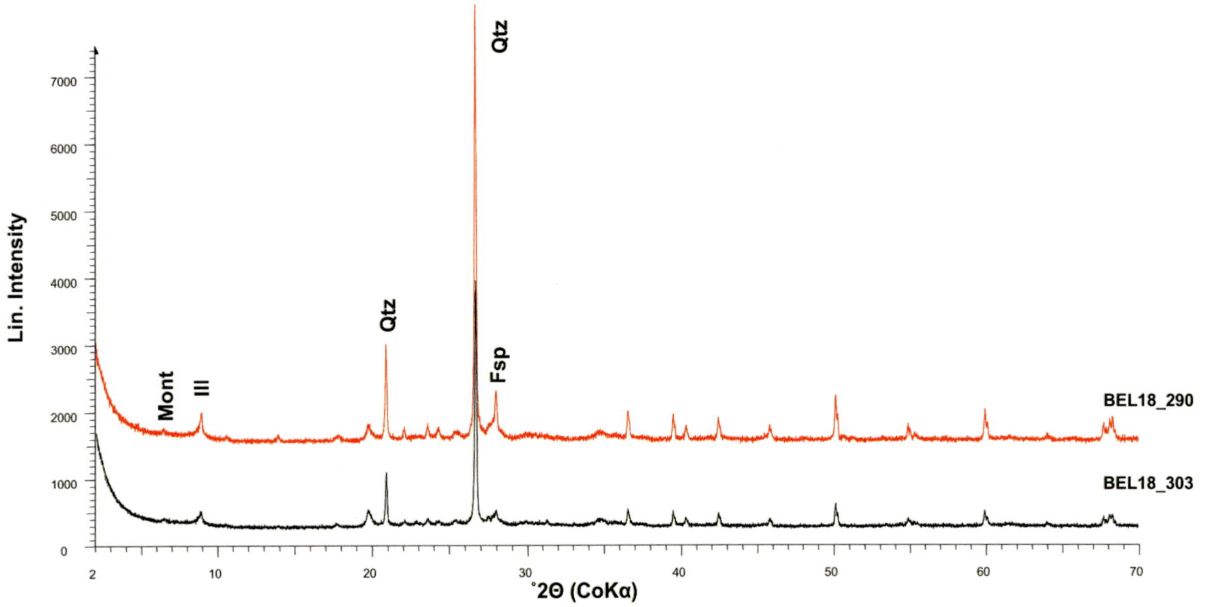

Figure 7. Diffractograms of samples BEL 303 (Starčevo style pottery) and BEL 290 (barbotine decorated vessel). Mnt: montmorillonite; Ill: illite; Qtz: quartz; Fsp: feldspar.

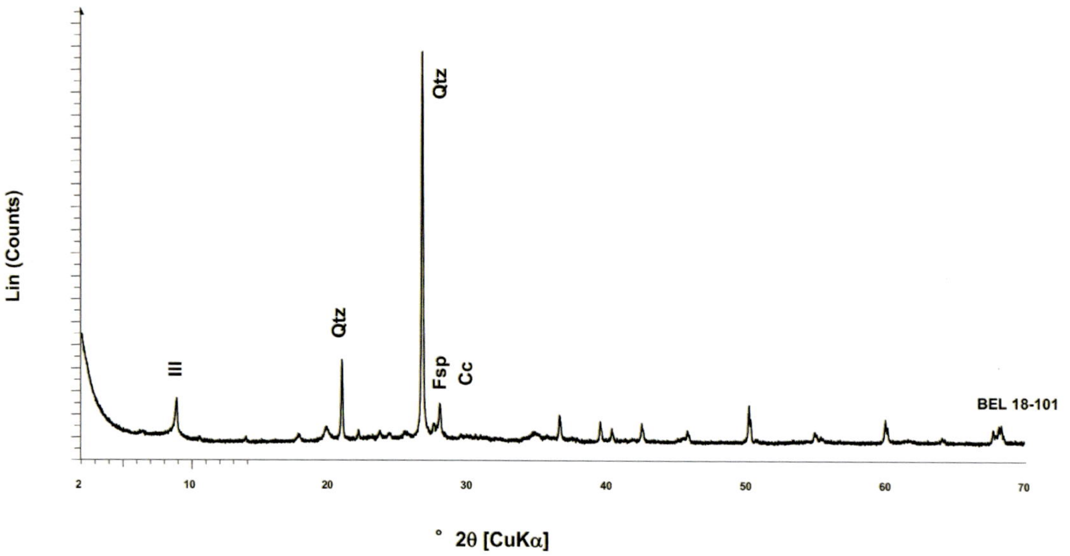

Figure 8. Diffractogram of sample BEL 101 (dark-burnished ware). Ill: illite; Qtz: quartz; Fsp: feldspar; Cc: calcite.

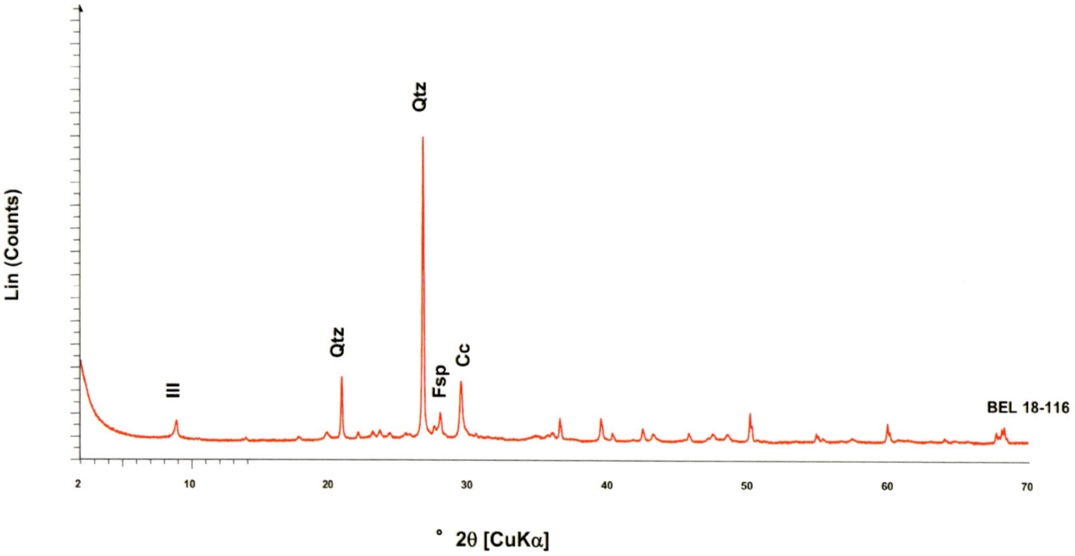

Figure 9: Diffractogram of sample BEL 116 (cooking pot). Ill: illite; Qtz: quartz; Fsp: feldspar, Cc: calcite.

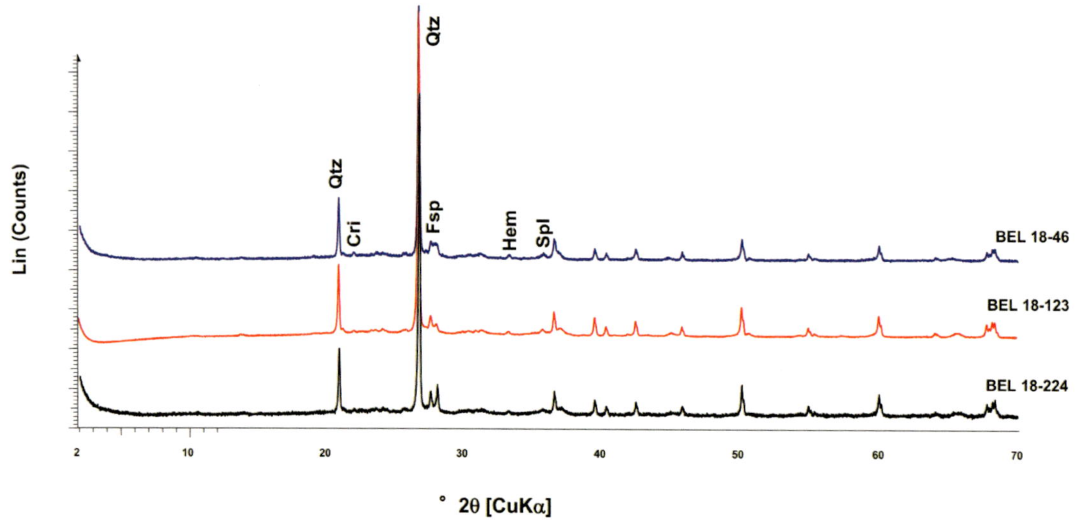

Figure 10. Diffractograms of samples BEL 46 (pithos), BEL 123 (chimney), and BEL 224 (chimney). Cri: cristobalite; Qtz: quartz; Fsp: feldspar; Hem: hematite; Spl: spinel.

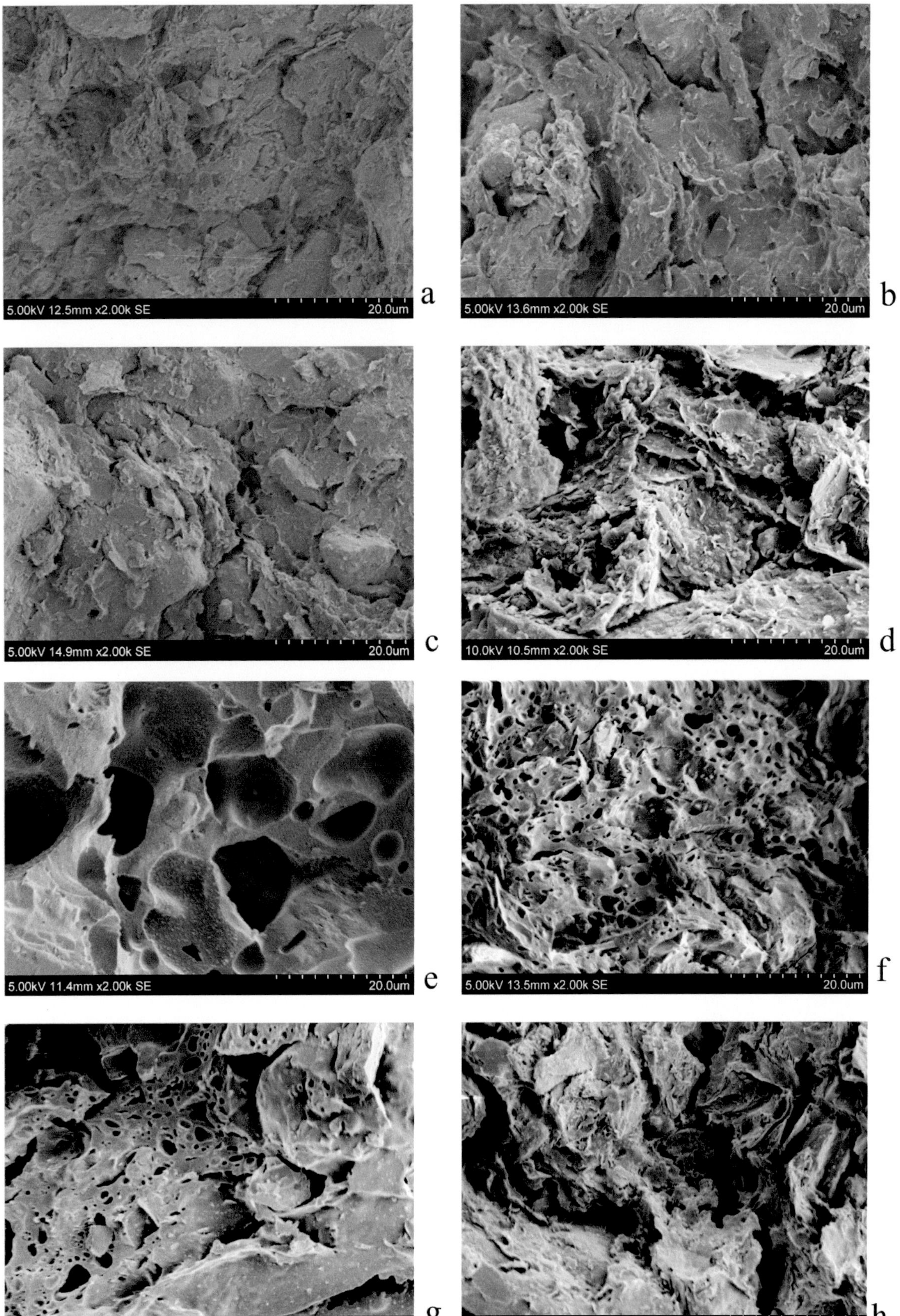

Figure 11. SEM photomicrographs of ceramic samples from Belovode: a) BEL 68: bowl, dark-burnished pottery; b) BEL 95: bowl, dark-burnished pottery; c) BEL 118: amphora, dark-burnished pottery; d) BEL 116: cooking pot; e) BEL 46: amphora; f) BEL 123: chimney; g) BEL 224: chimney; h) BEL 334: bowl, Starčevo style.

Table 3. Summary of the results of the SEM analysis (IV= initial vitrification 750–800°C; V= extensive vitrification 900–950°C; C= continuous vitrification 1000–1050°C; DB=dark-burnished).

Sample	Chronological Horizon	DB	Refiring	Degree of Vitrification
BEL 46	1 (C-D)			C
BEL 68	1 (C-D)	X	X	IV
BEL 94	1 (C-D)	X	X	IV
BEL 95	1 (C-D)	X	X	IV
BEL 115	1 (C-D)			IV
BEL 116	1 (C-D)			IV
BEL 118	1 (C-D)	X	X	IV
BEL 123	1 (C-D)			C
BEL 132	1 (C-D)		X	IV
BEL 224	2 (Gradac-C)			C
BEL 288	4 (A)			IV
BEL 295	4 (A)			IV
BEL 303	5 (Starčevo/A)			IV
BEL 334	5 (Starčevo/A)			IV

Discussion: raw material selection and processing

The analysis different technological aspects of the production of pottery at Belovode. The petrographic fabrics (A, B and C) detected in the assemblage contained quartz, plagioclase, muscovite, amphibole, epidote and metasedimentary rocks. Samples of Neogene sandy-clay collected close to the site (Figures 12a,b,c) and a specimen from a sandy layer of the Neogenic formation that outcrops 10 km away from Belovode (Figure 12d) have a composition comparable with the coarser fraction identified in the thin section analysis of Fabric Groups A and C and could have been used for ceramic manufacturing at the site. Neogene limestone sampled nearby had a shelly oolithic composition (Figure 12e). This weathered to a sandy clay with shell, microfossils and calcite (Figure 12f), found in the same area and matching well the Fossiliferous Fabric Group B.

Given this scenario of geological variability, the diversity among the petrographic groups could be explained by the use of different raw material sources available in the surrounding environment. Fabric Group B could have been produced from a mixture of the sandy clay used for Fabric Group A (Figures 12a–d) together with material rich in shells (Figure 12f) that was not plastic enough to be used on its own. Another explanation for the presence of inclusions of different lithology in Fabric Group B could be the existence of mixed alluvial clay sources not yet sampled by fieldwork.

These results suggest that potters deliberately selected different raw materials and manipulated them in order to produce a specific set of recipes suited to the vessels they were producing. For instance, when producing dark-burnished pottery bowls, they seem to have refined the material by sieving and levigation. On the other hand, when manufacturing pithoi and amphorae, they used coarser pastes to which they occasionally added tempers. The fabric of the (cooking) pots is notable, being characterised by the presence of shells and large fragments of limestone. While this type of fabric is very common in cooking pots, few dark-burnished vessels, which appear to be fired to higher temperatures (around 800°C), contain shells or limestone. The rare choice of a clay rich in shells and limestone for the manufacturing of dark-burnished pottery at Belovode could be explained by the behaviour of calcite (the mineral component of limestone and shells) during the firing process. Above 750–800° calcite starts to lose CO_2, which is then recovered during the cooling phase. This results in an initial decrease and subsequent increase in calcite volume. Since clay shrinks during firing, the re-formed calcite no longer has enough space within the ceramic material and the vessel desegregates (Picon 1995). Organic-tempered vessels from the earliest phases at Belovode have an oxidised surface and a reduced core, and their typology is connected to the Starčevo phenomenon (Spataro 2014). Organic tempering is not present in any of the Vinča style pottery analysed from the same horizon at Belovode. This suggests the existence of several distinct pottery recipes in the earliest phases of the site. Organic tempering disappears in the later phases, while the other traditions (Fabric Group A) are more persistent.

The presence of organic-tempered fabric within the inner part of the dark section and the outer, red part deserves a separate discussion. Vessels characterised by this kind of fabric are stylistically connected to Starčevo; organic tempering is recognised as one of the characteristics of this Middle Neolithic phenomenon (e.g. Spataro 2014). This kind of technological choice does not appear to be used in any of the pots of the Vinča style found at Belovode. This is remarkable considering that sherds of the Starčevo tradition, with organic tempering, are found in the same horizon – n. 5 (see Mirković *et al.*, Chapter 12, this volume) as the Vinča A pottery. It is possible to hypothesise the existence of different pottery recipes in the earliest phases of the site. Organic tempering disappears in later horizons, whilst other recipes are transmitted through to the latest phases. Notably, some of the outliers (especially

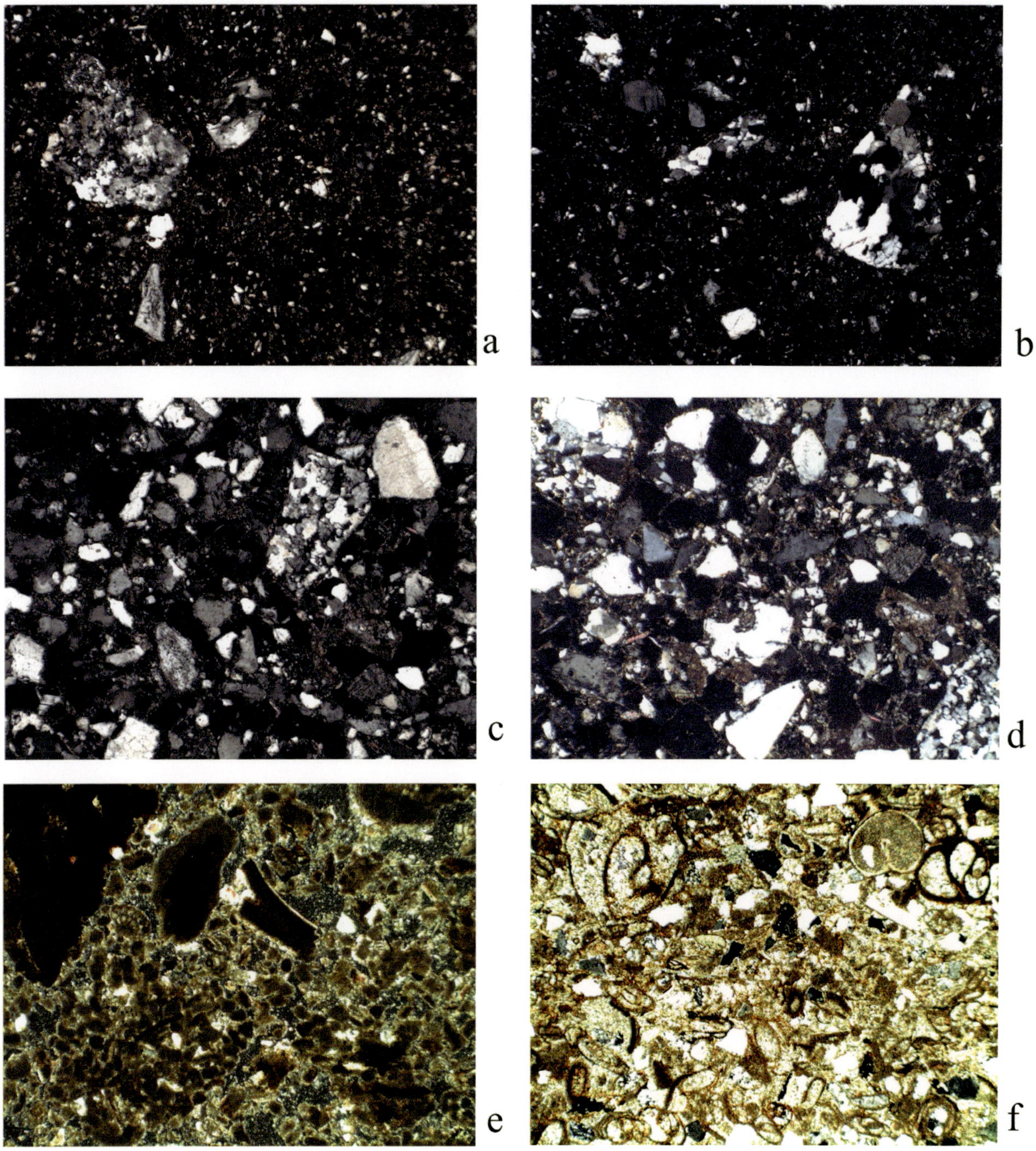

Figure 12. Thin section photomicrographs of selected geological samples from Belovode: a) sandy clay (Point 2 Sample 3) XP; b) sandy clay (Point 12 Sample 13) XP; c) sandy clay (Point 14 Sample 15) XP; d) sand (Point 24 Sample 26); e) shelly-limestone (Point 1 sample 2) XP; and f) sand (Point 1 sample 1) XP. Image width = 3 mm.

BEL 155 containing serpentinite and BEL 325 with mica-schist) may represent non-local production, made with raw materials that are incompatible with the geology of the area around Belovode.

Discussion: firing technology

Thin section petrography and XRPD analysis have enabled us to interpret aspects of pottery pyrotechnology at Belovode. The mineral suite detected by XRPD suggests that most of the sherds analysed were fired to temperatures greater than 750–850°C. This is confirmed by the optical activity in the clay matrix in thin section (Quinn 2013) and the SEM results. The results show that potters at Belovode were able to achieve high temperatures (at least *c.* 750°C), especially when producing dark-burnished pottery. Only three samples (BEL 46, BEL 123, and BEL 224) underwent temperatures of around 1000°C, however sample BEL 46 is from a destruction layer and could have been re-fired in a destruction event. BEL 123 and BEL 224 are fragments of so-called 'chimneys'. These are elongated, cylindrical ceramic forms that are open at both ends and have been connected to the smelting process, but analysis of these objects showed no contamination with metallic elements (Radivojević and Kuzmanović Cvetković 2014). Nevertheless, it is possible to connect the high firing temperatures undergone by these samples with their use in some form of pyrotechnological process. Variations in the surface and fabric colour of many of the sherds suggests that the potters were not always capable of controlling the redox conditions of the firing. During the production of dark-burnished pottery, however, the firing atmospheres were more controlled, resulting in a homogeneous, black, shiny surface and almost complete reduction of the fabric.

The firing temperature estimates revealed by our analysis contradict those published by Kaiser *et al.* (1986) but are in accordance with results of Maniatis and Tite (1981: 73), Goleanu *et al.* (2005), and Spataro (2018). Kaiser employed the thermal expansion technique which could lead to overestimates of firing temperatures (Cuomo di Caprio 2007: 614). We cannot, however, exclude the possibility that potters in different Vinča communities could have employed different firing procedures. The link between Vinča pottery technology and early metallurgy rests especially on the argument that potters employed temperatures above 1000°C in reducing conditions to produce dark-burnished pottery. The firing temperatures estimated by Kaiser *et al.* (1986) seem to support this. Nevertheless, our analysis at Belovode shows that pottery, including dark-burnished pottery, was most likely fired at temperatures below 1000°C and not always in a controlled atmosphere. This is an issue that will be explored more deeply in Chapter 43, this volume, where we compare this scenario at Belovode to that reconstructed for Pločnik.

Conclusions

Our analysis for the Neolithic settlement of Belovode highlights the plethora of choices made during the production of pottery at the site. The analysis of sherds across the chronological phases shows patterns of continuity and change. Our results elucidate aspects of the procurement and processing of raw materials and show a connection between fabric and vessel type that is evidence of a set of persistent technological traditions transmitted over 700 years.

Fabric Group A is present throughout the periods observed; Fabric Group B seems to be absent in the earliest phases. These preliminary results point towards a scenario of strong continuity of pottery recipes, but with great variability of technological choices. We were also able to identify potential non-local production that allows discussion of pottery exchange on a regional and interregional scale.

Dark-burnished pottery is characterised by a specific paste recipe (Fabric Group A). It is notable that there is no significant change in pottery technology preceding or concomitant with the appearance of metallurgy, apart from a slight decrease in the use of burnishing surface treatment. Our results shed new light on the controversial question of pyrotechnology and its links with early metallurgy. Data from XRPD and petrography show that dark-burnished pottery was fired below 1000°C and not always in fully controlled atmosphere conditions.

Appendix

Appendix avaliable online as part of Appendix B at https://doi.org/10.32028/9781803270425/AppendixB_Ch14

The bibliographic reference for this chapter is:

Amicone, S. 2021. Belovode: technology of pottery production, in Radivojević, M., Roberts, B. W., Marić, M., Kuzmanović Cvetković, J., and Rehren, Th. (eds) *The Rise of Metallurgy in Eurasia*: 186–198. Oxford: Archaeopress.

Chapter 15

Figurines from Belovode

Julka Kuzmanović Cvetković

The subject of this chapter are the figurines that form a significant percentage of finds on all Vinča sites. At the time they were made, these 'trademark' handmade statuettes were a very important aspect of life within the Neolithic community. During the excavation campaign of 2012 and 2013, a total of 115 figurines were discovered in Trench 18, some fragmented, some whole and in largely original form. This is only one trench of 25 m^2 within a 40ha site (Šljivar and Jacanović 1996: 175), and therefore a very small sample. A further obstacle to understanding is the limited number of research publications available, giving the general impression that the site itself is incomplete. Despite these challenges, the following chapter presents a comprehensive analysis and attempts to answer at least some of the many questions surrounding the figurines.

The history of research into the Vinča figurines

At the very beginning of any story related to Vinča culture stands Miloje Vasić, the first researcher of Vinča and the first to point out the significance of the figurine form, the topic of one book of his seminal publications on Vinča (Vasić 1936b: 158). The categories he developed can almost be used even today. He divided the Vinča figurines into seven groups: I-standing figures; II-sitting figures; III-kourotrophic (child-nurturing) female figures; IV-standing male figures; V-figures of different shapes; VI-animal figures; and VII-amulets (Vasić 1936b: 1). A further typological categorisation of figurine forms, based on that of Vasić, was created by Milutin Garašanin (1951: 16) and Vasić remains entirely applicable. The categorisation is also based on the way in which the body was modelled: A-standing figures; B-sitting figures; C-kourotrophic figures; D-amulets; and E-animal figures.

Further research on the sculpture of the Neolithic period was undertaken by Dragoslav Srejović (1968: 191) who categorised Vinča figurines according to their geographical area: A-Podunavlje area; B-Kosovo; and C-Eastern Pomoravlje. In making his categorisations, Srejović took into account the stratigraphy of the settlements at Vinča. The oldest figures, from Starčevo, had a 'columnar' form; later developments included the modelling of the arms, legs, abdomen and breasts, which Srejović categorised within his 'realistic' style (Srejović 1968: 193).

At the site of Vinča, figurines recovered from between the relative archaeological depths of 8.5 and 8.0 m represent a phase of stabilisation, with details that closely correspond to the anatomy of a human body. Between 7.8 and 6.6 m there is a powerful predominantly 'realistic' trend characterised by depictions of anatomic details and clothes. Between 6.5 and 5.9 m, figurines demonstrate the greatest variation of Vinča figurines in terms of number, shape, and monumentality (Srejović 1968: 194). In the subsequent, final phase of Vinča culture, there are statuettes with crudely fashioned bird faces, lacking the anatomical characteristics of the preceding realistic style. Srejović was, however, primarily concerned with the aesthetics of the figurines and provided only a basic outline of their chronological development.

Belovode figurines

At the Belovode site, a significant number of small finds were discovered in Trench 18 during the two excavation seasons. A total of 115 clay figurines were found in several different places. They were discovered mostly in a shattered or damaged condition, and rarely ever in their original state. These breakages date to the Neolithic period, raising questions with regards to when and how they were broken (Porčić and Blagojević 2014). They clearly played an important role in the lives of Neolithic communities. Eighty (69.9%) of the figurines (i.e. the majority) are anthropomorphic, 17 (14.8%) are zoomorphic, and 18 (15.6%) are amulets (Figure 1).

The largest fragment of figurine body, a torso, is 84 mm long; the widest span of extended arms is 110 mm. Both miniature and large statuettes are present, but no significantly larger examples.

All the figurines were handmade out of clay with varying levels of execution. Fifty were crudely made with a rough surface, 27 were smoothed, and just two were polished. Thirty-nine figurines are black and grey in colour, 29 range from ochre to brown, and 12 are reddish orange, typical of clay baking in an oxygen rich atmosphere (i.e. open fire).

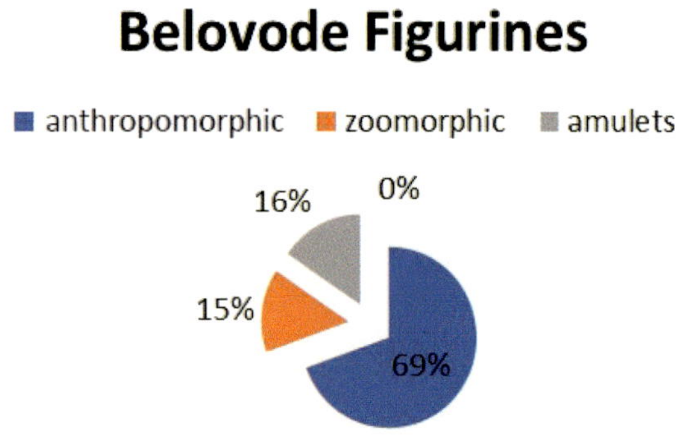

Figure 1. Figurine types in Trench 18.

Taphonomy (anthropomorphic figurines BAf 1-80)

Of the 80 anthropomorphic figurines discovered at Belovode, only two (2.5%) were found undamaged, in their original state (Figure 2). These were amongst the smaller sized of the assemblage, having a complete height of 68 mm; the fragments measured between 8 and 60 mm. The fact that most of the figurines were fragmented has led many authors to conclude that they were deliberately broken during rituals.

The most frequent type of damage is a missing head (22 figurines, i.e. 27.5%). It seems that the head was the easiest part to break off from the body. With some figurines, the head appears to have been produced separately and fixed into a hole in the neck – perhaps so that it could be detached more easily. Eleven figurines (13.75%) have only the head missing (i.e. 70–80% of the body is present). Two fragments (2.5%) comprise the head, shoulders and torso – almost one third of the body. These are broken in the waist area. In seven cases (7.5%), only the torso (20% of the whole figurine) was recovered. Only six fragments (7.5%) had both the head and the legs broken off with 50% of the figurine (body, chest and abdomen area) remaining. A further six fragments (2.5%) consisted of just the chest area and abdomen (i.e. no legs or heads), amounting to 30–40% of the figurine body. In four cases (5%), only the abdomen area (20–30% of the figurine body) was found.

Thirteen fragments (16.25%) represented the lower part of the figurine bodies, i.e. from the chest down (50–60% of the figurine body). These were mostly figurines with dresses or with their legs joined. There were four fragments (5%) which were firmly identified as parts of the arms (10–15% of the figurine body). Nine fragments (11.25%) representing legs were also found.

Anatomy

Thirty-eight of the 80 figurines (both whole or in the form of fragments) are 'standing'; only five are 'sitting'. The remaining fragments cannot be categorised with complete certainty. Nineteen figurines have a typical posture with arms, in the form of stumps, extended to the side; two statuettes have their hands on their hips; one has the hands on the abdomen; and three have their arms pointed towards the ground. For a comprehensive file with data for all figurines, see Appendix B_Ch15 (see below). Each figurine has a unique identifier, e.g. BAf 21, comprising a B (Belovode) followed by a category (Af – anthropomorphic; BA – amulet; BAn – animal) and a number.

In terms of gender, there is insufficient detail available to provide a complete picture. Seven statuettes have breasts, suggestive of the female gender; 15 have protruding abdomens suggestive of pregnancy and female gender; 12 statuettes have protruding steatopygic buttocks, a further characteristic of female gender. Only one has a vulva marked as a primary female sexual characteristic, and only one has phallus and scrotum clearly represented and is undoubtedly a male.

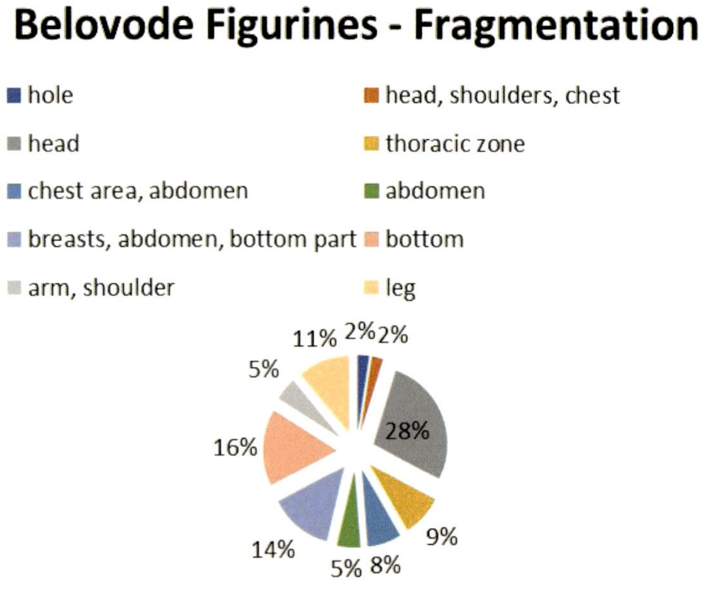

Figure 2. Figurine fragmentation in Trench 18.

Head shape varies and greater attention was frequently paid to the heads of figurines with triangular and pentagonal shaped faces, with fewer details added to the round and square shaped heads. The back of the head is often more profiled and emphasised, and six figurines have an array of perforations on their occipital bones; these must have contained some form of ornament. Hair is represented on 18 figurines. This is mostly straight and shoulder-length. Some figurines have a form of 'bun' and others have defined hairstyles with parted hair.

Eyes are represented on 20 figurines in total. Eleven have incised eyes, seven have moulded, protruding eyes; in three cases these are framed with incised lines. Sixteen of the figurines have almond-shaped eyes, two have round eyes, and three have rather slanted eyes; eyelashes, represented by incised lines, are depicted on four.

Clothes and ornaments

Vinča figurines are commonly represented wearing a dress with a v-neckline. On five of the Belovode figurines a single neckline is depicted with an incised line; on four there is a double neckline and on four others the neckline is emphasised with several lines. On these figurines, lines also depict dresses sewn together from vertically aligned strips.

A significant number of figurines have a loose dress that is gathered into a belt beneath the waist. Despite being tied up, it still follows the body outline perfectly and then drops to the ground. A variation of this dress, the so-called 'siren' model, becomes wider beneath knee area forming a bell shape.

Jewellery, in the form a medallion around the neck, is noticeable on five figurines. Figurine BAf29 has a necklace of round, flat pearls.

Typology based on body shape

The typology developed by Garašanin (1951: 16) was applied with a few necessary adjustments (Figure 3). The complete typology of figurines is set in the Appendix B_Ch15.

A Standing figurines

A1 Cylindrical/pillar like figurines

These were crudely made, in a cylindrical shape with a flat base. Five such fragments were recovered (BAf41, BAf69 (Figure 3.8.), BAf72, BAf76, BAf80). On two figurines (BAf41 and BAf69), a male reproductive organ is modelled; one figurine (BAf 80) has a protruding abdomen typical of a pregnant woman.

A2 Flat figurines

Flat figurines are straight, mostly with a head connected directly to the body, with arms in the form of limb stumps. Ten fragments were recovered (BAf14 (Figure 3.1), BAf15, BAf18, BAf19 (Figure 3.2), BAf22, BAf29 (Figure 3.3), BAf37 (Figure 3.4), BAf45, BAf65, BAf66), mostly representing the chest areas. Clothes are represented on two fragments (BAf14 (Figure 3.1), BAf15) and on one (BAf29/Figure 3.3)) there is a necklace of round pearls.

A3 Mixed figurines

A3a Figurines with a flat upper part and a cylindrical lower part and flat base

Twelve such figurines were identified (BAf20, BAf23, BAf40, BAf42, BAf46, BAf48, BAf50, BAf55 Figure 3.5), BAf56 (Figure 3.6), BAf57 (Figure 3.7), BAf62, BAf77). All were crudely made with few details. Arms consist of short limb remnants. The head and short neck are formed from a single piece of clay, connected without much differentiation. On several figurines there is a noticeable attempt to represent clothes with incised lines, mostly around the neck area or the hemline, with some parallel lines carved in the lower part of the base.

A3b Figurines with belts under the waist

All the figurines of this subtype (BAf3, BAf5, BAf6, BAf7, BAf43, BAf47, BAf49, BAf52, BAf78, BAf79) have the lower part of the body modelled in a similar way, beneath a wide, protruding abdomen. A line is depicted, both in the front and on the sides, most likely a belt. All the figurines become narrower towards the bottom, reducing to almost the half of the width of the narrow, flat base. All figurines have an emphasised gluteal area—they are almost steatopygic—with the lower part resembling the 'siren' type. They appear to have a very loose dress, gathered into the belt beneath the abdomen. The dress falls casually over the belt and narrows towards the hemline, closely following the body outline.

A3c Figurines with bell-shaped dresses

Only two fragments of this group were recovered (BAf34 and BAf44). Both depict the lower body, narrowing noticeably towards the knee area, and then widening into a bell shape. On figurine BAf34, the dress is represented by vertical lines and toward the lower area, around the base, there are two parallel, horizontal lines. On figurine BAf44, the dress widens into a bell shape.

A3d Figurines with modelled legs

There is only one, exquisitely made figurine with modelled legs (BAf9) but only the lower part of the body (i.e. from the waist to the knees) was recovered. It has a very realistically modelled gluteal area, a knee-length dress represented with lines, and separately modelled legs that are unfortunately broken. Further fragments of the legs (BAf17, BAf31, BAf32) were also recovered extending to the feet. The legs in these figurines were modelled separately with very defined feet.

3.1. B Af 14

3.2. B Af 19

3.3 B Af 29

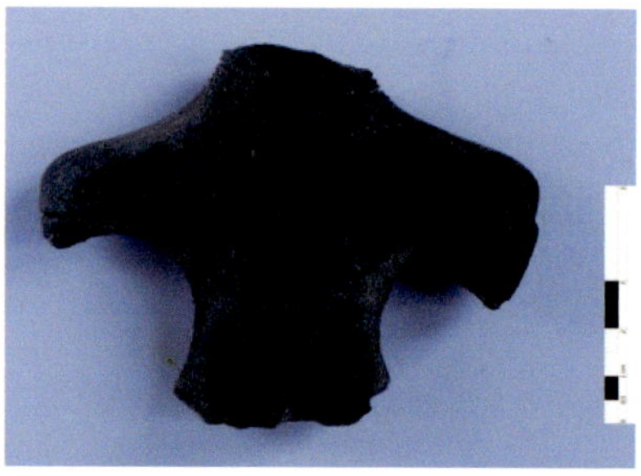

3.4. B Af 37

3.5. B Af 55

3.6. B Af 56

3.7. B Af 57

3.8. B Af 69

Figure 3. Belovode: anthropomorphic figurines.

B Sitting figurines

B1 Figurines sitting on the ground

Fragment BAf36 is no more than a left leg but clearly represents part of a figurine sitting on the ground. The leg is bent at the knee, with its lower part alongside the thigh. The foot is very realistically depicted.

B2 Figurines sitting on a pedestal

The fragments of figurines comprising just the abdomen area (BAf10 and BAf21) are likely to belong to the group of sitting figurines, although it is difficult to determine whether they are sitting on a pedestal or on a throne. Fragments BAf39 and BAf71 both represent the left side of the gluteal area, moulded and embellished with incised lines. These limited remains, however, make it difficult to claim with complete certainty what these figurines are leaning against. Figurine BAf70 comprises just the upper body part, but the way that the arms are laid on the abdomen is suggestive of a figurine on a throne.

C Kourotrophic (child-nurturing) figurines

The creation of kourotrophic (child-nurturing) figurines is likely to be a consequence of the increase in natality in the Neolithic period, as well as increased devotion to children and the family as the main social unit of a large Neolithic community. No such figurines were recovered from Trench 18.

Classification based on face shape

Rectangular faces are found on simple figurines with almost no detailing. Heads with rectangular faces are not separated from the neck, as can be seen on figurine BAf19. On the other three such figurines (BAf51, BAf55 (Figure 3.5), BAf73) the upper head shape is slightly rounded, with the sharp angles of the occipital bones depicted. Figurine BAf51 has a nose modelled in the shape of a snout and the eyes are represented by short horizontal lines. Figurine BAf55 (Figure 3.5) has a protruding nose and although crudely made, has a v-neck dress represented by incised lines. The head of figurine BAf73 has a somewhat longer neck and the back part of the head is modelled into a triangle.

Figurine BAf29 (Figure 3.3) has an oval face, modelled in parallel with the body, with an indistinct, moulded nose. Figurine BAf59 also has round head on a short neck. The whole figurine is made extremely crudely and there is only a moulded nose with two small holes.

Triangular faces are quite numerous; ten were present among the samples. Each figurine in the group is intrinsically different from the others but they have one thing in common: a triangular face with a protruding chin, usually on a short neck. Figurine BAf1 is very crudely made and has barely survived; it has a protruding nose. Figurines BAf2, BAf4 and BAf13 all have noses, slanted, almond-shaped eyes, and protruding ears that are horizontally perforated. Figurine BAf4 has a heavy line or 'stripe' above the forehead. Figurine BAf8 has a more emphasised profile and shoulder-length hair, represented by lines, at the back.

Its eyes are represented by heavily incised lines and its nose is realistically modelled. Figurine BAf58 has a heart-shaped face, a protruding nose and shoulder-length, straight hair represented by lines. The head of figurine BAf61 has no details but has a nose in a shape of a snout, and a rounded nape. Figurine BAf25 has an array of details that distinguish it from the others: on both sides the face is narrower at the temples and as a result the cheek bones are more protruding, the nose is larger, and the eyes are more deeply set. Hair is represented only on the upper part of the head, by parallel, incised lines with a central parting, with white incrustation. Figurine BAf27 has a damaged face and BAf38 is a small fragment with incised hair.

Nine figurines (BAf12, BAf16, BAf33, BAf35, BAf54, BAf67, BAf74, BAf75) have pentagonal faces. All have incised eyes, a protruding nose and a profiled nape with perforations. Some have hair represented: BAf12 and BAf16 have a form of 'bun' at the back of the head; BAf33 has straight incised hair; BAf67 has an emphasised, defined nape; and BAf75 also has straight hair represented by lines. Figurine BAf11 has a face shape somewhere between a triangle and pentagon. It has a sharp chin, a protruding nose and incised almond-shaped eyes with a lower line that extends to the temples. The nape is clearly defined with incised lines that represent the hair and which extend from the back of the head to the shoulders. The nape has three vertical perforations.

Amulets (BAm 1-18)

Some researchers consider these objects to be stylised idols; some simply called them amulets without trying to determine their purpose. Numerous other researchers consider them to be utilitarian objects that had a practical purpose in the sphere of rope making (Pantović 2014: 14).

Eighteen amulets were found in Trench 18 at Belovode. All are small and perforated along their length. The amulets have a simple shape, varying from cylindrical to slightly conical. Most have a flat base so they can stand independently and are vertically perforated with an opening that is 0.5 cm wide. They are mostly simply made, black or red coloured from clay baking, and have a glazed surface. Only two amulets, BAm10 (Figure 4.4) and BAm17, are embellished with incised parallel, slanting lines.

The amulets were classified using the typology developed by Pantović (2014: 15) after she analysed amulets from the south east Banat area and taking into account all the research of Vinča amulets carried out at the time. The full typology of amulets is in Appendix B_Ch15, Table 3.

A Two-armed amulets

Sixteen amulets found at Belovode could be included in this group.

A1- 'Pillar-like' body shape is characteristic for amulets BAm1, BAm5 (Figure 4.3), BAm6, BAm8, BAm9, BAm15, BAm16, and BAm17; their arms are small and short.

A2 - 'Shorter' amulets have very short bodies; amulets BAm10 (Figure 4.4) and BAm12 belong to this type.

A3 - 'Triangular' body shape with arms that get wider in a slantwise direction. Amulets BAm3 (Figure 4.4), BAm13, and BAm14 belong to this group.

Amulets BAm4, BAm7 and BAm8 are fragments and difficult to categorise.

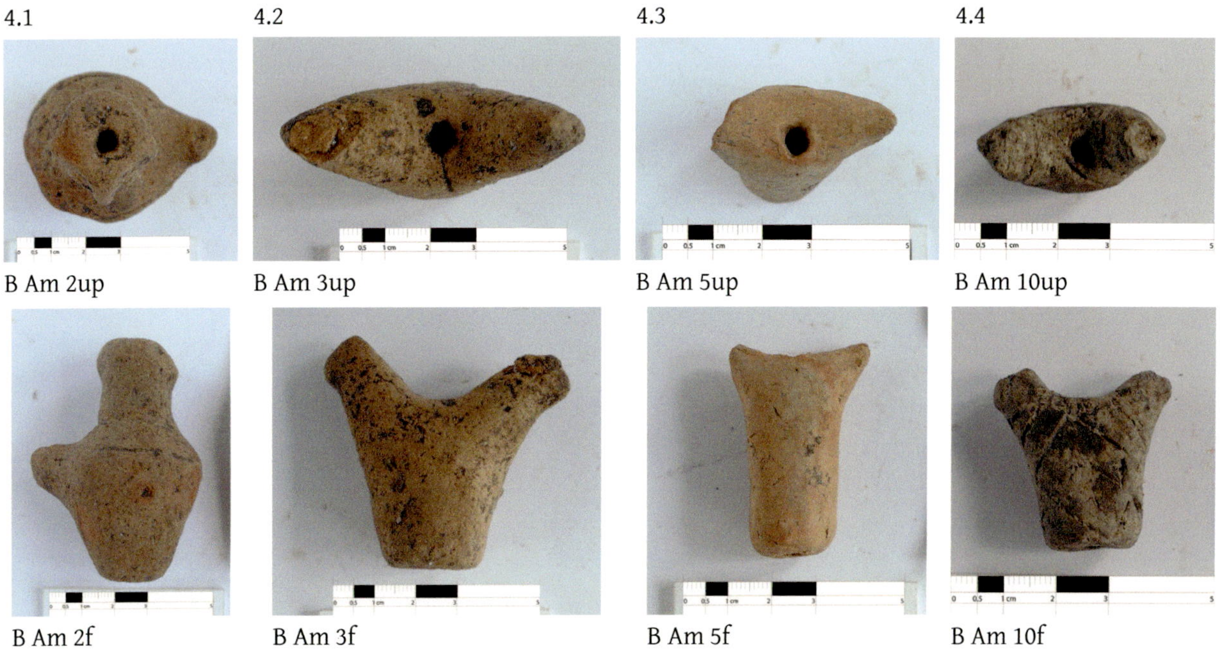

Figure 4. Belovode: amulets.

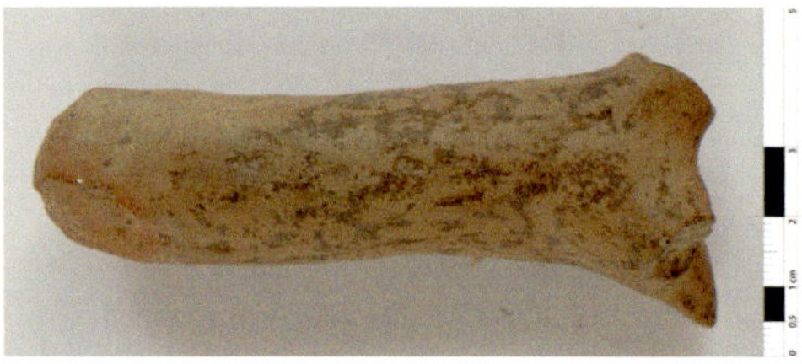

5.1 B An 1

5.2. B An 2

5.3. B An 3

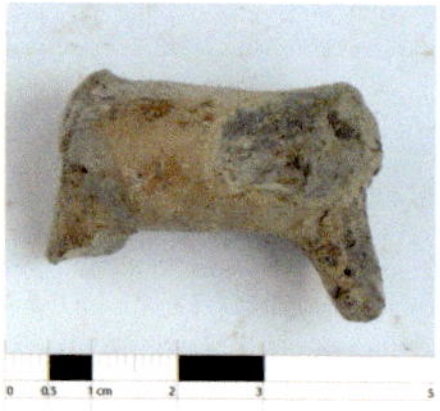

5.4. B An 5

5.5. B An 7

5.6. B An 8

5.7. B An 9

5.8. B An 12

Figure 5. Belovode: animal figurines.

C Multiple-armed amulets

Amulets BAm2 (Figure 4.1) and BAm11 belong to this group. Amulet BAm2 resembles a styled figurine with a rounded body and one shorter arm in the form of a stump, and it is only a single perforation along its length that defines it as an amulet. Amulet BAm11 is a damaged fragment and is difficult to categorise, but definitely belongs to the group of multiple-armed amulets.

C Multiple-armed amulets

Amulets BAm2 and BAm11 belong to this group. Amulet BAm2 resembles a styled figurine with a rounded body and one shorter arm in the form of a stump, and it is only a single perforation along its length that defines it as an amulet. Amulet BAm11 is a damaged fragment and is difficult to categorise, but definitely belongs to the group of multiple-armed amulets.

Animal figurines (BAn 1-17)

None of the 17 animal figures discovered at Belovode was complete and the animals they represent can be determined for only five of them with any certainty. Four figurines (BAn9, BAn11, BAn12 (Figure 5.8), BAn13) are perforated from the back to the base and, given that none are intact, there is a likelihood that they could actually be amulets or even figurines, which had some practical purpose. Most of the animal amulets were crudely made from clay but figurine BAn13, which could potentially represent a dog, and of which only the head was recovered, is made of white stone. They are all small in size – almost miniature, the largest being 7.5 cm in length.

Figurines BAn3 (Figure 5.8), BAn7 (Figure 5.5) and BAn8 (Figure 5.6) can be assumed to represent bulls. This can be particularly claimed for the figurine BAn3 (Figure 5.3), which represents a strong animal with a large front section, a short head and a protruding, realistically modelled phallus. In the case of the figurine BAn8 (Figure 5.6), only the front part with a large head and short legs was recovered.

Figurines BAn1 (Figure 5.1), BAn2 (Figure 5.2) and BAn6 have elongated bodies and are assumed to represent dogs. Figurine BAn13 has a realistically modelled, long, dog-like head with a thin snout, small ears and a thin, long neck; it closely resembles a greyhound.

Figurines BAn9 (Figure 5.7) and BAn11 both resemble a bear, an animal that was present in the Neolithic landscape. The remaining fragments are parts of legs, perhaps from animals, but it cannot be determined with certainty what species they represent, since many animals are depicted in a very similar manner with few differentiating features. The complete typology is in Appendix B_C15, Table 4.

Appendix

Appendix avaliable online as part of Appendix B at https://doi.org/10.32028/9781803270425/AppendixB_Ch15

The bibliographic reference for this chapter is:

Kuzmanović Cvetković, J. 2021. Figurines from Belovode, in Radivojević, M., Roberts, B. W., Marić, M., Kuzmanović Cvetković, J., and Rehren, Th. (eds) *The Rise of Metallurgy in Eurasia*: 199–204. Oxford: Archaeopress.

Chapter 16

Ground and abrasive stone tools from Belovode

Vidan Dimić and Dragana Antonović

Introduction

The analysis of ground and abrasive stone tools from Belovode encompassed an assemblage of 68 artefacts found in Trench 18, excavated in 2012 and 2013 and relatively dated to between the Vinča-Tordoš II (VinčaB1) and the Vinča-Pločnik I (Vinča C2) phases of the Vinča culture. The tools were categorised according to two criteria: the method of tool production and their functional-typological characteristics. 'Ground' and 'Abrasive' stone tools were further grouped into types and subtypes, according to their morphology and function.

The typological-functional analysis was conducted according to the morphological characteristics of the tools and traces of observable use. It was based on general observations and the correlation of the metric characteristics of certain tools and their position within a previously defined framework (Antonović 1992, 2003, 2014c), and was applied to all tools possessing the minimum preserved evidence necessary for analysis. The identification and definition of use-wear traces was conducted through comparison of data from the same category of material at other sites (Semenov 1976; Olausson 1983a, 1990; Adams 1988; 1989; 2002; Adams *et al.* 2009; Pritchard-Parker and Torres 1998; Dubreuil 2001; Plisson and Lompre 2008; Pawlik 2007; Antonović 1992: 20–23; Dimić 2013a, 2015). Tool traces caused by the production process were also recorded to provide indications of the techniques used by Belovode craftspeople in the processing of different types of rock. All analyses were carried out at the Institute of Archaeology in Belgrade using magnifying glasses of up to 16× magnification and the Celestron® USB microscope-camera (with up to 100× magnification), with a connecting camera.

Raw materials

At Belovode, as at many localities in the Vinča culture, various types of rocks (as determined by the macroscopic observation of their petrographic features) were used for the manufacture of ground and abrasive stone tools. Most of the artefacts derive from only three groups of rocks, while other sources were represented only by single object (Figure 1).

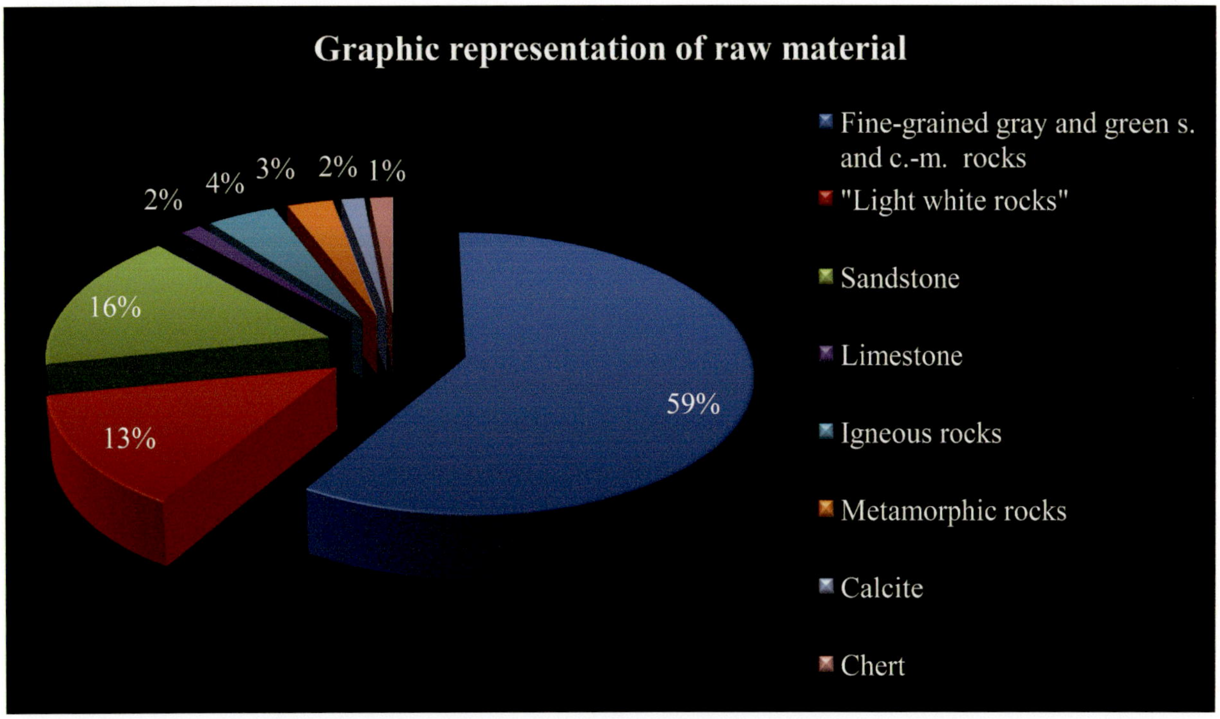

Figure 1. Graphic representation of raw material from Belovode, Trench 18.

The Rise of Metallurgy in Eurasia (Archaeopress 2021): 205–214

Fine-grained, grey-green sedimentary and contact metamorphic rocks were used in the production of more than a half of all tools collected during the Belovode 2012 and 2013 excavation campaigns. Implements made from these sources, and having a cutting edge, occur in all phases of the settlement. These rocks have a variety of geological origins (e.g. crystalline schists, cornite, metasandstone and metaalevrolite), but the same technical-physical features: they are fine-grained, shades of grey-green, have a conchoidal fracture, and are very hard. This hardness ensured that tools such as axes, adzes, chisels, and hammers could be used for procedures which required frequent impacts.

Tools made of 'light white stone' were used in the later phases of the settlement, at the end of Vinča Tordoš II Phase (Vinča B1) and in the Gradac Phase (Vinča B2–C1). The term 'light white stone' refers to a distinctive group of macroscopically similar rocks whose main characteristics are that they are light in weight, porous, relatively soft, and occur in various shades of off-white and yellowish-white. According to the sparse analyses available, these rocks are variously defined as magnesite, chert, tuff, diatomite, and porcelanite (Antonović 1997; Antonović 2003: 45–47; Antonović and Šarić 2011: 68; Šarić 2002). Tools made of this raw material were mostly used in central Serbia and east and central Bosnia during later the phases of the Vinča and Butmir cultures and represent a definable characteristic of these phenomena. In central Serbia, 'light white stone' was the main raw material not only for ground stone tools, but also for the chipped stone industry. It was not intensively used outside the territory of the classic variant of the Vinča culture. This descriptive definition of the group of raw materials in Vinča and Butmir cultures was first recognised in late 19th century (Radimsky and Hoernes 1895: 29, 53–54) and has since been widely used in archaeological literature (Antonović1997; 2003; 2011; Bogosavljević-Petrović 1992; Bogosavljević-Petrović *et al.* 2012; Dimić 2013a). In some assemblages, microscopic analyses revealed the 'light white stones' to be magnesite and chert (Antonović 2003: 45–46; Antonović and Šarić 2011: 68), which certainly cannot be termed 'light' due to their bulk mass but the porousness of the rock creates the false impression of lightness (see Antonović 1997; Šarić 2002). Partially silicified rocks with a hardness of *c.* 4.5 on the Mohs scale were the most commonly used raw material for production of the tools discovered in the earlier excavations at Belovode. Only rarely were raw materials with a lower hardness used, but since their characteristics were only macroscopically determined, the types of rock cannot be stated with total certainty.

Sandstone, a rock with abrasive properties, was a very common raw material at Belovode. Fine-grained, compact and well-silicified sandstones with high level of quartz were used to make all the abrasive tools. Other types of rocks were present in small amounts. One adze, one amulet, three abrasive tools, one fragment of perforated axe and two atypical fragments were made of diabase, aplite, gneiss, quartzite, chert, limestone, and calcite (Table 1).

The analysis of the stone raw materials used for tool production from Trench 18 at Belovode is completely compatible with earlier analyses of the stone tool assemblage from Trenches 1 to 6, which were excavated in the period 1994–1997 (Antonović 2000).

Typological-functional analyses

The 68 macrolithic artefacts from Trench 18 could be placed into two basic groups, the first characterised by mainly untreated tools with abrasive features and the second including tools shaped by grinding (Figure 2).

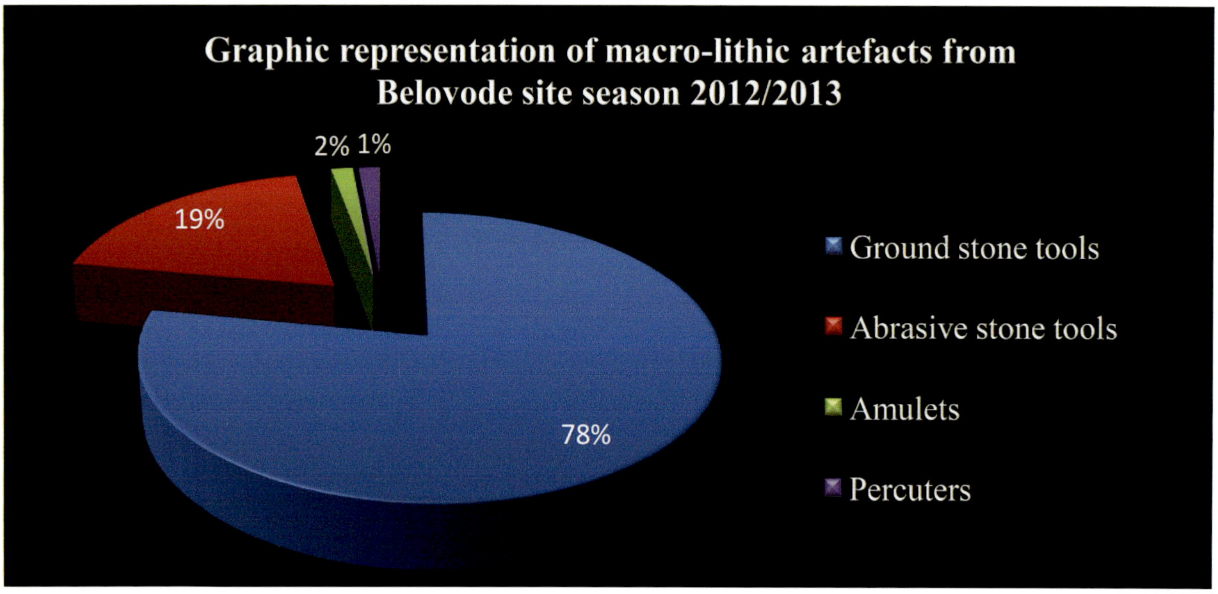

Figure 2. Graphic representation of macro-lithic artefacts from Belovode site, Trench 18.

Table 1. Types of ground and abrasive stone implements in Trench 18.

Type of tool	Number	Complete tool	Length in mm	Rock type
I/1b	3	0	64-90	Metamorphic sandstone, crystalline schist
I/3e	1	0	46	Crystalline schist
III/1a	5	1	65-90	Crystalline schist, metamorphic sandstone, metaalevrolite
III/1b	4	0	45-88	'Light white stone', crystalline schist, cornite
III/1c	1	0	66	'Light white stone'
III/1e	1	0	76	Crystalline schist
III/2a	2	0	65-96	Crystalline schist, metaalevrolite
III/3b	1	0	70	Crystalline schist
III/5a	1	0	52	Crystalline schist
III/5b	2	1	35-71	Crystalline schist
III/5d	2	1	47-113	Crystalline schist, limestone
V/1b	1	1	45	Crystalline schist
V/2b	4	2	47-50	Metaalevrolite, 'light white stone', crystalline schist
V/2e	1	1	73	Crystalline schist
V/4b	1	0	62	Crystalline schist
V/5b	3	0	37-91	Crystalline schist, cornite, metaalevrolite
V/6	2	1	46-47	Metamorphic sandstone, crystalline schist
XI/3a	1	1	123	Diabase
XI/3c	4	0	37-83	Coarse and fine-grained sandstone
XI/6a	1	0	74	Sandstone
XI/6b	2	0	90-147	Fine grained sandstone
XIII/3	1	0	109	Sandstone
Querns	2	1	250-290	Sandstone
Hammerstones	2	1	67-85	Aplite, crystalline schist
Amulets	1	1	30	Calcite
Preforms	3	2	62-118	'Light white stone', gneiss, crystalline schist
Pebble/pottery grinder	1	1	28	Quartzite
Chunks	3	0	37-42	'Light white stone', crystalline schist, metaalevrolite
Undetermined	12	0	34-102	'Light white stone', crystalline schist, diabase, metaalevrolite, sandstone, chert

Most (43) of the artefacts had serious damage, which had resulted in the tool being discarded. Somewhat smaller levels of damage were observed on 11 artefacts, while 15 were almost completely preserved. All tools, with even the slightest morphological evidence for typological determination, were included in the analysis (Table 1).

The abrasive tools from Belovode were less numerous than the ground stone artefacts. They encompassed several smaller static grinders (eight objects), two querns, a handstone, a whetstone, and a pebble pottery-grinder (Figure 3), and a hammer-stone/percussion tool. The static grinders have mid-range dimensions, with traces of use-wear and one to two clearly defined working surfaces. They were made from fine-grained sandstones with abrasive characteristics. A grinder from Feature 14 (C-1859), made from fine-grained sandstone, with lateral edges partially processed by grinding, is especially notable. The technical-physical features of the raw material made it suitable for fine-processing, most probably final-processing, of objects made from hard materials (e.g. rock, bone, horn, and pottery). It has two clearly defined working surfaces: after a large dent occurred on one side, the object was turned and the grinding was conducted on the other side until the stone became too thin, and fragmented (Figure 4). To produce querns, craftspeople used massive pieces of compact raw material with distinctive abrasive characteristics; they were partially modified by edge-pecking. Both querns from Belovode have visible burning traces on their lower, dorsal sides and clearly defined, smooth, concave working surfaces on their ventral sides. The traces of use-wear on the ventral side suggests that they were used through movement in various directions, but mostly longitudinally and transversely to the quern (Figure 5; see Dubreuil 2001: 73–87; Adams *et al.* 2009: 48–53).

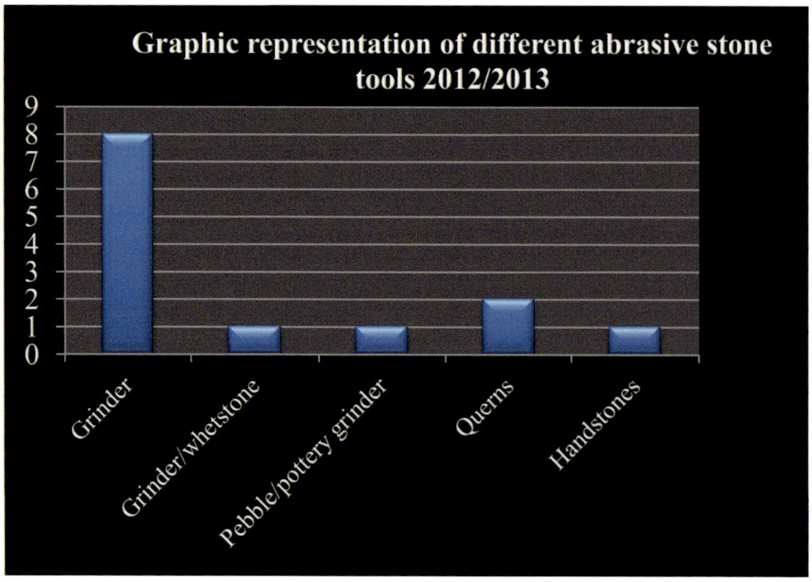

Figure 3. Graphic representation of different abrasive stone tools from Belovode, Trench 18.

Ground stone tools represent the majority of large stone tools at Belovode. Within this group, there are several types and subtypes including: axes (5), adzes (22), chisels (12), and three pre-forms and nine fragments that could not be classified with any certainty due to severe damage (Figure 6). Two types of adzes are represented: I/1 – three objects, and I/3 – one object. One axe could not be typologically determined due to severe damage. Adzes are the largest category of tool with 22 objects. Type III/1, the most common type in the Vinča culture, is the most dominant with

Figure 4. Grindstone (C-1859); grey coarse-grained sandstone. Working surfaces on both sides, dorsal and ventral, are clearly distinctive.

Figure 5. Querns from Belovode (Finds 224/2 and 225 in the ground stone database).

11 objects; Type III/5 includes five objects, while Types III/2 and III/1 are represented by two and one object respectively (Figure 7). A tool with a perforation could not be typologically attributed due to a high level of fragmentation. An amulet with a groove, made from calcite by pecking and grinding, is the only find of a cult object or item for personal adornment (Figure 8).

All ground tools with a cutting edge were made from fine-grained, hard and compact grey and grey-green rocks (crystallised schists, cornite, metasandstone and metaalevrolite), as well as from 'light white stones' of various hardness and silicification grades. The choice of raw material influenced the method of processing, i.e. the reduction technique used to create the desired size and shape. The traces observed on the ground tools indicate several stages of production (Antonović 2014b; Dimić 2015). The first of these was knapping. This was conducted mainly using the ventral side as the knapping platform, thus reducing the dorsal side and producing a recognisable semi-circular cross-section. This kind of knapping was usually practiced in the

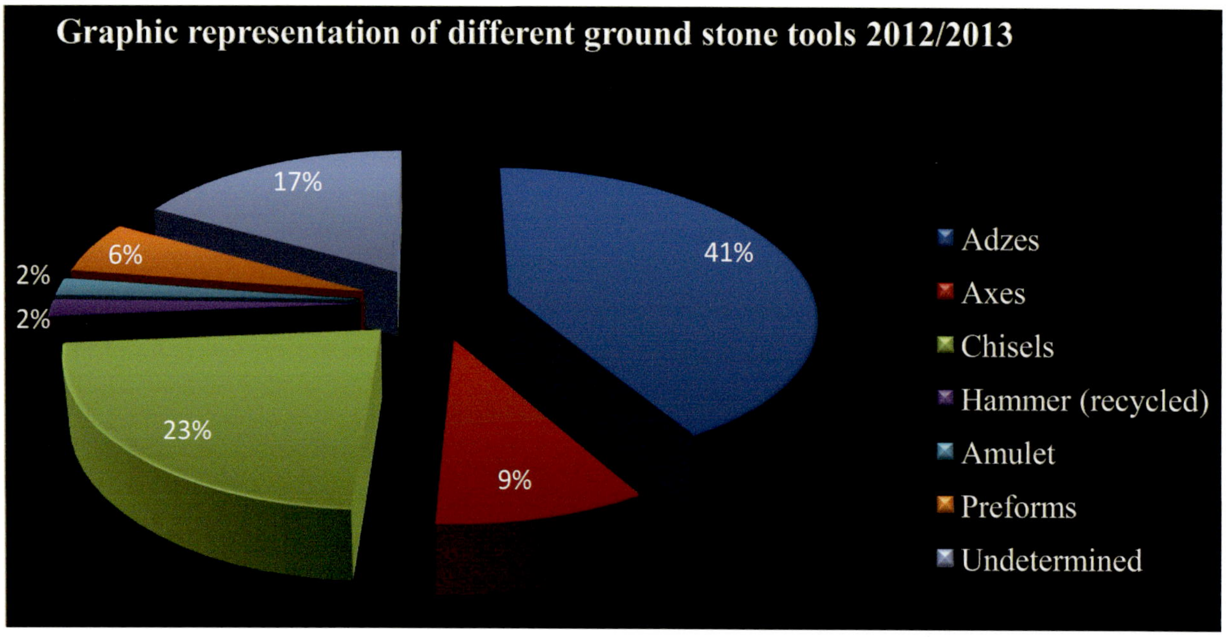

Figure 6. Graphic representation of different ground stone tools from Trench 18.

production of adzes and was a standard technique in the Vinča culture. By contrast, zig-zag knapping was used for producing axes, and can be observed on an example of a pre-form (Figure 9; Dimić 2013a). After knapping, the edges were retouched with a standard reduction of the butt, making it flat. The pre-form was finally processed by grinding on static grinders made from the fine-grained variety of sandstone. The grinding traces (rectilinear, parallel to the longitudinal axis of the object, and circular) indicate that this was conducted in several directions and that water was used. Particular attention was paid to processing the cutting edge.

When the use-wear traces of this tool category were investigated, different levels of destruction of edge and butt could be determined (Dimić 2015). Negatives of micro flakes, which occurred during intense use, can be observed on almost all cutting edges. Micro polish and rounded edges, produced by the pressure and friction that the tool head suffered from inside the haft, are visible on the butts (Figure 10). As previously mentioned, most of the tools suffered severe damage, although the position of the damage varied from the distal and proximal parts to the middle of the tool. After heavy damage, tools were rejected if they could not be recycled into a new object. Recycling was observed on three chisels, as well as on one larger tool. In this case, the tool (of unknown previous purpose) had suffered fragmentation of its distal part but had been transformed with minimal knapping into a hammer.

Conclusion

The analysis of the collection of ground and abrasive stone artefacts from the site of Belovode completely matches the results obtained through the examination of the same category of tools from Trenches 1–8 excavated between 1994 and 1997 (Antonović 2000). The production of stone tools on this Neolithic settlement appears in its fully developed form from the earliest occupation activity (Vinča–Tordoš I Phase, or Vinča A) and did not significantly change until the very end. The craftspeople of Belovode had clear affinities towards macroscopically similar grey and grey-green raw material, which is generally also the case in other Vinča settlements (Antonović 1992, 2000, 2003, 2014b). Fine-grained sediment and contact metamorphic hard rocks with a conchoidal fracture (e.g. crystallised schists, cornite, metasandstone and metaalevrolite) were used to produce woodworking tools (axes, adzes and chisels) in the same manner as throughout the entire Vinča culture (Antonović 2014b).

Primary processing began with the knapping of a suitable piece of raw material and its reduction to a desired shape. A retouch was then used to achieve fine modifications, to shape edges and create a cutting edge, while the final form was obtained by grinding on static grinders, using water. Polishing, as a special technique, was recorded on several objects. All polished edge cutting implements were made of green and grey fine-grained, hard rocks.

The quality and intensity of grinding depended on the quality of raw material. The best ground artefacts were those made of compact and hard rocks, indicating that the stone craftspeople of Belovode were familiar with the petrographic features of rocks and invested the most effort in producing implements from material that could endure intense work. The greatest attention was given to the process of grinding the cutting edge, while the rest of the tool was only partially ground.

Figure 7. Ground stone cutting implements from Belovode, Trench 18: 1) adze type III/5b; 2) adze type III/1a; 3) adze type III/5d and segment of dorsal side surface with manufacture traces of grinding.

The resulting tools were used for a variety of activities connected to wood processing, from tree felling and splitting, to the production of architectural elements and pieces of house furnishings.

We could not identify a clear change in the choice of raw materials or in the production techniques in the ground stone industry throughout the horizons of occupation in Trench 18. From the very beginning, stone tools produced in the settlement are not conspicuous for their attractiveness or the uniqueness of the manufacturing process, but for their consistency in the sizes, the types of tools, and the choices of raw material.

As in many other Vinča culture sites, Belovode populations also used objects made from 'light white stone', occurring from the Gradac Phase, although these are not as numerous as at some other sites (Antonović 1992, 2003, 2011:197, 2013: 26, 28; Bogosavljević-Petrović 2011: 215; Dimić 2013a). It seems that grey and grey-green rocks were abundant and available near the

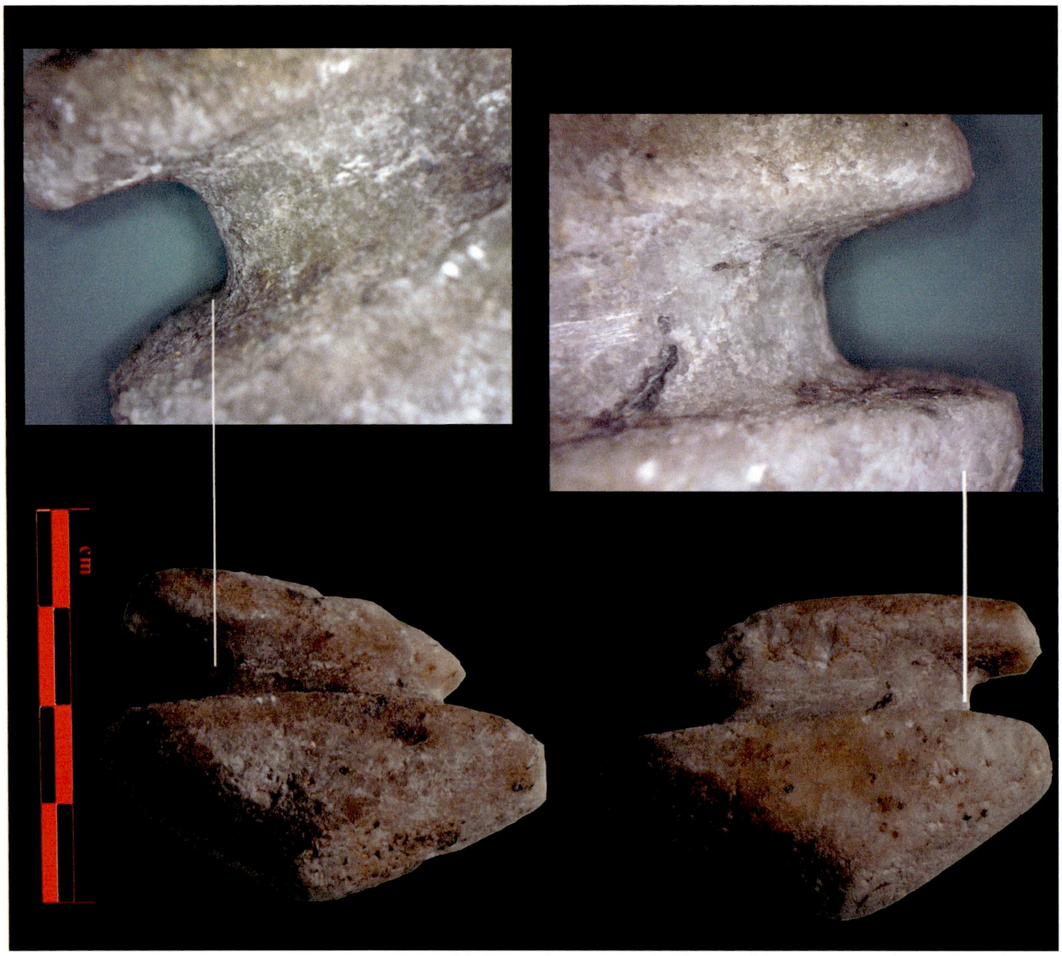

Figure 8. Ground stone amulet (C-1858) of calcite. Above: macro footage of groove and manufacture traces.

Figure 9. Preform of ground stone axe (Find 452); 'Light white stone'. Traces of knapping on both dorsal and ventral sides are clearly visible.

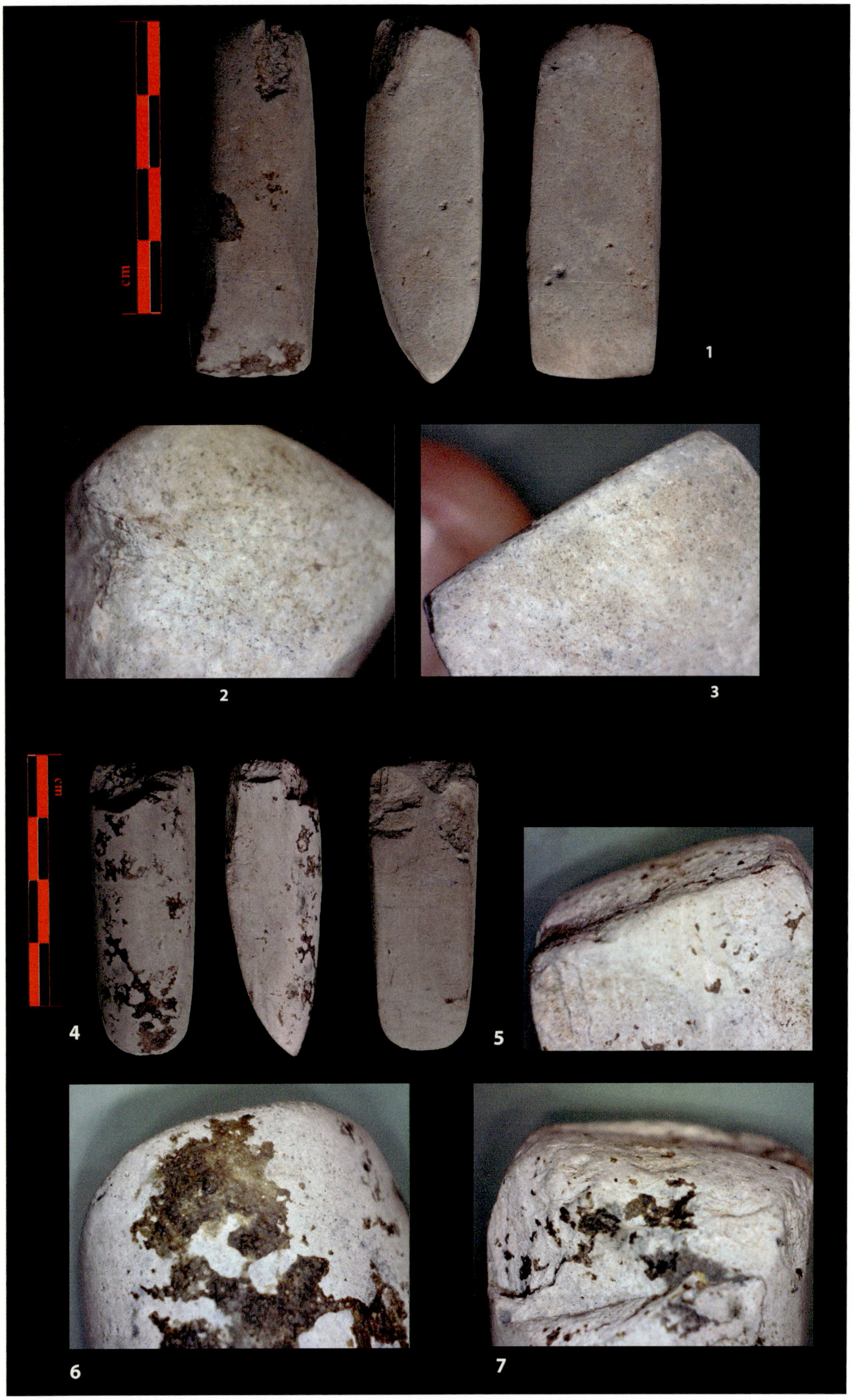

Figure 10. 1) Chisel (C-1817); 2) use-wear on butt 3); use-wear on cutting edge; 4) chisel (C-1823); 5) use-wear (rounded surfaces) on butt probably as an effect of hafting; 6) use-wear on the cutting edge is not very clear; 7) use-wear on butt.

settlement, so there was less necessity for 'light white stone' as elsewhere. Alternatively, it is possible that 'light white stone' deposits were limited in quantity at Belovode, perhaps occurring in smaller veins and interlayers, and not in whole geological layers as was the case with other areas in Serbia (Antonović 1997, 2003; Bogosavljević-Petrović *et al.* 2012; Dimić 2013a).[1]

Thus far, there is not enough data to identify the exact locations from which Belovode craftspeople could exploit suitable stone. Nevertheless, based on the processing method and the shape of tools, it can be concluded that the rocks were exploited from both primary and secondary deposits. Secondary rock deposits were most probably in the form of alluvium from nearby rivers and creeks; identification of primary deposits would require detailed geological prospection of the wider area around Belovode.

Stone tools from Belovode fit well within the framework of the ground and abrasive stone tool industry of the Vinča culture. They do not stand out for their production quality or attractiveness, but for their consistency over a long period with the production of tools of similar appearance, made from similar raw materials. Whether this was a result of defined production standards that completely fulfilled the needs of Belovode residents, or some other phenomena linked to the geological surroundings, or to social aspects remains to be explored in the future.

The bibliographic reference for this chapter is:

Dimić, V. and Antonović, D. 2021. Ground and abrasive stone tools from Belovode, in Radivojević, M., Roberts, B. W., Marić, M., Kuzmanović Cvetković, J., and Rehren, Th. (eds) *The Rise of Metallurgy in Eurasia*: 205–214. Oxford: Archaeopress.

[1] Geological analyses and prospection were not undertaken in the frame of the project. See Chapter 45, this volume for further details on stone raw materials from Belovode.

Chapter 17

Bone industry from Belovode

Selena Vitezović

Introduction

The bone industry encompasses all artefacts (tools, decorative items, manufacture debris) made from osseous raw materials (bone, antler, teeth, ivory, mollusc shells) (Averbouh 2000; Poplin 2004). Along with stone and flint, bone raw materials were very important for making everyday tools and other artefacts in all prehistoric societies. Their use, however, depended upon their availability as well as the economic and cultural preferences of a given community. Osseous raw materials had an important place in the Vinča culture, and they were frequently used for both everyday tools and decorative objects, representing a significant proportion of the material culture.

The osseous industry from Belovode was analysed from a technological perspective (cf. Inizan *et al.* 1995: 13 ff.), including the raw material choices, technology of manufacture, and typological data. The assemblage of about 60 artefacts includes those recognised during excavations as well as those separated during the post-excavation faunal analysis. Numerical data are not presented since they relate to only a segment of bone artefacts recovered so far at Belovode, and therefore are not statistically meaningful.

Techno-typological analysis

The raw material used within the Belovode settlement included large mammal ribs, metapodial bones from small ruminants, astragals from ovicaprines, wild and domestic pigs, red and roe deer antlers and occasionally teeth (only boar tusks and red deer canines). Bones were the predominant raw material, although antler artefacts also were found in small numbers.

Prehistoric craftspeople divided bones and antlers into blanks by chopping, breaking, or by direct and indirect percussion. The main technique used for final shaping of the artefacts was burnishing with various abrasive tools. The perforations on the bones were made by drilling with a flint borer.

The typological classification used here is based on the link between the supposed function and form of the active part of the objects, originally created by H. Camps-Fabrer (1966, 1979) and now used for numerous European prehistoric assemblages, with some modifications and improvements (e.g. Voruz 1984; Pascual Benito 1998; Beldiman 2007). The artefacts were classified into several groups: I pointed tools; II cutting tools; III burnishing tools; IV punching tools; V objects of special use; VI decorative items; VII non-utilitarian items; and VIII incomplete artefacts. Within these groups, further subtypes and variants were identified based on morphology, function, manufacturing technique, and raw material used (Vitezović 2007, 2011a, 2013a; 2016a: 79–98; see also Bačkalov 1979; Beldiman 2007).

I Pointed tools

Subtype I1. Awls or medium-sized pointed tools were the most common tool type. Two subtypes were recorded at Belovode: awls made from long bones (subtype I1A) and from flat bones, mainly ribs (subtype I1B).

For awls of subtype (I1A), ovicaprine metapodials were predominantly chosen. The bone was split longitudinally so that the artefact was shaped from a segment of the semi-circular cross-section, and then by grinding and burnishing (Figure 1). A proximal or distal epiphysis segment may be preserved at the basal part. Only one awl was made from an entire distal segment and has a complete distal epiphysis preserved at the basal part. There is a full cross-section of the epiphysis at the mesial section, and the distal end is shaped into the point by scraping and burnishing (Figure 1).

One completely preserved awl made from an ovicaprine metapodial bone should be highlighted. It was made from a longitudinally split bone and has one half of the distal epiphysis at the proximal end, a semi-circular cross-section, and a fine tip at the distal end (Figure 1). This example is of note because of its very small size and intensive traces of use (worn surfaces), indicating that it was in use for a long time, repeatedly re-sharpened, until reduced to its size at discovery.

Awls fashioned from small, ruminant metapodials were widespread across the whole of later prehistoric Europe, especially in the Neolithic and Chalcolithic periods (e.g. Schibler 1981: Table I/1; Voruz 1984: Figure 24; Sériadès 1992b: Plates 137/a–c and 191/c; Stratouli 1998: Tables 40/1 and 42/1–6; Pascual Benito 1998: 48,

Figure 1. Awls produced from sheep/goat metapodial bones.

Figure III.16; Deschler-Erb *et al.* 2002: Figure 507/1–4; Lang 2005, Table 187/1–25; Hüser 2005, Table 1). They are very common in the Vinča culture (cf. Срејовић and Јовановић 1959: Figure 1; Perišić 1984: Tables 2/10, 2/11, and 4/31; Russell 1990: Plate 14.1f; Vitezović 2007: 86–87, 105; Vitezović 2011b: Figure 5).

At Belovode, awls made from ribs (subtype I1B) were also relatively numerous, with the ribs coming mainly from large mammals (cattle- and red deer-sized animals). The ribs were split longitudinally, so that the blank for further shaping was a single bony plate. Ribs are very resilient in their fresh state and so not easy to work. In order to split a rib, it was necessary first to extract segments by breaking or by direct or indirect percussion. The obtained segments were then split with a flint burin (cf. Christidou 2001: 42). The final shape may have been obtained by scraping with a flint tool and/or by burnishing with an abrasive tool. On the Belovode awls, only traces of the final phase are visible, indicating burnishing and polishing on the lateral sides by a fine-grained, abrasive stone tool.

The final form of the awls is triangular, or they have straight edges in the proximal and mesial parts with the edges subsequently converging to the fine point. Only one-sided awls were present at Belovode (double awls were not recovered), some of which were particularly well made, with basal parts carefully cut and burnished (Figure 2). Traces of intensive use are visible on most of them: polish, striations, and worn outer surfaces, and smoothed and abraded spongy tissue on the inner surfaces.

This is a tool type that is common in the majority of Neolithic and Chalcolithic sites in Europe (e.g. Schibler 1981: Table 37/5–8; Voruz 1984: Figure 20; Deschler-Erb *et al.* 2002: Figure 10/3–7; Lang 2005: Tables 189/26–29, and 190/1–4; Hüser 2005: Table 6). Rib awls are also fairly characteristic for the Vinča culture, although their relative number within one assemblage may vary due to the method of recovery (used ribs are less conspicuous than, for example, ovicaprine metapodials) They are known in large numbers at, for example, Vinča-Belo Brdo (Срејовић and Јовановић 1959: Figure 2) or Motel Slatina (Vitezović 2007: Tables XXI, XXIII, XXIV and XXV).

Subtype I2. Heavy points were not numerous; one specimen, made from a large mammal (cattle- or red deer-sized animal) rib, should be mentioned. It was made from an unsplit segment (unlike awls), but one bone plate is only partially preserved due to intensive use. Its distal end is a massive, heavily worn point. In addition, one beam segment of roe deer antler was used as a heavy point, minimally modified, with only the crown tines being removed. The tip was smoothed and blunt from use.

Figure 2 Awls produced from ribs.

Figure 3. Fine pointed tool (needle) with broken perforation at the basal part.

Subtype I3. Three needles or fine-pointed tools were discovered. All three belong to the subtype of eyed needles (I3A), which are 'true' sewing needles with a small perforation placed near the base. They are all fragmented, broken exactly at the perforation, perhaps during use. They were made from small diaphyses segments, which were carefully burnished and polished. Perforations were made by drilling with a fine-pointed flint borer, and their diameter does not exceed 4 mm (Figure 3). Eyed needles are rarely encountered (perhaps due to fragmentation) in the Vinča culture bone industries. A few are known from Vinča-Belo Brdo (Perišić 1984: Tables 17/130 and 17/131), one from Selevac (Russell 1990: Figure 14.2), and one from Drenovac (Vitezović 2011b: Figure 16). They are known in somewhat larger numbers at the Neolithic site of Khirokhitia in Cyprus, where they are linked with the processing soft plant fibres (Legrand 2007: 76–83).

II Cutting tools

In this group, one particular artefact should be highlighted: an adze made from a red deer antler that had been shed, as indicated by traces of rodent gnawing covering a large part of one surface; traces of manufacture are superimposed on these. A crown segment was used, i.e. the segment of a beam and segments of the two upper tines. The end of the beam was obliquely cut and had a fine cutting edge which is partially preserved. A slightly ellipsoidal perforation was placed at the basal part of the tool. It has clear traces of manufacture where the outer cortex was carved out and then the hole was created by drilling (Figure 4). The working edge is, unfortunately, only partially preserved, so the use-wear traces are poorly visible (polished distal end). This tool is most likely to have been used as an adze in woodworking.

Figure 4. Large cutting tool made from red deer antler.

Antler is often used to make large cutting tools, especially those used in woodworking, since it is a material resilient to shock from impact (cf. Guthrie 1983; Christensen 2004). Most often, axes and adzes are made from lower segments of beam, sometimes with the actual base being used, although similar axes and adzes are found on some Vinča sites (cf. Bačkalov 1979: Table XXXI/8).

III Burnishing tools

Subtype III2. Two fragmented tools from split boar tusks were most likely used as scrapers. They were roughly triangular in shape with one of the edges worn and damaged from use. They were probably used for processing plant materials (cf. Maigrot 2003: 124–128).

V Objects of special use

Subtype V2: Spoons. One fragmented spoon made from an antler segment was discovered at Belovode. Only the spoon part is preserved, while the handle is broken. It was made from a longitudinally split segment of red deer antler beam, and shaped by cutting, scraping and burnishing with flint tools and abrasive stones. The spoon is rectangular and the fragmented handle had a square cross-section (Figure 5). The spoon part is almost completely flat and worn on both the outer and inner surfaces with the spongy tissue being almost completely abraded. The exact function of this artefact is difficult to establish but it may be assumed that it was used in contact with soft, organic materials, perhaps for applying grease in the working of leather or applying organic pigments to textiles.

Spoons of this type (with separate bowl and elongated handle) are rare in the Vinča culture. This contrasts with the earlier Starčevo culture period, when spoons that were carefully made from *Bos* metapodials were common (cf. Vitezović 2011a). Furthermore, a strict choice of raw material and careful manufacture are not characteristic of this artefact type, and spoons are encountered with a great variety of shapes, sizes, levels of curation, and raw material choices. Several spoons, also made from red deer antler, were discovered at the eponymous site, Vinča-Belo Brdo (cf. Васић 1932: 39–40; Bačkalov 1979: Tables XXVII/ 1–12 and XXIX/ 1–3, 5; Игњатовић 2008: 269, Catalogue 198); one antler spoon was recovered at Vitkovo (Витезовић 2012); and one unusual specimen made from a mandible was found at Pavlovac-Kovačke Njive (Vitezović 2014).

Figure 5. Fragmented spoon made from red deer antler.

Subtype V4: Used astragals. Four astragals were found at Belovode, two from ovicaprines and two from species not usually used in their manufacture – a wild pig and a domestic pig (Figure 6). The pig astragals were used in their natural shape with the entire prominent surfaces being used so intensively that they became completely flattened and worn down. The ovicaprine astragals have had their lateral sides worn down, with some loss in volume. They also have perforations, one through the central part and a second through the upper part. The perforations were made by drilling; polishing from use can be observed in their interiors.

The Late Neolithic and Chalcolithic periods in southeast Europe are characterised by the use of short bones, either unmodified or with minimal modification, i.e. the natural shape of the bone is largely preserved (e.g. Bolomey and Marinescu-Bîlcu 1988: 347, Figure 7/6; Lang 2005: Table 191/16–18; Zidarov 2005; Bacvarov and Vitezović 2014; Kogălniceanu *et al.* 2014). In the Vinča culture, only astragals were used (worked phalanges have not yet been discovered), in their original shape or with one or more perforations added. These were predominantly ovicaprine and cattle astragals and, more rarely, from red deer. Only at Belovode have pig astragals been reported (Jacanović and Šljivar 2001). Used astragals are known from Divostin (Lyneis 1988), Selevac (Russell 1990), Drenovac (Vitezović 2011b: 129–130, Figure 12), Motel Slatina (Vitezović 2007: 98–100, Tables IV and VII) and Pavlovac-Kovačke Njive (Vitezović 2014). They show great diversity in the position and degree of use of the perforations and the overall degree of wear.

The resemblance of the used astragals to gaming and/ or gambling pieces from ancient Greece or Rome has provided a strong analogy for some researchers, with others seeing the presence of perforations as relating to their use as pendants (cf. Russell 1990: 538–539; Jacanović and Šljivar 2001). However, their intensive use-wear suggests otherwise, although an exact method of their use is difficult to reconstruct (cf. experiments by Meier 2013 on astragals in Bronze Age Hungary). Experimental reconstructions also showed that it was possible to use astragals as loom weights

Figure 6. Used astragals.

(Grabundžija *et al.* 2016). Specimens from Draguşeni-Ostrov were interpreted as burnishers (for an overview of some of the hypotheses see Bolomey and Marinescu-Bîlcu 1988: 347 and Figure 7/6 and Kogălniceanu *et al.* 2014: 292–294). Used astragals have a large temporal and geographical distribution, and it is very likely that they had more than one function. For the Vinča culture objects, intensive polish and shine suggests contact with soft, organic materials (cf. Maigrot 2003; Legrand 2007 for interpretation of use-wear traces), while the loss of volume on some suggests contact with very abrasive, inorganic materials. It is most likely that these astragals were used in relation to the production of textiles and leather (Vitezović 2007; see also Grabundžija *et al.* 2016), although some of the most heavily used pieces may have been used on pottery (cf. experimental results in Meier 2013).

Subtype VI: Decorative items. Only two decorative items were discovered. One is a small, ring-shaped object with a prong at one end; it was most likely made from a large, long bone segment. It is perforated in the centre, finely made and polished, and was probably used as some sort of bead or other decoration on clothes (Figure 7). It has a partial resemblance to the type of decorative items known from Early Neolithic sites in the Near East (cf. Russell 2001: Figure 3) but closer analogies are not currently known so, for the moment, this is a unique find from the Vinča culture.

The other decorative artefact is a residual canine of a red deer. It has a perforation, approximately 3 mm wide, through the upper part and shows intensive polish and some damage to the lower part from use (Figure 7). The perforation itself is also polished, suggesting it was in use for some time (cf. d'Errico 1993) with the degree of wear being between stages 2 and 3 using the model proposed by Bonnardin (2008: 300).

Red deer canines were often used for decorations from early prehistory (cf. Barge-Mahieu and Taborin 1991; d'Errico and Vanhaeren 2002; Choyke 2001; Taborin 2004 and references therein) and they certainly had specific meanings, in addition to their aesthetic value (e.g. protection, symbols of identity and/or prestige). They were even imitated in other materials such as other osseous raw materials, stone, etc. (Choyke 2001). Both red deer canines and their copies occur in the Starčevo culture (cf. Vitezović 2011a) but are not frequently found at Vinča culture sites. One copy comes from Selevac (Russell 1990: Figure 14.7a) and two were identified from Vinča-Belo Brdo, one in bone and one probably in shell (Игњатовић 2008: cat. 222). The find from Belovode is therefore very important for the study of the geographical and chronological distribution of these ornaments.

Results and discussion

The osseous artefacts recovered from Belovode during the 2012 and 2013 excavations represent a cross-section of the entire bone industry from the Vinča culture settlement. Pointed tools dominate, followed by cutting and burnishing tools, and then objects of special use. The bone tools were used for small crafts, and most likely for processing leathers, hides and plant materials (wood and plant fibres). Only a few heavier tools were recovered, and some tool types are completely absent, including hammers and other punching tools, as well as fishing and hunting equipment (e.g. harpoons, fishhooks). Very few decorative objects were found.

Figure 7. Ornaments: perforated tooth and ring-shaped ornament.

The raw materials were obtained mainly from domestic animals that were available within the settlement. Antlers were at least partly collected (i.e. shed antlers), and a few artefacts were made from skeletal elements obtained from hunted wild animals, for example, the used astragal from a wild pig and the pendant made from a red deer canine. Manufacture debris was not present, suggesting that the activity areas for processing osseous raw materials took place elsewhere in the settlement.

Conclusion

The techniques used and the typological repertoire produced fits well within the Vinča culture bone industry (cf. Bačkalov 1979; Russell 1990; Vitezović 2007). The overall bone industry demonstrates high technological know-how and familiarity with raw materials. The specific traits of the Belovode assemblage are the use of pig metapodials, since pig bones were mainly avoided (cf. Vitezović 2013b), and a previously unknown type of decorative item, in the shape of a small ring with prong at one side.

The bibliographic reference for this chapter is:

Vitezović, S. 2021. Bone industry from Belovode, in Radivojević, M., Roberts, B. W., Marić, M., Kuzmanović Cvetković, J., and Rehren, Th. (eds) *The Rise of Metallurgy in Eurasia*: 215–220. Oxford: Archaeopress.

Chapter 18

Chipped stone industry at Belovode

Elmira Ibragimova

The excavations in Trench 18 at Belovode during the 2012 and 2013 field seasons yielded evidence for a chipped stone industry comprising a total of 529 individual artefacts. Chipped stone items were recovered and recorded both during the excavations and during the flotation of soil samples.

The analysis of the chipped stone industry aimed to determine the raw materials exploited, investigate the production techniques and methods used, and identify the types of tools being made. Where possible, the material, technological and typological characteristics were compared throughout the sequence of occupation at Belovode in order to examine the evolution of chipped stone industry. Some questions connected to raw material on the Belovode site have been investigated previously (Bogosavljević-Petrović and Marković 2012), but complex analysis was conducted for the first time. A comparison with chipped stone industries both at Pločnik and within the broader context of other Vinča culture sites is presented in Chapter 47 of this volume.

Field methodology and phasing

Over the two excavation seasons at Belovode, 529 chipped stone artefacts were recovered, primarily through excavation, with 276 (52% of the whole collection) recorded to the spatial resolution of a stratigraphical spit. The precise three-dimensional location was recorded using the EDM for 242 artefacts (46%) from flotation samples and for 10 artefacts (2%) recorded as a 'special find'. Each chipped stone artefact is therefore recorded to the minimum level of a stratigraphical spit; 104 artefacts (20%) were recovered from archaeological features.

The stratigraphic sequence revealed by the excavations of Trench 18 is summarised in a Harris matrix (see Chapter 7, this volume). Three distinctive phases in the evolution of the site can be defined: spits 1–8 (mainly Horizon 1), spits 9–15 (Horizons 2–3) and spits 16–20 (Horizons 4–5). Within the periodisation scheme based upon the detailed typo-chronological analysis of the pottery and an extensive radiocarbon dating with Bayesian modelling (see Chapter 37, this volume), the following Vinča culture phases can be assigned: Horizons 4–5 relate to Vinča Tordoš I-II (5420–4987 cal. BC); Horizons 2–3 are associated with a period from the Gradac Phase to Vinča Pločnik I-IIa Phase (4985–4615 cal. BC); and Horizon 1 refers to the final phase, Vinča Pločnik II (4615–4560 cal. BC) (Chapter 37, Table 4, this volume).

The chipped stone items were presented in all horizons therefore the analysed chipped stone industry spans the entire Vinča culture sequence at Belovode.

Database structure and methodology

The recording of the chipped stone finds encompassed the excavation context which always included the spit or feature number but, depending upon the circumstances of recovery, could also include the EDM measurement and find number, and the reference for the flotation sample. Each chipped stone artefact was measured for length, width and thickness. The raw material of each chipped stone artefact was examined macroscopically to identify the main physical characteristics: colour, natural surface information (pebble or tabular), and transparency. This enabled the identification of several geological groups: various sedimentary rocks (varieties of flint, limestone), igneous rocks (e.g. obsidian, chalcedony) and metamorphic rocks (e.g. quartzite, schist) (see Inizan *et al.* 1999: 21). Other parameters also noted were calcination, visible use-wear traces and cortex location, blank morphology and attribution to chipped stone category. The blade debitage was differentiated based upon the width of the blanks resulting in groups of 'blades' (>12 mm), 'bladelets' (8–12 mm), and 'microblades' (< 8 mm). This division is based primarily on the observation of debitage metrics derived from the experimental usage of different techniques of flint knapping (Pelegrin 1998).

The morphological characteristics of both blades and cores were recorded in detail. The blades were examined for the dorsal pattern, platform and metrics and morphology of bulbar parts; technical stigmata[1]; and individual impact zone preparation. The cores were examined for the number and location of flaking

[1] This term was used by J. Pelegrin in the evaluation of blade production techniques and refers to 'the character of the butt determined by the platform preparation (dimensions. aspect, edge angle)' and 'discrete details determined by the detachment itself (cracks, lip, ripple on the bulb, aspect of the bulb)' (Pelegrin 2006: 42).

surfaces and striking platforms, and the impact zone preparation. All the tools were described separately in a text document while processing the finds.

The analysis of the chipped stone industry concentrated upon establishing the *chaîne opératoires* from the selection and use of the raw materials to the techniques and methods used in the production, use and eventual discard of the final object (cf. Bar-Yosef and Van Peer 2009; Boëda 1990; Inizan *et al.* 1999: 14;). In order to identify the complex *chaîne opératoires* underpinning the chipped stone industry, all by-products of the activities connected to each stage were also analysed. The morphological types and terminology follow Šarić's (2006) classification of the Early and Middle Neolithic chipped stone tool assemblages in Serbia.

Raw material

The raw material of the Belovode chipped stone industry is highly varied, though flint was the main raw material used. Examination of the natural surfaces of the flint artefacts revealed two distinct raw materials: 'pebble flint' - small and medium-sized river gravels, all having a rolled surface and, in some cases, specific double patination; and 'tabular flint', with a chalky surface which came probably from an outcrop. The colour of pebble flint varies from black to light brown, with a texture ranging from fine-grained to medium-grained. Tabular flint has a cream colour and a fine-grained texture. The pebble to tabular flint ratio shows a clear chronological change. A predominance of brown and light-brown pebble flint is characteristic of Horizon 5, whilst in Horizons 2–3 the number of tabular cream flint suddenly increases until it becomes predominant in Horizon 1 (Table 1, Figure 1). Several chipped stone artefacts were made using honey brown flint with a rolled natural surface. This type of flint could be attributed to the 'Balkan flint' identified across the region though further analysis is required (Bogosavljević-Petrović and Marković 2012; Bogoslavljević-Petrović and Starović 2013; Bonsall *et al.* 2010; Antonović *et al.* 2005). Small amounts of obsidian (see Chapter 49, this volume), chalcedony (see Chapter 48, this volume) and limestone were also used throughout the full sequence of the site. The physical properties of the different raw materials could well have influenced the techniques of production and the types of objects being produced. Within the Belovode assemblage, tabular and pebble flint and obsidian were used for microblade production, whilst tabular flint and occasionally fine-grained pebble flint were used in the production of blades and bladelets; flake production was based on pebble flint.

One of the characteristics of the Belovode industry is the occurrence of a small number of obsidian items: flakes, microblades and a rejuvenation flake. The sources were defined as Carpathian 1 (Chapter 49, this volume), which means it was a long-distance import. Chalcedony items are also present in the Belovode collection: a blade, two fragments of bladelets, a scraper and an exhausted core. One flake was made from coarse-grained material, probably limestone.

Techniques and methods of blade production

The distribution of the main chipped stone categories (Table 2, Figure 2) clearly shows a gradual change in the chipped stone industry at Belovode, reflecting changes in production systems on the site. During Horizons 5–2, the proportion of waste products (e.g. fragments, flakes, chips), flake, and fragment tools is distinctly

Table 1. Distribution of chipped stone raw material type from Belovode, by horizon.

Raw material	Number of items (% of total)		
	Horizon 1	Horizons 2-3	Horizons 4-5
Cream tabular flint	121 (54)	101 (42)	7 (11)
Honey coloured flint	4 (2)	9 (4)	1 (1.75)
Other colours pebble flint	56 (25)	69 (28)	36 (57)
Brown transparent flint	6 (3)	6 (2)	2 (3)
Quartzite	0	1 (0.5)	1 (1.75)
Chalcedony	1 (0.5)	2 (1)	1 (1.75)
Obsidian	1 (0.5)	4 (2)	2 (3)
Cream pebble flint	2 (1)	2 (1)	1 (1.75)
Limestone (?)	0	1 (0.5)	0
Other	33 (14)	47 (19)	12 (19)
Total	**224**	**242**	**63**

Table 2. Distribution of main chipped stone categories from Belovode, by horizon.

Chipped stone category	Number of items (% of total)		
	Horizon 1	Horizons 2-3	Horizons 4-5
Microblades	25 (11)	49 (20)	12 (19)
Bladelets	38 (17)	26 (11)	6 (10)
Blades	38 (17)	32 (13)	5 (8)
Flakes and chips	30 (13)	19 (8)	8 (13)
Fragments	6 (3)	2 (1)	10 (16)
Blade cores	8 (4)	2 (1)	2 (3)
Flake cores	1 (1)	44 (18)	0
Blade tools	57 (26)	7 (3)	9 (14)
Flake and fragment tools	13 (5)	15 (6)	9 (14)
Microtools	7 (3)	6 (2)	2 (3)
Total	**224**	**242**	**63**

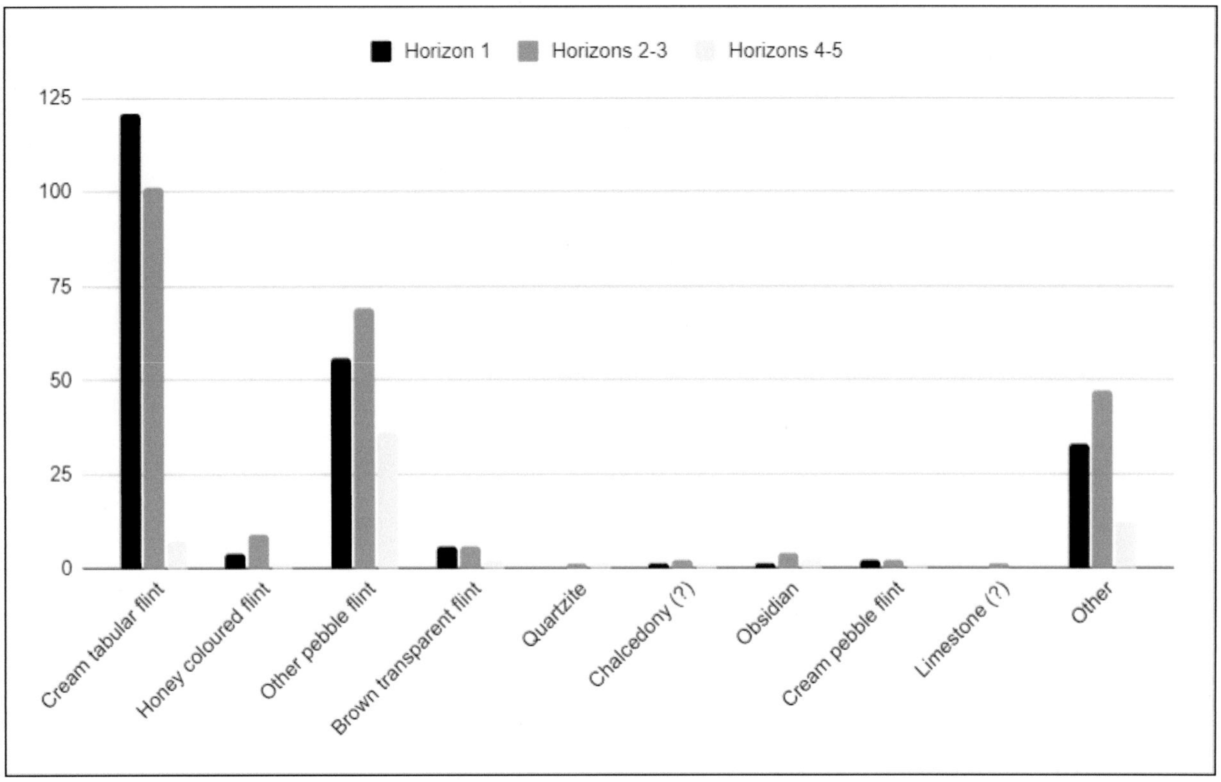

Figure 1. Distribution of chipped stone raw material types from Belovode site by horizon showing number of items.

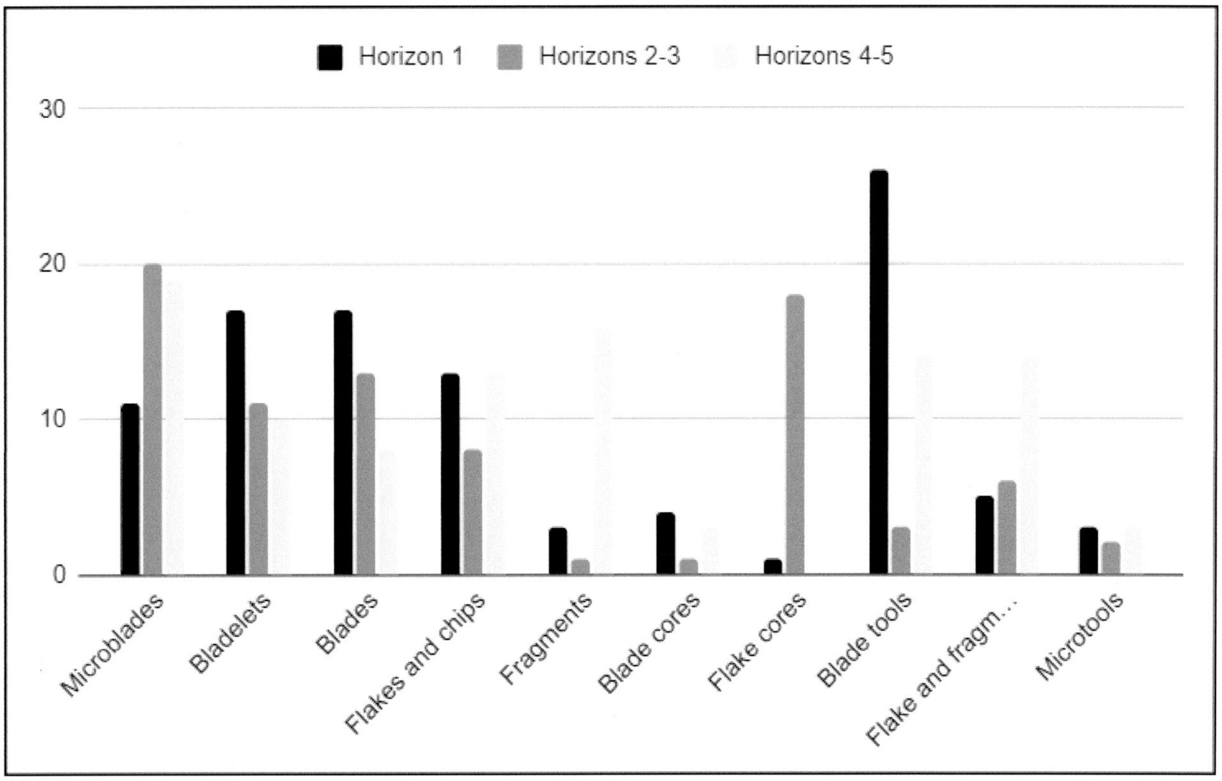

Figure 2. Distribution of main chipped stone categories from Belovode site by horizon, showing percentages of total assemblage.

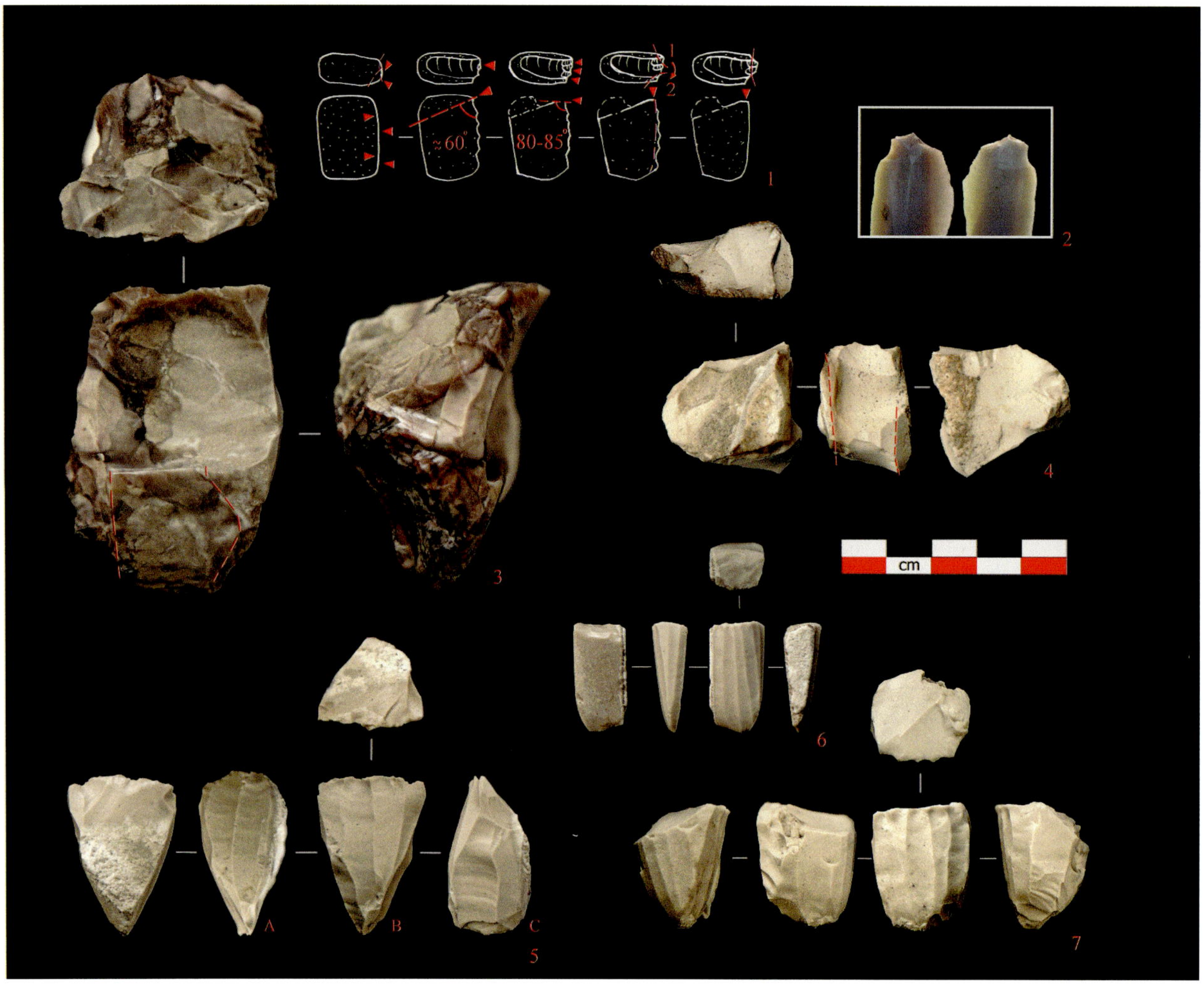

Figure 3. Belovode tabular flint reduction. 3.1 - reduction sequence scheme; 3.2 - dihedral bulbar part of the blade with slightly abraded overhang; 3.3–3.7 - blade, bladelet and microblade cores.

higher than in Horizon 1. There is a high proportion of microblades and microtools in Horizons 2 and 3. In Horizon 1, the proportion of blades increases and that of waste products (e.g. flakes, fragments) decreases.

A detailed technological analysis identified two groups of reduction products: microblades and bladelets/ blades. The two groups had distinct sets of reduction methods and techniques and utilised several different types of raw materials.

Microblade production

Microblade production played a significant role in the Belovode chipped stone industry although the amount of debitage connected to that production is minimal. Microblades form 19–20% of the whole collection in Horizons 2–5 and 11% in Horizon 1 (Table 2) but three cores were found in Horizon 1 and two cores in Horizons 2–5. The small number of waste products from the microblade production is not sufficient evidence to determine whether the cores for microblades were formed in a distinct way or simply that the cores from the last stages of reduction for blades or bladelets were reused to obtain the microblades.

Fine-grained tabular flint was the preferred raw material for microblade production (Table 3). Three single-platform cores of tabular flint with fluted flaking surfaces were discovered (Figure 3.4, 3.6 and 3.7), all found in Horizon 1, and 69 microblades from all site horizons. Pebble material was used less commonly for microblade production (Table 4). There are 14 microblades and one secondary microcore made on a large flake (Table 5, Figure 4.2) connected to Horizons 2–6.

Microblade reduction could be realised on the large frontal side of the core (e.g. flat exhausted microcore, Figure 3.6) or on a narrow flaking surface (Figure 3.4). The dorsal scar pattern of the microblades (with one cortical side having an angle of more than 50º) demonstrates that reduction could occasionally expand to the lateral faces of the core.

The products of this reduction were regular, thin-sectioned microblades with a straight profile. The

Figure 4. Belovode pebble flint reduction: 4.1–4.3 - blade, bladelet and microblade cores; 4.4 - proximal part of a large blade made of 'Balkan flint'; 4.5 - bladelet with flat butt.

Table 3. Main chipped stone categories of tabular flint items from Belovode.

Category	Number of items (%)
Blades (+tools)	63 (27)
Bladelets (+tools)	54 (23)
Microblades (+tools)	68 (29)
Flakes (+tools)	30 (13)
Fragments	12 (5)
Cores	7
Total	**234**

Table 4. Main chipped stone categories of pebble flint items from Belovode.

Category	Number of items (%)
Blades (+tools)	52 (28)
Bladelets (+tools)	38 (21)
Microblades (+tools)	15 (8)
Flakes (+tools)	53 (29)
Fragments	19 (10)
Cores	6 (3)
Core rejuvenation flakes	1 (1)
Total	**184**

whole piece microblades are 19–32 mm long. The mean ratio of width to thickness for 55 unretouched microblades is 4.4 (Figure 5). The microblades have trapezoid (93%) or triangular (7%) sections. A regular dorsal pattern is typical for 72%, and 24% had cortex still present on one side (Table 6). Flat or facetted butts predominate.

The extremely regular character of core debitage surface (Figure 3.4, 3.6 and 3.7) and the morphology of microblades gives us reason to assume that a pressure technique was used for microblade production.

Bladelet and blade production

Bladelet and blade production is strongly correlated with the use of tabular flint (Table 3). Its waste products occur mostly within Horizons 1 and 2–3 in the form of fragments, blade cores and flakes. Tabular flint flakes

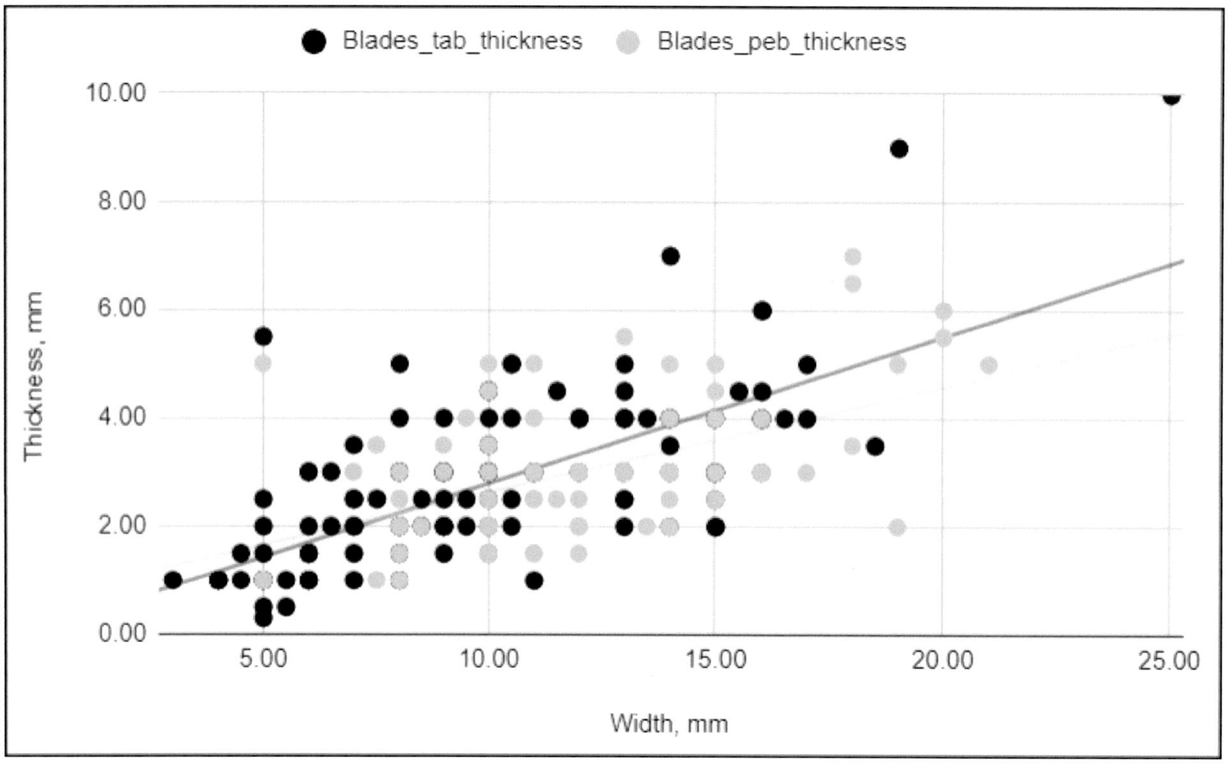

Figure 5. Width and thickness of blades, bladelets and microblades on tabular and pebble flint from Belovode.

Table 5. Types of cores from Belovode and their distribution across site horizons.

Raw material	Blank type	Tabular flint			Pebble flint			Other raw material			Total
	Horizon	1	2-3	5	1	2-3	5-6	1	2-3	4-5	
Pyramidal	Blades, microblades	1	0	0	0	0	0	0	0	0	**1**
Pyramidal	Blades	1	0	0	2	0	0	1	0	0	**4**
Pyramidal, narrow flaking surface	Microblades	1	0	0	0	0	0	0	0	0	**1**
Pyramidal, change -of orientation	Bladelets, microblades	1	1	0	0	0	0	0	0	0	**2**
Prismatic	Bladelets	1	0	0	0	0	0	1	0	0	**2**
Prism-type	Flakes	0	0	0	1	0	0	0	0	0	**1**
Secondary core	Microblades	0	0	0	0	0	1	0	0	0	**1**
N/A	Bladelets	2	0	0	0	0	1	0	0	0	**3**
Total		**7**	**1**	**0**	**4**	**0**	**2**	**2**	**0**	**0**	**15**

should be regarded as by-products of that production process. They are small (usually not more than 21 mm wide, see Figure 6) and the largest part is covered with a cortex (Table 7), indicating the connection of the flakes with the core preparation process.

Reduction was mostly performed on tabular flint slabs with a thickness of no less than 25 mm. Pre-cores and cores in the initial stage of reduction are almost absent in the assemblage (excluding one prismatic core with one negative of blade removal on Figure 3.3), so the initial length of cores was reconstructed on the stepped-ended whole piece blades (up to 61 mm). The modes of pre-core preparation were reconstructed on the basis of blade morphology: the presence of three crested blades in the collection suggests a frontal crest preparation with a series of transversal flake removals (Figure 3.1).

In the first stages of reduction, the flaking platform was formed with an angle of about 60º between the platform and the flaking surface. Before each blade removal, the

Table 6. Location of cortex on tabular blades, bladelets and microblades from Belovode.

Cortex location	Blades	Bladelets	Microbladelets
Primary	0	0	0
Left edge	7	5	5
Right edge	4	3	8
Both edge	2	1	1
Distal end	1	0	2
Plain	42	32	41
Chalky	3	4	0
Total	**59**	**45**	**57**

edge of the platform was facetted to keep the angle at about 80–85º, and subsequently abraded (Figure 3. 1).

Reduction in most cases would start at the narrow side of the slab and expand to the large frontal side when the reduction front had insufficient curvature (Figure 3.7). This trend is reflected in the morphology of the blades themselves. Some blades and bladelets in the collection have a cortical side (or even two sides) (Table 6), which often has a semi-abrupt or abrupt angle (average 50º, but up to 80º). Such blades could be the result of reduction of lateral faces. Two secondary crested flakes could also be evidence of the lateral crest formation in the course of blade reduction. After one front was exhausted, reduction could, in some cases, switch to

Table 7. Cortex location on tabular flint flakes from Belovode.

Cortex location	Number of items (%)
Primary	2 (8)
Secondary	9 (38)
Full	13 (54)
Total	**24 (100)**

the ex-flaking platform and, after its exhaustion, to some other side of the core (Figure 3.5).

Most blades and bladelets connected to that reduction correspond to plain debitage. Blades and bladelets without any trace of the cortex are present in 71% and 64% of the assemblage (Table 6) and have a unidirectional dorsal pattern.

Several features indicate that the technique of indirect percussion could be used to produce bladelets in the Belovode industry. Indirect percussion 'involves the application of an intermediary tool, called a punch, which can be of wood, antler, bone or metal' (Inizan *et al.* 1999: 32). Bladelets and blades have massive sections with the average width/thickness ratio estimated at 3.82 for bladelets and 4.16 for blades. The average flaking angle of bladelets and blades is estimated at 87.6º (ranging from 75º to 95º). Butts are flat (64%), facetted (24%) or dihedral (12%) (Figure 3.2). Four items have a

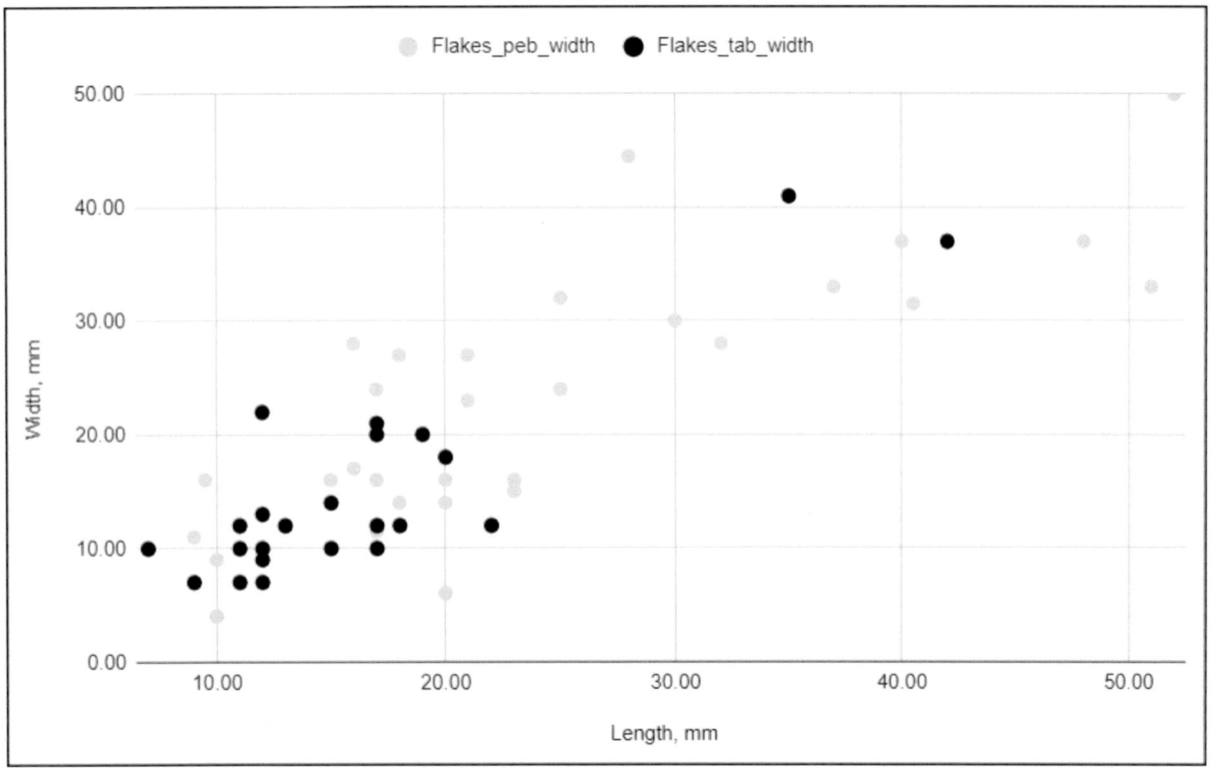

Figure 6. Width and length of pebble and tabular flint flakes from Belovode.

small lip; only one has a mesial belly. Overhangs were slightly abraded but not reduced completely as would be the case for direct percussion with a soft hammer (Pelegrin 2000).

A separate group of debitage products is obtained in the course of blade/bladelet reduction of pebble flint. Waste products of the reduction sequence include cores, fragments, and flakes. The technique of hard mineral percussion could be used in the course of a frontal reduction of pyramidal and prismatic cores (Figure 4.1 and 4.3). In some cases, points of impact are clearly visible on the platform of the core (Figure 4.1). This resulted in blades that are different from those made on tabular flint: they have thinner sections on average, with width to thickness ratio of around 4.5 (Figure 5). The width of the blades does not exceed 20 mm, while whole blades are between 28 and 55 mm long. Blades and bladelets of this group are mainly irregular, with deep striking platforms (e.g. Figure 4.5).

Most pebble flint flakes were probably produced in the course of blade and bladelet reduction sequences; the same raw material was used (Table 4) and there is just one flake core in the whole collection (Table 5).

All the operations connected to blade, bladelet and microblade production were performed locally. Two exceptions are relatively small groups of artefacts that lack a technological context. The first consists of obsidian items: four microblade fragments, two flakes and one rejuvenation core flake (Table 1). There is no other evidence to presume that the reduction of obsidian bladelets took place somewhere on the site except the rejuvenation core flake. (Tablets appear only in the course of blade reduction; the examined item did not appear to be reused). The second group includes items of honey-coloured flint (Table 1) amongst which a group of relatively large blades (up to 21 mm in width) should be mentioned. They were probably made from so-called 'Balkan flint' and must have been produced somewhere outside the examined area (Figure 4.4).

Typology

The blanks which were the result of the production processes described above were, in some cases, retouched to become tools. Blades, bladelets and their fragments (Figure 8.15, 8.16, 8.17, 8.18, 8.19, 8.20, 8.21, 8.22) were predominantly used for tool production, with flakes and microblades being used more rarely.

Many of the blades and bladelets have specific use-wear traces: a bright, mirror-like polish often connected with usage as sickles (Figure 8.1–8.9). This attribute was not considered in the present classification in order to prevent the mixing of morphological and functional approaches, with the latter being conducted by N. Skakun and V. Terekhina (Скакун *et al.* 2015) and to be published soon. The basic modes of tool fashioning were semi-abrupt, abrupt regular and irregular retouch, the latter most probably a result of tool usage. Flat retouch and burin blow techniques were used only marginally.

The tool assemblage from Belovode is characterised by two main tool groups: retouched blades and scrapers made on blades and flakes. All other tool type groups are represented by only a few examples and include drills, microdrills, burins, geometric microliths, tranchets, and retouched flakes (Table 8, Figure 7).

Retouched blades are the most commonly represented throughout all horizons (43–32% of tool collection; Table 8). Mainly proximal (34%) or mesial (33%) fragments of blades and bladelets were used, which is probably related to a distal curvature of the blanks. Most blades and bladelets have only irregular retouch or notches (Figure 8.1, 8.3, and 8.7). Often these tools are truncated (Figure 8.5 and Figure 8.8) or backed with abrupt or semi-abrupt retouch (Figure 8.6). Only two tools were retouched with a flat lamellar retouch.

Scrapers were made on blades, bladelets, or flakes. Usually, the distal ends of the blades or bladelets were retouched with a semi-abrupt retouch forming a symmetrical convex (Figure 9.5, 9.6, 9.9) or slightly straightened (Figure 9.4) edge. The number of blade end-scrapers increases continually throughout Horizons 5–1. Flake end-scrapers are particularly characteristic for Horizons 4–5, though they exhibit a large variety of working edge forms (Figure 9.10). Two microscrapers made on small flakes were also noted (Figure 9.1 and 9.2), whilst side-scrapers are only present in Horizon 1 (Figure 9.8).

Tools that are less represented at Belovode (15–16% in Horizons 1 and 18% in Horizons 2–5) include microdrills, geometric microlith, tranchets, burins, drills, and retouched flakes. Microdrills were made on tabular flint microblades with a width of 3–8 mm (Figure 8.12 and 8.13). The length of the whole tool could be up to 36 mm. Alternate or direct retouch was used to form a symmetric square-sectioned working tip, although most tools recorded in the collection have had their tips broken off as a result of use. These tools appear in Horizons 2–3 but decrease in Horizon 1. Tranchets were made on large flakes or fragments. The wide edges of these tools were retouched with semi-abrupt or abrupt direct or bifacial retouch. They are found in Horizons 1 and 4–5. Geometric microliths are represented in the tool assemblage by one trapezoid tool found in Horizon 5 (Figure 9.3). This is the mesial part of a pebble flint bladelet which was retouched with an alternating abrupt retouch. One burin was made on the transversal break of a large flake. A drill was retouched in the same manner as microdrills—with alternate retouch, but a

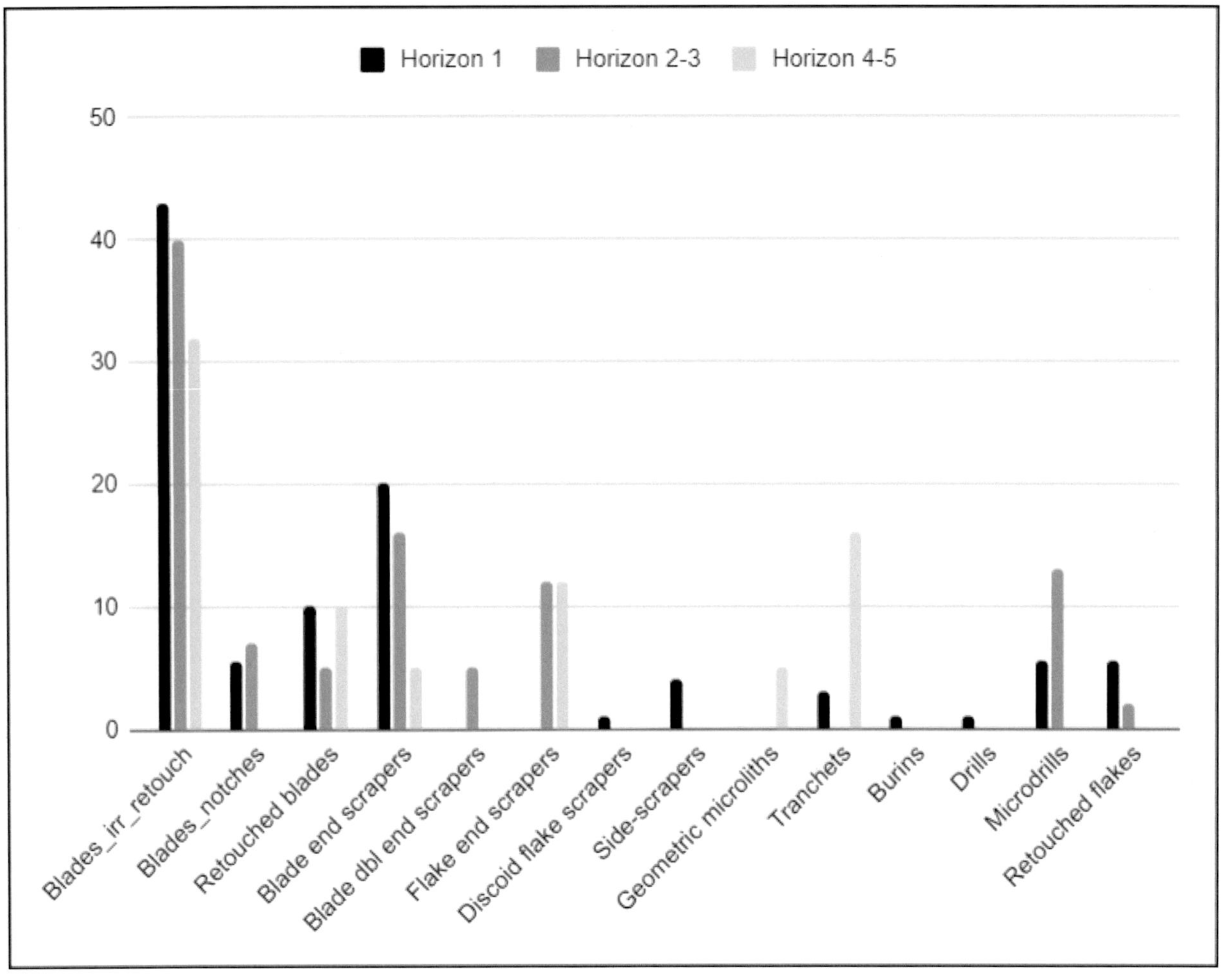

Figure 7. Basic tool categories at Belovode, showing percentages of total assemblage.

Table 8. Basic tool categories from Belovode.

		Number of items (% of total)			Total
		Horizon 1	Horizon 2-3	Horizon 4-5	
Blades	Irregular retouch	32 (43	27 (40)	6 (32)	**65**
	Notches	4 (5.5)	5 (7)	0	**9**
	Retouched blades	7 (10)	3 (5)	2 (10)	**12**
Blade scrapers	Blade end scrapers	15 (20)	11 (16)	1 (5)	**27**
	Double end scrapers	0	3 (5)	0	**3**
Flake scrapers	Flake end scrapers	0	8 (12)	6 (12)	**14**
	Discoid	1 (1)	0	0	**1**
	Side-scrapers	3 (4)	0	0	**3**
Others	Geometric microliths	0	0	1 (5)	**1**
	Tranchets	2 (3)	0	3 (16)	**5**
	Burins	1 (1)	0	0	**1**
	Drills	1 (1)	0	0	**1**
	Microdrills	4 (5.5)	9 (13)	0	**13**
	Retouched flakes	4 (5.5)	1 (2)	0	**5**
Total		**74**	**67**	**19**	**160**

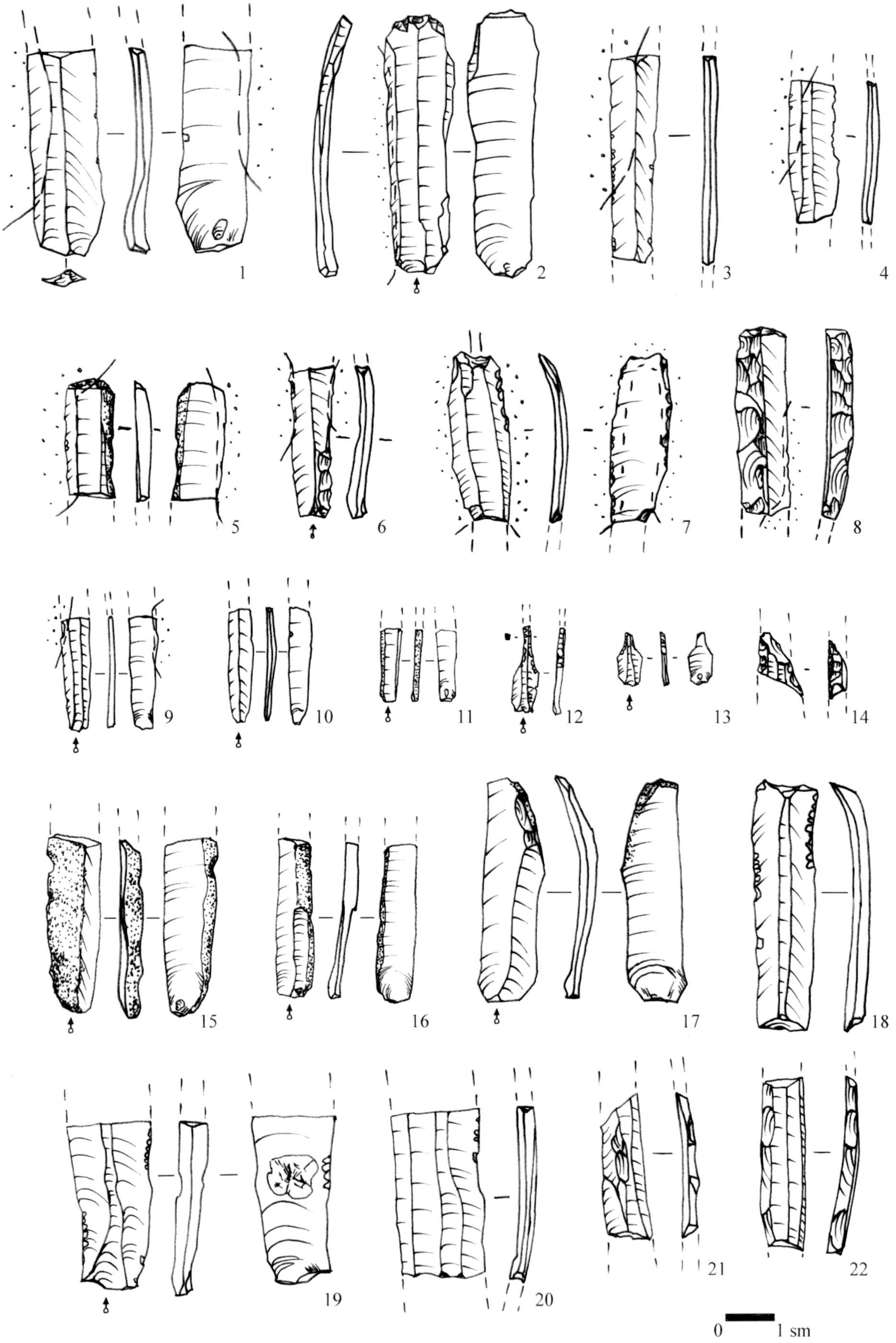

Figure 8. Belovode lithic tools, blades, bladelets and microblades.

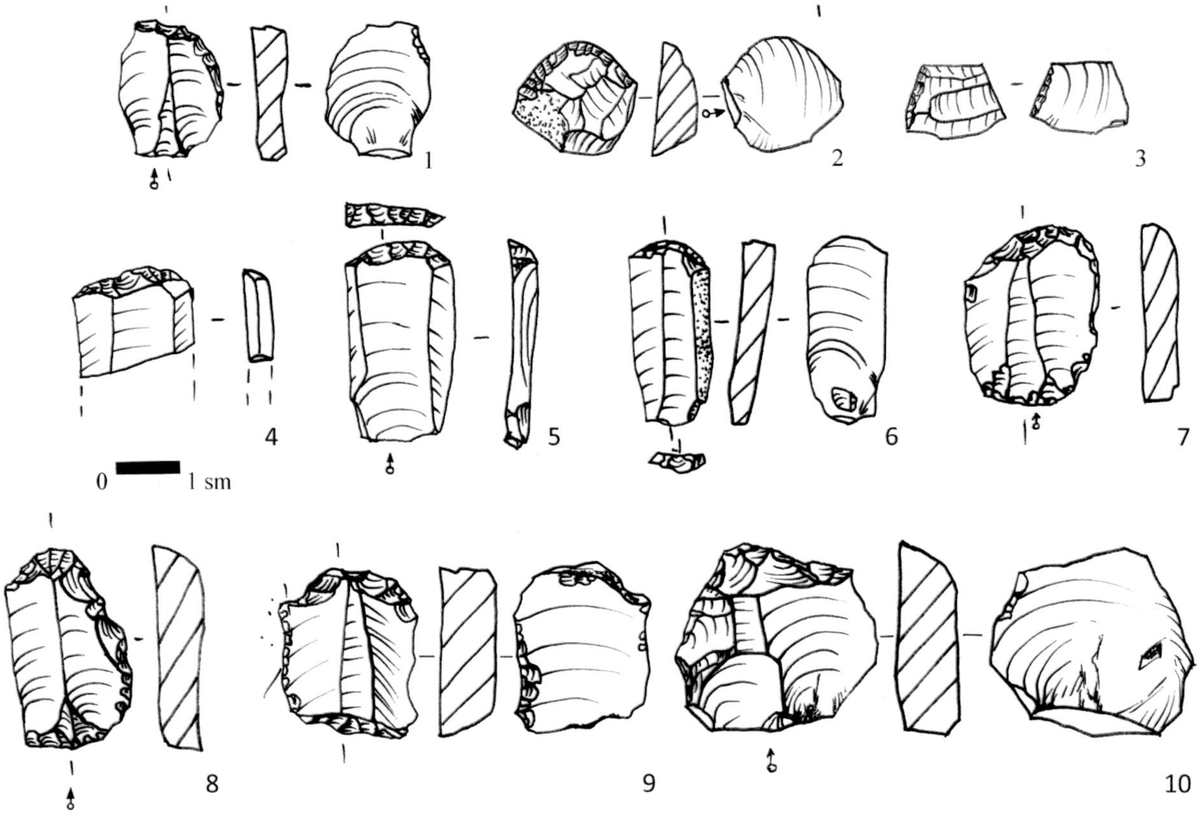

Figure 9. Belovode lithic tools. 2.2 - microscrapers; 2.3 - trapeze; 2.4 - end- and side-scrapers.

bladelet was used as a blank (Figure 8.14). Retouched flakes include those with semi-abrupt irregular retouch. These tools are more numerous in Horizon 1.

Conclusion

The analysis of the Belovode chipped stone industry enabled the identification of several *chaîne opératoires* relating to blade, bladelet and microblade production (Figure 10). The first of these involved the production of microblades from cream-coloured, tabular flint. The blade reduction was performed on single-platform cores, from the narrow flaking surface or from the frontal part, and a pressure technique was probably used. The resulting fine microblades were typically used with minimal retouching or were subsequently transformed into microdrills. The second *chaîne opératoire* involved the production of bladelets and blades from cream-coloured tabular flint using a punch technique. The reduction started in the frontal part of the core and would then expand to lateral parts or change its orientation. The resulting bladelets and blades were extensively used and, due to their large sections, the blades and bladelets were retouched to form scrapers and drills etc. However, a large proportion of the fragments of blades and bladelets could have been used with minimal retouch (or without it). The third *chaîne opératoire* involved the production of blades, bladelets and flakes from pebble flint using frontal reduction and direct percussion techniques. The resulting blades, bladelets, and flakes were used for various tasks, apparently without any strong preselection. Whilst all three *chaîne opératoires* took place at Belovode, the obsidian bladelets and core rejuvenation flake, as well as the blades made from honey flint, lack a clear technological context.

The main trajectory of chipped stone industry evolution throughout Horizons 1–5 could be characterised as moving from microblade production towards a blade production (blade width 11–25 mm). Flake production plays a marginal role in all horizons, while a pronounced 'microlithic component' is characteristic in Horizons 2–3.

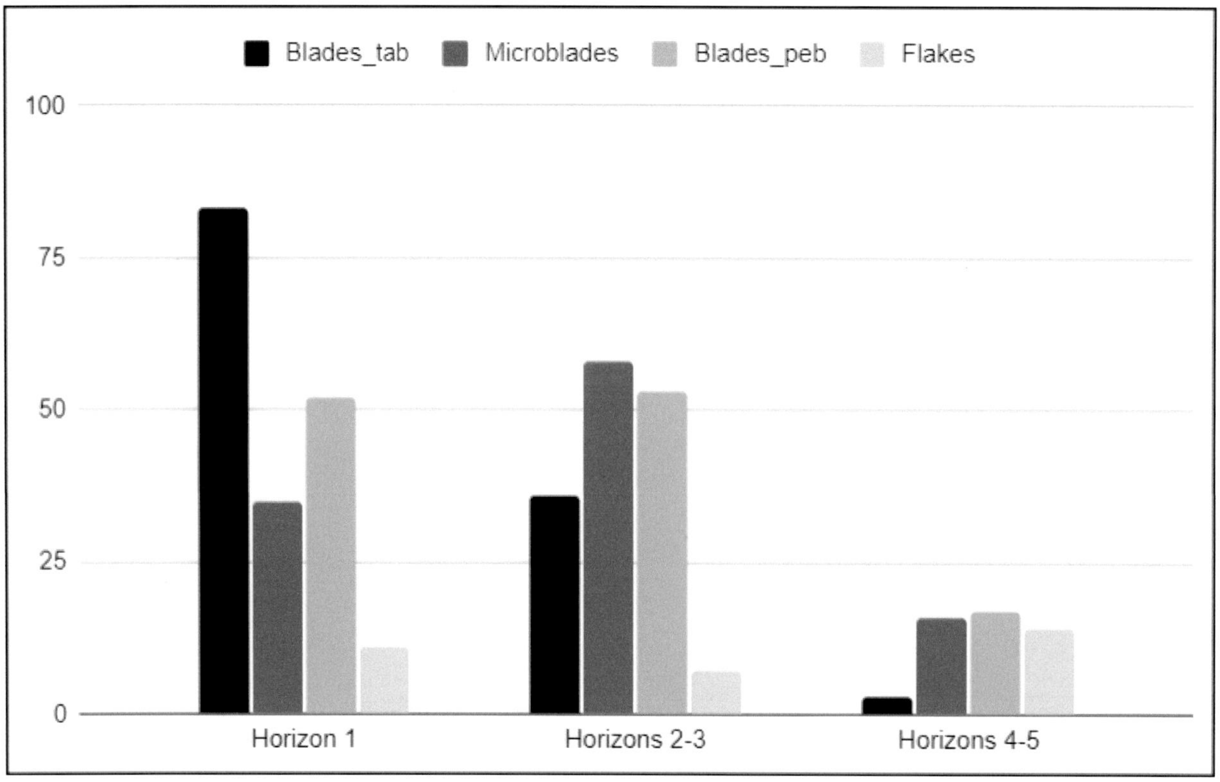

Figure 10. The evolution of the chipped stone industry at Belovode. Ratio of three *chaîne opératoires* (tab. - tabular flint; peb. - pebble flint) with marginal flake production, showing number of items.

The bibliographic reference for this chapter is:

Ibragimova, E. 2021. Chipped stone industry at Belovode, in Radivojević, M., Roberts, B. W., Marić, M., Kuzmanović Cvetković, J., and Rehren, Th. (eds) *The Rise of Metallurgy in Eurasia*: 221–232. Oxford: Archaeopress.

Chapter 19

Chemical and technological analyses of obsidian from Belovode

Marina Milić

Introduction

The Neolithic period of the central Balkans is marked by dynamic lifestyles that included long-distance exchange of resources, stimulating contact between diverse groups. Obsidian is a highly distinctive raw material that was procured and used by these peoples from the earliest Neolithic. Obsidian sources are not known from the central Balkans, so this material had to be imported from distant quarries such as those in the Carpathians, the Aegean, or the islands of the central Mediterranean (Milić 2014; Tripković 2003a). At the site of Belovode, only seven pieces of obsidian were found during the 2012–2013 excavation seasons, representing less than 1% of the chipped stone assemblage. In order to identify the origin of the obsidian from Belovode, we used a portable X-Ray Fluorescence (pXRF) spectrometer. The methods utilised are outlined in Milić (2014). Using the results of technological and sourcing studies of obsidian, the aim of this chapter is to examine the origin and nature of the obsidian artefacts that were found at the Belovode settlement. The exchange of obsidian amongst Vinča communities in the central and southern Balkans is addressed in Chapter 50 of this volume.

Characterisation of obsidian in the Balkans: the background

Obsidian is volcanic glass formed when lava rapidly cools on the edges of flows. It is a hard but fragile and razor-sharp material, that is easy to knap and has reflective surfaces that can be aesthetically attractive. Perhaps due to such properties, in prehistory obsidian was exchanged over hundreds of kilometres. It was used for the production of chipped stone tools by the prehistoric populations of central and southeast Europe. From an archaeometric perspective, the homogenous composition of this material enables accurate matching of archaeological artefacts to a specific geological source, provided the fingerprint of the source is known (Pollard and Heron 2008).

The sources of good quality obsidian that were exploited by prehistoric communities in Eurasia are located in several different environmental settings. In Anatolia, two large volcanic regions are known. The first, located in eastern Anatolia, includes sources at Bingöl and Nemrut Dağ that were exploited from the Palaeolithic period by groups living in southeastern Anatolia, Mesopotamia and the Zagros region. The second volcanic region, located in central Anatolia (Cappadocia), includes Göllü Dağ and Nenezi Dağ, the most extensively exploited obsidian sources by the Neolithic populations (Chataigner 1998). Obsidian from these locations was transported east to the Levant and west to the Aegean.

In the Aegean Sea area, obsidian from the Melian sources of Adamas and Demenegaki was also widely exploited. These raw materials were transported by maritime routes from as early as *c.* 11,000 BC in the Upper Palaeolithic (Carter 2009). In the central Mediterranean region, sources of obsidian are known from the islands of Panteleria, Lipari, Palmarola, and Sardinia, where several quarries were located (Tykot 2011). Finally, in continental Europe, obsidian is found at numerous places in the foothills of the Carpathian Mountains in present-day Hungary, Slovakia, Romania, and Ukraine (Biró 2014; Williams-Thorpe *et al.* 1984). This material is known as Carpathian obsidian; only Carpathian 1 (Slovakia) and Carpathian 2 (Hungary) were widely used in prehistory.

Obsidian characterisation studies were developed in the 1960s, initially to analyse the distribution of Anatolian obsidian in the Near East and Anatolia, and that of the Melian obsidian in the Aegean basin (Cann and Renfrew 1964; Renfrew *et al.* 1965; 1966; 1968). The first analyses of obsidian artefacts from sites in Serbia (Vinča culture) have shown that the Carpathian Mountains were the sources for this raw material.[1] Further work on obsidian assemblages from Serbia (e.g. Chapman 1981; Tripković and Milić 2008; Williams-Thorpe *et al.* 1984) has confirmed that all the obsidian analysed originated from two main source areas: Carpathian 1 in the Zemplin Hills of eastern Slovakia, and Carpathian 2 in the Tokaj Mountains of northeastern Hungary.

[1] Cann and Renfrew (1964) analysed three pieces from Vinča-Belo Brdo and assigned them to Group Ia – Carpathian source

Results of pXRF analyses

This chapter focuses on the results of chemical and technological analyses of seven obsidian artefacts that were found during the excavation seasons of 2012 and 2013. Elemental characterisation was conducted using a pXRF: a hand-held Olympus Innov-X Delta XRF device.[2] This method is non-destructive, fast and enables analyses of large obsidian sample, in this case 100% of recovered finds (Milić 2014). For the purpose of obsidian sourcing, the pXRF was set to the 'Soil setting' using three-beam mode which records heavy metals, transitional metals and light elements. The levels of nine elements (Ti, Mn, Fe, Zn, Rb, Sr, Zr, Ba, Pb) were measured, assessed and compared to geological samples collected at six sources including Carpathian 1 and Carpathian 2, Melos and Giali in the Aegean, and Göllü Dağ and Nenezi Dağ in central Anatolian. The discrimination of the sources was conducted by plotting three trace elements, Rb, Sr and Zr, which are commonly used in obsidian provenancing for clustering the source groups. The composition of archaeological assemblages was plotted against the geologically derived data (e.g. Liritzis 2008; Poupeau *et al.* 2010). PXRF results identified the Carpathian 1 source as the origin of these artefacts (Figure 1). This is not surprising as this source region was the major supplier of obsidian in the Balkans in the Neolithic (further details in Chapter 50, this volume). This raw material is characterised by its shiny, glossy surface, often with a completely transparent or semi-transparent appearance (Figure 2). These distinctive properties make it easily recognisable in archaeological contexts, not only in relation to other raw materials used for chipped stone manufacture, but also in relation to other obsidian types (Milić 2014; Milić *et al.* 2013).

Technological characteristics of Belovode obsidian

The obsidian assemblage consists of four blades (three are prismatic) and three flakes (Figure 2):

Sample 1 - Inv. 26204 (B3/13): distal fragment of a regular unipolar prismatic blade with parallel edges (2.35 x 0.7 x 0.2 cm) and almost completely transparent in appearance. There is no visible use-wear or modification of the edges on this blade.

Sample 2 - Inv. 26203 (B2/13): medial fragment of a unipolar prismatic bladelet (1.9 x 0.7 x 0.2 cm) with no traces of use-wear or edge modification; almost completely transparent in appearance.

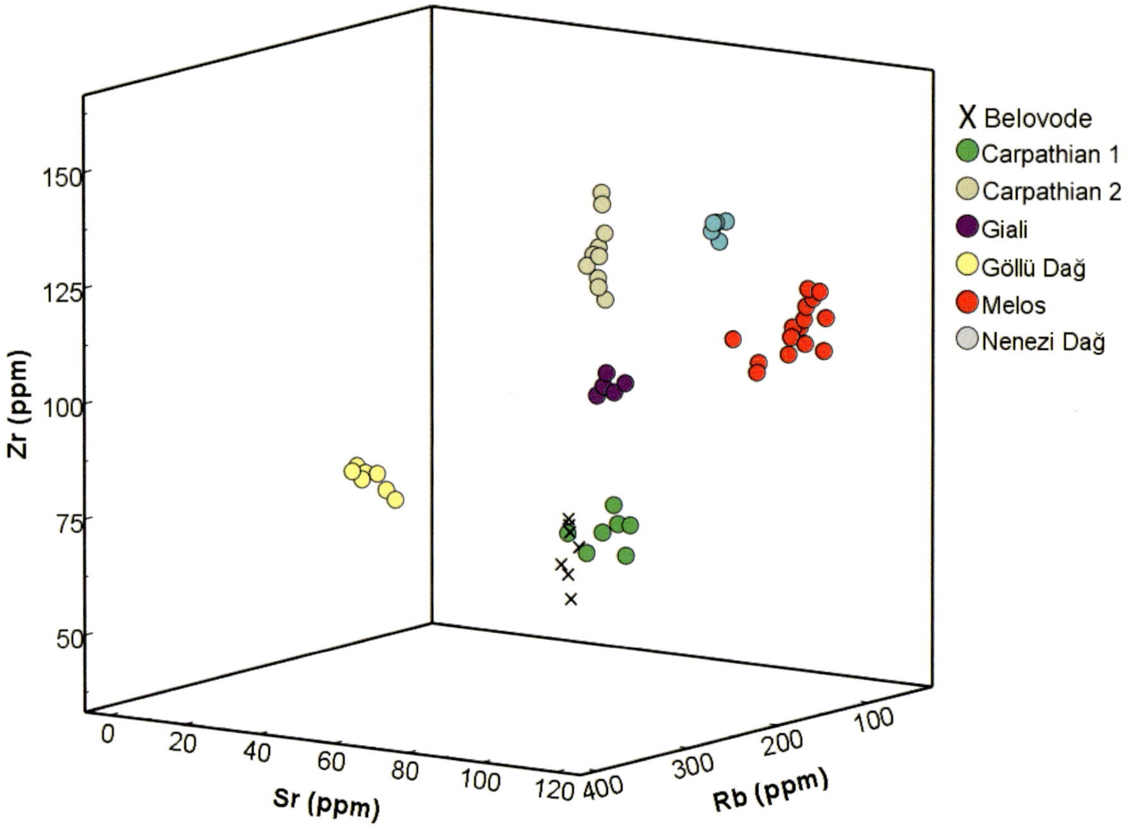

Figure 1. 3D scatter plot of Zr, Sr and Rb discriminating obsidian artefacts from Belovode against source material (analysed with pXRF).

[2] The instrument is owned by the Institute of Archaeology, UCL.

Figure 2. Obsidian artefacts from Belovode.

Sample 3 - Inv. 26205 (B4/13): medial fragment of a unipolar prismatic bladelet (1.5 x 0.8 x 0.2 cm) with some traces of use-wear; semi-transparent in appearance.

Sample 4 - Inv. 26210 (B9/13): proximal fragment of a unipolar prismatic bladelet (1.5 x 0.8 x 0.2 cm) with no traces of use-wear or edge modification; completely transparent in appearance.

Sample 5 - Inv. 26202 (B1/13): fragment of a chip (0.9 x 0.8 x 0.15 cm) with small amount of cortex preserved on the dorsal surface; almost completely transparent in appearance.

Sample 6 - Inv. 26209 (B8/13): medial fragment of a blade-like flake (2.1 x 1.5 x 0.3 cm); semi-transparent in appearance.

Sample 7 - Inv. 26207 (B6/13): a complete core tablet (3 x 2.6 x 1.1 cm) rejuvenated from a faceted platform of a core used for bladelet manufacture; small amount of cortex is also preserved on the dorsal surface; semi-transparent in appearance.

Chronology and contexts of obsidian artefacts

On the basis of the current excavation record, it appears that obsidian artefacts were not knapped within the settlement. Four blades and a chip (samples 1–5) were found within the later Vinča culture phases (sample 2 is from Vinča C2 while the other four belong to Vinča D), dated to the first half of the 5th millennium BC (see Chapter 50). Sample 6 was found within the Gradac Phase (beginning of the 5th millennium BC, Vinča B2/C), while sample 7 belongs to the earliest horizon, Vinča B1 (late 6th millennium BC). Contextually, only two artefacts, sample 2 and sample 6, were found within enclosed pits, F21 and F38 respectively, although there was no apparent special treatment of these pieces. The rest of the assemblage was found within layers that are not allocated to any specific features, although they could be dated to the later phases of the Vinča culture, i.e. the first half of the 5th millennium BC (see Chapter 50). The rare and isolated occurrences of obsidian in these regions are not unusual and it could be argued that all of the artefacts were brought to Belovode in finished form, while their production has most likely taken place at different locales. Such places often contain larger assemblages that include pieces from different productional stages such as cores, débitage (e.g. cortical and non-cortical flakes) and finished artefacts (e.g. regular end-blades). Nevertheless, it should be stressed that the obsidian pieces described in this chapter were found within only a small excavation area and future excavation of a larger area of the site might reveal obsidian knapping deposits. The wider production and exchange of obsidian artefacts in Vinča settlements is further discussed in Chapter 50.

The bibliographic reference for this chapter is:

Milić, M. 2021. Chemical and technological analyses of obsidian from Belovode, in Radivojević, M., Roberts, B. W., Marić, M., Kuzmanović Cvetković, J., and Rehren, Th. (eds) *The Rise of Metallurgy in Eurasia*: 233–235. Oxford: Archaeopress.

Chapter 20

Archaeobotanical evidence of plant use at the site of Belovode

Dragana Filipović

Introduction

The first application of systematic archaeobotanical sampling and recovery by flotation at Belovode yielded charred plant materials that offer important insights into plant and land use practices at this early farming settlement. The analysis of plant macro-remains from Belovode was aimed at: determining the range of plant species represented in the archaeological deposits; exploring the potential role and importance of different crop and wild plants for the inhabitants during the Late Neolithic–Early Chalcolithic; and detecting possible changes in plant use through time. The results contribute to current knowledge of plant production and consumption at the end of the 6th and throughout the 5th millennium BC in southeast Europe and, particularly, the plant economy of sites associated with the Vinča culture phenomenon of the central Balkans.

Methods and materials

Field methodology – sampling and recovery

Over the two excavation seasons (2012–2013), 49 archaeobotanical samples were collected from archaeological contexts selected by the excavators. A standard minimum of 5 litres of soil was aimed for, while a greater amount of soil was selected for flotation from contexts that showed traces of burning or contained immediately visible charred plant material. The sample provenance and volume data are provided in Appendix B_Ch20.1. Charred plant remains, and a few mineralised examples, were separated from the soil through hand- or machine-flotation. The light (i.e. floating) fraction of the samples was collected in pieces of cloth of 0.3 mm aperture, whilst the heavy residue (i.e. material that sinks in water) was retained in a 1 mm mesh. The total soil volume processed was 299 litres.

Laboratory procedure – sorting, identification and quantification

Macroscopic plant remains (excluding wood charcoal) were extracted from the entire light fraction of 40 archaeobotanical samples. The light residue of the remaining nine samples contained large amounts of modern vegetal material and the sorting proved highly time-consuming. Therefore, at this stage of the analysis, smaller portions (i.e. 25%) of the samples were fully sorted for non-wood macro-botanical material. The heavy fraction of all 49 samples was completely sorted and archaeological materials such as bone, (mainly chipped) stone, pottery, worked clay, malachite, beads and plant remains were removed for further analysis.

The majority of plant remains from Belovode have been charred, indicating exposure to high temperatures as the main route to preservation (e.g. Jacomet and Kreuz 1999; Pearsall 2000). A single mineralised *Solanum* seed was encountered in one sample, and a mineralised dung pellet in another. The extracted plant remains were observed under a low-power (10 x–40 x) stereomicroscope. A personal reference collection and relevant botanical literature (e.g. Bojňanský and Fargašová 2007; Cappers *et al.* 2006; Schoch *et al.* 1988) facilitated the taxonomic identification of plant remains, in particular the seeds of wild taxa. The names of crop taxa follow Zohary *et al.* (2012); *Flora Europea* (Tutin *et al.* 1964–1980) was used as the source of nomenclature for wild taxa.

Initial identification resulted in *c.* 120 taxa, which include intermediate, 'cf.' and some descriptive categories (see Appendix B_Ch20.1). Intermediate and broad categories (such as *Triticum monococcum/dicoccum*; *T. monococcum/dicoccum*/'new type' hulled wheat) refer to the remains that have been distorted and not preserved well enough to allow more precise determination. This is especially valid for glume bases of which the majority are highly eroded and lack elements for a more specific attribution (i.e. to a particular hulled wheat type). 'Cf.' categories represent poorly preserved specimens of the corresponding definite identifications (e.g. cf. *Lens culinaris*).

All items were quantified. Each complete seed/grain was counted as one item, as were fragments of fruit stones, nutshell, cereal chaff, pods, stalks and stems/culms. If seed cores or seed coats of the same taxon were found in the same sample, the seed part occurring in a greater number was scored. Fragmented cereal grains were recorded in the following way: a) apical or embryo end, whichever was more numerous, was counted as one grain; b) two 'halves' of longitudinally broken grain were given a score of one. Hulled wheat spikelet fork was counted as two glume bases. The terminal spikelet fork (topmost chaff element of hulled wheat ear) was counted as one; free-threshing cereal rachis was also scored as a single item. In the case of food ('bread') remains, possible nutmeat and indeterminate vegetal matter, the volume was measured (in millilitres). Nutshell fragments of *Corylus avellana* and fruit stone fragments of *Cornus mas* were not converted into the minimum number of whole items (MNI). The number of these remains is generally low across the assemblage, and no complete nut or fruit respectively were found in the samples. Hence it was impossible to estimate (per count, volume or weight) the amount of fragments representative of the whole item. The fragments were therefore simply taken as an indicator of the presence of fruits of these two taxa and were routinely counted as a single whole specimen regardless of the number of fragments in the sample.

The taxa counts obtained through the analysis of light and heavy fractions of each sample were combined to provide the totals. In the calculations, 'cf.' categories were amalgamated with the respective precise identifications (for example, counts of *Lens culinaris* combine *Lens culinaris* and cf. *Lens culinaris* scores). This reduced the number of taxa to a more manageable total of 90. Item scores for the sub-samples representing 25% of the light fraction of the nine samples were multiplied by four to obtain counts for the whole (100%) light fraction. Because this method may introduce some inaccuracies in the item counts, sub-samples of one of these samples were successively sorted (i.e. two sub-samples of 25%, then a 50% sub-sample). Plant remains from each of the sub-samples were then counted separately and compared, revealing only slight differences between the actual and the multiplied totals. Therefore, the multiplied totals are used in the quantitative analysis (both 'original' and extrapolated counts are given in Appendix B_Ch20.1). The botanical richness of the samples is expressed as abundance (absolute number of counted items) and density (number of counted items per litre of soil).

In order to provide a general overview of the botanical composition of the samples, and to explore and compare the frequency and abundance of different plant types, further amalgamation of the taxa was conducted. Einkorn, emmer and 'new type' wheat were combined in the 'hulled wheat' group. Free-threshing wheat and barley were included in the 'free-threshing cereal' group, whilst large-seeded pulses form the 'legumes' group. Remains of wild taxa with potentially edible fruits represent the 'wild-edible taxa' group but this does not exclude the possible dietary, medicinal etc. role of (some) plants included in the 'wild/weed taxa' group.

Results and discussion

Overview of the assemblage

Table 1 lists the identified taxa and gives the absolute counts of the remains and the frequency of taxa (i.e., number and percentage of samples in which they occur). The botanical assemblage appears taxonomically diverse, with 10 crop types and 52 wild taxa. Its overall

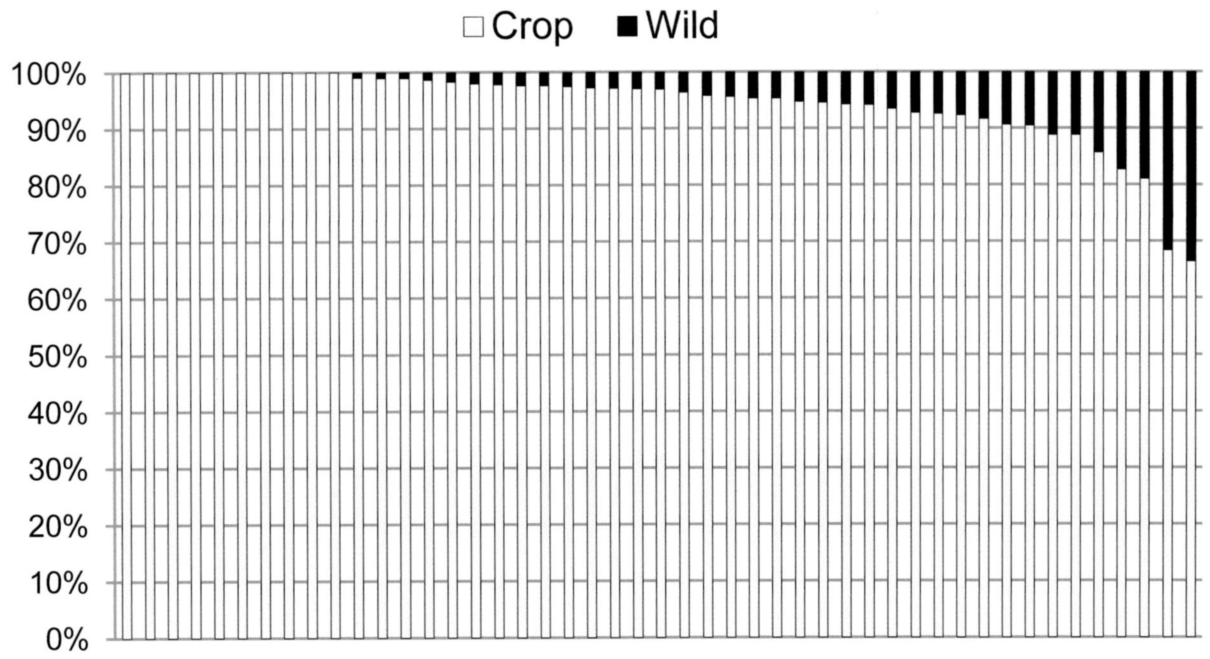

Figure 1. Proportions of crop and wild material in the samples from Belovode.

Table 1. Absolute number of remains and frequency of individual and groups of taxa at Belovode.

CROPS	plant part	common name	ubiquity		abundance	
			count	%	count	% of all crop
HULLED WHEATS						
Triticum monococcum	grain	einkorn	33	67	129	2
Triticum monococcum	glume base		41	84	650	12
Triticum dicoccum	grain	emmer	6	12	11	0.2
Triticum dicoccum	glume base		35	71	371	7
Triticum monococcum/dicoccum	grain		16	33	43	1
cf. *Triticum*, 'new type'	grain	'new type' hulled wheat	1	2	2	0.04
Triticum, 'new type'	glume base		32	65	172	3
Triticum dicoccum/'new type'	grain		1	2	1	0.02
Triticum dicoccum/'new type'	glume base		35	71	473	8
T. monococcum/dicoccum/'new type'	grain		13	27	32	1
T. monococcum/dicoccum/'new type'	glume base		47	96	3256	58
HULLED WHEAT	GRAIN		40	82	218	4
HULLED WHEAT	GLUME BASES		47	96	4922	87
HULLED WHEAT	ALL ITEMS		**47**	96	**5140**	**91**
FREE-THRESHING CEREALS			count	%	count	% of all crop
Triticum aestivum/durum	grain	free-threshing wheat	1	2	1	0.02
Triticum aestivum	rachis	bread wheat	9	18	22	0.39
Hordeum vulgare	grain	barley	7	14	10	0.18
Hordeum vulgare	rachis		4	8	6	0.11
FREE-THRESHING CEREAL	GRAIN		8	16	11	0.20
FREE-THRESHING CEREAL	RACHIS		12	24	28	0.50
FREE-THRESHING CEREAL	ALL ITEMS		**17**	35	**39**	**1**
Triticum sp.	grain	wheat	18	37	46	1
Cerealia indeterminata	grain	indeterminate cereals	29	59	144	3
ALL CEREAL	GRAIN		45	92	419	7
ALL CEREAL	CHAFF		47	96	4950	88
ALL CEREAL	ALL ITEMS		47	96	5369	95
Cerealia indeterminata	culm node	indeterminate cereals	1	2	1	0.02
LEGUMES			count	%	count	% of all crop
Lens culinaris	seed	lentil	28	57	81	1
cf. *Lathyrus sativus/cicera*	seed	grass pea	1	2	4	0.07
Pisum sativum	seed	pea	8	16	18	0.32
Vicia ervilia	seed	bitter vetch	21	43	86	2
Leguminosae sativae	seed	indeterminate pulses	18	37	35	1
ALL LEGUME	SEED		**39**	80	**224**	4
? Leguminosae sativae	stalk fragments	indeterminate pulses	1	2	13	0.2
OIL/FIBRE PLANTS						
Linum usitatissimum	seed	flax/linseed	5	10	40	1

Table 1 continued. Absolute number of remains and frequency of individual and groups of taxa at Belovode.

	plant part	common name	ubiquity		abundance	
			count	%	count	% of all crop
ALL CROP	SEED+CHAFF		**47**	96	**5633**	100
FRUIT/NUT TAXA	**plant part**		count	%	count	
Cornus mas	stone	Cornelian cherry	11	22	11	
Corylus avellana	shell	hazel	13	27	13	
Fragaria vesca	seed	wild strawberry	2	4	2	
Physalis alkekengi	seed	chinese lantern	5	10	26	
Prunus cf. *domestica* var. *insititia*	stone fragments	European plum	2	4	2	
Prunus cf. *spinosa*	stone fragments	sloe/buckthorn	3	6	5	
cf. *Prunus* sp.	stone fragments	plum	1	2	1	
cf. *Pyrus/Malus* sp.	seed	pear/apple	1	2	4	
Rubus idaeus/fruticosus	seed	raspberry/blackberry	3	6	9	
Rubus sp.	seed		5	10	20	
Sambucus ebulus	seed	elderberry	2	4	2	
fruit stone/nutshell	fragment count		13	27	37	
cf. nut meat	volume (ml)		11	22	4.71	
? fruit flesh	fragments		2	4	7	
WILD/WEED TAXA						
cf. *Avena* sp.	fruit	(wild) oat	2	4	5	
Bromus cf. *arvensis*	fruit	field brome	5	10	5	
Bromus secalinus	fruit	rye brome	1	2	4	
Bromus sp.	fruit	bromegrass	4	8	5	
Campanulaceae, small-seeded	seed	bellflower family	1	2	1	
Capsella bursa-pastoris type	seed	shepherd's purse	1	2	4	
Caryophyllaceae	seed	pink family	1	2	4	
Cerastium sp.	seed	mouse-ear chickweed	1	2	4	
Chenopodiaceae	endosperm	goosefoot family	3	6	9	
Chenopodiaceae/Caryophyllaceae	endosperm		1	2	1	
Chenopodium album type	seed	fat hen	8	16	25	
Chenopodium sp.	seed	goosefoot	5	10	9	
Crucifereae, small-seeded	seed	crucifers	1	2	4	
Echinochloa crus-galli	fruit	cockspur	2	4	2	
Fallopia convolvulus	seed	black bindweed	6	12	13	
Galium aparine	seed	cleavers	3	6	5	
Galium sp.	seed	bedstraw	1	2	1	
Galium verum type	seed	yellow bedstraw	1	2	1	
Graminae	culm node	grasses	2	4	5	
Graminae	culm fragment	grasses	1	2	9	
Graminae, large	fruit	grasses	3	6	7	
cf. *Hordeum* sp. (wild)	fruit	wild barley	1	2	1	
Lamiaceae	seed	mint family	3	6	3	
Lapsana communis	seed	nipplewort	2	4	5	

Table 1 continued. Absolute number of remains and frequency of individual and groups of taxa at Belovode.

	plant part	common name	ubiquity		abundance	
			count	%	count	% of all crop
Leguminosae, small-seeded	seed	wild legumes	1	2	1	
cf. *Medicago* sp.	seed	medick	1	2	4	
cf. *Medicago* sp.	pod fragments	medick	1	2	4	
Phleum type	fruit	catstail	1	2	1	
Phragmites communis	culm node	common reed	1	2	1	
Polygonaceae	endosperm	knotweed family	2	4	5	
cf. *Polygonum aviculare*	seed	common knotgrass	1	2	1	
Polygonum persicaria type	seed	red leg type	2	4	8	
Polygonum sp., trigonous	endosperm	knotweed	2	4	3	
Schoenoplectus mucronatus	seed	bog bulrush	1	2	1	
Scirpus/Schoenoplectus sp.	seed	clubrush/bulrush	2	4	2	
Secale sp.	fruit	(wild) rye	1	2	1	
Setaria viridis/verticillata	fruit	green/bristly foxtail	3	6	6	
Solanaceae	seed	nightshade family	5	10	10	
Solanum nigrum	seed	black nightshade	5	10	5	
Solanum sp.	seed	nightshade	2	4	5	
Solanum sp.	mineralised seed	nightshade	1	2	1	
Sisymbrium type	seed	mustard/rocket	1	2	1	
Teucrium sp.	seed	germander	1	2	4	
Trifolium arvense type	seed	haresfoot clover	1	2	3	
Trifolium repens type	seed	white clover	4	8	42	
Trifolium sp.	seed	clover	5	10	9	
INDETERMINATE and OTHER	**plant part**		count	%	count	
unknown seed, endosperm or fruit			17	35	50	
unknown vegetal matter	volume (ml)		2	4	0.09	
? fruit skin/pod	fragments		5	10	130	
pod	fragments		4	8	9	
stalk	fragments		6	12	9	
food ('bread') remains	volume (ml)		1	2	0.04	
dung pellet	mineralised		1	2	1	
mouse pellet	charred		1	2	1	

composition is rather homogeneous, and crop remains largely dominate the content of the samples, as shown in Figure 1. The sample-by-sample data are provided in the Appendix B_Ch20.1. Images of a selection of the remains are provided in Appendix B_Ch20.2.

Crops

The crop assemblage includes different cereals and legumes and an oil/fibre plant (flax/linseed, *Linum usitatissimum* L.). The identified cereal types are: hulled wheats – einkorn (*Triticum monococcum* L.), emmer (*T. dicoccum* Schrank), and 'new type' hulled wheat (Jones *et al.* 2000; Kohler-Schneider 2003); free-threshing wheat (*T. aestivum* L./*durum* Desf.); and barley (*Hordeum vulgare* L.).

Based on the abundance and frequency, hulled wheats dominate by far the crop dataset (Figure 2). They were found in the vast majority of the samples (96%) and in large numbers (91% of the total crop content). It is fair to suggest that they were the main crops consumed at the site. The greatest portion of the hulled wheat content is comprised of hulled wheat glume bases (96% of the hulled wheat component, 87% of the total crop content), and they are present in 96% of the samples. They can be understood as the evidence of crop processing activity at the site and perhaps also of the routine disposal of crop processing by-products into domestic fires (cf. Green 1982; Hillman 1981, 1984a; van der Veen 2007). Free-threshing wheat and barley are far less common than hulled wheats. That this is not a result of the potential over-representation of hulled wheats (hulled wheat glume bases) in the assemblage (due to different processing and storage procedures for the two crop groups – see Hillman 1984b, 1985) can perhaps be demonstrated through the comparison of the (absolute and percentage) abundance of *grain* of the three cereal types, as shown in Table 1 and Figure 3. According to this, free-threshing cereals seem to have been of minimal importance at the site.

Among precisely identified hulled wheat grain and chaff, einkorn remains appear most frequently and in greatest numbers, followed by emmer and 'new type' hulled wheat. Several einkorn grains possibly belong to the two-seeded type (Kreuz and Boenke 2002). The relatively large number of indeterminate hulled wheat grain and glume bases (as well as indeterminate wheat and cereal grains), however, prevents any conclusions on the preference for a particular hulled wheat type using the frequency/abundance figures. For example, for a number of glume bases it could not be determined whether they belong to emmer or 'new type' hulled

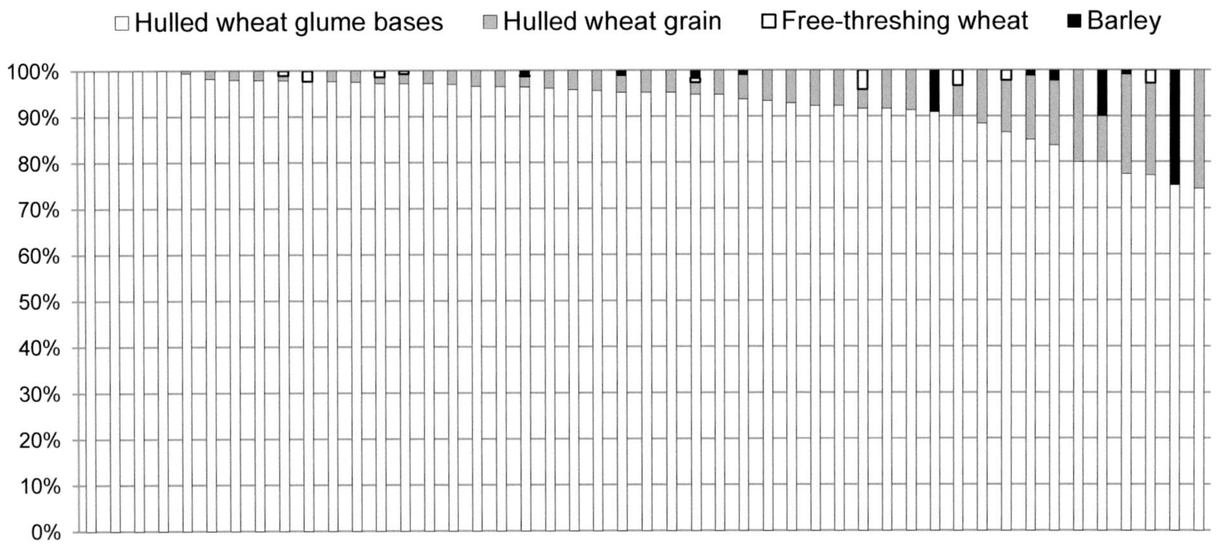

Figure 2. Proportional representation within the cereal content of the three cereal types identified at Belovode; proportions of hulled wheat grain and chaff shown separately.

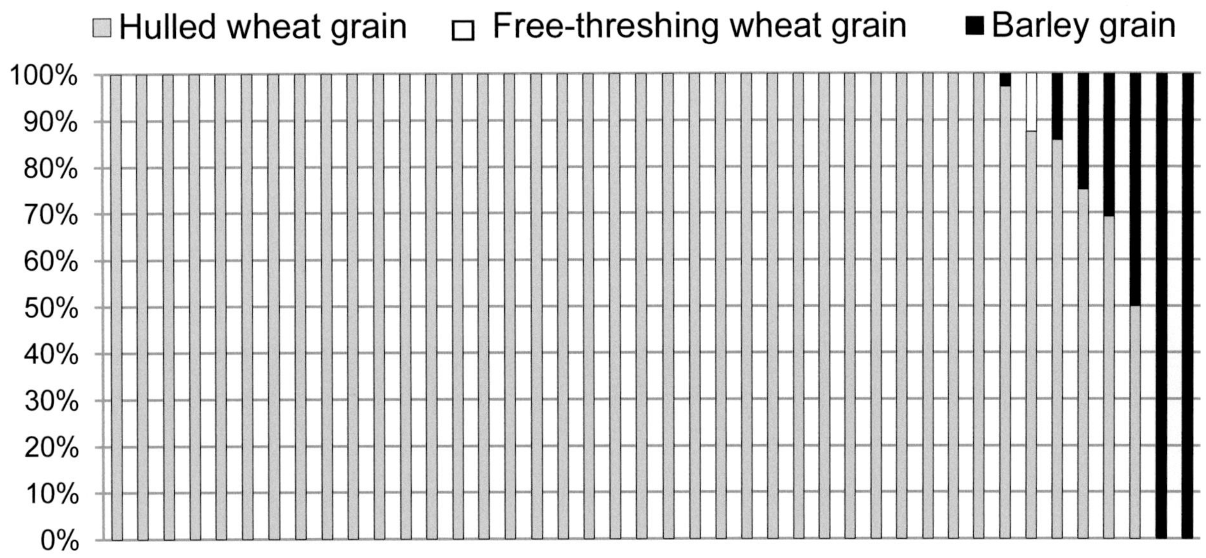

Figure 3. Proportions of grain of the three cereal types from Belovode.

wheat; for an even greater quantity of heavily eroded glume bases it could only be established that they derive from one of the three hulled wheat types. A similar observation applies to cereal grains. The very poor condition of the remains perhaps suggests that they were exposed to trampling and/or were displaced (and perhaps more than once) from the location of charring. There is also a possibility that at least some of the cereal chaff originates from daub used in house and feature construction, as it is evident from the observation of the fragments of daub found in the archaeological layers that cereal chaff was abundantly used as temper in building materials.

Large (cultivated) legumes are represented by seeds of lentil (*Lens culinaris* Medik.), pea (*Pisum sativum* L.), and bitter vetch (*Vicia ervilia* (L.) Willd.); few grass pea seeds (*Lathyrus sativus* L./*cicera* L.) were found in one sample. Lentils and bitter vetch occur in high frequency and abundance, suggesting their more important role compared to the other pulse crops. A number of large pulse seeds were too fragmented for identification beyond the family level.

Figure 4 shows the proportions of cereal (grain, chaff) and legumes. Cereal remains dominate the overall crop content of the samples; in a few cases pulse seeds outnumber cereal grains (in two of these, bitter vetch is dominant among crop *seeds* – see Appendix B_Ch20.1).

Wild-edible taxa

Several wild taxa with edible fruits were registered in the samples and could represent the remains of the processing and consumption of fruits gathered from the wild. The most common are finds of Cornelian cherry and hazelnut and these may have been more important than other wild fruits in the diet of Belovode inhabitants. A number of samples contained unidentified fragments of fruit stone or nutshell; also, in some samples, possible remains of nut cotyledons and fruit flesh were discovered.

Arable/ruderal taxa

The potential arable weed record is relatively diverse; the most frequent and abundant taxa being fat hen (*Chenopodium album* L.) and black bindweed (*Fallopia convolvulus* (L.) Á. Löve); followed by field brome (*Bromus arvensis* L.), black nightshade (*Solanum nigrum* L.) and clovers (*Trifolium* L.). According to Ellenberg *et al.*'s (1992) observations, these taxa seem to grow on soils of moderate to high fertility resulting from high nitrogen content such as that of soils in areas under (heavy) anthropogenic influence, e.g. in cultivation plots, waste areas and trampled ground (Šumatić *et al.* 1999). They are annuals and, if they represent arable weeds, would indicate highly disturbed crop growing conditions characteristic of well-tilled and weeded fields (cf. Jones 1992). Some of these and other taxa in the wild/weed assemblage are edible for both humans and animals, e.g. fat hen, black bindweed, small-seeded legumes and grasses; some could have been used as medicinal plants, such as cleavers (*Galum aparine* L.) and black nightshade (Tucakov 1986; Behre 2008). Fragments of reed stems perhaps derive from their use as a building material.

Botanical content and archaeological context

Table 2 gives the list of archaeological context types from which the analysed samples were collected; the contexts are grouped according to the respective occupation horizon identified in the excavated sequence. The total number of plant remains and the

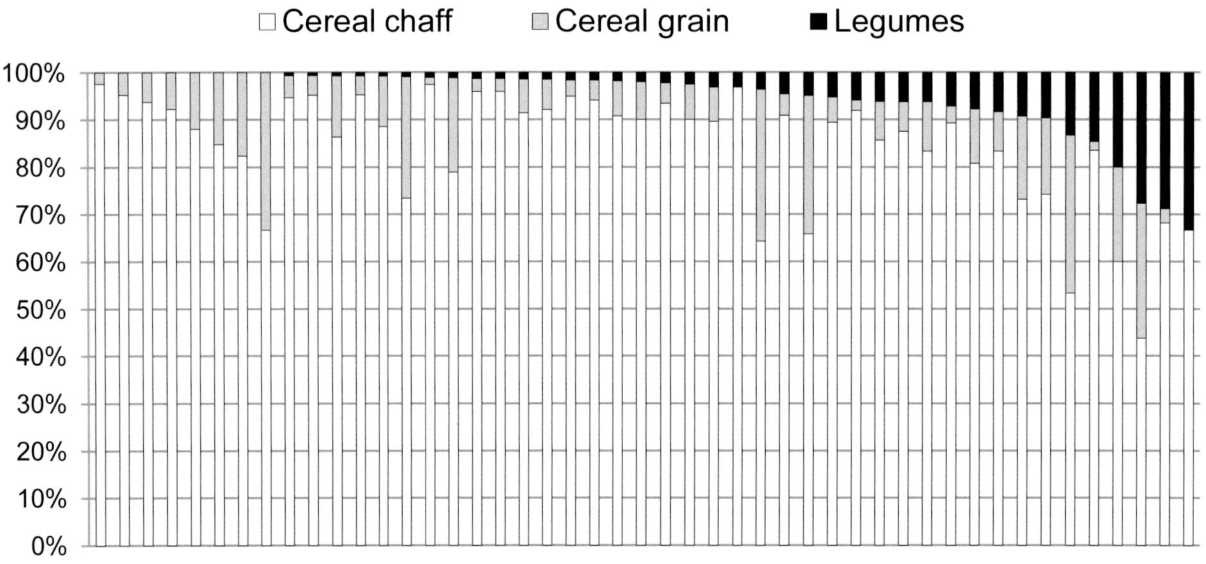

Figure 4. Comparison of the proportions of cereals (grain and chaff) and legumes within the crop assemblage from Belovode.

Table 2. List of archaeological contexts at Belovode from which the analysed samples derive; the contexts are grouped according to the settlement occupation horizons.

Sample number	Occupation horizon	Context type	Feature number	Spit number	Sample volume (litres)	*Total plant remains*	*Botanical density*	Hulled wheat grain	Hulled wheat glume bases	Free-threshing wheat	Barley	Lentil	Bitter vetch	Fruit/nut	Wild/weed
20		Arbitrary layer		5	5	*9*	*1.6*	1	4					1	
2		Arbitrary layer		5	5	*0*	*0.0*								
24		Arbitrary layer		6	6	*21*	*3.5*	1	14					1	2
1	1	Arbitrary layer			10	*14*	*1.4*		10		1			1	1
3		House rubble/ floor	3	6	5	*3*	*0.6*		2			1			
12 (2)		Pit	14	14	5	*127*	*25.4*	3	108	1	2	2		6	2
12 (1)		Pit			5	*149*	*29.8*	3	140	1		1			2
27		Arbitrary layer		8	3	*24*	*8.0*	3	108	1	2	2		2	2
13		Arbitrary layer		8	5	*21*	*4.2*	1	12					1	6
6 (2)		Arbitrary layer		8	15	*35*	*2.3*	2	27	1		1	1	1	1
34		Arbitrary layer		12	5	*219*	*43.8*		183			4	24		
17		Daub feature	20	10	5	*58*	*11.6*	1	49				1		4
4	2	Pit	8	7	15	*32*	*2.1*	3	23			1	1		
7		Rubbish pit		8	12.5	*32*	*2.4*	2	21			1	1	4	
6 (1)		Rubbish pit		10	5	*5*	*1.0*		3		1	1			
9		Rubbish pit	9	11	5	*73*	*14.6*	2	64					1	2
10		Rubbish pit		12	5	*439*	*86.6*	9	383	1	4	1	1	4	25
11		Rubbish pit		13	5	*157*	*31.0*	2	141	2			1	1	3

Table 2 continued. List of archaeological contexts at Belovode from which the analysed samples derive; the contexts are grouped according to the settlement occupation horizons.

Sample number	Occupation horizon	Context type	Feature number	Spit number	Sample volume (litres)	*Total plant remains*	*Botanical density*	Hulled wheat grain	Hulled wheat glume bases	Free-threshing wheat	Barley	Lentil	Bitter vetch	Fruit/nut	Wild/weed
23	3	Arbitrary layer		13	6	*196*	*32.7*	2	185	2		1	1	2	2
29		Ash bin	25	13	17	*143*	*8.4*	10	120				1		4
30		Ash bin	27	14	3	*18*	*6.0*	1	8		1	1	1	1	2
Total 2		Ash bin	39	13	4	*32*	*7.0*	1	25				2		
35		Burnt deposit	31	13	5	*306*	*52.4*	4	88	4		5	29	42	68
36		Burnt deposit	33	13	4	*90*	*22.0*	3	68			3		4	8
22		Daub feature	15	11	8	*281*	*22.9*	17	49			5	5	15	24
14 (2)		Fire pit	12	8	6	*46*	*7.7*	7	27	1			1	1	2
15		Oven	18	9	3	*110*	*36.7*	22	79		1		1		1
14 (1)		Oven		9	2.5	*46*	*18.4*	2	36			1		2	4
28		Pit	19	9	3	*23*	*7.7*	1	20						1
18		Pit		10	12	*231*	*19.3*	4	198			5		2	8
19		Pit		11	9	*172*	*18.9*		152			5		5	4
25		Pit	21	13	5	*109*	*21.6*	3	80		1	4	1	5	4
26		Pit		13	4	*59*	*14.8*	1	56				1		
21		Pit		13	7	*46*	*6.6*	1	40					3	2
Total 1		Post hole	35	13	4	*23*	*4.5*		15						2
31		Post hole	26a	14	6	*397*	*65.7*	2	354				5	5	20
32		Post hole	26b	15	13	*383*	*29.5*		336	8		13	5	1	8
33		Post hole	26d	14	2	*50*	*25.0*	3	39						4

Table 2 continued. List of archaeological contexts at Belovode from which the analysed samples derive; the contexts are grouped according to the settlement occupation horizons.

Sample number	Occupation horizon	Context type	Feature number	Spit number	Sample volume (litres)	*Total plant remains*	*Botanical density*	Hulled wheat grain	Hulled wheat glume bases	Free-threshing wheat	Barley	Lentil	Bitter vetch	Fruit/nut	Wild/weed
39		Burnt deposit	32	17	4	*118*	*29.0*	10	77	2		6	2	1	7
41	4	Pit	43	16	4	*49*	*12.3*	1	42			2	1		
42		Pit		20	4	*35*	*7.3*	2	22					2	2
40	5a or 5b	Burnt deposit	32	19	10	*901*	*90.1*	29	798			9		9	24
Total 4	5	Hearth	45	19	8	*114*	*13.1*	3	85			1		4	10
43	5	Hearth	46	20	1	*274*	*274.0*	34	208		3	2		1	6
Total 3	5a	Oven	44	18	1	*0*	*0.0*								
38	5a	Pit	37	18	4	*102*	*25.5*	5	89		1	1		1	1
37	5a	Pit	38	19	4	*59*	*14.8*	6	36		1	1			3
44	5b	Pit	51	18	8	*128*	*16.0*	5	108			1		1	5
45	unknown	Post hole	48	21	6	*303*	*50.3*	8	278			2		2	7

sample density are also shown, as well as the number of items within major plant categories. Samples originating from the same context in most cases appear different in terms of the botanical richness and they were therefore not amalgamated. In addition, most of the samples collected from pit features come from distinctive excavation layers (spits) and may represent discrete depositional episodes.

Abundance and density of the remains

The botanical density of the assemblage is 21 items on average, although two samples did not produce any seed/fruit/chaff remains. In reference to this, the density of individual samples can be described as low to average, with the majority containing less than 30 items per litre of processed soil (*c.* 76% = 37 samples, see Figure 5). There are, however, several samples (12%) that yielded more than 50 items per litre of soil; one had a relatively high density of 274 items per litre (Sample 43 from Feature 46). In terms of the absolute counts of the remains, 45% (22) of the samples contained at least 100 items (Figure 6) and can be described as relatively rich, while another 31% (15 samples) produced more than 30 items each.

Among the analysed contexts, the burnt deposits and fire-related features (hearths in particular) yielded the greatest quantities of plant remains, especially those discovered in the earliest occupation phase (e.g. Feature 46 and Feature 32 in Horizon 5). Their density is generally high, suggesting a steady build-up and good preservation of fire residue or perhaps intact evidence of the final use of these features. Unlike the hearth (Feature 46), where the highest botanical density was recorded, the oven (Feature 44) in Horizon 5 did not contain any plant remains. It may have been cleaned on abandonment or perhaps the temperatures were too high to allow survival of the plant material (cf. Boardman and Jones 1990).

Burnt deposits are of variable richness and density; that in Horizon 5 (Feature 32), for example, produced a very large number of remains. A single identified fire pit was relatively poor in plant material, as were two of the ash features in Horizon 3. These may have been emptied in the past. Another ash feature (Feature 25) produced >100 remains and may have preserved residue from the fire-related contexts registered within this horizon, if not the corresponding excavation layer (Spit 13). In fact, a random, small sample from Spit 13 yielded almost 200 plant remains. However, it is possible that the charred material originating from, for instance, the relatively rich burnt deposit (Feature 35) belonging to the same level, was spread over the surrounding area. The content of another botanically rich spit (Spit 12, Horizon 2) may have accumulated in a similar way.

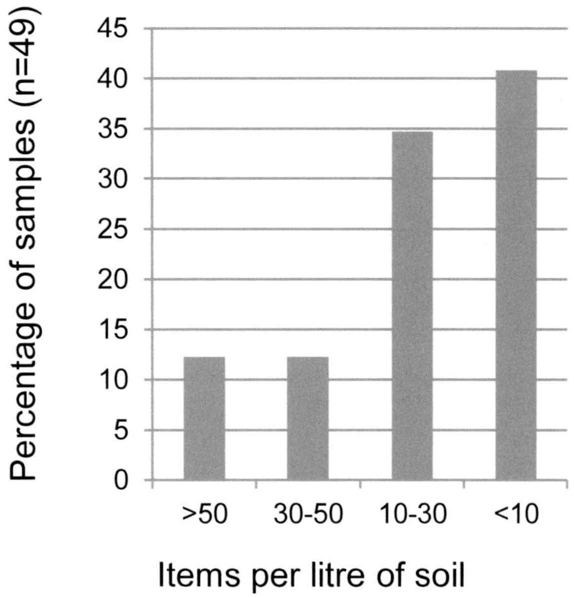

Figure 5. Botanical density values across the samples.

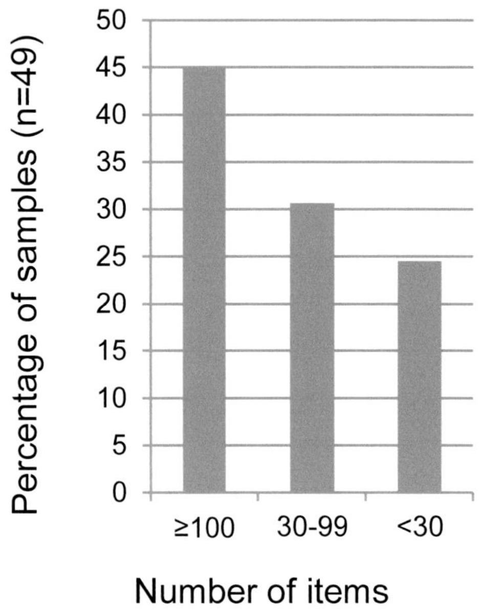

Figure 6. Botanical abundance figures across the samples.

Pit features were found in all occupation horizons and the samples from these contexts contained relatively large quantities of plant material (in most cases more than 100 items per sample). The botanical density of different layers of the pit fills was consistently low (up to or around 30) and could indicate regular deposition of small amounts of plant material through the use-life of the pits. An exception is the rubbish pit in Horizon 2 (Feature 9) where one of the lower layers of the fill had a high density and may reflect concentrated disposal of rubbish rich in charred plant remains, such as from cleaning of a fire feature. There was no evidence of *in situ* burning within the pits so it is clear that the plant material derived from elsewhere.

Features identified as post-holes are, based on the quantity and density of botanical material, comparable to some fire-related features and pits. One of the post-holes (Feature 26a) discovered in Horizon 3 is particularly rich, as is another post-hole (Feature 48) which has an unclear stratigraphical association. Given that plant remains found here have not been charred *in situ*, the pits made for inserting wooden posts may have been re-used for disposing of rubbish (floor sweepings, fire and food residue).

The feature representing house rubble and/or floor (Feature 3) contained very little plant material. The function of daub concentrations (Features 15 and 20) is not clear; the plant material found in them may have derived from plant temper used in the construction of the features but also from the use of the features (e.g. if they represented oven/hearth structures).

Sample composition

The composition of the samples from various contexts and occupation phases is largely uniform (Table 2). As noted, hulled wheat glume bases dominate all samples/contexts; other taxa occur only in very small numbers. There are few samples with more than 10 (some with more than 30) remains other than hulled wheat glume bases – i.e. hulled wheat grains, pulses or seeds of wild taxa (see Table 2), but these are still minor components. Based on the composition, there are no significant differences between contexts/deposits. The spectrum of plant categories is somewhat wider in certain deposits, for instance in pits and some of the spits (e.g. Spit 12 in Feature 9; see Table 2). It is thus clear that most of the analysed contexts received charred material from the processing/consumption of a similar range of crops and wild plants and discarding of the waste into fires. The sample from a burnt deposit (Feature 31) is of interest as it is a good indicator of the range of plants procured and consumed at Belovode. In addition to the evidence of hulled wheat de-husking and sieving (hulled wheat glume bases and weed seeds), this sample also contains a relatively high number of remains of fruit and nut taxa (Cornelian cherry stone, common hazelnut shell, seeds of Chinese lantern and probably blackberry), as well as bitter vetch, some lentil and several grains of free-threshing wheat.

The spectrum of plant taxa in the samples from different occupation horizons remains, for the most part, unchanged through time, as shown in Table 3. The apparent differences, such as the absence in some of the horizons of the generally rare free-threshing cereals and bitter vetch, could perhaps be explained by the small number of analysed samples and/or small sample volume from these horizons. The lack of free-threshing wheat and bitter vetch in the earliest occupation horizon at Belovode may, however, be

Table 3. Presence/absence of the main plant categories across the settlement occupation horizons at Belovode.

Occupation horizon	Number of samples	Hulled wheat grain	Hulled wheat glume bases	Free-threshing wheat	Barley	Lentil	Bitter vetch	Fruit/nut	Wild/ weed
1	7	X	X	X	X	X		X	X
2	11	X	X	X	X	X	X	X	X
3	20	X	X	X	X	X	X	X	X
4	3	X	X	X		X	X	X	X
5	7	X	X		X	X		X	X

indicative as it echoes the generally low visibility or absence of these two crop types at Early Neolithic (Starčevo culture, *c.* 6200–5400 BC) sites in the north-central Balkans (present-day Serbia), at least according to the evidence available so far (Filipović 2014). They seem to gain importance later, in the Late Neolithic–Early Chalcolithic and are common at Vinča culture sites (*c.* 5400–4500 BC). Similarly, grass pea, here found in Horizon 3, has in this region currently been found only in the Vinča culture layers and in very small numbers.

As illustrated in Figure 7, the representation of all the different plant categories seems to be the greatest in Horizon 3. This may simply be a result of the largest number of samples deriving from this particular occupation phase, and perhaps a considerable archaeobotanical potential of the sampled contexts (i.e. more fire-related deposits than in earlier/later horizons). Based on the presently available data, it appears that the major crop and wild-gathered types have been in use to a similar degree over time and no variations in their consumption could be identified.

Conclusions

The first archaeobotanical sampling and flotation at the site of Belovode resulted in the recovery of a diverse collection of crop and wild plant remains preserved through charring. Several cereal and pulse crop types were identified in the assemblage. Hulled wheat glume bases were found in all analysed contexts; they show that cereal processing was a widespread practice at the site and testify to the importance of cereal production. It is possible that einkorn was the main cultivated crop, but this observation remains tentative as most of the hulled wheat glume bases and grain were preserved below the level of possible identification to a wheat type. Gathering of wild-edible fruits was another aspect of plant economy at Belovode. In addition to the by-

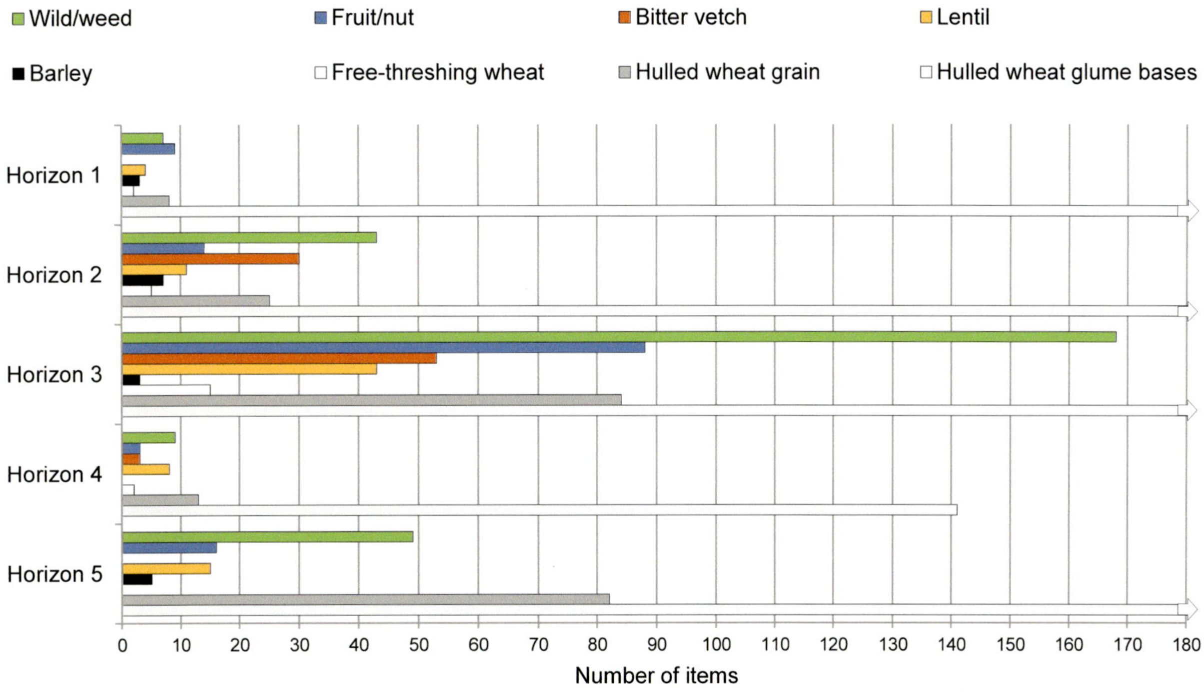

Figure 7. Absolute counts of the main plant categories within the occupation horizons.

products of cleaning of crops, fruit processing residue is also ubiquitous and co-occurs in the samples with crop remains.

Fire-related contexts (i.e. features and deposits showing traces of burning) and pit fills yielded most of the remains. The evident similarity in the composition (and, in some instances, density) of these two botanically productive context types could indicate that the rich deposits from pits represent episodes of dumping of fire residue, which included diverse food processing/ consumption waste.

There seems to have been no major changes in the range and degree of use of the identified cultivated and collected plants over time, although the crop spectrum appears to have been narrower in the earliest phase of occupation and was later enlarged through the addition of free-threshing wheat and bitter vetch. Hulled wheats remain the staple crop for the whole duration of the settlement. Small quantities of lentil are found throughout the sequence and may point to the continued, perhaps minor, reliance on this pulse crop. Overall, the crop husbandry appears stable and unchanged over time, but one cannot exclude possible changes in the methods of crop field management (e.g. tilling, weeding, fallowing), the size and permanence of fields and the intensity of cultivation. Detailed analysis of the weed record, combined with stable isotope analysis, would shed some light on these aspects of plant production at Belovode.

Appendix

Appendix avaliable online as part of Appendix B at https://doi.org/10.32028/9781803270425/AppendixB_Ch20

The bibliographic reference for this chapter is:

Filipović, D. 2021. Archaeobotanical evidence of plant use at the site of Belovode, in Radivojević, M., Roberts, B. W., Marić, M., Kuzmanović Cvetković, J., and Rehren, Th. (eds) *The Rise of Metallurgy in Eurasia*: 236–248. Oxford: Archaeopress.

Chapter 21

Animal remains from Belovode

Ivana Dimitrijević and David Orton

Introduction

The animal remains from Belovode are significant in several ways. Firstly, they extend coverage of Vinča period zooarchaeology to the otherwise unstudied Mlava valley of east-central Serbia. Indeed, they represent only the second Vinča assemblage to be studied from anywhere in eastern Serbia – the first having been described in a very brief report on Korbovo (Blažić in Babović 1986), by the Danube on the other side of the Serbian Carpathians. Secondly, along with Pločnik, they add to a small but growing number of central Balkan Neolithic sites at which faunal data are available for successive phases of occupation. Finally, they provide a rare glimpse of early Vinča animal use, although sadly the sample sizes from the earliest layers are very small, rendering the assemblage less significant in this specific respect than that from Pločnik. This report provides only an empirical overview of the fauna, but the results are discussed in wider context elsewhere in this volume (see Chapter 51), alongside those from Pločnik (see Chapter 35).

The faunal study from Belovode includes all material excavated from Trench 18 during the 2012 and 2013 seasons. The vast majority of specimens were excavated by hand from defined cultural layers, with a smaller number coming from various types of features (pits, ovens, etc.). A small quantity of animal bone from wet-sieved soil samples was retained but has yet to be studied and is not described here.

Previous archaeozoological work at Belovode involved a large sample of 3487 bone fragments from the 1995–2002 excavation campaigns, spanning seven trenches (Jovanović *et al.* 2003). Perhaps due to the authors' backgrounds in veterinary science, this study focused on the morphology of domestic species, with no separation by stratigraphical/chronological phases. This analysis showed the dominance of cattle (42%), followed by pig (27%), sheep/goat (25%) and dog (6%), while the remains of wild species were found in very small numbers (*c.* 1%). The findings from the new excavations revise this picture quite substantially.

Methods

Specimens were recorded using the EUROFARM recording protocol, as also employed at Pločnik (see Chapter 35) and at Kočićevo in northern Bosnia (Orton 2014). Animal bone fragments from each context were divided into 'diagnostic' and 'un-diagnostic' based on criteria designed to minimise bias in favour of more readily identifiable taxa and elements. 'Un-diagnostic' thus includes: skull fragments that cannot be identified to the skull element; all ribs; vertebrae other than atlas, axis and sacrum; blade fragments of pelvis and scapula; mandible fragments that cannot be assigned a side; long bone shaft fragments with less than half the circumference of the shaft and without any portion of articular or metaphyseal surface; tooth fragments that could not be assigned to a jaw and tooth class; and fragments that could not be identified to element. Specimens with butchery marks or pathologies were automatically considered 'diagnostic'. 'Un-diagnostic' specimens were only counted by body size category and element type. Their general size range, weight, surface condition and distribution were recorded at context level.

'Diagnostic' specimens were recorded individually to the lowest possible taxonomic group. The following information was recorded where appropriate: maximum length, weight, taxon, element, symmetry, element part, diagnostic zones (Dobney and Reilly 1988), epiphyseal fusion, tooth eruption/wear, sex, surface condition, burning, gnawing, and metrics. For specimens with butchery marks and pathological changes, the location and description of these features was recorded and at least one photograph or scan taken.

Taxonomic identification was carried out using the reference collection of the Laboratory for Bioarcheology of the Faculty of Philosophy in Belgrade, aided by published morphological criteria (Boessneck 1969; Boessneck *et al.* 1964; Halstead *et al.* 2002; Payne 1985; Prummel 1988; Prummel and Frish 1986; Schmid 1972; Zeder and Lapham 2010; Zeder and Pilaar 2010). Identification of wild and domestic species of cattle and pig was also checked using metrics. Measurements were taken wherever possible following von den

Driesch (1976). Number of identified specimens (NISP) and diagnostic zones (DZ)—following Watson's (1979) system as modified by Bogucki (1982)—were used as quantification measures.

Kill-off data for the main domestic animals were based on mandibular tooth eruption/wear and on epiphyseal fusion. For caprines, tooth eruption/wear was recorded according to both Payne's (1973) and Grant's (1982) systems, while Grant's system was followed for domestic cattle and pig. Payne's mandibular wear stages (A–I) were applied and scaled using suggested ages from Payne (1973) for sheep and goats, from Halstead (1985) for cattle, and from Hambleton (1999) for pigs. In a deviation from Payne's system, fragmentary mandibles that could not be assigned to a single age class were attributed to possible classes in proportion to the suggested length of those classes in months, rather than by reference to the numbers of mandibles already definitively assigned to each class. We consider this approach more robust for fragmentary material since it does not give undue weight to the potentially very small number of specimens that can be assigned to a single class.

Zeder's (2006) fusion stages were used for caprines, while for domestic cattle and pig, fusion stages (early, middle and late) from Reitz and Wing (2008) were followed, based on Silver's (1969) suggested ages for the fusion of epiphyses. The sex of pigs was assessed according to canine morphology (Schmid 1972), while the sex of cattle and caprines was based on pelvic morphology following Grigson (1982) and Boessneck *et al.* (1964).

Overall preservation

In general, fragmentation and bone preservation are moderate. In most contexts a very small percentage of fragments show evidence of burning, the exception being the remains from a 'refuse pit' (F43) which are severely fragmented and consist of 90% burnt bones. Carnivore/pig gnawing is fairly frequent but rarely severe: around half of all contexts show some evidence of gnawing, but only a handful had marks present on more than around 10% of specimens.

Relative proportions of taxa

Taxonomic frequencies are shown by NISP and DZ by horizon in Table 1 (see Chapter 37 for details of phasing and absolute chronology). In all phases, domestic animals are much more numerous than wild specimens, contributing 80-90% of NISP. Overall, cattle are the most abundant species (46-48%, depending on interpretation of some borderline wild/domestic specimens), followed by sheep and goat (collectively 17%), pig (16-19%) and dog (2%). Wild species are represented by red deer (7%), roe deer (2%), wild pig (3%), and very small numbers of remains from aurochs, bear, lynx, fox, wild cat, beaver, badger and hare. Five bird bones and a single tortoise specimen are not included in these calculations. No fish remains were found, at least in the hand-collected fraction. Taxonomic diversity seems to be a function of sample size within this sample: the more bones recovered from a given horizon, the more species are present (cf. Pločnik; see Chapter 35).

There are some changes in the relative abundance of species over time at Belovode (Figure 1 and Figure 2), though the samples for the earlier phases are small and should be treated with some caution, especially those from Horizons 4 and 5 (merged here for that reason). The contribution of cattle declines from around two-thirds of remains (over 80% of domestic specimens) in Horizons 5–4 to only around 45% (*c.* 50% of domesticates) in Horizon 2, before increasing again slightly in Horizon 1. Both pigs and caprines show the reverse trend, increasing across Horizons 5–4 to 2 and then decreasing in the final phase. Meanwhile, the contribution of wild taxa roughly halves between Horizons 5–4 and 3, from *c.* 16% to *c.* 8%, before increasing to *c.* 12% again in the final phase.

Oddly, the overall decline over time in relative abundance of cattle is the opposite of the trend seen at several other large Vinča sites, including Gomolava, Selevac, and Pločnik (see Orton 2012; Legge 1990; Chapter 35, this volume). Given the relatively small sample sizes in all but the latest phases, however, along with the limited excavation area and the possibility of spatial variation within the site, this result should be treated with a degree of caution.

Domestic cattle (*Bos taurus*)

Cattle are the most abundant species in the faunal sample from Belovode in all phases. All body parts are present, with mandibles, metapodials, and phalanges being more numerous than other elements by NISP. When DZ counts by anatomical region are normed by the number of DZ in each region of a complete carcass, however, anatomical representation appears fairly even, albeit with a slight underrepresentation of cranial and other axial elements (Figure 3a). A broadly similar pattern was seen at Pločnik (see Chapter 35), albeit with considerable variation between phases. No such change over time is observed at Belovode, with the Vinča D (Horizon 1b) and earlier (Horizon 5–Horizon 2) layers having similar anatomical profiles. Spatial element distribution cannot be discussed here since the faunal remains derive from just one trench.

Although a large number of cattle teeth are present in the sample, relatively few provided useful data for age estimation. A total of 19 mandibles and loose teeth

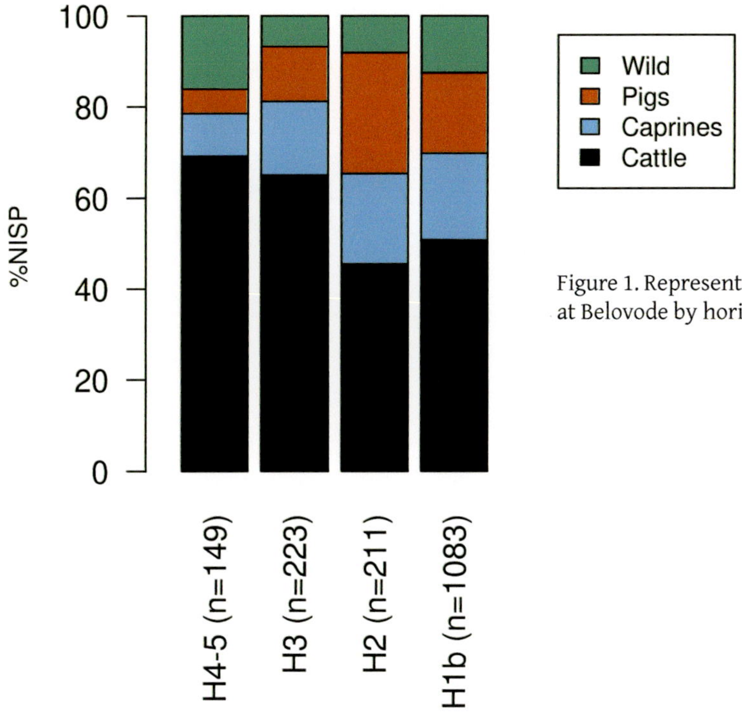

Figure 1. Representation of main domestic taxa and of wild mammals at Belovode by horizon, in terms of % NISP.

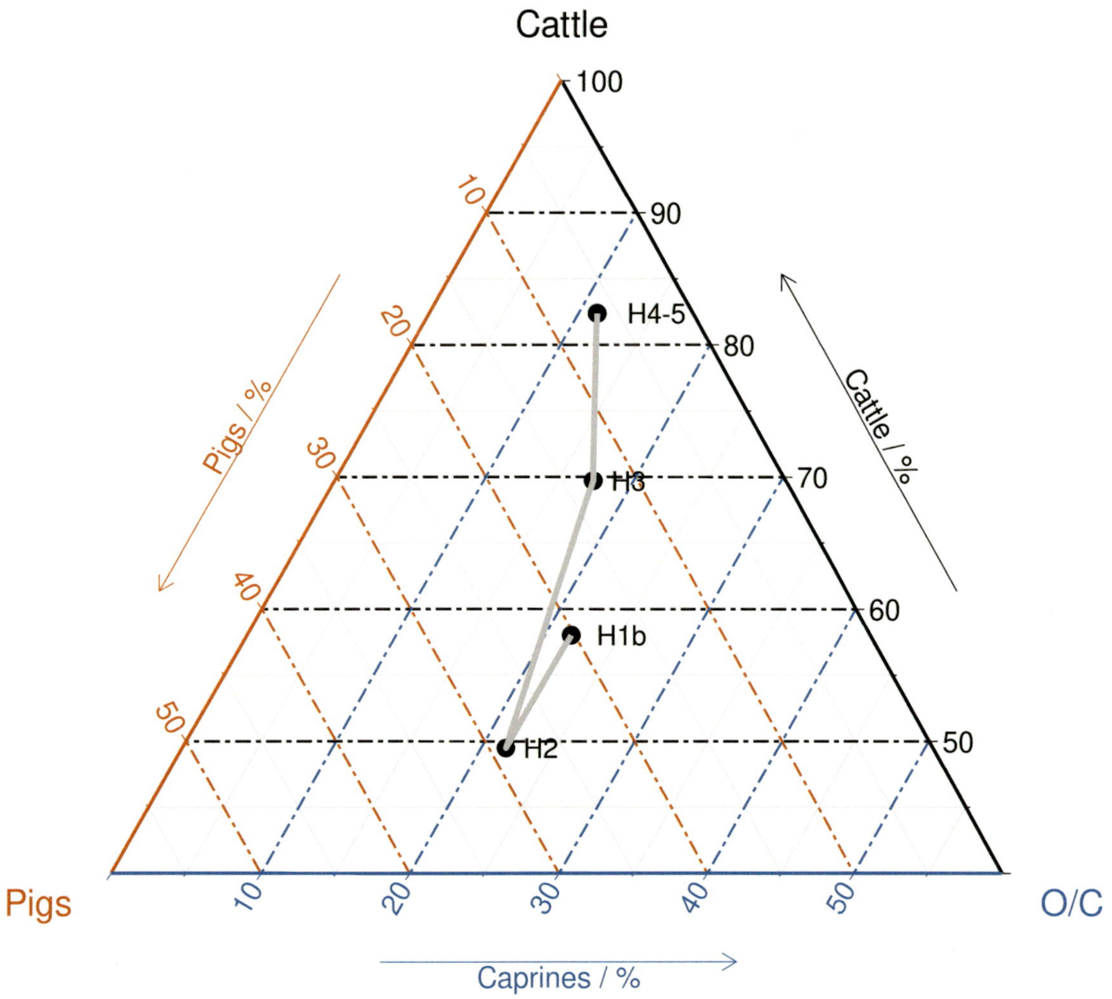

Figure 2. Ternary plot showing contributions of main domestic taxa to NISP at Belovode, by horizon.

Table 1. Distribution of taxa at Belovode by horizon, as NISP and as Diagnostic Zones (DZ; see text for explanation). *For deer, numbers i

		Horizon 5 (Vinča A1-A2)				Horizon 4b (Vinča A2-B1)				Horizon 4a (Vinča B1-B2)			
Common name	**Latin name**	NISP	%	DZ	%	NISP	%	DZ	%	NISP	%	DZ	%
Domestic cattle	*Bos taurus*	8		5		53	68.8	20	65.6	42	59.2	10.5	48.8
Aurochs	*Bos primigenius*												
Cattle (indet.)	*Bos sp.*	2		2		5	6.5	1.5	4.9	2	2.8	2	9.3
Domestic pig	*Sus domesticus*	1				1	1.3	1	3.3	6	8.5	2	9.3
Wild pig	*Sus scrofa*					4	5.2	2.5	8.2	2	2.8	2	9.3
Pig (indet.)	*Sussp.*	1		1		2	2.6						
Sheep	*Ovis aries*												
Goat	*Capra hircus*												
Sheep/goat	*Ovis/Capra*					5	6.5	1	3.3	9	12.7	2	9.3
Dog	*Canis familiaris*					1	1.3	1	3.3				
Red deer	*Cervus elaphus*	2		1		3	3.9	1.5	4.9	8(1)	11.3	3	14.0
Roe deer	*Capreolus capreolus*					2	2.6	2	6.6	1	1.4		
Brown bear	*Ursus arctos*									1	1.4		
Fox	*Vulpes vulpes*												
Badger	*Meles meles*												
Beaver	*Castor fiber*					1	1.3						
Wildcat	*Felis silvestris*												
Lynx	*Lynx lynx*												
Hare	*Lepus europaeus*												
	Total identified	**14**		**9**		**77**	**100**	**30.5**	**100**	**71**	**100**	**21.5**	**100**
Cattle or red deer		3		1		4				2			
Sheep/goat/roe deer						3				1			
Dog family													
Deer family													
Carnivore (indet.)													
Large mammal		22				129				190			
Medium mammal		1				40				20			
Small mammal													
Mammal		1				53				62			
Tortoise													
Bird						1							
	Total unidentified	**27**		**1**		**230**				**275**			
	Total	**41**		**10**		**307**		**30.5**		**346**		**21.5**	

brackets represent the number of antler specimens, while numbers outside the brackets **exclude** antlers.

Horizon 3 (Gradac-Vinča C)				**Horizon 2 (Vinča C)**				**Horizon 1b (Vinča D1)**				**TOTAL**			
NISP	%	DZ	%	NISP	%	DZ	%	NISP	%	DZ	%	NISP	%	DZ	%
145	62.0	27.5	46.6	96	41.7	30.5	36.7	550	47.4	144	39.0	894	50.2	237.5	41.4
								6	0.5	2.5	0.7	6	0.3	2.5	0.4
1	0.4	1	1.7	4	1.7	2.5	3.0	20	1.7	7.5	2.0	34	1.9	16.5	2.9
27	11.5	8	13.6	56	24.3	24	28.9	192	16.8	68	18.5	283	15.9	103	18.0
2	0.9	1	1.7	8	3.5	4	4.8	40	3.5	22.5	6.1	56	3.1	32	5.6
2	0.9	2	3.4	11	4.8	4	4.8	29	2.4	12	3.1	45	2.5	19	3.3
2	0.9							22	1.9	4	1.1	24	1.3	4	0.7
								4	0.3	1.5	0.4	4	0.2	1.5	0.3
34	14.5	8	13.6	42	18.3	11	13.3	180	15.7	47	12.8	270	15.2	69	12.0
8	3.4	4	6.8	4	1.7	3	3.6	22	1.9	13.4	3.6	35	2.0	21.4	3.7
4(24)	1.7	1.5	2.5	5(3)	2.2	4	4.8	57(7)	5.0	30	8.1	79(35)	4.4	41	7.2
7	3.0	4	6.8	3	1.3			17(2)	1.5	10	2.7	30(2)	1.7	16	2.8
								7	0.5	2.7	0.7	8	0.4	2.7	0.5
1	0.4	1	1.7					1	0.1			2	0.1	1	0.2
								3	0.2	3	0.5	3	0.2	3	0.5
				1	0.4							2	0.1		
								1	0.1	1	0.3	1	0.1	1	0.2
								1	0.1	0.4	0.1	1	0.1	0.4	0.1
1	0.4	1	1.7					2	0.2	0.6	0.2	3	0.2	1.6	0.3
234	**100**	**59**	**100**	**230**	**100**	**83**	**100**	**1154**	**100**	**370.1**	**100**	**1780**	**100**	**573.1**	**100**
32		3		12		5		57		4		110		13	
5				8				29		4		46		4	
								1				1			
								(6)				(6)			
1		1						1				2		1	
502		1		338				1271		1		2452		2	
107				115				464				747			
								8				8			
325				111				302				854			
								1				1			
				3				1				5			
972		**5**		**587**		**5**		**2135**		**9**		**4226**		**20**	
1206		**64**		**817**		**88**		**3289**		**379.1**		**6006**		**593.1**	

were used for the final analysis (Table 2a). Considering the small number of elements, they are presented as a whole, not by phase. To the extent that the small sample can be relied upon, the kill-off data (Figure 4a) indicate that around one quarter of individuals was killed by eight months, and an additional third between 8 and 30 months. Fewer than 20% of individuals appear to have survived beyond three years, and almost no specimens from our small sample fall definitively into the 'old adult' or 'senile' categories. However, a group of loose teeth (n = 8) could be assigned to only a very broad Payne stage (E+) and are hence excluded from the survivorship curve, suggesting that the number of individuals surviving to beyond three years may be somewhat underestimated.

Unlike dental specimens, epiphyseal fusion data are sufficiently abundant to permit comparison between Horizon 1b and the earlier levels (Figure 4a). In both cases, survivorship appears significantly higher than for the dental data – presumably due to poor preservation of the younger unfused postcranial specimens, *vis-à-vis* mandibles – with almost all 'early fusing' specimens fused. The horizons then diverge, with 60% survivorship to c. 42 months ('middle fusing') in Horizons 5–2 but 85% in Horizon 1b. For the latest-fusing group the Horizon 1b data show a marked drop to *c.* 60% fused. The paradoxical up-tick between middle- and late-fusing groups in the earlier levels can perhaps be put down to relatively small sample sizes, particularly for the former. This being the case, one cannot confidently conclude any difference between the two periods.

Sex could only be determined for two cattle pelvises, both of which were male.

Forty-three fragments (*c.* 6% of all cattle specimens) had butchery marks, mostly various cuts, with one case of impact also detected in the form of a compression fracture. These were most commonly observed on humerus and pelvis fragments but were also present on all body parts, excluding the skull.

Pathologies are not numerous in the Belovode assemblage. Only four cattle specimens displayed some pathological malformations. These took the form of varying degrees of exostosis and abnormal depressions, observed on a distal humerus, proximal radius, carpal 2+3 and first phalanx.

Sheep/goat (*Ovis/Capra*)

Sheep and goat remains are discussed together, since the majority of specimens could only be assigned to a combined sheep/goat category. Sheep outnumber goats around 6:1 by NISP amongst positively identified specimens in the assemblage as a whole, although only around 10% could be identified to species. Since there are virtually no such specimens from horizons prior to 1b, however, it is impossible to say whether this was consistent over time. Taken together, caprines make up around 9% of domestic remains in the Vinča A–B levels (Horizons 5–4), rising to between 15 and 18% in later phases. Again, this is a reversal of an overall trend *away* from caprines over the course of the Neolithic in the central Balkans (Orton 2012).

All areas of the body are represented in the sheep/goat assemblage, though mandibles and metapodials are noticeably more abundant than other elements. Very few phalanges are present despite their high preservation potential, but recovery bias may be responsible. Plotting weighted DZ by anatomical region (Figure 3b) reveals some possible differences in anatomical representation between horizons, although limited data are available from the earlier horizons. In particular, there appears to be an increase in frequency of hindlimb specimens in Horizon 1b. The sample sizes and excavation area are too limited to read much into this phenomenon, but it presumably reflects changes over time in the kinds of deposits represented within the trench, suggesting that the horizontal stratigraphy at the site was not homogeneous in terms of bone deposition.

Dental and fusion data for sheep/goat are not separated by phase, due to small sample sizes, but are presented as a whole. Only 30 mandibular specimens, including loose teeth, could be assigned to a specific Payne wear stage (Table 2b, Figure 4b). Based upon these, the majority of sheep and goats (approximately two thirds) were slaughtered at between two months and two years old, i.e. while still juvenile – an unusually high number for the region as a whole (cf. Chapters 35 and 51). The third of animals surviving beyond two years appear to have been killed off gradually over the remainder of their potential lifespan.

Fusion data are based on 48 specimens, excluding tiny samples of two specimens from fusion group C and one from group F (Figure 4b). Despite small samples, data from the first two fusion groups are broadly consistent with the dental data. Beyond this, however, the data become hard to interpret: around 56% of group D specimens are fused, representing almost twice the apparent survival rate to 30 months suggested by the dental data. This then rises to over 75% for group E. Such paradoxical upticks are common with the kind of small samples seen here but are rarely so pronounced.

Sex was determined for three specimens—all sheep—of which one was male and two female.

Butchery marks are present on just two specimens, a radius and a pelvis. On both elements they consist of shallow cuts of different lengths, located away

Table 2. Distribution of cattle, caprine, and pig mandibles and loose lower teeth into relative age stages following Payne (1973), as adapted by Halstead (1985) and Hambleton (1999). See text for explanation of how uncertain mandibles/teeth were assigned to stages.

	Stage	Suggested age (months)	Raw count				Corrected
A. Cattle	A	0-1	0				**0**
	B	1-8	1		1		**1.25**
	C	8-18	3				**3.35**
	D	18-30	0	2	2		**2.95**
	E	30-36	2	1			**3.59**
	F	young adult	3				**4.87**
	G	adult	1	1	1		**1.83**
	H	old adult	0				**0.83**
	I	senile	0				**0.33**
			10	4	4		**19**
B. Caprines	A	0-2	0				**0.16**
	B	2-6	2	3	1		**4.64**
	C	6-12	2	4		6	**7.26**
	D	12-24	2				**7.94**
	E	24-36	0				**0.75**
	F	36-48	1		3		**1.75**
	G	48-72	0	4			**3.5**
	H	72-96	1	1			**3.5**
	I	96-120	0				**0.5**
			8	12	4	6	**30**
C. Pigs	A	0-2	0	1			**0.42**
	B	2-7	0		1		**1.08**
	C	7-14	5	1			**6**
	D	14-21	5	1			**6.04**
	E	21-27	6				**6.46**
	F	27-36	1				**1**
	G	adult	0				**0**
	H	old adult	0				**0**
	I	senile	0				**0**
			17	3	1		**21**

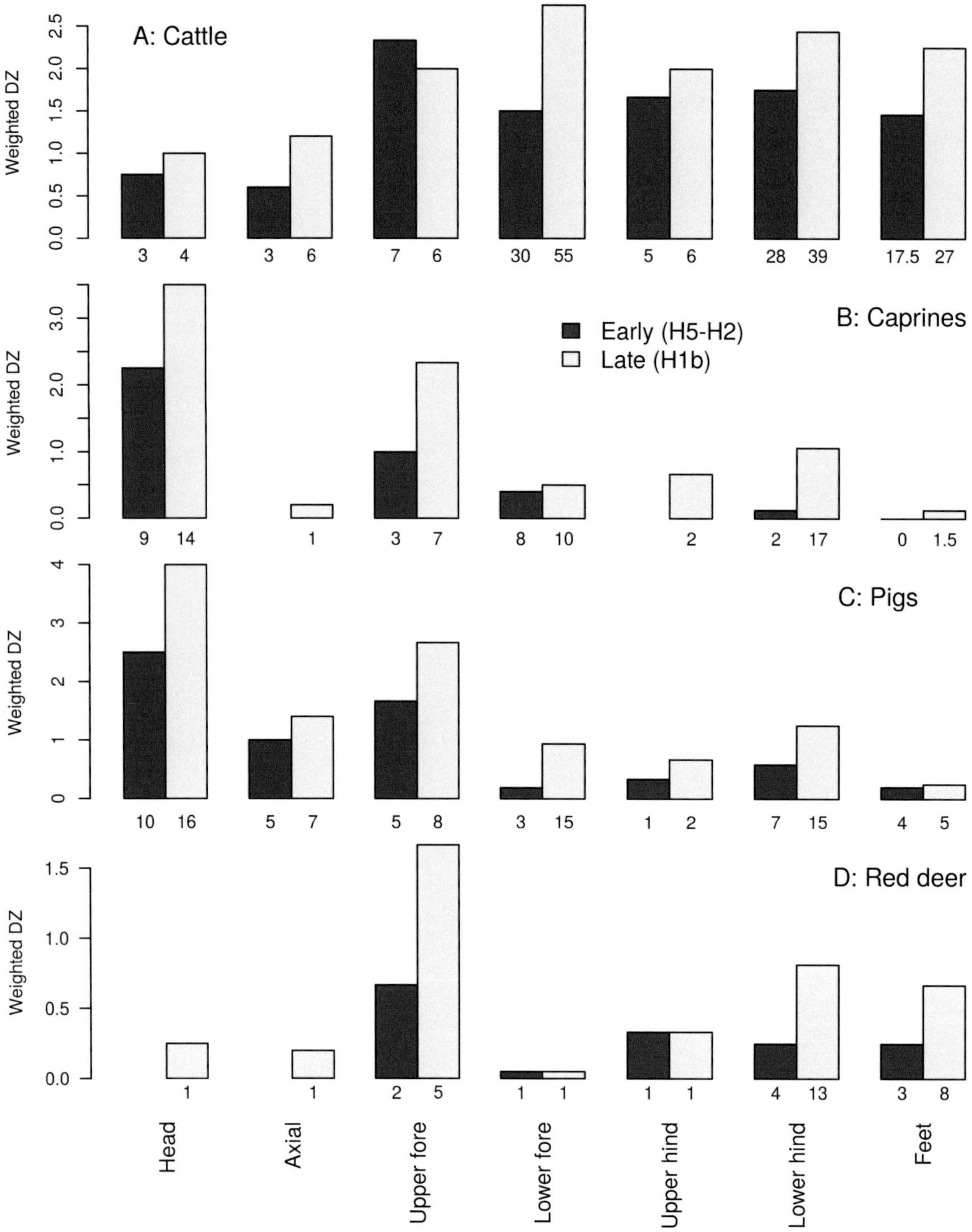

Figure 3. Body part representation (by percentage weighted Diagnostic Zones) for major taxa at Belovode, by broad period. See text for explanation of DZ system. Weighting involved dividing the number of observed DZ in each body region by the number of DZ from that region in a complete carcass. Numbers beneath bars indicate total DZ before weighting.

from the articulations. A single mandible exhibited some irregularities of tooth wear but otherwise no pathologies were found among the sheep/goat remains.

Domestic pig (*Sus domesticus*)

The relative frequency of domestic pig increases dramatically over the earlier horizons, from *c.* 5-7% (depending on the indeterminate *Sus* specimens) by NISP in Horizons 5–4 to 24-28% in Horizon 2, before dipping slightly to *c.* 17-20% in Horizon 1b. All body parts of the pig are present, with mandibles and skull fragments being the most numerous by NISP. Weighted DZ counts indicate over-representation of cranial elements and also upper forelimbs (Figure 3c). There is no noticeable difference in the element representation

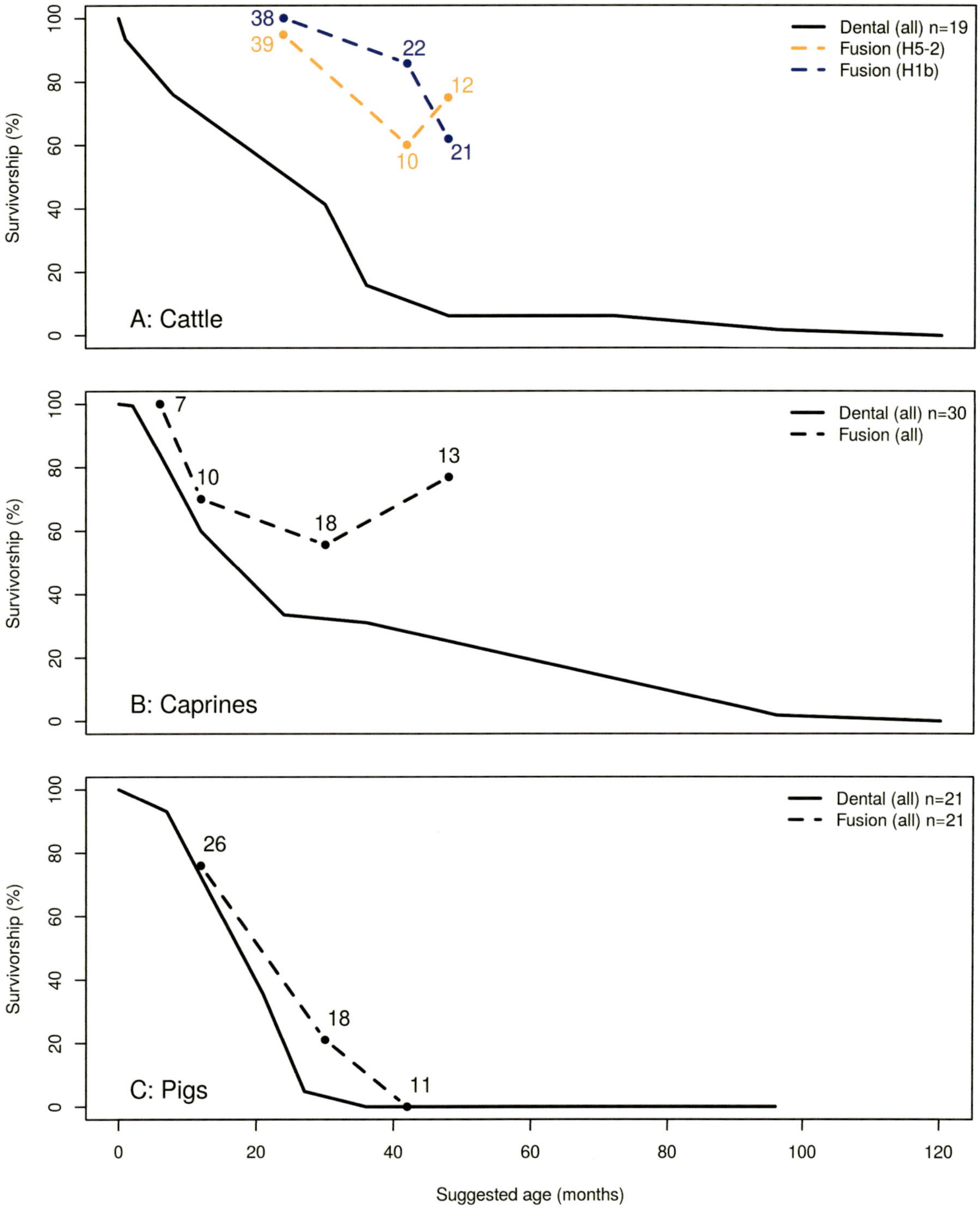

Figure 4. Age-at-death data for major domestic taxa at Belovode. Each plot overlays survival estimates based on epiphyseal fusion (dotted lines) on survivorship curves constructed using Payne's (1973) mandibular age stage system (solid line). See text for sources of fusion groupings and suggested ages.

between phases, nor any pattern that might indicate specific depositional practices. Indeed, the very close similarity to the pig element profiles from Pločnik (Chapter 35) suggests that the variation between body parts is likely to be largely taphonomic.

Dental data are based on a sample of 21 mandibles and loose teeth (Table 2c, Figure 4c). The vast majority of specimens (*c.* 95%) appear to have been killed fairly evenly between 7 and 27 months. A small number survived until 27-36 months (*c.* 5%), but no definite old adult or senile individuals were recognised in the sample.

Fusion stages are also presented in Figure 4c with all phases combined and fit very well with the dental results. Again, no specimens older than about three years are apparent.

Based on canine morphology, sex could be determined for nine specimens, of which two were female and seven male.

Cut marks were observed on seven domestic pig specimens, representing a variety of elements (atlas, long bones, pelvis, scapula). Two specimens had pathological changes. One scapula featured a possible healed break on the posterior part of the blade, close to the collum scapulae, while a mandible displayed an abnormality in tooth eruption.

Red deer (*Cervus elaphus*)

Red deer is the most abundant wild species in the Belovode assemblage (4.4% of total NISP, or 6.4% including antlers). Almost all body parts are present. The most numerous are antler fragments, followed by phalanges, metatarsals, humeri, tibiae, pelvises, calcanei and scapulae. By weighted DZ (Figure 3d) there is a clear over-representation of upper forelimbs—including both scapulae and humeri—which is hard to explain and sits in stark contrast to the very balanced element representation seen for red deer at Pločnik (Chapter 35). The near absence of cranial elements, excluding antlers, at Belovode, again in contrast to Pločnik, might indicate that the heads were removed from the carcasses of hunted dear before transport to the site.

The red deer brought to the site seem to have been overwhelmingly mature: of 33 specimens with fusion zones present, all are fused although only one specimen is from the latest-fusing group. No sex data is available.

Only three specimens – a pelvis, a scapula and a humerus, featured cut-marks, while a single third phalanx showed pathological change in the form of probable exostoses.

Summary

Although an assemblage from Belovode has previously been published (Jovanović *et al.* 2003), the present study revises the picture considerably while also allowing analysis by phase for the first time, along with consideration of anatomical representation and age at death. Perhaps surprisingly, the overall taxonomic distribution within the present sample differs substantially from that previously reported: cattle are more clearly dominant here, at 50% of NISP (cf. 42%), while domestic pigs and caprines are less frequent (16-18% and 17% respectively; cf. 27% and 25%). Dogs make up only 2% of NISP in the present sample compared to 6% in the earlier study. Most stark, however, is the difference in representation of wild species: at least ten times higher here than previously reported.

One possible explanation for these discrepancies involves change over time and differential chronological coverage between the two studies: our data from Horizon 2, for example, shows a taxonomic composition fairly close to that reported by Jovanović *et al.* (2003). Yet there is little reason to believe that the earlier sample (drawn from seven trenches) is heavily focused on a single period of occupation. Alternatively, we may be seeing spatial variability across the site – a possibility that should encourage caution regarding the results of both studies. Finally, differences in recovery, identification, or quantification methods may be responsible.

One of the most significant things about the new Belovode assemblage is that, like Pločnik, it presents an unusual opportunity to look at animal use in the earlier part of the Vinča period, and at change and continuity across the metallurgical horizon. The paucity of data from the earliest levels at the site does limit this somewhat but there are, nonetheless, some interesting trends. The overall increase in frequency of caprines at the expense of cattle is the opposite of that seen both in the region as a whole (Orton 2012) and at a number of major multi-phase sites, including Pločnik (Chapter 35), although the increase in numbers of pigs (at least up to Horizon 2) fits better with the regional picture.

Age-at-death results are generally inconclusive due to limited data, the need to combine horizons, and some inconsistency between dental and epiphyseal data. They suggest a relatively high degree of juvenile mortality for both cattle and caprines compared to Pločnik, which might suggest more specialised production, but the epiphyseal data are actually very similar between the two sites.

While this report has presented only a very brief overview of the data, the results are discussed alongside those from Pločnik elsewhere in this volume (Chapter 51).

The bibliographic reference for this chapter is:

Dimitrijević, I. and Orton, D. 2021. Animal remains from Belovode, in Radivojević, M., Roberts, B. W., Marić, M., Kuzmanović Cvetković, J., and Rehren, Th. (eds) *The Rise of Metallurgy in Eurasia*: 249–258. Oxford: Archaeopress.

Chapter 22

Belovode: past, present and future

Benjamin W. Roberts and Miljana Radivojević

Introduction

The 2012 and 2013 excavations and subsequent post-excavation analyses by *The Rise of Metallurgy in Eurasia* project team at the site of Belovode built upon two decades of earlier work led by the National Museum of Belgrade and the Museum in Požarevac (Jacanović and Šljivar 2003; Šljivar 2006; Šljivar and Jacanović 1996b, 1996c, 1997c; Šljivar *et al.* 2006). This earlier work across 17 trenches had identified four building horizons (Belovode A–D), the presence of the entire Vinča culture ceramic sequence from Vinča Tordoš (A–B1) to the Gradac Phase (I–III) as well as stone tools, figurines, obsidian blades, animal bone and, most importantly for the current research, evidence for the smelting of copper ores. As detailed in Chapter 5, it was the archaeometallurgical analysis of five small copper slags from Trench 3 together with the radiocarbon dating of the excavated horizon in which they were found that provided evidence for copper smelting at *c.* 5000 BC (Radivojević *et al.* 2010a) and the foundation for *The Rise of Metallurgy in Eurasia* project. However, in the absence of any detailed publication on these earlier excavations at Belovode, further questions relating to broader context of the earliest evidence for copper smelting could not be explored.

The geophysical and aerial survey of the site (see Chapters 9 and 39), the systematic, detailed and digitised methodological approach to excavation of Trench 18 (see Chapter 10), and the extensive programme of post-excavation analyses were all designed to enable such questions to be addressed as comprehensively as possible. It is with this aim in mind that the data encompassing the entire excavation archive at Belovode has been made freely available online to accompany this monograph (see Appendix A). Current and future generations of scholars can therefore easily access, use and analyse the data to evaluate our own interpretations as well as for their own research. It is hoped that this initiative, which we believe to a first for a prehistoric site in the Balkans, will provide a model for future projects and publications.

On metallurgy

The evidence for copper metallurgy at Belovode excavated in Trench 18 during 2012 and 2013 encompasses each stage in the *chaîne opératoire* of metal production from the black and green copper ore selection to the working of copper metal artefacts. The evidence is not only far more extensive than had previously been discovered but, due to the excavation methodology and post-excavation programme, the context for the metallurgy is far more precisely recorded. As detailed in Chapter 11, the metal technologies revealed either confirmed or developed the results and interpretations from earlier archaeometallurgical research (Radivojević 2007, 2012, 2013, 2015; Radivojević *et al.* 2010a; Radivojević and Rehren 2016).

The excavation of *c.* 1300 malachite and azurite minerals, including malachite beads but excluding sherds with traces of malachite throughout the Belovode site sequence not only provided detailed evidence for the careful selection of copper minerals and ores (see Chapter 11), but also for their ubiquity. It is difficult to consider copper minerals and ores to have been either a rare, precious or restricted resource as has been considered in debates over early metallurgy. The radiocarbon dating and Bayesian modelling of the entire stratigraphic sequence at Belovode (Chapter 37) demonstrates that, despite this level of early exploitation of copper minerals at Belovode, the evidence for copper metal production dates to the 49th century BC with two metal droplets found in a sealed refuse pit (Feature 21) and confirms rather than pre-dates the original *c.* 5000 BC date (Radivojević *et al.* 2010a). However, given that these original slags were dated according to their excavated horizon in the adjacent trench rather than directly according to their excavated pit feature, the new dates provide both further confirmation and greater chronological security for the earliest evidence for metallurgy in the world.

The process involved in smelting copper ores to metals is at least partially, if not fully, revealed in Feature 6: we identified slagged sherds, which had been exposed to a high temperature process reaching *c.* 1100°C, copper slags and metal droplets in association with the presumed (and ephemeral) 'hole-in-the-ground' installations for the early smelting of manganese-rich black and green copper ores. There is no evidence to suggest that the smelting process was undertaken away from any other areas of activity. Its location outside of any structures as well as being potentially within an enclosed residential area of grouped wattle and daub structures implies that access was neither visually nor physically restricted. The experimental replication of the copper smelting process as evidenced at Belovode demonstrates the fundamental necessity of small groups of people working together throughout the preparation and execution. The widely held ideas taken from later Classical mythology and selected African ethnographies of early metallurgy being practiced in secretive isolation by appointed specialist individuals are directly contradicted by the open and communal evidence at the Belovode settlement site.

On communities

The scale and duration of the community who lived at Belovode is now far better understood. The geophysical survey encompassed *c.* 26 ha and revealed *c.* 550 anomalies identified as burnt houses and several linear anomalies identified as ditches encompassing an area over *c.* 33 ha (see Chapter 9). A clearly higher area of settlement density in the southern part of Belovode can be contrasted with a clustered area of settlement density in the northern area. There are structures in the eastern area that are widely spaced and enclosure ditches in the northern and western areas of the site. The density and positioning of the structures in the settlement area can also identify the point of origin of the settlement and indicate how it spread over time, even though most of it remains unexcavated. Earlier and far larger estimations regarding the size of the Belovode settlement site can now be substantially reduced in the light of these results. Estimations of the maximum population size of the community Belovode based on the house groupings (see Chapters 9 and 38) and mathematical modelling (see Chapter 40) suggest *c.* 1000–1500 people, which is likely to have been in the later Vinča phases of the site.

The excavation and radiocarbon dating of the entire stratigraphic sequence at Trench 18 provides a far more precise chronology for the life of the community at the site of Belovode spanning *c.* 5350–4650 BC. This chronology is sub-divided into five horizons according to radiocarbon dated and stratified ceramic typo-chronologies drawn from the identification of *c.* 14,500 stylistically and typologically indicative ceramic fragments (e.g. rims, handles, bottoms, and ornamented belly fragments) excavated during the 2012 and 2013 excavations of Trench 18. In addition, there were a further *c.* 35,500 non-diagnostic ceramic fragments also recorded. No previous discussions of the chronology of the site of Belovode draw on this level of detailed analysis and neither do they integrate the entire ceramic and stratigraphical sequence with absolute radiocarbon dates (see Chapters 12 and 37).

The identification and analysis of craft production beyond metallurgy by the community at Belovode is fundamental to understanding how and why metallurgy emerged. Within this cross-craft framework, the scale, material preparations and high temperature processes underpinning ceramic production are especially important. The stratified and radiocarbon dated sequences, in addition to the detailed recording of the pottery in Trench 18, meant that the samples used to analyse the technology of ceramic production could be carefully chosen (see Chapter 14). The analysis demonstrated that the potters in the Belovode community had a specific set of recipes depending on the vessels that they were seeking to produce, involving the selection and manipulation of different raw materials in different pyrotechnological conditions. What the analysis did *not* reveal is a straightforward connection between pottery and metal production with evidence instead indicating that the pottery, including dark-burnished pottery, was most likely fired at temperatures below 1000°C and that the ability of the potters to control the atmosphere was variable. This still does not exclude the possibility of reaching the temperatures required to smelt copper prior to the Gradac Phase, as evidenced in previous analyses that detected smelted copper debris that pre-dated the copper slag from this phase (Radivojević 2015). In addition, the presence of a well-contextualised lead-based slag in horizons dated to the early Vinča culture (see Chapter 3) speaks of metal extraction practices that did reach temperatures in excess of 1000°C. The question, therefore, is not about the capacity to reach these high temperatures but rather how the demand for either highly burnished pots or copper metals shaped the technological parameters of their production (see also Chapters 43 and 52).

Craft production in other materials by the community at Belovode appears fairly consistent throughout the occupation of the site. The creation of pointed, cutting and burnishing tools, largely from domestic animal bones, throughout the occupation of the site was most likely for processing leathers, hides and plant materials (wood and plant fibres). No manufacturing debris was recovered indicating that use and/or disposal occurred in the contexts excavated (see Chapter 17). Similarly, the production of woodworking tools such as axes, adzes and chisels from locally sourced grey and grey-green raw stone remained largely unchanged throughout the occupation of the site with the only temporal distinction being the introduction of light white coloured ground stone tools during the Gradac Phase (see Chapter 16). Three different *chaînes opératoires* relating to flint blade, bladelet and microblade production could be identified at Belovode (see Chapter 18). In addition, there are several obsidian blades likely produced elsewhere (see Chapter 19).

Understanding the subsistence strategies of the community at Belovode had been previously limited by the absence of archaeobotanical sampling and recovery by flotation. The archaeobotanical results from the charred remains recovered (see Chapter 20) showed the presence of hulled wheat glume bases in all analysed contexts demonstrating the widespread practice and importance of cereal processing, tentatively identified as einkorn, throughout the duration of the settlement. Alongside the cereal crops, there is also consistent evidence for the gathering and processing of wild-edible fruits. The evidence for crop husbandry and wild edible fruit collection is stable and largely unchanged, with only the later addition of free-threshing wheat and bitter vetch to distinguish the passing of time. Whilst zooarchaeological evidence had been collected and reviewed from several of the earlier excavations at Belovode, there had been no chronological control and an (over-) emphasis on domestic species. The current zooarchaeological results (see Chapter 21) not only contradict the proportions stated in the earlier report, highlighting the dominance of cattle over pigs and caprines, but also reveal the far higher presence of wild species who clearly continued to be hunted by the community at Belovode.

Further work

The excavations at Belovode in 2012 and 2013 by *The Rise of Metallurgy in Eurasia* project comprised only a single trench – initially measuring 5 x 5 m (subsequently extended by 2 x 3 m) and 2.3 m deep. The positioning of the trench was deliberately focussed on the eastern area of the site near to two earlier trenches (Trenches 3 and 17) where metallurgical remains had been identified. The project sought to excavate and analyse a complete material, structural and environmental sequence at Belovode that would include further metallurgical remains in order to understand the organisation of metal production in context. It uncovered 51 features including a wattle and daub structure, pits, hearths and ash bins, pottery concentrations and a circular structure comprising six sub-oval post holes and a wide range of material and environmental evidence, all of which were analysed in the post-excavation process. These results enabled the project to largely achieve its original aims. However, the results have also generated several new avenues of investigation that would constitute a programme for further work.

1. Whilst the detailed relative and absolute Vinča chronology of occupation at Belovode has been well established by the project, the placing of Trench 18 in the eastern area and away from the central plateau where the archaeological layers from earlier excavations have been up to 4 m in depth. Further work would see excavations to characterise the differences in the dating, intensity and nature of this occupation on the central plateau.
2. The pre- and post-Vinča activity at the site remains poorly understood – whether the underlying Starčevo pit features and the transition from Starčevo ceramic forms and technology to Vinča or the presence of the Late Chalcolithic Kostolac culture. Within this context, the apparently abrupt end of the Vinča settlement at 4571–4482 cal. BC remains largely unexplained but fits well with the disappearance of other Danubian Vinča sites, including Belo Brdo in Vinča. Further targeted excavations would be able to evaluate continuities or changes in activity at the site.
3. Whilst the geophysical and aerial survey by the project has provided invaluable insights into the broader spatial patterning of settlement at Belovode, the chronology of the different spatial settlement phases has yet to be determined by absolute dating. It can be assumed that the settlement construction spread outwards with a gradual increase in the overall size but there is no radiocarbon dating for the creation of different settlement phases or the construction of the different enclosure ditches.
4. The organisation of craft production and subsistence across the community at Belovode remains poorly understood within a cross-craft perspective. The excavation of different areas and structures might provide further insights into debates over craft specialisation, craft intensification and community activities – and whether these change through time.
5. The organisation of subsistence strategies is also not well understood beyond the crop and animal types present in the sequence throughout Trench 18. The role and nature of garden plots and fields as well as the intensity and management of cultivation could be illuminated by targeted excavations of those larger areas between the settlement structures revealed in the geophysical survey as well as beyond.

The bibliographic reference for this chapter is:

Roberts, B.W. and Radivojević, M. 2021. Belovode: past, present and future, in Radivojević, M., Roberts, B. W., Marić, M., Kuzmanović Cvetković, J., and Rehren, Th. (eds) *The Rise of Metallurgy in Eurasia*: 259–261. Oxford: Archaeopress.

Photo by Jugoslav Pendić

Part 3
Pločnik

Chapter 23

Pločnik: landscape and settlement perspectives

Miroslav Marić

In this chapter, the Late Neolithic landscape and known settlement patterns surrounding the site of Pločnik are discussed. Archaeologically known for almost a hundred years, Pločnik is the only Late Neolithic site in the Toplica region that has been excavated over several campaigns spanning several decades. Surveys and chance finds in the area enable us to create a broad picture of the adjacent contemporary landscape and its communities. Aspects of their locations are examined and presented here.

The Toplica region

The site of Pločnik is in the Toplica region (Figure 1), where the modern administrative centre is the city of Prokuplje. The region lies in southwestern Serbia and encompasses the valley of the Toplica river. Its western border lies on the foothills of the Kopaonik Mountains, whilst the eastern edge is the South Morava Valley and the Niš region. In the north, the Toplica region ends in the area of Veliki and Mali Jastrebac, bordering Župa, whilst in the south the boundary is represented by the mountains Vidojevica and Pasjača. The area is 2.231 km^2 in size, or about 60 km from east to west and about 30 km north to south. Although a significant portion of the region is mountainous, the central part around the valley of the Toplica River and its tributaries is predominantly flat or mildly undulating. The highest peak is found on Mount Jastrebac (1492 m), whilst the eastern edge of the region reaches heights of around 200 m above sea level.

The modern climate of the area is moderately continental with warm summers and moderately cold winters in the valley of the Toplica, and long and severe winters on the surrounding mountains. The average rain fall is about 690 mm per m^2 and this region is among the driest in Serbia. The area is, however, abundant with smaller streams and mineral waters that originate on the slopes of the surrounding mountains and empty into the Toplica. There is no evidence to support the notion of a different climate in the Late Neolithic, but studies of the palaeoclimate are still to be undertaken in the region.

The geology of the Toplica region is rather diverse, ranging from the very old Precambrian rock to the Quaternary alluvium, diluvium and proluvium

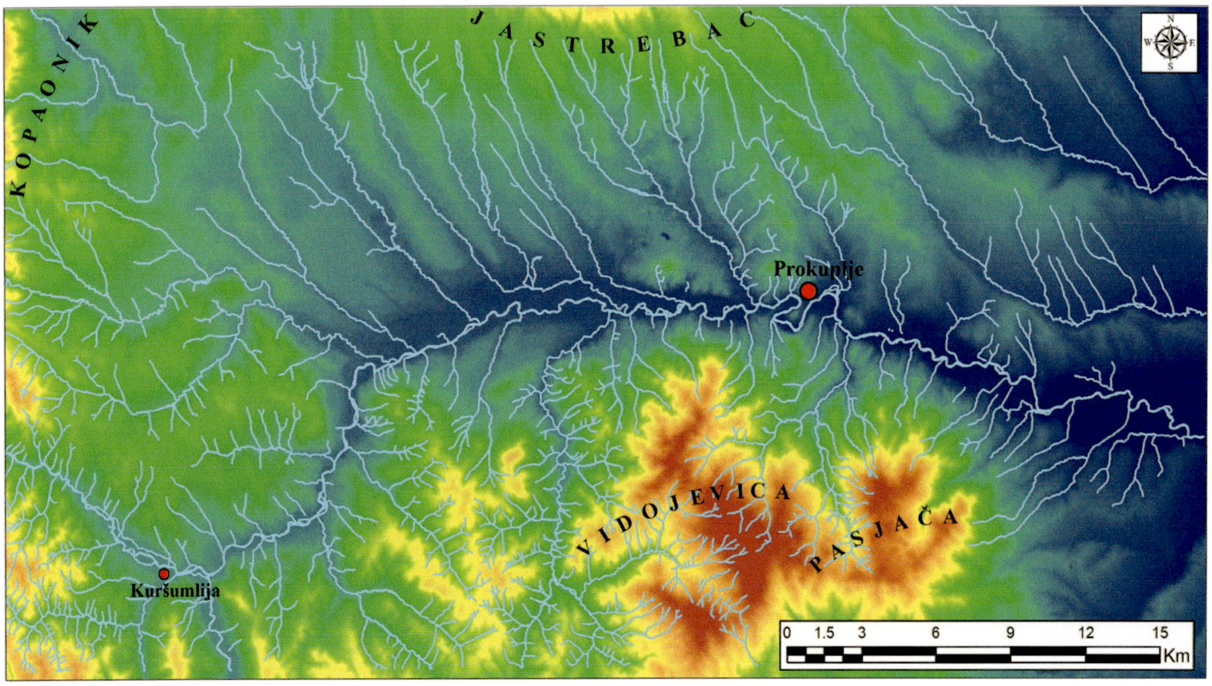

Figure 1. The Toplica region.

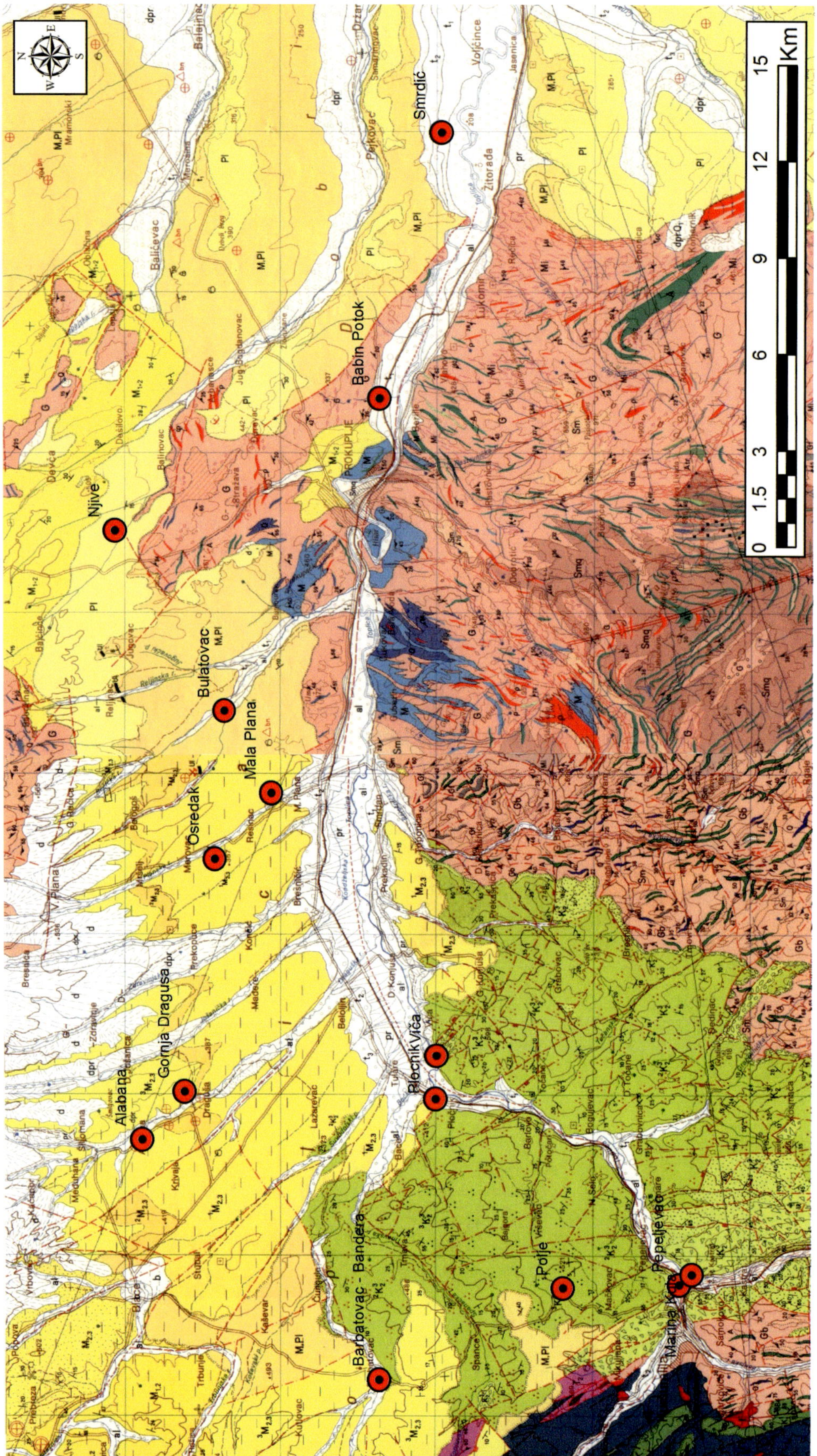

Figure 2. Geological map of the Toplica region.

sediments. The oldest geological formations can be found to the south and southwest of the modern city of Prokuplje, consisting mainly of Precambrian small-grained biotitic and bi-mical gneiss, leptinolite, mica-schists and mica rocks found on Mount Vidojevica and Mount Pasjača (Figure 2, marked with letters Gb, Sm and Smq). These sediments are infused with outcrops of amphibolite, amphibolite schists and marble. West of Mount Vidojevica, towards Kuršumlija, the geology of the area consists of Cretaceous sediments of aleurolites, marlstone and various sandstone sediments (Figure 2, marked as $^{1-3}K_2^3$. West of Kuršumlija, the Jurassic sediments consist of diabase rocks, occasionally mixed with Trias outcrops of sericite schists and limestones (Figure 2, marked with bb and T_2). These sediments are cut through by the Quaternary alluvial river valley of the Toplica and its tributaries (Figure 2, marked with pr, al, dpr and d). Although limited in meandering by their size and the seasonal character of their flow, all the tributary rivers managed to carve their valleys through the Miocene clay, sand and gravel deposits, creating several river terraces in the process. It is exactly within this contact zone, between the Miocene clay sediments and Quaternary alluvial/diluvial deposits, that the majority of known Late Neolithic settlements are located. Six of the 14 (43%) such sites considered in this chapter are located on Quaternary deposits whilst seven (50 percent) are found on the Miocene sediments. Only one settlement site, Marina Kula, is known to be located on older sediments, but others could exist in the unsurveyed tributaries between Kuršumlija and Mount Vidojevica. The northern bank of Toplica River opens towards the mildly undulating landscape of north-south oriented tributaries originating from the southern slopes of Mount Jastrebac and Mount Kopaonik. This specific geography makes possible that many settlements could be located in this region. This is further supported by the close proximity of several known settlements (Alabana and Gornja Draguša, Osredan, Mala Plana, and Bulatovac), which could indicate a higher density of settlement due to the beneficial geographic and soil characteristics.

The Quaternary river terraces which developed along the valleys of Toplica, Kosanica and Banjska reka can be divided into three types, according to their relative heights (Malešević *et al.* 1980: 36). The upper terrace (t3) lies at relative height of about 35 m and is only partially preserved in the lower reaches of the Toplica. The middle terrace (t2) is about 10 m lower than t3 and is preserved almost entirely in the lower reaches of the Toplica, whist the lower terrace (t1) is preserved in full through the whole course of the Toplica, and also in the valleys of the Kosanica and Banjska rivers. Its relative height is about 10 m, and the deposits are 2–3 m thick. These terraces are formed mostly of gravel, sands, and clayish materials. In the vertical profiles of these terraces the lower layers are made up of gravel formations, whilst upper layers are formed mostly by sands and clayish materials (Malešević *et al.* 1980: 37).

The soil in the region is diverse and, although soil acidity is predominantly high throughout, the soil fertility is also moderately high. Larger portions of extremely acidic soil can be found south of Mount Jastrebac, although the largest quantity of these soils is located in mountain areas. The soil acidity can be attributed to the dissolution of limestone from the mountain rocks by surface and subsurface water, and its deposition throughout the landscape. Around 70% of the soil in the Toplica area is Cambisol in numerous variants and states; about 12% are Vertisols; 5% are alluvial deposits; and 13% various other soils. It could be assumed that the economy of the Late Neolithic communities of the area was a mix of agriculture and cattle herding, with possible specialised transhumance settlements located in the surrounding hills.

The Neolithic of the Toplica region

Being on one of the river valley corridors that link the Kosovo region with central Serbia and the Morava Valley, the Toplica region was settled in the Early Neolithic, and a human presence was most likely established even before. Early and Middle Neolithic sites have been known since the beginning of the 20th century (Fewkes 1936; Garašanin and Garašanin 1951; Kuzmanović Cvetković 1988) but have never been excavated in detail and knowledge of these periods is limited at best. The Late Neolithic period is far better studied, largely because of Pločnik, where research began in 1927 (Grbić 1929) and, albeit with several interruptions, has continued until present day. In the wider Toplica region there are 17 known sites dating to the Late Neolithic period (Figure 3 shows the position of 14), but the close proximity of some of these sites allows for the existence of more in the area. Besides Pločnik, only Kremen, near Mačina (Đurić 1986; Šljivar and Antonović 1996) has been published in some detail.

Based on the position of known Late Neolithic sites (Figure 3), several observations can be made. The sites are set on the banks of local rivers, often on an elevated terrace near the confluence of two smaller streams. The closest recorded distance between two sites is about 350 m between the sites of Pepeljevac and Marina kula, but these may actually represent the remains of a single site. The second smallest recorded distance is 2.5 km, between the sites of Alabana and Gornja Draguša. This distance is slightly larger than the examples found in the Braničevo region, but it should be noted that the poor state of research in that area does not allow for a precise analysis of inter-settlement distances. It is also possible that some of the sites along the meandering Toplica have been destroyed over the millennia. Today, the river mostly erodes its left bank, moving northwards

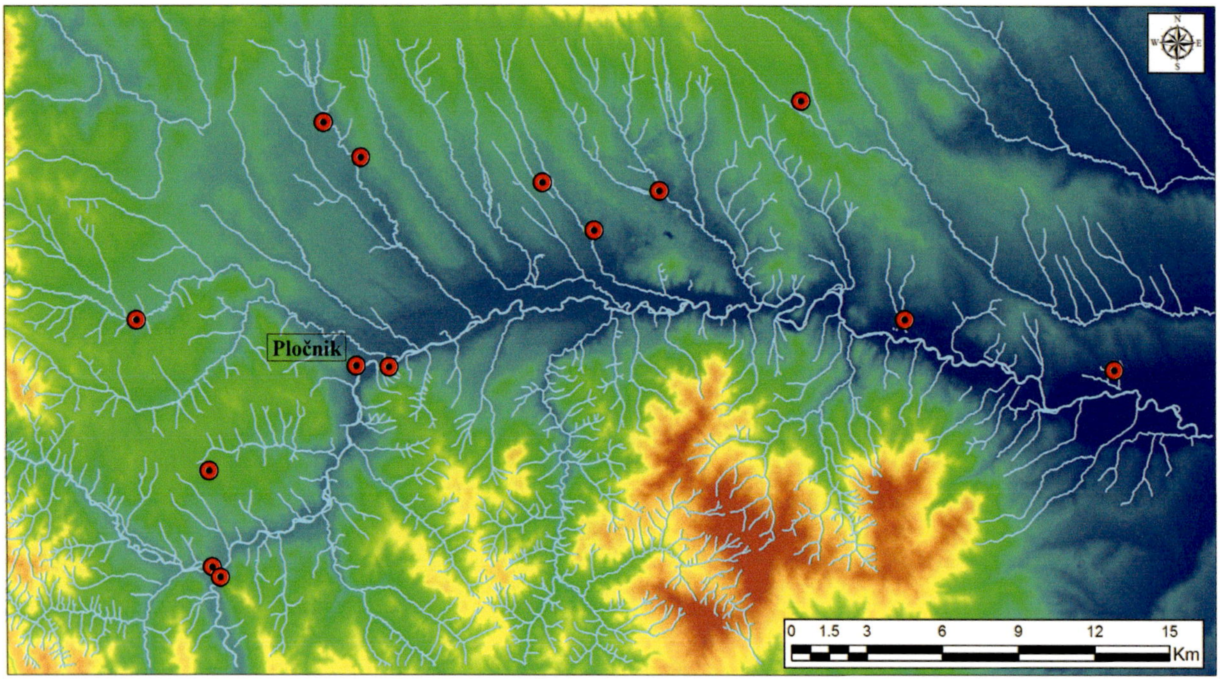

Figure 3. The locations of known Late Neolithic sites in the Toplica region.

towards the gentle plains beyond, and a large bend has formed, even in the area of Pločnik. The proximity of the sites to water is also reduced compared to the Braničevo region (Table 1), where sites tend to be further away from major rivers, especially in the Velika Morava valley. For sites in the Toplica region, the vertical distance from water ranges between 2 and 20 m (Marina Kula near the site of Kuršumlija) and this does not differ much between regions. The terraces of the Toplica valley are not as flat as those in the Braničevo region, and their inclination can reach up to 17°, but is predominantly below 5° (Figure 4). The known sites are largely found on terrain with a southern or southwestern aspect, but they are not sufficiently numerous to be statistically representative, or to allow the identification of preferences in the settlement pattern.

Most of the sites (8 of 14) are found on Eutric Cambisols, the most common soil type in the Toplica region although not the most fertile. This kind of soil is intensively used around the world for agriculture and can be reasonably productive when properly tilled. Worldwide, they cover an estimated 15 million km^2, and are well represented in the temperate and boreal regions that were under

Table 1. Landscape characteristics of Late Neolithic sites in the Toplica region.

Site no.	Slope degree	Aspect	Soil type	Distance to water (m)
1	5.3931	SW	Eutric Cambisols	175.096193
2	1.926039	SE	Calcareous Aluvial deposit	89.451039
3	2.452959	S	Eutric Cambisol	446.583932
4	3.000104	S	Non Calcareous Vertisol on Neogene sediments	98.346026
5	2.272111	NE	Eutric Cambisol on Neogene sediments	146.172191
6	5.592044	SW	Eutric Cambisol	46.040752
7	16.744549	NW	Eutric Cambisol	269.872728
8	1.828303	NW	Non Calcareous Vertisol on Neogene sediments	164.606246
9	1.080327	SW	Eutric Cambisol	132.565328
10	5.62231	N	Calcareous Vertisol	231.349852
11	4.689361	SE	Non Calcareous Cambisol on Neogen sediments	137.956684
12	6.209617	SW	Eutric Cambisol	378.509889
13	3.05944	S	Eutric Regosol	95.100291
14	1.02741	SW	Non Calcareous Vertisol on Lake sediments	263.34905

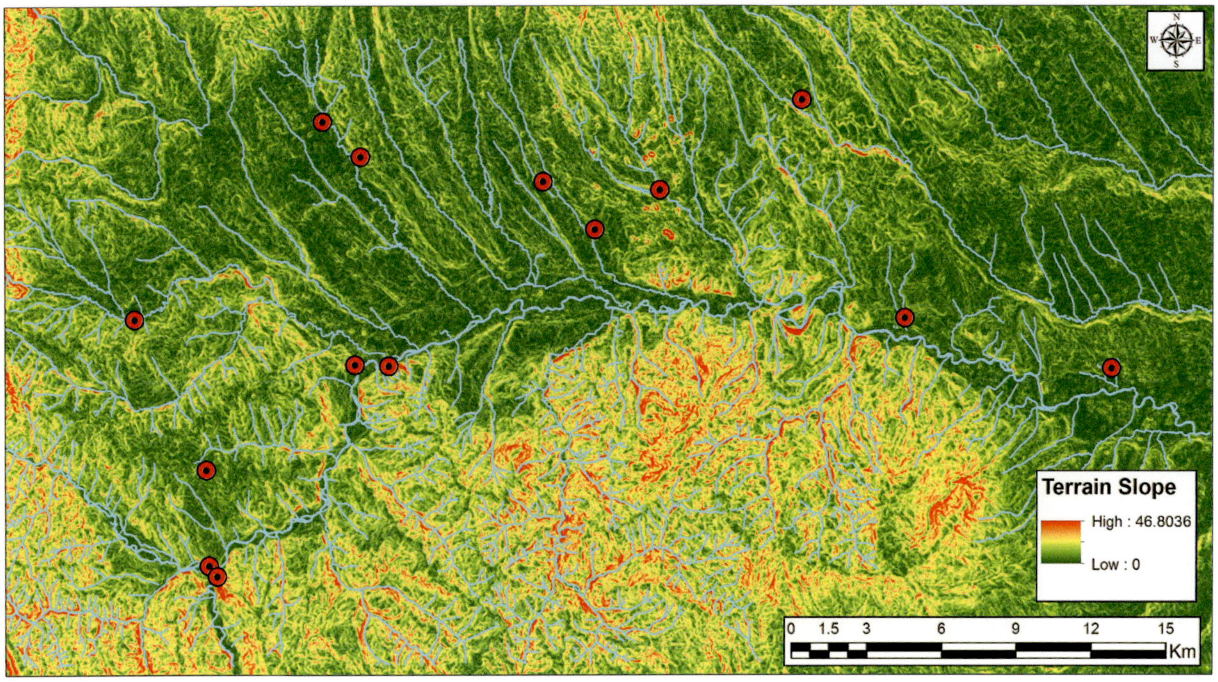

Figure 4. Terrain slope values in the Toplica region.

the influence of glaciation during the Pleistocene (FAO website). The horizon differentiation is weak as this represents the beginning of soil formation. In the Toplica region, Eutric Cambisols account for almost 70% of the arable soil, with Vertisols being the second most prevalent at 11.5%. Four Late Neolithic sites (28.5% of the total) are found on Vertisols, although more may still be uncovered.

If the number of known sites is compared with the total soil type for the area (Table 2) it can be seen that the Eutric Cambisols are, in fact, less densely populated than the Vertisols, which are chosen three times more often for settlement formation.

It is surprising that the Vertisols appear to be more popular settlement sites as the physical properties of this soil type are hardly ideal for the primitive agriculture technologies of the Late Neolithic. Vertisols have a high clay content and, when wet, tend to be too sticky and unworkable without modern equipment; when devoid of moisture, they tend to harden and crack (Vimani *et al.* 1982). Perhaps, however, it was the ability of Vertisols to store water for long periods when fully

Table 2. Site density according to soil types in the Toplica region.

Number of sites	Soil type	Soil coverage (%)	Site density
8	Eutric Cambisol	70.0	0.114
4	Vertisol	11.5	0.347
2	Other	18.5	0.108

charged (Vimani *et al.* 1982: 83) that made them suitable for an agricultural cycle that was heavily dependent on rainfall, as the Late Neolithic tradition must have been.

The number of known sites in both regions is still relatively small so these primary observations regarding the settlement-landscape interaction may change significantly with future research in both regions.

Pločnik settlement and its surroundings

The Late Neolithic site of Pločnik is located about 20 km west of Prokuplje, on an elevated terrace on the left bank of the Toplica river (Figure 5). The position of the settlement was almost perfectly chosen. Immediately to the southwest, the broad Toplica valley, which widens several km west of Prokuplje, starts to narrow again as it enters the mountainous region formed by the northwestern slopes of Mount Radan and the southern foothills of the Kopaonik mountains to the northwest. In this location, the settlement of Pločnik overlooked one of the major routes between the Kosovo plain and the Morava valley. Like Belovode, Pločnik had several water sources in the immediate vicinity, the principal being the Toplica, supplemented with waters of the River Backa which most likely represented the eastern edge of the settlement. Unfortunately, the proximity to the Toplica caused significant damage to the settlement remains, as the river creates a wide arc at this point, its left bank undermining archaeological remains. The current vertical height difference between the land surface and the riverbed is, at places, up to 10 m, but in the Neolithic, it was less, perhaps around 5 m (archaeological deposits reach over 4 m in certain areas

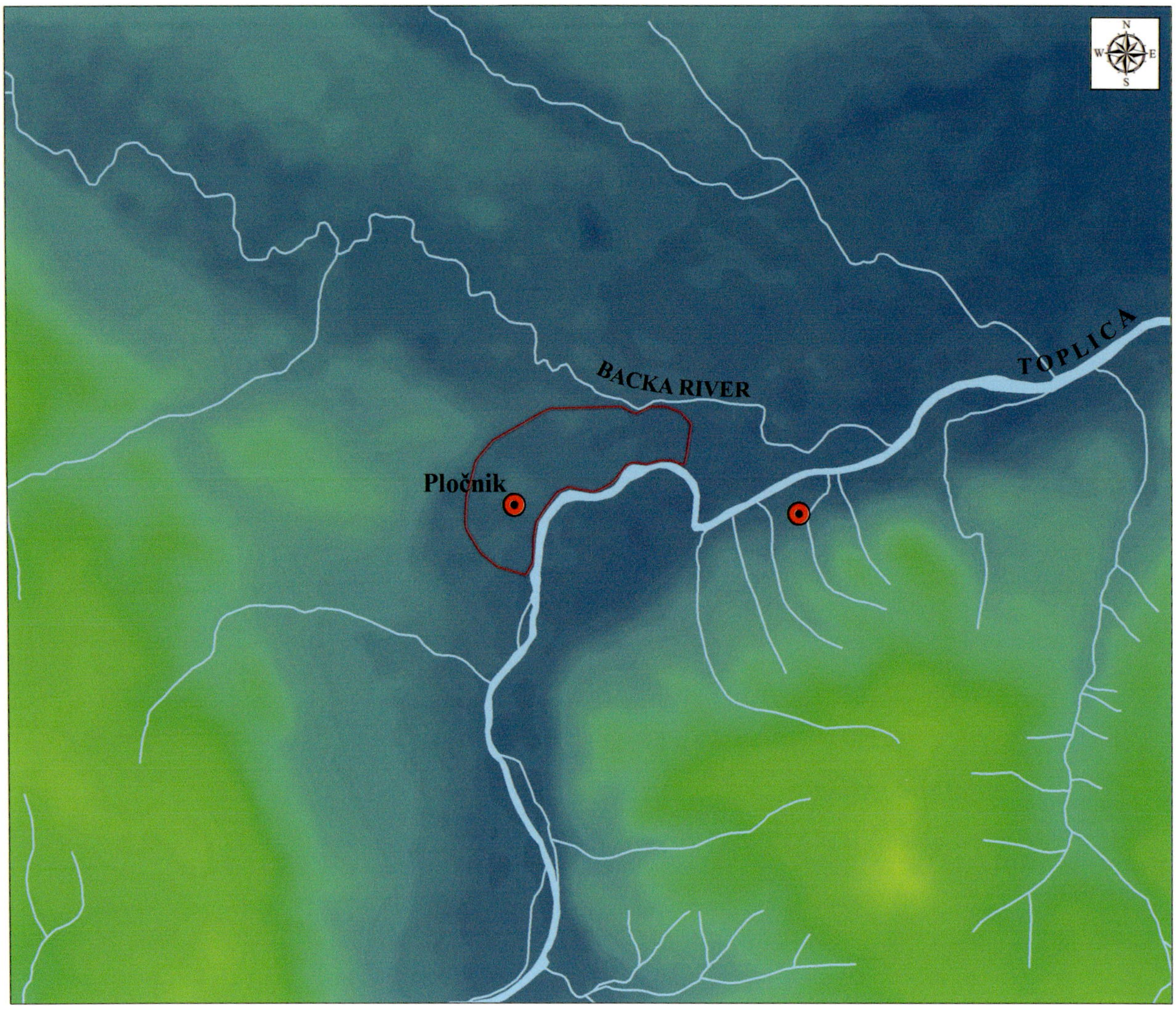

Figure 5. The position of the site of Pločnik.

of the site), and this is still visible in the vertical section of the bank. This difference made the settlement safe from seasonal flooding and enabled long-term habitation. Although the archaeological deposits contain several metres of prehistoric remains, it does not appear that Pločnik was a tell-type settlement, as it was not limited by its surrounding landscape, which would lead to the pronounced vertical accumulation of archaeological remains in certain areas. Rather, it would appear that the site was located on a sloping terrace that rises away from the river towards the west and the nearby hills of Bandera and Ravan. The slope is gradual, being about 10 m per 100 m of distance. The thickest archaeological deposits are found closest to the Toplica, becoming thinner towards the hilly background west of the river, where only about 80 cm of deposit is present (Šljivar and Kuzmanović Cvetković 1998b: 81)

The structures (Chapter 24, Rassmann *et al.* this volume) appear to be clustered with wide, open spaces in between that could represent estates, gardens or backyards of individual households, all indicating a spread-out, horizontal settlement. However, it is not certain whether these groups of structures represent the concurrent households of extended families, consisting of dwellings and commercial/storage objects, or whether each structure is an individual household of several families related by blood and could have been built sequentially in the same area of the settlement, which was the 'property' of a particular family.

The bibliographic reference for this chapter is:

Marić, M. 2021. Pločnik: landscape and settlement perspectives, in Radivojević, M., Roberts, B. W., Marić, M., Kuzmanović Cvetković, J., and Rehren, Th. (eds) *The Rise of Metallurgy in Eurasia*: 265–270. Oxford: Archaeopress.

Chapter 24

Pločnik: geomagnetic prospection data as a proxy for the reconstruction of house numbers, population size and the internal spatial structure

Knut Rassmann, Roman Scholz, Patrick Mertl, Jugoslav Pendić and Aleksandar Jablanović

Introduction

The Vinča culture settlement of Pločnik was surveyed during two campaigns in 2012 and 2013. The site is partly covered by a modern village. Representative data are available for the northern settlement area, which contains *c.* 300 house anomalies. Additional areas were surveyed to the west of this (west of the Prokuplje-Kuršumlija road and the railway line). While the surveys provide valuable results, they were constrained by the area available for investigation. Despite gaps around the southern end of the modern settlement (also caused by the unavailability of certain areas), the data allows a general reconstruction of the ancient settlement boundary which enclosed an area of *c.* 26 ha. Around 60% (16 ha) of this was surveyed, revealing many archaeological and geological features. The distribution of houses, pits, and open areas (i.e. spaces without architectural features) suggest the presence of at least three principal house groups in the northern settlement area.

Due to the limited availability of survey areas for the southern settlement, this can be reconstructed only approximately. This part of the site occupies the lower terrace, closer to the river, where only smaller areas would have been available as potential settlement areas. It should be noted that a significant portion of the site may have been destroyed by the fluctuations of the Toplica river. Given the visible outline from the geomagnetic data, it is likely that a large section of the site is missing in the southeast, having been eroded away by processes that are visible even today, along the banks of the river bordering the site.

When the sum of collected data is examined from all segments of the surveyed site areas, we could argue for the possible existence of four to five principal house structure groups, a valuable starting point for the reconstruction of the spatial layout of the Vinča settlement of Pločnik. The availability of data from a wide geophysical survey such as this allows the development of an optimised framework for future field research campaigns.

Survey methodology

The survey was conducted using two systems: a 16-channel magnetometer (SENSYS MAGNETO®-MX ARCH) and a 5-channel magnetometer (SENSYS MAGNETO®-MX ARCH). For a description of the technical details see Chapter 9, this volume). The current series of surveys at Pločnik started in November 2012 with the 16-channel magnetometer and continued in small areas in Spring 2013 with the 5-channel system (Figure 1). As at Belovode, our initial aim was to gain insights into architectural features and to determine the spatial layout of the settlement. Data from both the micro and macro levels are fundamental to situating the archaeometallurgical research into a broader settlement context.

Large portions of the modern village that partly covers the Vinča site at Pločnik are used as farmland and gardens. These, however, pose even greater survey constraints than at Belovode, and were not suitable for the 16-channel device, which demands an extra vehicle and trailer and is more appropriate for non-urbanised zones. Despite these difficult conditions, the survey nonetheless managed to cover an area of *c.* 26 ha (Figure 2). The addition of the 5-channel system was of significant value in the second campaign as it facilitated survey of confined areas such as orchards and narrow strips of terrain that could not be traversed with the larger 16-channel configuration. This allowed us to examine an area in the centre of the present village, near the bank of the Toplica river. Although this small promontory in the central village appeared to be promising, no evidence was found for building remains (Figure 3). This is in direct contrast with the results for the areas covered by orchards in the southern portion of the northern settlement area. These were surveyed with the single aim of providing enough ground coverage to facilitate at least a partial footprint of any possible large anomalies (e.g. burnt house structures). Survey transects were made at an average of 2–3 m intervals, both to allow for the tree line and based on our awareness that the features

Figure 1. Overview of the areas surveyed, and the instruments used in the 2012 and 2013 campaigns.

detected in previous campaigns could easily span this gap. Our assumptions proved accurate and the 5000 m^2 surveyed in this manner yielded indications of at least seven houses.

The survey revealed more than 300 houses, mainly on the northern periphery of the settlement. Detailed analysis of the more significant anomalies at both the individual and site-wide level revealed more general patterns of spatial distribution. The results of the survey, combined with the implementation of a GIS database of identified structures, allowed the organisation, classification and appraisal of the entire dataset. The geomagnetic data was processed as for Belovode (Chapter 9, this volume).

Results

Western periphery

An area of approximately 10 ha was surveyed in the western periphery of the modern village. Although the data did produce some dipolar indications, the area was not very heavily contaminated by modern objects. A smaller survey area revealed some geological features and weak linear structures that were too small be more precisely classified. We suggest that this area be closely monitored in the future as there are some indications that this may represent an entrance feature (Figure 3A). Ideally, the entire zone between the present-day village of Pločnik and the swell of the hills to the west (an area

Figure 2. Overview of geomagnetic data.

of around 20 ha) would be targeted for geomagnetic survey. Although the western periphery does not seem to cover any clusters of settlement activity (it appears likely that the settlement did not extend to the plateau), the information gleaned from such as survey would potentially allow further refinement of our understanding of the ancient landscape.

Southern area

As with the western periphery, we found no clear house structures in this area however the survey did reveal a cluster of rounded anomalies, each with a diameter of approximately 1–2 m (Figure 3B). These have a maximum geomagnetic response of at least 50 nT. Similar values are known for compact layers of burnt clay, such as those known from house anomalies (Hofmann *et al.* 2007: 58) so it is unlikely that these objects represent ancient pits. Of more importance, perhaps, is their proximity to the modern road and an abandoned construction site; these anomalies could easily be of recent date.

Northern area

The northern settlement area shows a remarkable number of burnt houses (*c.* 290), which vary in orientation and spatial order (Figure 4). Some are grouped in rows and are organised in orientation and spaced at regular distances (Figure 4.1). In other areas, orientation within the house group varies (Figure 4.2–3). The most frequently observed orientation is NW–SE. This is different to that noted at Belovode, where a NE–SW orientation dominates (Chapter 9, this volume).

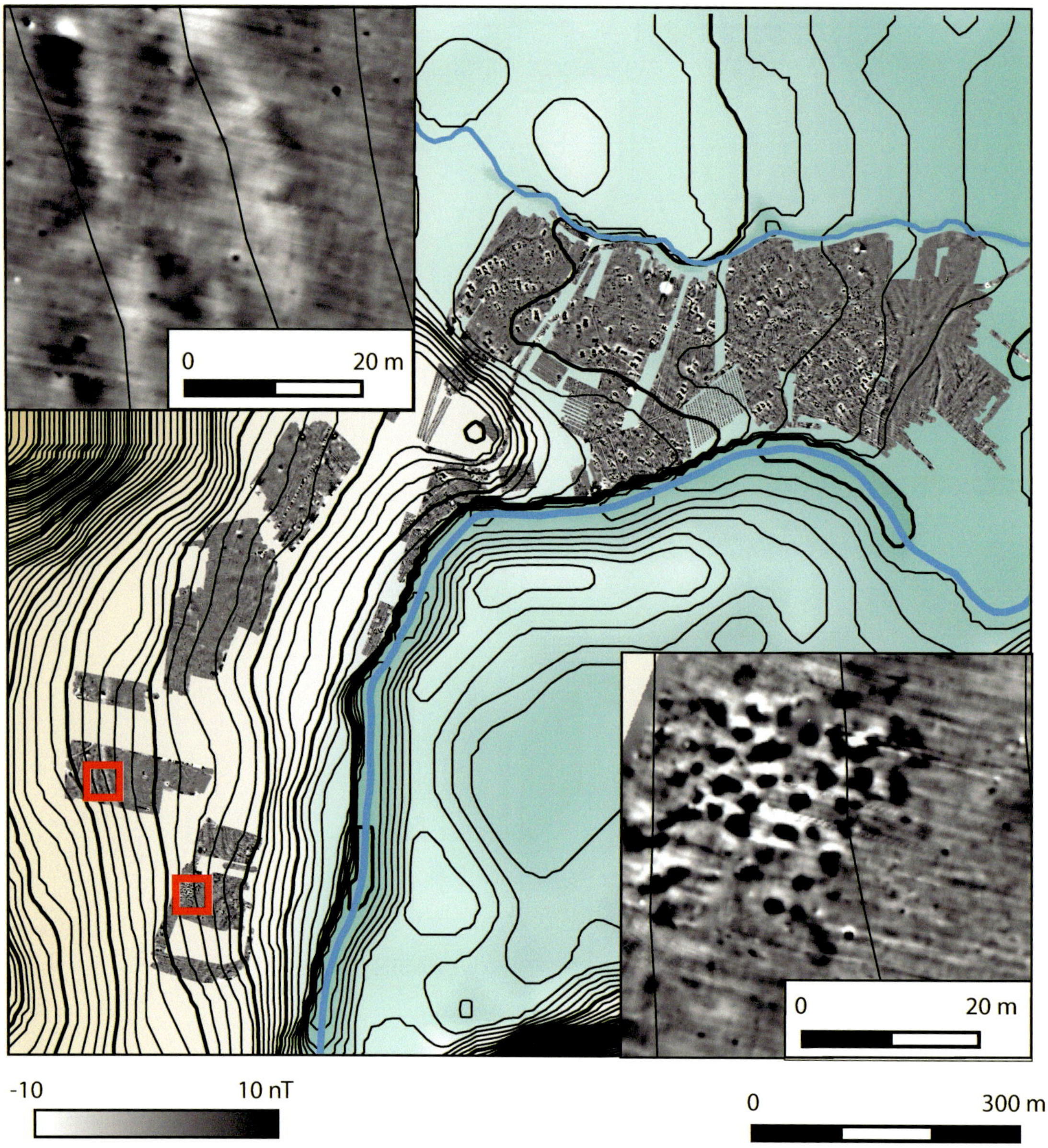

Figure 3. Overview of geomagnetic data on the western periphery of the settlement area.
A) Linear anomalies: possible entrance? B) Cluster of pits.

A small number of houses in the central part of the northern area have a N–S orientation (Figure 5.4–6).

The individual house anomalies vary widely in size and shape (Figure 5). In some cases, there are clear indications of architectural details such as post-holes (Figure 5.7). These indicate the internal space was divided into two rooms, with the entrance placed on the southwest side (Figure 5.1–3).

Discussion

The methodological approach to this study was much the same as that for Belovode; we began the spatial analysis by digitising the house anomalies and found the results to be comparable. The different house patterns seen in Figure 4 illustrate the limitations of digitisation, particularly the difficulties in reconstructing the size and shape of the houses in detail. Despite these limitations, the results provide clear and quantifiable indications of variation in the size and shape of house features. As with Belovode, we compared the size of the houses to the size of the area of daub (generated by a 6 nT contour line which helped with converting objects to polygons). Our analysis considered only those house objects with compact agglomerations of daub larger than 2 m x 2 m.

The mean value of the daub area surface was calculated to be 18.9 m^2 per feature, a figure significantly higher than that at Belovode (13.8 m^2). Less clear, however, is the difference in house sizes. The mean house size

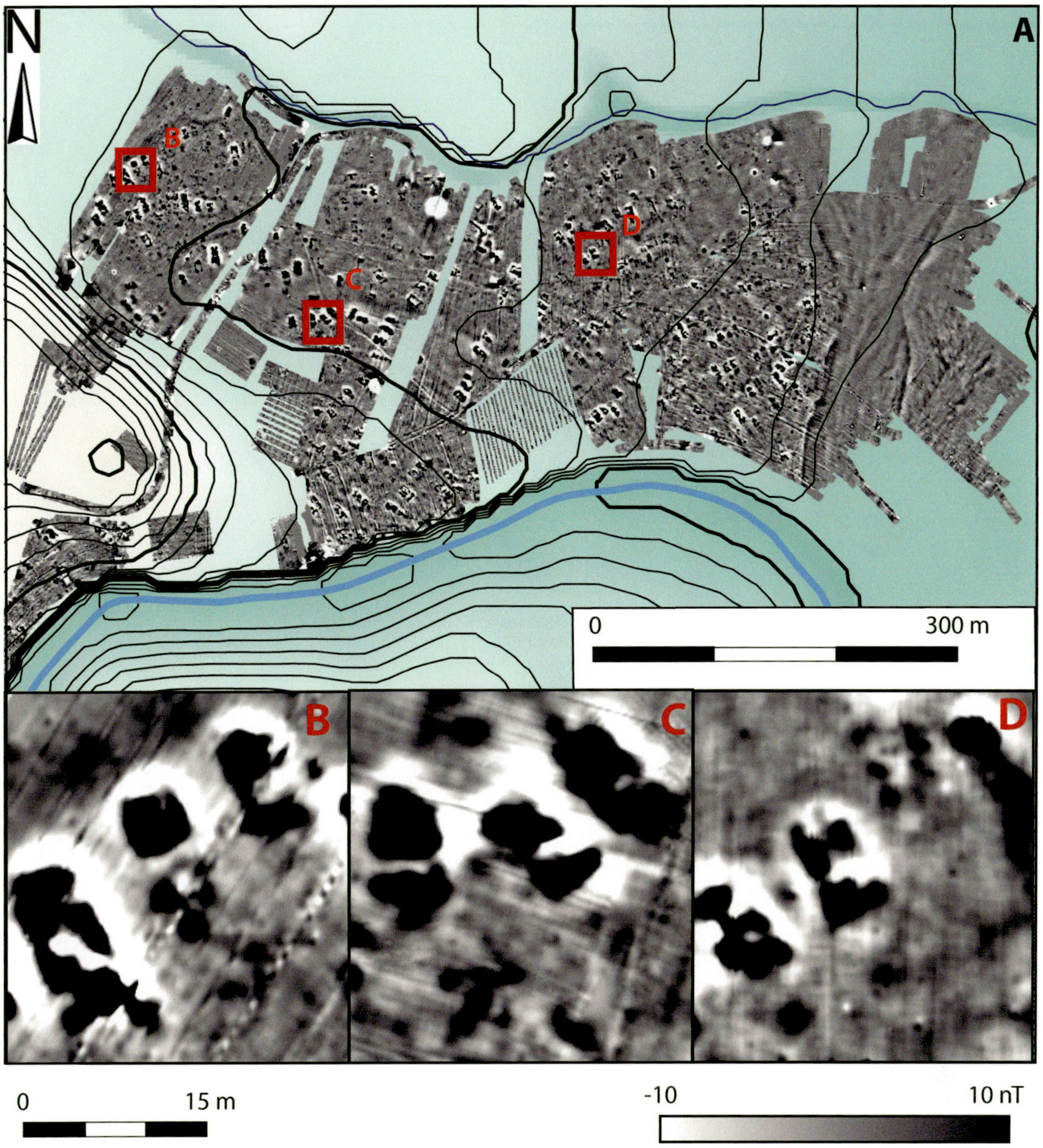

Figure 4. Northern settlement area. Overview of the geomagnetic data. 1) Burnt houses with the same orientation; 2) Burnt houses with varying orientations; 3) Burnt houses with varying orientations and degrees of preservation.

at Pločnik was 56.2 m²; at Belovode it was 52.6 m². It is possible that this comparison is simplistic, as there are multiple peaks in the size of houses and daub areas but there is a strong indication of a trend for larger houses and a greater degree of variation in house size at Pločnik than at Belovode (Figure 6).

As noted, the establishment of the southern and western boundaries of the settlement was challenging. In direct contrast, data from the northern part of the surveyed area provided a clear indication of settlement activities, indicating that these zones were outside the settlement proper. In this direction, the settlement boundary appears to have been either very close to the areas excavated in 2012–2013 or to have been destroyed or otherwise rendered invisible by the railway, train station and the construction of the asphalt road. Nonetheless, the geomagnetic survey data from Pločnik offers a tentative approximation of the total settlement territory and, as was mentioned above, our estimates of its size approach the upper limit at 26 ha (Figure 7). One could feasibly argue that the settlement area was actually somewhat smaller than this reconstruction. There are clear indications that earlier estimations of around 100 ha were unrealistic (Šljivar and Kuzmanović Cvetković 1998a).

Only data from the northern settlement area were considered reliable for a more detailed spatial analysis. As with Belovode, we used Kernel Density Estimation (KDE) of the house centroids and the 6 nT-polygons. Both results illustrate the same spatial trend: there

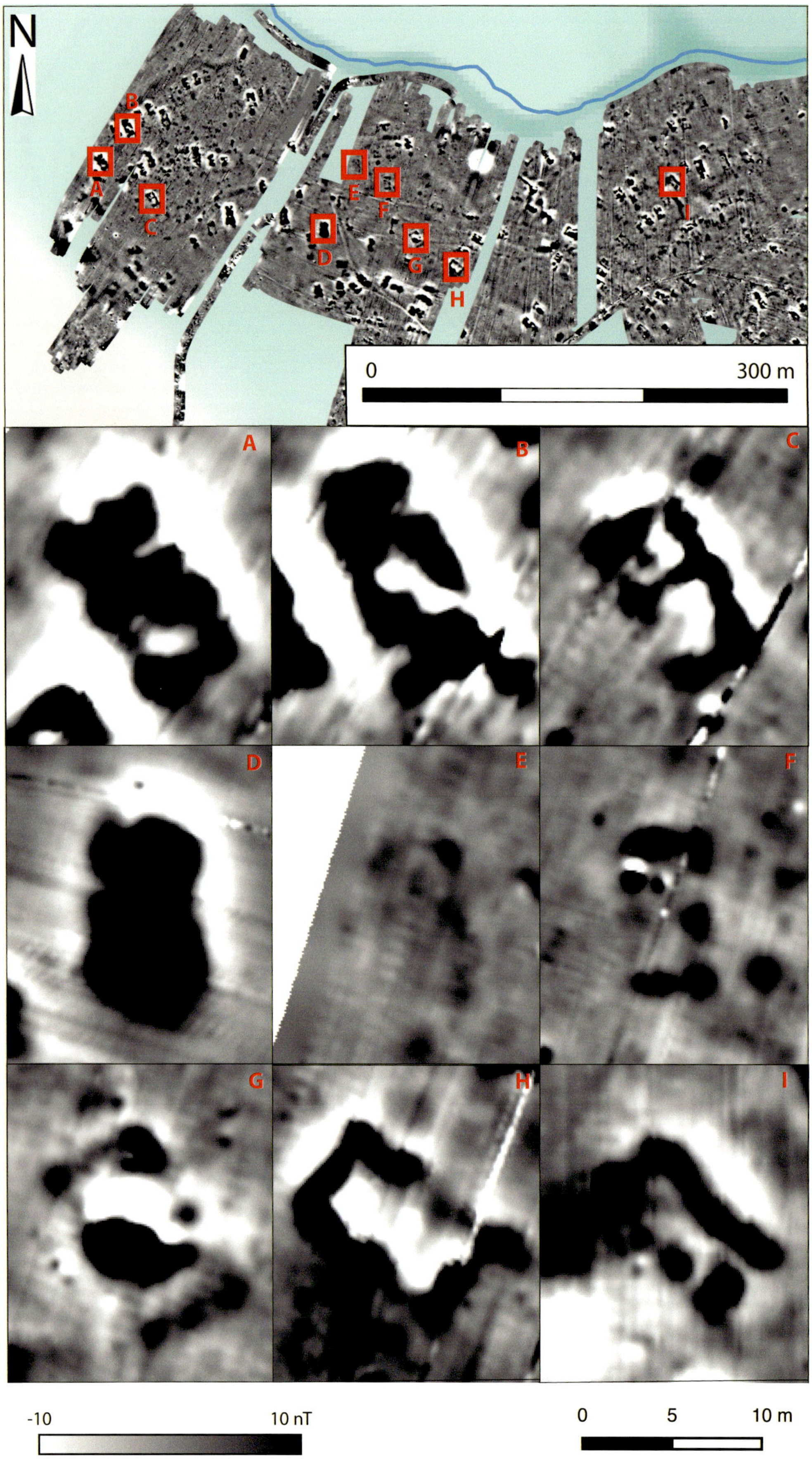

Figure 5. Northern settlement area. 1–9: Houses of different sizes, with diverse orientations, differences in their destruction by fire and stage of preservation.

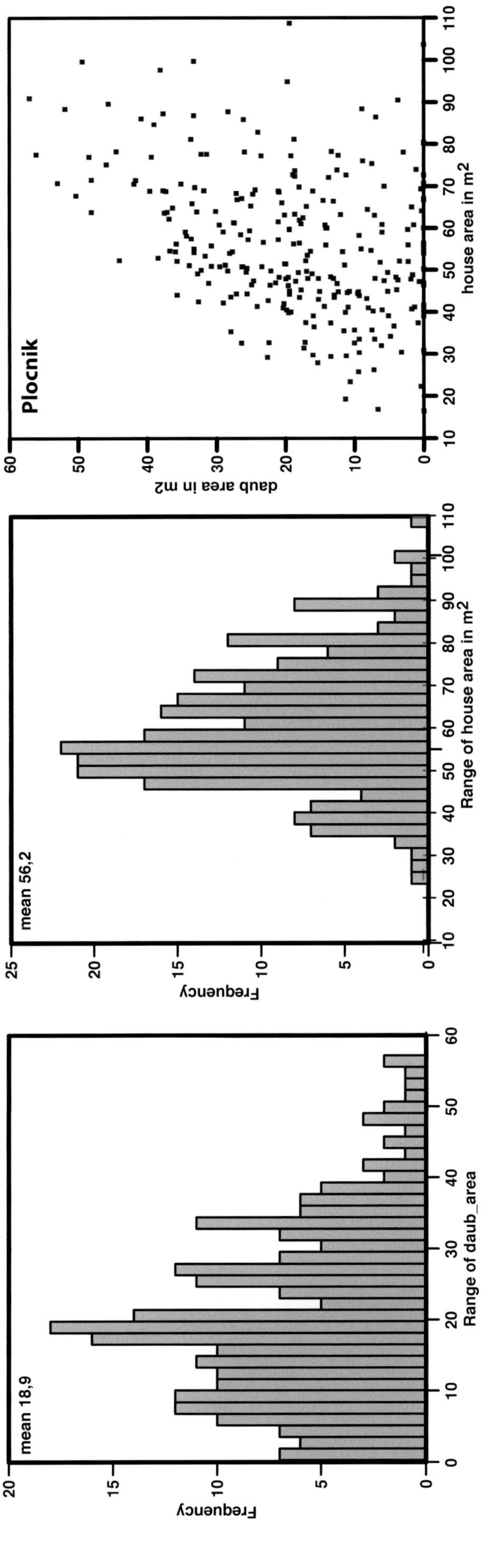

Figure 6. Analysis of the size of the daub area (6nT line) in relation to the size of reconstructed houses. A) Histogram of the size of daub area; B) Histogram of the size of the reconstructed houses; C) Scatter plot of the size of houses and the underlying daub area.

are areas with a higher density of houses or areas with daub that were surrounded by zones without any evidence for dwellings. These analyses allowed us to reconstruct three principal house groups within which were smaller houses or building clusters that we termed 'subsidiary house groups'. The system of principal house groups and the minor units of subsidiary house groups were the main spatial elements through which the settlement space was structured (Figure 8).

Based on the entire results of the Pločnik survey, the existence of four to five principal house clusters can be proposed, if the area to the far south is included (although this area was unavailable for survey, it can still be cautiously considered part of the Neolithic settlement). The number of subsidiary house groups is less easily established however the upper limit is presented in Figure 9, based on the data currently available. Three of the reconstructed principal house clusters vary in size from 3–4 ha, closely in line with Belovode (4–5 ha), with a maximum of seven subsidiary groups within the central and eastern principal clusters (looking only at the data from northern areas); the western group is much smaller, with approximately three subsidiary house groups.

Spatial layout of the settlement

As for Belovode, the KDE calculation marks only one phase for Pločnik, when the settlement reached its largest extent. Further fieldwork is essential for this model to be better defined. We assume that the settlement varied in size throughout the different stages of the Vinča culture. It is likely that for most of its existence it was much smaller than the hypothetical reconstruction illustrated in Figure 9.

While the problematic geomagnetic results for the southern areas have been described, there is a possibility that the settlement was limited to the lower terraces; our data indicate no settlement remains on higher ground (Figure 3). The lower area provides enough space for a maximum of only one to two house groups. The specific terrain and landscape situation of Pločnik, in combination with geomagnetic data, enable us to reconstruct the upper limit of the zone covered by the settlement as well as the maximum number of house groups. However, due to the presence of the modern village it is likely that some more obvious site/settlement boundaries remain unavailable

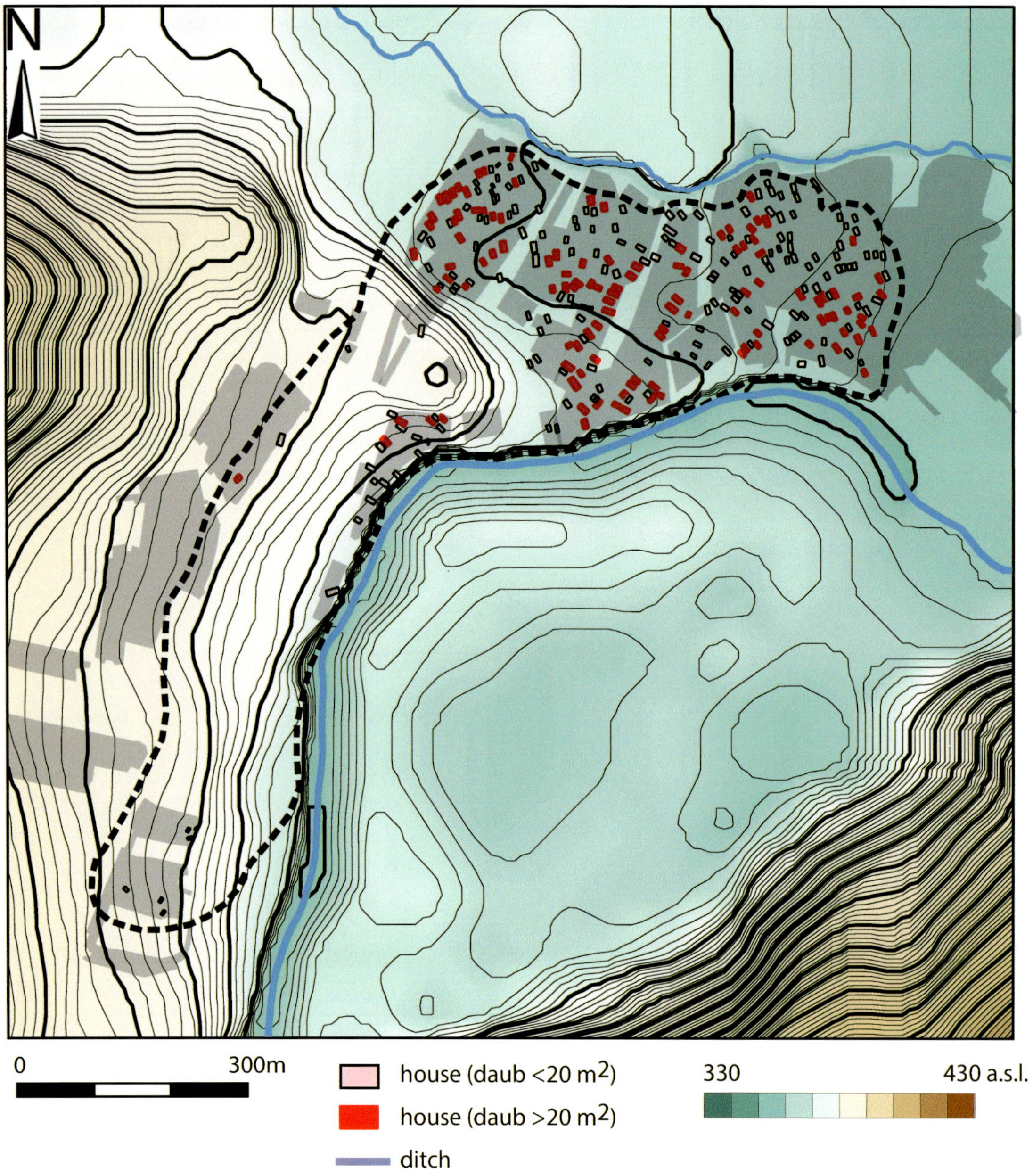

Figure 7. Interpretation of the geomagnetic data.

for appraisal, at least by geomagnetic survey. Further archaeological investigations could profit from using different geophysical methods more suited to narrow or non-traversable surfaces possibly contaminated with the metal junk and other debris which characterise the fenced-in yards in the immediate vicinity of the modern houses.

The scale of the settlement seems to have been similar to that of Belovode. If we begin from our upper estimate for the size of whole settlement of *c.* 26 ha, we might estimate a maximum of four to five house groups, each of *c.* 3–4 ha, comprising a total of *c.* 16 ha, together with the structure-free spaces inside the settlement area (Figure 9). If one calculates that ten houses can accommodate 50 people, then the four to five house groups, each with a minimum of 25 houses, can be expected to have housed a population of *c.* 1250 residents.

Conclusions and future directions

The reconstruction of the spatial distribution and the upper limits of the population size at the site of Pločnik is only an approximation. The weakest point is currently the static character of the data, which is without chronological differentiation. Despite these obvious deficiencies, the conclusions drawn here are helpful insofar as they sharpen our perspective on the question of population size. The results of the survey are also valuable for the optimisation of the design of future fieldwork, which should prioritise analysis of the

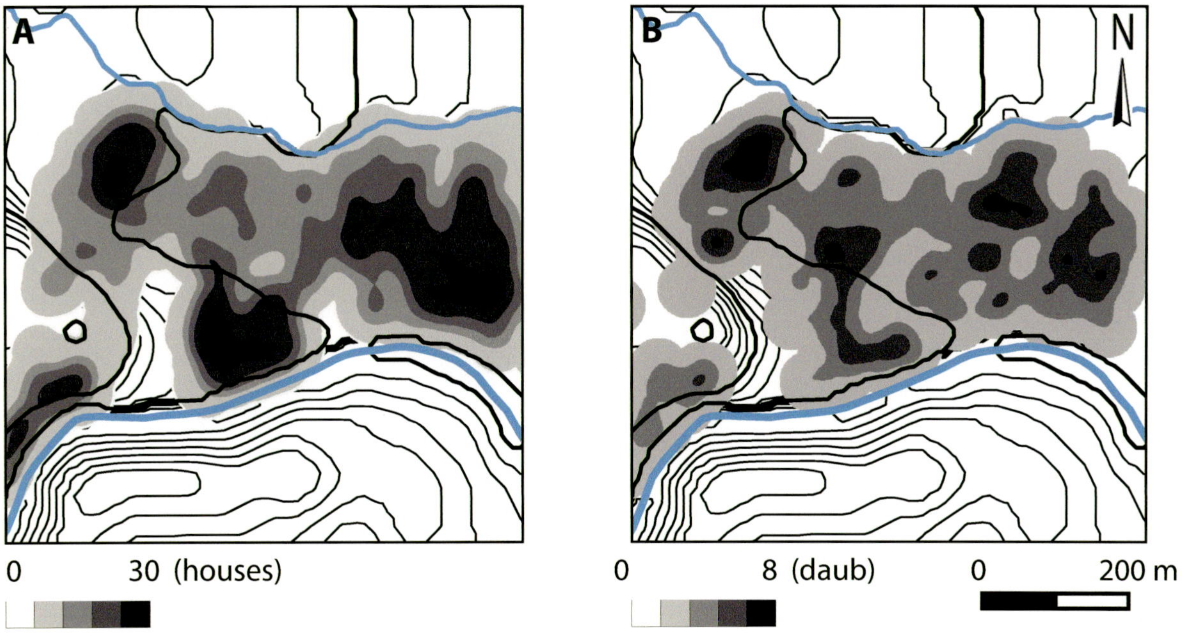

Figure 8. Visualisation of the house cluster by Kernel Density Estimation (biweighted interpolation, radius 40 m). A) Interpolation based on the centroids of the reconstructed houses; B) Interpolation based on the centroids of the daub areas.

chronological differences between the principal house groups and their interrelations. A promising starting point would be more detailed documentation of the long settlement section along the lower edge of the river terrace by 3-D photogrammetry, and a systematic sampling strategy including archaeobotanics and soil chemistry analysis. An additional geomagnetic survey to include areas in the upper north, beyond the Backa river that borders the site, would be helpful in determining whether the settlement also spread in that direction (Figure 2).

A systematic augering program would potentially interpolate this information in the settlement. Auger data should aim to sample all principal house groups. These results would provide the thickness of the settlement layer. The river has sharply eroded large areas of the settlement, opening a window onto the site's stratigraphy for a distance of over 300 m, a serendipitous opportunity for research and further sampling. Samples for geochemical and archaeobotanical analyses should be extracted from both auger drillings and the river profile. The chemical data (especially phosphorus and strontium) may deliver indications of human impact along different chronological horizons and in principal house groups. Areas with precise geomagnetic data could be easily sampled by focussing on a specific type of anomaly and taking into account variation in the collected data in terms of anomaly size, orientation, compactness and shape, etc. In addition to houses, other anomalies like pits and ditches should be sampled along with the specific cluster of anomalies in the southern periphery (Figure 3) in order to verify its source. The open question regarding the exact western and southern border of the Vinča settlement could also be answered with the auger and would complement further geophysical research. The reconstruction of the changes at the site of Pločnik should be set in relation to changes in the landscape surrounding the site, and future fieldwork would ideally also take into account other Vinča sites in the surrounding area.

The bibliographic reference for this chapter is:

Rassmann, K., Scholz, R., Mertl, P., Pendić, J. and Jablanović, A. 2021. Pločnik: geomagnetic prospection data as a proxy for the reconstruction of house numbers, population size and the internal spatial structure, in Radivojević, M., Roberts, B. W., Marić, M., Kuzmanović Cvetković, J., and Rehren, Th. (eds) *The Rise of Metallurgy in Eurasia*: 271–280. Oxford: Archaeopress.

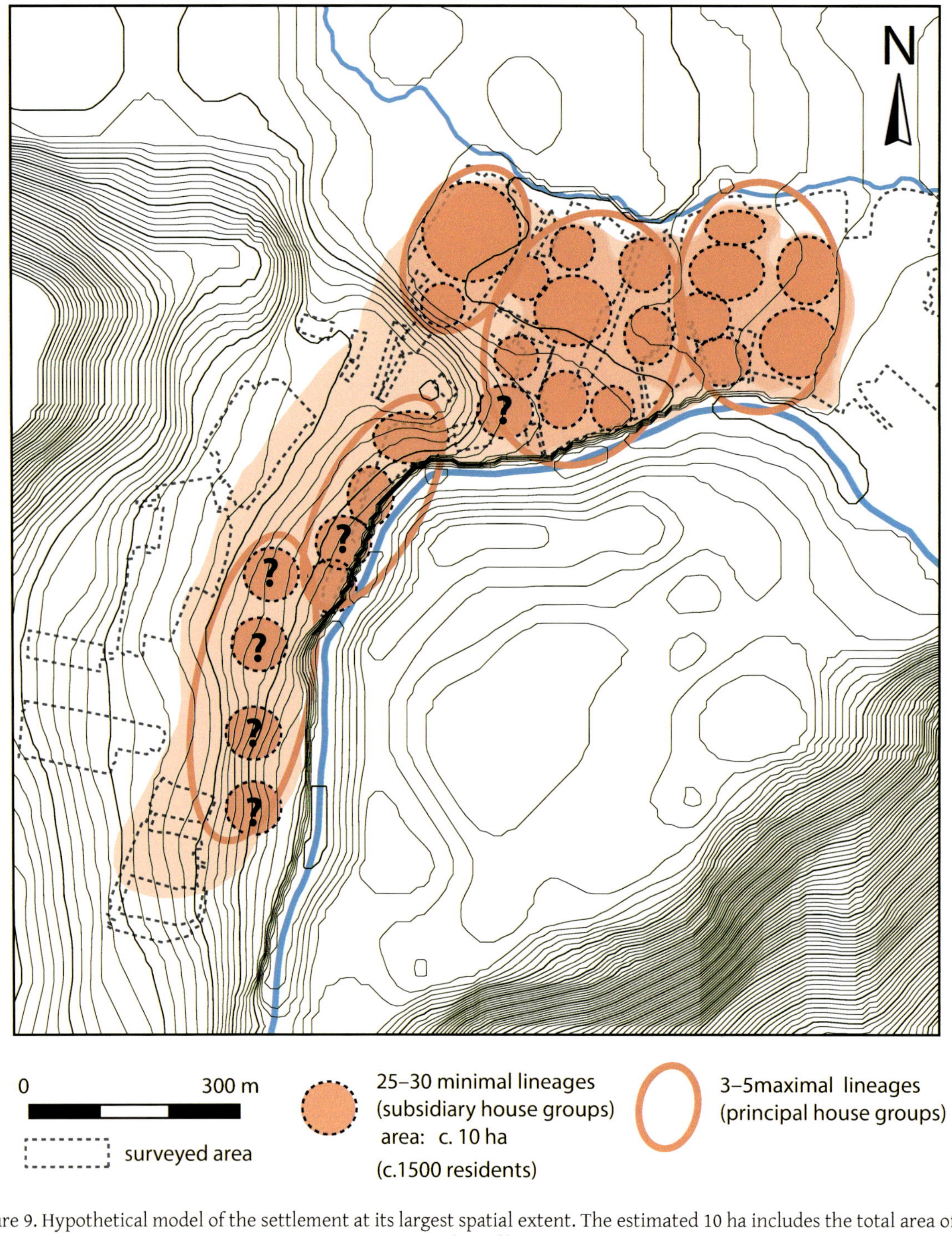

Figure 9. Hypothetical model of the settlement at its largest spatial extent. The estimated 10 ha includes the total area of the minimum number of house groups.

Chapter 25

Pločnik: excavation results

Miroslav Marić, Jugoslav Pendić, Benjamin W. Roberts and Miljana Radivojević

The results of the geophysical survey conducted at the site of Pločnik in 2012 and 2013 indicate that the total area of the settlement can be estimated at 35 ha (Figure 1, see Chapter 38 this volume), although it may have been larger during the Neolithic-Chalcolithic periods. The meandering of the Toplica river may have destroyed a significant portion of the site through erosion, a process that is visible even today after the seasonal swelling of the river during the spring and late autumn. The archaeological excavations were undertaken in the southwest part of the site since copper implements were previously discovered in this area of the settlement (e.g. Grbić 1929; Šljivar and Kuzmanović 1997a; Šljivar *et al.* 2006; Stalio 1964; Stalio 1973). The single trench, Trench 24, was placed between two previously excavated trenches, that is Trenches 20 (to the north) and 21 (to the south), which had yielded metallurgical finds, including both the earliest copper and tin bronze production, anywhere (Radivojević *et al.* 2013; Radivojević and Kuzmanović 2014). It was also deliberately placed over an area which, in an earlier survey, had registered geophysical anomalies indicating

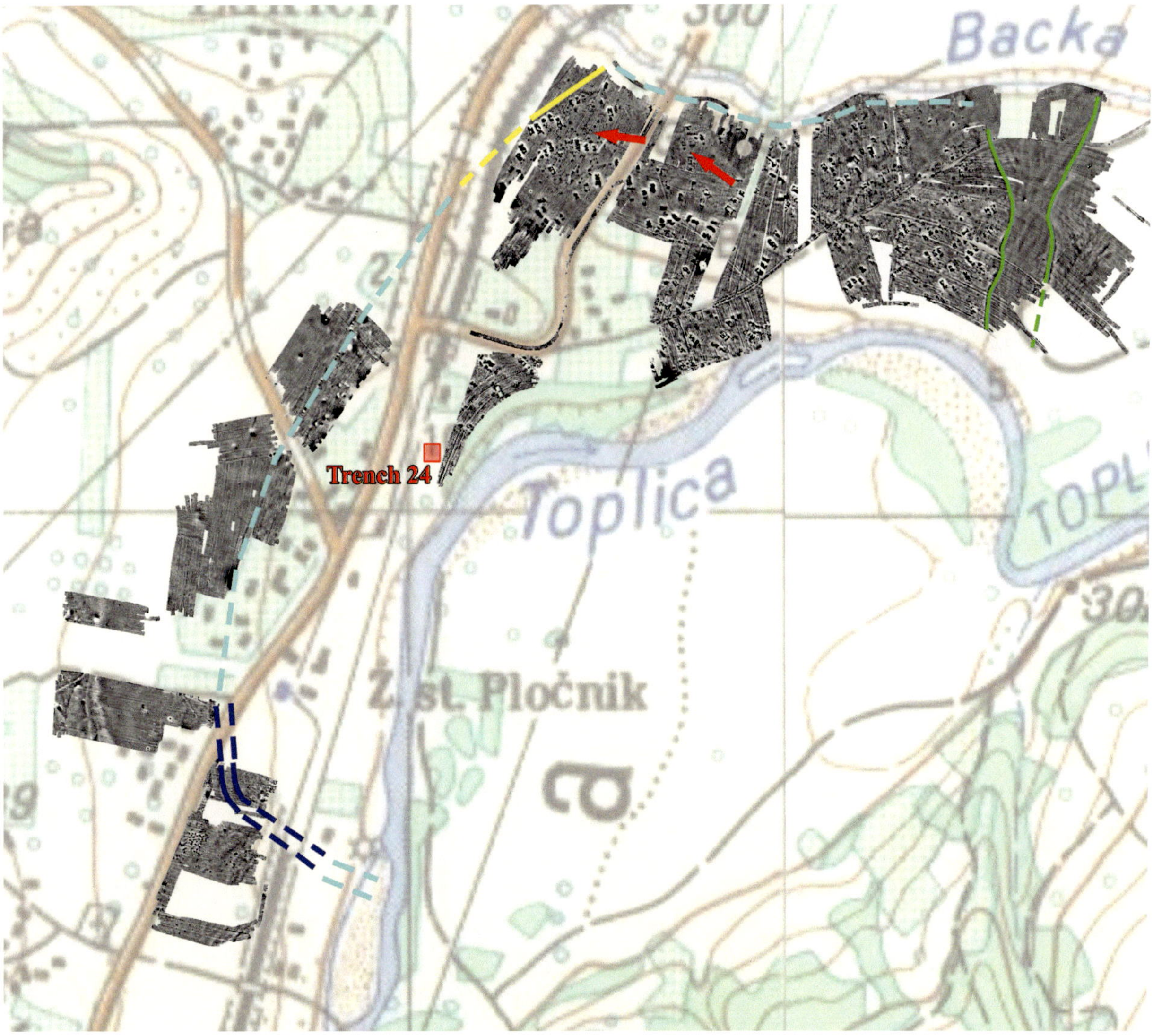

Figure 1. The geophysical survey of Pločnik archaeological site

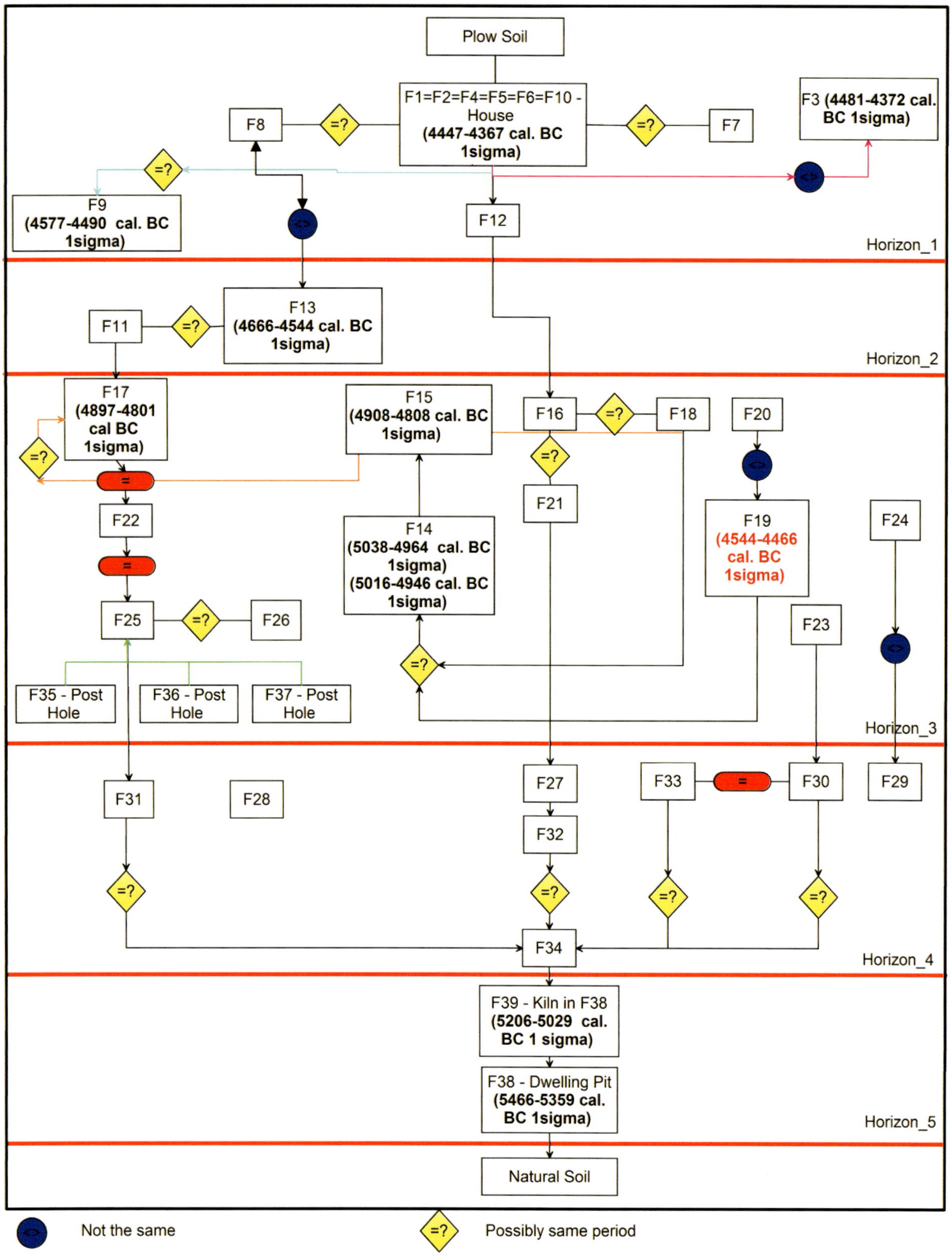

Figure 2. Relative stratigraphy of Trench 24.

archaeological features (currently unpublished). As at Belovode, the initial size of the trench was planned to be 5 x 5 m but was quickly extended towards the south in order to encompass the full extent of Feature 1, a burnt wattle and daub structure found in the final occupational horizon at the site. In the 2013 excavation season, the trench was shortened to its originally planned extent due to financial and time constraints.

The two archaeological field seasons in Trench 24 showed that there is approximately 3.7 m of stratified archaeology in this area of Pločnik, which is not unusual for the site when compared to excavated trenches from earlier campaigns (Šljivar 1996: 85). It appears that the thickness of archaeological deposits increases with proximity to the Toplica river, which could imply that the meandering of the river has destroyed the core of the settlement over the millennia. It can be assumed that the settlement gradually expanded from its core area (as is also very probable at Belovode) towards the hilly terrain in the background. This core area would have been to the east of Trench 24 and has now been destroyed by the Toplica.

The same single context excavation and recording methodology applied at Belovode was used at Pločnik (see Chapter 724, this volume). A sequence of five phases (Figure 2) was established based upon the relative stratigraphy of archaeological features found in Trench 24. As at Belovode, these horizons were labelled 1 to 5, with the first being the latest and the last being the earliest settlement occupation horizon. However, beyond the relative stratigraphy, these divisions are dated absolutely by radiocarbon dating (see Chapter 37, this volume). The relative horizons do not reflect a specific chronological phasing based upon finds, even though the finds corroborate the proposed divisions (see Chapter 37). In this chapter we will describe the characteristics of features discovered and, where applicable, try to elaborate upon their possible function. The features will be presented following the relative chronology established by the excavations with the more important features outlined in detail. Where possible, 3D models of features are also provided, with interactive models published on the project website and in the Appendices (see Chapter 7).

Structural features in Trench 24

The excavations during 2012 and 2013 in Trench 24 revealed and recorded 39 archaeological features with the majority being fully excavated. These features can be divided into several different types based upon their form and function, including wattle and daub rectangular structures; kilns; finds concentrations; pits; and dwelling dugouts. The following catalogue of excavated features begins with the most recent settlement horizon.

Horizon 1

Features 1, 2, 4, 5, 6 and 10 form the remains of a rectangular wattle and daub structure which was discovered at the bottom of spit 6 and extended down to spit 10. The excavated dimensions of the structure are 6.3 x 3.5 m which is not uncommon for Vinča structures, albeit that their dimensions tend to be larger towards the end of the period (cf. Bogdanović 1988: 47–48, Tables 5.3 and 5.4; Marić 2011: 72, Tables 3 and 4). The larger part of the wall debris was not found, and it seems to have been cleared out in the past or destroyed by later ploughing. In spits 1–5 there were extremely low quantities of architectural debris or archaeological finds, so it remains unclear how the wall debris disappeared. The remaining daub is orange baked and well made, although damaged by later activities in certain places (Figure 3). The occurrence of larger stone blocks on the sides of the daub and in the central area indicates the possible existence of pedestals for load-bearing beams that comprised the main construction of the walls (Figure 4). This building technique seems indigenous to the site, as at least one other feature was discovered in previous excavations with the same evidence for construction, although unfortunately it remains unpublished. The finds discovered in the debris were rather scarce and consist of a small number of orange burnt vessel fragments, one polished stone axe which was found in the central area, a small handful of fragments of metal artefacts, and a metal droplet, the latter of which indicates potential association of this large rectangular structure with metal extraction activities (see Chapter 26, this volume).

Immediately to the northwest of the daub, a concentration of stones and pottery fragments was discovered on the same level as the debris and may represent the original house inventory that was removed from the daub following the destruction of the structure. It is interesting that no heating or cooking installations were identified in the debris.

Upon removing the floor level of the structure, a substructure made of split timbers arranged in parallel rows following the longer axis of the daub outline was discovered (Figures 5 and 6). An identical floor construction method was discovered in Grivac, in the Late Vinča culture horizon (Bogdanović 2004: 180, Figure 8.15, 197, Figures 8.48–8.49). The individual timbers had diameters of between 7 and 11 cm and were placed flat side down, with the curved side forming the actual substructure of the floor upon which the clay was placed and compacted to produce the floor level. At the widest preserved part of the structure, 34 or 35 parallel rows of timber could be counted. The floor construction is somewhat different to that of Feature 3 at Belovode where laid wooden planks were flat on both sides. Unfortunately, the wood used for this

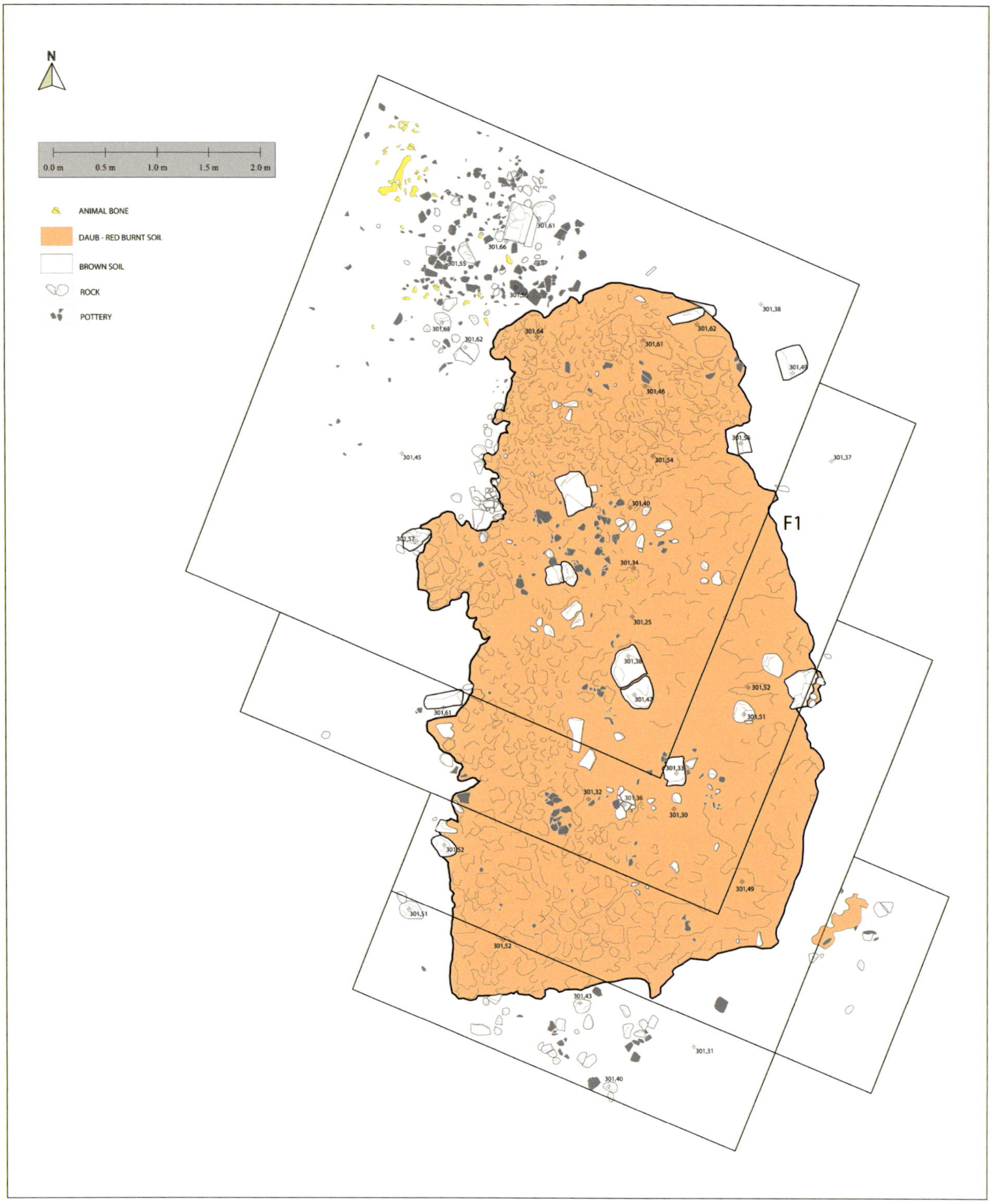

Figure 3. Scale drawing of Feature 1. Regularly spaced larger stones indicate the position of load bearing posts.

substructure could not be identified based solely on the remaining imprint.

Feature 3 was detected next to the west profile of the trench, close to corner A (Figure 7). It is a small feature consisting of several large stones mixed with the remains of a burnt structure that could have been a kiln, or possibly a furnace, due to the find of a copper metal ring/band (see Chapter 26, this volume and Figure 7) in its vicinity. The shape of the feature is unusual: like a portion of a slightly elevated wall of an almost rectangular structure, or at least an edge of such a structure. Its remit is unknown as it remains incompletely excavated, however it has a striking resemblance to the excavated rectangular firing structures in Trenches 20 and 21 (see Figure 8, Chapter 6, this volume) (Radivojević *et al.* 2013: 1033, Figure 2; Šljivar and Kuzmanović Cvetković 2009a: 61), which

Figure 4. Section drawing of stone pedestal used for load bearing wall-posts

also had metal artefacts and casting debris associated with them. What separates Feature 3 from the previous finds is the fact that it was possibly located just outside the rectangular wattle and daub (dwelling?) structure (see Figure 5). Yet, it would not be unusual to have the fireplace for metallurgical activities in an economic area, as we have seen at Belovode (Chapter 11). Fragments of metal artefacts found inside the rectangular wattle

Figure 5. Orthogonal image of Features 1 and 7.

and daub structure in Trench 24 further enforces the argument of the potential association of this structure with metallurgical activities (i.e. casting, working).

Feature 7 is a sub-oval concentration of finds discovered to the northwest of Feature 1 in spit 8 (Figure 3). These included numerous fragments of worked and unworked stone, pottery fragments and animal bones. It is not completely clear whether this concentration represents the context of Feature 1 as there are almost no fragments of wall debris amongst the finds, and only some of the discovered pottery fragments revealed traces of intensive burning (Figure 4) which is usually connected with the destruction of daub structures.

Feature 8 is a smaller concentration of pottery and stone fragments extending from the eastern edge of Feature 1 towards the eastern profile of Trench 24 (Figure 5, upper left corner). The exact shape of the concentration is uncertain as it extends under the profile of the trench. It can be assumed that the finds do not originate from Feature 1 as some of the fragments extend under the substructure of the floor (Feature 2). It is possible that these represent material deposited between structures in order to harden the walking surfaces. Similar cases are known at other sites of the period.

Feature 9 was located to the southwest of Feature 1 and is a concentration of large ground stone axes (Antonović 2003: 53; see Chapter 31). These were made from white stone and mixed with several grinding stones (Figure 8); they possibly represent the remains of a workshop for the production of these implements that was located close to the burnt wattle and daub structure (Feature 1).

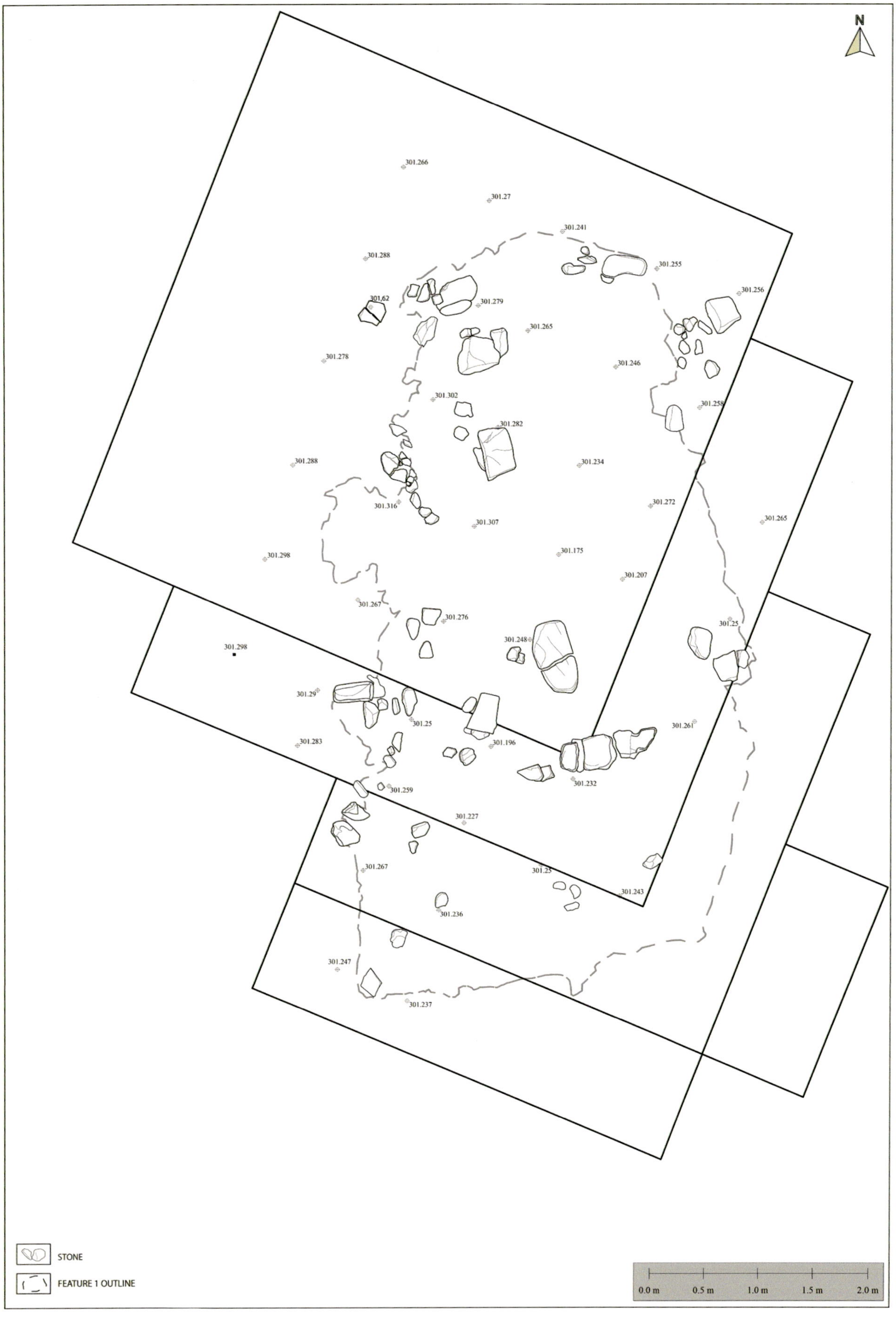

Figure 6. Orthogonal view of floor substructure (Feature 2), Feature 3 and Feature 8.

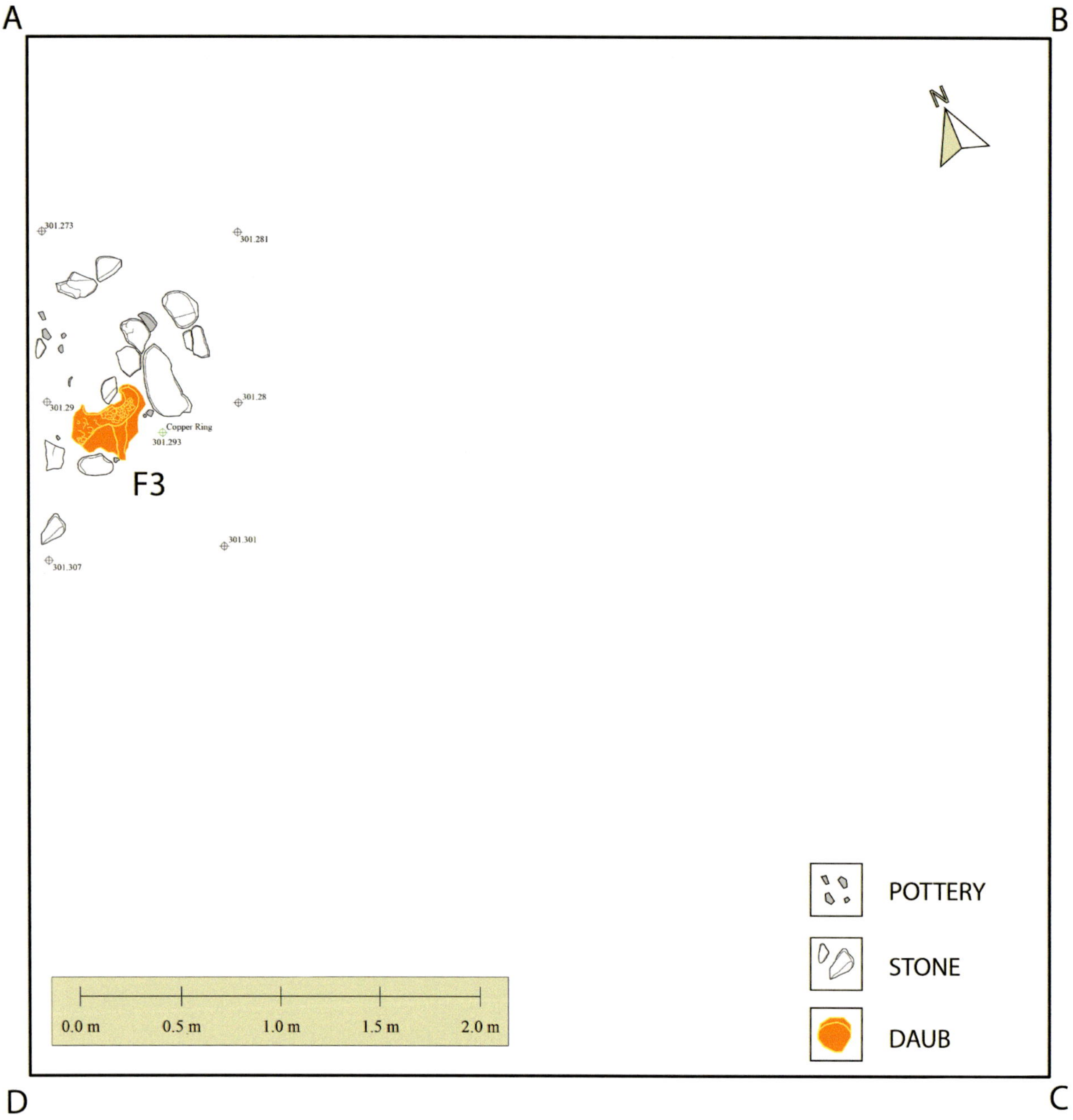

Figure 7. Feature 3 and the position of the copper ring

These white stone axes are a common find in the Late Vinča period on central Serbian sites (Antonović 1997: 34). It is notable that these white stone axes are the only types of non-metal artefacts found associated with massive copper implements at Pločnik (see Chapter 6) (Šljivar 1999; Šljivar *et al.* 2006: 261–265).

Horizon 2

Feature 11 represents the remains of a kiln found close to corner D of Trench 24 in spit 11. The kiln was only partially excavated as a significant portion remained under the west profile of the trench, but it was a typical horseshoe shaped kiln, as found on almost all Vinča culture sites. No dome or wall remains were detected and four kiln floors were excavated, all of which were damaged in the southeast section. Each kiln floor had been constructed upon foundations of a mixture of pottery fragments and stones (Figure 9). It is interesting to note that the earliest foundation level of the kiln consisted exclusively of large stone fragments (Figure 10). The kiln does not appear to be within a structure made of wattle and daub, and the number of floor renovations indicates that it was in use for a long period.

Horizon 3

Features 14 and 15 were detected in spit 14 and represent the *in situ* remains of a kiln as well as the

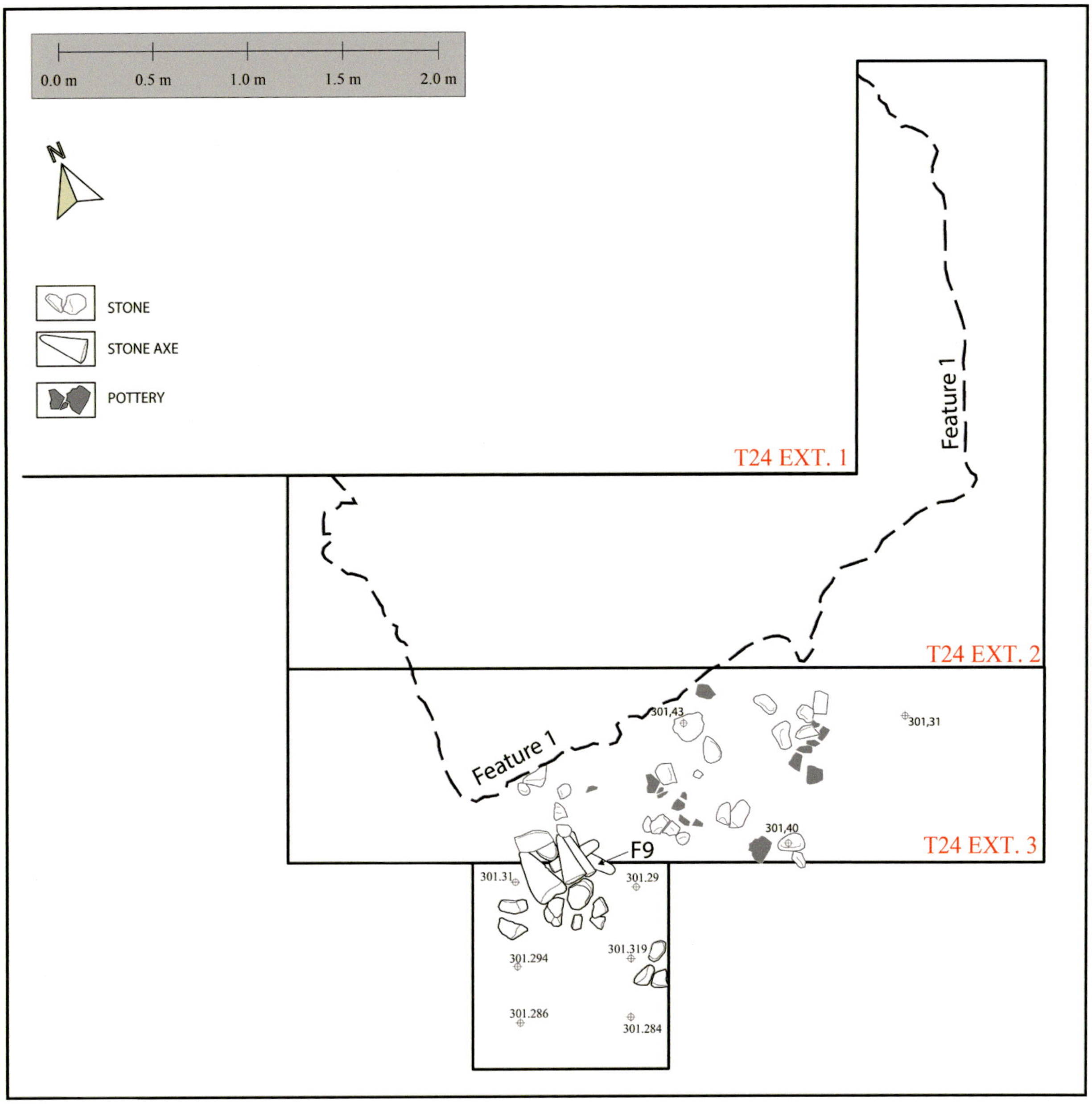

Figure 8. Feature 9 and its relative position to Feature 1.

surrounding ash (Figure 11). Feature 14 is the kiln and is located next to the eastern profile of the trench. As with Feature 11, only the remains of two kiln floors were found, whilst the dome and the walls were dismantled prior to its abandonment. A large pile of white ash with charcoal was found around it, most likely the by-product of kiln operation. It is interesting to note that the second kiln floor had a foundation only of pebbles (Figure 12) whilst the first kiln floor had pebbles tightly packed with pottery fragments in a basin constructed of brown, compact soil (Figures 13 and 14). During the renovation, the kiln also seems to have been enlarged. It appears to have been free standing, without a structure around it, for the entire duration of its use. Both floors were somewhat tilted towards the south, which would indicate that the firebox opening was on that side.

Features 16 and 21 comprise a large concentration of finds consisting predominantly of pottery fragments. The concentration was detected in spit 14 (Feature 16) and extended into spit 15 (Feature 21). It has a roughly elliptical form, and could have been a part of a pit, although the edges were not detectable. The dimensions measure 1.7 x 1.5 m. Besides pottery, several broken grindstones were found in the concentration (Figure 15).

Features 17, 22, 25 were found in spit 14 in the southwest part of Trench 24 and comprise an approximately triangular area consisting of lumps and patches of yellow and red soil (Figure 15 and 16). These features most likely represent the daub of a destroyed structure, the majority of which is under the southwest

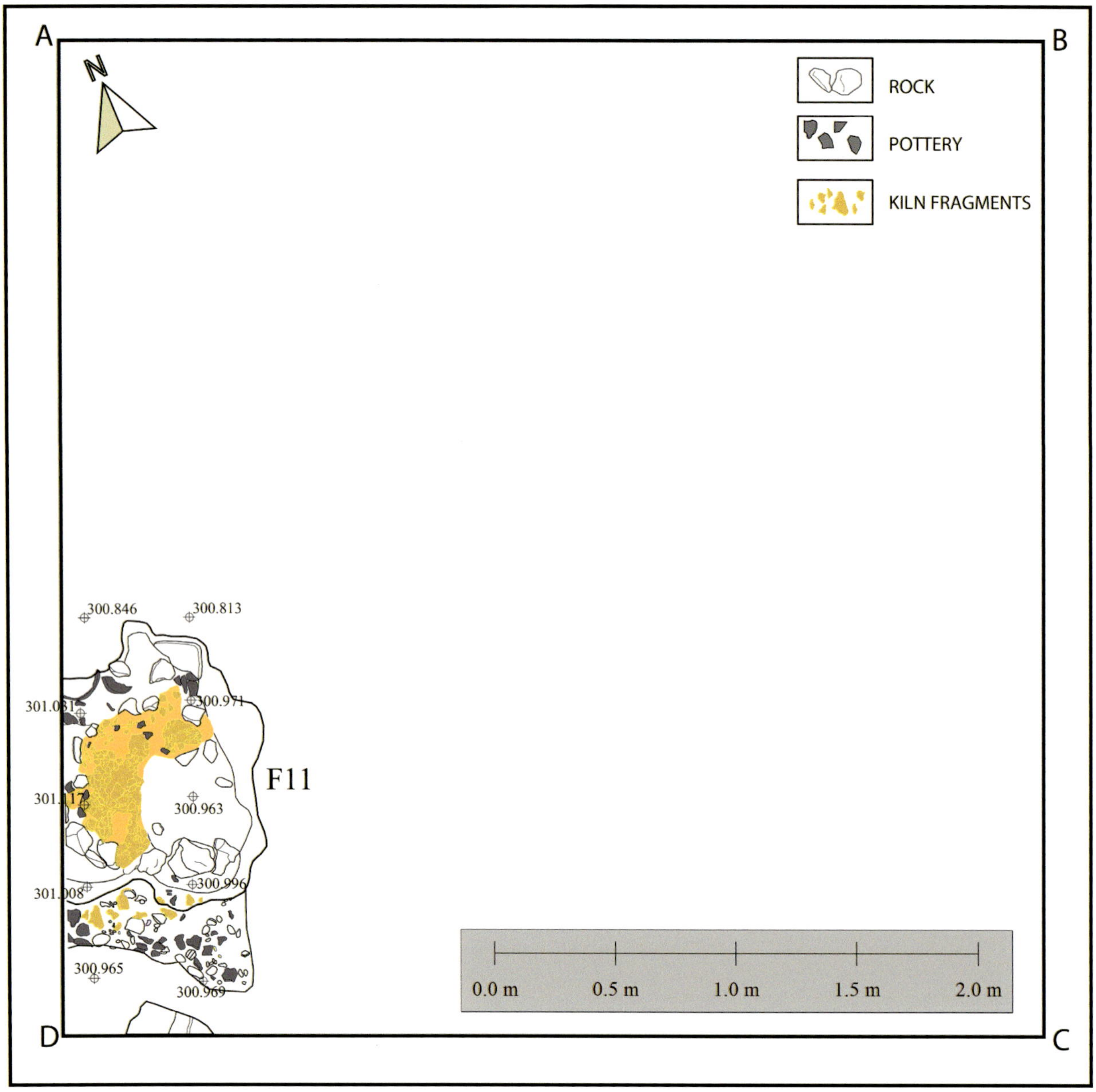

Figure 9. Feature 11 upon discovery.

profile of the trench. Its general orientation is similar to that of wattle and daub structures found in this area of the site.

Features 35, 36, and 37 are three post holes which were found underneath Features 17, 22 and 25 (Figure 16). An additional post hole was also visible in the west profile of the trench but was not excavated as a separate feature as the majority was under the trench profile. The post holes were constructed by digging large individual holes, which were round in the case of Features 35 and 36 (diameter 70–74 cm), and elliptical for Feature 36 (diameter 85 cm). A post was then placed vertically, and the soil backfilled and compacted to keep it in place. The presence of an entire post hole detected in the west profile (A-D profile), we can say with certainty that the holes were dug to a depth of about 1.15 m. This technique is known from other Vinča period sites in the region (Bogdanović 1988: 42, Figure 5.6; Bogdanović 2004: 169, Figure 8.4, 177, Figure 8.12, 178, Figure 8.13, 179, Figure 8.14), and appears parallel to the construction of foundation trenches.

Feature 19 is a sub oval concentration of finds (Figure 15) which was detected in spit 14. It is close to 1.2 m in diameter and is located close to corner C. The content of the feature comprised largely broken grind stones and unworked stones, mixed with pottery fragments. The feature was most likely the infill of a pit whose outline could not be identified with certainty.

Figure 10. The primary foundation of Feature 11 comprising damaged ground stone implements.

Figure 11. Features 14, 15 and 19 (pottery concentration).

Figure 12. The substructure of floor 2 in Feature 14.

Horizon 4

Feature 29 is a roughly rectangular feature in the southeast corner (C) of Trench 24 which was detected in spit 20 (Figure 17). The feature comprises different layers of burnt soil, ash and charcoal mixed with finds. It resembles a pit that extended into the eastern and southern profiles. Unfortunately, due to time and money constraints it was not completely excavated, and the excavation was ended at the bottom of spit 22.

Feature 30 consists of a very compact, dark brown and red fired soil with charcoal found in spit 20 (Figure 17). It is several centimetres thick and contained almost no finds. Its lower boundary is Feature 34.

Feature 31 is a layer of several centimetres of white and grey ash deposit which was found in spit 20 (Figure 17). It is about 1.5 x 0.9 m in dimensions and its thickness varies from 3 to 10 cm in places. Its lower boundary is Feature 34.

Figure 13. The substructure of pebbles and pottery in Feature 14.

Figure 14. The basin used to hold the substructure of floor 1 in Feature 14.

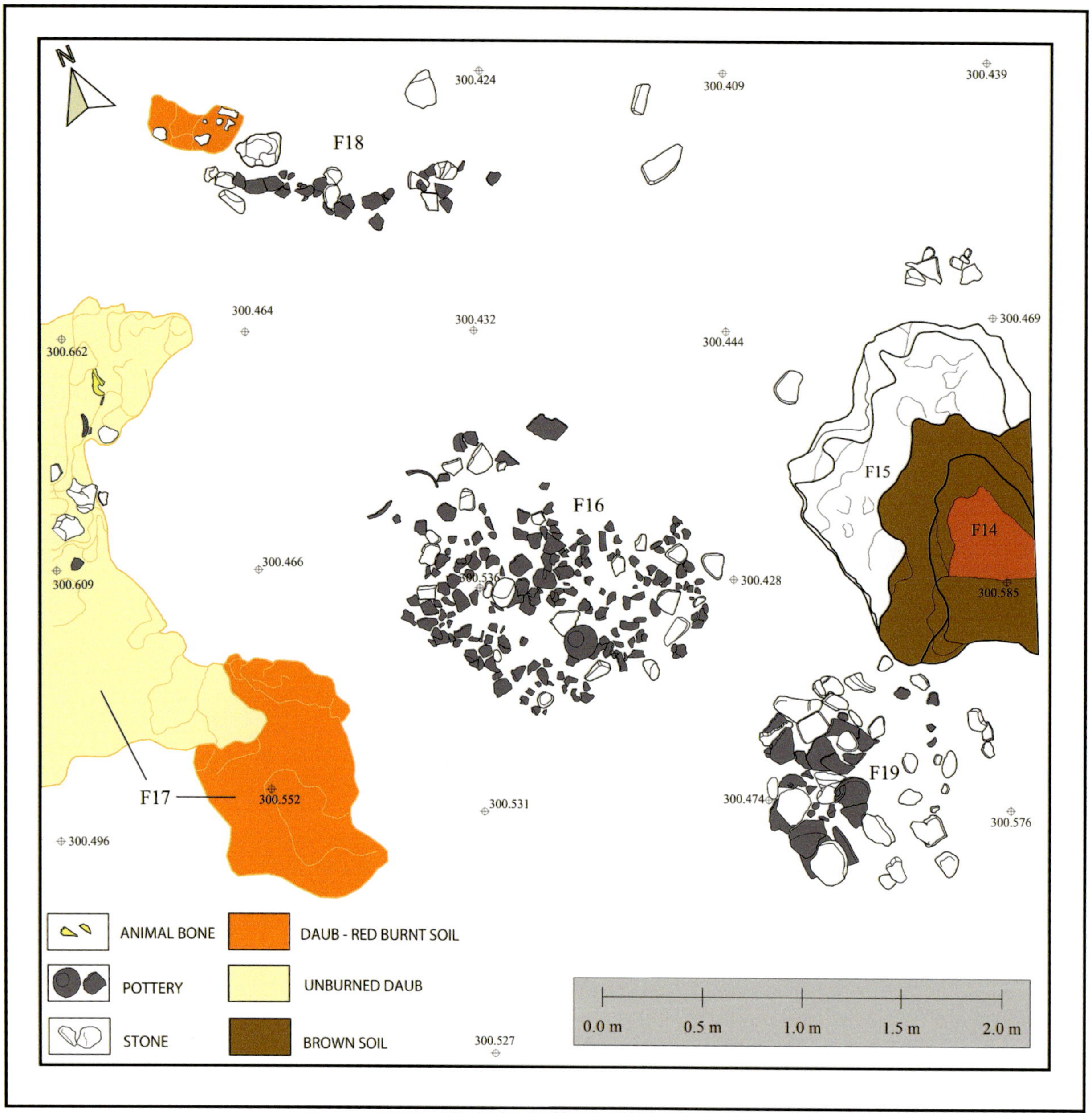

Figure 15. Features 14, 15, 16, 17, 18 and 19 in spit 15.

Feature 32 was found in spit 20 and is a thick (10–15 cm) crescent shaped feature consisting of orange fired daub, charcoal and grey ash (Figure 17). The largest span of the feature is 1 x 0.65 m. The feature represents the remains of a destroyed kiln, but the remains were not found *in situ*.

Feature 34 is a large, irregular rectangular area of white ash, several centimetres thick (Figures 18 and 19). The feature covered most of the trench surface in spit 21. It consists of several finely deposited layers of white ash, occasionally mixed with charcoal, daub fragments and orange or red baked soil. It was damaged by later activities on the site, including by the cuts of the holes (Features 35, 36 and 37) dug for the construction of the structure defined through units 17, 22 and 25 (Figure 18). It is unclear what the exact function of this feature may have been, but its rather regular dimensions and thickness may indicate the remains of a structure made of light materials (e.g. hay or straw) that was burnt in a fire. The direct dates set this feature right at the beginning of the 5th millennium BC (Chapter 37, Table 1), which is about the time when the earliest copper metal artefacts appear at the site. It is therefore possible that this massive ash deposit feature was associated with metallurgical activities.

Horizon 5

Due to monetary, safety and time constraints, towards the end of 2013 field campaign it became necessary to cease excavations on an area of Trench 24. It was

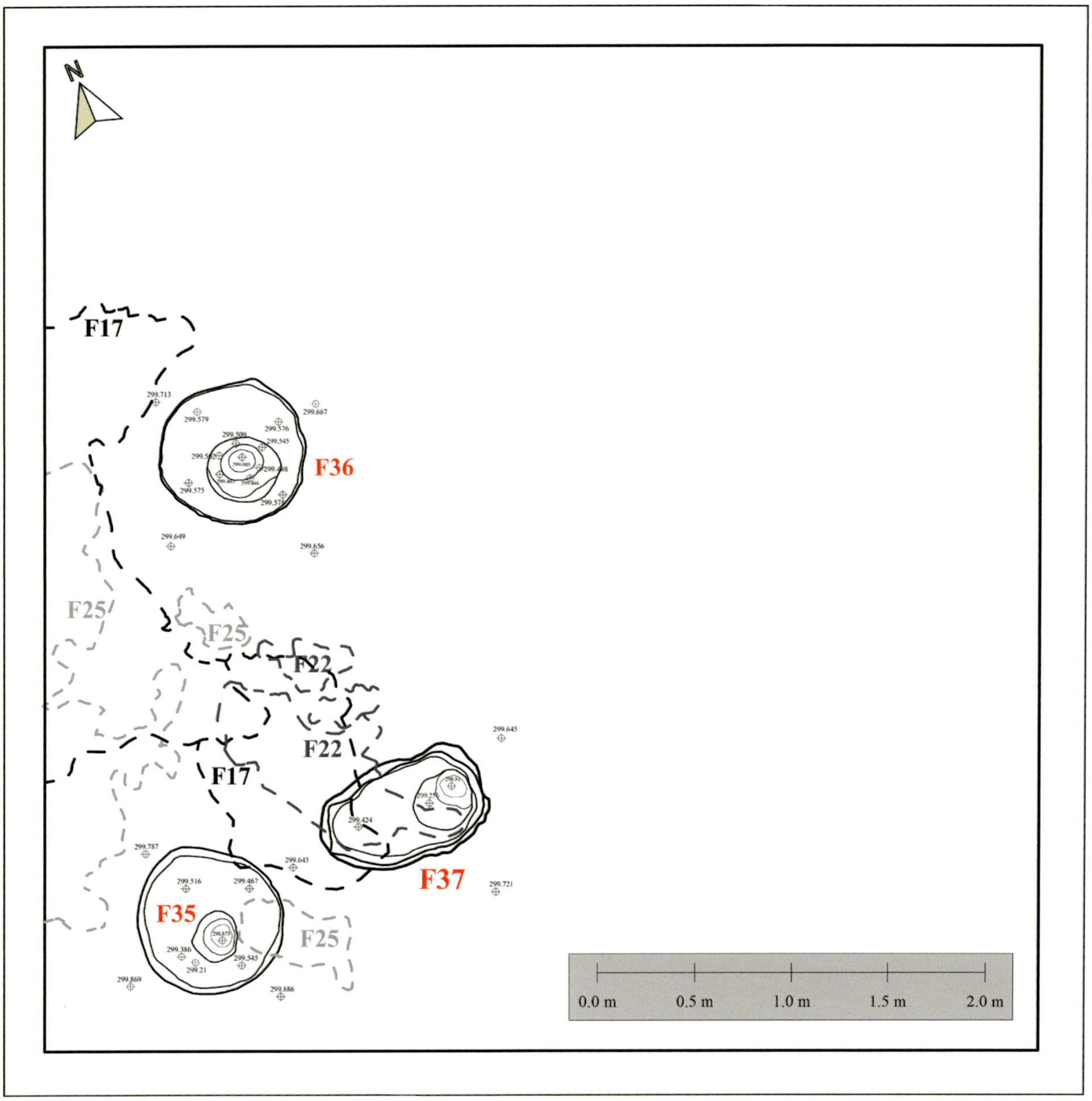

Figure 16. Features 35, 36, and 37, the post-holes for a structure comprising Features 17, 22 and 25 shown in dashed lines of different colour.

decided that just the western part of the trench would be excavated until the natural soil was reached. During this process Feature 38 was detected between spits 22 and 25. It soon became clear that it was a pit structure with several separate cells (Figure 20) and was found in its secondary role as a refuse pit. This was clearly evidenced by the highly fragmented finds and the diversity of layers comprising the infill of the cut. Layers of ash were separated by thin charcoal and soil layers until the natural soil which occurred at the base of spit 25. The feature extended beyond the excavated part towards the west and east, whilst towards the north and south, the edges of the pit were clearly visible. At its deepest the feature was over 80 cm deep and consisted of at least two cells, the first to the southeast and the second to the northwest. Based on this evidence, it was possible to deduce that it may have had a figure eight shape, which is not uncommon for late Starčevo and early Vinča period pits in the central Balkans (Bogdanović 1988: 38, Figure 5.2; Petrović 1999/2000: 8, Figure 5; Marić 2013b: 20, Figure 3, 25, Figure 8)

Feature 39 was detected in spit 24 and within the extent of Feature 38 at the very northeast edge of the excavated area in Trench 24. It consisted of orange baked daub which had an elliptical shape and extended under the eastern profile (Figure 20), so that only about half of the feature could be excavated. The inside of the baked daub wall consisted of charred soil filled with carbonised seeds lying on a red baked floor (Figure 21).

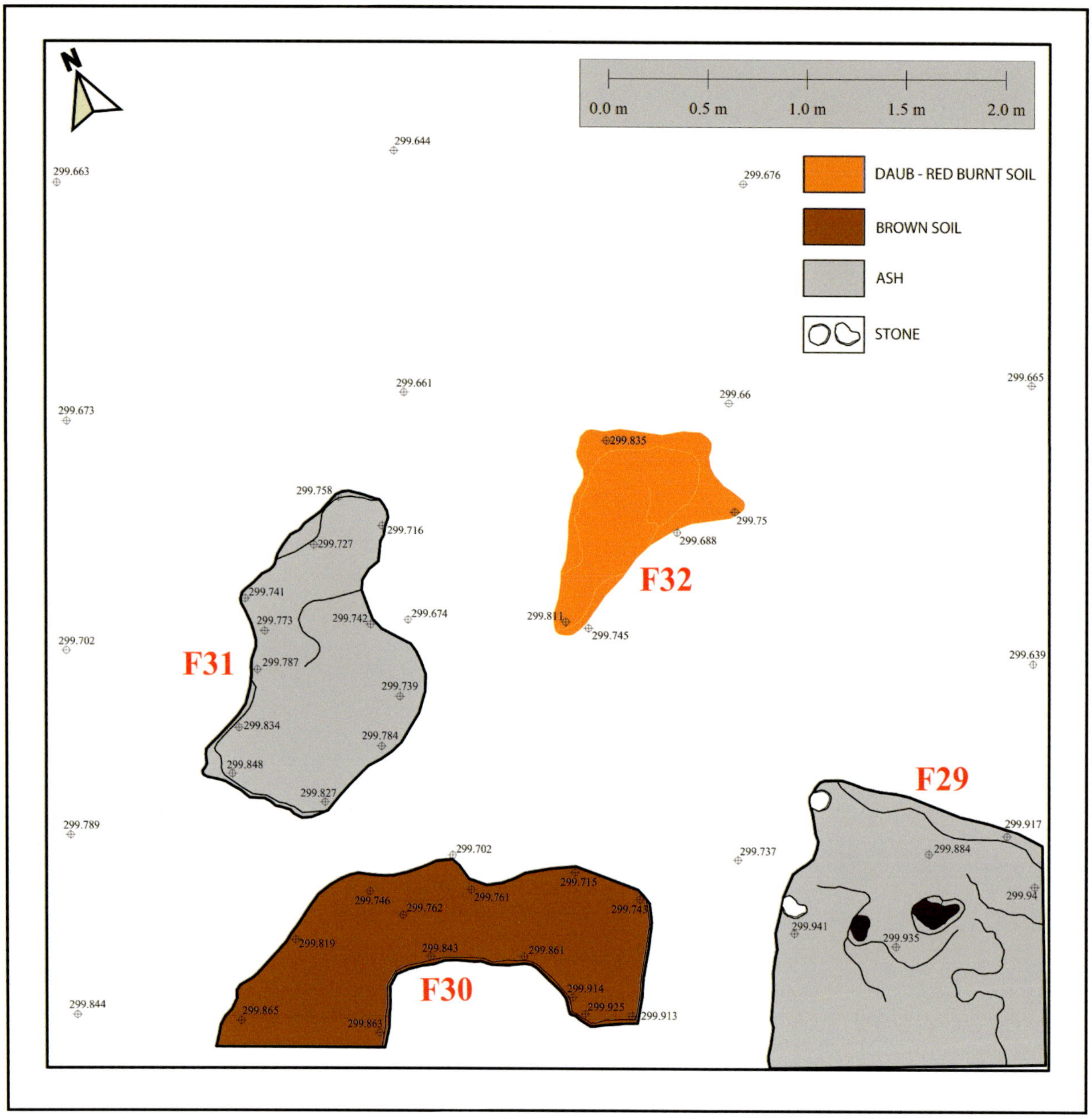

Figure 17. Features 29, 30, 31 and 32.

Based upon parallels from the late Starčevo-early Vinča site of Jaričište 1 (Marić 2013b: 20, Figure 4) it could be concluded that the feature represented the remains of a kiln constructed in the side wall of the pit house by digging a hole in the soil which was then strengthened through repeating firing.

The bibliographic reference for this chapter is:

Marić, M., Pendić, J., Roberts, B. W. and Radivojević, M. 2021. Pločnik: excavation results, in Radivojević, M., Roberts, B. W., Marić, M., Kuzmanović Cvetković, J., and Rehren, Th. (eds) *The Rise of Metallurgy in Eurasia*: 281–300. Oxford: Archaeopress.

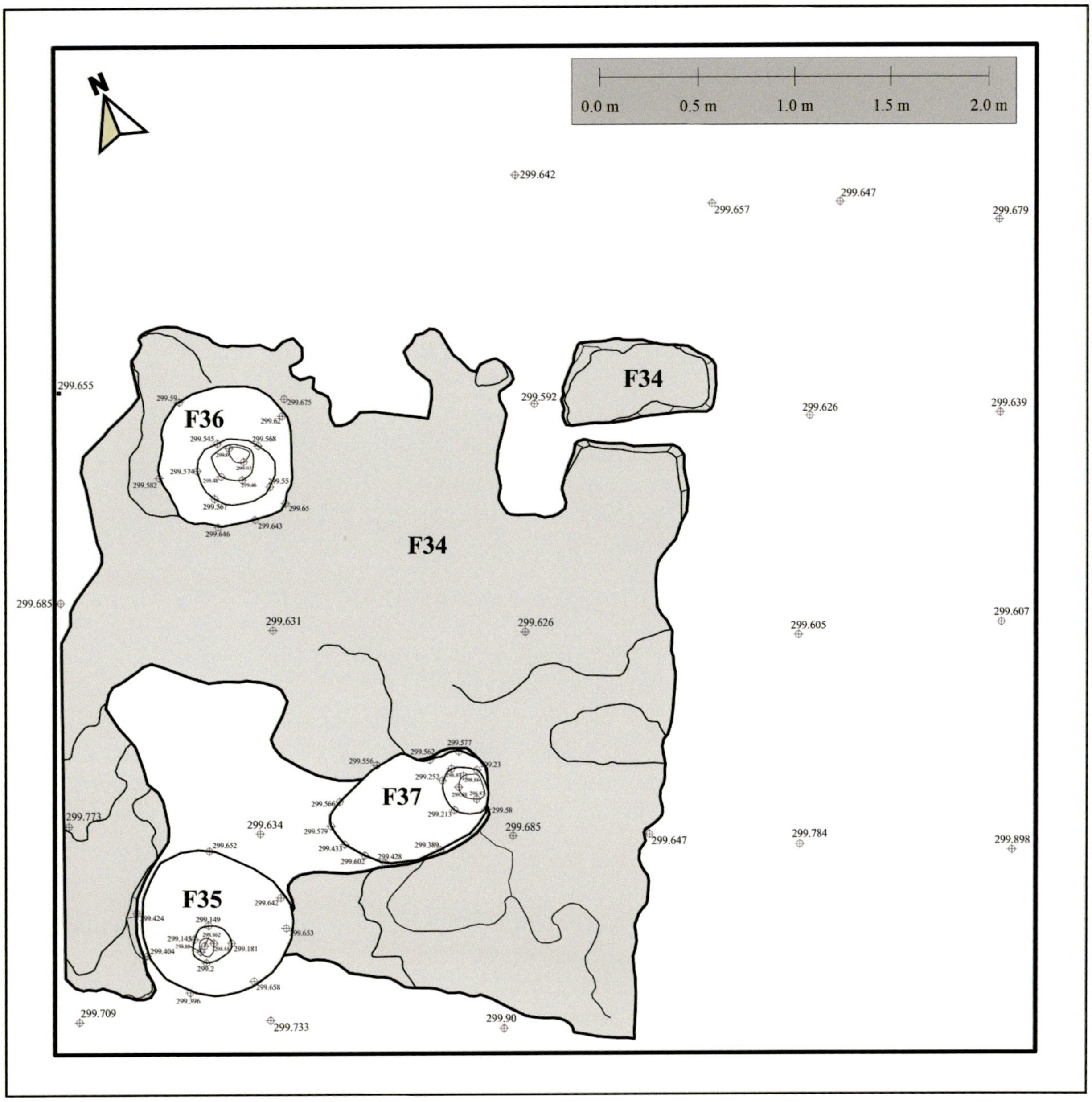

Figure 18. Feature 34 which was damaged by Features 35, 36, and 37 cutting it.

Figure 19. Feature 34 upon discovery in spit 21. Pits (Features 35, 36 and 37) visible in the ash.

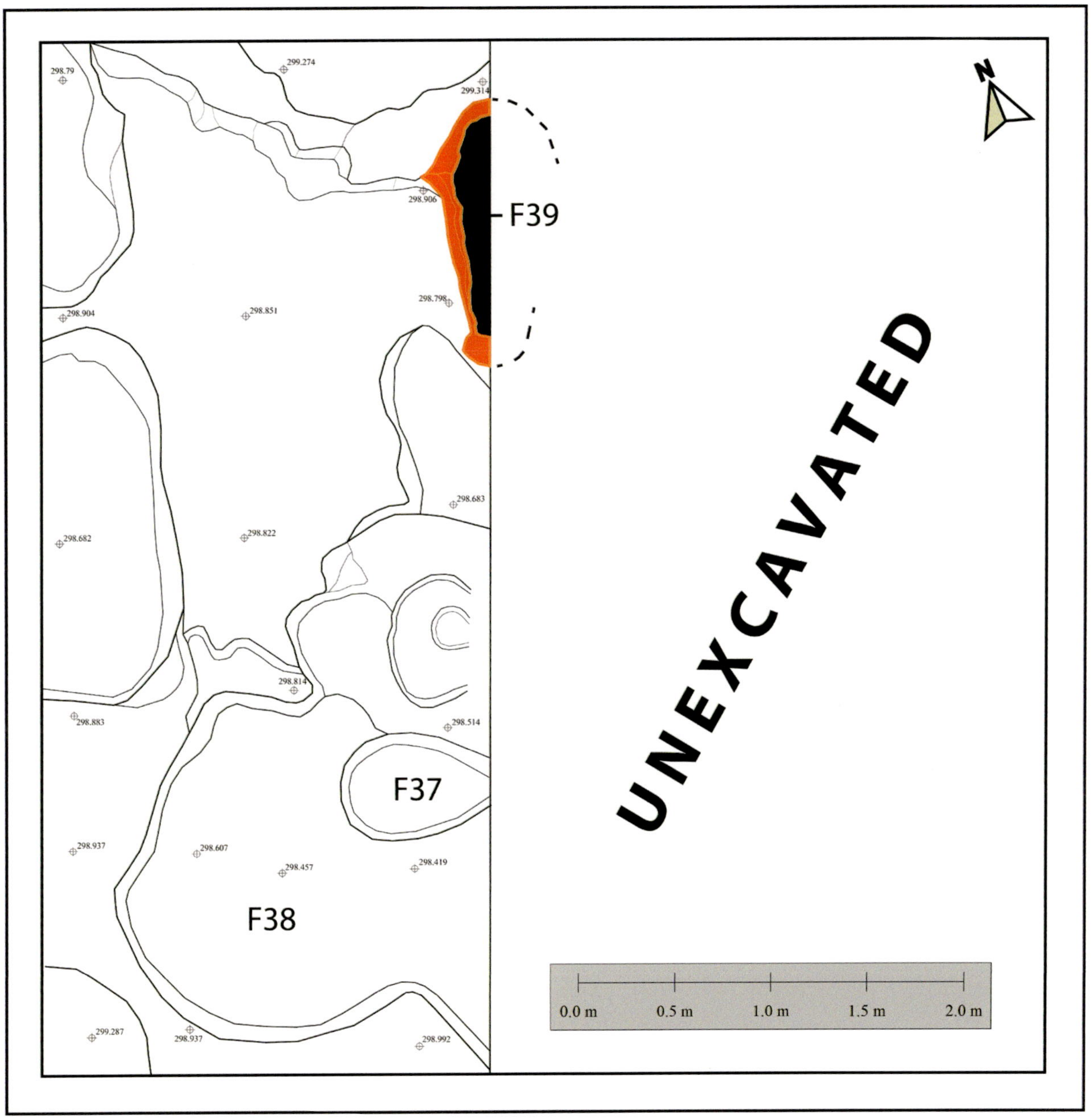

Figure 20. Feature 38 and 39, spit 25.

Figure 21. Cross-section of Feature 39. Photo taken from the west.

Chapter 26

Pločnik: technology of metal production

Miljana Radivojević and Thilo Rehren

Metallurgical materials recovered during the excavation campaigns of 2012 and 2013 in Pločnik show similar characteristics to samples already studied and published previously (Radivojević 2012, 2015; Radivojević and Kuzmanović Cvetković 2014; Radivojević and Rehren 2016; Radivojević *et al.* 2013). They include, as for Belovode (Chapter 11), predominantly malachite minerals and ores (Table 1), that occur as roughly beneficiated pieces and without a distinct spatial patterning in Trench 24. In comparison to Belovode, they occur less frequently across all five horizons, partially explained by the fact that most of Trench 24 is a large rectangular feature – a house (F1=F2=F4=F5=F6=F10), and there is very little economic area surrounding it (Chapter 25, Figure 4).

The uncovered copper mineral samples are macroscopically similar to those already characteristic of both Pločnik and Belovode: dominated by green (malachite) minerals with black/dark specs (Figure 1b), and some that are more purely green, with an occasional occurrence of blue (azurite) (see Table 1). In addition to the expected malachite, Trench 24 produced an abundance of green-yellow minerals (Figure 1a) that reacted to the magnet used during the excavations (hence termed 'magnetic' in our archives). It is notable that these were found exclusively in Horizon 1 (see below). They were analysed along with the other metallurgy-related materials and the results are presented below.

Copper finds from Trench 24 can be separated into decorative minerals (malachite beads) and metal artefacts. The latter derive mostly from inside the large house in Horizon 1 (Chapter 25, Figure 4), but also include a copper metal ring (C_P2/13) found within Horizon 1, in Feature 3, external to the house (Figure 2).

A copper metal bead (C_P4/13) was found spatially associated with Feature 15, which is interpreted as a typical kiln in Horizon 3. Feature 3 is the only feature potentially related to metallurgical activities. It was detected adjacent to the west profile of the trench, close to corner A (Figure 2). This small feature consists of several large stones mixed with the remains of a burnt structure that could have been a kiln, or more likely a furnace, due to the find of a copper metal ring (Table 1) in its vicinity. The shape of Feature 3 is unusual (Figure 3): it appears to be a portion of a slightly elevated wall of an almost rectangular structure, or at least the edge of such a structure (see also Chapter 25, Figure 7). Its extent is unknown as it was not fully excavated (see Figure 3), however, it bears a striking resemblance to the excavated rectangular firing structures in Trenches 20 and 21 (Radivojević *et al.* 2013: 1033, Figure 2; Šljivar and Kuzmanović Cvetković 2009a: 61), which also had

Figure 1. a) Example of green-yellow 'magnetic' mineral (P8/12); b) Example of black and green copper rich mineral (P125/13).

Table 1. The list of minerals and metallurgical materials from excavation campaigns 2012 and 2013 at Pločnik. Note an indicated subset analysed in depth with various analytical instruments.

No	no. finds	trench	spit	find no.	EDM	type of material	OM	metall-ogra-phy	SEM-EDS	EPMA	LIA	NAA	LA-ICP_MS	weight (g)	reasoning for provenance analysis
P6/12		24	S6	20	89	magnetic	X		X						
P14/12		24	S5	40	129	magnetic	X		X						
P58/12	1	24				bloated pottery	X								
P61/12		24	S8	96	283	metal droplet	X		X						
P117/12		24/F4	Sec 3	181	738	magnetic									
P8/13	2	T24	S10	214	977	magnetic					X	X			the earliest magnetic mineral
P10/13	2	T24	S10	220	988	fragmented metal	X	X	X	X		X	X	0.22	metal artefact
P13/13	1	T24	S11	233	1017	fragment of a bracelet/wire	X	X	X			X		0.15	metal artefact
P14/13	1	T24	S12	242	1054	malachite					X	X			malachite near P13/13
P18/13	1	T24	S14	272	1192	azurite									
P19/13	1	T24	S14	273	1197	malachite									
P55/13	1	T14/F15	S16	366	1606	malachite					X	X			ore choice consistency
P108/13	1	T24	S21	515	1904	malachite bead									
P117/13	2	T24	S21	530	1951	malachite bead x 2									
P121/13	1	T24	S21	537	1958	malachite					X	X			the last spit with minerals
P175/13	1	T24	S21	no	no	ceramic nozzle?									
P125/13	1	T24	S21	546	1967	malachite	X		X						

Table 1 continued. List of studied materials from Pločnik and analytical techniques applied

No	no. finds	trench	spit	find no.	EDM	type of material	OM	metall-ogra-phy	SEM-EDS	EPMA	LIA	NAA	LA-ICP_MS	weight (g)	reasoning for provenance analysis
C_P1/13	1	T24	S9	155	587	metal loop / ring	X	X		X	X	X	X	0.47	metal artefact
C_P2/13	1	T24/F2 (north)	S9	195	908	Metal band / ring	X	X		X	X		X	1.93	metal artefact
C_P3/13	1	T24	S10	217	985	Malachite bead									
C_P4/13	1	T24	S14	276	1200	Metal bead								0.05	
C_P5/13	1	T24	S18	405	1705	Malachite bead									
C_P6/13	1	T24	S21	516	1905	Malachite bead									
C_P7/13	1	T24	S21	524	1913	Malachite Bead									
C_P8/13	1	T24	S21	531	1952	Malachite bead									
C_P9/13	1	F32	S21	573	2078	Malachite bead									
C_P10/13	1	T24	S21	529	1950	Malachite bead flot									
C_P11/13	1	T24	S21	529	1950/2	Malachite bead flot									
C_P12/13	1	T24	S21	529	1950/3	Malachite bead flot									
Pf19/12		24 ext2		75	189	malachite bead									
Pf33/12		24	S8	96	283	magnetic	X		X					0.4	
Pf35/12		24	S8	102	297	magnetic	X		X					1.93	

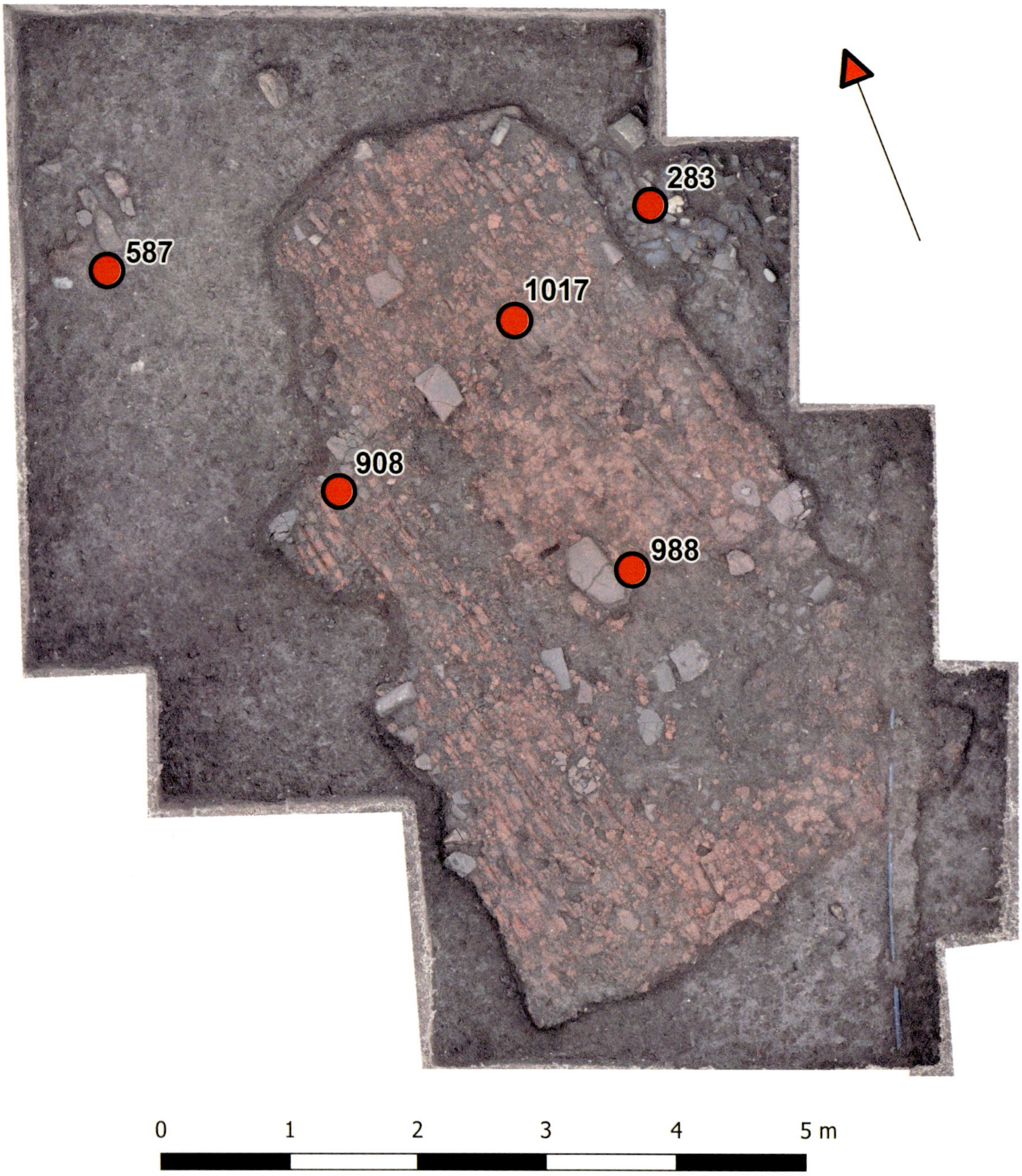

Figure 2. Overview of Horizon 1 house and spatial distribution of metal artefacts. All numbers are EDM identifiers for objects analysed in this chapter: P61/12 (283), C_P1/13 (587), C_P2/13 (908), P10/13 (988), P13/13 (1017) (prepared by M. Marić).

copper metal artefacts and casting debris associated with them. Feature 3 is distinct from the previous finds in that it was possibly located just outside a house but it would not be unusual to have the fireplace for metallurgical activities in an economic area, as seen at Belovode (Chapter 11). The excavated assemblage did not include any slag or slagged sherds, although the copper metal droplet (P61/12) may be indicative of what has previously been characterised as a 'slagless' metallurgy at Pločnik (Radivojević 2012; Radivojević and Rehren 2016: 220), with only one other previous find with a similar structure.

The direct radiocarbon dates for the house (F1=F2=F4=F5=F6=F10) and Features 3 and 15 frame the chronology of metallurgical finds from Trench 24 (see Table 2). In Chapter 37 (this volume), Marić *et al.* model the site chronology using the Bayesian statistical method, which combines both the radiocarbon dates and the relative stratigraphy recorded during the

Figure 3. Feature 3 in Trench 24. Note low rise red walls of what could have been a rectangular structure (centre).

excavation. These modelled dates are presented in Table 2, and we will refer to these when discussing the dating of metallurgical samples from Pločnik.

The modelled dates for Horizon 1 metallurgical samples (Table 2): P61/12, P10/13, P13/13, C_P1/13 and C_P2/13 indicate that they start in layers dated to 4631-4462 cal. BC to 4446-4231 cal. BC (95.4% prob.), or possibly 4576-4491 cal. BC to 4431-4324 cal. BC (68% prob.). This dating framework is also valid for all 'magnetic' minerals (see Table 1). The metal bead sample C_P4/13 belongs to Horizon 3, which starts at 5013-4968 cal. BC and ends at 4894-4747 cal. BC (95.4% prob.), or possibly 5036-4951 cal. BC to 4927-4621 cal. BC (68% prob.). This is consistent with the beginning of the Gradac Phase on the site (Chapters 25 and 37, Table 3), and is amongst the earliest securely dated copper metal objects from Pločnik, making it the earliest secure date for the beginning of metallurgy at the site.

Several fragments of a ceramic object from spit 21 (Horizon 4) could potentially resemble a tuyère (Figure 4), however a pXRF scan of the object did not show any contamination with potential ore elements either insides or outside, and the diameter of *c.* 2 cm is too large for the expected type of on-site activities. Interestingly, it was found in association with Feature 34, a large irregular rectangular area comprising fine white ash mixed with charcoal, daub fragments and baked soil. A radiocarbon date directly associated with this feature sets it at around 5003-4987 cal. BC (68% prob.), which is around the time that copper metallurgy emerges in Pločnik.

Methodology

The sampling and analytical strategy is identical to that applied at Belovode (see Chapter 11, Table 3). Polished blocks of copper metal artefacts C_P1/13, C_P2/13 and P10/13 were prepared for metallographic examination using, as an etchant, ammonia hydrogen peroxide made from equal proportions of ammonia (NH_4OH), water and 3% H_2O_2.

Results: technology of metal making and working in Pločnik

Ten samples were selected for in-depth microstructural and compositional analysis: five minerals/ores and five pieces of copper metal making and working evidence (a droplet and four fragments of copper metal artefacts). All provide evidence for copper-based metallurgy on the site which, in terms of the nature and volume of the process, is consistent with previous studies on archaeometallurgical materials both from this settlement and beyond, across the Vinča culture (Radivojević 2012, 2015; Radivojević and Rehren 2016).

Table 2. Direct dates for metallurgical

No	trench	spit	find no.	EDM	Horizon	type of material	associated feature
P61/12	24	S8	96	283	1	metal droplet	F1=F2=F4=F5=F6=F10
C_P1/13	T24	S9	155	587	1	Metal earing	F1=F2=F4=F5=F6=F10
C_P2/13	T24/F3	S9	195	908	1	Metal ring	F3
P10/13	T24	S10	220	988	1	metal foil corroded	F1=F2=F4=F5=F6=F10
P13/13	T24	S11	233	1017	1	metal bracelet (?)	F1=F2=F4=F5=F6=F10
C_P4/13	T24	S14	276	1200	3	Metal bead	F15

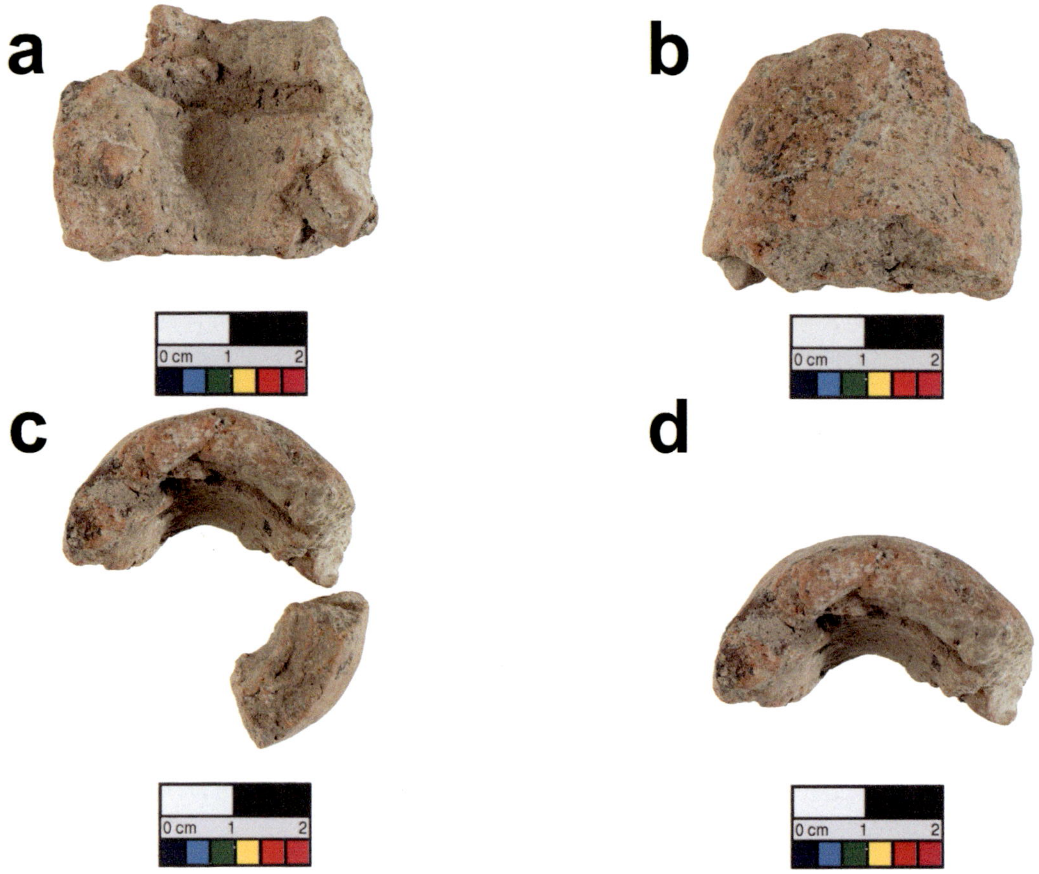

Figure 4. A cylindric ceramic artefact from Horizon 4, Pločnik.

Processing: archaeological minerals

Of the five minerals/ores analysed, one was malachite (P125/13, see Figure 1b, Table 3) and four belonged to the category of 'magnetic' minerals (Figure 6), that were more unusual and clearly different from the typical black and green copper-rich minerals that regularly occur at both Pločnik and Belovode. The black and green copper mineral, P125/13, contains both copper and manganese phases, with Mn content in the latter close to 50 wt% (Table 3). The so-called 'magnetic' minerals, P6/12, P14/12, Pf33/12 and Pf35/12 are a mixture of iron oxides and members of the olivine family of minerals (Table 4). The lightly coloured phase (pale yellow) is iron oxide, and the darker phase (see Figure 6) includes forsterite (Mg-end member of the olivines) and monticellite ($CaMgSiO_4$). In all cases, the texture of the samples identified them as natural rock fragments rather than metallurgical products. Their distinctive colour (mostly dark green) may explain their collection by the Pločnik communities. We do not currently know whether they were somehow related to the smelting process, since our evidence is for 'slagless' smelting processes (see below), or whether they are unrelated to metallurgy. Their presence does, however, seem to be anthropogenic, since such finds are not known from other excavations in Pločnik.

activities in the site of Pločnik

modelled C14-date 1σ	modelled C14-date 2σ	directly dated feature/spit (2σ)
(start) 4576-4491 cal BC to (end) 4431-4324 cal BC	(start) 4631-4462 cal BC to (end) 4446-4231 cal BC	MAMS22083 4454-4356 cal BC
(start) 4576-4491 cal BC to (end) 4431-4324 cal BC	(start) 4631-4462 cal BC to (end) 4446-4231 cal BC	MAMS22083 4454-4356 cal BC
(start) 4576-4491 cal BC to (end) 4431-4324 cal BC	(start) 4631-4462 cal BC to (end) 4446-4231 cal BC	MAMS22081 4493-4365 cal BC
(start) 4576-4491 cal BC to (end) 4431-4324 cal BC	(start) 4631-4462 cal BC to (end) 4446-4231 cal BC	MAMS22083 4454-4356 cal BC
(start) 4576-4491 cal BC to (end) 4431-4324 cal BC	(start) 4631-4462 cal BC to (end) 4446-4231 cal BC	MAMS22083 4454-4356 cal BC
(start) 5013-4968 cal BC to (end) 4894-4747 cal BC	(start) 5036-4951 cal BC to (end) 4927-4621 cal BC	MAMS22090 4937-4796 cal BC

Figure 5. 'Magnetic' minerals. a) P6/12; b) P14/12; c) Pf33/12; d) Pf35/12.

Table 3. SEM-EDS compositional data for copper (green) and manganese (black) rich phases in P125/13. All values are averages of six to fourteen analyses of each sample / phase and normalised to 100%.

	MgO	Al_2O_3	SiO_2	P_2O_5	K_2O	CaO	MnO	FeO	NiO	CuO	ZnO	PbO
	wt%	wt%	wt%	wt%	wt%	wt%	wt%	wt%	wt%	wt%	wt%	wt%
P125/13 green	0.0	0.0	0.0	0.0	0.0	0.0	0.7	0.8	0.0	97.2	2.2	0.0
P125/13 black	0.0	2.4	6.1	0.9	0.0	1.0	47.6	7.2	0.0	12.8	4.0	18.4

Table 4. SEM-EDS compositional data for 'magnetic' minerals, with iron oxide (light) and olivine (dark) rich phases. All values are presented as stoichiometrically calculated at% and averages of two to nineteen analyses of each sample / phase and corrected with factors based on CRM analysis; the uncorrected data is reported in the Appendix B_Ch26.

	Mg	Al	Si	P	Ca	V	Cr	Mn	Fe	Ni	O
	at%	at%	at%	at%	at%	at%	at%	at%	at%	at%	at%
P6/12 light	0.0	0.0	0.0	0.0	0.0	0.2	0.0	0.0	49.7	0.0	50.1
P6/12 dark	11.9	9.9	12.0	0.0	0.1	0.0	0.0	0.1	8.7	0.0	58.5
P14/12 light	9.7	1.8	19.0	0.0	9.6	0.0	0.3	0.0	0.7	0.0	60.0
P14/12 dark	22.4	1.0	17.0	0.0	0.2	0.0	0.0	0.0	3.2	0.0	58.7
Pf33/12 light	0.0	0.0	2.6	0.0	0.1	0.0	0.0	1.0	44.9	0.0	51.3
Pf35/12 light	1.4	0.0	1.4	0.0	0.0	0.0	0.0	0.0	46.6	0.1	50.7
Pf35/12 dark	24.1	0.4	16.8	0.0	0.1	0.0	0.0	0.0	2.6	0.0	58.5

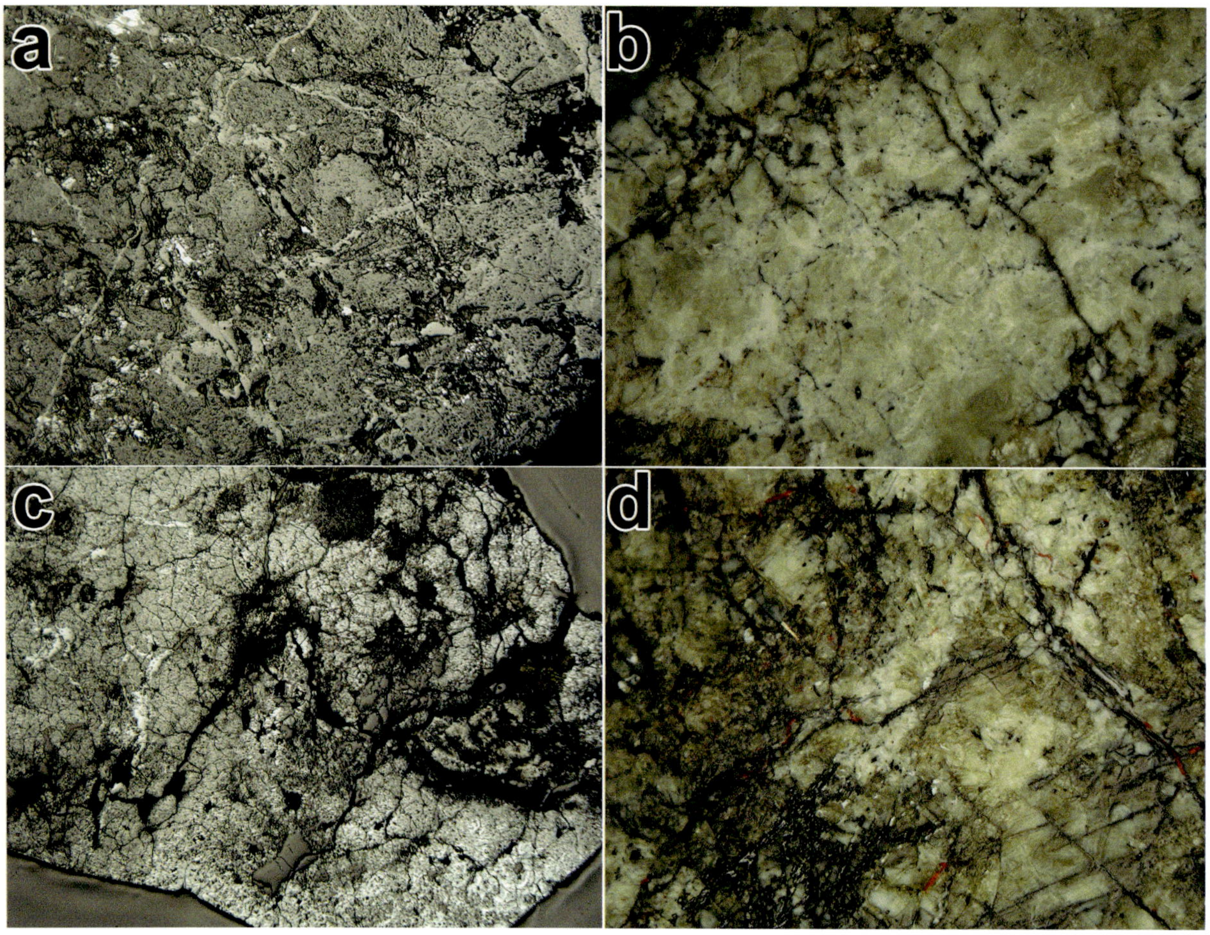

Figure 6. a) Photomicrograph of P6/12 taken under plain polarised light, 25x magnification, 6.4 mm width. Note that the lighter phase is iron oxide, while darker belongs to olivine family; b) photomicrograph of P14/12 taken under cross polarised light, 50x magnification, 3.2 mm width; c) photomicrograph of Pf33/12 taken under plain polarised light, 25x magnification, 6.4 mm width; d) photomicrograph of Pf35/12 taken under cross polarised light, 50x magnification, 3.2 mm width.

Making and working: copper minerals and metal artefacts

The total assemblage of copper-related artefacts from Pločnik includes thirteen mineral ornaments, one copper metal droplet, and fragments of five finished copper metal artefacts. Almost all are produced from copper-rich minerals or directly linked to metallurgy, demonstrating local metal smelting and working activities.

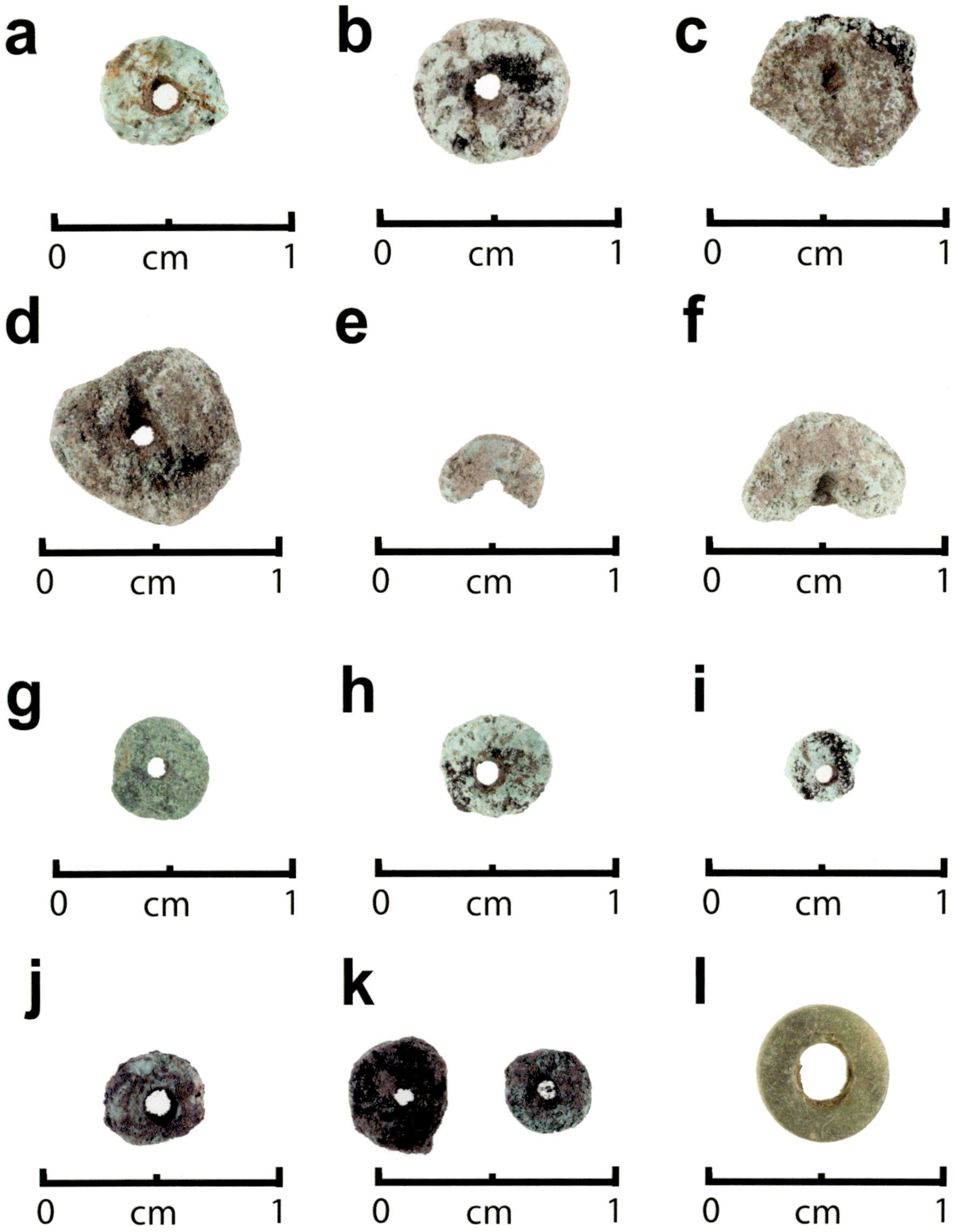

Figure 7. Mineral beads from Trench 24, Pločnik. a) C_P3/13; b) C_P5/13; c) C_P6/13l; d) C_P7/13; e) C_P8/13; f) C_P9/13; g) C_P10/13; h) C_P11/13; i) C_P12/13; j) P108/13; k) P117/13; l) Pf19/12.

Copper mineral artefacts (ornaments)

Thirteen mineral ornaments (beads) were found, of which all but one (Pf19/12, green stone ring bead, Figure 7/l) are made of malachite of varying purity (Figure 7)., The malachite beads can be roughly divided into three distinctive typological categories: circular or cylindrical (C_P3/13, C_P5/13, C_P7/13, C_P8/13, C_P9/13, C_P11/13, P108/13, P117/13); flat disc (C_P6/13, C_P10/13); and ring beads (C_P12/13) (cf. Wright *et al.* 2008; Wright and Garrard 2003). Visually, all beads show a thick, light green corrosion layer, while seven of the thirteen (C_P5/13, C_P7/13, C_P9/13, C_P11/13, P117/13; Figure 7b, 7d, 7f, 7h, 7k) have visible black (likely manganese-rich) areas. All but one are finished objects, fragmented after use or during the post-depositional processes. The exception is C_P6/13 (Figure 7c), which is a blank with an unfinished central hole indicating local bead manufacture.

Copper metal droplet

As noted in Chapter 11 on Belovode metallurgy, the 'Droplet' category of metallurgical debris can include semi-molten and fully molten pieces of ore/metal that could be produced by a wide range of activities. These droplets can represent anything from an attempt to smelt metal through to accidental loss of debris from smelting, melting or casting; most importantly, they are not worked any further, and this distinguishes them from the category of copper 'Artefacts'. One object, P61/12, recovered from Trench 24 in 2012, fits

Figure 8. Metal droplet P61/12. Note green staining on top.

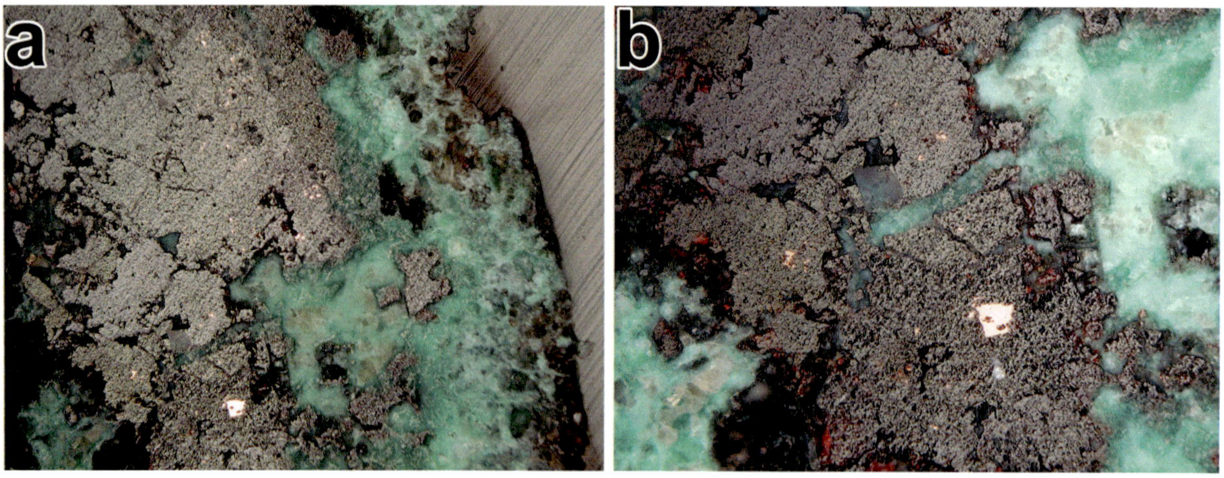

Figure 9. a) Photomicrograph of P61/12 taken under cross polarised light, 50x magnification, 3.2 mm width. Note bright yellow residual copper metal in a matrix of copper oxide (grey) and green corrosion products; b) photomicrograph of P61/12 taken under cross polarised light, 100x magnification, 1.6mm width.

Table 5. SEM-EDS compositional data for metal droplet P61/12, both metal and dross phase. All values are averages of four to fifteen analyses of each sample / phase.

	O	S	Cu	Zn
	at%	at%	at%	at%
P61/12 metal	1.7	0.0	98.3	0.0
stdev.s	*0.3*	*0.0*	*0.3*	*0.0*
P61/12 dross	33.6	0.7	65.7	0.0
stdev.s	*2.7*	*2.8*	*2.9*	*0.0*

the 'Droplet' criteria (Figure 8). It was found within the house in Horizon 1 and is not associated with any firing feature. The droplet has an amorphous shape, is covered in thick green patina, and is less than 1 cm in length. As with previous, similar, examples from Belovode (Chapter 11; see also Radivojević and Rehren 2016), this sample was initially collected and labelled as a mineral, its partially metallic structure only being revealed when it was cut with a saw (Figure 9). The droplet contains predominantly dross (copper oxides) with small specs of bright yellow copper metal that indicates what must initially have been a fully developed metallic phase. Optically, the dross is characterised by bright red internal reflections under cross polarised light, typical for cuprite (Cu_2O) (Figure 9). The presence of sulfur (Table 5) implies that the ore used contained both primary and secondary copper mineralisations. Overall, P61/12 can be interpreted as deriving from a 'slagless' melting event as there is no iron detected (see Table 5), or from a smelting event using copper ore of high purity. Experiments with copper smelting and refining have yielded similar material, retained on the crucible wall due to insufficiently high temperatures (i.e. only just above the melting point), or the presence of impurities (S. Timberlake, personal communication). Since it was found inside the dwelling, we assume that this droplet was in a secondary location, away from the location of high temperature processing where it was produced.

Copper metal fragments

Four of the five copper metal artefacts were analysed in detail with the results presented here, while the fifth was used entirely for provenance analysis (see Chapter 41). All four examined artefacts originate from Horizon 1 (Figure 2) and are associated with the house (F1=F2=F4=F5=F6=F10) and fireplace (Feature 3) in its vicinity (C_P2/13). Two artefacts appear to have been uncovered in their final form: an earring/loop (C_P1/13) and a (band) ring (C_P2/13); the other two are potentially fragments of unspecified decorative objects.

C_P1/13 is a loop or a ring, almost 1 cm in diameter (Figure 10a). It has a pale yellow, bright (copper) metal body, with green corrosion products developing on its edges (Figure 11). The object has been cast, and then worked with a combination of techniques carefully designed to respond to the desired function of the object. It has a fully recrystallised microstructure that was initially cold worked and then annealed, as evidenced by the presence of annealing twins (Figure 11b). The object must have been subject to several cycles of cold working and annealing, given the small size of the grains, the shape of the object, and also the noticeable folding line, implying that it began as a sheet of copper metal that was worked by folding onto itself. The final stage of production is preserved as deformed annealing twins, which commonly originate from low temperature mechanical or thermal stressing (Rostoker and Dvorak 1990: 23): the presence of twinned slip lines suggests that the final stage in the making of this artefact was cold working.

C_P2/13 also exhibits a cold worked microstructure, with the $Cu+Cu_2O$ eutectic deformed into distinct layers (Figure 12). The fully recrystallised grain structure presents traces of several cycles of annealing (annealing twins) and cold working and distortion (slip lines, mechanical deformation, deformed annealing twins). These repeated cycles are further indicated by the reduction in grain size by *c.* 20% (Rostoker and Dvorak 1990: 16), however the reduction in thickness did not cause intolerable brittleness for this object. The final stage of manufacture of this piece of folded metal was also cold working.

Based on its shape, P10/13 was most likely once a decorative item, or potentially a piece preserved as metal stock (Figure 10d). The fully recrystallised grain structure developed within a regularly distributed copper–copper oxide eutectic (Figure 13). Annealing twins, slip lines, and deformed twin lines all indicate that this object was worked in the same manner as the rest of the assemblage: cold working followed by annealing, in several cycles. The microstructure seems fully recrystallised and the reduced grain size (from working) corroborates the evidence for cycles of annealing and cold working, or recrystallisation at temperatures low enough to prevent further grain growth (American Society for Metals 1979: 60). Judging by the elongation of grains and the separation of the copper–copper oxide eutectic into distinct layers, as well as the greater presence of slip lines, it appears that one side of the artefact was worked more intensively than the other (Figure 13a). The concentration of slip lines towards the working surface (Figure 13a) indicates that the final stage of manufacture was cold finishing, or they may have resulted from use of the finished object. The other side of the object preserves the dendritic structure of the initial cast, next to a significant amount of copper oxide inclusions, segregated towards the

Figure 10. Copper metal artefacts from Trench 24, Pločnik. a) C_P1/13; b) C_P2/13; c) C_P4/13; d) P10/13; e) P13/13.

surface. This side bears some traces of annealing twins (Figure 13b), suggesting that it was also hammered, although not as extensively as the other side. Of note are two distinct layers of corrosion in the structure of the artefact. These could have results from it being folded onto itself, either during production, or when in use.

A fragment of a bracelet or wire (P13/13) is a completely corroded form of what was once a copper metal body (Figure 14). Specks of bright yellow copper metal in the predominantly oxidised copper (dross) matrix confirm this. It is unclear what the object might originally have been, however judging by its shape it may have been a

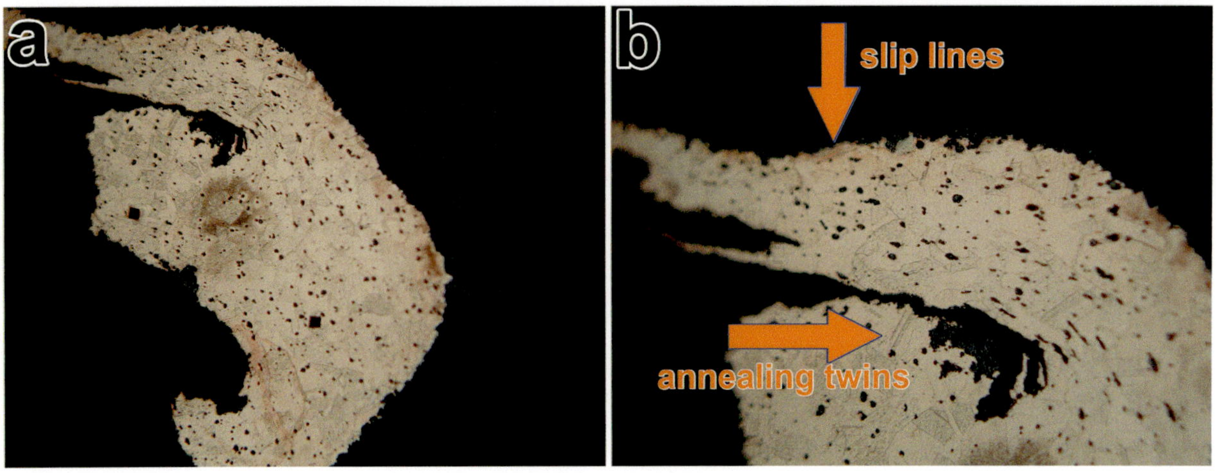

Figure 11. a) Photomicrograph of C_P1/13, taken under plain polarised light, 100x magnification, 1.6 mm width, etched with ammonia hydrogen peroxide; b) photomicrograph of C_P1/13, taken under plain polarised light, 200x magnification, 0.85 mm width, etched with ammonia hydrogen peroxide. Note the folding line in the middle.

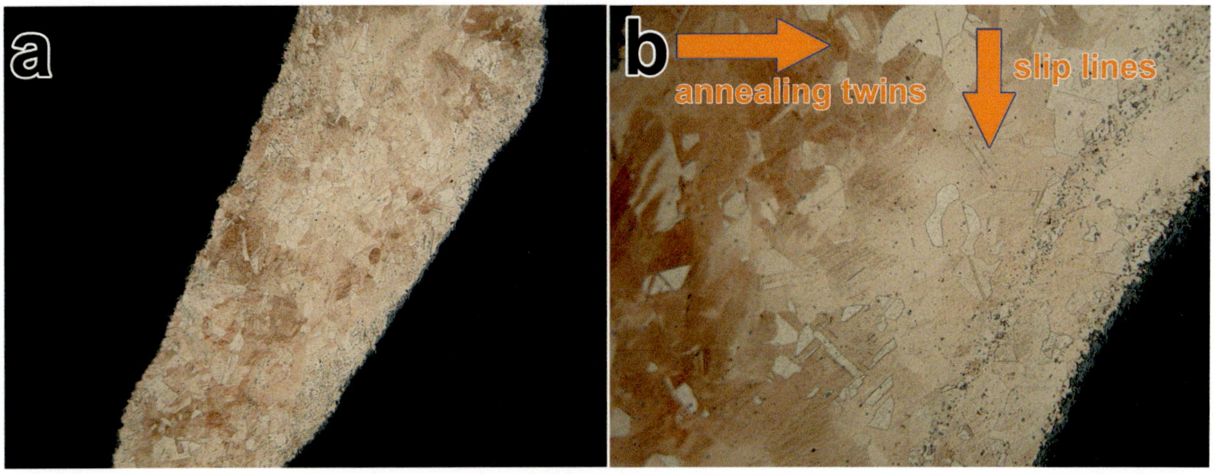

Figure 12. a) Photomicrograph of C_P2/13, taken under plain polarised light, 50x magnification, 3.2 mm width, etched with ammonia hydrogen peroxide; b) photomicrograph of C_P2/13, taken under plain polarised light, 200x magnification, 0.85 mm width, etched with ammonia hydrogen peroxide. Note annealing twins and slip lines.

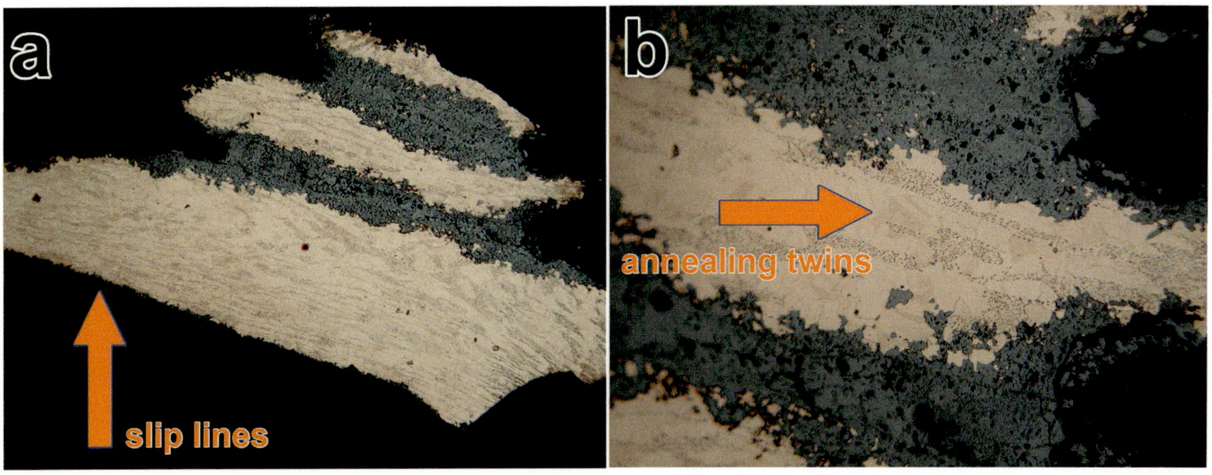

Figure 13. a) Photomicrograph of P10/13, taken under plain polarised light, 50x magnification, 3.2 mm width, etched with ammonia hydrogen peroxide. Note one side (bottom) worked harder than the other; b) photomicrograph of P10/13, taken under plain polarised light, 200x magnification, 0.85 mm width, etched with ammonia hydrogen peroxide.

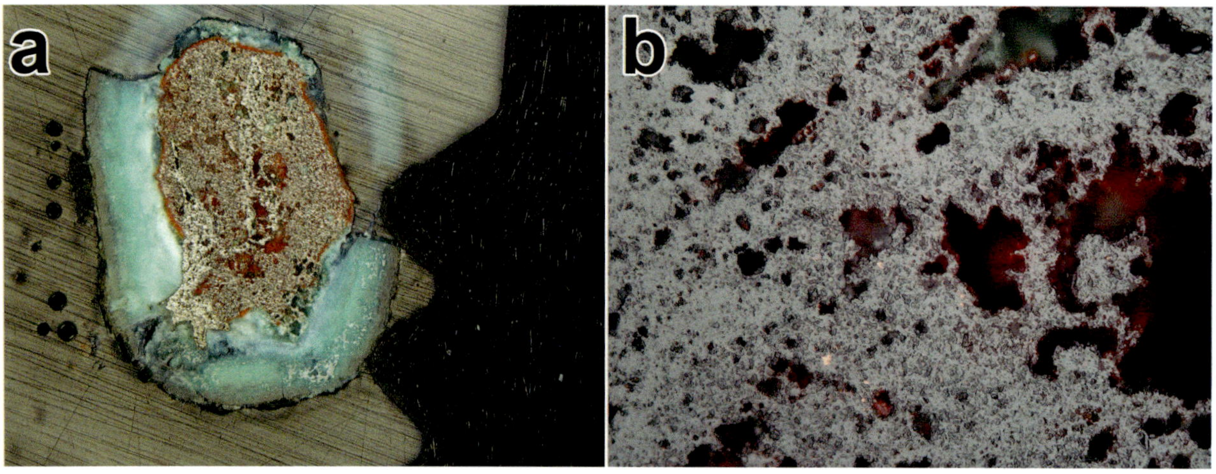

Figure 14. a) Photomicrograph of P13/13, taken under plain polarised light, 50x magnification, 3.2 mm width, etched with ammonia hydrogen peroxide; b) photomicrograph of P13/13, taken under plain polarised light, 500x magnification, 0.34 mm width, etched with ammonia hydrogen peroxide.

decorative. As such, it fits into the overall picture of an assemblage of exclusively decorative metal items from Trench 24.

While all the objects are made from copper of high purity, electron probe examination offers insight into the significantly different trace element signatures of three of them: P10/13, C_P1/13 and C_P2/13 (Table 6). We use ppm (parts per million) to discuss µg/g (microgram per gram) concentrations in this text. P10/13 is made of very pure copper with readings of Ni and possibly S; C_P1/13 is dominated by Pb (0.2 wt% on average) and Ag (*c.* 200 ppm on average) as main impurities; and C_P2/13 shows a significant reading of S (*c.* 160 ppm on average) and Sn (*c.* 100 ppm). The iron content is low in all three, suggesting either that the copper was highly purified (by multiple melting episodes), or that they all stem from a copper ore of high purity. The difference in trace element signatures would imply that each artefact derives from a different ore batch. This conclusion is strengthened with LA-ICP-MS results (Table 6a), with largely consistent results when it comes to the dominant Ni content in the copper ore used to make these artefacts. The Pb content, however, comes out as significant (c. 1.8 wt%) in C_P1/13. While it clearly shows lead content in the smelted copper ore, it is possible that the laser ablation picked up a phase rich in lead in the analysed metal section (copper and lead do not mix well).

Discussion and conclusion

The metallurgical evidence from Pločnik fits well within the broader picture developed in previous work and publications (Radivojević 2012, 2015; Radivojević and Kuzmanović Cvetković 2014; Radivojević and Rehren 2016; Radivojević *et al.* 2013). The copper metal droplets from previous research have already established the 'slagless' metallurgy principle at the settlement, and while evidence for primary production of copper may still be present, it remains to be found in future excavations. However, the new finds from Pločnik confirm it as a place for secondary processing (such as melting/refining), and we have been able to directly date a firing structure (Table 2) that, potentially, looks similar to several others discovered in its vicinity, and is associated with finished and semi-finished copper and tin bronze artefacts (Radivojević *et al.* 2013; Šljivar and Kuzmanović Cvetković 2009a).

That Pločnik belonged to what we have termed a multi-consumer metallurgical network is not novel information; Radivojević *et al.* (2010) previously established the link between metal production at Belovode and some artefacts from Pločnik. Prior to that, Pernicka *et al.* (1993; 1997) identified at least three different sources of copper that supplied this settlement from across the Balkans. In their application of the complex networks modularity method, Radivojević and Grujić (2018) were able to trace, in high resolution, the shifting nature of copper metal supply to Pločnik, which adjusted twice during the course of the Vinča culture: 1) when copper mineral use shifted to extractive metallurgy, in *c.* 5000 BC; and 2) when Vinča culture sites in the north were abandoned and new economic relationships forged with KGK VI cultural complex communities in Bulgaria (see also Figures 9-11 in Chapter 3, this volume). We can see, therefore, that the Pločnik communities were able to shift their subsistence economy towards the east of the Balkans, which may have contributed to extending the lives of these communities well into the 44th century BC (see Chapter 37, Table 3), almost 200 years longer than Belovode, and other key Vinča sites north from Pločnik. In Chapter 41 we will synthesise the metal technologies of both Belovode and Pločnik, and present a detailed exploration of new provenance data for both sites, feeding into a comprehensive

Table 6. EPMA compositional data of metal artefacts P10/13, C_P1/13 and C_P2/13 (selected significant trace element values), given in wt%. Values above *c.* 0.01 wt% (100 ppm) are considered reliable based on CRM measurements; values below this are indicative only. All data are corrected for values obtained from the reference material, using a procedure reported in the methodology section of Belovode metallurgy (Chapter 11). Values sought but not found at levels above *c.* 0.01 wt% were indicated as not detected (n.d.).

	S	Mn	Fe	Co	Ni	As	Ag	Sn	Te	Au	Pb	Bi	Analytical Total
	wt%	wt%	wt%	wt%	wt%	wt%	wt%	wt%	wt%	wt%	wt%	wt%	
P10/13	0.005	0.001	0.002	n.d.	0.008	n.d.	n.d.	0.030	n.d.	n.d.	0.008	n.d.	99.93
P10/13	n.d.	0.003	0.001	n.d.	0.016	n.d.	n.d.	n.d.	0.014	n.d.	n.d.	n.d.	99.95
P10/13	0.002	0.005	n.d.	n.d.	0.011	n.d.	n.d.	n.d.	0.015	n.d.	n.d.	n.d.	99.93
P10/13	0.001	0.011	n.d.	0.005	0.012	0.051	n.d.	0.016	n.d.	0.023	n.d.	0.017	99.65
P10/13	0.001	n.d.	0.001	0.009	0.012	n.d.	n.d.	n.d.	n.d.	n.d.	n.d.	n.d.	99.58
P10/13	0.001	n.d.	0.002	n.d.	0.019	0.005	n.d.	n.d.	0.004	n.d.	0.001	0.023	99.44
P10/13	0.005	0.001	n.d.	n.d.	0.015	n.d.	n.d.	0.002	0.003	n.d.	n.d.	n.d.	99.61
P10/13	0.004	n.d.	0.001	0.002	0.010	n.d.	0.011	n.d.	0.002	n.d.	n.d.	0.078	99.73
P10/13	0.005	0.005	n.d.	n.d.	n.d.	0.022	0.003	0.014	n.d.	n.d.	n.d.	0.017	99.59
P10/13	n.d.	n.d.	0.003	n.d.	0.004	n.d.	0.007	0.002	n.d.	n.d.	0.038	n.d.	99.77
C-P1/13	0.001	0.003	n.d.	n.d.	0.007	n.d.	0.024	n.d.	0.028	n.d.	0.280	n.d.	99.61
C-P1/13	n.d.	n.d.	0.002	0.004	0.006	n.d.	0.007	0.005	n.d.	n.d.	0.419	n.d.	99.34
C-P1/13	0.001	n.d.	0.008	n.d.	0.005	n.d.	0.019	0.006	n.d.	n.d.	0.260	n.d.	99.62
C-P1/13	0.005	n.d.	n.d.	0.002	n.d.	n.d.	0.024	0.026	n.d.	n.d.	0.445	n.d.	99.59
C-P1/13	0.002	n.d.	0.001	n.d.	0.004	n.d.	0.019	0.001	0.003	n.d.	0.087	0.031	100.17
C-P1/13	0.002	n.d.	n.d.	n.d.	0.004	n.d.	0.024	n.d.	0.015	n.d.	0.254	n.d.	99.57
C-P1/13	n.d.	n.d.	0.007	n.d.	0.013	n.d.	0.011	n.d.	0.024	n.d.	0.066	n.d.	99.57
C-P1/13	0.002	n.d.	n.d.	0.004	n.d.	n.d.	0.010	0.016	0.014	n.d.	0.211	0.010	99.91
C-P1/13	n.d.	n.d.	0.006	n.d.	0.014	n.d.	0.029	0.026	0.007	n.d.	0.343	0.017	100.03
C-P1/13	n.d.	0.002	n.d.	0.003	0.009	0.034	0.014	n.d.	0.012	n.d.	0.469	n.d.	99.68
C-P2/13	0.005	n.d.	n.d.	0.002	n.d.	0.016	0.002	0.025	0.037	n.d.	n.d.	0.014	99.99
C-P2/13	0.005	n.d.	n.d.	0.001	n.d.	n.d.	n.d.	0.006	n.d.	n.d.	0.007	n.d.	99.41
C-P2/13	0.005	0.015	0.005	0.009	0.006	0.011	n.d.	0.013	0.026	n.d.	0.016	n.d.	99.76
C-P2/13	0.006	n.d.	0.003	0.002	0.001	n.d.	0.002	0.023	0.007	n.d.	0.014	n.d.	99.52
C-P2/13	0.005	n.d.	0.003	0.003	0.007	0.020	0.002	n.d.	0.015	n.d.	0.003	n.d.	99.73
C-P2/13	0.004	0.007	0.005	n.d.	0.015	n.d.	n.d.	n.d.	n.d.	n.d.	0.008	0.019	99.68
C-P2/13	0.004	0.009	0.005	0.002	0.014	n.d.	0.007	0.004	n.d.	n.d.	0.034	0.002	99.65
C-P2/13	0.007	n.d.	0.001	0.009	n.d.	n.d.	n.d.	0.005	n.d.	n.d.	n.d.	n.d.	99.82
C-P2/13	0.005	0.004	n.d.	0.011	0.008	n.d.	n.d.	0.018	n.d.	n.d.	n.d.	n.d.	99.79
C-P2/13	0.001	0.004	0.002	0.004	n.d.	0.003	n.d.	0.011	n.d.	n.d.	n.d.	n.d.	99.90

Table 6a. LA-ICP-MS analysis of copper metal phases in production debris and artefacts. Values sought but not found above the indicated detection limit were treated as not detected (n.d.).

	Cu	Co	Ni	Zn	As	Ag	Sb	Te	Pb	Bi
	%	*μg/g*	*μg/g*	*μg/g*	*μg/g*	*μg/g*	*μg/g*	*μg/g*	*μg/g*	*μg/g*
detection limit		*1*	*20*	*5*	*37*	*4*	*5*	*2*	*2*	*1*
P10/13	100.0	23	84	31	bdl	7	bdl	bdl	bdl	bdl
C_P2/13	100.0	41	21	27	bdl	6	bdl	bdl	bdl	bdl
C_P1/13	98.2	bdl	82	131	bdl	230	19	bdl	17800	130

picture of the evolution of metallurgy in the Balkans, from the core Vinča sites to the entire region, and beyond.

The bibliographic reference for this chapter is:

Radivojević, M. and Rehren, Th. 2021. Pločnik: technology of metal production, in Radivojević, M., Roberts, B. W., Marić, M., Kuzmanović Cvetković, J., and Rehren, Th. (eds) *The Rise of Metallurgy in Eurasia*: 301–316. Oxford: Archaeopress.

Appendix

Appendix avaliable online as part of Appendix B at https://doi.org/10.32028/9781803270425/AppendixB_Ch26

Chapter 27

Pottery from Trench 24 at Pločnik

Neda Mirković-Marić, Marija Savić and Milica Rajičić

Introduction

The 2012 and 2013 excavations of Trench 24 at Pločnik revealed a total of *c.* 53,600 ceramic fragments encompassing *c.* 37,000 non-diagnostic and 16,500 stylistically and typologically indicative ceramic fragments (e.g. rims, handles, bottoms and ornamented belly fragments). The analysis of the ceramic assemblage was conducted with the aim of establishing the relative ceramic sequence excavated at the site which, together with the stratigraphic sequence and the radiocarbon dating programme, would enable the creation of a detailed typo-chronological framework. This is fundamental, not only for temporal resolution, but also for investigating patterns of ceramic production and consumption at Pločnik and their potential interconnections to metal production and consumption.

The analysis of ceramics previously excavated at the site by Šljivar and his associates has been published only in short summary and then according to nine vessel type classes Arsenijević and Živković 1998: 281–291) (Figure 1). For the new analysis conducted on the pottery material from Trench 24 excavated within the frame of *The Rise of Metallurgy in Eurasia* project, the methodological approach using the old pottery type classes was applied, but the recording characteristics in the processing model (containing fabric, colour, surface treatment, and decoration) were modified with metrical characteristics concerning rim percentages and rim diameter, weight of the fragments and total belly weight (cf. Orton 2013). The new pottery database also encompassed all the excavated sherds from the trench, in contrast to earlier work on the previous trenches which was influenced by traditions in which the ceramic assemblage was previously selected, and some fragments discarded in the field—an approach which could alter the final data.

The problem of the Pločnik periodisation within the Vinča culture and within its South-Morava variant has been a matter of disagreement between many authors. Previously published material from Pločnik dated Horizon I and II to the early or developed stage of Vinča Tordoš II phase (Šljivar 1996: 90). This contributed to the interpretation that the Vinča culture developed in the core area of the classical variant (central Serbia), and that the South Morava variant settlements developed in late Vinča Tordoš phase I (Nikolić 2004: 220). The periodisation of the subsequent Gradac Phase and the place of Pločnik within this has also been a matter of disagreement. According to the published material from the eastern part of the site, the Vinča culture at Pločnik ended with the Gradac Phase (Horizon III, Šljivar 1996: 90–94, Šljivar and Kuzmanović Cvetković 1997a, 1997b). Garašanin considered that the Vinča Pločnik phase in the South Morava variant ended with Vinča Pločnik I under the expansion of the Bubanj-Hum Chalcolithic culture (Garašanin 1973: 102; Garašanin 1979: 189). Jovanović considered that the Gradac Phase was tripartite and that the Neolithic in this region lasted longer than in the classic region, stating that Pločnik was a representative of Gradac Phase III (Jovanović 1994: 1–11; Perić 2006: 238). In this paper, we will present the main representations and characteristics of all the vessel types in specific horizons defined during excavation of the Trench 24 at Pločnik in order to help

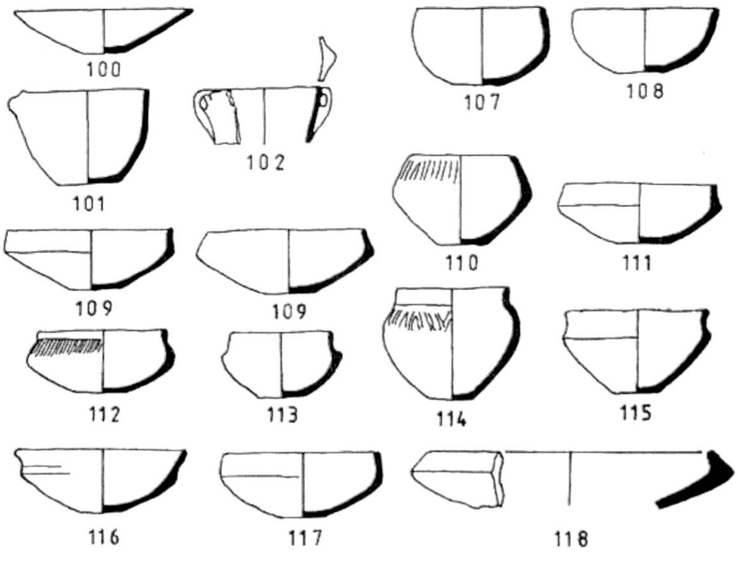

Figure 1. Bowls types according to Арсенијевић and Живковић 1995.

clarify the stratigraphic picture of the site itself. Code numbers for vessel types are indicated throughout in parentheses.

Horizon 1

The horizon is characterised by highest prevalence of bowls, fragments of which account for 64% of the whole assemblage (Figure 2). The most dominant are those from bowls with a turned-in rim (117), comprising 62% of the bowl assemblage in the horizon (Figure 3; 12.1–3). Conical bowls are well represented, both with slightly curved walls (type 101; 12%) and straight walls (100), and with thickened rims (104b, Figure 12.4; 9%); Biconical bowls are represented sporadically with only a few types: with a low upper cone (109, Figure 12.6), with a massively profiled angular shoulder (118), with a high cylindrical (113) or concave neck (115, Figure 12.5) and with a rounded shoulder and high cylindrical neck (114).

When compared to other horizons, sherds from amphorettae (320) and beakers with handles (220) make up a significant percentage of the whole assemblage (7% and 5% respectively), while those from jugs (340) account for only 1%. Among amphorettae, those with a concave upper and convex lower portion (325) are of note. Beakers represented include those that are biconically shaped with either a conical neck and ribbon handles (223), with conical neck and handles on the rim (221) or with a cylindrical neck and ribbon handles (222, Figure 12.7–8; 14.1%).

In the storage vessel class, amphorae fragments comprise 17% of the whole assemblage of the horizon, the most numerous being from vessels with a biconical body with a funnel (307, Figure 14.2; 7%) or cylindrical neck 305 (4%). Containers for the storage of solids, i.e. pithoi (400), are represented by 3% of the assemblage; the type with conical neck (404) dominates.

Among the cooking vessels, pots sherds are not frequent in this horizon, at only 4% of the assemblage. The most commonly represented pots are biconical with a conical upper part (422), biconical with a concave upper part (424), and globular (421). Lids (700) are represented to a lesser degree; those with a handle are most prevalent. So-called 'Gradac cups' (240, Figure 14.3) are also present.

Vessel ornamentation in the horizon is dominated by channelling, usually horizontal, arched and slanting. Polished and incised ornamentation are rare, as are tongue and wart-shaped protuberances in the form of pseudo-handles. The most frequently represented handles are band-like; button, tongue and band-shaped handles with the extended ends are less common.

Surface treatment is limited to smoothing; polished bowls are scarce. The same treatment also dominates in the amphorae and beakers.

The most prominent features of this horizon are 1, 2, 4, 5, 6 and 10.

Features 1, 2, 4, 5, 6 and 10 form the remains of a rectangular wattle and daub structure which was discovered at the bottom of spit 6 (Marić *et al.*, Chapter 25, this volume) and it should be noted that a good portion of the material from the house has secondary burning. The diagnostically significant pottery material from Feature 1 includes fragments of handles and bellies, and rim fragments, mostly of bowl type 117, but also types 100, 104. There are also fragments of biconically profiled beakers with a conical neck and ribbon shaped handle (223), of amphorae with a biconical body and a funnel-shaped neck (307); plates sherds (600) are also common in this phase, for example from plates with a rounded profile and thickened rim (601).

Feature 2 contains rim fragments belonging mostly to bowl type 117, but also types 100, 101, 104a, 104b, 104c, 106, 108, 109, 113, and 118. Beakers (220) are represented, including types those that are biconically shaped with a conical-neck and rim-handles (221), with a cylindrical neck and ribbon handles (222) and biconically profiled beakers with a conical neck and ribbon shaped handle (223). Gradac cups (240) are also present. Amphorae are represented, primarily with sherds from a biconical type with a funnel-shaped neck (307), as well as types with a mildly biconical profile and cylindrical neck (305) and biconically-shaped with a conical neck (306), with even upper and lower parts (303). Amphorettae (320) are present, especially a type with concave upper and convex lower portions (325), and jugs also occur (340). Pithoi (400) are represented with sherds from a biconically-shaped type with a conical neck (404); cooking pots with a rounded conical profile (421) and cone-shaped profile (425) are also represented. Lids (700) are present in a flat form with handle (704).

Diagnostically significant pottery material from Feature 4 includes rim fragments belonging to bowl types 100, 101, 104b, 107, 117, and sherds from amphorettae (320), biconically profiled beakers with conical necks and ribbon shaped handles (223), from biconically-shaped amphorae with conical necks (306), and from amphorae with cylindrical necks (305) and funnel-shaped necks (307). Pithoi (400) with biconically-shaped body and conical neck (404) are also represented.

Feature 5 contains fragments of bottoms, bellies and handles, and rim fragments belonging to bowl types 100, 101, 104b, 104c, 109, 117, to amphorettae (320),

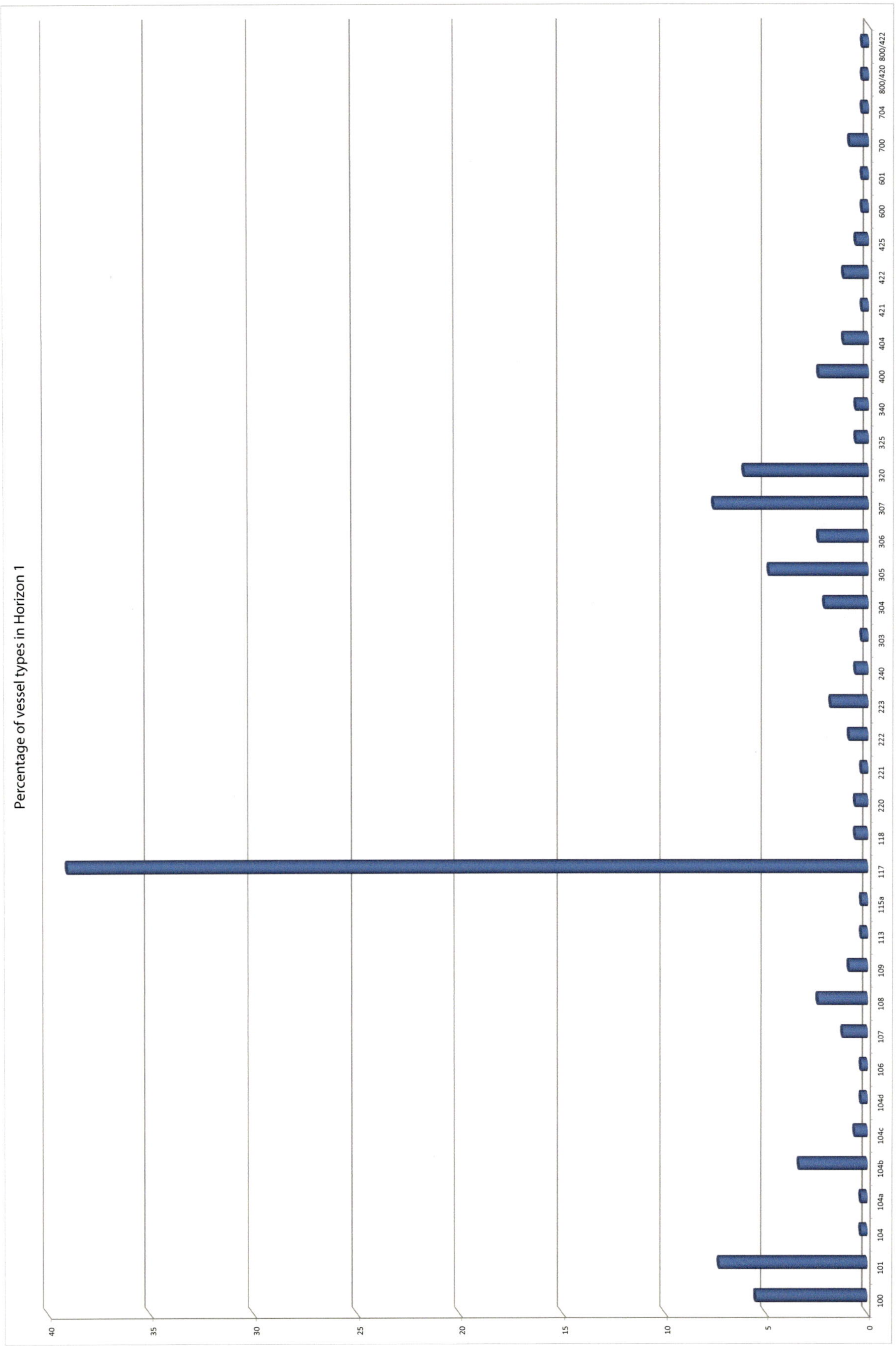

Figure 2. Representations of vessel types in Horizon 1.

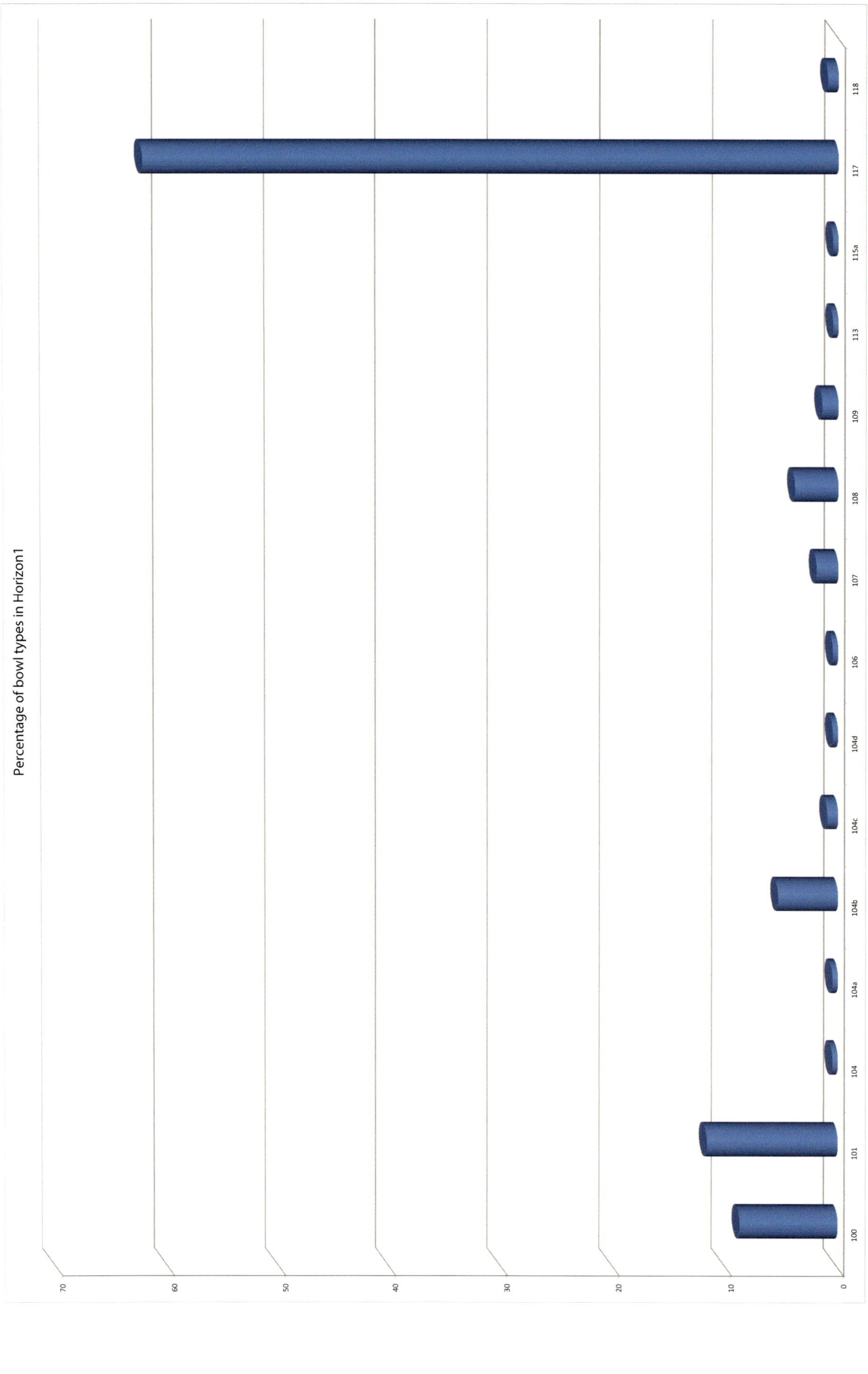

Figure 3. Representations of bowl types in Horizon 1.

biconically profiled beakers with conical necks and ribbon shaped handles (223), amphorae with pear-shaped bodies and elongated conical necks (304), pithoi (400), and biconically-shaped cooking pots with conical upper parts (422) and cone-shaped ones (425).

Feature 8 is a smaller concentration of pottery and stone fragments (Marić *et al.*, Chapter 25, this volume). The diagnostically significant pottery includes fragments of handles and bellies, several rim fragments belonging to bowl type 117 (in larger numbers), as well as bowl types 100, 101, and 104b. Beakers with cylindrical neck and ribbon handles (222) are also represented, as are amphorae with cylindrical necks (305), with pear-shaped bodies and elongated conical necks (304) and those with carinated profiles (301), and biconically-shaped pithoi types with conical necks (404).

Horizon 2

The most common vessels of this horizon are bowls, accounting for 58% of the whole horizon assemblage (Figure 4). The most numerous are conical with straight walls (100, Figure 13.7; 9%); slightly curved walls (101; 14%) and followed by those with profiled rim (104b, Figure 5; 13.6, 8; 17%). The next most common are sherds from biconical bowls, the majority being from bowls with a turned-in rim (117; 10% of all bowls, Figure 13.1–2) and with a massive profiled angular shoulder (118, Figure 13.4; 16%). Biconical bowls with a concave upper cone (115; 10%), and a high upper concave cone (115a; 14%) are also well represented, as are biconical bowls with a low upper cone (109; 7%). Other types are less common, e.g. Type 112 (Figure 13.5) and Type 110 (Figure 13.3).

Beakers (220) are not very numerous and are represented by just 1% of the assemblage, while jugs were not detected. In the spits that correspond to Horizon 2, biconical beakers with conical-necks and rim-handless (221), Gradac cups (240), and jugs (340) are all present. Of the storage vessels, amphorae are represented by 10% of the assemblage, equally distributed between those with a pear-shaped body with elongated conical (304) or cylindrical (305, Figure 14.5) neck, a biconical body and conical neck (306), a funnel-shaped (307, Figure 14.6–7) and cone shaped (308, Figure 14.8) varieties. Pithoi (400) are represented by 3% of the assemblage, specifically the biconically-shaped type with a conical neck (404). Food preparation vessels are represented by pot fragments (420), which make up 7% of the assemblage; those from biconically-shaped pots with a conical upper part (422) and concave upper parts (424) are the most numerous.

The most common ornamentation technique is channelling: horizontal, slanting and arched (Figure 15.7). Frequent ornaments include polished strips and lines. Incised ornamentation occurs on an insignificant percentage, as does red colour. Although it does not appear in the features, graphite ornamentation is represented by individual samples (a total of four, Figure 15.8) in the spits corresponding to the horizon. Band-shaped handles are frequently represented, then tongue, elbow, button, fan-shaped, protoma-handles and turban, wart like, cylindrical, vertical, oval, and strip with expanded ends. Most of the bowls are polished, with fewer being smoothed or highly polished. The same treatments also dominate the surfaces of amphorae.

Horizon 3

In this horizon, there is a significant representation of bowls (79% of the assemblage) (Figure 6). The most common are conical with a profiled rim (104b, Figure 15.3–5; 12%) and with slightly curved walls (101, Figure 7; 35%). Among the biconical bowls, those with a concave upper cone (115, Figure 15.1; 7%) are most represented, especially those with an oversized concave upper cone (115a, Figure 15.2; 22%). There is a significant presence of biconical bowls with a low upper cone (7%). Sherds from other forms of biconical bowls are few in number but present in a large number of simultaneous variations. These span types with a turned-in rim (117, Figure 16.1–2), with a low (111, Figure 16.7) or high cylindrical neck and angular shoulders (113), biconical bowls with a low (112, Figure 16.5) or high cylindrical neck and rounded shoulder (114, Figure 16.6), biconical bowls with a massive square profiled shoulder (118, Figure 16.3–4), and with cones of a equal height (110), with a funnel-shaped upper cone (116). Globular forms (106) are also present (Figure 16.8). A small percentage (1%) of other vessels are represented, including biconical beakers with a cylindrical neck and rim handles, Gradac cups (240, Figure 17.7–8), and square-shaped vessels and jugs.

In the storage vessel category, amphorae (300) are the most represented, accounting for 20% of the total horizon assemblage. Types with a biconical body and a funnel-shaped neck (307, Figure 17.3–4) and a biconical body and cylindrical (305, Figure 17.1) or conical (304, Figure 17.2) are the most common (both with 4%). There are also examples with cone shaped neck (308, Figure 17.6). Pithoi are not very numerous. Of the cooking vessels, pots are represented by 4% of the assemblage, and biconically profiled pots with conical upper part (422) by 3%; fragments of pots with a rounded conical profile (421) and fragments of cone-shaped pots (425) are found only in small numbers. Amphorettae (320, Figure 17.5) are present, as are cylindrical vessels—'chimneys' (450)—which are represented by 1% of the assemblage.

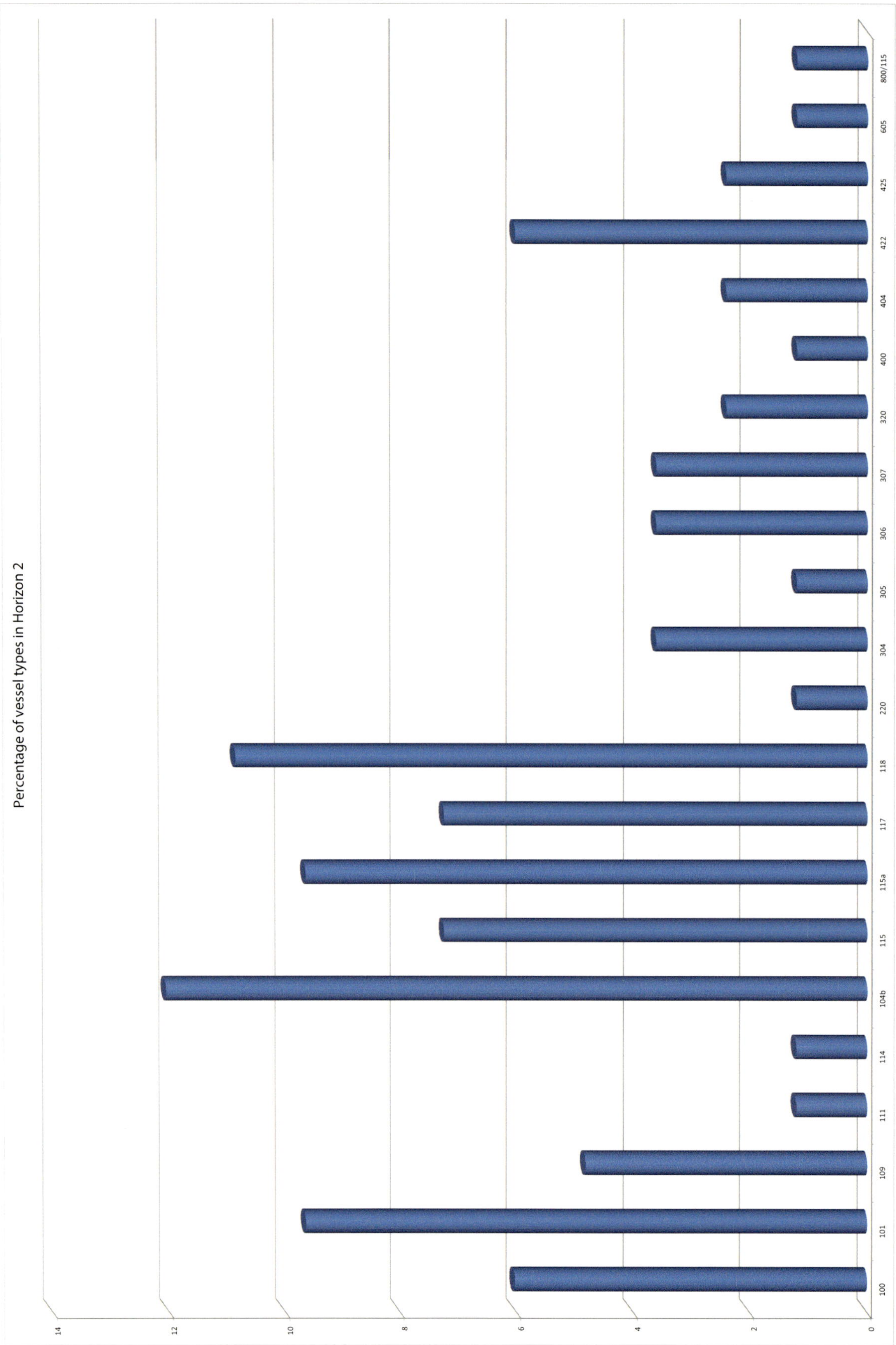

Figure 4. Representations of vessel types in Horizon 2.

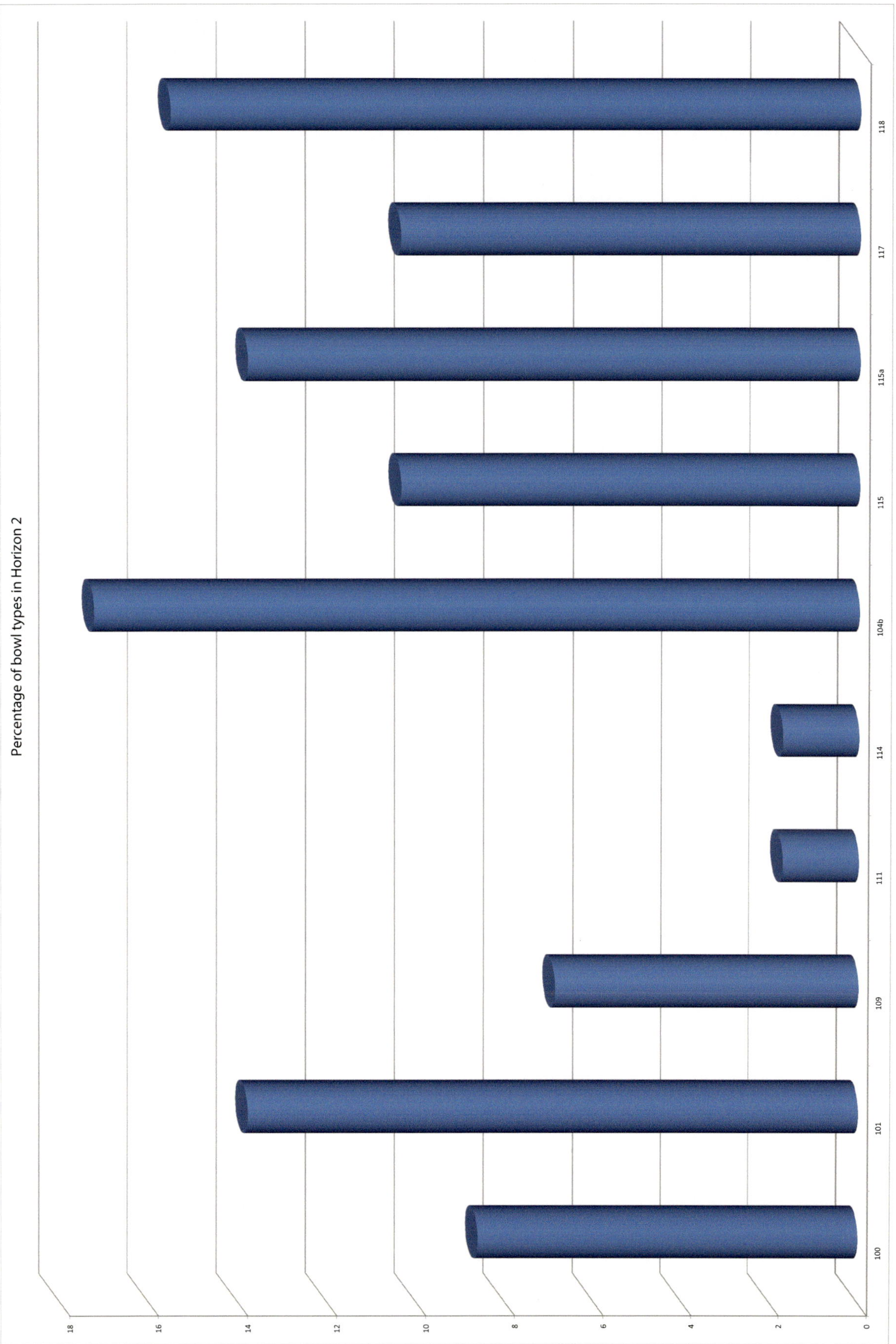

Figure 5. Representations of bowl types in Horizon 2.

The most common ornamentation in the horizon is channelling with horizontal forms being by far the most common, followed by spiral, and arched; channelled garlands are also found. Polished ornamentation in the form of lines, strips, and 'nets' are more numerous than incised ornamentation, which consists of lines and bands filled with encrustation. Graphite painting occurs on individual samples (within Features 16, 18 and 19). In a few cases, there are examples of the black topped decoration technique and of red painting (Figure 15.6). Handles are frequently band-shaped, and there are also button, tongue, tunnel, and oval, striped with extended ends at the junction with the body of the vessel. On the shoulder of the vessels, crescent and pinched, wart-like protuberances in the form of pseudo-handles occur, predominantly in bowls with a small flattened ellipsoid. Most bowl sherds are polished, with fewer being highly polished or smoothed. The same treatment dominates the amphorae fragments.

The most important features of this horizon are 14-15, 16, 21, 17, 22, 25, 35, 36, 37 and 19 (Marić *et al.*, Chapter 25, this volume). Feature 16, 21 is a large concentration of finds comprising predominantly pottery fragments (Marić *et al.*, Chapter 25, this volume). Diagnostically significant pottery material from Feature 16 includes fragments of handles and bellies decorated with channelling and polishing, and several rim fragments belonging to bowl types 100, 101, 118, 109, 111, 113, 115a, 115, 117, and 104b. Gradac cups (240) and amphorettae (320) are also present. Amphorae are represented by fragments from types with a mildly biconical profile and cylindrical neck (305), with a pear-shaped form with an elongated conical neck (304) and from biconical amphorae with a funnel-shaped neck (307); cooking pots with a biconical-shape and conical upper part (422) are also represented.

Diagnostically significant pottery from Feature 21 includes fragments of handles and bellies decorated with channelling and polishing, and several rim fragments belonging to bowl types 104a, 109, and 115a. There also fragments of pithoi (400) and from biconical amphorae with a funnel-shaped neck (307).

Feature 19 is a sub-oval concentration of finds (Figure 14) detected in spit 14 (Marić *et al.* Chapter X, this volume). Diagnostically significant pottery from Feature 21 contains fragments of handles and bellies decorated with channelling, and several rim fragments belonging to bowl types 109, 110, 115a, and 117. There also fragments of pithoi (400) and of biconical amphorae with a pear-shaped body with elongated conical neck (304) and of those with cylindrical neck (305). Pithoi are represented by fragments from the biconically-shaped type with a conical neck (404); cone-shaped cooking pots (425) and cylindrical 'chimneys' (450) are also present.

The fragmentation and character of the pottery material indicates that these features could be described as refuse pits.

Horizon 4

In this Horizon, we see a wide representation of bowls (up to 53% of the whole Horizon 4 assemblage) (Figure 8). By far the most represented are conical bowls with a profiled rim (104b) (Figure 9; 18.1–2), which account for 42% of all bowls. Other variations of the conical bowls are represented in smaller numbers, including those with slightly rounded walls (101, Figure 18.4; 10%), with straight walls (100; 3%) and with a thickened rim (104a, Figure 18.3; 2%). A large percentage of fragments are from biconical bowls with a rounded shoulder and low cylindrical neck (112, Figure 18.5; 12.5%) and those with high cylindrical neck and rounded shoulder (114, Figure 18.6; 8%). Other biconical bowls are represented in smaller numbers: with unequally high upper and lower part (110), with a high cylindrical neck and square shoulder (113, Figure 18.7) and with a concave upper part (115, Figure 18.8). The same is true for closed spherical bowl with a ring neck (106) and hemispherical bowls (108). Pedestal beakers (200) are present, but not in large numbers (1%). Plates (600) are common in this phase, those with conical profile and massively thickened rim (605) accounting for 4%.

Storage vessels are represented by amphorae and pithoi. Amphorae (300) are represented by 9% of the assemblage, the most common being those with a mildly biconical profile and cylindrical neck (305, Figure 19.1–4) and biconically-shaped amphorae with a conical neck (306, Figure 19.7). Pear-shaped amphorae with an elongated conical neck (304) and biconical amphorae with a funnel-shaped neck (307, Figure 19.5–6), which were pre-dominant in Horizon 3, are present in only very small numbers. Pithoi (400) are represented by 5% of the assemblage.

Of the cookware, pots (420) are represented by 8% of the Horizon 4 assemblage, the most common being those with a biconical profile and conical upper part (422, Figure 19.8). Lids (700) are present in a small percentage, with or without a handle.

The most common ornamentation of the horizon is channelling - horizontal or vertical and intertwining. In many cases there is also incised ornamentation including lines filled with encrustation, ribbons, geometric ornaments and punctuation, as well as lines and bands, and bundles of lines, sometimes also filled with encrustation (Figure 20.1–4, 6–8). There are a small number of examples of a black topped decoration technique, as well as the red colouring paste and polished ornamentation. Impressed ornamentation on vessels of the intermediate fabric and barbotine (Figure

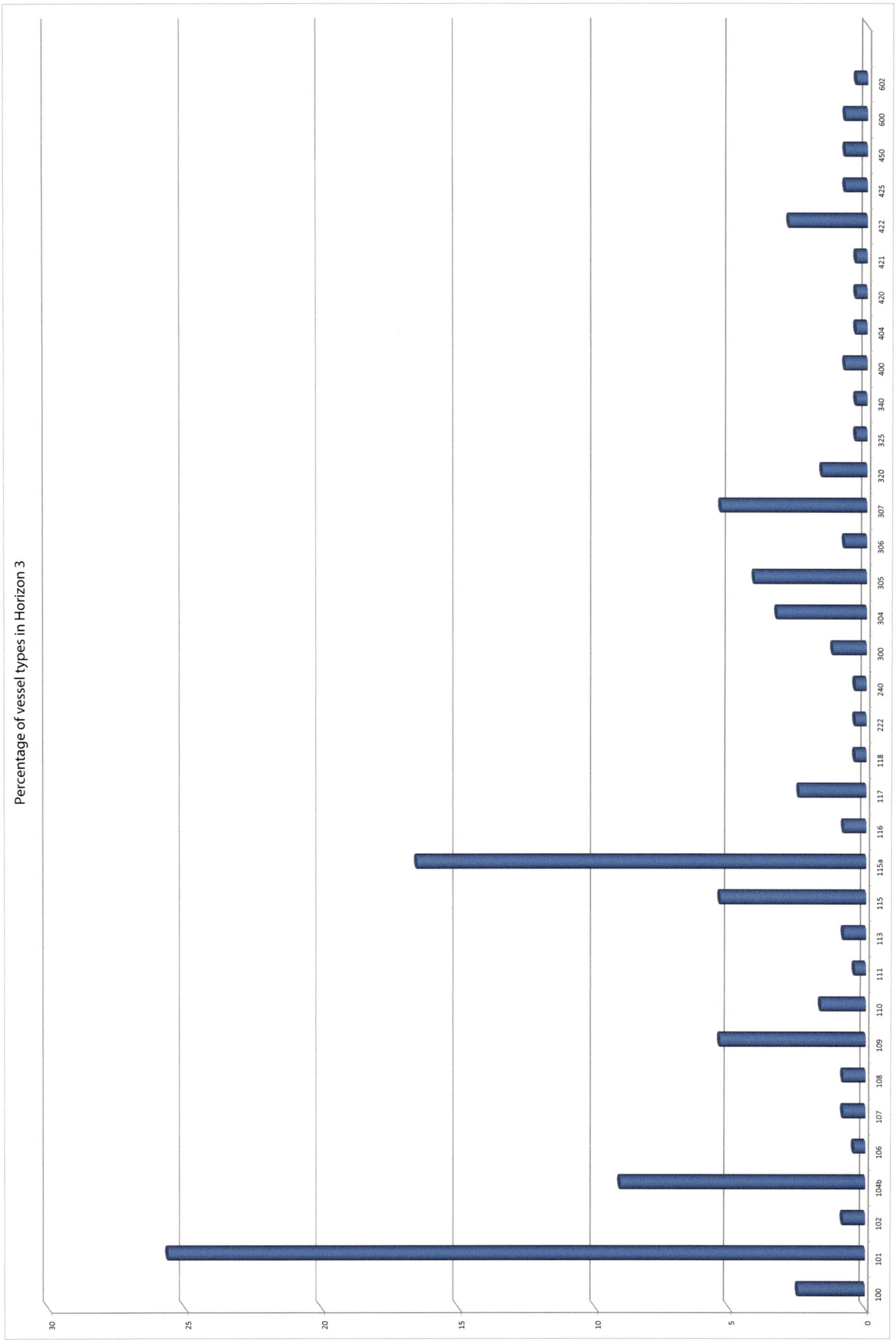

Figure 6. Representations of vessel types in Horizon 3.

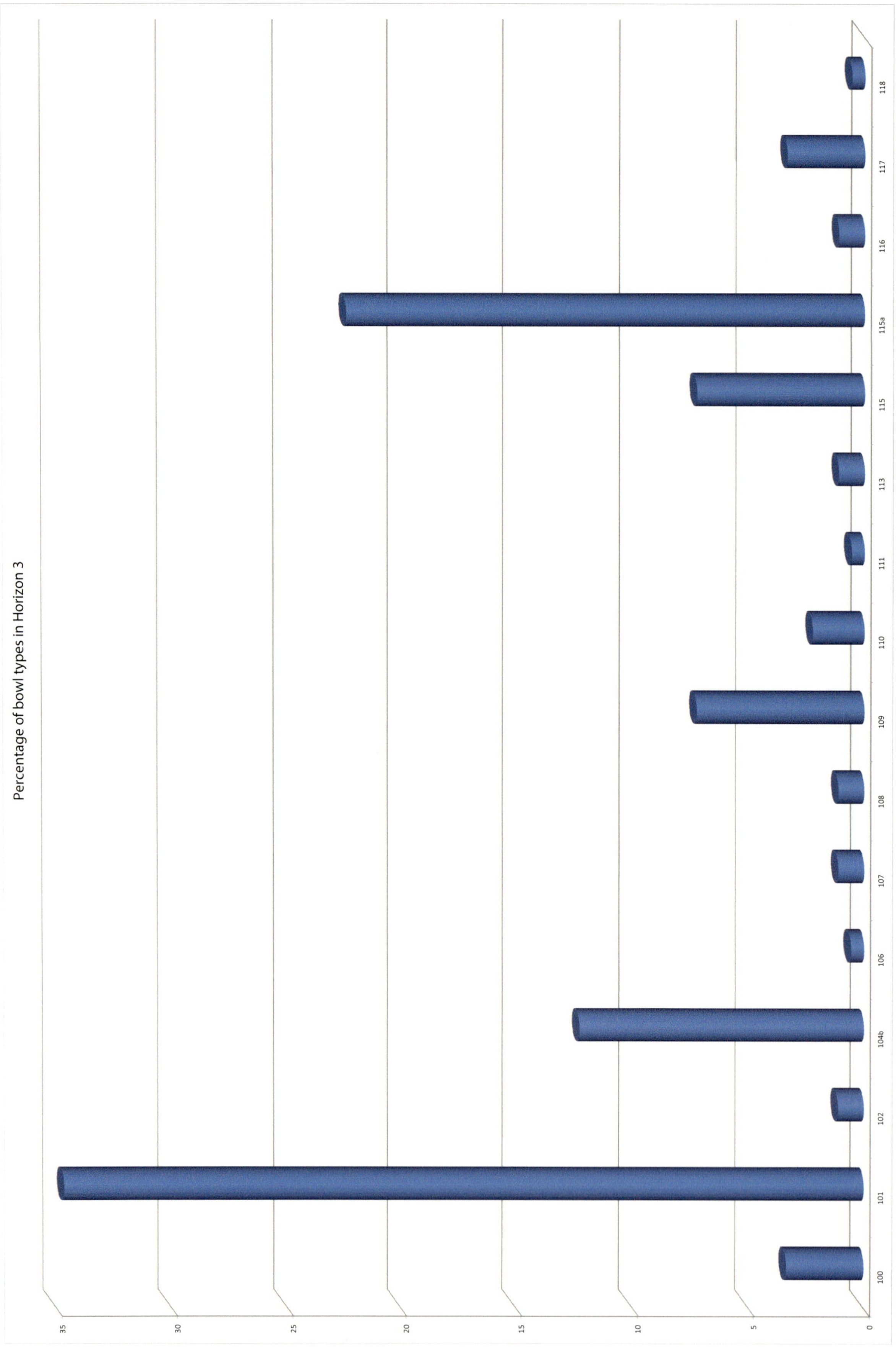

Figure 7. Representations of bowl types in Horizon 3.

20.5) is present in small numbers. Strap handles are present in most cases, but also forms also occur: striped forms with button-endings, cylindrical, tongue-shaped, striped with extended ends, tunnelled, crescent and, in particular the flap, then articulated, button shaped, striped with a protoma. There are also projections, in the form of pseudo-handles: crescent, nipple and ellipsoid flattened of the shoulder. Most bowl fragments are polished, with smaller numbers of highly polished and smoothed examples. Amphorae and cups are also polished and smoothed.

The most important features of this horizon are 29, 30, 31, 32, 34 (Marić *et al.*, Chapter 25, this volume). Feature 34 is a large, irregular rectangular area of white ash (Marić *et al.*, Chapter 25, this volume). Diagnostically significant pottery material from this feature includes fragments of handles and bellies decorated with channelling, polishing and incisions, several rim fragments belonging to bowl types 100, 101, 104b, 106, 107, 108, 109, 110, 112, 113, and 114, pedestal bowls (200), amphorae with a mildly biconical profiles and cylindrical necks (305) and pear-shaped amphorae with elongated conical necks (304). Cooking pots (420) are represented by fragments from types with biconical profiles and conical upper parts (422) and cone-shaped pots (425). Pithoi (400) fragments are also present, specifically the biconically-shaped type with a conical neck (404). The fragmentation and character of the pottery material indicates that this feature can be described as a refuse pit.

Horizon 5

There was relatively little material in Horizon 5, but a high representation of bowls (Figure 10) fragments of which account for 87% of the whole horizon assemblage. The most common fragments are from conical bowls with a profiled rim (104b, Figure 21.1–2; 33%), with a slightly curved walls (101) (10%) and with straight walls (100, Figure 11; 5%). Rounded forms include spherical bowls with a ring/cylindrical-shaped neck (106; 7% of bowls), spherical bowls (107; 5%), and hemispherical bowls (108; 3%). Several forms of biconical bowls are represented with the majority having high cylindrical necks and rounded shoulders (114; 14%) and low cylindrical neck and rounded shoulder (112, Figure 21.3–6; 10%). Other biconical bowl types are less numerous but still encountered, including those with a low upper cone (109), with an equally high upper and lower part (110), with a high cylindrical neck and squared shoulder (113, Figure 21.7) and with a concave upper part (115).

Pedestal beakers (200) and amphorettae (320) are represented by a very small percentage of fragments. Containers for storing food and liquid, i.e., amphorae (300) are represented by 9% of the fragments, including mild biconical profiles and a cylindrical neck (305) and, to a lesser extent, pear-shaped amphorae with elongated conical neck (304).

In the first two spits of the horizon, channelling is the most common ornamentation (Figure 22.1–2), especially slanting forms, while incised ornamentation also occurs. Polishing is present on only a small number of fragments, while in the final spits (24–25 in Feature 38), curved, vertical and arc channels and barbotine (Figure 22.8) are the most common decorations. The latter are present in large quantities, as is incised ornamentation in the form of bundles of lines, and incised strips filled with punctuation (Figure 22.3–7). There are also pseudo-handles that act as decoration with types including: wart; tang; button; lunate; fan-shaped; and horn-shaped. The most common are strap handles, button-shaped, ribbon, with extended ends at the junction with the vessel, tunnel-shaped, but also tongue-shaped, prismatic, and horn shaped with button closures. The handles on the bowls are pseudo handles, often in rudimental form; they are smaller and have decorative character, but also functional, making the slippery burnished or polished vessel easier for handling. Most bowl fragments are polished; some are highly polished or smoothed. Amphorae and beakers are smoothed and polished.

Although a small amount of pottery is present, and it cannot make a significant difference to the typology, the presence of spherical bowls, the domination of barbotine and different temper of the vessels with intermediate and coarse fabrics (cf. Amicone, Chapter 29, this volume) suggests that we can separate material from spits 24, 25 as being earlier than materials in other spits of this horizon. Though there are few fragments that are typologically definable, barbotine is a decoration also present on bowls with a short cylindrical neck and rounded shoulder (cf. Garašanin and Garašanin 1979: Supska spit 9, Table 36).

The most important features of this horizon are 38 and 39 (Marić *et al.*, Chapter 25, this volume). Feature 38 was a pit house with several separate cells but was found in its secondary role as a refuse pit, evidenced by the highly fragmented finds and the diversity of layers comprising the infill of the cut (Marić *et al.*, Chapter 25, this volume). The amount of pottery fragments from Feature 38 is substantial. Diagnostically significant pottery material includes fragments of bellies decorated with channelling and polishing technique, rim fragments belonging to bowl types 104b, 104a, 112, and 114 (in large quantity), as well as types 100, 101, 106, 107, 108, 109, 110, 112, 113, and 115a. Cooking pots (420) are represented by fragments from types with biconical profiles and conical upper parts

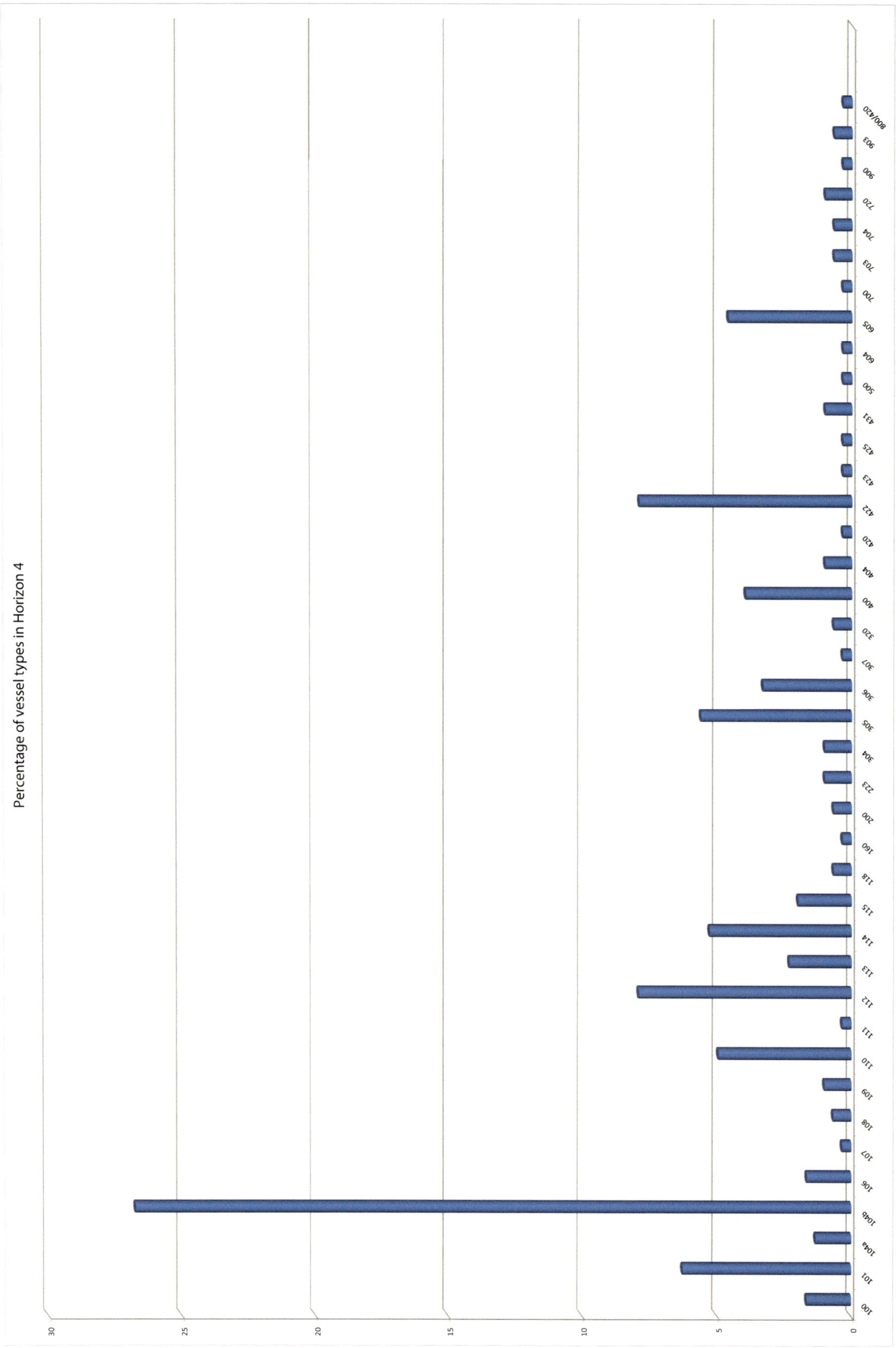

Figure 8. Representations of vessel types in Horizon 4.

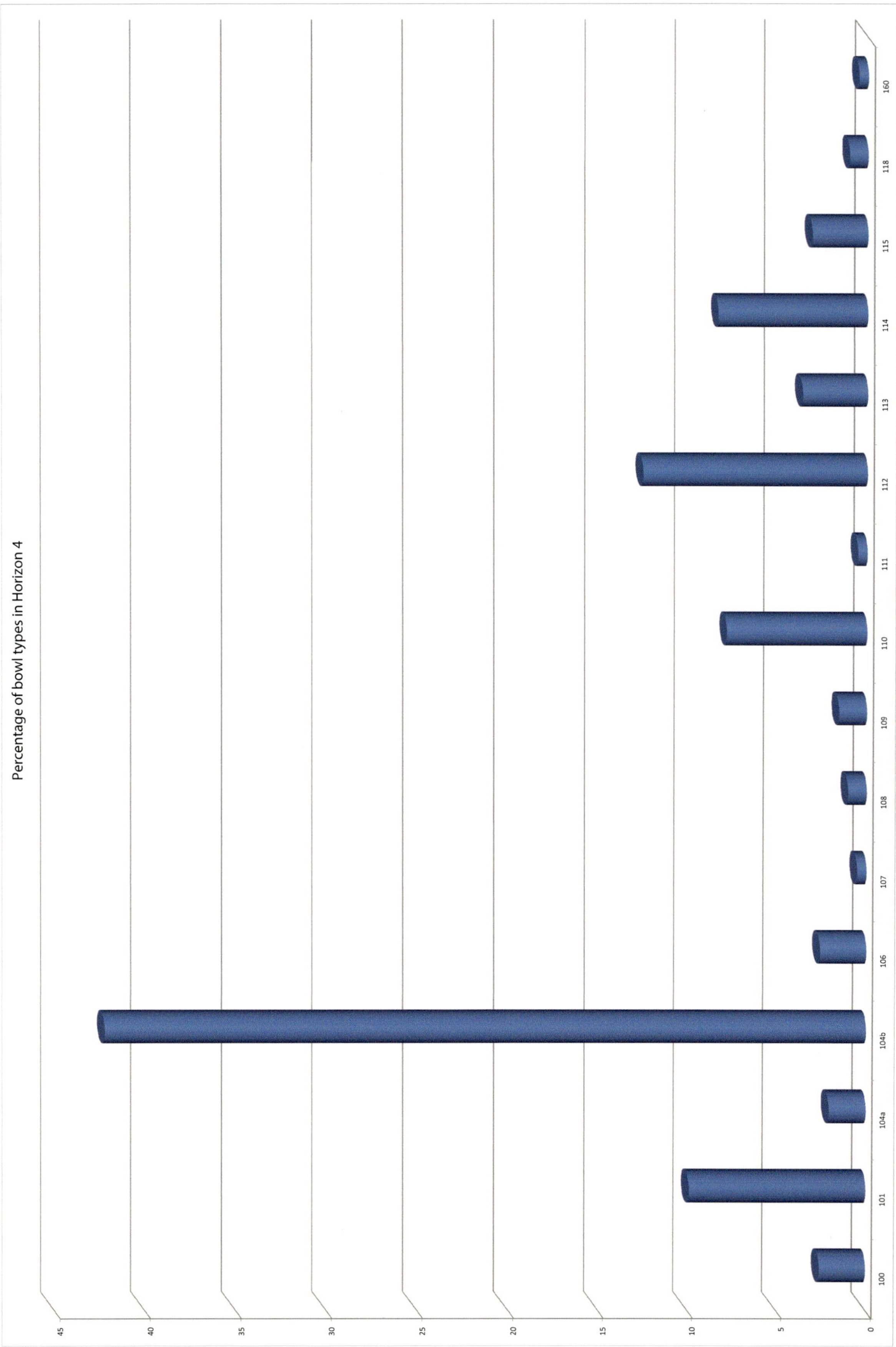

Figure 9. Representations of bowl types in Horizon 4.

(422) and those with cylindrical profiles (423). Beakers (220) are also represented with sherds from biconically profiled forms with conical necks and ribbon shaped handle (223), also amphorae with pear-shaped forms with elongated conical necks (304), with a mildly biconical profile and cylindrical necks (305), and from biconically-shaped amphorae with conical necks (306), biconical amphorae with funnel-shaped necks (307), amphorettae (320). Biconically-shaped pithoi (400) with conical neck (404) are also present, as are cooking pots (420) with biconical profiles with a conical upper part (422). There are few fragments from cooking pans (500) with sporadic fragments from plates with conical a profile and massively thickened rim (605), lids (700, 703), prosopomorphic lids (720), and plastic vessels (900).

The fragmentation and character of the pottery material indicates that this feature could be described as a refuse pit.

Relative chronology

Horizon 1

The material of this horizon has the characteristics of late Vinča Pločnik II phase (according to Garašanin 1979) in the classical variant with elements distinctive of the South-Morava and Kosovo variants: conical bowls with rim handles, beakers with two handles, and jugs. Although the presence of these types, as well as type 117 has analogies with the material of Horizon 2, the low representation of the other bowl types, complete degradation in the production, and also a higher percentage of beakers with rim handles separates this horizon from the previous one and fits Gradac III Phase as proposed by Jovanović (1994: 10).

Horizon 2

There is a large representation of bowls with a concave upper cone, and those with oversized concave upper cone, as well as of bowls with turned-in rims and with massively profiled angular shoulders, which have analogies in the material from the Vinča Pločnik phase II according to Garašanin (1979: 175–181).

The final phase of the Vinča culture at the Vinča site is represented by the material from the spits above 4.5 m (Vasić 1936a, Garašanin 1979) and corresponds to other sites of classical variants: Crkvine in Stubline, Crkvine in Mali Borak (Spasić 2011: 101–146), Jakovo-Kormadin (Glišić and Jovanović 1960: 113–139), Obrež-Beletinci (Brukner 1962: 89–122, Tables IV–VI), as well as Gomolava Ib in Vojvodina (Brukner 1980: 33–37, Table IV).

There are analogies in late Vinča culture settlements in the Grivac VI horizon (Nikolić 2004: 213–216), Divostin II b horizon (Madas 1988: 143–171, Figure. 6.2–6.21, Bogdanović 1990: 99–107), in spits 3 and 2 in Supska (Garašanin and Garašanin 1979: 40, 41, Tables I–VII). The material of this horizon has analogies in the spits 4–1 excavated in 1962/1963 and 1985/86, and from the last dwelling horizon excavated during 2001–2003 on the site of Motel Slatina and the last residential horizon of Trench XV at Drenovac (Perić 2006: 243). For the Kosovo variant there are analogies in later horizons at the sites of Predionica and Valač (Garašanin 1998: 80–84).

Horizon 3

The characteristics of this horizon are a wide variety in terms of vessel types, primarily bowls, dominated by a few major types. We see the abandonment of old types, characteristic of Horizons 4 and 5, and the emergence of new ones that will be dominant in the following, younger horizons. Ornamentation in all the techniques is extensive. Especially dominant are spiral channels, which are very numerous on amphorae, occurring on the shoulder and belly, and sometimes with elongated neck, which is decorated with horizontal channels.

Horizon 3 can be defined as Vinča Tordoš II—Vinča Pločnik I phase, or Gradac Phase in the classical sense. According to Jovanović, a feature of the Gradac I stage are: jugs with ribbon handles, biconical bowls with button-shaped handles, altars with a protoma in the form of bulls heads, plates with a thickened rim, and amphorae with a high funnel-shaped neck, with channelling and incision as decorative techniques (Jovanović 1994: 22; Perić 2006: 238).

The horizon between 7.5 and 4.5 m at Vinča can be considered as the height of the Vinča culture (Tasić and Tomić 1969: 57). On the eponymous site, this phase is characterised by bowls with cylindrical necks and reinforced rounded shoulder; bowls with a concave upper cone, and deep bowls with a conical neck and much accentuated shoulder (Schier 1996: 146, Figure 7 S242, S243, S52, Figure 8 S244, S54). Within the classical variant there are analogies in the Gomolava Ia, a–b phase at Vojvodina (Brukner 1980: 33–37, Tables I–III).

Within the South Morava variant, an analogy is found within the Grivac V horizon (Nikolić 2004: 209–213, 220–226), the older Vinča horizon at Crnokalačka bara (Tasić and Tomić 1969: 60), spits 4, 5 and 6 at Supska (Garašanin and Garašanin 1979: 39, Tables XIV–XX), horizons V–VII in Selevac (Vukmanović and Radojčić 1990: 289–316), Gradac (Stalio 1972: Tables I, II, IV, and V), in the material within the spits 5–7 excavated in 1985/1986 at the Motel Slatina site, as well as in the material from spits 5 and 6 from Trenches I–III from 1962/1963 (Perić 2006: 243). Amphorae with channelled spirals are occur in Rudna Glava hoards (Jovanović 1994: Figure 3).

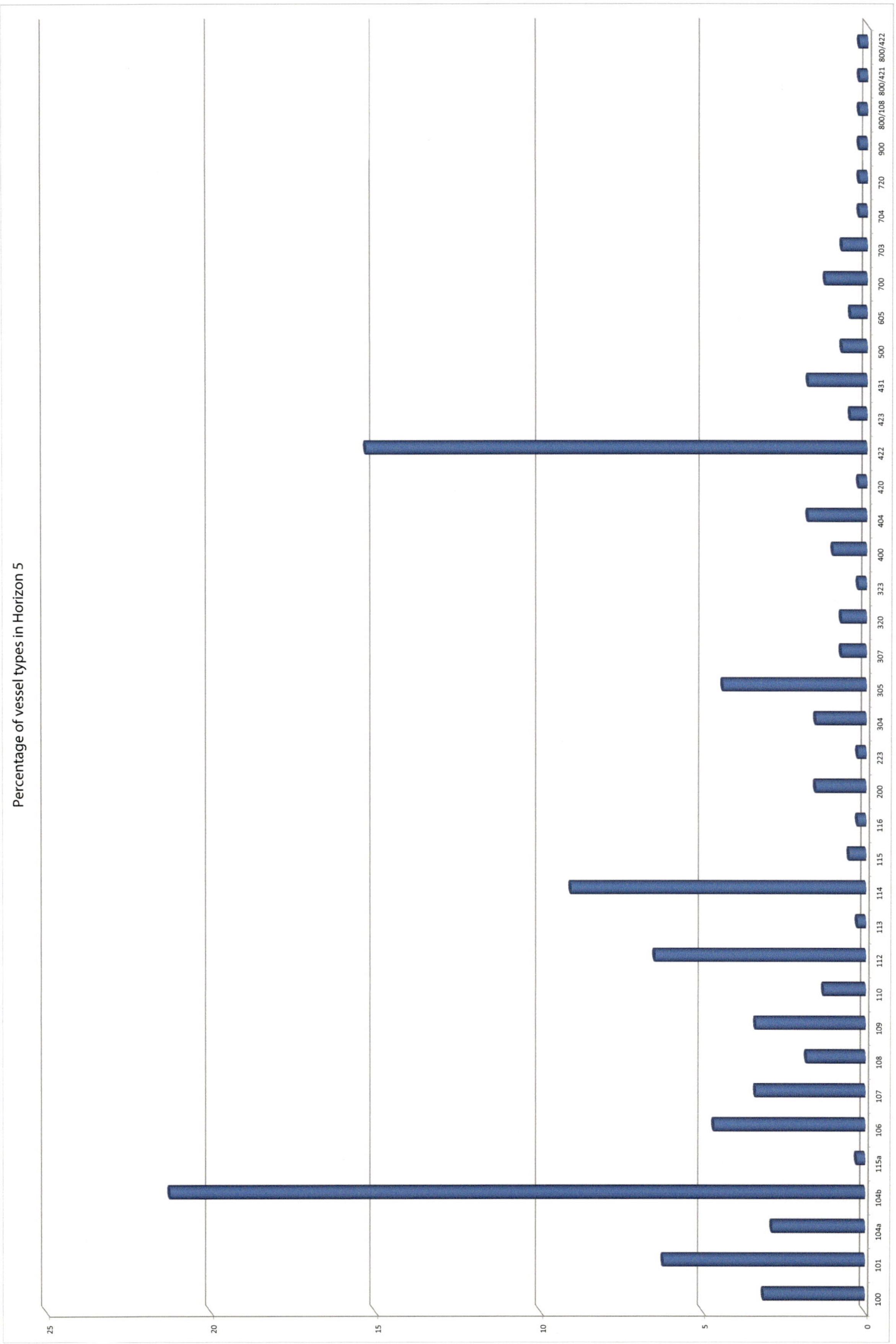

Figure 10. Representations of vessel types in Horizon 5.

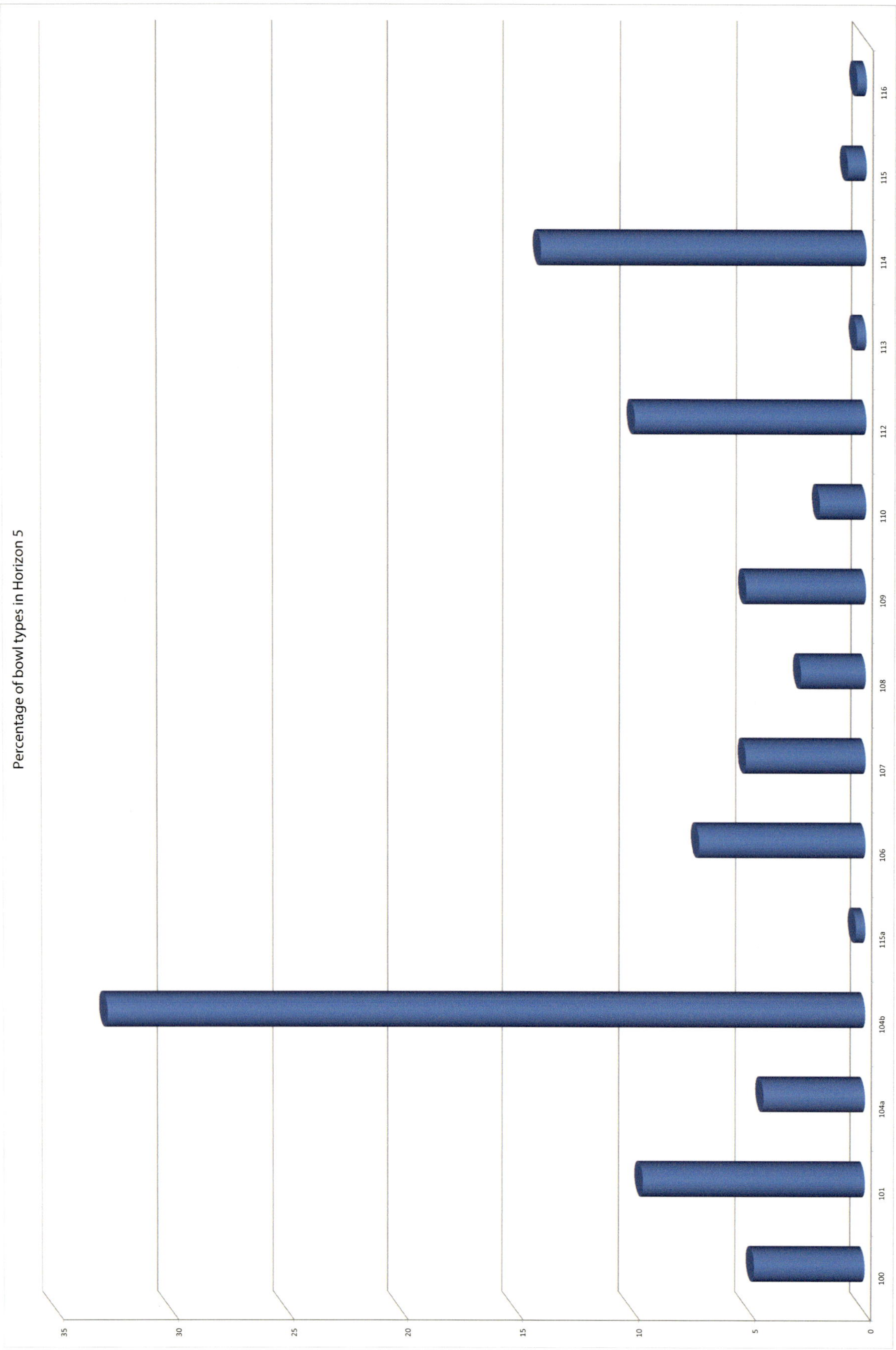

Figure 11. Representations of bowl types in Horizon 5.

The house horizon at Predionica, and the settlement of Fafos I in Kosovo variant (Garašanin 1998: 80–82) also belong to this stage.

Horizon 4

There is a large percentage of bowls with short and long cylindrical neck and rounded shoulders. There are no characteristics of the Gradac Phase present (types 117, 118, 102, 240, and 340); this material fits in an earlier stage of the Vinča culture. Bowls with a short upper part, characteristic of the early stages of the Vinča culture are scarce. This material can best be defined as belonging to the Vinča Tordoš II phase.

In Vinča, the Vinča Tordoš II phase belong spits between 8.5 and 6.5 m (Garašanin 1993: 7–20). According to Schier (1996), this includes Phases 4–5c in which commonly includes rounded bowls with short necks (Schier 1996: S135, S136), biconical bowls with rounded shoulders and vertical necks (Schier 1996: Figure 6 S153, S163) and those with curved or angular shoulders with very long, sometimes concave neck (Schier 1996: Figure 7 S178, S169, S185).

The material has analogies in Supska spit 7 (Garašanin and Garašanin 1979: XXI 38, 1–7, Table XXII, 1–7, XXIII, 1–8, XXIV, 1–5, 1 XXV, 1–5) and construction horizons I–IV at Selevac (Vukmanović and Radojčić 1990: 295–296, Figure 9.3, Type 314–316).

Horizon 5

The material has analogies with that in spits 7, 8 and 9 at Supska, which have been dated to the Vinča Tordoš phase I and I / II (Garašanin и Garašanin 1979: Table XXVI). The material at Vinča to 8 m represents the three forms of bowls: shallow with a very short upper cone (Vasić 1936a: Figure 30, 31, No. 62–65, Schier 1996: Figure S21, S28); with a short-rounded shoulder (Vasić 1936a: Figure 14, 18, No. 87, 33c, 67a, Schier 1996: Figure 5 S124, S126); and deep bowls of spherical shape with a short neck ring (Vasić 1936a: Figure 18 Nos. 93-98, 31, 32; Schier 1996: Figure 5 S141, S146, S135, Figure 6 S137, S132, S143, S135).

In the material currently published from Belovode, the oldest phase (Belovode A) is attributed to a developed stage of Vinča Tordoš I, while Phase B Belovode is attributed to Vinča Tordoš II (cf. Šljivar and Jacanović 1995, 1996, 1997b; Šljivar *et al.* 2006: 251).

The typical forms of Tordoš early stages, which dominate the classical variant—primarily bowls with a short upper cone (the earliest strata of the Belovode, Vinča)—are present in a small numbers, so this horizon can be typologically dated to the Vinča Tordoš II / I with the possibility that the two lowest spits belong to the earlier period.

Discussion and conclusions

Based on the statistical analysis of material from Trench 24 and analogies from other Vinča culture sites, confirmed by new radiocarbon dates (Chapter 37, this volume), we can say that the complete span of Vinča culture is represented. The material from the lower spits indicates the existence of the older phase in this trench and this requires additional attention. Specifics of the South Morava Vinča variant associated with the development of metallurgy can also be followed in Trench 24 (Garašanin 1979: 188), in the so-called Gradac Phase. Some characteristics of these regional variations, especially jugs and beakers with handles, are present in the assemblage (Jovanović 2006: 221–235). A larger sample would allow better typological analysis of the specific types of cups and pitchers specific for this variant (Jovanović 2006: 221–235), however the stratigraphy of Trench 24 fits best within the tripartite division of the Gradac Phase proposed by Jovanović (1994: 1–11) as confirmed by absolute dates from individual horizons.

The representation of vessels traditionally considered serving and consumption class of vessel, namely bowls, is hugely disproportionate in relation to those for cooking and storage, as at Belovode, and may lead to the conclusion that these vessels had other functions, for example for the short term storage of food or for mechanical preparation of food and beverages (Vuković 2010: 12). The very small number of cooking pans compared to other cooking vessels could indicate some preferences concerning food preparation in different regions with Vinča period settlements.

Surface treatment of the Pločnik vessels varies, sometimes affected by post-depositional processes. The surface of the vessel could be smoothed, burnished or polished. While cooking pots and pans, and pithoi (i.e. vessels for storage of food) have smoothed surfaces, burnishing and polishing appear on vessel classes for serving and consumption (bowls, beakers, jugs), but also on vessels for the storage of liquid (amphorae and amphorettae). Burnishing and polishing of the bowls and amphorae have a practical purpose, making the vessel less porous. These processes compact particles on the surface, making it harder and more resistant to abrasion.

The bibliographic reference for this chapter is:

Mirković-Marić, N., Savić, M. and Milica Rajičić, M. 2021. Pottery from Trench 24 at Pločnik, in Radivojević, M., Roberts, B. W., Marić, M., Kuzmanović Cvetković, J., and Rehren, Th. (eds) *The Rise of Metallurgy in Eurasia*: 317–344. Oxford: Archaeopress.

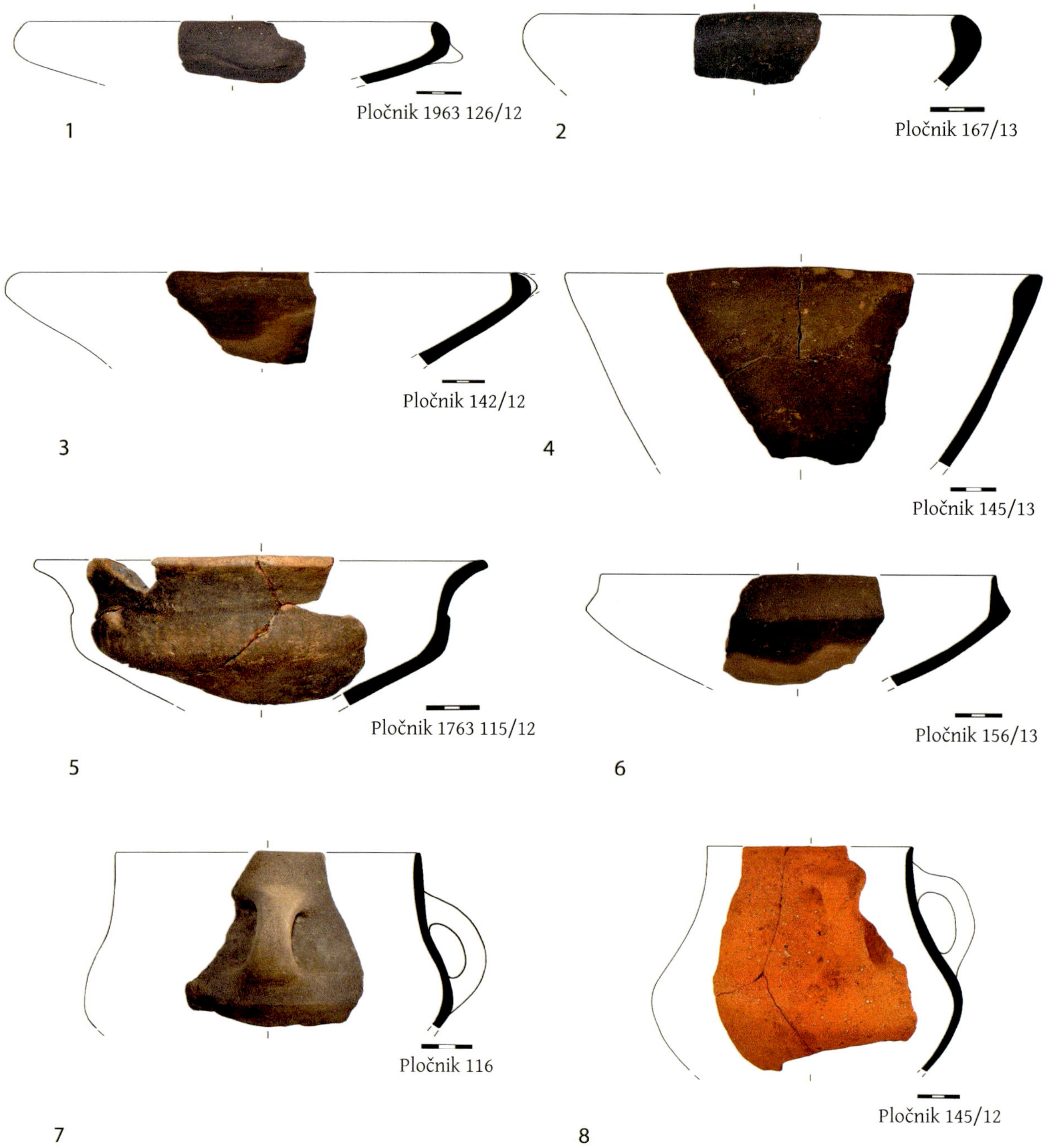

Figure 12. Most characteristic vessel types of Horizon 1.

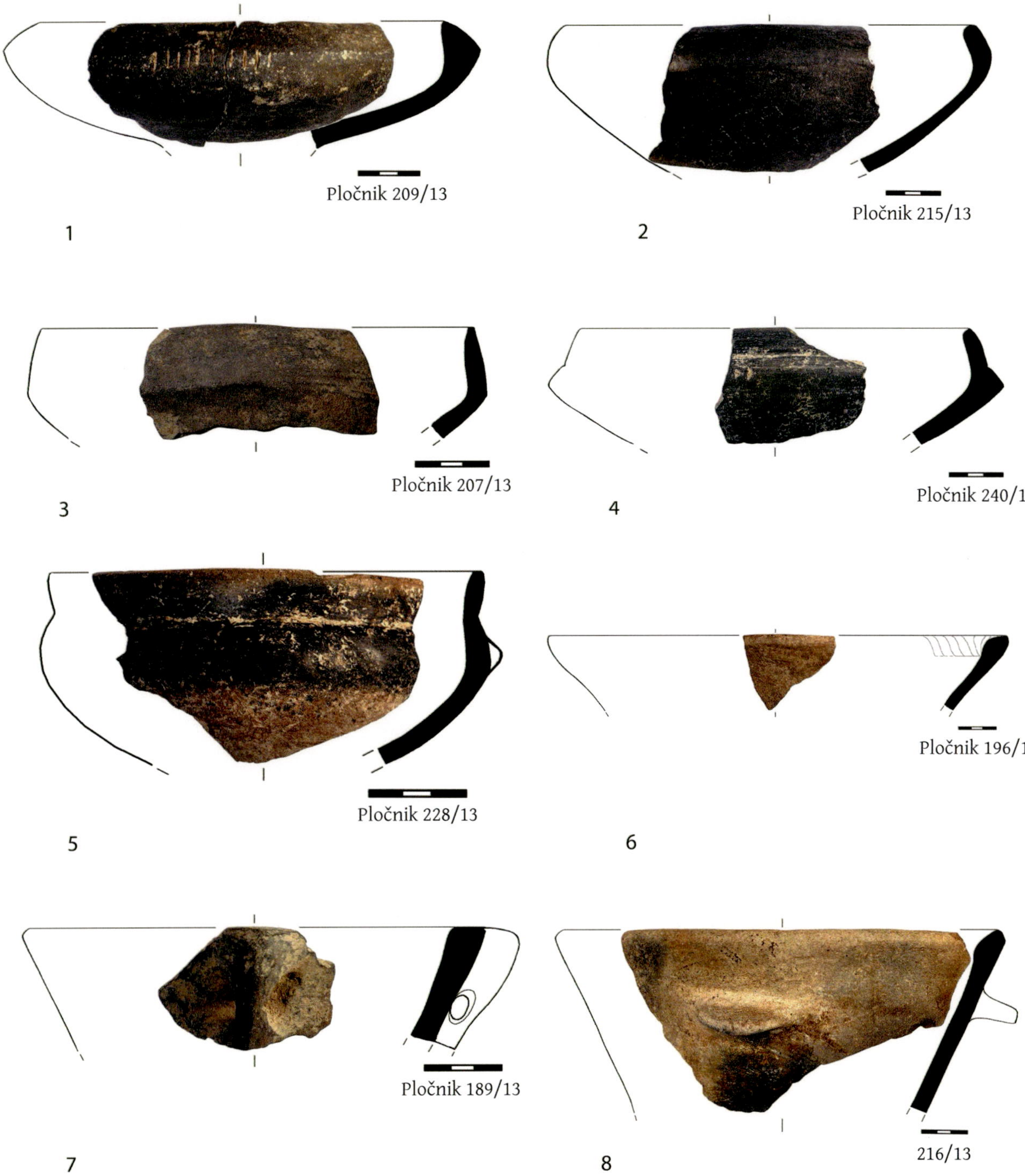

Figure 13. Most characteristic vessel types of Horizon 1.

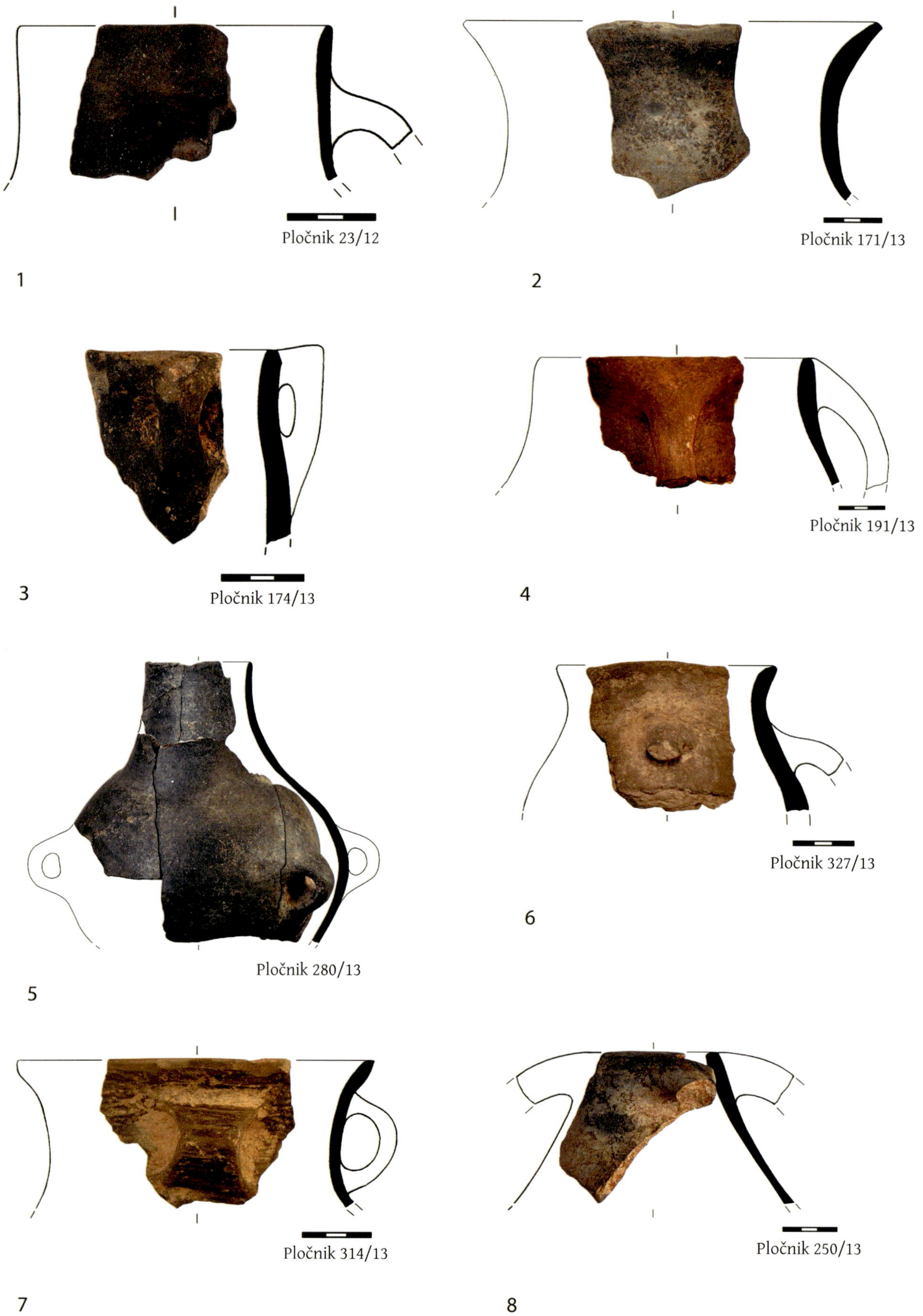

Figure 14. Most characteristic vessel types of Horizons 1 and 2.

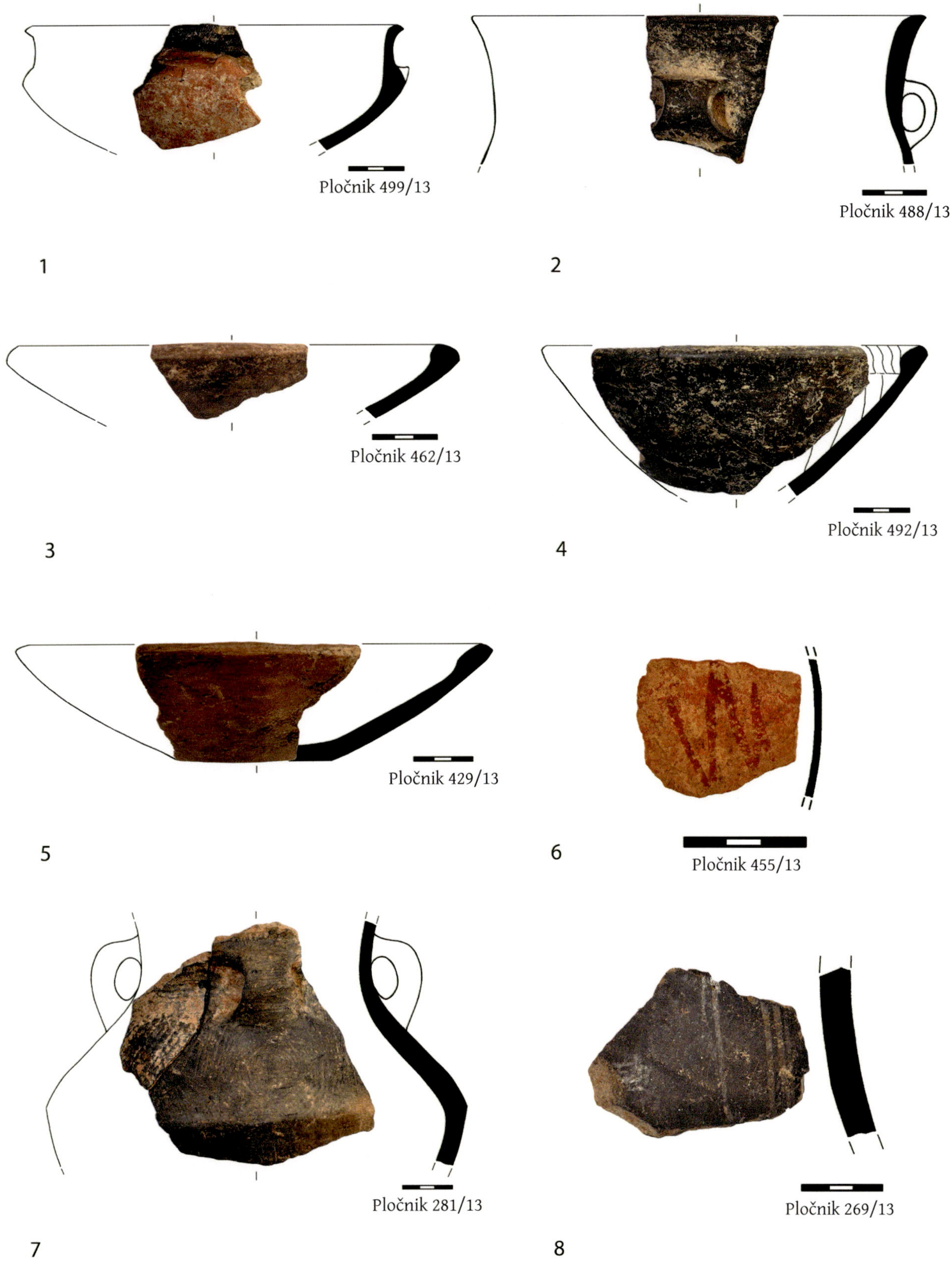

Figure 15. Most characteristic vessel types and ornamentation of Horizons 2 and 3.

Figure 16. Most characteristic vessel types of Horizon 3.

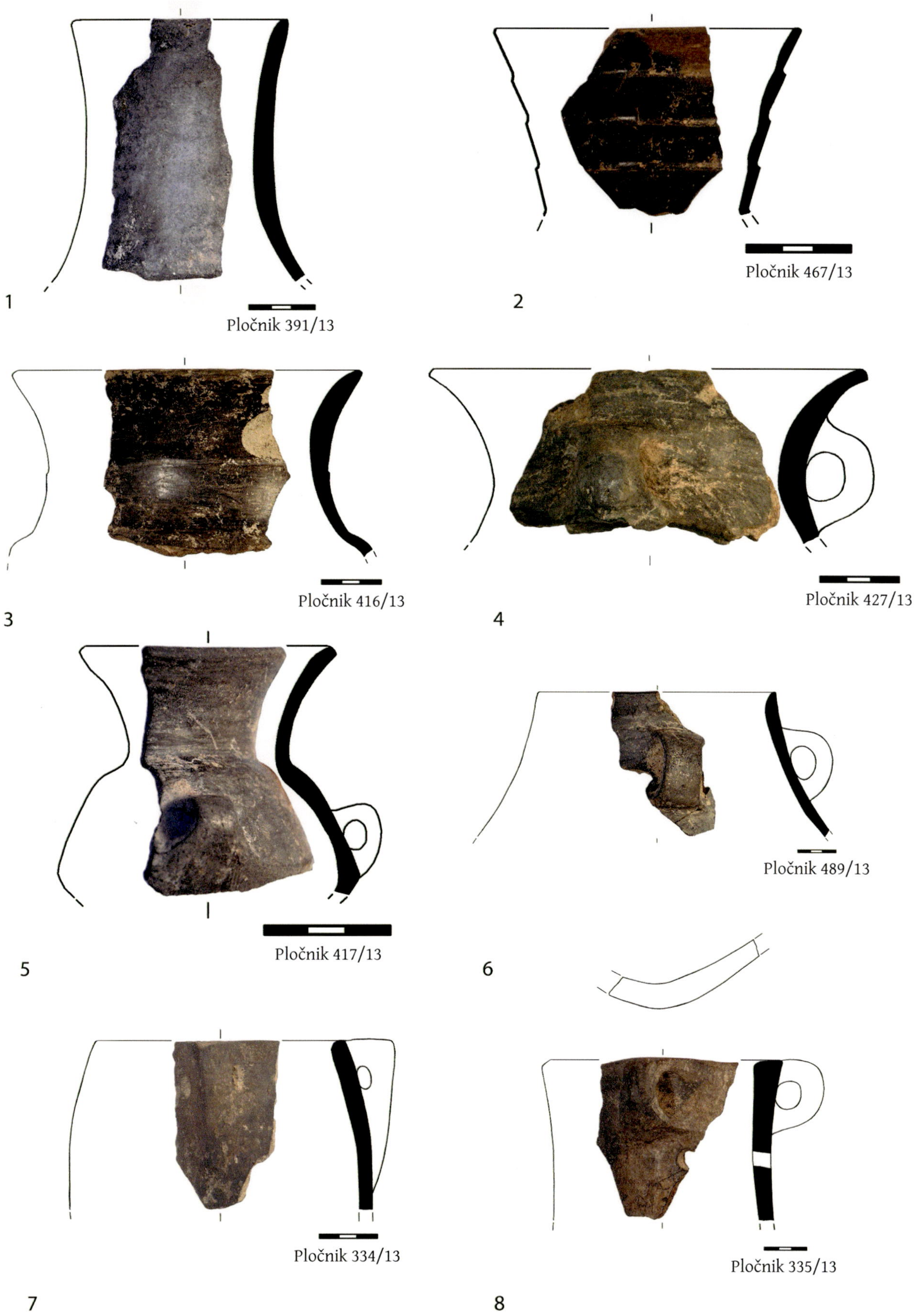

Figure 17. Most characteristic vessel types of Horizon 3.

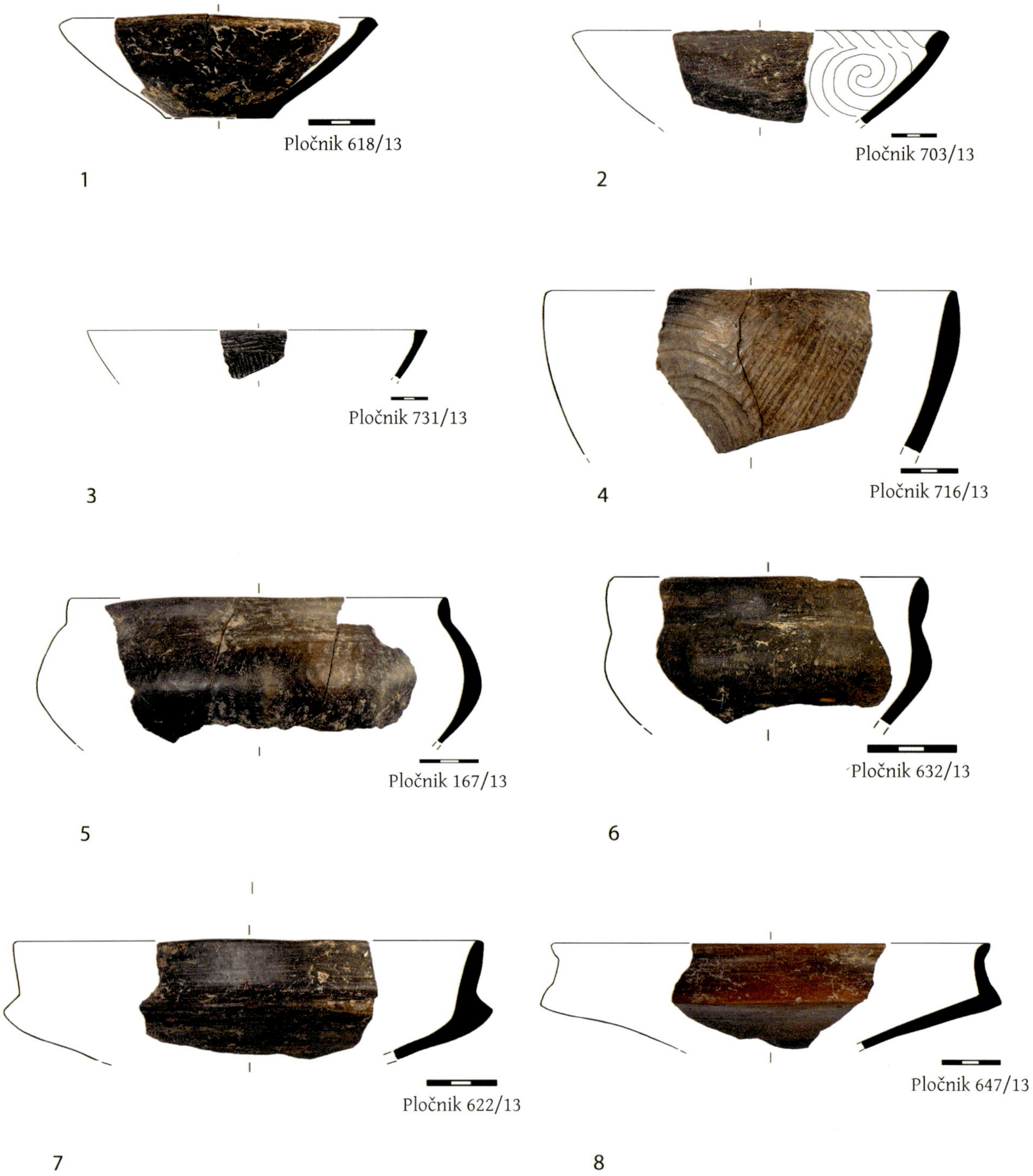

Figure 18. Most characteristic vessel types of Horizon 4.

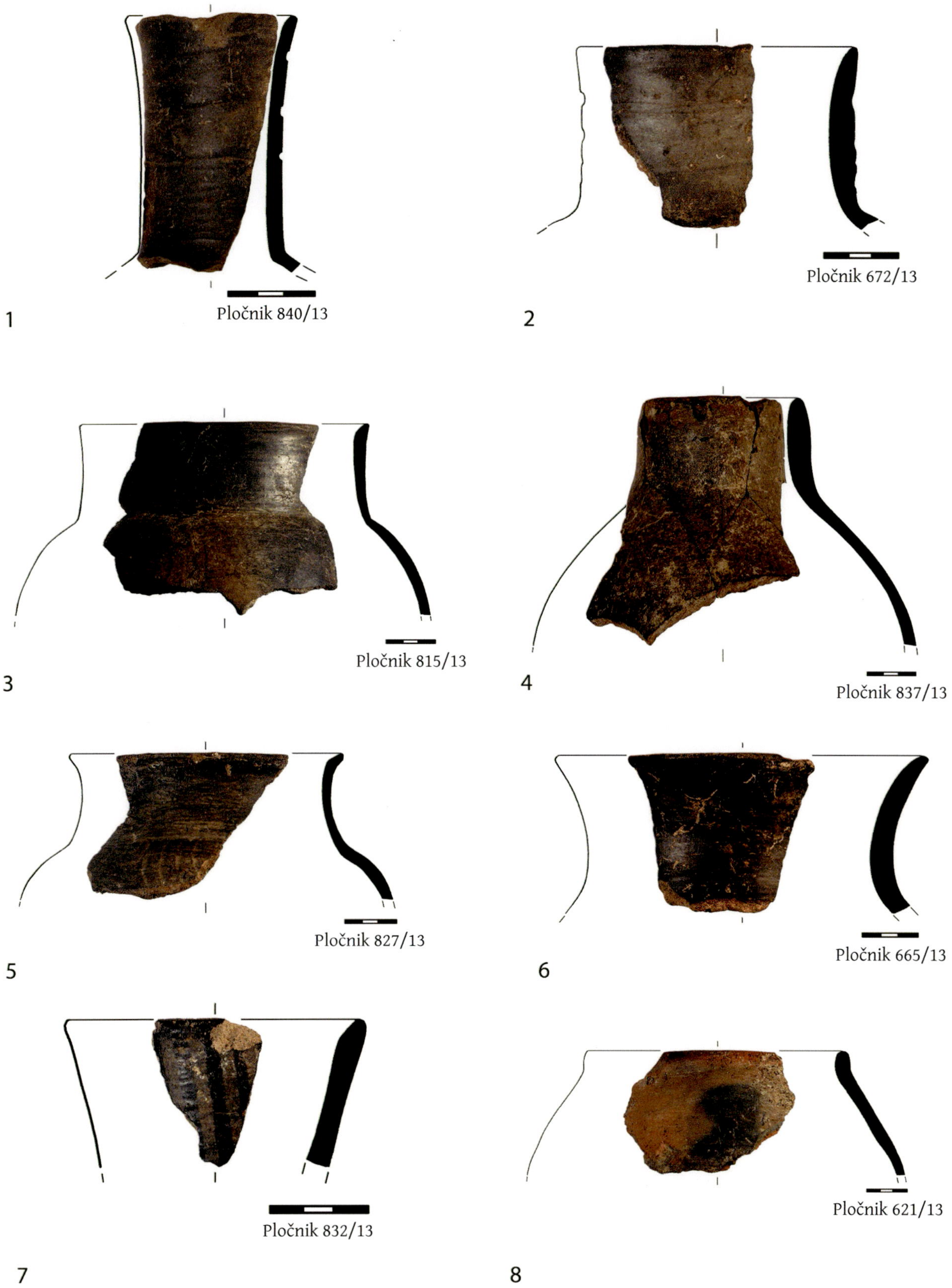

Figure 19. Most characteristic vessel types of the Horizon 4.

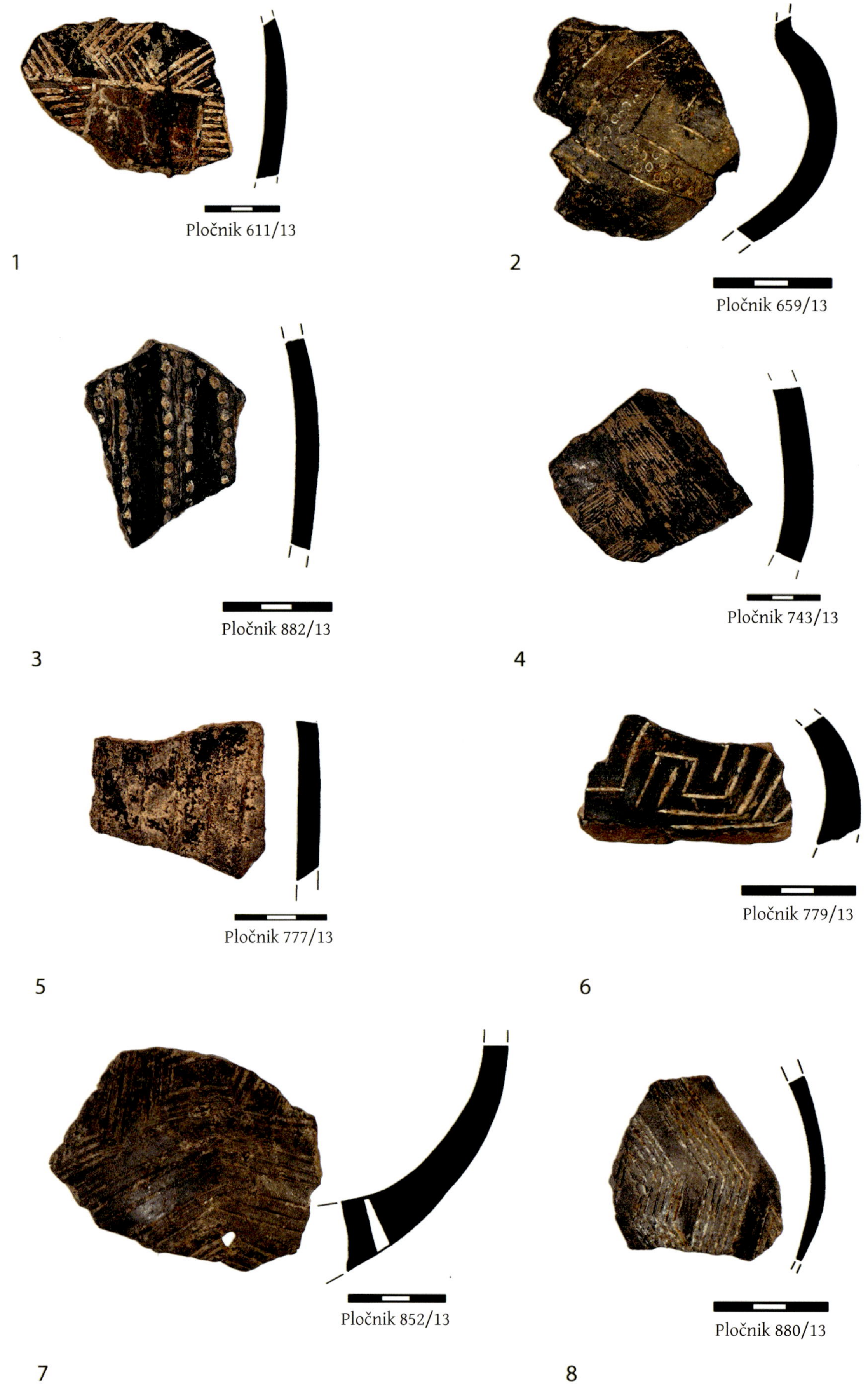

Figure 20. Most characteristic ornamentation of Horizon 4.

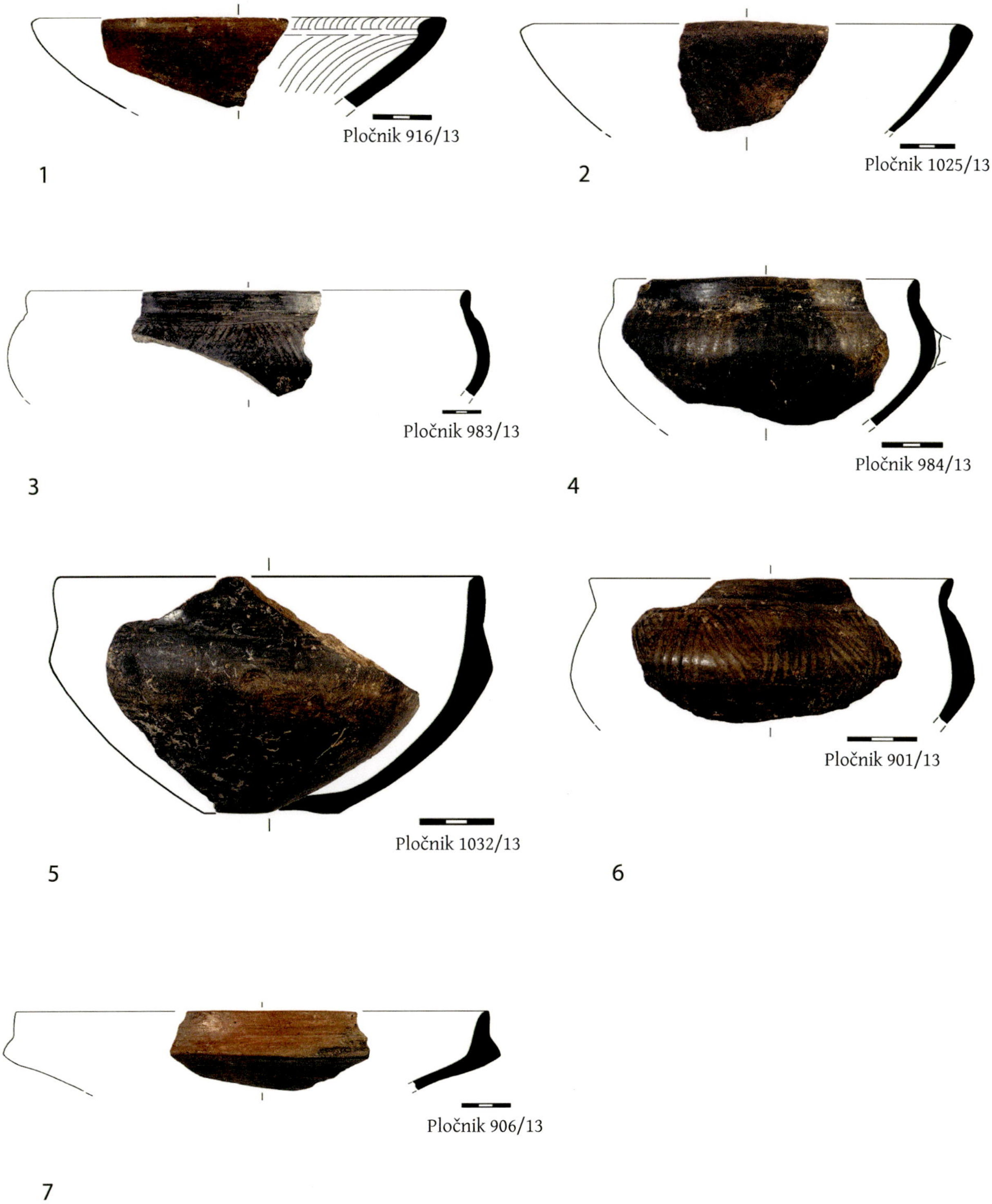

Figure 21. Most characteristic vessel types of Horizon 5.

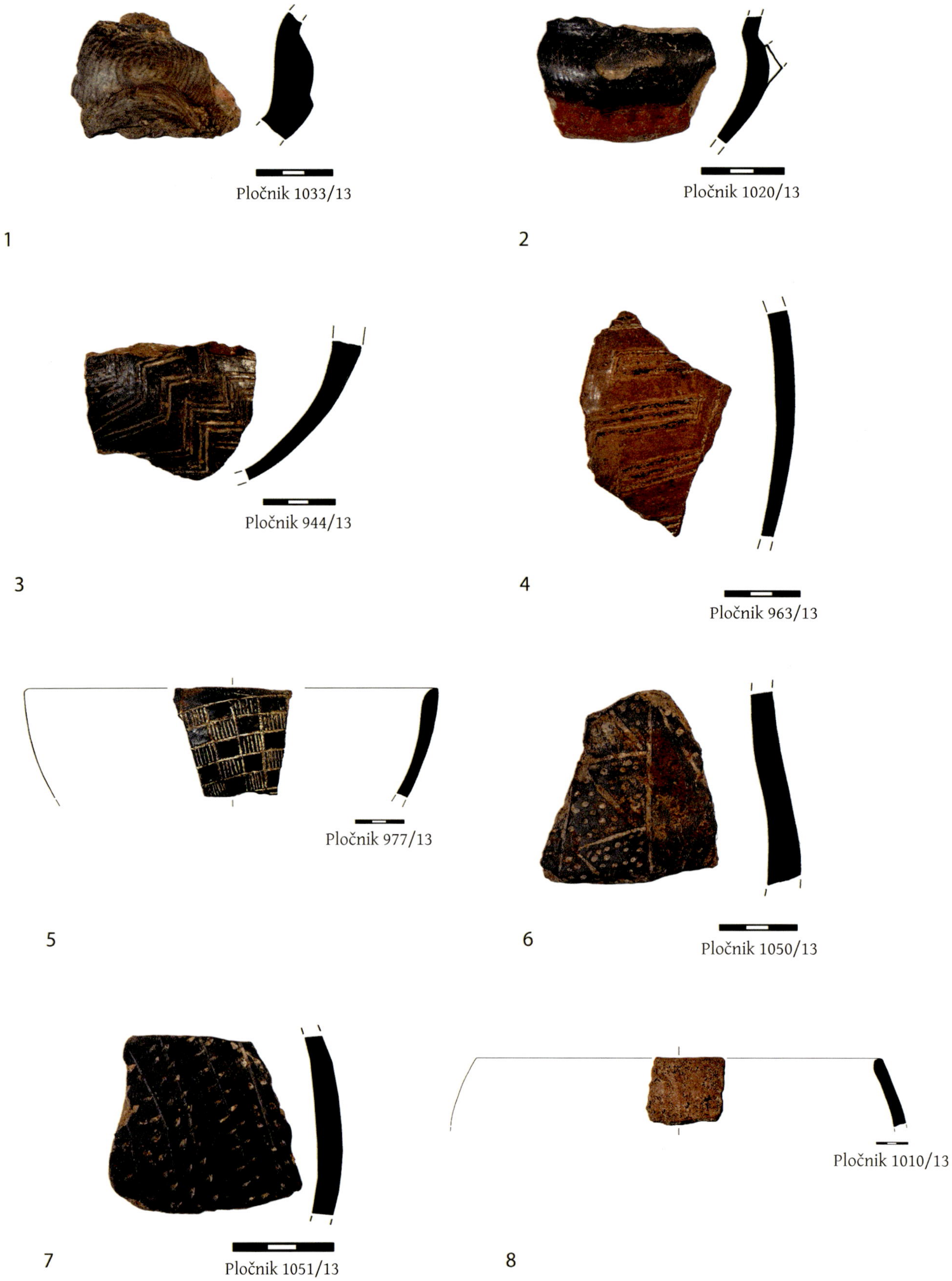

Figure 22. Most characteristic ornamentation of Horizon 5.

Chapter 28

Chronological attribution of pottery from Trench 24 at Pločnik based on correspondence analysis

Neda Mirković-Marić and Miroslav Marić

In this chapter, we use correspondence analysis (CA) to analyse the pottery data in form of a table in which the rows contain data from individual features found in Trench 24 at Pločnik during archaeological excavations between 2012 and 2013. The columns contain quantities of individual types of vessels, specifically those bowls deemed to be the most sensitive, typologically, in correlation to the chronology of the period.

It is our goal to determine the existence of correspondence between certain features and the types of vessels found within them; we ask whether the types are equally distributed amongst the features or whether a pattern of association exists between certain features and bowl types. The analysis is based on the R Script developed by Gianmarco Alberti (2013a), available on his website 'Correspondence Analysis in Archaeology' (http://cainarchaeology.weebly.com). The first step in the analysis to determine the strength of association between the rows and columns of the table being analysed. This is achieved using a correlation coefficient (valued between 0.0 and 1.0) equal to the square root of the table variability. The correlation coefficient in the observed situation is 1.446, whilst the coefficient of the table is 0.822 (Figure 1), which makes the correlation of data statistically significant (Healey 2013: 289–290).

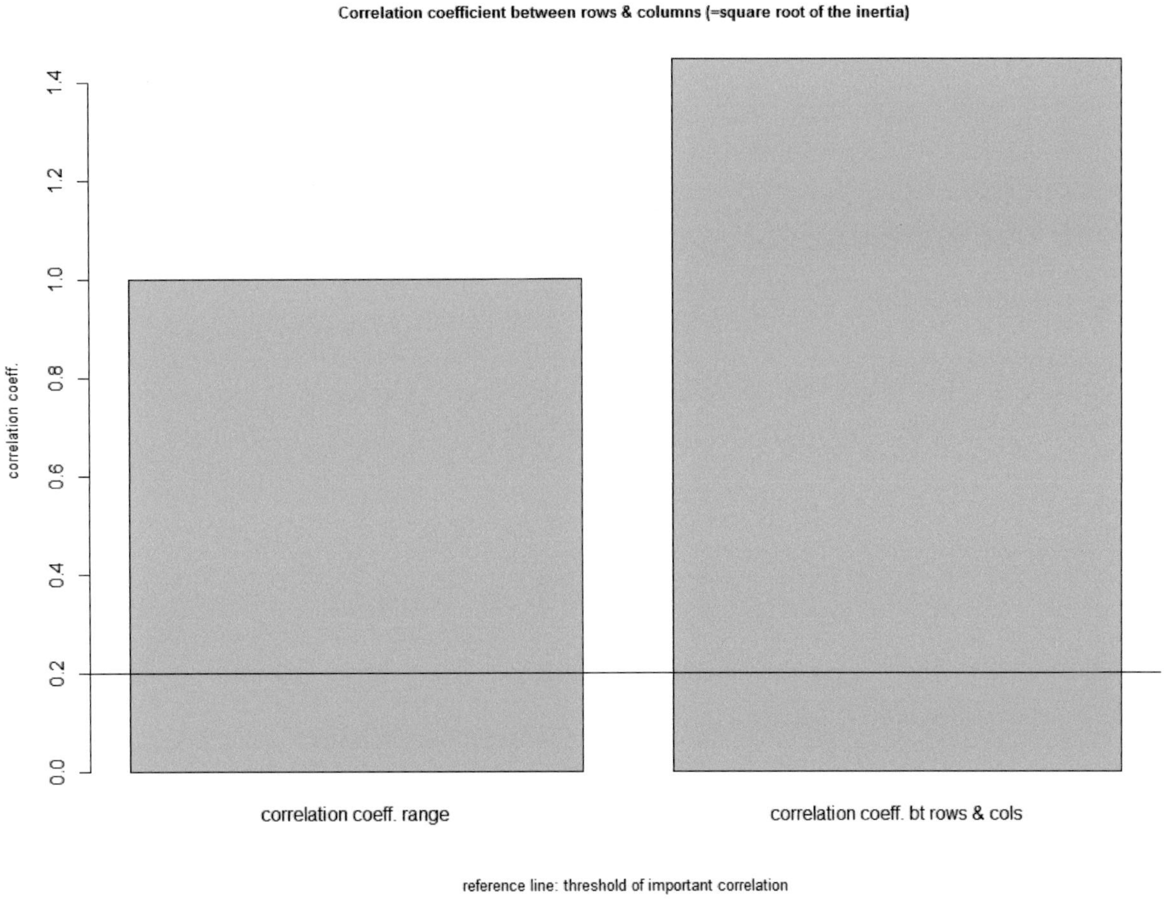

Figure 1. The correlation coefficient of data.

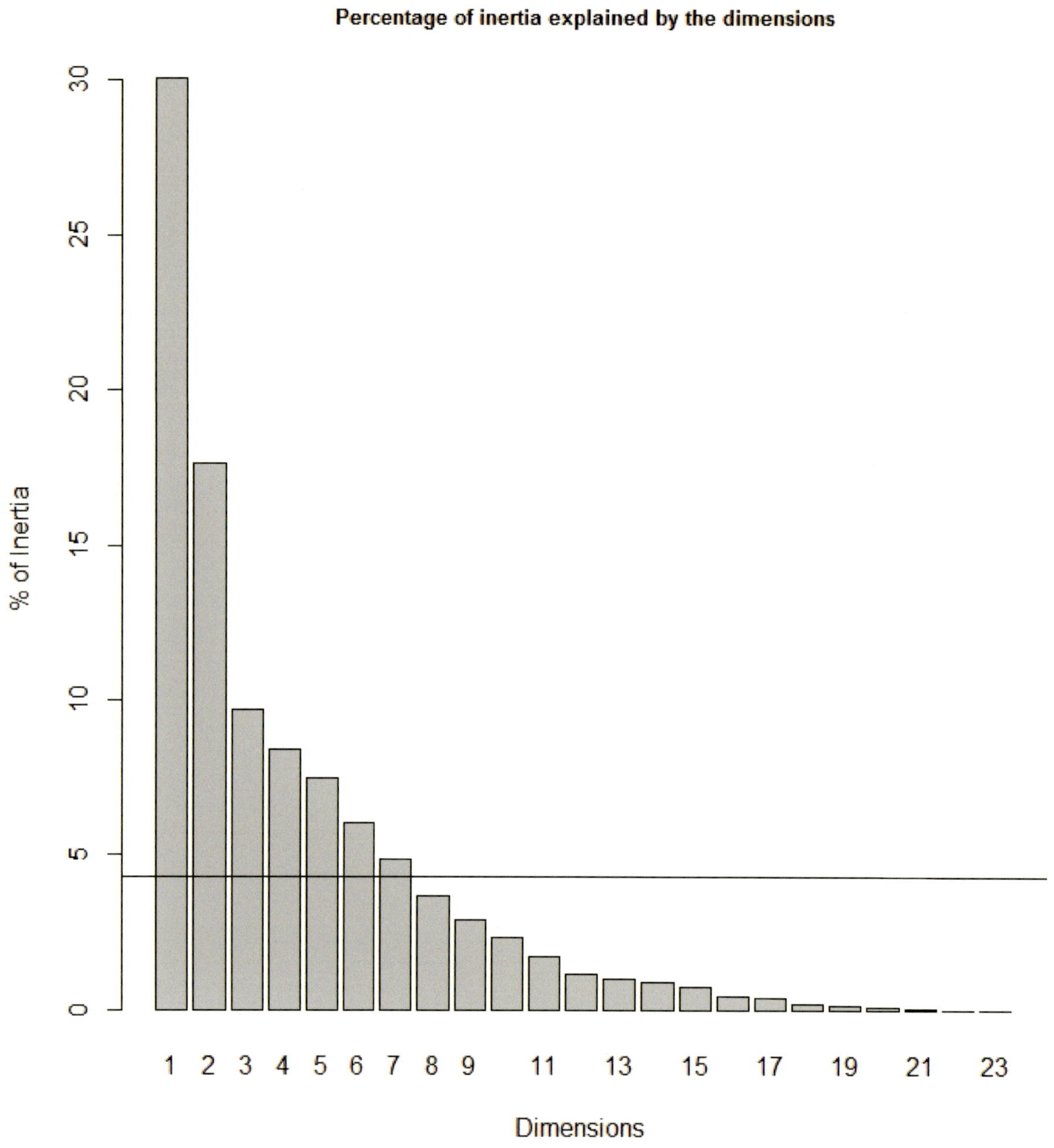

Figure 2. The dimensions that have larger average value than the percent of the average inertia.

Having established the existence of correlation of data, we need to determine the number of dimensions (i.e. groups of data) in order to interpret the patterns. Some authors consider that the number of retained dimensions should reflect the arbitrarily set majority value of total variation (e.g. 90%), or that all values with inertia greater than 0.2 (Eigen value) should be included (Hair *et al.* 2009: 591). It is also possible to calculate a scree plot (similar to principal component analysis) to identify proper dimensions (Drennan 2009: 286–288). A fourth method, used here, is based on the calculation of the average inertia, where all the dimensions that explain more than the average inertia are kept for analysis. The data being analysed here indicate that all dimensions which have average inertia above 4.34% for columns and 3.7% for rows (Figure 2) should be included in the analysis, i.e. there are 7 groups of bowls in the assemblage that need to be analysed in order to achieve 84.2% explanation of the data variability. The same can be seen in the scree plot given below (Figure 3), which shows a cut-off point at the 8th dimension (cut-off shown in red dashed line), having a value of 3.7% of the average inertia.

Before proceeding with the analysis, it is necessary to illustrate which bowls contribute to which dimension

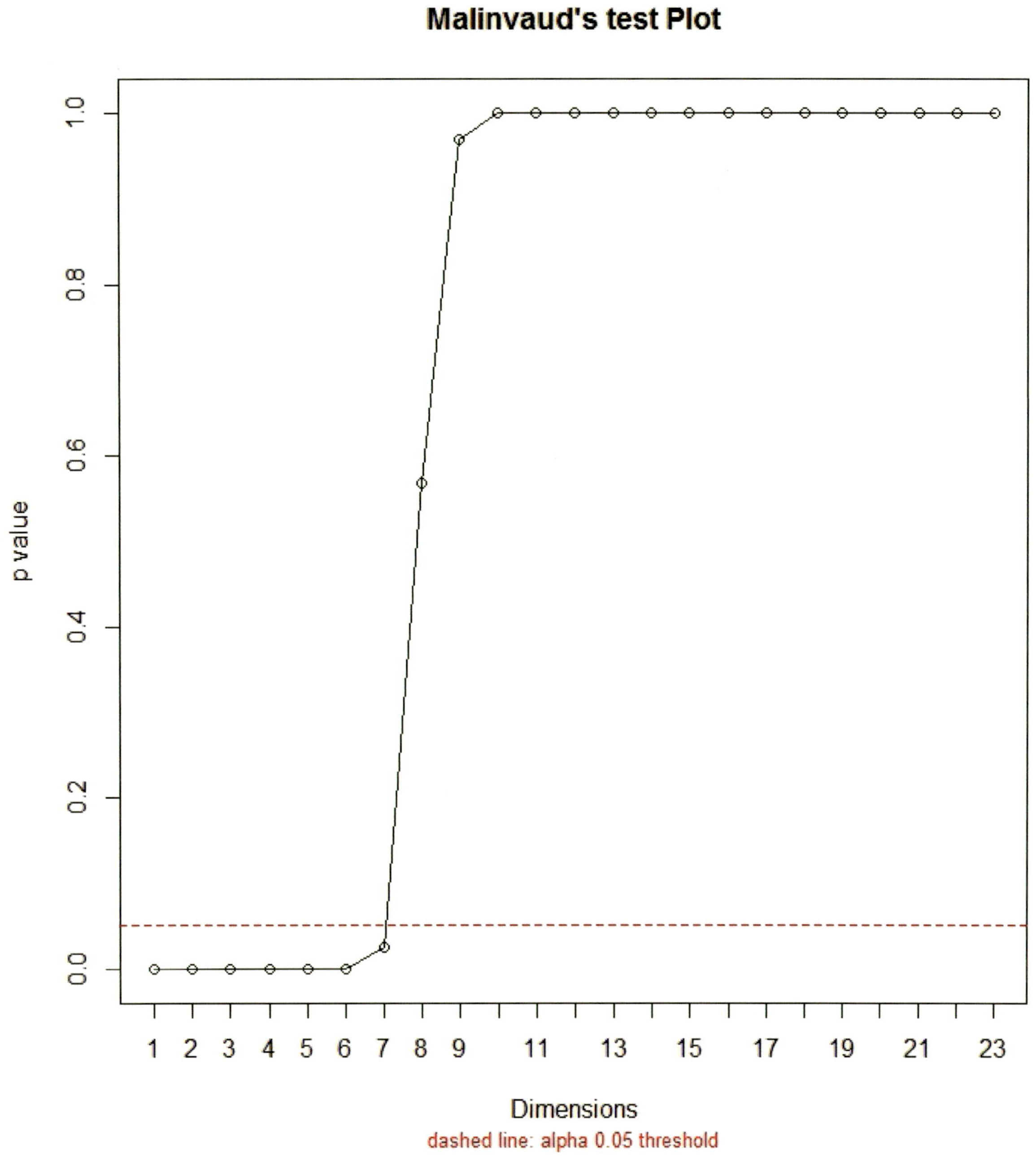

Figure 3. Scree plot of dimension included in the analysis.

and how observed features contribute to the individual dimensions to be analysed. Figure 4 identifies the types of bowls that contribute to each of the seven dimensions in permills using a bar plot, whilst Figure 5 illustrates the contribution of individual features to each dimension. The contribution in permills is given on the vertical axis, while the reference line (horizontal) indicates the threshold of the average contribution beyond which any contribution is considered significant in the definition of the dimension (Greenacre 2007: 82).

The bar plots in Figure 4 clearly indicate that some of the bowl types, such as 101 or 102, contribute significantly to more than one dimension (for example type 101 contributes to Dimensions 4, 5 and 6 while type 102 contributes significantly to all dimensions but the first two), whilst other types, like 117, contribute significantly only to Dimension 1. The bar plots of the contribution of individual features in Trench 24 to certain dimensions show another interesting occurrence in Figure 5. Comparing bar plots it is easy to see that certain features contribute significantly to only one dimension (for instance, features F1, F2, F4, F5, F8 and F9 contribute exclusively to Dimension 1), whilst others contribute significantly to more than one dimension (F38 contributes to Dimensions 1, 2, 5, 6, and

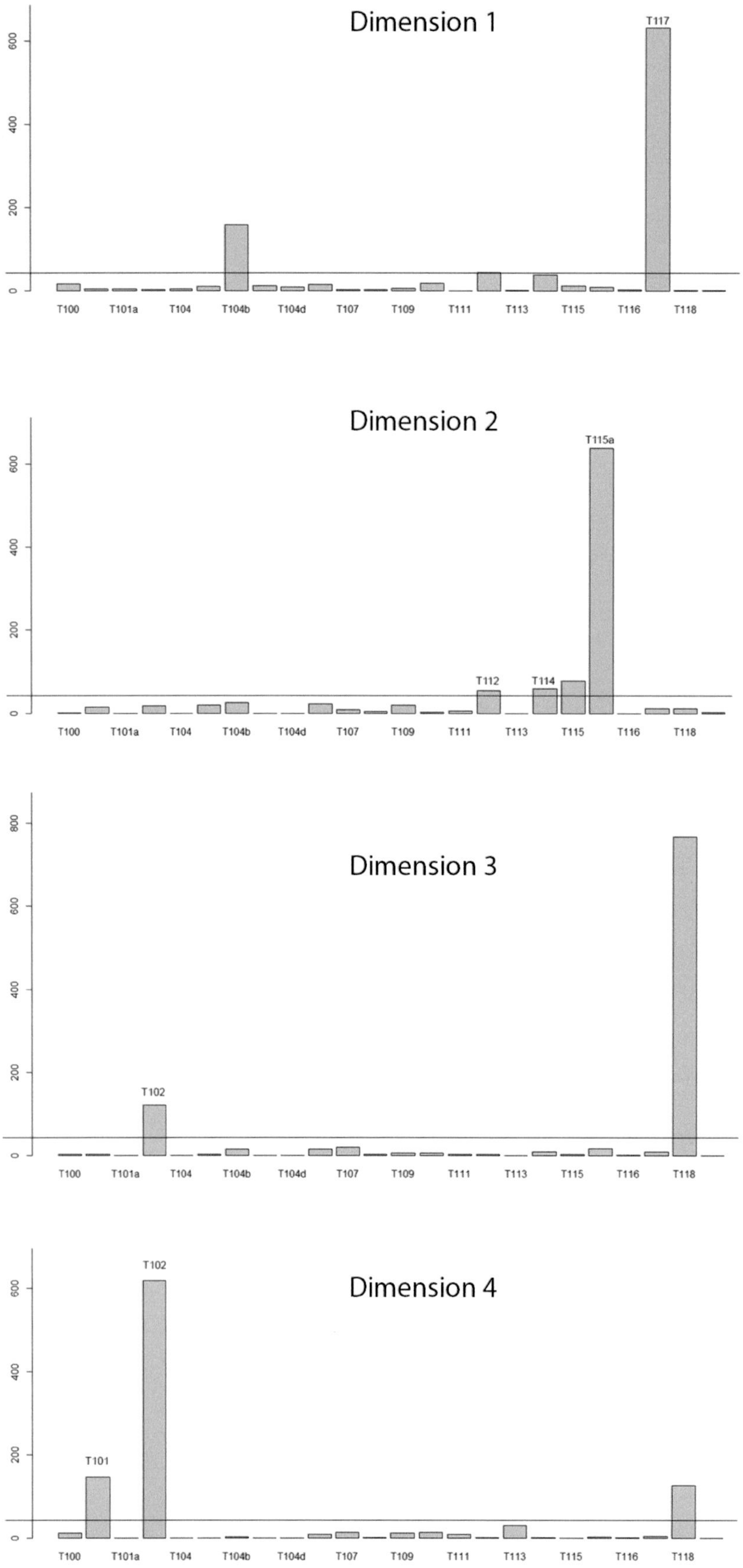

Figure 4. Contribution of bowl types to individual dimensions.

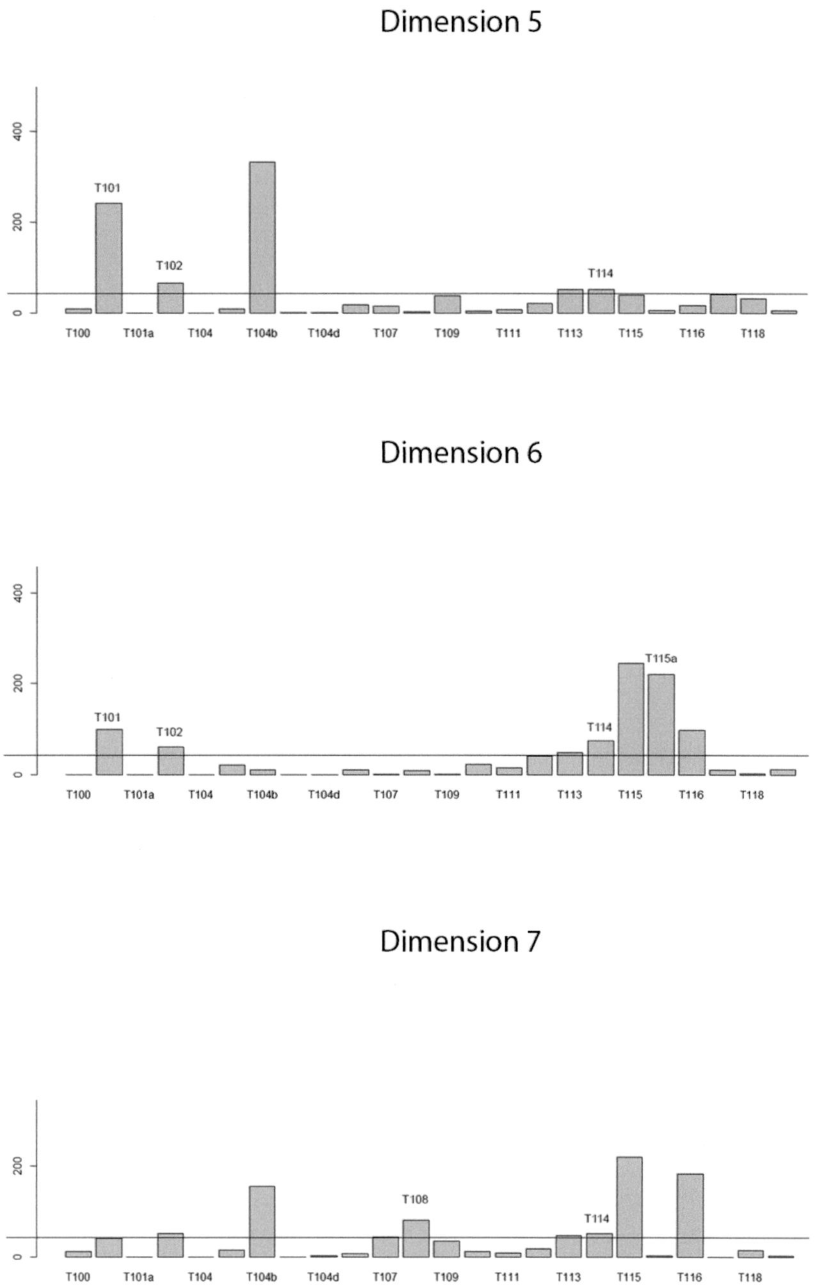

Figure 4 continued. Contribution of bowl types to individual dimensions.

7). This indicates that certain types of bowls are more common throughout the sequence of Trench 24, while others appear in and dominate only certain parts of the same sequence. The one-dimensional types of bowls are clearly indicative of certain chronological periods.

The bar plot of feature contribution to individual dimensions also shows that certain features are contributing more to some dimensions, whilst others have no statistically significant contribution, which would reflect their chronological positioning in the sequence of the trench. Having established this, we can now turn the focus of our analysis to the correspondence of the statistically significant dimensions. The most analysis is performed against the most significant dimension, Dimension 1. The results are plotted on standard biplots (Figure 6) which contain an axis for each of the dimensions being analysed. The centroid of the biplot represents the average profile of the data and can be thought of as a place in which there is no difference between the profiles (i.e. data), and there is a homogeneity in data (Greenacre 2007: 32), i.e. the bowl types are equally distributed in terms of features. The more diverse the profiles, the more the profile

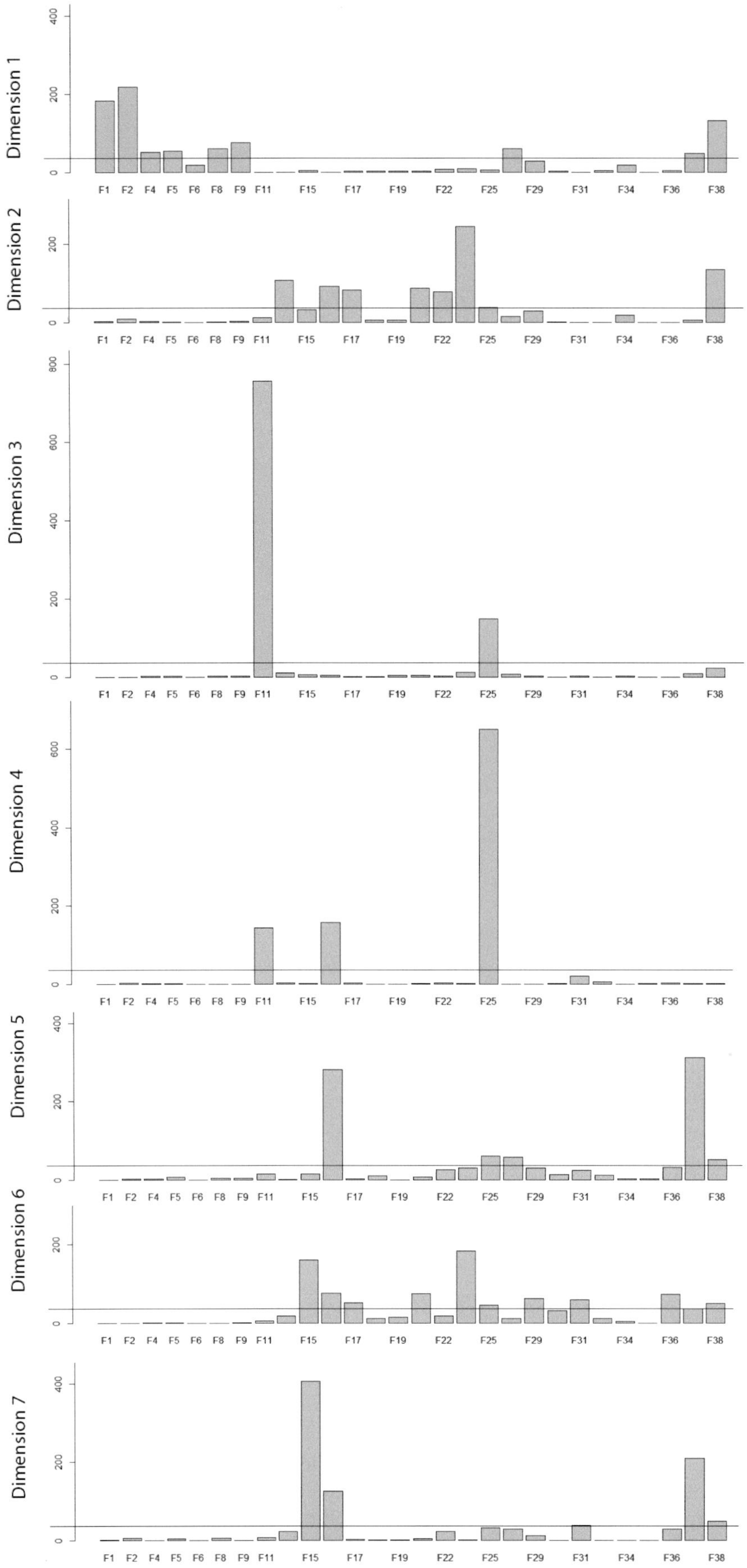

Figure 5. Contribution of individual features to dimensions.

points will be distributed away from the centroid. When the relative distances of points of different type (column versus row data) are compared, then the correspondence between the categories making up the table is obtained. If the datapoint for the row data is closer to that of the column data, the proportion of this column category towards the row profile is greater (i.e. further away from the average value). Figure 6 shows the CA biplots with the analysis of each of the dimensions. It can be said that the first two dimensions, explaining 47.7% variability, have as the dominant and significant opposite poles, pottery types 104b and 117 for Dimension 1, whilst Dimension 2 is defined by several types including 112, 114, 115 and 115a (Figure 6, top left). If we look to the absolute contribution to the total inertia of the data (Figure 6, top left) we can see that the most intensive colour is on the vectors of 104b, 112 and 114, indicating that, while these types have a significant contribution to the dimensions, they have a low contribution to the variability of the table overall.

Looking at the CA analysis biplot for Dimensions 1 and 3, which explain an additional 9.7% of variability/inertia (Figure 6, middle left), we can clearly identify that the dominant and opposite poles for Dimension 3 are pottery types 102 and 118. The angular proximity of these types to the axes of their dimensions indicates that there is virtually almost no contribution, positive or negative to Dimension 1, i.e. that these types define only Dimension 3. If we look at the absolute contribution of bowl types (Figure 6, middle left), type 118 has a very low contribution to the table variability, whilst type 102 has some contribution, but not nearly as great, evidenced in Figure 3, where type 102 can be seen as statistically significant in Dimensions 3, 4, 5, 6 and 7, but not to the same extent.

The next CA biplot, for Dimensions 1 and 4, explains 8.4% of inertia (Figure 6, bottom left) and shows two major pottery types that define this dimension, 101 and 102, the latter being a more important contributor to Dimension 4, but in its negative pole. Type 118, even though present in the dimension is not a major contributor. No other types present in the distribution are statistically important (Figure 7, bottom left).

The situation is somewhat different when analysing CA biplot for Dimensions 1 and 5, which explain 7.5% of inertia. Pottery types 101 and 104b are major contributors to Dimension 5, albeit to its opposite poles (Figure 6, top right), as is type 116. It is quite clear from the biplot that type 101 contributes almost entirely to Dimension 5, whilst type 104b, with a somewhat different angular proximity, contributes to both dimensions, although more to Dimension 5 than Dimension 1. If we compare this to the biplot for absolute contribution to the variability of the table (Figure 7, top right) we see that pottery types 102 and 114 have little contribution to the variability of table overall.

The penultimate CA biplot refers to Dimensions 1 and 6 (Figure 6, middle right). The principal and opposite pottery types are 115, 115a and 116 for Dimension 6, and also types 101, 102, 113 and 114. Looking at the absolute contribution to the inertia, those contributing least are types 116, 104b and 114 (Figure 7 middle right). Looking at the last CA biplot for Dimensions 1 and 7 (Figure 6, bottom right) the principal pottery types for Dimension 7 include 104b, 108, 115, and 116, while low contributors to the inertia of Dimension 7 are types 108 and 114 (Figure 7, bottom right).

If we look at the CA scatterplots for the first two dimension (Figure 8, top left) we can see that the main difference detected for the bowl types is that between types 104b and 117 in Dimension 1, and types 112, 115 and 115a in Dimension 2. The main difference thus lies between features F1, F2, F4, F5, F6, F8 and F9 where pottery type 117 dominates and features F28, F32, F36 and F37 where type 104b is dominant, against features F21 and F23 where type 115a appears, features F15 and F25 where type 115 dominates, and features F29, F34 and F38 where type 112 occurs. The other features occurring nearer to the centre of the plot have little differentiation as far as the principal types are concerned, so the distribution of these types in these features does not depart significantly from the average. The CA scatterplot for Dimensions 1 and 3 shows a large concentration around the centroid which would indicate features without a significant differentiation of principal type, i.e. the average distribution. The only two different cases are features F25, where there is a proportion of type 102 present (not too large as they are quite separate), and feature F11 in which there is a large proportion of type 118 (Figure 8, middle left).

Similar is true for the CA scatterplot of Dimensions 1 and 4 in which pottery again T102 type 102 again stands out in feature F25, although not with a large proportion, and type 118 in feature F11, but with much higher proportion (Figure 8, bottom left). All other types appear clustered around the centroid, indicating insignificant departure from the average distribution. The next scatterplot shows a different image for Dimensions 1 and 5 (Figure 8, top right). The principal pottery types of Dimension 5, 101 and 104b, plot slightly off the centroid and have large proportions in features F11 and F31 (for type 101), whilst type 104b is present in somewhat higher proportions in features F22 and F23. The other contributing type, 113, can be found in small proportions in features F16 and F25. In Dimension 6 (Figure 8, middle right), the principal type of bowl, 115, can be found in larger proportion in features 15, 30 and 36, whilst type 115a is dominant in feature 17 and there are some found in features 21 and 23. Type 101 is found

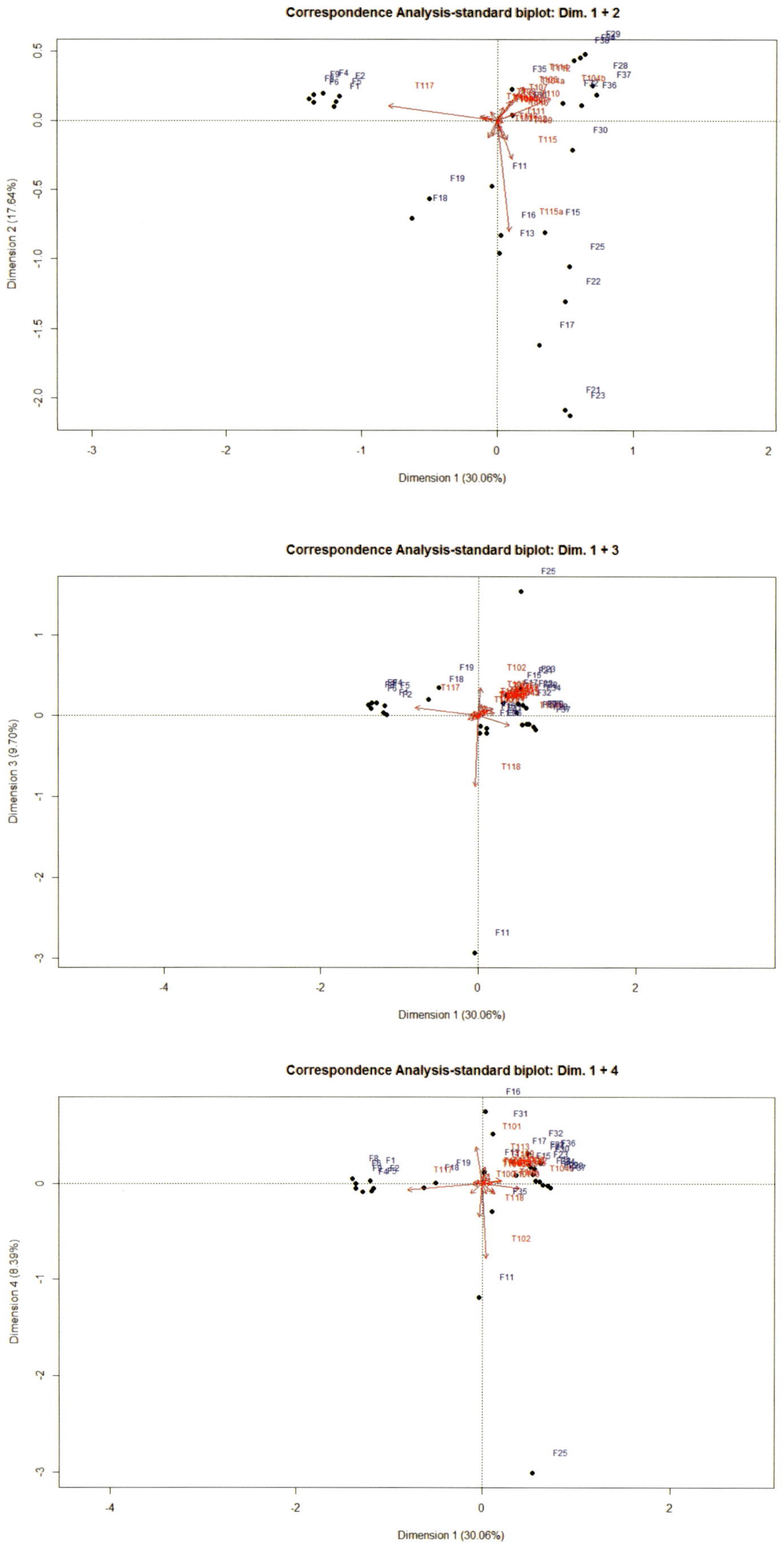

Figure 6. Standard CA biplots with dominant types of vessels.

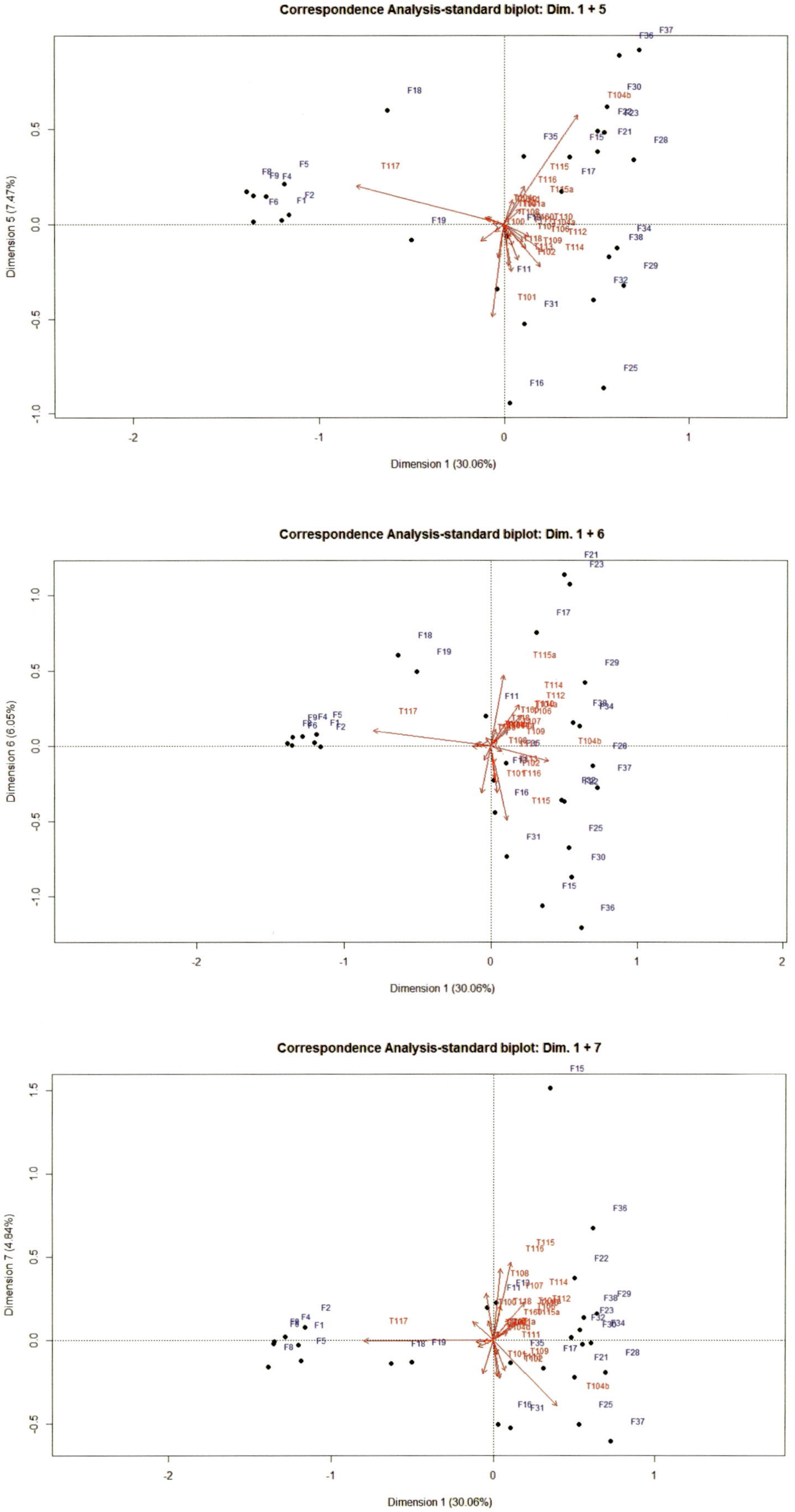

Figure 6 continued. Standard CA biplots with dominant types of vessels.

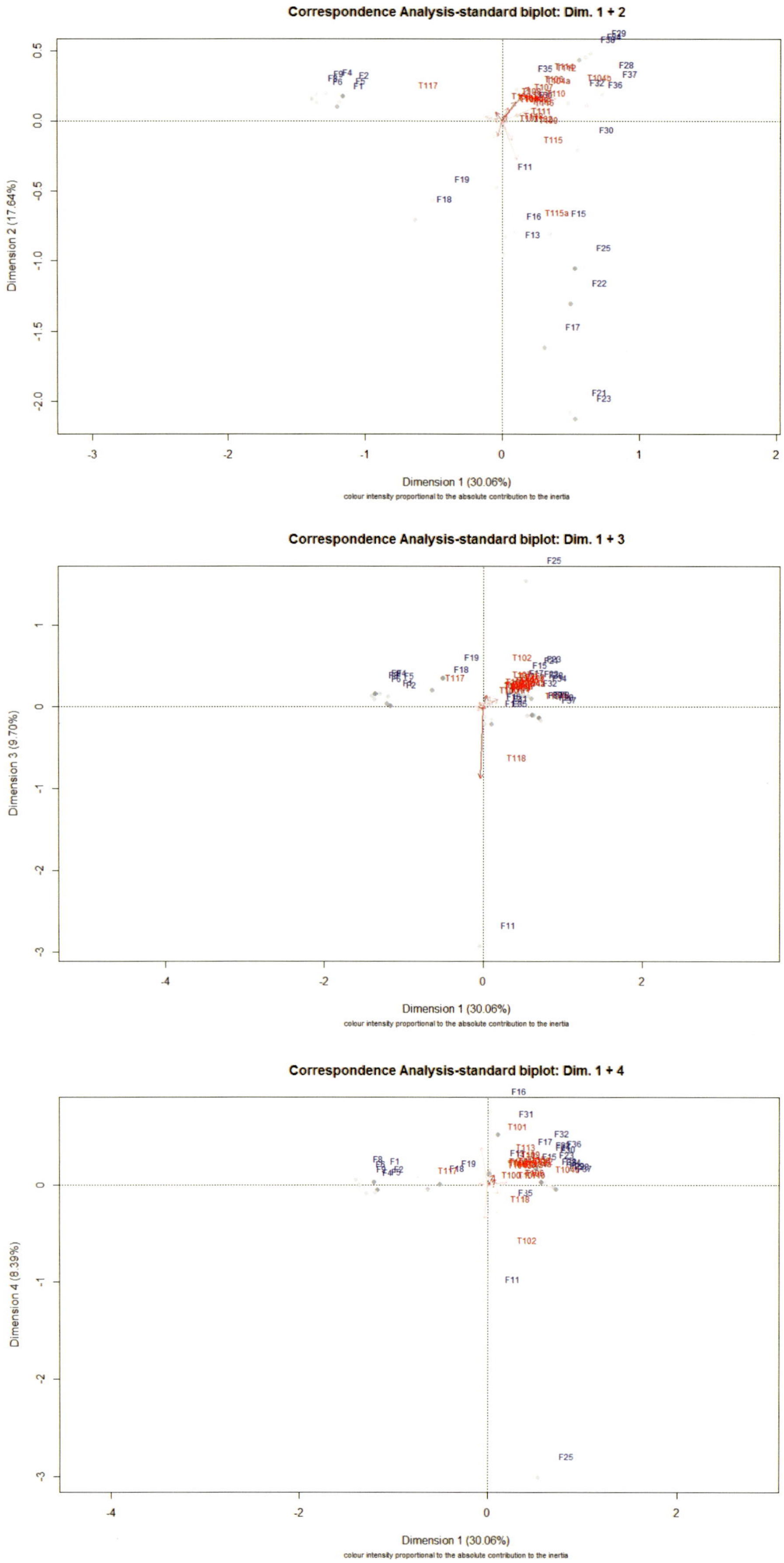

Figure 7. Standard biplot showing absolute contribution to the inertia of bowl types.

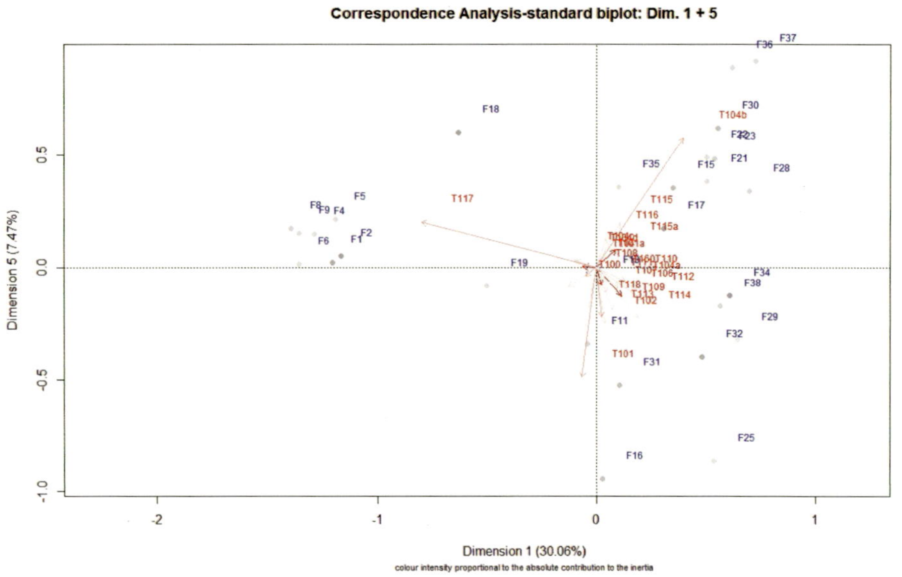

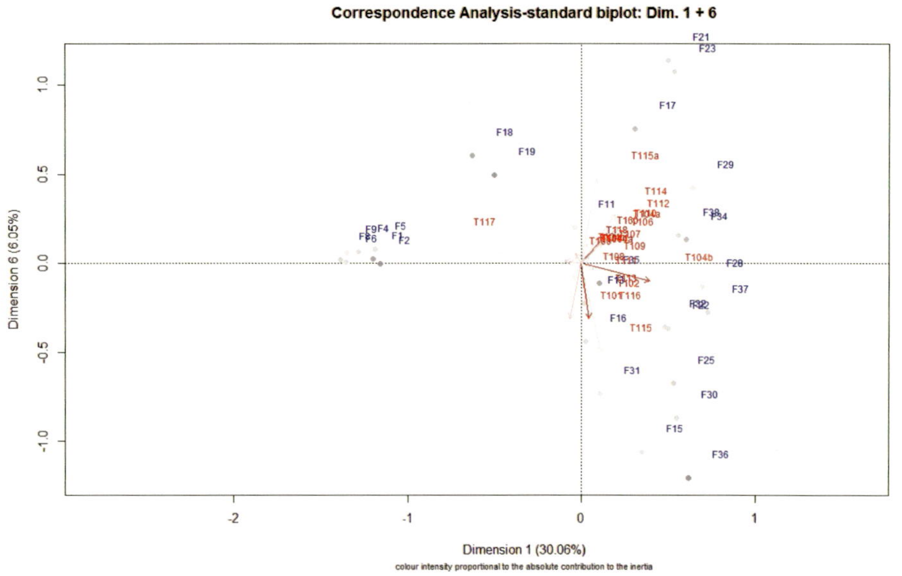

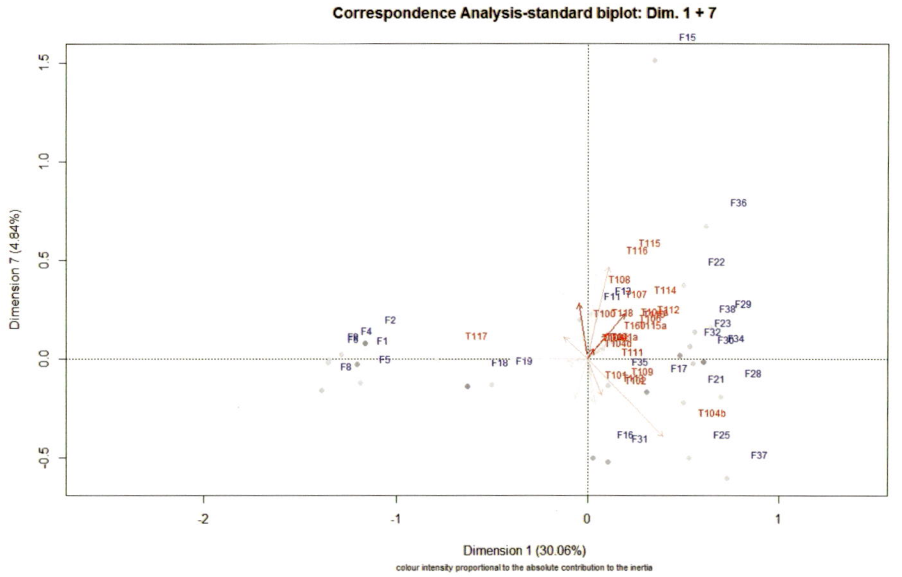

Figure 7 continued. Standard biplot showing absolute contribution to the inertia of bowl types.

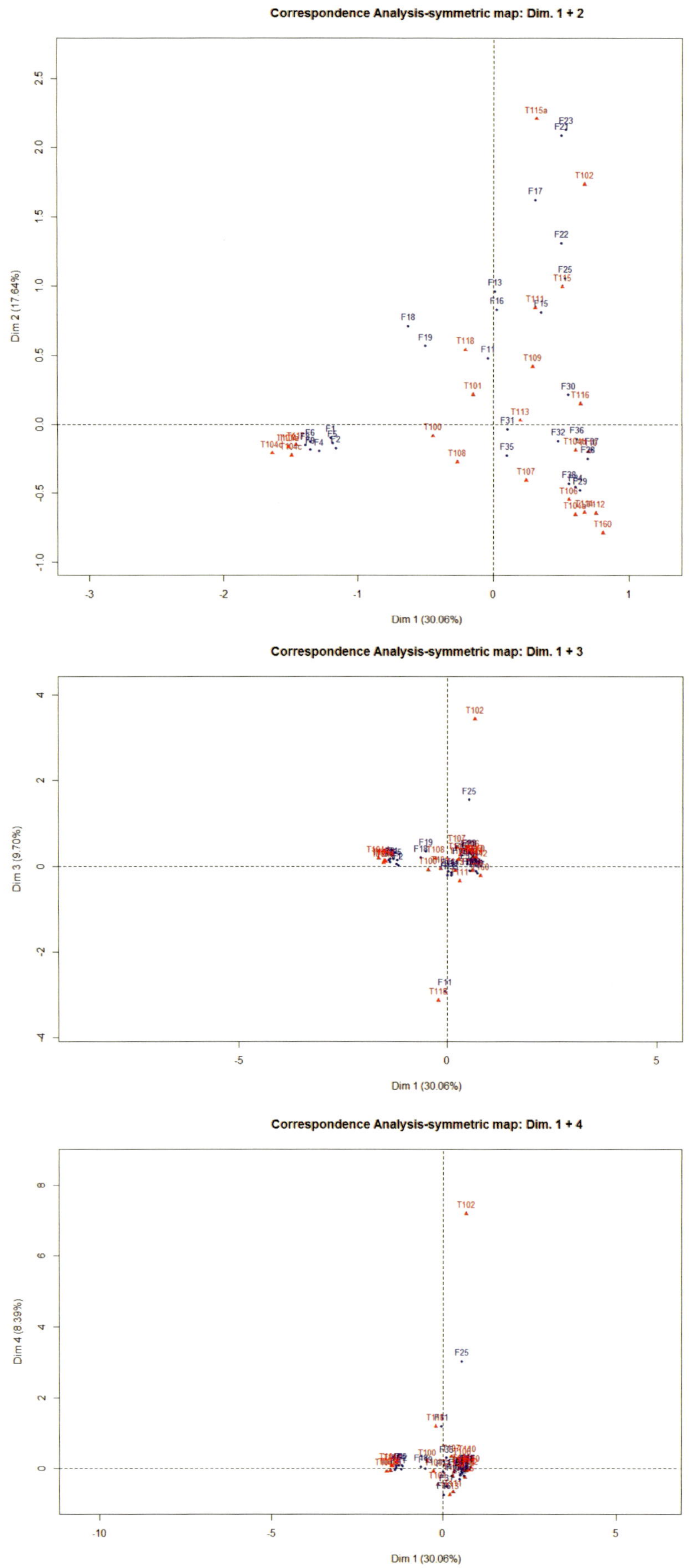

Figure 8. Scatterplot representations of correspondence analysis.

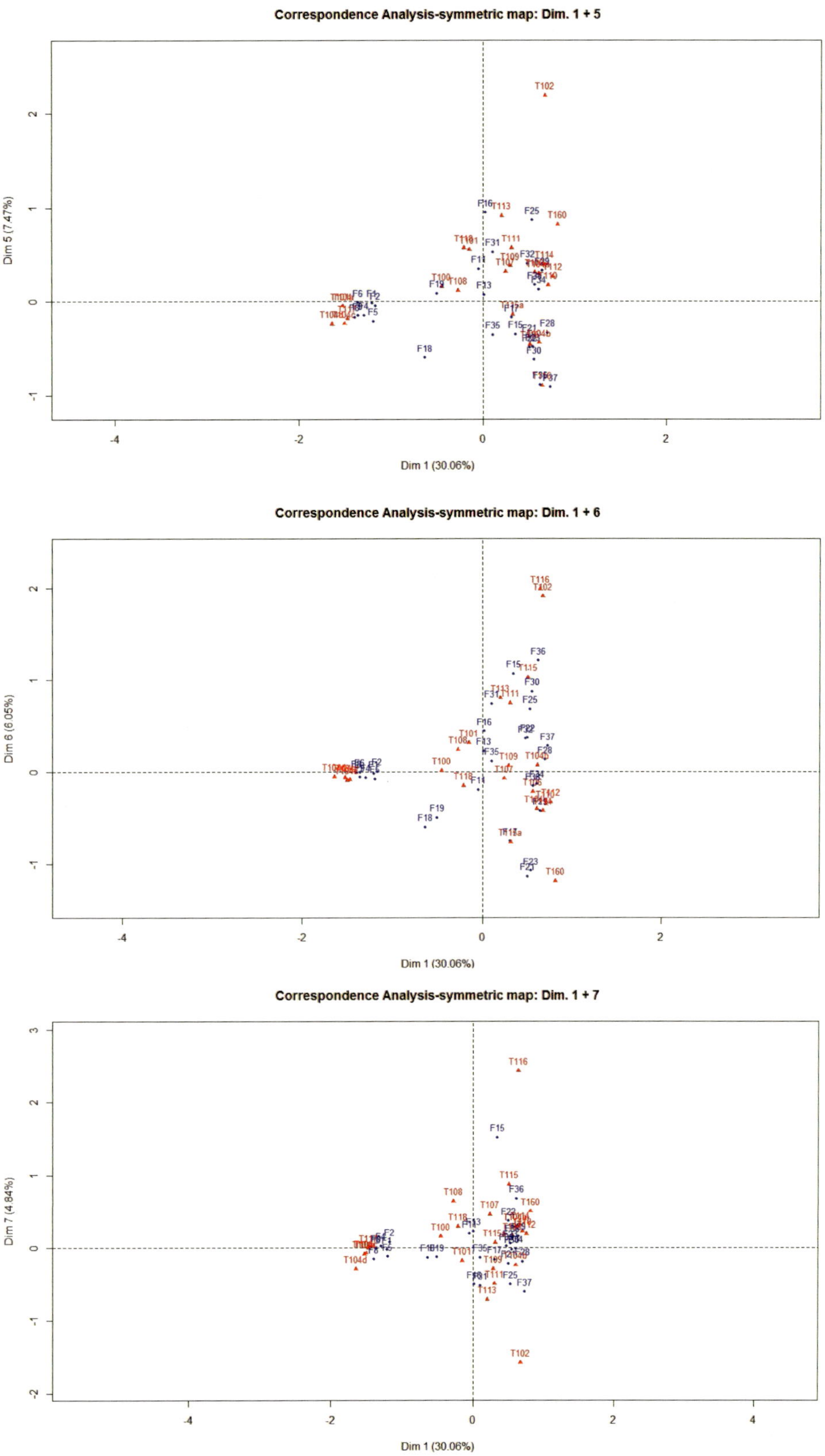

Figure 8 continued. Scatterplot representations of correspondence analysis.

in significant proportions in features 13 and 16. Type 102 has no strong proportion in particular features and type 116 can be found in feature 36 in a very small proportion. Finally, the CA scatterplot for Dimension 7 (Figure 8, bottom right) shows that one of the principal types, 115, appears in larger than average proportions in feature 36, while the other principal type, 116, has a very small proportion in feature15. Type 104b appears in high proportion in features F17 and F28, whilst type 108 has no clear association with any particular feature.

At this point it is necessary to consider the degree of correlation between the row profiles and the dimensions. This will enable understanding of how features relate to the established dimensions (i.e. groups of bowls). It is best viewed using a bar plot where the correlation is expressed with a coefficient ranging from 0.0 to 1.0 (the latter being the ideal) (Greenacre 2007: 86). The bar plots show that most features have a very strong correlation with the Dimension 1 (Figure 9, top left), whilst those that do not, correlate strongly with Dimension 2. The Dimension 3 bar plot (Figure 9, left, third from top) is particularly interesting as it shows that the only strong correlation occurs in features F11 and F25. The Dimension 4 correlation plot (Figure 9, bottom left) shows the strongest correlations with features F11 and F25 as in the previous bar plot, with the addition of features F16 and F31. The situation changes with the correlation bar plots of Dimensions 5 and 6 (Figure 9, top and middle right) where a clearer correlation is visible in the deeper features. Finally, the bar plot of Dimension 7 indicates correlation in the middle section of the trench, with features F13, F15 and F16 and a cluster comprising features F36, F37 and F38 (Figure 9, bottom right).

Finally, the quality of representation should also be considered, i.e. whether all points are well displayed in the chosen dimensions, or whether some suffer from under-representation, making them stand out in a different pattern from the rest of the analysed assemblage. The bar plots in Figure 10 show the representation of points in all retained dimensions, showing that some of the features are constantly under-represented in the assemblage. Feature F11, for example, is only adequately represented in Dimension

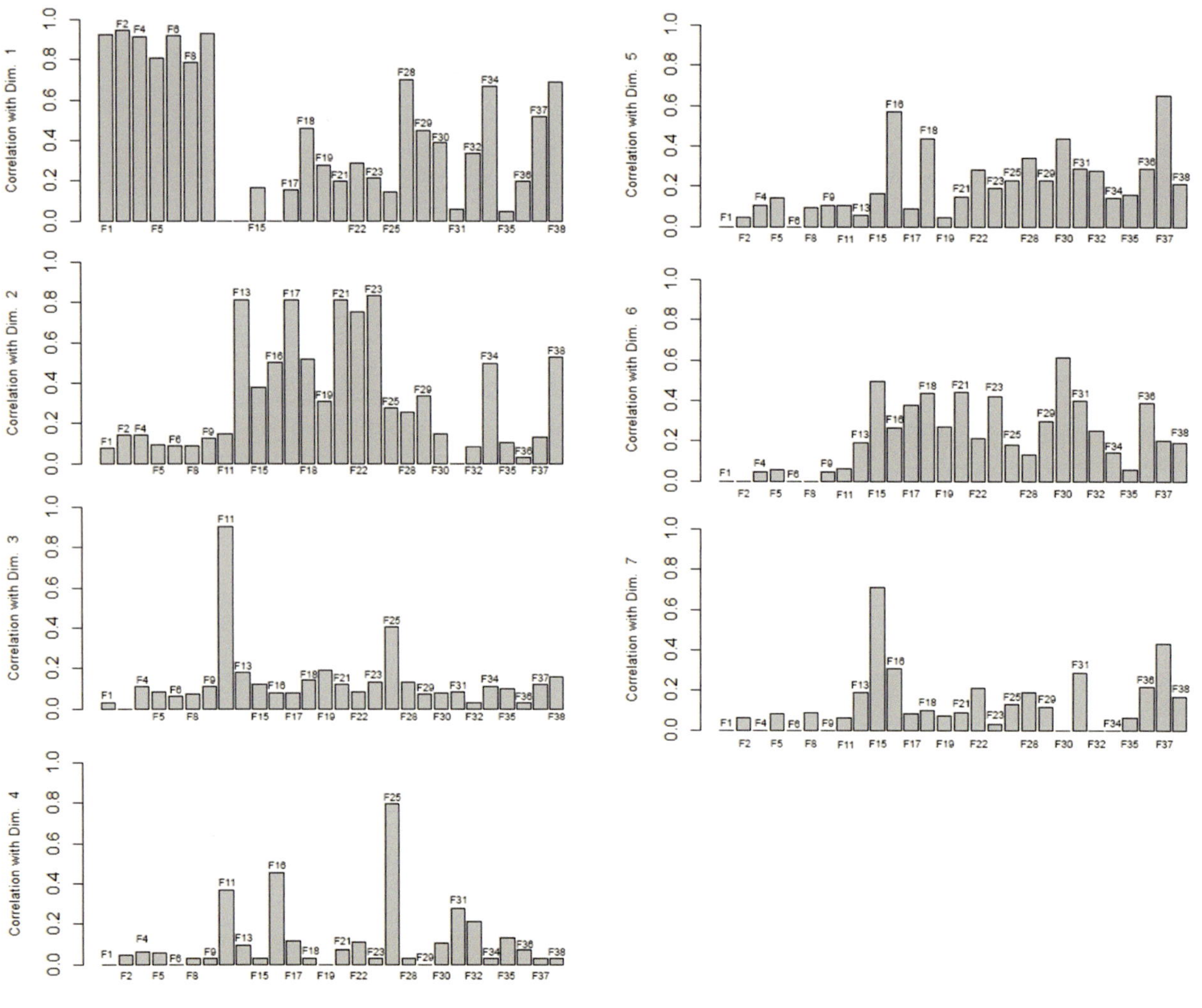

Figure 9. Correlation between row profiles and observed dimensions.

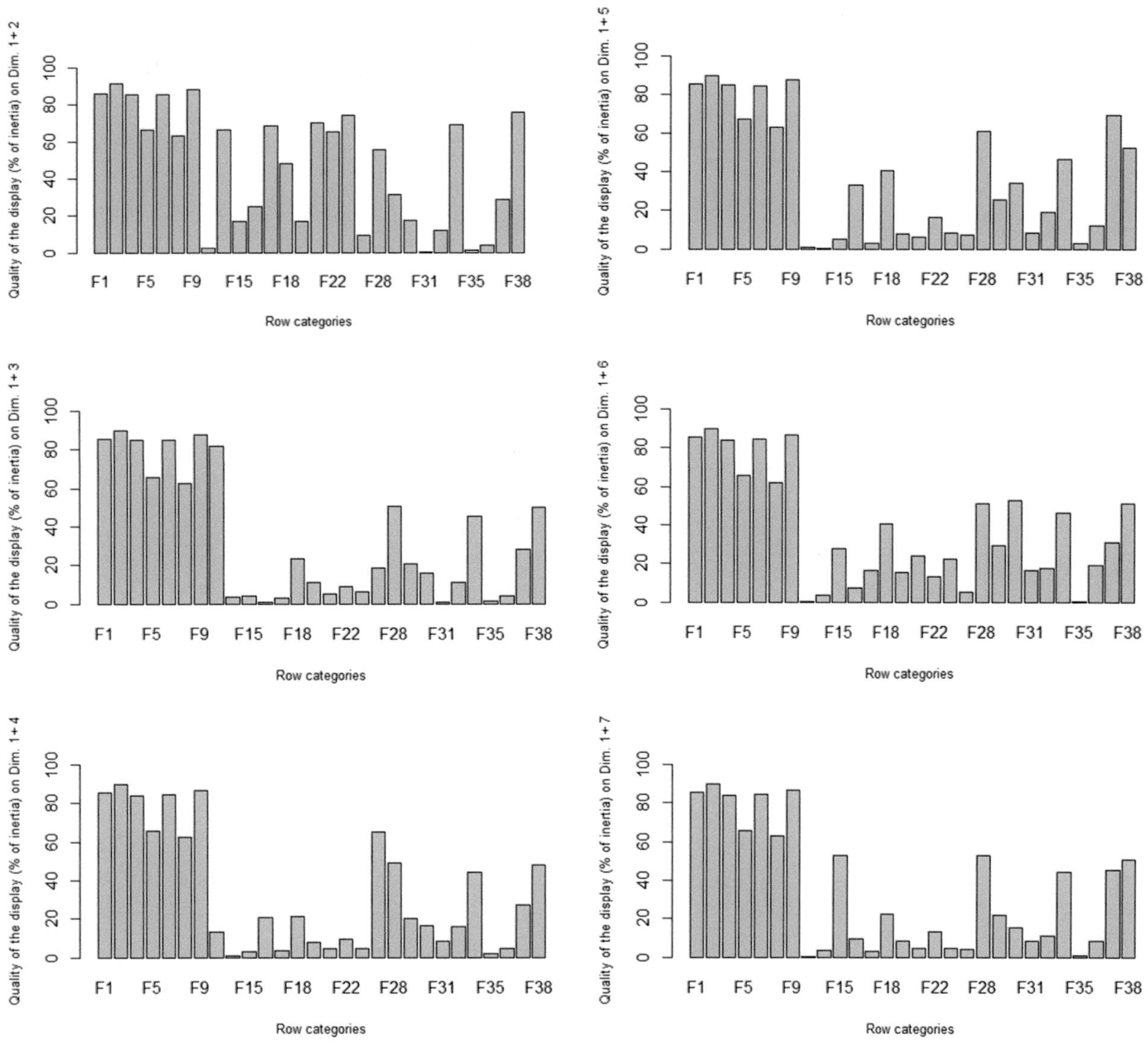

Figure 10. Quality of representation of dimensions per feature.

3, whilst in all others it is well below statistically significant representation. The same is true for features F13 and F35. Several other features also show under-representation (e.g. F17, in certain dimensions). Their points on the scatter plot should also be evaluated with caution, as they may be misleading. It is worth mentioning that the three most under-represented features are a kiln (F11), an ash deposit (F15), and a post hole (F35), none of which are strongly associated with large quantities of pottery.

Conclusions

The results of the CA enabled us to distinguish seven main trends of variation in the dataset. Dimensions 1 and 2 accounted for most of the variation, explaining almost 50% of the inertia. Though this is far from a simple explanation, it enables us to indicate that Dimension 1 is determined by the opposition of types 117 (the negative pole) and type 104b (the positive pole) which explains over 30% of variability in the dataset, whilst Dimension 2 can be defined as the opposition of types 115 and 115a (the negative pole), and types 112 and 114 (the positive pole). But what does this mean?

To understand this, the scatterplot or the biplot (Figure 5, top left) should be examined from which it can be concluded that the more the features of Dimension 1 lie on the right side, closer to 104b, the more they will be associated with this type of pottery, i.e. this type will make up a higher proportion of their assemblage. Of course, this does not mean that the assemblage will not contain other types, but rather that the proportion of this particular type will be significantly greater than that of others. On the opposite pole is type 117, which has such a high proportion in features associated with it (F1, F2, F4, F5, F6, F8, F9), that almost no other types will appear it those assemblages, or that their proportions will be diminishingly smaller. In Dimension 2, it is clear that types 115 and 115a have such large proportions on

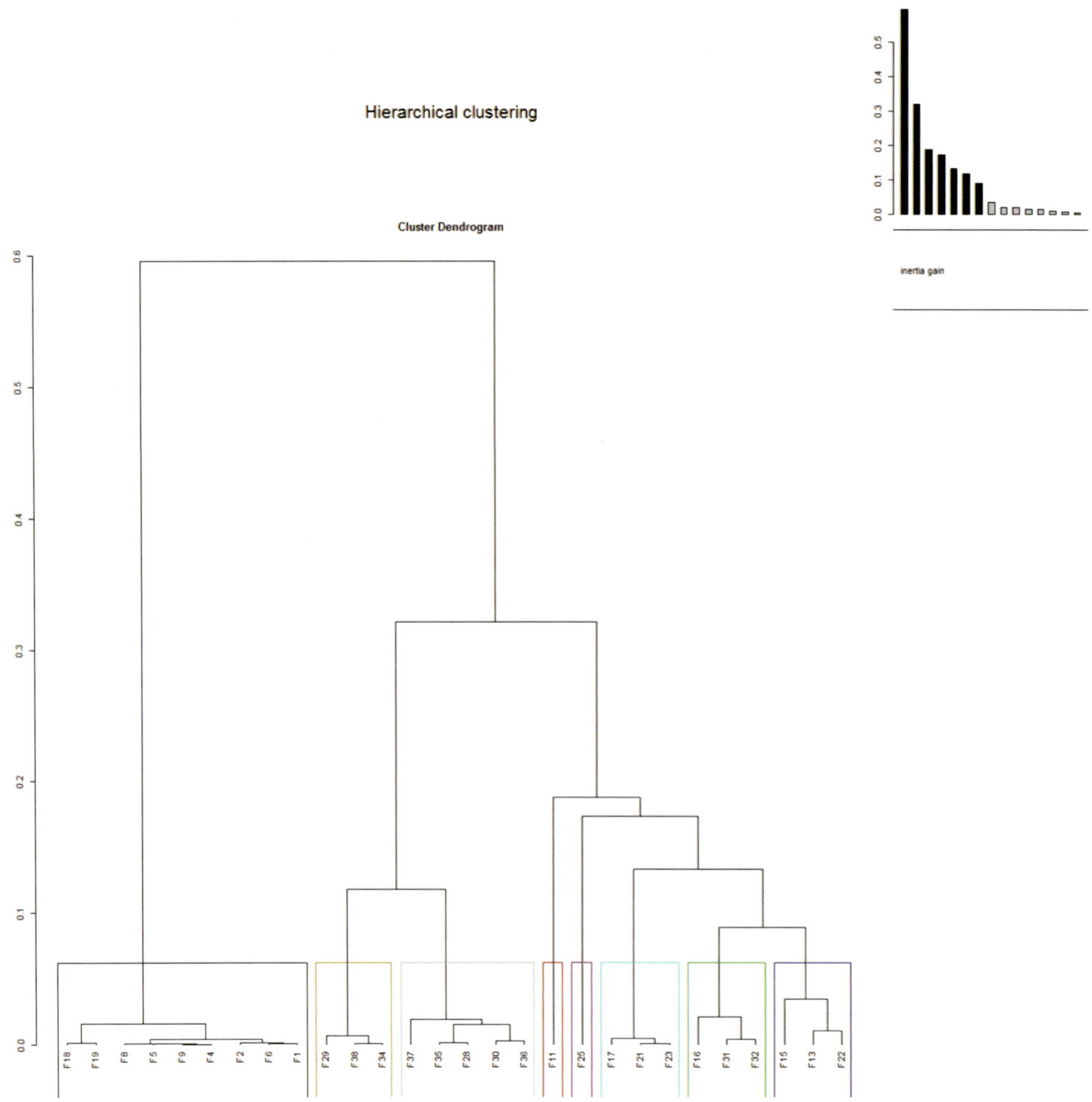

Figure 11. Cluster analysis on row data showing similarity in features based on the bowl types.

their pole that almost no other types will occur with them in features such as F13, F17, F21, F22 and F23. On the opposite, positive pole, types 112 and 114 appear but their contribution to Dimension 2 is rather limited and confined to just three features (F29, F34 and F38).

The standard biplot of Dimensions 1 and 3 (Figure 6, middle left) shows that almost 10% of inertia can be explained by types 102 and 118, the first being on the positive pole of the dimension, the second on the negative. However, in this situation the determination is so strict that no other type contributes to this dimension, and it can be seen that the higher-than-average proportion of both types is limited to just one feature (25 in the case of type 102 and feature F11 in the case of type 118) both of which seem to be under-represented in most dimensions (Figure 9). These results should therefore be viewed with caution, as they show that the higher proportion of these two types is most likely the result of a bad sample, and that the two features are best not included in the final interpretation.

A similar situation is evident on the biplot of Dimensions 1 and 4, which amount to 8.4% of the inertia. Again, it is the case that pottery types 102 and 118 the define Dimension 4, with the additional introduction of type 101. As in the previous case, this is most likely due to a bad sample, and it is best to avoid using the two features F11 and F25 for CA in which these types dominate. However, type 101 is somewhat better represented in the assemblage, even though almost no other types show association with it. The symmetric scatterplot (Figure 87, bottom left) shows a very centric plot of points, indicating very limited deviation from the average distribution in most features.

The biplot of Dimensions 1 and 5, amounting to 7.47% of inertia, show the dominance of pottery types 101 and 104b (Figure 6, top right), but also the opposition of types 113 and 114, which are found opposite to 101 and 104b. As type 113 can be mostly associated with features F16 and F25 (Figure 8, top right) which are poorly represented in most dimensions, it is questionable whether this type should be taken into consideration (Figure 9). Conversely, type 114 is well represented in features F29 and F32 (Figure 8, top right). Feature F29, although not extremely well represented, is always above the statistically significant limit unlike feature F32 which is constantly underrepresented in all dimensions. and should be considered with care.

In the biplot of Dimensions 1 and 6, covering 6.05% of inertia, the dominant types are 115 and 115a, which slow a clear opposition to each other (Figure 6, middle right) meaning that the higher proportion of the first will result in a lower proportion of the second in the features and vice versa. There is also a visible opposition in other, less dominant bowl types: type 114 occurs in larger proportions with type 115a, whilst types 101, 102, 113 and 114 occur in larger proportion with type 115. Aside from features F1–F9, this dimension is well correlated with most other examined features and is also well represented in the same features (Figure 9, middle right).

Finally, the biplot of Dimensions 1 and 7 (4.84%) shows the dominance of pottery types 104b and 114, with 115 and 116 as the opposites. Also present is type 108 which occurs in larger proportions with the latter three types and type 102 with 104b. This dimension is well correlated with several features (F15, F16, F22, F29, F31, F36, F37, F38) and well represented in features F15, F37 and F38 (Figure 10, bottom right).

Further analysis

Having analysed the profiles in the dataset, it is possible to do further piece of analysis with the CA results which will enable us to distinguish clusters of points that are similar in their assemblage profiles (Greenacre 1988: 41). The rows (or columns) are progressively aggregated such that every successive merging results in the smallest change in the inertia of the table until it is reduced to just one row containing marginal columns of the original table (Greenacre 2007: 116–117). This implies that rows or columns which, in merging, produce small changes to the inertia of the table, have similar profiles; this merging can be represented as a dendrogram as shown in Figure 10. Four major groups of bowls appear in similarly profiled assemblages. The first group consist of types 101a, 104 (with subtypes c and d) and type 117. The second consists of two sub-clusters, the first containing types 104a, 112, 114 and 160 and the second consisting of types 106, 107 and 110. The third major group also has two sub clusters; types 100 and 108, and types 101, 109, 111 and 113. The last major cluster consists of two types, 115 and 116. All other types are presented as individual clusters as they appear in several different profiles (e.g. type 104b or type 118) and cannot be distinctly associated with specific assemblage profiles. Based on the dendrogram it can be said that four major groups of bowls dictate the similarity of features.

If the same analysis is applied on to row data (Figure 11) the resulting dendrogram identifies several major groups with the most similar bowl profiles. The first group consists of two sub-clusters with the larger containing features associated with structure of wattle and daub (F1, F2, F4, F5, F6, F9) and a concentration of movable finds found in interaction with this structure (F8). The second sub-cluster contains features F18 and F19, both discovered underneath the wattle and daub structure. The second large cluster consists of three features, two of which (F29 and F34) are in the same construction horizon whilst feature F38 lies immediately below feature F34. A cluster with similar profile to this comprises features F28, F30, F35, F36 and F37. It is interesting to note that F35, F36, and F37 are post holes from the same structure, and are similar to the other cluster due to the fact that they have been cut through the infill of feature F38. Feature F28 is a large, partially excavated refuse pit detected at the same level as F30.

The next cluster consists of three features, F17, F21 and F23, all belonging to the same construction horizon. Similar to these features in profiles are the next two clusters, which show more similarity with each other. These include features F16, F31 and F32 in one cluster, and features F13, F15 and F22 in a second. The first cluster contains two features (F31 and F32) from the same construction horizon and one (F16) from a horizon above. The second cluster also comprises two features (F15 and F22) from the same horizon and one (F13) from a horizon above. However, as previously established, the data from feature F13 should be treated with special attention as it shows clear signs of under-representation in the analysis. Finally, two features, F11 and F25, stand apart as they are too specific or under-represented to fit with the rest of the assemblage, as already seen during the previous steps of the analysis.

The bibliographic reference for this chapter is:

Mirković-Marić, N. and Marić, M. 2021. Chronological attribution of pottery from Trench 24 at Pločnik based on correspondence analysis, in Radivojević, M., Roberts, B. W., Marić, M., Kuzmanović Cvetković, J., and Rehren, Th. (eds) *The Rise of Metallurgy in Eurasia*: 345–361. Oxford: Archaeopress.

Chapter 29

Pločnik: technology of pottery production

Silvia Amicone

Introduction

This chapter presents results from the analysis of ceramic assemblages found in Trench 24 from the Late Neolithic/Early Chalcolithic site of Pločnik (*c.* 5200–4400 BC). It will elucidate aspects of the development of pottery production during the transition to the Chalcolithic, particularly around the time of the emergence of metallurgical activities at this site. We aim to identify changes in pottery production and potential pyrotechnological associations, both in the phases preceding the introduction of metallurgy and during the phases of metal production. Emphasis is placed on aspects related to raw material selection and preparation, and on the pyrotechnology behind the so-called 'dark-burnished pottery' and 'graphite-painted pottery'. Vessels decorated with graphite are found across the 5th millennium BC in the Balkans (e.g. Leshtakov 2005; Martinon 2017; Todorova 1986: 107). The first documented use of graphite to decorate pottery comes from Promachon-Topolnica in the Struma valley and is dated to the beginning of the 5th millennium (Vajsov 2007). Within the Vinča culture, it appears for the first time during the Gradac phase (Perić 2006: 238), and this is also the case at Pločnik (cf. Mirković *et al.*, Chapter 27, this volume). It has been proposed that the use of graphite decoration was connected to the emergence of early metal production. The metallic sheen that is produce by the light-reflective qualities of graphite may have been aesthetically appealing to prehistoric communities (Todorova 1981). Other have drawn a technological link and proposed that the high temperatures necessary for copper metallurgy (around and exceeding *c.* 1100ºC) could also have been needed to produce graphite-painted pottery (Renfrew 1969). Finally, the acquisition of graphite would have required

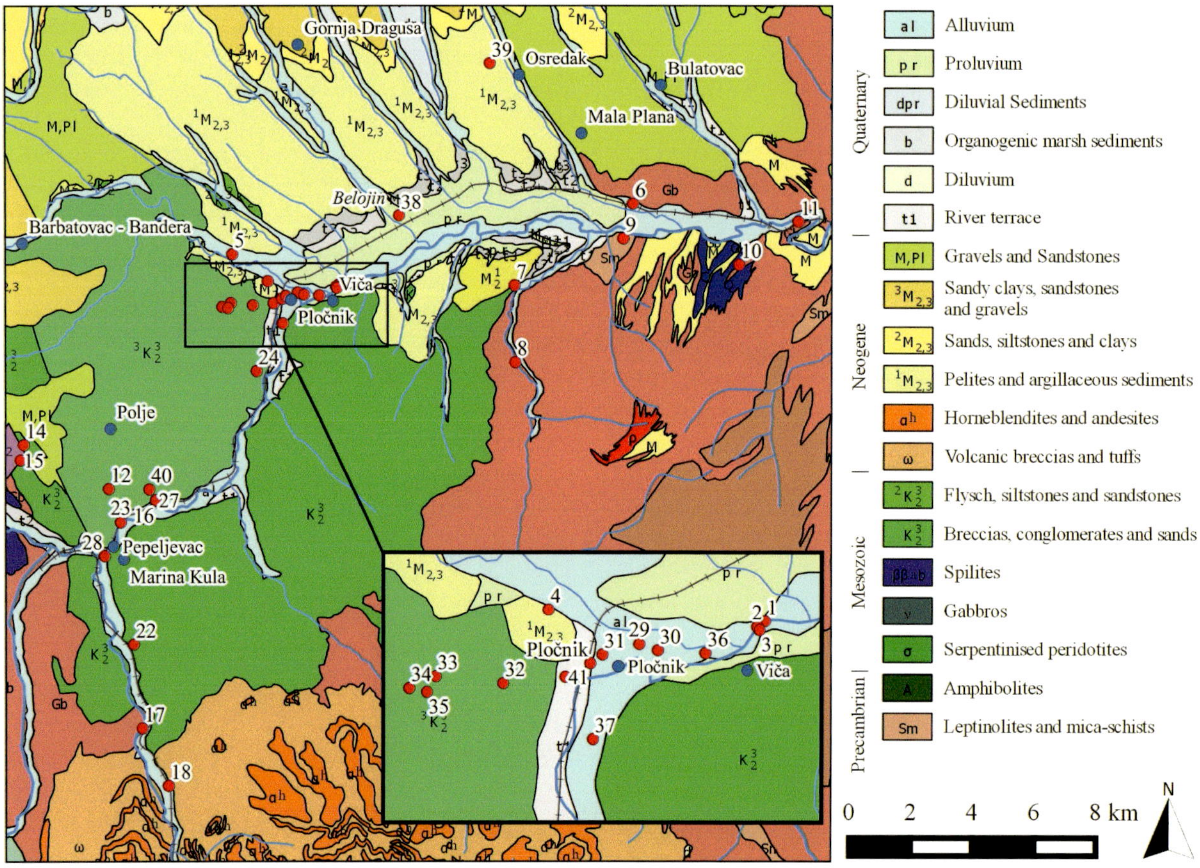

Figure 1. Geological map of the area surrounding Pločnik. Red dots indicate the location of specimen samples in the area. (Map by Enrico Croce)

The Rise of Metallurgy in Eurasia (Archaeopress 2021): 362–374

Table 1. Number of samples analysed with petrography and XRPD according to building horizons.

Petrography				XRPD		
Pločnik 24				Pločnik 24		
Number of samples	Horizon	Absolute Chronology		Number of samples	Horizon	Absolute Chronology
34	1	4631-4231 cal. BC		11	1	4631-4231 cal. BC
22	2	4927-4621 cal. BC		8	2	4927-4621 cal. BC
45	3	5036-4951 cal. BC		7	3	5036-4951 cal. BC
29	4	5121-4976 cal. BC		9	4	5121-4976 cal. BC
17	5	5389-5003 cal. BC		13	5	5389-5003 cal. BC

participation in specialist trade networks comparable to those required for copper exploitation (e.g. Leshtakov 2005; Radivojević and Grujić 2018). However, the relationship between the emergence of graphite-painted pottery within Vinča culture and extractive metallurgy has never been thoroughly investigated.

Methods

A detailed technological characterisation of the ceramic assemblages was undertaken, using both macroscopic and microscopic techniques. Macroscopic analysis was conducted by a team of pottery specialists working at Pločnik (Mirković *et al.*, Chapter 27, this volume), with pottery classified according to the same methods employed at Belovode (cf. Chapter 12, this volume).

Following this initial classification, 147 of ceramic samples were analysed by thin section petrography and a smaller selection of specimens by X-ray powder diffraction (XRPD) analysis and scanning electron microscopy (SEM), in order to characterise the raw materials and the manufacturing processes employed at Pločnik (Table 1 and Appendix B_Ch29/Table 1). In addition, a prospection of raw materials for potter making was carried out by the authors in the area around the site with the aim of exploring the nature of the raw materials available for pottery making.

Pločnik is situated on a fertile floodplain characterised by alluvial quaternary deposits related to the activity of the Toplica river (Figure 1), which flows from the Kopaonik mountain range *c.* 50 km from the site. The area surrounding the site shows the presence of Cretaceous deposits of flysch, sandstone, marl, and olistostrome, covering the Precambrian formations constituted by leptinolite and mica-schist, fine-grained biotite and gneisses, and amphibolite. The latter outcrops are located east of the site. West of Pločnik, the Kuršumlija area is characterised by gabbro dolerites, rare basaltic pillow lavas, and small isolated outcrops of mainly serpentinised hazburgites.

Results of the petrographic analysis

The selected samples could be divided into fabric groups according to the presence of different minerals and rocks and the characteristics of matrices and voids (see Appendix B_ Ch29/Table 2).

Group A: Sedimentary Rock Fabric

Fabric Group A is the dominant petrographic group (Figure 2a–e) and includes 128 samples. It is characterised by heterogeneous, coarse (subgroup A1) to fine-grained samples (subgroup A2) and is marked by generally poorly sorted sub-angular to sub-rounded inclusions of quartz and plagioclase that seem to derive from quartz-arenite or arkosic-arenite sandstones. Few samples contain fragments of metarenite rocks, but iron-rich inclusions and clay pellets are common. Other inclusions comprise small amounts of muscovite, amphibole, perthitic feldspar, microcline chert and, more rarely, calcite and biotite. This evidence suggests that the raw material used for the first petrographic group was a primary clay containing clasts of sandy rocks. Additionally, the matrix for this group is non-calcareous and it is light yellow to bright orange to dark red in plane-polarised light (PPL) and light brown to dark reddish to grey in cross-polarised light (XP). It is quite homogenous; inhomogeneity is caused by distinct firing horizons. It ranges from moderately active to weakly active to inactive. In several samples, relict coils occur, related to the use of the coiling technique.

Group B: Mica-schist Rock Fabric

Fabric Group B is a heterogeneous group (Figure 2f). It includes eight samples and is characterised by a coarse- to fine-grained fabric and by the presence of poorly sorted, angular to sub-angular inclusions of quartz and muscovite. The dominant mineral inclusions comprise quartz and muscovite. Plagioclase, microcline, feldspar, biotite, iron-rich inclusions and clay pellets are frequent; foliated rock fragments rich in quartz and muscovite (probably mica-schist) are also common, while rock fragments composed of quartz (probably metarenite) are rare. Based on these mineral inclusions and the rock fragments, the raw material used to produce these objects appears to have come from a foliated metamorphic rock composed of quartz and muscovite. This composition is indicative of mica-schist. The bimodality of the coarse fraction observed in some samples suggests that these sherds could have been tempered with a sand derived from the erosion of a metamorphic rock. Additionally, the matrix in this group is non-calcareous; it ranges from light brown to bright yellow to light orange in PPL, and light brown to reddish yellow in XP. It also appears to be quite homogenous with inhomogeneity caused by distinct firing horizons; it is moderately active to weakly active. As with Fabric Group B, relict coils observed in some samples are related to the use of the coiling technique.

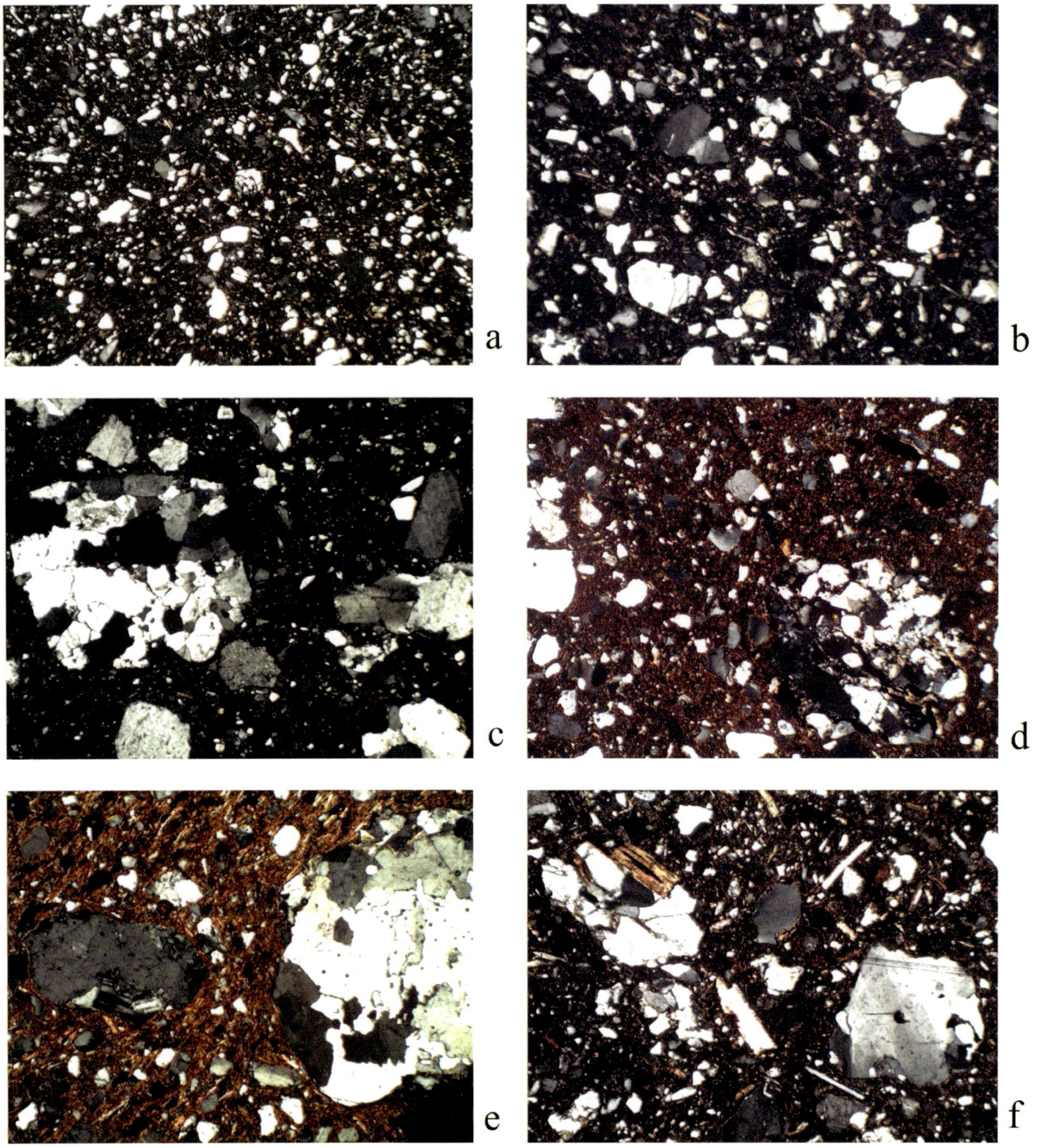

Figure 2. Thin section photomicrographs of selected ceramics from Pločnik: a) Fabric Group A with sedimentary rocks, fine XP; b) Fabric Group A with sedimentary rocks, medium-fine XP; c) Fabric Group A with sedimentary rocks, coarse XP; d) Fabric Group A with arkosic arenite XP; e) Fabric Group A with sedimentary rocks and abundant muscovite XP; f) sample PL-B (21-19) with mica-schist XP. Image width = 3 mm.

Two samples, PL 24-23 and PL 24-211 (Figure 3a) are dominated by the presence of poorly sorted sub-angular to sub-rounded inclusions of quartz, plagioclase, muscovite and biotite; these minerals appear to have derived from weathering of medium-grained igneous rocks (andesite). Other common inclusions include amphibole and smaller amounts of pyroxene. The angularity of the inclusions and the poor degree of sorting suggest that this clay was residual in origin or only minimally transported. The matrix is non-calcareous and is light brown in PPL and brown in XP; it is also quite homogenous. Sample PL 24-23 is optically inactive, while PL 24-211 is moderately active.

In addition to the groups detailed above several outliers were identified. Sample PL 24-32 (Figure 3b) is a medium- to fine-grained sample characterised by the presence of serpentinite. Other common minerals comprise inclusions of quartz, poorly sorted sub-angular to sub-rounded muscovite, serpentinite, clay pellets and opaque minerals. The matrix is non-calcareous, homogenous and optically active. It is light brown in both PPL and XP.

Sample PL 24-319 (Figure 3c) shows an abundance of epidote. Other common minerals include poorly sorted sub-angular to sub-rounded inclusions of quartz, feldspars, muscovite, amphibole as well as more metamorphic rocks (probably psammites) and opaque minerals. These petrographic characteristics suggest that the raw material used to produce this sherd was a secondary clay containing clasts of metamorphic rocks. The matrix is non-calcareous, homogenous and optically inactive, and is dark red in both PPL and XP.

Sample PL 24-113 (Figure 3d) stands out for its high levels of amphibole and possible fragments of amphibolite. Other frequent inclusions comprise quartz muscovite and opaque mineral. The raw material used to produce this vessel could be a primary clay formed from the erosion of metamorphic rocks (probably amphibolite). The matrix is non-calcareous, homogenous and weakly active. It is yellow in PPL and brown in XP.

Sample PL 24-208 (Figure 3e) is a very coarse specimen characterised by the presence of fragments of metamorphic rocks, probably gneiss and phyllite. Other common inclusions are quartz, polycrystalline quartz, feldspars, and opaque minerals. The matrix is non-calcareous, homogenous, and optically inactive. It is orange in PPL and dark orange in XP.

Samples PL 24-82 and PL 24-315 (Figure 3f) are dominated by abundant, well-sorted and rounded fragments of schist and phyllite. Other common minerals are quartz, epidote and, more rarely, fragments of mudstone and clay pellets. The strong bimodality of the coarse fraction suggests that this sherd might have been tempered with a sand derived from the erosion of a metamorphic rock. The matrix is non-calcareous and is light brown in PPL and dark brown in XP. It is quite homogenous and weakly active.

Samples PL 24-202 and PL24-258 (Figure 3g) are fragments of loom weights. These are characterised by tiny fragments of quartz, muscovite and opaque minerals. Grog has probably been added as temper. The evidence suggests that the raw material used to produce these loom weights was a calcareous, secondary clay. The matrix is optically active and is yellow in PPL and light orange in XP.

Sample PL 24-336 (Figure 3h) is a very coarse floor fragment. It is made from highly calcareous clay (probably a marl) mixed with a clay rich in sedimentary rocks that seems similar to that used to produce vessels classified in Group A. The matrix is active and its colour is yellow in PPL and light orange in XP.

Results of the XRPD and SEM analyses

The XRPD analysis of 48 samples (see Table 2, Figure 4 and 5) showed that most of the samples contain illitic clay, which can be distinguished by a main diffraction peak at 10 Å d spacing and further peaks at increasing 2θ° with variable intensities (Deer *et al.* 1992). Overall, the main mineralogical assemblages that were detected through XRPD analysis are constituted by quartz, felspars, calcite and illite. Several samples (Figure 5) exhibit a weak diffraction peak corresponding to a d-value of approximately 14 Å which points to either chlorite or montmorillonite. The presence of amphibole has been verified in some samples (PL 24-23, PL 24-113, PL 24-211, and PL 24-315) by its main characteristic peak at about d ~ 8 Å. In addition, cristobalite (d = 4.08 Å) was detected in sample PL 24-23 (Figure 6). The presence of illite peaks suggests that the maximum firing temperature of most of the samples must have been below 850–900°C (Kulbicki 1958; Maggetti 1982). Only a few samples show the presence of calcite in their diffraction patterns, which means that these vessels could not have been fired at temperatures above 850°C. Alternatively, calcite might occur as a secondary mineral (Maggetti 1982; Maritan 2004). Overall, the XRPD results suggest that the maximum temperature to which most of the ceramic specimens analysed had been fired was not above 900°C, and probably around 750–800°C. This includes all the dark-burnished pottery samples with or without graphite painting decoration. One sample however, PL 24-23, in which cristobalite was detected, was probably exposed to temperatures over 1000°C. Notably, this same sample, characterised by mineral phases that form at high temperatures, is most likely a

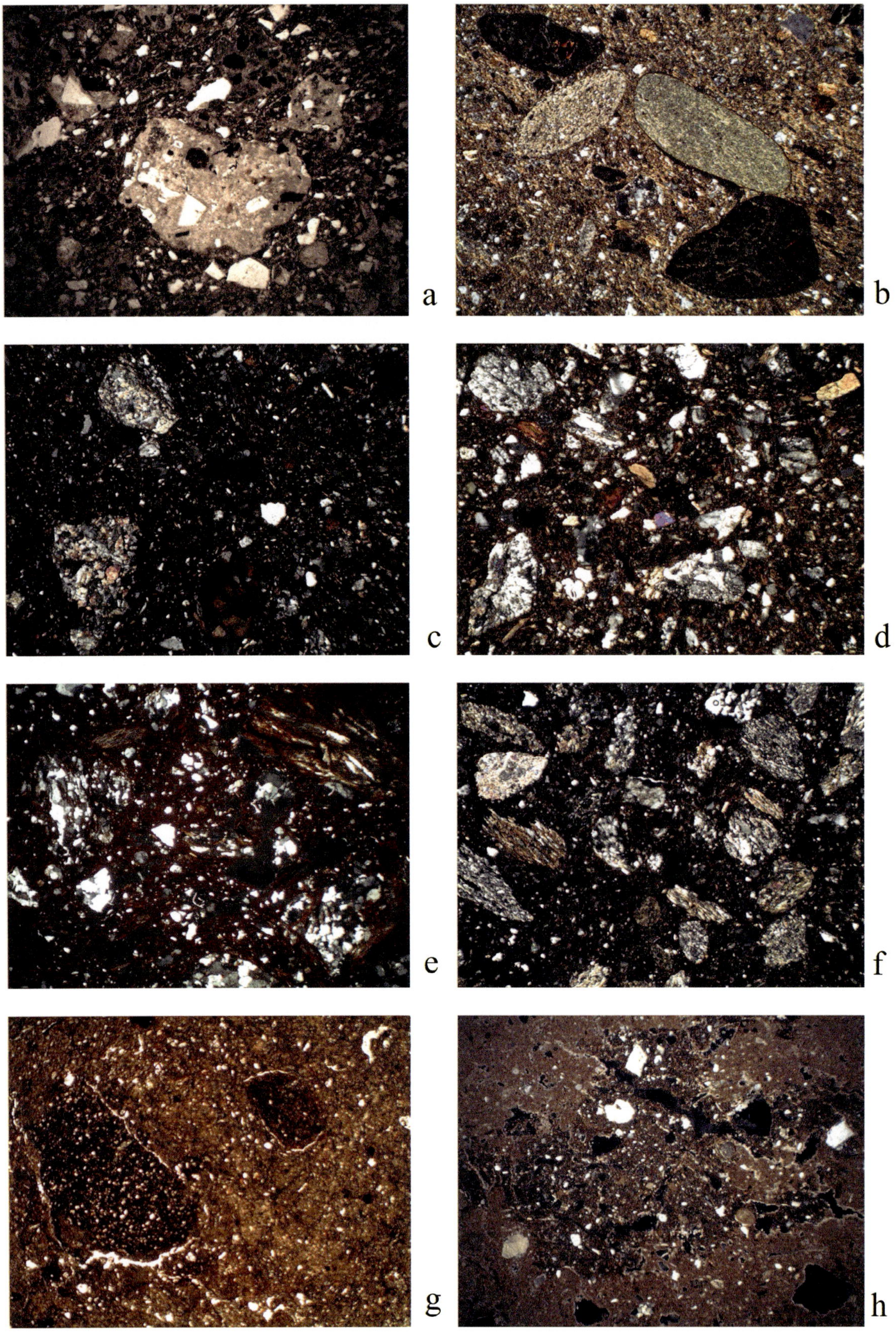

Figure 3. Thin section photomicrographs of selected ceramics from Pločnik: a) PL 24-23 with volcanic rocks XP; b) PL 24-32 with serpentinite XP; c) PL 24-319 with abundant epidote XP; d) PL 24-113 with amphibolite XP; e) PL 24-208 with gneiss XP; f) PL 24-315 with phyllite XP; g) PL 24-202 with grog tempering XP; h) PL 24-336 made from calcareous clay XP. Image width = 3 mm, except e = 6 mm.

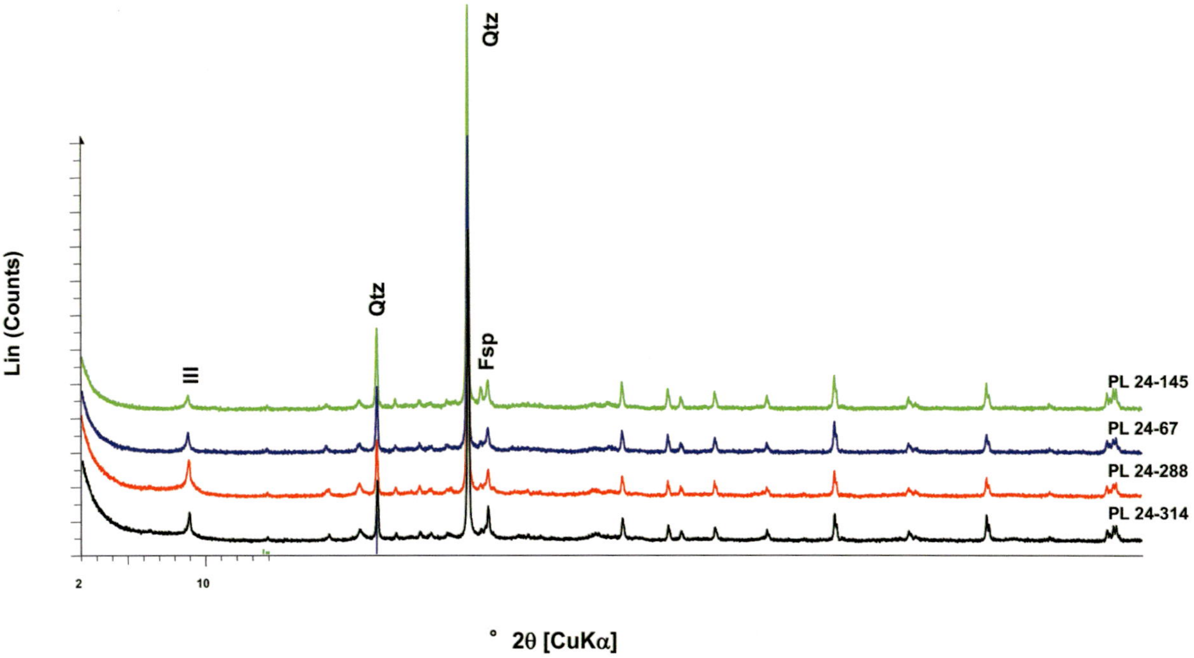

Figure 4. Diffractograms of samples PL 24-314, PL 24-288, PL 24-267, and PL 24-145 (dark-burnished pottery). Mnt: montmorillonite; Ill: illite; Qtz: quartz; Fsp: feldspars.

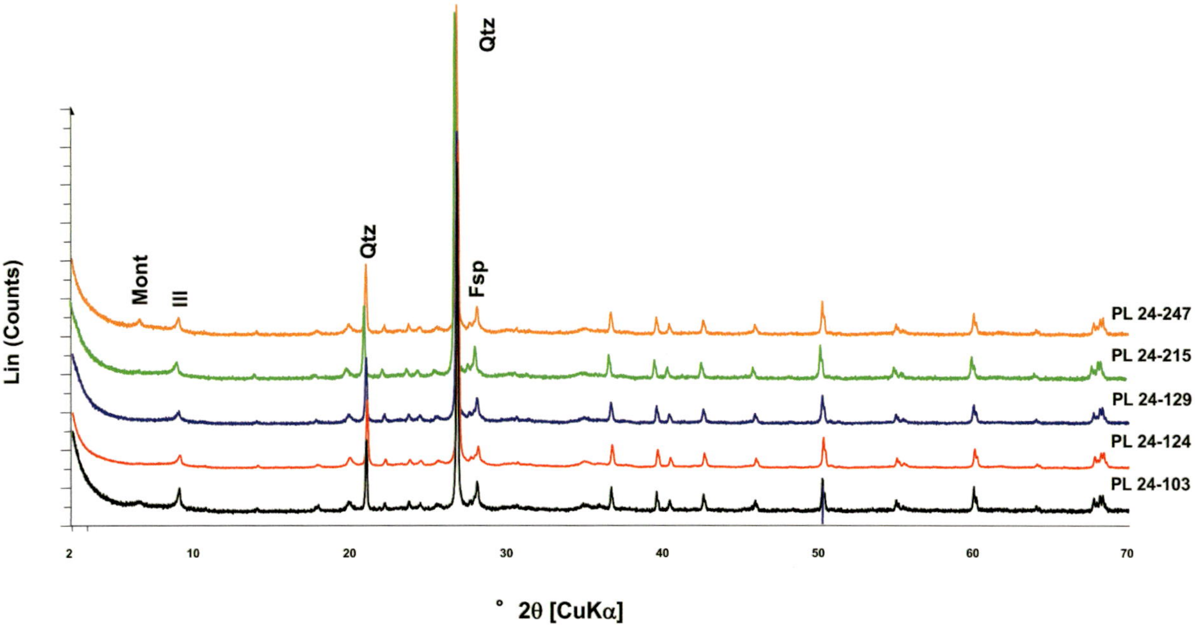

Figure 5. Diffractograms of samples PL 24-103, PL 24-124, PL 24-129, PL 24-215 and PL 24-247 (graphite-painted pottery). Ill: illite; Qtz: quartz, Fsp; feldspars.

of non-local production. The original vessel fragment from which the sample derives has a perforated wall. While its function remains uncertain, it could be a fragment of a 'chimney'.

The fresh fractures of eleven samples of pottery from different phases of the settlement were also analysed under scanning electron microscope (SEM) to observe their degree of vitrification (Maniatis and Tite 1981). Two dark-burnished samples were re-fired in reducing conditions at set interval temperatures ranging from 700–1100°C. Most of the samples show an initial vitrification that is compatible with temperatures between 750°C and 850°C (Figure 7; Table 3). The re-firing experiment (Figure 8) shows that the degree of vitrification of the five samples started to change within this temperature interval. However, PL 24-247, decorated with graphite, shows a microstructure

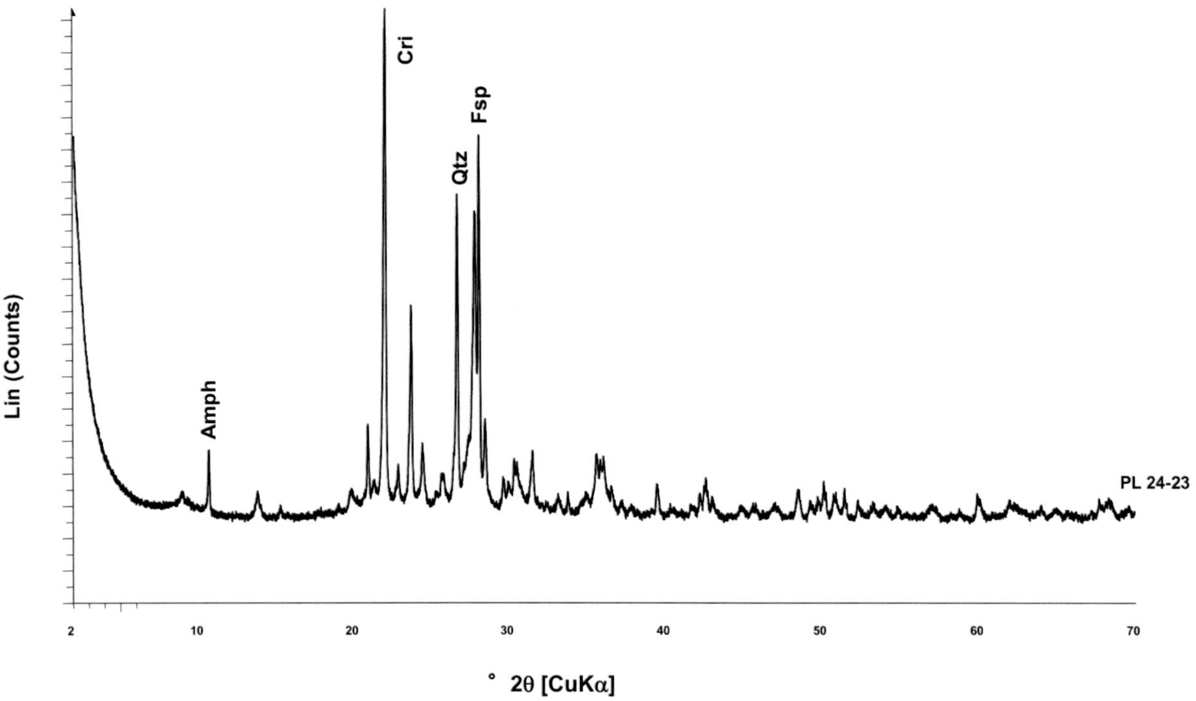

Figure 6. Diffractogram of sample PL 24-23 (wall fragment, fabric with volcanic inclusions). Amph: amphibole; Cri: Cristobalite; Qtz: Quartz; Fsp: feldspars.

(Figure 7f) compatible with higher temperatures of around 900°C. This sample also shows the presence of bloating pores, which are normally formed under reducing firing conditions (Maniatis and Tite 1981: 61).

Discussion: raw material selection and processing

The petrographic analysis of the ceramic samples from Pločnik and the geological specimens collected in the field showed that sherds belonged to Fabric Group A and can be associated with clay sources available in the area surrounding the site. These sherds could represent local production. The petrographic characteristics of some geological samples, i.e. sample 6 from site 5, sample 7 from site 6, and sample 38 from site 35 (Figure 1), and a sample of modern daub collected in the modern nearby village, closely match those of the specimens of Fabric Group A. These sandy clays contain quartz, muscovite, and various sedimentary rocks (Figures 9 a–d).

It is difficult, perhaps even impossible, to localise the exact outcrops of clay exploited by the potters to produce the ceramic material characterised by Fabric Group A. We can probably, however, exclude the alluvial clay deposited by the Toplica river. Clay and sand samples from the riverbank were analysed (Figures 9e–f) and the nature of the sandy fraction was revealed to be polymictic (Maggetti 1994), with fragments of rocks of different origins (e.g. sedimentary, volcanic). The samples attributed to Fabric Group A are compositionally homogeneous with a narrow range of minerals and sedimentary or meta-sedimentary rocks. The raw material could have come from the Neogenic and Mesozoic formation which outcrops around Pločnik (Figure 1) and contains sandy clays suitable for pottery making.

Some differences were observed in the grain size, sorting and distribution of the inclusions. It is not clear if this variability is related to different practices of paste preparation adopted by the potters when producing different vessel types, as it was also observed in Belovode (Chapter 14, this volume). It was possible, however, to observe a general tendency towards more abundant and coarser inclusions in vessels with thicker walls and a generally finer fabric amongst samples with thinner walls. It was not always clear whether the Pločnik ceramics were tempered or not, although this was more likely for some specimens than for others (Appendix B_Ch29/Table 2). As at Belovode, tempered pottery appears more common in the earliest phases of the settlement. In general, it seems that once the suitable raw material was collected, it was processed only minimally. It is notable that the fabric of the dark-burnished pottery is not particularly fine-grained and it is not possible to associate its production with any specific paste recipe.

While the provenance of the samples belonging to Fabric Group A—the majority of the sherds from Pločnik—is fairly clear, the samples belonging to the

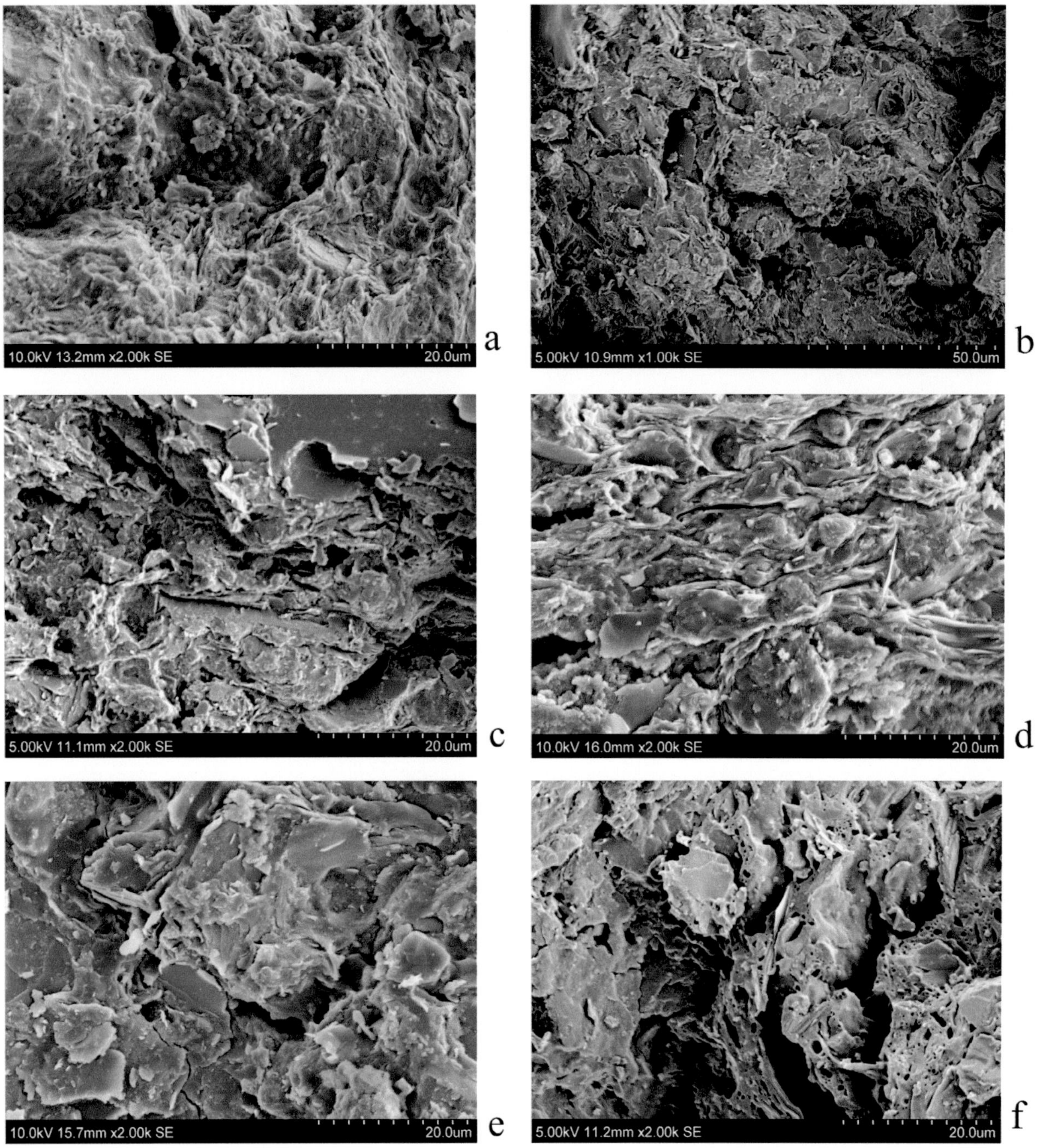

Figure 7. SEM photomicrographs of ceramic samples from Pločnik: a) PL 24-83 (bowl); b) PL 24-103 (bowl, dark-burnished ware with graphite decoration); b) PL 24-124 (dark-burnished ware with graphite decoration); b) PL 24-129 (dark-burnished ware with graphite decoration); c) PL 24-215 (dark-burnished ware with graphite decoration); and d) PL24-247 (dark-burnished pottery with graphite decoration).

other attested fabric groups and the outliers require a more detailed discussion. Specimens of Fabric Group B are characterised by an abundance of mica and mica-schist fragments that seem to be added as temper, and they also contain fragments of sedimentary rocks. It is possible to hypothesise that these specimens were made from the same clay used for production of Fabric Group A, to which mica-schist was later added as temper. This tradition seems to be very common in the last horizon of the settlement, when dark-burnished pottery production is considerably reduced. Formations with mica-schist outcrops are located *c.* 10 km east of Pločnik (Figure 1 and Figure 10a,b) and fragments of this type of rock were also found in the excavated trench.

Table 2. Summary of the XRD results (DB=dark-burnished; GP=graphite-painted).

Sample	Chronological Horizon	DB	GP	Optical Activity	Qtz	Fsp	Cc	Am	Ill	Msc	Hem	Cri	Spl	Temp
PL 24-15	1 (Gradac II-III)	X		moderate	X	X			X					< 900 °C
PL 24-23	1 (Gradac II-III)			moderate	X	X		X		X		X		> 1000 °C
PL 24-32	1 (Gradac II-III)			weak	X	X			X					< 900 °C
PL 24-34	1 (Gradac II-III)			moderate	X	X			X?	X?				< 900 °C
PL 24-54	1 (Gradac II-III)			absent	X	X			X					< 900 °C
PL 24-70	1 (Gradac II-III)	X		weak	X	X			X					< 900 °C
PL 24-73	1 (Gradac II-III)			absent	X	X			X					< 900 °C
PL 24-74	1 (Gradac II-III)			absent	X	X			X					< 900 °C
PL 24-75	1 (Gradac II-III)			absent	X	X			X					< 900 °C
PL 24-83	1 (Gradac II-III)			absent	X	X	X		X					< 900 °C
PL 24-101	2 (Gradac I)	X		weak	X	X			X					< 900 °C
PL 24-103	2 (Gradac I)		X	moderate	X	X			X					< 900 °C
PL 24-107	2 (Gradac I)	X		weak	X	X			X					< 900 °C
PL 24-113	2 (Gradac I)			high	X	X		X	X					< 900 °C
PL 24-124	2 (Gradac I)		X	weak	X	X			X					< 900 °C
PL 21-129	2 (Gradac I)		X	weak	X	X			X					< 900 °C
PL 24-132	2 (Gradac I)	X		weak	X	X			X					< 900 °C
PL 24-145	2 (Gradac I)	X		weak	X	X			X					< 900 °C
PL 24-157	3 (Gradac I)	X		weak	X	X	X		X					< 900 °C
PL 24-161	3 (Gradac I)	X		weak	X	X			X					< 900 °C
PL 24-179	3 (Gradac I)			weak	X	X			X					< 900 °C
PL 24-186	3 (Gradac I)	X		moderate	X	X			X					< 900 °C
PL 24-204	3 (Gradac I)	X		weak	X	X			X					< 900 °C
PL 24-209	4 (B2)	X		weak	X	X			X					< 900 °C
PL 24-211	4 (B2)	X		moderate	X	X		X	X					< 900 °C
PL 24-215	2 (Gradac I)		X	weak	X	X			X					< 900 °C
PL 24-247	3 (Gradac I)		X	weak	X	X			X					< 900 °C
PL 24-263	3 (Gradac I)	X		weak	X	X			X					< 900 °C
PL 24-267	3 (Gradac I)	X		high	X	X			X					< 900 °C
PL 24-275	4 (B2)	X		absent	X	X	X		X					< 900 °C
PL 24-287	4 (B2)			moderate	X	X			X					< 900 °C
PL 24-288	4 (B2)	X		high	X	X			X					< 900 °C
PL 24-299	4 (B2)			high	X	X			X					< 900 °C
PL 24-303	4 (B2)	X		weak	X	X			X					< 900 °C
PL 24-307	4 (B2)	X		weak	X	X			X					< 900 °C
PL 24-313	5 (A2-B1)			weak	X	X			X					< 900 °C
PL 24-314	5 (A2-B1)	X		weak	X	X			X					< 900 °C
PL 24-315	5 (A2-B1)	X		weak	X	X		X	X					< 900 °C
PL 24-318	5 (A2-B1)			weak	X	X	X		X					< 900 °C
PL 24-319	5 (A2-B1)	X		moderate	X	X		X	X					< 900 °C
PL 24-320	5 (A2-B1)	X		weak	X	X			X					< 900 °C

PL 24-323	5 (A2-B1)			moderate	X	X			X					< 900 °C
PL 24-324	5 (A2-B1)			absent	X	X			X					< 900 °C
PL 24-328	5 (A2-B1)			moderate	X	X			X?	X?				< 900 °C
PL 24-329	5 (A2-B1)	X		moderate	X	X			X					< 900 °C
PL 24-331	5 (A2-B1)	X		weak	X	X			X					< 900 °C
PL 24-332	5 (A2-B1)	X		weak	X	X			X					< 900 °C
PL 24-333	5 (A2-B1)			moderate	X	X			X					< 900 °C

Table 3. Summary of the results of the SEM analysis. (IV= initial vitrification 750–800 °C; V= extensive vitrification 900–950 °C; C= continuous vitrification 1000–1050 °C; DB=dark-burnished; GP=graphite-painted.)

Sample	Chronological Horizon	DB	GP	Refiring	Degree of Vitrification
PL 24-23	1 (Gradac II-III)				V
PL 24-83	1 (Gradac II-III)				IV
PL 24-103	2 (Gradac I)		X		IV
PL 24-124	2 (Gradac I)		X		IV
PL 24-129	2 (Gradac I)		X		IV
PL 24-157	3 (Gradac I)	X		X	IV
PL 24-161	3 (Gradac I)	X		X	IV
PL 24-215	2 (Gradac I)		X		IV
PL 24-247	3 (Gradac I)		X		IV
PL 24-303	4 (B2)	X			IV
PL 24-313	5 (A2-B1)				IV

The petrographic characteristics of the specimens PL 24-23 and PL 24-211 point to a non-local production of these vessels. The sampled sherds are characterised by the presence of fragments of andesite that might be related to an area extending from Rudare (*c.* 20 km from Pločnik) to Kosovo (Niševac), which has volcanic breccia, tuff, and andesite formations (Figures 1 and 10c). The outlier sample PL 24-32 could be from another non-locally produced vessel. This specimen is dominated by the presence of serpentinite, which is characteristic of the area surrounding Kuršumlija and therefore a likely candidate for its place of origin. Sample PL 24-113 show an abundant presence of amphibole and amphibolite and is therefore compatible with an area south of Prokuplje which is marked by the presence of amphibolite (formation A; Figure 1). Samples PL 24-208, PL 24-315 and PL 319 were made with material from an area characterised by metamorphic rocks (e.g. gneiss and phyllite). These types of metamorphic rocks outcrop west of the site (formation Sm; Figure 1). Finally, two of the analysed loom weights (PL 24-202 and PL 24-258) and one floor fragment (PL 24-336) were produced with a very calcareous clay, probably a marl (Figure 10d). In the geological prospection we were able to find clays with these characteristics in the vicinity of Pločnik (Figure 1; sample 5, site 4).

Discussion: firing technology

The comparison between the results of petrographic analysis, XRPD and SEM shows that the potters at Pločnik must have been able to achieve relatively high firing temperatures (at least *c.* 750°C) but tended not to exceed temperatures of *c.* 850°C. The variations in colour of many of the surfaces and fabrics indicate that they were not always able to perfectly control the firing atmospheres. This observation is also valid for the dark-burnished pottery, however the five sherds characterised by graphite-painted decoration were probably fired in more controlled reducing firing temperatures. Graphite burns off at relatively low temperatures in oxidising conditions (Kreiter *et al.* 2014), so it is possible that the potters were aware of this and were in possession of a relatively advanced pyrotechnology that enabled them to adequately control redox conditions in order to produce this type of decoration.

The firing temperatures estimated for Pločnik match the results of the analysis carried out on the material from Belovode. The only sample that could have been fired to temperatures exceeding 1000°C is of non-local production and is probably a fragment of a 'chimney'.

Conclusions

The range of analyses conducted on the Pločnik samples provided important information about pottery production at this site. Raw material procurement and processing traditions do not seem to change significantly during the occupation of the site. This points to a scenario of strong continuity in the transmission of knowledge of pottery making. During the last phase of the settlement, however, Fabric Group B became more commonly used and there was significant simplification in the range of surface treatments. The social implications of this scenario, with both strong continuity and change, will be further discussed in Chapter 43 in light of the results of the analysis carried out on samples from two other trenches excavated at Pločnik (Amicone 2017). By drawing on a systematic geological prospection of available clay sources, the present analysis also allowed

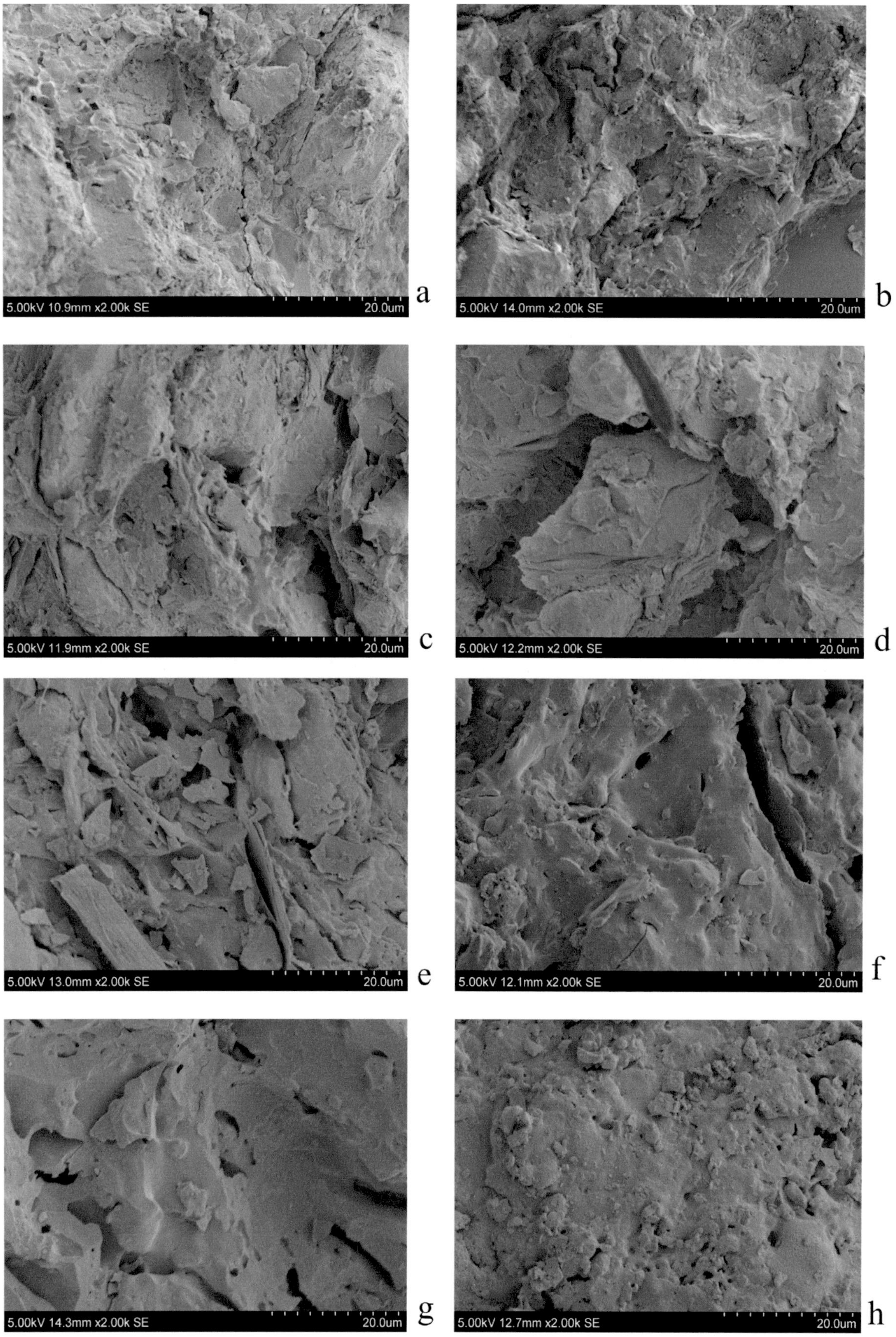

Figure 8. SEM photomicrographs of BEL 157 (bowl, dark-burnished pottery, re-fired in reducing atmosphere at different temperatures): a) as received; b) 700°C; c) 750°C; d) 800°C; e) 850°C; f) 900°C; g) 950°C; h) 1000°C.

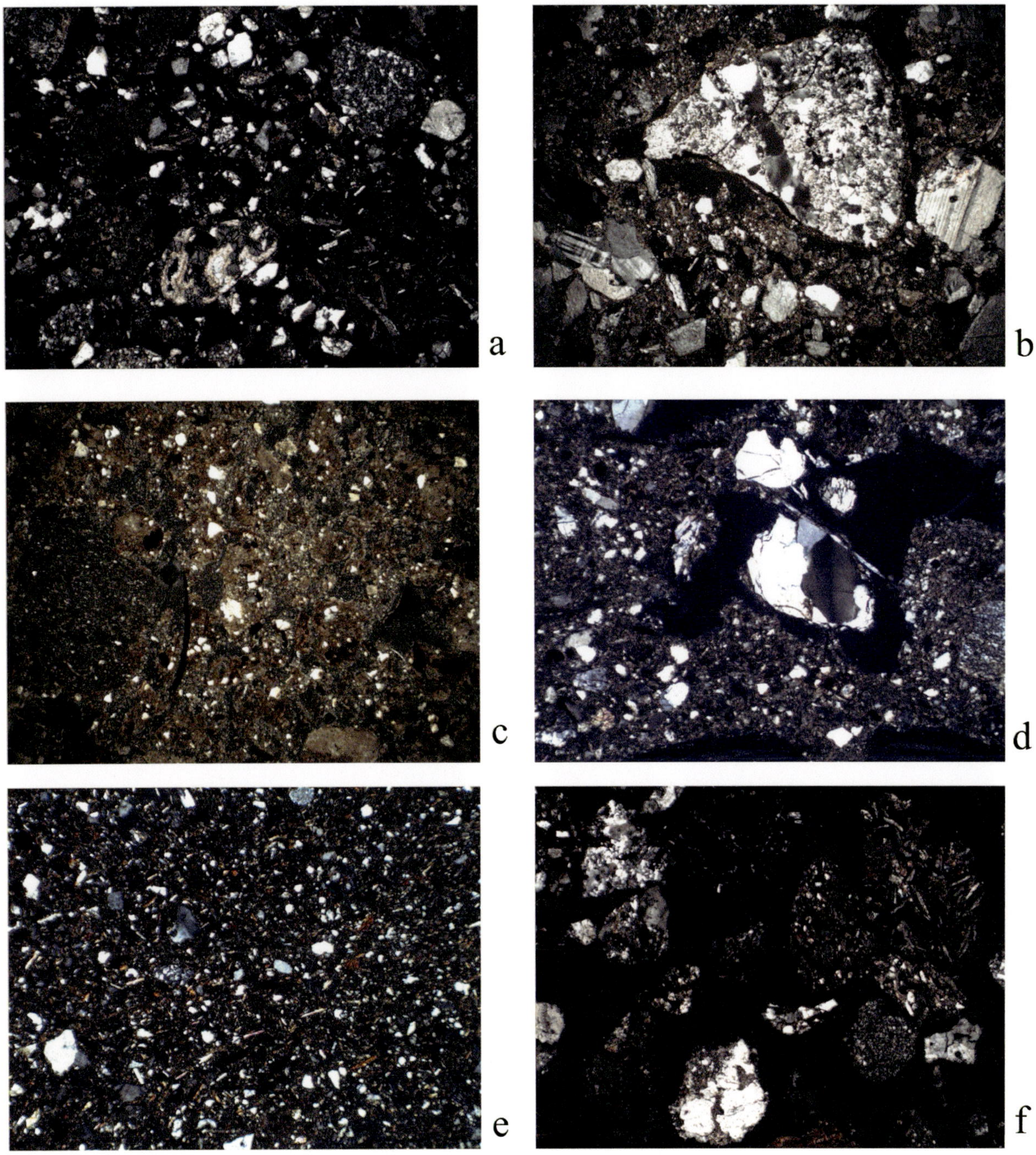

Figure 9. Thin section photomicrographs of selected geological samples from Pločnik: a) sandy clay (point 5, sample 6) XP; b) sandy clay (point 7, sample 7) XP; c) clay (point 35, sample 38) XP; d) modern daub; e) Toplica river clay XP; f) Toplica river sand XP. Image width = 6 mm, except d and f = 3 mm).

consideration of pottery importation and exchange, on both a regional and interregional scale. Finally, the results of XRPD and SEM analysis contributed to the ongoing discussion regarding the pyrotechnological link between pottery and metallurgy. Our analysis demonstrated that, as claimed in previous literature (e.g. Kaiser *et al.* 1986), Vinča potters were indeed in possession of advanced pyrotechnological skills, although these abilities may have been overemphasised by researchers in order to claim an association with metallurgy.

Appendix

Appendix avaliable online as part of Appendix B at https://doi.org/10.32028/9781803270425/AppendixB_Ch29

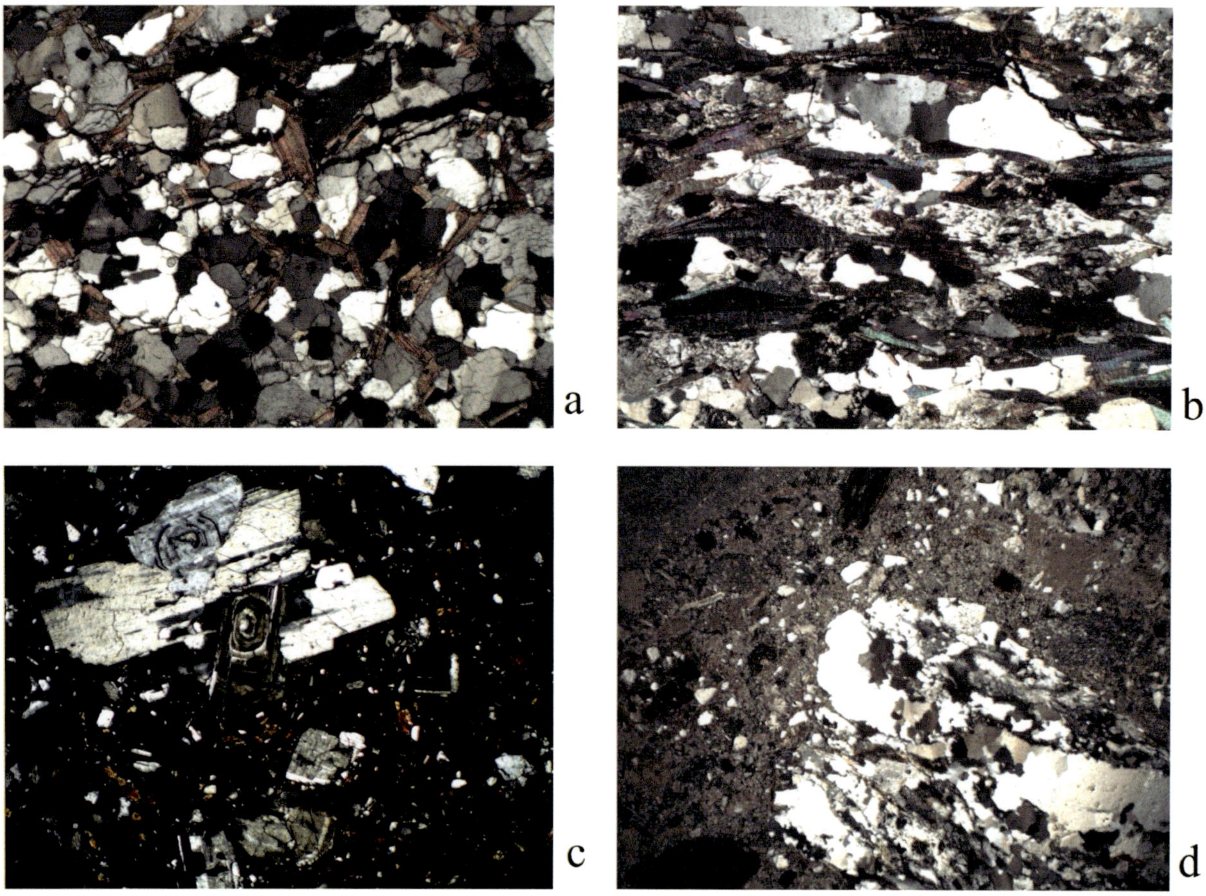

Figure 10. Thin section photomicrographs of selected geological samples from Pločnik: a) rock fragment (point 8, sample 10) XP; b) rock fragment (point 9, sample 11) XP; c) rock fragment (point 18, sample 21) XP; d) (point 4, sample 5). Image width = 6 mm, except d and f = 3 mm.

The bibliographic reference for this chapter is:

Amicone, S. 2021. Pločnik: technology of pottery production, in Radivojević, M., Roberts, B. W., Marić, M., Kuzmanović Cvetković, J., and Rehren, Th. (eds) *The Rise of Metallurgy in Eurasia*: 362–374. Oxford: Archaeopress.

Chapter 30

Figurines from Pločnik

Julka Kuzmanović Cvetković

Although the archaeological site of Pločnik was discovered over 90 years ago, and even though the early phase of Vinča culture was named after it, there is little literature about it, especially in relation to figurines. A publication by Miodrag Grbić (1929: 13) contains only lists of figurines presented in tables. In some excavation reports the findings of figurines were only briefly mentioned (Šljivar and Kuzmanović Cvetković 1998a: 1). The only dedicated text (Kuzmanović Cvetković and Šljivar 1998: 173) relates to the finding of the monumental head of a statuette with a pentagonal face; it is that text that the present authors address with regard to the way the statuette was made and its meaning for the Neolithic community.

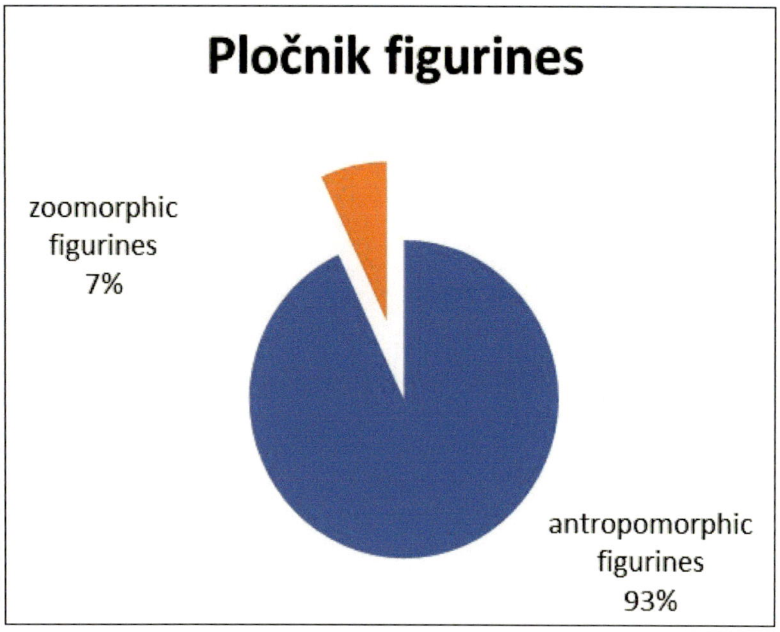

Figure 1. Proportion of anthropomorphic and zoomorphic figurines found in Trench 24.

Pločnik figurines

A total of 88 figurines were discovered in Trench 24 at Pločnik during the 2012/2013 campaigns. Almost all were in pieces, some broken on purpose and some unintentionally. Eighty-two (93%) of the figurines were anthropomorphic, and six (7%) zoomorphic, representing animals (Figure 1). The figurines were all small in size—the largest fragment was 102 cm long—which is typical for Vinča culture. One piece of head (PlAf19) was considerably larger: part of the cheek of the pentagon shaped face was 10 cm long. This figurine could be considered 'monumental'. The detailed measurements of all the figurines can be found in Table 1. A comprehensive file with data for all figurines is in Appendix B_Ch30 (see below). Each figurine has a unique identifier, e.g. Pl Af 21, comprising a Pl (Pločnik) followed by a category (Pl Af – anthropomorphic; PlAm – amulet; BAn – animal) and a number.

The figurines were made of refined clay, with colours ranging from dark ochre to nearly black. Fifty-two (63.5%) of the figurines were black-grey in colour; 14 (17%) were shades varying from ochre to brown, a similar number vary from red to orange, and two (2.5%) remain unclassified.

Thirty-nine figurines (47.5%) were crudely made; 36 (45%) have smooth surfaces, and only four (5%) are highly polished and can be considered as examples of exceptional production. Only two figurines (2.5%) remain unclassified due to their size and state of conservation.

Taphonomy (anthropomorphic figurines)

Of the 82 anthropomorphic figurines found in Trench 24, only six (7%) were recovered whole with minor damage, i.e. with 95–100% conservation (Figure 2). Eight (10%) were missing only the head, so that 70–90% of the figurine was conserved. Sixteen (21%) heads were found, each representing 20–30% of the whole figurine by length or height, which could possibly testify that the heads were intentionally broken off. Eight fragments (10%) comprised a head with a part of the bust (30–45% of the whole figurine); six fragments (7%) comprised the bust and abdomen (40–50% of the figurine); and three fragments (4%) comprised only the abdomen (25–20% of the figurine). In two fragments (2%) the upper part of the figurine was lost, leaving just the abdomen and below (60% of the figurine conserved);

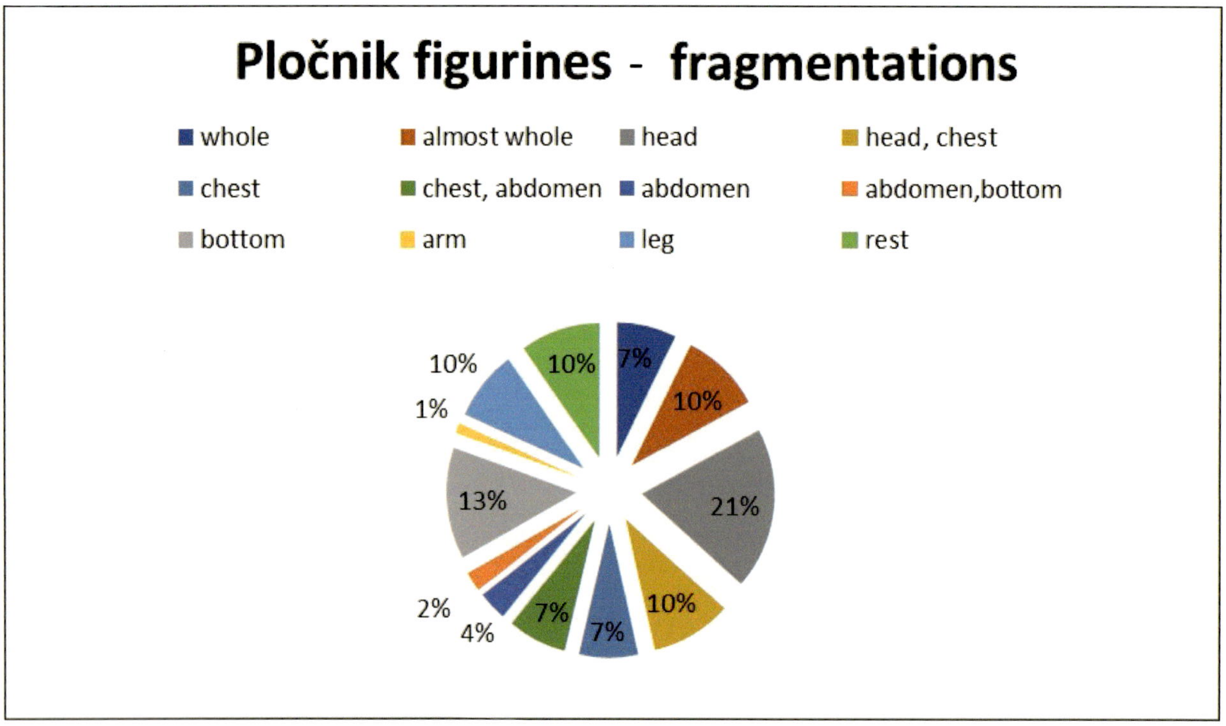

Figure 2. Figurine fragmentation in Trench 24.

in 11 figurines (13%), only 20–50% of the lower body was conserved. There was only one piece of an arm (5–10% of the whole figurine), but seven examples (10%), of legs, either together or in fragments (10–40% of the figurine). There were a further ten fragments for which we could not, with certainty, claim which part of the body they represented.

Anatomy

Since not all the figurines found are conserved to the same degree, all analyses must be conducted with caution. We can, with certainty, claim that 38 figurines were made in a standing pose and only four were sitting. Twenty-three figurines were made with arms: 15 with their arms spread, one with arms on hips, one with arms lifted up, five with their arms down, and one with each arm in a different pose. Anatomical details were often emphasised: 14 of the figurines had their bottoms represented and 15 had their stomachs represented – probably pregnant women. We can also confidently claim that two figurines represent men, both having the penis represented. Two of the figurines had the vulva represented, and breasts were depicted on nine figurines, indicating that these are representations of women.

Heads were present on 27 examples. Traces of red dye were conserved on 21 of these: one had amber traces and five had traces of white incrustation. Perforations were found on two heads. The eyes were represented in 18 cases. For 15 figurines these were moulded and in three cases they were incised. Eyes were round on eight figurines, almond-shaped on eight, and upturned on two. Eyelashes were represented on only one face. The mouth was also depicted on one of the faces. One of the figurines had a mask with sheep horns. Hair was represented on only two figurines.

Clothes and ornaments

The figurines from the 2012/2013 campaign at Pločnik include extraordinary fragments that allow us to imagine the wealth of figural sculpture that existed. One completely conserved figurine, PlAf55, has a representation of a long dress, without many details. Long dresses are uncommon but are usually straight, with a V-neck, and sewn out of strips of cloth that may be wide or narrow. Ten figurines had dresses with a V-neck; on three examples this is depicted with a single incised line; on four figurines the line is double, and on three figurines it is a triple line.

Two figurines have representations of hair and also have a triple line passing over the shoulders, diagonally across the breast and underneath the other arm like a 'sash'.

Two figurines that have the lower parts conserved have clothing represented by lines incised both horizontally and vertically, forming a net. Two figurines have notches on the shoulders, probably representing folds on the dresses. The fragment of the sitting figurine (PlAf48), that was only partly conserved, has unusual clothing: a thin dress is draped over the legs in folds, with a small belt that has a medallion suspended between the legs.

Typology based on body shape

The basis for this typology that of Garašanin (1951: 16), with amendments necessary because of the nature of the material from Pločnik. The complete typology of figurines is provided in Appendix B_Ch30.

A Standing figurines

A1 Pillar figurines

Only two pillar figurines, PlAf55 and PlAf70, were found in the lowest levels; they had few details.

A2 Flat figurines/sketches

(PlAf2 (Figure 4.1), PlAf11, PlAf13 (Figure 4.2), PlAf24, PlAf49, PlAf53 (Figure 3), PlAf74 (Figure 4.8)).

Flat figurines are straight and flattened, as though carved out of thin plates of clay. The first was revealed during the excavations in the river profile as a chance find: PlAf2 (Figure 3.1), named 'Janko', is very special because it has a double face. On the front side there is a small head on a strong neck, with round, plastically modelled eyes. The left arm of the figurine is bent at the elbow and raised upwards; the right arm is in fragments. On the reverse, the face is made in almost the same way.

Figurine PlAf11 is made with little detail. The arms are spread, and at the base of the neck there is a dent created for a specially made head, which is missing in this case.

Figurine PlAf13 (Figure 4.2) is completely flat, in a simple dress. Figurines PlAf24, PlAf35, PlAf49 and PlAf33 (Figure 3) are almost without details, and figurine PlAf61 has very small arms in the form of stumps, and a nose and upturned eyes.

A3 Mixed/combined figurines

With this group the upper part is flat, but the lower part is cylinder-shaped and flattened so the figurines can stand on a base. The lower part often represents a long dress, sometimes with some detail represented with incised lines.

A3a Upper part flat, the lower part flat and cylinder-shaped

(PlAf15, PlAf16, PlAf20, PlAf25 (Figure 4.3), PlAf28, PlAf33 (Figure 3), PlAf35, PlAf69, PlAf72)

Figurine PlAf15 was crudely made with small, spread arms shaped as stumps. Figurines PlAf16, PlAf20 and PlAf69 have only their lower parts conserved, in the form of cylinders, with incised lines that probably represent dresses. Figurine PlAf25 (Figure 4.3), almost a miniature, has a cylinder-shaped body with highly emphasised bottom, and fist resting on the right thigh. Figurine PlAf81 also has an emphasised stomach and bottom, and a dress made with vertical lines.

Figurine PlAf72 has many details, although only the top part was conserved. It has a plastically modelled nose, a v-neck and four horizontal perforations on each arm stump. There are traces of red dye and white incrustation on the head.

A3b Figurines with a belt under the waist

(PlAf23, PlAf26, PlAf39, PLAf58)

With this type of figurine, the upper part is flattened and elongated, as though they had a strap or a ribbon on the line under the stomach, underneath which the skirt becomes narrow. Figurine PlAf23 is the true example of the type. Only the lower part below the waist is conserved; the stomach and bottom are emphasised. From figurines PlAf26, PlAf39 and PlAf58 only the upper parts were conserved. It appears that the line of the belt was where the figurine broke; judging by the profile, they belong to this type.

A3c Figurines with bell-shaped dresses

None were found in this trench.

A3d Figurines with modelled legs

(PlAf5, PlAf6, PlAf12, PlAf27, PlAf34, PlAf42, PlAf44 (Figure 4.5), PlAf46, PlAf56, PlAf60, PlAf64 (Figure 3), PlAf65 (Figure 3 and Figure 4.6), PlAf68, PlAf73, PlAf77)

With this type of figurine, the legs were carefully modelled with the line that separates them emphasised; this is also the area where the figurine can often break. The bottom and legs are decorated with carvings that probably represent clothing. Fragments PlAf5, PlAf6, PlAf51 and PlAf73 are parts of legs; PlAf12 and PlAf56 are representations of feet. Figurine PlAf44 (Figure 4.5) is particularly finely made and is polished. It has a profiled stomach and bottom with strong hips and legs separated by a line, and arms spread but unfortunately the head is missing. On the body there are traces of red dye. Figurines PlAf42, PlAf64 (Figure 3), and PlAf73 have defined and carefully modelled bottoms decorated with spiral and diagonal lines. Figurine PlAf65 (Figure 3 and Figure 4.6) is completely conserved. It has a mask with modelled sheep horns on the head, arms in the form of stumps, spread horizontally, a defined bottom, and legs separated by a line. There are traces of red dye on the body. A modelled penis indicates that the figurine is male.

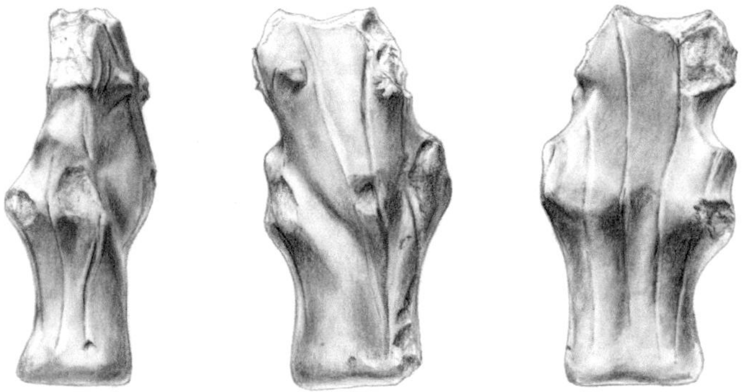

Pl Af 33

Pl Af 43

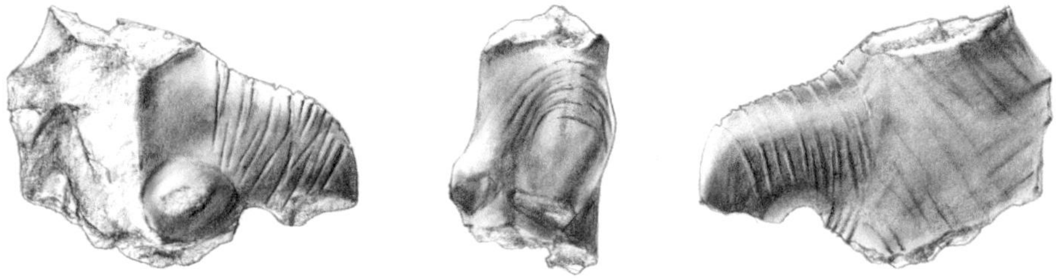

Pl Af 50

Pl Af 53

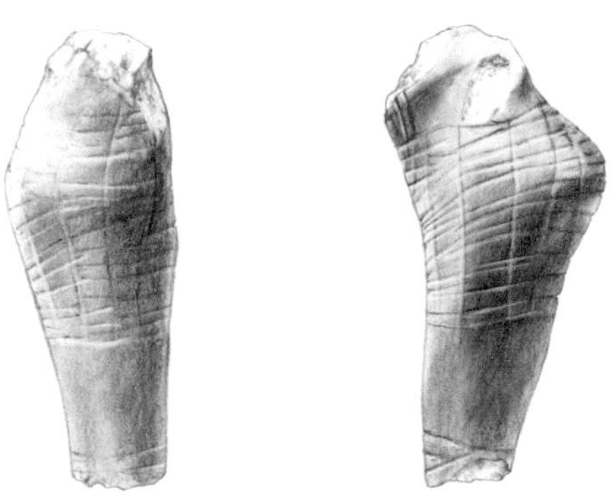

Pl Af 64

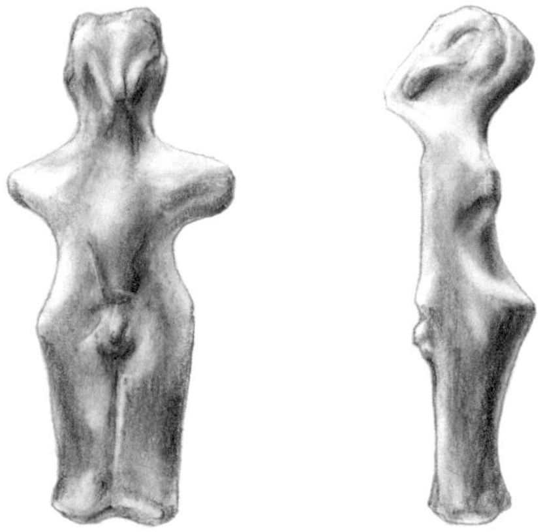

Pl Af65

Figure 3. Pločnik: drawings of anthropomorphic figurines.

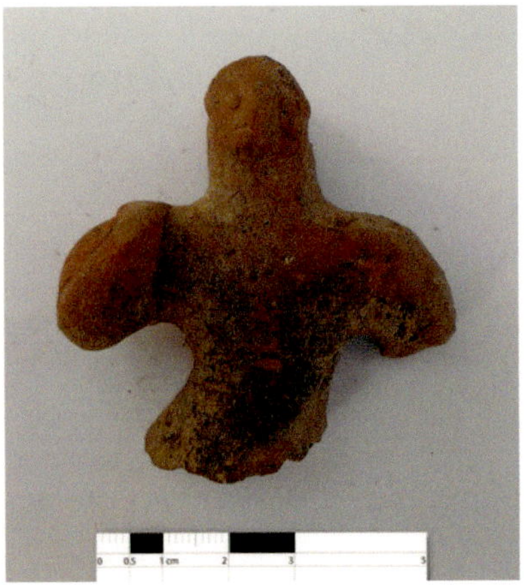

4.1. PlAf 2

4.2. PlAf 13

4.3. PlAf 25

4.4. PlAf 37

4.5. PlAf 44

4.6. PlAf 65

4.7. PlAf 71

4.8. PlAf 74

Figure 4. Pločnik: photographs of anthropomorphic figurines.

A4 Head figurines

A special group of figurines comprises heads and necks that are flattened at the bottom. Each figurine in the group has a unique shape: PlAf36 has an oval face on a long neck; PlAf38 is totally oval; PlAf40 consists of an oval face with a pointed nose, and a representation of the hair in locks; and PlAf41 has a pentagonal head on a short neck, almond eyes and a shaped nose.

B Sitting figurines

B2 Sitting on a base

The only part of figurine PlAf22 that is conserved is part of a bench decorated with diagonal, carved lines that form a meander ornament; there are traces of red dye. The lower part of figurine PlAf31 is conserved and is decorated with diagonal carvings.

B3 Sitting on a throne

Figurine PlAf48 is truly special. It is represented sitting on a base made together with the figurine. Only the seat and legs below the knees were conserved. There is a simple v-neck garment with a belt around the waist and a medallion between the legs. There are grooves on the feet. Figurine PlAf59 is conserved from the knee down, with a realistic foot.

C Kourotrophic figurines

There are no kourotrophic figurines from Trench 24.

Classification based on face shape

1. Oval faces

Figurines with oval faces are often made very coarsely with little detail. They have only indicated noses and

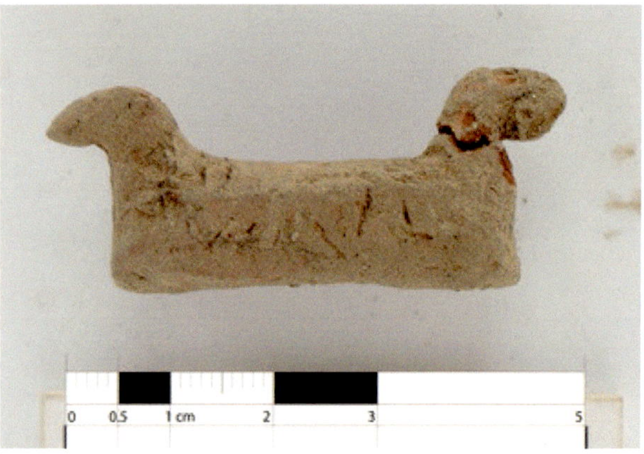

5.1. P An 4 p

5.1. P An 4 up

5.2. P An 5 p

5.2. P An 5 up

Figure 5. Pločnik: photographs of animal figures.

carved eyes represented by diagonal, straight or round lines. The neck is not differentiated; everything was made in one piece. Of the figurines from Trench 24, ten have oval-shaped heads (PlAf2 (Figure 3.1), PlAf24, PlAf32, PlAf35, PlAf36, PlAf38, PlAf40, PlAf57, PlAf75 and PlAf80). Figurine PlAf2 stands out, having two identical faces; figurine PlAf40 has carvings representing hair that goes down to the shoulders.

2. Rectangular faces

Only one figurine, PlAf61, can be put in this group. Almost completely conserved, it has a flat body with arms like stumps, spread horizontally, with an almost rectangular-shaped head that is continued straight to the torso without a neck. On the face there is a plastically modelled nose and diagonal carvings representing eyes.

3. Triangular faces

A face shaped like a triangle with a strong chin distinguishes this group. The upper line of the forehead is mildly curved on many of them. Figurines PlAf8, PlAf37 (Figure 4.4), PlAf43 (Figure 3), PlAf47, PlAf58 were all made in a very similar way, with an elongated face with a plastically modelled, pointed nose and eyes. These figurines usually have a strongly profiled back of the head. Figurines PlAf4 and PAf30 stand out because of their strongly curved forehead line and emphasised ears. Figurine PlAf65 (Figure 3 and Figure 4.6) has a mask with sheep horns covering the face, and figurine PlAf1 has a heart-shaped face, the forehead line dented in the middle to create the heart shape.

The bibliographic reference for this chapter is:

Kuzmanović Cvetković, J. 2021. Figurines from Pločnik, in Radivojević, M., Roberts, B. W., Marić, M., Kuzmanović Cvetković, J., and Rehren, Th. (eds) *The Rise of Metallurgy in Eurasia*: 375–381. Oxford: Archaeopress.

4. Pentagonal faces

Figurines in this group were made with a lot of care and with many details so that each has a special feature, even though they are a part of one group.

Figurines PlAf9, PlAf14 and PlAf29 are made in a similar way with a curved forehead line and dents near the temples that represent cheekbones.

5. Figurines with bird-like faces

No figurines of this type were found in Trench 24.

Amulets

No amulets were found during the 2012/2013 campaign.

Animal figures

Only six animal figures were found, which is unusual for this site based on observations by the author over several years at Pločnik. The most interesting is a figurine of a bird with two heads on either side of a cylinder (PlZf4; Figure 4.1). One more bird head, PlZf3, was found, this one with a long beak and neck. A highly stylised head, PlZf5 (Figure 4.2), that could represent a bear was also found. It cannot be said with certainty what other animals were represented by the remaining fragments.

Appendix

Appendix avaliable online as part of Appendix B at https://doi.org/10.32028/9781803270425/AppendixB_Ch30

Chapter 31

Ground and abrasive stone tools from Pločnik

Vidan Dimić and Dragana Antonović

Introduction

This analysis of ground and abrasive stone tools from Pločnik is based on the examination of assemblage of artefacts found during 2012 and 2013 in Trench 24. The assemblage is very characteristic of the Vinča culture and spans Vinča Tordoš I (Vinča A) to the Gradac Phase (Vinča B2–C1) when occupation at Pločnik terminated in a great destructive fire.

We analysed more than 100 artefacts but selected for detailed study only those finds with a clear context; 72 artefacts are discussed here. A large number (41) of large abrasive tools (static grindstones, grinders and querns made of various types of sandstone of local origin) were not included in this study because of their unclear context.

As with the Belovode assemblage, the ground and abrasive stone tools were classified according to the production method of the tools and their typological and functional features. The typological analysis was based on general observations and the correlation of metric characteristics of certain tools and their place within the methodological framework established by Antonović (1992, 2003, 2014c). Tool function was examined through the correlation of morphological characteristics and visible traceological markers for all tools with minimum preserved evidence (Semenov 1964; Olausson 1983a, 1990; Adams 1988; 1989; 2002; Adams *et al.* 2009; Pritchard-Parker and Torres 1998; Plisson and Lompre 2008; Pawlik 2007; Lunardi 2008; Antonović 1992: 20–23; Dimić 2013a, 2015). In addition to specific use-wear traces, other production marks were also recorded, providing indications of the

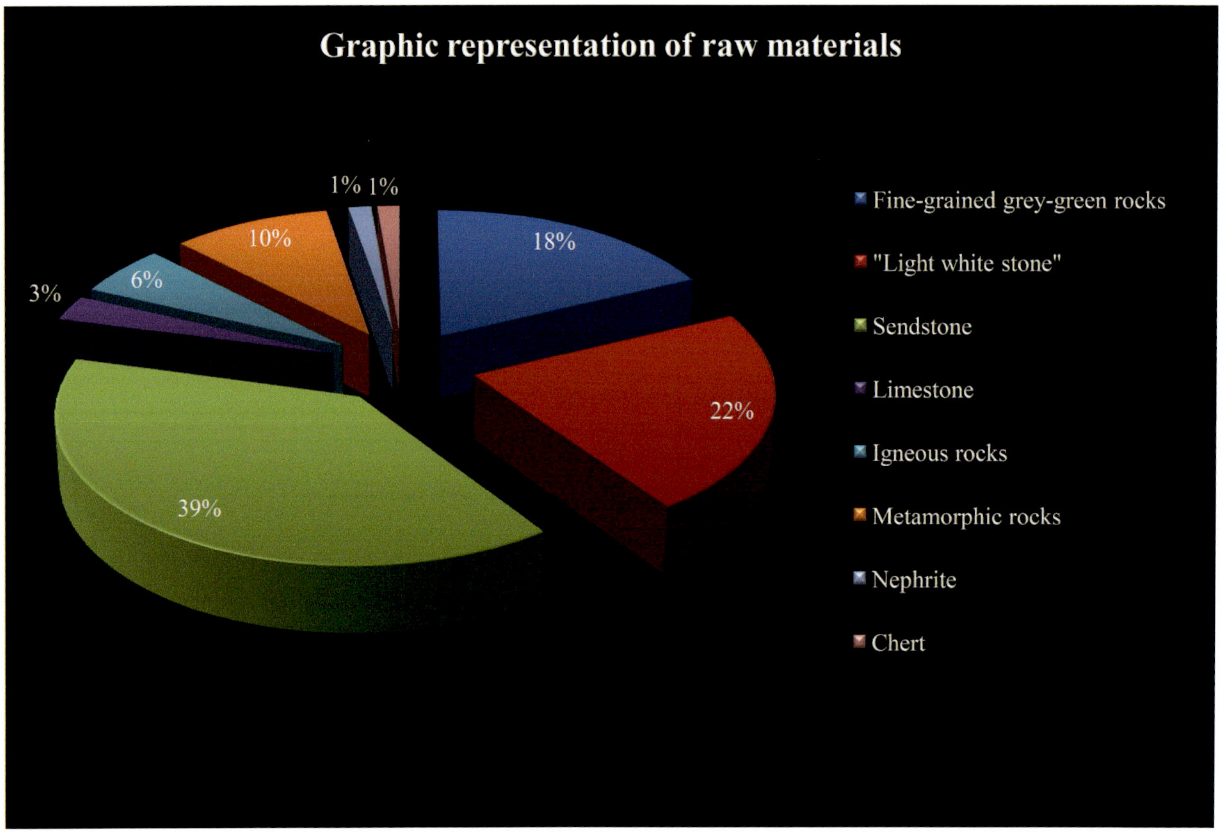

Figure 1. Graphic representation of raw materials from Pločnik.

methods and processing techniques used by Pločnik craftspeople for different types of rock. All analyses were carried out at the Institute of Archaeological in Belgrade using magnifying glasses with up to 16× magnification and a stereo microscope (Olympus®) with up to 100× magnification with a connecting camera.

Raw materials

The stone raw materials from the site of Pločnik are typical of the Vinča culture (Figure 1). Like all other Vinča culture sites, Pločnik had ground-edge implements (e.g. adzes and chisels) produced from fine-grained compact rocks of different geological origins (sedimentary and contact-metamorphic rocks) but with the same technical characteristics: they produce a distinct conchoidal fracture that enables processing by knapping, and a toughness to prevent them breaking during repeated impacts.

Two large groups of these rocks can be distinguished by colour: green-grey and white. Based on the macroscopic analysis, the group of green-grey rocks consisted of cornite, crystalline schist, and metaalevrolite. The white rocks could not be precisely defined by macroscopic analyses. These so-called 'light white stones' were the most popular stone raw material in later Vinča culture phases (Antonović 1997, 2003). In Pločnik, they were present in different degrees of silification from completely soft and powdery to those which were significantly harder, similar to chert (see Belovode report, Chapter 16, this volume). The abrasive tools at Pločnik were made from various types of sandstone. Fine-grained, very compact and hard quartz sandstones represent the main type of raw material. In addition, rocks and minerals such as silicified limestone, some igneous rocks (granite, andesite and aplite) and metamorphic rocks (marble, quartzite and chert) were also used. Nephrite was used to make one miniature adze. Most of the nephrite artefacts from the Neolithic and Chalcolithic cultures of the Balkans come from Bulgaria (Kostov 2013). Deposits of nephrite have not been found in Serbia, although areas geologically suitable for its occurrence do exist (Antonović 2003: 34–37, 139). Nephrite is found in Austria, Bulgaria, Italy, Poland, Russia, Ukraine, and Turkey (Kostov 2013). A study of the provenance of samples of this mineral found at Neolithic localities in Serbia might indicate trade relations with such remote areas.

Typological classification

During the excavation campaign of 2012–2013 at Pločnik, more than a hundred large stone tools were discovered, of which 72 were macro-lithic artefacts with detailed contextual information and therefore included in this analysis (Figure 2). Almost half (47.2%) of these were fragmented or partially damaged, while 38 objects (52.8%) were completely preserved. The tools can be classified into two categories:

1. artefacts with abrasive features, which were not formed by intentional grinding in the process of production, but subsequently during their use; and
2. ground stone tools, which obtained their final form in the manufacturing process by grinding or polishing.

The abrasive stone tools included 28 complete and partially damaged artefacts (Figure 3). The most numerous are 15 whetstones, produced from fine-grained sandstones of different shades of grey and red. They are of various shapes and sizes, from flat ellipsoid to elongated, irregular rectangular in cross-section (Type XII/1: twelve examples, Type XII/2: three examples; types according Antonović 2003: 52–60). Pestles are also represented (four examples). These are made from hard, compact sandstone, as are two tools considered to be used for the thinning of metal objects (Freudenberg 2009: 343). Three hand grinders and three pebble grinders, possibly used for pottery polishing, were also present (Type XI/6: one example, Type XI/3: two examples). Five hammerstones of almost spherical shape were identified; few anvils and handstones were found.

Ground stone tools are present in slightly more significant numbers than the abrasive tools (Figure 2). Out of 34 artefacts with different degrees of preservation the following stand out: two decorative objects (pendants?); one fragment of a marble vessel; and one fragmented mace head (Type X/2). The remaining 30 objects belong to the group of edge-ground implements (Figure 4).

As at other Late Vinča sites in the Balkans, adzes from Pločnik are the most common ground stone tools. Twenty-three were found at the site, manufactured mostly from fine-grained rocks (e.g. crystalline schists, cornite, metaalevrolite) and 'light white stone' of different grades of hardness. The adzes have different degrees of preservation, from complete specimens (C-563, C-570, C-609, C-610, C-649, C-727, C-728, C-729, C-730, C-731, C-732, C-733, C-669) to fragments of the distal or proximal end. The dimensions of complete tools vary from 60–210 mm, most commonly being from 120–140 mm (Figure 5). The most dominant type are adzes have a wider distal end and a slightly convex edge (Type III/1: eleven examples). Next are elongated adzes with the slightly wider distal end (Type III/3: five examples), and other forms of adzes (Type III/2 with three specimens, Types III/4 and III/5 with two specimens and Type III/7 is represented by one artefact).

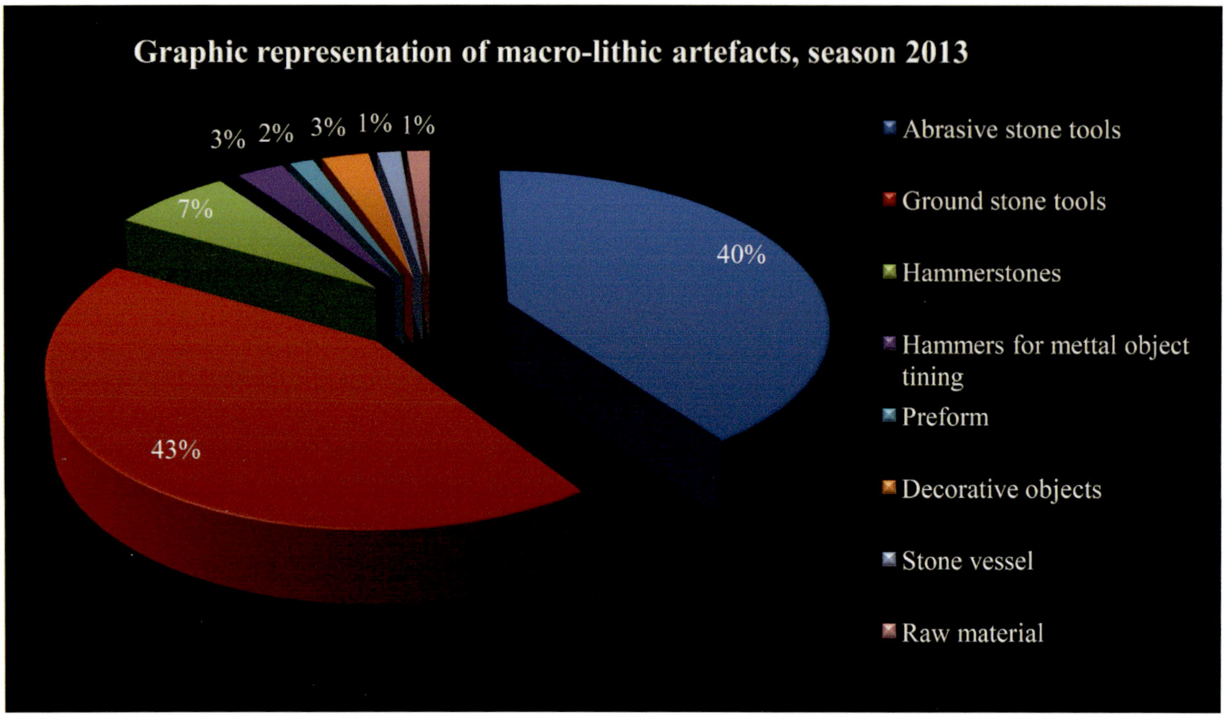

Figure 2. Graphic representation of macro-lithic artefacts from Pločnik.

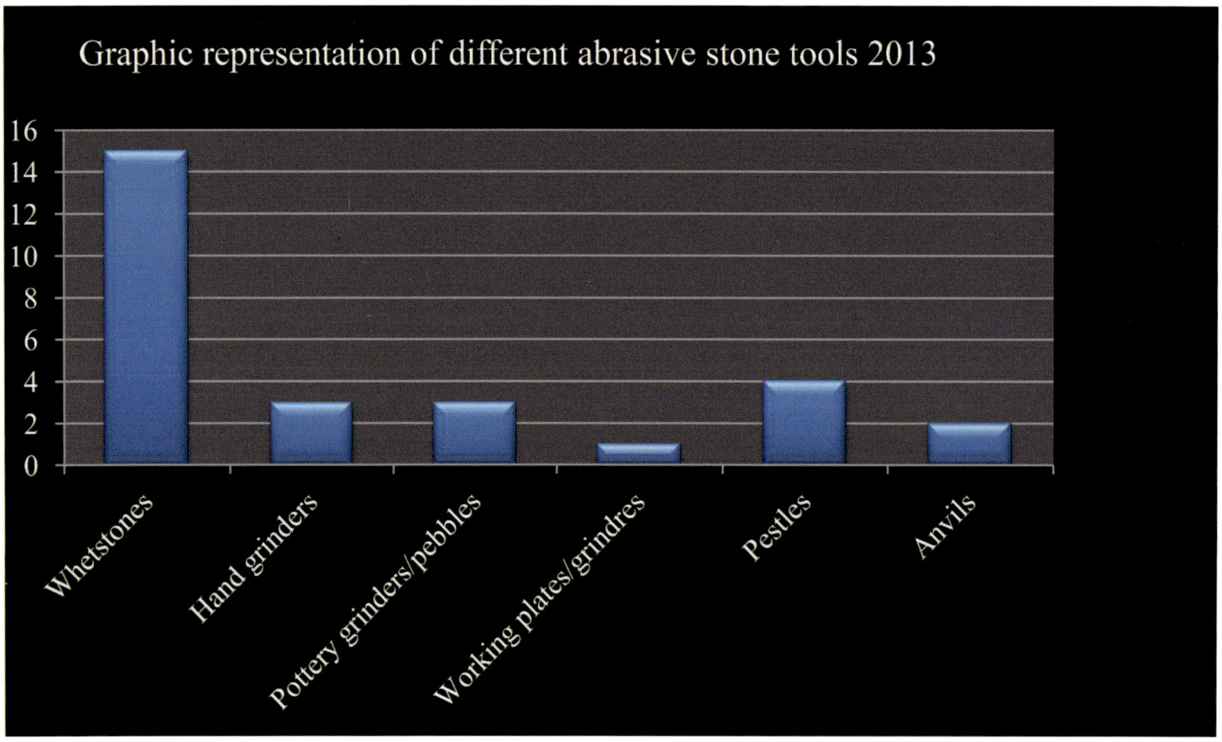

Figure 3. Graphic representation of different abrasive stone tools from Pločnik.

Particularly noteworthy is the hoard of eight adzes manufactured from the 'light white stone'. These belong mostly to the type of elongated adzes with a narrower proximal and slightly wider distal end with convex edge (Types III/1 and III/3). Microscopic examination has determined that four of the adzes carry distinctive use-wear traces on their cutting edge and butt, which most likely occurred during their use in woodworking (see Figures 10 and 11).

Four chisels were also found, two of which were manufactured by the recycling of flakes produced

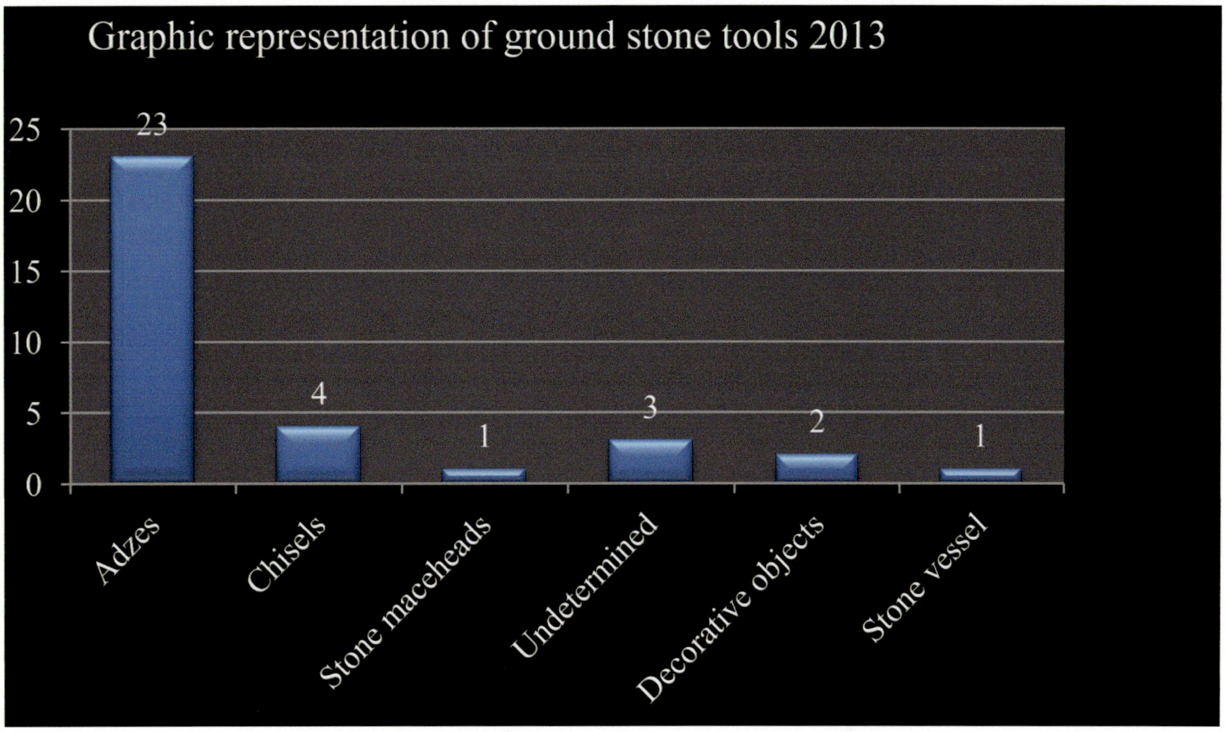

Figure 4. Graphic representation of ground stone tools from Pločnik.

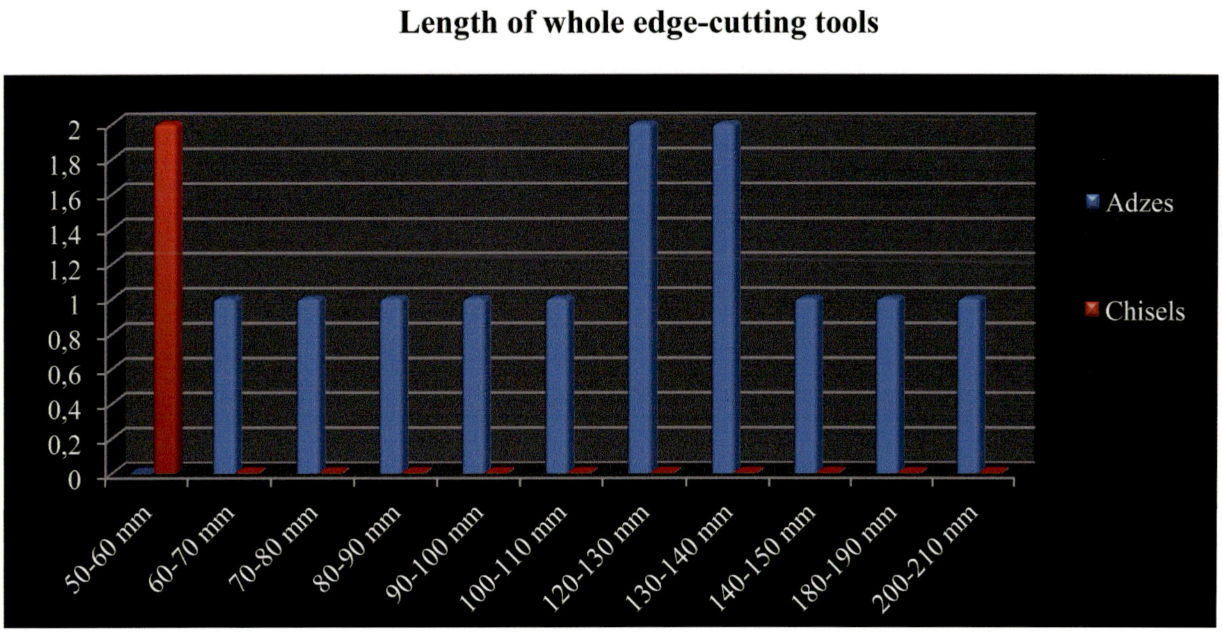

Figure 5: Length of complete tools with a cutting edge from Pločnik.

during the fragmentation of larger ground stone tools. The chisels and the adzes were produced from the raw materials of metaalevrolite, cornites, and 'light white stone'. From a morphological perspective, three types are noted: chisels with parallel sides, and the convex edge which is not in the plane of symmetry of the tool (Type V/3 – two examples); chisels with a wider proximal end, and a convex edge that is not in the plane of symmetry of the tool (Type V/2 – one example); chisels with a wider distal end, and a slightly convex edge that is not in the plane of symmetry of the tool Type V/5 – one example).

Due to their high degree of fragmentation, three objects are classified as 'undetermined'. No examples of axes were found in the excavations at Pločnik between 2012 and 2013.

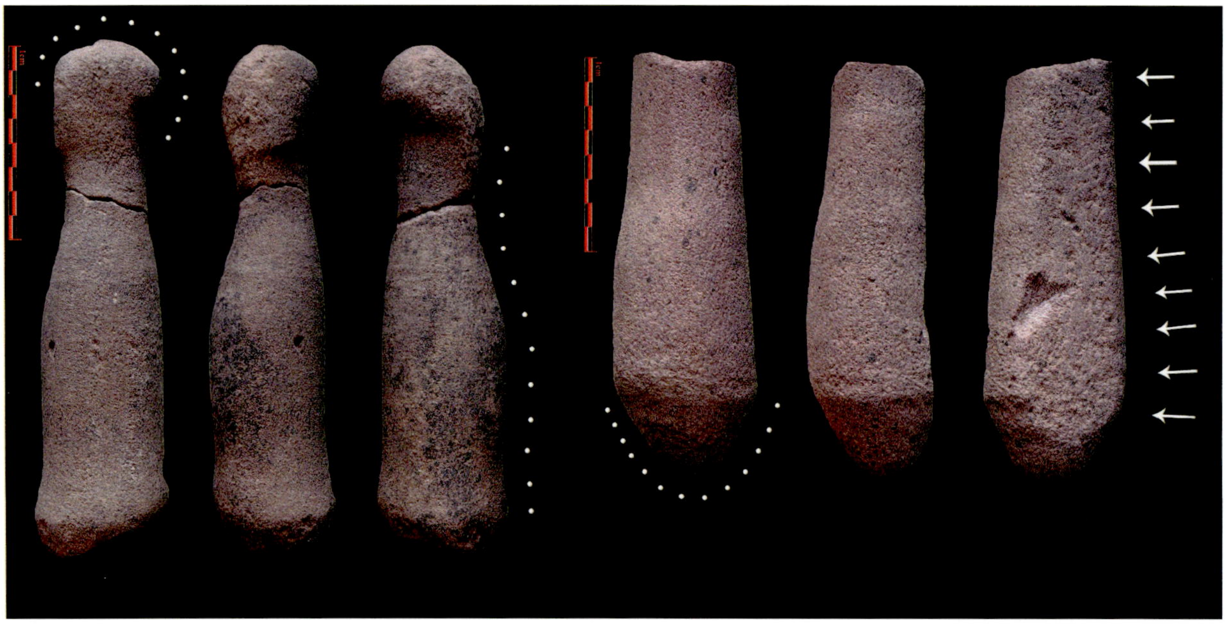

Figure 6. Pestle C-574 (left), pestle C-574e (right).

Figure 7. Pestle C-574c with clearly distinctive working surface. Traces of use are visible as parallel cylindrical furrows produced by circular working motion.

Traceology: production, function and use-wear traces

There is insufficient data to reconstruct the complete *chaîne opératoire* of ground and abrasive stone tools from Pločnik but examination of specific traces on the tools indicates some of the sequences. The traces relate to the techniques used, from the modification of a piece of stone to the fashioning of a complete tool, to the treatment of the tool during its use, fragmentation, re-sharpening, re-use, and deposition.

The abrasive stone tools are characterised by distinctive abrasive features related to the raw material from which they were made i.e. various types of sandstone of different hardness, granularity and colour. Most of these tools (whetstones and grinders) were not produced by the communities of Pločnik. Rather, they were used in their natural form as pebbles or slabs, found in rock deposits or in the gravel of the Toplica river and in nearby creeks.

One group of abrasive tools from Pločnik that were modified through primary treatment are the pestles. All have visible signs of pecking and subsequent grinding on the surface. Traces of pecking are noticeable in the form of small dents and notches that occurred as grains were removed from the rock mass in order to reduce the piece of stone to the desired shape. Grinding was then used to remove edges and make the object easier to grip. The pestles from Pločnik are of different shapes, from elongated to irregular spheres, and are characterised by a hemispherical protrusion on the distal end (e.g. C-574, C-574b, C-574c, C-574e). The protrusion was produced by pecking and served to provide better adherence to the concave surface of the mortar (see Figure 6).

Each of the use-wear traces on these tools comprises clearly visible concentric circles, extending from the top to the base of the hemispherical protrusions, produced by exerting pressure on the surface in circular movements (Figures 6 and 7). These tools were probably used for the pulverising and preparation of certain foods (e.g. cereals and other grains) but may also have been used for crushing pigments and copper ore such as malachite.

Two tools that resemble similar objects from Germany (Freudenberg 2009: 343) were supposedly used for the thinning and hammering of metal objects (Figure 8). One is a pebble of suitable dimensions and weight to be used in its natural form, and the other is a stone of roughly cylindrical shape that was modified by pecking and grinding (C-621). Both have clearly defined, highly smoothed working surfaces. Based on use-wear traces, the second object was used as a multifunctional tool.

For the other abrasive tools, traces of use vary depending on function. On whetstones and grindstones, very smooth and partially concave surfaces occurred as a result of extended use (Figure 9). As use of metal at Pločnik is unequivocally confirmed, it could be assumed that the whetstones, usually used in stone and bone tool processing, were probably also used for sharpening metal objects.

Very hard, sizeable, and compact pebbles of quartzite, sandstone or granite were often used as pounders. It can be presumed that these were collected from the alluvial deposits of the Toplica river. Traces of use appear on them in the form of shallow and concentrated dents, a consequence of pounding resulting in the loss of grains and micro flakes.

Figure 8. Tools which probably have been used for hammering and thinning metal objects. Traces of use are visible as rounded and smoothed surface on the distal end (left), and distal and proximal ends (right).

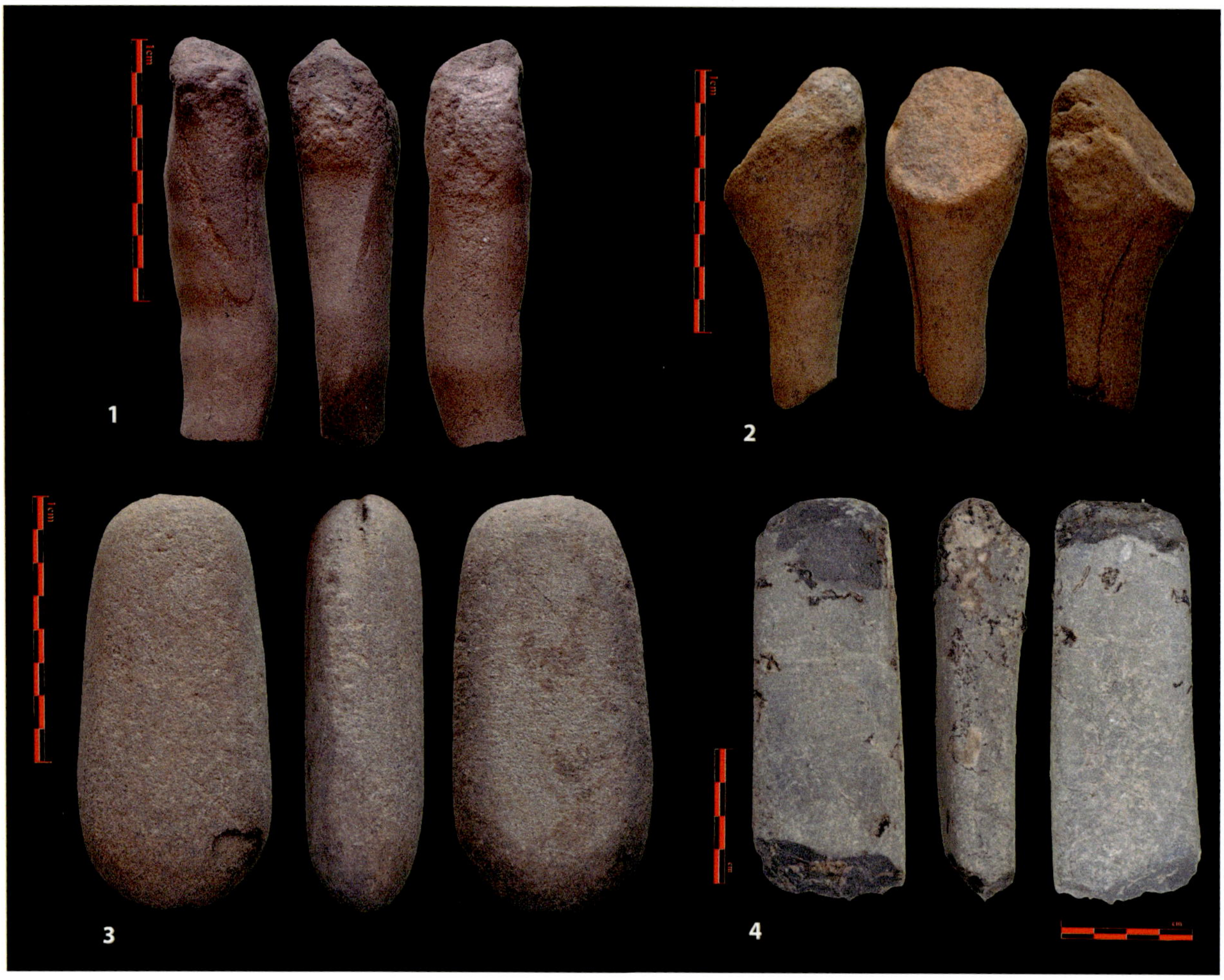

Figure 9. Whetstones from Pločnik.

Several production stages can be identified for the ground stone tools from Pločnik, depending upon the characteristics and quantity of available raw material. These include knapping, retouching, grinding, sharpening, retouching, and regrinding. The pecking technique was probably used to produce object C-566. The basic reduction technique was knapping, which left negative scars on almost all ground stone tools. The reduction of the dorsal side was often achieved by flaking, using a platform on the ventral side. The flaking technique was used on almost all ground tools, mainly due to the raw material used—fine-grained rocks with a distinct conchoidal fracture ('light white stones', crystalline schist, cornite, metaalevrolite).

Based on production traces, after flaking and retouching of tools, grinding was carried out on static grindstones made from fine-grained sandstone, with the use of water.[1] This was performed with both circular and rectilinear movements. Special attention was paid to the cutting edge, while the rest of the object was, in most cases, only superficially ground. This kind of processing is characteristic of the end period of Vinča culture (Antonović 2003: 132–133). In most cases, tools were knapped and retouched, and only the edge was meticulously ground. The rest of the tool was only partially worked in order to reduce irregularities and ridges produced during knapping, that might cause the tool to break due to uneven spreading of the impact force. Some of the most representative objects of this approach are the 'deposit' of adzes made from 'light white stones' (Figure 10). These tools are made of more silicified and compact stone. Higher quality grinding was also noticed on one adze (C-649) made of metaalevrolite. Polishing, the finest processing technique, was partially applied only to the adze made of nephrite; there is no evidence of it on other ground stone tools from this assemblage.

The recycling of fragmented tools, which has been observed on three artefacts (C-573, Find-464, Find-488), was achieved by retouching and regrinding of flakes and chunks obtained following the fragmentation of larger ground tools.

[1] A 'skim' can be seen on the surface of certain tools due to thickening of the dust produced during grinding and water. Skim resembles a smear, with grinding marks macroscopically visible, e.g. on C-729 and C-730, produced from 'light white stone'.

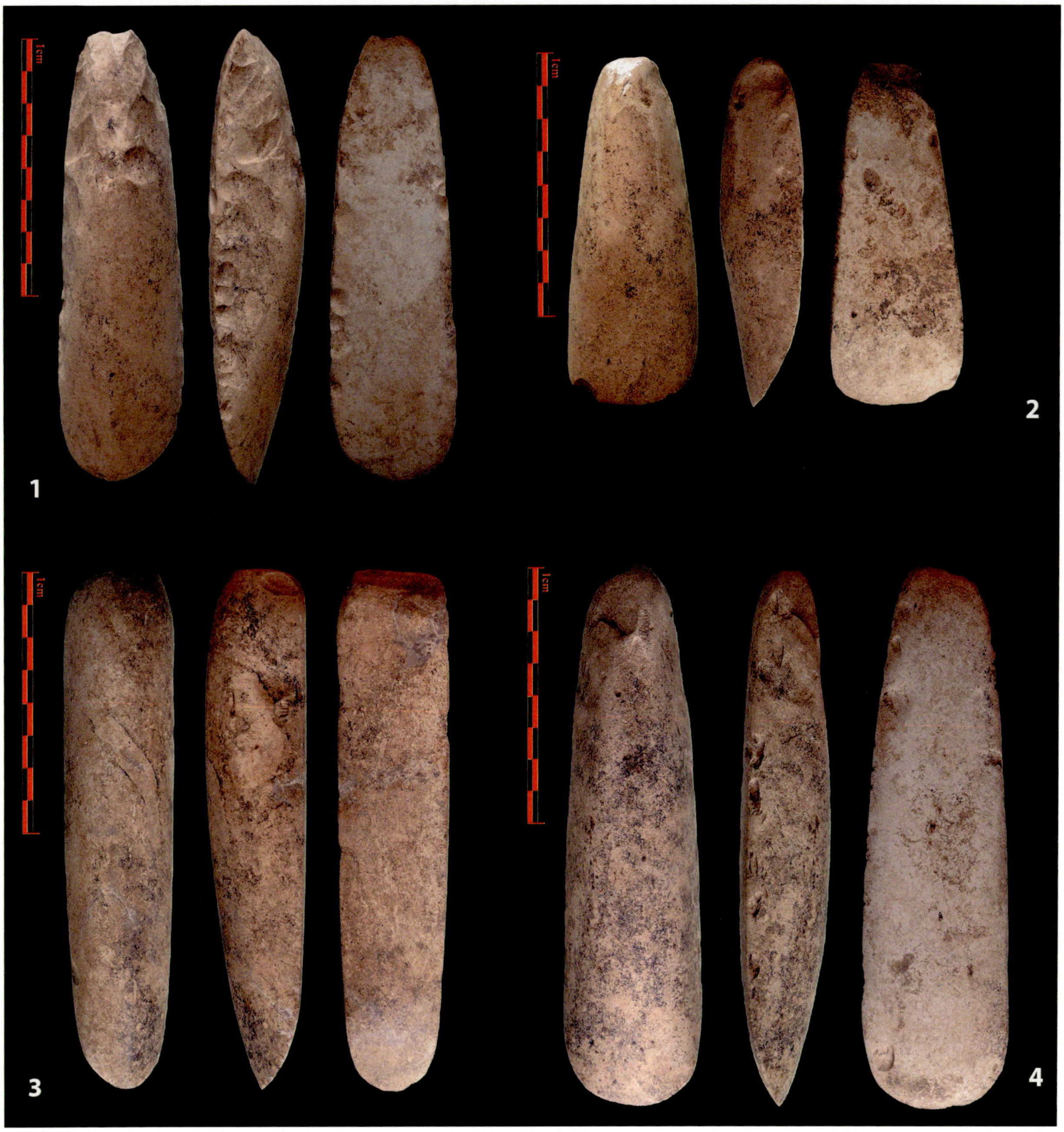

Figure 10. Adzes from deposit of 'light white stone' adzes from Pločnik: 1) C-728; 2) C-732; 3) C-733; 4) C-731.

Ground stone tools with a cutting edge (adzes and chisels) could have been used for a wide spectrum of woodworking activities: felling of trees, adzing and gouging of trunks, stripping bark and adzing of branches, as well as for processing wood for house-building elements and house furniture. All these activities left recognisable use-wear traces on cutting edges and butts (Semenov 1976: 126; Lunardi 2008; Pawlik 2007; Dimić 2013a, 2015). Traces were defined on ten adzes, including four from the adze hoard (Figures 11 and 12; C-570, C-600, C-609, C-610, C-649, C-728, C-729, C-731, C-732). Artefact C-699 is classified as a miniature adze but was used as a chisel. Use-wear traces could be observed on all other adzes that were microscopically examined.

Traces of use similar to those recorded on the adzes are also visible on one of the four chisels found at Pločnik. Micro-polished and rounded edges are clearly visible on the proximal end, while the cutting edge is slightly damaged by intense use, with furrows that occur perpendicular and inclined to the cutting edge. These are more intense on the dorsal side but can also be clearly distinguished on the ventral side. From traces on the butt it can be concluded that the chisel was hafted to a handle made of wood, bone, or horn.

Figure 11. Use-wear traces on adze C-732.1) Adze C-732; 2) polished surface and rounding occurred on butt; 3) furrows/striations on the edge (dorsal side) under magnification 32x; 4) edge (ventral side) at magnification 48x.

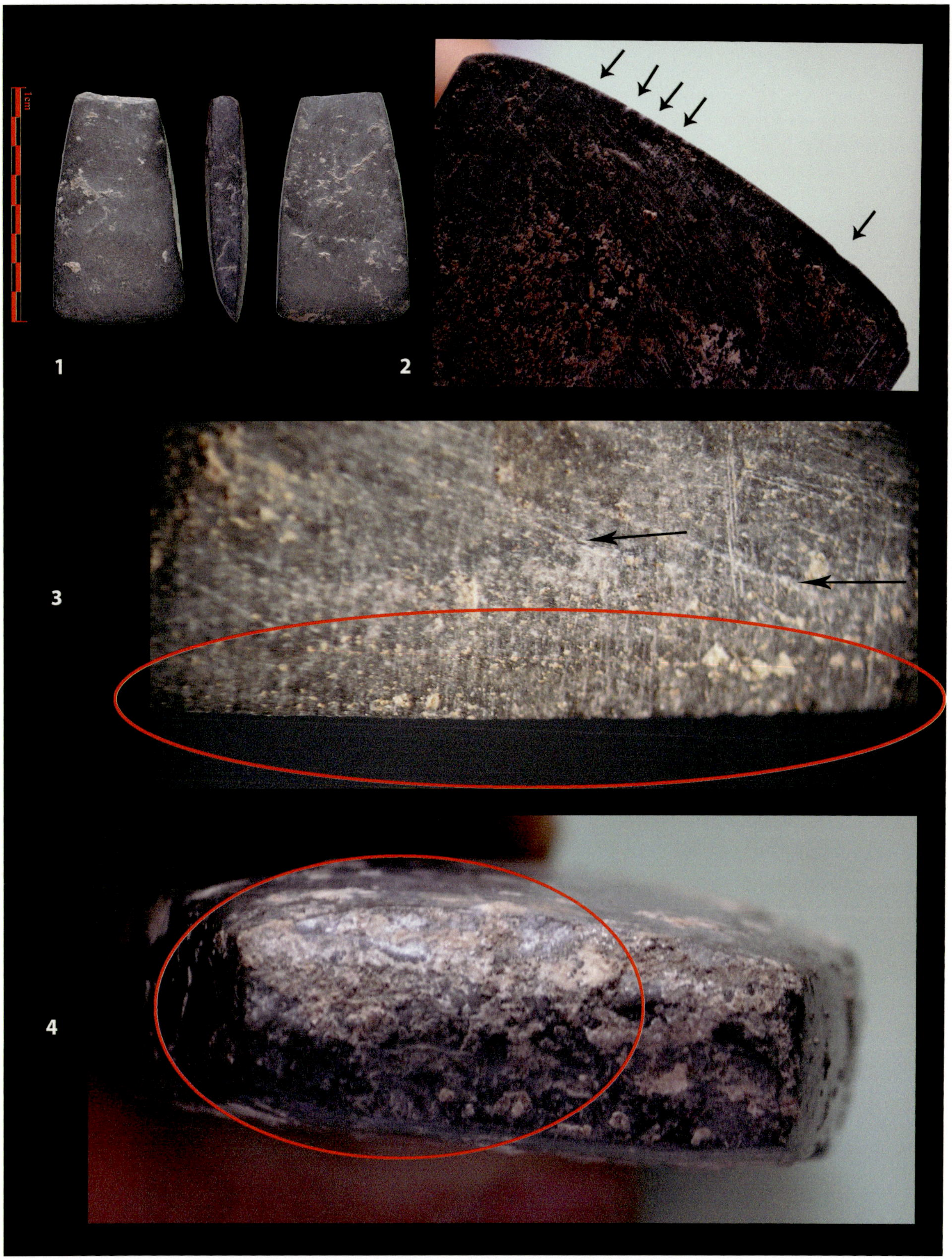

Figure 12: Use-wear traces on adze C-649. 1) Adze C-649; 2) edge (dorsal side) macro; 3) furrows/striations on edge (dorsal side) under magnification 40x; 4) polished surface, rounding and gloss on butt.

Conclusion

The ground and abrasive stone industry from the site of Pločnik is characterised by raw material selection and tool types typical of the Vinča culture. From the time of the earliest settlement in the Vinča Tordoš I Phase (Vinča A), the method of stone processing did not change significantly until the end of occupation at the site. Hard and compact, grey and green-grey rocks were used for manufacture of tools with a cutting edge. After reduction of the raw material by knapping, the whole tool surface was ground, with special attention paid to the cutting edge. Smaller abrasive tools were not (or only minimally) manufactured and were commonly left in their natural form. In the earlier layers of the settlement at Pločnik, maximal use of raw material through recycling and modification is evident, as well as the use of secondary (recycled) tools. Stone tools were valuable, and craftspeople from Pločnik were using them judiciously.

The first changes in the stone industry are visible from the end of Vinča Tordoš II (Vinča B) Horizon and the beginning of the Gradac Phase, when the so-called 'light white stones' came into use. The quality of tools with cutting edges decreased; raw materials were still primarily reduced by knapping but the whole tool surface was ground only on certain tools such as the hoard of adzes. Overall, the cutting edges were meticulously ground, while the remainder of the object only partially. This practice is also a mark of later phases of Vinča culture and most probably occurred as a consequence of the beginning of the use of copper (Antonović 2003: 132–133). Alternatively, it may have been due to demographic expansion and intensified building activities, with increased need for woodworking tools (e.g. axes, chisels, adzes), resulting in an emphasis on quantity rather than quality.

It should be noted that a large quantity of massive abrasive tools (41 objects) were not included in this analysis. The highest proportion are typologically connected to massive static grinders. with clearly defined working surfaces. Given their context, distribution and quantity in a small research area—Trench 24—the logical conclusion is that there was a workshop for the manufacture of stone tools in this part of the settlement.

There is currently insufficient data regarding possible locations from which Pločnik craftspeople may have obtained raw materials. It can be concluded with a high degree of certainty however, that the alluvial deposits of the Toplica river and/or the nearby streams provided one source. Sandstones could also have been gathered from river deposits, but it is more likely that they were exploited from primary deposits in the nearby hills. Based on the results of analysis from other Vinča sites, fine-grained, green-grey and 'light white stones' were certainly exploited from such sources (Antonović *et al.* 2005: 63; Šarić and Cvetković 2013: 42).

The ground stone industry at Pločnik fits entirely within established Vinča culture technological frameworks. Nevertheless, it has certain particularities including: the quantity and types of abrasive tools; the hoard of elongated ground adzes with a high degree of processing; and tools thought to be used in the hammering and thinning of metal objects. Based on the number of stone tool finds and the quality of their production, it is no exaggeration to state that Pločnik was a large economic centre in the Vinča culture, where the production of unique, massive copper tools probably derived from earlier, well-developed stone tool production practices.

The bibliographic reference for this chapter is:

Dimić, V. and Antonović, D. 2021. Ground and abrasive stone tools from Pločnik, in Radivojević, M., Roberts, B. W., Marić, M., Kuzmanović Cvetković, J., and Rehren, Th. (eds) *The Rise of Metallurgy in Eurasia*: 382–392. Oxford: Archaeopress.

Chapter 32

Bone industry from Pločnik

Selena Vitezović

Introduction

The bone industry encompasses all artefacts (tools, decorative items, manufacture debris) made from osseous raw materials (bone, antler, teeth, ivory, mollusc shells) (Averbouh 2000; Poplin 2004). Along with stone and flint, bone raw materials were very important for making everyday tools and other artefacts in all prehistoric societies. Their use, however, depended upon their availability as well as the economic and cultural preferences of a given community. Osseous raw materials had an important place in the Vinča culture, and they were frequently used for both everyday tools and decorative objects, representing a significant proportion of the material culture.

The osseous industry from Pločnik was analysed from a technological perspective (cf. Inizan *et al.* 1995: 13 ff.), including the raw material choices, technology of manufacture, and typological data. The assemblage of about 60 artefacts includes those recognised during excavations as well as those separated during the post-excavation faunal analysis. Numerical data are not presented since they relate to only a segment of bone artefacts recovered so far at Pločnik, and therefore are not statistically meaningful.

Techno-typological analysis

Predominant among the osseous raw material used within the Pločnik settlement are long bones and ribs from large and medium-sized mammals. Antlers are rare, and only from red deer, and *Bos* astragals and *Spondylus* shell were each found only once. Prehistoric craftspeople divided the bones into blanks by chopping, breaking, and direct and indirect percussion. The final shape of the artefacts was obtained by burnishing and polishing with various abrasive tools.

The typological classification used here is based on the link between the supposed function and form of the active part of the objects, originally created by H. Camps-Fabrer (1966; 1979) and now used for numerous European prehistoric assemblages, with some modifications and improvements (e.g. Voruz 1984; Pascual Benito 1998; Beldiman 2007). The artefacts were classified into several groups: I pointed tools; II cutting tools; III burnishing tools; IV punching tools; V objects of special use; VI decorative items; VII non-utilitarian items; and VIII incomplete artefacts. Within these groups, further subtypes and variants were identified based on morphology, function, manufacturing technique, and raw material used (Vitezović 2007, 2011a, 2013a; 2016a: 79-98; see also Bačkalov 1979; Beldiman 2007).

I. Pointed tools

Subtype I1. Several awls (medium-sized pointed tools) were discovered at Pločnik. They were made from longitudinally split segments of diaphyses of long bones (subtype I1A) or from longitudinally split ribs (subtype I1B). Most were fragmented with just tips or mesial parts preserved; very little information could be extracted on their manufacture and use.

Two awls made from large mammal (cattle- or red deer-sized animals) ribs (subtype I1B) were somewhat better preserved with basal and mesial parts, and only their pointed distal ends are missing. These were made from longitudinally split ribs, then shaped by grinding and burnishing. The base was carefully cut and burnished. The awls are heavily worn; the distal end was probably broken during use and the spongy tissue on the lower (inner) surfaces is smoothed and abraded (Figure 1).

Subtype I2. One heavy point was made from a cattle ulna. The epiphysis was preserved (although now fragmented) and the diaphysis was shaped into a large point. Polish from use is visible on most of the tool's surface. Ulnae points are not common in the Vinča culture, although they do occur occasionally at Neolithic and Chalcolithic sites in southeast Europe (e.g. Elster 2001: Figure 20; Gál 2011: Figure 3; Beldiman *et al.* 2012: Plate 6).

Subtype I3. Several fine-pointed tools (needles) were also discovered. These were made from split ribs or from very small segments of long bones. They are well made, polished and have very fine, sharp tips (Figure 2). These small, slender tools may have been used in fibre processing, or for various tasks such as net-making, basketry or similar. Fine-pointed tools are common in Vinča culture bone industries (cf. Russell 1990; Vitezović 2007) and similar artefacts are widely found across

Figure 1. Awls produced from ribs.

Figure 2. Fine pointed tools (needles) produced from ribs.

Neolithic and Chalcolithic sites in Europe (e.g. Hüser 2005: Table 5; Deschler-Erb *et al.* 2002: Table 511/4–26). No examples of eyed needles were recovered at Pločnik.

III Burnishing tools

Subtype III2. Scrapers at Pločnik were made from longitudinally split ribs (Figure 3). They have straight or slightly curved working edges and vary in shape and size due to fragmentation. Manufacturing traces are not preserved because of the intensive use. The bone is heavily worn at the distal part, the spongy tissue at the inner surfaces is completely abraded in the active part, and the outer surfaces are worn, showing polish and with striations.

Scrapers are a simple tool type that is encountered on various sites across Neolithic and Chalcolithic Europe (cf. Hüser 2005: Table 8; Lang 2005: Table 183/1–5). Like rib awls, these they can be overlooked if the faunal record is not carefully examined, so it is difficult to assess how common they truly were. In the Vinča culture, they are noted at several sites, although in varying quantities at Vinča-Belo Brdo (Срејовић and Јовановић 1959: Figure 6; Perišić 1984: Table 12/98, 101; and Pavlovac-Kovačke Njive (Vitezović 2014: Table I/6).

Subtype III4. Several artefacts from Pločnik are classified as spatula-chisels. Their main characteristic

Figure 3. Scraper produced from split rib.

is a relatively sharp working edge that is straight or curved from use. They were used for tasks that combined cutting and scraping. One example is made from a large, proximal segment of cattle radius, with a segment of proximal epiphysis at the base and a large portion of diaphysis shaft. It is fragmented, but probably comprised the entire epiphysis and diaphysis in the proximal and mesial part. In the distal portion it was obliquely cut. On the distal end, it has a curved working edge, smoothed and worn from use. This is an expedient, *ad hoc* tool, and the entire form and the raw material are very unusual.

Another spatula-chisel was made from a cattle ulna, from the epiphysis segment (fragmented) and the diaphysis. The diaphysis is cut from both sides in order to obtain a sharp, straight cutting edge. The basal part is fragmented, and the distal end is worn from use (Figure 4). Ulna cutting and scraping tools are more common in the Chalcolithic period (cf. Beldiman *et al.* 2012: Plates 140–145; Lang 2005: Table 194/1) and are previously unknown from Vinča culture sites.

The third specimen is the most carefully worked, from a flat segment of long bone. It is trapezoidal in shape, with a small, straight and sharp working edge that was intensively used (Figure 5). It imitates the shape of fine, small-sized stone chisels (cf. Antonović 2003). The working edge is only 1 cm wide; it is heavily worn, polished, and slightly chipped from use. Its small dimensions and intensive polish suggest the use on soft, organic materials (cf. Maigrot 2003; Legrand 2007) however, it is not possible to determine precisely the contact material at this stage of analysis. Similar artefacts are known, for example, at Arbon Bleiche 3 (Deschler-Erb *et al.* 2002; Table 514/3–7).

Figure 4. Spatula-chisel made from cattle ulna.

V Objects of special use

Subtype V4. Used astragals. Only one *Bos* astragal with traces of use was recovered at Pločnik. It does not have any traces of deliberate modification, but the prominent parts are completely flattened and abraded from intensive use. (For a discussion on the possible use of astragals and references see Chapter 17, this volume, on the bone industry at Belovode.)

VI Decorative objects

Only one decorative object was discovered, but it is an extraordinary find: a large bead fashioned from *Spondylus* shell. It has an elongated, cylindrical shape and vertical perforation (Figure 6). The traces of manufacture have not been preserved, but polish and wear from use can be seen. The outer surface is also slightly damaged, but it is not clear what caused this wear. A further stray find in the vicinity of the Pločnik settlement should also be mentioned: over 300 *Spondylus* shell beads (also elongated, but with much smaller dimensions), probably originating from the same context and possibly of Vinča culture date. These are currently stored at the National Museum in Niš (Stojić and Jocić 2006).

Very large beads made from mollusc shells are not usually found within the Vinča culture area. In the central Balkan region, the most common objects made from mollusc shells are bracelets (i.e. curved fragments,

Figure 5. Spatula-chisel made from long bone segment.

originally in the shape of an open or closed circle – cf. Срејовић and Јовановић 1959; Dimitrijević and Tripković 2002, 2006; Игњатовић 2008: 227, Catalogue 221). Only from the cemetery of Botoš-Živanića Dolja are several large, elongated beads known (Marinković 2010).

Spondylus ornaments are a pan-European phenomenon (cf. Comşa 1973; Willms 1985; Siklósi 2004; Séfériadès 2010; Ifantidis and Nikolaidou 2011) and appear in numerous prehistoric cultures and in diverse contexts. They caught the attention of researchers very early, and are usually associated with status, prestige, and luxury (cf. Séfériadès 2010, with references). There are, however, still many questions that remain unanswered, including the extent of their distribution in time and space and the modes by which they were exchanged. In the central Balkans region, *Spondylus* items were previously thought to be concentrated in the Danube valley (cf. maps in Willms 1985; Dimitrijević and Tripković 2006; Séfériadès 2010). New finds, however, and the careful examination of previously published data, demonstrated that they were also present in the Pomoravlje region, for example, at Divostin (McPherron *et al.* 1988); Drenovac (Vitezović 2007); and Vitkovo (Vitezović 2013c) (see also Vitezović 2016b).

The find of large *Spondylus* beads with contextual data at Pločnik is very important, since it demonstrates the presence of these ornaments within the settlement. Furthermore, large beads are rare in the modern Serbian territory with no identical specimens known. This find suggests a much wider typological repertoire of mollusc shell ornaments within the Vinča culture than was previously thought.

Results and discussion

The osseous artefacts recovered from Pločnik during the 2012 and 2013 excavation seasons represent a cross-section of the entire bone industry from the Vinča culture settlement. Most are heavily fragmented, suggesting that this was a refuse area (at least for osseous industry), although it may have also been an abandoned activity area, where bone objects were exploited and then broken, and tools that were no longer usable were discarded.

Figure 6. Cylindrical bead from *Spondylus* shell.

The predominant objects are pointed and burnishing tools, used mainly on organic materials such as leathers, hides or plants. Only a few objects of special use and decorations were discovered. Some tool types are not represented, such as heavy-duty tools such as axes or hammers, or hunting and fishing gear. Manufacture debris is also absent, suggesting that the working area where the osseous artefacts were produced was outside of the excavated area. Ribs were most commonly used, followed by variety of long bone segments, including some that do not often occur in Vinča culture industries, such as ulnae or radii. Antlers were found only in fragments although earlier finds from Pločnik suggest it had a rich antler industry (Grbić 1929: Figures 123–125).

The techniques used and typological repertoire evidenced fits well with current understanding of the Vinča culture bone industry (cf. Bačkalov 1979; Russell 1990; Vitezović 2007). The specific traits of the Pločnik assemblage are the use of unusual skeletal elements, e.g. ulnae used for heavy points and spatula-chisels, and the unique find of a cattle radius (for raw material choices in the Vinča culture, cf. Vitezović 2013b). The *Spondylus* bead is very important for the analysis of exotic raw materials within the Pločnik settlement, but also for the wider study of their distribution and the study of trade and exchange routes.

The bibliographic reference for this chapter is:

Vitezović, S. 2021. Bone industry from Pločnik, in Radivojević, M., Roberts, B. W., Marić, M., Kuzmanović Cvetković, J., and Rehren, Th. (eds) *The Rise of Metallurgy in Eurasia*: 393–396. Oxford: Archaeopress.

Chapter 33

Chipped stone industry at Pločnik

Elmira Ibragimova

The excavations in Trench 24 at Pločnik during the 2012 and 2013 field seasons yielded evidence for a chipped stone industry comprising a total of 1046 individual artefacts. The larger part of the chipped stone items was recovered and recorded during the excavations, with a smaller part recovered during the flotation of soil samples.

The analysis aimed to determine the raw materials exploited, investigate the production techniques and methods used, and identify the types of tools being made. Where possible, the material, technological and typological characteristics were compared throughout the sequence of occupation at Pločnik in order to examine the evolution of the industry. A comparison of chipped stone industries, both at Belovode and within the broader context of at other Vinča culture sites is presented in Chapter 47 of this volume.

Stratigraphic context and principles of field documentation

Over the two excavation seasons at Pločnik, 1018 chipped stone artefacts were recovered, primarily through excavation, with 718 (70% of the whole collection) recorded to the spatial resolution of a stratigraphical spit. The precise three-dimensional location was recorded using the EDM for 286 artefacts (28%) from soil flotation samples and for 42 artefacts (4%) which were gathered as 'special finds'. Each chipped stone artefact was therefore recorded to the minimum level of a stratigraphical spit. Part of the collection—245 items or 23%—was recovered from archaeological features.

The stratigraphic sequence revealed by the excavations of Trench 24 at Pločnik was summarised in a Harris matrix (see Chapter 7, this volume). Two distinctive phases in the evolution of the site can be defined: spits 1–9 (Horizon 1), and spits 10–25 (Horizons 2–5). Within the periodisation scheme based upon the detailed typo-chronological analysis of the pottery and an extensive radiocarbon dating with Bayesian modelling (see Chapter 37, this volume), the following Vinča culture phases can be assigned: Horizon 1 equates to the late Vinča Pločnik II Phase (4616–4385/4360 cal. BC) and Horizons 2–5 equate to Vinča Tordoš I Phase up to Vinča Pločnik II Phase (5210–4616 cal. BC) (Chapter 37, Table 3, this volume). It is suggested that the part of the settlement excavated in Trench 24 was not settled for over a hundred years (Chapter 37, this volume), between Horizons 2 and 3.

Database structure and methodology

The recording of the chipped stone finds encompassed the excavation context which always included the spit/feature number but, depending upon the circumstances of recovery, could also include the EDM measurement and find number, and the reference for the flotation sample. Each chipped stone artefact was measured for length, width and thickness. The raw material of each chipped stone artefact was examined macroscopically to identify the main physical characteristics: colour, natural surface information (pebble or tabular) and transparency. This enabled the identification of several geological groups: various sedimentary rocks (varieties of flint, limestone), igneous rocks (e.g. obsidian, chalcedony) and metamorphic rocks (e.g. quarzite, schist) (see Inizan *et al.* 1999: 21). Other parameters also noted were calcination, visible use-wear traces and cortex location, and blank morphology and attribution to chipped stone category. The blade debitage was differentiated based upon the width of the blanks resulting in groups of 'blades' (>12 mm); 'bladelets' (8–12 mm); and 'microblades' (< 8 mm). This division is based primarily on the observation of debitage metrics derived from the experimental usage of different techniques of flint knapping (Pelegrin 1998).

The morphological characteristics of both blades and cores were recorded in detail. The blades were examined for the dorsal pattern, platform and metrics and morphology bulbar parts'; technical stigmata[1]; and individual impact zone preparation. The cores were examined for the number and location of flaking surfaces and striking platforms and the impact zone preparation. All the tools were described separately in a text document while the finds were processed.

[1] This term was used by J. Pelegrin in evaluation of blade production techniques and refers to 'the character of the butt determined by the platform preparation (dimensions. aspect, edge angle)' and 'discrete details determined by the detachment itself (cracks, lip, ripple on the bulb, aspect of the bulb)' (Pelegrin 2006: 42).

Figure. 1. Pločnik tabular flint reduction. 3.1 and 3.2 - blades with facetted (1) or flat (2) blade butts; 3.3–3.7 - bladelet cores. Core 7 has visible traces of heat treatment.

The analysis of the chipped stone industry concentrated upon establishing the *chaînes opératoires* from the selection and use of the raw materials to the techniques and methods used in the production, use and eventual discard of the final object (cf. Bar-Yosef and Van Peer 2009; Boëda 1990; Inizan *et al.* 1999: 14). In order to identify the complex of *chaînes opératoires* underpinning the chipped stone industry, all by-products of the activities connected to each stage were analysed. The morphological types and terminology follow Šarić's (2006) classification of the Early and Middle Neolithic chipped stone tool assemblages in Serbia.

Raw material

The lithic industry at Pločnik is characterised by the wide variety of raw material used, including sedimentary (various types of flint), igneous (chalcedony) and metamorphic rocks (shale, schist, quartzite) (Table 1; see also Chapter 48, this volume). Two major flint varieties with distinct proveniences could be distinguished: white-cream, fine-grained tabular flint with a chalky, natural surface (Figure 1)

Table 1. Distribution of chipped stone raw material types of Pločnik, by horizon.

Raw material	Number of items (% of total)	
	Horizon 1	Horizons 2-5
Cream tabular flint	247 (50)	51 (10)
Dark brown pebble flint	41 (8)	50 (9)
Other colours pebble flint	68 (14)	150 (28)
Brown transparent flint	26 (8)	51 (10)
Quartzite	4 (1)	2 (0)
Chalcedony	1 (0)	4 (1)
Brown schist with white lines	4 (1)	9 (2)
Cream pebble flint	5 (1)	26 (5)
Grey banded schist	2 (0)	31 (6)
Siliceous shale (?)	14 (3)	4 (1)
Other	83 (17)	145 (28)
Total	**495**	**523**

Figure 2. Pločnik pebble flint reduction. 4.1 and 4.3. - large blocks of poor-quality flint (EDM 1447); 4.2 - pebble flint fragments; 4.4. and 4.5 - blade and bladelet with flat butts; 4.6 - pebble flint blade core; 4.7 and 4.8 - fragment of blade and core made on banded grey-greenish schist.

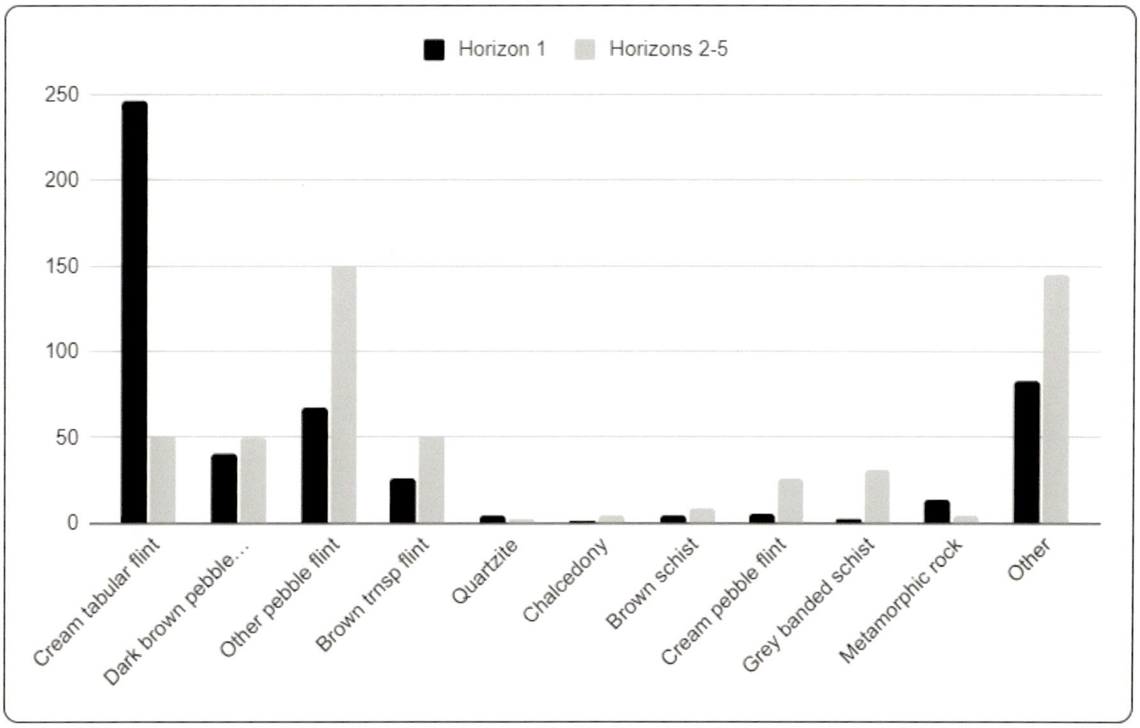

Figure 3. Distribution of chipped stone raw material types from Pločnik site by horizon showing number of items.

and a large group of pebble flints with a rolled, natural surface, sometimes with bronzing (Figure 2.1, 2.2 and 2.3). It is assumed that the white tabular flint variety is related to sources of silicified magnesite (Antonović *et al.* 2005), which were actively used for ground stone tool production at Pločnik ('light-white stone'). This observation, however, needs to be proven by chemical and petrographic analyses and further comparative work. The acquisition of pebble flint might be connected to gravels from the nearby Toplica river or its drainage network where small pebbles, as well as large blocks of poor-quality material, were probably obtained (Figure 2.1, 2.2, 2.3).

The use of different raw materials has a clear chronological pattern. In Horizon 1, tabular flint comprises up to 50% of the whole excavated assemblage (Table 1 and Figure 3). In contrast, the earlier horizons exhibit a wide variability of raw materials including metamorphic greenish stone, brown schist with white lines, banded grey-greenish schist (Figure 2.7 and 2.8), chalcedony and quartzite.

Techniques and methods of blade production

The analysis of the main chipped stone categories shows a change from the lower to the upper horizons, which reflects a shift in the chipped stone industry at Pločnik (Table 2). The industry of the lower horizons is characterised by the intensive production of flakes and bladelets. In Horizon 1, flake production decreases whilst the use of blades and blade tools (produced on site or elsewhere) increases (Table 2 and Figure 4). We do not observe the sudden end of one tradition and the beginning of the other, rather the process changed gradually. Technological study could give more detailed information on the evolution of the chipped stone industry.

Table 2. Distribution of main chipped stone categories from Pločnik site, by horizon.

Chipped stone category	Number of items (% of total)	
	Horizon 1	**Horizons 2-5**
Microblades	18 (3)	20 (4)
Bladelets	44 (9)	46 (9)
Blades	66 (13)	32 (6)
Flakes	152 (31)	201 (39)
Fragments	103 (21)	96 (18)
Preforms	1 (0.15)	11 (2)
Blade cores	9 (1.75)	7 (1)
Flake cores	10 (2)	36 (7)
Flake, fragment tools	23 (5)	38 (7)
Blade tools	68 (14)	34 (7)
Microtools	1 (0.15)	2 (0)
Total	**495**	**523**

Reduction at Pločnik aimed to produce flakes, blades, and bladelets. The debitage connected to their production is present throughout all the horizons of the site but was more significant in Horizons 1 and 2–5. Two main raw material groups were widely used, each employing specific techniques and methods.

Bladelet (and blade?) production

Two main raw materials were used: tabular cream flint and pebble flint of different colours. Tabular flint was preferred throughout the occupation of Pločnik. The distribution of the main categories of tabular flint items (Table 3) demonstrates that this raw material was preferred for blade production and blade tools (36%), bladelets, microblades and tools also being made (26%). A range of waste products from blade production is found at the settlement, including fragments of blocks of tabular flint, pre-cores, blade cores, rejuvenation core tablets, and flakes. This indicates that production of bladelets and microblades—and probably also blades—was conducted on site.

Several flakes made using tabular flint could have been connected to flake production but there are no cores connected with flake production made on such flint (Table 4). Moreover, primary and secondary flakes (Table 5) are usually of small size (Figure 5) and most are probably connected with blade pre-core preparation and blade reduction. It is likely that these flakes are a by-product of blade, bladelet or microblade reduction.

Reduction was performed on the flanks of the tabular flint slabs with a thickness of 10–25 mm (Figure

Table 3. Main chipped stone categories of tabular flint items from Pločnik.

Category	Number of items (%)
Blades (+tools)	114 (36)
Bladelets (+tools)	47 (16)
Microblades (+tools)	32 (10)
Flakes (+tools)	74 (24)
Fragments	27 (9)
Nodules	3 (1)
Preforms	1 (0.25)
Cores and core fragments	8 (3)
Core rejuvenation flakes	2 (0.75)
Total	**308**

1.3, 1.4 and 1.5). The length of these slabs could be reconstructed on the stepped-ended full blades (up to 61 mm). The pre-core preparation would consist of a ridge formation or even be performed without any ridges if the shape of the nodule was convenient. The presence of one fully crested blade, as well as a blade wholly covered with a cortex ('*entame*'), might suggest such an explanation.

The flaking platform was formed afterwards with special attention given to keeping the angle between the platform and the flaking surface at more than 70º. Before every blade reduction, the edge of the platform was facetted to keep the angle at about 80–85º. The reduction method included bladelet and microblade reduction from the lateral cortical sides of the core to keep a convenient curvature of the flaking surface. This assumption is supported by occurrences of blades, bladelets, and microblades having a cortical side (or even two sides), which often have a semi-abrupt or abrupt angle (average 50º, but up to 80º). Only 44% of bladelets and 46% of microblades have no trace of the cortex, whilst this is the case for 73% of the blades (Table 6).

During blade reduction, the lateral ridges could have been formed to keep convenient curvature of the flaking surface. The by-products of this process are two secondary crested blades found in the trench in Horizons 19 and 25.

Several features might indicate that the technique of indirect percussion was used to produce blades, bladelets, and microblades on tabular flint. Indirect percussion 'involves the application of an intermediary tool, called a punch, which can be of wood, antler, bone or metal' (Inizan *et al.* 1999: 32). This technique affects the morphology of the blades, bladelets, and microblades of *plain débitage*. They are all quite thick in section with the average width:thickness ratio being 3.9 for microbladelets, 4.2 for bladelets and 4.4 for blades. Their flaking angle is often close to 85º (ranging from 80º to 95º). Striking platforms can be deep (on average 3 mm, but up to 6 mm). Butts of the blades are flat or facetted (Figure 1.1 and 1.2). Overhangs were slightly abraded (Figure 1.1) and are rarely reduced (as would have been the case if the soft hammer technique was used).

Of particular interest is a small group of artefacts made on a variety of flint which has strong similarities to tabular cream flint but a grainier structure (Figure 1.6 and 1.7). It is suggested that this was exploited in a similar way to tabular flint though the quality of the material could be deliberately improved with heat treatment. One core on this raw material bears a set of traits probably indicative of heat treatment (see Inizan *et al.* 1999: 24). The striking front and facetted part of platform display greasy patches in contrast to the matt surface of the rest of platform (Figure 1.7).

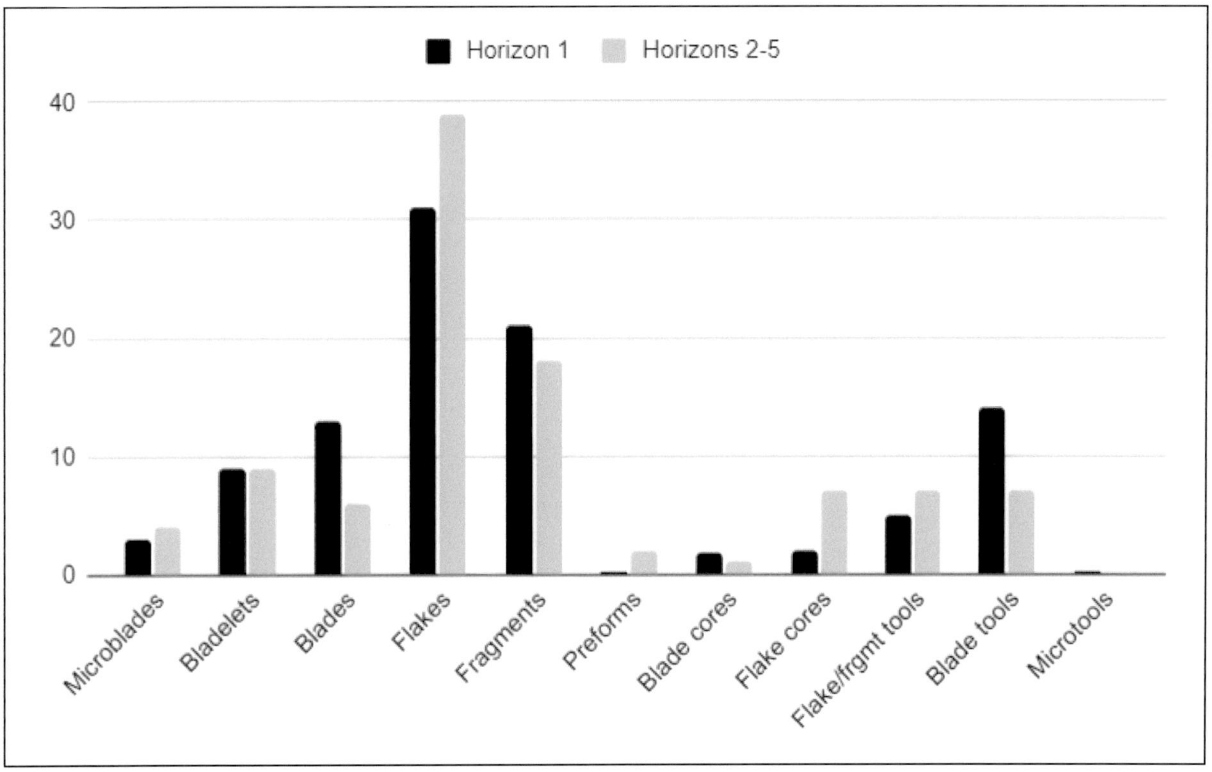

Figure 4. Distribution of main chipped stone categories from Pločnik site by horizon showing percentages of total for each horizon.

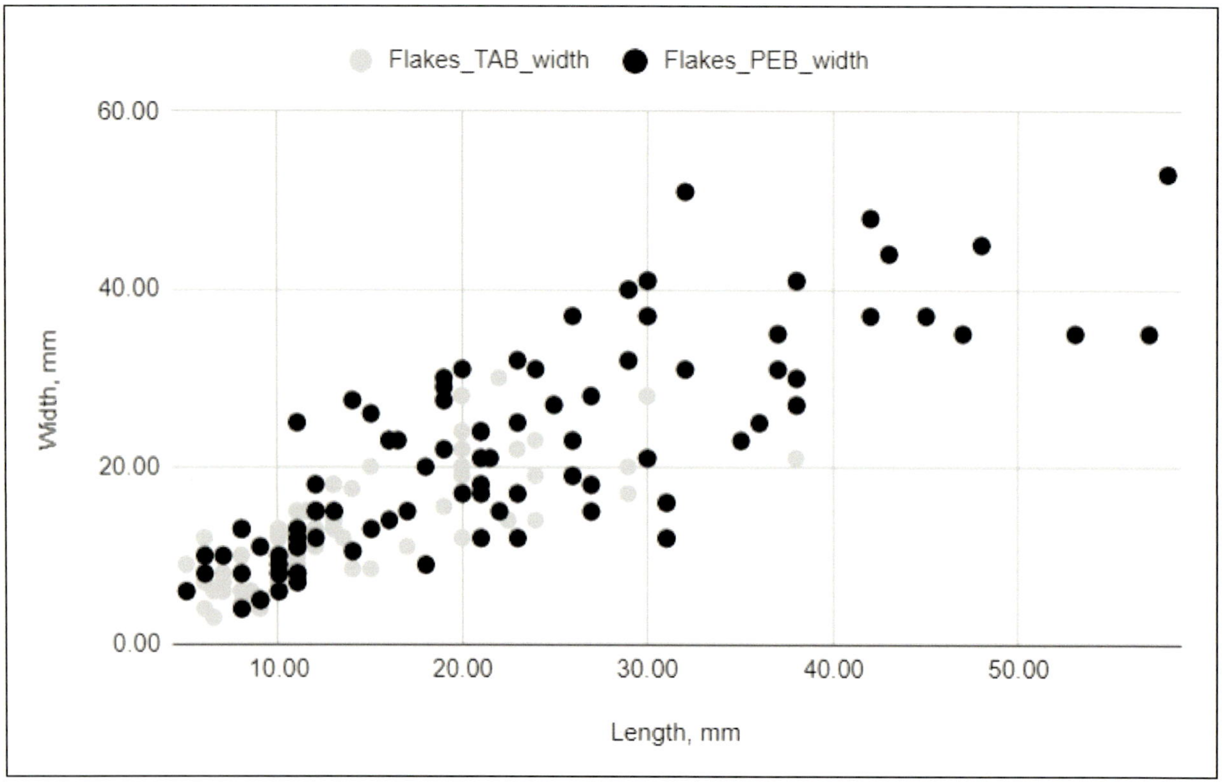

Figure 5. Width and length of flakes from Pločnik.

Table 4. Types of cores from Pločnik and their distribution across site horizons.

Raw material	Blank type	Tabular flint		Pebble flint		Other raw material		Total
	Horizon	1	2-5	1	2-5	1	2-5	
Pyramidal	Blades, bladelets	1	0	4	2	1	2	**10**
Pyramidal with narrow flaking surface	Blades, bladelets, microblades	5	0	0	4	0	3	**12**
Pyramidal, change of orientation	Blades, bladelets	0	0	0	1	0	0	**1**
Globular	Flakes	0	0	1	11	0	1	**13**
Prism-type	Flakes	1	0	2	6	1	3	**13**
Flat with convergent negatives	Flakes	0	0	1	0	0	0	**1**
Precore	?	1	1	0	7	0	4	**13**
N/A	?	1	0	1	5	0	4	**11**
Total		**9**	**1**	**9**	**36**	**2**	**17**	**74**

Table 5. Location of cortex on the tabular flakes from Pločnik.

Cortex location	Number of items (%)
Primary	7 (11)
Secondary	20 (29)
Full	42 (60)
Total	**71**

Table 6. Location of cortex on the tabular blades, bladelets and microblades from Pločnik.

Cortex location	Blades	Bladelets	Microbladelets
Primary	1	2	1
Left side	5	10	12
Right side	5	8	2
Both sides	3	0	0
Distal end	6	0	0
Plain	64	16	14
Chalky	3	0	1
Total	87	36	30

Table 7. Main chipped stone categories of pebble flint items from Pločnik.

Category	Number of items (%)
Blades (+tools)	19 (6)
Bladelets (+tools)	15 (5)
Microblades (+tools)	10 (3)
Flakes (+tools)	173 (54)
Fragments	61 (19)
Nodules	4 (1)
Preforms	5 (2)
Cores and core fragments	29 (9)
Core rejuvenation flakes	3 (1)
Total	**319**

Table 8. Cortex location for pebble flint flakes from Pločnik.

Cortex location	Number of items (%)
Primary	8 (5)
Secondary	36 (25)
Full	103 (70)
Total	**147**

Pebble flint was also used for blade reduction, but its products are not numerous (Table 7). This sample is relatively small, compared with tabular flint blades. The mean ratio width:thickness estimate is 4.8 and the width of the blades does not exceed 16 mm. Blades and bladelets of this group are mainly irregular and have a large variance of flaking angles. Their butts are flat, facetted, or linear (Figure 2.4 and 2.5). These products could be attributed more to frontal, rather than narrow-surface reduction (Table 4 and Figure 2.6). Hard and/or soft mineral hammers were probably used.

Flake production

Flake production is strongly correlated with pebble flint and was conducted on the site as indicated by numerous flake cores (Table 4) and a wide range of cortical waste in the collection (Table 8). Flakes were produced and used in an expedient fashion: globular cores with minimal preparation are widespread, though several nodules and cores discovered in one location (EDM 1447, spit 16) show evidence of heat exposure. This occurred before a series of flake removals, after which specific shiny surfaces were exposed (Figure 2.1and 2.3). The large size of the blocks as well as poor quality of the material (with caverns as on Figure 2.2) makes it improbable that a deliberate heat treatment was applied. The main technique used for flake production was probably direct percussion using a hard mineral hammer.

Typology

The blanks which were the result of the production processes described above (Figure 7.13–7.18) were often retouched to become tools. Abrupt and semi-abrupt retouch was preferred with burin blow and notches being much rarer. Irregular retouch was most probably caused by the usage of tools.

The tool assemblage from Pločnik is characterised by a predominance of retouched blades and scrapers (Table 9, Figure 6). Only a few examples of drills, microdrills, burins, geometric microliths, retouched flakes and tranchets were represented.

One group of tools has specific use-wear traces: a bright, mirror-like polish often connected with usage as sickles (Figure 7.1–7.6 and 7.9). This group was not considered in this classification in order to prevent the mixing of morphological and functional approaches, with the latter being conducted by N. Skakun and V. Terekhina (Скакун *et al.* 2015) and to be published soon.

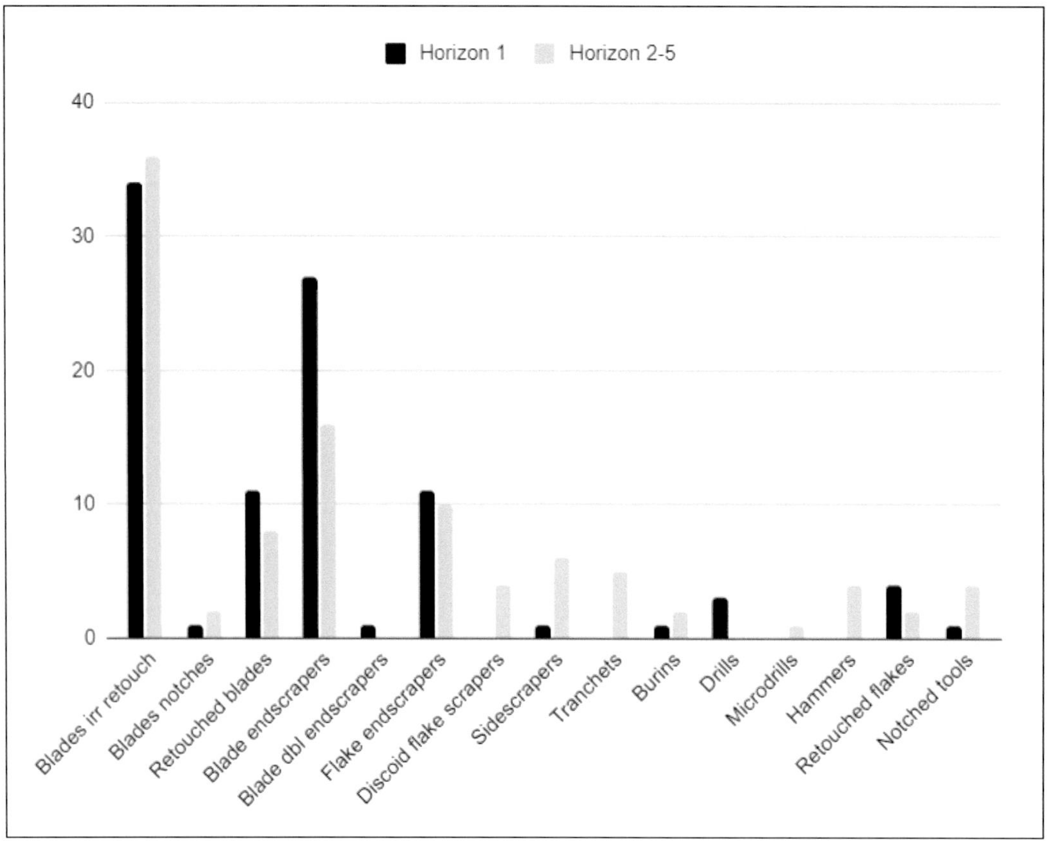

Figure 6. Basic tool categories at Pločnik site showing percentage of total assembly.

Table 9. Basic tool categories from Pločnik

		Number of items (% of total)		Total
		Horizon 1	Horizon 2-5	
Blades	Irregular retouch	32 (34)	30 (36)	**62**
	Notches	1 (1)	2 (2)	**3**
	Retouched blades	12 (11)	7 (8)	**19**
Blade scrapers	Blade endscrapers	25 (27)	13 (16)	**38**
	Double endscrapers	1 (1)	0	**1**
Flake scrapers	Flake endscrapers	10 (11)	8 (10)	**18**
	Discoid	0	3 (4)	**3**
	Sidescrapers	1 (1)	5 (6)	**6**
Others	Tranchets	0	5 (5)	**4**
	Burins	4 (1)	2 (2)	**4**
	Drills	3 (3)	0	**3**
	Microdrills	0	1 (1)	**1**
	Hammers	0	4 (4)	**4**
	Retouched flakes	4 (4)	2 (2)	**6**
	Notched tools	1 (1)	3 (4)	**4**
Total		**94**	**85**	**178**

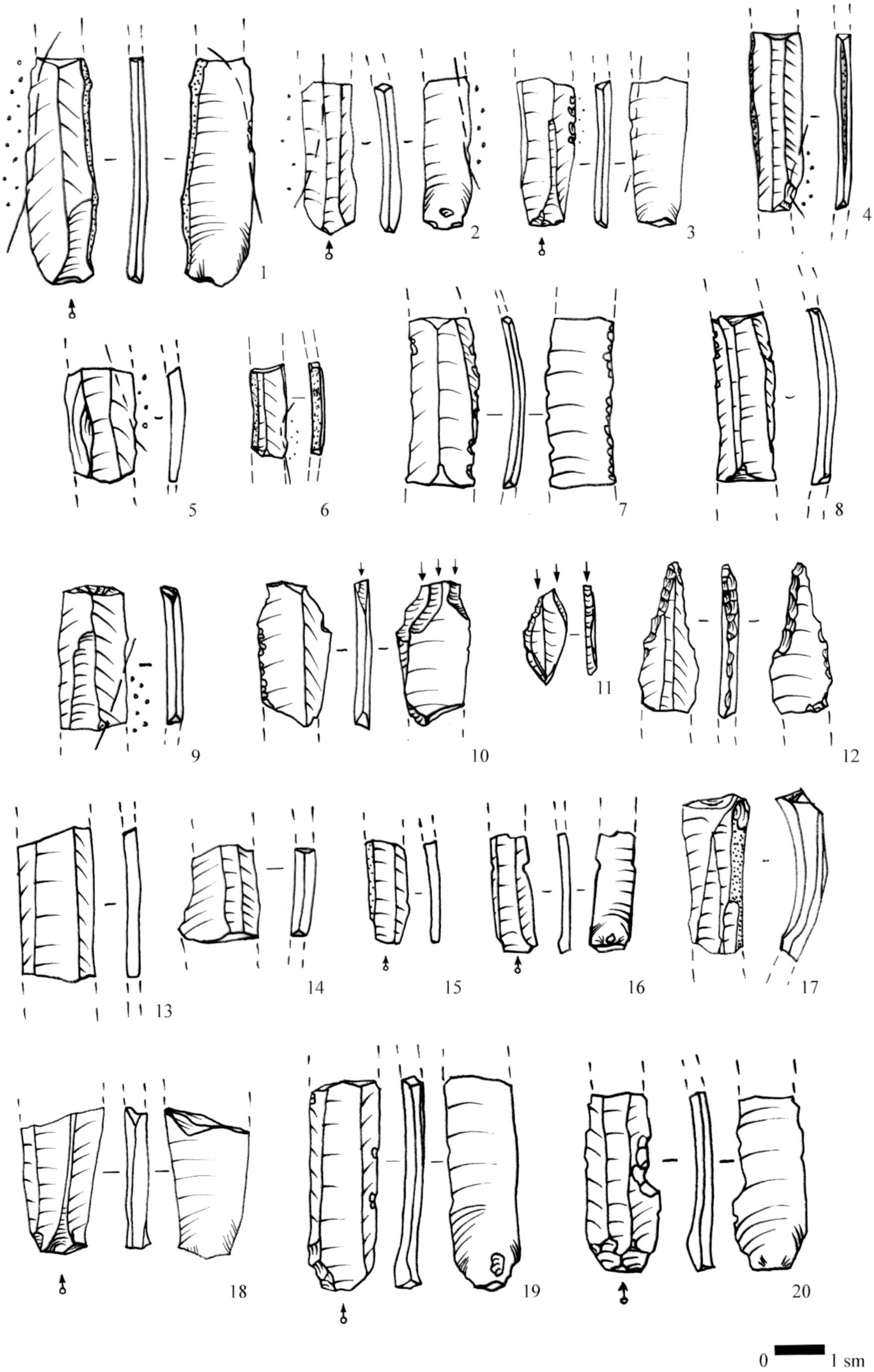

Figure 7. Pločnik lithic blades and bladelets, tools.

Retouched blades are the best represented tools in both periods (34% in Horizon 1 and 36% in earlier horizons). Mainly mesial fragments are present, but proximal and distal parts of the blades and bladelets were also used. Most blades and bladelets only have an irregular retouch (Figure 7.3, 7.7, 7.8 and 7.19) with notches being rare (Figure 7.20). In several cases, they have abrupt and semi-abrupt retouch (flat retouch was recorded just once). Truncation or backing with abrupt and semi-abrupt retouch (Figure 7.4 and 7.9) were used for the accommodation of the tool handle. Often, these tools have a mirror-like polish.

Scrapers are the second largest tool category (40% in Horizon 1 and 36% in Horizons 2–5) and include blade scrapers (endscrapers and double endscrapers) and flake scrapers (endscrapers, sidescrapers, and discoid scrapers). For blade endscraper production, large regular blades with a width varying from 14 to 22 mm were mainly used and, in Horizon 1, were mostly made from tabular flint. Usually, the distal ends of the blanks in these cases were retouched with semi-abrupt retouch forming a symmetrical convex (Figure 8.1 and 8.4) or slightly straightened edge (Figure 8.2). In some cases, the edge was asymmetrical (Figure 8.5). The share of blade scrapers increased from 16% in Horizons 2–5 to 28% in Horizon 1. Flake sidescrapers (6%) (Figure 8.6 and 8.7) and discoid scrapers (4%) (Figure 8.8) are typical for Horizons 2–5.

Tools that are less well represented at Pločnik (9% in Horizon 1 and 18% in Horizons 2–5) include tranchets, burins, drills, retouched flakes and notched tools. Tranchets (Figure 8.10 and 8.11) are rough tools made on large flakes or fragments with wide working edges covered with semi-abrupt retouch. Tools of this category are particularly common in the earliest horizons, occurring between the 25th and the 19th spits. For early horizons, blade burins are typical (burins on transversal break and one dihedral burin) (Figure 7.10 and 7.11). In Horizon 1, flakes were predominantly used (three burins on transversal break). All drills (Figure 7.12) were attributed to the Horizon 1 and were produced in a similar way. The tabular flint blades were selected, and the distal end was retouched by alternate abrupt retouch forming a symmetrical working tip. The resulting drills were square in section and 3–5 mm thick. Retouched flakes (Figure 8. 9) include flakes with semi-abrupt irregular retouch (four tools attributed

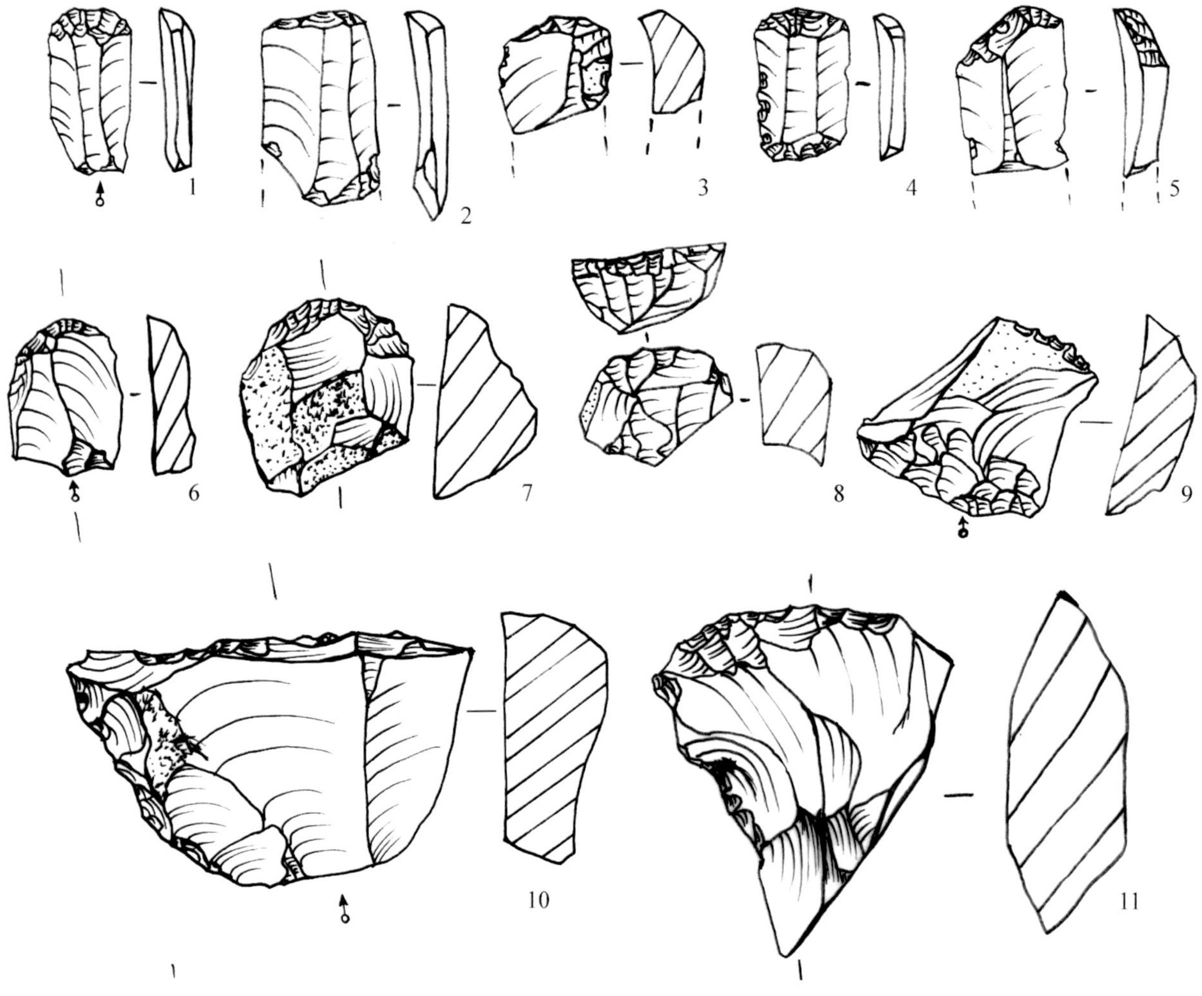

Figure 8. Pločnik lithic tools. 8.1–8.8 – scrapers; 8.9 - retouched flake; 8.10. and 8.11 - tranchets.

to Horizon 1; two other tools were found in the earlier Horizons). Notched tools were made on flakes, and have a concave edge formed with semi-abrupt retouch. They are mostly specific to early horizons, as are the hammers made from pebbles or exhausted flake cores that were used for flake production.

Conclusion

The analysis of the Pločnik chipped stone industry enabled the identification of *chaîne opératoire* relating to blade production using cream, tabular flint and a pebble flint, and also to flake production (Figure 9).

The first *chaîne opératoire* strongly correlates to a specific raw material: the cream tabular flint occurring in the form of the small slabs. The blade reduction was performed from the narrow flaking surface and a punch technique was probably used. The reduction products (blades and bladelets) were subsequently used without any strict selection process. However, the thick blades were retouched to produce end-scrapers or drills, and the fine blades and bladelets were used in composite tools with minimal retouching.

The second *chaîne opératoire* is strongly connected to pebble flint. Percussion techniques were probably used to obtain the bladelets and blades. They were predominantly used as inserts, since they have finer section. Their usage pattern is similar to that involving cream tabular flint.

The third *chaîne opératoire* is linked to flake production and is based on local river gravel sources. These flakes and flake cores are not standardised. The flakes were made with direct percussion and were used without retouch or were retouched to form scrapers or tranchets.

The reduction processes relating to the three outlined *chaîne opératoires* took place locally. Heat treatment was probably used in the course of blade production to improve the raw material knapping characteristics.

The evolution of the chipped stone industry is observable with an increased production of tabular blades and bladelets from 10% in Horizons 2–5 to 35% in Horizon 1. Both the flake industry and blade production were based on a large variety of pebble flint and other raw material with more simplified methods and techniques occurring in Horizons 2–5.

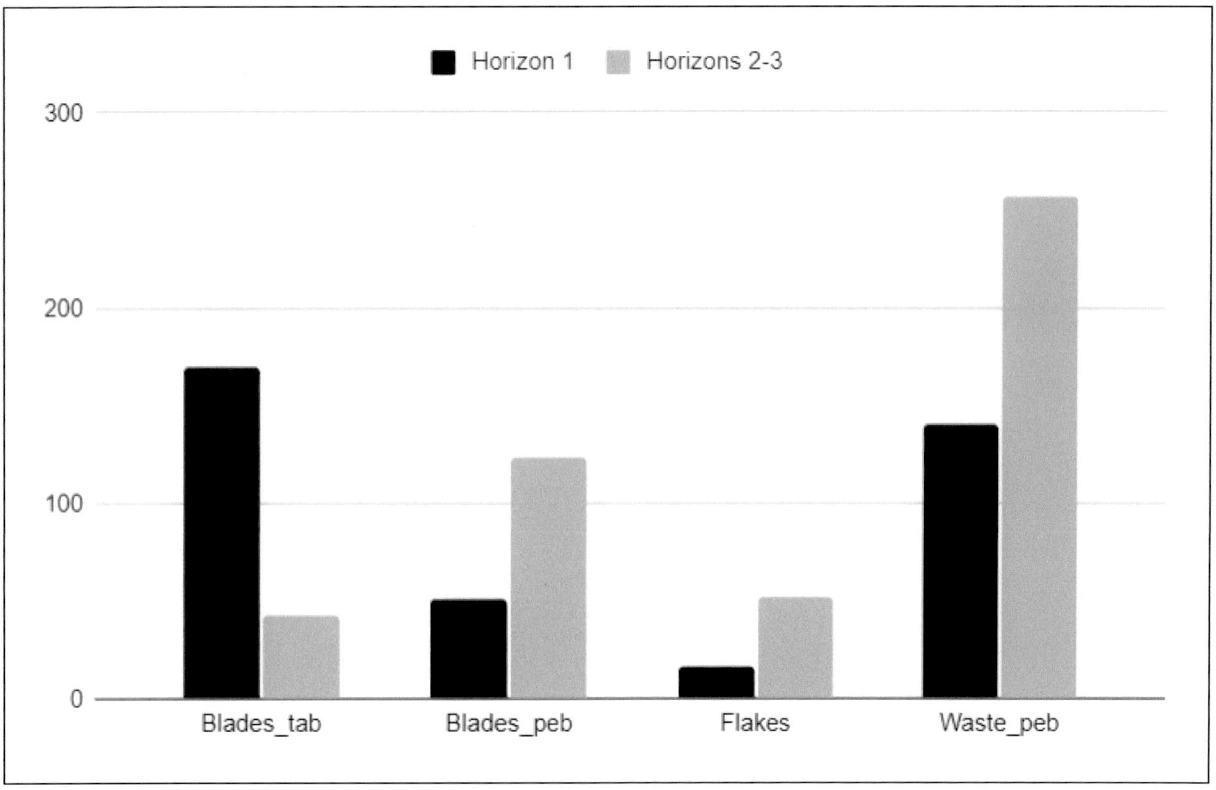

Figure 9. Evolution of the chipped stone industry at Pločnik. Ratio of three *chaîne opératoires* (tab. - tabular flint; peb. - pebble flint) and unattributed waste products, showing number of items.

The bibliographic reference for this chapter is:

Ibragimova, E. 2021. Chipped stone industry at Pločnik, in Radivojević, M., Roberts, B. W., Marić, M., Kuzmanović Cvetković, J., and Rehren, Th. (eds) *The Rise of Metallurgy in Eurasia*: 397–407. Oxford: Archaeopress.

Chapter 34

Plant use at Pločnik

Dragana Filipović

Introduction

The recent archaeological investigations at Pločnik included for the first time the study of charred macro-plant remains (seeds, chaff, fruit etc). The main goals of the sampling and analysis were to explore the potential for, and mode of, preservation of plant remains at the site; to determine the taxonomic composition of crop and wild assemblages; and to investigate the distribution of different plant taxa across the archaeological deposits and occupation horizons and assess their potential role in the economy. The archaeobotanical results from Pločnik represent an important contribution to the growing body of archaeobotanical evidence from Neolithic–Early Chalcolithic sites in Serbia and the Balkans as they offer new and systematically collected data on plant use at a Vinča culture site.

Methods and materials

Field methodology - sampling and recovery

A total of 68 archaeobotanical samples were collected from the majority of contexts excavated in the seasons 2012–2013. A minimum sample volume agreed with the excavators was 5 litres; the actual soil volume ranged from as little as 200 millilitres collected from inside ceramic vessels, up to 14 litres from contexts that contained plant material immediately visible in the field (e.g. charcoal concentration) or appeared burnt (fire-related features). The archaeological context and soil volume information for all the samples is given in Appendix B_Ch34.1. Flotation was applied for the extraction of charred and non-charred (mineralised) plant remains - manual flotation in the case of small samples and machine-aided for larger ones. The material that floated (light fraction) was captured and air-dried in pieces of mesh-like cloth with 0.3 mm openings; the heavy fraction was collected in a 1 mm mesh (mosquito net) placed inside a bucket or the flotation tank. Altogether, 479 litres of soil were floated.

Laboratory procedure - sorting, identification and quantification

Archaeological materials including pottery shards, bone, stone, beads and malachite were extracted from the heavy fraction of the samples; there was no plant material in the heavy residue. Plant remains other than wood were removed from the light fraction. For 60 samples, 100% of the light fraction was sorted. Seven samples were mainly composed of modern vegetal material and lumps of soil, and this slowed down the sorting; these samples were, therefore, split to 50% (1 sample) or 25% (6 samples) and the 50% and 25% sub-samples respectively were sorted. Further, the light fraction of Sample 66 (from Feature 39) contained a much greater quantity of charred material compared to the other Pločnik samples (established simply by 'scanning' the sample bag). This sample was also divided to 25%; over 600 crop items were encountered in the sorted 25% sub-sample.

Sample sorting and observation of the remains were carried out under a low-power stereomicroscope (magnification range 10 x–40 x). In the process of identification, the remains were compared for their morphology and, sometimes, anatomy with modern specimens from the personal reference collection and with the drawings, photographs and descriptions in the available literature (e.g. Beijerinck 1947; Cappers *et al.* 2006; Schoch *et al.* 1988). The names of wild taxa were taken from *Flora Europea* (Tutin *et al.* 1964–1980); crop names follow Zohary *et al.* 2012.

The full list of crop and wild taxa is provided in Appendix B_Ch34.1. and includes some 140 identifications among which are precise as well as intermediate, 'cf.', and some descriptive categories. The level of determination depended on the state of preservation of the remains (e.g. complete or incomplete, distorted) and, for wild taxa, the availability of the reference material. Thus, a number of wild taxa were identified to a species/genus 'type' and require further work towards more accurate attribution. Intermediate or amalgamated categories, such as *Triticum monococcum/dicoccum*/'new type' hulled wheat, include remains that are too poorly preserved to be recognised as belonging to a particular taxon within the category. Remains that bear most of the characteristics of a precisely identified taxon, but are fragmented or eroded, were labelled 'cf.' (e.g. cf. *Pisum sativum*).

In order to assess quantitative representation of the taxa, all remains were counted. This was done in the following manner: a complete cereal grain was recorded as one item; apical and embryo ends of fragmented cereal grains were counted separately and the higher score was included in the grain total; cereal grain longitudinal 'halves' were paired, and each pair counted as one; hulled wheat spikelet fork was counted as two glume bases; terminal spikelet fork was scored as one; free-threshing cereal rachis segment was recorded as one; pulse seed was counted as one, as was wild seed. The following were also counted as a single item: culm node, fragment of culm/stalk/pod, fragment of fruit stone/nutshell. The amount of cf. nut meat, cf. parenchyma, food/bread remains, fruit (?) flesh, indeterminate vegetal matter and dung was measured as volume (in millilitres).

Each fragment of nutshell and fruit stone was scored as one and included as such in the total number of fruit/nut items. In the case of *Cornus mas,* both complete and fragmented fruits and fruit stones were discovered. Each complete fruit/fruit stone was scored as one; fragments, regardless of their number in a sample, were together scored as one item (and, where appropriate, added to the number of complete fruits/fruit stones to obtain the minimum number of items (MNI)). In Sample 52 (Feature 37), however, *Cornus mas* fruit stone fragments were numerous (compared to their number in the rest of the samples); their volume was measured here, as well as the volume of the complete fruit stone found in the same sample (*c.* 0.5 ml) which was used as a reference. The volume of the fragments was then converted to the number of complete stones and added to the *Cornus mas* MNI for this sample.

Finally, the number of remains of different taxa in the sorted sub-samples (50% or 25%) of the eight samples (see above) were multiplied up to acquire the totals for the whole samples (both counts for the sub-samples and the whole samples are given in Appendix B_Ch34.1).

In order to provide a general overview of the sample composition, the number of initial identifications was reduced to 94 through amalgamation of the definite and 'cf'.' categories (see Table 1). Further, hulled wheat types were considered both separately and within the 'hulled wheat' group. The wild taxa producing potentially collected (edible) fruits were included in the 'fruit/nut' group. 'Wild/weed taxa' include wild taxa that may have represented arable weeds, although some possess medicinal properties or may have been valued for other reasons (e.g. common reed possibly used as crafting/construction material – e.g. Stevanović 1997). The botanical richness of the samples was assessed based on the total number of items per sample (abundance) and the number of items per litre of floated soil (density) (Table 2). The types of remains included in these calculations are indicated in Appendix B_Ch34.1 (i.e., the remains recorded by volume were omitted). The potential degree of use of different crop and fruit/nut taxa was evaluated using as parameters the number of remains (abundance) and the number of occurrences (ubiquity) across the assemblage (Table 1).

Table 1. Absolute number of remains and frequency of individual and groups of taxa at Pločnik.

CROPS	**plant part**	common name	**ubiquity**		**abundance**	
HULLED WHEATS			count	%	count	% of all crop
Triticum monococcum	grain	einkorn	42	62	269	3
Triticum monococcum	glume base		44	65	1341	15
Triticum dicoccum	grain	emmer	6	9	36	0.4
Triticum dicoccum	glume base		40	59	825	9.4
Triticum monococcum/dicoccum	grain		2	3	8	0.09
cf. *Triticum,* 'new type'	grain	'new type' hulled wheat	3	4	3	0.03
Triticum, 'new type'	glume base		44	65	871	10
Triticum dicoccum/'new type'	glume base		38	56	851	10
T. monococcum/dicoccum/'new type'	grain		17	25	79	0.9
T. monococcum/dicoccum/'new type'	glume base		58	85	3437	39
HULLED WHEAT	GRAIN		46	68	395	5
HULLED WHEAT	GLUME BASES		58	85	7325	84
HULLED WHEAT	ALL ITEMS		**59**	87	**7720**	**88**

Table 1 continued. Absolute number of remains and frequency of individual and groups of taxa at Pločnik.

CROPS	**plant part**	common name	**ubiquity**		**abundance**	
FREE-THRESHING CEREALS			count	%	count	% of all crop
Triticum aestivum/durum	grain	free-threshing wheat	3	4	21	0.2
Triticum aestivum/durum	rachis	free-threshing wheat	16	24	312	4
Hordeum vulgare	grain	barley	11	16	21	0.2
Hordeum vulgare	rachis		2	3	2	0.02
FREE-THRESHING CEREAL	GRAIN		14	21	42	0.5
FREE-THRESHING CEREAL	RACHIS		17	25	314	4
FREE-THRESHING CEREAL	ALL ITEMS		**25**	37	**356**	**4**
Triticum sp.	grain	wheat	34	50	203	2
Cerealia indeterminata	grain	indeterminate cereals	47	69	336	4
ALL CEREAL	GRAIN		53	78	976	11
ALL CEREAL	CHAFF		58	85	7639	87
ALL CEREAL	ALL ITEMS		59	87	8615	99
Cerealia indeterminata	culm node	indeterminate cereals	1	1	4	0.05
LEGUMES			count	%	count	% of all crop
Lens culinaris	seed	lentil	30	44	77	0.9
cf. *Lathyrus sativus/cicera*	seed	grass pea	2	3	3	0.03
Pisum sativum	seed	pea	5	7	10	0.1
cf. *Vicia ervilia*	seed	bitter vetch	1	1	1	0.01
Leguminosae sativae	seed	indeterminate pulses	18	26	29	0.3
ALL LEGUME	SEED		**39**	57	**120**	1
OIL/FIBRE PLANTS			count	%	count	% of all crop
Linum usitatissimum	seed	flax/linseed	1	1	1	0.01
ALL CROP	SEED+CHAFF		**59**	87	**8736**	100
FRUIT/NUT TAXA	**plant part**		count	%	count	
Cornus mas (MNI)	fruit+stone	Cornelian cherry	16	24	41	
Corylus avellana (MNI)	shell	hazel	2	3	2	
Fragaria vesca	seed	wild strawberry	7	10	12	
Physalis alkekengi	seed	chinese lantern	8	12	63	
Prunus cf. *domestica* var. *insititia*	stone fragments	European plum	2	3	2	
Prunus cf. *spinosa*	stone fragments	sloe/buckthorn	2	3	5	
Prunus type	stone fragments	plum	1	1	2	
cf. *Pyrus/Malus* sp.	seed	pear/apple	3	4	6	
Rubus idaeus/fruticosus	seed	raspberry/ blackberry	6	9	30	
Rubus sp.	seed		9	13	10	
Sambucus ebulus	seed	elderberry	1	1	1	

Table 1 continued. Absolute number of remains and frequency of individual and groups of taxa at Pločnik.

	plant part	common name	**ubiquity**		**abundance**	
FRUIT/NUT TAXA			count	%	count	% of all crop
fruit stone/nutshell	fragments		5	7	7	
cf. nut meat	volume (ml)		4	6	0.07	
? fruit flesh	volume (ml)		7	10	0.89	
WILD/WEED TAXA			count	%	count	
cf. *Androsace* sp.	seed	rock jasmine	1	1	2	
cf. *Asperula* sp.	seed	(woodruff)	1	1	4	
cf. *Avena* sp.	fruit	(wild) oat	1	1	1	
Avena sp.	awn fragments		3	4	6	
Bromus cf. *arvensis*	fruit	field brome	3	4	14	
Bromus secalinus	fruit	rye brome	1	1	8	
Bromus sp.	fruit	bromegrass	7	10	12	
Buglossoides arvensis	mineralised seed	field gromwell	3	4	8	
cf. Caryophyllaceae	endosperm	pink family	1	1	1	
Centaurea sp.	seed	starthistle	1	1	1	
Chenopodiaceae	seed/endosperm	goosefoot family	3	4	3	
Chenopodium album type	seed	fat hen	8	12	13	
Chenopodium sp.	seed	goosefoot	6	9	11	
Chenopodiaceae/Caryophyllaceae	endosperm		1	1	1	
cf. Compositae, small-seeded	seed	daisy family	1	1	1	
Cynodon type	fruit	Bermuda grass	3	4	4	
Echinochloa crus-galli	fruit	cockspur	1	1	1	
Fallopia convolvulus	seed/endosperm	black bindweed	15	22	40	
Galium aparine type	seed	cleavers	1	1	1	
Graminae	fruit	grasses	1	1	1	
Graminae	culm fragment	grasses	3	4	4	
Graminae, large	fruit	grasses	6	9	6	
cf. *Hordeum* sp. (wild)	fruit	wild barley	2	3	5	
Hypericum sp.	seed	St. John's wort	1	1	1	
Lamium type	mineralised seed	deadnettle	1	1	1	
cf. Lamiaceae	seed	mint family	1	1	1	
Lolium type	fruit	ryegrass	1	1	1	
Phleum type	fruit	catstail	1	1	1	
Phragmites communis	culm node	common reed	1	1	1	
Poa sp.	fruit	meadow-grass	2	3	2	
cf. *Polygonum aviculare*	endosperm	common knotgrass	2	3	2	
cf. *Polygonum* sp.	endosperm/seed coat	knotweed	5	7	9	
Polygonaceae	seed	knotweed family	2	3	2	
cf. *Raphanus raphanistrum*	pod fragment	wild radish	1	1	1	
Rumex sp.	keeled seed	dock	1	1	1	
Scrophularia type	seed	figwort	1	1	3	
Setaria viridis/verticillata	fruit	green/bristly foxtail	6	9	9	

Table 1 continued. Absolute number of remains and frequency of individual and groups of taxa at Pločnik.

	plant part	common name	ubiquity		abundance	
WILD/WEED TAXA			count	%	count	% of all crop
Solanum nigrum	seed	black nightshade	13	19	31	
Solanum sp.	seed	nightshade	3	4	10	
Solanaceae	seed	nightshade family	10	15	25	
Stachys annua type	mineralised seed	woundwort	1	1	1	
Teucrium sp.	seed	germander	3	4	7	
Trifolium arvense type	seed	haresfoot clover	1	1	2	
Trifolium repens type	seed	white clover	1	1	1	
Trifolium sp.	seed	clover	1	1	1	
cf. *Vaccaria pyramidata*	seed	cowherb	1	1	1	
Verbena officinalis	seed	common vervain	1	1	4	
Vicia/Lathyrus sp.	seed	wild vetch/ vetchling	1	1	1	
INDETERMINATE and OTHER			count	%	count	
unknown seed or seed core			12	18	39	
unknown vegetal matter	volume (ml)		3	4	0.5	
? fruit skin/pod	fragments		8	12	55	
bud			4	6	7	
stalk	fragments		5	7	45	
cf. culm base	fragments		1	1	1	
cf. parenchyma	volume (ml)		5	7	0.2	
food ('bread') remains	volume (ml)		3	4	0.3	
cf. dung	volume (ml)		4	6	1.1	
mouse pellet	charred		1	1	1	

Results and discussion

Overview of the assemblage

The absolute counts of the identified crop and wild taxa and the frequency of their occurrence across the assemblage are provided in Table 1. Detailed data for each analysed sample are given in Appendix B_Ch34.1. The remains are mostly charred, though a certain number of seeds of wild taxa and fruit stone/nutshell fragments were preserved through mineralisation.

As shown in Figure 1, the vast majority of the samples are mainly composed of crop remains. The three samples that appear dominated by wild material (Samples 8, 10 and 11 – see Appendix B_Ch34.1) in fact yielded only few remains (for example, as few as one in Sample 11). Also, a number of samples contained very small amounts of botanical material, e.g., less than 30 remains. All samples are, however, included in this initial evaluation.

Crops

A wide range of crop types were registered including einkorn (*Triticum monococcum* L.), emmer (*T. dicoccum* Schrank), 'new type' hulled wheat (Jones *et al.* 2000; Kohler-Schneider 2003), free-threshing wheat (*T. aestivum/durum/turgidum*), barley (*Hordeum vulgare* L.), lentil (*Lens culinaris* Medik.), pea (*Pisum sativum* L.), bitter vetch (*Vicia ervilia* (L.) Willd.), grass pea (*Lathyrus sativus* L./*cicera* L.) and flax/linseed (*Linum usitatissimum* L.). Hulled wheat types are most abundant (88% of all crop content) and frequent (found in 87% of the samples) and they greatly exceed the occurrence of free-threshing cereals (free-threshing wheat and barley) and legumes. They were likely the most important crops at Pločnik. In the largest number of samples, the hulled wheat component is represented almost exclusively by hulled wheat glume bases (Figure 2) testifying to the practice of regular on-site crop cleaning and disposal of crop processing by-products. The dominance of hulled wheat glume bases in the Pločnik assemblage may potentially

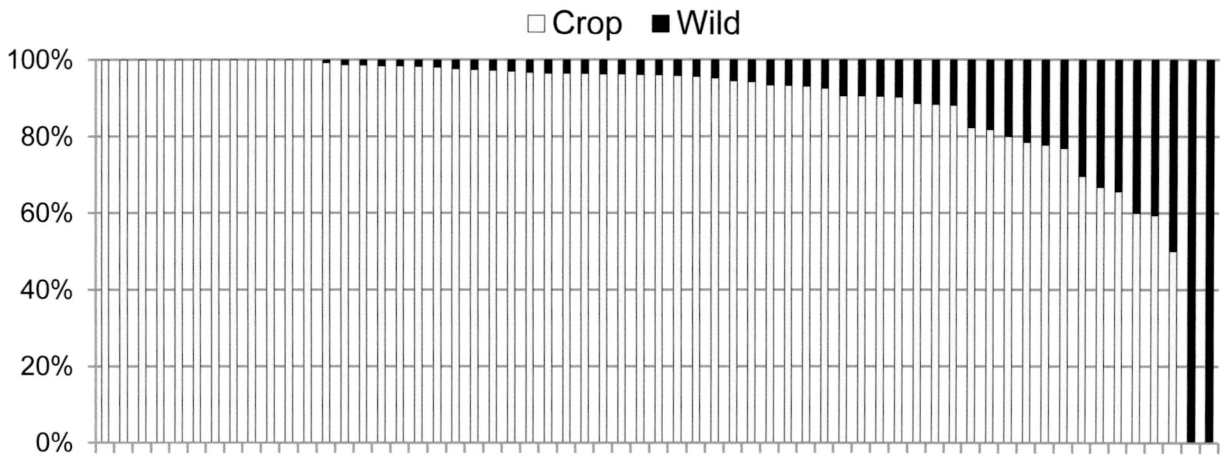

Figure 1. Proportions of crop and wild plant material in the samples from Pločnik.

obscure and/or diminish the role of other cereal types in the plant economy, given that the comparable evidence of free-threshing cereals (i.e. cleaning waste) has smaller chances of being preserved among charred material due to differences in the processing and storage of these crops (cf. Hillman 1984b, 1985). Still, as shown in Table 1 and Figure 3, the comparison of abundance and frequency across the samples of cereal *grain* also suggests the overall higher significance of hulled wheat. In the few samples with seemingly large proportions of barley grain, only a few cereal grains were found and are not considered statistically representative.

A closer examination of the hulled wheat record reveals that einkorn remains, both grain and glume bases, outnumber emmer and 'new type' wheat. It is, however, important to note that a considerable number of glume bases were too distorted to be precisely identified (indeterminate glume bases + emmer/'new type' hulled wheat glume bases = 64% of all crop content – see Table 1). The percentages of glume bases identified to each of the three hulled wheat types are not too far apart: einkorn – 15%, emmer – 9.4%, 'new type' hulled wheat – 10%; this may suggest a similar degree of use of the different hulled wheats. On the other hand, the number of einkorn grain is significantly larger than the number of emmer and 'new type' hulled wheat grain combined, perhaps indicating greater importance of einkorn.

The quantity of free-threshing wheat and barley grain and rachis is generally small. Some rachis segments of free-threshing wheat may belong to bread wheat (*Triticum aestivum* L.) and it is possible that some of the grain also derives from the hexaploid type. The majority of free-threshing wheat rachis (85% of the total number) comes from a discrete, undisturbed context (Sample 66) – oven Feature 39 located in a pit-house (Feature 38) representing the earliest occupation level of the site (Appendix B_Ch34.1). The pit-house was radiocarbon-dated to the early phases of the Late Neolithic/Vinča culture (see Chapter 37, this volume). Based on their morphology (see Appendix B_Ch34.2), these particular rachis segments belong to a tetraploid type of free-threshing wheats (such as e.g., *T. durum*). The absolute

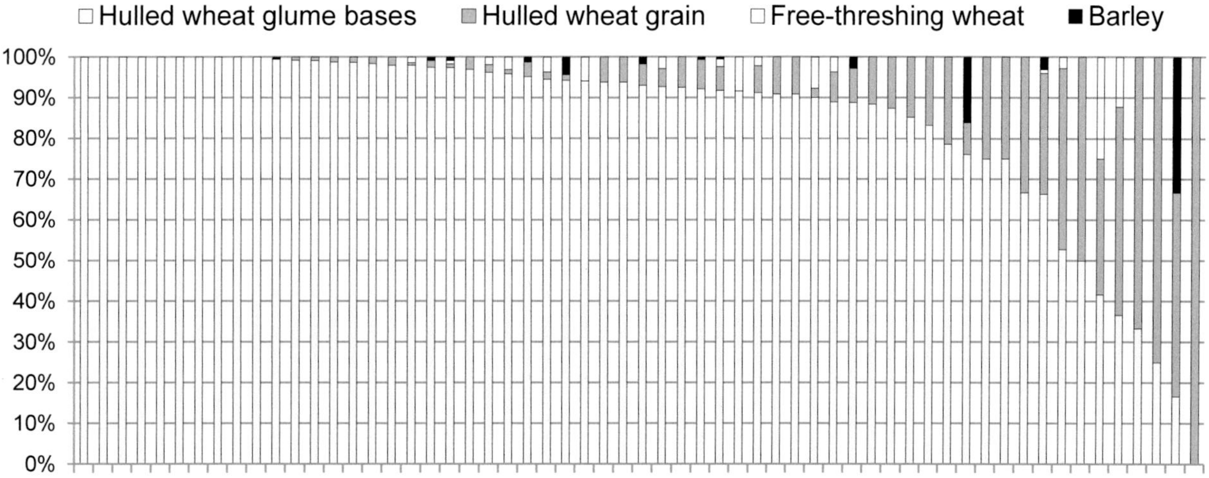

Figure 2. Proportional representation within the cereal content of the three cereal types identified at Pločnik; proportions of hulled wheat grain and chaff shown separately.

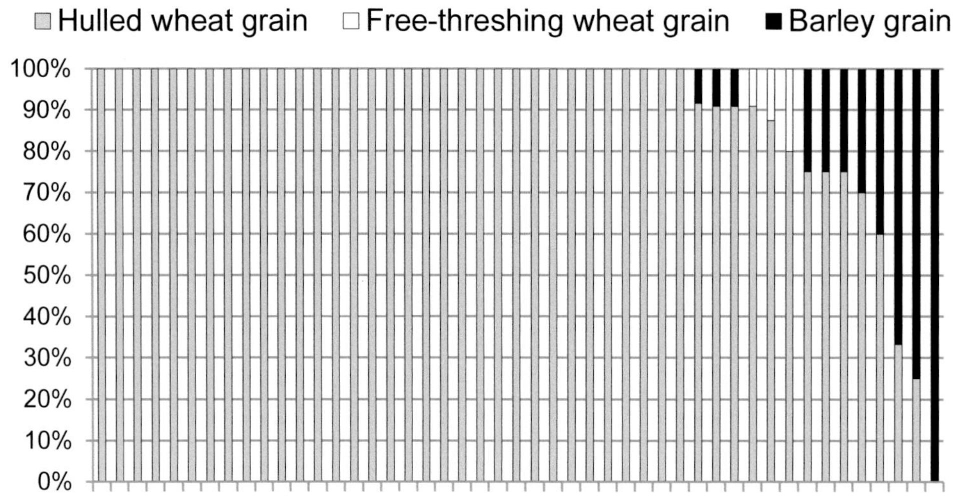

Figure 3. Proportions of grain of the three cereal types from Pločnik.

date for the pit-house is taken as reflecting the date of the *in situ* burnt rachis remains from the oven. These are, for now, the earliest dated finds of free-threshing wheat in Serbia. Several grains of *T. aestivum* were reported for the site of Belotić in western Serbia attributed to the Early Neolithic/Starčevo culture (Borojević 1990), but the absolute date of the specimens was not provided (Sue Colledge pers. comm.). Radiocarbon dates of free-threshing wheat grain from another Starčevo culture site, Mesarci (Borojević 1990), revealed that the material is of modern age (Sue Colledge pers. comm.).

At Pločnik, the remains of free-threshing wheat and barley rarely co-occur in the samples. This could potentially mean that they have been grown and processed separately, despite their identical cleaning requirements (Hillman 1985; Jones 1990). Their abundance is, however, too low for any conclusions to be drawn.

The presence of legumes is minor by both abundance and frequency (Table 1). In Figure 4, the proportions of cereals and legumes in the samples are compared. Lentils are more ubiquitous relative to other pulses; this potentially signifies their more important or perhaps major role amongst pulse crops. Lentil is followed by pea, which is represented by only ten seeds in total. Very small quantities of grass pea were found at Pločnik and at few other Neolithic sites in the central Balkans; it may have represented accidental inclusion among cultivated legumes. Bitter vetch could have had a similar, 'non-crop' status at Pločnik, although it is frequently found in the region as part of crop assemblages from the Late Neolithic (Filipović 2014).

Fruit/nut taxa

Remains of a number of wild edible taxa were discovered in the samples among which Cornelian

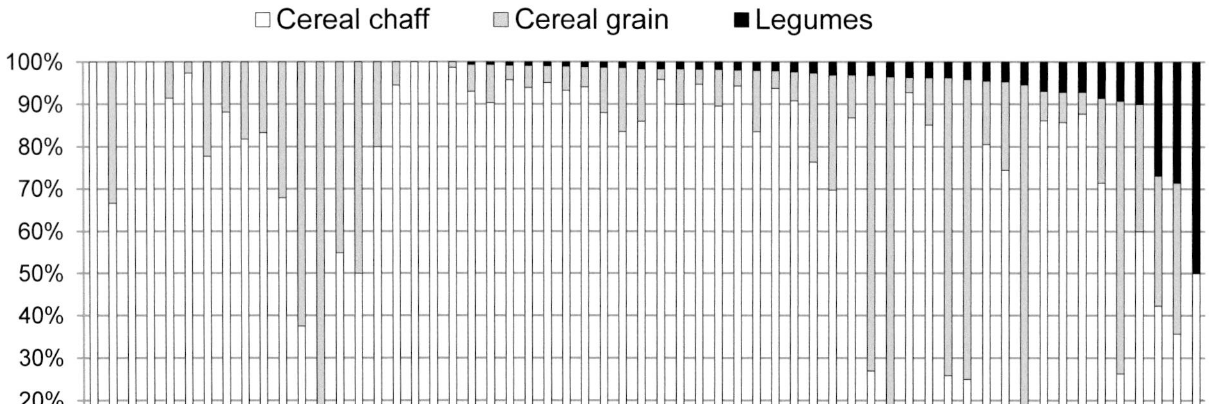

Figure 4. Comparison of the proportions of cereals (grain and chaff) and legumes within the crop assemblage from Pločnik.

cherry (*Cornus mas* L.) appears most prominent. Also noteworthy are finds of Chinese lantern seeds (*Physalis alkekengi* L.), in particular a 'cache' of *c.* 30 seeds in a burnt deposit (Sample 51). Seeds that are most likely to be wild blackberry (*Rubus ideaus/fruticosus* L., *Rubus* sp.) are also relatively frequent, as are seeds of wild strawberry (*Fragaria vesca* L.). Fragments of stone of wild plum types and of hazelnut shell (*Corylus avellana* L.) were occasionally encountered.

Wild/weed taxa

Many wild taxa were documented that could have represented weeds of arable fields. Most are present in very small numbers and in only one or few samples. The exceptions are brome types (*Bromus*), goosefoot (*Chenopodium*), black bindweed (*Fallopia convolvulus* (L.) Á. Löve) and black nightshade (*Solanum nigrum* L.), which are more frequent and abundant than other taxa. Other than *Bromus*, three of the more archaeologically visible wild taxa at Pločnik are annual weeds, which were recognised as probable indicators of the small-scale intensive crop cultivation that characterised Neolithic agriculture in central and southern Europe (Bogaard 2004; Bogaard and Halstead 2015). This form of cultivation may have been practiced at Pločnik, but further analysis of the weed record is needed to test this impression.

Some of the wild/weed plants have edible seeds or leaves (e.g. fat hen – *Chenopodium album* L., black bindweed, dock – *Rumex*, clovers – *Trifolium*) and/or medicinal qualities (e.g. common vervain – *Verbena officinalis* L., black nightshade) (Beeston *et al.* 2006; Behre 2008; Tucakov 1986). They would have grown in disturbed habitats such as arable and ruderal areas and, if of interest to the Neolithic inhabitants of Pločnik, would have been easily accessed.

Botanical content and archaeological context

Summarised data regarding the archaeological context of the samples are given in Table 2. The contexts located within the same occupation horizon are shown together and are grouped based on the feature or spit number. The overall abundance of plant remains and the botanical density are also shown, and absolute counts of the remains of some major plant categories indicated. These are used as a basis for the discussion of the differences and similarities between the contexts. Multiple samples deriving from the same context, or the same arbitrary layer within a context, were not amalgamated as they sometimes appear different in terms of botanical density and composition (e.g. samples from Feature 15).

Table 2. List of archaeological contexts at Pločnik from which the analysed samples derive; the contexts are grouped according to the settlement occupation horizons.

Sample number	Occupation horizon	Context type	Feature number	Spit number	Sample volume (litres)	*Total plant remains*	*Botanical density*	Hulled wheat grain	Hulled wheat glume bases	Free-threshing wheat	Barley	Lentil	Bitter vetch	Fruit/nut	Wild/weed
2	1	Arbitrary layer		5	5	*2*	*0*		2						
4		Arbitrary layer		7	5	*0*	*0*								
6		Arbitrary layer (NW corner)		8	5.5	*0*	*0*								
7		Arbitrary layer		9	5	*3*	*1*	1	2						
3		House rubble	1	6	5	*0*	*0*								
5		House rubble		8	5	*0*	*0*								
8		House floor substructure	2	9	5	*1*	*0*								1
9		House floor substructure		9	5	*1*	*0*		1						
10		House floor (Section 6)	4	9	5	*8*	*2*		2			2		1	3
11		House floor (Section 3)		9	4.5	*1*	*0*							1	
12		Stone-lined feature (hearth?)	3	9	5	*0*	*0*								
1		Burnt daub	6	4	7	*0*	*0*								

Table 2 continued. List of archaeological contexts at Pločnik from which the analysed samples derive; the contexts are grouped according to the settlement occupation horizons.

Sample number	Occupation horizon	Context type	Feature number	Spit number	Sample volume (litres)	*Total plant remains*	*Botanical density*	Hulled wheat grain	Hulled wheat glume bases	Free-threshing wheat	Barley	Lentil	Bitter vetch	Fruit/ nut	Wild/ weed
13	2	Arbitrary layer		10	8	*3*	*0*		2					1	
14		Arbitrary layer		11	7	*36*	*5*	1	32					1	
15		Arbitrary layer		12	14	*192*	*14*		180						
16		Rubbish pit	13	12	4	*77*	*19*	1	66		3	1		1	1
17	3	Arbitrary layer		14	8	*120*	*15*	1	109			1		3	2
19		Arbitrary layer		15	14	*152*	*11*		148						
26		Arbitrary layer		16	11	*94*	*9*	3	79		1	2			7
32		Arbitrary layer		17	10	*201*	*20*		180		1	1			7
18		Vessel contents		15	0.05	*0*	*0*								
Vessel 1		Vessel contents		16	0.4	*68*	*170*		64	4					
20		Concentration of pottery	16	15	0.5	*9*	*18*	1	7						
21		Oven floor	14	15	3	*18*	*6*		11	1				1	3
23		Oven floor		15	5	*86*	*17*	1	77			2		1	2
36		Oven floor		16	5	*12*	*2*	1	10						
24		Rake-out from oven Feature 14	15	15	9	*140*	*16*	1	122						1
25		Rake-out from oven Feature 14		15	11	*32*	*3*		25						3
39		Rake-out from oven Feature 14		15	8	*60*	*8*		50						7
40		Rake-out from oven Feature 14		15	8	*87*	*11*	1	80						2
30		Rake-out from oven Feature 14		16	9	*14*	*2*		9					1	2
31		Rake-out from oven Feature 14		16	8	*126*	*16*	1	114	1	1	1		2	3
28		Displaced oven (burnt daub)	23	16	6	*61*	*10*	1	52	2		1			1
37		Displaced oven (burnt daub)		17	9	*266*	*30*	4	247			1		3	5
33		Displaced oven (burnt daub)	26	18	5	*56*	*11*	3	46			1		1	1
29		Ash lens	24	16	9	*108*	*12*	1	94	3				4	2
22		House rubble (burnt and unburnt)	17	15	7	*59*	*8*	1	51	1					
27		Burnt daub	22	16	7	*160*	*23*	3	146			1		1	2
34		Burnt daub (house rubble?)	25	18	7	*74*	*11*	4	61			1		1	
54		Post hole	35	22	10	*362*	*36*	14	293	9		5	1	3	22
57		Post hole		22	7	*17*	*2*	1	5			2		2	1
58		Post hole		22	7	*10*	*1*	1	3						4
56		Post hole	36	22	10	*271*	*27*	16	219	5		6		2	11

Table 2 continued. List of archaeological contexts at Pločnik from which the analysed samples derive; the contexts are grouped according to the settlement occupation horizons.

Sample number	Occupation horizon	Context type	Feature number	Spit number	Sample volume (litres)	*Total plant remains*	*Botanical density*	Hulled wheat grain	Hulled wheat glume bases	Free-threshing wheat	Barley	Lentil	Bitter vetch	Fruit/ nut	Wild/ weed
41	4	Arbitrary layer		19	12	*478*	*40*	7	432		4	3		4	14
44		Arbitrary layer		20	11	*177*	*16*	11	141		1	1		1	20
55		Arbitrary layer		21	6	*58*	*10*	5	29			1		4	16
38		Unburned daub (fill of a rubbish pit?)	27	18	8	*46*	*6*	3	32		1	1			3
42		Pit fill	28	19	5	*74*	*15*	3	54		1	3		1	6
43		Pit fill		19	2	*30*	*15*	2	19		4			1	1
53		Pit fill		21	7	*159*	*23*	10	122	5		3		1	6
48		Ashy layer	29	21	11	*8*	*1*	3	3						
49		Ashy layer		21	3	*8*	*3*	3	1		2				
47		Burnt deposit/ashy layer	30	21	14	*210*	*15*	10	157	3	1	2		13	11
45		Burnt deposit/ashy layer	31	21	14	*27*	*2*	3	11			3			1
46		Burnt deposit/ashy layer		21	12	*13*	*1*	2	6					2	1
50		Burnt deposit/ashy layer	32	21	7	*256*	*37*	84	60	20		8			4
51		Burnt deposit/ashy layer		21	9	*178*	*20*	30	67	1	3			37	17
59		Burnt deposit/ashy layer	34	22	2	*26*	*13*	2	20					1	
Vessel 2		Vessel in the burnt deposit/ashy layer		22	0.2	*5*	*25*		5						
52		Post hole	37	22	11	*236*	*21*	8	100			12		52	44
Vessel 3	5	Vessel in the pit-house	38	23	0.15	*152*	*1013*	1	146	2				2	
60		Pit-house		23	5	*33*	*7*	4	5	3		1		4	2
61		Pit-house		25	10	*140*	*14*	12	92					4	4
62		Pit-house		25	8	*84*	*11*	16	19	1		7			8
63		Pit-house		25	4	*30*	*8*	12	6			1		6	
64		Pit-house		25	8	*41*	*5*	15	5			2		2	2
65		Pit-house		25	7	*30*	*4*	7				1			1
66		Oven in NE room of Feature 39	39	25	14	*3732*	*267*	80	3204	272				16	48

Abundance, density and composition of the remains

The largest number of plant remains were recovered from Horizon 5 (the earliest). In contrast, Horizon 1 (the latest) yielded a negligible amount of plant material. Although the average density of the assemblage is 31, in most samples it does not exceed 30 (90% or 61 samples), as illustrated in Figure 5. Only three (4%) of the samples produced more than 50 items per litre of soil. Among them, the sampled content of Vessel 3 (from Feature 38) had a particularly high density of over 1000; also of note is the sample from inside Vessel 1 (found within Spit 16)

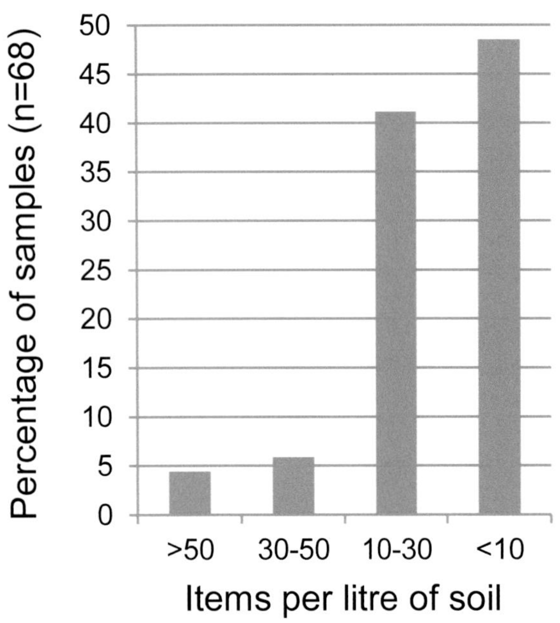

Figure 5. Botanical density values across the samples from Pločnik.

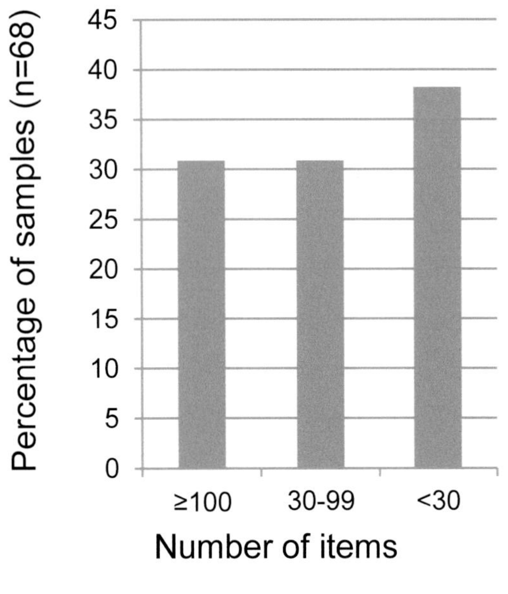

Figure 6. Botanical abundance across the samples from Pločnik.

with a density of 170. Almost 40% (26) of the samples contained fewer than 30 remains (Figure 6). Of these, seven samples did not yield any plant material. Around 30% of the samples were relatively rich, with a hundred or more remains, and another 30% contained between 30 and 100 items.

The overall picture is one of uniformity; different context types are very similar in the number of remains and density, and their taxonomic composition does not vary much either. This is most likely a consequence of mixing and/or re-deposition of the material from different activity areas within the individual occupation horizons, and possibly also between the different occupation levels, for instance, through the re-use of space (e.g. pits) for practices such as disposal of household waste.

The greatest quantities of plant material were found in fire-related features (ovens and rake-outs) and burnt deposits or ashy layers. The density of these contexts, however, is generally low perhaps due to slow, gradual deposition of moderate amounts of charred material, or maybe as a result of disturbance and mixing of the material with the surrounding matrix. With respect to the latter, it is maybe indicative that, in Horizons 4 and 3, arbitrary layers (presumably non-burnt) appear as rich as the ovens or burnt deposits found within the same spits (Spits 15, 16, and 17). The possible effect of mixing is visible in the generally uniform composition of all these contexts, and the vast predominance of hulled wheat glume bases within them. Even where the samples are slightly more taxonomically diverse, their general composition shows similarities. For example, in Horizon 4, the presence of material other than hulled wheat glume bases is somewhat more visible both in the arbitrary layers and in the burnt deposits (and includes hulled wheat grain, barley, lentil, wild plants – see Figure 7). It looks as though the charred material from burnt features and deposits was re-deposited over wider areas (as may also have been the case with entire structures, given the traces of displaced ovens, e.g. Features 23 and 26). Perhaps this happened in the course of house destruction. It is also possible that, in some instances, large quantities of hulled wheat bases originate from burnt construction material (i.e. daub tempered with vegetal matter), accumulations of which were discovered in several locations (e.g. Feature 22 in Horizon 3).

Oven Feature 14 appears to have been regularly cleaned, with crop processing residue thrown into the fire; the sweepings may have been deposited in the associated rake-out (Feature 15) and/or elsewhere – for example an ash lens (Feature 24). Another oven (Feature 39) has the fill discovered *in situ* in one of two rooms of the pit-house (Feature 38) in Horizon 5. This is of interest since the remains of (last) spent fuel and

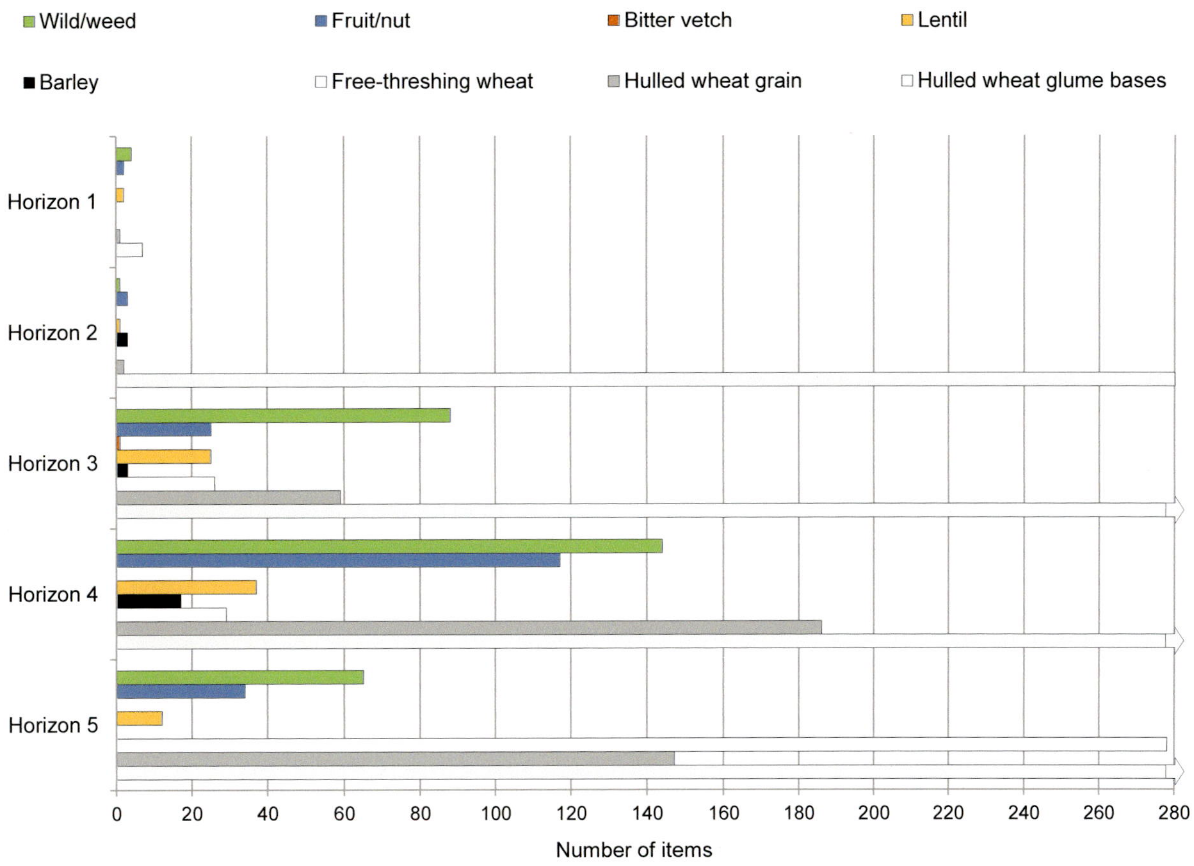

Figure 7. Absolute counts of the main plant categories within the occupation horizons at Pločnik.

perhaps also food spills seem to be preserved here. This context yielded the largest number of crop remains at the site; the proportions of crop types are shown in Figure 8. They include a significant number of glume bases from all three identified hulled wheat types, some einkorn and emmer grain, several free-threshing wheat grains, high quantity of tetraploid free-threshing wheat rachis segments, and a small amount of fruit/nut and wild/weed remains (see Appendix B_Ch34.1). The composition of the context most likely reflects the practice of cleaning (fine-sieving – cf. Hillman 1981, 1984b) of small amounts of hulled wheat grain on a daily or piecemeal basis. It further implies that the three hulled wheats may have been combined in processing and consumption, while they could have also been grown and stored together (e.g. Jones and Halstead 1995). Free-threshing wheat rachis segments may also have resulted from small-scale cleaning of grain prior to consumption. They could suggest that the crop was stored unthreshed (i.e. grain in ears) given that the early stages of processing (generally carried out outdoors) of free-threshing cereals release clean ('free') grain and remove most of the chaff (Hillman 1981, 1985). However, the ethnoarchaeological study conducted by Jones (1984, 1990) of present-day processing of free-threshing cereals demonstrated that by-products of fine-sieving of free-threshing wheat or barley still tend to contain low proportions of rachis nodes/internodes, in addition to weed seeds (Jones 1990: 93, Table 6). Thus, the rachis found in the oven (Feature 39) could represent discard from fine-sieving of clean-stored free-threshing wheat grain, combined in burning with hulled wheat fine-sieve by-product.

As noted, burnt deposits and ashy layers generally have low density and some contain more abundant plant remains (e.g. Features 30 and 32), while some yielded few. In Feature 32 (Samples 50 and 51), more hulled wheat grain is found than in any other context, similar in number to hulled wheat glume bases. A number of grains here could not be assigned to a particular wheat or cereal type (see Appendix B_Ch34.1). The deposit could potentially represent the remains of whole charred spikelets mixed with free-threshing grain and rachis, and the remains of edible fruits (*c.* 30 seeds of Chinese lantern, Cornelian cherry stone, hazelnut shell).

Non-burnt deposits such as pits and post-holes are similar in their composition and they all have low density suggesting gradual build-up of the material. Post-holes yielded relatively high quantities of the remains, potentially resulting from the secondary use of these features as rubbish pits. The contents of post-hole Feature 37 (Sample 52) are interesting as they include complete fruit and fruit stones of Cornelian

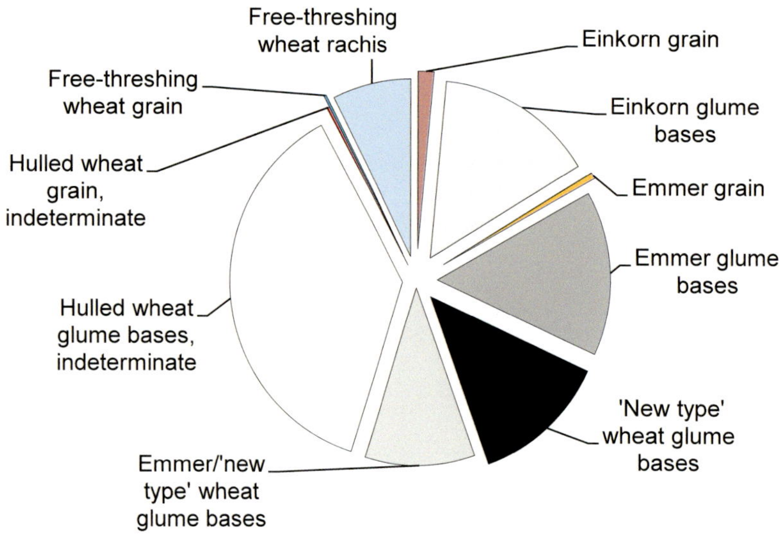

Figure 8. Relative proportions of the crop remains in oven Feature 38 (Sample 66) at Pločnik.

cherry and likely fragments of the fruit skin and stalk, probably reflecting a single episode of waste-dumping. Given the preservation of highly fragile parts of the fruit, the deposit seems to have remained undisturbed.

Of the four ceramic vessels whose contents were analysed, two yielded relatively high (Vessel 1) and very high (Vessel 3) amounts of hulled wheat glume bases per litre of soil; the vessels themselves do not show traces of burning. In Spit 16 (Horizon 3), from which Vessel 1 was retrieved, there were several burnt contexts, some of which were also rich in this type of remains (though they had lower density) – for instance, rake-out (Feature 15), burnt daub from oven (Feature 23) and ash lens (Feature 24). Since, as already observed, it appears that charred material was 'smeared' over the whole area, it is possible that some of it ended up in the vessel. Based on the high density of almost pure, hulled wheat chaff in the vessel, it is possible that all of it was deposited in a single episode. The high botanical density of the sample from Vessel 3 (from pit-house Feature 38) may also point to the deposition of the remains as a single event. This deposit is also almost entirely composed of hulled wheat chaff, in contrast to other samples from the pit-house which, in addition to hulled wheat chaff, contain some hulled wheat grain as well as small amounts of lentil, fruit/nut and wild/weed taxa, and so appear more 'mixed'. Tentatively, the residue in the vessel may have derived from a discrete episode of hulled wheat processing and burning of by-products.

Plant use through time

With the exception of Horizon 1, in which plant remains where almost absent (see Figure 7), Horizons 2–5 look similar in terms of the overall abundance of different taxa, and the range of the taxa. Table 3 shows presence/absence of crop types (and the two groups of wild plants) through the occupation levels. All three types of hulled wheats are present throughout the sequence. Their place in the plant economy of the site seems to have been constant, and the activities related to their processing (and consumption) also appear unchanged over time.

The absence of free-threshing wheat from Horizon 2 (and Horizon 1) is most likely apparent rather than real and is probably a product of the small number of analysed samples and the small quantity of remains from this level. The lack of barley in Horizon 5 could, however, be genuine. A considerable amount of plant material was recovered from this horizon, particularly from Feature 39, which seems to be the only analysed context where most of the remains were preserved at the 'original' charring location rather than being scattered over the occupation area as with fire-related features and burnt deposits in the later horizons (see above). It is thus possible that free-threshing wheat was the only free-threshing cereal used in the earliest phase of the settlement. In Horizon 4, barley remains are relatively common; in Horizon 3 (and later) they are almost invisible (see Table 2). The overall abundance of barley is too low to indicate whether its relative presence/absence through the horizons reflects variation in its use over time. Indeed, the role and significance of barley in the Neolithic crop husbandry of the region is ambiguous since, although present from the Early Neolithic onwards, it is consistently found in very low frequencies and numbers (Filipović 2014). In the wider region, however, large concentrations of hulled and naked barley were discovered, for example from the Early Neolithic in Bulgaria (Marinova 2007). The low presence of barley at Pločnik and elsewhere in Serbia could also be due to the limited archaeobotanical record from the area and/or a result of taphonomy.

The near absence of bitter vetch throughout the occupational sequence at Pločnik contrasts with the relatively frequent occurrence of this crop at other Late Neolithic sites in Serbia (Filipović 2014). This perhaps demonstrates that there were intra-regional differences in the crop choice and use between Late Neolithic Vinča culture sites.

Table 3. Presence/absence of the main plant categories across the settlement occupation horizons at Pločnik.

Occupation horizon	Number of samples	Hulled wheat grain	Hulled wheat glume bases	Free-threshing wheat	Barley	Lentil	Bitter vetch	Fruit/ nut	Wild/ weed
1	12	X	X			X		X	X
2	4	X	X		X	X		X	X
3	27	X	X	X	X	X	X	X	X
4	17	X	X	X	X	X		X	X
5	8	X	X	X		X		X	X

Conclusions

The archaeobotanical sampling and analysis at Pločnik provided first insights into the plant use and crop husbandry at the site. The charred plant assemblage is largely composed of crop remains, the vast majority of which comprise hulled wheat glume bases. The main agricultural activity was probably the cultivation of einkorn, emmer and 'new type' hulled wheat. Although no storage deposits were detected, hulled wheat glume bases, as evidence of crop cleaning, suggest that harvested wheat was stored in spikelets and processed on a regular basis. Free-threshing wheat (both tetra- and hexaploid) and barley are also present in the assemblage, but it is difficult to assess their status compared to hulled wheats. Free-threshing wheat appears more frequent and abundant in the samples than barley and, unlike barley, is present from the earliest occupation phase; it may have been more important. Several legume types were recorded, though probably only lentil and pea were actually grown, while grass pea and bitter vetch may have been accidental inclusion. Wild fruits contributed to the diet; their remains serve as evidence of exploitation of local wild resources.

By-products of food processing and preparation were thrown into fires. The accumulated debris in fire-related features and burnt deposits seems to have been spread and trampled in across the occupation/use area, perhaps in the process of building destruction and/or infilling. This practice resulted in highly indistinct botanical composition and density across different archaeological contexts. Nonetheless, in a few cases, the general characteristics of the context and slight differences in the botanical content enabled recognition of activities that could have produced the remains, for example, burning of hulled wheat spikelets (burnt deposit Feature 32); possible fruit consumption discard (post hole Feature 37); fine-sieving of hulled and free-threshing wheat and disposal of by-products (oven Feature 39).

The current results indicate that a similar spectrum of crop and wild plants was procured and consumed throughout the duration of the settlement at Pločnik. The only possible change relates to barley, which was not found in Horizon 5, appears relatively common in Horizon 4, and then drops to minor occurrences from Horizon 3 onwards. The overall quantity of barley is, however, too low to allow firm conclusions. A much larger dataset is needed, both from Pločnik and from elsewhere in the region, in order to fully explore possible variations in the use of crops and other plants through the Neolithic.

Appendix

Appendix avaliable online as part of Appendix B at https://doi.org/10.32028/9781803270425/AppendixB_Ch34

The bibliographic reference for this chapter is:

Filipović, D. 2021. Plant use at Pločnik, in Radivojević, M., Roberts, B. W., Marić, M., Kuzmanović Cvetković, J., and Rehren, Th. (eds) *The Rise of Metallurgy in Eurasia*: 408–421. Oxford: Archaeopress.

Chapter 35

Animal remains from Pločnik

Jelena Bulatović and David Orton

Introduction

The faunal assemblage from Pločnik is significant in that it is amongst the larger Vinča collections to have been studied, provides a rare glimpse into animal use in the earlier part of the Vinča period—with continuity into and through the Gradac Phase—and is one of very few faunal assemblages from any part of the Neolithic thus far reported from southern Serbia (although see Bulatović 2011). This report provides only an empirical overview of the fauna, but the results are discussed in wider context in J. Bulatović's PhD thesis (2018), as well as elsewhere in this volume (Chapter 51) alongside those from Belovode (see Chapter 21).

The study encompasses all animal remains excavated from Trench 24 during the 2012 and 2013 seasons. All animal bones found during the excavations were recovered and recorded. The vast majority were excavated by hand from defined cultural layers, with a smaller number coming from various types of features such as pits, house structures and post-holes. Some smaller animal bone fragments were also recovered and recorded from soil samples that were subsequently wet-sieved.

We present relative distributions of taxa found at the site and changes in their proportions through time, with the focus on the main domestic species (cattle, sheep, goat and pig), plus red deer as the most abundant wild species. We discuss their body-part representation, age-at-death data, pathological changes and butchery marks. Although animal remains were collected during previous excavations at Pločnik, those from Trench 24 are the first to be analysed and reported.

Methods

Specimens were recorded using the EUROFARM faunal recording profile, as also employed at Belovode and at Kočićevo in northern Bosnia (Orton 2014). An overview of methodology can be found in the chapter on Belovode zooarchaeology (Chapter 21) and will not be repeated here. Results are reported by stratigraphic horizons as set out in Chapter 10 of this volume.

Relative proportions of taxa

The total number of animal remains from the Pločnik assemblage studied is 15,840. Of these, 12,868 were recovered by hand, while the remainder (2972) were collected by wet-sieving. Of the hand-collected specimens, exactly 4000 (31.1%) were identified to species or at least to the genus level, and thus were used for quantification purposes.

The relative frequency of taxa based on NISP (Number of Identified Specimens) and DZ (Diagnostic Zones) is given in Table 1 for the main hand-collected sample. Results from the wet-sieved samples are shown in Table 2. Domestic cattle dominate the assemblage with sheep and goat (taken together) the second most common taxon, followed by domestic pig, red deer, and wild pig.

Relative frequencies of taxa show some trends over time (Figures 1 and 2). The contribution of domestic cattle increases overall between Horizon 5 and Horizon 1 with a concomitant overall decrease in caprines and pigs. None of these trends are entirely continuous, however, with pig frequency peaking in Horizon 3 and caprines in Horizon 2.

The contribution of hunted species increases slightly during the first part of the occupation, peaking at 13.7% by NISP in Horizon 3, before dropping markedly to below 5% in the later Gradac horizons. Taxonomic richness also peaks in Horizons 4 and 3, which feature remains of fur-bearing animals such as fox, bear, and various mustelids as well as five bird remains in Horizon 4. Lower taxonomic richness in Horizon 5 may simply reflect a much smaller overall sample size, but the same cannot be said for the later Gradac phases. Other wild animals present in the assemblage include aurochs, roe deer, wolf, beaver and hare.

Given the apparent hiatus in occupation—at least in this part of the site—between Horizon 3 and Horizon 2 (Chapter 10, this volume), it is notable that the importance and breadth of hunting shifts quite dramatically at this point (Figure 1, Table 1). On the other hand, the biggest shift in relative proportions of the two most frequent taxa—cattle and caprines—occurs between Horizons 2 and 1 (Figure 2, Table 2).

Table 1. Taxonomic distribution of hand-collected animal remains at Pločnik by horizon, as NISP and as Diagnostic Zones (DZ; see in Chapter 21 for explanation). *for deer, numbers in brackets represent the number of antler specimens while numbers outside the brackets exclude antlers, as do row/column totals.

		Horizon 5 (Vinča A2-B1)				Horizon 4 (Vinča B2)				Horizon 3 (E. Vinča C / Gradac I)				Horizon 2 (Vinča D2 / Gradac II)				Horizon 1 (L. Vinča D2/ Gradac III)				TOTAL			
Common name	Latin name	NISP	%	DZ	%	NISP	%	DZ	%	NISP	%	DZ	%	NISP	%	DZ	%	NISP	%	DZ	%	NISP	%	DZ	%
Domestic cattle	*Bos taurus*	146	50.3	48	39.7	673	58.3	217	52.8	337	52.1	102.5	44.7	492	59.0	116.5	44.0	838	78.0	226.5	73.7	2486	62.2	710.5	53.3
Aurochs	*Bos primigenius*	1	0.3	1	0.8	2	0.2	2	0.5	2	0.3	1	0.4	4	0.5			1	0.1			10	0.3	4	0.3
Cattle (indet.)	*Bos* sp.	2	0.7	2	1.7	13	1.1	6	1.5	2	0.3			1	0.1	2	0.8	8	0.7			26	0.7	10	0.7
Domestic pig	*Sus domesticus*	31	10.7	13	10.7	114	9.9	46.5	11.3	90	13.9	33.5	14.6	62	7.4	15.5	5.9	55	5.1	21	6.8	352	8.8	129.5	9.7
Wild pig	*Sus scrofa*	1	0.3			50	4.3	20.5	5.0	31	4.8	15	6.5	11	1.3	1.5	0.6	14	1.3	5.5	1.8	107	2.7	42.5	3.2
Pig (indet.)	*Sus* sp.	8	2.8	1	0.8	19	1.6	6	1.5	16	2.5	4	1.7	11	1.3	3	1.1	4	0.4	2	0.7	58	1.5	16	1.2
Sheep	*Ovis aries*	20	6.9	20	16.5	29	2.5	27	6.6	13	2.0	12	5.2	41	4.9	43	16.2	15	1.4	13	4.2	118	3.0	115	8.6
Goat	*Capra hircus*	5	1.7	4	3.3	16	1.4	13	3.2	5	0.8	5	2.2	5	0.6	5	1.9	5	0.5	3	1.0	36	0.9	30	2.2
Sheep or goat	*Ovis/Capra*	44	15.2	11	9.1	154	13.3	38	9.2	79	12.2	24	10.5	165	19.8	60.5	22.8	93	8.7	22.5	7.3	535	13.4	156	11.7
Dog	*Canis familiaris*	6	2.1	7	5.8	13	1.1	8.2	2.0	19	2.9	11.2	4.9	17	2.0	9	3.4	6	0.6	3	1.0	61	1.5	38.4	2.9
Red deer*	*Cervus elaphus*	17	5.9	8	6.6	51(5)	4.4	17	4.1	35(7)	5.4	9	3.9	20(5)	2.4	6.5	2.5	30(8)	2.8	7.5	2.4	153(25)	3.8	48	3.6
Roe deer*	*Capreolus capreolus*	6	2.1	3	2.5	12	1.0	7	1.7	6(1)	0.9	3	1.3	3	0.4	1	0.4	2	0.2	2	0.7	29(1)	0.7	16	1.2
Brown bear	*Ursus arctos*					1	0.1	0.2		2	0.3	0.2	0.1	1	0.1	0.4	0.2	1	0.1	1	0.3	5	0.1	1.8	0.1
Wolf	*Canis lupus*									1	0.2			1	0.1	1	0.4	1	0.1			3	0.1	1	0.1
Fox	*Vulpes vulpes*									2	0.3	3	1.3									2	0.1	3	0.2
Badger	*Meles meles*					1	0.1															1			
Beaver	*Castor fiber*									1	0.2	1	0.4									1		1	0.1
Wild cat	*Felis silvestris*									1	0.2	1	0.4									1		1	0.1
Otter	*Lutra lutra*					1	0.1	1	0.2													1		1	0.1
Marten	*Martes* sp.	1	0.3	2	1.7					5	0.8	4	1.7									6	0.2	6	0.4
Hare	*Lepus europaeus*	2	0.7	1	0.8	6	0.5	1.8	0.4									1	0.1	0.4	0.1	9	0.2	3.2	0.2
	Total identified	**290**	**100**	**121**	**100**	**1155**	**100**	**411.2**	**100**	**647**	**100**	**229.4**	**100**	**834**	**100**	**264.9**	**100**	**1074**	**100**	**307.4**	**100**	**4000**	**100**	**1333.9**	**100**

Table 1 continued. Taxonomic distribution of hand-collected animal remains at Pločnik by horizon, as NISP and as Diagnostic Zones (DZ; see in Chapter 21 for explanation). *for deer, numbers in brackets represent the number of antler specimens while numbers outside the brackets exclude antlers, as do row/column totals.

		Horizon 5 (Vinča A2-B1)				Horizon 4 (Vinča B2)				Horizon 3 (E. Vinča C / Gradac I)				Horizon 2 (Vinča D2 / Gradac II)				Horizon 1 (L. Vinča D2/ Gradac III)				TOTAL			
Common name	**Latin name**	**NISP**	**%**	**DZ**	**%**	**NISP**	**%**	**DZ**	**%**	**NISP**	**%**	**DZ**	**%**	**NISP**	**%**	**DZ**	**%**	**NISP**	**%**	**DZ**	**%**	**NISP**	**%**	**DZ**	**%**
Cattle or red deer		19				49				38				31				72				209			
Sheep/ goat/roe deer		16				1				2								2				21			
Carnivore (indet.)						1												1				2			
Large mammal		197				1898				1169				1957				1787				7008			
Medium mammal		123				355				298				327				117				1220			
Small mammal		5																				5			
Mammal		82				18				29				42				227				398			
Bird	Aves					5																5			
Total unidentified		**442**				**2327**				**1536**				**2357**				**2206**				**8868**			
Total		**732**		**121**		**3482**		**411.2**		**2183**		**229.4**		**3191**		**264.9**		**3280**		**307.4**		**12868**			

Table 2. Taxonomic distribution of wet-sieved animal remains at Pločnik by horizon, as NISP.

Latin name	Horizon 5	Horizon 4	Horizon 3	Horizon 2	Horizon 1	Total
Bos taurus				2		**2**
Sus domesticus	1	2	1			**4**
Sus scrofa			1			**1**
Ovis aries				2		**2**
Ovis/Capra		4	6	6	2	**18**
Cervus elaphus				1		**1**
Large mammal	2	55	62	32	30	**181**
Medium mammal	80	79	108	103	16	**386**
Small mammal	9	22	39	4	1	**75**
Mammal	42	538	976	479	263	**2298**
Pisces		3	1			**4**
Total	**134**	**703**	**1194**	**629**	**312**	**2972**
No. of samples	3	11	16	3	3	3
Avg. frags/sample	44.7	63.9	74.6	209.7	104.0	990.7

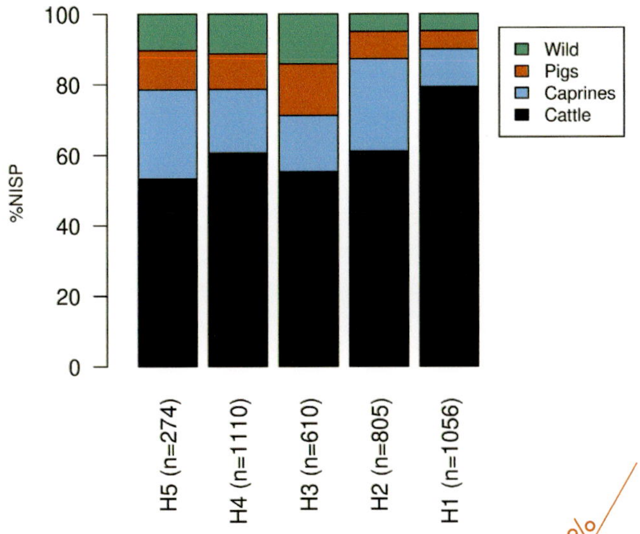

Figure 1. Representation of main domestic taxa and of wild mammals at Pločnik by horizon, in terms of %NISP.

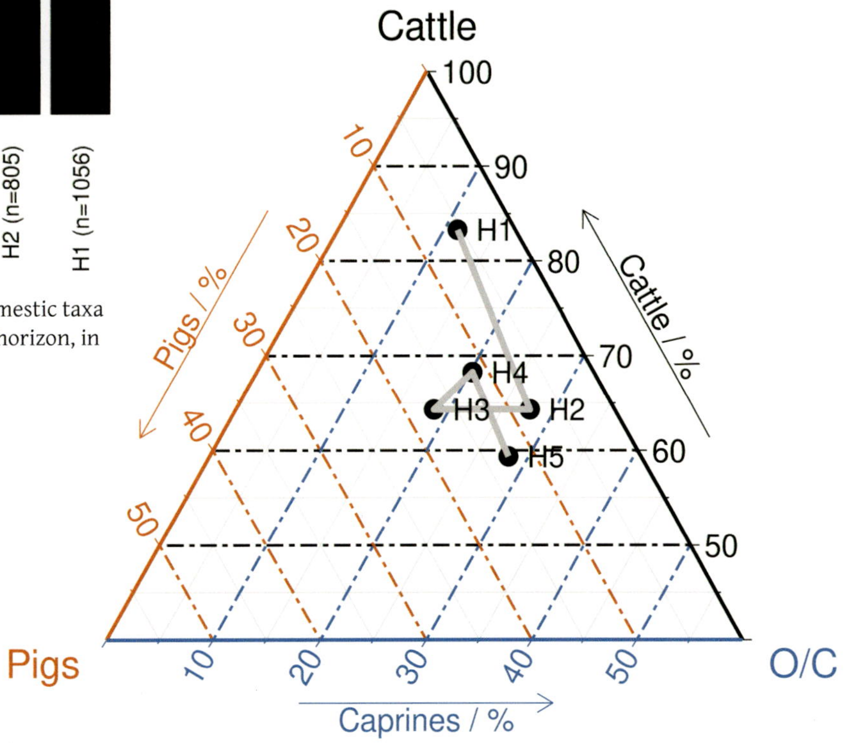

Figure 2. Ternary plot showing contributions of main domestic taxa to NISP at Pločnik, by horizon.

Taxa by context type

Studies of animal bones from some other Vinča sites have shown relationships between taxonomic frequencies and feature types (Orton 2008; Russell 1993), raising the possibility that trends observed in the Trench 24 assemblage might be linked to shifts in the representation of features within the trench over time. To assess this, a breakdown of taxa by horizon and context type was undertaken and is presented in Table 3 and Figure 3. Bones from the cultural layer clearly dominate in all horizons apart from the earliest, which mostly consists of pits (Figure 3a). The proportion of bones from pits decreases sharply over time, with none at all in the final horizon, while house remains only contribute bones in Horizons 3 and 1. Unfortunately, only Horizon 4 has sufficient bones from both pits and the cultural layer to allow a direct comparison Figure 3b), and even this is based on a small sample (n=77) from the pits. In this case the pits feature fewer cattle and more pigs than the cultural layer - a pattern that also holds for all horizons combined (Table 3). Where bones are available from house remains, these consistently have more pigs and wild taxa and fewer cattle and sheep, but the differences between Horizons 3 and 1 are clearer than those within them.

Despite these contextual differences, the overwhelming dominance of bones in all cultural layers apart from Horizon 5 limits any impact from changing representation of context types within Trench 24. Indeed, the key changes between the earlier and later horizons—increased cattle, reduced wild fauna—are apparent from the cultural layer assemblages alone.

Domestic cattle (*Bos taurus*)

Domestic cattle is the most abundant species in each horizon, representing more than half of the overall assemblage both in terms of NISP and DZ (Table 1). The presence of one complete metacarpal bone allowed the estimation of withers height (using the factor of Motolsci 1970), which was 115.6 cm. All body parts of cattle were represented, although certain elements were noticeably more abundant than others (Figure 4a). Interestingly, this varies between phases: while the earlier horizons (5–3) show a relatively even body-part representation by corrected DZ, apart from an over-representation of upper forelimbs, the post-hiatus levels (2–1) feature markedly fewer cranial bones and an abundance of phalanges. This pattern is hard to explain in terms of differential preservation, recovery, or overall site function. Slaughter and primary butchery further from the site might explain the paucity of cranial specimens, but in this case selection for transport would be expected also to reduce the number of specimens from the feet - the opposite of what is observed. The pattern is thus more likely to relate

Table 3. Representation of main domestic taxa and of wild mammals by horizon and context type, as NISP.

Horizon	Context	n	Cattle		Pigs		Caprines		Wild	
			NISP	%	NISP	%	NISP	%	NISP	%
1	Layer	**891**	716	80.4	41	4.6	93	10.4	41	4.6
	Pits									
	Houses	**154**	112	72.7	13	8.4	20	13.0	9	5.8
2	Layer	**787**	480	61.0	62	7.9	205	26.0	40	5.1
	Pits	**14**	8	57.1		0.0	6	42.9		
	Houses									
3	Layer	**479**	268	55.9	65	13.6	84	17.5	62	12.9
	Pits	**21**	15	71.4	2	9.5	3	14.3	1	4.8
	Houses	**80**	39	48.8	14	17.5	8	10.0	19	23.8
4	Layer	**1033**	635	61.5	97	9.4	187	18.1	114	11.0
	Pits	**77**	38	49.4	17	22.1	12	15.6	10	13.0
	Houses									
5	Layer	**17**	11	64.7	2	11.8	3	17.6	1	5.9
	Pits	**257**	135	52.5	29	11.3	66	25.7	27	10.5
	Houses									
Total	Layer	**3207**	2110	323.5	267	47.2	572	89.8	258	39.5
	Pits	**369**	196	230.5	48	42.9	87	98.4	38	28.3
	Houses	**234**	151	121.5	27	25.9	28	23.0	28	29.6

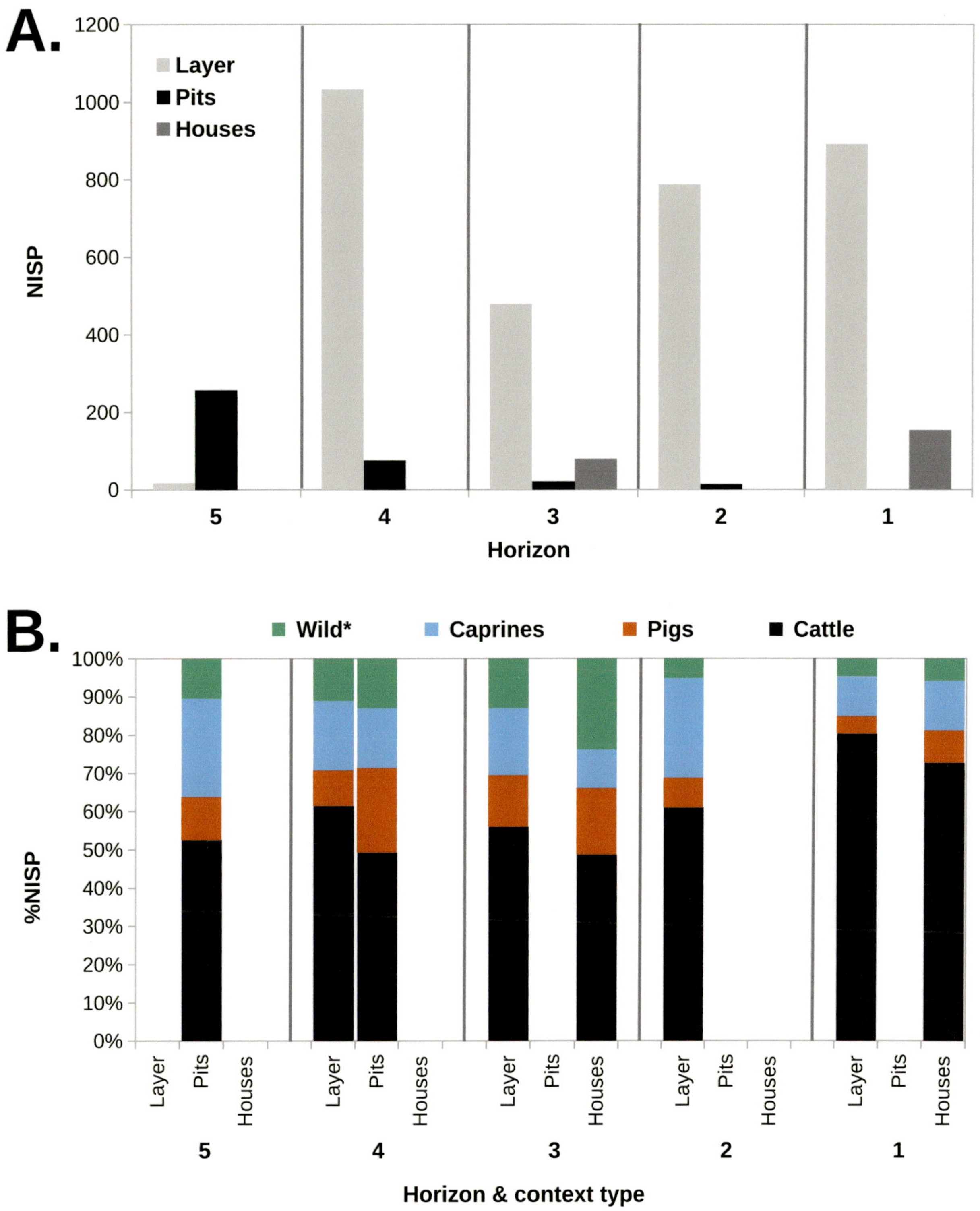

Figure 3. (A) total NISP by main context types in each horizon at Pločnik; (B) representation of main domestic taxa and of wild mammals by horizon and context type (%NISP), where total NISP > 50.

to changes in the specific deposit types represented within the trench.

Figure 5a shows cattle survivorship curves based on dental eruption/wear and on epiphyseal fusion, plotted on the same scale according to the standard suggested ages. The dental data (see Table 4a) for Horizons 5–3 show relatively low juvenile mortality, with around 70% of individuals surviving their first year and *c.* 55% reaching three years, followed by a slightly sharper drop to *c.* 30% at four years. In the later horizons (2–1) juvenile mortality appears to decrease still further, with remarkably continuous mortality throughout the (sub)adult years. Given the small samples, however, the difference between phases should be treated with some caution. These data, and those from Belovode, are placed in regional context in Chapter 51 of this volume.

Samples for epiphyseal fusion were also large enough to be plotted separately for the earlier and later horizons. Due to the nature of the cattle fusion groups, these data only provide information on survivorship to between *c.* 18 and *c.* 40 months. Both fusion curves appear to show rather higher survivorship within this age range than do their respective dental counterparts. While the fusion curve for Horizons 2–1 is almost

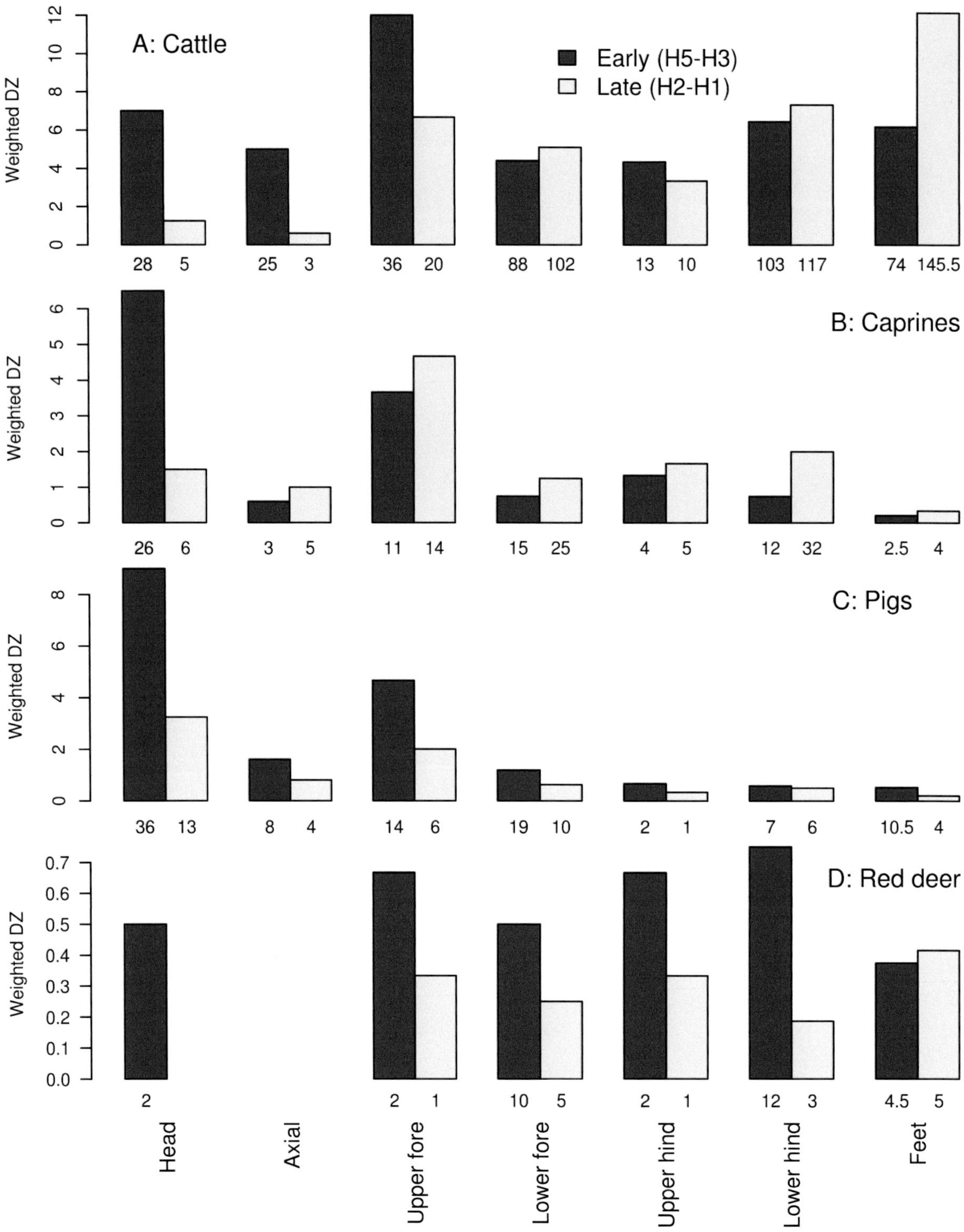

Figure 4. Body part representation (by percentage weighted Diagnostic Zones) for major taxa at Pločnik, by broad period. See text in Chapter 21 of this volume for explanation of DZ system. Weighting involved dividing the number of observed DZ in each body region by the number of DZ from that region in a complete carcass. Numbers beneath bars indicate total DZ before weighting.

perfectly parallel to the corresponding dental curve, that for Horizons 5–3 features a marked uptick in percentage fusion from 67% in the mid-fusing group to 84% amongst late-fusing bones. The fact that an *increase* in survivorship is theoretically impossible serves to emphasise that fusion data do not constitute true survivorship curves, but rather a series of separate survival estimates for discrete samples. Nonetheless, this result is hard to explain: the sample sizes for late-fusing elements are the smallest but, at 45, that for the early horizons is still very healthy.

The overall discrepancy between the dental- and fusion-based curves may be taphonomic. Although differential preservation favours more mature specimens in both cases, this effect is typically more pronounced for unfused postcranial elements than for relatively robust juvenile mandibles. Nonetheless, both sets of results suggest surprisingly little mortality in the first three years of life.

There were 11 cattle pelvises that could be sexed; based on their morphology, nine are assigned to female and two to male.

Butchering marks were found in 4.5% of cattle specimens. Cut and chop marks were most frequently found on the metapodial bones, navicular-cuboids and astragali, followed by humeri, mandibles and pelvises.

Pathological changes were observed in 39 cattle specimens. The most frequent were articular depressions observed in the phalanges (n=26), mostly on the first phalanx (n=16), but also on the carpal 2+3, navicular-cuboid, patella and scapula. These depressions are interpreted as a result of osteochondrosis, hereditary and/or environmental factors (Thomas and Johannsen 2011). Other pathological changes include bone proliferations, possible bone tissue necrosis, malformation, and irregular tooth wear. There is a slight increase in the rate of pathology over time, but this is not statistically significant (χ^2 (df = 1) = 0.918, p = 0.338 (Bulatović 2018: 152)).

Table 4. Distribution of cattle, caprine, and pig mandibles and loose lower teeth into relative age stages following Payne (1973), as adapted by Halstead (1985) and Hambleton (1999).

			Early (Horizons 5-3)			Late (Horizons 2-1)		
	Stage	Suggested age (months)	Raw count		Corrected	Raw count		Corrected
A. Cattle	A	0-1	0	1	**0.13**	0		**0**
	B	1-8	0	4	**2.56**	3	1	**3.42**
	C	8-18	4	2	**7.26**	0		**0.58**
	D	18-30	1		**3.3**	3		**3**
	E	30-36	1	3	**1.6**	2		**2**
	F	young adult	6		**7.16**	2		**2**
	G	adult	0	4	**2**	3	7	**6.5**
	H	old adult	0		**2**	0		**3.5**
	I	senile	6		**6**	5		**5**
			18	14	**32**	18	8	**26**
B. Caprines	A	0-2	1		**1**	0		**0**
	B	2-6	0	2	**0.8**	0		**0**
	C	6-12	8		**9.2**	1		**1**
	D	12-24	3		**3.33**	0		**0**
	E	24-36	2	1	**2.33**	0		**0**
	F	36-48	10		**10.34**	3		**3**
	G	48-72	9	2	**10**	3	3	**4.5**
	H	72-96	0		**1**	3		**4.5**
	I	96-120	2		**2**	1		**1**
			35	5	**40**	11	3	**14**
C. Pigs	A	0-2	0	1	**0.29**	1	1	**1.29**
	B	2-7	0		**0.71**	0	1	**1.11**
	C	7-14	13		**13**	1		**1.6**
	D	14-21	4		**4**	1		**1**
	E	21-27	4		**4**	2		**2**
	F	27-36	0		**0**	3		**3**
	G	adult	0		**0**	0		**0**
	H	old adult	0		**0**	0		**0**
	I	senile	0		**0**	0		**0**
			21	1	**22**	8	2	**10**

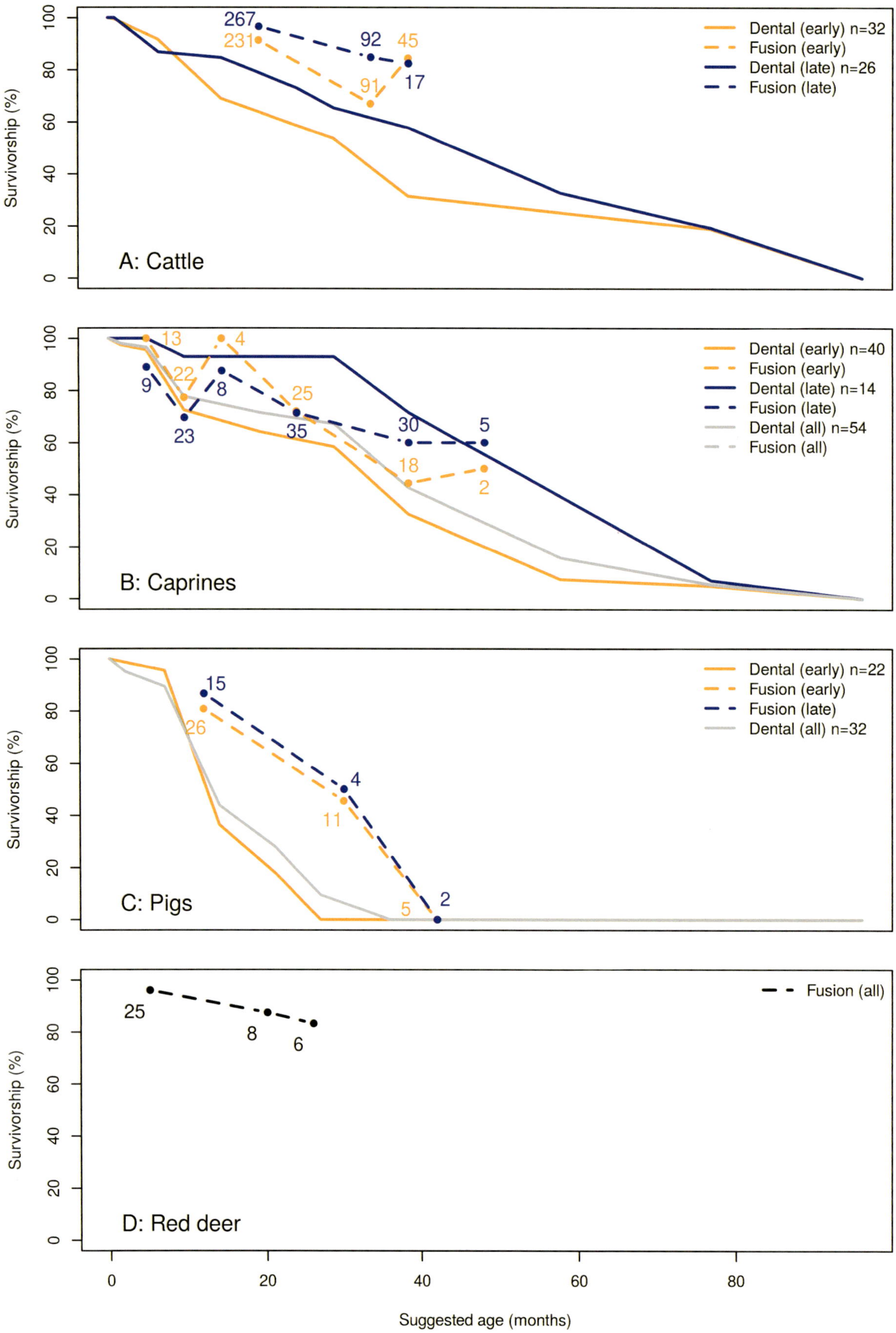

Figure 5. Age-at-death data for major domestic taxa at Pločnik. Each plot overlays survival estimates based on epiphyseal fusion (dotted lines) on survivorship curves constructed using Payne's (1973) mandibular age stage system (solid line). See text in Chapter 21 for sources of fusion groupings and suggested ages.

Sheep and goat (Ovis/Capra)

After cattle, caprines are the next most abundant taxon in each phase, at least by NISP. Positively identified sheep outnumber goats by 3:1 overall and between 2:1 and 8:1 (by NISP) in individual phases, although these figures should be treated with caution given that only 22% of the caprine remains were identified to species. The DZ data give very similar ratios. The average withers height—using Teichert's (1975) factor—was 53.3 cm in both earlier (5–3) and later (2–1) horizons, indicating that sheep size did not change through time. Overall, element representation is biased towards heads and upper forelimbs (Figure 4b). Phalanges, carpals, and the smaller tarsals are barely represented, presumably due to hand-collection. There is a marked reduction in frequency of cranial elements between the earlier and later horizons (as for cattle) but the relative proportions of postcranial elements change very little.

Kill-off data based on tooth eruption and wear is shown in Figure 5b and Table 4b. A total of 54 mandibles and loose lower teeth could be assigned to Payne (1973) stages, including 17 positively identified sheep and 10 goats. Separate survivorship curves were not constructed for the two species due to small sample sizes plus the risk of age-correlated identification bias (Mallia 2015). Curves are plotted for all horizons combined, as well as for both the early (5–3) and late (2–1) groups, though the latter should be treated with extreme caution given the sample size of just 14.

The dental curve for the early horizons is not dissimilar to that for cattle, with *c.* 70% of individuals surviving to one year, followed by little change over the second year, then an apparent renewed kill-off in the third bringing survivorship down to *c.* 30%. An apparent decrease in juvenile mortality to almost none during the later horizons can almost certainly be put down to the vagaries of small samples.

The fusion data (following Zeder 2006) support this assumption, since the curves for both phases track each other closely over the first few years, only diverging after fusion group D (18–30 months). The uptick seen in group C (*c.* 12–18 months) for both phases probably represents a recovery bias. This group consists only of proximal and intermediate phalanges, small bones that are often missed by hand collection, as appears to have been the case at Pločnik (see Figure 1b). Apart from explaining the small sample size for this group (a total of 12 across all phases) hand-collection may also have selected against the smaller, unfused specimens, resulting in the anomalous high percentage of fusion observed here. If this group is ignored, both fusion curves match quite closely with the overall and Horizon 5–3 dental curves up until 30 months, after which they part company; the curve for the early horizons continues to match the dental data while the late horizons curve does appear to show greater survivorship to *c.* 48 months. Fusion data for the final group, F (48–60 months) are extremely scarce and should probably be discounted.

It was possible to determine sex only for one sheep pelvis, which was assigned to male.

Evidence of butchery, in the form of cut and chop marks, are observed on three radii, two pelvises, a metatarsal, and an axis.

Pathological changes were observed on four specimens. Besides one goat horn core with a circular depression, most likely due to later calcium resorption under nutritional or lactation stress (Abarella 1995), all others are dental problems (i.e. abnormal tooth wear). Non-metric traits are also observed in two sheep/goat mandibles which have double *foramen mentale*.

Domestic pig (Sus domesticus)

Domestic pig is the third most frequent species representing 9–10% NISP and 10–11% DZ of the whole assemblage (depending on the true status of '*Sus* sp.' specimens). Withers height was estimated for one whole calcaneum using Teichert's (1969) factor, and it was 68.3 cm. Body part representation is broadly similar to that for sheep and goats, but with no clear change between phases (Figure 4c). The pattern of over-representation of cranial elements and upper forelimbs matches very closely that observed at Belovode (Chapter 21, this volume), and is likely to be primarily taphonomic.

The kill-off data based on tooth eruption and wear are presented in Figure 5c and Table 4c. A total of 32 mandibles and loose third molars were assigned to Payne's stages, scaled according to Hambleton's (1999) suggested ages. Due to small sample sizes, separate survivorship curves could not be created for pigs in each phase, but the combined curve indicates very high mortality between 6 and 12 months, with around 40% of individuals surviving their first year. Almost all appear to have been slaughtered by the end of stage F, at around three years.

The corresponding fusion curves are remarkably similar between the two phases, despite small sample sizes. Overall, the fusion results suggest rather later culling than the dental data, although, as for the cattle, this may well be an artefact of differential preservation. Both techniques indicate almost no survivorship beyond three years.

Based on the canine morphology, sex is assigned to 29 pig specimens, of which 26 are female and three male.

Butchering marks are evident on only five pig bones – two mandibles, a humerus, a radius and an ulna. There is one pathological condition in one pig mandible: a rotated third molar.

Red deer (Cervus elaphus)

Red deer is the most commonly represented wild species, even though its remains comprise *c.* 4% of specimens identified in the whole assemblage based on NISP or DZ. Given the small sample size, all parts of the red deer skeleton are present in remarkably even numbers—postcranial axial skeleton aside—compared to the domestic species (Figure 4d): the opposite of what one might expect due to selective carcass transport if deer were hunted at reasonable distances from the site.

Kill-off data based on long bone fusion (n=39) indicate that mature animals were primarily hunted, although unfused epiphyses of proximal radius, distal metatarsal bone, and femur indicate the presence of a small number of immature individuals (Figure 5d).

Evidence of butchery in the form of short and long cuts was observed on only two specimens: a basioccipital skull fragment and a radius. Pathological changes were also recorded in two specimens: an articular depression in a third phalanx and bone proliferation in a radius.

Conclusions

The results from Pločnik make a small but significant contribution to the zooarchaeology of the Vinča period, not least by extending coverage to the previously unstudied Toplica Valley, and to southern Serbia more generally. The overall trend of increasing cattle and decreasing wild fauna in the latest levels fits with trends seen at some—although not all—more northerly sites, such as Gomolava and Selevac (Legge 1990; Orton 2008, 2012). This is an interesting finding given that the only previously studied Vinča faunal assemblage from the Late Neolithic of southern Serbia, Vitkovo (Bulatović 2011), showed a very different pattern with a sheep-dominated domestic fauna much closer to that seen in Anzabegovo in Macedonia (Bökönyi 1976) than to the main cluster of Vinča assemblages to the north. The trends observed at Pločnik are not continuous, however, with the main decrease in importance of hunting coinciding with the apparent hiatus between Horizons 3 and 2, while the biggest shift towards cattle occurs between Horizons 2 and 1.

Age-at-death data for both cattle and caprines at Pločnik show an apparent generalist strategy, with some signs of increasing survivorship to maturity in later horizons – the opposite of what would be expected in the case of a shift towards more intensive milk use. Pigs appear to have been mostly killed young, as is typical, with no appreciable difference between phases.

Anatomical representation at the site is hard to interpret. There is a clear drop in relative frequency of cranial elements for both cattle and caprines between the earlier and later horizons, with no such trend for pigs. Taken at face value this might be interpreted in terms of slaughter of cattle and caprines further from the site followed by selective transport, but this is undermined by an *increase* in relative frequency of phalanges from both taxa. The true explanation may lie in the realm of *intra*-site use of space.

The bibliographic reference for this chapter is:

Bulatović, J. and Orton, D. 2021. Animal remains from Pločnik, in Radivojević, M., Roberts, B. W., Marić, M., Kuzmanović Cvetković, J., and Rehren, Th. (eds) *The Rise of Metallurgy in Eurasia*: 422–432. Oxford: Archaeopress.

Chapter 36

Pločnik: past, present and future

Benjamin W. Roberts and Miljana Radivojević

Introduction

The 2012 and 2013 excavations and subsequent post-excavation analyses by *The Rise of Metallurgy in Eurasia* project team at the site of Pločnik built upon nearly a century of discoveries and excavations led initially by the National Museum of Belgrade (Grbić 1929; Stalio 1960, 1962, 1964, 1973) and co-led latterly by the National Museum Belgrade and Museum of Toplica, Prokuplje (Kuzmanović Cvetković 1998; Šljivar 1996, 1999, 2006; Šljivar and Kuzmanović Cvetković 1997a, 1998a, 1998b; Šljivar *et al.* 2006). This later phase of work across 23 trenches (see Chapter 6) has, as at Belovode, yet to be fully published with the only detailed analysis being done on the metallurgical remains (Radivojević 2012; 2015; Chapter 6) together with a small programme of radiocarbon dating (Radivojević and Kuzmanović Cvetković 2014: 17–18). The evidence for copper metal production at Pločnik comprises only two droplets of smelting or melting activity (Radivojević and Rehren 2016: 220; see Chapter 6) and rectangular firing structures with copper minerals, metal artefacts and casting debris in association, as excavated in Trenches 20 and 21 (see Chapter 6, Figure 8) (Radivojević *et al.* 2013: 1033, Figure 2; Šljivar and Kuzmanović Cvetković 2009a: 61). However, the importance of selecting the site of Pločnik for *The Rise of Metallurgy in Eurasia* project lay primarily in the extensive evidence for metal artefacts from the moment of its discovery and the potential to explore questions around Vinča metal consumption practices. The metal artefacts known from the site ranged from copper beads to the famous discovery of massive copper metal implements which have been found from1928 onwards (Grbić 1929; Šljivar 1996, 1999; Šljivar *et al.* 2006; Šljivar and Kuzmanović Cvetković 1996–2009; Stalio 1964; see Chapter 6). Most recently, archaeometallurgical analyses revealed a tin-bronze foil from an undisturbed context at Pločnik dated to an occupation horizon of *c.* 4650 BC, making it the earliest known tin-bronze artefact anywhere in the world (Radivojević *et al.* 2013). As at Belovode, the absence of a detailed publication meant that further questions relating to evidence for early metal primary or secondary production and metal consumption could not be explored. The same methodological approach used at Belovode to investigate in detail the archaeological context of early metallurgy and metal at Pločnik, encompassing geophysical and aerial survey (see Chapters 24 and 39), systematic excavation and sampling, followed by extensive post-excavation analyses, was employed. As at Belovode, the entire excavation archive is made available online for current and future scholars (see Appendix A).

On metallurgy

As detailed in Chapter 25, Trench 24, was placed between two previous Trenches (20 and 21) which had produced rectangular firing structures and metallurgical finds including the earliest known tin bronze metal (Radivojević *et al.* 2013; Radivojević and Kuzmanović 2014). The evidence for copper metallurgy excavated in Trench 24 encompasses several stages in the *chaîne opératoire* of metal production including ore selection and melting and/or refining. The archaeological context of the metallurgical evidence is far more precisely documented, radiocarbon dated and contextualised than in earlier excavations at the site and serves to build upon the results and interpretations of earlier archaeometallurgical research.

The recovery of mainly green coloured malachite minerals and ores from throughout the stratigraphic sequence of Trench 24—albeit at a lower frequency than at Belovode—highlights the ubiquity of copper bearing minerals and ores throughout both of these Vinča culture settlement sites. The presence of green-and-black and green-yellow minerals in Horizon 1 and occasional occurrence of blue azurite provides further evidence for the careful selection of minerals and ores by their colours for specific uses at different times in the occupation of the site. Whilst neither copper smelting slag nor slagged sherds were excavated at Pločnik by *The Rise of Metallurgy in Eurasia* project, a copper metal bead was found in association with a kiln (F15) in Horizon 3 that provides the earliest secure radiocarbon dated evidence for metal at the site, contemporary with the start of the Gradac ceramic phase (see Chapter 26). In addition, a copper metal ring was found in association with a stone rectangular-shaped burnt structure (F3) in Horizon 1, thought to be a kiln or furnace, but this could only be partially excavated due to its location in the corner of the trench. Furthermore, the extensive concentration of ground stone tools found at Pločnik, which has been interpreted as a workshop/production area (F9) in Horizon 1, also encompassed two tools thought to have been used in the hammering and thinning of metal objects (see Chapters 31 and 45). The new evidence at Pločnik not only confirms the presence of secondary metal production activities at the site such as melting and/or refining and possibly also hammering/thinning but also provides secure radiocarbon dates for a copper metal object and subsequently a potential firing structure, confirming the contemporary appearance of the Gradac Phase and metallurgy at the site.

On communities

Understanding of the scale and duration of the community who lived at Pločnik has been significantly advanced by *The Rise of Metallurgy in Eurasia* project. However, as one area of the site is occupied by a village and the site has also been partially destroyed by the construction of a railway line and the erosion processes of the river Toplica, even the extensive geophysical and aerial surveys undertaken could only cover *c.* 60% (16 ha) of the site. Hence, whilst the geophysical survey revealed *c.* 300 anomalies identified as burnt houses in at least three major groups, this was only in the northern area of the settlement. The overall reconstruction of the Vinča settlement area is thought to be *c.* 26 ha and spans at least four to five major house groupings (see Chapter 24). There are potentially larger houses which demonstrate more variation in size than at Belovode. As at Belovode, earlier and much larger estimations of the Vinča settlement at Pločnik must be revised downwards (*contra* Šljivar and Kuzmanović Cvetković 1998a). The estimation of the population of the community living at Pločnik, based on house groupings (see Chapter 24) and mathematical modelling (see Chapter 38 and 40), suggests *c.* 600–1250 people during the later Vinča phases.

The excavation and radiocarbon dating of the entire stratigraphic sequence at Trench 24 identified 39 features across activity Horizons 1–5, which were radiocarbon dated and subsequently modelled (see Chapter 37), as well as detailed sub-divisions according to stratified ceramic typo-chronologies based on *c.* 14500 diagnostic ceramic fragments (Chapter 42). This provides a far more precise relative and absolute chronology for the occupation of the site spanning *c.* 5200–4400 BC than had previously been achieved. It not only spans the Vinča culture but enables a temporal and typological refinement of the tripartite Gradac ceramic phase (see Chapters 37 and 42) and reveals the late dating of the final phases of activity at Pločnik, overlapping with the Bubanj-Salçuta-Krivodol (BSK) communities of the Middle Chalcolithic in the Central and Eastern Balkans, potentially confirming continuities rather than collapses in the southern Vinča culture communities (cf. Radivojević and Grujić 2018).

The recovery and analysis of the charred plant assemblage provides the first archaeobotanical evidence of plant use and crop husbandry at the site (see Chapter 34). The excavations revealed the by-products of food preparation and plants following their burning, discard and probable spreading and re-distribution. The overall assemblage demonstrated that the community at Pločnik consistently cultivated primarily einkorn, emmer and 'new type' hulled wheat. Whilst lentils and peas may also have been grown, the continued presence of wild fruits highlights the importance of food sources beyond the domesticated crops. In certain excavated contexts, it was possible to identify the residues of specific activities such as the fine-sieving of hulled and free-threshing wheat and disposal of by-products (F39). The recovery and analysis of the animal bone assemblage represents the first archaeozoological evidence from the site of Pločnik (see Chapter 35). As with the archaeobotanical assemblage, it is comparable to evidence found at other—admittedly more distant—Vinča culture settlement sites. The excavations revealed an increasing shift towards cattle, a decreasing emphasis on the hunting of wild fauna, and no clear evidence for an age-at-death pattern that would indicate intensive milk production. The subsistence evidence revealed at Pločnik compares well with that at other Vinča culture sites and provides the foundations for further research in this formerly neglected area (see Chapters 50 and 51).

A large rectangular wattle and daub house structure (F1, 2, 4, 5, 6 and 10) in Horizon 1 provided a major focus for the excavations at Pločnik. Beyond radiocarbon dates that place it at the very end of the Vinča culture sequence (see Chapter 37), the structure revealed evidence of a construction technique of pedestals for load-bearing beams that is currently specific only to the site of Pločnik. In contrast, the preservation in the daub outline of a sub-structure of parallel rows of split timbers demonstrated that the floor construction is evidenced elsewhere (see Chapter 25). The assemblage from inside the house structure was relatively limited, comprising a small number of ceramic vessels, a polished stone axe and metal fragments/droplet, however evidence of a concentration of stone and ceramic debris found to the northwest of the structure, suggests that the latter objects were removed before and/or after the destruction and burning of the house (see Chapter 25).

The presence of the partially excavated stone rectangular-shaped burnt structure (F3), thought to be a kiln or furnace, and the concentration of large ground stone tools (F9), thought to be a workshop (Chapter 31), also in Horizon 1 in close proximity to the wattle and daub house structure can provide potential insights into the organisation of craft production. Whilst the metallurgical interpretation of the former (F3) is discussed above and in Chapters 26 and 41, the quantity and varied typology of the ground stone tools and the group of elongated and carefully polished stone axes in the latter (F9) is strongly suggestive of a workshop or production area. The connections spanning ground stone and metal production and use have previously been highlighted with white stone axes also only appearing during the Gradac Phase (see Chapter 45) and being the only non-metal artefacts found with the massive copper implements at Pločnik (Šljivar 1999; Šljivar *et al.* 2006: 261–265). However, the excavations undertaken by *The Rise of Metallurgy in Eurasia* project revealed not only a ground stone workshop or production area in close proximity to a potential metalworking area but also stone tools thought to be used for hammering and/or thinning metal objects, as highlighted above. This close spatial association of stone and metal production activities occurred in the open and outside of the house structure.

The production of ceramics is extensively evidenced at Pločnik with the repeated repairing and re-use of kilns (F11, F14, F15 and F32) in Horizons 2, 3 and 4 and potentially in Horizon 1 (F3), in certain cases potentially over a long period. The analysis of the pottery production techniques demonstrated that the craftspeople were not always able to control the firing atmospheres and whilst they were able to achieve *c.* 750°C, they rarely exceeded *c.* 850°C (see Chapters 29 and 43). The association of a copper bead with a kiln (F15) in Horizon 3 (see Chapter 26) highlights that whilst the pyrotechnological conditions of the Vinča ceramics might not have been easily transferable for copper smelting, they were certainly sufficient for copper melting, refining or manipulating, and that this may well have been performed by the same craftspeople in the same place. The production of chipped stone tools occurred across three different *chaîne opératoires*, in cream tabular flint, pebble flint and local river gravel sources. The thick blades were retouched to produce end-scrapers or drills, the fine blades and bladelets were used in composite tools and the flakes were made into scrapers or tranchets (see Chapters 33 and 47). The evidence for bone tool production is limited as the excavated area may well have been a place where bone objects, mainly pointed and burnishing tools for leathers, hides or plants, were used and subsequently broken and abandoned (see Chapters 32 and 46). The networks revealed by the different raw materials being exploited and artefacts being made at Pločnik are evidenced by the copper ore and copper metal objects (see Chapter 41), ground stone tools (see Chapter 45) and ceramics (see Chapters 42 and 43). However, the presence of a rare, large and elongated *Spondylus* bead not only contextualises the earlier stray find of over 300 *Spondylus* shell beads from the vicinity of the site but highlights the very distant connections involved, spanning the spatial extent of the Vinča culture and beyond (see Chapter 32 and 46).

Further work

The excavations at Pločnik in 2012 and 2013 by *The Rise of Metallurgy in Eurasia* project comprised only a single trench, initially measuring 5 x 5 m and subsequently extended to encompass the large rectangular wattle and daub burnt structure (F1, F2, F4, F5, F6 and F10) in Horizon 1. The project aimed to excavate and analyse a complete material, structural and environmental sequence at Pločnik that would include further metal artefacts and metallurgical remains in order to understand metal production and consumption in context. The 39 features spanned wattle and daub rectangular structures, kilns, finds concentrations, pits, and dwelling dugouts. These results enabled the project to largely achieve the original aims but also created new avenues of investigation for further work.

1. Whilst the project has made significant contribution to establishing the spatial scale of settlement at Pločnik, it is now clear that there are at least four major groupings of burnt houses whose chronologies have yet to be determined. Further targeted excavations and sampling would enable a far clearer sense of where and when increases and decreases and settlement activity and demographics occur. The western and southern borders of the settlement could be more precisely defined with further geophysical survey and excavation.

2. The complete sequence of activity at Pločnik, including pre- and post- Vinča culture activity requires further definition. Only the western area of Trench 24 could be excavated to the natural soil to expose the complete sequence of occupation at the site, revealing in the process a complex pit structure (F38), potentially comparable to late Starčevo and early Vinča period pits in the central Balkans (see Chapter 25). Given the broader debates around the Starčevo-Vinča transition and the subsequent Vinča- Bubanj-Salçuta-Krivodol (BSK) transition, the evidence for activities and their dating at Pločnik would make the site a priority for further investigation.

3. The organisation of subsistence activities is not well understood with the archaeobotanical and archaeozoological sampling and analyses representing a major development at the site as well as a significant development in the region. Further targeted excavations for additional samples and, in particular

taking advantage of the extensive stratigraphy exposed for over 300 m along the river Toplica, would enable a far more detailed picture of subsistence practices throughout the site.

4. The organisation of craft production at Pločnik—and in particular the interconnections of stone, ceramic and metal production spatially and temporally across the site—require far more detailed excavation and post-excavation analyses. Whilst metallurgical remains and metal artefact fragments are indeed associated with the partially excavated burnt rectangular stone structure (F3), a feature type also found in earlier largely unpublished excavations, the activities relating to the feature (type) remain incompletely defined.

5. There has been relatively little survey and fieldwork in the landscape surrounding the site of Pločnik, whether to explore the potential sources of different materials used at the site (stone, copper minerals and ores, graphite etc.), the management of the land for arable or pastoral agriculture or the presence of smaller and potentially contemporary Vinča sites.

The bibliographic reference for this chapter is:

Roberts, B. W. and Radivojević, M. 2021. Pločnik: past, present and future, in Radivojević, M., Roberts, B. W., Marić, M., Kuzmanović Cvetković, J., and Rehren, Th. (eds) *The Rise of Metallurgy in Eurasia*: 433–436. Oxford: Archaeopress.

Part 4
The Rise of Metallurgy in Eurasia: a view from the Balkans

Photo by Marko Djurica

Chapter 37

Relative and absolute chronology of Belovode and Pločnik

Miroslav Marić, Miljana Radivojević, Benjamin W. Roberts and David C. Orton

The chronology of the sites of Belovode and Pločnik has been discussed several times in the past two decades since excavations were renewed at each site (Arsenijević and Živković 1998; Šljivar 1996; Šljivar and Jacanović 1996a; Šljivar and Kuzmanović Cvetković 1997a), though not in detail and rarely integrating the relative and the absolute chronological sequences. Chronologies from both sites have been published (Whittle *et al.* 2016), drawing upon eight radiocarbon dates from the two trenches then existing for Belovode (Trenches 7 and 8) and nine radiocarbon dates from three trenches for Pločnik (Trenches 14, 15 and 16). These dates were allegedly focussing on the dating of metallurgical finds, however, the nature of those finds was not known at the time, so the best guess was to date materials from the spits, which only adds to the doubt on the chronological precision achieved.

The seriation and ceramic sequences of both sites resulting from this project are discussed in detail elsewhere in this volume (see Chapters 12, 13, 27 and 28) and in this chapter will be used only to illustrate relative chronology compared to other chronologies used for Vinča culture (see Chapter 4). The identical spit and context excavation methodology employed on both sites enables us to compare chronologically sensitive pottery forms to other relevant sites of the period. This chapter presents 29 new radiocarbon dates from throughout the excavated sequences: 17 from Trench 18 at Belovode and 12 from Trench 24 at Pločnik. The systematic excavations, relative ceramic sequences and the intensity of radiocarbon dating enables a far more precise modelling of dates for the identifiable activity horizons at both sites.

Stratigraphy

A detailed discussion of features and the stratigraphic situation in each excavated trench is provided in Chapters 10 and 25 of this volume; a short summary is given here to facilitate understanding of the chronology. Both sites were excavated using 10 cm spits, with definable features being recorded and excavated separately but within a framework of relative stratigraphy, providing secure units for selecting samples and dating the sites.

Trench 18 at Belovode comprised approximately 2.2 m of cultural deposits starting directly from the modern plough zone just 10 cm bellow the present-day surface. Based on stratigraphic observations of defined contexts, five activity horizons (labelled Horizons 1 to 5 from surface to bottom) were defined within the trench, with four possible sub-horizons (Horizons 1a and 1b, and 4a and 4b). The general thickness of anthropogenic deposits is similar to that of previous partially published trenches, e.g. Trench 2 (2.4 m) and Trench 5 (2.3 m), but the thickness of deposits can reach up to 4 m when pits are present at the bottom of a trench (e.g. Šljivar and Jacanović 1996a: 55). As no detailed analysis of the ceramic forms in their stratigraphic sequence was published by the previous excavator, with only bowls and goblets featuring in Arsenijević and Živković (1998), it is difficult to compare the finds from Trench 18 with other excavated trenches. We can, however, claim with certainty that the site of Belovode contains the whole sequence of the so-called Vinča 'culture'.

Trench 24 at Pločnik yielded somewhat thicker anthropogenic deposits than Belovode, ending in archaeologically sterile soil at about 3.7 m below the present surface in the excavated portion of the trench. It should be noted that approximately 80 cm before the occurrence of the yellow clayish subsoil at the bottom of the trench, an infill of a single feature (F38) was detected, which is most likely an early phase pit-dwelling dug into the natural soil and subsequently re-used as a refuse pit. Since the last feature of Trench 24 at Pločnik belongs to an early habitation phase, the thickness of the cultural deposits in the trenches at both sites can be broadly equated, especially if we acknowledge about 60 cm of archaeologically sterile soil between the modern surface and Feature 1.

The detailed study of the finds from the cultural deposits at both sites shows that they correspond with the established phases of the Vinča culture and its gradual transformation from the Neolithic towards the Chalcolithic, as known from other sites of the central Balkan region (see Garašanin and Garašanin 1979; Jovanović 1994; Stalio 1972). In our limited area of excavations, we were able to gather enough archaeological information to reconstruct the entire Late Neolithic occupation sequences at both Belovode and Pločnik.

Relative and absolute chronology of Vinča culture

The basis for the relative chronology of the Vinča culture is the rich and diverse evidence for the production and consumption of pottery. Ceramic vessel fragments are found in vast numbers as refuse in pits or as *in situ* assemblages within destroyed wattle and daub structures throughout the settlements of the period. The reference assemblage used for the creation of the Vinča culture ceramic chronology sequence is that of Belo Brdo in Vinča, excavated by Miloje Vasić for more than 20 years and published largely in the last of his four-volume 'Prehistoric Vinča' books (Vasić 1936c). However, it was not Vasić but rather the German archaeologist Friedrich Holste who first tried to develop a relative chronology of the Late Neolithic period in the central Balkans based on Vasić's finds from Belo Brdo (Holste 1939). Holste correctly suggested a four-part division of the Vinča culture, a framework that was further supported a decade later by Vladimir Milojčić (1949) and Milutin Garašanin (1951). The latter would become a crucial contributor to the Late Neolithic relative chronology of the central Balkan area, dedicating much of his fruitful scientific career to the matter. He divided the Vinča culture into two major periods, early and late (the so-called Tordoš and Pločnik Phases respectively) and defined a finer sub-division of both the early and late phases (Garašanin 1951). Over the following decades, he further refined the principal chronological divisions with the addition of the transitional Gradac Phase (Garašanin 1979), which marked the end of Neolithic Vinča culture and the beginning of copper processing in the region. Towards the end of his career, he made one more revision to the chronological system (Garašanin 1993) by further sub-dividing the early (Tordoš) phases of the culture.

Influenced by Garašanin's work, other authors established their own relative chronological systems for the period. The best-known examples include the work of Stojan Dimitrijević on the chronology of Late Neolithic Croatia (Dimitrijević 1968) and Berciu and Lazarovici for the Late Neolithic relative chronology of Romania (Berciu 1961; Lazarovici 1979, 1981). An all-encompassing chronology for the whole of the territory of Vinča culture was proposed by Chapman in his seminal work on the southeast European Late Neolithic (Chapman 1981) but was not widely used or accepted by authors in the region. Further schemes were made at the beginning of 1990s, but were more specific site cases, rather than all-encompassing pottery based relative chronology sequences (Bogdanović 1990; Vukmanović and Radojčić 1990). The work of Bora Jovanović on the southern, Pločnik variant of the Vinča culture led to a refinement of the pottery sequence and instigated the notion of the longer lasting Vinča sites in the south of Serbia, indicated by discernible differences in the pottery assemblages of the late phases (Jovanović 1994). Finally, with the broader application of absolute dating based on radiocarbon samples, a new wave of papers on the Vinča culture started to appear from the late 1980s (Breunig 1987; Schier 1996; Tasić *et al.* 2015; Tasić *et al.* 2016b; Whittle *et al.* 2016) further refining existing divisions and periodisation. No radical change has occurred, however, demonstrating that the old periodisation system based on relative chronology devised from pottery sequences remains fundamentally correct despite all the adjustments made over the decades.

Incorporating the relative chronology of two archaeological sites found over a hundred kilometres away from the type site was somewhat challenging. Regional variations (as established by Chapman 1981: 19–31) tend to increase with distance from the type assemblage, necessitating revision of the examined site assemblage, including the comparison and analysis of pottery types common to both. We therefore propose a relative chronological system (Figure 1) adapted to provide both a relative chronology based purely on pottery types and their relative depths (Garašanin 1979) and a second based on pottery seriation and absolute dates (Schier 1996), both based upon the type site of Belo Brdo. The proposed relative chronological framework is an attempt to draw upon the strengths of both research traditions, i.e. the long-established chronological scheme of Garašanin (1979) and the ^{14}C dates underpinning the chronology of Schier (1996) that lacks the analysis of the final four metres of the Belo Brdo pottery. In order to accommodate the slight differences in the phasing of the two systems, we introduce transition boundaries (denoted as blurred white regions in the Belovode/Pločnik column in Figure 1). These boundaries nullify the slight relative depth differences between the Garašanin and Schier phases and represent the short time difference between the appearance of certain types of pots in the reference assemblage based on pure pottery observation and the pottery seriation combined with absolute dating. A note of caution is required here. Although we accept Schier's sub-division of certain phases, we found it difficult to identify such minute differences in the limited assemblage excavated from the trenches of Belovode and Pločnik during just two seasons of excavations. The reference assemblage from Belo Brdo, however, consists of a significantly larger collection of pottery sherds that were collected over 11 excavation campaigns, covering a large area and spanning 20 years of field research.

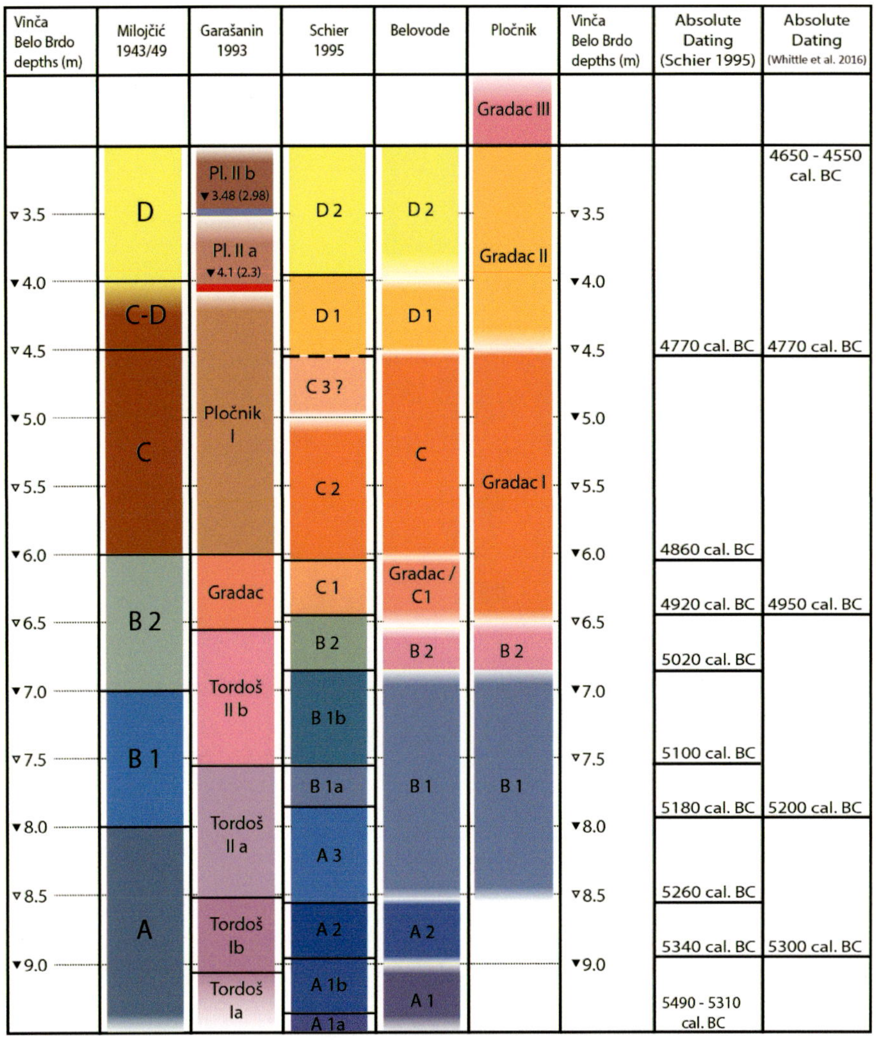

Figure 1. Selected chronological schemes for the Vinča culture based on the Vinča Belo Brdo pottery assemblage (adapted from Schier 1996, Figure 9).

Radiocarbon dates from Belovode and Pločnik

A total of 37 samples taken from trenches at both sites were collected for ^{14}C dating, including 22 from Belovode and 15 from Pločnik (Table 1). In order to minimise the residual effect of samples, a strategy was adopted of selecting short-lived samples such as grains and animal bone where possible, although this proved difficult to achieve in the field. Whenever possible, samples were also taken from secure contexts in order to minimise the risk of residual or intrusive measurements. The samples chosen from spits were associated with confirmed metallurgical finds in order to help date them by proxy. More detailed descriptions of the individual features from Table 1 can be found in Chapters 10 and 25 of this volume.

From the 37 samples chosen, four (MAMS 22072, 22080, 22085 and 22094) were withheld as reserves in case the original samples failed. The extracted collagen from animal bones was purified by ultrafiltration and freeze dried, then combusted in an Elemental Analyzer. The resulting carbon dioxide was converted catalytically to graphite. Sampled charcoal was pre-treated using an Acid-Base-Acid sequence, whilst the seeds were washed in HCl (hydrochloric acid) before analysis. Only two samples of animal bones (out of 19) failed to produce enough collagen for dating.

The dating results were initially calibrated using IntCal13 (Reimer *et al.* 2013) and SwissCal 1.0 by the Manheim Laboratory, whilst the ^{14}C ages were normalised to $\delta^{13}C$=-25 (Stuiver and Pollach 1977). The $\delta^{13}C$ value was obtained from the isotope determination in the AMS system with a typical uncertainty of 0.2%.

Discussion

The integration of relative and absolute chronologies was performed after the ^{14}C results were obtained from the Curt-Engelhorn-Centre of Archaeometry in Mannheim. As the archaeological excavations were conducted using a hybrid single context recording, the relative chronology of each excavated site is best illustrated by a diagram showing relationships between individual contexts found within the trenches (Figures 2 and 3). These were supplemented with absolute dates to illustrate the absolute age. The relative phasing based on pottery finds was subsequently compared to the absolute dating in order to establish phasing at both sites.

Belovode

At Belovode, five horizons were established during the excavations, marked 1–5 from the surface down. To obtain absolute chronological ranges for the horizons, several Bayesian models were created from the available dates.

A total of 17 radiocarbon results were available for the whole Belovode sequence (Table 1). This allowed a formal estimation of the site chronology using an explicit statistical method, combining both the radiocarbon dates and the relative stratigraphy recorded during the

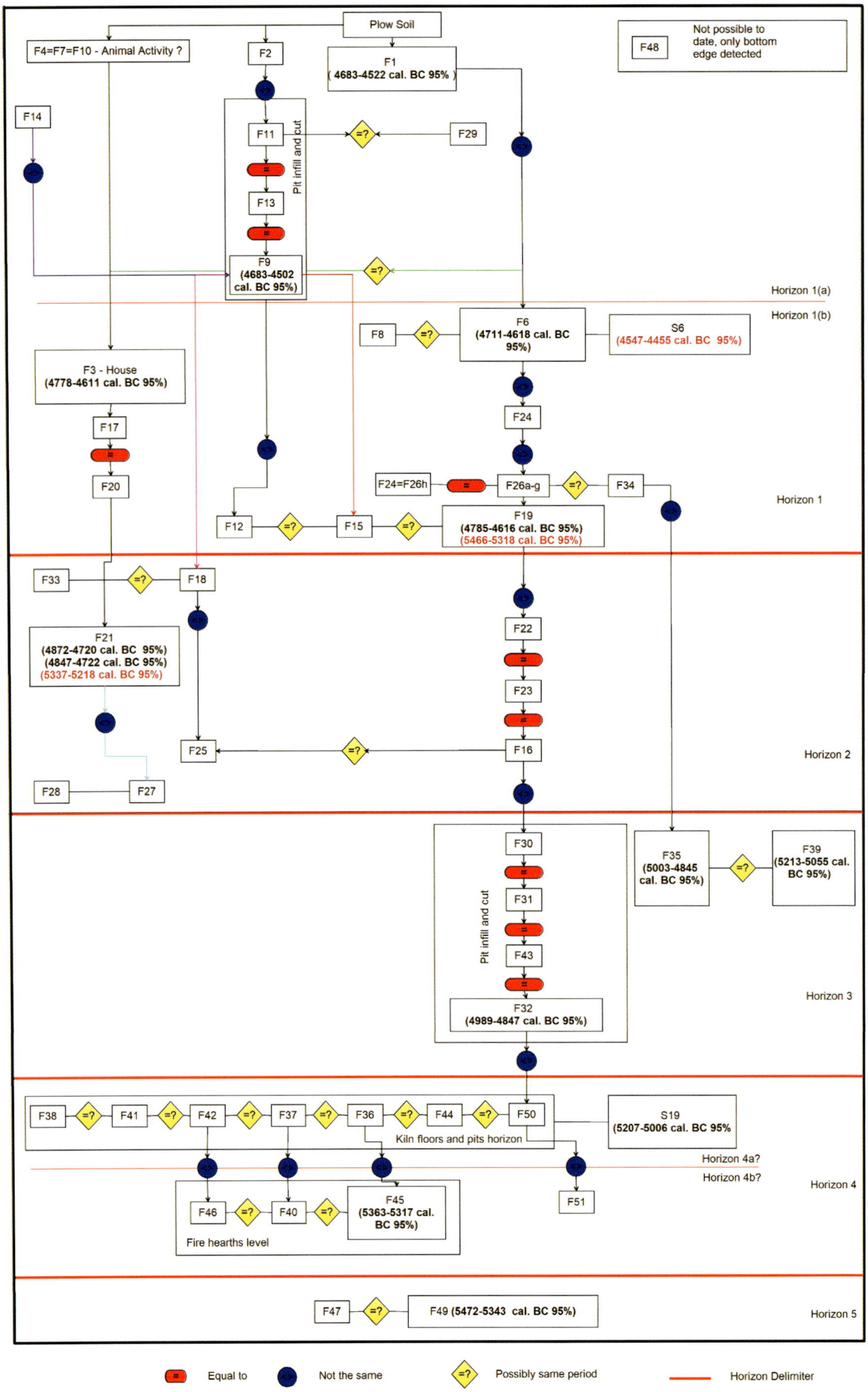

Figure 2. Schematic diagram of stratigraphic relations between features in Belovode Trench 18.

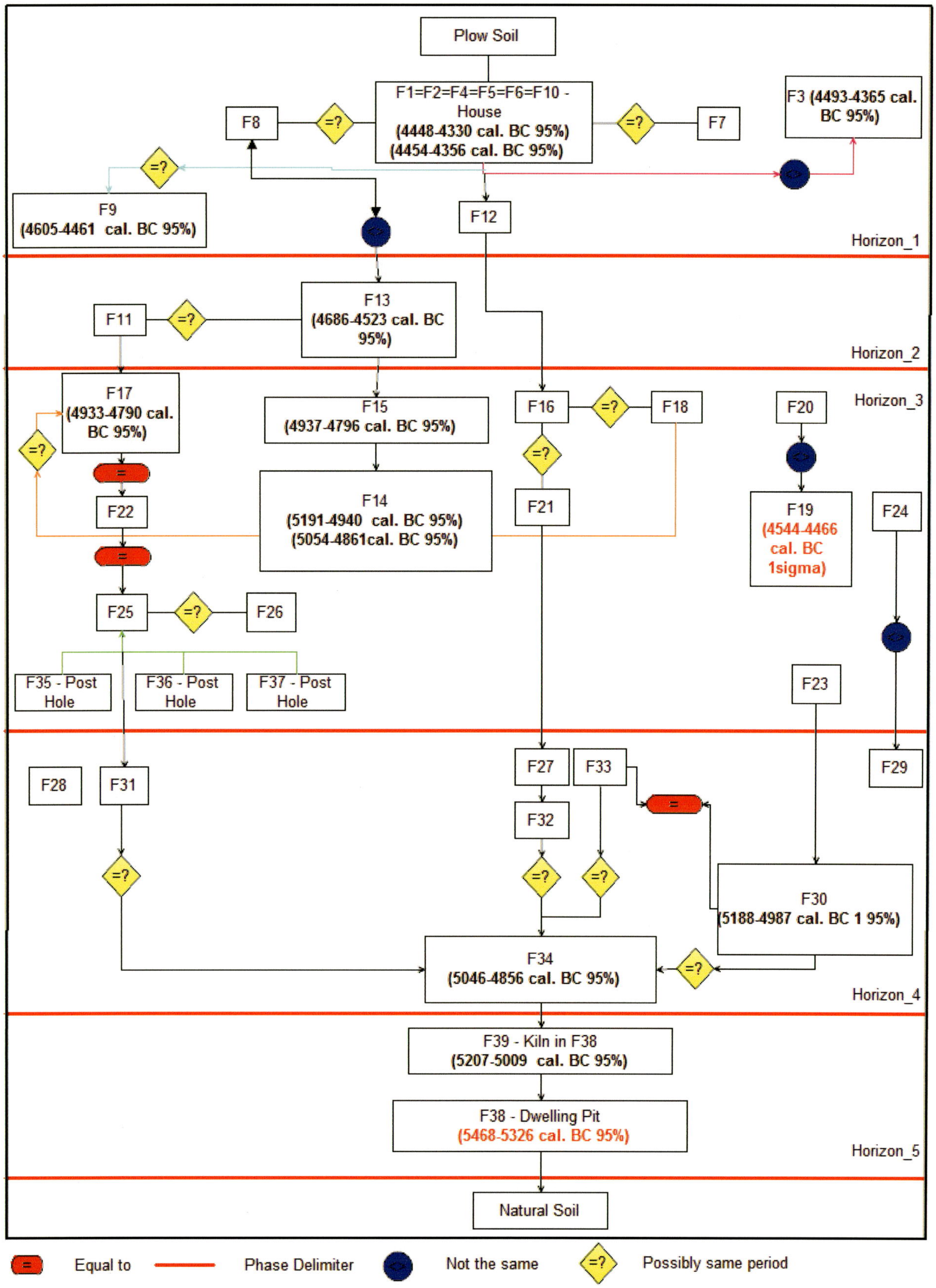

Figure 3. Schematic diagram of stratigraphic relations between features in Pločnik Trench 24.

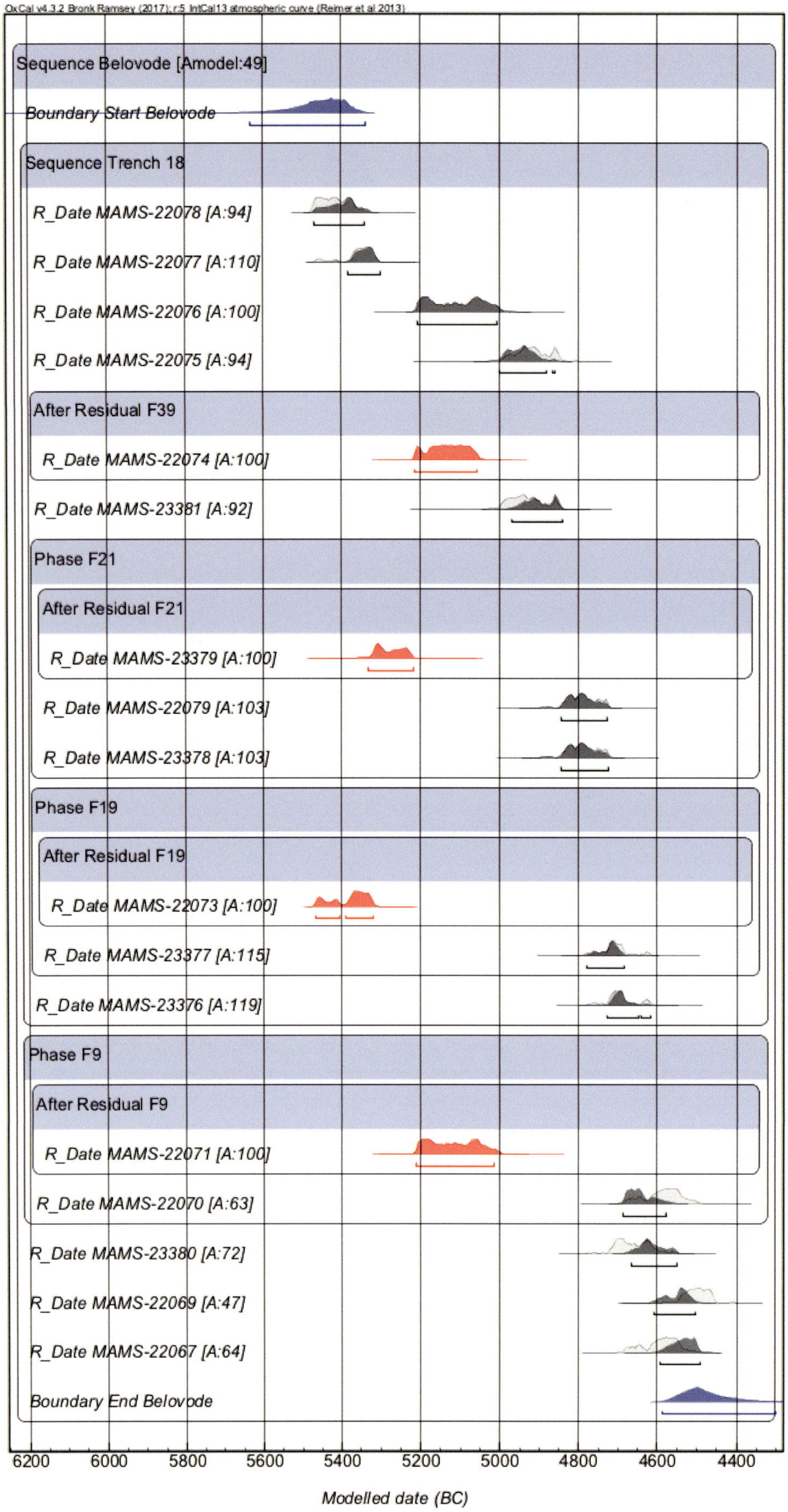

Figure 4. Stratigraphic model with corrected residual samples for Belovode Trench 18.

excavation. Such an approach, combining two types of chronological information, provides date estimates which can be more precise and more robust than the individual chronological data that generates the model.

Modelling was performed in OxCal v4.2 using IntCal13 calibration curve (Bronk Ramsey 2009; Reimer *et al.* 2013). All absolute date measurements were placed into the first model based on their spit numbers (i.e. relative depths). When more than one sample existed from the infill of the same cut feature, the same principle was applied, on the presumption that deeper samples represent earlier deposits. Using such an approach, four dates were identified as statistically inconsistent with the rest of the Trench 18 sequence and were partially removed from the modelling, using the 'After' function of OxCal. These four dates (MAMS-22071, MAMS-22073, MAMS-22074 and MAMS-23379) are assumed to represent residual samples and cannot be included fully in the model. Two of the excluded dates originate from charcoal samples, which could indicate an old wood effect (Ward and Wilson 1978), whilst the others, measured on animal bone samples, could indicate reused or residual bone which was found out of its original context (articulated bone specimens having unfortunately not been available). The final relational model is presented in Figure 4. Upon the removal of these problematic samples the overall agreement of the model was Aoverall: 65.5.

In addition to modelling individual dates, our model contained the dating of the span of each horizon, in an attempt to establish the transitionary periods between relative phases as well as the beginning and the end of the Neolithic occupation of the site (cf. Tasić *et al.* 2015). In order to achieve this, the dates of each horizon were modelled separately and a value for the start and the end of the horizon was obtained (Figure 5).

The start of Horizon 5 can be modelled at 5648–5338 cal. BC (95.4% prob.), which corresponds with the use span of ossuary pit Z on Belo Brdo in Vinča (Tasić *et al.* 2016b: 128, Table 4), i.e. the end of Starčevo burials, which corresponds with the 68% probability modelled at 5491–5375 cal. BC. It must be noted that the sole dating sample from Horizon 5 in Trench 18 originates from a hearth and is a charcoal fragment. This could imply old wood effect, as the ceramic finds in the spit surrounding the feature show trademarks of Vinča A pottery style, but examples with chaff found in inclusions, indicative of Starčevo pottery technology, were also recovered from the same spit (Chapters 12 and 13, this volume). The hearth itself yielded three non-distinct ceramic fragments, which cannot be dated more precisely. It is our belief that the hearth itself is probably linked with the end of Starčevo occupation, whilst the spit immediately above it may be linked with the Vinča A Phase of the settlement. The transition boundary to Horizon 4b is modelled at 5452–5318 cal. BC (95.4% probability), which would correspond to very early Vinča A found in the Pannonian plain (Whittle *et al.* 2016: 12). The ceramic finds from this sub-horizon, although limited in numbers, show clear analogies with Vinča A pottery and the number of pot fragments containing chaff is negligible. At Belo Brdo, this date corresponds to the very end of the Starčevo occupation of the site. Horizon 4a, comprising a series of discarded kiln floors, is dated with one sample from the immediate vicinity of Feature 36 (MAMS-22076). The modelled boundary between Horizons 4b and 4a is at 5366–5054 cal. BC (95.4%) which corresponds with the start of Vinča occupation at Belo Brdo itself, defined by pits at around 9.3 m relative depth (Tasić *et al.* 2016b: 136–7, Table 8). The subsequent boundary between the end of Horizon 4a and Horizon 3 start is modelled at 5139–4860 cal. BC (95.0%) which compares to layers between 7.5 and 6.5 m relative depth at Belo Brdo in Vinča (Tasić *et al.* 2016b: 136–7, Table 8), i.e. the Vinča B1–B2 period (Figure 1). The modelled transition of Horizon 3 to Horizon 2 is 4951–4760 cal. BC (95.4%), corresponding to layers between 6.2 and 5.6 m at Belo Brdo (Figure 1), or the Vinča Gradac–Vinča C period (Tasić *et al.* 2016b: 136–7, Table 8). The end of Horizon 2 and the start of Horizon 1b boundary in Belovode Trench 18 is modelled at 4818–4692 cal. BC (95.4%) and directly corresponds to layers between 4.9 and 4 m, or the Vinča C–D1 period (Figure 1). The transition to the last phase of Neolithic life on the site, and the boundary between Horizon 1b and 1a, is modelled at 4701–4540 cal. BC (95.4%) corresponding to layers below 4.0 m relative depth at Belo Brdo, or the Vinča D2 period. It should be stated that although the sub-division of Horizon 1 was not clearly detectable in the trench while excavating due to the lack of overlapping features, the modelling strongly suggests its existence as visible in Figure 4, when compared to the type site of Belo Brdo in Vinča (Schier 1996). Thus, Horizon 1a would comprise Features 1 and 9 whilst Horizon 1b would contain Features 3, 6 and 19. Finally, the end of Neolithic life in Trench 18 at Belovode is modelled to 4601–4383 cal. BC (95.4% probability) or 4571–4482 cal. BC (68.2%), which corresponds to modelled results from Belo Brdo itself at 4570–4460 cal. BC (Tasić *et al.* 2015: Figure 8; Tasić *et al.* 2016b: 128, Table 4), but also on a wider scale to the majority of the Vinča world (Whittle *et al.* 2016: 38, Figure 35), especially in the Danube-Sava-Tisza region.

The model also clearly illustrates that the whole span of the Late Neolithic Vinča culture is present at Belovode (Table 2), although specific phases may be absent for certain areas of the site that may not have been permanently occupied during all periods of the Late Neolithic settlement. It seems that the Late Neolithic settlement of Belo Brdo (Schier 1996: 160, Figure 11)

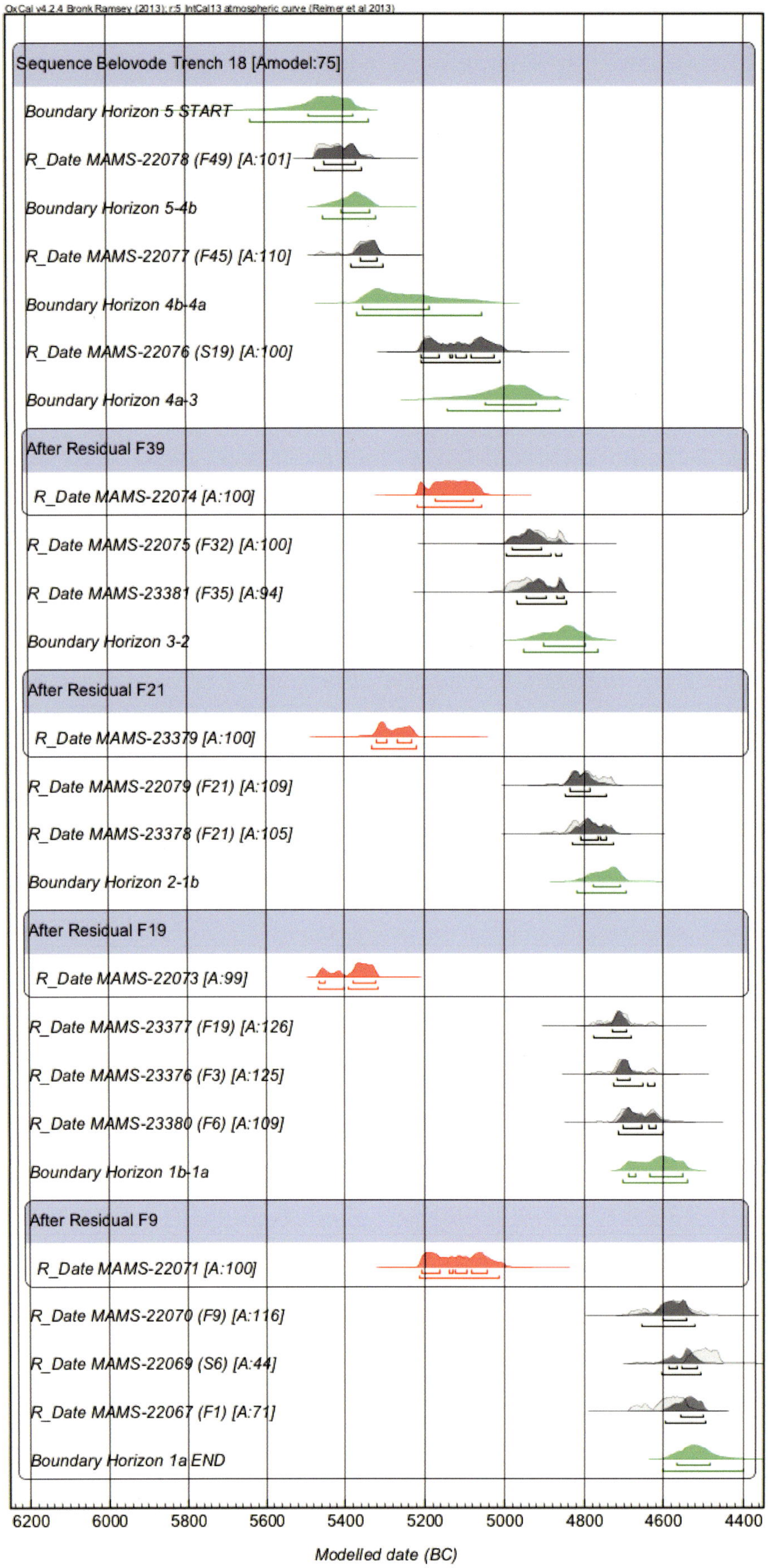

Figure 5. Final Bayesian model for Belovode Trench 18 (green distributions represent horizon boundaries).

in Vinča begins earlier than Belovode but not by very much. The occasional find of Late Starčevo pottery in some of the earlier trenches at Belovode (Šljivar *et al.* 2015) indicates that the location was not unknown in the Early Neolithic, which is typical of many sites occupied by Vinča communities (Belo Brdo being the best known). Based on the absolute dates presented here, but also compared to the pottery typology, the earliest Vinča settlement in Belovode was established early in Vinča A1 Phase (Figure 5). The horizon, defined by a series of small, elliptical hearths and discarded kiln floors in Trench 18, is (like the previous Horizon 5) dated by a single sample, preventing us from undertaking more detailed modelling of this phase, though still providing an insight into continuous occupation of this part of the settlement in the Late Neolithic. The Vinča B Phase at Belovode (Horizon 3) shows evidence of an increase in activity in Trench 18, with numerous pottery fragments recovered, a large refuse pit (Feature 32), a hearth with an ash deposit immediately next to it, indicating economic activity, although it is unclear whether this was connected with food preparation or raw material resource preparation and transformation.

The transitional, traditionally non-metallic to metallic Vinča Gradac Phase at Belovode, and the beginning of the Vinča C period, are contained within Horizon 2, defined by several pits, one of which (Feature 21) shows strong traces of fire use and contains multiple lumps of malachite ore in its infill.

The beginning of the subsequent Vinča D1 Phase at Belovode (Horizon 1b) is defined by the appearance of a wattle and daub structure (Feature 3), surrounded by a pit to the south (Feature 19) and a hearth installation with metallic copper droplets (Feature 6) to the east. Unfortunately, Schier's study, although comprehensive, does not include the last phase of the settlement in Vinča (Schier was forced to leave Serbia in 1992 due to UN sanctions before he could finish his work), but a recent study (Tasić *et al.* 2016b: 136–7, Table 8) fills

Table. 1. Radiocarbon dates from Belovode and Pločnik

Lab No. MAMS	Site	EDM No.	Feature / Spit	Horizon	Sample type	14C age	±	13C	Cal. 1σ (cal BC)	Cal. 2σ (cal BC)
22067	Belovode	38	F1	1(a)	Ch. wood	5741	23	-27.5	4648-4542	4683-4522
22068	Belovode	368	F3	1(b)	Animal bone (Medium ungulate)				No collagen	No collagen
22069	Belovode	272	S6	1(a)	Animal bone (Large mammal)	5669	25	-24.0	4522-4463	4547-4455
22070	Belovode	667	F9	1(a)	Animal bone (Bos Taurus)	5735	26	-23.6	4648-4535	4683-4502
22071	Belovode	790	F9	1(a)	Animal bone (Bos/Cervus sp.)	6151	28	-15.5	5206-5046	5210-2015
22072	Belovode	1212	F18	2	Charcoal				Reserve sample	Reserve sample
22073	Belovode	1379	F19	1(b)	Charcoal	6389	23	-25.7	5461-5322	5466-5318
22074	Belovode	2429/2446	F39	3	Charcoal	6179	23	-24.2	5208-5074	5213-5055
22075	Belovode	2585	F32	3	Charcoal	6026	23	-19.9	4960-4851	4989-4847
22076	Belovode	2639	S19	4	Charcoal	6143	27	-26.4	5206-5024	5207-5006
22077	Belovode	2677	F45	4	Charcoal	6365	25	-33.2	5363-5317	5464-5303
22078	Belovode	2899	F49	5	Charcoal	6422	23	-29.1	5466-5372	5472-5343
22079	Belovode	1694	F21	2	Ch. Grain (Einkorn/Emmer)	5923	25	-20.7	4835-4731	4847-4722
22080	Belovode	1570	F21	2	Ch. Grain (Wheat)				Reserve sample	Reserve sample
22081	Pločnik	587	F3	1	Animal bone (Bos Taurus)	5611	25	-20.9	4481-4372	4493-4365
22082	Pločnik	588	F1	1	Animal bone (Bos Taurus)				No collagen	No collagen
22083	Pločnik	743	F2	1	Animal bone (Bos Taurus)	5573	25	-19.9	4447-4367	4454-4356

Table. 1. Radiocarbon dates from Belovode and Pločnik

Lab No. MAMS	Site	EDM No.	Feature / Spit	Horizon	Sample type	14C age	±	13C	Cal. 1σ (cal BC)	Cal. 2σ (cal BC)
22084	Pločnik	974	F9	1	Animal bone (Bos Taurus)	5701	26	-16.6	4577-4490	4605-4461
22085	Pločnik	974	F9	1	Animal bone (Sus scrofa)				Reserve sample	Reserve sample
22086	Pločnik	1121	F13	2	Animal bone (Medium mammal)	5745	26	-23.0	4666-4544	4686-4523
22087	Pločnik	1240	F14	3	Charcoal	6089	22	-22.2	5038-4964	5191-4940
22088	Pločnik	1385	F19	3	Animal bone (Bos Taurus)	5686	25	-23.5	4544-4466	4583-4457
22089	Pločnik	1440	F17	3	Charcoal	5970	25	-23.5	4897-4801	4933-4790
22090	Pločnik	1462	F15	3	Charcoal	5981	25	-23.4	4908-4808	4937-4796
22091	Pločnik	1683	F14	3	Charcoal	6076	25	-21.8	5016-4946	5054-4861
22092	Pločnik	2453	F38	5	Charcoal	6408	26	-23.3	5466-5359	5468-5326
22093	Pločnik	2663	F39	5	Grain	6145	26	-31.6	5206-5029	5207-5009
22094	Pločnik	2663	F39	5	Grain				Reserve sample	Reserve sample
23373	Pločnik		F1	1	Animal bone (Bos Taurus)	5517	30	-28.9	4438-4336	4448-4330
23374	Pločnik		F30	4	Animal bone (Bos Taurus)	6104	25	-23.3	5188-4987	5205-4943
23375	Pločnik		F34	4	Animal bone (Ovis/Capra sp.)	6064	25	-20.4	5003-4987	5046-4856
23376	Belovode		F3	1(b)	Animal bone (Ovis/Capra sp.)	5829	24	-19.5	4725-4619	4778-4611
23377	Belovode		F19	1(b)	Animal bone (Ovis/Capra sp.)	5839	26	-21.9	4768-4626	4785-4616
23378	Belovode		F21	2	Animal bone (Ovis/Capra sp.)	5922	26	-16.3	4834-4730	4872-4720
23379	Belovode		F21	2	Animal bone	6305	27	-20.4	5317-5228	5337-5218
23380	Belovode		F6	1(b)	Animal bone	5808	26	-15.2	4711-4618	4724-4555
23381	Belovode	2420	F35	3	Grain	6038	27	-27.4	4987-4857	5003-4845

Table 2. Absolute dates of Belovode settlement phases based on final model (Figure 5).

Posterior density interval (2σ)	Posterior density interval (1σ)	Horizon	Relative Chronology (Belo Brdo)
4600-4400 cal BC	4564-4479 cal BC	Horizon 1a end	End Vinča D2
4702-4540 cal BC	4689-4669 cal BC (10.6%) or 4634-4549 cal BC (57.6%)	Horizon 1a start	Start Vinča D2
4817-4692 cal BC	4776-4709 cal BC	Horizon 1b start	Vinča C-D1
4951-4762 cal BC	4899-4797 cal BC	Horizon 2 start	Gradac-Vinča C
5140-4859 cal BC	5139-4859 cal BC	Horizon 3 start	Vinča B1-B2
5366-5054 cal BC	5351-5185 cal BC	Horizon 4a start	Start Vinča A
5452-5318 cal BC	5404-5335 cal BC	Horizon 4b Start	End Starčevo
5648-5338 cal BC	5491-5375 cal BC	Horizon 5 Start	Starčevo

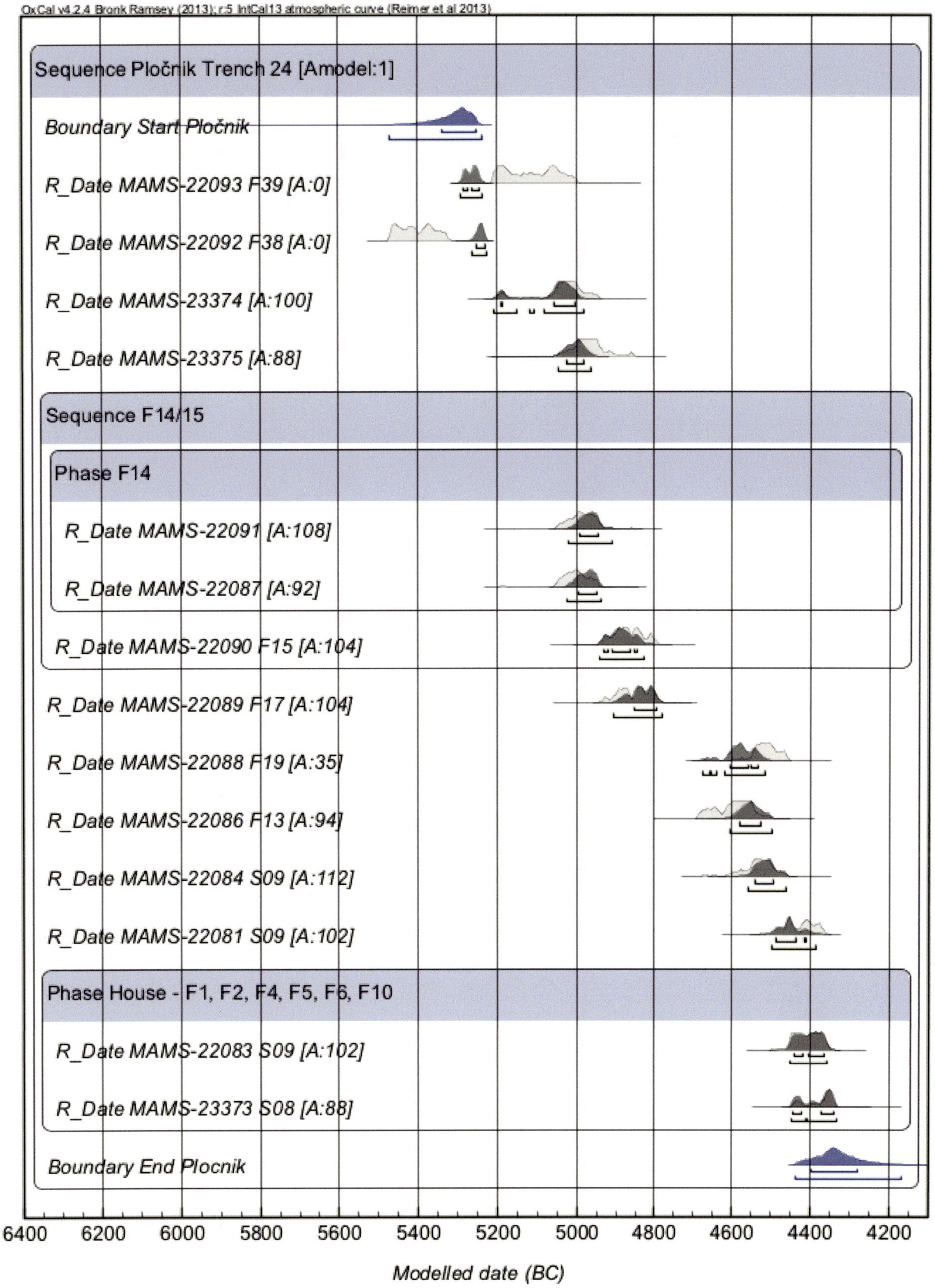

Figure 6. Stratigraphic model without corrected residual or intrusive samples for Pločnik Trench 24.

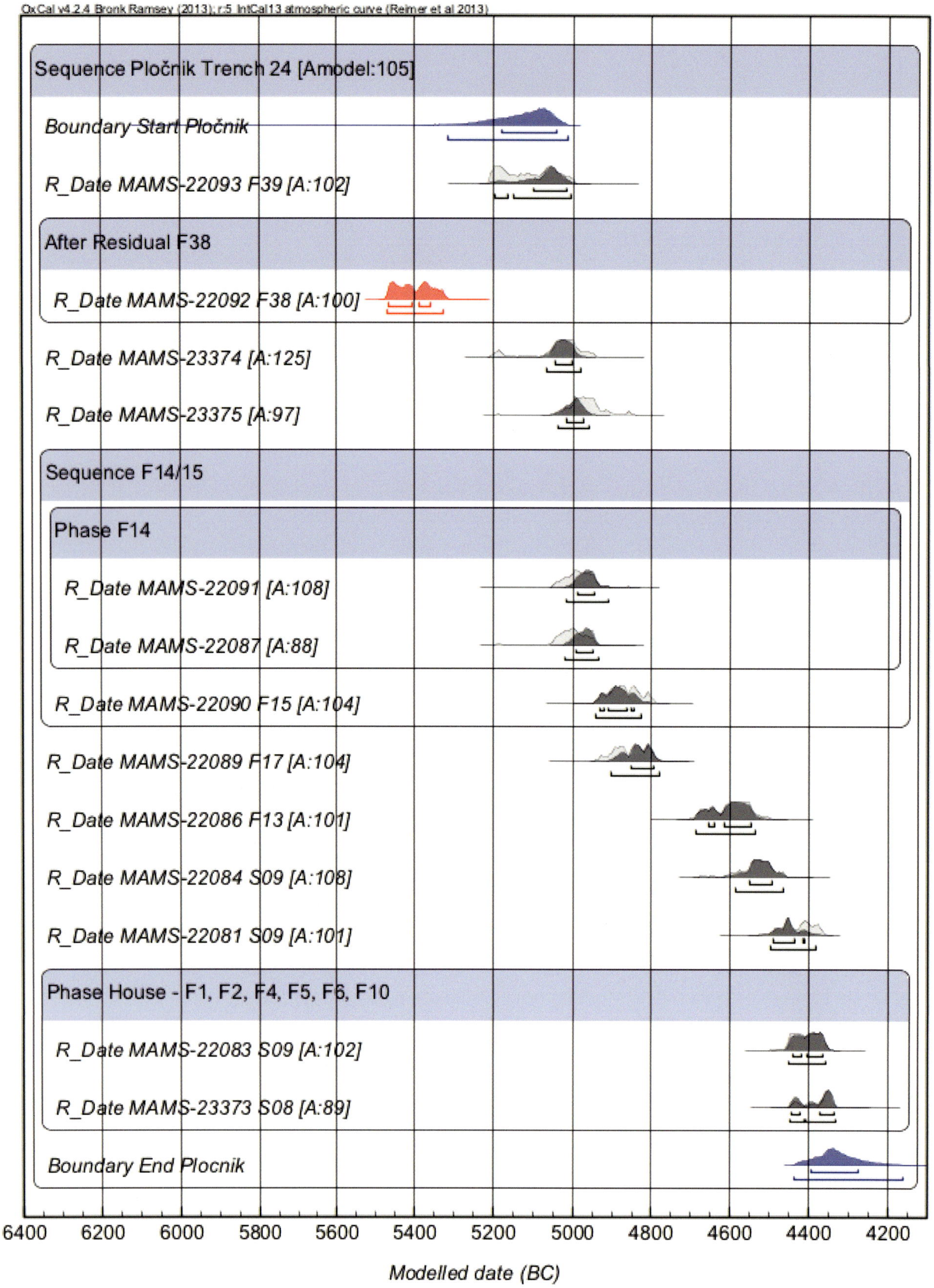

Figure 7. Stratigraphic model with corrected residual or intrusive samples for Pločnik Trench 24.

this gap with new dates. The Vinča D1 Phase at Belo Brdo in this study has absolute dating values similar to those obtained on Belovode samples, which is not surprising due to the proximity of the two sites. Finally, the modelled date for the Vinča D2 period in Belovode is equivalent to that for the Belo Brdo type site in Vinča, as is the end of Neolithic occupation on both sites (Tasić *et al.* 2015: 1077), indicating a larger occurrence that engulfed the whole of the Danubian Vinča at the turn of the 46th century BC, resulting in a complete abandonment of well established, permanently occupied settlements.

Pločnik

Like Trench 18 at Belovode, Trench 24 at Pločnik comprised five activity horizons, numbered 1 to 5 from the topsoil to the bottom of the trench (Figure 3). The sample list contained 17 samples in total, but one (MAMS-22082) did not yield collagen and two additional samples (MAMS-22085 and MAMS-22094) were kept in reserve and were not dated (Table 1). The same method used to model the absolute dates from Trench 18 at Belovode was used for Trench 24 at Pločnik. The 14 samples were first modelled stratigraphically based on their spit number in order to identify residual or contaminated samples. In the process, two samples (MAMS-22088 and MAMS-22092) were identified as intrusive and residual respectively (Figure 6, upper portion). The remaining 12 samples were built into a slightly altered model (Figure 7) using the 'After' and 'Before' functions of OxCal to identify the residual and intrusive samples for further modelling, and to keep them distinct (red coloured). The altered model (Figure 7) with high agreement (Amodel:104), corroborates the relative stratigraphy recorded during the excavations in the trench.

The stratigraphic model was then enhanced by adding boundaries between identified horizons in order to define the start and the end of each horizon, i.e. its duration. The final model (Amodel:90), presented in Figure 8, indicates that the start of Neolithic occupation of the southern part of Pločnik settlement occurred in 5389–5003 cal. BC (95.4% prob.), perhaps in 5180–5028 cal. BC (67.4% prob.) or 5189–5186 cal. BC (0.8% prob.), which corresponds to layers between 9.3 and 7.0 m relative depth in Vinča, or the Vinča A–Vinča B2 period (Tasić *et al.* 2016b: 136–7, Table 8). If the 67.4% probability posterior density estimate is considered, it could narrow the starting period to between 8 and 7 m at Belo Brdo, coinciding with the end of Vinča A and the beginning of the Vinča B1 Phase (Figure 1). The boundary at the end of Horizon 5 and the beginning of Horizon 4 is modelled at 5121–4976 cal. BC (95.4%), which corresponds to the layers between 7.05 and 6.55 metres at Vinča, or the Vinča B2 period in relative chronology (Figure 1). The end of Horizon 4 and beginning of the subsequent Horizon 3 (Figure 8) is modelled at 5036–4951 cal. BC (95.4% prob.). This period corresponds to the Vinča B2–C transition, or the so-called Gradac Phase (Garašanin 1979) at Belo Brdo (Tasić *et al.* 2016b: 136–7, Table 8). A single sample (MAMS-22086) dates the span of Horizon 2. The end of Horizon 3 and the beginning of Horizon 2 in Pločnik is modelled at 4927–4621 (95.4% prob.), which would correspond to the layers between relative depths of 6.0 and 4.0 m at Belo Brdo (Tasić *et al.* 2016b: 136–7, Table 8), or the Vinča C–D1 span in relative chronology (Figure 1). Taking into account the 68.2% probability for the posterior density estimate of this boundary, modelled at 4894–4746 cal. BC, the start of Horizon 2 would fall between 6.05 and 4.95 m relative depth at Belo Brdo (Tasić *et al.* 2016b: 136–7, Table 8), or the Vinča C Phase. The end of the penultimate horizon in Pločnik, and the beginning of Horizon 1 is modelled at 4631–4462 cal. BC (95.4% prob.), corresponding to the layers between 3.4 and 1.3 m at Belo Brdo (Tasić *et al.* 2016b: 136–7, Table 8), or the latter half Vinča D2 Phase and beyond. Although it may seem so, this is not surprising for the southern variant of the Late Vinča (the metallic Vinča) culture, as some authors have hypothesised that it has a prolonged duration compared to the Danubian Vinča (Jovanović 1994) and even direct contact and overlap with the early Bubanj-Salçuta-Krivodol (BSK) communities in the Central and Eastern Balkans area (Tasić 1979; Tasić 1995) which, in its earliest phase, contains pottery types closely resembling those of the Late Vinča. Other authors (Srejović 1984b) have claimed that Vinča communities of the southern variant disintegrated under the influence of the BSK complex, melting into the new societies that started occupying the Balkans area from the 45th century BC. It appears that the new dates presented here, extrapolated from a strict and constrained Bayesian statistical framework can further corroborate the hypotheses of the previous researchers regarding the transition from 'metallic' Vinča to the Middle Chalcolithic proper. They also illustrate that, in Trench 24 at Pločnik, there appears to be no activity linked with the Vinča D1 period, either because this part of the site was temporarily abandoned during that specific period, or because the limited scope of the trench prevented us from detecting such features.

Finally, the end of the Neolithic occupation of Pločnik and the end of Horizon 1 is modelled at 4446–4231 cal. BC (95.4% prob.) or 4430–4326 cal. BC (68.2% prob.). This posterior density interval, obtained on two dates (MAMS-22083 and MAMS-23373) originating from the burnt daub structure discovered *in situ* in Trench 24 has no direct comparison at either Belovode or Belo Brdo in Vinča, as by this period both these sites were long abandoned after a fiery end that occurred towards the end of the 46th and the beginning of the 45th century BC and enveloped a host of Late Neolithic sites in the wider

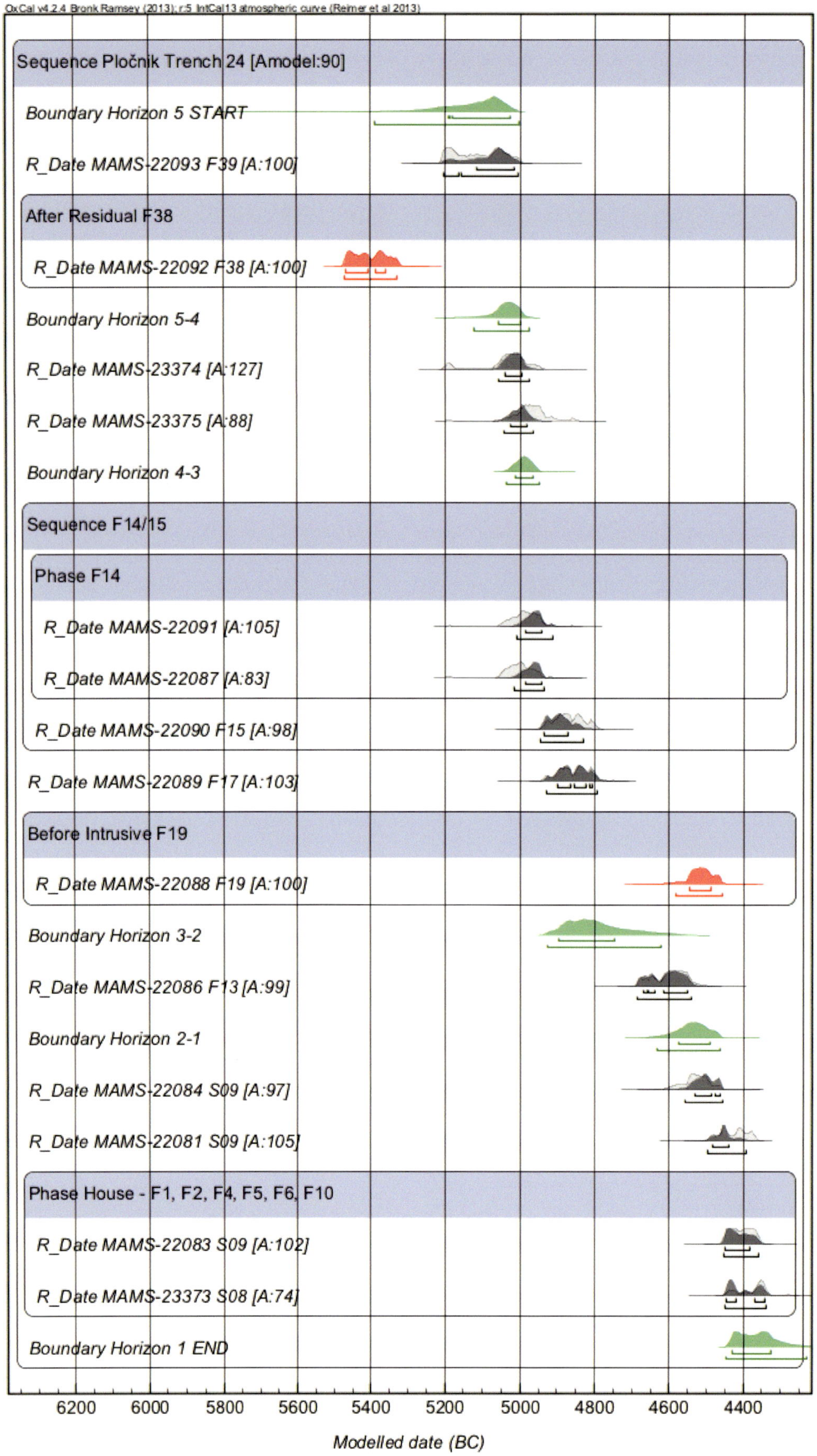

Figure 8. Final Bayesian model for Pločnik Trench 24 (green distributions represent horizon boundaries).

area of the Danube and its tributaries (Tasić *et al.* 2015). However, in Belo Brdo itself, there is evidence for Middle Chalcolithic (or Early Copper Age in the terminology of the original scholarship) occupation, with four burials identified as belonging to Bodrogkeresztúr phase of the Middle Chalcolithic (Jevtić 1986: Figure 2). Of the four burials discovered, two were dated: Burial 1 at 4354–4244 cal. BC (95.4% prob.) and Burial 2 at 4314–4084 cal. BC (95.4% prob.). These postdate the modelled abandonment of Late Neolithic Belo Brdo by about 150–250 years (Tasić *et al.* 2015: 1077) but are significantly less distant when compared to the modelled dates of Trench 24 in Pločnik, where this margin is between 80 and 100 years.

Another, albeit single, AMS date from the Middle Chalcolithic site of Bodnjik, located about 160 km northwest of Pločnik in western Serbia, is a much better illustration of the possibility—and probability—of contemporaneity between late Vinča culture in the south of Serbia and the emerging BSK communities. The site of Bodnjik, located on a hilltop near the town of Koceljeva in western Serbia, was researched over several campaigns in the mid-1990s (Palavestra *et al.* 1993; Palavestra *et al.* 1996), yielding two rectangular burnt daub structures with pottery material typical of early BSK complex in a single layered settlement. Of four samples submitted for AMS dating, only one had sufficient collagen (OxA-26309, 5579BP, +/-35), calibrating to 4466–4347 cal. BC at 95% probability, or 4448–4369 cal. BC at 68.3% probability (Živanović 2013: 54). This value overlaps with the modelled values obtained from dates for the end of Horizon 1 in Trench 24 at Pločnik (Table 3), indicating a strong possibility of contemporaneous life of late Vinča and early BSK communities in the central Balkans.

A similar set of early radiocarbon dates was obtained in northeast Bulgaria, at the site of Lîga (about 180 km northeast of Pločnik). Located 1 km north of the modern village of Telish in the Cherven Briag municipality of Bulgaria (Merkyte 2005: 9), this site has three horizons, two of which (Lîga 2 and 3) are comparable to the Bubanj-Hum Ia phase (Merkyte 2005: 16, Figure I.5) in Serbia (Lîga 1, datable to Vinča D period by the lead author (MM), was heavily damaged by the construction of later settlements). Excavations between 2000 and 2002 yielded six AMS dates, three from the settlement of Lîga 2 and three from the later cemetery established over a part of the settlement denoted as Horizon Lîga 3 (Merkyte 2005: 34, Figure II.12). Modelled together, according to the context descriptions (Merkyte 2005: 33–36), the modelled highest posterior density for the start of Lîga 2 is 4837–4357 cal. BC (95% prob.) or 4566–4386 cal. BC (68% prob.), placing it safely in the very late Vinča D2 period, i.e. the period associated with the conflagration of Danubian late Vinča sites. However, the end of this horizon indicates that it extends well into the 44th century BC, modelled at 4416–3996 cal. BC (95% prob.) or 4387–4176 cal. BC (59.4% prob.) or 4068–4028 cal. BC (5.8% prob.), illustrated in Figure 9. Based on these results, it is clear that the Middle Chalcolithic Lîga 2 settlement existed in parallel to the final phase of Late Neolithic occupation at Pločnik, further corroborating the possibility of contact between the two communities, already established by the ^{14}C date from Bodnjik.

From the final Pločnik Trench 24 model (Figure 8) it is easily discernible that the Late Neolithic occupation begins later than at Belovode and Belo Brdo, midway through the Vinča A2 Phase. Trench 24 was located towards the southern edge of the Late Neolithic settlement, so that this need not be the case for the central part but the contemporary village of Pločnik prevents us from large scale excavation in this area (see Chapter 24). A noticeable increase in activity is present in the following Horizon 4, dated to the Vinča B2 Phase, with several notable features, most distinct being the rectangular ash deposit (Feature 34). The transitional Gradac Phase of Horizon 3 has an abundance of activity, including the edge of a burnt daub structure (Feature 17) opposite a dismantled kiln surrounded by a large ash deposit originating from its use (Features 14/15). The latter two indicate a longer lasting economic activity. Unfortunately, they were

Table 3. Absolute dates of Pločnik settlement phases based on final model (Figure 8).

Posterior density interval (2σ)	**Posterior density interval (1σ)**	**Horizon**	**Relative Chronology (Belo Brdo)**	**Relative Chronology (Jovanović 1994)**
4446-4231 cal BC	4431-4324 cal BC	Horizon 1 end	-	Gradac III
4631-4462 cal BC	4576-4491 cal BC	Horizon 1 start	Vinča D2	Gradac II
4927-4621 cal BC	4894-4747 cal BC	Horizon 2 start	Vinča C-D1	Gradac I
5036-4951 cal BC	5013-4968 cal BC	Horizon 3 start	Gradac phase	
5121-4976 cal BC	5057-5001 cal BC	Horizon 4 start	Vinča B2	Vinča B2
5389-5003 cal BC	5199-5190 cal BC (2.5%) or 5182-5028 cal BC (65.7%)	Horizon 5 start	Start Vinča A2-B1	Start Vinča A2-B1

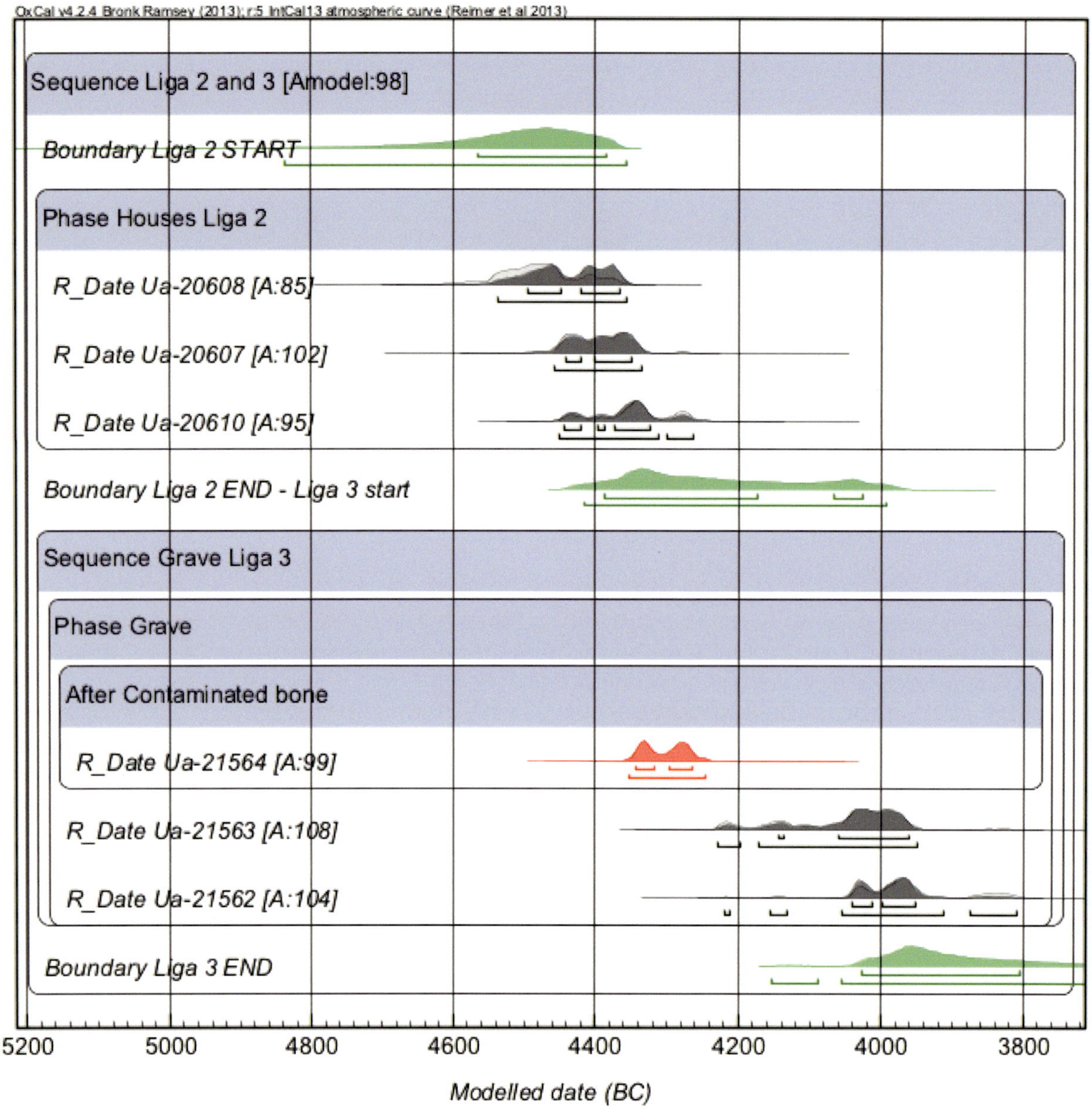

Figure 9. Bayesian model for Lîga 2 and 3 radiocarbon dates (adapted from Merkyte 2005: 33–36).

located next to the eastern profile of the trench, so it was not possible to establish the exact nature of the activity as the kiln was only partially excavated. Horizon 2, defined by another dismantled kiln next to the western profile and a partially excavated pit in the northeast corner of the trench, seems to mimic the economic activity of the previous horizon. The Late Vinča C and Early D1 Phase of Horizon 2 indicates that the area was still not exclusively residential, but rather had a mixed role in the late settlement. In the final horizon of the Neolithic occupation of Pločnik, well past the onset of the Middle Chalcolithic in the rest of the Balkans and Pannonian plain, the residents of the settlement erected a large rectangular daub structure, which was completely burnt at the end of the settlement occupation, well into the second half of the 45th or the first half of the 44th century cal. BC. This abrupt change in the nature of the space may signify a new purpose given to the previously predominantly economic area, but also some other changing aspect, such as an increase in the number of households due to yet unexplained reasons, or similar. However, these questions extend far beyond the limits of both this chapter and this volume.

The bibliographic reference for this chapter is:

Marić, M., Radivojević, M., Roberts, B. W. and Orton, D. C. 2021. Relative and absolute chronology of Belovode and Pločnik, in Radivojević, M., Roberts, B. W., Marić, M., Kuzmanović Cvetković, J., and Rehren, Th. (eds) *The Rise of Metallurgy in Eurasia*: 439–454. Oxford: Archaeopress.

Chapter 38

The social organisation of the Vinča culture settlements. New evidence from magnetic and archaeological excavation data

Knut Rassmann, Martin Furholt, Nils Müller-Scheeßel and Johannes Müller

New large-scale magnetic surveys of Vinča period settlements can provide fresh insights into the social organisation of Late Neolithic communities. In the following chapter we compare the results of such surveys of a large region of southeastern Europe with the regional archaeological study conducted in the Bosnian Visoko valley (Müller *et al.* 2013a) in order to correct previous estimations of settlement sizes and population numbers and to discuss the internal social composition of Vinča period settlements.

Around three decades ago, Clemens Lichter published a compilation of houses and their architecture in the Neolithic and Copper periods in southeast Europe (Lichter 1993). Despite numerous excavations in the 20th century, Lichter's overview illustrated general limitations in our knowledge of these archaeological phenomena, particularly concerning the spatial layout of settlements and their internal structure, which are crucial for understanding the social, functional context of houses and the lives of their inhabitants.

Today we are better situated to discuss spatial organisation in settlements and related topics thanks to ambitious settlement research and large-scale magnetic surveys centred around sites like Okolište in central Bosnia (Müller *et al.* 2013a), Uivar in Romania (Schier and Draşovean 2004: 151, Figure 3), Polgár-Csöszhalom (Racky and Anders 2006), Tölna-Möcs in Hungary (Rassmann *et al.* 2015), Vráble in Slovakia (Furholt *et al.* 2014), Bordjoš in Serbia (Medović *et al.* 2014) and Cucuteni-Tripolje-Settlements (Müller *et al.* 2016) (Figure 1). Our new research on the settlements at Pločnik and Belovode should be viewed in this context. The large-scale and high-resolution magnetic surveys at these sites provide general information on the settlements, such as their size and intra-site spatial structure, analysed here alongside the excavation results and using comparisons with data from other Late Neolithic and Copper Age settlements.

In this chapter, we explore the spatial organisation of Belovode and Pločnik in a broader southeast European context. Central to our arguments are the excavation in Okolište and new magnetic surveys in Serbia, e.g. at Drenovac (Perić *et al.* 2016), Crkvine/Stubline (Crnobrnja 2011: 131), and Bordjoš (Medović *et al.* 2014). The value of the research at Okolište lies in the high-resolution excavation data, insights into houses, their chronology, and connected social spheres, and the extensive exploration of the surrounding landscape, the Visoko Basin (400 km^2). With a combined analytical strategy with archaeological data at different scales and an interdisciplinary approach, Okolište can be seen as paradigmatic for modern settlement archaeology (Müller *et al.* 2013a; Hofmann 2013b).

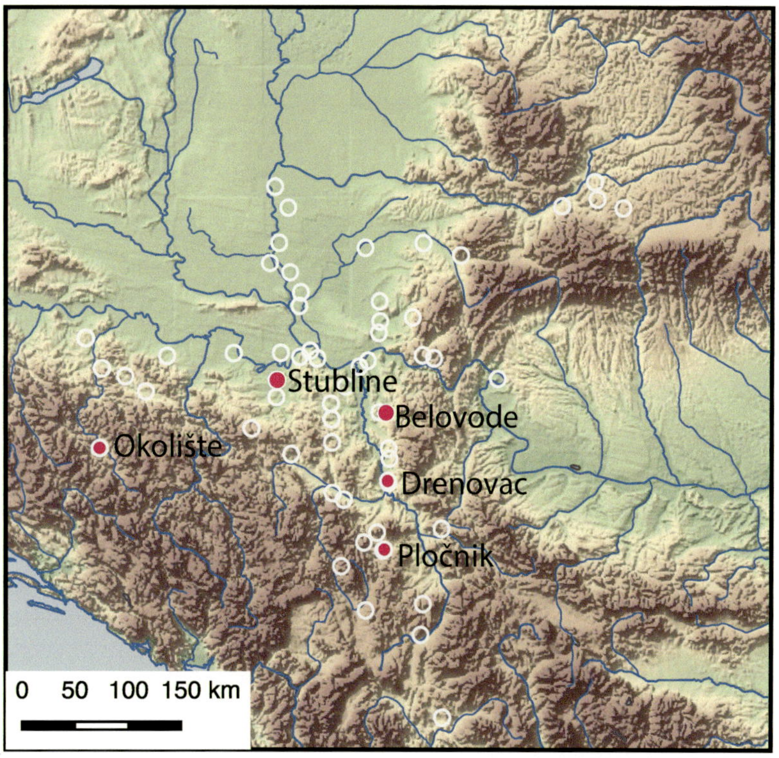

Figure 1. The distribution of Vinča sites (after Whittle *et al.* 2016: 32, Figure 28). Settlements with geomagnetic surveys are marked: 1 Belovode; 2 Crkvine/Stubline; 3 Drenovac; 3 Pločnik; and 5 Okolište.

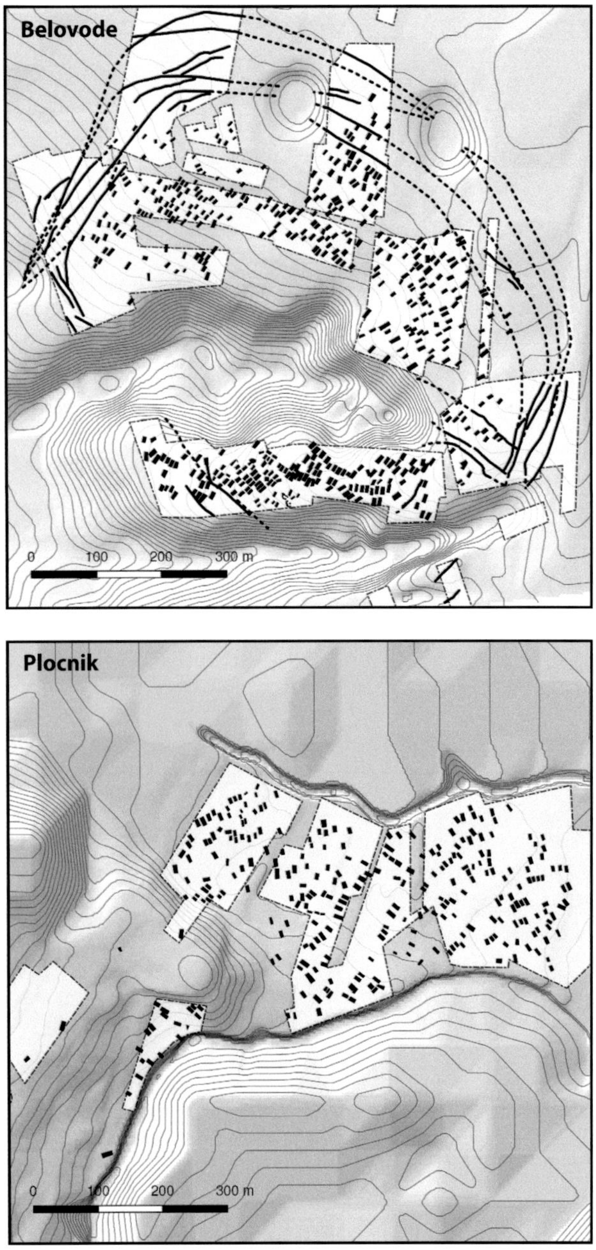

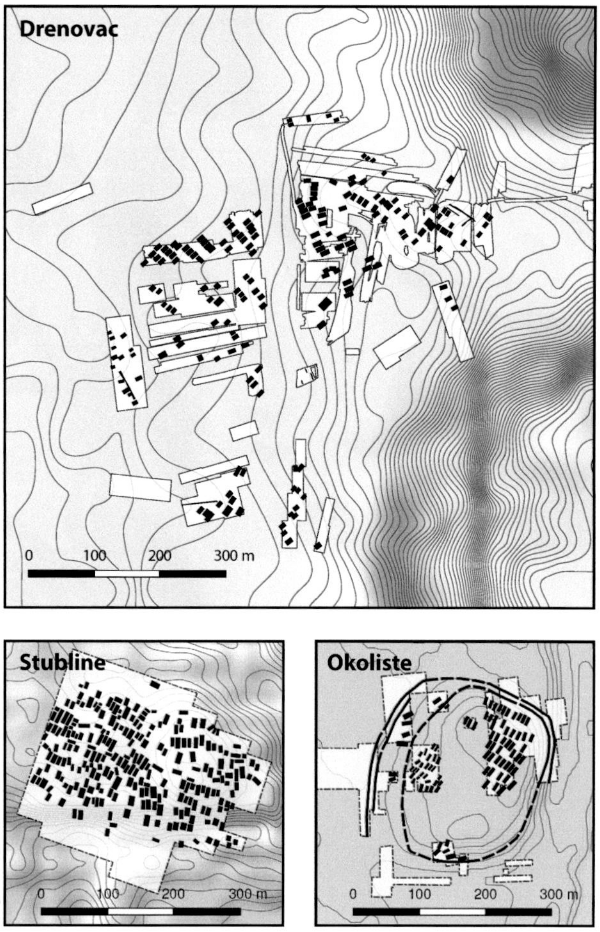

Figure 2. Overview of magnetic surveys of Vinča settlements.

Settlement size comparisons (Figure 2)

The new magnetic surveys at Belovode and Pločnik confirmed settlement areas in the range of 25–35 ha and corrected former estimates that assumed up to 100 ha. A similar correction is now provided for the Vinča settlement of Drenovac, where the new magnetic survey revealed a settlement size of around 35 ha (Perić *et al.* 2016). The new magnetic survey at Drenovac covers 10 ha, that is less than 35% of the settlement area. Based on the 198 burnt houses in the magnetic data, an estimate of more than 600 houses for the complete settlement was discussed by Perić *et al.* (2016: 22). This is much higher than our reconstruction for Belovode and Pločnik, perhaps because Perić *et al.* neglected variations in building density within the settlement. As in Belovode and Pločnik, the settlement at Drenovac includes areas with relatively low building density. A linear interpolation of the house density (accounting for 10 ha or 30% of the settlement area) from the 198 houses to a corresponding value for the 35 ha of the entire settlement is not a reasonable assumption. Further, it is unknown how many of the houses existed contemporaneously. A figure of 300–350 houses (1500–1750 inhabitants; 5 persons per house), as we reconstructed for Belovode and Pločnik, seems more likely. It remains to be seen which estimates future surveys will confirm.

Settlement size and house number

In comparison to Pločnik and Belovode, a remarkably different settlement pattern was revealed by the magnetic survey at the site of Crkvine in the village of Stubline (Crnobrnja 2011). The size of the survey area was around 7.7 ha. The settlement is situated on an elongated, narrow ridge framed to the north by an erosion ravine and by a small stream in the south. The northern, eastern and southern boundaries can be reliably reconstructed by combining magnetic data and topography; the western border of the settlement is less clearly determined. Still, we estimate a settlement area not larger than 11–12 ha. Inside the prospected area of 7.7 ha, around 250 burnt houses are clearly visible. The houses are densely positioned in rows, resembling the pattern at Okolište or the central areas at Belovode or Drenovac.

In contrast to the varying building densities at Belovode, Drenovac, and Pločnik, the magnetic plan at Crkvine indicates an even density of house placement, at around 32 houses per ha. The estimated house number is thus 300–350 for the complete settlement (*c.* 11 ha), which is close to our calculation for Pločnik and Belovode.

Published estimations of the building density for Okolište in Bosnia are significantly higher (Müller *et al.* 2013a: 43 Table 2). For the early settlement phase at Okolište, 7 ha in extent, 500–650 houses were estimated; in the later phases, 100–75 houses are estimated within the then smaller settlement area of 1.2 ha (Müller *et al.* 2013a: 53 Table 2). Comparing this to the observation for Crkvine, the building density estimations for Okolište should perhaps be corrected. The house total for Okolište was reconstructed using the assumption that the settlement layer would mainly consist of house debris. However, if we include general settlement debris and materials used for pathway construction in the calculation of debris volume, the estimated number of houses would be considerably lower. Further, we have new indications of a more complex taphonomic process behind the formation of the settlement layer, indicated by excavation data from all trenches, drilling profiles (Hofmann and Müller-Scheeßel 2013: 79, Figure 10) and the resistivity data (Erkul *et al.* 2013: 108–109, Tables 4–5). The wide range in the resistivity data indicates a varying composition of the settlement layer and is evidence against a homogeneous agglomeration of clay from house remains. We cannot go into detail here, but it seems more likely to us that only 50% of the settlement layer derives from houses. These theoretical considerations lead us to reduce the estimate of house numbers to about 300 in the earliest phase. This is in line with our estimations for Belovode, Pločnik, Drenovac, and Crkvine.

Orientation of houses

Recently, new investigations of Neolithic house orientations have been published (Müller-Scheeßel *et al.* 2020; Hofmann and Müller-Scheeßel 2020), so it seems worth considering this matter from the perspective of Belovode, Pločnik, and Crkvine. We assume that the houses vary around north, and we compute their orientation based on their two longest sides as explained in Müller-Scheeßel *et al.* (2020). For a detailed insight into the dataset, see Appendix B_Ch38.

The orientation of the 1033 houses from these three sites varies greatly. While the mean direction is 25.8°, i.e. north-northeast, a significant proportion of the houses is also aligned to the northwest. The variation within each of the settlements is much narrower. The houses at Belovode (n = 487) are mainly orientated towards the northeast with a mean direction of 41.8°; those in Crkvine (n = 251) face north-northeast with a mean direction of 16.9°; and those in Pločnik (n = 295) face towards the northwest (mean direction = 329.2°). Thus, in the terminology of Hofmann and Müller-Scheeßel (2020), Belovode and Crkvine follow the 'Balkan' orientation, while Pločnik, the furthest south of the three, is in line with the 'Upper Tisza' orientation.

Whereas the house orientation at Belovode seems uni-modal, the houses in Pločnik show a bi-modal distribution, with a large proportion of the houses orientated northwest (*c.* 310°) and another group facing north-northwest (*c.* 340°). In contrast, at Crkvine, and to a lesser extent also in Pločnik, several houses are aligned orthogonally to the others, as is visible on the magnetic plan, creating a minor second group facing either west-northwest (*c.* 290°) in the case of Crkvine, or northeast (*c.* 40°) in the case of Pločnik. A similar observation has been made for Okolište (Hofmann 2013b). Why some houses so blatantly violate the common orientation remains open to question.

Assuming a counterclockwise change in orientation, the settlement of Crkvine seems to be quite large in its earliest phase and then gradually shrink in size, whereas the opposite can be deduced for Belovode. In the case of Pločnik, the bi-modal distribution of house orientation could indicate the existence of several distinct settlement phases, separated by a hiatus.

Elsewhere, an average change in orientation between 3.9° and 9.5° with an emphasis on 5° per 100 years was determined (Hofmann and Müller-Scheeßeel 2020; Müller-Scheeßel *et al.* 2020). Excluding outliers, the change of 50° for the Belovode houses presumably indicates the presence of different chronological phases in the settlement history. In comparison with Crkvine, with the bulk of its data falling within a range of 40°, this might indicate a wider chronological range for the revealed houses.

The Okolište research clearly indicates a chronological range for the houses revealed in the magnetic survey, with the *c.* 45 houses in the magnetic map belonging to different settlement phases between 5100 and 4700 BC (Hofmann 2013b). The fact that the houses are not automatically contemporaneous has to be taken into account in dealing with all magnetic data, not only that from Vinča sites. Visible houses are mainly burnt, with less clear, unburnt houses in the upper settlement layer. The opportunity to gain insights into the early stages of a settlement is limited by the superimposition of older houses by material of younger settlement phases. Okolište might be an exception since the oldest settlement gets gradually smaller, with a chronological shift from the periphery to the centre, with the youngest houses in the centre of the tell and above this.

The Okolište settlement shows differences in spatial organisation. Okolište and Stubline are similar in size and building density. Pločnik, Belovode and Drenovac are clearly different, being much larger with a varying building density. Belovode follows two patterns. The central part on the promontory is similar to Stubline and Okolište with a high building density and a strict structuration in house rows whereas the periphery, outside of the promontory, has a lower building density. In Drenovac, some parts of the western and northern areas also have a high building density but between these there are areas with clearly lower building density. The high-density areas are similar to those on the promontory at Belovode. There might be a similar spatial order at Belovode as at Drenovac. Also similar are the house orientations and the size of settlements. So far, Pločnik appears most divergent, being a little smaller at 26 ha and with a more regular distribution of house groups. The clearest difference is the house orientation to the northwest. However, such chronological considerations should be considered as hypotheses to be proven (or disproven) by further investigations through excavations or drillings (see Müller-Scheeßel *et al.* 2020).

The relation between open space and built space

The magnetic data for Crkvine allows for a further valuable observation concerning the relationship between the building areas and the open space in the settlement (Crnobrnja 2011: 131). A total of 250 burnt houses in the survey area (7.7 ha) were aligned in rows, covering *c.* 3.5 ha. This leaves 4 ha of open space for communal activities (pathways, communication, meeting places, etc.). An initial hypothetical reconstruction of built space in Okolište (Hofmann and Müller-Scheeßel 2013: 100, Table 32) failed to consider the very likely existence of unbuilt, communal areas, therefore adjustment is necessary. Using the smaller figure of *c.* 300 houses for Okolište, the house density of *c.* 40 houses per ha is close to the *c.* 32 houses per ha at Crkvine. If we assume for Okolište a similar ratio of built to unbuilt space for communal activities as at Crkvine (3:4), this would further reduce the estimated house number at Okolište.

Okolište and the Visoko Basin as a model for the size of social groups in the Late Neolithic

The research program in Okolište and the surrounding Visoko Basin provides data for the reconstruction of both regional and local settlement patterns, and it is possible to connect the chronological dynamics of settlement patterns at both scales. The history of the settlement of Okolište is characterised by a stepwise reduction in settlement size from the early 7 ha village in around 5200 BC to a relatively small settlement of 1.2 ha in around 4600 BC (Müller *et al.* 2013a). On the other hand, during this time in the Visoko Basin we can observe the development of an increasing number of small settlements with a strikingly uniform size of around 0.3 ha (Hofmann 2013b: Table 170). At one of these, Kundruci, which was founded in around 4900 BC (Furholt 2012, 2013), the excavation data allowed for an estimation of house numbers. The first settlement phase at Kundruci consisted of around 4–10 houses (about 20–50 persons). Based on the similarity in material culture, house types and the organisation of space, as well as the synchronous settlement size reduction in Okolište, we propose that people formerly living in Okolište founded the settlement at Kundruci.

In the settlement of Crkvine, most house rows consist of a similar number of houses. Unfortunately, the magnetic data from Okolište is much less detailed, but given the similar overall structure, it seems plausible to assume a similar social composition, and we thus propose that social sub-units of about 20–50 persons also existed in Okolište. In the light of the data form Kundruci and the other small settlements in the Visoko Basin covering a similar settlement area of around 0.3 ha, we propose that the decline of Okolište was a result of the departure of such sub-units consisting of 20–50 persons from the site. This would represent a process of social fission, where social groups, constituting house rows in a larger village like Okolište, left in a kind of dioecism and founded new, separate small hamlets in the vicinity.

Social units of 20–50 persons are often described in anthropological contexts as minimal for a lineage group (Hahn 2012: 33 ff.) We propose that lineage groups of this size formed the social backbone of the larger Vinča sites, and that they had sufficient political independence to separate from the larger social units.

Conclusions

The comparison of different settlements assigned to the Vinča culture (and Butmir culture) has revealed both differences and similarities. The magnetic surveys corrected previous estimations of size, reducing previous estimates of 100 ha sites to sites of less than 30 ha. Despite this correction, the large Vinča settlements of between 25 and 35 ha indicate agglomerations with more than 1000 inhabitants. This is clearly different to the dispersed settlements of the Linear Pottery and the younger Lengyel-Culture. The complex Vinča settlements are closer in form to the settlement mounds in the Eastern Pannonian Basin.

The demographic dimension of the Vinča settlements is unique within the Late Neolithic and Copper Age in central and southeastern Europe. Only the large Tripolje settlements in Moldova and Ukraine are larger in size. The existence of social sub-units is likely, and we proposed the existence of such groups comprising 20 to 50 people, who occupied house rows in larger settlements like Okolište and Crkvine.

Large unbuilt areas and the existence of large enclosure systems around these extensive sites indicate the presence of settlement-wide social institutions connected to the construction and maintenance of communal features. Nevertheless, social fission, the breaking out of a social sub-unit from the overall settlement and the founding of a new, smaller site, was a viable and frequent phenomenon during the Late Neolithic. One model to explain this involves changes in inheritance practices (Müller 2017).

The bibliographic reference for this chapter is:

Rassmann, K., Furholt, M., Müller-Scheeßel, N. and Müller, J. 2021. The social organisation of the Vinča culture settlements. New evidence from magnetic and archaeological excavation data, in Radivojević, M., Roberts, B. W., Marić, M., Kuzmanović Cvetković, J., and Rehren, Th. (eds) *The Rise of Metallurgy in Eurasia*: 455–459. Oxford: Archaeopress.

A deeper understanding of these phenomena needs a broadening of the comparative perspective to include settlements from a wider geographical range in a larger diachronic scope. The advances in the possibilities to conduct large-scale geophysical investigations has stimulated—and will continue to stimulate—further research in this field, which will also help to refine or reject the models proposed here.

Appendix

Appendix avaliable online as part of Appendix B at https://doi.org/10.32028/9781803270425/AppendixB_Ch38

Chapter 39

Belovode and Pločnik: site visibility and remotely sensed data

Jugoslav Pendić

Introduction

This chapter discusses the application of remote sensing approaches for the study of the Belovode and Pločnik sites and their respective surroundings. Some of research methods have been known and used for decades (e.g. aerial photography, archive aerial imagery, satellite reconnaissance) while others have only recently become established as routine field tools (e.g. structure-from-motion photogrammetric survey). The purpose of the study was to improve understanding of the location of the sites and to complement research undertaken in the field.

Belovode is unique in that the probable extent of the site is readily identifiable through aerial and low earth orbit imagery (Figure 1, a and b). A roughly circular shape of dark-toned surface sediment, encompassing *c.* 60 ha can be viewed throughout the year (with visibility varying in rhythm with vegetation cycles). The site envelops two parallel, shallow valleys, 300 m apart (Figure 1, e and g). The northern valley contains an intermittent stream that is seasonally active; the southern valley is largely dry throughout the year. The footprint of the settlement covers both the central and southern plateaux, widening towards the north, and appears to include areas beyond the boundaries established through near-surface geophysics (as seen in Chapter 9, this volume).

Unlike Belovode, Pločnik (Figure 1, c and d) is not so readily outlined. Its limits are better known through research history and interaction with industrial and infrastructural development of the area, that have both affected the total extents of the site. Whereas the site of Belovode is relatively isolated and removed from human agency, Pločnik lies beneath a modern settlement and is located on a major regional routeway and near to railroad structures (Figure 1, f and h). This introduces a few problems. Just as modern debris, activities, housing and industry affect some methods of near-surface geophysical survey, so do they affect remotely sensed datasets: earth removal and redeposition, agricultural activities, construction and material dumping create effects that must be taken into account when interpreting remotely sensed information.

Despite this, a continuous darker toned sediment, distributed in a way that largely supports the outlines of the site proposed by magnetic survey, can be tracked throughout the year; any ambiguities in the footprint are related to intrusive activities of various magnitudes. For both Pločnik and Belovode, the appearance of the site on remote imagery depended largely on the season at the time of the acquisition and type of sensor used. This was of particular importance for the unmanned aerial vehicle (UAV) surveys and the follow-up photogrammetric terrain reconstruction, as this required low vegetation and/or open fields for the on-site element (i.e. maximum visibility of bare ground).

Overview of data used

A selection was made from open access and commercially available remotely sensed imagery (Table 1). The aim was to obtain a wide temporal perspective of the site, providing a historicity of the location, and to exploit the existing visible record acquired by the multispectral array of orbital sensors.

The Landsat Mission Data Repository is a readily available data source for landscape studies. In orbit since 1972, Landsat sensors have been continuously acquiring imagery of the Earth's surface, with variation across the individual missions in terms of spectral and spatial resolution (Lasaponara and Masini 2011; US Geological Survey 2020a). The project is the joint effort of the US Geological Survey (USGS) and the National Aeronautics and Space Administration (NASA). It can be accessed and appraised through the USGS online database explorer (US Geological Survey 2020b). Designed as a civilian observation satellite for collecting medium-to-high spatial resolution multispectral data (Deroin *et al.* 2011), Landsat has been through eight mission series, with the ninth iteration announced for March 2021. The current active missions are Landsat 7 and 8. Ease of access to the data has improved considerably since the earlier missions (cf. survey results and methodology employed by Montufo 1997). An open and free data policy established in 2008 (Zhu *et al.* 2019) saw a steep increase in user engagement with the Landsat collections.

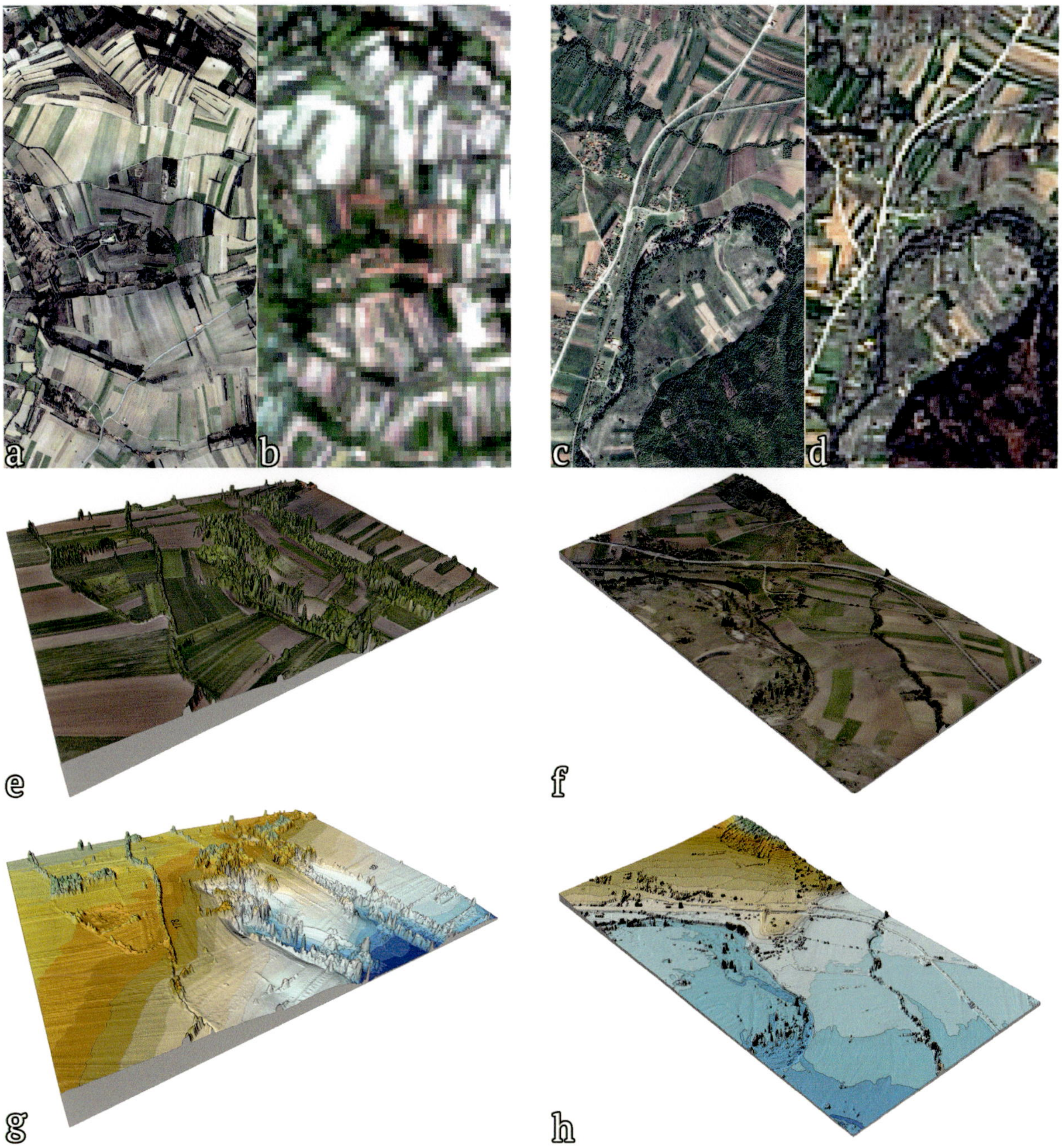

Figure 1. Overview of both sites: a-b) Belovode, in high resolution/low resolution satellite imagery; c-d) Pločnik, the same; e) isometric view of Belovode, orthoimage; f) isometric view of Pločnik, orthoimage; g) Belovode, terrain model, isometric view; h) Pločnik, the same.

Table 1. Datasets used for the study.

Description	Data access	Data host	Product type
Landsat imagery	Open access	US Geology Survey (https://earthexplorer.usgs.gov/)	Medium-to-high resolution multispectral imagery from low orbit
Declassified satellite imagery	Open/Restricted access	US Geology Survey (https://earthexplorer.usgs.gov/)	Medium resolution black and white panchromatic imagery from low orbit
Historical Aerial Photography (HAP) imagery	Restricted access	MGI (http://www.vgi.mod.gov.rs/)	High resolution black and white panchromatic imagery, high altitude aerial survey
Unmanned Aerial Vehicle (UAV) survey	/	/	High resolution RGB imagery from low altitude aerial survey

An historical overview of Pločnik and Belovode was achieved using national archives, but also through use of external sources including the satellite surveillance and reconnaissance missions maintained by the US Government during the Cold War period. This imagery has proven invaluable for archaeological study in the Near East (Beck and Philip 2012; Bitelli and Girelli 2009; Menze and Ur 2012; Ur 2003), notably where little or no systematic mapping projects had been undertaken (Cassana and Cothren 2008) but it has also seen application beyond this region (Goossens *et al.* 2006; Watanabe *et al.* 2017).

Looking at national sources, Historical Aerial Photographs (HAPs) are available for the whole territory of the Socialist Federal Republic of Yugoslavia (SFRY)[1] and are held by the Military Geographic Institute (MGI) in Belgrade as film rolls and/or digital scans. The archive has been formed through systematic reconnaissance by the army through multiple decades. The first missions were flown from Pleso Airport, near Zagreb (Croatia) using the Petlyakov Pe-2 bomber, adapted to take imagery with a large-format aerial camera (Buder 1984). The following decades saw the constant build-up of this collection by the SFRY Army and its inheritors, to its current (vast) proportion.

Finally, to obtain a contemporary view of both sites, a photogrammetric survey approach was deployed. This method has been increasingly used in field applications during the past decade for its affordability, its wide range of applications, and its rich outputs (for comparison see Carvajal-Ramirez *et al.* 2019; Dubbini *et al.* 2016; McCarthy 2014; Pakkanen *et al.* 2020; Remondino 2013). It allows the use of a custom set of image data for the creation of accurate spatial plans, orthographic maps and terrain models of high detail. This is the only approach used that required a substantial field presence. A three-member team was involved, using an industrial grade hexarotor platform with RGB sensor.

Data processing

Landsat 8 data

When dealing with Landsat datasets, researchers can typically expect a spatial resolution of 60 m pixel size for Landsat 1-5 (Multi Spectral Scanner), and up to a high resolution of 30 m for Landsat 4-5 TM (Thematic Mapper Plus) and Landsat 7-8 (ETM+ and Operational Land Imager respectively), with possibilities for performing a resolution merge with panchromatic imagery, which cuts image pixel size by half (15 m) for all products Landsat 7 and later. The detectable event on the ground thus needs to be sufficiently large not to be engulfed by the resolution of the image. This acts as limiting factor for detection applications: features smaller than the pixel size will be unnoticed. Despite this, Landsat imagery has been employed as a tool for predictive location modelling, by delineating favourable environmental settings (Custer *et al.* 1986) or as a more direct solution for detecting (Bloom *et al.* 1997) or accounting for the evolution of landscape and changes over time in the vicinity of archaeological content (Deroin *et al.* 2011).

This chapter considers only Landsat 8 output. Using Landsat 7 could have reduced the number of times we revisited the locations, but this was non-essential. Landsat 8 also has different characteristics to Landsat 7 including improved signal-to-noise ratio and calibration, narrower spectral bands, and higher 12-bit radiometric resolution (Roy *et al.* 2016). L1 products were selected from the Landsat 8 collection, in order to create a time series across the yearly vegetation cycles. Pre-processing was performed using PCI Geomatica Focus software (Banff ver.2019-11-25; www.pcigeomatics.com) and included haze removal (Zhang *et al.* 2002), pansharpening (Zhang 2002a,b) and clipping the scenes to a more manageable area of 25 km radius around Belovode. False colour composites (Parcak 2009) were created for enhanced visualisation of the dataset.

Historical Aerial Photographs (HAP)

Regions of the SFRY have been surveyed from various altitudes at different scales on a roughly ten-year basis. The potential of this imagery is largely untapped in Serbian archaeology and published examples of its use are rare (Бабовић 1992; Бугарски and Иванишевић 2014a,b; Ivanišević and Bugarski 2015; Ivanišević *et al.* 2015; Ivanišević *et al.* 2018). There are further references to purposefully acquired (now historical) aerial imagery by contemporary researchers (Грбић 1936, 1950, 1951; Дероко 1950; Дероко 1951a,b; Васић 1987).

Table 2. Declassified imagery sources and corresponding site coverage.

Label	Acquisition date	Camera ground resolution	Site coverage
DS1022-2104DA040	26.07.1965	(2.7–7.6 m)	Belovode
DS1103-2139DA092	30.05.1968	(2.7–7.6 m)	Belovode
DZB1206-500023L00400	19.07.1973	(6–8 m)	Belovode, Pločnik
DS1022-2104DA047	26.07.1965	(2.7–7.6 m)	Pločnik
DS1042-2105DA048	23.06.1967	(2.7–7.6 m)	Pločnik

[1] The country was known as the SFRY between 1963 and 1992, but the collection of HAPs began earlier, in the 1950s, when the country was the Federal People's Republic of Yugoslavia (1945–1963).

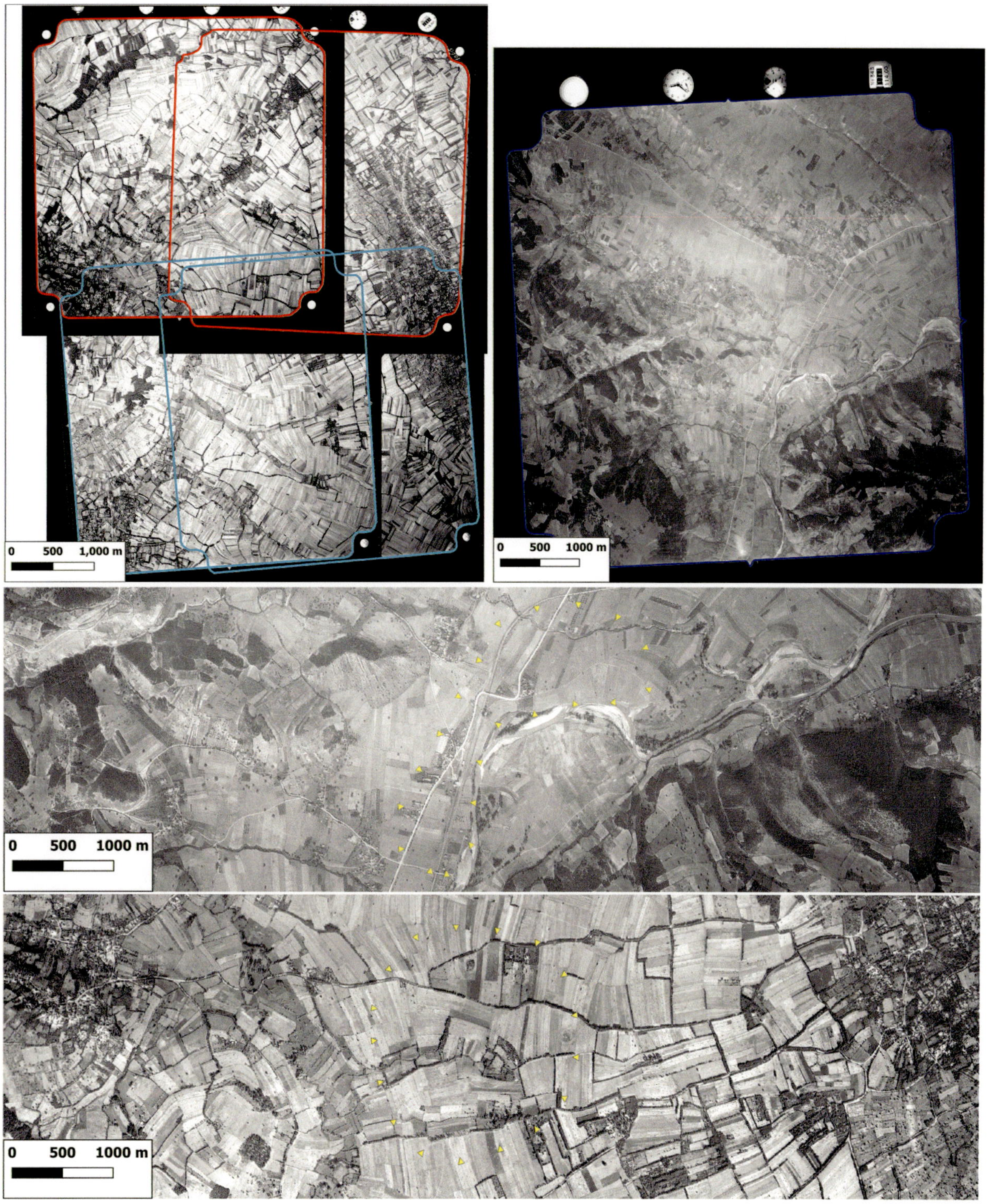

Figure 2. Historical aerial photography of both sites. Upper-left: MGI image coverage of Belovode, forward and lateral. Upper right: Pločnik, single frame covering the whole site. Middle image: Pločnik, site area overimposed on HAP. Lower image: Belovode, HAP mosaic, site area overimposed on HAP.

Selecting imagery for a given region is not a straightforward task, especially in rural surroundings with few or no permanent features. The main challenge is to recognise a top-down view of landscape that is several decades old and probably very much altered. The current MGI mission catalogue is not fully digitised (apart from the image data itself) and relies upon hardcopy documents. Once the area of interest (AOI) is determined, it can be provided as a hard copy printout or digital scan of the whole or partial negative.

For the Belovode area, imagery from 1953, 1968, and 1981 (at 1:20000, 1:32500, and 1:50000 scales, respectively) was identified. Images from both 1953 (see Table 2) and

1981 were of good quality but that from 1968 was over exposed (or poorly developed) and of limited use. The site was covered with four frames, with site extents visible in the corner of the frames from two adjacent rows (Figure 2, a and c).

Images had a declared overlap of 60% forwards and 30% sideways. Adjacent photographs on a roll (or from adjacent rows) can therefore be used for stereo reconstruction of terrain geometry, if accurate ground control points can be obtained for area coverage (Cantoro 2015). However, the overlap found for the Belovode area was insufficient, the horizontal overlap between forward rows being significantly below 30%.

For Pločnik we selected the earliest recorded imagery, taken in 1959, at 1:32500 scale. A single image covered the extent of the site and a wide area around it, hence there was no need to introduce multiple frames (Figure 2, b and d).

Data was provided in 8-bit tiff format at 1270 x 1270 dpi. To accurately synchronise this imagery with all other data included in the study, an initial attempt was made with automatic registration of the raster images on Google Earth and Bing aerial base layer, using AutoGR software (Cantoro 2012). This did not produce the desired results. The SIFT (Scale Invariant Feature Transform) algorithms of AutoGR, while quite powerful for building a base of common points between two overlapping images, require a substantial degree of similarity between them. This was impossible to achieve as both sites have undergone significant change through multiple decades of land use. Damage to the negatives from which digital scans were made was also detrimental to this approach. For registering and orthorectifying images, we again employed the PCI Geomatica OrthoEngine (Ortho Banff Edition 2019-11-25) module, using Shuttle Radar Topography Mission (SRTM) 1-ArcSecond Digital Elevation Model (DEM) as a terrain correction layer. Once registered, each individual frame could be studied individually in a GIS environment.

Declassified low orbit satellite imagery

Declassified satellite data owe their good reputation for use in scientific study due to their very high spatial resolution of 2–3 m (US Geological Survey 2008). Further, just as for the HAP imagery, some of the missions predate industrial/urban development that has affected the landscape and, potentially, any archaeology present.

Some of the output was delivered in stereo pairs that could be used for DEM generation (Altmaier and Kany 2002). This meant that terrain composition lost due to natural processes or human activity could still be accessed with a level of detail even surpassing open access terrain data (Cassana and Cothren 2008). Another benefit was that the films and filters used in combination were partially sensitive to the invisible spectrum of light, so that features detectable in infra-red wavelengths would stand out on images from certain missions (Fowler 2012).

The USGS database of declassified historical imagery (US Geological Survey 2008) offers a clear overview of the available scenes. The task of accurately positioning image strips across the close-to-accurate extents within the database can be challenging, and slight offsets in presented coverage are unavoidable. It is recommended that, when selecting frames that enclose the AOI, those where the AOI is closest to the central axis of the image are given priority. Distortion increases towards the frame edges where the surface viewing angle can be severe, depending on the position and orientation of the satellite at the instance of image acquisition.

For the purpose of this research, the mission data analysed (Table 2) was collected on 26 July 1965 - DS1022-2104DA040 (KH-4A), 30 May 1968 - DS1103-2139DA092 (KH-4A), and 19 July 1973 - DZB1206-500023L004001 (KH-9 HEXAGON lower resolution mapping). The first two of these (KH-4A) cover the Belovode site; the third (KH-9) extends over both sites. Pločnik is seen on several other frames: 26 July 1965 - DS1022-2104DA047 (KH-4A) and 23 June 1967 - DS1042-2105DA048 (KH-4A) (Figure 3).

Unmanned Aerial Vehicle (UAV) photogrammetry

Photogrammetric survey has developed from a rarely used (in archaeology) and complex methodology demanding considerable resources, expert personnel, equipment and mobilisation protocols, into an operation that can be conducted by a very small teams, or even individuals without specialist training (Greenshaw 2018). The processing software and hardware, and the single or multiple sensor platforms used are accessible to research and educational institutions, and available from small vendors. The process relies on the collection of a great number of images from a non-metric, consumer camera, aimed at covering the documented surface from multiple viewing angles; the data can be used to obtain ortho rectified plans of areas, digital surface models (DSM) and, in limited cases, digital terrain models (DTM).

For the Belovode survey, an Aibotix hexarotor platform, fielding an Olympus E-PL5 camera was used. The flight path was projected with 80% forward overlap of photographs (forward motion of the vehicle) and 60% side overlap (sideways overlap of traversed lines). The

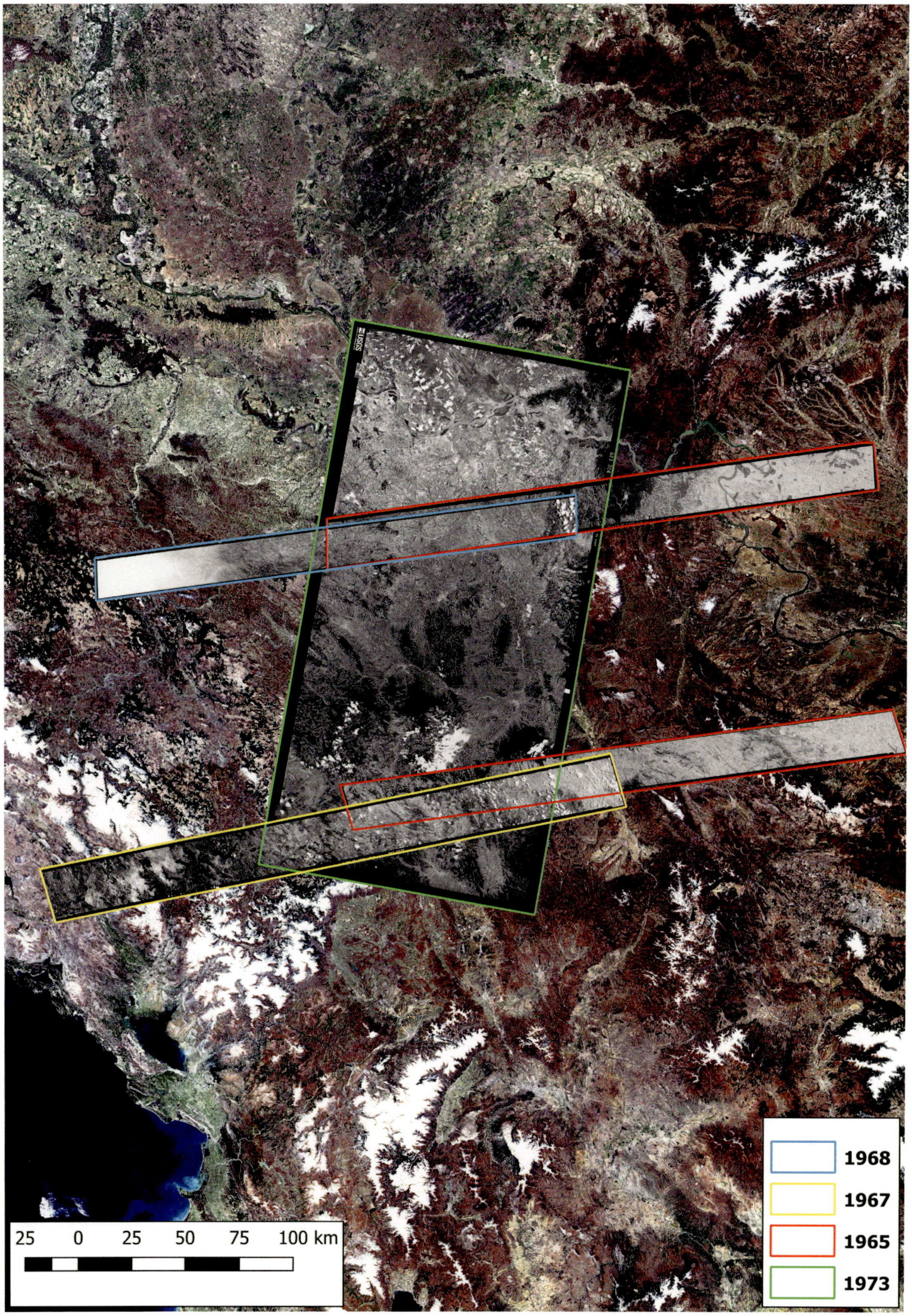

Figure 3. Area covered by declassified imagery individual frames, between various missions

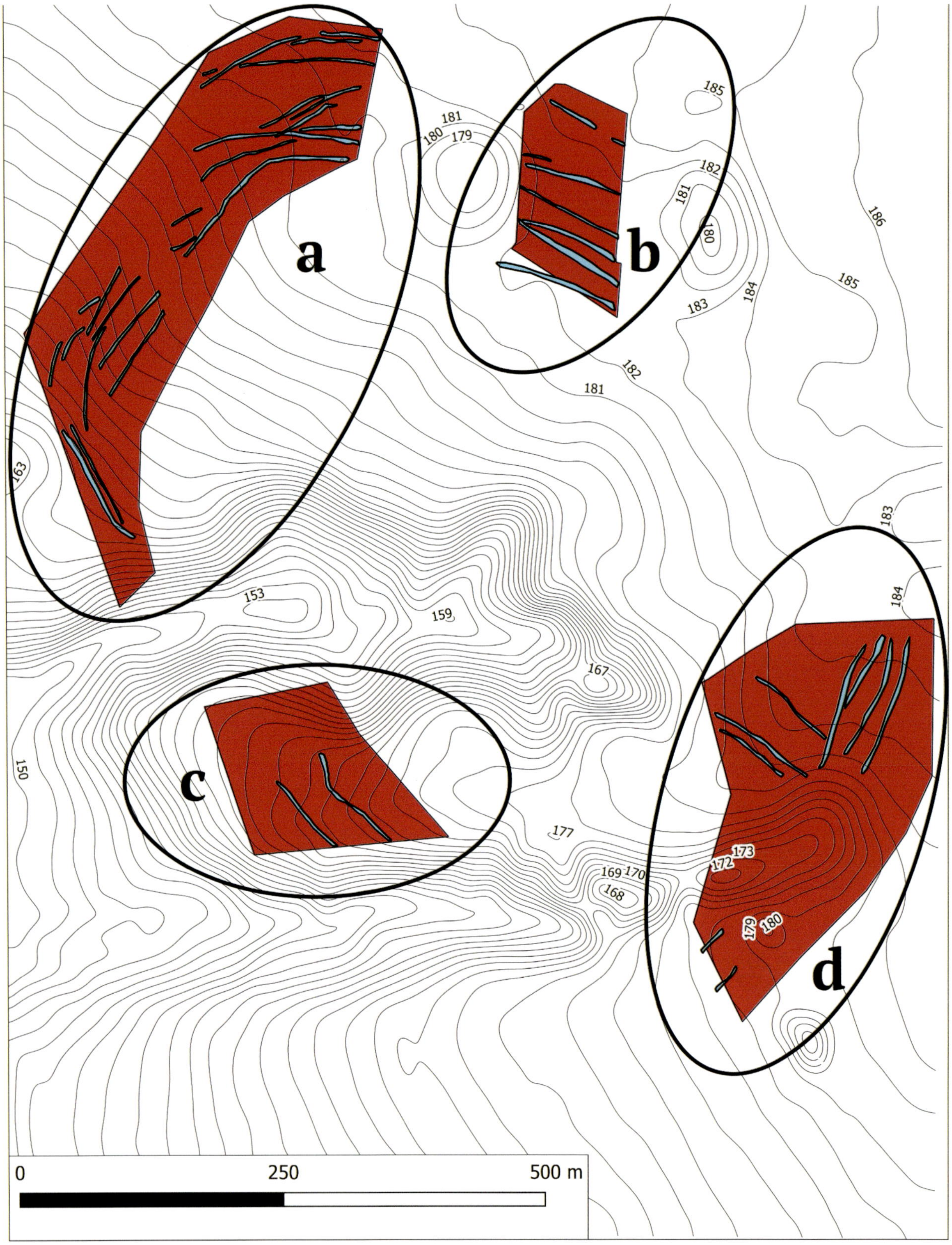

Figure 4. Spatial constraints applied to processing UAV imagery of Belovode, introduced to speed up the process and allow efficient management of data.

AOI was a 1 km^2 rectangular footprint encompassing the outline of the site (as seen on the historical and modern remotely sensed data). Photographs were taken in strictly nadir mode between midday and late afternoon. This introduced significant variations in image brightness that were corrected in post-acquisition stage, before processing with the photogrammetric reconstruction software.

The height of the flight was set to 100 m above ground in order to target ground sampling distance of approximately 2.5 cm. A series of ground control points was set across the AOI, with positions measured by the differential GPS unit. The low flight plan and sensor combination required a total of 1550 images to conform to the project specification; pixel size was chosen to produce a high-resolution terrain model of the site.

The camera network and orientation were generated using Agisoft Photoscan (now Metashape) software but the reconstruction of a dense and accurate point cloud through multi-view stereo process was completed using Nframes Sure (Rothermel *et al.* 2012), exploiting the more robust and customizable pipeline for production. The same software was used to create DTMs and a dense point cloud. The DTMs were later processed in LiDAR visualisation software, RVT (Zakšek *et al.* 2011; Kokalj and Somrak 2019) and reclassified in QGIS (QGIS.org 2020, 3.12.3).

Localised analysis was performed on areas where a prominent presence of subsurface structures was documented (Chapter 9, this volume). The proposed dimensions of ditches surrounding the habitation area, alongside a nearby concentration of household features, were a promising set of targets, despite the low possibility to observing any effects on the surface. Dense point clouds were created by using maximum pixel resolution of the nadir UAV images, with spatial constraints included (Figure 4, a-d) to speed up calculations and increase manageability of the data. Further to this, a rigid export model was adopted by using only reconstructed points from multiple (3+) stereo models (for details see nFRAMES SURE 1.3). Since most of the site was covered by photographs from a minimum of seven and maximum of nine positions, only weak reconstructions were excluded. Micro shifts in terrain were thus studied on a DTM generated from 66 million points.

The operation at Pločnik differed from that at Belovode in the amount of data collected. A time series of UAV surveys was conducted over the northern and southern halves of the site during low vegetation periods (first mission in November 2015, the second in December the same year). The southern survey (November) focussed on the more urbanised portion of the site mixed with arable fields and the excavation zone. The northern survey (December) concentrated on the confluence of the Backa and Toplica rivers and potential traces of past riverbed movement. A complete site plan was produced in early April 2016, spanning a little over 1 km^2. For all surveys, an identical platform was used, with different sensors (Nikon Coolpix A), and a total of 1807 images acquired (Figure 5, a-f). For both sites, further steps included the use of LAStools (LAStools ver.200509) employed to filter, de-noise and classify the photogrammetry-generated point cloud, and to create a high-resolution, bare ground surface model.

Discussion

Belovode

The footprint of Belovode owes its clear outline to a soil composition extensively changed by human agency in the past, through actions tied to the everyday life of a settlement. The intensity of the change is evident: the physical boundary of the site presumably corresponds to the outer rim of the ditches, but the soil colourisation conforms to the space close to and around the households, as seen on the magnetometry plan and the colour response (Figure 6). Where the magnetometry survey had yet to cover areas of ground (Figure 6, c and d), it is possible to plot the extent of the sediment from actively used areas (at least one phase of it) using satellite imagery and data obtained from UAV flights. To understand this result and its reliability, it is important to account for the effect of human-induced changes to the topography and land cover in the modern period.

Land use and site encroachment

At the time of the massive development in national agricultural capacities (mid 1950s to 60s) only a few areas were covered with mature woods or low growth (Figure 7a). A notable distinction is seen in the northern portion of the site where field orientation is shifted compared to the contemporaneous situation, and land plots are grouped into a single system, most likely to support the use of mechanised equipment (Figure 7 a-1). The northern trenches and household concentration lie in this area; any near-surface archaeology would surely have been affected. The date of the change falls in a period between two images: MGI (1953) and Corona archive (1965), when the introduction of the mechanised farming saw an increase in the use of massive agricultural machines across the country (Simonović 1967). The rest of the site has seen little change in this respect. Nearly all the fields are regularly maintained and cultivated, and although the land plot system is drifting it can be tracked and referenced to the present organisation. Several active farmsteads can be seen within site boundaries, both on the northern and southern sides of Belovode from as early as the first aerial image available. Together with access roads, and the earthwork in the pond area, these mark the only other direct encroachment onto the site, and the most significant impact on the local stratigraphy (Figure 7a 3,4,8). As the temporal series of images advances, the number of overgrown fields and the extent of medium-

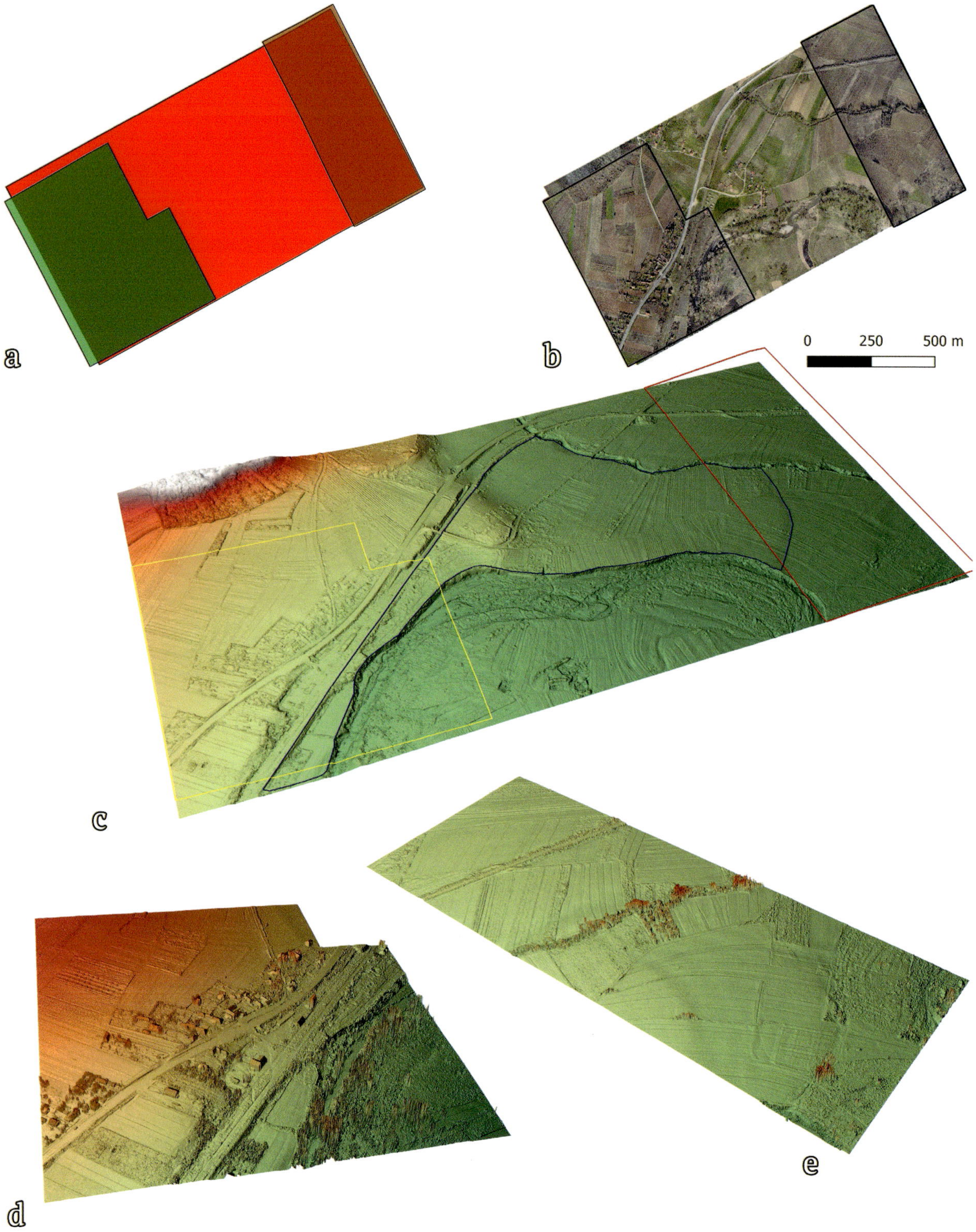

Figure 5. Pločnik, overview of UAV mission: a) ground coverage of several UAV flights (Nov, Dec 2015 in brown and green, respectively, April 2016 in red); b) the same, with included orthoimagery; c) April 2016, isometric view of DTM of Pločnik, with emphasis on Toplica river, and marked Nov 2015 UAV ground coverage (in red) and Dec 2015 UAV ground coverage (in yellow); d) Dec 2015 DSM terrain model; f) Nov 2015 DSM terrain model.

to-high growth vegetation increases, underlining the trend of land abandonment (Figure 7, b-g). This trend can be corroborated throughout the whole country by cross-referencing present-day images, HAP frames and declassified images. As regular maintenance of the land declines, low to medium growth recaptures the fields, with a gradual introduction of mature, high growth in parts of the site (Figure 7a 5,7). The

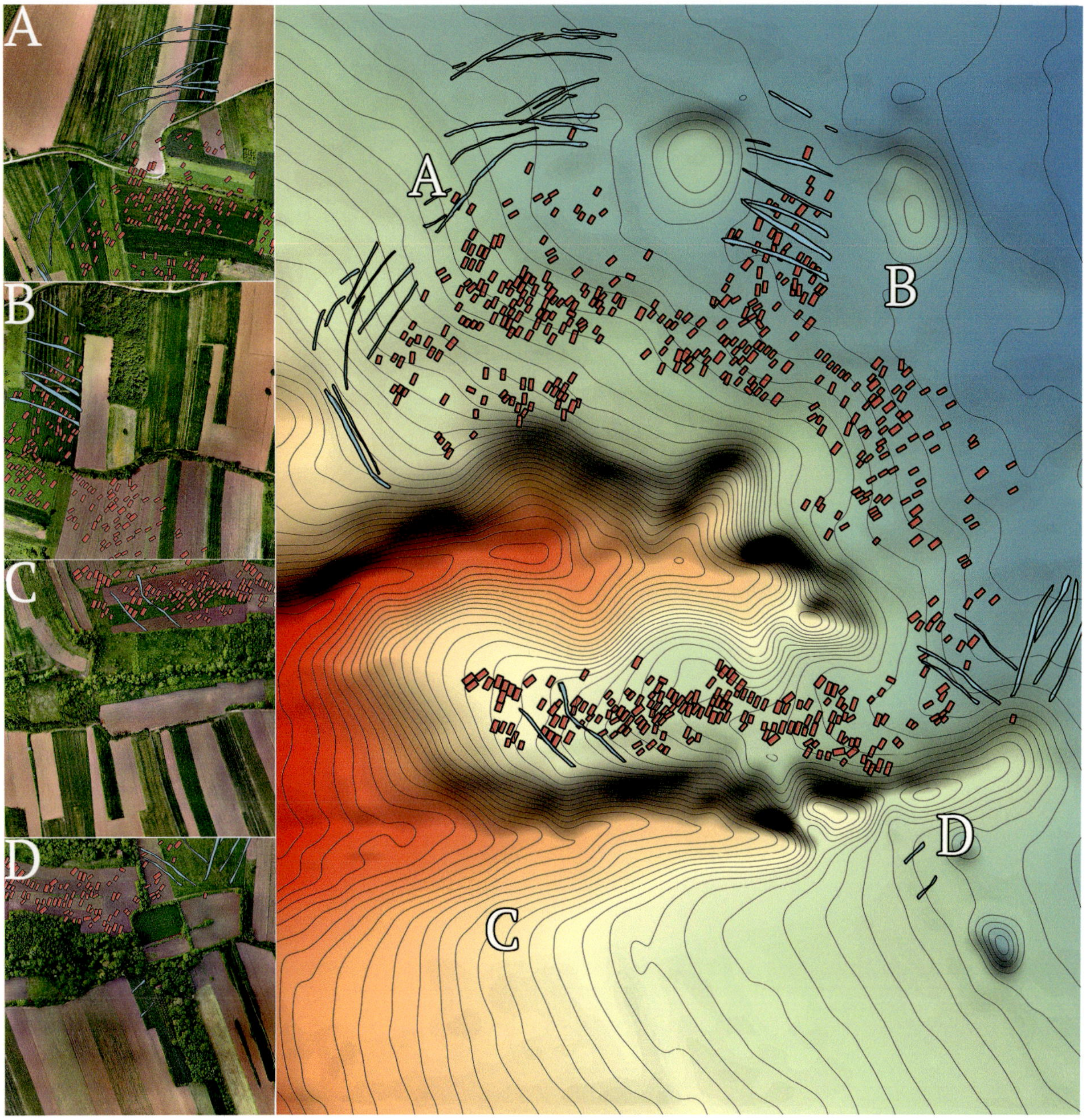

Figure 6. Belovode, overview of site: a-d) boundaries of soil colourisation and their respective positions related to the magnetometry results (in blue: system of ditches surrounding the site; in dark red: interpreted positions of household anomalies)

impact of land cultivation on archaeological sites has been discussed previously (Darvill 1987; Mantellini and Berdimuradov 2019; Spandle *et al.* 2010; Yorston *et al.* 1990) but no detail is available on the subject of the abandonment of previously (modern) cultivated land and the impact of the returning uncultivated plant growth on subsurface remains and sediment shifting processes. These revolve around changes in erosion processes, bioturbation effects (rooting and burrowing), but also the potential impact of the re-introduction of cultivation (repeated land cleaning processes).

One aspect of potential interest is the presence of two ponds in the northern and northeastern parts of the site respectively. The magnetogram shows that ditch structures become more complex and extend to encompass these areas into the general settlement footprint. This is clear for the far-north pond, but only indicated for the eastern one (labelled 'Bull's pond' on MGI archival maps; Figure 7 a6). The pond in the far north (Figure 7 a2) was actively used for watering as early as the mid-20th century, as evidenced by the multiple tracks leading to and from it; its use decreased somewhat towards the modern period (visible overgrowth increases). The potential significance of these two seasonal water resources to the Belovode settlement has been unexamined so far as the date of their formation and thus their availability to the ancient population, are unknown.

Figure 7. Historical aerial photos of Belovode: a) left: Belovode, HAP, 1953; right: same ground coverage in UAV orthomosaic from 2015; b–g) Belovode site in 1953 (MGI), 1965 (declassified), 1968 (declassified), 1973 (declassified), 1981 (MGI), and orthomosaic from 2007 (Google Earth).

Site boundaries and visibility

HAP imagery provides mixed output on this subject. Visibility is dependent on access to the bare ground The MGI image was recorded in September 1953 when most of the fields would be still occupied with cultivated plants. The same is true of the 1965 July Corona KH-4A image, however the record from May 1968 provides a strikingly different appearance, and it is possible to map out the boundaries of Belovode in terms of the affected sediment. The bare ground colour visibility is highest in the early months of the year in the early winter months. The images from MS satellite sources (see Figure 8) indicate that the 1968 image represents the seasonal limit for remote sensing surveys, with the optimal start being in late February.

Landsat 8 data shows a sequence of increasing and decreasing access to the surface soil throughout the year (Figure 8). The false colour composite, with SWIR and NIR channels in red provide the best control images over the study area. The coarse resolution is a constraint but is non-detrimental to the study. Geophysics survey shows few-to-no traces of settlement on the southernmost plateau however survey of this area was limited to only few land plots; it can now be supplemented by the satellite data.

The orthorectified plan of the site from UAV data represents the most detailed and accurate resource available for remote soil mapping. Even seen in only RGB, the colourisation of the impact zone is clearly visible (Figure 9). The eastern boundary, which is not clear in either HAP or declassified data, can be accurately delineated and conforms to the area of the site, encompassing the households and nearby ditches. It is interesting to note that the colour signal seems to stop short of the ponds (with the ditches encompassing more ground in the north; it is presumably similar in the eastern portion of the site). This might be indicative of the organisation of space being focussed on—or at least being affected by—the access to and use of the ponds. The western boundary is only partially visible, opening into the intermittent stream and drainage valleys leading towards the Busur river, but can be followed through several open fields. Where information on the southern plateau was completely lacking, the timing of the UAV survey allows us to propose an extension of the interest zone further southward: future magnetic and sediment sampling efforts should focus on this area.

Potential for detection of sub-surface structures via remote sensing

The geometric response was less positive: overall visualisation of the site through the LiDAR Relief Visualisation Toolbox (RVT) offered no direct evidence of subsurface structures (Figure 8). Attempting to measure elevation differences through profiles also gave no consistent output; we tried bare ground profiling, but there were multiple vegetation-covered surfaces, and we had not documented changes in plant height. In some respects, the photogrammetry-derived point cloud data (e.g. density and response to obstacles) are very similar to Airborne Laser Scanning data packages, minus some (highly) positive traits, such as multiple echoes and the consequent ability to create DTMs under thick canopy. The method is guaranteed to work on readily identifiable ground features and as complementary data to a LiDAR survey (Corns and Shaw 2013; Иванишевић and Бугарски 2015). Its usefulness for the current study or similar applications should be assessed according to the occasion. At Belovode, traces of underground features are simply not seen on the ground or in the photogrammetric record (Figure 10).

Pločnik

The Belovode UAV data showed remote, rural scenery, despite the relative proximity of the village of Veliko Laole, and state roads within walking range. The site of Pločnik is set in a very different locality: the primary excavation area is only few metres away from the railway (the field house for the remote sensing team and excavation team was a nearby—now derelict—railway station); and the archaeological site lies beneath the contemporary village of Pločnik. Earth used in the construction of some of the auxiliary houses in the existing village was sourced from local fields and sometimes includes pottery fragments in the plastered composition. While the landscape remains rural, it is a zone where contemporary and ancient living collide with force.

To add to this complexity, the re-deposition of naturally occurring sediment is documented at the site (Marić, Chapter 23, this volume). The Toplica river is cutting into the eastern part of the site in a long-term process. We can also document another occurrence, similar in origin, at the northern portion of site, where the Toplica and Backa rivers once affected the site boundaries and possibly removed a portion of the settlement.

Land use and site encroachment

Pločnik is actively used for agriculture and very little overgrowth is present. The only non-active fields are the excavation zones active since middle of the 20th century, under the protection of the museum in Prokuplje. The historical overview of the land use shows no changes.

Few infrastructural changes are notable: the state road between Prokuplje and Kuršumlija has been expanded from the existing route and extended northbound to

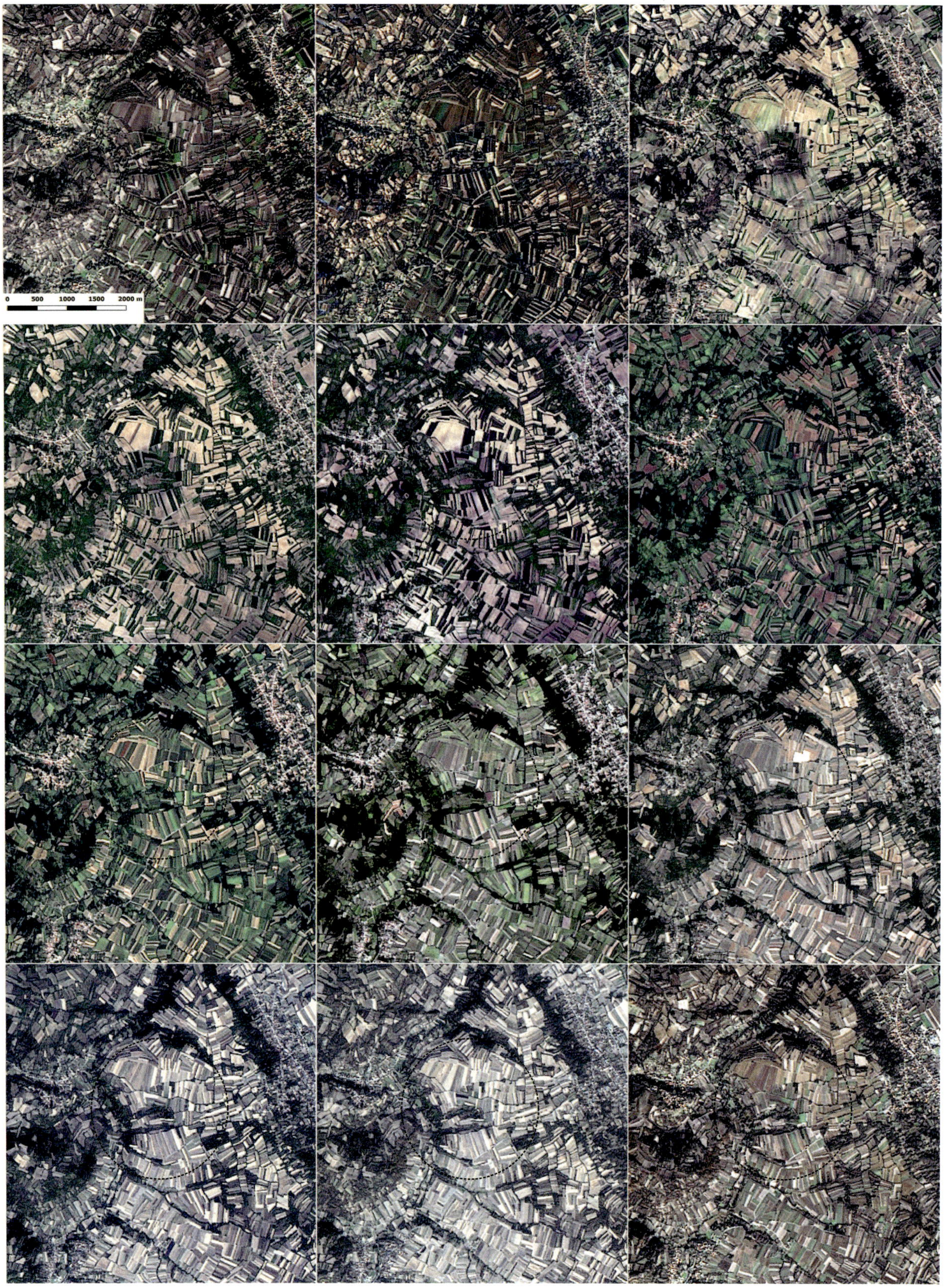

Figure 8. Belovode, Landsat 8, from Jan to Dec 2018 (left to right, in declining order).

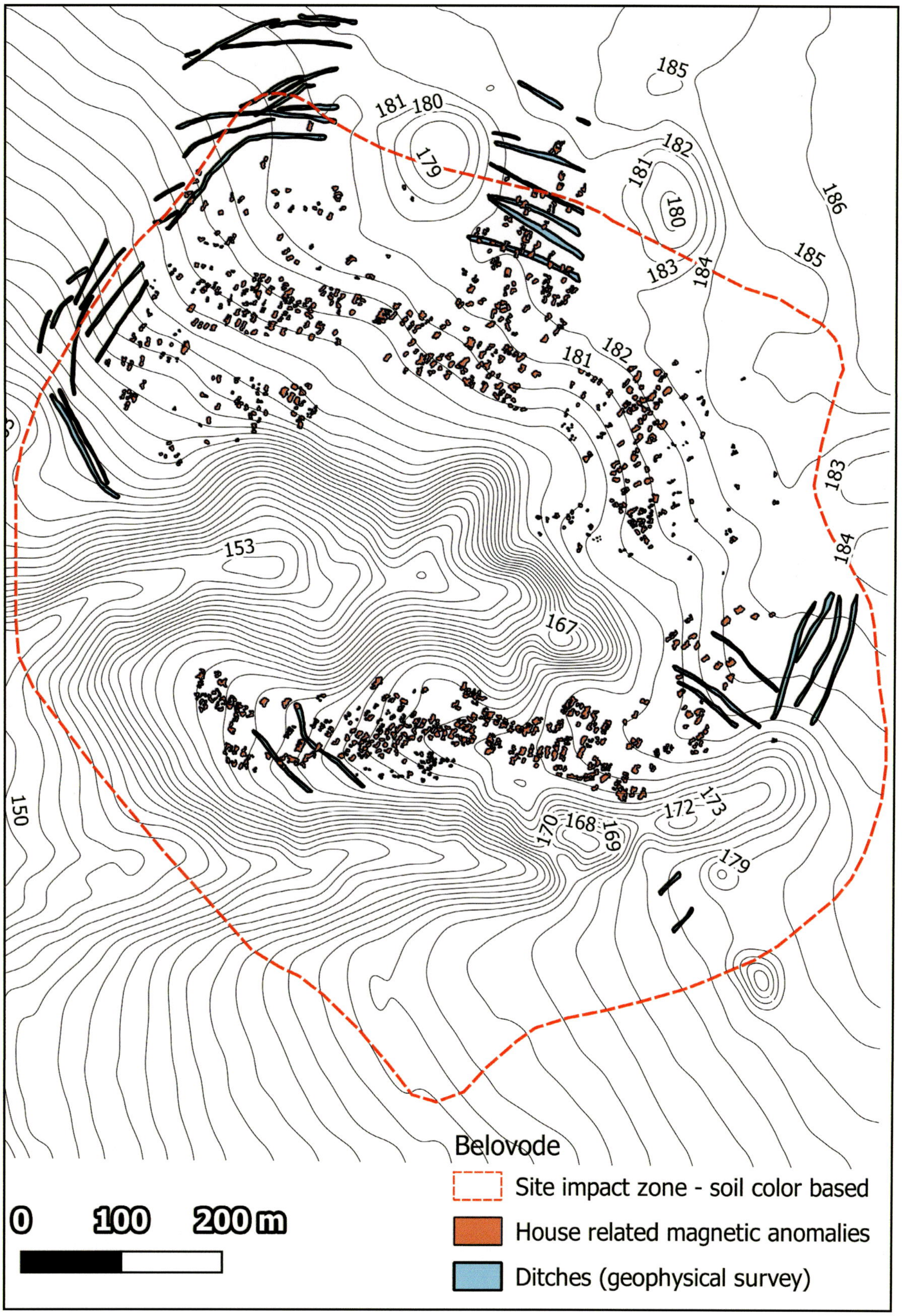

Figure 9. Estimated coverage of compositionally changed sediment, as seen from the remote sensing sources, with underimposed structure plan of magnetometry mapped ditches (light blue) and household related anomalies (dark orange).

Figure 10. Belovode, orthomosaic with overimposed visualisations of DTM (Sky View visualisation reported here), clearly demonstrating absence of detectable features related to ditches surrounding the settlement.

Prokuplje. The western part of the modern Pločnik settlement has also seen expansion: the overall amount of the encroachment onto the archaeological site by house building is much lower at the start of the image time-series. Between the historical imagery of 1959, 1965, and 1973 and the more recent imagery from 2015, a number of new buildings have been erected directly over the ancient settlement (Figure 11, a and b) adding to the existing disturbance: the number of structures has nearly doubled in the area that can be identified as

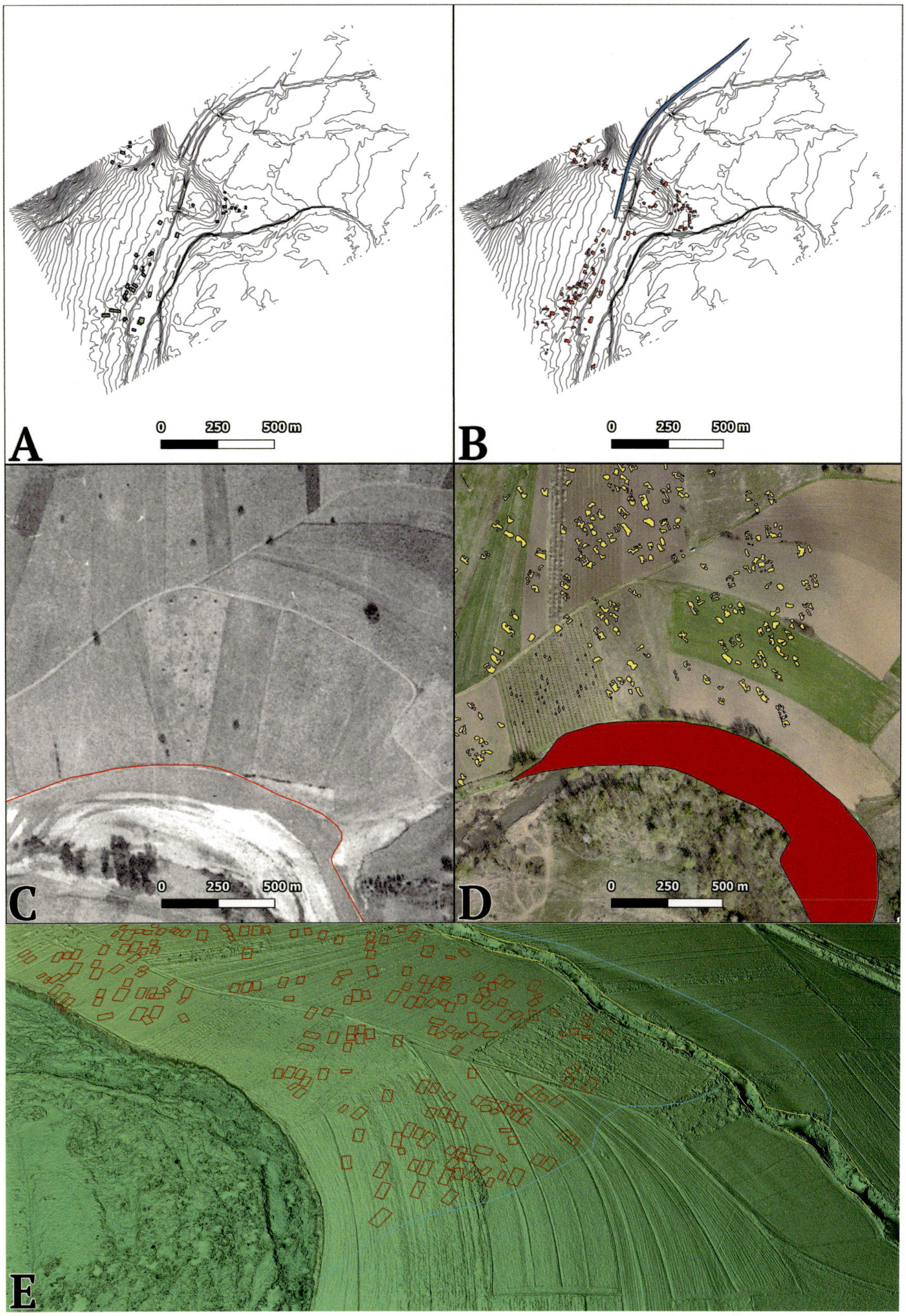

Figure 11. Pločnik. a) vectorised outines of structures of modern settlement in 1959; b) the same, in 2015/2016 with noticeable increase of structures in the middle area of the site; c) Pločnik, showing Toplica river bank in 1959 at the eastern edge of the site, with shoreline marked in red; d) the same position, showing area (shape marked in dark red) that was removed by the riverflow; e) the eastern boundary of the site: isometric view of the DTM with overimposed interpetation of household positions in red, and start of the terrain decline/boundary of distinctive soil colour related to site premises marked in blue.

central to ancient Pločnik (see magnetogram, Rassmann *et al.*, Chapter 24, this volume).

It is difficult to estimate the full extent of the Pločnik site at its zenith (or in its early stages) but the intensity of the process of erosion in modern times is indicative of how it may have been affected in the past. In the northeastern part of the site, fluctuations in the flow of the Toplica river have caused significant damage over a period of 65 years. Between the 1959 and 2015 surveys, a massive segment of the settlement (over 2 ha of surface) has been carried away by the river (Figure 11, c and d). The Toplica flow moved a little over 40 m northwards, cutting into a zone with groups of households.

Site boundaries and visibility

The outline of the site is detectable through the remotely sensed imagery only in part. Even the coarser resolution of the pansharpened Landsat 8 scene clearly delineates a darker patch of sediment present at the join of the lower planes of the Backa and Toplica rivers. It is unclear whether this can be attributed to the impact of the settlement alone, rather than to the colour of the alluvial sediment of the Toplica, which takes on a darker tone throughout the valley. The near-surface geophysics detects the abrupt end of the site at this location, as well as a massive change in the magnetic response, resembling a riverbed trace extending from in the direction of the Backa river. The terrain model shows a steep (although not so abrupt) decline at this spot, even more pronounced through visualisation tools (Figure 11e). At the same time, shifts in soil colourisation and terrain, seen on the fine resolution orthoimage of the UAV mission, suggests that additional survey with geophysical methods and core-augering would cover the site extent towards the north and across the current riverbed of the Backa (markedly different to that suggested by the magnetogram and digital surface model).

The site does not seem to extend westwards beyond the railroad and state road but the southern boundary of archaeological structures potentially extends almost 400 m beyond the primary excavation zone, along the western bank of the Toplica; unfortunately this cannot be corroborated by either the UAV or historical imagery.

Acknowledgments

We would like to extend our gratitude to the Belgrade Vekom team for great cooperation and help rendered during fieldwork operations and preparation of UAV material; to Konrad Wenzel and Mathias Rothermel, and the whole team at Nframes, for making available for testing their dense image reconstruction software (SURE) with our survey data; and to PCI Geomatica for making available satellite image processing software.

The bibliographic reference for this chapter is:

Pendić, J. 2021. Belovode and Pločnik: site visibility and remotely sensed data, in Radivojević, M., Roberts, B. W., Marić, M., Kuzmanović Cvetković, J., and Rehren, Th. (eds) *The Rise of Metallurgy in Eurasia*: 460–476. Oxford: Archaeopress.

Chapter 40

Population size and dynamics at Belovode and Pločnik

Marko Porčić and Mladen Nikolić

Demography of the Late Neolithic populations in Central Balkans

The reconstruction of the demographic aspects of Late Neolithic societies in the central Balkans is important for understanding two fundamental cultural processes: 1) social complexity (Alberti 2014; Carneiro 1986, 2000; Dunbar 1993; Feinman 2011; Johnson 1982; Kosse 1990; Porčić 2012a, 2019a); and 2) technological evolution (Collard *et al.* 2013; Derex *et al.* 2013; Henrich 2004; Kline and Boyd 2010; Shennan 2001). The intensity and scale of demographic archaeological research into the Late Neolithic in the central Balkans has been rather limited, largely due to little interest among traditionally oriented local archaeologists and a lack of relevant data; information comes from a small number of studies (e.g. Hofmann *et al.* 2009; Müller 2006, 2007; Porčić 2011, 2012a, 2019a, 2020).

The first attempt to systematically estimate the population size of the Late Neolithic Vinča culture settlements was made by John Chapman (1981: 48) who estimated that sites with an area up to 1.9 ha had population in the range between 50 and 300 people, sites with areas between 4 and 4.9 ha 200–500 people, and large sites with 20–29 ha had a population of between 1000 and 2500 people. The most comprehensive effort to reconstruct the Late Neolithic population size and population dynamics was undertaken by J. Müller, R. Hoffman and colleagues for the Okolište settlement and Visoko Basin in Bosnia (Hofmann *et al.* 2009; Müller 2006, 2007). According to the reconstructions, population size in Okolište initially increased from around 750 to around 2000 people between Phase 1 (*c.* 5200–5000 cal. BC) and Phase 2 (*c.* 5000–4500 cal. BC), and then decreased to around 200 people during Phase 3 (*c.* 4850–4700 cal. BC) (Müller *et al.* 2013b). In addition to reconstructing population size, an attempt was made to reconstruct population dynamics in terms of population growth rates throughout the Bosnian Neolithic (Hofmann 2013a). The rates for the Early Neolithic in Bosnia are rather low compared to the values expected for the Neolithic Demographic Transition (hereafter NDT). The usual NDT rates are between 1 and 2% (Bocquet-Appel 2002; Guerrero *et al.* 2008), while those reconstructed for Starčevo-Impresso and Kakanj are 0.25% and 0.48% respectively (Hofmann 2013a). The Late Neolithic growth rates reported by Hofmann are even lower at 0.13% and 0.05% for Butmir 1 (*c.* 5200–5000 cal. BC) and Butmir 2 (*c.* 5000–4750 cal. BC) respectively, whereas in the Butmir 3 phase (*c.* 4750–4500 cal. BC), population growth rate becomes negative (-0.63%) suggesting a population decrease. This pattern seems to resemble the boom and bust scenario reconstructed for Western Europe (Shennan *et al.* 2013; Timpson *et al.* 2014), but with much longer periods of population growth.

Porčić also attempted to reconstruct population size on several Late Neolithic sites in Serbia and Romania (Porčić 2009, 2011, 2012a, 2019a). For tell sites such as Uivar and Gomolava, the reconstructed final population size was 100–200 people, while for the flat site of Divostin the estimates ranged between 1000 and 2500 people. Porčić also made an attempt to estimate growth rates on the Late Neolithic sites of Parca and Gomolava at 0.7% and 0.8% respectively (Porčić 2009: 344–345). The macroregional reconstruction of the population dynamics suggests that the population size increased in the Late Neolithic and probably decreased at the end of the Vinča period (Porčić 2020).

This short review suggests that population size could have been relatively high at settlements such as Okolište, Divostin and probably Stubline (Crnobrnja 2012; Porčić 2011, 2012a, 2019a). It also suggests that the growth rates were still relatively high in the Late Neolithic, and only slightly lower than growth rates characteristic of the NDT. The aim of this chapter is to use the settlement data from Belovode and Pločnik to reconstruct the population size and dynamics of the Late Neolithic societies in the central Balkans.

Data and method

Estimating house numbers

The site of Belovode has been the subject of extensive geophysical survey (Rassmann *et al.*, Chapter 9, this volume) and the size of the settlement area (i.e. the area enclosed by ditches) can be estimated at *c.* 25 ha. On the promontory to the south, rows of houses are visible in the geomagnetic data. This contrasts with the northern part of the settlement, which has a lower density of houses (Rassmann *et al.*, Chapter 9, this volume, Figure

10). Analysis of the geophysics suggests that there were around 550 houses, with an average floor area of *c.* 50 m^2 (Rassmann *et al.*, Chapter 9, this volume, Figures 9 and 10). Given that almost 70% of the settlement was surveyed, this number can be extrapolated to 785 houses for the entire site[1]. Radiocarbon dates suggest that the final settlement horizon lasted from *c.* 4650–4470 cal. BC. It is likely that most of the houses visible in the geophysics plan came from this horizon, but it is possible that some came from earlier periods (e.g. as at Okolište, see Hofmann 2013a; Müller *et al.* 2013b). Moreover, Rassmann *et al.* (Chapter 9, this volume) suggest that, based on their spatial relations with the ditches, not all the houses visible in the geophysics plan were contemporaneous. Based on the geophysics, however, we assume that most houses belonged to the latest stage of the settlement, which reached its largest extent during that phase.

Pločnik has also been extensively geophysically surveyed (Rassmann *et al.*, Chapter 9, this volume). It is roughly estimated that around two thirds of the total site area (*c.* 26 ha) was surveyed with a magnetometer. It should be emphasised that part of the site was destroyed by the river Toplica. Analysis of the geophysics plan suggests that there was a total of 351 houses in the surveyed area, with an average house floor area equal of around 60 m^2 (Rassmann *et al.*, Chapter 9, this volume, Figure 6). When this number is extrapolated to the unsurveyed area, the resulting estimate is 526 houses. As for Belovode, this includes only burnt houses. Radiocarbon dates suggest that the final settlement horizon lasted from *c.* 4530–4366 cal. BC and, as at Belovode, we assume that most houses belonged to the latest stage of the settlement.

Estimating population size and dynamics

The estimation of population size from house-related data is not straightforward (Cameron 1990; Porčić 2011; Porčić and Nikolić 2016), especially if the duration of the settlement or settlement phase is longer than the average use-life of a house (for detailed and thoughtful discussions of the contemporaneity problem see Dewar 1994; Kintigh 1984, 1994; Schacht 1984). Any such archaeological estimation depends on assumptions concerning population dynamics and is only as good as these assumptions. Porčić and Nikolić (2016) present a method that takes into account the issue of non-contemporaneity of houses and uses a model-based approach to estimate the parameters of population dynamics such as intrinsic growth rate, and initial and final population size. This method is probabilistic (Approximate Bayesian Computation) in the sense that it does not rely on fixed point estimates of input parameters but allows the analyst to define a prior distribution of possible values. It is based on simulating a large number of population dynamics and house accumulation scenarios by sampling from the prior distribution of parameters and keeping only those parameter combinations which satisfy constraints. In the case of Belovode and Pločnik, the constraints pertain to the total amount of accumulated houses; only demographic parameters for which the simulations produce a number of houses close to the empirically observed figure are considered as possible. The final result is the posterior probability distribution of demographic parameters, that is: growth rate, initial population size, and final population size.

Population growth and house accumulation models

In order to apply the method, a population growth model first needs to be defined. Two general patterns are possible:

1) Rapid abandonment model: the population was growing or was close to constant throughout the final phase of the settlement, which was abandoned rapidly; all remaining houses were abandoned at the same time (e.g. an extensive fire destroyed the settlement). This scenario is shown schematically in Figure 1a.

2) Gradual abandonment model (Schacht 1984): the population was first growing or was close to constant, and then declined and the settlement was gradually abandoned. This scenario is shown schematically in Figure 1b.

Given that it is not possible with current data to determine which of the two is more probable, both models are used as premises for population reconstruction.

Following Porčić and Nikolić (2016) and for reasons described therein, the logistic growth model is taken as a population dynamics model. The house accumulation model is also adopted from Porčić and Nikolić (2016). In the original paper published by Porčić and Nikolić, however, only house accumulation formula for a growing population is presented. Here, we present a formula for the number of accumulated houses when the population is decreasing.

Let P_i be the size of the considered population in step i, where $i \subset \{0,1,...,T-1\}$ for some number of years, T. Let S denote the average number of individuals per house and let L denote the average duration of the house in years. The expected number of houses in the archaeological deposit is given by

[1] In principle, it is also possible to estimate the number of unburnt houses (Hofmann 2013a; Müller *et al.* 2013b) however, in order to calculate such figures, one would need to be aware (or have a reasonably accurate estimate) of the total volume of sediment on the site as well as the average volume of sediment associated with a single house. As these figures are not available, we have limited the present analysis to burnt houses alone.

$$\sum_{i=0}^{T/L-1} \frac{P_{i \cdot L}}{S}$$

The justification for this expression is as follows. We assume that the houses built in step 0 are built simultaneously. If there is no previous history of the site, that assumption is correct. Otherwise, it is an approximation, the error introduced becoming less important the longer the duration of the site. All the houses last for L years, subsequently becoming part of the deposit. Every L years, new houses are built due to the deterioration of the existing ones, the exact number being the size of population in that year divided by the number of individuals per house. Note that, since the population size is decreasing, there is never any need to build houses, except for the deterioration of the existing ones. Since the houses are built every L years, only the population sizes in years that are multiples of L are relevant. Hence the term $P_{i.L}$. The number of L-year periods in T years is T/L, but since the houses are built at the beginning of the period, we round the quotient upwards and subtract 1 because of 0 based indexing.

Consider a simplified example. Let the population size sequence be: 100, 80, 60, 50, 40, 20; let L be 2; and S be 4. T is therefore 6. In the given expression, i should go from 0 to 2 (6/2 – 1 = 2) and population sizes included in the summation are P_0, P_2, P_4,i.e., 100, 80, 40. Therefore, the number of houses that will enter the archaeological deposit is 25 + 20 + 10 = 55.

Prior distributions of parameters

Rapid abandonment model. Growth rate is sampled from a prior uniform distribution (0, 0.06). Given that the settlement phase visible on the geophysics survey plan is not the founding phase of the settlement, the initial population size might have been relatively high. Therefore, the initial population size value is sampled from a uniform prior ranging between 200 and 5000 people. Average use-life of the house was sampled from a prior normal distribution with a mean of 40 years and a standard deviation of 5 years, based on use-life estimates of Neolithic houses from the literature ranging from 20–50 years (see Gerritsen 2008; Whittle 2003: 140–141). The conversion constant which transforms average house floor area into average household size (Brown 1987; Porčić 2012b) is sampled from a uniform distribution ranging between 6 and 10 m² per person as most ethnographically recorded conversion constants for sedentary communities are in this range (Porčić 2012b). The occupation span of houses visible in the geophysics plan is sampled from a uniform distribution ranging between 200 and 400 years. Carrying capacity is sampled from a prior distribution ranging between 500 and 10000 people.

Gradual abandonment model. This model is more complicated since we can have different growth rate values for the increase and decrease component and the duration of the increase and decrease phase may also vary. For these reasons we sample the increase component growth rate from a uniform distribution ranging from 0 to 0.06, and the decrease component is sampled from a uniform distribution ranging from -0.1 to 0. It is assumed that the most probable scenario is that where both growth and decline have equal durations. Therefore, the increase component duration is sampled from a normal distribution with mean equal to one half of the total final settlement phase duration and a standard deviation equal to 20 years.

All other input parameters are sampled from the same distributions as in the rapid abandonment model.

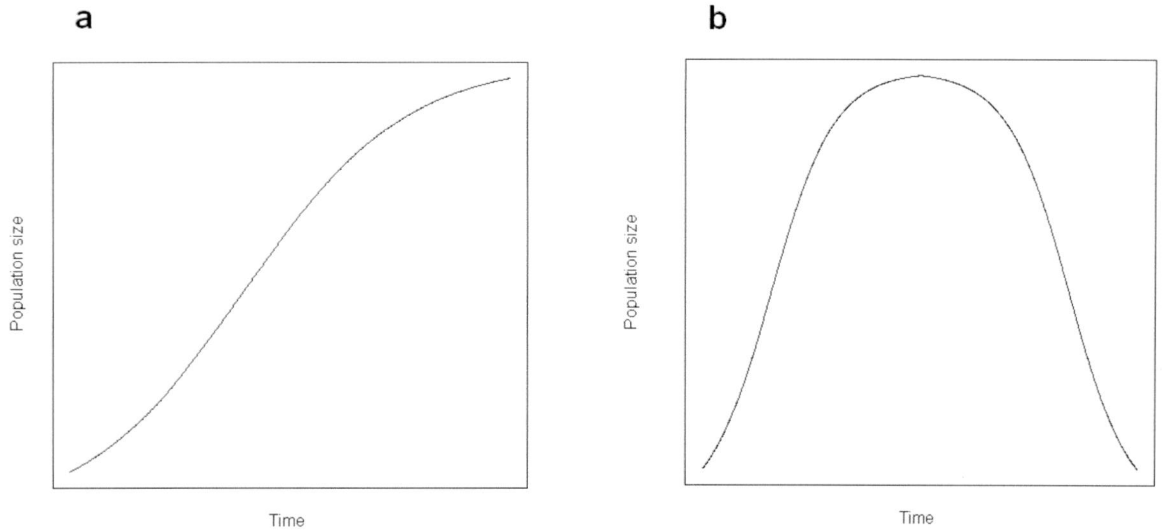

Figure 1. Schematic representation of a) rapid abandonment model; and b) gradual abandonment model.

Constraints

Belovode. The constraints for the rapid abandonment model are: 1) predicted number of accumulated houses must be between 783 and 883 houses (based on the count of houses from the geophysics plan); 2) the number of houses at the beginning of the phase must not be over 753 houses (estimated number of houses within ditches). The first two constraints for the gradual abandonment model are the same as for the rapid abandonment model with two additional constraints: 3) difference between carrying capacity and maximum population size is less than 10 people; 4) final population size must be less than 5 people and greater than 0.

Pločnik. The constraints for the rapid abandonment model are: 1) predicted number of accumulated houses must be between 450 and 550 houses (based on the count of houses from geophysics plan). The first constraint for the gradual abandonment model is the same as for the rapid abandonment model with two additional constraints: 2) difference between carrying capacity and maximum population size is less than 10 people; 3) final population size must be less than 5 people and greater than 0.

Results

Belovode

There are 1269 parameter combinations out of 500,000 that satisfy the constraints of the rapid abandonment model. The histogram in Figure 2a summarises the posterior distribution of initial population size. The mean is 509, the median is 476, and the interquartile interval is 333–647. The final population size posterior distribution histogram is presented in Figure 2b. The mean is 995, the median is 943 and the interquartile interval is 759–1173. In Figure 2c the histogram of the growth rate posterior distribution is presented. The mean is 0.014 (1.4%), the median is 0.006 (0.6%) and the interquartile interval is 0.002–0.022.

There are 486 parameter combinations out of 500,000 that satisfy the constraints of the gradual abandonment model. The histogram in Figure 3a summarises the posterior distribution of initial population size. The mean is 728, the median is 688, and the interquartile interval is 483–934. The maximum population size posterior distribution histogram is presented in Figure 3b. The mean is 1110, the median is 1073 and the interquartile interval is 892–1300. In Figure 3c histograms of the positive and growth rate posterior distributions are presented. The mean positive growth rate (population increase phase) is 0.042 (interquartile interval 0.034–0.052) and the mean negative growth rate is -0.36 (interquartile range -0.46 to -0.24, Figure 3d).

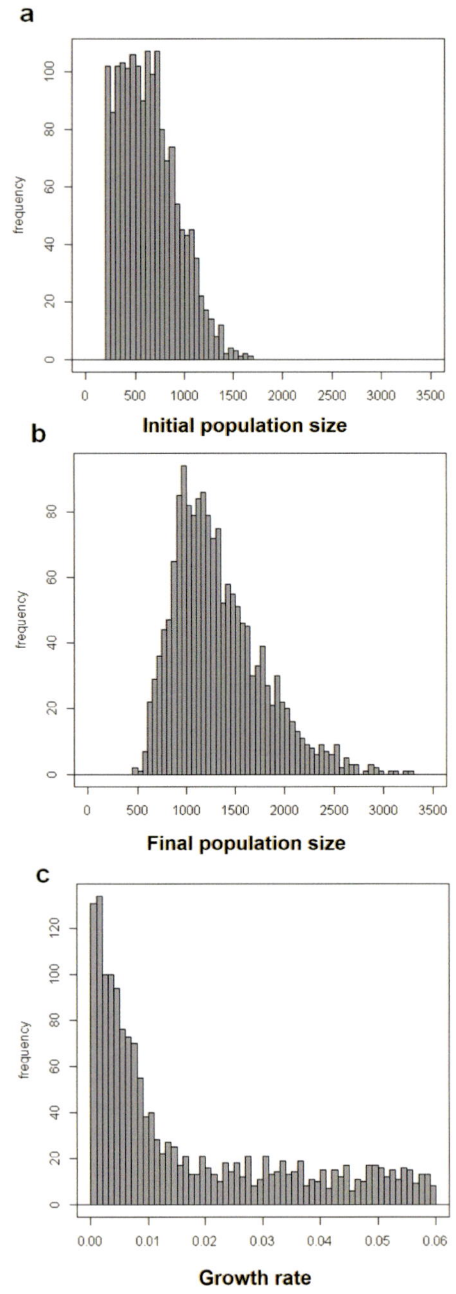

Figure 2. Results for the Belovode rapid abandonment model. Posterior distributions of a) initial population size; b) final population size; and c) intrinsic growth rate.

Pločnik

There are 910 parameter combinations out of 500,000 that satisfy the constraints of the rapid abandonment model. The histogram in Figure 4a summarises the posterior distribution of initial population size. The mean is 421, the median is 397, and the interquartile interval is 303–508. The final population size posterior distribution histogram is presented in Figure 4b. The mean is 793, the median is 747 and the interquartile interval is 608–916. In Figure 2c the histogram of the growth rate posterior distribution is presented. The mean is 0.016 (1.6%), the median is 0.01 (1%) and the interquartile interval is 0.002–0.011. There are 261 parameter combinations out of 500,000 that satisfy the constraints of the gradual abandonment model. The histogram in Figure 5a summarises the posterior distribution of initial population size. The mean is 477, the median is 457, and the interquartile interval is 333–477. The maximum population size posterior distribution histogram is presented in Figure 5b. The mean is 691, the median is 653 and the interquartile interval is 585–773. In Figure 5c histograms of the positive and growth rate posterior distributions are

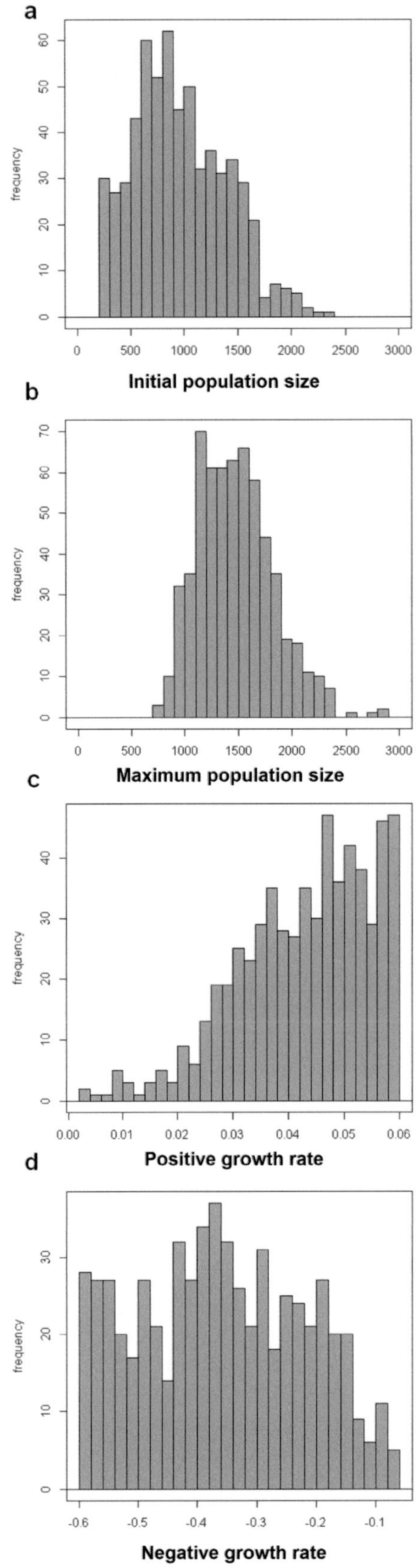

Figure 3. Results for the Belovode gradual abandonment model. Posterior distributions of a) initial population size; b) final population size; c) positive growth rate; and d) negative growth rate.

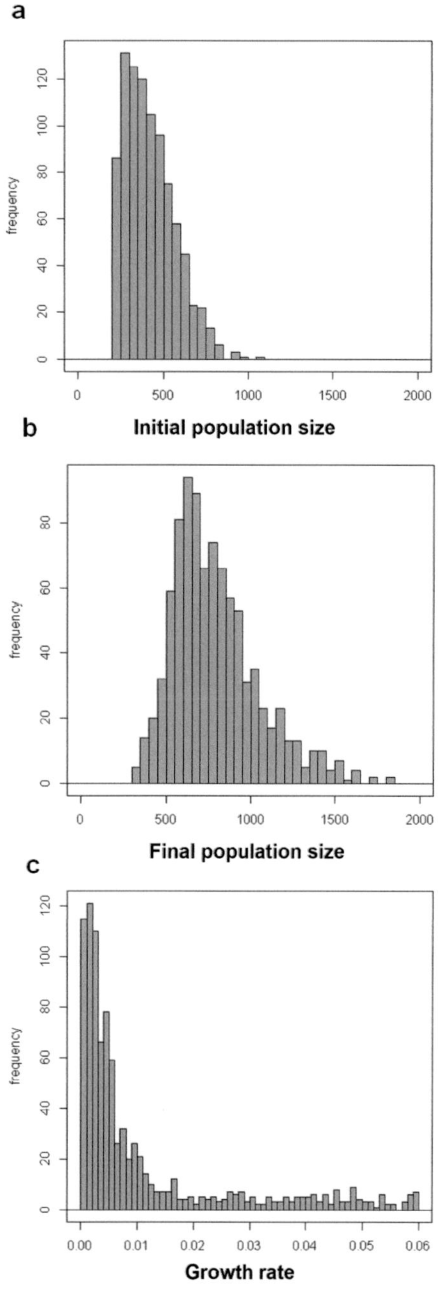

Figure 4. Results for the Pločnik rapid abandonment model. Posterior distributions of a) initial population size; b) final population size; and c) intrinsic growth rate.

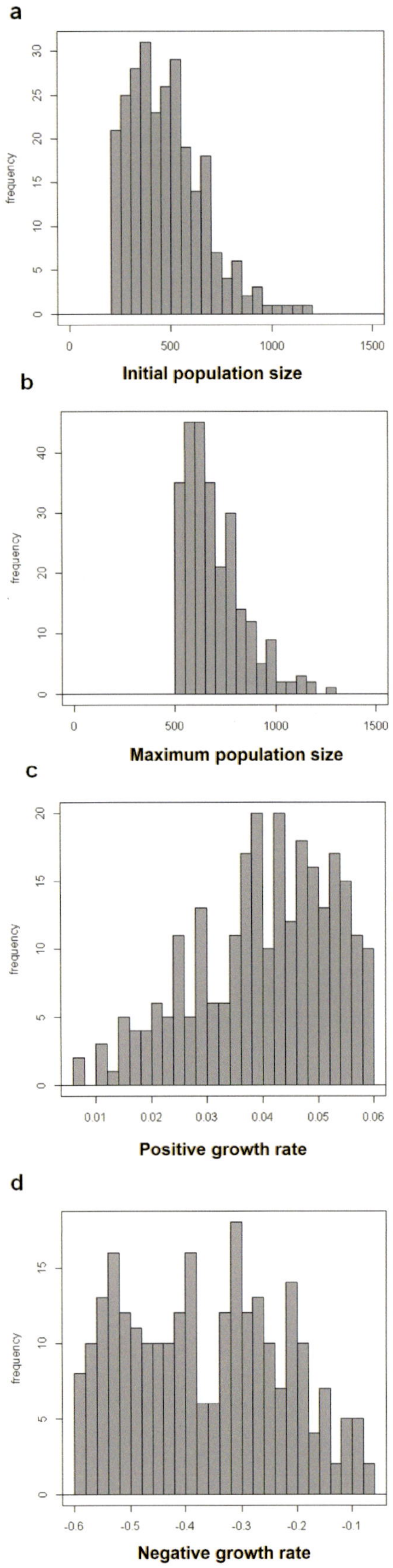

Figure 5. Results for the Pločnik gradual abandonment model. Posterior distributions of a) initial population size; b) final population size; c) positive growth rate; and d) negative growth rate.

presented. The mean positive growth rate (population increase phase) is 0.04 (interquartile interval 0.033–0.05) and the mean negative growth rate is -0.36 (interquartile range -0.48 to -0.26, Figure 5d).

Discussion and conclusion

Maximum population estimates for Belovode and Pločnik are within the limits of estimated population size of other Late Neolithic settlements in the central Balkans such as Okolište, Divostin, and Stubline, i.e. around 1000 people.

The extremely high intrinsic growth rate related to the gradual abandonment model at both sites should not be interpreted as evidence for population boom. The estimated initial population size was large to begin with in most cases, and for the underlying logistic model of growth the actual growth rate (the first derivative of the logistic growth curve) is very low in the region where current population size is close to the carrying capacity. For example, when population size is higher than half the carrying capacity, the growth starts to decelerate and turns from an exponential-like mode into the logarithmic-like model (for details see Porčić and Nikolić 2016). For example, Belovode population dynamics are illustrated in Figure 6a by randomly sampling 50 parameter sets from the set of simulation results for the gradual abandonment model. This illustration presents that, according to gradual abandonment model, Belovode population slowly increased by around 200–300 people in the first half of settlement duration. This also applies to growth rate estimates for the rapid abandonment which are significantly lower than gradual abandonment model estimates but are still relatively high (in the order of magnitude of the NDT) – the mean initial population size estimate is around one half of the final population size which means that most of the time the growth is quasi-logarithmic so the actual growth rates are much smaller than the intrinsic growth rate (see Figure 6b). The pattern at Pločnik is identical (Figure 7).

It is interesting to note that the results of the gradual model for both Belovode and Pločnik actually suggest that the abandonment was rapid, as the expected values for the decrease phase growth rate are extremely low (which means that the decrease was rapid). To what extent this result is the artefact of the model assumptions (e.g. prior distribution of the length of growth/decline in the last phase) remains to be seen, but both models predict similar population size values for both sites, and are of the same order of magnitude as population estimates for similar Late Neolithic sites in the central Balkans.

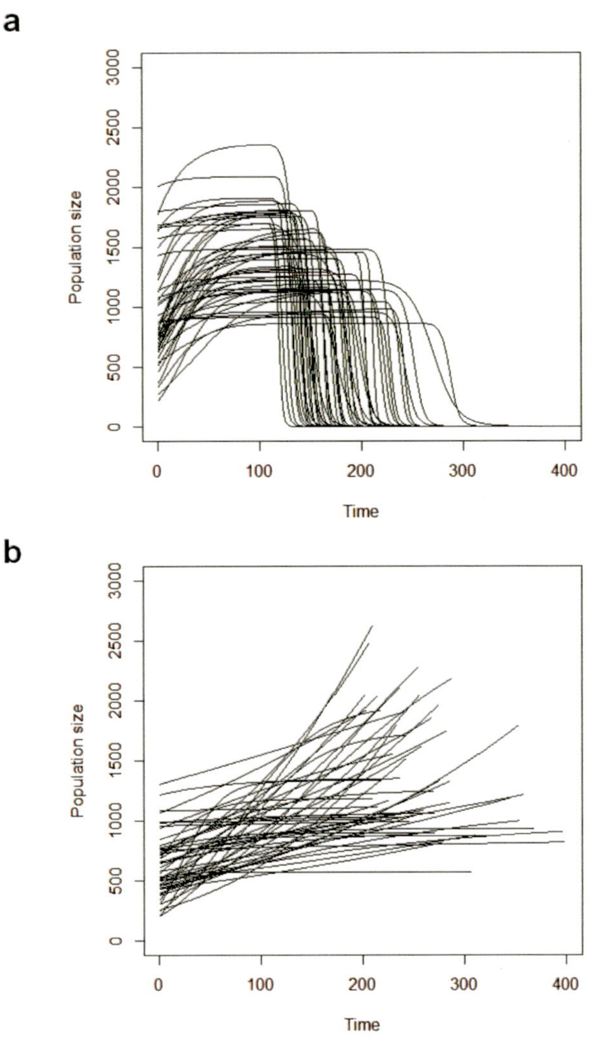

Figure 6. Illustration of population dynamics for Belovode based on 50 randomly sampled posterior parameter sets for a) gradual abandonment model; and b) rapid abandonment model.

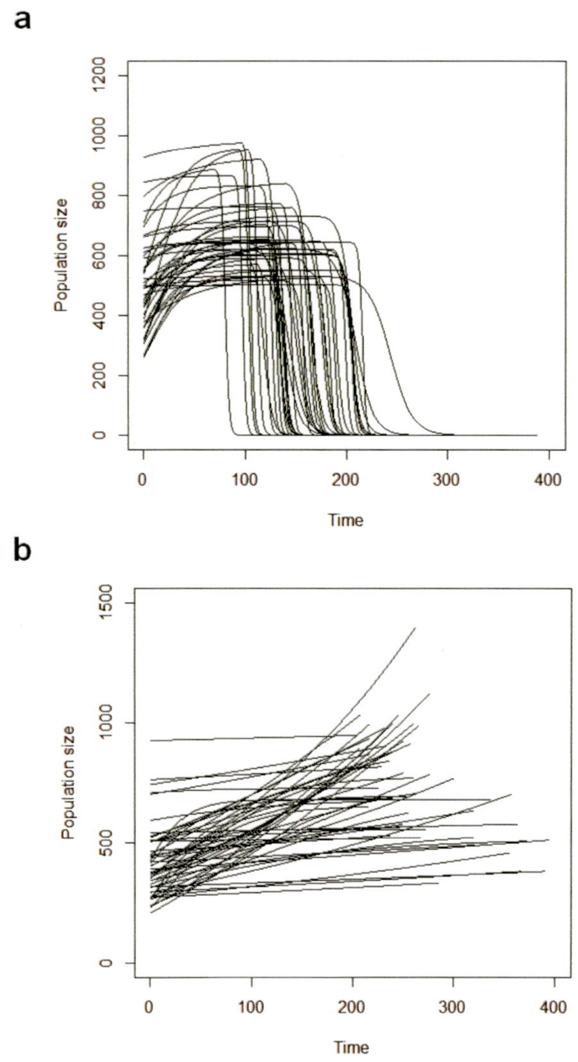

Figure 7. Illustration of population dynamics for Pločnik based on 50 randomly sampled posterior parameter sets for a) gradual abandonment model; and b) rapid abandonment model.

The bibliographic reference for this chapter is:

Porčić, M. and Nikolić, M. 2021. Population size and dynamics at Belovode and Pločnik, in Radivojević, M., Roberts, B. W., Marić, M., Kuzmanović Cvetković, J., and Rehren, Th. (eds) *The Rise of Metallurgy in Eurasia*: 477–483. Oxford: Archaeopress.

Chapter 41

Metallurgical knowledge and networks of supply in the 5th millennium BC Balkans: Belovode and Pločnik in their regional context

Miljana Radivojević, Thilo Rehren and Ernst Pernicka

The recent set of excavations (campaigns 2012 and 2013) at the sites of Belovode and Pločnik (see Chapters 11 and 26) have shown the use of copper minerals and metallurgical activities to be highly consistent with results from previous analytical research (Radivojević 2007, 2012, 2013, 2015; Radivojević *et al.* 2010a; Radivojević and Rehren 2016). Specific aspects to emerge so far include: persistent selection of black and green manganese-rich copper ores for metal extraction; similar engineering parameters involved in the early copper smelting technology; field evidence from Belovode supporting the presence of pottery-lined hole-in-the-ground installations; consistent metal making and working technology remains at both sites; and direct absolute dating evidence that leaves no doubt for *c.* 5000 BC as the beginning of copper metallurgy in the Balkans. In this chapter we will synthesise the evidence for this *c.* 7,000 years old copper production technology within the Vinča culture, in its local and regional perspective, including data and debates on the provenance of copper ores.

Tainted ores as the main trait of the Vinča culture copper smelting

The persistent selection of black and green manganese-rich copper ores at both sites speaks, yet again, of shared practices for copper smelting technology between these two sites, and within the wider Vinča culture phenomenon. Previous research has already revealed the enduring practice of the selection of these minerals in the Early Neolithic in the Danube Gorges (*c.* 6200 BC), together with a parallel set of criteria applied to the selection of copper minerals with largely green appearance (free of impurities) (Radivojević 2015). While we cannot claim that these black and green minerals were already used as ores for smelting as early as *c.* 6200 BC, this particular preference for these distinctively coloured 'precious stones' implies either a potentially long experimentation phase or an alignment with ideology and rituals of which these may have been a significant part; a discussion of these hypotheses is beyond the scope of this chapter. Importantly, the 'purer' green minerals, more commonly used for bead making, still presented black impurities (most likely manganese oxide) in some of the 25 studied examples (see Figure 16 in Chapter 11, and Figure 7 in Chapter 26, both this volume), which moderates our initial assumption of an exclusive selection of black and green minerals for metal extraction (Radivojević *et al.* 2010a), and reinforces the close link between the preceding bead making linked to lithic technology and the emerging metallurgical technology.

Oxidic copper minerals were, however, often mixed with primary (sulfur-rich) copper minerals at both sites, as confirmed through the analysis of copper metal droplets (Bf56/13 and P61/12) with chalcocite inclusions (Bf56/13, see Figure 21d in Chapter 11, this volume) in both rounded (heated) and more angular (natural) form (also see Figure 9b, Table 5 in Chapter 26, this volume). Another type of similarly (dark)coloured minerals appeared consistently alongside black and green manganese rich copper ores at Pločnik; these were preliminarily called 'magnetic' minerals, as they were highly responsive to a magnet. Compositional analysis identified them to be a mixture of iron oxides and members of the olivine family of minerals (see Figure 6, Table 4 in Chapter 26, this volume). In the absence of any slag from this site, we cannot say whether these formed as part of the smelting process; however, their distinctive colour (dark green) could have played a role in their initial selection by the prospectors at that time.

Another indication for the types of copper ores selected for metallurgical activities at Belovode and Pločnik came from EPMA and LA-ICP-MS analyses of metal phases in production evidence and artefacts. Close scrutiny of the data confirms the consistent selection of manganese-rich copper ores that most commonly also contained nickel and cobalt, some remnants of primary copper minerals (iron, sulfur) and a potential association with polymetallic deposits that contain arsenic, tin, lead and bismuth (see Tables 8, 8a, 11, 13, 15 in Chapter 11 and Tables 6 and 6a in Chapter 26, this volume). Polymetallic deposits that could potentially match sources used to make these metals exist in the

Table 1. EPMA compositional data of metal phases in copper production evidence, droplets and finished artefacts from the sites of Belovode and Pločnik, given as µg/g. Values above *c.* 0.01 wt% (100 µg/g) are considered reliable based on CRM measurements; values below this are indicative only. All data are corrected for values obtained from the reference material, using a procedure reported in the methodology section of the Belovode metallurgy chapter (11). Values sought but not found at levels above *c.* 0.01 wt% were indicated as not detected (n.d.). Bold figures stand for data reported in this volume (Chapters 11 and 26) and normal figures for previously conducted analyses in Belovode and Pločnik (Radivojević 2012; Radivojević and Rehren 2016). Belovode and Pločnik data were analysed with the same Electron Probe Micro-Analyser (EPMA) located at the UCL Institute of Archaeology (see Table 3a, Chapter 11 in this volume).

	site	type	S	Mn	Fe	Co	Ni	Zn	As	Ag	Sn	Sb	Te	Au	Pb	Bi
			µg/g	µg/g	µg/g	µg/g	µg/g	µg/g	µg/g	µg/g	µg/g	µg/g	µg/g	µg/g	µg/g	µg/g
B23/12	**Belovode**	**slagged sherd**	**67**	**46**	**15**	**14**	**84**	**n.d.**	**115**	**4**	**70**	**n.d.**	**50**	**n.d.**	**70**	**67**
B47/12/2	**Belovode**	**slag**	**96**	**26**	**117**	**32**	**64**	**n.d.**	**46**	**32**	**46**	**n.d.**	**82**	**n.d.**	**82**	**57**
Belovode 30a	Belovode	slagged sherd	n.d.	n.d.	10	n.d.	70	n.d.	20	n.d.	n.d.	n.d.	n.d.	20	n.d.	n.d.
Belovode 31a	Belovode	slagged sherd	n.d.	n.d.	n.d.	n.d.	60	n.d.	10	n.d.	n.d.	n.d.	n.d.	90	n.d.	n.d.
Belovode 31b	Belovode	slagged sherd	20	n.d.	20	n.d.	70	n.d.	0	n.d.	n.d.	n.d.	n.d.	110	n.d.	n.d.
Belovode 131	Belovode	slag	150	8150	36100	10600	960	3000	220	n.d.	n.d.	235	n.d.	150	n.d.	n.d.
Belovode 134	Belovode	slag	80	3000	10200	500	45	60	20	n.d.	n.d.	n.d.	n.d.	210	n.d.	n.d.
Bf21/12	**Belovode**	**droplet**	**45**	**39**	**15**	**27**	**135**	**n.d.**	**220**	**3**	**91**	**n.d.**	**40**	**n.d.**	**57**	**86**
Belovode M6	Belovode	droplet	34	n.d.	n.d.	n.d.	n.d.	n.d.	n.d.	n.d.	n.d.	n.d.	42	n.d.	n.d.	n.d.
Pločnik 52	Pločnik	droplet	52	n.d.	n.d.	n.d.	37	n.d.	11	n.d.	n.d.	n.d.	34	120	n.d.	n.d.
B71/12	**Belovode**	**fragment**	**91**	**23**	**27**	**15**	**83**	**n.d.**	**112**	**90**	**98**	**n.d.**	**40**	**n.d.**	**62**	**162**
P10/13	**Pločnik**	**fragment**	**73**	**26**	**10**	**16**	**107**	**n.d.**	**78**	**21**	**64**	**n.d.**	**38**	**23**	**47**	**134**
C_P1/13	**Pločnik**	**loop**	**43**	**5**	**24**	**13**	**61**	**n.d.**	**34**	**181**	**80**	**n.d.**	**103**	**n.d.**	**2800**	**58**
C_P2/13	**Pločnik**	**band**	**160**	**40**	**23**	**43**	**50**	**n.d.**	**50**	**12**	**105**	**n.d.**	**85**	**n.d.**	**81**	**36**
Pločnik 67	Pločnik	tool (?)	96	n.d.	n.d.	n.d.	41	n.d.	n.d.	n.d.	n.d.	11	0	n.d.	n.d.	n.d.
Pločnik 73	Pločnik	bracelet	56	n.d.	n.d.	n.d.	n.d.	n.d.	n.d.	n.d.	n.d.	n.d.	0	107	n.d.	n.d.
Pločnik 75	Pločnik	sheet	31	n.d.	47	n.d.	49	n.d.	n.d.	n.d.	n.d.	n.d.	0	77	n.d.	n.d.
Pločnik 143	Pločnik	chisel	25	n.d.	n.d.	n.d.	53	n.d.	n.d.	n.d.	n.d.	n.d.	23	75	n.d.	n.d.
Pločnik 145	Pločnik	chisel	n.d.	n.d.	75	n.d.	28	n.d.	n.d.	n.d.	n.d.	n.d.	12	61	n.d.	n.d.

Bor mining region in eastern Serbia and include massive copper sulfide deposits dominated by pyrite and different copper sulfides (enargite, bornite, covellite or chalcocite) and accessory chalcopyrite, amongst others (Janković 1967; Antonijević and Mijatović 2014; Jelenković 1999; Neubauer and Heinrich 2003; Monthel *et al.* 2002; Sillitoe 1983). While provenance data will be addressed in detail below, in Table 1 we synthesise electron microprobe compositional analyses of copper prills in production evidence (slag and slagged sherds) and copper metal artefacts analysed thus far from the sites of Belovode and Pločnik (including data for copper metal artefacts from Radivojević 2012).

Although the data come from metallurgical materials from the same sites, and in places from the same occupational horizons. For instance, the freshly analysed copper metal phases from the production evidence from Belovode present only slightly more elevated readings of As than the Pločnik metal artefacts, but significantly less Pb (which is *c.* 0.3 wt% in C_P1/13), Ag (c. 180 ppm in C_P1/13) and S (c. 160 ppm in C_P2/13), being used at each site, although such a generally low variability may also occur within the same mineralisation (Table 1). Ni remains consistently present in all data, confirming the previously assumed wide availability of this element in the copper ores exploited by the Vinča communities (Radivojević 2012; Radivojević and Rehren 2016). We shall use ppm (parts per million) to discuss µg/g (microgram per gram) concentrations throughout this text.

In conclusion, the low variability of trace elements such as As, Pb, Ag and S between these two sites is more likely to reflect the use of different batches of copper ores than the use of different copper ore sources. This assumption is possibly corroborated by the fact that the observed artefacts cluster in different 'workshop' areas at both sites. For example, the four samples, Pločnik 67/73/75/145, come from the same dwelling (Radivojević 2012, also Chapter 6 in this volume), and Belovode 23/12, 47/12/2, f21/12, 71/12 from the wider area of Feature 6 ('workshop area', see Chapter 11 in this volume). Hence, we could interpret this as different craftsmen or workshops using different copper ores or ore batches for metal making within these settlements.

On the whole, the Vinča culture communities of Belovode and Pločnik exploited copper ores with relatively pure copper carbonate (malachite) associated with black minerals (manganese oxide) and residual primary ore minerals (e.g. chalcopyrite, chalcocite) and consistent (albeit at low concentrations) nickel content. The variability of the concentrations of As, Ag, Pb or S is not so large as to imply the use of different copper sources, and we are inclined to propose the use of different batches of copper ore from a single deposit. The connectedness of these two sites when it comes to shared copper resources has been indicated before (cf. Pernicka *et al.* 1993, 1997; Radivojević *et al.* 2010a); however, the provenance study offers more data and therefore a more complex picture of copper supply to these two settlements, as will be shown in the next section.

The most precious ten grams of early Balkan metallurgy

The research conducted thus far has confirmed the consistently ephemeral nature of early copper smelting technology in the first half of the 5th millennium BC in the central Balkans. To the previously analysed 8.3 g of (free) slag samples (Radivojević 2007, 2012), this research has now added a further 0.82 g (see Table 1, Chapter 11 in this volume), taking the whole assemblage to slightly short of 10 g of available direct metal production evidence for the stated time frame. Although this excludes the volume of slag adhering to the ceramic sherds from Belovode, it is clear that the early copper smelting technology in the Vinča culture is consistent in its near invisibility, as recognised in the results of both macro- and micro-analytical approaches conducted here and in previous research.

A common feature of all slag finds (free samples and slagged masses on sherds from Belovode) is that they solidified quickly from being almost fully liquefied at the time of the smelt, resulting in a high proportion of glassy phase. The slag matrix areas are mostly located towards the surfaces of the samples, in contact with copper-rich phases such as metal droplets, 'dross' and corrosion products. The latter also fill the pores scattered across visible sections. Hence, copper containing compounds are the dominant phases in all samples; these are followed by various phases which are evenly distributed throughout a particular area of crystallisation (e.g. delafossite, spinels; see Tables 7 and 10 in Chapter 11, this volume).

The bulk composition of the glassy matrices of all slag samples predominantly consists of silica, alumina, lime, and iron and copper oxides, followed by fuel components: phosphorus oxide, potash and magnesia, and other ore elements (besides copper): manganese, cobalt and zinc (see Tables 6, 7 and 9 in Chapter 11, this volume). The ternary plot of components understood to represent typical pottery ($SiO_2/Al_2O_3/TiO_2$), fuel ash ($CaO/MgO/P_2O_5/K_2O$) and ore ($FeO/MnO/ZnO/NiO/CoO/As_2O_3/SnO_2/Sb_2O_3$) contamination in the glassy slag matrices (re-cast as Cu-free) in Figure 24, Chapter 11 of this volume, illustrates the consistency of metal smelting practices across several Vinča culture sites: Belovode, Vinča-Belo Brdo and Pločnik. The strong concentration of all glassy matrices in the silica+alumina+titania corner implies that they were predominantly formed by mostly acidic oxides, which explains their enhanced viscosity. A small number of copper metal droplets: B29/12 and P61/12 (see Chapters 11 and 26, this volume), together with Belovode M6 and Pločnik 52 (Radivojević and Rehren 2016) may be seen as the exceptions to the 'slagging' rule: they exhibit a smelting attempt together with the presence of primary copper minerals, but not much slag. Yet, given the huge discrepancy between the number of implements (estimated at *c.* 4,300 items, or 4.7 tonnes in the 5th millennium BC Balkans) (Pernicka *et al.* 1997; Ryndina 2009) and production evidence (such as <10 g slag), the 'slagless' metallurgy may well have been the 'rule' rather than the exception. At this point it is important to emphasise that the identified slag from the Vinča culture sites covers the first half of the 5th millennium BC, and that the overwhelming majority of the 4,300 artefacts are from the second half of the 5th millennium BC. This leaves many open questions about the nature of copper extraction technology in the second half of that millennium, including its location and identification during the excavations (cf. Radivojević *et al.* 2010a).

Nevertheless, although the early slagging process in the Vinča culture is characterised as heterogeneous, almost fully liquefied and resulting in highly viscous debris that needs to be crushed in order to extract metal, when compared to similar evidence throughout the Near East, dated between the mid-5th to the 3rd millennium BC, this fits well within the general picture of borderline stable metal extraction processes (cf. Bourgarit 2007). Of particular interest for comparison here is the recently discovered pit (no. 28) at the site of Akladi Cheiri on the southern Bulgarian Black Sea coast (see Figure 3, Chapter 3, this volume) (Rehren *et al.* 2016, 2020). This context, broadly dated to the middle of the 2nd half of the 5th millennium BC, yielded around 100 kg of copper ores, tailings, and production debris, including some 300 fragments of slagged sherds and one largely preserved (melting) crucible. Microstructural analyses of the slag on these sherds reveal striking similarities with the *c.* 700–800 years older examples from Belovode, with the main components being copper-rich phases, delafossite and magnetite (Rehren *et al.* 2020: 144, Figures 9–2). The regular occurrence of copper and copper-iron sulfide phases, however, implies that the copper ore was a mixture of carbonate and sulfide minerals (strong iron presence suggests chalcopyrite or bornite), which fits well with the estimated age of this exceptional late-5th millennium BC metallurgical assemblage. The ternary plot of pottery (SiO_2/Al_2O_3/TiO_2), fuel ash (CaO/MgO/P_2O_5/K_2O) and ore (FeO/MnO/ZnO/NiO/CoO/As_2O_3/SnO_2/Sb_2O_3) contamination in the glassy slag matrices (re-cast as Cu-free) of Vinča culture and Akladi Cheiri slag (Figure 1) exhibits broadly consistent smelting conditions, modified by a stronger iron and fuel ash component for Akladi Cheiri.

The Akladi Cheiri assemblage also confirmed another connection with the Vinča culture metallurgy: the hole-in-the-ground smelting installations lined with fragmented pottery sherds (Rehren *et al.* 2020). This is well exemplified with hundreds of Akladi Cheiri sherds presenting a pattern of localised heat impact and slag cover that goes around broken sections, clearly implying that these were not crucibles, but lined smelting installations similar to those hypothesised for the sites of Belovode and Gornja Tuzla (Radivojević and Rehren 2016). This is yet another point that strongly indicates the transmission of metal smelting technology from the Vinča culture sites to the later KGK VI settlements on the Bulgarian Black Sea coast.

The excavation campaigns of 2012 and 2013 at Belovode yielded more support for the shape and nature of these 'hole-in-the-ground' installations: the charred and burnt soil in the bowl-shaped feature, F6, and its clear association with the slagged sherds and other production debris in Trench 18 (Table 1, Figures 2 and 3 in Chapter 11, this volume) was a highlight of this project. Taken together with the initial research at this site (Radivojević and Rehren 2016) and the growing number of examples from Bulgaria (Akladi Cheiri) and Romania (Foeni) (Rehren *et al.* 2016; Radivojević *et al.* in preparation), the new evidence confirms that the pottery-lined hole-in-the-ground installation was the earliest type of a smelting 'furnace' in this part of the world.

Another highlight of the excavation campaign was the thorough dating programme for organic samples directly associated with the metallurgical remains from both Belovode and Pločnik (see Table 1, Chapter 37, this volume). As a result of the former, metallurgical activities were firmly placed within the period from the 49th to the 47th centuries cal. BC, respectively, confirming conclusions from previous research (Radivojević *et al.* 2010a). In the latter, a metal bead (C_P4/13) was directly dated (with associated charcoal) to the beginning of the 5th millennium BC, which makes it the earliest secure date for the beginning of metallurgy at Pločnik (see Chapters 11 and 26, this volume). The end date for Belovode occupation (and metallurgical activities) can be placed at the turn of the 46th century cal. BC, in alignment with what we know thus far about the (fiery) end of the Vinča culture settlements in the northern (Danubian) area. Pločnik, on the other hand, preserved evidence for prolonged site activities, well into the second half of the 45th or the first half of the 44th century cal. BC, with metallurgical activities following through to the very end of its occupation (see Table 2 in Chapter 26, this volume). These dates are also in agreement with the dating of metallurgical activities at the site of Gornja Tuzla in Bosnia (Radivojević and Rehren 2016). The Vinča culture metallurgy was therefore actively practiced for *c.* 600 years (5000–4400

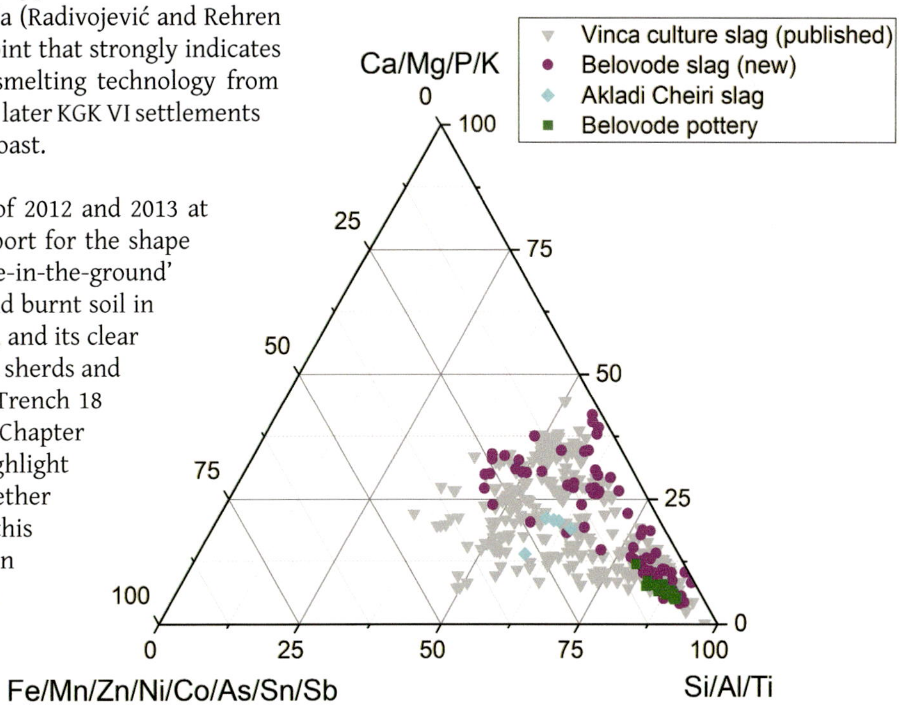

Figure 1. SiO2/Al2O3/TiO2-CaO/MgO/P2O5/K2O-FeO/MnO/ZnO/NiO/CoO/As2O3/SnO2/Sb2O3 ternary plot (cast as Cu-free) for all Vinča culture metallurgical slag matrix data and 'cold' pottery analysis (from Radivojević 2007; 2012 and Chapter 11, this volume), with Akladi Cheiri slag (Rehren *et al.* 2020).

BC) and technological knowledge was transmitted across the Balkans during the 5th millennium BC, confirming the conclusions of previous research (e.g. Radivojević 2015).

Acquisition and circulation of copper minerals, ores and artefacts: lead isotope analysis

A total of 19 artefacts were analysed for provenance (lead isotope and trace element analysis, see Table 2) within the scope of this project. The rationale for their selection ranged from association with production debris and metal artefacts, to minerals of various qualities marking top, bottom and middle stratigraphic points during the excavations (Table 2). The methodology for sample preparation and analysis has been extensively reported elsewhere (cf. Nørgaard *et al.* 2019)

The lead isotope abundance ratios in all investigated objects are summarised in Table 3 and illustrated in Figures 2–11 with comparative datasets (datasets are also available as Appendix to this Chapter). Sample P8/13 (an iron-based mineral) is excluded from further consideration as it plots far from the copper-based group of artefacts and is not relevant to the discussion on copper metallurgy in this chapter. The clustering of samples is evident in the plots of two sets of isotope abundance ratios of lead: $^{207}Pb/^{206}Pb$ vs $^{208}Pb/^{206}Pb$ (Figure 2) and $^{206}Pb/^{204}Pb$ vs $^{207}Pb/^{204}Pb$ (Figure 3). All samples fit broadly into four categories:

Table 2. List of samples selected for provenance analysis (lead isotope- LIA and trace elements- NAA) and rationale for selection.

No	trench	spit	find no.	EDM	type of material	LIA	NAA	reasoning for provenance analysis
B21/13	18ext	S4	1378	420	malachite (green)	X	X	top layer malachite
B23/12	18	S5	77	182	slagged sherd (metal phase)	X	X	slagged sherd
Bf21/12	18	S6	112	272	metal droplet	X	X	metal droplet
Bf22/12/2	18	S6	116	276	malachite	X	X	workshop area mineral
B47/12/3	18	S6	112	272	metal prill associated with the slagged sherd	X	X	production evidence / metal
Bf43/13	18/F21	S12	1492	477	metal droplet	X	X	metal droplet
B108/13	18	S13	1691	550	malachite (black and green)	X	X	associated with Bf43/13
B350/13	18ext/F39	S13	2429	881	malachite	X	X	Near ash feature 39
B155/13	18	S14	1785	612	copper mineral	X	X	a different type of a copper mineral
B385/13	18	S19	2596	937	malachite	X	X	earliest malachite occurrence
P8/13	T24	S10	214	977	(magnetic) iron mineral	X	X	the earliest magnetic mineral
P10/13	T24	S10	220	988	fragmented metal (foil/stock)		X	metal artefact
P13/13	T24	S11	233	1017	fragment of a metal bracelet/wire		X	metal artefact
P14/13	T24	S12	242	1054	malachite	X	X	malachite near P13/13
P55/13	T14/F15	S16	366	1606	malachite	X	X	ore choice consistency
P121/13	T24	S21	537	1958	malachite	X	X	earliest malachite occurrence
C_P1/13	T24	S9	155	587	Metal loop / ring	X	X	metal artefact
C_P2/13	T24/F2 (north)	S9	195	908	Metal band / ring	X		metal artefact

Table 3. Lead isotope abundance ratio for selected copper-based materials from the sites of Belovode and Pločnik. The isotope ratios measured were 208Pb/206Pb, 207Pb/206Pb, and 206Pb/20Pb, with relative uncertainties of less than 0.01 for the first two ratios and less than 0.03% for the last. The ratios 208Pb/204Pb, and 207Pb/204Pb, were calculated from the other ratios.

Lab no.	**original label**	**description**	$^{208}Pb/^{206}Pb$	$^{207}Pb/^{206}Pb$	$^{206}Pb/^{204}Pb$	$^{208}Pb/^{204}Pb$	$^{207}Pb/^{204}Pb$
MA-140907	B23/12	Slagged sherd (metal)	2.0686	0.82887	18.891	39.077	15.658
MA-140908	B108/13	Malachite (black and green)	2.0729	0.84099	18.544	38.440	15.596
MA-140909	B47/12/3	Slagged sherd (metal)	2.0634	0.83062	18.855	38.904	15.661
MA-140910	B21/13	Malachite (green)	2.0348	0.82165	19.071	38.806	15.670
MA-140911	B155/13	Copper mineral	2.0676	0.83777	18.663	38.586	15.635
MA-140913	B385/13	Malachite	2.0749	0.84209	18.511	38.409	15.588
MA-140914	Bf21/12	Metal droplet	2.0690	0.83813	18.662	38.610	15.641
MA-140915	Bf22/12/2	Malachite	2.0764	0.84264	18.505	38.425	15.593
MA-140916	Bf43/13	Metal droplet	2.0762	0.84260	18.506	38.421	15.593
MA-140917	P8/13	(Magnetic) iron mineral	2.1666	0.91460	16.934	36.689	15.488
MA-140920	P14/13	Malachite	2.0725	0.83601	18.749	38.857	15.674
MA-140921	P55/13	Malachite	2.0766	0.84262	18.508	38.435	15.595
MA-140922	P121/13	Malachite	2.0750	0.84241	18.500	38.387	15.584
MA-140923	C_P1/13	Metal loop / ring	2.0805	0.84766	18.440	38.363	15.630
MA-140924	C_P2/13	Metal band / ring	2.0628	0.83206	18.812	38.805	15.653

Group 1 is a tight cluster predominantly comprising minerals, both from Belovode (Bf22/12/2, B108/13, B385/13) and Pločnik (P55/13, P121/13), accompanied by a single copper metal droplet (Bf43/13). The $^{206}Pb/^{204}Pb$ isotope abundance ratios of these objects cluster around 18.544–18.500;

Group 2 is a pair of samples consisting of copper metal droplet Bf21/12 and copper mineral B155/13. The $^{206}Pb/^{204}Pb$ isotope abundance ratios of these objects cluster around 18.663–18.662;

Group 3 consists of artefacts that form a dispersed agglomeration away from Groups 1 and 2. It includes metal from the production process at Belovode (B47/12/3), a slagged sherd (B23/12) and a copper metal band (or a ring) C_P2/13, from Pločnik. These artefacts show a clear association with metal droplets / prills in slags (B47/12/3 and B23/12) from F6 (workshop) at Belovode, as well as confirming the contemporaneous links to the Pločnik community and the metal band / ring (C_P2/13), all dated to the 46th century BC (see Chapter 26). Their $^{206}Pb/^{204}Pb$ isotope abundance ratios vary between 18.855 and 18.812.

The three 'outlier' samples do not belong to any of the mentioned groups and include the copper metal loop / ring from Pločnik C_P1/13 and two malachite samples from Belovode (B21/13 and P14/13).

The grouping of these objects could indicate consistency with three or more distinctive ore deposits or may reflect a single geologically complex source. In order to locate potential sources exploited by the Vinča culture communities, these data were plotted against the existing database of lead isotope ratios of Balkan ores and artefacts (metal and mineral-based), which largely come from the same laboratories that analysed the samples in Table 3 (the Max-Planck-Institute for Chemistry in Mainz and the Curt-Engelhorn-Centre for Archaeometry in Mannheim, Germany, with labels HDM and CEZA, respectively) (Pernicka *et al.* 1993, 1997; Kunze and Pernicka 2020). Additional lead isotope data by Gale (1991) (label GALE) and the mean of two samples from Majdanpek by Amov (1999) (label AMOV) as well as Amov and Văkova (1994) (label A&V) for Bulgarian deposits are also included. The data presented in Table 4 is the Balkan copper

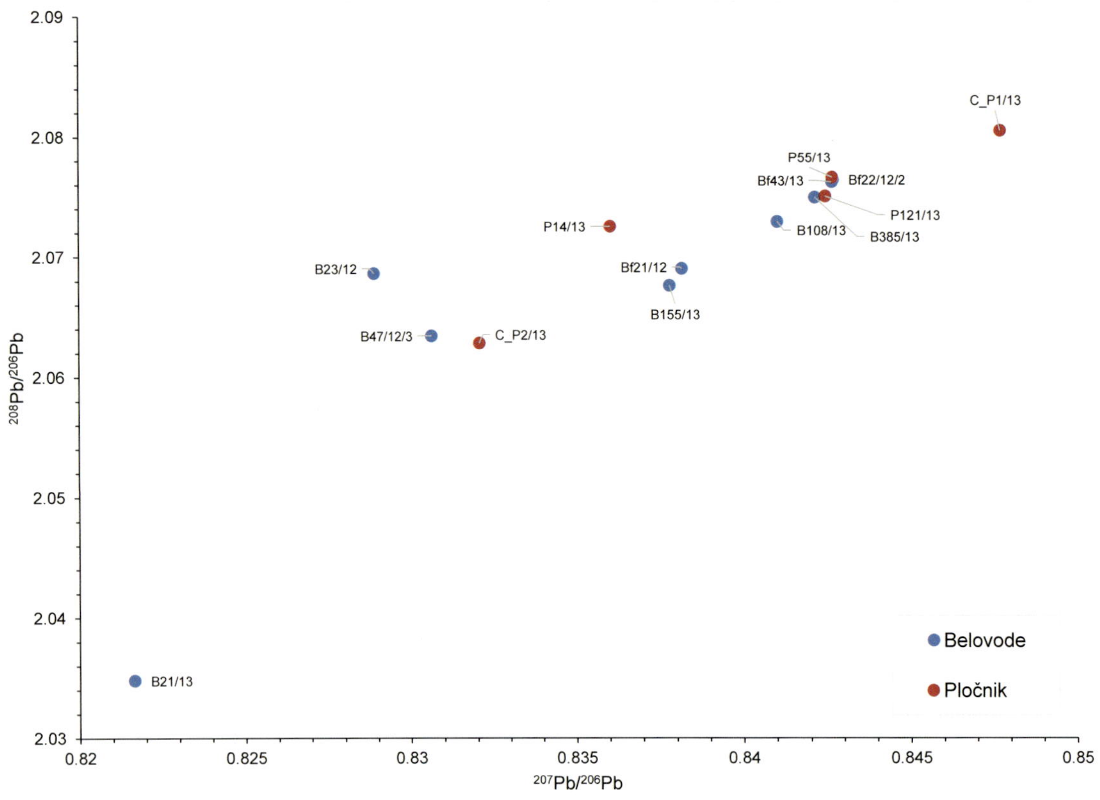

Figure 2. Lead isotope abundance ratios of Belovode and Pločnik artefacts presented in a $^{207}Pb/^{206}Pb$ vs $^{208}Pb/^{206}Pb$ diagram. Error bars are smaller than the symbol size.

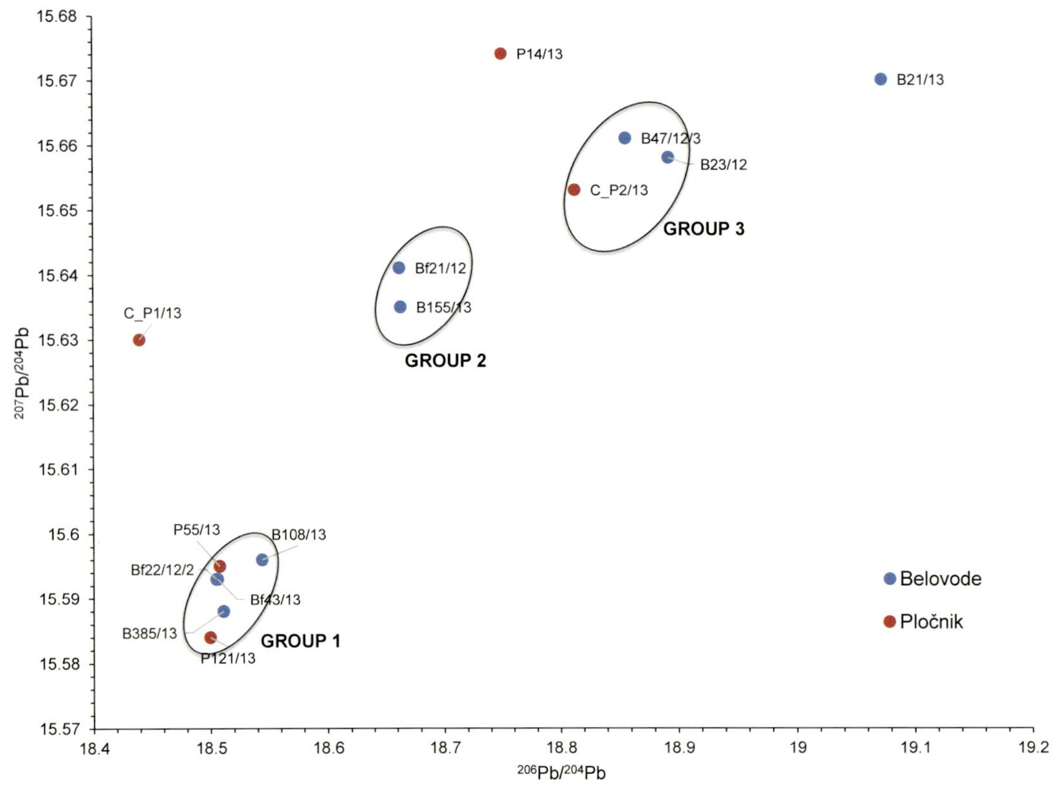

Figure 3. Lead isotope abundance ratios of Belovode and Pločnik artefacts presented in a $^{206}Pb/^{204}Pb$ vs $^{207}Pb/^{204}Pb$ diagram.

deposits database used for comparison, however only the sources consistent with the lead isotope abundance ratios in archaeological samples are included in the accompanying figures. The same applies to Tables 5–8 which, combined, include 177 copper-based archaeological artefacts (malachite, copper production and metal implements) from the Balkans, spanning the mid-6th to mid-4th millennium BC, with the largest concentration of items in the Early Chalcolithic (EC, 5000–4600 BC) and Middle Chalcolithic (MC, 4600–4450 BC). This unique dataset originated from the Curt-Engelhorn-Centre for Archaeometry in Mannheim, Germany. The additional dataset from Bulgaria (http://oxalid.arch.ox.ac.uk/Bulgaria/Bulgaria.html) was not used due to the difficulties in matching the exact level of contextual information available for the data in Tables 4–11.

From ore to metal: the pathways of lead isotope abundance ratios

A key feature of the tight clustering of copper minerals from Belovode (Bf22/12/2, B108/13, B385/13) and Pločnik (P55/13, P121/13) in *Group 1* is that these minerals come from top, middle and bottom spits, and have been chosen to check the consistency of ore choice from the beginning until the end of the occupation at both settlements. Interestingly, one of these (Bf22/12/2, from Belovode Horizon 1b), is an exact match for the earliest metal smelted at Belovode (Bf43/13, Belovode Horizon 2), while copper mineral B108/13 in this cluster originated from the same spit and horizon as Bf43/13 (see Table 2). The whole cluster shows more clearly that there was a consistent supply of copper minerals / ores from a source (or a group of sources) throughout the occupation of these two settlements, from *c.* 5200 BC until 4600 BC, or for *c.* 600 years. Looking at the direct dates provided (see Table 1, Chapter 37), both settlements joined the same supply network at almost the same time: the copper mineral from Belovode (B385/13) comes from spit 19, directly dated to the 51st century BC, which correlates well with the probability distribution of ^{14}C dates for P121/13 (associated with spit 21 and F30/F34), set also at around the 51st century BC. Also, the copper mineral samples B108/13, Bf43/13 and P55/13 show similar dating to the 49th century BC, which again reinforces the existence of the same supply networks between these sites and the ore deposits in eastern Serbia. This further supports the likely transmission of a shared metallurgical knowledge between the communities of Belovode and Pločnik, from the very beginning until the end of the Belovode occupation (46th century BC, see Chapter 37).

Table 4. Lead isotope abundance ratios of copper deposits in the Balkans. Regions are abbreviated as follows: BSC=Black Sea Coast, SRB = Serbia, W= West Bulgaria, STR= Strandza, THR = Thracia (data from Amov and Văkova 1994; Gale *et al.* 1991; Kunze and Pernicka 2020; Pernicka *et al.* 1993, 1997). Ore minerals are abbreviated as follows: chalc=chalcocite, cov=covelline, pyr=pyrite, enarg=enargite, chpyr=chalcopyrite, born=bornite, gal=galena, sphal=sphalerite, mal=malachite, fahl=fahlore, magn=magnetite.

Label	Sample	Location	Region	description	$^{208}Pb/^{206}Pb$	$^{207}Pb/^{206}Pb$	$^{204}Pb/^{206}Pb$	$^{208}Pb/^{204}Pb$	$^{207}Pb/^{204}Pb$	$^{206}Pb/^{204}Pb$
HDM	TG 197	Bor	SRB	Cu ore (chalc)	2.0900	0.85100	0.054820	38.127	15.524	18.242
HDM	TG 197 A	Bor	SRB	Cu ore (cov, pyr, enarg)	2.0866	0.84910	0.054680	38.158	15.528	18.288
HDM	TG 197 B	Bor	SRB	Cu ore (chalc)	2.0915	0.85050	0.054750	38.203	15.536	18.266
HDM	TG 197 C	Bor	SRB	Cu ore (cov, pyr, enarg)	2.0864	0.84930	0.054720	38.128	15.520	18.275
HDM	TG 197 E	Bor	SRB	Cu ore (enarg, pyr, cov)	2.0895	0.84930	0.054650	38.234	15.541	18.298
HDM	TG 197 F	Bor	SRB	Cu ore (chpyr, pyr)	2.0882	0.84990	0.054750	38.137	15.522	18.263
HDM	TG 197 G	Bor	SRB	Cu ore (cov, pyr, enarg)	2.0891	0.84980	0.054660	38.217	15.546	18.294
HDM	TG 197 I	Bor	SRB	Cu ore (cov, pyr)	2.0887	0.84970	0.054730	38.163	15.525	18.271
HDM	TG 197 J	Bor	SRB	Cu ore (enarg, pyr, cov)	2.0909	0.85050	0.054710	38.219	15.547	18.279
HDM	TG 197 K-1	Bor	SRB	Cu ore (born, pyr, chpyr)	2.0894	0.85070	0.054790	38.138	15.527	18.253
HDM	TG 197 K-4	Bor	SRB	Cu ore (cov, pyr)	2.0920	0.85120	0.054730	38.222	15.552	18.271
HDM	TG 197 L	Bor	SRB	Cu ore (gal, pyr, cov)	2.0915	0.85090	0.054760	38.193	15.538	18.261

Table 4 continued. Lead isotope abundance ratios of copper deposits in the Balkans. Regions are abbreviated as follows: BSC=Black Sea Coast, SRB = Serbia, W= West Bulgaria, STR= Strandza, THR = Thracia (data from Amov and Văkova 1994; Gale *et al.* 1991; Kunze and Pernicka 2020; Pernicka *et al.* 1993, 1997). Ore minerals are abbreviated as follows: chalc=chalcocite, cov=covelline, pyr=pyrite, enarg=enargite, chpyr=chalcopyrite, born=bornite, gal=galena, sphal=sphalerite, mal=malachite, fahl=fahlore, magn=magnetite.

Label	Sample	Location	Region	description	$^{208}Pb/^{206}Pb$	$^{207}Pb/^{206}Pb$	$^{204}Pb/^{206}Pb$	$^{208}Pb/^{204}Pb$	$^{207}Pb/^{204}Pb$	$^{206}Pb/^{204}Pb$
HDM	TG 197 COV	Bor	SRB	Cu ore (cov)	2.0873	0.84930	0.054660	38.185	15.537	18.294
HDM	TG 197 ENARG	Bor	SRB	Cu ore (enarg)	2.0854	0.84880	0.054710	38.115	15.515	18.278
HDM	HDM 1395	Bor	SRB	native copper	2.0860	0.84540	0.054030	38.608	15.646	18.508
HDM	HDM 1396	Bor	SRB	native Cu	2.0901	0.85040	0.054770	38.160	15.526	18.258
HDM	TG 240	Velika Brestovica (near RG)	SRB	Cu slag	2.0587	0.83200	0.053140	38.742	15.656	18.818
HDM	TG 253 A-1	Majdanpek	SRB	Cu slag, metallic Cu	2.0783	0.84400	0.054170	38.367	15.581	18.461
HDM	TG 253 A-2	Majdanpek	SRB	oxidic Cu ore	2.0753	0.84430	0.054170	38.310	15.567	18.460
HDM	TG 253 B	Majdanpek	SRB	Cu slag, met. Cu	2.0763	0.84430	0.054140	38.349	15.576	18.470
HDM	TG 253 C	Majdanpek	SRB	Cu slag	2.0781	0.84370	0.054080	38.425	15.601	18.491
HDM	TG 253 E-1	Majdanpek	SRB	native Cu	2.0832	0.84510	0.054140	38.478	15.609	18.470
HDM	TG 253 E-2	Majdanpek	SRB	Cu ore (cupr)	2.0750	0.84280	0.054100	38.351	15.578	18.483
HDM	TG 253 F	Majdanpek	SRB	Cu ore (mal, azur)	2.0780	0.84400	0.054250	38.306	15.560	18.435
HDM	TG 253 G	Majdanpek	SRB	Cu ore (chpyr)	2.0752	0.84270	0.054150	38.326	15.563	18.489
AMOV	/	Majdanpek	SRB	Cu ore	2.0807	0.84440	2.464000	38.511	15.629	18.509
HDM	BG-17	Ai Bunar	THR	malachite	2.0829	0.84410	0.054000	38.572	15.631	18.519
HDM	BG-17e	Ai Bunar	THR	malachite	2.0836	0.84440	0.053980	38.599	15.643	18.525
HDM	BG-17f	Ai Bunar	THR	malachite	2.0836	0.84450	0.054000	38.585	15.639	18.519
HDM	BG-17g	Ai Bunar	THR	malachite	2.0823	0.84420	0.053990	38.568	15.636	18.522
HDM	BG-17h	Ai Bunar	THR	malachite	2.0830	0.84420	0.054000	38.574	15.633	18.519
HDM	BG-17i	Ai Bunar	THR	malachite	2.0824	0.84410	0.054030	38.542	15.623	18.508
CEZA	MA-14991	Ai Bunar	THR	malachite	2.0819	0.84380	0.054018	38.541	15.621	18.512
GALE	AB1	Ai Bunar	THR	malachite	2.0811	0.84831	0.054051	38.502	15.695	18.501
GALE	AB1a	Ai Bunar	THR	malachite	2.0871	0.84597	0.054086	38.589	15.641	18.489
GALE	AB2	Ai Bunar	THR	malachite	2.0791	0.84368	0.054115	38.419	15.590	18.479
GALE	AB2a	Ai Bunar	THR	malachite	2.0797	0.84360	0.054022	38.496	15.616	18.511
GALE	AB4	Ai Bunar	THR	malachite	2.0805	0.84398	0.054051	38.492	15.614	18.501
GALE	AB5	Ai Bunar	THR	malachite	2.0812	0.84396	0.054075	38.487	15.607	18.493

Table 4 continued. Lead isotope abundance ratios of copper deposits in the Balkans. Regions are abbreviated as follows: BSC=Black Sea Coast, SRB = Serbia, W= West Bulgaria, STR= Strandza, THR = Thracia (data from Amov and Vǎkova 1994; Gale *et al.* 1991; Kunze and Pernicka 2020; Pernicka *et al.* 1993, 1997). Ore minerals are abbreviated as follows: chalc=chalcocite, cov=covelline, pyr=pyrite, enarg=enargite, chpyr=chalcopyrite, born=bornite, gal=galena, sphal=sphalerite, mal=malachite, fahl=fahlore, magn=magnetite.

Label	Sample	Location	Region	description	$^{208}Pb/^{206}Pb$	$^{207}Pb/^{206}Pb$	$^{204}Pb/^{206}Pb$	$^{208}Pb/^{204}Pb$	$^{207}Pb/^{204}Pb$	$^{206}Pb/^{204}Pb$
GALE	AB5a	Ai Bunar	THR	malachite	2.0819	0.84437	0.053990	38.561	15.639	18.522
GALE	75	Ai Bunar	THR	malachite	2.0811	0.84421	0.054133	38.443	15.595	18.473
GALE	77D	Ai Bunar	THR	malachite	2.0816	0.84391	0.054037	38.522	15.617	18.506
GALE	78C	Ai Bunar	THR	malachite	2.0813	0.84398	0.054133	38.449	15.591	18.473
GALE	79	Ai Bunar	THR	malachite	2.0857	0.84581	0.054198	38.483	15.606	18.451
HDM	BG-11a	Radka	W	native copper	2.0703	0.83660	0.053390	38.777	15.670	18.730
HDM	BG-11b	Radka	W	native copper	2.0819	0.84200	0.053760	38.726	15.662	18.601
HDM	BG-11c	Radka	W	chpyr, pyr, gal, fhl, born	2.0845	0.84470	0.054060	38.559	15.625	18.498
HDM	36	Radka-Panagjuriste	W	ore?	2.0866	0.84560				
HDM	BG-6a	Zidarovo	BSC	malachite	2.0664	0.83660	0.053320	38.755	15.690	18.755
HDM	BG-6b	Zidarovo	BSC	malachite	2.0754	0.84210	0.053950	38.469	15.609	18.536
HDM	BG-6c	Zidarovo	BSC	born, chpyr	2.0666	0.83770	0.053600	38.556	15.629	18.657
CEZA	MA-141987	Zidarovo	BSC	malachite	2.0656	0.83593	0.053427	38.662	15.646	18.717
A&V	42.1	Zidarovo, Urta	BSC	galena	2.0757	0.84110	0.053903	38.509	15.605	18.552
A&V	42.2	Zidarovo, Urta	BSC	galena	2.0677	0.83750	0.053671	38.525	15.605	18.632
A&V	42.3	Zidarovo, Urta	BSC	galena	2.0718	0.83850	0.053576	38.671	15.650	18.665
HDM	BG-12d	M. Tarnovo-Bradceto	STR	malachite	2.0646	0.83550	0.053360	38.692	15.658	18.741
HDM	BG-12a	M. Tarnovo-Bradceto	STR	native Cu	2.0733	0.83690	0.053410	38.819	15.669	18.723
HDM	BG-12b	M. Tarnovo-Bradceto	STR	native Cu	2.0689	0.83600	0.053540	38.642	15.614	18.678
HDM	BG-12c	M. Tarnovo-Bradceto	STR	native Cu	2.0671	0.83410	0.053160	38.884	15.690	18.811
HDM	BG-12e	M. Tarnovo-Bradceto	STR	native Cu	2.0731	0.83510	0.053330	38.873	15.659	18.751
HDM	ZDRelo 1 C1	Ždrelo	SRB	malachite	2.0735	0.83991	0.053706	38.601	15.639	18.620
HDM	ZDRelo 1 C2	Ždrelo	SRB	malachite	2.0713	0.83931	0.053714	38.602	15.625	18.617
HDM	ZDRelo 2	Ždrelo	SRB	malachite	2.0730	0.83988	0.053697	38.574	15.641	18.623
HDM	Mali Sturac 152	Mali Sturac	SRB	malachite	2.0810	0.83903	0.053565	38.701	15.664	18.669
HDM	TG196	Rudna Glava	SRB	Cu ore mal	2.0374	0.83190	0.053250	38.261	15.622	18.780
HDM	TG196-1	Rudna Glava	SRB	Cu ore (magn, chalc, mal)	1.5448	0.63460	0.039670	38.937	15.996	25.206

Table 4 continued. Lead isotope abundance ratios of copper deposits in the Balkans. Regions are abbreviated as follows: BSC=Black Sea Coast, SRB = Serbia, W= West Bulgaria, STR= Strandza, THR = Thracia (data from Amov and Văkova 1994; Gale *et al.* 1991; Kunze and Pernicka 2020; Pernicka *et al.* 1993, 1997). Ore minerals are abbreviated as follows: chalc=chalcocite, cov=covelline, pyr=pyrite, enarg=enargite, chpyr=chalcopyrite, born=bornite, gal=galena, sphal=sphalerite, mal=malachite, fahl=fahlore, magn=magnetite.

Label	Sample	Location	Region	description	$^{208}Pb/^{206}Pb$	$^{207}Pb/^{206}Pb$	$^{204}Pb/^{206}Pb$	$^{208}Pb/^{204}Pb$	$^{207}Pb/^{204}Pb$	$^{206}Pb/^{204}Pb$
HDM	TG 196-2	Rudna Glava	SRB	Cu ore mal	2.0466	0.83720	0.053590	38.190	15.623	18.660
HDM	TG 196-2g-1	Rudna Glava	SRB	Cu ore (mal, azur)	1.0507	0.45250	0.027300	38.493	16.578	36.635
HDM	TG 196-2g-2	Rudna Glava	SRB	Cu ore (chalc, mal)	1.9548	0.79740	0.050840	38.447	15.684	19.669
HDM	TG 196-2g-3	Rudna Glava	SRB	Cu ore (mal)	1.5461	0.60670	0.037920	40.774	15.999	26.372
HDM	TG 196-2g-4	Rudna Glava	SRB	Cu ore (mal)	2.0493	0.83880	0.053590	38.239	15.652	18.660
HDM	TG 196-3	Rudna Glava	SRB	Cu ore (mal)	2.0409	0.83270	0.053220	38.345	15.646	18.789
HDM	TG 196-4	Rudna Glava	SRB	Cu ore (mal)	1.8599	0.75150	0.047640	39.039	15.774	20.990
HDM	TG 196-4.3	Rudna Glava	SRB	Cu ore (chpyr, mal, azur)	1.8791	0.75520	0.047930	39.207	15.756	20.864
HDM	TG 196 A	Rudna Glava	SRB	Cu ore (chpyr, mal, azur)	1.6961	0.68300	0.042970	39.469	15.895	23.271
HDM	TG 196 B	Rudna Glava	SRB	Cu ore (magn, chalc)	1.8239	0.74830	0.047490	38.410	15.758	21.059
HDM	RUDNA 1	Rudna Glava	SRB	Cu ore (mal)	1.8846	0.76650	0.048520	38.844	15.800	20.612
HDM	RUDNA 2	Rudna Glava	SRB	Cu ore (mal)	1.9339	0.78460	0.049970	38.703	15.701	20.013
HDM	RUDNA 3	Rudna Glava	SRB	Cu ore (mal)	1.9935	0.81190	0.051810	38.477	15.672	19.302
HDM	RUDNA 4	Rudna Glava	SRB	Cu ore (mal)	0.6723	0.30890	0.017500	38.425	17.655	57.156
HDM	RUDNA 5	Rudna Glava	SRB	Cu ore (mal)	1.9535	0.78010	0.049710	39.300	15.693	20.118
GALE	RG1	RUDNA GLAVA 1	SRB	Cu ore	2.0757	0.84975	0.054526	38.069	15.584	18.340
GALE	RG2	RUDNA GLAVA 2	SRB	Cu ore	2.0780	0.85069	0.054564	38.084	15.591	18.327
GALE	RG3	RUDNA GLAVA 3	SRB	Cu ore	2.0835	0.85226	0.054576	38.175	15.616	18.323
GALE	RG4	RUDNA GLAVA 4	SRB	Cu ore	2.0795	0.85223	0.054538	38.130	15.626	18.336
HDM	TG 249 A	Rudnik	SRB	Cu ore (mal)	2.0770	0.83840	0.053550	38.784	15.655	18.673
HDM	TG 249 AA	Rudnik	SRB	Pb-Zn ore (gal, sphal, pyr, chpyr)	2.0783	0.83870	0.053520	38.834	15.671	18.686
HDM	TG 249 B	Rudnik	SRB	Cu slag	2.0783	0.83830	0.053560	38.806	15.652	18.672
HDM	TG 249 BB	Rudnik	SRB	Pb-Zn ore (gal, sphal, pyr, chpyr)	2.0793	0.83870	0.053510	38.861	15.674	18.689
HDM	TG 249 D	Rudnik	SRB	Cu slag	2.0788	0.83880	0.053520	38.843	15.674	18.685
HDM	TG 249 EE	Rudnik	SRB	Cu ore (chpyr)	2.0783	0.83870	0.053540	38.815	15.663	18.677
HDM	TG 249 F	Rudnik	SRB	Pb-Zn slag	2.0788	0.83840	0.053500	38.858	15.672	18.692
HDM	TG 249 FF	Rudnik	SRB	Pb-Zn ore (gal, sphal, pyr, chpyr)	2.0770	0.83850	0.053590	38.759	15.647	18.661

Table 4 continued. Lead isotope abundance ratios of copper deposits in the Balkans. Regions are abbreviated as follows: BSC=Black Sea Coast, SRB = Serbia, W= West Bulgaria, STR= Strandza, THR = Thracia (data from Amov and Văkova 1994; Gale *et al.* 1991; Kunze and Pernicka 2020; Pernicka *et al.* 1993, 1997). Ore minerals are abbreviated as follows: chalc=chalcocite, cov=covelline, pyr=pyrite, enarg=enargite, chpyr=chalcopyrite, born=bornite, gal=galena, sphal=sphalerite, mal=malachite, fahl=fahlore, magn=magnetite.

Label	Sample	Location	Region	description	$^{208}Pb/^{206}Pb$	$^{207}Pb/^{206}Pb$	$^{204}Pb/^{206}Pb$	$^{208}Pb/^{204}Pb$	$^{207}Pb/^{204}Pb$	$^{206}Pb/^{204}Pb$
HDM	TG 249 G	Rudnik	SRB	Pb-Zn slag	2.0796	0.83860	0.053520	38.853	15.668	18.683
HDM	TG 249 O-1	Rudnik	SRB	Cu ore (chpyr, mal)	2.0770	0.83830	0.053540	38.792	15.657	18.677
HDM	TG 249 O-2	Rudnik	SRB	Cu slag	2.0785	0.83880	0.053550	38.818	15.666	18.676
HDM	TG 249 T	Rudnik	SRB	Cu slag	2.0786	0.83870	0.053530	38.830	15.668	18.681
HDM	TG 249 Y	Rudnik	SRB	Cu slag	2.0801	0.83910	0.053450	38.919	15.699	18.710
HDM	Mali Sturac 152	Mali Sturac	SRB	malachite	2.0810	0.83903	0.053565	38.701	15.664	18.669
HDM	BG-1b	Medni Rid	BSC	chpyr, pyr	2.0645	0.83590	0.053410	38.654	15.651	18.723
HDM	BG-1c	Medni Rid	BSC	chpyr, born, mal, pyr, fhl	2.0745	0.84500	0.054110	38.339	15.616	18.481
CEZA	MA-135454	Rosen, Medni Rid region	BSC	malachite	1.8253	0.74540	0.047404	38.505	15.724	21.095
CEZA	MA-141990	Rosen, Medni Rid region	BSC	malachite	1.8196	0.73895	0.046903	38.795	15.754	21.320
CEZA	MA-145532	Rosen, Medni Rid region	BSC	malachite	1.8180	0.74178	0.047142	38.564	15.735	21.213
CEZA	MA-141989	Propadnala Voda, Medni Rid region	BSC	malachite	1.7136	0.68114	0.042754	40.080	15.932	23.390
CEZA	MA-145533	Propadnala Voda, Medni Rid region	BSC	malachite	1.8215	0.74430	0.047296	38.513	15.737	21.144
CEZA	MA-145534	Propadnala Voda, Medni Rid region	BSC	surface slag	1.5612	0.64458	0.040514	38.535	15.910	24.683
CEZA	MA-147718	Propadnala Voda, Medni Rid region	BSC	malachite	1.1363	0.48083	0.029397	38.654	16.356	34.017
CEZA	MA-152442	Propadnala Voda, Medni Rid region	BSC	malachite	1.6644	0.68392	0.043203	38.525	15.830	23.147
CEZA	MA-147715	Propadnala Voda, Medni Rid region	BSC	malachite	1.8286	0.74644	0.047415	38.566	15.743	21.090
CEZA	MA-147716	Propadnala Voda, Medni Rid region	BSC	malachite	2.0422	0.82842	0.052989	38.540	15.634	18.872
CEZA	MA-147722	Propadnala Voda, Medni Rid region	BSC	malachite	1.9733	0.80186	0.051187	38.551	15.666	19.536
CEZA	MA-147719	Kyumyur-lake, Medni Rid region	BSC	malachite	1.3509	0.56261	0.034962	38.639	16.092	28.603
CEZA	MA-147721	Kyumyur-lake, Medni Rid region	BSC	malachite	1.7921	0.72624	0.045977	38.978	15.795	21.750
CEZA	MA-147723	Surneshko Kladenche, Medni Rid region	BSC	malachite	2.0298	0.82333	0.052643	38.558	15.640	18.996
CEZA	MA-135453	Varli Brjag	BSC	malachite	2.0637	0.83611	0.053490	38.581	15.361	18.695
CEZA	MA-141988	Varli Brjag	BSC	malachite	2.0638	0.83396	0.053287	38.730	15.650	18.766
CEZA	MA-135455	Atiya	BSC	malachite	2.0394	0.82900	0.052948	38.517	15.657	18.886
CEZA	MA-141992	Zelenata Kanara	BSC	malachite	2.0613	0.81434	0.051801	39.793	15.721	19.305

Table 5. Lead isotope abundance ratios of Vinča culture malachite and copper metal implements (data from Pernicka *et al.* 1993; Radivojević *et al.* 2010a).

rel period	abs period	Cultural attribution	Site	Type	Region	$^{208}Pb/^{206}Pb$	$^{207}Pb/^{206}Pb$	$^{204}Pb/^{206}Pb$	$^{208}Pb/^{204}Pb$	$^{207}Pb/^{204}Pb$	$^{206}Pb/^{204}Pb$
EC	5000-4600 BC	Vinča culture	Medvednjak	malachite bead	SRB	2.0760	0.84290	0.054080	38.388	15.586	18.491
EC	5000-4600 BC	Vinča culture	Medvednjak	malachite bead	SRB	2.0554	0.83990	0.052940	38.825	15.865	18.889
EC	5000-4600 BC	Vinča culture	Selevac	malachite	SRB	2.0328	0.82360	0.052560	38.676	15.670	19.026
EC	5000-4600 BC	Vinča culture	Selevac	malachite	SRB	2.0552	0.83110	0.053290	38.566	15.596	18.765
EC	5000-4600 BC	Vinča culture	Selevac	malachite	SRB	2.0739	0.84200	0.054050	38.370	15.578	18.501
EC	5000-4600 BC	Vinča culture	Selevac	malachite	SRB	2.0749	0.84220	0.053990	38.431	15.599	18.522
EC	5000-4600 BC	Vinča culture	Selevac	malachite	SRB	2.0734	0.84200	0.053880	38.482	15.627	18.560
EC	5000-4600 BC	Vinča culture	Selevac	malachite	SRB	2.0762	0.84290	0.054070	38.398	15.589	18.495
EC	5000-4600 BC	Vinča culture	Selevac	malachite	SRB	2.0672	0.83400	0.053400	38.712	15.618	18.727
EC	5000-4600 BC	Vinča culture	Selevac	malachite	SRB	2.0735	0.84160	0.054010	38.391	15.582	18.515
EC	5000-4600 BC	Vinča culture	Selevac	malachite	SRB	2.0615	0.83320	0.053360	38.634	15.615	18.741
EC	5000-4600 BC	Vinča culture	Selevac	malachite	SRB	2.0731	0.84400	0.054010	38.384	15.627	18.515
EC	5000-4600 BC	Vinča culture	Selevac	malachite	SRB	2.0657	0.83500	0.053340	38.727	15.654	18.748
EC	5000-4600 BC	Vinča culture	Selevac	malachite	SRB	2.0681	0.83490	0.053330	38.779	15.655	18.751
EC	5000-4600 BC	Vinča culture	Selevac	malachite	SRB	2.0718	0.84010	0.053790	38.516	15.618	18.591
EC	5000-4600 BC	Vinča culture	Selevac	copper prill	SRB	2.0422	0.82440	0.052700	38.751	15.643	18.975
EC	5000-4600 BC	Vinča culture	Selevac	copper prill	SRB	2.0441	0.82530	0.052650	38.824	15.675	18.993
EC	5000-4600 BC	Vinca culture	Belovode	Belovode 12 mineral	SRB	2.0693	0.83947	0.053807	38.457	15.602	18.585
EC	5000-4600 BC	Vinca culture	Belovode	Belovode M13 mineral	SRB	2.0771	0.84338	0.054113	38.385	15.586	18.480
EC	5000-4600 BC	Vinca culture	Belovode	Belovode M17 mineral	SRB	2.0754	0.84249	0.054080	38.377	15.579	18.491
EC	5000-4600 BC	Vinca culture	Belovode	Belovode M20 slag	SRB	2.0610	0.83512	0.053548	38.490	15.596	18.675
EC	5000-4600 BC	Vinca culture	Belovode	Belovode M21 slag	SRB	2.0550	0.83089	0.053104	38.699	15.647	18.831
EC	5000-4600 BC	Vinca culture	Belovode	Belovode M22a slag	SRB	2.0546	0.83065	0.052994	38.769	15.675	18.870
EC	5000-4600 BC	Vinca culture	Belovode	Belovode M23 slag	SRB	2.0088	0.81254	0.051824	38.760	15.681	19.296
MC	4600-4450 BC	Vinča culture	Gomolava	bracelet	SRB	2.0777	0.84570	0.054150	38.369	15.618	18.467
MC	4600-4450 BC	Vinča culture	Pločnik	chisel	SRB	2.0812	0.84330	0.054000	38.541	15.617	18.519

Table 5 continued. Lead isotope abundance ratios of Vinča culture malachite and copper metal implements (data from Pernicka *et al.* 1993; Radivojević *et al.* 2010a).

rel period	abs period	Cultural attribution	Site	Type	Region	$^{208}Pb/^{206}Pb$	$^{207}Pb/^{206}Pb$	$^{204}Pb/^{206}Pb$	$^{208}Pb/^{204}Pb$	$^{207}Pb/^{204}Pb$	$^{206}Pb/^{204}Pb$
MC	4600-4450 BC	Vinča culture	Pločnik	flat axe	SRB	2.0836	0.84420	0.054010	38.578	15.630	18.515
MC	4600-4450 BC	Vinča culture	Pločnik	flat axe	SRB	2.0825	0.84390	0.054020	38.551	15.622	18.512
MC	4600-4450 BC	Vinča culture	Pločnik	hammer axe	SRB	2.0518	0.82680	0.052890	38.794	15.632	18.907
MC	4600-4450 BC	Vinča culture	Pločnik	hammer axe	SRB	2.0574	0.83040	0.053050	38.782	15.653	18.850
MC	4600-4450 BC	Vinča culture	Pločnik	chisel	SRB	2.0657	0.83310	0.053320	38.742	15.625	18.755
MC	4600-4450 BC	Vinča culture	Pločnik	chisel	SRB	2.0629	0.83360	0.053330	38.682	15.631	18.751
MC	4600-4450 BC	Vinča culture	Pločnik	chisel	SRB	2.0775	0.84030	0.053650	38.723	15.663	18.639
MC	4600-4450 BC	Vinča culture	Pločnik	chisel	SRB	2.0748	0.83860	0.053510	38.774	15.672	18.688
MC	4600-4450 BC	Vinča culture	Pločnik	chisel	SRB	2.0840	0.84430	0.054020	38.578	15.629	18.512
MC	4600-4450 BC	Vinča culture	Pločnik	chisel	SRB	2.0844	0.84430	0.053980	38.614	15.641	18.525
MC	4600-4450 BC	Vinča culture	Pločnik	chisel	SRB	2.0751	0.84050	0.053720	38.628	15.646	18.615
MC	4600-4450 BC	Vinča culture	Pločnik	chisel	SRB	2.0836	0.84420	0.054010	38.578	15.630	18.515
MC	4600-4450 BC	Vinča culture	Pločnik	chisel	SRB	2.0585	0.83070	0.053040	38.810	15.662	18.854
MC	4600-4450 BC	Vinča culture	Pločnik	chisel	SRB	2.0831	0.84390	0.053990	38.583	15.631	18.522
MC	4600-4450 BC	Vinča culture	Pločnik	chisel	SRB	2.0650	0.83520	0.053330	38.721	15.661	18.751
MC	4600-4450 BC	Vinča culture	Pločnik	hammer axe	SRB	2.0824	0.84360	0.054010	38.556	15.619	18.515

Table 6. Lead isotope abundance ratios of 'Group of 16' artefacts (data from Pernicka *et al.* 1997).

Sample	rel period	abs period	Cultural attribution	Site	Type	Region	208Pb/206Pb	207Pb/206Pb	204Pb/206Pb	208Pb/204Pb	207Pb/204Pb	206Pb/204Pb
HDM2004	EC	5000-4600 BC	Hamangia II	Durankulak	malachite bead	BSC	2.0513	0.82980	0.052990	38.711	15.660	18.871
HDM1559	MC	4600-4450 BC	Vinča culture	Pločnik	axe	SRB	2.0574	0.83040	0.053050	38.782	15.653	18.850
HDM1568	MC	4600-4450 BC	Vinča culture	Pločnik	axe	SRB	2.0585	0.83070	0.053040	38.810	15.662	18.854
HDM1982	MC	4600-4450 BC	Varna II	Durankulak	bracelet	BSC	2.0556	0.83060	0.053080	38.726	15.648	18.839
HDM1987	MC	4600-4450 BC	Varna I	Durankulak	bracelet	BSC	2.0567	0.83260	0.053140	38.703	15.668	18.818
HDM2016	MC	4600-4450 BC	Varna II	Durankulak	bracelet	BSC	2.0551	0.82960	0.053010	38.768	15.650	18.864
HDM1920	MC	4600-4450 BC	KGK VI?	Ruse	axe	NE	2.0538	0.83220	0.053210	38.598	15.640	18.793
HDM2043	MC	4600-4450 BC	KGK VI?	Ruse	borer	NE	2.0514	0.83010	0.053170	38.582	15.612	18.808
HDM2059	MC	4600-4450 BC	KGK VI?	Ruse	borer	NE	2.0581	0.83160	0.053020	38.817	15.685	18.861
HDM2068	MC	4600-4450 BC	KGK VI?	Ruse	borer	BSC	2.0510	0.83000	0.052920	38.757	15.684	18.896
HDM2089	LC	4450-4100 BC	KGK VI?	Ruse	pin	NE	2.0586	0.83170	0.053140	38.739	15.651	18.818
HDM2101	LC	4450-4100 BC	KGK VI?	Ruse	borer	NE	2.0529	0.82810	0.052920	38.793	15.648	18.896
HDM2102	LC	4450-4100 BC	KGK VI?	Ruse	borer	NE	2.0571	0.83190	0.053130	38.718	15.658	18.822
HDM2103	LC	4450-4100 BC	KGK VI?	Ruse	borer	NE	2.0598	0.83250	0.053060	38.820	15.690	18.847
HDM1303	FC	4100-3700 BC	Bodrogkeresztur	Urovica	axe	SRB	2.0551	0.83070	0.053120	38.688	15.638	18.825
HDM2736	PB	3700-3200 BC	KSBh IV?	Malorad	dagger	W	2.0610	0.83150	0.053010	38.879	15.686	18.864

Table 7. Lead isotope abundance ratios of Akladi Cheiri and related Middle to Late Chalcolithic sites on the Black Sea Coast (data from Kunze and Pernicka 2020; Rehren *et al.* 2020).

Sample	rel period	abs period	Cultural attribution	Site	Type	Region	208Pb/206Pb	207Pb/206Pb	204Pb/206Pb	208Pb/204Pb	207Pb/204Pb	206Pb/204Pb
MA-135452	LN	5500-5000 BC	Karanovo III/IV	Alepu	malachite	BSC	2.0137	0.81748	0.052293	38.508	15.633	19.123
MA-152458	LN	5500-5000 BC	Karanovo III/IV	Hadzidimitrovo	malachite	BSC	1.9778	0.79940	0.050966	38.806	15.685	19.621
MA-152459	LN	5500-5000 BC	Karanovo III/IV	Hadzidimitrovo	malachite	BSC	2.0640	0.83408	0.053309	38.718	15.647	18.759
MA-152460	LN	5500-5000 BC	Karanovo III/IV	Dana Bunar 2	malachite	BSC	2.0611	0.83394	0.053263	38.697	15.657	18.775
MA-133042	LN	5500-5000 BC	Karanovo III/IV	Akladi Cheiri	malachite	BSC	1.8394	0.75056	0.047712	38.552	15.731	20.959
MA-133043	LN	5500-5000 BC	Karanovo III/IV	Akladi Cheiri	malachite	BSC	1.4452	0.59936	0.054075	26.726	16.018	26.726
MA-133044	LN	5500-5000 BC	Karanovo III/IV	Akladi Cheiri	malachite	BSC	2.0196	0.81944	0.052367	38.566	15.648	19.096
MA-133045	LN	5500-5000 BC	Karanovo III/IV	Akladi Cheiri	malachite	BSC	2.0541	0.83240	0.053208	38.605	15.644	18.794
MA-133046	LC	4450-4100 BC	KGK VI	Akladi Cheiri	limonite/malachite	BSC	1.9520	0.79405	0.050682	38.515	15.667	19.731
MA-133047	LC	4450-4100 BC	KGK VI	Akladi Cheiri	limonite/malachite	BSC	2.0372	0.82633	0.052840	38.554	15.638	18.925
MA-133049	LC	4450-4100 BC	KGK VI	Akladi Cheiri	crucible/ceramic	BSC	2.0586	0.82916	0.052938	38.887	15.663	18.890
MA-135447	LC	4450-4100 BC	KGK VI	Akladi Cheiri	furnace wall with mal	BSC	2.0487	0.82598	0.052778	38.817	15.650	18.947
MA-151990	LC	4450-4100 BC	KGK VI	Akladi Cheiri	copper prill/crucible	BSC	2.0541	0.83184	0.053150	38.647	15.650	18.814
MA-133048	LC	4450-4100 BC	KGK VI	Akladi Cheiri	crucible/slag	BSC	2.0560	0.83383	0.053390	38.509	15.618	18.730
MA-151990	LC	4450-4100 BC	KGK VI	Akladi Cheiri	slag on pottery	BSC	2.0326	0.82476	0.052758	38.527	15.633	18.814
MA-151990	LC	4450-4100 BC	KGK VI	Akladi Cheiri	slag on pottery	BSC	2.0450	0.82926	0.053047	38.551	15.633	18.852
MA-151990	LC	4450-4100 BC	KGK VI	Akladi Cheiri	slag on pottery	BSC	2.0604	0.83429	0.053315	38.646	15.649	18.757
MA-151990	LC	4450-4100 BC	KGK VI	Akladi Cheiri	slag on pottery	BSC	2.0656	0.83695	0.053527	38.590	15.636	18.682
MA-135448	LC	4450-4100 BC	KGK VI	Akladi Cheiri	malachite	BSC	0.9262	0.40165	0.024028	38.546	16.715	41.616
MA-135449	LC	4450-4100 BC	KGK VI	Akladi Cheiri	malachite	BSC	2.0075	0.81549	0.052162	38.486	15.634	19.171
MA-135450	LC	4450-4100 BC	KGK VI	Akladi Cheiri	malachite	BSC	1.9927	0.80684	0.051556	38.651	15.650	19.396
MA-135451	LC	4450-4100 BC	KGK VI	Budzhaka	malachite	BSC	2.0457	0.83001	0.053128	38.505	15.622	18.822

Table 8. Lead isotope abundance ratios for EC and MC copper metal artefacts (data from Pernicka *et al.* 1993, 1997; Radivojević *et al.* 2010a).

LABEL	period	absolute dating	cultural attribution	Site	type of site/context	Region	Type	Axe type	$^{208}Pb/^{206}Pb$	$^{207}Pb/^{206}Pb$
HDM 1904	EC	5000-4600 BC	Dikilitash Slatino Gradesnica	Slatino	settlement	W	chisel	unk	2.0835	0.84470
HDM 1912	EC	5000-4600 BC	Marica III	Marica	settlement	THR	heavy axe	unk	2.0714	0.83690
HDM 1975	EC	5000-4600 BC	Hamangia III	Durankulak	cemetery	BSC	finger ring	unk	2.0824	0.84400
HDM 2005	EC	5000-4600 BC	Hamangia III	Durankulak	cemetery	BSC	finger ring	unk	2.0339	0.82410
Belovode M34	MC	4600-4450 BC	Vinča culture	Belovode	settlement	SRB	copper ingot	unk	2.0773	0.84348
HDM 1308	MC	4600-4450 BC	KSBh / Vinča?	Sumrakovac	stray	SRB	hammer axe	Pločnik	2.0615	0.83390
HDM 1496	MC	4600-4450 BC	Vinča culture	Gomolava	cemetery	SRB	bracelet	unk	2.0777	0.84570
HDM 1555	MC	4600-4450 BC	Vinča culture	Pločnik	settlement/hoard?	SRB	chisel	unk	2.0812	0.84330
HDM 1556	MC	4600-4450 BC	Vinča culture	Pločnik	settlement/hoard?	SRB	flat axe	unk	2.0836	0.84420
HDM 1557	MC	4600-4450 BC	Vinča culture	Pločnik	settlement/hoard?	SRB	flat axe	unk	2.0825	0.84390
HDM 1558	MC	4600-4450 BC	Vinča culture	Pločnik	settlement/hoard?	SRB	hammer axe	Pločnik	2.0518	0.82680
HDM 1559	MC	4600-4450 BC	Vinča culture	Pločnik	settlement/hoard?	SRB	hammer axe	Pločnik	2.0574	0.83040
HDM 1560	MC	4600-4450 BC	Vinča culture	Pločnik	settlement/hoard?	SRB	chisel	unk	2.0657	0.83310
HDM 1561	MC	4600-4450 BC	Vinča culture	Pločnik	settlement/hoard?	SRB	chisel	unk	2.0629	0.83360
HDM 1562	MC	4600-4450 BC	Vinča culture	Pločnik	settlement/hoard?	SRB	chisel	unk	2.0775	0.84030
HDM 1563	MC	4600-4450 BC	Vinča culture	Pločnik	settlement/hoard?	SRB	chisel	unk	2.0748	0.83860
HDM 1564	MC	4600-4450 BC	Vinča culture	Pločnik	settlement/hoard?	SRB	chisel	unk	2.0840	0.84430
HDM 1565	MC	4600-4450 BC	Vinča culture	Pločnik	settlement/hoard?	SRB	chisel	unk	2.0844	0.84430
HDM 1566	MC	4600-4450 BC	Vinča culture	Pločnik	settlement/hoard?	SRB	chisel	unk	2.0751	0.84050
HDM 1567	MC	4600-4450 BC	Vinča culture	Pločnik	settlement/hoard?	SRB	chisel	unk	2.0836	0.84420
HDM 1568	MC	4600-4450 BC	Vinča culture	Pločnik	settlement/hoard?	SRB	chisel	unk	2.0585	0.83070
HDM 1569	MC	4600-4450 BC	Vinča culture	Pločnik	settlement/hoard?	SRB	chisel	unk	2.0831	0.84390
HDM 1570	MC	4600-4450 BC	Vinča culture	Pločnik	settlement/hoard?	SRB	chisel	unk	2.0650	0.83520
HDM 1571	MC	4600-4450 BC	Vinča culture	Pločnik	settlement/hoard?	SRB	hammer axe	Pločnik	2.0824	0.84360
HDM 1903	MC	4600-4450 BC	KSBh	Radlovci	stray	W	hammer axe	Pločnik	2.0770	0.84300
HDM 1910	MC	4600-4450 BC	KSBh	Dragoman	hoard	W	hammer axe	Pločnik	2.0771	0.84280
HDM 1911	MC	4600-4450 BC	KSBh	Dragoman	hoard	W	hammer axe	Pločnik	2.0751	0.84190

Table 8 continued. Lead isotope abundance ratios for EC and MC copper metal artefacts (data from Pernicka *et al.* 1993, 1997; Radivojević *et al.* 2010a).

LABEL	period	absolute dating	cultural attribution	Site	type of site/context	Region	Type	Axe type	$^{208}Pb/^{206}Pb$	$^{207}Pb/^{206}Pb$
HDM 1940	MC	4600-4450 BC	Karanovo VI?	Ai Bunar	mine	THR	hammer axe	Pločnik	2.0795	0.83840
HDM 1976	MC	4600-4450 BC	Hamangia IV	Durankulak	cemetery	BSC	bracelet	unk	1.9927	0.80690
HDM 1977	MC	4600-4450 BC	Hamangia IV	Durankulak	cemetery	BSC	bracelet	unk	2.0825	0.84390
HDM 2011	MC	4600-4450 BC	Varna I	Durankulak	cemetery	BSC	bracelet	unk	2.0679	0.83250
HDM 2025	MC	4600-4450 BC	Hamangia IV	Durankulak	cemetery	BSC	bracelet	unk	2.0675	0.83250
HDM 2026	MC	4600-4450 BC	Hamangia IV	Durankulak	cemetery	BSC	bracelet	unk	2.0679	0.83250
HDM 2041	MC	4600-4450 BC	Boian-Spantov	Tell Ruse	settlement	NE	borer	unk	2.0786	0.84090
HDM 2049	MC	4600-4450 BC	Boian-Spantov	Tell Ruse	settlement	NE	borer	unk	2.0801	0.84150
HDM 2064	MC	4600-4450 BC	Boian-Spantov	Tell Ruse	settlement	NE	borer	unk	2.0776	0.84420
HDM 2080	MC	4600-4450 BC	Boian-Spantov	Tell Ruse	settlement	NE	borer	unk	2.0772	0.84280
HDM 2124	MC	4600-4450 BC	Sava IV (Hamangia IV)	Goljamo Delcevo	settlement	BSC	borer	unk	2.0772	0.84040
HDM 2702	MC	4600-4450 BC	KSBh	Darzhanitsa	hoard	W	heavy axe	Gumelnica	2.0447	0.82580
HDM 2703	MC	4600-4450 BC	KSBh	Darzhanitsa	hoard	W	Heavy axe	Salcuta	2.0777	0.84310
HDM 1422	MC	4600-4450 BC	KSBh I	Bubanj	settlement	SRB	Chisel	unk	2.0764	0.84250
HDM 1431	MC	4600-4450 BC	KSBh I?	Stari Kostolac	stray	SRB	chisel	unk	2.0763	0.84280
HDM 1905	MC	4600-4450 BC	KSBh I	Djakovo	settlement	W	flat axe	unk	2.0753	0.83910
HDM 1942	MC	4600-4450 BC	Varna II	Durankulak	cemetery	BSC	Finger ring	unk	2.0794	0.84150
HDM 1943	MC	4600-4450 BC	Varna I	Durankulak	cemetery	BSC	Finger ring	unk	2.0744	0.84490
HDM 1944	MC	4600-4450 BC	Varna II	Durankulak	cemetery	BSC	bracelet	unk	2.0749	0.84480
HDM 1945	MC	4600-4450 BC	Varna II	Durankulak	cemetery	BSC	bracelet	unk	2.0754	0.84490
HDM 1947	MC	4600-4450 BC	Varna II	Durankulak	cemetery	BSC	bracelet	unk	2.0104	0.81540
HDM 1948	MC	4600-4450 BC	Varna II	Durankulak	cemetery	BSC	bracelet	unk	2.0789	0.84670
HDM 1949	MC	4600-4450 BC	Varna II	Durankulak	cemetery	BSC	bracelet	unk	2.0481	0.82760
HDM 1967	MC	4600-4450 BC	Varna I	Durankulak	cemetery	BSC	Finger ring	unk	2.0839	0.84410
HDM 1968	MC	4600-4450 BC	Varna I	Durankulak	cemetery	BSC	bracelet	unk	2.0191	0.81940
HDM 1972	MC	4600-4450 BC	Varna I	Durankulak	cemetery	BSC	bracelet	unk	2.0658	0.83260
HDM 1974	MC	4600-4450 BC	Varna II	Durankulak	cemetery	BSC	Finger ring	unk	1.8189	0.74210
HDM 1979	MC	4600-4450 BC	Varna II	Durankulak	cemetery	BSC	bracelet	unk	2.0730	0.83820

Table 8 continued. Lead isotope abundance ratios for EC and MC copper metal artefacts (data from Pernicka *et al.* 1993, 1997; Radivojević *et al.* 2010a).

LABEL	period	absolute dating	cultural attribution	Site	type of site/context	Region	Type	Axe type	$^{208}Pb/^{206}Pb$	$^{207}Pb/^{206}Pb$
HDM 1980	MC	4600-4450 BC	Varna II	Durankulak	cemetery	BSC	bracelet	unk	1.9644	0.79780
HDM 1981	MC	4600-4450 BC	Varna I	Durankulak	cemetery	BSC	bracelet	unk	2.0583	0.83440
HDM 1982	MC	4600-4450 BC	Varna II	Durankulak	cemetery	BSC	bracelet	unk	2.0555	0.83060
HDM 1984	MC	4600-4450 BC	Varna II	Durankulak	cemetery	BSC	bracelet	unk	2.0827	0.84350
HDM 1985	MC	4600-4450 BC	Varna II	Durankulak	cemetery	BSC	bracelet	unk	2.0762	0.83810
HDM 1986	MC	4600-4450 BC	Varna I	Durankulak	cemetery	BSC	bracelet	unk	2.0766	0.84140
HDM 1987	MC	4600-4450 BC	Varna I	Durankulak	cemetery	BSC	bracelet	unk	2.0567	0.83260
HDM 1989	MC	4600-4450 BC	Varna I	Durankulak	cemetery	BSC	bracelet	unk	2.0833	0.84420
HDM 1991	MC	4600-4450 BC	Varna II	Durankulak	cemetery	BSC	bracelet	unk	2.0831	0.84410
HDM 1992	MC	4600-4450 BC	Varna II	Durankulak	cemetery	BSC	bracelet	unk	2.0819	0.84390
HDM 1993	MC	4600-4450 BC	Varna II	Durankulak	cemetery	BSC	bracelet	unk	1.9036	0.77490
HDM 1994	MC	4600-4450 BC	Varna II	Durankulak	cemetery	BSC	bracelet	unk	2.0394	0.82480
HDM 1995	MC	4600-4450 BC	Varna II	Durankulak	cemetery	BSC	bracelet	unk	2.0482	0.83030
HDM 1996	MC	4600-4450 BC	Varna I	Durankulak	cemetery	BSC	bracelet	unk	2.0833	0.84390
HDM 1998	MC	4600-4450 BC	Varna I	Durankulak	cemetery	BSC	bracelet	unk	2.0823	0.84410
HDM 2000	MC	4600-4450 BC	Varna I	Durankulak	cemetery	BSC	Finger ring	unk	1.7460	0.71350
HDM 2001	MC	4600-4450 BC	Varna II	Durankulak	cemetery	BSC	Finger ring	unk	2.0794	0.84270
HDM 2002	MC	4600-4450 BC	Varna I	Durankulak	cemetery	BSC	Finger ring	unk	2.0198	0.81920
HDM 2003	MC	4600-4450 BC	Varna I	Durankulak	cemetery	BSC	Finger ring	unk	2.0613	0.83900
HDM 2006	MC	4600-4450 BC	Varna I	Durankulak	cemetery	BSC	spiral ring	unk	2.0699	0.83520
HDM 2009	MC	4600-4450 BC	Varna II	Durankulak	cemetery	BSC	spiral ring	unk	2.0589	0.83630
HDM 2012	MC	4600-4450 BC	Varna I	Durankulak	cemetery	BSC	bracelet	unk	2.0795	0.84750
HDM 2013	MC	4600-4450 BC	Varna I	Durankulak	cemetery	BSC	finger ring	unk	2.0799	0.84120
HDM 2014	MC	4600-4450 BC	Varna I	Durankulak	cemetery	BSC	bracelet	unk	1.8107	0.73960
HDM 2015	MC	4600-4450 BC	Varna I	Durankulak	cemetery	BSC	bracelet	unk	2.0649	0.83470
HDM 2016	MC	4600-4450 BC	Varna II	Durankulak	cemetery	BSC	bracelet	unk	2.0551	0.82960
HDM 2017	MC	4600-4450 BC	Varna II	Durankulak	cemetery	BSC	bracelet	unk	2.0097	0.81680
HDM 2018	MC	4600-4450 BC	Varna I	Durankulak	cemetery	BSC	bracelet	unk	2.0428	0.82850

Table 8 continued. Lead isotope abundance ratios for EC and MC copper metal artefacts (data from Pernicka *et al.* 1993, 1997; Radivojević *et al.* 2010a).

LABEL	period	absolute dating	cultural attribution	Site	type of site/context	Region	Type	Axe type	$^{208}Pb/^{206}Pb$	$^{207}Pb/^{206}Pb$
HDM 2019	MC	4600-4450 BC	Varna II	Durankulak	cemetery	BSC	bracelet	unk	2.0622	0.83860
HDM 2023	MC	4600-4450 BC	Varna II	Durankulak	cemetery	BSC	bracelet	unk	1.9966	0.81020
HDM 2024	MC	4600-4450 BC	Varna II	Durankulak	cemetery	BSC	bracelet	unk	1.9977	0.81030
HDM 2027	MC	4600-4450 BC	Varna II	Durankulak	cemetery	BSC	Finger ring	unk	2.0825	0.84410
HDM 2028	MC	4600-4450 BC	Varna I	Durankulak	cemetery	BSC	Finger ring	unk	2.0366	0.82870
HDM 2029	MC	4600-4450 BC	Varna I	Durankulak	cemetery	BSC	Finger ring	unk	2.0831	0.84410
HDM 2031	MC	4600-4450 BC	Varna I	Durankulak	cemetery	BSC	Finger ring	unk	1.7515	0.71690
HDM 2032	MC	4600-4450 BC	Varna II	Durankulak	cemetery	BSC	borer	unk	2.0818	0.84330
HDM 2034	MC	4600-4450 BC	Varna II	Durankulak	cemetery	BSC	pin	unk	2.0745	0.83900
HDM 2035	MC	4600-4450 BC	Varna II	Durankulak	cemetery	BSC	spiral ring	unk	2.0827	0.84420
HDM 2036	MC	4600-4450 BC	Varna I	Durankulak	cemetery	BSC	spiral ring	unk	2.0063	0.81450
HDM 2137	MC	4600-4450 BC	Varna I	Durankulak	cemetery	BSC	Finger ring	unk	2.0292	0.82300
HDM 2138	MC	4600-4450 BC	Varna I	Durankulak	cemetery	BSC	Finger ring	unk	2.0775	0.84610
HDM 1920	MC	4600-4450 BC	KGK VI?	Tell Ruse	settlement	NE	axe	Kamenar	2.0538	0.83220
HDM 1921	MC	4600-4450 BC	KGK VI?	Tell Ruse	settlement	NE	flat axe	Kamenar	2.0835	0.84430
HDM 1922	MC	4600-4450 BC	KGK VI?	Tell Ruse	settlement	NE	chisel	unk	1.9217	0.78140
HDM 1923	MC	4600-4450 BC	KGK VI	Tell Ruse	settlement	NE	axe	Gumelnica	2.0847	0.84460
HDM 2039	MC	4600-4450 BC	KGK VI?	Tell Ruse	settlement	NE	borer	unk	2.0794	0.84680
HDM 2040	MC	4600-4450 BC	KGK VI?	Tell Ruse	settlement	NE	borer	unk	2.0808	0.84150
HDM 2043	MC	4600-4450 BC	KGK VI?	Tell Ruse	settlement	NE	borer	unk	2.0514	0.83010
HDM 2044	MC	4600-4450 BC	KGK VI?	Tell Ruse	settlement	NE	borer	unk	2.0831	0.84420
HDM 2045	MC	4600-4450 BC	KGK VI?	Tell Ruse	settlement	NE	borer	unk	2.0831	0.84390
HDM 2047	MC	4600-4450 BC	KGK VI?	Tell Ruse	settlement	NE	double spiral headed pin	unk	2.0807	0.84500
HDM 2050	MC	4600-4450 BC	KGK VI?	Tell Ruse	settlement	NE	borer	unk	2.0774	0.84320
HDM 2051	MC	4600-4450 BC	KGK VI?	Tell Ruse	settlement	NE	borer	unk	2.0809	0.84690
HDM 2052	MC	4600-4450 BC	KGK VI?	Tell Ruse	settlement	NE	borer	unk	2.0814	0.84700
HDM 2053	MC	4600-4450 BC	KGK VI?	Tell Ruse	settlement	NE	chisel	unk	2.0829	0.84400

Table 8 continued. Lead isotope abundance ratios for EC and MC copper metal artefacts (data from Pernicka *et al.* 1993, 1997; Radivojević *et al.* 2010a).

LABEL	period	absolute dating	cultural attribution	Site	type of site/context	Region	Type	Axe type	$^{208}Pb/^{206}Pb$	$^{207}Pb/^{206}Pb$
HDM 2054	MC	4600-4450 BC	KGK VI?	Tell Ruse	settlement	NE	borer	unk	2.0786	0.83720
HDM 2055	MC	4600-4450 BC	KGK VI?	Tell Ruse	settlement	NE	borer	unk	2.0800	0.84660
HDM 2056	MC	4600-4450 BC	KGK VI?	Tell Ruse	settlement	NE	borer	unk	2.0732	0.84040
HDM 2057	MC	4600-4450 BC	KGK VI?	Tell Ruse	settlement	NE	borer	unk	2.0817	0.84360
HDM 2058	MC	4600-4450 BC	KGK VI?	Tell Ruse	settlement	NE	borer	unk	2.0782	0.84300
HDM 2059	MC	4600-4450 BC	KGK VI?	Tell Ruse	settlement	NE	borer	unk	2.0581	0.83160
HDM 2060	MC	4600-4450 BC	KGK VI?	Tell Ruse	settlement	NE	borer	unk	2.0667	0.83670
HDM 2061	MC	4600-4450 BC	KGK VI?	Tell Ruse	settlement	NE	borer	unk	2.0628	0.83360
HDM 2063	MC	4600-4450 BC	KGK VI?	Tell Ruse	settlement	NE	borer	unk	2.0465	0.82770
HDM 2065	MC	4600-4450 BC	KGK VI?	Tell Ruse	settlement	NE	borer	unk	2.0760	0.84270
HDM 2067	MC	4600-4450 BC	KGK VI?	Tell Ruse	settlement	NE	borer	unk	2.0838	0.84230
HDM 2068	MC	4600-4450 BC	KGK VI?	Tell Ruse	settlement	NE	borer	unk	2.0510	0.83000
HDM 2070	MC	4600-4450 BC	KGK VI?	Tell Ruse	settlement	NE	borer	unk	2.0849	0.84440
HDM 2071	MC	4600-4450 BC	KGK VI?	Tell Ruse	settlement	NE	borer	unk	2.0843	0.84420

The Group 1 cluster in both diagrams $^{207}Pb/^{206}Pb$ vs $^{208}Pb/^{206}Pb$ (Figure 4) and $^{206}Pb/^{204}Pb$ vs $^{207}Pb/^{204}Pb$ (Figure 5) implies strong consistency with Majdanpek, a large copper deposit in eastern Serbia (Figure 12), and to a lesser extent with Zidarovo, an ore field located on the western slope of the Medni Rid region in southeast Bulgaria. The partial overlap of Majdanpek with the mixed sulfide copper ores from Zidarovo has already been discussed by Pernicka *et al.* (1997: 139) in relation to the likely origins of the Serbian Chalcolithic copper metal artefacts. The authors argue that it is unlikely that Zidarovo was the source for the kind of copper metal that circulates only in Serbia and not anywhere in the Black Sea coast region. Further consistencies with the previously published copper minerals from Belovode (Figures 4 and 5) and Selevac (Figures 6 and 7) strengthen the argument that Majdanpek was one of the main copper deposits exploited during the Vinča culture (cf. Pernicka *et al.* 1993, 1997). Also, the Zidarovo lead isotope abundance field is clearly distinguished from Majdanpek in the $^{206}Pb/^{204}Pb$ vs $^{207}Pb/^{204}Pb$ diagram (see Figure 5, for instance).

A slight overlap of the Majdanpek field with the lead isotope abundance ratios of another prolific Chalcolithic copper deposit, Ai Bunar in Bulgaria, does not compromise the likelihood of Majdanpek being the major source for the mentioned artefacts (also easily distinguishable by their trace element patterns, see Table 12); Majdanpek is also consistent with a portion of the Selevac malachite assemblage and previously published copper metal implements from Pločnik (Figures 7–9, Tables 4 and 5), confirming arguments that Vinča culture communities were utilising copper ores from this mine towards the end of the Vinča culture, particularly at the prolonged end at Pločnik (Pernicka *et al.* 1993; Radivojević and Grujić 2018).

In Figure 11, a plot of Early Chalcolithic (EC) and Middle Chalcolithic (MC) metals from the Balkans shows a handful of objects with high consistency with the Majdanpek field, and an exact match with copper metal droplet Bf43/13 (such as a copper chisel HDM1422 from the MC period in Bubanj, Krivodol-Sălcuţa-Bubanj I culture, abbreviated as KSBh I/II, see Table 8). Further consistencies with copper implements from the MC occupation of the sites of Gomolava or Ruse (Figure 11, Table 8) indicate a wide network of copper supply that extended mainly along the lower Danube but also across eastern Serbia / western Bulgaria which, at the

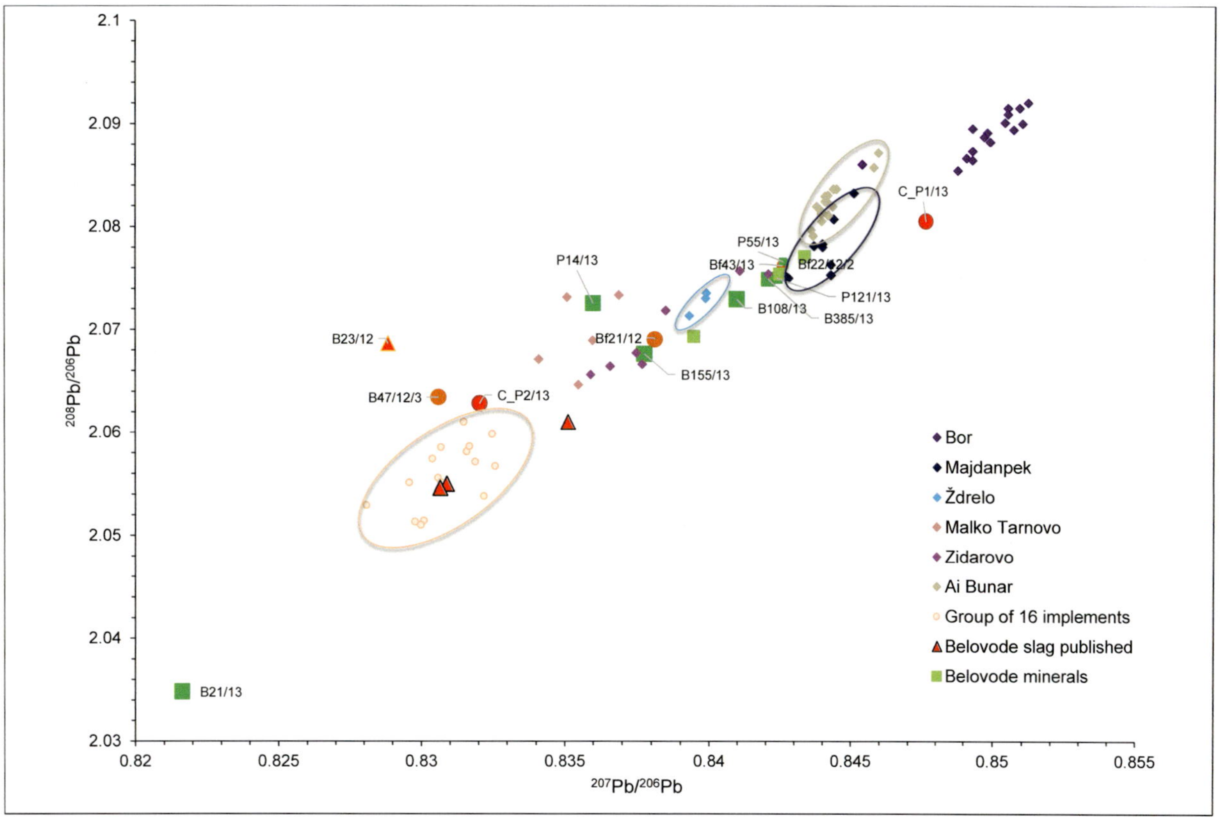

Figure 4. Lead isotope abundance ratios of Belovode and Pločnik artefacts presented in ^{207}Pb/^{206}Pb vs ^{208}Pb/^{206}Pb ratio against the published Belovode data and a selection of Balkan copper deposits and a group of artefacts (Group of 16). Error bars are smaller than symbols

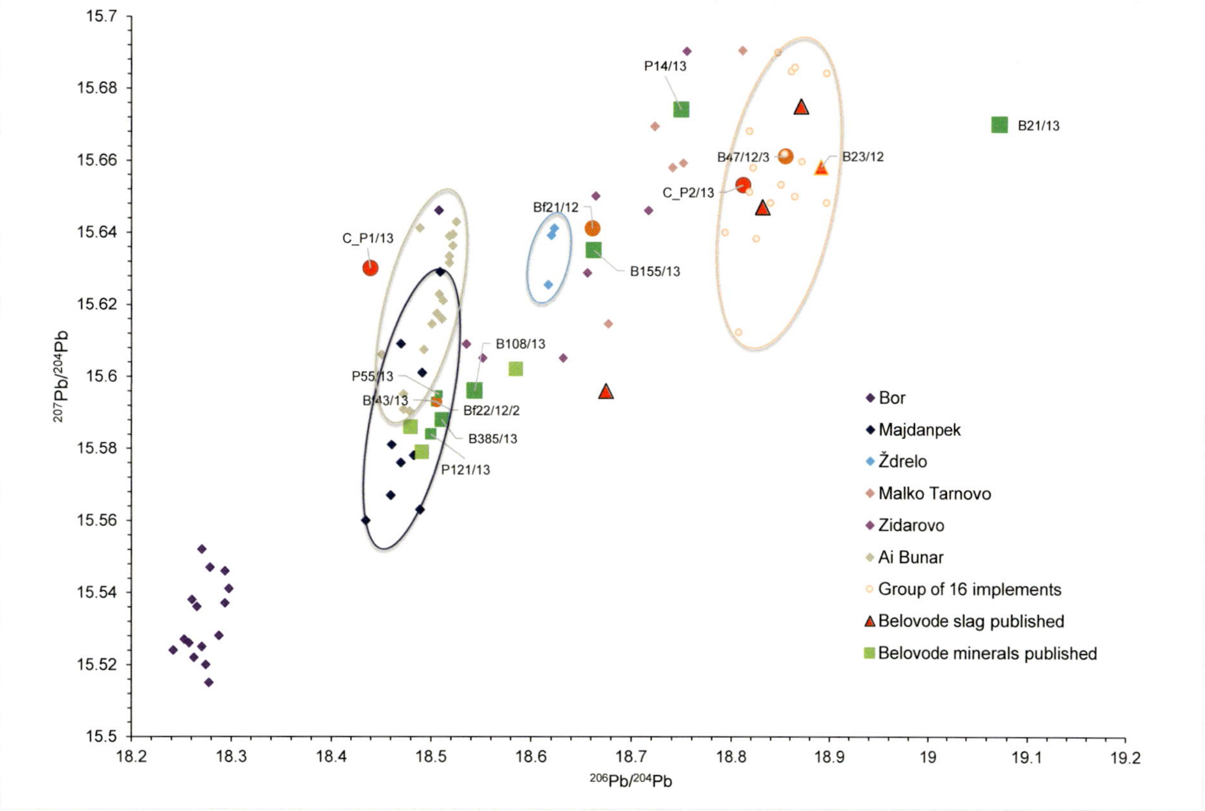

Figure 5. Lead isotope abundance ratios of Belovode and Pločnik artefacts presented in ^{206}Pb/^{204}Pb vs ^{207}Pb/^{204}Pb ratio ratio against the published Belovode data and a selection of Balkan copper deposits and a group of artefacts (Group of 16).

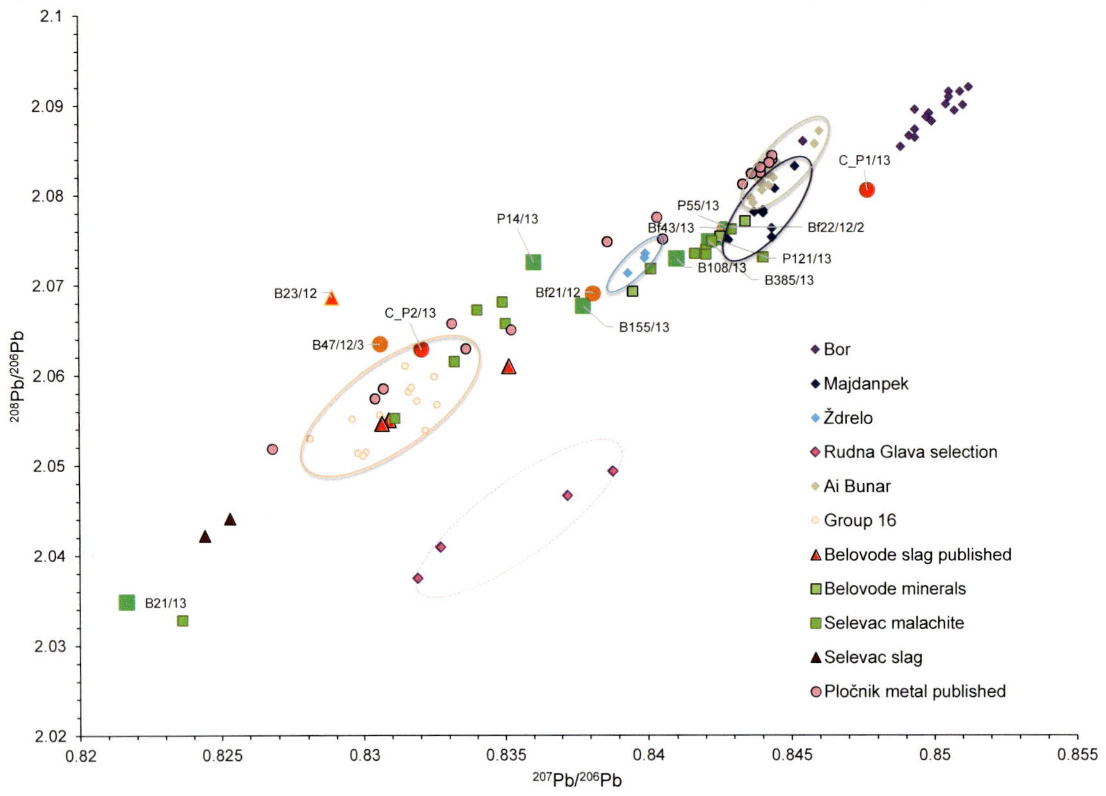

Figure 6. Lead isotope abundance ratios of Belovode and Pločnik artefacts presented in $^{207}Pb/^{206}Pb$ vs $^{208}Pb/^{206}Pb$ ratio against the published Belovode, Selevac and Pločnik data and a selection of Balkan copper deposits and a group of artefacts (Group of 16). Error bars are smaller than symbols.

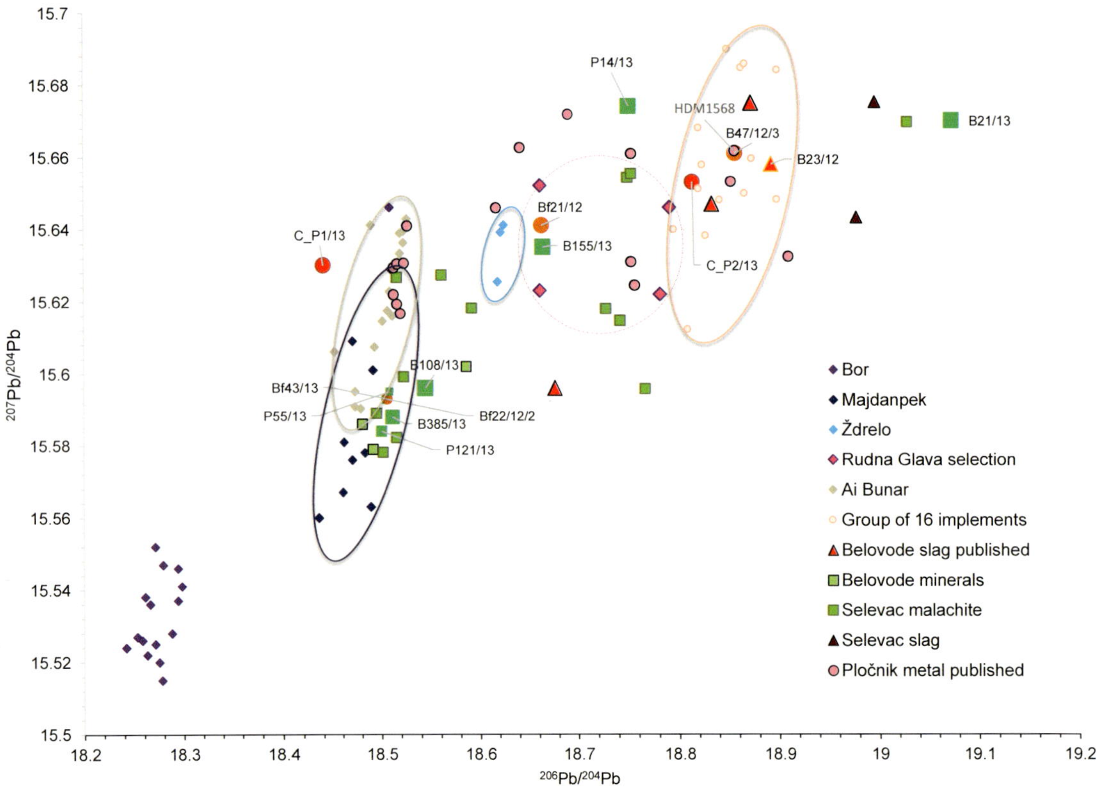

Figure 7. Lead isotope abundance ratios of Belovode and Pločnik artefacts presented in $^{206}Pb/^{204}Pb$ vs $^{207}Pb/^{204}Pb$ ratio against the published Belovode, Selevac and Pločnik data and a selection of Balkan copper deposits and a group of artefacts (Group of 16).

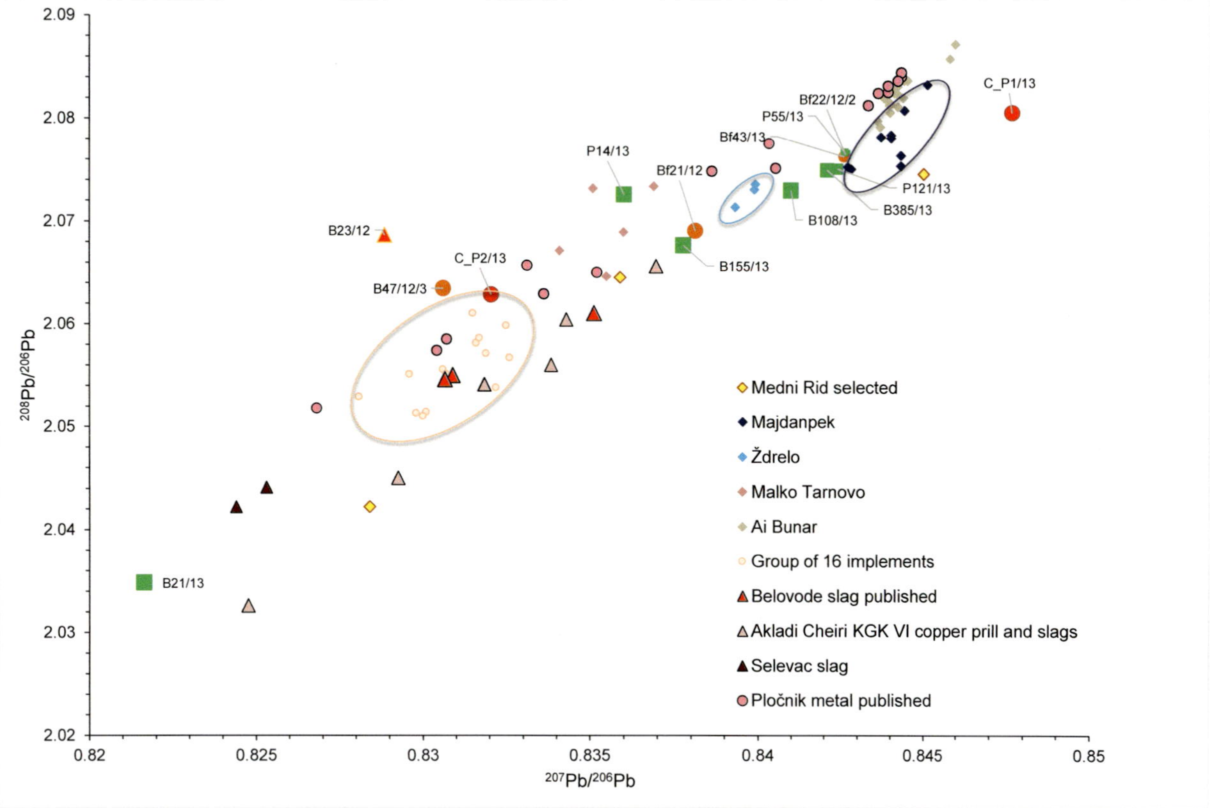

Figure 8. Lead isotope abundance ratios of Belovode and Pločnik artefacts presented in a $^{207}Pb/^{206}Pb$ vs $^{208}Pb/^{206}Pb$ diagram against the published Belovode, Selevac and Akladi Cheiri metal production data, Pločnik artefacts and a selection of Balkan copper deposits and a group of artefacts (Group of 16). Error bars are smaller than symbols.

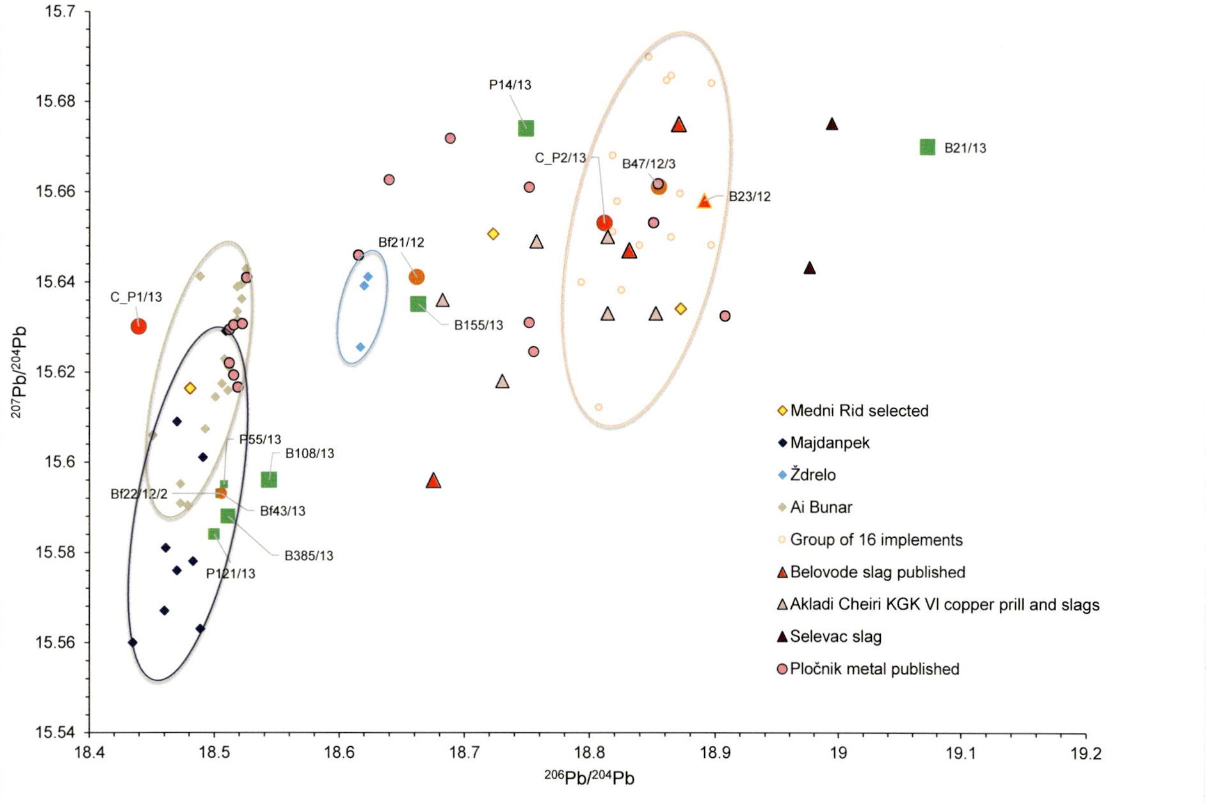

Figure 9. Lead isotope abundance ratios of Belovode and Pločnik artefacts presented in a $^{206}Pb/^{204}Pb$ vs $^{207}Pb/^{204}Pb$ diagram against the published Belovode, Selevac and Akladi Cheiri metal production data, Pločnik artefacts and a selection of Balkan copper deposits and a group of artefacts (Group of 16).

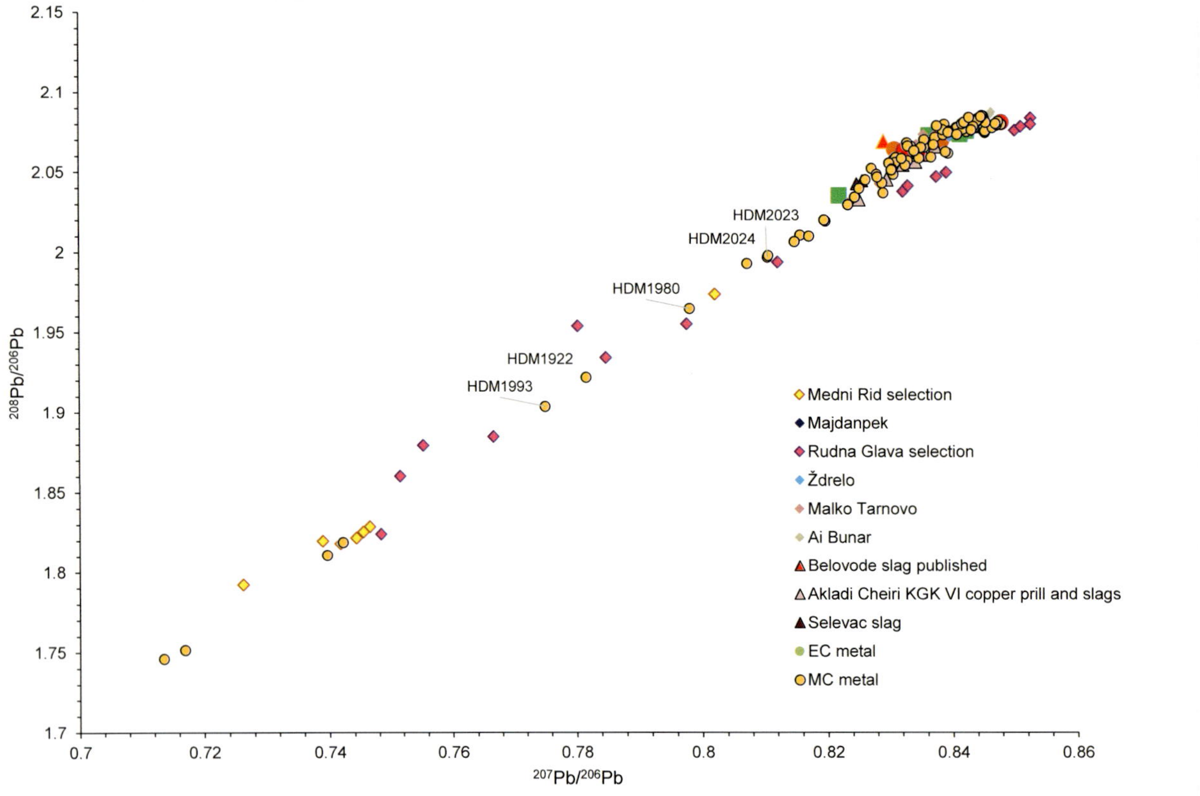

Figure 10. Lead isotope abundance ratios of Belovode and Pločnik artefacts presented in a $^{207}Pb/^{206}Pb$ vs $^{208}Pb/^{206}Pb$ diagram against the published Belovode, Selevac and Akladi Cheiri metal production data, EC/MC copper metal artefacts and a selection of Balkan copper deposits. Error bars are smaller than symbols.

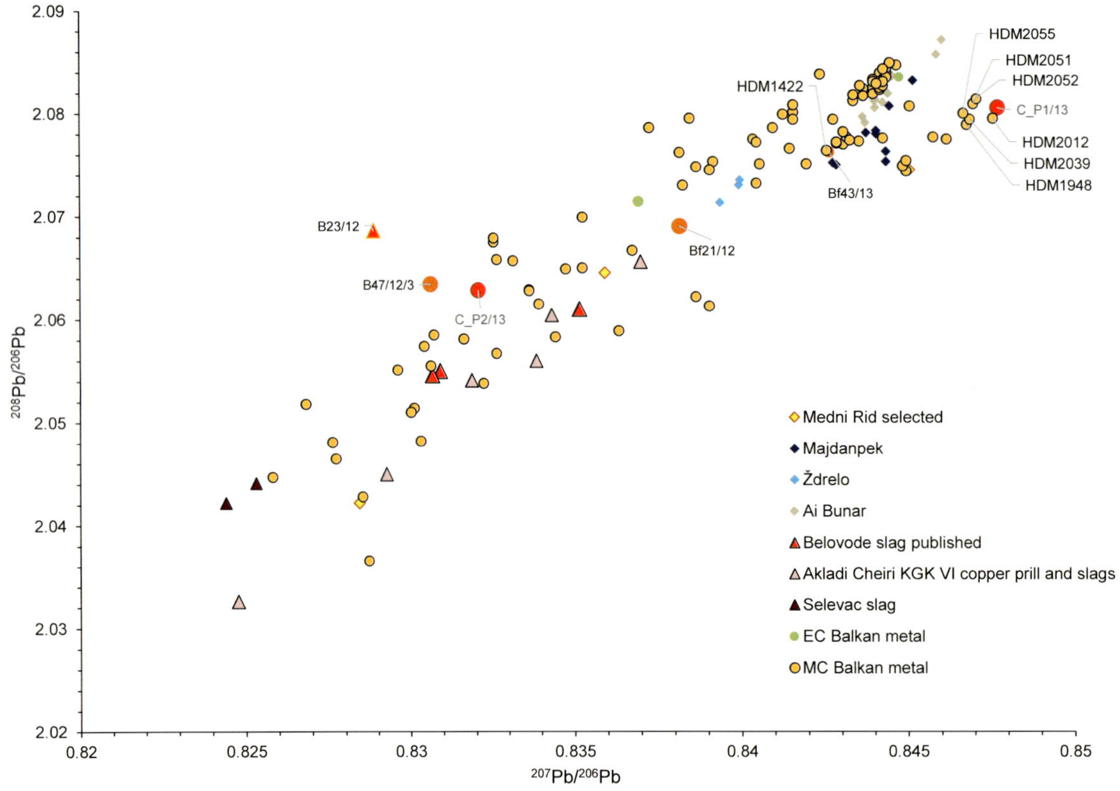

Figure 11. Lead isotope abundance ratios of Belovode and Pločnik artefacts presented in a $^{207}Pb/^{206}Pb$ vs $^{208}Pb/^{206}Pb$ diagram against the published Belovode, Selevac and Akladi Cheiri metal production data, EC/MC copper metal artefacts and a selection of Balkan copper deposits. This is an enlarged section of Fig. 10. Error bars are smaller than symbols.

time, was occupied by the Krivodol-Sălcuţa-Bubanj I communities. It is noteworthy that copper minerals in Group 1 are both pure green and black-and-green, hence could have been used for both smelting and malachite bead making (cf. Radivojević *et al.* 2010a). There is, however, no production evidence in this group of finds, barring the copper metal droplet (Bf43/13), which originated from a 'slag-less' process (Chapter 11, this volume).

Group 2 consists of a copper metal droplet (Bf21/12) and a mineral sample (B155/13) that come from Belovode Horizons 1b and 2 respectively (see Chapters 11 and 37). These further support the continuity of copper exploitation and production practices between these two horizons as indicated by the Group 1 artefacts. The ore field of Ždrelo, despite being currently defined by only three ore samples (Figures 4 and 5, and Table 4) appears to be more closely related to Bf21/12 and B155/13, particularly in the $^{206}Pb/^{204}Pb$ vs $^{207}Pb/^{204}Pb$ projection (Figure 5). The possible significance of this copper mine for metal production at Belovode has already been emphasised in Radivojević *et al.* (2010), particularly as it is less than 10 km away from the site, easily accessible on foot and visible from this settlement. These samples are also broadly consistent with malachite previously published from Belovode and Selevac (Figure 6 and Table 5). Although the latter appear to show relative consistency with the Rudna Glava ore field in the $^{206}Pb/^{204}Pb$ vs $^{207}Pb/^{204}Pb$ projection (Figure 7), including two more implements from Pločnik, the $^{207}Pb/^{206}Pb$ vs $^{208}Pb/^{206}Pb$ diagram excludes Rudna Glava as a possible source. Most data of this mineralisation are outside both diagrams, and two samples from Gale (1991) were excluded because they have model ages that are completely different from all the other samples from Rudna Glava and may have been contaminated as suggested by Pernicka *et al.* (1993).

Rudna Glava is a well-documented ancient mine that traditionally has been at the centre of discussions on the beginnings of metallurgy in southeastern Europe (Jovanović 1971; Jovanović and Ottaway 1976), most prominently in relation to the origins of copper for the Vinča culture metal artefacts (see opposing views in Pernicka *et al.* 1993; Jovanović 1993). However, lead isotope analysis produced a different conclusion. The radiogenic nature of lead from this copper mine clearly distinguishes it from Vinča culture archaeological ores and artefacts (see Pernicka *et al.* 1993: 31, Fig. 13, 14). This is despite the clear evidence for exploitation of Rudna Glava during the Vinča culture period, starting from *c.* 5500 BC and intensifying around the Gradac Phase (*c.* 5000 BC), which has been demonstrated both through material culture and direct ^{14}C dating (e.g. Jovanović 1971, 1995; Radivojević and Rehren 2016; Pernicka *et al.* 1993: 40).

The Group 3 artefacts are inconsistent with any currently identified copper deposits in the Balkans. However, they are consistent with the isotope field of sixteen 5th millennium BC copper artefacts (implements and a malachite bead), excavated mainly from the MC occupations at the sites of Pločnik, Durankulak and Ruse (Table 6). This assemblage of artefacts has been previously identified as Grouplet #7 (Pernicka *et al.* 1997: 112, Table 4), whose distinctive grouping outside the known Serbian and Bulgarian copper ore deposits (Pernicka *et al.* 1997: 106) indicates an independent assemblage with a source yet to be identified.

Figures 4 and 5 demonstrate the consistency of this 'Group of 16' with copper metal embedded in production evidence (B47/12/3, and previously published Belovode slag M21 and M22a, see Table 5), while Figures 6 and 7 show further variety in copper supply from this as yet unidentified source, adding a few Pločnik implements and Selevac malachite. Of note is the exact consistency (Figure 7) of two contemporaneous artefacts: production evidence (B/47/12/3, metal droplet) from the Belovode workshop and a copper metal chisel from Pločnik, HDM 1568, once again clearly demonstrating the consistent supply and cooperative links between these two communities. It is therefore reasonable to assume a scenario where copper metal produced in Belovode's workshop F6 ended up in the form of a copper chisel used at Pločnik.

Further comparisons with the copper production evidence from the site of Akladi Cheiri in Bulgaria reveal only a minor overlap with this 'Group of 16' artefacts field (Figure 9), while it has been convincingly shown that copper ores smelted at this site originate from the Medni Rid mining region (Rehren *et al.* 2020: 150–151, Figure 15). The ore samples from Akladi Cheiri show the same radiogenic lead 'wide' scatter as Medni Rid ores in southeast Bulgaria. Figure 10 shows only a selected reading of the lead isotope abundance ratios from Medni Rid, which largely matches Akladi Cheiri production, as well as good consistency with a selection of MC copper artefacts from Bulgaria. Another deposit with radiogenic lead is that at Rudna Glava, the wide scatter of which also appears consistent with several copper metal objects, such as bracelets and a chisel from the MC period in Durankulak (HDM 1980, 1993, 2023, 2024) and Ruse (HDM 1922) (Figure 10, Table 8). However, this impression could be due to the expanded scale of the diagram. Rudna Glava samples are systematically lower in the $^{208}Pb/^{206}Pb$ ratio and can thus be clearly distinguished in the upper right part of the diagram. This means that the Th/U ratios at Rudna Glava and at Medni Rid are different, which is only apparent at higher resolution. In the region of radiogenic lead isotope ratios this difference is less pronounced. There is also some similarity in the trace element signature of Medni Rid and Rudna Glava (cf. Kunze and Pernicka 2020: 416–417, Fig. 13 and 15), which will be discussed in more detail below.

The outliers, two minerals (B21/13 and P14/13) and a copper metal loop/ring (C_P1/13) do not appear consistent with any known lead isotope abundance ratios of Balkan copper deposits, although they are broadly consistent with other archaeological artefacts. Malachite sample B21/13 clusters closely with one of the Selevac malachite samples and one of the two slags from this settlement (Figure 7). Pločnik malachite (P14/13) is also broadly consistent with some of the Selevac malachite as well as Pločnik implements (Figure 7). The metal loop / ring from Pločnik (C_P1/13) groups well with six other copper metal implements (Figure 11) from the MC occupation (4600–4450 BC) of Durankulak (HDM: 1948, 2012) and Ruse (HDM: 2039, 2051, 2052, 2055). Incidentally, these two sites are in the lower Danube area (Figure 12), which features prominently in interpretations of the movement of ores or metals from the Vinča culture sites of Belovode and Pločnik.

Looking at the entire lead isotope abundance ratio range of all EC and MC Balkan metals analysed thus far (124 in total, Table 8 and Figure 10), a great majority of them (104 artefacts, Figure 11) are largely consistent with the copper deposits exploited by the Belovode and Pločnik communities. This applies equally to cases where the lead isotope abundance ratios of copper sources are known, as well as to cases where they are only indicated by a distinctive cluster of metal artefacts, as is the case with the 'Group of 16' (Figure 9) or an artefacts cluster surrounding C_P1/13 (Figure 11). It may also be assumed that the 'Group of 16' represents the exploitation of multiple copper deposits with similar lead isotope abundance ratios, or that a portion of these artefacts clustered closely due to the homogenisation effect resulting from mixing of different ore samples. Both scenarios could be equally probable in the absence of deposit(s) that are consistent with the 'Group of 16' artefacts range.

Overall, the lead isotope abundance ratios of Belovode and Pločnik artefacts investigated in detail here, together with the contemporary objects from across Serbia and Bulgaria (or Vinča culture and other coeval phenomena) reveal complex dynamics of copper exploitation between *c.* 5200 BC and *c.* 4450 BC in the Balkans. The earlier stages of copper mining activities (*c.* 5200–4600 BC) suggest the use of at least three major deposits: Majdanpek, Ždrelo and the 'Group of 16' source. In this context, the possibility of Rudna Glava exploitation (potentially limited to MC here, Figure 10) cannot be confirmed given the current state of assembled lead isotope abundance data. The *c.* 4600–4450 BC (MC) period introduces Ai Bunar and Medni Rid, with supply networks heavily entangled between all five (or six) major deposits by that time, as well as the main production / consumption sites, such as Pločnik, Belovode, Ruse and Durankulak. This hypothesis is further explored below with trace element analysis, which was conducted on largely the same set of samples (Table 9).

Acquisition and circulation of copper ores and artefacts: trace element analysis

Seventeen artefacts were analysed for their trace element contents (Tables 2 and 9) using the methodology reported by Kuleff and Pernicka (1995). The iron oxide sample P8/13 is excluded here from further consideration, as for the previous section.

Table 9 summarises results of NAA of the predominantly copper based objects, showing varying ranges of ten trace elements (As, Sb, Co, Ni, Ag, Au, Zn, Sn, Se, Te) and iron. All trace elements are in the lower range, with occasional spikes of Ni at 1600 ppm and Sn at 3800 ppm (metal artefact P13/13) and significant Zn readings at 1.2 wt% and 3 wt% in minerals P121/13 and B385/13, respectively, which happen to be the earliest copper minerals from both Belovode and Pločnik that belong to the distinctive Group 1 cluster in Figures 2 and 3. It is of note that other minerals from Group 1 (including the metal droplet Bf43/13) also present relatively high levels of Zn, varying between *c.* 1300 and 6000 ppm.

A comparison of seven trace elements commonly used for provenance analysis: As, Sb, Co, Ni, Ag, Au, Se (sensu Pernicka 1990) across the published dataset for the Chalcolithic copper artefacts (Tables 10 and 11) (Pernicka *et al.* 1993, 1997; Radivojević and Grujić 2018) and newly acquired data on copper metal (Table 9) offers a varied picture of the quality of copper ores used at Belovode and Pločnik, the Vinča culture in general and EC/MC copper metal in the Balkans more widely. The generally low values across the seven trace elements in metal production samples from Belovode (Bf43/13, B23/12, B47/12/3 and Bf21/12) (Table 9) are largely consistent with the readings of previously published similar samples from this site (Table 11), apart from the elevated Ni and Co readings (also see Table 1).

In contrast, the newly acquired dataset for Pločnik metal has a different trace element signature from the previously published data (Table 11). The contrasting values of As, Sb, Ag, Au and Se, albeit with huge fluctuations within the dataset, speaks to the use of different sources or different types of copper ores for producing metal artefacts from this settlement. This accords with the observation that Pločnik copper

Table 9. Trace element data for Belovode and Pločnik copper-based artefacts. Note seven trace elements used for provenance analysis as bold, and artefacts paired with either deposits or another cluster of Chalcolithic artefacts as shaded.

Site	Artefact type	Cu	Fe	As	Sb	Co	Ni	Ag	**Au**	**Zn**	**Sn**	**Se**	**Te**	**Au**	Zn	Sn	**Se**	Te
							%	%	**μg/g**	**μg/g**	**μg/g**	**μg/g**	**μg/g**	**μg/g**	μg/g	μg/g	**μg/g**	μg/g
Belovode	Slagged sherd	1.61	2.30	9.9	1.16	55	47	2.1	**0.01**	**176**	**< 100**	**4.4**	**< 7**	**0.01**	176	< 100	**4.4**	< 7
Belovode	Malachite (black and green)	33.0	0.63	46.6	8.66	65	340	3.1	**0.071**	**5990**	**< 440**	**< 2**	**14.9**	**0.071**	5990	< 440	**< 2**	14.9
Belovode	Metal from slagged sherd	40.2	1.30	15.7	2.02	68	160	2.3	**0.012**	**145**	**< 75**	**2.1**	**< 11**	**0.012**	145	< 75	**2.1**	< 11
Belovode	Malachite (green)	46.2	1.26	14.1	0.84	5.8	69	0.8	**< 0.01**	**1330**	**< 49**	**< 0.7**	**< 4**	**< 0.01**	1330	< 49	**< 0.7**	< 4
Belovode	Copper mineral	35.1	2.70	530	12.2	4.0	61	< 1	**0.031**	**3060**	**< 110**	**0.78**	**14.4**	**0.031**	3060	< 110	**0.78**	14.4
Belovode	Malachite (green)	45.3	3.20	252	12.1	0.62	41	< 1	**0.271**	**555**	**< 90**	**< 1**	**4.1**	**0.271**	555	< 90	**< 1**	4.1
Belovode	Malachite (green)	47.1	0.53	30.5	1.37	2.98	100	< 2	**0.014**	**30800**	**< 140**	**0.97**	**3.3**	**0.014**	30800	< 140	**0.97**	3.3
Belovode	Metal droplet	103	0.04	4.0	0.2	15.5	180	9.9	**0.044**	**44**	**< 280**	**3.2**	**13**	**0.044**	44	< 280	**3.2**	13
Belovode	Malachite (green)	61.9	0.30	434	4.12	0.9	19	152	**0.103**	**1300**	**< 81**	**159**	**2.9**	**0.103**	1300	< 81	**159**	2.9
Belovode	Metal droplet	47.3	0.62	50.8	0.86	1.94	20	36.2	**0.085**	**6030**	**< 29**	**26.8**	**4.3**	**0.085**	6030	< 29	**26.8**	4.3
Pločnik	(magnetic) iron mineral	0.10	12.1	30.2	2.09	204	1870	< 2	**0.01**	**425**	**< 170**	**< 2**	**< 16**	**0.01**	425	< 170	**< 2**	< 16
Pločnik	metal foil/stock corroded	89.9	0.01	0.92	0.28	14.9	86	6.4	**0.079**	**25.8**	**19**	**0.8**	**< 6**	**0.079**	25.8	19	**0.8**	< 6
Pločnik	Metal wire/bracelet	46.3	2.20	38	7.3	64	1600	30	**0.7**	**1360**	**3800**	**53**	**290**	**0.7**	1360	3800	**53**	290
Pločnik	Malachite (green)	37.7	4.60	3360	172	48.7	303	21.3	**0.02**	**1150**	**58**	**20.1**	**36**	**0.02**	1150	58	**20.1**	36
Pločnik	Malachite (green)	21.9	7.50	513	25.1	3.3	121	< 2	**0.948**	**4810**	**< 150**	**2.9**	**2.5**	**0.948**	4810	< 150	**2.9**	2.5
Pločnik	Malachite (black and green)	25.2	2.50	228	12.3	25	214	11.7	**1.34**	**12300**	**< 160**	**< 2**	**7**	**1.34**	12300	< 160	**< 2**	7
Pločnik	Metal loop / ring	85.7	0.04	10.6	6.26	2.9	74	230	**0.41**	**61**	**< 280**	**6.2**	**15**	**0.41**	61	< 280	**6.2**	15

Table 10. Trace element data for EC and MC copper production and implements. Regions are abbreviated as follows: BSC=Black Sea Coast, 2012; Radivojević

LABEL analytical	relative period	absolute period	cultural attribution	Site	Sample label	Type of site/context	Region
MA-110610	EC	5000-4600 BC	Vinča culture	Belovode	Belovode M6	settlement	SRB
MA-110620	EC	5000-4600 BC	Vinča culture	Belovode	Belovode M14	settlement	SRB
MA-071498	EC	5000-4600 BC	Vinča culture	Belovode	Belovode M21	settlement	SRB
MA-071499	EC	5000-4600 BC	Vinča culture	Belovode	Belovode M22a	settlement	SRB
MA-114275	EC	5000-4600 BC	Vinča culture	Pločnik	Pločnik145	settlement	SRB
L 355	EC	5000-4600 BC	Vinča culture	Pločnik	Pločnik 217	settlement	SRB
MA-110617	EC	5000-4600 BC	Vinča culture	Belovode	Belovode 131	settlement	SRB
MA-110609	EC	5000-4600 BC	Vinča culture	Belovode	Belovode 134	settlement	SRB
L 354	EC	5000-4600 BC	Vinča culture	Pločnik	Pločnik 216	settlement	SRB
MA-114274	EC	5000-4600 BC	Vinča culture	Pločnik	Pločnik 143	settlement	SRB
MA-110622	EC	5000-4600 BC	Vinča culture	Vinča	Vinča 79	settlement	SRB
MA-110616	EC	5000-4600 BC	Vinča culture	Vinča	Vinča 91	settlement	SRB
HDM 1483	EC	5000-4600 BC	Vinča culture	Selevac	HDM 1483	settlement	SRB
HDM 1494	EC	5000-4600 BC	Vinča culture	Selevac	HDM 1494	settlement	SRB
HDM 1904	EC	5000-4600 BC	Dikilitash Slatino Gradesnica	Slatino	HDM 1904	settlement	W
HDM 1912	EC	5000-4600 BC	Marica III	Marica	HDM 1912	settlement	THR
HDM 1975	EC	5000-4600 BC	Hamangia III	Durankulak	HDM 1975	cemetery	BSC
HDM 2005	EC	5000-4600 BC	Hamangia III	Durankulak	HDM 2005	cemetery	BSC
MA-103713	MC	4600-4450 BC	Vinča culture	Gornja Tuzla	Gornja Tuzla 190	settlement	SRB
MA-103719	MC	4600-4450 BC	Vinča culture	Gornja Tuzla	Gornja Tuzla 183	settlement	SRB
MA-103721	MC	4600-4450 BC	Vinča culture	Gornja Tuzla	Gornja Tuzla 185	settlement	SRB
MA-103723	MC	4600-4450 BC	Vinča culture	Gornja Tuzla	Gornja Tuzla 195	settlement	SRB
MA-103715	MC	4600-4450 BC	Vinča culture	Gornja Tuzla	Gornja Tuzla 193	settlement	SRB
MA-103716	MC	4600-4450 BC	Vinča culture	Gornja Tuzla	Gornja Tuzla 192a	settlement	SRB
MA-103718	MC	4600-4450 BC	Vinča culture	Gornja Tuzla	Gornja Tuzla 198a	settlement	SRB
MA-103720	MC	4600-4450 BC	Vinča culture	Gornja Tuzla	Gornja Tuzla 186	settlement	SRB
MA-103724	MC	4600-4450 BC	Vinča culture	Gornja Tuzla	Gornja Tuzla 196	settlement	SRB
HDM 1308	MC	4600-4450 BC	KSBh / Vinča?	Sumrakovac	HDM 1308	stray	SRB
HDM 1496	MC	4600-4450 BC	Vinča culture	Gomolava	HDM 1496	cemetery	SRB
HDM 1555	MC	4600-4450 BC	Vinča culture	Pločnik	HDM 1555	settlement/hoard?	SRB
HDM 1556	MC	4600-4450 BC	Vinča culture	Pločnik	HDM 1556	settlement/hoard?	SRB
HDM 1557	MC	4600-4450 BC	Vinča culture	Pločnik	HDM 1557	settlement/hoard?	SRB
HDM 1558	MC	4600-4450 BC	Vinča culture	Pločnik	HDM 1558	settlement/hoard?	SRB
HDM 1559	MC	4600-4450 BC	Vinča culture	Pločnik	HDM 1559	settlement/hoard?	SRB
HDM 1560	MC	4600-4450 BC	Vinča culture	Pločnik	HDM 1560	settlement/hoard?	SRB
HDM 1561	MC	4600-4450 BC	Vinča culture	Pločnik	HDM 1561	settlement/hoard?	SRB
HDM 1562	MC	4600-4450 BC	Vinča culture	Pločnik	HDM 1562	settlement/hoard?	SRB
HDM 1563	MC	4600-4450 BC	Vinča culture	Pločnik	HDM 1563	settlement/hoard?	SRB
HDM 1564	MC	4600-4450 BC	Vinča culture	Pločnik	HDM 1564	settlement/hoard?	SRB
HDM 1565	MC	4600-4450 BC	Vinča culture	Pločnik	HDM 1565	settlement/hoard?	SRB

SRB = Serbia, W= West Bulgaria, NE= Northeast Bulgaria, ROM = Romania, THR = Thracia. Data from Pernicka *et al.* 1993, 1997; Radivojević and Grujić 2017.

Artefact type	Axe type	Category	Base	Cu (%)	As (µg/g)	Sb (µg/g)	Co (µg/g)	Ni (µg/g)	Ag (µg/g)	Au (µg/g)	Se (µg/g)
copper metal droplet	unk	production	copper	55	9.4	1.83	0.68	40	58	0.089	270
copper metal droplet	unk	production	copper	73	7.1	1.21	0.68	21	0.7	0.02	43
copper slag	unk	production	copper	86	1.1	0.3	9.3	65	9.7	0.03	1.6
copper slag	unk	production	copper	94	1.6	0.2	5.4	22	6	0.03	1.3
copper chisel	unk	metal implement	copper	100	2	0.19	10.4	27	5.3	0.067	2.6
copper ingot	unk	metal implement	copper	100	1	5.0	2.0	43.3	0.0	30.4	0.0
copper slag	unk	production	copper	68	5.4	0.91	450	130	2.3	0.016	2.4
copper slag	unk	production	copper	54	12.6	2	2990	580	4	0.039	9
copper chisel	unk	metal implement	copper	100	1	1	7.4	29.1	0.0	0.1	0.0
copper chisel	unk	metal implement	copper	100	3	0.68	8	83	7.9	0.3	6.3
copper slag	unk	production	copper	76	12.2	2.74	62	51	3.9	0.037	2.5
copper slag	unk	production	copper	49	16.3	3.7	860	320	2.6	0.04	4.3
copper prill	unk	production	copper	90	2	0.5	56	36	3.1	0.67	0.8
copper prill	unk	production	copper	97	1.1	0.3	61	108	5.8	0.019	1.9
chisel	unk	metal implement	copper	100	1	68	1	52	540	13.2	67
heavy axe	unk	metal implement	copper	100	2	14.7	1.5	130	530	10.2	70
Finger ring	unk	metal ornament	copper	100	770	770	2	39	420	8.1	83
Finger ring	unk	metal ornament	copper	88	6.3	0.44	11.1	20	370	4.6	27.8
copper slag droplet	unk	production	copper	88	6	0.48	45	50	5.3	0.087	3.5
copper awl	unk	metal implement	copper	94	8.7	1.93	10.8	130	11.2	2.28	19.3
copper hook	unk	metal implement	copper	92	4.3	5.9	7.3	50	173	6.8	29.9
copper bracelet	unk	metal ornament	copper	83	93.8	23.2	1.33	140	65	3.7	8.3
copper wire	unk	metal implement	copper	94	5.7	0.85	18.2	240	10.3	1.16	9.8
metal droplet	unk	production	copper	85	2.25	0.33	4.2	50	3.5	0.097	1.8
copper awl	unk	metal implement	copper	88	4.6	0.43	5.9	45	6.3	0.16	5.2
copper metal droplet	unk	production	copper	86	8	3.8	186	120	7.6	0.169	3.5
copper wire	unk	metal implement	copper	89	4.1	1	13.8	150	8.1	0.37	6.4
hammer axe	Pločnik	metal implement	copper	100	4790	100	3.7	78	71	4.2	44
bracelet	unk	metal ornament	copper	100	1	0.1	3.2	33	13.8	0.07	6
chisel	unk	metal implement	copper	100	5.9	2.6	2.7	31	12.9	0.1	6.7
flat axe	unk	metal implement	copper	100	770	540	0.8	10	650	14	250
flat axe	unk	metal implement	copper	100	1120	730	0.9	11	610	10.6	175
hammer axe	Pločnik	metal implement	copper	100	0.5	0.06	13.4	42	4.9	0.058	4
hammer axe	Pločnik	metal implement	copper	100	0.4	0.05	42	30	5.5	0.017	3
chisel	unk	metal implement	copper	100	0.4	0.23	3	31	5.8	0.21	5.1
chisel	unk	metal implement	copper	100	0.4	0.18	42	32	5.4	0.019	2
chisel	unk	metal implement	copper	100	0.7	51	3.1	41	235	4.6	76
chisel	unk	metal implement	copper	100	2	52	4.2	26	800	15.9	64
chisel	unk	metal implement	copper	100	2	63	0.8	5	660	11.3	106
chisel	unk	metal implement	copper	97	3	400	0.3	16	700	10.8	128

Table 10 continued. Trace element data for EC and MC copper production and implements. Regions are abbreviated as follows: BSC=Black 2012; Radivojević

LABEL analytical	relative period	absolute period	cultural attribution	Site	Sample label	Type of site/context	Region
HDM 1566	MC	4600-4450 BC	Vinča culture	Pločnik	HDM 1566	settlement/hoard?	SRB
HDM 1567	MC	4600-4450 BC	Vinča culture	Pločnik	HDM 1567	settlement/hoard?	SRB
HDM 1568	MC	4600-4450 BC	Vinča culture	Pločnik	HDM 1568	settlement/hoard?	SRB
HDM 1569	MC	4600-4450 BC	Vinča culture	Pločnik	HDM 1569	settlement/hoard?	SRB
HDM 1570	MC	4600-4450 BC	Vinča culture	Pločnik	HDM 1570	settlement/hoard?	SRB
HDM 1571	MC	4600-4450 BC	Vinča culture	Pločnik	HDM 1571	settlement/hoard?	SRB
HDM 1903	MC	4600-4450 BC	KSBh I	Radlovci	HDM 1903	stray	W
HDM 1910	MC	4600-4450 BC	KSBh I	Dragoman	HDM 1910	hoard	W
HDM 1911	MC	4600-4450 BC	KSBh I	Dragoman	HDM 1911	hoard	W
HDM 1940	MC	4600-4450 BC	Karanovo VI?	Ai Bunar	HDM 1940	mine	THR
HDM 1976	MC	4600-4450 BC	Hamangia IV	Durankulak	HDM 1976	cemetery	BSC
HDM 1977	MC	4600-4450 BC	Hamangia IV	Durankulak	HDM 1977	cemetery	BSC
HDM 2011	MC	4600-4450 BC	Varna I	Durankulak	HDM 2011	cemetery	BSC
HDM 2025	MC	4600-4450 BC	Hamangia IV	Durankulak	HDM 2025	cemetery	BSC
HDM 2026	MC	4600-4450 BC	Hamangia IV	Durankulak	HDM 2026	cemetery	BSC
HDM 2041	MC	4600-4450 BC	Boian-Spantov	Tell Ruse	HDM 2041	settlement	NE
HDM 2049	MC	4600-4450 BC	Boian-Spantov	Tell Ruse	HDM 2049	settlement	NE
HDM 2064	MC	4600-4450 BC	Boian-Spantov	Tell Ruse	HDM 2064	settlement	NE
HDM 2080	MC	4600-4450 BC	Boian-Spantov	Tell Ruse	HDM 2080	settlement	NE
HDM 2124	MC	4600-4450 BC	Sava IV (Hamangia IV)	Goljamo Delcevo	HDM 2124	settlement	BSC
HDM 2702	MC	4600-4450 BC	KSBh I	Darzhanitsa	HDM 2702	hoard	W
HDM 2703	MC	4600-4450 BC	KSBh I	Darzhanitsa	HDM 2703	hoard	W
MA-140926	MC	4600-4450 BC	KSBh I?	Kmpije Bor	Kmpije 2	settlement	SRB
MA-140927	MC	4600-4450 BC	KSBh I?	Kmpije Bor	Kmpije 3	settlement	SRB
MA-140929	MC	4600-4450 BC	Foeni Petresti (~ Late Vinča)	Foeni Cimitir Foeni	Foeni Petresti 6	settlement	ROM
HDM 1422	MC	4600-4450 BC	KSBh I	Bubanj	HDM 1422	settlement	SRB
HDM 1431	MC	4600-4450 BC	KSBh I?	Stari Kostolac	HDM 1431	stray	SRB
HDM 1905	MC	4600-4450 BC	KSBh I	Djakovo	HDM 1905	settlement	W
HDM 1942	MC	4600-4450 BC	Varna II	Durankulak	HDM 1942	cemetery	BSC
HDM 1943	MC	4600-4450 BC	Varna I	Durankulak	HDM 1943	cemetery	BSC
HDM 1944	MC	4600-4450 BC	Varna II	Durankulak	HDM 1944	cemetery	BSC
HDM 1945	MC	4600-4450 BC	Varna II	Durankulak	HDM 1945	cemetery	BSC
HDM 1947	MC	4600-4450 BC	Varna II	Durankulak	HDM 1947	cemetery	BSC
HDM 1948	MC	4600-4450 BC	Varna II	Durankulak	HDM 1948	cemetery	BSC
HDM 1949	MC	4600-4450 BC	Varna II	Durankulak	HDM 1949	cemetery	BSC
HDM 1967	MC	4600-4450 BC	Varna I	Durankulak	HDM 1967	cemetery	BSC
HDM 1968	MC	4600-4450 BC	Varna I	Durankulak	HDM 1968	cemetery	BSC
HDM 1972	MC	4600-4450 BC	Varna I	Durankulak	HDM 1972	cemetery	BSC
HDM 1974	MC	4600-4450 BC	Varna II	Durankulak	HDM 1974	cemetery	BSC

Sea Coast, SRB = Serbia, W= West Bulgaria, NE= Northeast Bulgaria, ROM = Romania, THR = Thracia. Data from Pernicka *et al.* 1993, 1997; Radivojević and Grujić 2017.

Artefact type	Axe type	Category	Base	Cu (%)	As (µg/g)	Sb (µg/g)	Co (µg/g)	Ni (µg/g)	Ag (µg/g)	Au (µg/g)	Se (µg/g)
chisel	unk	metal implement	copper	100	0.6	15.4	1.7	10	450	7.1	44
chisel	unk	metal implement	copper	100	81	310	0.9	12	680	14.8	98
chisel	unk	metal implement	copper	100	0.6	0.1	16.3	20	2.9	0.02	1.7
chisel	unk	metal implement	copper	99	0.8	32	0.7	34	240	6.4	67
chisel	unk	metal implement	copper	100	17.3	12.7	1.2	44	155	11.9	33
hammer axe	Pločnik	metal implement	copper	100	840	160	2.8	36	240	5.9	35
hammer axe	Pločnik	metal implement	copper	100	0.9	4.9	1	85	12	0.212	0.5
hammer axe	Pločnik	metal implement	copper	100	0.8	7.6	1.5	72	39	0.79	3.1
hammer axe	Pločnik	metal implement	copper	100	0.8	2.68	1	84	12	0.35	0.8
hammer axe	Pločnik	metal implement	copper	100	1.3	1.55	1.5	41	850	12	320
bracelet	unk	metal ornament	copper	100	1	29	5.6	150	150	29.2	38
bracelet	unk	metal ornament	copper	99	2490	590	1	15	620	20.8	177
bracelet	unk	metal ornament	copper	98	3400	38	5	73	244	0.248	56
bracelet	unk	metal ornament	copper	94	340	15.2	30	284	450	0.61	100
bracelet	unk	metal ornament	copper	100	360	15.2	32	310	470	0.65	102
borer	unk	metal implement	copper	96	127	16.7	2	153	3400	53	204
borer	unk	metal implement	copper	83	15.3	6	5.5	23	147	14.8	135
borer	unk	metal implement	copper	100	2	1.8	3.5	125	54	0.83	42
borer	unk	metal implement	copper	100	10.9	3.1	39	890	11	0.46	1
borer	unk	metal implement	copper	100	92	2.75	11.2	67	360	2.94	274
heavy axe	Gumelnica	metal implement	copper	103	0.8	0.11	12.6	34	9.3	0.083	2.5
Heavy axe	Salcuta	metal implement	copper	102	10.9	2.15	0.1	52	11.9	0.275	1
Slagged sherd	unk	production	copper	4.76	53.8	2.36	30.1	123	2	0.033	3
metal	unk	metal implement	copper	84.9	2.9	1.13	1.34	66	7.6	0.215	2
Metal droplet?	unk	production	copper	46.5	964	42.7	5.2	105	666	3.87	42.3
Chisel	unk	metal implement	copper	98	3	8.9	1.9	90	129	0.69	45
chisel	unk	metal implement	copper	100	3.5	4.5	0.7	94	12.8	0.26	1.6
flat axe	unk	metal implement	copper	100	2	4.9	1.5	104	1350	7.8	110
Finger ring	unk	metal ornament	copper	96	2.3	6.3	1.5	22	1210	2.26	750
Finger ring	unk	metal ornament	copper	82	39	5.8	9.9	77	19	44	32
bracelet	unk	metal ornament	copper	100	0.7	0.97	9.7	82	104	19.4	21.3
bracelet	unk	metal ornament	copper	100	0.6	1.1	8.6	71	111	20	26.4
bracelet	unk	metal ornament	copper	95	188	9.2	98	152	233	4.78	102
bracelet	unk	metal ornament	copper	100	5.9	5.7	15.3	162	191	10.3	27.6
bracelet	unk	metal ornament	copper	100	132	3.4	21	68	142	19.5	12.6
Finger ring	unk	metal ornament	copper	100	173	320	5.6	75	520	13.1	60
bracelet	unk	metal ornament	copper	100	5	1.96	90	2400	180	27.2	43
bracelet	unk	metal ornament	copper	100	2240	72	1	18	171	0.075	17.8
Finger ring	unk	metal ornament	copper	100	820	7.5	660	250	230	12.9	22.5

Table 10 continued. Trace element data for EC and MC copper production and implements. Regions are abbreviated as follows: BSC=Black 2012; Radivojević

LABEL analytical	relative period	absolute period	cultural attribution	Site	Sample label	Type of site/context	Region
HDM 1979	MC	4600-4450 BC	Varna II	Durankulak	HDM 1979	cemetery	BSC
HDM 1980	MC	4600-4450 BC	Varna II	Durankulak	HDM 1980	cemetery	BSC
HDM 1981	MC	4600-4450 BC	Varna I	Durankulak	HDM 1981	cemetery	BSC
HDM 1982	MC	4600-4450 BC	Varna II	Durankulak	HDM 1982	cemetery	BSC
HDM 1984	MC	4600-4450 BC	Varna II	Durankulak	HDM 1984	cemetery	BSC
HDM 1985	MC	4600-4450 BC	Varna II	Durankulak	HDM 1985	cemetery	BSC
HDM 1986	MC	4600-4450 BC	Varna I	Durankulak	HDM 1986	cemetery	BSC
HDM 1987	MC	4600-4450 BC	Varna I	Durankulak	HDM 1987	cemetery	BSC
HDM 1989	MC	4600-4450 BC	Varna I	Durankulak	HDM 1989	cemetery	BSC
HDM 1991	MC	4600-4450 BC	Varna II	Durankulak	HDM 1991	cemetery	BSC
HDM 1992	MC	4600-4450 BC	Varna II	Durankulak	HDM 1992	cemetery	BSC
HDM 1993	MC	4600-4450 BC	Varna II	Durankulak	HDM 1993	cemetery	BSC
HDM 1994	MC	4600-4450 BC	Varna II	Durankulak	HDM 1994	cemetery	BSC
HDM 1995	MC	4600-4450 BC	Varna II	Durankulak	HDM 1995	cemetery	BSC
HDM 1996	MC	4600-4450 BC	Varna I	Durankulak	HDM 1996	cemetery	BSC
HDM 1998	MC	4600-4450 BC	Varna I	Durankulak	HDM 1998	cemetery	BSC
HDM 2000	MC	4600-4450 BC	Varna I	Durankulak	HDM 2000	cemetery	BSC
HDM 2001	MC	4600-4450 BC	Varna II	Durankulak	HDM 2001	cemetery	BSC
HDM 2002	MC	4600-4450 BC	Varna I	Durankulak	HDM 2002	cemetery	BSC
HDM 2003	MC	4600-4450 BC	Varna I	Durankulak	HDM 2003	cemetery	BSC
HDM 2006	MC	4600-4450 BC	Varna I	Durankulak	HDM 2006	cemetery	BSC
HDM 2009	MC	4600-4450 BC	Varna II	Durankulak	HDM 2009	cemetery	BSC
HDM 2012	MC	4600-4450 BC	Varna I	Durankulak	HDM 2012	cemetery	BSC
HDM 2013	MC	4600-4450 BC	Varna I	Durankulak	HDM 2013	cemetery	BSC
HDM 2014	MC	4600-4450 BC	Varna I	Durankulak	HDM 2014	cemetery	BSC
HDM 2015	MC	4600-4450 BC	Varna I	Durankulak	HDM 2015	cemetery	BSC
HDM 2016	MC	4600-4450 BC	Varna II	Durankulak	HDM 2016	cemetery	BSC
HDM 2017	MC	4600-4450 BC	Varna II	Durankulak	HDM 2017	cemetery	BSC
HDM 2018	MC	4600-4450 BC	Varna I	Durankulak	HDM 2018	cemetery	BSC
HDM 2019	MC	4600-4450 BC	Varna II	Durankulak	HDM 2019	cemetery	BSC
HDM 2023	MC	4600-4450 BC	Varna II	Durankulak	HDM 2023	cemetery	BSC
HDM 2024	MC	4600-4450 BC	Varna II	Durankulak	HDM 2024	cemetery	BSC
HDM 2027	MC	4600-4450 BC	Varna II	Durankulak	HDM 2027	cemetery	BSC
HDM 2028	MC	4600-4450 BC	Varna I	Durankulak	HDM 2028	cemetery	BSC
HDM 2029	MC	4600-4450 BC	Varna I	Durankulak	HDM 2029	cemetery	BSC
HDM 2031	MC	4600-4450 BC	Varna I	Durankulak	HDM 2031	cemetery	BSC
HDM 2032	MC	4600-4450 BC	Varna II	Durankulak	HDM 2032	cemetery	BSC
HDM 2034	MC	4600-4450 BC	Varna II	Durankulak	HDM 2034	cemetery	BSC
HDM 2035	MC	4600-4450 BC	Varna II	Durankulak	HDM 2035	cemetery	BSC
HDM 2036	MC	4600-4450 BC	Varna I	Durankulak	HDM 2036	cemetery	BSC

Sea Coast, SRB = Serbia, W= West Bulgaria, NE= Northeast Bulgaria, ROM = Romania, THR = Thracia. Data from Pernicka *et al.* 1993, 1997; Radivojević and Grujić 2017.

Artefact type	Axe type	Category	Base	Cu (%)	As (µg/g)	Sb (µg/g)	Co (µg/g)	Ni (µg/g)	Ag (µg/g)	Au (µg/g)	Se (µg/g)
bracelet	unk	metal ornament	copper	100	4	0.16	7.9	26	4	0.078	0.7
bracelet	unk	metal ornament	copper	98	5.4	4.8	12.5	82	330	6.8	68
bracelet	unk	metal ornament	copper	99	5	8	1	107	910	36	16
bracelet	unk	metal ornament	copper	100	4.4	7.1	1	26	640	9.5	40
bracelet	unk	metal ornament	copper	100	234	1550	3	47	880	20	310
bracelet	unk	metal ornament	copper	100	6	7.4	2.5	56	590	11.6	62
bracelet	unk	metal ornament	copper	97	66	120	7.5	57	330	100	50
bracelet	unk	metal ornament	copper	100	2.6	7.6	6.1	61	460	35	181
bracelet	unk	metal ornament	copper	100	3.9	122	0.5	41	440	3.4	235
bracelet	unk	metal ornament	copper	100	56	370	0.5	63	720	16.9	81
bracelet	unk	metal ornament	copper	100	5100	830	0.5	61	580	14.5	256
bracelet	unk	metal ornament	copper	100	12.8	3.4	35	350	144	56	11.8
bracelet	unk	metal ornament	copper	100	3	0.6	4	156	580	34	33
bracelet	unk	metal ornament	copper	100	53	7.8	107	64	180	3	96
bracelet	unk	metal ornament	copper	100	240	1500	2	105	650	7.9	78
bracelet	unk	metal ornament	copper	100	3.6	112	0.5	34	420	3	221
Finger ring	unk	metal ornament	copper	83	162	2.2	7.2	60	99	25	20.1
Finger ring	unk	metal ornament	copper	90	620	136	24.7	53	350	11.6	9.9
Finger ring	unk	metal ornament	copper	95	640	12.4	78	98	500	13.2	66
Finger ring	unk	metal ornament	copper	84	3.7	0.24	3	15	164	22.6	29
spiral ring	unk	metal ornament	copper	84	9	23.4	0.5	48	1640	12.9	33
spiral ring	unk	metal ornament	copper	92	1.3	2.91	1.5	96	240	33	47
bracelet	unk	metal ornament	copper	97	2.5	0.33	1.5	5	21	0.218	1.7
finger ring	unk	metal ornament	copper	95	1	0.21	1.5	27	720	2.63	400
bracelet	unk	metal ornament	copper	95	266	3.4	36	74	167	17.4	27.8
bracelet	unk	metal ornament	copper	100	13.5	11	2	27	600	7.7	53
bracelet	unk	metal ornament	copper	100	0.9	1.01	2	12	390	9.8	55
bracelet	unk	metal ornament	copper	100	0.9	0.44	2.5	33	181	11.4	31
bracelet	unk	metal ornament	copper	100	20.3	16.6	26.2	93	390	18.8	61
bracelet	unk	metal ornament	copper	100	1.1	2.73	1.5	152	350	27.1	37
bracelet	unk	metal ornament	copper	100	3	2.45	6.4	118	280	40	52
bracelet	unk	metal ornament	copper	100	2	2.31	3.5	100	276	37	55
Finger ring	unk	metal ornament	copper	100	6300	1310	1	25	860	9.8	279
Finger ring	unk	metal ornament	copper	88	9.2	1.89	10.3	174	158	39	63
Finger ring	unk	metal ornament	copper	100	1100	320	3.5	95	1980	15.4	33
Finger ring	unk	metal ornament	copper	95	22.6	1.21	5	105	205	55	26.8
borer	unk	metal implement	copper	98	2	5.5	1.5	294	630	7.9	80
pin	unk	metal ornament	copper	98	0.7	1.61	2	40	225	4.6	227
spiral ring	unk	metal ornament	copper	97	2	27.2	1	22	330	8.7	61
spiral ring	unk	metal ornament	copper	97	9.9	9.3	13	132	215	100	72

Table 10 continued. Trace element data for EC and MC copper production and implements. Regions are abbreviated as follows: BSC=Black 2012; Radivojević

LABEL analytical	relative period	absolute period	cultural attribution	Site	Sample label	Type of site/context	Region
HDM 2137	MC	4600-4450 BC	Varna I	Durankulak	HDM 2137	cemetery	BSC
HDM 2138	MC	4600-4450 BC	Varna I	Durankulak	HDM 2138	cemetery	BSC
HDM 1920	MC	4600-4450 BC	KGK VI?	Tell Ruse	HDM 1920	settlement	NE
HDM 1921	MC	4600-4450 BC	KGK VI?	Tell Ruse	HDM 1921	settlement	NE
HDM 1922	MC	4600-4450 BC	KGK VI?	Tell Ruse	HDM 1922	settlement	NE
HDM 1923	MC	4600-4450 BC	KGK VI	Tell Ruse	HDM 1923	settlement	NE
HDM 2039	MC	4600-4450 BC	KGK VI?	Tell Ruse	HDM 2039	settlement	NE
HDM 2040	MC	4600-4450 BC	KGK VI?	Tell Ruse	HDM 2040	settlement	NE
HDM 2043	MC	4600-4450 BC	KGK VI?	Tell Ruse	HDM 2043	settlement	NE
HDM 2044	MC	4600-4450 BC	KGK VI?	Tell Ruse	HDM 2044	settlement	NE
HDM 2045	MC	4600-4450 BC	KGK VI?	Tell Ruse	HDM 2045	settlement	NE
HDM 2047	MC	4600-4450 BC	KGK VI?	Tell Ruse	HDM 2047	settlement	NE
HDM 2050	MC	4600-4450 BC	KGK VI?	Tell Ruse	HDM 2050	settlement	NE
HDM 2051	MC	4600-4450 BC	KGK VI?	Tell Ruse	HDM 2051	settlement	NE
HDM 2052	MC	4600-4450 BC	KGK VI?	Tell Ruse	HDM 2052	settlement	NE
HDM 2053	MC	4600-4450 BC	KGK VI?	Tell Ruse	HDM 2053	settlement	NE
HDM 2054	MC	4600-4450 BC	KGK VI?	Tell Ruse	HDM 2054	settlement	NE
HDM 2055	MC	4600-4450 BC	KGK VI?	Tell Ruse	HDM 2055	settlement	NE
HDM 2056	MC	4600-4450 BC	KGK VI?	Tell Ruse	HDM 2056	settlement	NE
HDM 2057	MC	4600-4450 BC	KGK VI?	Tell Ruse	HDM 2057	settlement	NE
HDM 2058	MC	4600-4450 BC	KGK VI?	Tell Ruse	HDM 2058	settlement	NE
HDM 2059	MC	4600-4450 BC	KGK VI?	Tell Ruse	HDM 2059	settlement	NE
HDM 2060	MC	4600-4450 BC	KGK VI?	Tell Ruse	HDM 2060	settlement	NE
HDM 2061	MC	4600-4450 BC	KGK VI?	Tell Ruse	HDM 2061	settlement	NE
HDM 2063	MC	4600-4450 BC	KGK VI?	Tell Ruse	HDM 2063	settlement	NE
HDM 2065	MC	4600-4450 BC	KGK VI?	Tell Ruse	HDM 2065	settlement	NE
HDM 2067	MC	4600-4450 BC	KGK VI?	Tell Ruse	HDM 2067	settlement	NE
HDM 2068	MC	4600-4450 BC	KGK VI?	Tell Ruse	HDM 2068	settlement	NE
HDM 2070	MC	4600-4450 BC	KGK VI?	Tell Ruse	HDM 2070	settlement	NE
HDM 2071	MC	4600-4450 BC	KGK VI?	Tell Ruse	HDM 2071	settlement	NE

metal must have come from three, or possibly four, different deposits, most prominently from Ai Bunar (see Figures 6 and 7) and was possibly not directly produced on the site. The latter notion is based on the initial analysis of Belovode production evidence (slags M21 and M22a, see Table 10) showed that it matched the trace element pattern of copper implements from Pločnik (Radivojević *et al.* 2010a: 2785, Fig. 11), as well as the fact that the lead isotope abundance ratio of another sample of Belovode production evidence (B47/12/3) matched with one of the Pločnik copper chisels (HDM1568, see Figure 7). On the other hand, given that the same copper ores from identical sources in eastern Serbia were reaching both Belovode and Pločnik (e.g. Figures 4–9), and that copper smiths at both sites had equal opportunity and knowledge to extract metal (see for instance Radivojević *et al.* 2013), it is very likely that we may still discover a smelting workshop at Pločnik.

The variation in the trace element levels in the Vinča culture and EC / MC copper metal assemblages becomes

Sea Coast, SRB = Serbia, W= West Bulgaria, NE= Northeast Bulgaria, ROM = Romania, THR = Thracia. Data from Pernicka *et al.* 1993, 1997; Radivojević and Grujić 2017.

Artefact type	Axe type	Category	Base	Cu (%)	As (μg/g)	Sb (μg/g)	Co (μg/g)	Ni (μg/g)	Ag (μg/g)	Au (μg/g)	Se (μg/g)
Finger ring	unk	metal ornament	copper	72	11.5	0.65	17.9	132	145	20.5	58
Finger ring	unk	metal ornament	copper	54	69	2.13	9.9	99	170	8.7	18
axe	Kamenar	metal implement	copper	100	4.8	1.09	11.6	230	79	12.9	30
flat axe	Kamenar	metal implement	copper	100	246	1210	1.5	19	690	14.2	430
chisel	unk	metal implement	copper	100	4.5	1.12	19.9	93	141	20.2	14.6
axe	Gumelnica	metal implement	copper	100	2	169	1	32	620	5.1	106
borer	unk	metal implement	copper	96	3500	870	3.5	58	5800	0.96	14.2
borer	unk	metal implement	copper	100	1	2.68	2.5	22	300	4.6	282
borer	unk	metal implement	copper	100	2	14.3	3	143	520	0.36	117
borer	unk	metal implement	copper	77	4	750	1.5	25	340	13.8	113
borer	unk	metal implement	copper	100	330	450	2.5	18	670	26.3	104
double spiral headed pin	unk	metal ornament	copper	100	4	6.8	1.5	154	2600	14.5	87
borer	unk	metal implement	copper	100	32	87	5.5	117	126	0.97	42
borer	unk	metal implement	copper	93	8600	197	2	214	2840	15.9	18.8
borer	unk	metal implement	copper	98	9400	236	3.5	271	2560	14.4	19.6
chisel	unk	metal implement	copper	99	117	440	2.5	17	430	9.6	490
borer	unk	metal implement	copper	100	5	3.1	1.5	81	600	17.8	59
borer	unk	metal implement	copper	100	12	860	1.5	510	13100	16.2	8.2
borer	unk	metal implement	copper	99	12.7	1.75	5.9	105	25	0.34	38
borer	unk	metal implement	copper	100	2430	197	3	17	650	9.2	140
borer	unk	metal implement	copper	100	1.3	3.1	2	64	11	0.29	0.7
borer	unk	metal implement	copper	99	4.1	0.78	1.5	41	64	1.15	152
borer	unk	metal implement	copper	100	2	1.53	2	17	852	5.3	15
borer	unk	metal implement	copper	100	3	0.63	7.3	204	42	0.67	38
borer	unk	metal implement	copper	98	4.5	0.58	16.4	210	22	0.04	94
borer	unk	metal implement	copper	100	138	19.6	18.7	109	39	1.4	92
borer	unk	metal implement	copper	100	1.4	2.92	1.5	57	2210	8.5	129
borer	unk	metal implement	copper	97	7.1	1.58	2	8	340	12.8	370
borer	unk	metal implement	copper	99	1.7	122	1.5	34	600	10.2	134
borer	unk	metal implement	copper	98	4.7	249	3.5	11	560	8.7	98

more evident as the number of artefacts analysed grows (Table 11). The greatest variations are in As, Sb, Co, Ni and Ag content, while Au remains at low levels overall. This complements the complex copper supply picture created by the lead isotope abundance ratios (Figure 11), indicating the exploitation of five (or six) major copper deposits at the time.

The immense work of building a trace element dataset for copper metal artefacts and main copper deposits in the Balkans conducted by Pernicka and collaborators (1993, 1997) laid the foundations for further comparisons. More recently, this dataset has been enriched with more samples that pushed the origins of metallurgy in the Balkans back to the beginning of the 5th millennium BC (Radivojević *et al.* 2010a; Radivojević 2012), which are also fully published as an open access document (Radivojević and Grujić 2017). Here, we use the nine chemical clusters of 335 copper based early metal artefacts from Serbia and Bulgaria, obtained through average-link cluster analysis (excluding malachite and samples younger than Proto Bronze Age)

Table 11. Trace element data comparison of newly analysed and previously published data for Belovode, Pločnik, Vinča culture, EC and MC copper metal, from production debris and implements.

label	period	description / number	As (ppm)	Sb (ppm)	Co (ppm)	Ni (ppm)	Ag (ppm)	Au (ppm)	Se (ppm)
Bf43/13	EC	Metal droplet	50.8	0.86	1.94	20	36.2	0.085	26.8
B23/12	MC	Slagged sherd (metal)	9.9	1.16	55	47	2.1	0.01	4.4
B47/12/3	MC	Slagged sherd (metal)	15.7	2.02	68	160	2.3	0.012	2.1
Bf21/12	MC	metal droplet	4	0.2	15.5	180	9.9	0.044	3.2
P10/13	MC	metal foil/stock corroded	0.92	0.28	14.9	86	6.4	0.079	0.8
P13/13	MC	metal wire/bracelet	38	7.3	64	1600	30	0.7	53
C_P1/13	MC	Metal loop / ring	10.6	6.26	2.9	74	230	0.41	6.2
Belovode average	EC/MC	n= 5 copper metal artefacts	**4.9**	**0.9**	**93.2**	**55.6**	**15.3**	**0.03**	**63.7**
Belovode min	EC/MC		1.1	0.2	0.68	21	0.7	0.000	1.3
Belovode max	EC/MC		9.4	1.83	450	130	58	0.089	270
stdev	EC/MC		*3.6*	*0.7*	*199.5*	*45.3*	*24.1*	*0.0*	*116.7*
Pločnik average	EC/MC	n= 21 copper metal artefacts	**135.8**	**113.2**	**7.8**	**29.2**	**260.5**	**6.9**	**52.7**
Pločnik min	EC/MC		0.4	0.05	0.3	5	0.000	0.017	0.000
Pločnik max	EC/MC		1120	730	42	83	800	30.4	250
stdev	EC/MC		*329.6*	*205.9*	*12.2*	*17.2*	*299.6*	*7.9*	*67.3*
Vinča culture average	EC/MC	n= 41 copper metal artefacts	**76.2**	**60.6**	**49.1**	**60.4**	**146.7**	**4.0**	**38.2**
Vinča culture min	EC/MC		0.4	0.05	0.3	5	0.000	0.000	0.000
Vinča culture max	EC/MC		1120	730	860	320	800	30.4	270
stdev	EC/MC		*244.8*	*157.7*	*151.7*	*64.3*	*248.0*	*6.5*	*65.4*
EC average	EC	n= 17 copper metal artefacts	**49.6**	**51.3**	**91.1**	**71.5**	**115.6**	**4**	**34.4**
EC min	EC		1.0	0.2	0.7	20.0	0.000	0.000	0.000
EC max	EC		770	770	860	320	540	30	270
stdev	EC		*185.7*	*185.9*	*225.3*	*73.5*	*203.4*	*8*	*67*
MC average	MC	n= 132 copper metal artefacts	**452**	**123.6**	**16.3**	**109.0**	**564.3**	**12.4**	**83.1**
MC min	MC		0.4	0.1	0.100	5	2	0.02	0.5
MC max	MC		9400	1550	660	2400	13100	100	750
stdev	MC		*1446.7*	*292.4*	*61.4*	*227.7*	*1329.7*	*16.3*	*112.0*

(Pernicka *et al.* 1997: 91, Table 2) as the comparative database for the newly analysed Belovode and Pločnik artefacts (Table 9). Importantly, the majority of artefacts analysed by Pernicka *et al.* (1993, 1997) are from Middle, Late and Final Chalcolithic, while only a handful represent the Early Chalcolithic period; the research presented here adds more artefacts to the EC period and takes the debate deeper into the beginnings of the 5th millennium BC.

The chemical clusters from Pernicka *et al.* (1997) are presented as the 10% percentile, median and 90% percentile of the distribution of elemental concentrations (Figures 12–14, see relevant selection in Table 12). The clusters of Ai Bunar, Majdanpek, Rudna Glava and Medni Rid in Table 12 are derived from analysis of copper ores, while clusters #8 and #2 come from cluster analyses of copper implements. For the comparison of metal objects with ore samples, the seven trace elements listed above were normalised to 100% copper based on the assumption that during smelting they behave in largely the same way as copper (Pernicka 1999) so that their ratios with respect to copper are not substantially altered. This approach was also used for the comparison of slag and copper ore samples from archaeological contexts with copper ore samples from mineralisations / deposits, because these are usually mixed with silicates that do not contain

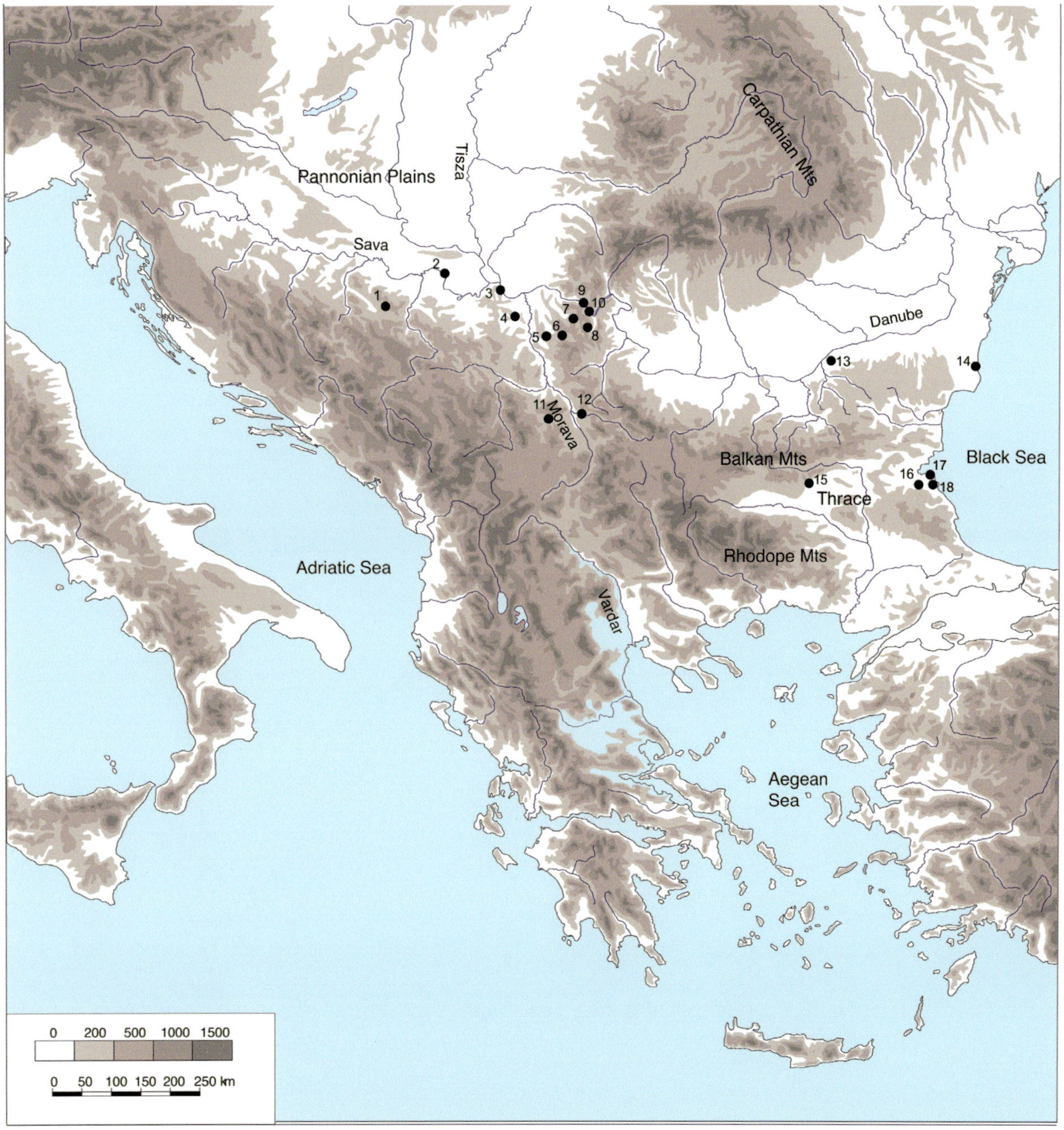

Figure. 12. Map of sites and copper deposits mentioned in this chapter: 1.Tuzla; 2.Gomolava; 3.Vinča; 4.Selevac; 5.Belovode; 6.Ždrelo; 7.Majdanpek; 8.Rudna Glava; 9.Vlasac; 10.Lepenski Vir; 11.Pločnik; 12.Bubanj; 13.Ruse; 14.Durankulak; 15.Ai Bunar; 16.Zidarovo; 17.Akladi Cheiri; 18.Medni Rid.

those chalcophile and siderophile elements and thus function only as a diluent. A similar principle was previously applied to compare artefacts from Belovode and Pločnik against the trace element pattern of copper deposits, such as those from Ai Bunar, Majdanpek and Rudna Glava (Pernicka *et al.* 1993: 32–33 and Table 9, 1997: 158–160 and Tables A2 and A3a). These copper deposits, being confirmed as sites of ancient exploitation, were selected to test the assumptions derived from the discussion of the lead isotope data, although all chemical clusters and copper ore datasets published thus far were also probed for their consistency with the trace element data in Table 9.

Figure 13 presents the unique trace element pattern of the following copper minerals (malachite) from Belovode: B21/13, B108/13, B155/13, B385/13, and a copper metal wire / bracelet, P13/13 from Pločnik. This pattern is characterised by significant levels of Ni (up to 1600 ppm), Co (up to *c.* 60 ppm) and As (up to *c.* 150 ppm), while other values are much lower (e.g. Au). Since half of these artefacts fall into Group 1 (see Figure 3), this is consistent with the lead isotope abundance field for Majdanpek (e.g. Figures 4 and 5), the whole assemblage is plotted against the trace element signature of this copper ore field. While the consistency is not immediately apparent due to the prominent Ni peak that extends outside the Majdanpek ore trace element field, the rest of the trace element string is within the compositional envelope of this ore field. These artefacts are also consistent with previously published minerals and malachite beads from Lepenski Vir, Vlasac, Belovode, Pločnik and Gomolava (see Table 10) (Radivojević and Grujić 2017).

We deem these corresponding trace element patterns as indicative of the likely origins of this group of artefacts from Majdanpek, reinforcing the consistencies indicated in the discussion of the lead isotope data above. It is noteworthy that the trace element field for Majdanpek comes from 11 ore samples of native copper and copper ore, and from a deposit that has very poor preservation of ancient mining, being one of the largest Balkan copper deposits exploited in modern times. It seems very likely that the mineralisation that was exploited in the past could have been more enriched in Ni, however, it has long been bulldozed away and the opportunity to acquire more relevant samples has been lost.

The distinctive trace element pattern of a large body of metal production samples, including the freshly analysed B47/12/3 and B23/12, besides other Belovode, Vinča and Gornja Tuzla slags, resembles those of Rudna Glava ores (Figure 14, see Table 12 for data) but also those of the

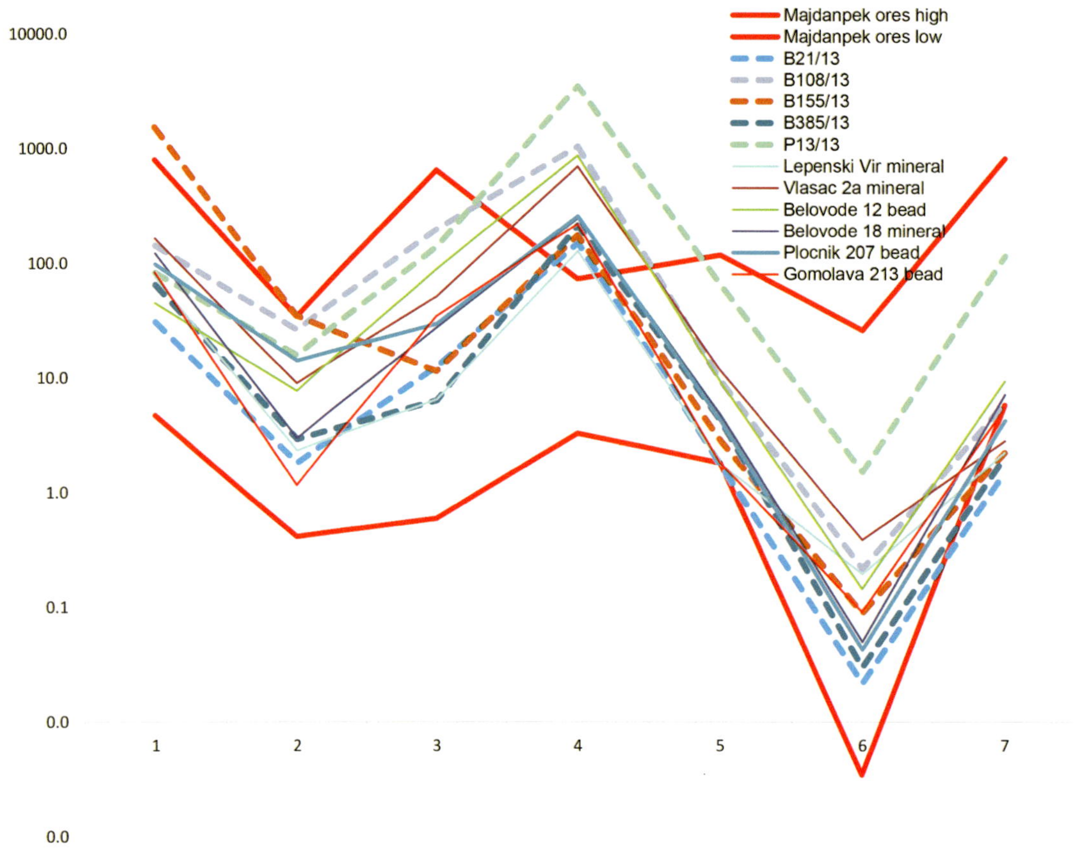

Figure 13. Trace element signature comparison of Late Neolithic, EC and MC artefacts against the signature of Majdanpek copper ore field.

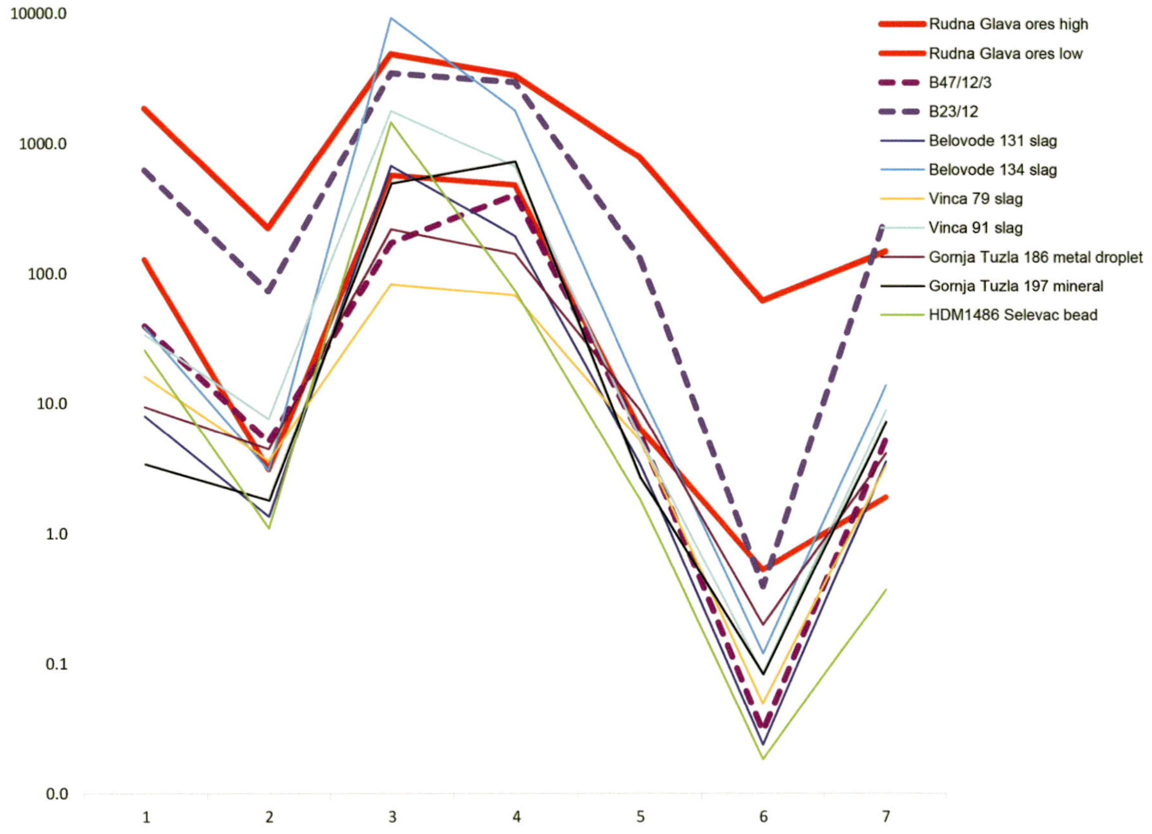

Figure 14. Trace element signature comparison of EC and MC artefacts against the signature of Rudna Glava copper ore field.

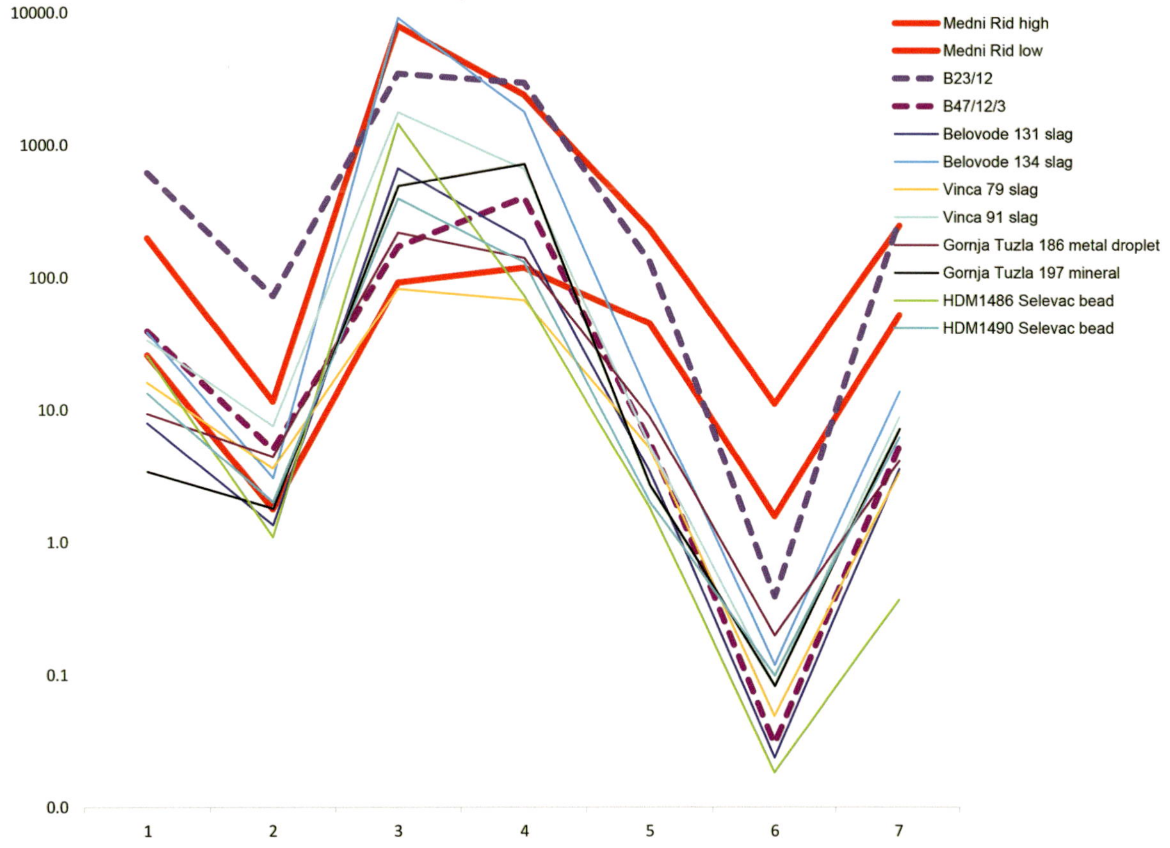

Figure 14a. Trace element signature comparison of EC and MC artefacts against the signature of Medni Rid ore field.

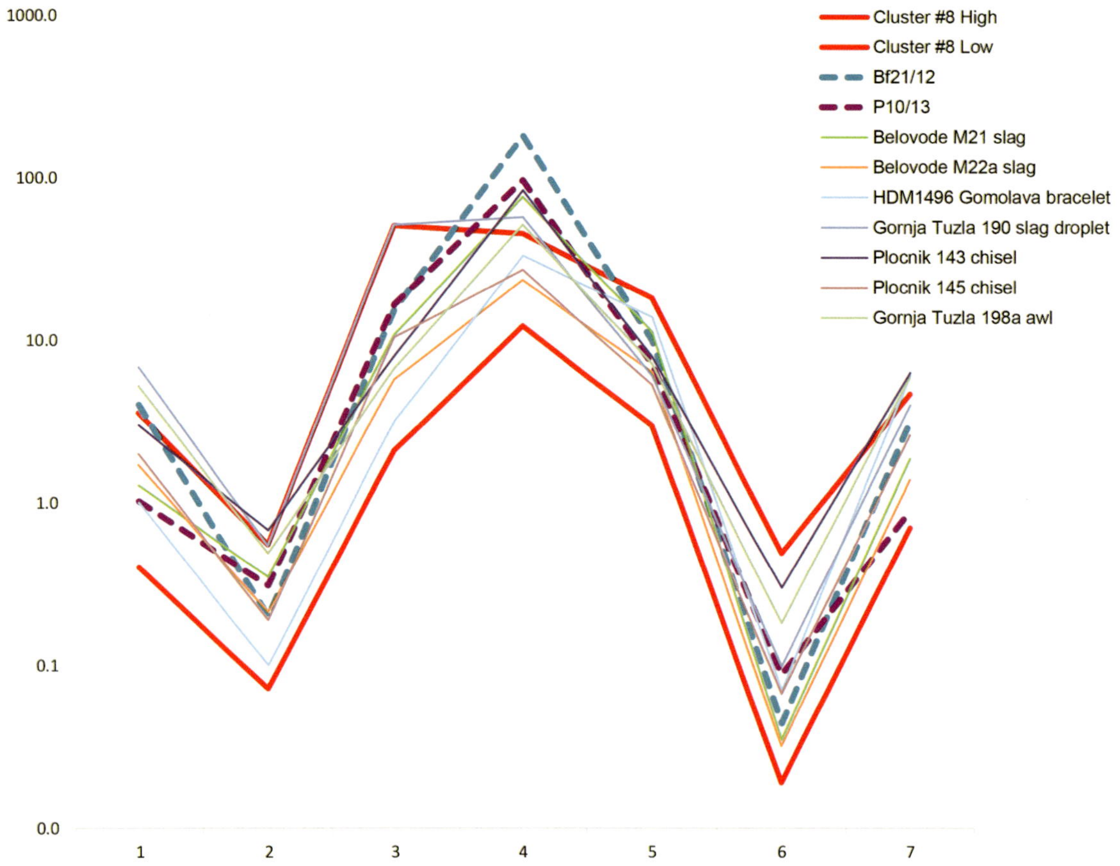

Figure 15. Trace element signature comparison of EC and MC artefacts against the signature of chemical cluster #8.

Medni Rid ore field in southeast Bulgaria (see Kunze and Pernicka 2020: 413–414, Table 3) (Figure 14a). The significant Co and Ni levels, as well as lower Sb and Au readings, characterise this group. The major difference is the lower Au concentrations in the ore samples but this may be due to the fact that the objects were most likely produced from ores originating from the oxidised zone of the copper deposits, which are usually somewhat enriched in Au. These parts of the ore deposits are no longer available; at least one of them, Medni Rid, was heavily worked in modern times. Both deposits also contain radiogenic lead (Figure 10) but so far, copper metal objects with such highly radiogenic signature have been identified only in eastern Bulgaria, from the Bulgarian MC period and later. This suggests that they may be related to the Medni Rid region, especially since the exploitation of these deposits in the later Chalcolithic is, at least indirectly, confirmed by the evidence for Chalcolithic copper smelting at nearby Akladi Cheiri (Rehren *et al.* 2020). Yet, no copper metal artefact analysed thus far from the EC or MC period from Serbia is isotopically consistent with Medni Rid. Rudna Glava, on the other hand, does not show optimal consistency with MC copper implements from Bulgaria (Figure 10); however, the samples in Figure 14 and 14a are largely a mix of EC/MC production evidence, or rather analyses of copper metal prills embedded in this material.

Seven out of 10 artefacts that produce this distinctive pattern are copper metal prills suspended in slag matrix, or droplets, and hence subjected to a set of variable conditions that would result in depletion of elements such as Ag, Au and Ni (depending on the copper ore content) (cf. Tylecote *et al.* 1977). The key to the provenance of these samples is their Co and Ni content, elaborated above, since Belovode production evidence (Table 11) exhibits increased readings of Co and Ni. In the extreme, the newly formed inclusions in Belovode slagged sherds and free slag samples (Tables 7 and 10, and Chapter 11 this volume) contain up to 60 wt% Co and a few percent of Ni, while EPMA analysis also points to high levels of Co and Ni in the previously published production data (Table 1). Co and Ni content are therefore crucial information for designating the provenance of ores used in copper smelting at the sites of Belovode, Vinča and Gornja Tuzla.

The Vinča culture metal producing sites are also geographically located within close reach of the river valleys of the Sava and the Danube, which (amongst others) possibly served as communication routes. Rudna Glava is a mining complex in the hinterlands of the Danube Gorges in eastern Serbia, with several shafts discovered during excavations by B. Jovanović in the 1960s (Jovanović 1971), and which also served

Table 12. A selection of chemical clusters made of Chalcolithic copper deposits and artefacts (data from Kunze and Pernicka 2020; Pernicka *et al.* 1993, 1997)

		As	Sb	Co	Ni	Ag	Au	Se
		µg/g	µg/g	µg/g	µg/g	µg/g	µg/g	µg/g
Cluster #2 (58 samples)	median	15	5	1.2	53	19	0.4	1.4
	10% percentile	4	2	0.4	23	9	0.2	0.5
	90% percentile	99	14	4	94	59	0.8	5
Cluster #8 (15 samples)	median	1.1	0.18	13.4	32	5.5	0.07	1.7
	10% percentile	0.4	0.072	2.1	12.2	2.98	0.02	0.7
	90% percentile	3.56	0.554	50.4	45	18.1	0.5	4.7
Cluster Ai Bunar (9 samples)	median	50000	3500	37	192	458	6.3	193
	10% percentile	10053	1397	12	97	229	1.7	101
	90% percentile	90889	10691	60	407	2329	54.3	991
Cluster Majdanpek (11 samples)	median	39	3.7	21	22	10	0.08	19
	10% percentile	5	0.4	1	3	2	0.003	6
	90% percentile	793	34.3	645	73	117	25.91	810
Cluster Rudna Glava (19 samples)	median	525	27	2611	805	132	3.2	23
	10% percentile	127	3	560	472	6	0.5	2
	90% percentile	1846	221	4792	3280	768	60.6	146
Cluster Medni Rid region (13 samples)	median	74	6	1574	609	120	5	76
	10% percentile	26	2	91	117	45	2	52
	90% percentile	197	12	7897	2361	230	11	242

as an iron mine during Roman times, and into the 20th century (Pernicka *et al.* 1993: 41; Jovanović 2001), due to the abundance of magnetite ore besides malachite. Given the relatively small scope of this mining complex, it had been—unlike Majdanpek—mostly preserved at the time of excavation and is now part of the certified European Union Industrial Heritage Tour (Graf 2013; Muhi *et al.* 2018).

All of this is important in weighing the trace element consistency of the Vinča culture production evidence with Rudna Glava. The site has a confirmed activity from the mid-6th millennium BC and most notably around 5000 BC and could have been a larger complex in the past or included several adjacent outcrops. The most significant aspect is the compositional structure that bears a unique Co and Ni signature, found widely in metal production evidence dated between 5000 BC and 4400 BC (Radivojević and Rehren 2016). Crucially, Rudna Glava is only *c.* 100 km away from Belovode, a short distance in comparison to Medni Rid, which is *c.* 800 km away. Finally, a crucial to this discussion is the fact that the radiogenic lead contained in Rudna Glava ores contradicts the trace element consistency of this mine with Vinča culture metal production evidence (see example in Figure 7). On the other hand, not all Group 1 artefacts that are isotopically consistent with Majdanpek show the same consistency for trace elements. While pairing of both methods is clearly a more credible approach for designating the copper source, with the current state of the database we can consider Majdanpek as a major source of copper in the Chalcolithic but by no means the only one. Ideally, matching isotope and trace element patterns would be complemented by field observations of ancient mining activities combined with absolute dating. In this combination, a relationship with contemporary archaeological finds could be considered as proven.

Two of these three prerequisites (lead isotope and trace element consistencies in ore vs. artefacts comparisons) are presently fulfilled for Majdanpek, and possibly for

Ždrelo in the EC period, and Ai Bunar and Medni Rid for MC and later periods. The exploitation of the latter is at least indirectly indicated by a substantial number of artefacts that can be geochemically related. For Ždrelo, however, more field work and more geochemical analyses would be required to properly characterise this deposit. Rudna Glava remains as a special case where there is a copper mineralisation with abundant evidence for Chalcolithic mining but without matching archaeological artefacts that could be associated with the mine with high probability, especially with lead isotope analysis. It remains to be seen whether other mineralisations with radiogenic lead will be identified in eastern Serbia in the future that may match the Vinča culture assemblage better in their combination of lead isotope and trace element analysis. The possibility cannot, of course, be excluded that eventually, Chalcolithic copper artefacts may be found which are geochemically comparable with the ore samples from Rudna Glava, or that more ore samples from this mining complex may clarify this situation. Our assumption is that those artefacts that potentially came from the Rudna Glava ores are more likely to belong to the period between *c.* 5500–4600 BC.

The final group of artefacts with a particular trace element pattern in Figure 15 represents two artefacts analysed here, a metal droplet from Belovode (Bf21/12) and a metal foil from Pločnik (P10/13). Their patterns are very similar to those of two published slag samples from Belovode, to production evidence from Gornja Tuzla and to a handful of copper metal artefacts from Belovode, Pločnik and Gomolava. This is a purely production / artefact cluster and is highly consistent with cluster no. 8 identified by Pernicka *et al.* (1997: 89), formed by 13 copper metal implements and two malachite beads from the sites of Selevac, Pločnik, Durankulak, Gomolava, Daržanica, Hotnica and Zlotska pecina. It had been recognised earlier as the only trace element pattern to match the Belovode slags M21 and M22a (Radivojević *et al.* 2010a: 2785, Fig. 11), as well as yet another (in addition to 'Group of 16' or grouplet #7) to indicate an as yet unidentified copper deposit (Pernicka *et al.* 1997: 89). Spikes in Ni content in Bf21/12 and P10/13 again show that this deposit was rich in Ni and, as such, could also have been located in eastern Serbia, possibly even in the vicinity of Majdanpek or Rudna Glava (which are only about 24 km apart).

While this will form part of future research, the main outcome of the trace element comparison is that it indicates that at least three different copper deposits were exploited during the Vinča culture period (Figures 13–15), most likely located in the same geographic area, eastern Serbia. Importantly, these were not the only sources used by the Belovode and Pločnik communities, as further corroborated by the remaining seven artefacts (out of 16, see Table 9), which individually display a unique trace element signature that is yet to match a deposit or a published artefact assemblage.

Discussion and conclusion

The detailed analysis of technology and provenance data highlight the most important aspect of Belovode and Pločnik metallurgy in the wider context: its super-connectedness across the Vinča culture, as well as to important trade and exchange nodes along the known communication routes, such as the lower Danube. The shared access to copper sources and knowledge of metal making reveals a wide and complex network of interactions between metal producing and consuming communities in modern day Bosnia, Serbia and Bulgaria. While multiple datasets were employed to dissect this connectedness on a micro-level (e.g. engineering parameters for metal extraction, redox conditions, viscosity, close consistencies for individual mineral, slag and metal implements), a macro-level of data comparison was previously achieved using a robust approach stemming from complexity science (Radivojević and Grujić 2018). The research presented here not only underlines, but also complements the gaps in data and interpretations from the complex networks research, thereby opening avenues for more fine-grained research in the future.

In addition to providing direct absolute dates for metallurgical assemblages, the micro-level approach to data interpretation has both emphasised the connectedness of metal producing and consuming sites along important communication routes and highlighted the importance of several east Serbian copper deposits for the early phase of metallurgy evolution in the Balkans (Early Chalcolithic, *c.* 5000–4600 BC).

Direct ^{14}C dating of materials associated with metallurgical materials at Belovode and Pločnik provided, for the first time, high-resolution evidence of the beginning, evolution and end of metallurgy at these sites and, most importantly, data regarding the cooperation between these communities concerning access to copper ores. For example, the close consistency of Belovode production evidence with copper implements from Pločnik, and similar correspondence of copper minerals from the earliest levels at both sites, dated to the 51st century BC, reinforces assumptions about their close connectedness and involvement in metallurgical knowledge sharing. Equally, the matching data from smelted copper in one horizon and malachite associated with another at Belovode implies consistent supply networks throughout the occupation of this settlement. Pločnik, on the other hand, exhibits a tight connection with Ai Bunar copper supply shortly after 4600 BC, based on information drawn from both provenance and complex networks data (see also Figure 9, Chapter 3 this volume).

The consistency of copper supply links between Belovode, Pločnik, Gornja Tuzla, Gomolava, Selevac, Vinča, Ruse and Durankulak also accords with the conclusions of the complex networks approach and underlines the super-connectedness of these nodes and their likely communication routes. Direct dating again offers a clearer picture of the development of these relationships, placing Belovode, Vinča and Pločnik at the heart of these links, having early communication with the population buried in Durankulak, and potentially being the community who migrated to Gornja Tuzla, a settlement which demonstrates the same knowledge of supply and technology as held by the Vinča and Belovode communities, soon after their villages were abandoned in haste. While the current state of information for the latter is limited, our hope is that the analysis of metal making practices and supply networks presented here opens more avenues for research into the lives of Vinča communities after most of the villages disappeared in around 4600 BC.

Finally, the complex debate on the origins of copper ores used for metal making in the Vinča culture and beyond shows the limitations of the current approach and the available datasets, despite the fact that this is currently one of the richest in the field of early metallurgy. Our hope is that new perspectives for more research and analysis will open, based on the directions we offer in this interpretation.

Overall, the Balkan Chalcolithic communities were utilising copper from at least six (or seven) copper deposits, two of which remain unidentified. These are: Majdanpek, Ždrelo, Ai Bunar, Medni Rid, 'Group of 16', 'Cluster #8' and potentially Rudna Glava (or an associated Co/Ni rich mineralisation). Of these, Vinča culture communities were not using metal from Medni Rid, while copper from Ai Bunar was only used during the extended occupation of Pločnik, or beyond *c.* 4600 BC. This list of deposits is based on the analysis of consistency of studied artefacts with known copper resources; it is important to emphasise that it is not exhaustive. Further, it appears that consistencies based on trace element analysis for two deposits, Rudna Glava and 'Cluster #8', are not reflected in the lead isotope abundance ratio plots, which may point to the need to obtain more samples, identify novel deposits, or look for alternative explanations. It would be interesting to see whether the trace element field of Ždrelo, or lead isotope analysis of an expanded sample set, if and when available, could potentially offer additional explanations for some of the sampled artefacts.

Another important point is that the set of data presented here adds *c.* 500 years to the age of previously analysed artefacts (Pernicka *et al.* 1993, 1997), which brings the analysed assemblage closer to the date of established activity at the ancient mine of Rudna Glava (*c.* 5000 BC). This site has been the subject of contentious debates, mainly revolving around the radiogenic lead in the isotope signature and the lack of metallurgical materials that show consistency with this source. The newly acquired trace element data from the unique set of Serbian copper slags dating between *c.* 5000 and 4400 BC might now have filled this gap (Figure 14), although the question remains of why this Co and Ni abundant source shows good consistency in the trace element and not on the lead isotope analysis front.

The current consensus is that Rudna Glava provides the closest elemental match with the most important metal production evidence, which, given the unpredictable nature of early copper smelting (and hence varied depletion / enrichment patterns in the newly formed slag phases), and taken alongside corresponding ^{14}C dates and, most notably, spatial proximity to the metal production sites, makes it (or an associated Co/Ni rich mineralisation) a potential candidate for copper exploitation from the beginning of the Vinča culture until its end. However, the radiogenic lead isotope abundance ratios of Rudna Glava will continue to present a problem for linking the deposit to any Chalcolithic metallurgical artefacts. The most important point is that confirmation of the exploitation of copper ores with elevated Co and Ni levels gives direction for further research for more ancient copper deposits in eastern Serbia, which might potentially bear a signature that exhibits stronger consistency with the archaeological artefacts analysed here. The abundance of copper mineralisations in this region was mentioned at the beginning of this chapter, and is reinforced here once again, particularly as the next chemically corresponding mining region (that is rich in Co and Ni), Medni Rid, lies between 800 and 1000 km away from the Vinča culture metal producing sites in Serbia.

Our research is far from complete and opens more questions than it provides answers, despite the enlarged dataset since the research in the 1990s. Our hope, however, is to inspire future investigations in this region that will help us understand the intricate web of connectedness of the world's first metal making communities.

Appendix B_Ch41

Appendix avaliable online as part of Appendix B at https://doi.org/10.32028/9781803270425/AppendixB_Ch41

The bibliographic reference for this chapter is:

Radivojević, M., Rehren, Th. and Pernicka, E. 2021. Metallurgical knowledge and networks of supply in the 5th millennium BC Balkans: Belovode and Pločnik in their regional context, in Radivojević, M., Roberts, B. W., Marić, M., Kuzmanović Cvetković, J., and Rehren, Th. (eds) *The Rise of Metallurgy in Eurasia*: 484–527. Oxford: Archaeopress.

Chapter 42

The pottery typology and relative chronology of Belovode and Pločnik: concluding remarks

Neda Mirković-Marić and Miroslav Marić

Introduction

The periodisation and the relative chronology of the Vinča culture has been the subject of many authors, such as Holste (1939), Milojčić (1949), Garašanin (1979), Chapman (1981), Schier (1996) and, more recently, Tasić *et. al.* (2015) and Whittle *et. al.* (2016). The widely accepted framework comprises two early and two late phases and these exist in almost all proposed chronological schemes (except Schier 1996). The key, middle period, known as the Gradac Phase, is associated with the appearance of early copper metallurgy.

Previously, it was observed that the Gradac Phase was hard to define for the eponymous and the classical variant sites but was easily discernible for the south Morava and Kosovo variants of the Vinča culture. Milutin Garašanin distinguished several variants: 'Classical' (Vojvodina and Serbia until the confluence of South and West Morava), 'South Morava', 'Kosovo', 'East-Bosnian', 'Transilvanian', 'Oltenian' and later, 'Srem-Slavonian' (Garašanin 1979: 184–191). It was subsequently recognised that even sites in central Serbia and Šumadija regions (Selevac, Medvednjak, Supska, Crnokalačka bara) showed characteristics similar to the South Morava variant and differed from the Danubian variant.

Garašanin considered that the Vinča Pločnik Phase of the South Morava variant ended with Vinča Pločnik I 'under the expansion of the Bubanj-Hum Chalcolithic culture' (Garašanin 1973: 102; Garašanin 1979: 189). Perić concluded, in 2006, that the Vinča culture persevered for the longest period in the middle Morava valley and the Šumadija region of central Serbia, where settlements like Divostin, Supska, Drenovac, Motel Slatina, and perhaps others such as Sevojno near Paraćin, continued to exist even after the settlement of Belo Brdo was abandoned (Perić 2006: 238).

Jovanović considered that the Gradac Phase in the south was tripartite and that the Neolithic in this region lasted longer than in the Classic region; he believed that Pločnik represented the late Gradac Phase III (Jovanović 1994: 1–11, Perić 2006: 238). He places the Gradac I Phase between Vinča B2 (or Vinča Tordoš II) and C1 phases (Vinča Pločnik I), followed by the Gradac II Phase which is comparable with the end of the Vinča D Phase (Jovanović 1994: 6).

According to previous AMS dating of several Vinča culture sites, the Vinča A Phase could be confined to the period between 5400/5300 and 5200 cal. BC; Vinča B between 5200 and 5000 cal. BC; Vinča C between 5000/4950 and 4850; and Vinča D between 4850 and 4650/4600 cal. BC. The Gradac Phase is dated to *c.* 5000 cal. BC (Whittle *et al.* 2016; Chapter 37 this volume). The Gradac Phase of the Vinča culture is parallel with: Gradeshnitsa III A-B-Poljanica Sava-Vidra (Boian III)-Pre-Cucuteni-Maritsa V, Dikili Tash II-Paradimi IV-Classic (Late) Dimini-Sakalhat-Tisza (transition)-Sopot B-Zeleziovice (Jovanović 2006: 223).

Both the Belovode and Pločnik sites are crucial to the understanding of the Vinča culture and its internal development, especially in the broader context of the social and economic changes during the early Chalcolithic and the rise of metallurgy in the Balkans. Based on the statistical analysis of material from Trench 24 at Pločnik and analogies from other Vinča culture sites, further confirmed by new C14 dates (Chapter 37, this volume), we can say that almost the complete span of Vinča culture is present on the site. Although not providing a sufficiently large sample to allow better typological analysis of types of cups and pitchers specific for this variant (Jovanović 2006: 221–235), the stratigraphy of Trench 24 fits more in the tripartite division of the Gradac Phase as proposed by Jovanović (1994: 1–11). This is confirmed by absolute dates from individual horizons. Certain characteristics of the material link the site of Belovode with the South Morava regional variant of the Vinča culture (Garašanin 1979: 188) more than with the Classical variant dominant in the Danube area nearby, although some features of this regional variation, especially jugs and beakers with handles, are well represented in the assemblage (Jovanović 2006: 221–235).

Gradac Phase pottery characteristics in the Vinča culture

According to Garašanin, the Gradac and Vinča Pločnik phases are characterised by intensive changes such as the development of early metallurgy and an increase in the number of elevated settlements. Like Garašanin, Jovanović linked the Gradac Phase with the start of major changes in all aspects of Vinča culture and social life, as reflected in the material culture of the period (Perić 2006: 235–250). Garašanin identified the main forms of pottery and figurines for this phase as plates with thickened rims (known in the South Morava and Kosovo variants from the earliest Vinča period); biconical jugs with one handle; amphorae with rounded shoulders and elongated flat necks; conical bowls on short conical or solid pedestals (already known in the Vinča B (Tordoš II) Phase in the South Morava and Kosovo variants); tripods (characteristic of the Vinča C (Pločnik I) Phase in the Classical variant); altars or lamps (?) with stylised deer heads; Vidovdanka type figurines (large standing, mostly female figurines with outstretched arms); spiral and meander ornamentation; bundled incised lines; and motifs that mimic metopic division.

In his own periodisation, Jovanović (2006) associates each phase with basic types of ceramic vessels as follows:

The Gradac I Phase (Rudna Glava, Layer 5 in Supska, Horizons V–VIII in Selevac, earlier horizon of Predionica, Crnokalačka bara) is identifiable by jugs with ribbon handles, biconical bowls with button shaped handles, and altars with deer head protomas. There are also plates with thickened rims, amphorae with elongated funnel shaped necks, and amphorae with short necks and emphasised shoulders. Other characteristic vessel types include a form with a narrow, concave neck decorated with spirals, and single handled jugs.

The Gradac II Phase (Horizons 3–4 in Supska, final stages in Predionica and Valač) is identifiable by the appearance of goblets with two handles in addition to the previously mentioned jugs.

The Gradac III Phase (Pločnik, Predionica, elevated settlements) is characterised by the occurrence of graphite ornamentation and the absence of altars with deer head shaped protomas. Two handled goblets and jugs with cylindrical handles are still present.

Later, Jovanović (2006: 224) summarises the Gradac types as:

Vessels with rounded shoulders and pronounced funnel shaped disproportionate neck and flaring rims. The recipient and the neck are generally decorated with parallel rows of shallow channels, sometimes polished; jugs with single ribbon handle, connecting upper part of the neck and the shoulder, but not over the rim. The lower part is predominantly angular, but rounded examples are known to exist; altars with four legs and shallow rectangular recipient with one or two deer heads; and voluminous figurines with lines that follow natural shape of the body with polygonal faces, Vidovdanka type.

Pottery characteristics at Belovode and Pločnik

The sites of Belovode and Pločnik represent two extraordinary windows onto the Vinča culture and, as new research and absolute dates indicate, each shows a complete range of the culture in each specific region. Aside from being linked with the phenomenon of the earliest copper metallurgy and the appearance of finished copper implements, the position of both sites within the framework of the Gradac Phase is significant and specific.

During the two campaigns of excavation of Trenches 18 and 24 at Belovode and Pločnik respectively, analyses of the ceramic material were performed on *c.* 50,000 fragments from each site, including 14,500 stylistic and typologically indicative fragments per site.

Pločnik

The development of pottery types at Pločnik follows the typical development of shapes during the life of the Vinča culture, with some characteristic elements distinctive of the South Morava and Kosovo variants, such as conical bowls with rim handles, beakers with two handles, and jugs. Also noticeable are conical bowls, and plates with thickened rims, which are present throughout all phases of Neolithic Pločnik. In the following summary, type numbers are indicated in parentheses.

In Horizon 5 (the earliest), biconical bowls are well represented; the majority having high cylindrical necks and rounded shoulders (114) and low cylindrical necks and rounded shoulders (112). Interestingly, bowls and plates with thickened, profiled rims (104b) are present in Pločnik from the oldest phases, unlike other sites where this occurrence has not yet been observed.

The same trends can be seen in Horizon 4 where bowls with thickened rims (104b) are the most numerous. A large percentage of the pottery in this horizon consists of biconical bowls with a rounded shoulder and low cylindrical neck (112) and with a high cylindrical neck and rounded shoulder (114), accompanied by other forms such as bowls with unequally high upper and lower parts (110), with a high cylindrical neck and square shoulder (113), and those with a concave upper part (115).

Horizon 3 is characterised by a wide variety of pottery types, abandonment of old types characteristic of Horizons 4 and 5, and the emergence of new forms that become dominant in the following, younger horizons. Among the biconical bowls, the majority have a concave upper cone (115), many with an oversized, concave upper cone (115a).

Other forms of biconical bowls are present in many variations but are few in number. These include types with a turned-in rim (117), with a low (111) or high (113) cylindrical neck and angular shoulders, biconical bowls with a high cylindrical neck and rounded shoulder (114), biconical bowls with a massive, square shoulder (118), and with cones of an equal height (110), and with a funnel-shaped upper cone (116). A small percentage of vessels in the horizon assemblage comprises biconical beakers with a cylindrical neck and rim handles, Gradac cups, square-shaped vessels, and jugs.

In Horizon 2 the biconical bowls are present in large quantities, the highest percentage being those with a turned-in rim (117), with a massive profiled angular shoulder (118), with a concave upper cone (115), and with a high upper concave cone (115a). Beakers (220) are not very numerous in the features and no jugs were detected. In the spits that correspond to Horizon 2, biconical shaped beakers with a conical neck and rim handles (221), Gradac cups (240), and jugs (340) were also present.

In Horizon 1, the most dominant vessels are bowls with a turned-in rim (117). The number of vessel shapes decreases, as does the extent of surface treatment and decoration. Compared to other horizons, amphorettae (320) and beakers with handles (220) are significantly represented in the whole assemblage, while jugs (340) are less numerous. Beakers are biconically shaped with a conical neck and ribbon handles (223), or a conical neck and handles on the rim (221), and with a cylindrical neck and ribbon handles (222).

Belovode

In Horizon 5 at Belovode, bowls with a short upper part (109), and those with equally high upper and lower parts (110) form the majority of the total bowl types in the assemblage, together with bowls with a short cylindrical neck and angular shoulder (111).

In Horizon 4, conical bowls are the most common, while those with a thickened rim (104a) or a profiled rim (104b) are less frequent. Rounded bowls are numerous and represented by spherical forms with a short neck (106) and with a spherical (107) or semi-spherical (108) body. Amongst the biconical bowls, those with a short biconical neck and rounded shoulder (112) are most numerous and dominate other biconical types.

In Horizon 3, conical bowls are again numerous and include the straight walled type (100), the slightly curved walled type (101), those with rim handles (102), with a thickened rim (104a), and those with a profiled thickened rim (104b). Rounded forms are present in small numbers including spherical bowls with a short cylindrical neck (106), bowls with hemispherical (107) and globular (108) receptacle. Bowls and plates with thickened rims are represented in Horizons 3 and 4. The highest proportion of biconical bowls are of type 117 with a turned-in rim, however there are many other types, such as the bowls with a short (112) and long (114) cylindrical neck and rounded shoulders, and those with funnel-shaped walls (116). Others appear in smaller proportions, e.g. those with a short upper cone (109), and with a short (111) or long (113) neck. A small percentage of biconical beakers with a conical neck and peripheral handles (221) also occur. Jugs are not represented, either in features or in horizon spits.

In Horizon 2, there are many bowls with a funnel-shaped profile (116), the most numerous having a turned-in rim (117). Other biconical bowls are also represented including those with a short upper cone (109); with a short (112), or long (114) cylindrical neck and rounded shoulders, and those with short (111) and long (113) cylindrical neck and angular shoulders. Biconical bowls with a concave upper cone (115) are not numerous. Jugs are represented by a type with elongated biconical profile, long cylindrical neck and a strap handle (341) and by a type with a sharp biconical profile and a massive conical neck and rim strap handle (344). Biconical beakers with a conical neck and rim handles (221) occur but with only individual examples. In the spits of the horizon there are biconical beakers with a cylindrical neck and ribbon-shaped handles (222) as well as biconical beakers with a conical neck and band-shaped handle (223).

In Horizon 1, the most numerous bowls are those with turned-in rims (117). Other biconical bowls are also present but only in small numbers, with types including: short (111) or long (113) vertical necks and angular shoulders; short (112) or long (114) vertical necks and rounded shoulders; concave (115) and funnel-shaped (116) upper cones, and biconical bowls with a massive angular shoulder (118). Beakers with a conical neck and handles on the rim (221) occur sporadically. In addition, in the layers corresponding to the horizon, there are biconical and conical beakers with a conical neck and ribbon handles (223). Jugs (340) are not represented in the features but are present in a small number of spits corresponding to the horizon. Amongst these are distinctive types with elongated biconical profiles, a long cylindrical neck and strap handles (341); with a sharper biconical profile, a massive conical neck and a strap handle (343); and jugs with a sharper biconical profile and a massive conical neck and strap handle on the rim (344). Several fragments came from spits S05, S06, S07 of the original trench extent and spits S02, S03, S05, S06 from the eastern extension of the trench.

Correspondence analysis

If the pottery type frequency is aggregated in occupation horizons as described in Chapters 13 and 28 of this volume, correspondence analysis can be applied to determine whether there is dependence between the pottery types (column data) and horizons (row data) on both sites. This technique allows the decomposition of total inertia (i.e. the variability) of the table to define the smallest number of dimensions which capture the data variability. This information can be illustrated on scatterplots which illustrate the distance between data points of the same pottery type based on the degree to which individual rows (horizons) have similar profiles (the relative frequencies of pottery types). The same principle can be applied in reverse, analysing similarities between column data based on row profiles (i.e. pottery type similarities per horizon). The origin of the axis on a scatterplot (0,0 value) is the centroid value, or the average profile value, which can be thought of as the position where no difference exists between profiles, i.e. where the profiles are homogenous (Greenacre 2007: 32). Accordingly, the greater the difference between profiles, the more the profile points will be spread out, away from the centroid value.

One more concept requires further explanation: the distance between points of different types, i.e. the row to column distance. The closer a row point is to a column point, the more distant from the average is the proportion of that column category on the row profile (and vice versa).

Correspondence analysis also identifies and describes outlier values which are profiles that deviate substantially from the others, having either a very small or very large frequency and/or very few categories present. In graphical representations, such points lie far away from the rest of the cloud of points, causing these to cluster together.

The same R procedure (Alberti 2013a, 2015) was used as in previous chapters without any additional alterations to data or the script.

The data to be analysed were aggregated according to the frequency of bowl types, which are the most chronologically sensitive pottery type in Vinča assemblages. This resulted in a table of 10 rows and 21 columns. Belovode horizons were marked as BELHOR 1–5, whilst Pločnik horizons were marked PLOHOR 1–5. The square root of the total inertia (i.e. the correlation coefficient between the rows and columns) was calculated to be 1.087, indicating significant dependencies in the data, the threshold being set at 0.20 according to Bendixen (1995: 576) and Healey (2013: 289–290). The chi-square test (Drennan 2009: 192–188) is also significant (χ^2= 2486.791, df = 180, p-value < 2.2×10^{-16}).

Determining how many dimensions need to be retained for the analysis is more challenging. Whilst the average rule (Lorenzo-Seva 2011), which retains all dimensions that explain more than the average inertia (in percentage), suggests three dimensions, the Malinvaud test (Saporta 2006) indicates seven dimensions. Although no strict rule exists, the number of retained dimensions is a trade-off between the increasing explained data variability (through keeping multiple dimensions) and the increasing complexity of the interpretation (when having more than two retained dimensions). In this case, three dimensions were retained, together explaining 79.3% of the variability based on the fact that they each explain more than 11% of the average inertia (a value obtained when 100% is divided by the number of rows or columns, whichever is smaller, minus 1).

It should be clarified that our interest in the data lies in interpreting the row points in the space defined by the columns, i.e. we seek to understand similarities between the two sites on the basis of the proportion of pottery types in each occupational horizon. Towards that goal, a contribution bar plot (in permills) of pottery types to the three dimensions retained must be inspected (Figure 1). It can be seen that BELHOR1 and PLOHOR1, 3 and 4 each have a major role in the definition of Dimension 1; BELHOR4 and PLOHOR3 have the same role in Dimension 2; and BELHOR 4, and PLOHOR1, 2, 3 and 5 define Dimension 3.

If the absolute contribution of bowl types to the inertia of average profile is examined (Figure 2) certain types of bowls show above-average values for specific horizons (illustrated by the different colour intensity of the red vectors). If a symmetric map (Figure 3) is considered, it is apparent exactly which bowl types are more closely linked to which horizon (Figure 3). The relative position of blue dots (horizons) compared to the positions of red triangles (bowl types) indicates clearly which bowls can be more—or less—closely connected with which horizons. For instance, bowl types 104c and 117 can both be found in Pločnik and Belovode Horizon 1 but are much more closely positioned to Pločnik Horizon 1, indicating their more than average appearance in this assemblage. Both horizons are highly correlated with Dimension 1 which explains 35.53% of variability of data. However, the positioning of bowl types 100, 107, 108, and 116, closer to Belovode Horizon 1, indicates that these are also more common than average in the horizon assemblage. It must not be forgotten that certain horizons under-contribute in all dimensions

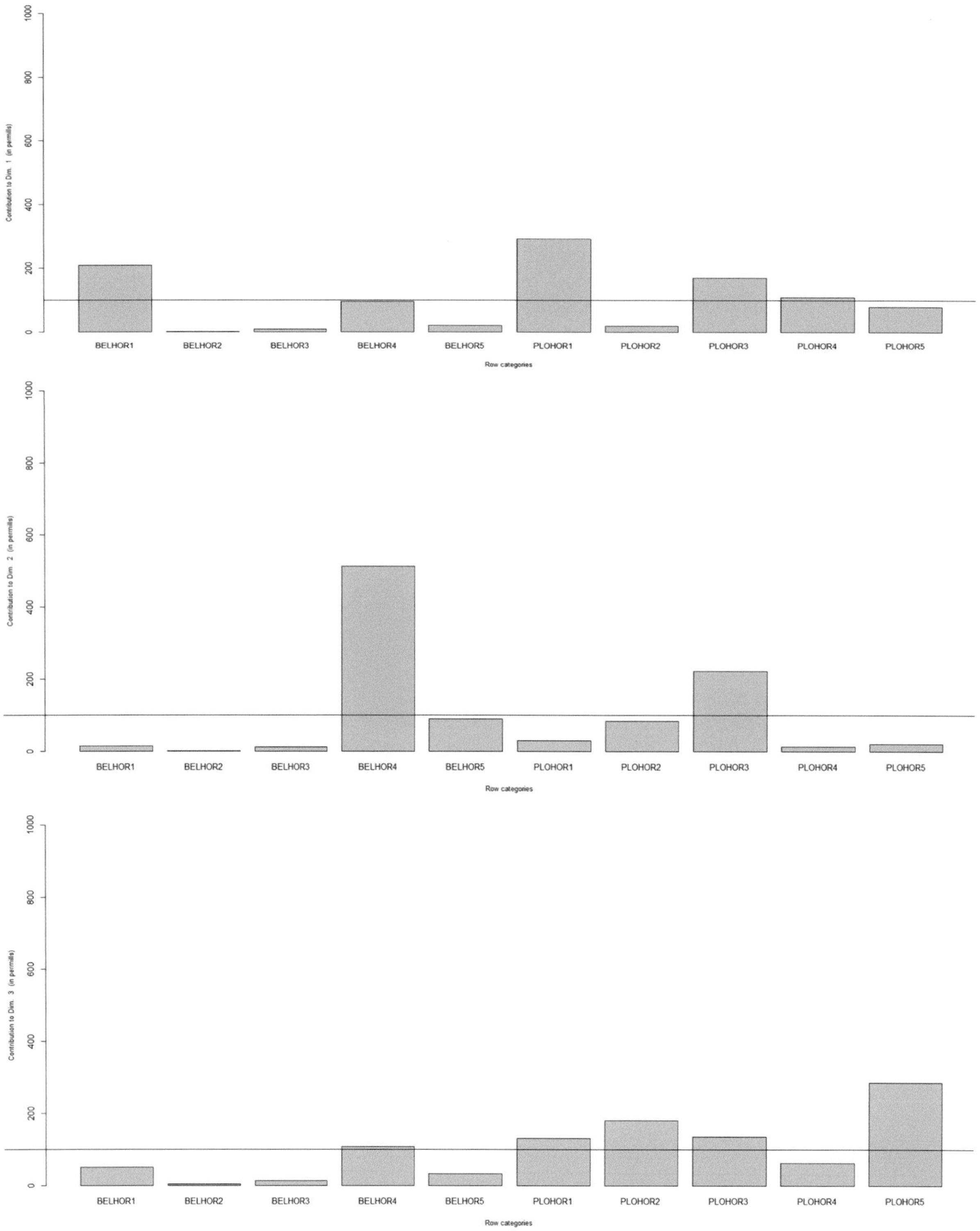

Figure 1. Contribution bar plot of three dimensions retained.

(Figure 1) and are also not equally represented (Figure 4) so their data should be examined with care.

Finally, if hierarchical clustering is examined (Figure 5), it is evident that a multiple cluster solution is present. The clustering detects that Horizon 1 for both Pločnik 1 and Belovode 1 belong to the same cluster, which seems to be the closest in profile to that of Belovode Horizons 4 and 5, most likely due to the above-average presence of conical bowls with both flat and curved walls (types 100 and 101). However, these bowls are not chronologically very sensitive, as they seem to appear in the Early Neolithic period and continue until the end. Another cluster comprises Pločnik Horizons 2 and 3, which appear mutually similar (and would fall into the Gradac I Phase according to Jovanović), but not as similar as

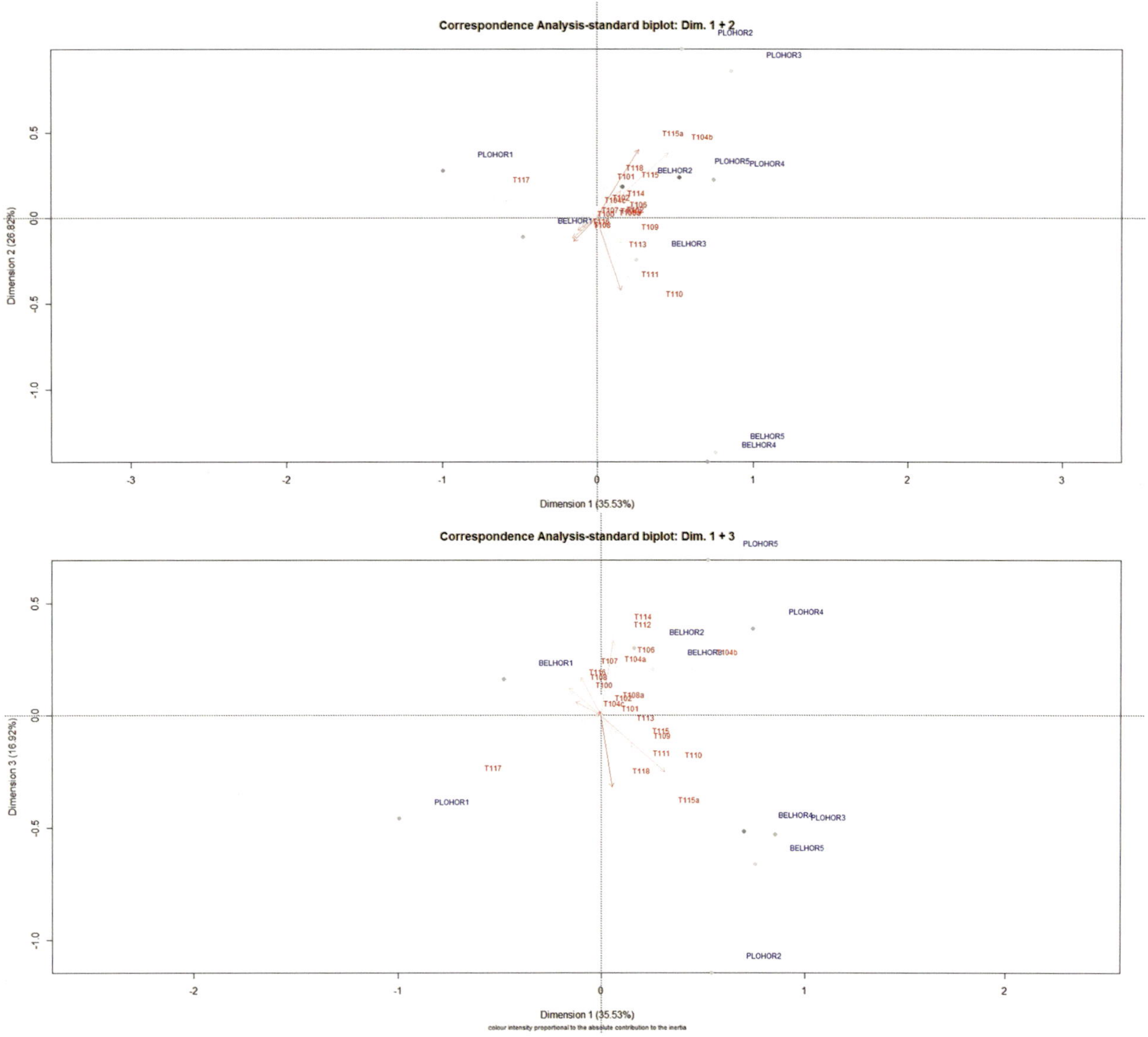

Figure 2. Absolute contribution of bowl types to the inertia of average profile.

the largest cluster to which they are related, consisting of Belovode Horizons 2 and 3 and Pločnik Horizons 4 and 5. This is not surprising, as these horizons are all absolutely dated somewhere between the Vinča B1 and Vinča Gradac phases, thus being slightly earlier in date.

Gradac Phase pottery characteristics

Towards the Gradac Phase we can identify no alteration in the established pottery technology that would indicate major changes during the transitional period from early to late Vinča phases (see Chapter 43, this volume). The main characteristic of this phase on both sites is the persistence of older bowl forms in parallel to the introduction of new types that become typical for the later phases of Vinča culture (bowls with turned in rims and bowls with concave upper parts). Specifically, on both sites we see the appearance of two-handled goblets and jugs with single handle. This kind of vessel is common in the South Morava and Kosovo variety of Vinča culture (Garašanin 1979: 188). Beakers with two handles and jugs are recorded at Crnokalačka Bara (Tasić and Tomić 1969: 41). There are also Vinča sites with these features in the Požarevac and Niš regions (Stojić and Jacanović 2008; Stojić and Jocić 2006), as well as in the Vranje region (Bulatović 2007) and the Leskovac region (Bulatović and Jović 2009: 129; Garašanin 1979: 188). Different types of jugs and goblets were recorded at the Divostin site in the Divostin II Phase (Madas 1988: 146–147) and at Rudna Glava (Jovanović and Ottaway 1976: Table XII). These are not common in the areas north of the Danube river, although there is a beaker with handles mentioned at the Gomolava site (Brukner 1980: 34). These vessel types are characteristic for the later period of the Vinča culture. Goblets characteristic of the Gradac Phase are detected in two graves at Durankulak in Bulgaria (Todorova 2002b: 30, Tables

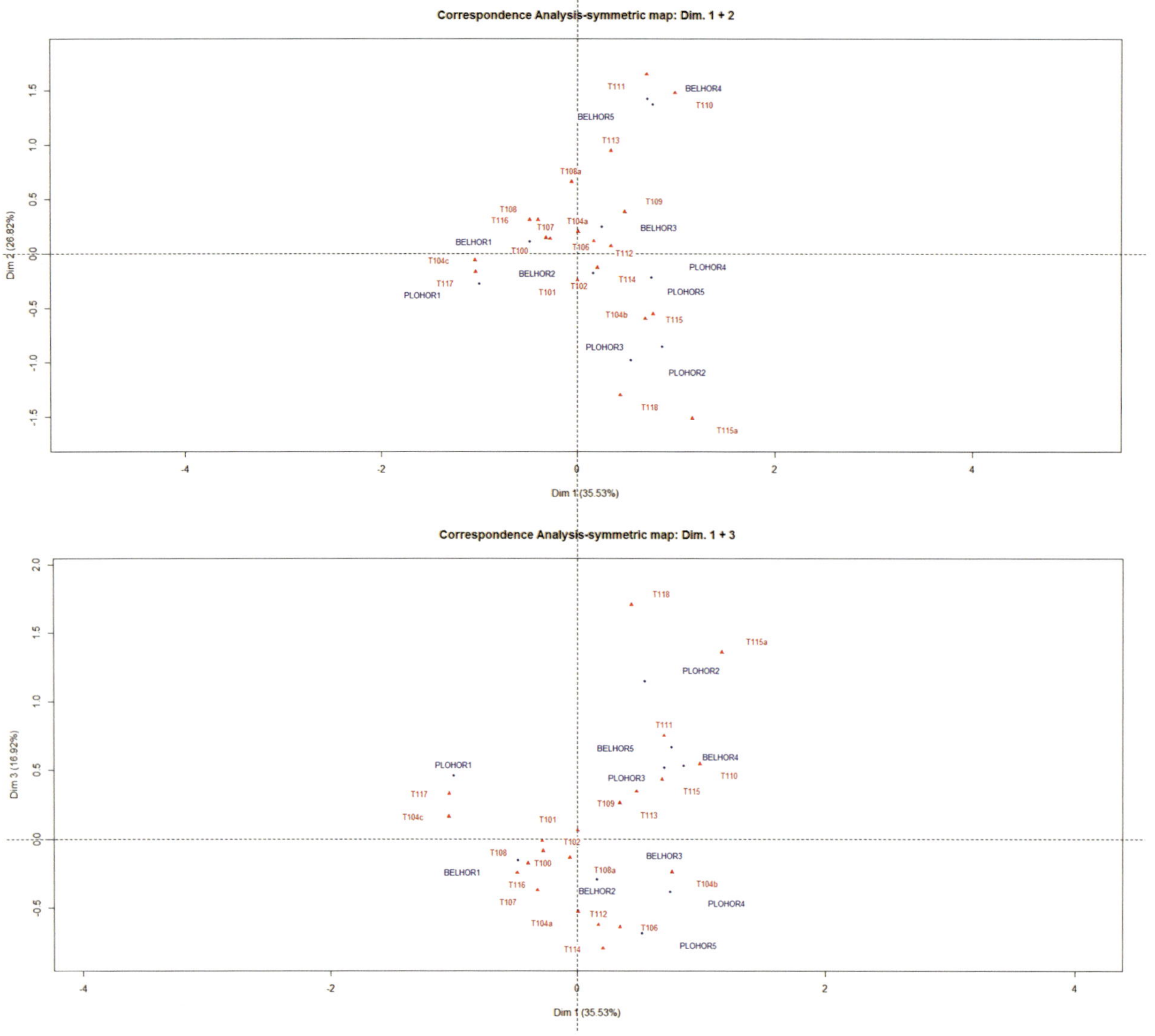

Figure 3. Correspondence analysis symmetric map of retained dimensions.

5, 12 and 35, Table 8, 10; Jovanović 2006: Figure 8). Jugs with handles are known in northeastern Turkey (Parzinger 2005: 62, Table 17; Jovanović 2006: Figure 8). There is also an increased presence of amphorae on both sites, particularly those with funnel-shaped necks. The presence of goblets, jugs and amphorae could indicate a change in drinking habits or liquid storage and management in the community during everyday life, as well as during communal feasts and events. It would be useful to perform residue analysis on these typical forms in order to clarify the matter. Bowls and plates with thickened rims occur in the older literature and are depicted as the main characteristic of the Gradac Phase and the transitional period between early and late Vinča culture. This is not highly visible at the eponymous site but is evident at other sites of the Classical variant of the Vinča culture such as Supska (Garašanin and Garašanin 1979). This phenomenon was considered proof that transition from the earliest to the later part (Gradac Phase) of the Vinča culture was more pronounced in the ceramic typology of the Morava valley than in that of the Danube and Sava valleys (Tringham and Krstić 1990b: 571). These vessel types are more common in the southern variant, e.g. Crnokalačka Bara (Tasić and Tomić 1969: 42), Gradac kod Zlokućana (Stalio 1972: Table XXVI 1; Garašanin 1979: 187), and Divostin (Madas 1988: 148) but are not exclusive to it, also being present in the Classical variant sites like Baranda-Trnovača or Vinča Belo Brdo in the Vinča B (Tordoš II) Phase (Jovanović 1965: 35; Garašanin 1979: 174), and at Selevac near Smederevska Palanka (Tringham and Krstić 1990b: 296, 571). The presence of plates/bowls with thickened rims is evident at both Belovode and Pločnik, with a slight difference: at Pločnik, they are present from the earliest phases until the end of the Neolithic life in the settlement whereas at Belovode they appear only from the onset of the Gradac Phase.

The absence (or very small number) of cooking pans on both sites deserves further mention. Cooking pans are not 'representative' vessels, nor typo-chronologically

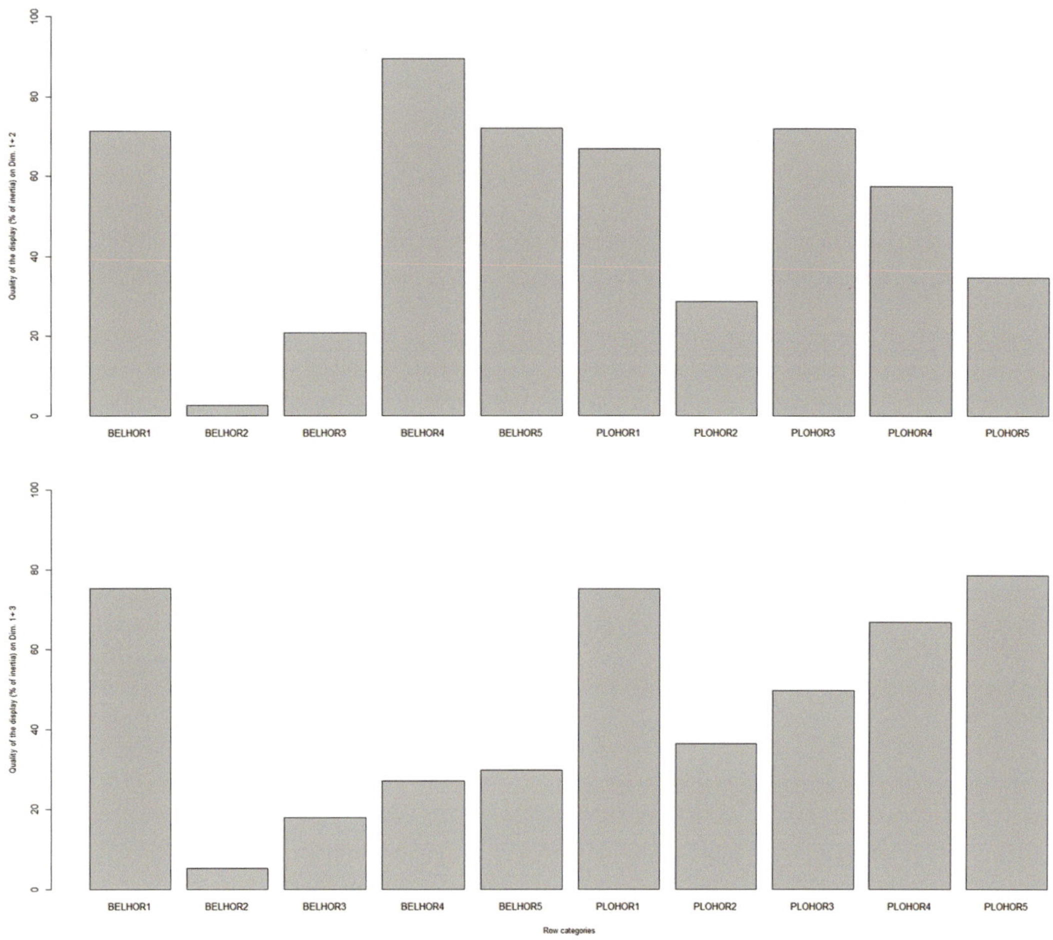

Figure 4. Quality of the display (percentage of inertia) of retained dimensions.

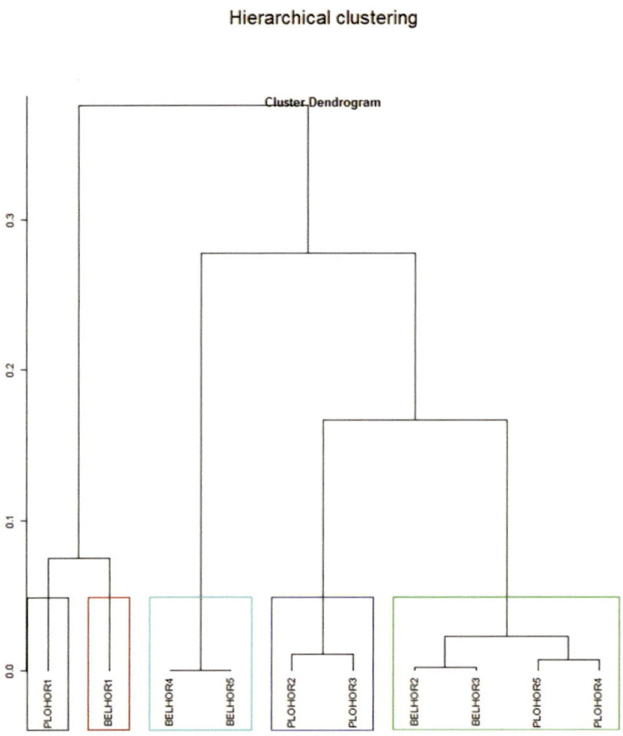

Figure 5. Hierarchical clustering dendrogram of Pločnik and Belovode horizons.

sensitive, so they were rarely mentioned in earlier reports. This type of vessel is very often recorded in the pottery assemblages of the Vojvodina Late Neolithic sites like Opovo (Tringham *et al.* 1992: 374), Gomolava (Brukner 1980: 20), Crna Bara (Garašanin and Garašanin 1958: Table 5), Trnovača-Baranda (Jovanović 1965: 33) as well as in the Classical variant of the Vinča culture, e.g. Selevac, near Smederevska Palanka (Tringham and Krstić 1990b: 313) throughout the whole duration of the Vinča culture. On the eponymous site of Belo Brdo in Vinča, this type of vessel is very common (Vuković 2013). At the site of Gruža near Grivac, Bogdanović notes 'shallow bowls with massive walls' (Bogdanović 1992: 42–43). For the late Vinča Pločnik Phase site in the village of Poljna, Spasić also mentions cooking pans (Spasić 1993: 183). There are no cooking pans from Grivac (Nikolić 2004), and just one from the Divostin houses of the Divostin IIb horizon (Madas 1988: 148). Still, the number of recorded cooking pans in the southern variants is smaller than in the northern Vinča region. We cannot exclude misinterpretations in the processing of this type of vessel, in that sometimes rim fragments of cooking pans can be interpreted as conical bowls (Vuković

2013: 130). Given that the cooking pots from Belovode have a specific fabric, however, it would be impossible to wrongly associate them with the conical bowls. In older excavations of the site, a small number of cooking pans were recorded but none were recovered in some horizons (Arsenijević and Živković 1998: 284). The fact that cooking pans were not so widely used south of the Danube may be related to the cooking habits of the specific topo-geographical regions.

Specific processes occur towards the end of occupation on both sites. Bowls with turned-in rims dominate the assemblage and goblet numbers increase, especially at Pločnik. The number of shapes present at this site also decreases, with just one type of vessel (bowls with thickened rim) being dominant. Standardisation of both shape and dimension is noticeable. This trend is also recorded on other Late Neolithic sites in the region, indicating changes in the organisation of production towards craft specialisation (see Chapter 43 this volume).

Ornamentation

During the early periods (Vinča A and B), the decorative techniques of barbotine and imprinted ornamentation occur mainly on the cooking ware, but other decoration styles like black topped, channelling and incised ornamentation are also present. Through all phases, surface treatment of serving vessels is dominated by burnishing and polishing. This is decorative but also has a functional purpose, reducing thermal shock resistivity and making vessel walls less permeable (Skibo 2013: 102).

Towards the Gradac Phase, a flourishing of style occurs with copious ornamentation, especially channelling. At Belovode this is dominated by channelling, predominantly slanted, but also horizontal and vertical. Polishing is well-represented in the form of lines but is only present in certain features. Black topped techniques and organised barbotine are both common; incised decoration is rare.

At Pločnik, ornamentation produced by all the techniques is widely found, but spiral channels are particularly common, especially on amphorae, with spirals on the shoulder and belly, while the neck is decorated with horizontal channels.

At Belovode, later phases of the Vinča settlement are dominated by polished lines, while the pottery from some features is dominated by channelling which may be horizontal, slanting, vertical, or in arcades. Engraved ornamentation rarely occurs.

Although at Belovode—as at some sites in central Serbia—the ornamentation is abundant throughout all phases of the Vinča culture, at Pločnik a decline can be observed towards the end, when surface treatment is limited to smoothing and polished bowls are rare. The same surface smoothing treatment also dominates the amphorae and beakers.

An exceptional decoration technique, graphite painting, is applied to the vessels from Horizons 3 and 2 at Pločnik.

Discussion and conclusion

The nature of the connections and relationships between the Starčevo and Vinča populations remains unknown and inadequately researched, remaining an open question even after a century of research. Are they different populations? Did Starčevo communities evolve into Vinča communities?

Dimitrijević (1974, 1979) explains the appearance of the Vinča culture as the consequence of migratory pressure within the Starčevo population. Other authors, e.g. Leković (1990), consider it the result of assimilation, whilst others still support the notion of a cultural continuity between the two populations. There are also theories of gradual migrations (see Garašanin 1979: 200–203). Some data indicates the existence of a much more frequent mixing of Starčevo and Vinča traditions than published references would suggest, such as at Crnokalačka Bara (Tasić and Tomić 1969: 23), Gornja Tuzla (Vetnić 1974: 148), Drenovac (Vetnić 1974: 137), and Pavlovac (Garašanin and Garašanin 1958; Vuković 2015: 664–667).

Although there is evidence of Starčevo occupation at Belovode, pottery from the earliest phase at Pločnik exhibits different technological characteristics with respect to paste recipes. However, questions surrounding the succession of communities and their relationships cannot be answered based solely on the finds recovered from Trenches 18 and 24. The specifics of the southern variant of the Vinča culture, evidenced through the different vessel repertoires in the site assemblages, will only be explained through larger sample and wider analyses.

Jugs and goblets are specific to the southern variant of the Vinča culture, particularly in its later stage. These appear at both Belovode and Pločnik from the onset of the Gradac Phase. The lack of (or minute quantities of) cooking pans can also be considered characteristic of this southern Vinča culture variant. This could be evidence of local, perhaps even geographical, differences in cooking and serving habits, which cannot be directly linked with the development of metallurgy. Standardisation of shape and dimension are hints of other changes in the organisation of production toward the craft specialisation during the final stages of Vinča culture (Chapter 43 this volume).

The relationship between Vinča communities and the Chalcolithic Bubanj Hum-Krivodol-Salçuta culture is also under-explained. Older theories of population migrations and Bubanj Hum influences (Garašanin 1979: 205; Perić 2006: 237) have been replaced by notions of evolution from the late Vinča foundations, marked by new economic strategies or technological innovations (Greenfield 2010; Sherratt 1981; Parkinson *et. al.* 2004). There are striking similarities between the repertoires of the late Vinča (predominantly south variant) and early Bubanj Hum Ia assemblages, including bowls with turned-in rims, pedestal bowls, and beakers. The same is true with respect to surface treatment (polishing) and the reduction atmosphere firing. Certain elements of similarity can also be evidenced in figurines (Stojić and Jocić 2006: 33). It is our firm belief that a more in-depth technological analysis, encompassing the whole of technological knowledge with regard to vessel-making, will provide greater insight and move us toward solving this very important archaeological question.

The bibliographic reference for this chapter is:

Mirković-Marić, N. and Marić, M. 2021. The pottery typology and relative chronology of Belovode and Pločnik: concluding remarks, in Radivojević, M., Roberts, B. W., Marić, M., Kuzmanović Cvetković, J., and Rehren, Th. (eds) *The Rise of Metallurgy in Eurasia*: 528–537. Oxford: Archaeopress.

Chapter 43

Pottery technology at the dawn of metallurgy in the Vinča culture

Silvia Amicone, Miljana Radivojević, Patrick Quinn and Thilo Rehren

This chapter summarises the macroscopic and microscopic analyses of pottery sherds from the sites of Belovode and Pločnik, presented in Chapters 14 and 31, and provides insight into different technological traits in order to aid reconstruction of pottery making recipes in these two Vinča culture communities. Using a multi-pronged scientific approach, we reconstructed routines of raw material acquisition and processing, techniques of forming and finishing vessels, firing conditions and organisational aspects of pottery production. The possible non-local production identified in this research is also considered in order to understand the dynamics that shaped pottery circulation in these prehistoric communities (e.g. Quinn *et al.* 2010). These results also contribute significantly to the previous technological studies carried out on Neolithic pottery from sites in the central Balkans (Figure 1) (e.g. Dammers *et al.* 2012; Kaiser 1984, 1989, 1990; Kaiser *et al.* 1986; Kreiter *et al.* 2009, 2011, 2013, 2017a, 2017b, 2019; Spataro 2014, 2017, 2018; Szakmány *et al.* 2019).

Combining the results of the pottery technological study together with archaeological information presented in the other chapters of this monograph (Chapters 14 and 29), this synthesis offers a contribution to the understanding of Vinča pottery across different phases of its development. Also, the results of this study help us explore the technological links between the dark-burnished and graphite-painted pottery with extractive metallurgy of copper that emerged in the Vinča culture (Amicone *et al.* 2020a). We test the hypothesis that the ability of potters to exert sufficiently close control

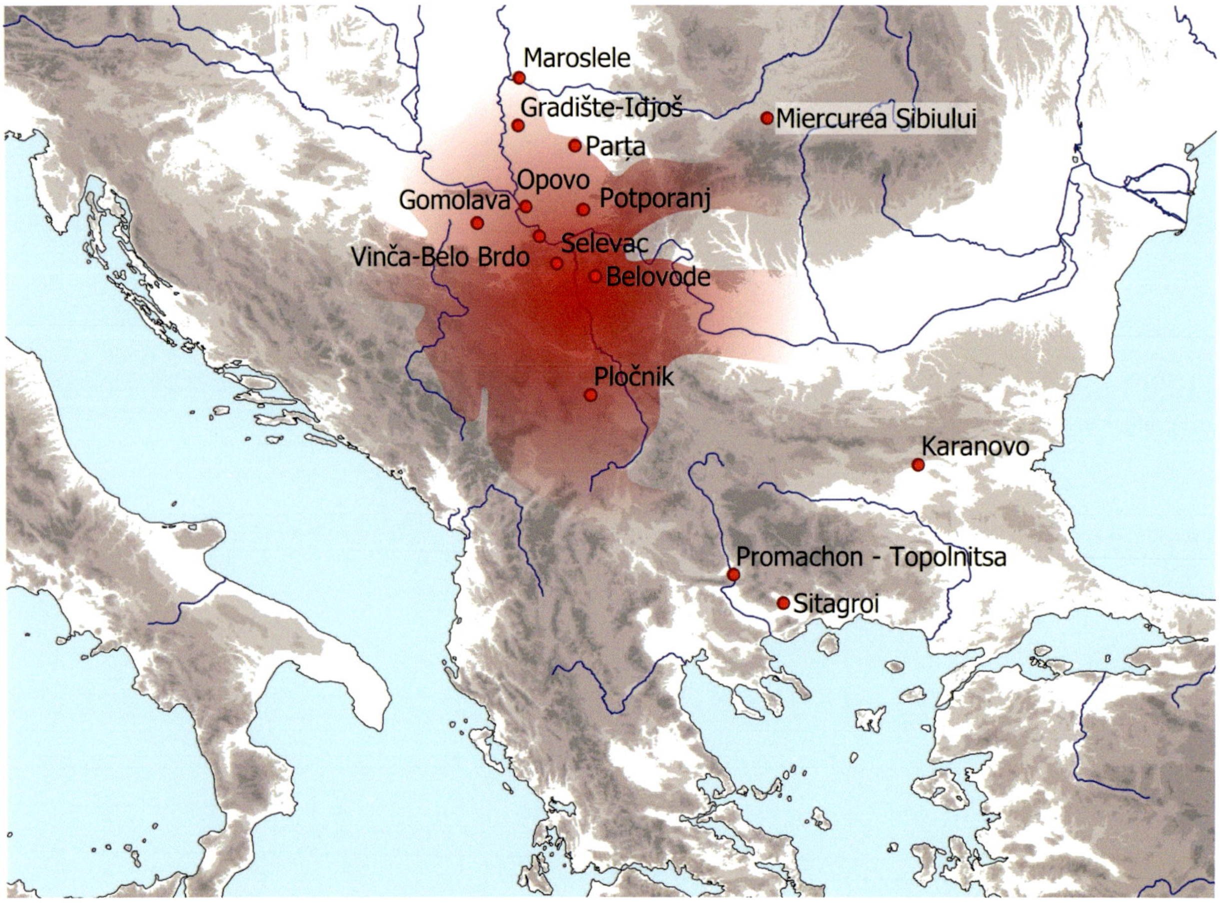

Figure 1. Distribution of the Vinča culture throughout all its periods (shaded) and location of sites mentioned in this study (map by Lars Heinze and Jugoslav Pendić).

over the redox atmosphere in a two-step firing process necessary to produce graphite-painted pottery could indeed provide a long sought-after link between pottery and metal making technologies.

Raw material selection

The first major theme highlighted by the combination of petrographic analysis on the pottery and geological samples from Belovode and Pločnik (Chapters 14 and 29) was raw material selection. The results revealed that potters at both sites selected clay immediately available from their surrounding landscapes. A worldwide study of distances travelled to acquire clays and temper sources (Arnold 2000: 343) suggests that the maximum threshold distance for raw material procurement lies at 7 km from the potter's production place (assuming that their own bodies are used for transport). Nevertheless, most potters did not travel more than 1 km to obtain clays and tempers under these conditions. At both sites, clay sources suitable for pottery making can be found within 1 km of the supposed boundaries of the settlements. These patterns have also been observed at other Vinča settlements (cf. Amicone *et al.* 2020a).

The potters of Belovode selected two different types of clay sources: the first was a non-calcareous Neogene secondary sandy-clay (see Geological Map in Chapter 14, points 2, 12, and 14) whose coarse fraction consists of minerals derived from the weathering of low-grade metamorphic rocks (probably metasedimentary rocks). This was used throughout the period in which the settlement was inhabited (Figure 2). The second type is a Neogene calcareous sandy clay source, rich in calcite and shells (see Geological Map, Chapter 14, point 1), exploited from Horizon 3 onwards (*c.* 5140–4859 cal. BC, 95% probability, Chapter 37). The Neogene formation also includes sandy layers that could have been a suitable source for temper.

Very good primary clay sources were available in Pločnik, deriving from Cretaceous formations rich in conglomerates, sandstones, and mudstone. These types of clay sources originate from the weathering of sedimentary rocks and therefore contain fragments of these types of clasts and related minerals. The geological sample showed that different outcrops of this clay source differ in coarseness, and both calcareous and non-calcareous clays were found. Throughout its occupation (Figure 3), the inhabitants of Pločnik selected non-calcareous clays which have good plasticity and need only minimal processing (e.g. cleaning or tempering). The calcareous clay sources were used only for loom-weights and plastered floors.

Alluvial clay from the Toplica river does not appear to have been used in pottery manufacturing. Given that this source of raw material was immediately available in the vicinity of the site, this is difficult to explain. One argument could be that the course of the river today is not the same as during the Neolithic period, as illustrated by the fact that a substantial portion of the site has been destroyed by its meandering. At the time the site was occupied, however, the river should have been close to the settlement and would have very likely bordered its southeastern extension.

Other factors could lie behind the decision not to use fluvial sediments for producing pottery: the clay sources from the Cretaceous formations may have been perceived to be more suitable or there may have been socio-political reasons that are today impossible to determine or reconstruct. It has been noted, however, that the distribution of resources in a given environment presents people with different possibilities and influences their movement within that environment (Michelaki *et al.* 2014). In this sense, the existence of multiple resources in single area allows people to simultaneously undertake a variety of tasks in the same general place. It is possible, therefore, that the inhabitants of Pločnik oriented themselves towards a specific clay source perhaps, for example, because that location also offered other important resources. Gosselain and Livingstone-Smith (2005) point out several cases in which potters discovered suitable clay sources by undertaking other tasks, especially those which require a close observation of soils (e.g. fetching water, digging foundations). In general, potters often selected raw material from locations that offered opportunities to undertake a variety of tasks at the same time. It is therefore important to note that the hills surrounding Pločnik are also rich in sandstone that was employed in the ground-stone industry. It is likely that the task of collecting raw materials for both pottery making and for the ground-stone industry were somehow merged, or at least linked, and took place in areas that offered opportunities to find suitable raw materials for both crafts. A similar logic could also explain the fact that some sherds are marked by a fabric rich in mica-schist (fabric PL-B, Figure 2f, Chapter 29). Inhabitants most likely exploited outcrops close to mica-schist quarries that were found about 6–7 km from the site (Geological Map in Chapter 29, Point 9). This mica-schist was also employed in the local ground stone industry (Dimić and Antonović Chapter45, this volume). In Horizon 1 (*c.* 4631–4231 cal. BC, 95% probability, Gradac II–III, see Chapter 37, this volume), the latest phase of the settlement, one third of the sherds are characterised by fabric PL-B, whilst previous horizons contain only a few such samples.

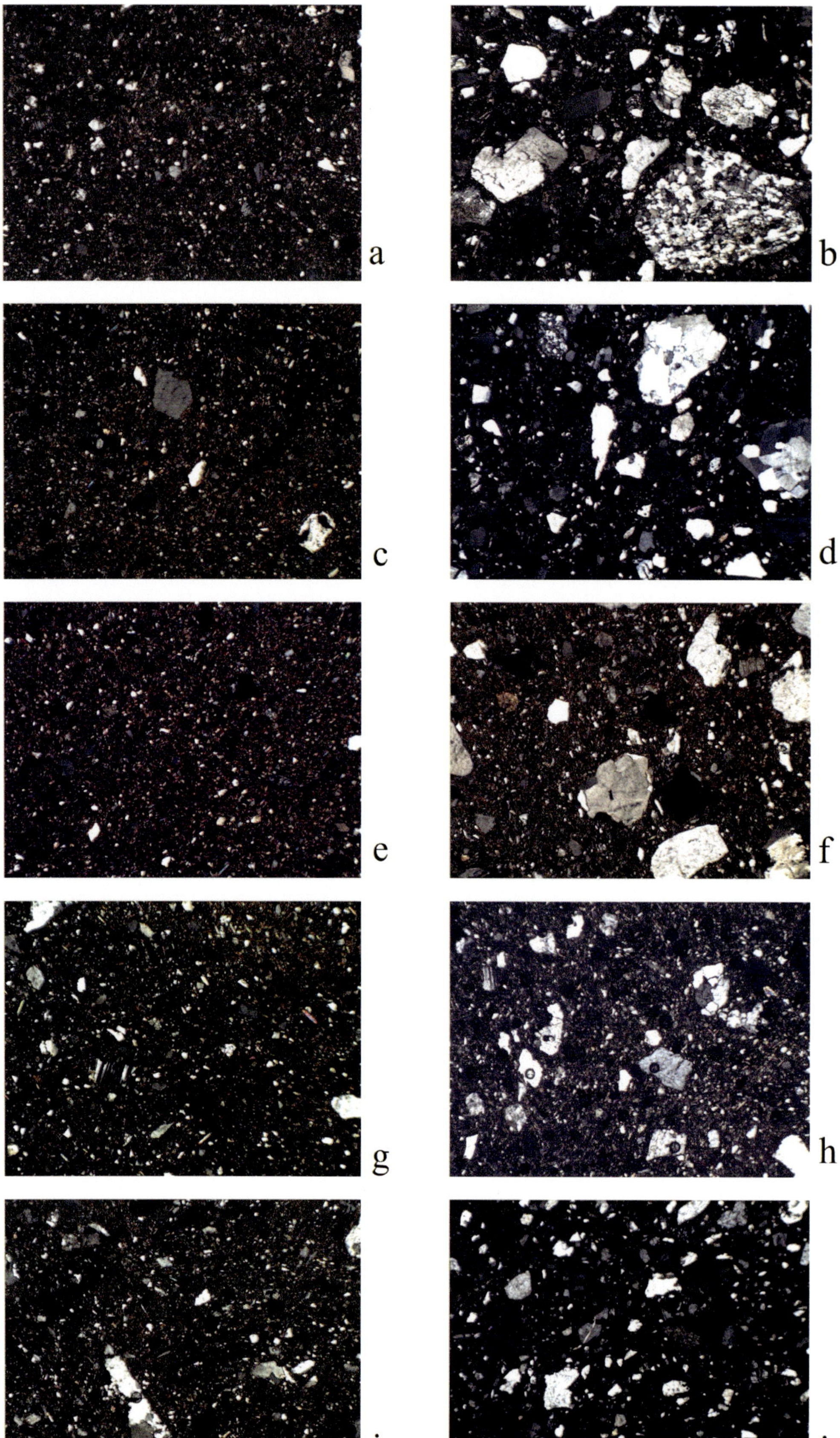

Figure 2. Thin section photomicrographs of fabric BEL-A in different Horizons. a) and b) Horizon 1; c) and d) Horizon 2; e) and f) Horizon 3; g) and h) Horizon 4; i) and j) Horizon 5 (image width = 3 mm).

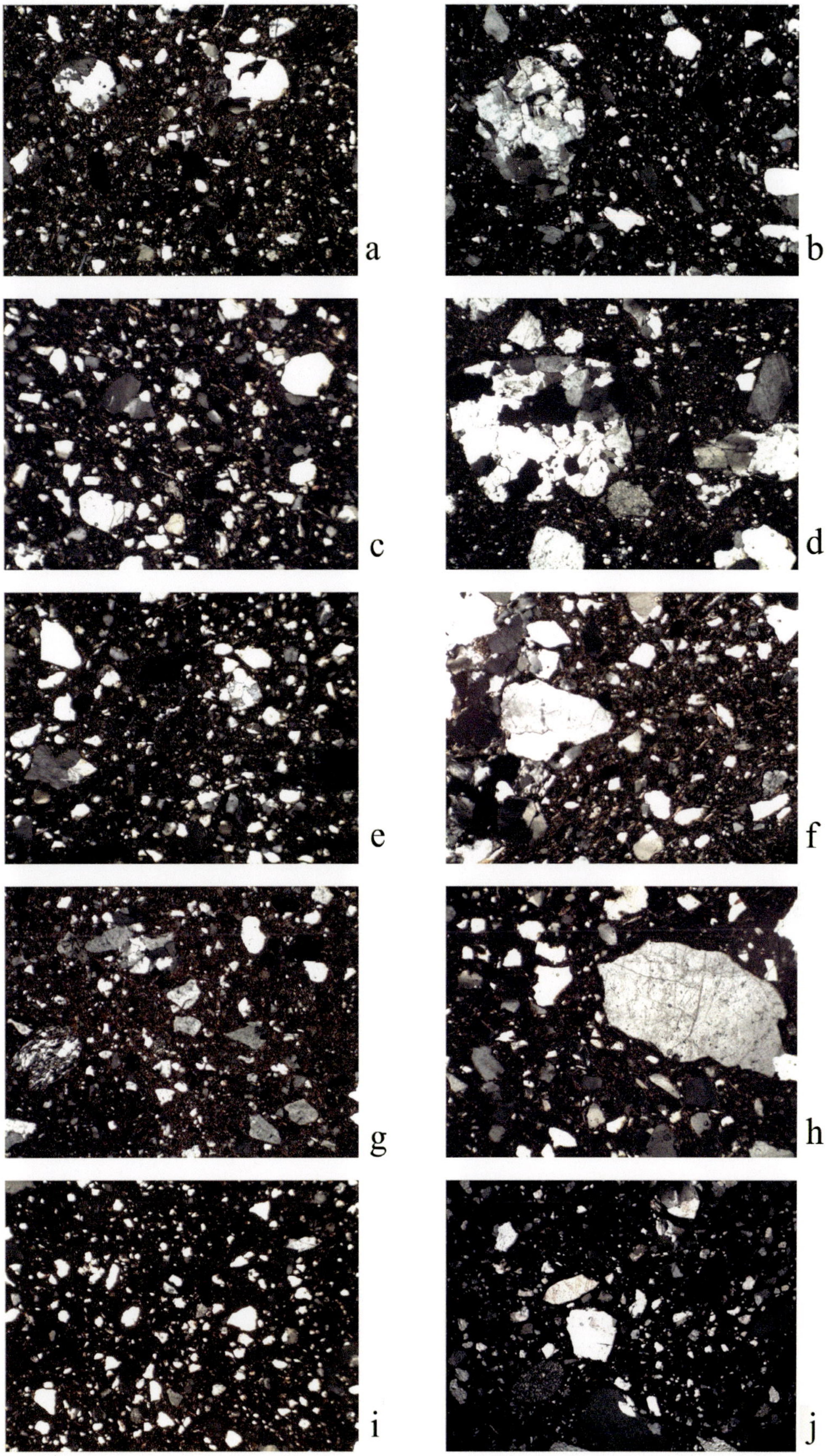

Figure 3. Thin section photomicrographs of fabric PL-A in different horizons. a) and b) Horizon 1; c) and d) Horizon 2; e) and f) Horizon 3; g) and h) Horizon 4; i) and j) Horizon 5 (image width = 3 mm).

Raw material processing

In both Belovode and Pločnik, clay processing seems to have been minimal. Tempering was not a very common practice and often barely recognisable. In Belovode this practice is attested to, especially in the earliest horizons, by the inclusion of rock fragments and minerals or organic tempering (e.g. chaff). In the earliest phases of the settlement, Horizon 5 and 4 (*c.* 5648–5054 cal. BC, 95% probability, Starčevo–Vinča A, Chapter 37, this volume), it is already possible to differentiate three different paste recipes which employ the same type of clay, but differ in terms of tempering: one is defined by mineral tempering, one by plant tempering and one is untempered. Plant tempering is associated with a specific pottery making recipe that is marked by barbotine decoration, oxidising firing conditions and, possibly, slightly lower firing temperatures. This pottery recipe is associated with material of the early/middle Neolithic Starčevo culture. In Belovode this pottery making tradition is only present in the earliest horizon, but probably co-existed and overlapped with other traditions marked by mineral tempering or the absence of tempering, which are associated with more typical Vinča culture pottery. From Horizon 3 onwards (*c.* 5140–4859 cal. BC, 95% probability, Vinča B1–B2, Chapter 37, this volume) mineral tempering is still used, but very sporadically.

In Pločnik, the bimodal grain size distribution noted in some of the coarse clay samples assigned to fabric PL-A (Figure 2, Chapter 29, this volume), from Horizons 5 and 4 (*c.* 5389–4976 cal. BC, 95% probability, Vinča A2–B2, Chapter 37) could be evidence of the intentional addition of aplastic material. Also, coarse samples assigned to fabric PL-B (with mica-schist) could have been tempered. Otherwise, there is no strong tradition for tempering observable at this site. Conversely, the results of the analysis carried out for other Vinča settlements (Amicone *et al.* 2020a) show that tempering is a quite common practice among potters who used a variety of different materials including river sand, plant material, rock fragments and grog.

The two sites also show differences in the relationships between pottery fabric, shape, and function. In the analysis carried out on the sherds from Pločnik it was possible to observe a slight tendency for coarser fabrics to be used for vessels characterised by thicker walls (e.g. amphora and pithoi), however a strict relationship between fabric and vessel type was excluded. Very interestingly, in terms of grain size distribution of the inclusions, the fabric used to produce dark-burnished bowls did not differ significantly from those used to produce other types of pots. At Belovode, however, we can observe a tendency towards a different selection and preparation of the paste according to the type and surface treatment of the vessels. Bowls, for instance, which are in most cases burnished or polished, have a fine fabric (fabric BEL-A2, Figure 2a, Chapter 14, this volume) that could indicate that a finer clay source was used to manufacture these vessels, or that the starting raw material was freed of its coarser fraction through sieving or levigation. By contrast, the majority of the cooking pots were characterised by a paste rich in shell fragments, microfossils and calcite (fabric BEL-B, Figure 4a,b, Chapter 14). The selection of clay sources rich in naturally occurring calcite or calcite tempering in connection with cooking pots is not surprising, since the presence of calcite could have a positive impact on the vessel's thermal shock resistance (Rice 2015: 324).

Similar patterns have been noticed at other Vinča sites (Figure 1) such as Opovo, Selevac, Gomolava (Kaiser 1984, 1990; Tringham *et al.* 1992), Parța, Miercurea Sibiului Petriş and Vinča-Belo Brdo (Spataro 2014, 2017, 2018). Kaiser, for example, noticed a correlation between pottery fabric, vessel form, and size, because each vessel type tended to be associated with a particular fabric and less strongly with others. In addition, at Potporanj (near Vršac) and Gradište-Iđoš (near Kikinda), which are both located in Banat (northern Serbia), a more systematic correlation between fabric and vessel shapes seems to be present (Amicone *et al.* 2020b). At Belovode, bowls which are normally dark and burnished have a very fine fabric that was probably obtained through cleaning (e.g. by levigation) of clay. Storage and cooking vessels (pots and pans), however, have a very coarse-tempered fabric.

Forming techniques, surface finishing and decoration

The combination of macro and micro analysis revealed that coiling was widely applied for the manufacturing of vessels at Belovode and Pločnik (Roux 2017). The use of this technique was recognised macroscopically by the presence of vessels with fractures preferentially aligned parallel to the rim plane and via the identification of joins between coils visible on the surface of numerous vessels. This was also confirmed by the distribution of inclusions and voids observed in the samples examined via thin section analysis (Quinn 2013: 179). Pinching techniques were employed at both sites, but only in the manufacturing of small vessels and sometimes in the forming of the bottoms of bowls. There is no evidence of the use of moulds, while the slab technique seems to be employed—at least at Pločnik—when producing vessels with a squared mouth or to produce the base of vessels. Forming techniques seem to be evenly employed across different horizons at Belovode and Pločnik. Similar patterns were also noticed at the sites of Selevac and Gomolava by Kaiser (1984: 208–217).

Different techniques of decoration were used at both sites including barbotine, incising (often filled with calcite and cinnabar), impressing, channelling, painting, and applied decorations. These were very common techniques of pottery decoration at that time in the Balkans (Bonga 2013). The development of decorative techniques follows similar trajectories at both sites, with an increasing presence of channelling corresponding with the emergence of the Gradac Phase (*c.* 5000 BC) and the gradual disappearance of barbotine decoration over time. Pločnik, however, shows some differences in the presence of graphite-painted pottery from Horizons 3 and 2 (*c.* 5036–4621 cal. BC, 95% probability, Gradac I, cf. Chapter 37, this volume).

Surface treatments such as smoothing, burnishing and polishing (Martineau 2010) occur at both Belovode and Pločnik. Burnishing and polishing are applied to different types of vessels, but they are predominant in the production of dark-burnished bowls. This surface treatment is very common and became widespread during the Middle and Late Neolithic of southern Europe, and in other material cultures such as Danilo, Hvar, Vinča and Karanovo (Spataro 2017). It is interesting to note that Pločnik Horizon 1 (*c.* 4631–4231 cal. BC, 95% probability, Gradac II–III, see Chapter 37, this volume) is marked by a significant decrease in burnishing, polishing and decoration techniques.

Pyrotechnology and colour

The reconstruction of aspects of pottery pyrotechnology at the studied sites is challenging since there is no clear evidence for the firing installations used. There is generally no conclusive evidence for pottery kilns in Vinča culture settlements. Features that could be compatible with these types of pyrotechnological installations were found, for example, in Trench 6 at Belovode (Šljivar *et al.* 2011), represented by three round firing structures, a fireplace and an assemblage of clay figurines (bulls), all of which the excavator ascribed to ritual-cult function. Yet, it is not common in Vinča households to find groups of round structures for firing next to each other, and these indeed could indicate the presence of kilns. Similar clusters of round structures were also identified from anomalies detected during the geophysical survey of Pločnik (see Rassmann *et al.*, Chapter 24, this volume); future excavations may show whether these are dwelling debris or, indeed, are much sought-after examples of kilns on a Vinča site.

In the absence of more secure evidence, it might be plausible that pots were simply (or additionally) fired in pits dug in the ground. As shown in a firing experiment carried out in Serbia (Svoboda *et al.* 2004/2005; Vuković 2018b), it is possible to produce the entire range of pottery found in Vinča sites without using a proper kiln. Despite the lack of clear evidence for firing installations in the archaeological record, it is still possible to reach preliminary conclusions based on results presented in Chapters 14 and 29, especially based on the colour of the vessels and the results of XRD and SEM analyses.

Most of the pottery from Belovode has a pale yellow to reddish yellow surface colour, indicating firing under oxidising conditions. Only the vessels with burnished or polished surfaces appear to be fired differently (in reducing conditions). By contrast, 60% of the sherds from all five horizons at Pločnik exhibit grey and black shades associated with burnishing or polishing and, more rarely, with graphite decoration (Amicone 2017: 190). This preference for dark pottery persists into Horizon 1 (Gradac II–III), although there is only occasional evidence for burnishing at this point. Some black-topped sherds occur at both sites, for example in Horizons 5 and 4 (Vinča A2–B2).

The colour of most of the examined sherds varies in cross-section (Figure 4) indicating variable atmospheric conditions during firing and/or a short firing duration. The fabric of dark-burnished vessels is never homogeneously grey. Rather, it is usually characterised by a reddish core and a darker grey matrix towards the surface; the boundary between them is not sharply delineated. Vessels fired completely in an oxidised atmosphere are also attested, with the cooking pots at Belovode being a good example. However, reddish sherds often come from the interiors of burned features and might therefore have turned red during destruction events. Vessels from other Vinča sites are more often grey in colour, and larger preserved fragments frequently display a range of shades from pale to dark grey (Chapman 2006; Kaiser 1984). Fragments from Pločnik decorated with graphite derived from building Horizon 3 (*c.* 5036–4951 cal. BC, 95% probability) and 2 (*c.* 4927–4621 cal. BC, 95% probability)(see Chapter 37), corresponding to the Gradac Phase I of the Vinča culture, exhibit a relatively homogeneous dark to light grey fabric in cross-section (Amicone *et al.* 2020b).

There are two common ways to achieve a grey or black colouration of pot surfaces during the firing process: iron reduction and carbon black (also called the smoking and smudging technique). In the first case, the black colour is connected with the nucleation of dark coloured iron-oxides (e.g. magnetite). In the second, the black colour derives from carbon particles that cover the vessel surface during firing and penetrate the pores (Jones 1986: 762–763). However, the XRD analysis carried out on samples from both sites (Table 2 in Chapters 14 and 29 respectively) did not reveal the presence of magnetite on the surface of black coloured pots, thus ruling out the iron reduction methods. However, carbon was detected on the surface of several dark-burnished sherds from both Pločnik and Belovode via μ-Raman analysis, suggesting that their lustrous black surfaces were the result of the carbon black smudging technique (Amicone *et al.* 2020b).

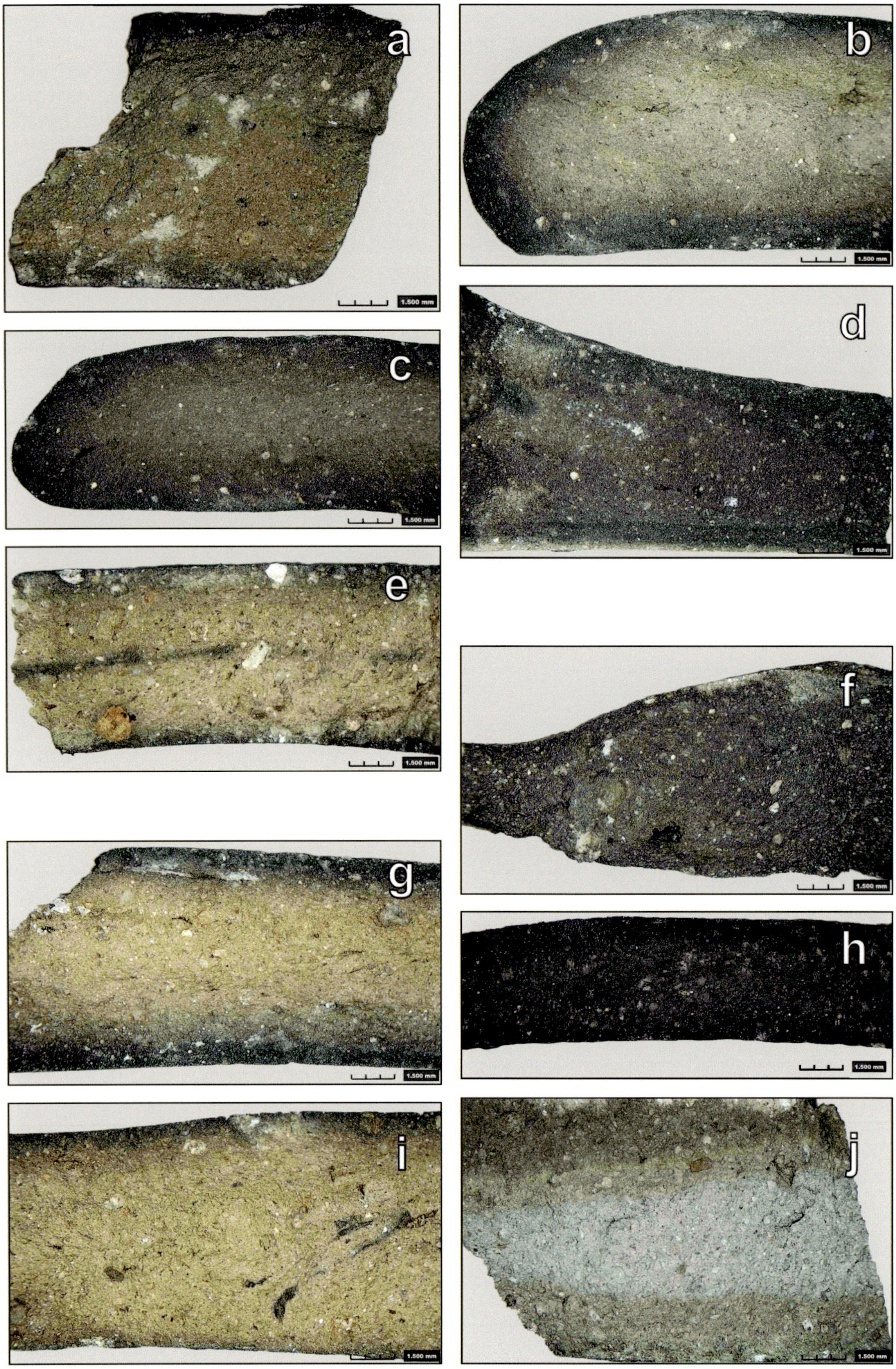

Figure 4. Selected dark-burnished and graphite-painted pottery sherds from the Vinča culture sites of Belovode and Pločnik seen in fresh break, revealing the fired colour of their fabric and the presence of firing horizons: a) BEL 289; b) BEL 94; c) BEL162; d) BEL 219; e) BEL 299; f) PL 24-129 (graphite-painted); g) PL 24-107; h) PL 24-124 (graphite-painted); i) PL 24-288; j) PL 24-145.

The nucleation of dark iron oxide with a lower oxidation state (e.g. magnetite) happens at temperatures above 700 °C in non-calcareous clays and above 800 °C in calcareous clays (Maritan 2004). Firing temperatures are not, however, of particular relevance in smudging, which is normally done towards the end of an oxidising firing process (when temperatures are lower) by smothering the burning fuel and pottery with substances that produce smoke, such as dry manure, green branches, or leaves. The lustre of the smudging process is less bright if temperatures are too high, so this technique is usually carried out at low temperatures (Shepard 1980: 88–90). However, the technique requires careful control of the presence of oxygen in the firing in order to avoid burning off carbon particles.

The results of thin section analyses, XRPD, and SEM on the selection of samples from Belovode and Pločnik all showed that pottery tended to be fired at temperatures between 750 °C and 900 °C, with most samples having been fired at around 750–800 °C. The only samples which were exposed to temperatures around and above 1000 °C are three possible fragments of the so-called 'chimneys' (BEL 18-124, BEL 18-224, and PL 24-23). Only one other fragment (BEL 18-46) showed microstructural and mineralogical characteristics compatible with very high temperatures, however this sample was retrieved from a destruction layer and could easily have been re-fired to a very high degree during this event.

Given these observations, it is possible to formulate some hypotheses about the firing procedures followed to produce dark-burnished pottery at Belovode and Pločnik. As most of the samples have a reddish core and dark margins, we can propose a two-step procedure in which vessels were initially fired in an oxygen-rich atmosphere during which the maximum firing temperature was achieved and sintering started to take place, giving the ceramics their rigidity. Then, with the temperature decreasing, a reducing phase was obtained by covering up the pit, possibly after organic material was added in order to increase the production of smoke. At the beginning of this reducing phase, the temperature was probably not high enough or sustained for long enough to allow the nucleation of magnetite. Only a few sherds of the analysed assemblages display a homogeneous black colour across their surface; the majority show surface colours that vary between light and medium grey to light red. This suggests that, even if potters had a certain awareness and ability in manipulating fire and obtaining reducing conditions, this knowledge was not completely mastered. This is probably only due to the lack of adequate firing installations that could have enabled more absolute control.

Graphite-painted pottery was probably fired in a similar two-step procedure but with a longer reducing process given the relatively homogeneous dark to light grey colour of most of these sherds in thin section. The reducing phase would have started earlier in the process than in the production of dark-burnished pottery, certainly before the initial oxidising phase had reached a temperature of about 700 °C, at which point the graphite would have begun to burn off relatively quickly (Kreiter *et al.* 2013: 176, 2014). It is possible that the maximum temperature was reached under reducing conditions, because the estimated firing temperature of the samples analysed is between 750 and 800 °C, a temperature at which graphite decoration would already have been severely damaged under oxidising conditions. Furthermore, the small bloating pores that characterise one of the analysed samples are normally formed under reducing conditions when vitrification starts (Maniatis and Tite 1975).

Analysis of the 'graphite' painted decoration of one out of five selected samples using µ-XRD[2] and µ-Raman revealed that it was actually coated with a type of artificial carbon black pigment which achieved a very similar metallic lustre to mineral graphite (the 'Glanzkohlenstoff' of Letsch and Noll 1978, 1983), and the distinction is difficult to make with the naked eye. This finding suggests that the same 'graphite' decorative effect on Vinča pottery could have been produced using several alternative processes, demonstrating the technological advancements achieved by the potters (Amicone *et al.* 2020b). Equally, it demonstrates an innovative means to decorate the pottery with metallic lustre, making the distinctive metal-like appearance that is one of the linking components between metal and pottery technologies.

All the evidence presented above seems to suggest that potters were able to exert a relatively close control over the atmosphere of the firing process. This is clearly demonstrated by the presence of vessels that are intentionally bi-coloured (the so-called black topped vessels), the dark-burnished pottery and, especially, the graphite-painted decorations. However, the results of the analyses conducted on samples from Belovode and Pločnik also show that the ability of potters to reach high temperatures was overemphasised in some early studies (e.g. Kaiser *et al.* 1986), while our results are in line with those of other studies (Maniatis and Tite 1981; Perišić *et al.* 2016; Spataro 2014, 2017, 2018). Perhaps too much attention has been paid to using firing temperatures to illustrate the Vinča potters' mastery of the pyrotechnological skills, rather than to their control over firing atmosphere conditions.

In order to produce a functional pot, temperatures of around 600–700 °C are more than enough. The common argument that the potters wanted a hard-fired vessel does not appear to be valid, since this would also have been achievable with a less sophisticated and more resource-efficient firing procedure. The question that remains, therefore, is not whether the potters were able to achieve and to control temperatures above 1000 °C, but rather why they would have preferred to do so in the first place. The impression is that this over-estimation of firing temperatures could have been influenced by the fact that these results matched too well with pre-existing models concerning the connections between pottery and copper pyrotechnology in the Balkans, such as that developed by Renfrew (1969) on the basis of the archaeometric analysis of Frierman (1969) on a sherd from Karanovo. Detailed study of copper smelting shows that this pyrotechnology is fundamentally different to that employed in dark-burnished production, because the sequence of redox conditions is reversed. In metallurgy, the process starts with relatively low temperatures in conjunction with a reducing atmosphere to form copper metal from the ore, followed by higher temperatures in less reducing conditions to melt and allow the metal to form larger prills. Yet, even if we conclude that pottery at Belovode and Pločnik—and more generally within the Vinča culture—was not fired in exactly the same conditions as those necessary for copper smelting, this would not exclude the possibility that certain advances in pottery pyrotechnology, such as the ability to manipulate firing conditions, could have laid the groundwork for further technological progress that was necessary for the smelting of copper. On the other hand, with regard to graphite-painted pottery we cannot exclude a reverse trajectory of transmission, or a parallel development, as this type of decoration was broadly contemporaneous with the emergence of metallurgy. It might be that the pyrotechnology of copper smelting could have evolved together with advanced pottery technology or may even have triggered the development of the graphite decoration technique (Amicone *et al.* 2020b).

Organisation of production

Vinča culture pottery has often been considered the product of specialised labour (for a discussion see Kaiser 1984: 282–287). It is worth remembering, however, that archaeologists and museum curators tend to select only the finest and most beautiful vessels for display and publication, frequently creating a bias in the material record.

Specialisation has been defined in different ways (e.g. Costin 1991: 4; Kaiser 1984: 280; Rice 1981, 1991: 259), but it may be said that it is the organisation of production in such a way that the number of the people involved in the manufacture of a commodity is smaller than the number of consumers. Archaeologists often use direct evidence such as production loci and debris to identify context, concentration, scale, and intensity of production (Costin 1991: 8) to describe the organisation of production. As previously mentioned, there is currently a lack of evidence for pottery production installations in Vinča culture settlements. This makes the task of defining the level of craft specialisation and organisation for the production of pottery in these settlements challenging. There is, however, indirect evidence that could give insights into this issue without knowledge of the exact location of the production areas.

The first is standardisation, which is considered to be positively correlated with specialisation because it is assumed that production with fewer producers will display less variability. An increase in routine activities connected with specialised production can also result in standardisation (Costin 1991: 34). Theoretically, the characteristics of the artefacts could therefore indicate the level of standardisation, efficiency and skill of production, and hence the level of specialisation. In practical terms, however, several problems arise when we try to define a strategy to assess these parameters.

Standardisation has often been measured by considering the metric variability of vessels (e.g. Roux 2003). Within the study of Vinča pottery, an attempt has been made by Vuković (2011; Vuković and Miloglav 2018) who studied the assemblages from Vinča and Montel Slatina. The statistical analysis of metric parameters of ceramic vessels from the two Late Neolithic sites identified some evidence of standardisation in the assemblages considered, despite the exceptional material fragmentation and the difficulties of distinguishing a relevant sample. According to the results of this study, Vinča pottery exhibits a relatively high level of standardisation that can be recognised through the values of the coefficients of variation for metric parameters (Vuković 2011). However, Vuković (2011: 97) also noted a different degree of standardisation between certain functional classes and within each individual class (e.g. carinated bowls versus bowls with incurved rims). This posed many questions that, as the author herself stated, required a more complex comparative investigation into the pottery from several additional sites.

The high fragmentation of the assemblages examined in the present research, together with the long period of deposition and the fact that the sample contains the products of a number of potters, makes an evaluation of standardisation very difficult. Despite this, an attempt was made using a different approach. Instead of considering metric variables, the focus was shifted towards aspects related to selection and processing of raw materials at both sites. At Pločnik, throughout the period considered, potters were using sandy clays characterised by the presence of quartz and different type of sedimentary rocks, accompanied by different mineral suites. Sometimes a few inclusions of muscovite are present and with others there is the rare inclusion of amphibole and epidote. Textural variation was noted in the grain size distribution and sorting of the inclusions. However, it was possible to observe a general tendency toward larger inclusions within thick-walled and large vessels, and smaller inclusions in thin-walled and small vessels. In most cases, coarse fabrics do not show clear evidence of tempering and it can be assumed that sandy clays of different coarseness were employed according to the size and wall thickness of the vessel produced. At Belovode, two main types of raw material sources were exploited: a non-calcareous clay source marked by metasedimentary rocks and a clay source rich in shells and microfossils. A weak association between fabric and shape was observed in that the dark-burnished bowls seem to have been produced with a clay that may have been well cleaned prior to its use; the clay source rich in shells and microfossils is mostly associated with pots that could have been used in cooking activities. However, at both sites the geological variability of the surrounding area is not very high, and it would therefore be unreasonable to expect locally made pots to contain a large variety of rocks and minerals (Arnold 2000). In addition, technological knowledge tends to be broadly distributed and shared among household producers (Foster 1965: 58). If other technological parameters are taken into consideration it is possible to observe that forming techniques at both sites seemed to be rather stable. However, the high fragmentation of the studied pottery did not allow for a full and detailed reconstruction of the *chaîne opératoire* of forming techniques.

More generally, this technological overview reveals that despite the fact that these assemblages show a relatively high degree of homogeneity, this could be the result of a strong and widespread technological tradition rooted in the tight learning networks that seem to characterise this material culture, rather than due to specialisation. However, the results of the technological study carried out by Kaiser (1984) on assemblages from Selevac and Opovo show that although there were similar notions of what constituted an appropriate pot and that certain aspects of pottery knowledge were shared across these sites, differences existed regarding the specific procedures followed in various steps of the production. For these reasons it is suggested that a large number of potters were active at these sites, which ultimately led Kaiser (1984) to conclude that Vinča ceramics were not made by highly specialised potters.

The discourse on firing techniques leads us to the final parameter to be considered, that of 'skill', a concept that refers to the 'proficiency with which activities are executed' (Bleed 2008: 156). Skilfully produced objects are expected to be well made, e.g. regularly shaped and complex in both form and decoration. Ethnoarchaeological research has focused on this topic (e.g. Roux and Corbetta 1989; Wallaert-Pêtre 2001). A few studies in archaeology have also considered the production sequence in term of skills (Budden 2008; Forte 2019; Michelaki 2008), with interesting results.

It is possible to draw some conclusions regarding 'skill' from what we have learned about the employment of pyrotechnics in Vinča culture. Our results have shown that potters were able to control, to an extent, the variables of atmosphere and temperature. However, differences in the colour of vessel surfaces show that potters were not always able to control firing conditions completely. The conclusion is that, even though people who were producing pottery at both sites show an awareness of how to manipulate fire in order to produce different colours, this knowledge is still not completely mastered. In practice, this is very likely because no conclusive kiln installations were employed and that conditions in a pit fire can never be fully controlled, even by highly skilled potters.

The presence of graphite-painted pottery at Pločnik deserves further consideration. Graphite-painted pottery has often been associated with the presence of craft specialisation (Evans 1973), as the unusually high degree of control that was necessary to exercise the firing of such decoration has been attributed to the skills of specialists. While dark-burnished pottery is abundant in pottery from all phases at Pločnik, examples decorated with graphite are rarely found. Only eleven small sherds of graphite-painted pots were found in Trench 24 at Pločnik, all derived from building Horizon 3 (*c.* 5036–4951 cal. BC, 95% probability) and Horizon 2 (*c.* 4927–4621 cal. BC, 95% probability), corresponding to the Gradac Phase I of the Vinča culture. These samples come from a single trench and therefore may not be fully representative given the current size of the settlement (*c.* 30 ha, see Rassman *et al.*, Chapter 24, this volume) or that assumed (possibly double the area) prior to erosion from the shifting Toplica river. The fragments belong to vessels that were all locally produced, as shown by the petrographic analysis of five of them.

Graphite could have been immediately available to the Pločnik communities. In Serbia, graphite deposits (Republic of Serbia Ministry of Mining and Energy n.d.). can be found in Donja Ljubata, Ibarski Rudnici, Jaram, Pasjača, Ušće, Veta and Vrška Čuka (Figure 5). It was not possible to find any reference in the literature for the exploitation of these deposits during the Neolithic and Chalcolithic, but it is very likely that some of these sources were used at the time. The deposit in Pasjača is about 34 km from the site of Pločnik and is easily accessible via the Toplica river valley, and there is no reason to assume that the access to this deposit was restricted. Equally, graphite could have been acquired through specialist trade networks from Bulgaria, via the same connections that brought metal from ore deposits in Bulgaria for the copper implements found in Horizon 1 at Pločnik (Pernicka *et al.* 1997; Radivojević and Grujić 2018).

As mentioned above, the manufacture of graphite-painted decoration required considerable mastery of

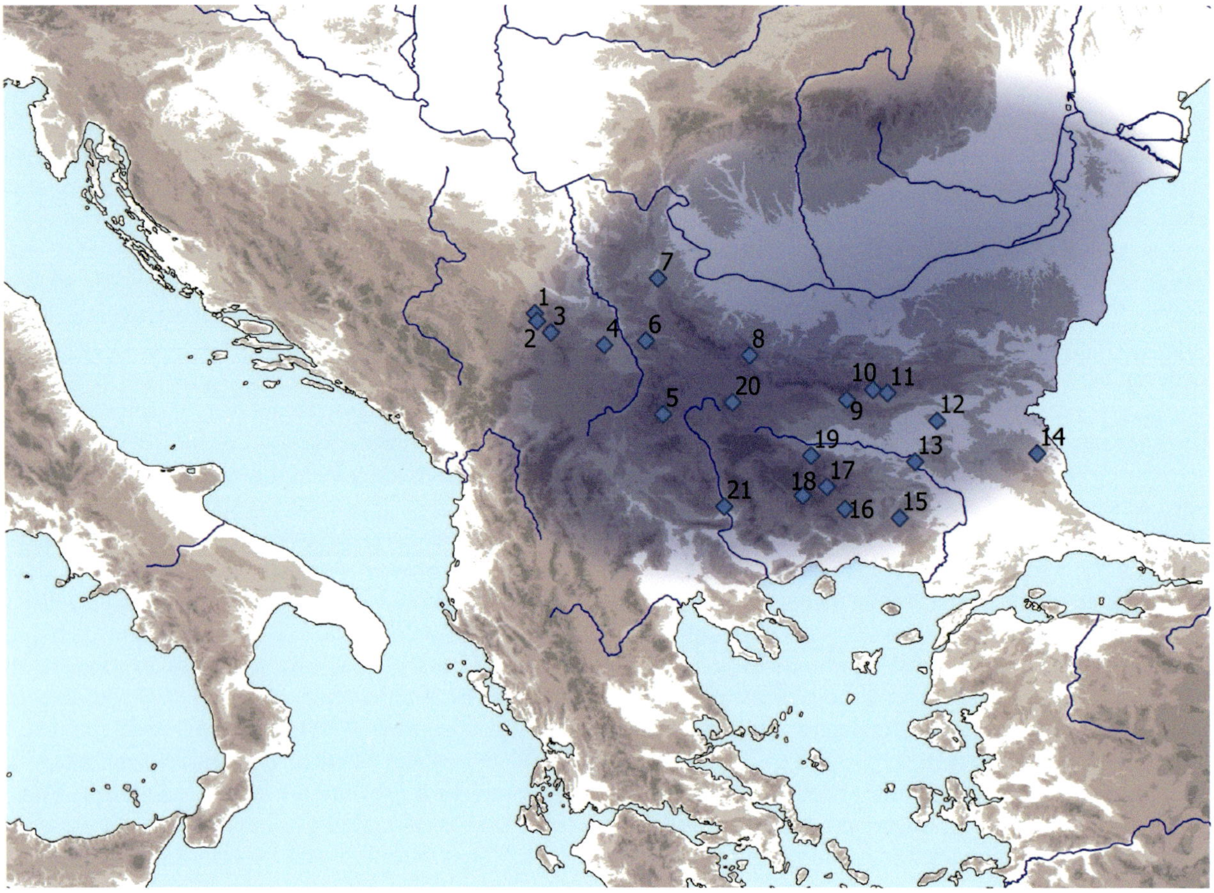

Figure 5. Map showing the main areas of distribution of graphite-painted pottery and deposits of graphite in Serbia and Bulgaria: 1 Ušće; 2 Ibarski Rudnici; 3 Jaram; 4 Pasjača; 5 Donja Ljubata; 6 Veta; 7 Vrška Čuka; 8 Ignatitsa, Vratsa District; 9 Kalofer-Samodivska chukara; 10 Shipka, Kazanlak District; 11 Seltse, Kazanlak District; 12 Sveti Ilija hills (Svetiiliyski Vuzvisheniya); 13 Shishmanovo, Harmanli District; 14 Gramatikovo, Burgas District; 15 Golyamo Kamenyane, Krumovgrad District; 16 Madan; 17 Chepelare; 18 Yagodina, Smolyan District; 19 Krichim; 20 Bistritsa, Dupnitsa District; 21 Lebnitsa, Sandanski District. (Map by Lars Heinze and Silvia Amicone).

redox conditions and this skill may not have been shared by all the potters at a site like Pločnik. Perhaps only a small number were able to produce this decoration at a given time, potentially marking the emergence of a degree of specialisation in pottery production but the overall evidence for this is currently insufficient to build a solid model.

In conclusion, the ceramic archaeological record considered here does not exhibit clear signs of high degree of craft specialisation. Within the limits imposed by the natural availability of clay sources in the area surrounding Pločnik and Belovode, the ceramic assemblages seem to be characterised by high homogeneity in paste recipes as well as other technological attributes. This could be the outcome of a tight learning network and strong technological traditions (that need to be discussed further) rather than specialisation. In addition, different functional shapes (serving vessels, storage vessels and cooking pots) show strong technological similarities both at the levels of the selection and procurement of raw material and of forming techniques. This evidence suggests the absence of separate production units specialised in the production of different functional shapes. Moreover, pottery circulation (see below) points to a similar scenario. There is, however, insufficient evidence to counter the hypothesis that the number of pottery producers was smaller than the number of consumers and therefore that a low degree of specialisation (*e.g.* household specialisation) existed. Further, graphite-painted vessels could have required specialised production with skills unlikely to have been shared by all people making pottery.

Pottery circulation

Recent analyses of Neolithic pottery in the Southern Balkans and the Aegean have challenged our view of pottery circulation at that time (see for example Pentedeka 2011; Quinn *et al.* 2010; Schneider *et al.* 1994; Tomkins and Day 2001). These studies have shown that pottery was circulating around routes and (long) distances comparable with those of other materials, such as obsidian and copper. This strongly contrasts with the traditional view that sees ceramics at this time as a commodity that was usually made and consumed locally, or at least was not transported over significant distances (e.g. Arnold 2000: 342; Vitelli 1993a, 1993b; Wijnen 1994). These new discoveries have encouraged us to rethink our previous models of craft production and circulation in the Neolithic. Understanding the circulation of pottery to and from Belovode and Pločnik is therefore key to the present study.

The results of thin section analysis carried out on the samples from Belovode and Pločnik allow us to discuss the possibility of non-local production for the two sites. The phenomena of importation and/or exchange of pottery on a regional and interregional scale can, to a degree, be addressed for the first time for both sites. However, the low geological variability of the area surrounding Belovode and the stylistic homogeneity that generally characterised Vinča material culture somewhat limit the identification of non-local production.

At Pločnik, several samples clearly show a petrographic profile that is not compatible with clay immediately available in the vicinity of the site, or even within the 7 km threshold suggested by ethnographic studies (Arnold 1985, 1993, 2000). In some cases, the comparison between the geological maps and the collected geological samples indicates possible areas of origin for some of the samples. Unfortunately, not enough is known about the settlement pattern of this area to formulate solid hypotheses about the provenance of the studied samples. The results do, however, show a very interesting network of exchange between Pločnik and the surrounding sites and beyond, and this seems to intensify towards Horizon 1 (*c.* 4631–4231 cal. BC, Gradac II–III, 95% probability) (see Chapter 37), the latest phase of occupation. The expansion of these trade networks matches well with the results of complex networks analysis that used the composition of copper artefacts from this and other Balkan sites to indicate supply networks (Radivojević and Grujić 2018). Pločnik changes from being part of a Vinča culture-dominated supply network to developing trading links with cultural complexes in east and central Bulgaria shortly after the massive and hasty abandonment of the northern Vinča culture sites in around 4600/4500 BC.

As mentioned above, the evidence for non-local production in Belovode is less clear. It was noted in Chapter 14 that within fabric BEL-A (Figure 2, Chapter 14), interpreted as being produced locally to Belovode, some technological variability exists in terms of tempering techniques, potentially reflecting different traditions. However, given the homogeneity of the geological environment around this site, it cannot be discounted that they originated from nearby contemporary villages that exploited similar raw material sources, yet had different tempering traditions. As discussed previously however, a few other samples show a petrographic profile that leaves no doubt about their non-local origin, and for some a presumed area of origin has been proposed (Figure 4, Chapter 14).

Kaiser (1984, 1990) mentioned the possible presence of non-local production at Selevac and Gomolava but does not elaborate on their possible provenance. The results of the ceramic thin section analysis (Tringham *et al.* 1992) carried out on specimens from Opovo, also seem to provide evidence of wider pottery circulation: fabrics are characterised by volcanic rock fragments and agate, which are not compatible with the area surrounding this site. Finally, Spataro (2014), in her work on the Starčevo/Vinča sites of Parţa and Miercurea Sibiului Ptriş, does not report the presence of any samples that could be potentially non-local.

In general, we can conclude that current evidence is not sufficient to allow us to draw a broader picture about pottery circulation in the Vinča culture during the Neolithic and Chalcolithic periods. The projects carried out so far are, however, starting to suggest a more complex picture than was previously supposed, also opening the possibility for long distance pottery exchanges. These could be explained in many ways. One likely scenario, for example, is that objects were regularly carried via supra-regional marriages, where women could have brought vessels and other ceramic material (e.g. loom weights) to their new homes (for a discussion on the mobility of women as opposed to more stationary men in prehistoric times, see Bentley *et al.* 2002; Knipper *et al.* 2017; Mittnik *et al.* 2019). But we should not exclude the possibility that certain vessels were exchanged for their contents rather than for the pot itself. So far, the (presumably imported) vessel repertoire from Pločnik and Belovode does not favour either of these models and could well be the result of both (or other) scenarios.

Conclusions

The results of the analyses carried out on samples from Belovode and Pločnik, considered in the context of other contemporary sites from the central Balkans, shed new light on various aspects of pottery manufacturing in the region at this time. Despite recognisable changes, both sites show a general pattern of continuity within their pottery making recipes over the observed periods. Kaiser observed a similar scenario at both Selevac and Opovo (Kaiser 1984: 292–297). This pattern of continuity in the technological tradition indicates that there was no significant change in the transmission of pottery making recipes related to the start of the Gradac Phase (*c.* 5000 BC), which denotes the appearance of metallurgy in Vinča culture. This evidence of continuity challenges the traditional vision of a profound change driven by the appearance of a novel technology in the Gradac Phase (e.g. Jovanović 1994; Garašanin 1994/1995) accompanied by an important social reconfiguration that would have impacted the learning networks responsible for the reproduction of technological traditions.

Despite the discontinuities noted in Horizon 1, this final stage of occupation at Pločnik may indeed reflect some dynamics of social change. The Horizon 1 (*c.* 4631–4231 cal. BC, 95% probability) (see Chapter 37) seems to correspond to the Gradac II-III Phases that, according to Jovanović (1994), is the final stage of development of Vinča culture in the South Morava Valley. The phase is characterised by an increased presence of metallurgy at the site and a reconfiguration of the network of metal supply, as mentioned above. Before Horizon 1, artefacts at Pločnik were made mostly from copper sourced from deposits in eastern Serbia (Radivojević 2012). However, metal from Horizon 1 shows consistency with Bulgarian copper ores (Pernicka *et al.* 1997). This is closely paralleled by the results of a study of pottery typology that blend the final developments of Vinča material culture with the Chalcolithic Bubanj-Hum phenomenon that seems to originate from east Bulgaria (see Mirković-Marić Chapter 42, this volume). As mentioned above, these newly forged links between Pločnik communities and Bubanj Hum (or further, Krivodol-Salcuta-Bubanj Hum I complex) in eastern and central Bulgaria seem to be supported by a sharing of both metal resources and pottery making technologies. Was this the connection that ended life in Pločnik as violently as at other Vinča culture sites? The results are inconclusive, although the arrival of newcomers in Pločnik is argued to represent the beginning of the end for this prehistoric settlement (Garašanin 1973). Based on the pottery technology study we can argue that throughout the occupation of Pločnik there was a general trend of continuation of pottery making recipes, with a diversification of resources in the latest stage, but no disruption of the potters' traditions. Could such gradual change have contributed to the abrupt ending for the Pločnik communities? There is no conclusive evidence to determine the exact cause of the site's ultimate destruction and abandonment.

The present research also provides an important contribution to the long prevailing theory that sees a link between pottery and metallurgy pyrotechnology (e.g. Gimbutas 1976; Kaiser *et al.* 1986; Renfrew 1969). It was clearly shown that dark-burnished pottery and graphite-painted pottery were not fired at the same temperature employed in copper smelting (up to around 1083 °C, the melting point of copper). These results are in agreement with a series of analyses already made on other assemblages of the Neolithic and Chalcolithic periods, that include dark-burnished pottery and graphite-painted pottery from the Balkans (e.g. Gardner 2003; Goleanu *et al.* 2005; Maniatis and Tite 1981; Youni 2000; Perišić *et al.* 2016; Spataro 2018), but they challenge the extensive analysis carried out by Kaiser which showed that pottery from Selevac and Gomolava was routinely fired at temperatures of around 1000 °C and beyond. Without denying the possibility of different firing routines employed in other Vinča culture communities, we suggest that previous studies have placed too much emphasis on the achievement of high firing temperatures as connection between pottery and copper metallurgy. Instead, we would highlight the mastery of fire control as the crucial link between these technologies. Vinča potters were able to achieve various decorative finishes to ceramics by skilled control of the firing conditions. For example, the production of both dark-burnished and graphite-painted pottery would have been achieved by opening and closing the firing installation and hence varying the air to fuel ratio in a two-step process. Potters also seem to have added relatively moist organic matter, such as straw and leaves which combusted quickly and used up free oxygen while depositing soot, as well as perhaps ash. This manipulation of the firing atmosphere, particularly the ability to obtain and sustain reducing conditions, could have been an important precursor to the development of early metallurgy, which also requires a two-step process but in reverse order: first a predominantly reducing atmosphere to extract copper metal from ores before more oxidising conditions are introduced to reach the higher temperatures needed to melt the copper metal (or in other words physical change is followed by chemical during this process). It is possible, even tempting, to envisage the transfer of this expert knowledge of manipulating redox conditions from ceramic production to another technological domain such as metallurgy.

There is no conclusive evidence that pottery making and copper smelting were highly specialist activities. Radivojević (2012) could not identify dwellings occupied by metal smiths in the Vinča culture, and neither could the present excavations at Belovode and Pločnik. Even if pottery were a specialist activity, it could have taken place seasonally, at times that would not require crafts people to engage in other subsistence activities. If the craft knowledge was not segregated within specialist groups, then the cross-fertilisation of pottery and metal making knowledge within the Vinča culture appears feasible.

Looking across the Balkans, the wider appearance of graphite decoration seems to be contemporary with the emergence of metallurgy during the start of the fifth millennium BC (Bailey 2000: 227; Vajsov 2007) and requires the same strongly reducing atmosphere for much of the firing cycle as is essential in copper smelting. Significantly though, given the current dating evidence that sets the early fifth millennium BC for the emergence of both metal making and graphite-painted decoration in Pločnik, it is more plausible to suggest a parallel development—or even reverse trajectory of transmission—in which the production of graphite-painted decoration was influenced by early metallurgy, and both were benefitting from the pre-existing experience with dark-burnished pottery (cf. Amicone *et al.* 2020b).

Yet, while graphite-painted decoration and metallurgy occur in the Balkans at broadly the same period it is important to bear in mind that the current evidence seems to suggest that the earliest emergence of the two are geographically unrelated, with graphite-painted pottery probably first appearing in the Struma Valley at the site of Promachon-Topolnitsa (Vajsov 2007), outside the Vinča culture (Figure 1) which was home to the earliest metallurgy (Radivojević *et al.* 2010). Importantly, the appeal of the black graphite correlates with the wider preference for black lustre in the Chalcolithic Balkan material culture, as well as the sought-after black component in black-and-green copper ores, found to have the most beneficial qualities to facilitate a successful smelt (Radivojević 2015; Radivojević and Rehren 2016). With this in mind, the two crafts are likely to have been generally linked, geographically, technologically and aesthetically, making them 'close cousins' rather than one being the direct precursor to the other.

The bibliographic reference for this chapter is:

Amicone, S., Radivojević, M., Quinn, P. and Rehren, Th. 2021. Pottery technology at the dawn of metallurgy in the Vinča culture, in Radivojević, M., Roberts, B. W., Marić, M., Kuzmanović Cvetković, J., and Rehren, Th. (eds) *The Rise of Metallurgy in Eurasia*: 538–551. Oxford: Archaeopress.

Chapter 44

Belovode and Pločnik figurines in their wider context

Julka Kuzmanović Cvetković

The Neolithic period of Central Balkan prehistory is characterised by, amongst other things, a large number of clay figurines. Based on numerous fragments and whole figurines, we propose several theories regarding the role they played in the lives of people living in Belovode and Pločnik.

Creation of figurines

The figurines that have survived into the present were normally made of baked clay, a very accessible and suitable material. There are, however, a few examples made of stone, and there are indications that figurines were also made of wood but have not been preserved (Naumov *et al.* 2009: 91). It is unlikely that all the figurines were 'home-made'; there are significant differences in the quality of production and there were certainly also specialist artists engaged in the process.

Figurines were made by sculpting clay into the desired shape and were not made in a mould (Vasić 1936a: 13–14). No figurine mould has ever been found at either Belovode or Pločnik. It is possible, however, that they were made of several pieces that were created separately. On some chest fragments, e.g. BAf22, there is a hole at the base of the neck, into which the head, made separately, was subsequently attached. This may have been because the head was more complicated to make, requiring many details such as hairstyle, eyes, nose, and mouth, which needed special attention. Alternatively, the figurines may have been created in a process where different specialists made different parts of the body.

During the Neolithic period, clay was easily accessible and frequently used for making objects intended for everyday use. It is evident from the high quality of the handmade pieces that the Vinča people had perfected both clay-shaping and fire-controlling skills (as proved by research conducted by Amicone (Chapter 43, this volume). These people also understood the qualities of materials very well and, clay being fragile, this determined the shape of the objects, especially the figurines. Often, a standard model was followed, defined by Vasić (1936a: 27) as a front-facing figurine with arms outstretched. This had a pillar-like, flat, or cylindrical body, and arms in the shape of short 'tentacles'. The hands and the fingers are depicted only when the arms lie alongside the body (e.g. BAf70).

The figurines were modelled in three dimensions in order to make them free-standing, whether within a domestic altar or simply as an ornament. Some were made with particular expertise, were very rich in details, of fine clay, and highly polished, presumably by specialised manufacturers or 'artists'.

What the figurines represent

The Vinča figurines depict the lifestyle and beliefs of the community, illustrating dress, jewelry, and activities. The female body is most often represented, with a protruding bosom and bottom, which were most likely the standards of beauty of the period, as well as symbols of fertility. Some pieces have a protruding belly, perhaps representing pregnant women, who played an important role in expanding the community. The significant number of pregnant figurines may be a consequence of a demographic expansion during the Neolithic period or may be related to a fertility cult (Srejović 1997: 711).

Belovode

The model of a figurine in a long dress made of a very thin fabric with a string tied up beneath the waist is extremely realistic. It is an archetypal model that can be found at almost all Vinča sites. Some such dresses were made of vertical stripes, either narrow or wide. Members of the Vinča community were familiar with weaving (Jovanović 2011: 28), and it is likely that they would dress up on special occasions. Examples of statuettes from Vinča (Vasić 1936a: Table LXXIX, 369, 370, 371; Tasić 2008b: 227, Catalogue 72), Vitkovac (Catalogue of the Exhibition 'The Field of Vitkovac', catalogue number 125) and Divostin (Srejović 1968: Figure 37) all have almost identical dresses. In his research, centred on the anthropomorphic sculpture of South Eastern Europe, Hansen provides a review of published material with similar statuettes having long dresses found in Romania and Bulgaria (Hansen 2007: Tables 206–373). A reconstructed version of a dress was displayed for visitors to the exhibition 'The Lords of Clay and Grain' at the Museum of Vojvodina in Novi Sad (Jovanović 2011: 80–81)

During the height of the Vinča culture, when communities were knowledgeable about the technology of copper processing, various pieces of jewellery were represented on many of the figurines. Previous excavations at Belovode have recovered female figurines wearing bracelets, belts with medallions made of stripes, or necklaces with medallions (Šljivar *et al.* 2011: 31). Figurine BAf29 has a necklace made of large, flat, probably copper pearls; figurine BAf50 has a circular medallion in its v-neck area and a medallion on each hip. Figurines with medallions were also found at Vitkovo, for example, a female wearing three medallions on the belt of her dress (Čađenović *et al.* 2003, catalogue number 125). Other examples were found also found at Vinča (Srejović *et al.* 1968: Figure 36).

There are also frequent representations of a sash over one shoulder, perhaps indicating status. Figurine BAf56 has a sash, with several medallions beneath the arms, in the centre of the stomach area.

The heads of the figurines received special attention. Triangular faces are often most realistic, with even the ears represented, but there are no mouths represented on any figurines. The heads are profiled and emphasised and they have perforations on their occipital bones that, it is assumed, were filled with ornaments made of grass or hair so as to make the figurines better resemble the individuals they were intended to represent (Tasić 2008b: 149). Several figurines have straight, shoulder-length hair depicted, which may also be related to the social status of the person represented. Some figurines have traces of colour on their faces and bodies. Some researchers (see Naumov 2009: 95) believe that this colouring emphasised the social status of the individual depicted. Perforations are also frequently found on the arms and hips suggesting that they might have been suspended. In some cases, the extent to which the bodies were realistically represented is remarkable. Two figurines have protruding gluteal areas. On BAf9 there is a light dress with several layers whereas on BAf71 there is only a thin stripe [of fabric?] depicted across the hips.

Pločnik

As much as the figurines from Trench 24 at Pločnik belong to Vinča culture in style and typology, they are significantly different from those found in Trench 18 at Belovode. No figurines from the Pločnik trench were wearing a long dress tied beneath the waist. Two figurines (PlAf23 and PlAf13), however, were modelled with long dresses that narrow towards the bottom. An almost identical long dress is represented on a Vinča figurine (Vasić 1936a: 48, Figure 278). As at Belovode, some of the Pločnik figurines have a sash falling across their chests (e.g. PlAf53), also likely representative of their status. Fragments of protruding gluteal area and legs were common, with incised lines that probably represent clothes (e.g. PlAf73). These kinds of fragments were also found in Fafos (Bulatović 2003: 18), Vinča (Vasić 1936a: Figure 130) and Tell Kirilevo (Pernicheva 2004: 464).

Several figurines from Pločnik are represented nude (e.g. PlAf44 and PlAf65), with protruding hips and waists, indicative of the fact that the artists were familiar with human anatomy but were limited by the material with which they worked. Nevertheless, they managed to show the layers of fat on the women's hips.

Figurine PlAf2 has two slightly different faces, a form embodied in Roman art as the god Janus. A similar head was found in Fafos and is held in the museum in Vranje (Vukanović 1969: 1; Bulatović 2003: 19).

The function of the figurines

What purpose did figurines have in the Vinča society? This question has no single, correct answer. Some authors believe that the figurines were deliberately broken after use in rituals (Srejović *et al.* 1984: 51; Bailey 2000: 285) and that figurine fragments were used to create and maintain societal networks (Chapman 2000; Chapman and Gaydarska 2007). Were they cult-related and thrown away following an act of magic (Srejović *et al.* 1984: 52)? Did their destruction enhance the magical effect? Or were they simply a domestic ornament, saved as a precious item throughout generations? Were they made to be kept, to be used for chiropractic purposes (Bailey 2005: 29), or were they supposed to represent gods?

Although Srejović supports the theory that figurines served cult purposes (Srejović 1984: 51), he also believes that they were part of an artistic opus (Srejović *et al.* 1984: 42) including 'masterpieces' intended to be exhibited and preserved, regardless of whether they represented a distant ancestor or a mythical creature. The figurine heads found at Pločnik (PlAf36, PlAf38, PlAf40, PlAf57) most likely had a completely prosaic, everyday function. Vasić interpreted their function as 'pawns' in table games (Vasić 1936a: 93), since they closely correspond to modern gaming pieces, being a couple of centimetres tall. The representation of a figurine on a throne is related to a goddess, 'Great Mother' (Garašanin 1973), the Neolithic period being considered a matriarchal era (Gimbutas 1974). The chair depicted in this type of figurine is usually richly ornamented (e.g. PlAf18). We are unlikely to ever determine whether these figurines represent the portraits of actual individuals, but it is more probable that they are pieces of art created according to contemporary standards.

Zoomorphic figurines

Zoomorphic figurines from Trench 18 in Belovode are realistic representations of domestic animals. Half of those found, i.e. six, depict a bull, a powerful animal with a strong body; in two figurines the male genitals are clearly emphasised. Bull (ox) figurines have been found previously at Belovode (Jovanović *et al.* 2004: 472; Šljivar and Jacanović 2005: 71) within a cult area dedicated to fertility. Analysis has shown that two species were represented: *Bos primigenius* and *Bos brachiceros* (Jovanović *et al.* 2004: 474).

Dogs were also well-represented with three made of baked clay and one exquisitely sculptured head made of alabaster. Since the dog was one of the first animals to be domesticated (Srejović 1997: 786), it is unsurprising that such figurines would be made. Animal heads made of stone and alabaster were also found at Vinča from the early Tordoš phase, i.e. between the depths of 7.8 and 4.1 m (Garašanin 1973: 89).

Two figurines resembling a bear indicate that the people of Belovode also created figurines that embodied the wild animals they encountered (Cermanović-Kuzmanović and Srejović 1992).

Only a few animal figurines were found in Trench 24 at Pločnik. The interpretation of Vasić (1936a: 93), that miniature heads of birds or animals actually represented pawns for playing table games, could be applied to a bird head that is missing its base. Figurine PlZf6, with two bird heads placed on a narrow pedestal, could have had a similar purpose.

Amulets

An amulet is 'an object made of different kinds of materials that a supernatural power is ascribed to' (Srejović 1997: 45). It was most frequently worn on a rope around the neck and was considered to have an apotropaic (i.e. protective) effect. Amulets could also be placed within houses, hung on a door panel or sometimes on the horns of animals, in order to protect home, family, or domestic animals (Pantović 2014: 39). Another view proposes that amulets were accessories for weaving, used as weights or for coiling yarn (Tringham and Stevanović 1990: 334–338). Two-tentacle (or Y-shaped) amulets, the most common form at Belovode, also represent a fertility symbol. Amulets that were highly decorated surely played a significant role in house interiors or were of importance to the individuals wearing them. Two-tentacle amulets that resemble a figurine with raised arms 'undoubtedly represent Deity' (Pantović 2014: 40). Ornithomorphic and zoomorphic amulets are assumed to indicate a connection with deities and a joining of symbolism such that the animal attributes become those of the deity. Nevertheless, the meaning of amulets still cannot be fully deciphered. We can only suggest, based on current findings, that the amulets are related to cult, religion, and belief.

Throughout the 2012 and 2013 campaigns, not a single amulet was found at Pločnik. This does not mean that none are present but rather that none could be found in the excavated part of the site.

Conclusion

Despite being only a small sample, and from areas of the sites with no defined function (e.g. living space or workshops), the figurines found at Belovode and Pločnik give us a more complete idea of life in these settlements. There remain, however, more questions than answers.

In an attempt to summarise the artistry and religious beliefs of the communities of Vinča culture, Srejović proposed that the creativity 'is defined by three components: local tradition (art and religion of Starčevo culture), a unique economic and social hierarchy (differentiating of economic work fields, distribution of work, creation of large families and their becoming independent) and the influences of the cultures of neighbouring territories, mostly from lower Danubia and Thrace' (Srejović *et al.* 1984: 42).

The oldest Vinča figurines, including those found at Belovode and Pločnik, are the pillar-like statuettes. Their further development was influenced both by the material from which they were made, and the development of the society as a whole, art representing a significant component of life. Although following a common process during the making of the statuettes, with a front-facing figurine with arms outstretched (Vasić 1936a: 27), the Neolithic artists were able to make each figurine unique. In addition to being pieces of art, the Vinča statuettes are distinct, both stylistically and typologically, and representative of the culture to which they belong.

Based on the material remains of their technological achievements, we can assume that there was a division of labour in the Vinča settlements. Men were probably hunters and metallurgists, and women made clothes, looked after children and prepared food. Some were probably skilled in making clay pottery and figurines, but towards the end of the period there may have already been professional pottery makers. Based on the figurines, we can also make observations about clothing. There was clearly a social code regarding women's clothing, since many of the figurines found throughout the area of Vinča culture are depicted wearing the same style of dress. Sculptors followed this standard but also added specific features to each piece.

Over time, in parallel with the development of society, there was a development of art. The shape and features of the figurines were very basic at first, later becoming more realistic. The quality however, always depended on the artist's talent, and probably also on the intended purpose of the figurine. One question that arises is the extent to which the figurines accurately represent Neolithic life. Those from Belovode feature several different dress styles, probably representing garments made from flax linen. These are full-length, flat dresses that narrow towards the hemline, and are tied around the waist with a piece of rope. In a few cases, these dresses widen near the base, almost resembling some modern ball gowns.

A few of the Pločnik figurines are almost completely naked and quite realistic, indicating a good understanding of human anatomy through very carefully sculpted legs, feet with separate toes, and very detailed depictions of the gluteal muscles.

The bibliographic reference for this chapter is:

Kuzmanović Cvetković, J. 2021. Belovode and Pločnik figurines in their wider context, in Radivojević, M., Roberts, B. W., Marić, M., Kuzmanović Cvetković, J., and Rehren, Th. (eds) *The Rise of Metallurgy in Eurasia*: 552–555. Oxford: Archaeopress.

Chapter 45

Ground and abrasive stone tools from Belovode and Pločnik: concluding remarks

Vidan Dimić and Dragana Antonović

Introduction

The Neolithic polished stone industry in Serbia appears as a fully developed operation, with clearly defined and formed types of tools; there is currently little evidence relating to its origins. An exception is the area of the Djerdap Gorge in northeastern Serbia, where the specificity of the populated area and immersion in different types of raw materials already in the Mesolithic, resulted in sedentary communities and the creation of an indigenous, totally unique industry of ground stone (Antonović 2003: 131, 142–143). In the rest of Serbia, it has not yet been possible to establish the beginning of the development of the polished stone industry. From the Early to the Late Neolithic in Serbia there is no substantial difference in processes, with tool types being the same for most of the raw materials, although there is some variation in the quantities of certain types of tools and materials depending upon the locality. The process for producing polished stone tools throughout the Neolithic remains utterly unchanged and the techniques used in the processing of raw materials to the finished tool are directly determined by the raw material used. Vinča craftspeople used various techniques, from primary flaking, retouching, and pecking, to grinding and polishing. The most common raw materials in the Vinča culture were grey and grey-green sedimentary and contact metamorphic rocks and later, during the Gradac Phase and, in particular, in the stage of Vinča–Pločnik (Vinča C), there is intensive use of raw materials known in the archaeological literature as 'light white stone'. A number of other rocks of similar macroscopic appearance were also in use, but in substantially smaller amounts. Research at the sites of Belovode and Pločnik in the seasons of 2012 and 2013, and analysis of assemblages of ground and abrasive stone tools (see Chapters 16 and 31 this volume) provide new data that contribute to the current knowledge of ground and abrasive stone production and, in the case of Pločnik, reveal unique insights.

Raw materials

Research concerning the exploitation of raw materials in the Neolithic has been conducted sporadically on a small scale and is usually associated with a thorough reassessment of certain micro-regions related to a particular site, such as terrain prospection of the environment of Belovode, Lojanik at Mataruška Banja, and modern the magnesite mine Lazac near Kraljevo (Bogosavljević-Petrović et al. 2012). These studies conclude that the Neolithic stone craftspeople could exploit the raw material from both primary (mines, quarries) and secondary deposits. It is not currently possible to state how the process of stone extraction from the parent rocks occurred during the Neolithic in Serbia due to the lack of excavated mining sites. Analogies can be drawn from ethno-archaeological studies (e.g. Burton 1984; Hampton 1999; Stout 2002, 2005) and well-researched Neolithic stone extraction sites in Central and Western Europe (Pétrequin and Jeunesse 1995; Pétrequin et al. 2013). The exploitation of copper ore from locations such as Rudna Glava (Jovanović 1982) or Ai Bunar (Chernykh 1978b) was achieved using a technique involving crushing rocks with heavy stone hammers. It can be assumed that similar mechanisms of exploitation were also applied to stone extraction. The collection of stones from secondary deposits—alluvial deposits in the beds of streams, rivers or their sandbanks—was also very important and potentially the main raw material source for some localities (Dimić 2013b). Since most Neolithic settlements were located in fertile river valleys intersected by small streams (see Marić, Chapters 8 and 23, this volume), this method of collecting raw materials is very rational. Raw materials from both primary and secondary deposits are evidenced in the archaeological record and are clearly observed in the analysis of polished stone tools. Indicators of each practice can be seen in the shape and size of tools, the existence or absence of the cortex, the techniques used in the processing of raw materials, as well as in the proportions of specific raw material at individual sites. It can be clearly concluded that (from the primary deposits in most cases) the exploited raw materials in the Vinča culture were 'light white stone' and grey-green tough rocks (Antonović 2003: 132, 143–144). There is a high probability that other stones were collected from secondary deposits where they already resembled the tools required.

Raw materials used to produce ground stone tools at Belovode and Pločnik follow this broader pattern.

The most common are grey-green sedimentary and contact metamorphic rocks and 'light white stone', while other stone types are used on a much smaller scale. Sandstones, which are very common were, in most cases, at both sites used to make abrasive tools and as tools for pounding or pulverising. Fine-grained sandstone was used for grindstones, while the coarser sandstones were used for static and hand grinders, pounders, and querns. In Pločnik, the use of sandstone dominates, and the uniformity of type and colour is clearly visible; it is highly likely that these sandstones originated from primary deposits.

Earlier analyses on assemblages from several Neolithic sites in Serbia (Antonović 2003: 37–45) demonstrated that the fine-grained, grey-green sedimentary and contact metamorphic rocks exploited are of various geological origins (crystallised schists, cornite, metamorphosed sandstone, metaalevrolite), but with the same technical-physical features (conchoidal fracture, hardness and, above all, toughness), which made them a suitable and preferred raw material for the production of tools whose usage involved frequent impacts (axes, adzes, chisels, hammers). At Belovode, this group of rocks comprises the raw material for 59% of stone tools, while at Pločnik it comprises only 18%. The transition from the earlier phase of the Vinča culture (Vinča B1) to the phase of Vinča–Gradac (Vinča B2–C1) and later Vinča–Pločnik (Vinča C2) is characterised by the same pattern with the exception of the use of 'light white stone'. At Belovode, this accounts for 13% of the raw material, while at Pločnik it comprises 22%. This material has only recently started attracting the attention of researchers (Antonović 1997), largely due to the research and excavation of Vinča sites in the Šumadija basin, where it is a major raw material for ground stone tools with a cutting edge. At all the Late Neolithic sites in central and western Serbia, 'light white stone' is present in significant amounts (60–70% of the total stone material) (Antonović 1997: 33; Antonović 2003; Antonović *et al.* 2005). The category covers all macroscopically similar rocks whose main characteristics are lightness, porosity, low hardness and white colour. Previous petrographic analyses revealed that most examples are magnesite, whose density is 3 g/cm^3 (Antonović 2003: 45–47; Šarić and Cvetković 2013), so that the descriptor 'light' is not entirely justified. Yet the term 'light white stone' is a very common in the literature and for that reason is used here. The most frequently found examples are fine-grained, compact, tough rocks with conchoidal fractures. Silicified varieties can be very hard but appear only rarely compared to the softer varieties. Previous analyses of this raw material determined it to be sedimentary magnesite, dolomite silicified micrite, porcellanite, cherts, silicified tuff and diatomaceous earth (Antonović 1997; Antonović 2003: 45–47; Antonović and Šarić 2011: 68; Šarić 2002). Studies have also shown that magnesite and tuff are the most commonly used 'light white stone'. Widespread magnesite deposits are known in Kosovo, the Kopaonik basin, the Zlatibor massif, in Šumadija, Fruška Gora and into Bosnia (Antonović 2003: 21–22; Figure 3). Microscopic and other precise petrographic analyses of raw materials from Belovode and Pločnik were not conducted but, based upon macroscopic examination—which cannot give a precise provision—the 'light white stone' from both sites, and particularly from Pločnik, can be conditionally defined as magnesite. Unfortunately, the exact source of the raw materials used for making stone tools at Belovode and Pločnik cannot currently be identified.

Tool production technology

Vinča culture sites have produced few assemblages or locations where there is a possibility of defining the entire operational sequence for the manufacture of ground stone tools. At both Pločnik and Belovode, only certain stages of the operational process are archaeologically recognisable, with no stone 'workshop' being evidenced as at Crkvine in Mali Borak (Antonović 2011). It can be assumed that primary flaking was not performed in the settlement, but near it or at the source of the raw materials, as evidenced by examples of samples of semi-worked raw materials from several sites in Mali Borak, near Lazarevac in Central Serbia and numerous ethno-archaeological analogies (Antonović 2014b). It appears that primary flaking was conducted to test the technical characteristics of the stone, including its reaction to impact and the possible presence of cracks that could lead to undesirable fragmentation during processing. If the raw material met requirements it underwent further processing, involving reduction by knapping or pecking, and retouching to form semi-finished items. Grinding or polishing was the final process. All of these stages can be seen as traces on the tools.

The assemblages of ground stone tools from Belovode and Pločnik do not show any features that are unusual for the Vinča culture ground stone industry. All ground tools with a cutting edge from each site were made of fine-grained, hard and compact rocks with distinctive conchoidal fractures. The choice of raw material determined the method of processing, and in the case of fine-grained rocks with conchoidal fractures, this was knapping. This was conducted mainly using the ventral side as a knapping platform to reduce the dorsal side, producing a characteristic semi-circular cross-section. This kind of knapping was usually practiced in the production of adzes, and was a standard technique in the Vinča culture; zig-zag knapping was used for producing axes, and can be seen on an example of a pre-form (Belovode Find 452, see Figure 9 in Chapter

16, this volume; also Dimić 2013b, 2015). After knapping and retouching of the edges, the pre-form was ground on static grinders made of fine-grained sandstone, probably using water. Significant attention was paid to the processing of the cutting edge which was usually more finely ground than the rest of the tool.

Ground stone tools with a cutting edge (axes, adzes, chisels) from both sites show a great deal of uniformity of shape and size. Axes and adzes rarely exceed a length of 120 mm while the chisels are smaller with lengths up to 60 mm. The only exception are adzes from the hoard at Pločnik, which are up to 220 mm in length. Typologically, these tools differ little from each other. In most cases they have an arched edge at the wider, distal end or have almost parallel sides and an arched edge at one end. Similarly, the chisels from both sites do not differ from classical Vinča forms (Antonović 2003: 55). They are small with parallel sides and a narrow cutting edge no longer than 25 mm.

The production process of the abrasive tools from both sites is much simpler, and a certain number of artefacts were not modified at all, the raw material being used unworked in its natural form, e.g. pebble or slab. If further modification were needed, it was performed by pecking since the raw material did not permit the application of other techniques. Depending on the desired shape of the tool, edges and roughness were reduced by pecking (less often for grindstones, polishers, and grinders) while in a few cases the entire surface was pecked, for example with the pestles at Pločnik, in order to make them fit better in the hand.

For most tools with abrasive properties, the final form was created during long-term use rather than as a result of production processes as seen for tools with a cutting edge. As a result, and in contrast to the ground stone tools, there is a diversity in the shapes and sizes of abrasive tools at both Belovode and Pločnik.

Use of ground and abrasive stone tools

In the Neolithic period, ground and abrasive stone tools had wide application in everyday activities. Ground stone tools with a cutting edge, such as axes, adzes, and chisels, were used in the processing of wood (felling trees, splitting wood, clearing various shrubs and bushes, making planks and elements for residential construction etc.). The usage of these tools can generally be determined from analysis of specific traces of use-wear and damage. Indicators of function(s) include furrows, scratches, negative of flakes/micro-flakes, micro-polish and various forms of more substantial damage. These traces were formed during the penetration of the stone tool into the material or object where, due to friction and high pressure, more porous parts of the stone were lost. The form and distribution of these traces vary from one tool type to another. The intensity of the traces may also vary depending upon the raw material involved. Earlier research on these traces by Semenov and others (Semenov 1976; Olausson 1980, 1983b), indicated that axes are defined by furrows, polished surfaces and scratches that extend obliquely in the direction of the cutting edge, while linear traces characteristic for polished stone adzes and chisels extend perpendicular to the cutting edge (Semenov 1976; Olausson 1983b; Dimić 2015). It is important to stress that, in addition to a high percentage of fragmentation, we identified two tools—one from Belovode (Find 381) and one from Pločnik (C-699)—that have mixed traces on their surfaces which suggests multi-functional use. This is not uncommon in the Neolithic (Antonović 2003; Dimić 2013a, 2013b, 2015) and occurs even today when one tool can be used to perform several actions. A stone axe or adze could be used in a wide range of activities including: the primary processing of wood; the clearing of an area of shrubs and plants; in combat as a weapon; and in cult and ritual acts.

During heavy use, damage could also lead to the fragmentation of tools. If a remaining piece of raw material was large enough it could be redesigned using reduction techniques to create a new tool, which would then have a secondary function. A large proportion (63%) of tools at Belovode had suffered severe damage but recycling was recorded on only four. At Pločnik, recycling was also observed on only a small scale. The recycling of fragmented tools is not rare and represents the most rational utilisation of raw materials in response to several factors, primarily the reduction of the territory and the unavailability of sufficient quantities of high-quality raw materials. The practice of recycling in stone industries was widespread in Late Vinča culture phases (Vinča-Pločnik-Vinča C). An example of such a maximisation of raw materials is evident at the site of Crkvine - Mali Borak where, following fragmentation, pieces of ground stone tools were used for making chipped stone tools including scrapers, blades, and small blades (Bogosavljević-Petrović 2011: 217–223). The same practice was observed at the site of Lađarište in Vrnjačka Banja where broken axes and adzes were flaked and retouched to form scrapers (Dimić 2013b; Dimić 2015: 64).

Abrasive tools of various types (grindstones, static grinding slabs of all sizes, pestles, hand grinders and small, fine whetstones) were found at both Belovode (19% of the total assemblage) and Pločnik (40%). At Pločnik, the widespread use of certain types of sandstone (fine-grained, compact yellow and reddish sandstone) is particularly notable, implying that these were exploited from the same primary deposits. At

Belovode, mainly fine-grained sandstones of grey colour are dominant and are used for making querns, but reddish and mid-grained sandstones were used for grinders which suggests that these sandstones were also bought from the extraction site. Traces of use on these tools are clearly defined, reflecting the extended wear that ultimately changed their shape. This can be clearly seen on the grindstones from Belovode (Find 224/2 and Find 225), in the slightly concave polished working surfaces with parallel marks that extend along the length of the objects. A fragment of a grinder from the same site (C-1859) has a distinct working surface on both sides. Grinders were used so intensely that both the ventral and dorsal sides became so eroded that they eventually fragmented.

The most interesting abrasive tools are the pestles from Pločnik (C-574, C-574b, C-574c, C-574e), made of reddish, compact sandstone by a pecking technique. The working surface, in the form of hemispherical protrusions, is clearly defined, as are the parallel, concentric grooves formed by gradual attrition of the tools as they are pressed against the mortar and substance it contained. Exactly what material was processed with these tools is not possible to determine due to the lack of residue analysis but it can be assumed through ethno-archaeological analogies that pestles were used for pulverising raw materials used for either food (e.g. grains, berries, nuts), for the preparation of pigments (e.g. shells of mussels and snails, minerals) or potentially for the grinding of copper ore.

The suggestion that stone tools were used for manufacturing metal is supported by the presence of two tools from Pločnik (one, without C number was found in Trench 24, layer 16; second is C-621/EDM 1442) that were supposedly used for the thinning and hammering of metal objects. One is a pebble in its natural form (of suitable dimensions and weight), and the other is a stone of approximately cylindrical shape that was intentionally modified. The proposed function is based on analogy with very similar pieces from Fargau-Schlesen, Kr. Plön in Germany (Freudenberg 2009: 343).

Ground and abrasive tools from Belovode and Pločnik fully reflect the characteristics of this category of Neolithic stone tools in Serbia. The polished stone industry at Belovode does not stand out for the quality of craftsmanship or the aesthetics of the tools but rather for its uniformity in terms of raw materials and the shapes and dimensions of tools, which suggests a degree of standardisation in the production process. In contrast, the quality of processing of stone tools at Pločnik varies from partially ground to well-polished objects. There are also individual stone tool types from Pločnik that have never previously been identified in the Neolithic of Serbia, including the pestles with hemispherical protrusions and the stone hammers used for thinning metal objects. The hoard of adzes made from 'white light stone' excavated at Pločnik represents a unique find. Based on the number of stone tools found and the quality of their production, it is no exaggeration to say that Pločnik was an important manufacturing centre.

The bibliographic reference for this chapter is:

Dimić, V. and Antonović, D. 2021. Ground and abrasive stone tools from Belovode and Pločnik: concluding remarks, in Radivojević, M., Roberts, B. W., Marić, M., Kuzmanović Cvetković, J., and Rehren, Th. (eds) *The Rise of Metallurgy in Eurasia*: 556–559. Oxford: Archaeopress.

Chapter 46

Bone tool technology at Belovode and Pločnik

Selena Vitezović

Introduction

Osseous raw materials had an important place in the Vinča culture and were used for everyday tools along with flint and stone, as well as for decorative objects. A technological approach to their study (cf. Inizan *et al.* 1995: 13 *ff.*), which involves the analysis of the raw material choices, technology of manufacture and typological repertoire, is used here to provide important data relating to methods of raw material acquisition and management and to craft production and technological know-how. It can also give indirect evidence on crafts that have left little or no trace in the archaeological record, i.e. 'perishable technologies' (cf. Choyke 2013; Vitezović 2013a; 2016a).

Despite its importance, and despite over one hundred years of Vinča culture studies, the bone industry is among the least known categories of Vinča technologies. Faunal remains were not carefully collected on early excavations and the assemblages of bone objects were published only as short reports or mentioned in summary in excavation reports. Most of the publications where bone tools do feature are merely descriptive; very few have an analytical approach (Bačkalov 1979; Russell 1990; Vitezović 2007; 2011b).

The assemblages from Belovode and Pločnik, excavated in 2012 and 2013, are not large, probably due to the limited area of excavations, however careful sampling and collecting has yielded some important data, especially regarding the raw materials used and the typological range of the objects produced.

Raw material management

At most Vinča culture sites, the preferred osseous raw material was ovicaprine metapodials, which would be used to make fine and medium-sized points such as needles and awls. Next in abundance are large mammal ribs, which were used to make awls and various burnishing tools. In addition, other long bones from cattle and ovicaprines such as tibiae and radii, and astragals were occasionally used. Pig bones were mainly avoided and the use of certain skeletal elements such as scapulae or cranial bones (such as mandibles) is extremely rare (cf. Vitezović 2013b).

The antlers used were mainly from red deer and only occasionally from roe deer. In most cases these were shed antlers. The ratio between antlers and bones varies from site to site and it is likely that a level of regional specialisation existed as some sites were more engaged in antler procurement, manufacture, and use than others (cf. Vitezović 2013b). With the exception of boar tusks, which were used for tools, teeth occur only occasionally and were used as ornaments. The distribution of shells at Vinča culture sites is uneven although it is not clear if this is related to the sampling and recovery procedures or to other factors.

The evidence indicates that bone technology at Pločnik and Belovode did not differ greatly from the typical Vinča culture bone industry, though they each have certain specific traits. At Belovode, it is interesting to note the presence of pig bones, particularly two astragals, one from a wild and one from a domestic pig. Pig astragals were also mentioned from the earlier excavation campaigns at Belovode (Jacanović and Šljivar 2001) and therefore may be a specific trait of the Belovode settlement. The reasons for the avoidance of pig bones in the Vinča culture are not clear, although ruminant bones have a more regular axis and are therefore more convenient for manufacturing tools such as awls or points (cf. Vitezović 2013b: 68). Pig bones are not entirely unusable, as may be seen by their presence on other sites and in other cultures, for example, in the Bronze Age in Hungary or in the historic periods (e.g. Gál 2011: 145; Struckmeyer 2011: 188; Choyke 2013; 8, Figure 1.6; Colominas 2013: 90). The reasons behind the absence of pigs may be connected to butchery techniques, the related availability of adequate, usable pieces, or they may be entirely cultural. At Belovode, astragals were the only pig bones that were shaped into artefacts.

At Pločnik, there are occurrences of some less common skeletal elements, such as ulnae or radii, but very few, such as a spatula-chisel made from a cattle ulna and another from a cattle radius.

Antlers were present on both sites, although in varying quantities. At Belovode, there is evidence for the collecting of shed antlers, although more data is needed to establish whether this practice was planned and a regular activity. This assemblage from Pločnik yielded only limited information on antler use despite them being found in previous excavation campaigns at this settlement site (Grbić 1929: Figures 123–125).

The presence of mollusc shells at Pločnik is very important for studying the geographical distribution of this raw material and the routes of trade and exchange (see Chapter 32, this volume, on the Pločnik bone industry for a full discussion). Shells were a preferred raw material for personal ornaments throughout the whole of prehistory (cf. Taborin 1993, 2004 and references therein). Their physical characteristics, such as durability and hardness, were important and their exotic origin also certainly contributed (cf. Trubitt 2003). However, there are also some 'less technical' traits that surely played a role in raw material choice; shells are smooth, bright, and white. Smoothness and brightness may be emphasised by manufacture, by burnishing and polishing. White is a bright and shiny colour and has a wide range of symbolic meaning in different cultures, from 'death' to 'the divine' (Vollmar 2009). The importance of colour in raw material choice for decorative items has been observed elsewhere (e.g. Wright and Garrard 2002; Thomas 2011) and particularly in relation to osseous raw materials (Luik 2007; Vitezović 2012). The significance of white colour is underlined by the presence of the same types of ornaments made in white stones (see Vitezović 2012 and references therein).

Osseous raw material choice is firmly linked with the mechanical properties of some of the skeletal elements (see also Guthrie 1983 and Christensen 2004 and references therein). For instance, long bones tend to split along their axes and are easily shaped into pointed tools, while antlers are resilient to shock and most often used for heavy cutting and percussion tools. The artisans of both the Pločnik and Belovode appear to have had a high level of knowledge regarding the properties of the raw materials, with some local, perhaps even personal, preferences towards certain specific skeletal elements.

Both sites fit well into the general model for the acquisition and use of osseous materials proposed for the Vinča culture (Vitezović 2013b). Within this model, osseous materials were obtained as products from animal husbandry and hunting, through selective collecting, and through exchange. Bones were the most common and easiest to obtain and therefore were the material predominantly used, although with careful choice of the most appropriate and/or most desired skeletal elements. A prevalence of bones from domestic animals is observed on most sites (cf. Perišić 1984; Vitezović 2007). Antlers were acquired in the vicinity of the settlement, probably collected in conjunction with other raw materials, and may have been objects of small-scale exchange, as either raw materials or finished products. The teeth used for decorative purposes were obtained mainly through hunting, while shells were acquired through exchange networks about whose mechanisms our knowledge is still very limited.

The presence of raw materials obtained by hunting (such as the astragal or wild boar and red deer tooth) is interesting in terms of the relationship between animal husbandry and hunting and general relationships between wild and domestically obtained raw materials, however at this point the data are insufficient for any conclusions.

Technology

One of the conceptual frameworks that enables a better understanding of osseous industries is the concept of the 'manufacturing continuum' (*sensu* Choyke 1997; Choyke and Schibler 2007). The manufacturing continuum is largely focused on the effort put into the manufacturing of individual objects and their life duration. Two main classes are distinguished using the following criteria:

1) the regularity in raw material choice (species and skeletal elements);
2) the number of stages used in their manufacture;
3) whether they have been repaired; and
4) their exploitation index, which measures the degree of working, as defined by the proportion of the surface covered by manufacturing marks, relative to the degree of use (Choyke and Schibler 2007: 57).

Artefacts from an assemblage at a single site, or multiple assemblages from several sites, can therefore be aligned on an imaginary axis: at one end are carefully planned, standardised tools (Class I) and at the opposite end are *ad hoc*, expedient tools (Class II) (Choyke 1997; Choyke and Schibler 2007). Such a concept enables the analysis of the overall character of the industry in question and may also provide indirect information on the contexts in which the artefacts were found. For example, the prevalence of *ad hoc* tools may suggest non-permanent settlement, or completely used tools may come from a rubbish pit.

Craft production in the Vinča culture, and particularly its level of standardisation and specialisation and its role within society, are still not adequately analysed. More detailed technological case studies and new theoretical frameworks are needed for a more thorough approach to the problem. Increased standardisation and an increase in the production of diverse types of artefacts are observed as a general trend in the Vinča culture (cf. Tringham and Krstić 1990c; Tripković 2007; Vuković 2011) but have not yet been fully analysed.

For the bone industry, only a few preliminary observations can be outlined. Standardisation is visible (in the Vinča culture, in general in respect to the earlier, Starčevo culture) and an increased consistency led to a somewhat simplified process of manufacture. Although the investment of time and skill are not always extremely high, the overall quality of the tools is generally excellent. *Ad hoc* tools do occur, but rarely. Unfortunately, it is not possible to define any diachronic changes in the Vinča culture bone industries given the current state of research.

At Belovode, all the tools incline towards the Class I category and were made using the same techniques and following virtually the same sequence of operations. A somewhat larger number of *ad hoc* tools is present at Pločnik, but this deviation is small and may be explained by sample bias. Differences in manufacturing techniques between the two sites were not observed.

Standardisation within the bone industry has been noted previously for the sites of Drenovac and Slatina–Motel, located in the Pomoravlje region (Vitezović 2007). The implications are that the increased standardisation in bone manufacture signifies that the flint and stone tools used for production of bone tools were also standardised, while standardised tool shapes reflect the high production of 'perishable crafts' such as the processing of hides and plant fibres. However, more detailed analyses of multiple technologies are needed to establish more clearly the relationship between ground stone, chipped stone, and bone industries and as well as other crafts practised by communities of the Vinča culture. Workshops or working places were not identified with certainty at either site, although the identification of such areas is generally very difficult, especially in a limited area of excavation.

The presence of trade and exchange can be confirmed with certainty only for artefacts made from exotic raw materials, such as *Spondylus* and *Glycymeris* shells. The find of the *Spondylus* bead from Pločnik, the first of this type from a Vinča culture site in a secure context (cf. Chapter 32, this volume on the bone industry from Pločnik for detailed references), is extraordinary and supports the argument that the *Spondylus* trade routes were by no means limited to the Danube valley. *Spondylus* trade and exchange is an important phenomenon involving numerous Neolithic and Chalcolithic cultures (see Sériadès 2010 and references therein). For a long time it was considered that the *Spondylus* ornaments were restricted to the Banat and the region immediately around the Danube, and that they did not exist in the areas south of the Danube and Sava rivers (see, for example, the distribution map in Willms 1985, and especially the discussion in Chapman 1981). Such distribution maps, however, reflect the state of research and available data at the time. Recent research (e.g. at Drenovac or Vitkovo, cf. Vitezović 2007, 2013c) has revealed new finds of decorative shell items, and future studies will undoubtedly bring more new finds and new information. The find from Pločnik indicates the need for a revised model of general trade and exchange in the Neolithic period.

Typological repertoire

Within the typological repertoire of bone artefacts in the Vinča culture, tools are predominant with a much smaller proportion of weapons and personal ornaments, while non-utilitarian items are extremely rare: only a few possible musical instruments have been identified and figurines or other objects of ritual function were not fashioned from osseous raw materials (cf. Bačkalov 1979; Perišić 1984; Russell 1990; Vitezović 2007, 2011b).

Most of the tools recovered from Belovode and Pločnik were used in everyday tasks and crafts such as plant processing, textile production, the working of leathers and hides, and woodworking. The tools include awls, eyed and plain needles, heavy points, scrapers, spatulas, spatula-chisels, axes, and auxiliary items such as worked astragals, most likely related to textile production. Finer tools (awls, needles, scrapers) dominate the assemblage while heavy duty tools were not found in large number (only one preserved axe and no preserved heavy percussion tools). This may be related to the activity areas excavated within each of the settlements.

There are only a few personal ornaments from Belovode and Pločnik and no non-utilitarian objects. Furthermore, some of the artefact types that were noted on some other Vinča culture sites, such as hunting and fishing equipment, or tools related to flint processing such as retouching and pressing tools, were also absent.

Discussion and conclusion

The bone industries from Belovode and Pločnik fit well within the broader Vinča culture bone industry, both in terms of the techniques used and the typological repertoire produced (cf. Bačkalov 1979; Russell 1990; Vitezović 2007, 2011b). They demonstrate a high technological know-how and strong familiarity with the raw materials. They also provide indirect evidence of the importance of crafts related to textile production and to leather and hide working.

Each Vinča culture bone industry has locally specific special traits. For the Belovode assemblage, it is the use of pig astragals, and the previously unknown type of decorative item in a shape of a small ring with prong at one side. For the Pločnik assemblage, it is the use of unusual skeletal elements: ulnae used to make heavy points and spatula-chisels, and the unique find of a cattle radius. It is for future studies on bone industries in the Vinča culture to determine whether these traits remain unique for these sites or are more generally characteristic for the region and/or period.

At this stage, it is not possible to make any hypotheses on the diachronic development of the bone industries in any of its aspects: raw materials, techniques of manufacture or typological repertoire, however the data provided by these two sites will have an important place in future analyses.

The bibliographic reference for this chapter is:

Vitezović, S. 2021. Bone tool technology at Belovode and Pločnik, in Radivojević, M., Roberts, B. W., Marić, M., Kuzmanović Cvetković, J., and Rehren, Th. (eds) *The Rise of Metallurgy in Eurasia*: 560–563. Oxford: Archaeopress.

Chapter 47

Chipped stone industries in the Vinča culture

Elmira Ibragimova

Introduction

The analyses conducted on chipped stone artefacts from the sites of Belovode and Pločnik revealed gradual changes in both assemblages with the most significant developments taking place during the transition to the Vinča Pločnik II Phase (see Chapters 18 and 33, this volume). This chapter will compare the chipped stone industries excavated in 2012 and 2013 at Belovode and Pločnik to other Vinča culture sites as well as to adjacent contemporary cultural groups. Trends specific to and common for several temporal or territorial units of Vinča culture will be described through analysis of variation in raw material exploitation, production systems, and tool usage patterns.

Raw material variability and usage

The procurement of raw materials at the sites of Pločnik and Belovode during the Vinča Tordoš to the Vinča Pločnik I Phases is distinctively different. At Pločnik, a wider variety of raw materials was used, including non-siliceous materials (e.g. shale, schists). The industry was based on sources close to the site, most probably the bank of the nearest river. The small number of pieces made of imported raw materials and the active use of local fluvial sources in Pločnik Horizons 2–5 find analogies in the Middle and Lat Neolithic chipped stone industries of Bulgaria, in Thrace and at the Maritza river basin (Gurova 2014).

At Belovode, a smaller variation of raw materials was used and, in contrast to Pločnik, this included stones that probably had a distant provenance, such as obsidian (cf. Chapters18 and 49, this volume) and a honey-coloured flint similar to the so-called 'Pre-Balkan Platform flint' or 'Balkan flint'. The few obsidian finds took the form of bladelets or flakes, following the main chronological trend (described in details in Chapter 49 of this volume), although one rejuvenation core tablet was found in spit 16 at Belovode, and could point to sporadic reduction taking place locally.

'Balkan flint' is the other imported raw material used in Belovode. There is a substantial discussion connected to its provenance, morphology and ways of exchange and reduction throughout Starčevo and Vinča cultures as well as in the Neolithic of Bulgaria (e.g. Antonović *et al.* 2005; Bogoslavljiević-Petrović and Starović 2013; Bonsall *et al.* 2010; Gurova 2014; Šarić 2002). This raw material was actively used during the Early Neolithic at the settlements of the Karanovo I—Starčevo—Criş—Körös group, mainly for the production of medium to long regular blades (Gurova 2014). Starting from the Late Neolithic period in Bulgaria, 'Balkan flint' declined in use and significance. Within the Vinča culture, it was still used consistently but in relatively small amounts. For example, at Vinča Belo Brdo, its frequency increases slightly in the upper levels (from 6 m), while at Selevac the ratio fluctuates between 2 and 10 % in each level (Kaczankowska and Kozlowski 1990).

The final Vinča Pločnik II Phase at Belovode and Pločnik sees a more unified set of raw materials with white (cream) tabular flint predominating for blade production at both sites. Raw material with the same macro-characteristics is recorded from a number of settlements in Central and Western Serbia (Kaczankowska and Kozlowski 1990). Petrographical analyses have attributed the flint to several varieties: from 'white opal' or 'organogenic white-coloured flint' and 'silicified limestone' (Bogoslavljiević-Petrović and Marković 2012). The composition of five artefacts of white (cream) tabular flint from Belovode and Pločnik were also analysed. Two groups were identified that could represent different sources of flint; they comprise samples from both Belovode and Pločnik (see Chapter 48, this volume). The widespread use of white (cream) tabular flint coincides with the use of 'soft white stone' in the ground stone industry at Vinča sites in the same region, in the Vinča Pločnik II Phase. This stone is presumed to be silicified magnezite (Antonović *et al.* 2005). Several shafts were discovered in the West Morava basin where silicified magnezites could have been procured in the Neolithic period (Bogosavljević-Petrović and Marković 2014); these shafts need to be further investigated.

Technology and tool morphology

The Vinča Tordoš–Vinča Pločnik I production systems at the sites of Pločnik and Belovode show significant differences. The Gradac Phase horizons of Belovode reveal the developed microblade industry where a pressure technique was probably used. The microblades could be used without retouch or be retouched to form microdrills. Several technological traits (e.g. reduction of microcores, including a secondary one) as well as the occurrence of microtools (microdrills, microscrapers, trapezes) form a distinct 'microlithic component' of the Belovode industry, similar to that found at sites of the so-called 'eastern group' of Vinča culture that includes the sites of Leu, Cleanov-Fiera, Verbiçioara, Potporanj, Opovo (Kaczankowska and Kozlowski 1990: 46). At Belovode, though, the 'microlithic component' does not form such a large proportion, especially in the subsequent Horizon 1.

The expedient flake industry of the early horizons at Pločnik may be compared to the materials of Selevac dated within Vinča–Tordoš Phase presenting 'a pebble-working tradition' (Voytek 1988). There is a distinct shift in production on both sites in the Pločnik II Phase. The regular blade industry, with a tabular cream flint as preferred raw material, predominates in later horizons at both Pločnik and Belovode. Similar reduction methods were used at both sites, connected to pyramidal core reduction performed from the narrow side of the slabs. Nevertheless, there are some differences in the first stages of core reduction as well as in the final reduction stages. The sample from Belovode presents a greater number of intensively used cores with a change of orientation in core reduction and a large proportion of totally exhausted cores. The large number of pyramidal cores has analogies in the Vinča and Gomolava sites (Kaczankowska and Kozlowski 1990). The assemblage interpreted as a workshop at Gomolava included fragments of antler tools and granite supports, highlighting the materials and equipment used in the process of blade reduction (Voytek 1988).

The morphology of the products of this *chaîne opératoire* at both Belovode and Pločnik, i.e. blades made on tabular flint, probably shows evidence of the punch technique. Similar traits are observed from the late phases of Vinča culture with flaking angles of around 85–90° and regularity of the blades (Kaczankowska and Kozlowski 1990; Voytek 1988).

The tool typology of the latest Horizon 1 at Belovode and Pločnik follows the general trends of Late Vinča culture. Tool assemblages consist of end-scrapers, simple retouched blades with 15–20 mm width and flakes (Kaczankowska and Kozlowski 1990). The prevalence of these tools is characteristic for Vinča Pločnik Phase sites as well as for a range of sites of the Karanovo III–IV Phase (Todorova and Vajsov 1993; Gurova 2014). Though a growth of the microlithic component that is also regarded as a trend in the Late Neolithic at Bulgarian sites (Gurova 2014) is more typical for earlier Vinča sites.

Conclusion

A preliminary comparison with Vinča culture sites shows that the chipped stone industry of Belovode has analogies with the so-called 'Central group' of settlements such as the eponymous site Vinča, Gomolava, and Divostin (Kaczankowska and Kozlowski 1990). The Late Vinča industry of Pločnik is consistent with those of the described traditions, but the early horizons of the site demonstrate several locally specific traits such as significant usage of pebble raw material, and the importance of the flake industry and flake tools.

The study of the lithic industry at both sites would benefit from further investigation including a more detailed raw material analysis (petrographic study and geological prospection), and a strict statistical evaluation of the functional groups following extensive use-wear analysis.

The bibliographic reference for this chapter is:

Ibragimova, E. 2021. Chipped stone industries in the Vinča culture, in Radivojević, M., Roberts, B. W., Marić, M., Kuzmanović Cvetković, J., and Rehren, Th. (eds) *The Rise of Metallurgy in Eurasia*: 564–565. Oxford: Archaeopress.

Chapter 48

Geochemical characterisation of chipped stones from Belovode and Pločnik

Enrica Bonato, Martin Rittner and Silvia Amicone

A selection of stone artefacts recovered from the sites of Belovode and Pločnik was sampled and analysed with Laser Ablation Inductively Coupled Mass Spectrometry (LA-ICP-MS). The assemblage consisted of eight samples (Figure 1 and Table 1): four finds from Belovode (B1, B2, B3, B4) and four from Pločnik (P5, P6, P7, P8). On a macroscopic scale, the samples exhibited different colours: B1 and P7 had a clear amber colour; B2 was colourless and opaque; B3, P5, and P6 were white; B4 was yellow with a layering of different yellow shades; and P8 was a very dark green/black colour. The analyses were performed in order to understand the nature of raw materials used for the production of chipped stones at these sites and to investigate if these materials come from different sources (e.g. Bogosavljević *et al.* 2014; Bonsall *et al.* 2010; Gurova 2012; Kaczankowska and Kozlowski 1990).

Geochemical analyses were conducted at the London Geochronology Centre (Department of Earth Sciences, University College London) on an ESI NWR193 laser ablation system coupled to an Agilent 7700x ICPMS. Instrument parameters were: RF: power 1340 W; Ar Carrier Gas flow: 1.04 l/min; He gas flow: 950 ml/min; Sweep Time: 2.79 s; Dwell Time: 50 ms for most masses, 5 ms on 29 and 221, 10 ms on 23 and 20 ms on 44. Eight small specimens were cut to a size of several mm (Figure 1) and then mounted in an epoxy resin, ground (with SiC papers from 1000 P to 4000 P) and polished with Al_2O_3 for analysis. NIST SRM 610 glass standard (Reed 1992; Jochum *et al.* 2011) was used as the external concentration standard, with NIST SRM 612 as an alternative and for accuracy testing. Six to ten spots were analysed across each sample, avoiding the rims. Before and after each sample measurement, two to three spots on NIST SRM 610 were analysed as the bracketing standards. The following 57 masses, covering relevant major and trace elements, were measured: ^{7}Li, ^{9}Be, ^{11}B, ^{23}Na, ^{24}Mg, ^{27}Al, ^{29}Si, ^{31}P, ^{39}K, ^{43}Ca, ^{44}Ca, ^{45}Sc, ^{47}Ti, ^{51}V, ^{52}Cr, ^{55}Mn, ^{56}Fe, ^{59}Co, ^{60}Ni, ^{63}Cu, ^{66}Zn, ^{75}As, ^{85}Rb, ^{88}Sr, ^{89}Y, ^{91}Zr, ^{93}Nb, ^{95}Mo, ^{107}Ag, ^{115}In, ^{118}Sn, ^{121}Sb, ^{133}Cs, ^{137}Ba, ^{139}La, ^{140}Ce, ^{141}Pr, ^{146}Nd, ^{147}Sm, ^{153}Eu, ^{157}Gd, ^{159}Tb, ^{163}Dy, ^{165}Ho, ^{166}Er, ^{169}Tm, ^{172}Yb, ^{175}Lu, ^{178}Hf, ^{181}Ta, ^{182}W, ^{197}Au, ^{208}Pb, ^{209}Bi, ^{232}Th, ^{238}U, and mass 221 as a background monitor. Due to the unknown nature of the samples, the oxide scaling method was applied,

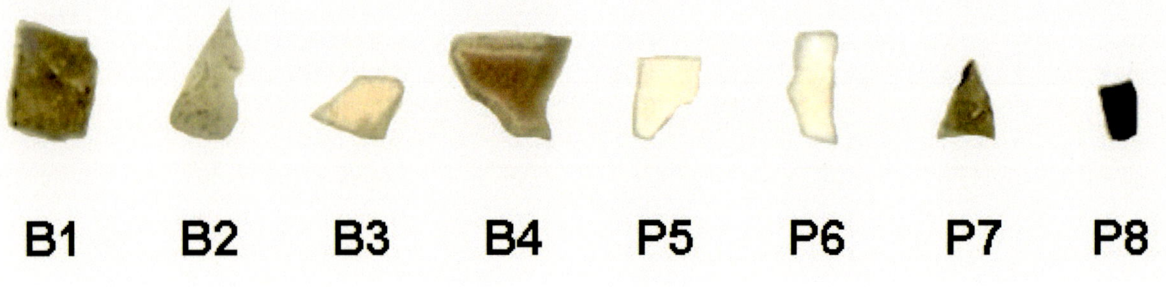

Figure 1. The eight samples from Pločnick and Belovode analysed with Laser Ablation Inductively Coupled Mass Spectrometry.

Table 1. List of samples from Belovode and Pločnik.

Sample	Site	Trench	Layer	Tool	Colour
1	Belovode	18	13	fragment	grey
2	Belovode	18	13	blade	grey
3	Belovode	18	7	blade	cream
4	Belovode	18	13	flake	cream
5	Pločnik	24	9	blade retouched	cream
6	Pločnik	24	3	blade	cream
7	Pločnik	24	5.6	flake	dark grey
8	Pločnik	24	5.6	blade	dark grey

The Rise of Metallurgy in Eurasia (Archaeopress 2021): 566–569

following Pitblado *et al.* 2013 (based on Gratuze 1999). ^{29}Si was used as the internal standard linking samples to NIST 610 standards. This approach assumes that all elements whose oxides make up the sample have been analysed.

Results and discussion

The results (Table 2) show that it is possible to identify one main group comprising of B1, B3, P5, P6 and P7. These samples show a chemical composition, which is typical of chert material (Olofsson and Rodushkin 2011) as can be clearly seen in Figure 2. Samples B2, B4, and P8 represent outliers. Sample B2 appears to be very pure in silica because of its low Aluminium (Al) content and significant levels of Uranium (U) (Table 2); this composition could be compatible with chalcedony (SiO_2). Sample B4 contains low Aluminium (Al) and high Calcium (Ca), Strontium (Sr) and Uranium (U) content (Figure 3 and Table 2), which is typical of silicified carbonaceous rocks. Sample P8 contains a significant amount of siliceous material (clay and silt), and its dark green/black colour indicates the possibility of significant content of organic matter. It presents high Aluminium (Al), Magnesium (Mg), Potassium (K) and Iron/Manganese (Fe/Mn) concentrations, which are the typical chemical components of clay minerals. The element concentration and element ratio discrimination plots clearly show this sample has a significantly different composition to the others (Figures 2 and 3). This sample, due to its high Silicon (Si) concentration can by identified as a siliceous shale. In summary, the outlying samples can be defined as chalcedony (B2), silicified carbonate (B4) and siliceous shale (P8).

The main group of chert samples can be further subdivided into two subgroups (Figure 4), one comprising samples B3, P5, and P6 and characterised by slightly lower concentration of (Strontium+Barium+Magnesium)/Calcium (Sr+Ba+Mg)/Ca and Aluminium/Calcium (Al/Ca) (Figure 2), and lower Iron+Manganese+Nickel (Fe+Mn+Ni) and Silicium (Si) content (Figure 4) compared to the second chert subgroup. The second subgroup includes samples B1 and P7, which show a high Aluminium (Al) content indicating a lower purity of the chert material compared to the other group, and a medium high Iron+Manganese+Nickel (Fe+Mn+Ni) content (Figure 4). It is interesting to notice that these two chert subgroups, which potentially represent different sources of chert, include samples from both Belovode and Pločnik. This could indicate the possibility of different sources of flint being exploited at both sites, or through the same supply network of chipped stone in the Vinča culture, however this hypothesis needs to be further investigated analysing a larger number of samples and using other techniques such as isotopic analysis.

The bibliographic reference for this chapter is:

Bonato, E., Rittner, M. and Amicone, S. 2021. Geochemical characterisation of chipped stones from Belovode and Pločnik, in Radivojević, M., Roberts, B. W., Marić, M., Kuzmanović Cvetković, J., and Rehren, Th. (eds) *The Rise of Metallurgy in Eurasia*: 566–569. Oxford: Archaeopress.

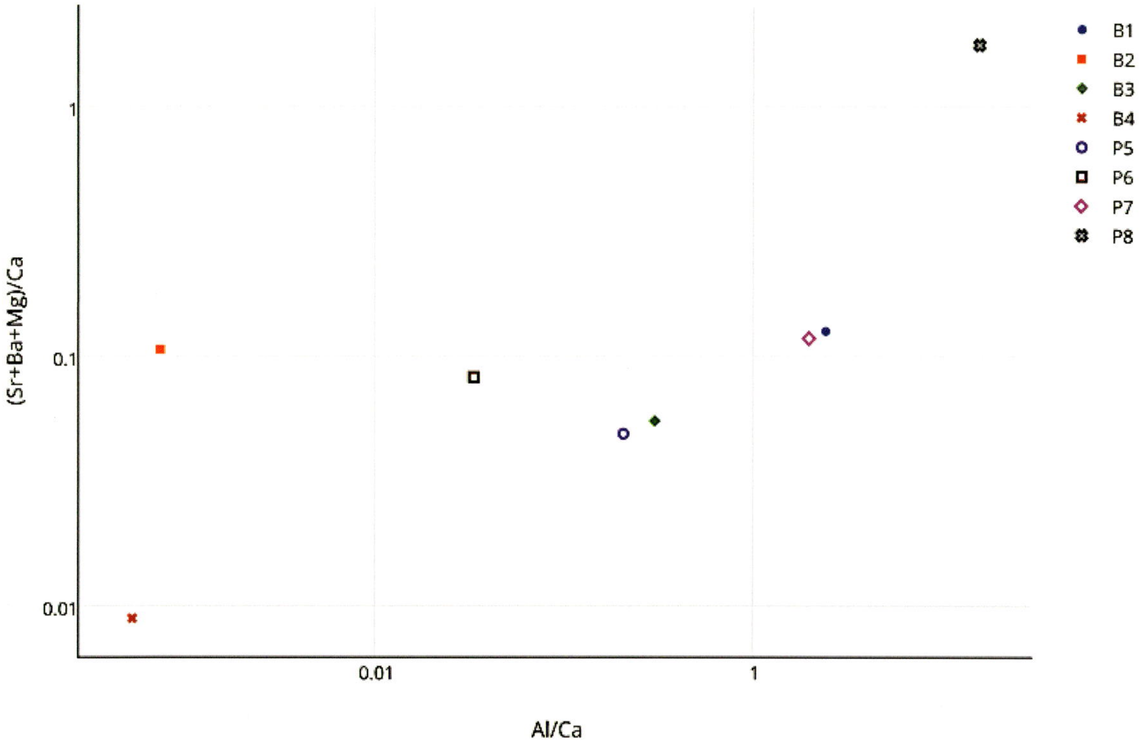

Figure 2. Sr+Ba+Mg vs Al, normalised to Ca. The plot displays the presence of a main group composed of samples B1, B3, P5, P6, and P7 and shows how samples B2, B4, and P8 plot separately from the main group.

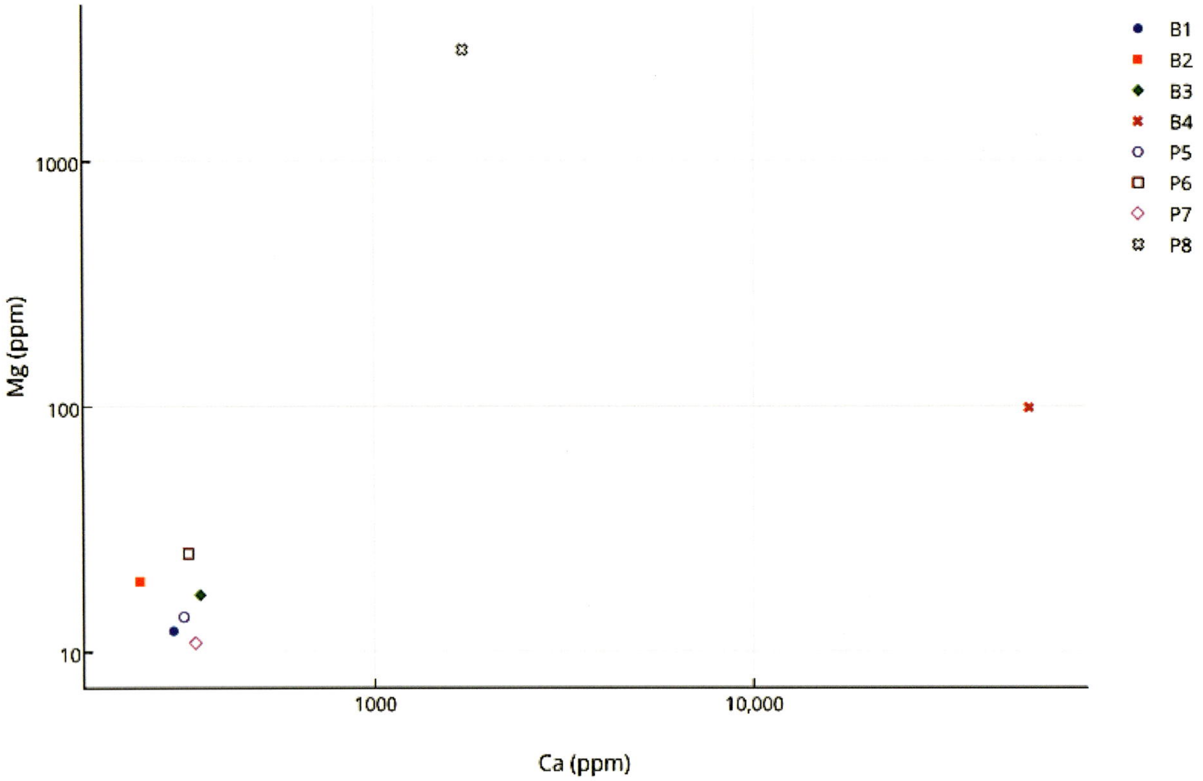

Figure 3. Bi-plot of Ca and Mg concentrations expressed in ppm. The graph highlights that the majority of the samples analysed have a low concentration of Ca and Mg.

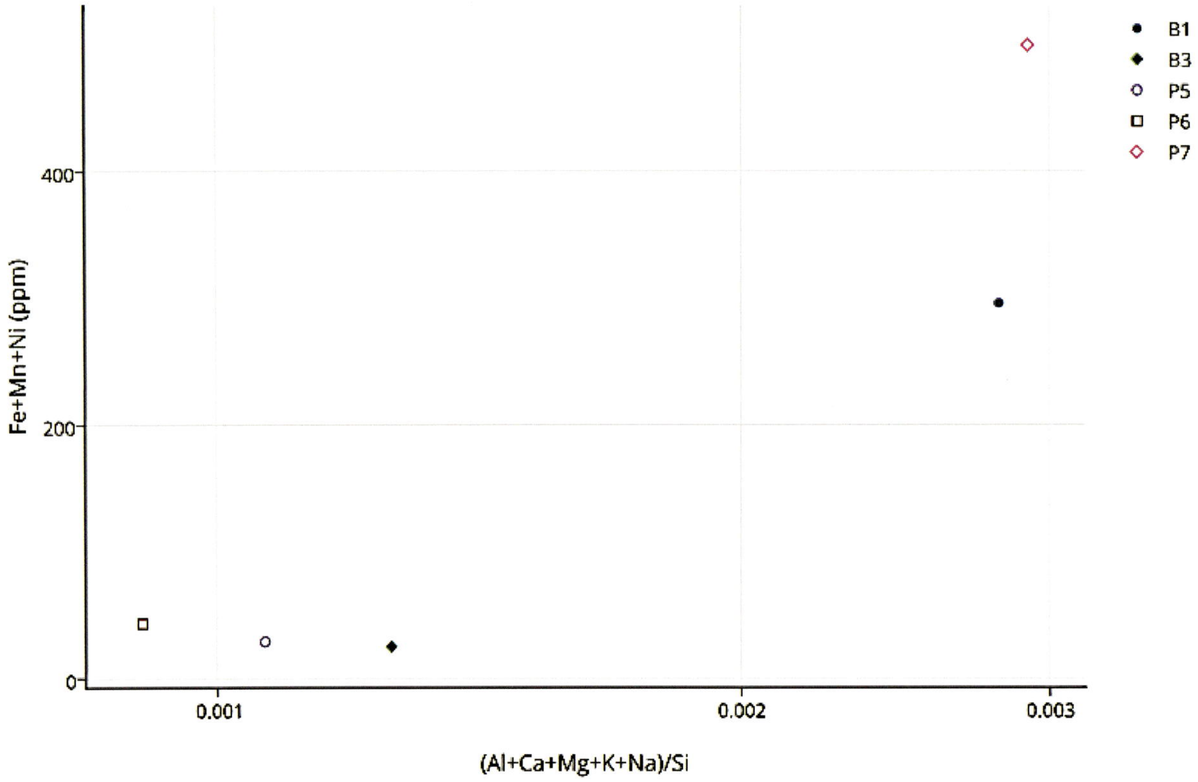

Figure 4. Fe+Mn+Ni (ppm) vs Al+Ca+Mg+K+Na (normalised to Si). The graph displays the presence of two main groups of samples, one characterised by low Fe+Mn+Ni and (Al+Ca+Mg+K+Na)/Si (samples B3, P5, and P6) and the other characterised by a high level of (Al+Ca+Mg+K+Na)/Si and a medium-high concentration of Fe+Mn+Ni (samples B1 and P7).

Table 2. Chemical values of the eight samples analysed through LA-ICP-MS. The values are expressed in ppm, with LOD: Limit of Detection.

	B1	B2	B3	B4	P5	P6	P7	P8
Li	5.480	3.472	0.139	0.586	0.074	0.061	34.856	39.170
Be	0.058	0.036	0.067	0.071	0.034	0.053	0.623	0.725
B	45.087	262.147	2.980	7.560	2.078	2.062	37.133	41.061
Na	105.457	276.060	18.935	181.335	26.782	17.055	1740.333	2019.705
Mg	12.706	19.462	17.026	98.339	13.856	25.092	2438.679	2843.195
Al	652.544	0.169	103.476	28.093	64.527	10.785	23626.582	27462.023
Si	464211.420	466414.493	466888.053	431936.455	466969.594	467021.094	427649.565	421414.206
P	105.006	102.802	101.234	123.042	89.011	86.084	449.819	509.383
K	183.544	114.125	103.209	60.995	78.925	48.068	6227.721	7236.798
Ca	287.032	244.957	345.761	53302.243	313.589	322.413	1526.181	1723.527
Sc	9.360	8.156	8.138	7.186	7.675	7.908	9.784	10.180
Ti	2410.357	3.989	7.642	4.620	5.383	4.774	1088.778	1055.075
V	5.251	0.082	0.565	0.745	0.801	0.647	40.214	45.904
Cr	3.222	2.008	0.719	1.146	0.919	0.641	19.739	22.296
Mn	1.047	0.112	0.197	0.891	0.151	1.706	61.577	71.692
Fe	270.246	32.606	24.573	34.729	28.235	41.446	16090.654	18715.906
Co	0.483	0.028	0.008	0.044	0.019	0.015	3.420	3.958
Ni	0.813	4.583	0.267	0.109	0.553	0.342	26.753	31.114
Cu	0.481	0.187	1.374	0.209	0.290	0.241	32.308	37.554
Zn	0.527	0.203	0.128	0.248	1.757	1.492	41.154	47.934
As	0.852	0.381	2.190	1.968	1.094	1.638	11.218	12.839
Rb	0.279	0.210	0.314	0.132	0.152	0.113	37.856	44.124
Sr	3.128	0.599	0.697	348.095	0.575	0.521	30.276	34.781
Y	0.637	0.008	0.019	0.068	0.012	0.010	5.104	5.775
Zr	15.974	0.155	0.328	0.392	0.619	0.421	28.130	12.502
Nb	6.442	0.025	0.035	0.086	0.044	0.044	4.200	4.093
Mo	0.178	0.025	0.011	0.035	0.013	0.015	0.125	0.071
Ag	0.006	2.196	0.007	0.011	0.007	0.008	0.031	0.032
In	0.010	59.136	0.008	0.017	0.007	0.007	0.022	0.025
Sn	1.080	61.043	0.374	0.380	0.282	0.374	1.333	1.339
Sb	0.531	3.106	0.717	0.849	0.471	0.908	0.350	0.329
Cs	0.209	1.780	0.023	0.019	0.008	0.009	2.645	3.081
Ba	20.386	5.691	1.349	30.137	0.943	1.001	161.726	184.292
La	2.448	0.003	0.034	0.027	0.059	0.008	5.276	5.833
Ce	3.736	0.003	0.074	0.042	0.177	0.027	9.933	11.212
Pr	0.411	0.007	0.010	0.005	0.026	0.003	1.159	1.316
Nd	1.439	0.004	0.042	0.020	0.139	0.014	4.678	5.359
Sm	0.232	0.002	0.007	0.011	0.027	0.010	1.037	1.194
Eu	0.038	0.003	0.003	0.003	0.007	<LOD	0.255	0.295
Gd	0.120	0.003	0.008	0.011	0.011	0.003	0.965	1.115
Tb	0.017	0.000	0.001	0.002	0.001	0.001	0.131	0.151
Dy	0.089	0.002	0.006	0.010	0.004	0.000	0.684	0.783
Ho	0.021	0.001	0.003	0.002	0.001	0.000	0.115	0.126
Er	0.093	0.008	0.010	0.010	0.003	<LOD	0.378	0.394
Tm	0.013	0.002	0.000	0.004	0.001	0.000	0.054	0.055
Yb	0.116	0.004	0.005	0.007	0.002	0.002	0.354	0.321
Lu	0.019	0.001	0.001	0.001	0.000	0.001	0.059	0.055
Hf	0.548	0.003	0.024	0.092	0.006	0.012	0.651	0.349
Ta	0.357	0.000	0.001	0.002	0.001	0.000	0.366	0.391
W	1.279	0.002	0.008	0.006	0.005	0.007	0.620	0.533
Au	0.001	0.004	0.001	0.003	0.002	0.002	0.007	0.006
Pb	1.611	0.009	0.031	0.040	0.017	0.028	11.628	13.245
Bi	0.015	0.002	0.003	0.009	0.011	0.006	0.087	0.098
Th	0.920	0.480	0.646	0.483	0.076	1.006	2.572	2.926
U	0.654	1.554	0.655	1.162	0.798	0.780	0.459	0.413

Chapter 49

Belovode obsidian in a regional context

Marina Milić

Introduction

Obsidian is known from numerous Neolithic settlements in the central Balkans (Figure 1). It is relatively common in the Starčevo period (*c.* 6000–5500 BC), and becomes most widely used throughout the early Vinča period (*c.* 5400–5000 BC), before a decline in the later phase of the Vinča culture (Tripković 2003a; Milić 2015). Even though the number of communities that engaged in the obsidian exchange network is significant, the overall percentage of this raw material in relation to other chipped stone types at each site is very small, often less than 1% of all lithics. The aim of this chapter is to give an overview of the consumption and exchange of obsidian between the Belovode community and contemporary settlement networks both within the Vinča culture and beyond, throughout the Balkans.

Obsidian in the Vinča culture settlements

Obsidian finds are documented at several settlements in the northern and central Balkans. These are interconnected through the Carpathian Mountains by a complex network of major rivers (the Danube, Tisza, Mureş, Körös) and their tributaries. The Morava river is one of the major tributaries of the River Danube, playing an important role in connecting the Carpathian and the Aegean basins, and this is particularly relevant for the distribution of obsidian throughout the central Balkans. This river, together with the Vardar (Axios) farther south, is believed to represent the main route-way between the Balkans and the Aegean (Chapman 1981), central to exchange networks in the Vinča period, as discussed below for the case of obsidian.

The earliest known occurrences of obsidian in the Balkans date to the Early and Middle Neolithic Starčevo communities that lived in the Danube regions from around 6000 BC, for example at the sites of Donja Branjevina, Starčevo, and Golokut. Current knowledge, from small excavated areas at several sites, shows that obsidian exchange only occasionally extended to the south of the Danube river, for example, to Livade, Popovića brdo, or Blagotin (Šarić 2002; Tripković 2003a). The artefacts that were exchanged in these early phases of the Neolithic were not a product of specialised manufacture. Obsidian is often found in the form of flakes, while the cores and regular blades are not known to have been a part of any excavated assemblages (pers. obs.).

The greatest expansion and use of Carpathian obsidian occurred during the Middle Neolithic and the early phase of Late Neolithic (middle Linearbandkeramik culture, with Bükk and Tisza variants in the Carpathians), which corresponds to the Vinča A–B1 phases (second half of the 6th millennium BC). During this period there was widespread consumption of Carpathian 1 obsidian in Balkan settlements located at great distances from the sources, sometimes up to 600 km 'as the crow flies'. Some of these communities used obsidian as a dominant raw material even though their distant location would not naturally promote such a pattern (that is, in accordance to projected fall-off curves relating to distance from source; see Renfrew *et al.* 1968). This is particularly visible at the site of Vinča-Belo Brdo, located 400 km from the sources. The obsidian assemblage here represented *c.* 70% of the overall lithic material in phase Vinča A–B1 (Radovanović *et al.* 1984) which can be dated to *c.* 5300–5000 BC. During this period, there were several settlements with larger concentrations of obsidian, seen as micro-regional centres for the redistribution of Carpathian obsidian. In addition to Vinča-Belo Brdo, there are also possible sites in the Vršac region (e.g. Potporanj-Kremenjak) where the estimated relative frequencies of obsidian are high at *c.* 30%[1] (Šarić 2002; Tripković 2003a). The larger amounts of obsidian that were in circulation in the Balkans within the Vinča A–B period are often linked to the existence of good relationships with communities that lived in the vicinity of the sources, i.e. the Bükk culture (Kaczanowska and Kozlowski 1990: 36; Radovanović *et al.* 1984; Voytek 1985: 249).

Characteristic of these early Vinča culture assemblages are micro-blade cores that were knapped into fine, regular bladelets. The cortical flake material discovered at Vinča-Belo Brdo and Potporanj-Kremenjak suggests that small, unprepared nodules were brought from the sources to sites where they were further transformed into micro-cores and blades. It is still uncertain whether they were directly procured or received through some exchange mechanisms (e.g. barter trade). In

[1] The quantities are still provisional as these sites were not systematically excavated and recorded.

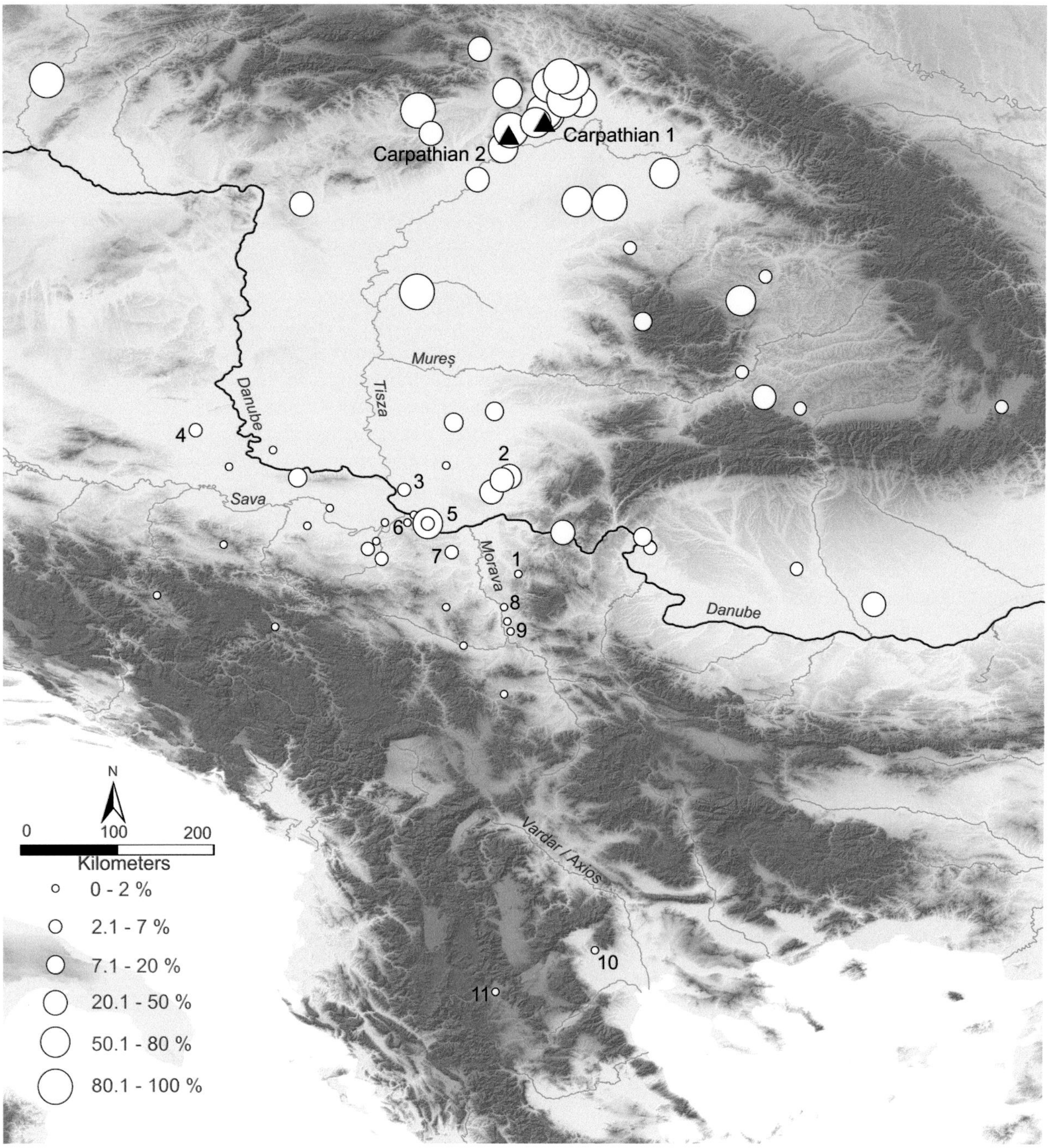

Figure 1. Map of the Late Neolithic Vinča sites mentioned in text with percentages of obsidian compared to other lithics: 1. Belovode; 2. Potporanj-Kremenjak; 3. Opovo; 4. Šamatovci; 5. Vinča-Belo brdo; 6. Banjica; 7. Selevac; 8. Supska; 9. Drenovac; 10. Mandalo; 11. Dispilio.

considering social relations, it is also significant that a rich assemblage is reported from Šamatovci in Slavonia (Croatia) dated to Vinča A–B1. Even though the overall obsidian percentage at Šamatovci is not as high as at Vinča-Belo-Brdo and the Vršac sites, the community at this site also produced many micro-cores and small, regular prismatic blades (Dimitrijević 1979).

Given the discrepancy of lithic assemblages at sites like Vinča-Belo Brdo, Potporanj-Kremenjak, or Šamatovci, one could presume that these early 'centres' possibly further served as redistribution places from which obsidian blades were exchanged with the settlements in the central Balkans, specifically during the late 6th millennium BC. In the Morava river region, obsidian finds are rare and currently only at the site of Selevac has a somewhat more significant assemblage been identified. Here, obsidian represents 5% of lithics in the Vinča A–B1 phases. Acquired cores were reduced on-site into micro-blades and waste (Voytek 1985). A similar picture is found at the sites of Vinča-Belo Brdo and Šamatovci however the overall quantity of obsidian found at Selevac is lower. Based on the current data, other assemblages in the region, including sample 7 from our campaigns at Belovode, (dated to Vinča A–B), represent only occasional occurrences of obsidian in lithics assemblages. It appears, therefore, that only certain centres were producing these micro-blades, while the others received them in finished form. As mentioned in Chapter 19, it should be emphasised that the obsidian artefacts from Belovode were recovered from a 5 x 5 m trench encompassing only the edge of the early settlement. We should, therefore, remain aware of the probability that a larger obsidian assemblage could be recovered if the early phases of the settlement are more fully explored.

The situation changes in the subsequent periods, in terms of the amount of obsidian at sites, the organisation of production and the geographical extent of circulation. At Vinča-Belo Brdo, which is rich in obsidian in the late 6th millennium BC, the material becomes rare and forms only *c.* 5% of the lithics assemblage in the early 5th millennium BC. Furthermore, from the onset of Vinča D2 period (after 4700 BC), obsidian appears only occasionally at this tell-site, representing less than 1% of the total chipped stone material (Radovanović *et al.* 1984; Tripković and Milić 2008). At this time, obsidian pieces are found mainly in the form of flakes or irregular blades and there is no indication that these were worked at the settlements, particularly those south of the Danube and Sava rivers (Milić 2015). To the north of Vinča culture territory, in Vojvodina, the picture is slightly different and there are still some possibilities that obsidian was worked on sites into regular prismatic blades. For example, at Opovo there are suggestions that obsidian cores were brought to the settlement in a prepared form and that they were treated carefully to produce micro-cores and blades (Tringham *et al.* 1985: 443).

Returning to the central Balkan sites and Belovode, we see that most obsidian artefacts (samples 1–6; Chapter 19) belong stratigraphically to the later Vinča phase (Vinča C–D2, dated to the period *c.* 4900–4500 BC). Similarly, at Selevac, in the final building horizon the percentage of obsidian decreases to less than 1%. Voytek (1985: 206–207) noted that several finished blades were recovered but there is no evidence for obsidian working in the settlement at this time. The other settlements south of the Sava and Danube follow this trend, including the Vinča-Belo Brdo community, which by this time seems to have lost its role as a potential regional 'distribution centre'.

Significant for this period, however, is that even though the overall obsidian presence at sites is low, Carpathian 1 obsidian circulation reaches its maximum spatial extent towards the south, overlapping with the Aegean zone in the southern part of the Balkans, in Greek Macedonia. Carpathian 1 obsidian is found at two settlements in western Greek Macedonia, at Mandalo (Kilikoglou *et al.* 1996; Milić 2015) and the lakeside site of Dispilio (Milić 2015). The site of Mandalo is dated to between *c.* 4600 and 4000 BC (Kotsakis *et al.* 1989), while the Dispilio Carpathian 1 obsidian finds could date to the Late Neolithic and Final Neolithic phases (between *c.* 5400 and 4500 BC) of the settlement however their precise contextual information is, unfortunately, not currently available.

Carpathian 1 pieces occur in very small amounts (less than 1% of assemblages) and in very random technological forms (mainly as flakes and irregular blades), mirroring the situation encountered within the central Balkan communities. These objects were most likely exchanged as finished artefacts, rather than nodules or even nuclei for blade production. In fact, their appearance seems to serve as a marker of novel exchange networks and technologies that developed in the central and southern Balkans in the 5th millennium BC.

Conclusion

Obsidian from the Carpathian sources, particularly from Slovakian Carpathian 1 localities, was exchanged within the Balkans from the Early Neolithic period, possibly from the late 7th millennium BC. Even though these territories can be considered as marginal for the distribution of Carpathian obsidian, its circulation in the central Balkan region can be followed throughout the following centuries, possibly until the end of the 5th or early 4th millennium BC. The peak of obsidian exchange is dated to the Mid to Late Neolithic periods (the second half of the 6th millennium BC). This phenomenon is documented in the areas closer to the sources (Bükk culture), but also in more distant regions, some 400 km south, where the Balkan Vinča culture communities lived. Some scholars argue for the existence of centres for obsidian re-distribution at the time of Vinča A–B1 (Williams-Thorpe *et al.* 1984). In these centres, obsidian nodules were brought, worked and obsidian blades were further exchanged. Such centres possibly existed at Vinča-Belo Brdo and Potporanj-Kremenjak, amongst other possible northern sites, where obsidian artefacts represent 30–70% of all lithics. At settlements south of the Danube and in the Morava river valley, Vinča communities would have been on the receiving end of these exchange activities, with obsidian forming only a small component (usually 1%) of the overall chipped stone assemblages. This would be the case with the one known example of obsidian flake of this date that we presently have at the site of Belovode.

In Vinča C2 and particularly Vinča D periods, the overall presence of obsidian in the circulation is low within the territory occupied by Vinča culture communities. The spatial extent of Carpathian 1 obsidian nevertheless extends beyond the realms of the Vinča culture, with examples recorded in the Aegean region at the sites of Mandalo and Dispilio. The system of exchange of these artefacts in the central and, possibly, southern Balkans is puzzling, since production and distribution centres are not known in the regions south of the Danube, as opposed to those presumed in Pannonia. Even the previously well supplied community at Vinča-Belo Brdo obtained only a relatively insignificant number of obsidian pieces in the later phases of the Vinča culture. Furthermore, micro-blade production (most likely performed by pressure-flaking on micro-cores), which was practiced in earlier Vinča periods, is not visible in this period. The exchanged obsidian artefacts are often present only in the form of irregular blades and flakes—only occasionally as prismatic blades—like the artefacts from Belovode in the Vinča D2 phase. A similar picture occurs at several sites south of the Danube and the Sava rivers and in the Morava river valley (e.g. Vinča-Belo Brdo, Banjica, Supska, Drenovac and Selevac).

The pieces of Carpathian obsidian that were exchanged in the later Vinča period were just remnants from the obsidian exchange networks that were established during the second half of the 6th millennium BC. To comment further, it would be necessary to examine a number of sites more extensively, with particular attention paid to the chronological, contextual and technological characteristics of obsidian assemblages. Based on currently available data, it is difficult to provide a more detailed image of obsidian exchange networks. For now, we might assume that the late Vinča obsidian exchange, at least in the regions south of the Danube, was carried out as a part of occasional or one-off events that were potentially piggybacking on the exchange of other less visible goods (e.g. salt; Urem-Kotsou 2016) or movement of people for other social reasons. This could indicate a degree of interaction amongst contemporary groups in the central and southern Balkans, along the Morava and Vardar (Axios) valleys. The dating of these finds to the Late Neolithic phase reminds us that these exchange networks could have been a part of novel social circumstances in which other materials became more widespread, with the occasional piece of exotic obsidian representing a part of that newly established exchange network. This is significant in considering interaction networks related to the emergence of the earliest metallurgy and the potentially new roles that communities such as that at Belovode came to possess in the central Balkans and wider region.

The bibliographic reference for this chapter is:

Milić, M. 2021. Belovode obsidian in a regional context, in Radivojević, M., Roberts, B. W., Marić, M., Kuzmanović Cvetković, J., and Rehren, Th. (eds) *The Rise of Metallurgy in Eurasia*: 570–573. Oxford: Archaeopress.

Chapter 50

Plant consumption at Belovode and Pločnik: a comparison

Dragana Filipović

Introduction

Plant remains recovered during new excavations at Belovode and Pločnik are a fresh contribution to the growing archaeobotanical archive from Neolithic sites in Serbia and the wider region (e.g., Borojević 2006; Filipović and Obradović 2013; Marinova 2007; Obradović 2020; Popova 2014; Reed 2015; Stojanova-Kanzurova and Rujak 2016). Until recently in Serbia, sampling for plant remains was employed at only a few Neolithic sites; at other sites concentrations of charred material were occasionally collected by the excavators but at a great majority of excavated sites, recovery of plant remains was not carried out. The detailed data now available for several Neolithic sites in Serbia, combined with those from sites in the neighbouring countries, will allow for both a site-specific and a more regional assessment of the plant evidence, and the study of different aspects of Neolithic crop cultivation and plant use.

As an initial step, the present chapter investigates the archaeobotanical evidence from Belovode and Pločnik side-by-side and highlights the similarities and differences between the two datasets that may shed more light on plant use in the region during this period. A general comparison of the charred crop and wild plant assemblages is conducted here on the basis of their composition, and on the number and frequency of the remains which potentially reflect the importance of different taxa in the plant economy. Evidence for crop and wild plant processing and use at the two sites is also discussed. Data patterning across the archaeological contexts and occupation horizons is explored using correspondence analysis.

Crop husbandry

Crop choice

The range of crops grown at Belovode and at Pločnik is identical. A similar array of crops has been identified at a further ten archaeobotanically sampled and analysed sites in Serbia that share elements of the Vinča culture (see Table 1). This crop package is characteristic of the Neolithic Balkans, but with differences across the region in terms of the presence or absence of some crops, e.g. absence of chickpea in the central and western part, based on the current evidence (Filipović 2014; Kroll 2013; Marinova and Popova 2008; Reed 2015); the scale of their presence, e.g. mass finds of grass pea in Greece and Bulgaria *vs.* only traces in the central and western Balkans (Filipović 2014; Marinova 2007; Valamoti and Kotsakis 2007; Valamoti *et al.* 2011; Reed 2013, 2015); and the timing of their appearance in various parts of the region, e.g. free-threshing wheat present from the Early Neolithic in Bulgaria and from the Late Neolithic in Serbia and Croatia (Filipović 2014; Marinova 2007; Reed 2015). The evidence from Belovode and Pločnik corroborates some previously observed trends in the Late Neolithic of the central and western Balkans: the general presence of einkorn, emmer, barley, lentil, pea and flax/linseed; slightly less common finds of bitter vetch and free-threshing wheat; low visibility of 'new type' hulled wheat (perhaps due to the relatively recent recognition as a separate wheat type) and grass pea; and the appearance of spelt wheat at some sites of this period (e.g. at Jagnjilo in Bosnia – Kroll, H. pers. comm.). The crop spectrum of the Late Neolithic is more diverse than that documented at Early Neolithic sites in Serbia (and elsewhere, e.g. Reed 2015). This trend coincides with a number of other developments associated with the emergence of the Late Neolithic Vinča culture phenomenon around 5300 cal. BC (cf. Bailey 2000; Chapman 1981; Garašanin 1984; Tringham 1971).

There are both similarities and differences in the level of occurrence of different crop types at Belovode and Pločnik. At both sites, einkorn, emmer and 'new type' hulled wheat are the most frequent and abundant crops (Figure 1 and Figure 2) and they are mainly represented by glume bases. Among them, einkorn appears more common although, as noted in the reports for the two sites, this may be a product of poor preservation of the majority of hulled wheat glume bases (of which a significant number were only broadly identified as emmer/'new type' or indeterminate glume bases). Free-threshing wheat (including tetraploid and hexaploid types) and barley occur in similar frequencies at the sites (Figure 1), but free-threshing wheat appears more abundant at Pločnik (Figure 2) where a large concentration of tetraploid free-threshing wheat rachis was discovered in an oven. Lentil is ubiquitous at both sites, although slightly less so at Pločnik; pea is much less common and is somewhat more frequent

Table 1. Crop types documented at Late Neolithic/Vinča culture sites in Serbia.

Archaeological site	Pavlovac	Drenovac	Motel Slatina	Belovode	Selevac	Jaričište 1	Pločnik	Gomolava	Vinča - Belo Brdo	Opovo
number of analysed flotation samples	185	440	2	41	53	2	68	41	195	267
einkorn	X	X	X	X	X		X	X	X	X
emmer	X	X	X	X	X	X	X	X	X	X
'new type' hulled wheat	X	X		X		X	X		X	(X)
free-threshing wheat	X	X	X	X			X	X	X	
barley	X	X		X	X		X	X	X	X
lentil	X	X	X	X	X		X	X	X	X
pea	X	X		X	X		X	X	X	
bitter vetch	X	X		X			(X)		X	
grass pea	X			X			X			
flax/linseed	X	X		X		X	X	X	X	X
reference	Obradović 2013, 2020	Filipović, Obradović 2013; Obradović 2013, 2020	Filipović, Obradović 2013	Filipović, this volume (Ch. 20)	Hopf 1974; Renfrew 1979; McLaren, Hubbard 1990	Borojević, Sheridan 2009	Filipović, this volume (Ch. 34)	van Zeist 2002	Filipović, Tasić 2012; Filipović *et al.* 2019	Borojević 2006

at Belovode than Pločnik. Bitter vetch is relatively frequent at Belovode (found in over 40% of the samples), whilst it is almost entirely absent at Pločnik (where a single cf. bitter vetch seed was recorded). The occurrence of grass pea is minimal at both sites. Flax/linseed at Belovode is not very frequent but is present in more samples than, for instance, free-threshing cereals and some of the pulses. At Pločnik, however, a single seed was found.

Based on the clear predominance of hulled wheats in the assemblages, it is evident that they formed the basis of cereal production at Belovode and Pločnik. Even when hulled wheat glume bases are disregarded, hulled wheat grains are more common than free-threshing cereal grains and rachis combined, especially at Belovode. This is shown in the site-specific reports in this volume and is here illustrated in Figure 3 (where proportions of different crops/crop parts are presented). Also shown in Figure 3 are the proportions of pulse at the two sites. The similar relative amounts of hulled wheat grain and pulse seed at Belovode perhaps indicate the high importance of pulse crops here (lentil and bitter vetch – see Figure 1 and Figure 2), in contrast to their apparently minor role at Pločnik.

Crop processing

The evidence, in the form of glume bases, of hulled wheat processing is remarkably ubiquitous at both sites. It implies storage in spikelets and piecemeal dehusking for preparation of food (cf. Hillman 1981; Jones 1984). Ethnographic studies of traditional crop cultivation reveal that earlier stages of crop processing (threshing, winnowing, coarse sieving) tend to be undertaken outdoors, perhaps near the fields or on settlement edges, whereas later stages (fine sieving, additional winnowing, hand picking) take place adjacent to or within domestic areas (Hillman 1984a; Jones 1984; Reddy 2003). Storage of crops often occurs following the early and prior to the late processing stages. In the Neolithic in the central Balkans, crops (and probably other plant food) seem to have been stored in a variety of ways: in large and small ceramic vessels, storage bins and pits, and perhaps

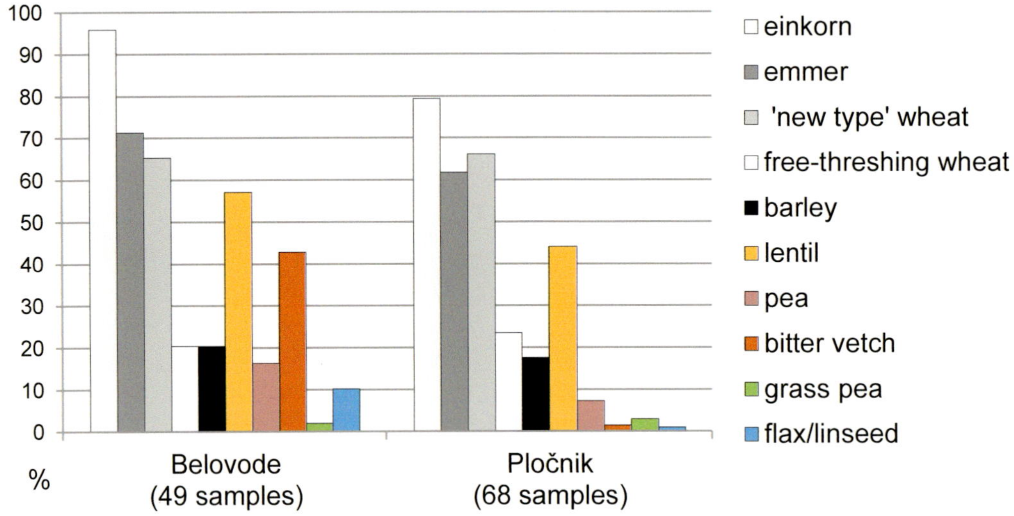

Figure 1. Percentage of samples (ubiquity) in which different crop types occur at Belovode and Pločnik.

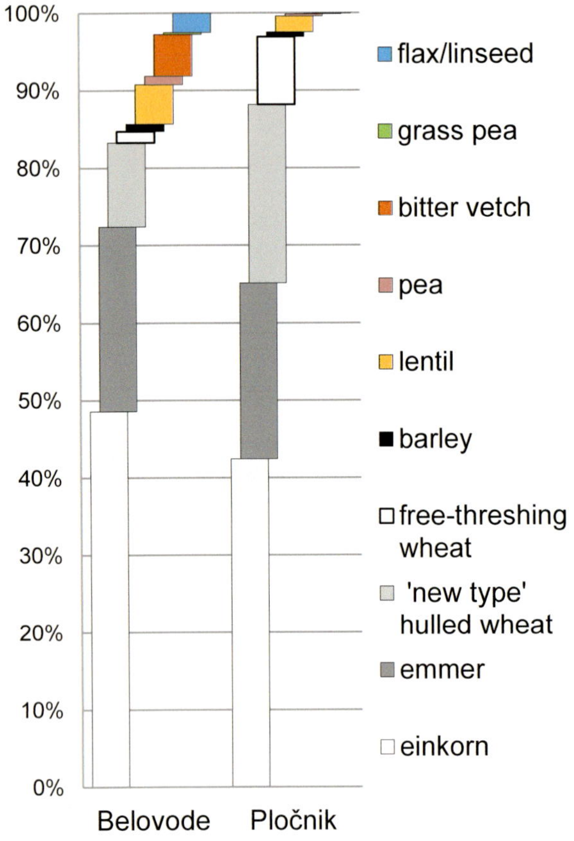

Figure 2. Relative proportions of crops in the assemblages from Belovode and Pločnik.

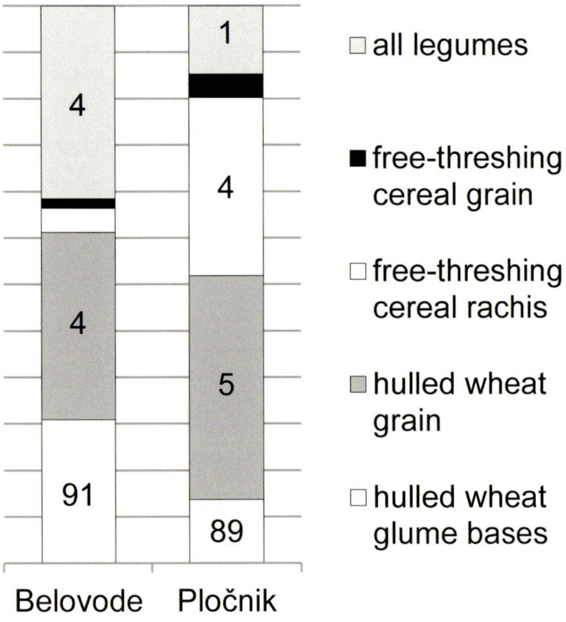

Figure 3. Comparison of proportional representation of selected crops/crop parts at Belovode and Pločnik (data labels indicate the percentage values).

also in bags made of leather, textile or other perishable material (e.g. Borojević 2006; Filipović *et al.* 2018; Hopf 1974; Stevanović 1997; Tripković 2011). Ethnographic research also shows that by-products of cleaning of crops taken from stores (in the case of hulled wheats this would comprise hulled wheat glume bases and weed seeds) are commonly thrown into domestic fires (Hillman 1981, 1984a; Jones 1984; van der Veen 2007) and this is a possible route of preservation of the large number of charred glume bases at these sites.

Arable weeds

Seeds of wild taxa (potentially representing arable weeds) within crop processing by-products are found in very small numbers, which is typical for the Balkan Neolithic (e.g., Borojević 2006; Filipović and Obradović 2013; Marinova 2007; Obradović 2020; Reed 2013, 2015; Valamoti 2004). Some taxa are more frequent than others (e.g. they occur in 5% or more samples). These include *Bromus* species and indeterminate large-grained Graminae, *Chenopodium* (primarily *album* aggregate), *Fallopia convolvulus*, *Polygonum* (*aviculare*), Solanaceae (mostly *Solanum nigrum*) and *Setaria viridis/verticillata*. A few other taxa are relatively common at Belovode whilst rare at Pločnik: *Galium aparine*, indeterminate member(s) of Lamiaceae family and *Trifolium* (mainly *repens* type). From a crop processing perspective, these include both large-seeded and small-seeded taxa (*sensu* Jones 1984, 1987), that is, seeds smaller than cereal grain that would have been removed together with hulled wheat glume bases in fine sieving, and seeds similar in size to cereal grains that would be picked out by hand (Hillman 1981, 1984a, 1985; Jones 1984, 1987). This supports the impression that, based on their composition, the crop assemblages reflect later stages of crop processing.

The wild/weed taxa documented at Belovode and Pločnik generally grow in disturbed habitats and in nutrient-rich conditions, such as those found in intensively cultivated, long-established plots (cf. Jones *et al.* 1999). The ecological composition of the wild/weed assemblage could point to a regime of intensive, small-scale crop cultivation – a practice demonstrated for Neolithic central Europe (Bogaard 2004) and inferred for Neolithic sites in the Great Hungarian Plain and the Teleorman Valley in south-central Romania (Bogaard *et al.* 2007; Walker and Bogaard 2011). The range of possible weed taxa from Belovode and Pločnik is very similar to the wild/weed record documented at other analysed Neolithic sites in the Balkans (Filipović and Obradović 2013; Marinova 2007; Reed 2013; Valamoti 2004), suggesting that intensive cultivation may have been a widely applied crop husbandry regime in the region (cf. Bogaard and Halstead 2015). The diverse crop spectrum characteristic of these sites could also be indicative of this, since extensively cultivated large fields are normally used for growing a narrow range of crops (van der Veen 2005). The possibly small scale of arable production is consistent with the lack of evidence of extensive anthropogenic influence on the vegetation, as shown by the regional pollen and wood charcoal analysis (Marinova and Thiébault 2008; Marinova *et al.* 2012, 2013; Willis 1994).

The potential weed evidence from the central Balkans is still too limited for detailed investigation into agricultural practices such as sowing and harvesting methods and timing, field size and location, field management (e.g. tilling, weeding) and so on. Some initial observations have been offered, such as the likely autumn sowing/summer harvest of cereal crops (Borojević 2006; Reed 2015); harvesting of cereal ears only, or ears separately from the straw (Kreuz *et al.* 2005); and possible intercropping (of einkorn and emmer) in some places to increase and secure yields (Borojević 2006; Filipović 2014; Reed 2015). Faunal evidence, including that from the wider region, offers some additional insight into arable cultivation. For instance, sheep (and goat) may have been an integral part of Neolithic crop husbandry by being kept in arable fields where they would have grazed on stubble and contributed to soil fertility (Bogaard and Halstead 2015; Gillis *et al.* 2020; Halstead 1996); cows may have been used to some degree for ard-ploughing and traction (Isaakidou 2006; see also Balasescu *et al.* 2006).

Collection of wild plants

Besides crops, plant production and consumption at Neolithic Belovode and Pločnik benefitted from the collection of various wild edible fruits and nuts which enriched the diet. Figure 4 shows the percentage ubiquity of the recorded wild-gathered taxa (*Prunus* sp. here includes *P. spinosa* and *P.* cf. *domestica* var. *insititia*; *Rubus* sp. combines *R. idaeus/fruticosus* and *Rubus* sp.). Figure 5 illustrates overall proportions of the taxa. Based on frequency, Cornelian cherry (*Cornus mas*) seems to have been highly valued at both sites, along with Chinese lantern (*Physalis alkekengi*) and most likely blackberry (*Rubus* sp.). The quantity of *Cornus mas* is lower at Belovode, where hazelnut (*Corylus avellana*) is found more frequently and in greater number than at Pločnik. Other taxa are present in similar proportions, but their frequency varies; for example, blackberry and wild strawberry (*Fragaria vesca*) are more common at Pločnik.

The differences in the level of occurrence of certain fruit and nut taxa at the two sites may potentially reflect preferences for particular fruits, or perhaps their varied availability in the surrounding landscape. They may also, to some extent, pertain to the preservation potential of the remains depending on the archaeological deposits from which they were recovered. For instance, in comparison to Belovode, Pločnik yielded more samples from primary contexts (*in situ* deposits) where charred remains could have had greater chances of being preserved in identifiable form than in secondary and tertiary contexts where the remains would have been trampled or re-deposited. However, the majority of Pločnik non-primary contexts also contained fruit/nut remains, some in relatively high quantities, thus the nature of the context may not be relevant to the observed differences. In any case, the overall number of remains of the collected taxa is low and this limits the basis for discussion of possible variations in the representation of fruit/nut taxa within and between the sites.

The fruit/nut taxa registered at Belovode and Pločnik were previously reported for a number of Early and Late Neolithic sites in Serbia and the wider region

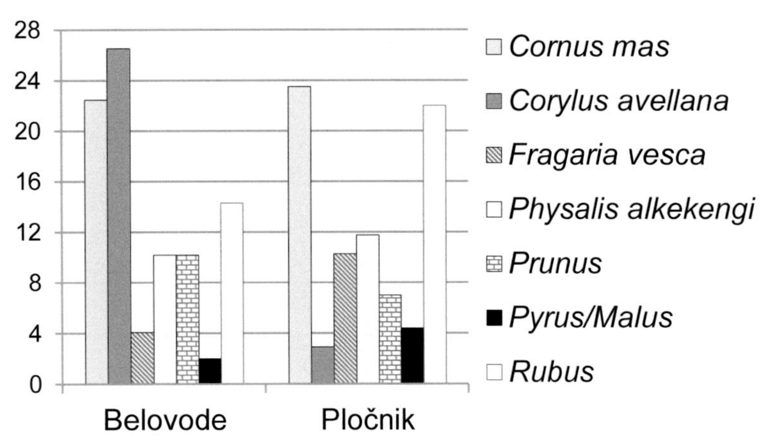

Figure 4. Percentage of samples (ubiquity) in which different fruit/nut taxa occur at Belovode and Pločnik.

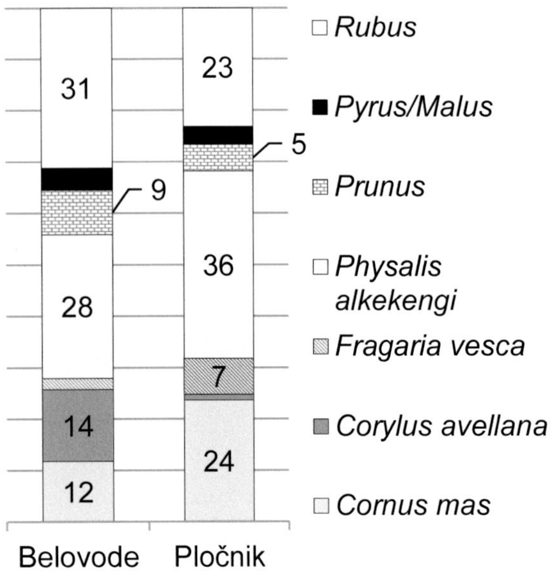

Figure 5. Comparison of proportional representation of fruit/nut taxa in the Belovode and Pločnik assemblages (data labels indicate the percentage values).

(Borojević 2006; Filipović and Obradović 2013; Marinova 2007; Obradović 2020; Reed 2013; Valamoti 2004). This selection of wild plant resources seems to have a long history of use and was, overall, consistent across the Neolithic Balkans. It also seems to have been widely available in the landscape. The palynological, anthracological and archaeobotanical records so far produced for the Neolithic of the region suggest that mixed oak forests with well-developed undergrowth were the main vegetation formation in eastern, central and western Balkans (e.g. Filipovitch *et al.* 1998; Marinova and Thiébault 2008; Marinova *et al.* 2012, 2013; Schroedter *et al.* 2011; Willis 1994). The generally wooded landscape, however, also included areas of open, light-demanding vegetation, perhaps in the form of wooded steppe or more open remnants of the Late Glacial steppe vegetation (e.g. Sümegi *et al.* 2012). Sparsely forested zones, along with oak forest edges and clearances, would have been the prime, easily accessible sources of these wild fruit. The remarkable geographic and temporal continuity in the choice of collected wild fruit in the Neolithic Balkans suggests that there may have been some investment in maintaining and expanding productive areas. This could have involved the reduction of forests, which would also have secured more land for pastoral and arable use. Thus, the fruit-collection areas with more open vegetation may have existed as natural elements but could equally have been a result of anthropogenic impact. The currently available palaeovegetation evidence for the Balkans does not, however, suggest major human impact on the vegetation during the Neolithic (e.g. Edwards *et al.* 1996; Marinova *et al.* 2012, 2013; Willis and Bennet 1994). The naturally existing zones of open vegetation during this period may have been sufficient to cover needs, and any potential management directed towards promotion and securing the growth of the desired fruit sources may have been strategic and carried out on a small scale (e.g. Filipović *et al.* 2017).

Patterns in the composition and distribution of plant remains

The archaeobotanical evidence from Belovode and Pločnik derives from a restricted excavation area and, for the most part, from contexts and deposits that seem to have been disturbed in prehistory or contain material of secondary origin coming from various, often inseparable sources (e.g. fills, construction materials, refuse deposits). This means that the datasets are of coarse resolution and of limited use for detailed investigation of plant-related activities and trends in plant consumption. Nonetheless, some general observations can be made on plant use and plant deposition through time (see archaeobotanical reports in this volume). Furthermore, some of the analysed deposits do seem to have preserved the 'original' content (i.e. *in situ* burnt plant material) whose composition potentially reflects distinct formation processes and can perhaps be correlated with the composition of the 'mixed' (secondary) deposits.

In order to explore the variability in the assemblage and identify potentially meaningful similarities and differences in the botanical composition within and between contexts/deposits and occupation horizons, correspondence analysis was carried out for the two datasets. The correspondence analysis was applied to combined crop and wild assemblages separately for the two sites using absolute counts of the remains. Prior to the analysis, all samples from the same context or arbitrary layer (spit) were amalgamated in order to reduce the number of samples in the analysis. Also, contexts/deposits containing fewer than 30 items were excluded, leaving 27 contexts from Belovode and 30 contexts from Pločnik. The computer software *Canoco for Windows 4.5* (ter Braak and Šmilauer 1997–1999) was used to run the analysis and *CanoDraw for Windows* (Šmilauer 1992) was used to produce the graphs, which show the arrangement of archaeological contexts along axis 1 (horizontal axis) and axis 2 (vertical axis). Botanical composition of the samples is illustrated using pie-charts based on item counts of different crop types/groups, combined fruit/nut taxa and combined wild/weed taxa.

Belovode

Figure 6 shows the distribution of the contexts classified according to the occupation horizon. As previously highlighted, there is virtually no variation in plant composition within and between different occupation phases. The vast majority of the contexts are dominated by hulled wheat glume bases, as shown in Figure 7; the exceptions are two outliers which contain much lower quantities of hulled wheat glume bases than the other contexts. One of these (burnt deposit Feature 31, Horizon 3, Sample 35) contains relatively high amounts of bitter vetch, fruit (primarily Chinese lantern and blackberry) and wild/weed (mostly clover seeds). The other (daub structure Feature 15, Sample 22) is relatively rich in

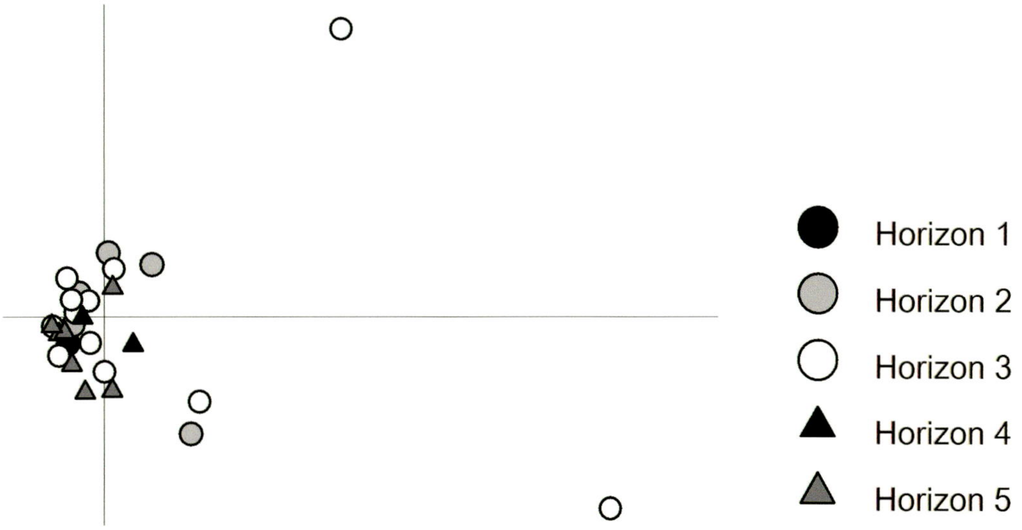

Figure 6. Correspondence analysis plot showing distribution of 27 Belovode archaeological contexts classified based on the occupation horizon.

Figure 7. Correspondence analysis plot showing relative proportions of main plant groups within the 27 archaeological contexts from Belovode.

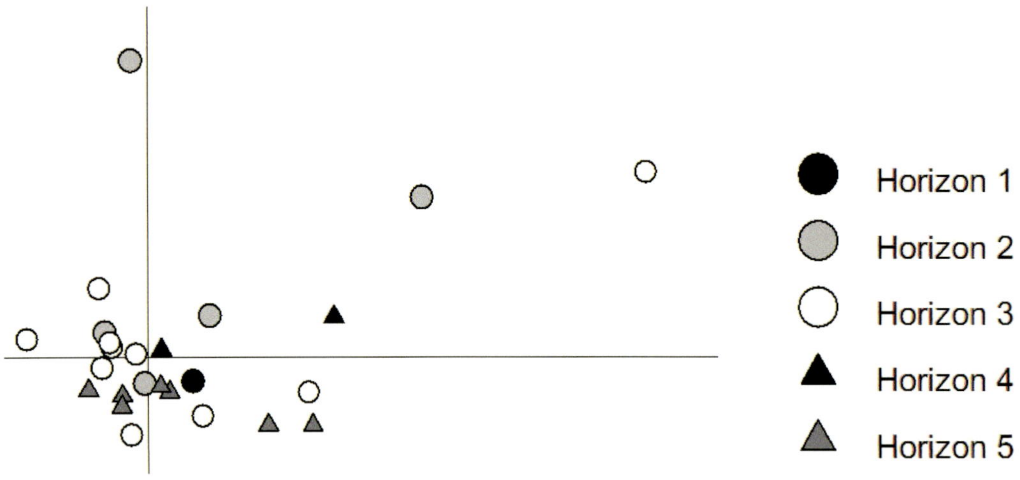

Figure 8. Correspondence analysis plot showing distribution of 25 Belovode archaeological contexts classified based on the occupation horizon.

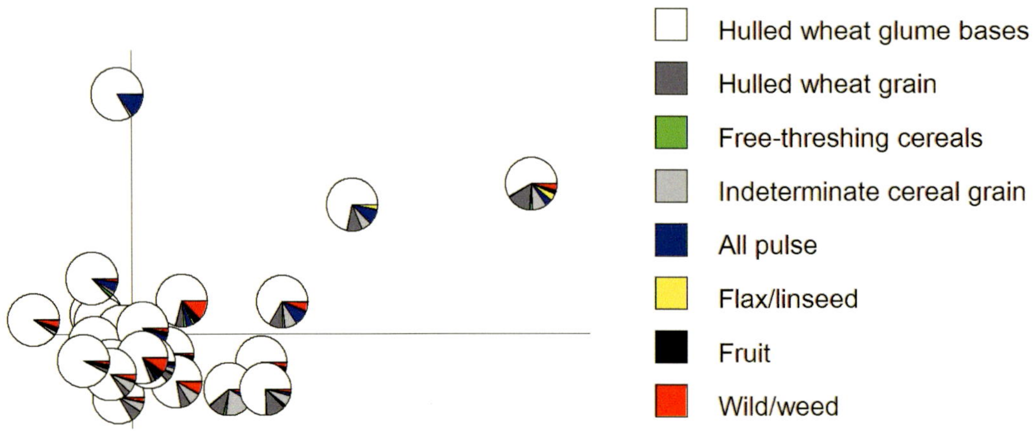

Figure 9. Correspondence analysis plot showing relative proportions of main plant groups within the 25 archaeological contexts from Belovode.

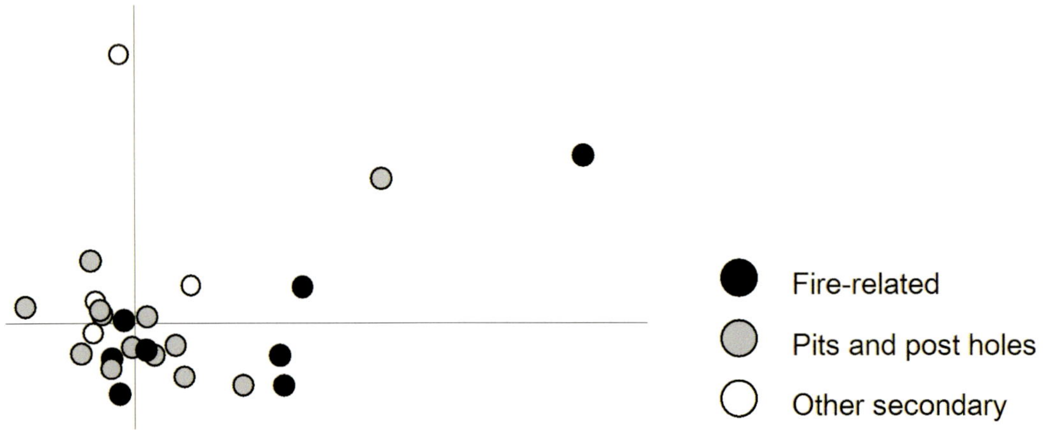

Figure 10. Correspondence analysis plot showing distribution of 25 Belovode archaeological contexts classified based on their summarised (depositional) category.

flax/linseed whilst it also contains small quantities of pulses. The distribution of the contexts in the plot does not significantly change after exclusion of the two outliers. Figure 8 shows the now 25 contexts classified by horizon. Most contexts still cluster around the origin; several contexts from different horizons are pulled away towards the right end of axis 1, and the top of axis 2. Figure 9 reveals that these comprise more pulse, flax and/or cereal grain than other contexts. For instance, the context at the top of axis 2 (arbitrary layer, Horizon 2, Sample 34) is another with relatively abundant bitter vetch seeds. Thus, the weak variability in the composition of the contexts relates to the quantity of hulled wheat glume bases *versus* the quantity of other admixtures. Figure 10 shows the summarised archaeological category of the contexts and, once again, indicates the overall compositional similarity between different (primary, secondary) contexts and horizons preventing the identification of specific processes or events by which the remains were (re-)deposited across and through the site. On the other hand, the uniformity in botanical composition suggests that the assemblage was produced by a constantly repeated, plant-related activity or set of activities (e.g. Fuller and Weber 2005; Fuller *et al.* 2014). Day-to-day crop processing and discard of crop processing by-products are the major activities relevant to the formation of the Belovode plant record. The consistency of the plant record through time perhaps reflects the unchanged tradition of these practices.

Pločnik

Figure 11 shows the distribution of the analysed contexts based on the occupation horizon from which they derive. Horizon 1 is not represented because the contexts contained fewer than 30 remains and were thus excluded. Figure 12 shows the proportions of main plant taxa/groups within the contexts. It was previously observed that most contexts are mainly composed of hulled wheat glume bases; in the plots, they form a group around the origin. Several contexts, however, are pulled away from the origin, towards the right end of axis 1. They are different in that they contain 50% or more of the material other than hulled wheat glume bases – for example, hulled wheat grain, indeterminate cereal grain and pulse, but also fruit and wild/weed seeds, such as the context located towards the top of axis 2 with relatively high amount of Cornelian cherry, Chinese lantern and blackberry (post-hole Feature 37, Horizon 4). As shown in Figure 13, these contexts are primary (in these cases – burnt deposits) or secondary (pits, arbitrary layers), as are contexts within the group at the origin. Their composition, which is somewhat different to that of the contexts displaying a clear crop processing by-product signature, indicates sources of material additional to crop cleaning: the cereal grain- and pulse-rich contexts may contain traces of burnt crop stores or food preparation accidents, and the fruit-rich context hints at fruit consumption. The species plot, Figure 14, confirms this as it reveals

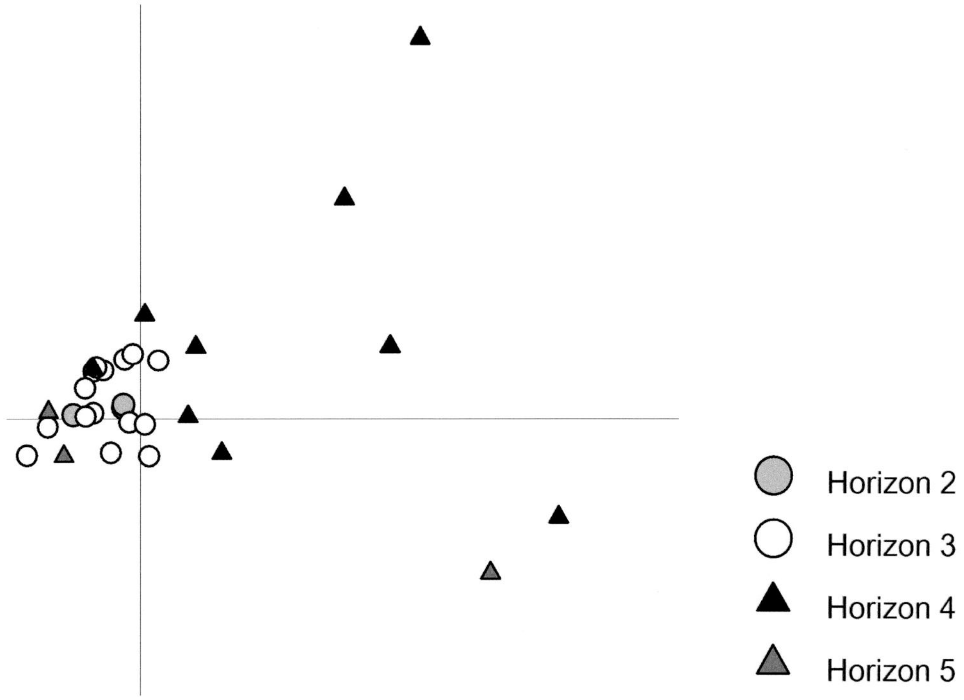

Figure 11. Correspondence analysis plot showing distribution of 30 Pločnik archaeological contexts classified based on the occupation horizon.

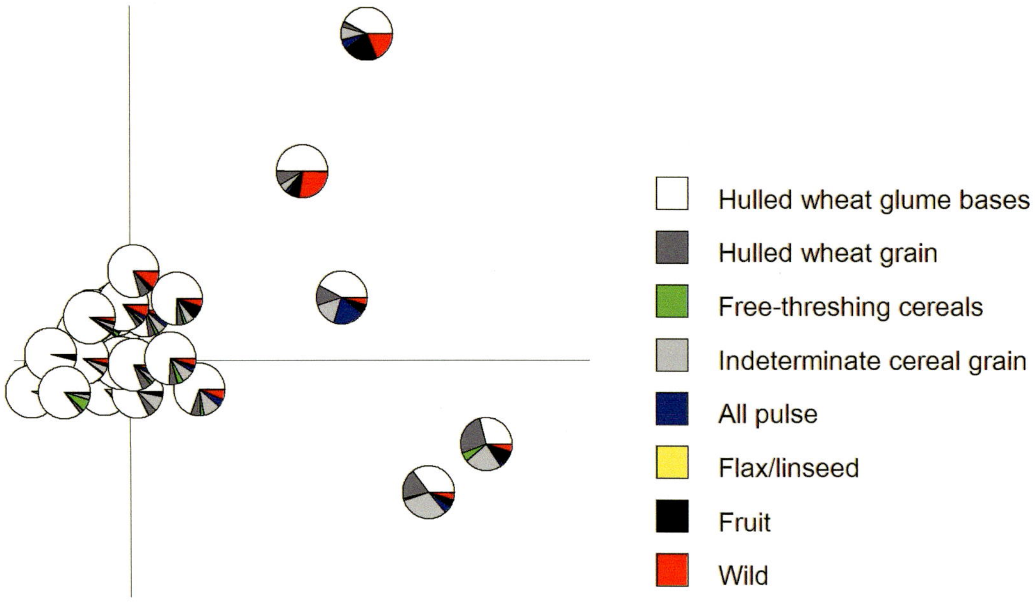

Figure 12. Correspondence analysis plot showing relative proportions of main plant groups within the 30 archaeological contexts from Pločnik.

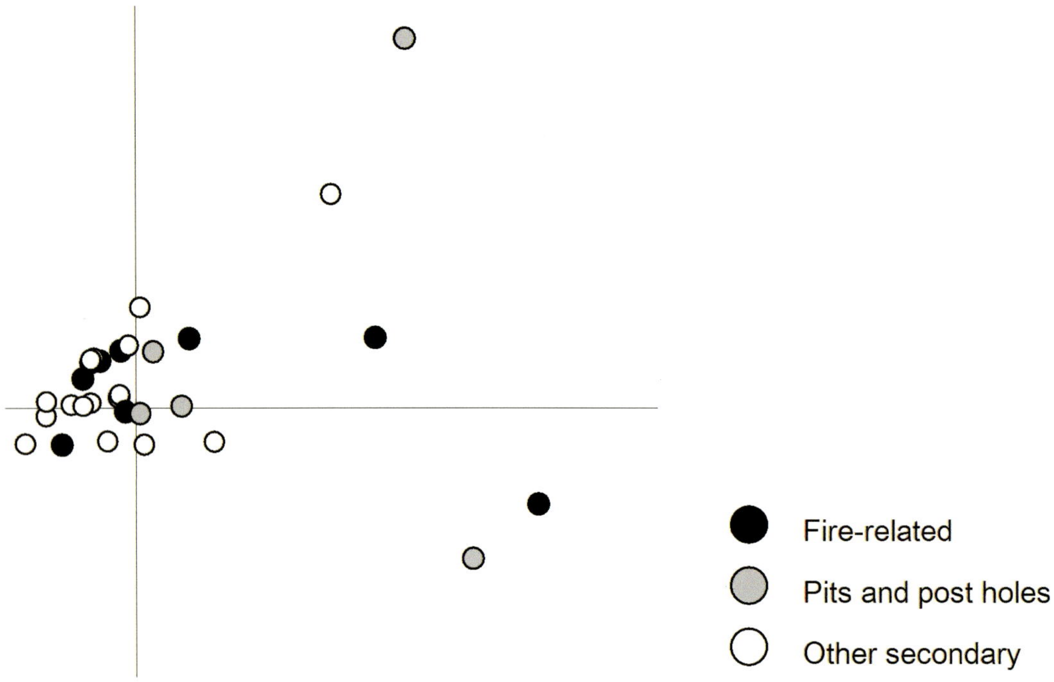

Figure 13. Correspondence analysis plot showing distribution of 30 Pločnik archaeological contexts classified based on their summarised (depositional) category.

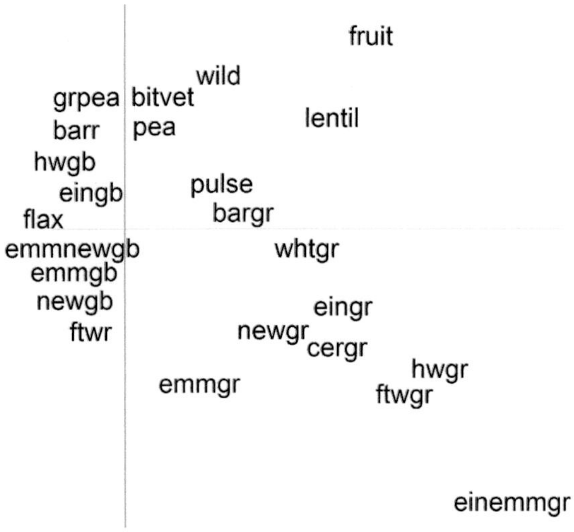

Figure 14. Correspondence analysis species plot showing distribution of main plant taxa registered in the 30 archaeological contexts from Pločnik; for cereal types, grain and chaff are shown separately.

a clear distinction between crop by-product (hulled wheat glume bases and free-threshing cereal rachis) concentrated near and at the origin, and cereal grain and other food (pulse, fruit) towards the right end of axis 1 (see Table 2 for descriptions of the codes used in the plot). The potential food-containing contexts derive specifically from occupation Horizons 4 and 5; this is not seen as significant in the temporal sense because contexts belonging to these phases are also present within the group dominated by crop processing by-product (and are of both primary and secondary character). Thus, in terms of activities and established practices producing the remains, the picture at Pločnik is similar to that at Belovode. Unlike Belovode, however, the composition of the Pločnik assemblage is somewhat more diverse as, besides crop processing, it preserved more visible evidence of plant storage, preparation and consumption.

Conclusions

Crop production at Belovode and Pločnik relied on a wide range of crops. Hulled wheats— einkorn, emmer and 'new type' wheat—probably played the most important part; einkorn appears more prominent than the other two hulled wheat types, but this could be due to identification issues. The evidence of hulled wheat processing in the form of glume bases is ubiquitous at the two sites and dominates the assemblages. It suggests storage of hulled wheat in spikelets, and piecemeal dehusking. Free-threshing wheat and barley are present in small quantities and were perhaps minor crops. Four pulse crops were recorded: lentil, bitter vetch, pea and grass pea. Their occurrence varies between the sites and they seem more important at Belovode, particularly bitter vetch which is virtually absent at Pločnik. A similar spectrum of crops is registered at other contemporary sites in the region and is wider than the range documented at Early Neolithic sites in the central Balkans. The appearance of free-threshing wheat, bitter vetch and grass pea in the Late Neolithic may be associated with the economic and technological innovations characterising the Vinča culture phenomenon.

Seeds of potential arable weeds are frequent but present in very small numbers and thus their use for inferring crop growing conditions and practices is limited. Even so, based on the similarity in the composition of the wild/weed assemblage with the ecologically analysed arable weed records from the neighbouring areas and central Europe, the possibility of small-scale intensive cultivation of crops is suggested.

The plant-based diet reliant on crops was supplemented by collected wild edible fruits and nuts. The ecology of fruit/nut species indicates exploitation of small trees and shrubs growing along forest-edges and in open

Table 2. Codes for taxa included in the correspondence analysis.

Code	Taxon	Code	Taxon	Code	Taxon
eingr	einkorn grain	hwgr	hulled wheat grain	lentil	lentil
eingb	einkorn glume base	hwgb	hulled wheat glume base	grpea	grass pea
emmgr	emmer grain	ftwgr	free-threshing wheat grain	pea	pea
emmgb	emmer glume base	ftwr	free-threshing wheat rachis	bitvet	bitter vetch
einemmgr	einkorn/emmer grain	bargr	barley grain	pulse	indeterminate pulses
newgr	'new type' grain	barr	barley rachis	flax	flax/linseed
newgb	'new type' glume base	whtgr	indeterminate wheat grain	fruit	fruit/nut taxa
emmnewgr	emmer/'new type' grain	cergr	indeterminate cereal grain	wild	wild/weed taxa
emmnewgb	emmer/'new type' glume base				

places. A similar choice of collected taxa was recorded at several other Neolithic sites in the Balkans. The regional uniformity and temporal continuity in the crop production are mirrored in the wild plant use.

Both assemblages provide some evidence, variably represented at the two sites, of plant *chaîne opératoire* – storage, processing, consumption and discard. The analysis of the composition of plant remains and their distribution within and across the occupation horizons confirms the previous observation that the same set of plant-related activities produced both assemblages. Moreover, those activities remained constant through a long period of time. It is, however, important to note that the observations are based on the evidence from relatively small excavation areas and may not apply on the settlement level.

Comparison of the evidence from Belovode and Pločnik and that from the wider region reinforces the impression of general continuity and consistency in the plant practice across the Neolithic central Balkans. Nevertheless, this does not exclude variations in plant use on a sub-regional or site-specific level. Indeed, some differences, though minor, were noted between Belovode and Pločnik plant datasets, and could now be compared with the results of the analysis of other archaeological materials from the two sites (e.g. faunal remains) and assessed against the charred plant archives from contemporary sites in the broader region.

The bibliographic reference for this chapter is:

Filipović, D. 2021. Plant consumption at Belovode and Pločnik: a comparison, in Radivojević, M., Roberts, B. W., Marić, M., Kuzmanović Cvetković, J., and Rehren, Th. (eds) *The Rise of Metallurgy in Eurasia*: 574–584. Oxford: Archaeopress.

Chapter 51

Evidence for animal use in the central Balkan Neolithic across the early metallurgical horizon: the animal remains from Belovode and Pločnik in context

David Orton, Jelena Bulatović and Ivana Dimitrijević

Introduction

The last decade has seen a considerable increase in the amount of data available on animal remains from Vinča sites (Blažić and Radmanović 2011; Bulatović 2012, 2018; Bulatović and Spasić 2019; El Susi 2011; Gaastra *et al.* 2019; Greenfield 2014, 2017; Orton 2012; Orton *et al.* 2016; see discussion in Stojanović and Bulatović 2013). The majority of these data, however, relate to sites in the latter part of the period, that is Vinča B2–D according to the periodisation of Milojčić (1949); Vinča-Gradac and Vinča-Pločnik following Garašanin (1973); or *c.* 5000–4600/4500 cal. BC in absolute terms (Radivojević and Rehren 2016; Whittle *et al.* 2016). Meanwhile, recent research on the origins of metallurgy in the region, as presented in this volume and elsewhere (Radivojević and Rehren 2016; Radivojević *et al.* 2010a, 2013; Whittle *et al.* 2016), places the threshold of systematic copper metallurgy in the region at approximately the same horizon, *c.* 5000 cal. BC.

Taken in this context, the animal bones from deep stratigraphic sequences at Belovode and Pločnik represent an opportunity (a) to bolster the available data from the earlier part of the period, making a comparison between late 6th and early 5th millennium assemblages more viable, and (b) to assess twin time-transects across the early metallurgical horizon, at sites which are known to have been involved in some of the earliest metallurgical activities.

The extent to which metallurgical innovations within the Balkan Neolithic were linked to, or implicated in, wider changes in social organisation, settlement forms, and landscape use remains an open question. The coincidence of the metallurgical horizon with the onset of the Gradac Phase is intriguing but may partly reflect our continuing reliance on relative dating alongside the small but growing number of available ^{14}C dates. The thresholds in the record will inevitably gravitate towards typo-chronological breakpoints where absolute dating is limited. Moreover, the transition between early Vinča (A–B1 or Tordoš) and late Vinča (B2–D or Gradac/Pločnik) is not characterised by any clear shift in the settlement evidence. Rather, it occurs against a backdrop of gradual trends towards increasingly large and permanent settlements with substantial architecture, a phenomenon that has been discussed at length elsewhere, along with its possible social significance (e.g. Chapman 1990; Kaiser and Voytek 1983; Orton 2008: 13–17; Porčić 2019; Tringham and Krstić 1990a; Tringham *et al.* 1985; Tripković 2003b, 2013; Whittle 1996: 105).

Suggestions of a late Vinča population decline and settlement dispersal (Brukner 2003; Chapman 1990; Tringham 1992) are undermined by very large final Vinča settlements at Stubline-Crkvine (Crnobrnja *et al.* 2009), Mali Borak-Crkvine (Arsić *et al.* 2011), and Divostin (McPherron and Srejović 1988) as well as continued occupation of the type-site at Belo Brdo (see dates in Tasić *et al.* 2015; Whittle *et al.* 2016). Nor is such a late-Vinča decline clearly evident when summed probability distributions of calibrated radiocarbon dates are used as a population proxy (Porčić 2020). Indeed, the most profound break visible in the settlement record of the region falls not at the turn of the 6th–5th millennium, nor even at the mid-to-late-6th millennium transition from the Starčevo-Körös-Criş complex to Vinča – which remains relatively obscure in terms both of dating and subsistence data – but rather at the very end of the Vinča period, around 4500 cal. BC, when the large village settlements disappear suddenly, giving way to a sparse settlement pattern of what appear to be small farmsteads (e.g. Palavestra *et al.* 1996; Trbojević 2005; Tripković and Penezić 2017; see also Bankoff and Winter 1990). It is acknowledged that the long-lived tell at Bubanj is an exception but that may point towards a different pattern in the south (see Bulatović and Milanović 2020; Trajković-Filipović *et al.* 2008). The development of metallurgical technology has sometimes been implicated in this change (e.g. Jovanović 1979) but, as indicated above, is now known to pre-date it by several centuries.

With regard to the extent that metallurgical innovations may have stimulated economic and societal change within late Neolithic communities –

or may themselves have been stimulated by ongoing processes – animal remains are one dataset that might be expected to reflect such change. Both herding and hunting represent socially situated practices, reflecting choices that are subject not only to environmental limitations but also to logistical considerations and to social incentives. Changes in social organisation, mobility patterns, or landscape use are all likely to have an impact on animal use, although the nature of this impact is hard to predict *a priori*.

This paper consists of two parts. Firstly, we discuss some of the current challenges facing zooarchaeology of the Neolithic in the central Balkan region, and how they might be mitigated. Secondly, we synthesise the available zooarchaeological data in order to assess any shifts in animal use across the early metallurgical threshold. This is, by choice, a top-down, exploratory affair, aiming to detect patterning – or lack thereof – on a regional scale rather than drilling down into processes of change at individual sites. We do, however, consider some common trends at multi-phase sites.

Part 1: Challenges for zooarchaeology of the Neolithic in the central Balkans

In this section we briefly discuss some of the factors which currently limit the potential of zooarchaeological research in the central Balkans.

Recovery methodology

It has long been understood within zooarchaeology (as with other areas of archaeology) that differences in recovery methodology can profoundly alter the make-up of faunal assemblages and hence the conclusions drawn from them (e.g. Payne 1972; O'Connor 2003). Since sieving appears to have been limited or non-existent at a substantial percentage of central Balkan Neolithic sites – probably the majority (see Orton 2012: Figure 2) – there is a substantial risk that much of our data is inherently biased towards larger taxa and larger skeletal elements. The new excavations at Belovode and Pločnik are no exception. Amongst other things, this renders problematic comparisons with contemporaneous sites on the Adriatic coast, where sieving is generally routine (Orton *et al.* 2016).

The impact of recovery differences has been demonstrated directly at several Balkan Neolithic sites where partial sieving regimes permitted comparison. In the 1991 season at Petnica for example, sieved fractions were bagged separately, making it possible to calculate that sieving increased yields by around 74% in terms of fragment counts. While many of the additional specimens were small, fragmentary pieces, the yield of diagnostic specimens was increased by 38% (Orton 2008: 210–211). At the Middle Neolithic site of Obre I in central Bosnia, comparison between the main 'Jugoslav' excavations and more carefully excavated 'American' trenches revealed a significant bias against smaller taxa in the former (Bökönyi 1974), while somewhat subtler size-based biases are apparent between different recovery methods at Opovo (Russell 1993: Tables 7.5 and 7.6).

In all these cases, it was only possible even to assess recovery biases because (a) the sieving regime was stated, and (b) at least some of the bones recovered by different methods were recorded and reported separately. Unfortunately, this is not always the case – indeed the frequent omission of explicit information on recovery methodology is as serious a problem as the lack of sieving itself. For example, a high frequency of sheep and goat remains at the Sopot site of Slavča, Nova Gradiška, compared to other Late Neolithic sites in the Posavina (Miculinić and Mihajlević 2003; see Orton 2014 for comparisons) probably partially reflects systematic sieving of one third of deposits (M. Mihajlević pers. comm.), but this is not mentioned in the zooarchaeological report.

While we understand that constraints of time and funding may simply make 100% sieving impractical, even a limited sieving programme can help to compensate for recovery biases – provided that it is thoroughly documented. Where 100% of sediment is sieved, the resulting bones can safely be combined with the hand-collected fraction, but in any other circumstances it is imperative that they be bagged separately and labelled as such, to permit statistical comparison (Baker and Worley 2014: 12–13).

On a brighter note, the growth of archaeobotanical research within central Balkan prehistory (see Filipović and Obradović 2013) opens up the possibility that heavy residues from flotation samples might routinely be used to assess the extent of recovery biases in faunal material. Such comparison has not yet been attempted for the central Balkan Neolithic but is briefly considered in the context of Pločnik (Chapter 35). Provisionally, comparison of all the wet-sieved material with its parent contexts shows significantly more caprine specimens (74% cf. 17% amongst hand-collected) and fewer cattle (7% cf. 63%), though some larger specimens may have been removed prior to sampling, exacerbating this contrast. Ideally, samples for sieving should be taken as 'whole-earth', without first removing larger pieces (Campbell *et al.* 2011).

Volume data

Related to recovery strategy and its documentation is the question of systematically recording volumes of excavated sediment. This is standard practice in many excavation systems, particularly those based around single-context recording. The latter is not common in Balkan archaeology, however, and the recording of unit volumes fits somewhat less naturally into the standard excavation protocols employed in the region. This is unfortunate, as where volume data are available, they allow the zooarchaeologist to calculate densities of recovered bone, in the sense of fragments, grams, or diagnostic zones per litre excavated. The same principle can, of course, also be applied to other finds categories. Averaged across similar contexts, such densities can enable the analysis to move from relative changes in taxonomic abundances to absolute increases or decreases in the deposition of remains from a given species (see Russell *et al.* 2013: 217–220 for an example where density data seriously impact upon interpretation). In the absence of such data, any analysis of trends over time at a given site will struggle to distinguish between an increase in the frequency of one species and a decrease in the frequency of its counterparts.

Excavation extent

For entirely practical reasons – time, funding, and contemporary land-use – the majority of excavations at central Balkan Neolithic sites over the last few decades have been fairly limited in extent. This greatly restricts the potential for understanding the processes by which animal bone deposits formed and precludes any serious analysis of spatial differentiation within settlements. This is unfortunate, as such analyses could potentially contribute much to our understanding of how settlements functioned. Of greater concern, however, is that this also raises the prospect that functional differentiation between deposits might pass undetected, with animal remains reported as 'the fauna' from a given site actually representing only a subset of depositional activities. At Gomolava, for example – a site with unusually extensive excavation – zooarchaeological study of Block I (Clason 1979) and of Block VII (Orton 2008: 179–182) produced rather different results, although differing methodology may also have played a role.

Apparent changes between phases might thus reflect changing use of space on a site rather than shifts in overall patterns of animal use. For example, the change in cattle element representation as one moves into the latest two horizons at Pločnik – phalanges becoming very frequent at the expense of cranial specimens (Chapter 35, Figure 4) – might represent a change in the types of deposits represented within Trench 24, or in the activities taking place in this part of the site, rather than an overall shift at the site. In the absence of evidence on the spatial distribution of activities, this is a risk we simply must accept.

To some extent, any call for large excavation areas is at odds with that for careful bone recovery, given the finite resources available for fieldwork. This is compounded when investigating deeply stratified sites as at Belovode and Pločnik, since considerable resources are required even to excavate a small area down to the natural. Achieving a good balance between extensive and intensive excavation is by no means an easy task.

The increasing application of geophysical survey to later Balkan Neolithic sites – as at Uivar (Schier and Draşovean 2004), Okolište (Hoffman *et al.* 2009), Stubline-Crkvine (Crnobrnja 2014) and Drenovac-Turska Česma (Perić *et al.* 2016) – presents a possible way forward, allowing for informed, targeted excavation of different settlement zones without necessarily exposing an extensive surface area. Unfortunately, zooarchaeological data are not yet available from any sites thus investigated, apart from a very basic report on the poorly preserved fauna from Okolište that does not consider spatial distribution or even phasing (Benecke 2009), and brief preliminary results from the Vinča site of Stubline presented in J. Bulatović's PhD thesis (2018). Animal remains from the Vinča site of Drenovac are also currently under study by one of the present authors (I. Dimitrijević) for PhD research. Should substantial spatial differentiation be revealed by such studies in future, it would raise doubts about the reliability and representativeness of zooarchaeological data from small-scale excavations at large Neolithic sites in the region.

Wild versus domestic forms of cattle and pigs

While sheep and goats in Balkan prehistory must necessarily have been descendants of near Eastern domesticates, both cattle and pigs had local wild forms present in the Neolithic. This immediately raises the methodological problem of distinguishing between the wild and domestic populations of each species, a problem which has traditionally been approached primarily from measurement data. For pigs, the situation seemed clear-cut until recently: bone and tooth measurements fall into two surprisingly clear groups which, since each includes both male and female specimens, are assumed to represent the domestic and wild populations. These are often present in broadly similar numbers. Figure 1 illustrates these groupings with particular reference to Belovode and Pločnik.

Research combining genetics with geometric morphometric analysis of pig tooth shape has suggested that the larger group includes both wild and domestic specimens, while the smaller group is, as previously assumed, exclusively domestic (Evin *et al.* 2015). In fact, the situation is likely to be quite complex given the possibility of various interbreeding and/or feralisation scenarios. Recent isotopic evidence from Gumelniţa sites indicates that the 'large domestic' pigs occupied a niche closer to 'wild' than to small domestic specimens, suggesting that they represent feral individuals (Balasse *et al.* 2016).

A similar dietary distinction between pigs identified as wild and as domestic is seen at Vinča-Belo Brdo, where domestic specimens have relatively enriched $\delta^{15}N$, while a rather different pattern is evident at Stubline (Gillis *et al.* 2020). Multivariate statistics (Orton *et al.* 2016; see also below for a more limited application) support this interpretation, revealing that pigs identified as wild (i.e. the large size group) systematically co-occur with deer, beaver, and dogs, while those recorded as domestic (i.e. the smaller group) place very close to domestic cattle. This being the case, we take the view that the large size group primarily represent functionally wild specimens, regardless of their ultimate or immediate ancestry. Ongoing collaborative work on a very large sample of pig remains from Gomolava may help to resolve these complexities.

Even accepting the size groupings at face value, the presence of two forms of pig within each zooarchaeological sample poses a problem. While some authors attribute all remains to one or the other population, many also report a subset of indeterminate specimens. In order to review published data, one must decide how to deal with these specimens: ignoring them would result in understating the overall contribution of pigs, while combining all pig specimens together would unjustifiably subsume animals that represent very different subsistence practices. Accordingly, the best approach (adopted in the review below) is probably to attribute the indeterminate specimens to wild and domestic groups *pro rata* based on the relative numbers firmly identified to each group. The final numbers of wild and domestic pigs calculated this way are an estimate but are probably the best approximation given the inherent uncertainty.

A similar situation exists for cattle, except that (a) there generally seem to be many more domestic than wild specimens, and (b) the metrical separation is much less clear-cut (see Rowley-Conwy 2003; Wright and Viner-Daniels 2015). The same recommendations regarding *pro rata* allocation of indeterminate specimens apply as for pigs, but the resulting numbers of aurochsen should be treated with caution. The fact that aurochsen place close to domestic cattle and domestic pigs, rather than to other large game, according to correspondence analysis (Orton *et al.* 2016, see also below) may indicate an interesting relationship between cattle-herding and aurochs hunting; alternatively, it may simply reflect widespread misidentification. It is hard to see how this problem could be resolved in the future within the scope of conventional zooarchaeological methods.

Identification of sheep and goat

The difficulty of distinguishing sheep and goat based upon their skeletal remains has attained an almost legendary status within zooarchaeology and is certainly not a Balkan- or Neolithic-specific problem. It does, however, pose particular problems for the region and

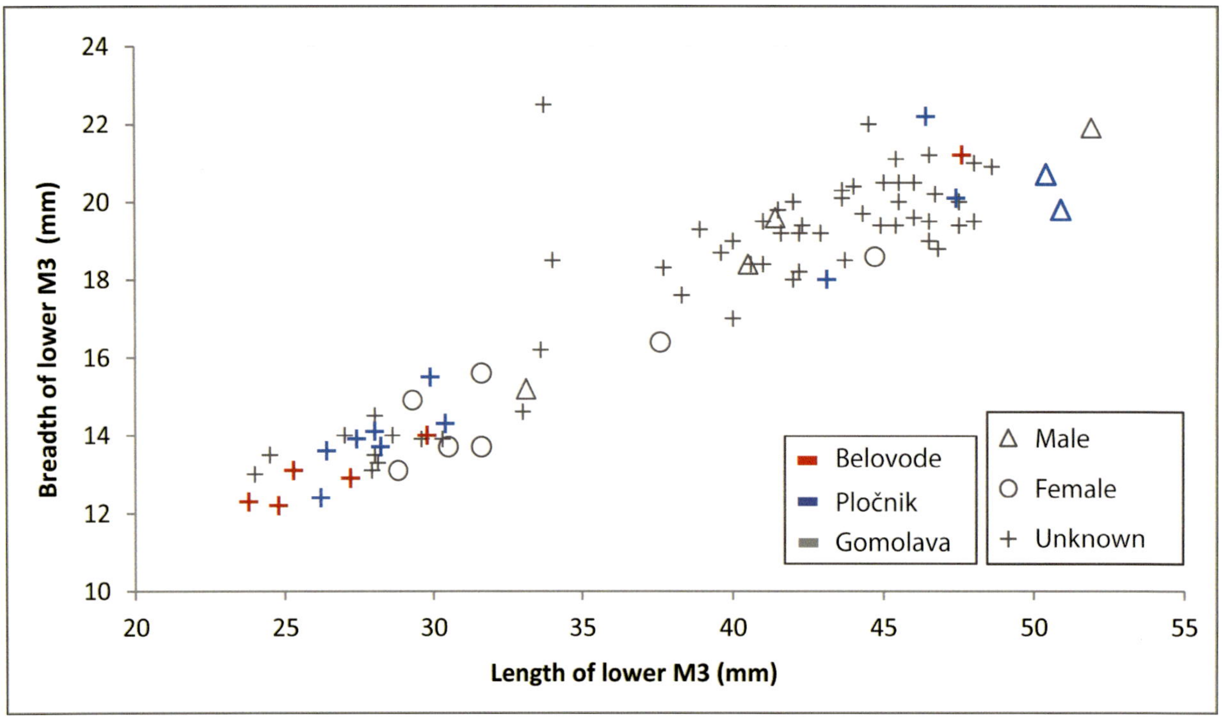

Figure 1. Dimensions of lower pig third molars from Belovode, Pločnik, and Gomolava, to demonstrate metrical separation into 'wild' and 'domestic' groups. Gomolava data are from Orton 2008.

period discussed here, e.g. in terms of recent arguments for changing management of sheep vis-à-vis goats over time (Greenfield and Arnold 2015). Sheep and goats can, in fact, be distinguished with reasonable confidence based on a wide range of morphological characters (e.g. Boessneck 1969; Halstead *et al.* 2002; Helmer 2000; Payne 1985; Prummel and Frisch 1986), but the ways in which these criteria are applied is not necessarily consistent. The number of different authors who have contributed to the available Balkan Neolithic data, and the fact that many of the relevant reports were published before some of the key identification references, mean that there is likely to be considerable variability in approaches to sheep/goat identification. Indeed, this can be illustrated by looking at the wide variation in overall rates of specific identification for caprines (Kassebaum 2019). Moreover, in some publications no attempt is made to distinguish between sheep and goats at all.

The upshot of this is that, while separation of sheep and goats for all aspects of analysis would be preferable due to their differing ecology and potential for human use, the only rational approach to a review of the varied publications available is to lump both species into a single category, which we assume, in fact, to be mostly sheep. To consider only positively identified specimens would be plainly unwise. Apart from excluding some assemblages altogether, it would inevitably understate the overall contribution of the two species compared to other taxa, seriously skewing results. A more viable alternative would be to estimate overall proportions of sheep and goat by extrapolating from the positively identified subset, following the same approach described above for wild versus domestic pigs and cattle, but given the asymmetrical nature of many caprine identification criteria (Zeder and Lapham 2010; Zeder and Pilaar 2010) and the typically small numbers identified to species, this is unlikely to yield reliable estimates. Moreover, identification rates for caprine mandibles may be correlated with age, such that older animals are under-represented in species-specific age-at-death analyses, seriously affecting likely interpretations (Mallia 2015).

Accordingly, sheep and goat are treated together in the review below. In future, the development of low-cost, minimally destructive collagen fingerprinting (ZooMS, Buckley *et al.* 2010) may help to detect sheep:goat ratios and to assess identification biases and errors in conventional methods. Application of ZooMS to mandibles from the Vinča site of Vitkovo revealed that 25 'sheep', 11 'goat', and 4 indeterminate caprine mandibles were in fact all sheep, with only one indeterminate result (Kassebaum 2019: 27). The routine application of such technology is, however, some way off at present.

The number and distribution of studied assemblages

It is something of a cliché for a review paper to make the observation that 'more work needs to be done'. Indeed, in archaeology, this is almost always the case. When it comes to zooarchaeological studies from the central Balkan Neolithic, however, the problem is not so much the number of studied assemblages – though more data is always welcome – but rather their geographical and temporal distribution.

The relatively limited amount of data available from early Vinča sites is noted in the introduction. There is a more acute lack of data from the post-Vinča Eneolithic, but this reflects the very sparse, almost ephemeral nature of the settlement record more than any research bias. Indeed, a number of assemblages from the period have been studied but the majority are extremely small (Blažić 1995; Bökönyi 1991; Bulatović 2010, 2018, 2020; Bulatović and Milošević 2015; Greenfield 1986, 2014; Kapuran and Milošević 2013; Lazić 1992; Radmanović *et al.* 2014; Vuković and Marković 2019).

Geographically, there remains a bias towards northern Serbia, Vojvodina, and the Romanian Banat as regards the Neolithic as a whole, with less data available for central and southern Serbia. This is starting to be rectified by recent publications (Bulatović 2011, 2012; Greenfield 2017; Greenfield and Greenfield 2014) and ongoing work at sites such as Drenovac and Pavlovac (I. Dimitrijević's PhD research), and the new data from Belovode and Pločnik. Data from Pavlovac, situated in the Southern Morava valley, will be an important first step in connecting our current knowledge of Neolithic animal use in central Serbia with a cluster of published assemblages from northern North Macedonia. Likewise, zooarchaeological research on some of the numerous sites in the Polimlje (see Derikonjić 1996) or the Kosovo plateau would be very welcome, extending our knowledge into areas that are presently unknown in terms of Neolithic subsistence.

One particular problem with the existing data is that geographical coverage shifts somewhat between periods, hindering temporal comparisons (see Orton 2012: Figure 10). Bačka, for example, is fairly well represented in terms of Starčevo-Körös-Criş assemblages but at present has no Later Neolithic data, while almost the reverse is true for the Belgrade region and Kolubara valley. Other areas, such as the Banat and Danube Gorges, have more continuity in representation.

Part II: Trends in animal use across the early metallurgical threshold

Sites and data

This review incorporates the new results from Belovode and Pločnik with existing data from Vinča sites within the phenomenon's core distribution (i.e. excluding Transylvania and North Macedonia but including the Banat group). In addition, Earlier Neolithic (i.e. Starčevo-Körös-Criş) data from within this same region are included for comparative purposes (n = 16), as are a handful of Eneolithic assemblages (n = 7 phases, at five discrete sites). The latter are from disparate typo-chronological phases with extremely limited absolute dating, so not a coherent group.

The Vinča data are divided into early (n = 7) and late (n = 16) groups, with the cut-off placed at *c.* 5000 cal. BC, i.e. approximately the start of Vinča B2 or Vinča-Gradac. Samples with NISP < 200 for a given site and period are excluded, with the exception of (a) the early phases at Belovode (NISP = 162) given their relevance to this volume; and (b) Humska Čuka (NISP = 185 (Bulatović 2018)) as one of only two Early Eneolithic (i.e. closely post-Vinča) assemblages. The important multi-layered site of Selevac is also omitted – with a degree of reluctance – for most purposes, since the

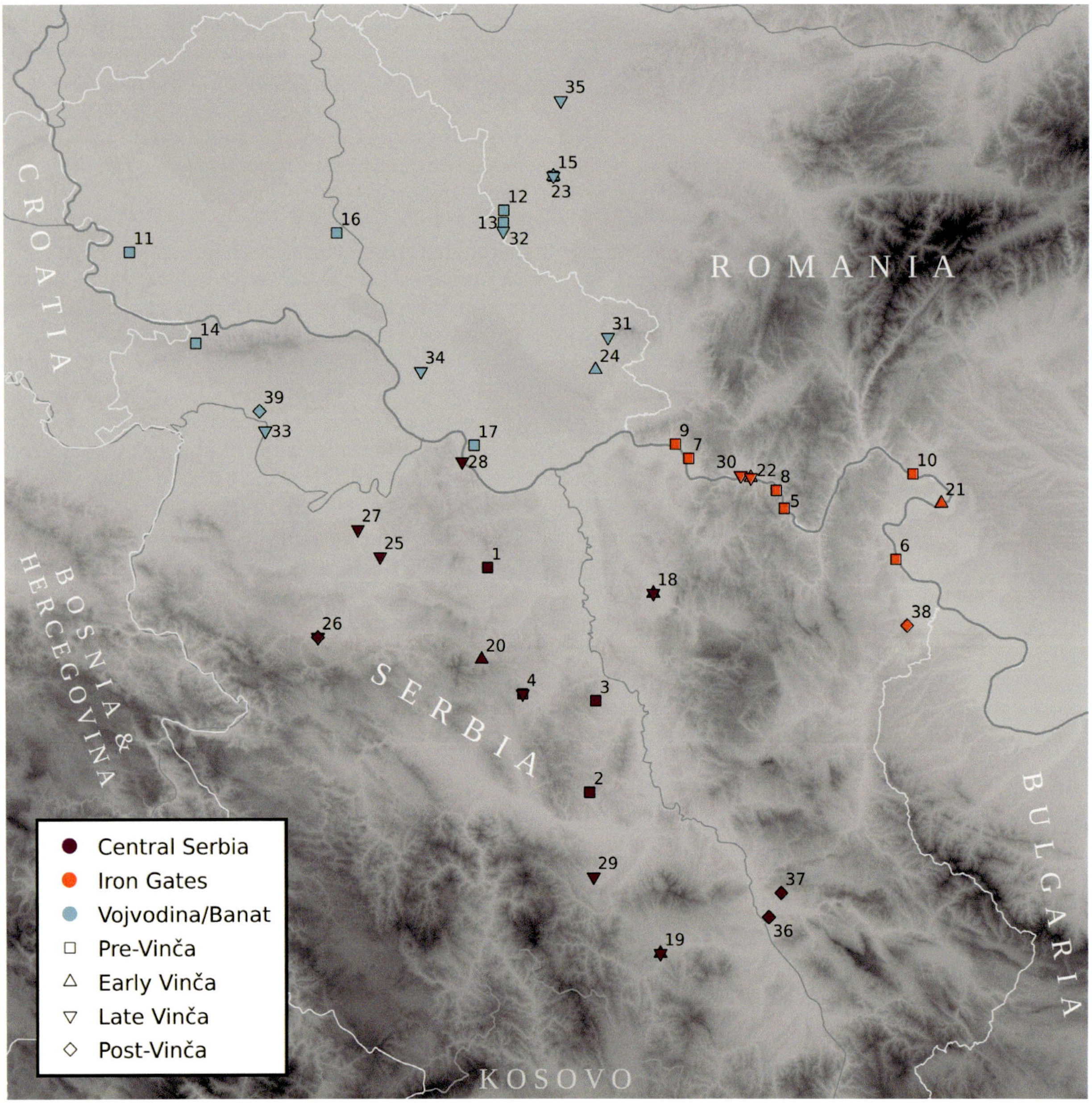

Figure 2. Map showing sites included in review, coloured by the regional groupings employed here. Blue: Vojvodina and Romanian Banat; red: Iron Gates region; purple: central Serbia. Numbers relate to Figure 3.

zooarchaeological report provided quantitative data only for the major taxa, without distinguishing between wild and domestic cattle or pigs on a phase-by-phase basis (Legge 1990). Figure 2 shows the sites included in the review.

Relative taxonomic abundances

Figure 3 shows (a) relative abundance of the main domesticates and (b) domestic:wild ratio at each site, divided by period and geographical region. Where data are reported from multiple phases at a site that fall into the same period category these have been combined.

The domestic versus wild plots reveal that wild species continue to be important at most sites in the region throughout the Neolithic, with no clear trends over time and a considerable degree of variability across the dataset as a whole. By the late Vinča period it is fairly clear that hunting played a larger relative role at sites on the southern Great Hungarian Plain and Iron Gates than in central Serbia. This may also be true for early Vinča, but there are, as yet, too few studied sites to be sure, while the wild-dominated assemblage from Stragari goes against this trend. Both Belovode and Pločnik fall towards the domesticate-dominated end of the spectrum in both periods.

Turning to relative percentages of the main domesticates, the initial impression is of an overall decrease in numbers of caprines between Early Neolithic and Vinča sites, with a concomitant increase in cattle and pigs. This pattern of decline in sheep and goat herding is particularly clear at the Great Hungarian

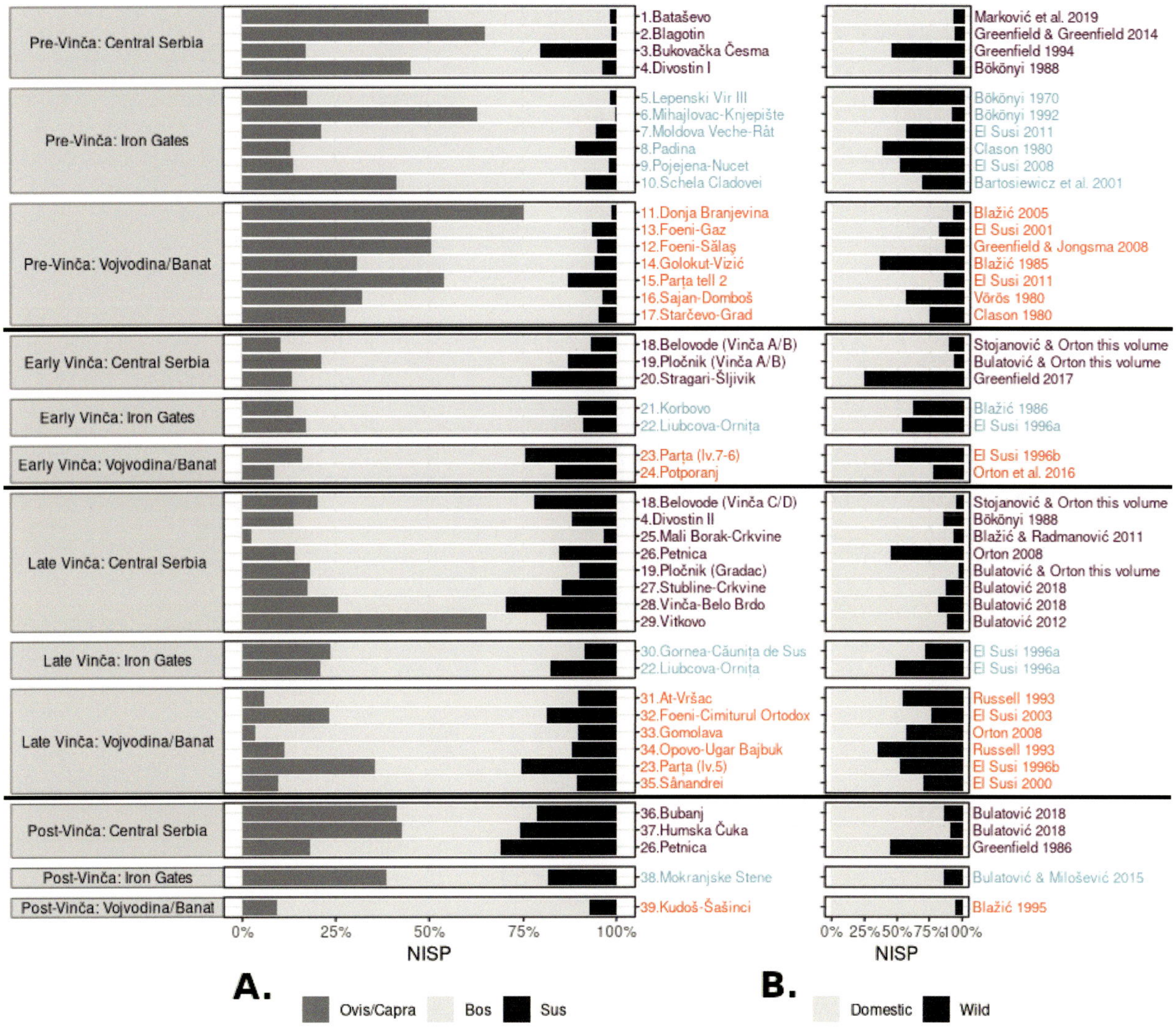

Figure 3. Overview of Neolithic zooarchaeological assemblages from the core Vinča region, in terms of (a) relative proportions of the main domesticates; and (b) relative contributions of wild and domestic species. Labels are coloured according to geographical regions, following Figure 1.

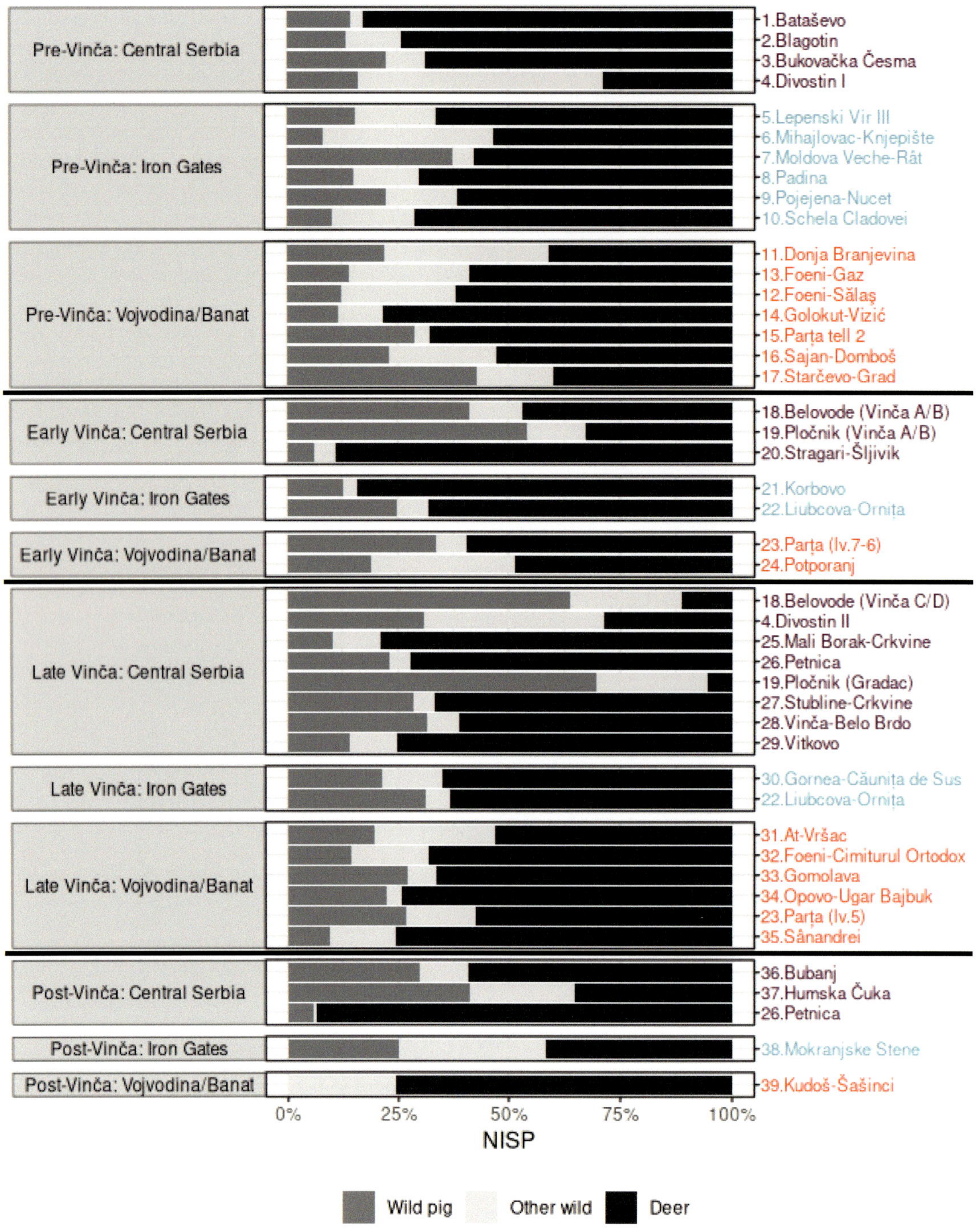

Figure 4. Major hunted taxa in central Balkan Neolithic assemblages. References as for Figure 3.

Plain sites, which have the greatest average percentage of caprines in the Early Neolithic and the lowest by Late Vinča. It is also evident between Early Neolithic and Early Vinča in central Serbia, albeit with considerable variation in the subsequent Late Vinča period. The increasing contribution of pig is more clear-cut, with the main shift again seeming to occur between the Early Neolithic and Early Vinča groups.

This means that there are no especially clear shifts visible between the Early and Late Vinča periods, unless one thinks in terms of inter-site variability. Given the relatively small number of sites in the former group this is a slightly tentative point to make, but there does seem to be a notable increase in variability from Early to Late Vinča, both within and between regions.

Although data are limited, there appears to be a shift back towards sheep and goat herding in the post-Vinča period. Three of five Eneolithic sites have a greater relative contribution of caprines than any of the Vinča assemblages apart from Vitkovo. These include the only two available Early Eneolithic (i.e. late 5th millennium BC, closely post-Vinča) assemblages: Humska Čuka and Bubanj. These caprine-rich assemblages were all collected overwhelmingly by hand, ruling out a recovery bias vis-à-vis Vinča-period sites.

Diversity of hunted species

In addition to considering the overall contribution of wild animals in relation to domesticates, it may also be worth examining the range of taxa hunted. In the Balkan Neolithic context, three species – red deer, roe deer, and wild pig – typically make up a very substantial percentage of the hunted fauna (Figure 4), but other species such as hare, beaver, and small carnivores may be under-represented due to the predominantly hand-collected nature of assemblages.

Taxonomic richness, i.e. the number of taxa represented, n_{TAXA}, can be a useful if basic measure of diet breadth, i.e. the diversity of resources exploited by a past community. In the context of a Neolithic settlement, a high diet breadth might be taken to represent more opportunistic hunting and/or use of a range of ecotypes; conversely, low diet breadth might represent targeted hunting of favoured species and/or use of a less varied landscape. Since n_{TAXA} is intimately related to sample size, however, it is necessary either to create an index variable or to plot results against sample size, typically transformed logarithmically. Figure 5 takes the latter approach to the Balkan Neolithic dataset, in this case including phase-by-phase results, where reported, rather than combining them by broad periods. The expected positive relationship between wild sample size and n_{TAXA} is

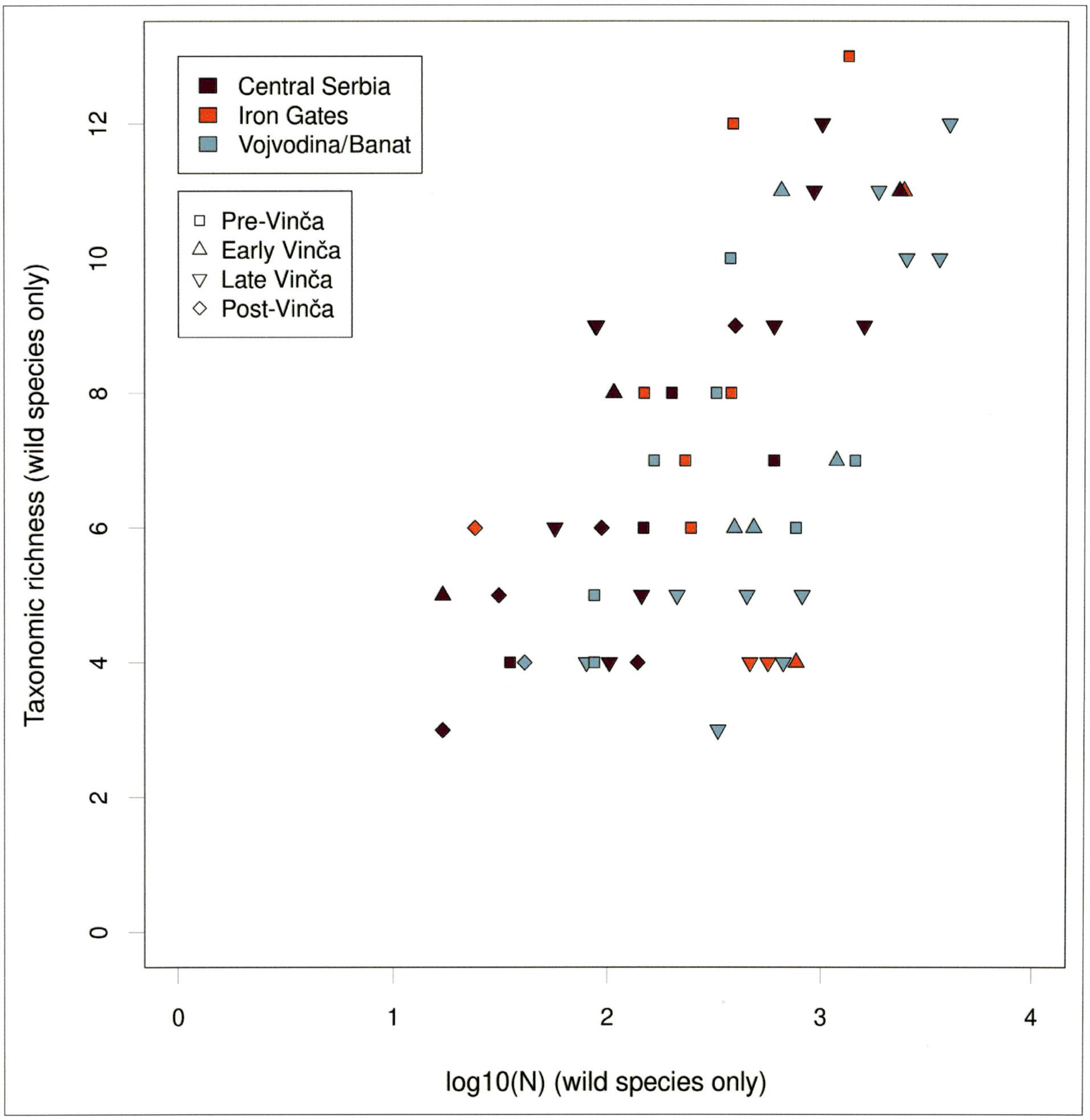

Figure 5. Relationship between sample size and taxonomic richness of wild taxa at central Balkan Neolithic sites.

observed, although there is considerable variation in n_{TAXA} for any given range of sample sizes. The Vojvodina/Banat assemblages show a slight tendency towards lower richness for a given wild sample size, particularly in the Vinča period – an observation which, if real, probably reflects environmental differences. Interestingly, the few Vinča sites representing the Iron Gates show a similar pattern, with surprisingly low taxonomic richness given substantial samples of wild remains: three of four assemblages have lower n_{TAXA} than any of the pre-Vinča assemblages in the region. Otherwise, taxonomic richness does not appear to change systematically over time.

Overall assemblage composition: correspondence analysis

As a final method of assessing overall taxonomic composition, we performed a correspondence analysis on data from individual assemblages (i.e. treating each phase at each site separately). Factor scores for major taxa fall into three groups: (a) sheep and goat (of necessity treated together; see Part I above); (b) cattle, domestic pigs, and aurochsen; and (c) a fairly diffuse cluster of other wild species plus dogs (Figure 6a). Plotting factor loadings for individual assemblages (Figure 6b) reveals some minor geographical trends: although there is considerable variation within each of the three regions, central Serbian sites tend more towards the cattle/pig pole, and the others more towards the wild cluster.

Temporal shifts are more clear-cut. Where pre-Vinča sites are spread primarily between the caprine and wild poles regardless of region, both Early and Late Vinča sites largely form an axis between the cattle/pig and wild poles. The exception is Vitkovo, which is unusual amongst Vinča assemblages in being dominated by sheep and goat remains despite being entirely hand-collected (Bulatović 2011, 2012). Post-Vinča assemblages – admittedly few in number – show a clear shift back towards the Early Neolithic pattern, particularly in central Serbia, with most assemblages plotting towards the caprine pole.

The main impression given by the correspondence analysis results is that, once again, the clearest changes are seen at the start of the Vinča period and after its end, with considerable continuity in between. There is perhaps more diversity amongst the Late Vinča assemblages, at least for sites on the Great Hungarian Plain, though this may partly just reflect the smaller number of Early Vinča samples.

Age-at-death of main domesticates

Figure 7 shows age-at-death results for (a) caprines and (b) cattle at all sites in the core Vinča region (including pre-Vinča assemblages) for which Payne (1973) format age data are reported, with sample sizes >15. The majority of these data have been discussed at great length – and often with considerable heat – in the context of ongoing debate on the validity of the Secondary Products Revolution model (Arnold and Greenfield 2006; Brochier 2013; Gillis *et al.* 2019; Greenfield 2005; Greenfield and Arnold 2014; Orton 2012; Vigne and Helmer 2007). We do not wish to revisit that debate in detail here. Rather, our aim is to place our new data from Belovode and Pločnik in context, and to comment upon any patterns that may be relevant to the question of changes – or lack thereof – across the early metallurgical horizon.

The available data are plagued by small sample sizes, and the new results from Belovode and Pločnik are unfortunately no exception. Once the data are broken down by period, no really reliable large samples remain; only five are worth plotting. From Belovode, only the caprines from the later phases qualify (n = 29), although Pločnik provided usable samples from early and late phases for both caprines (n = 34 and n = 20 respectively) and cattle (n = 24, n = 34). These numbers differ slightly from those in the site faunal reports (Chapters 21 and 35) since the natural breaks in the chronology at each site differed from the *c.* 5000 cal. BC threshold applied here. Small samples are a particular problem for Payne-format age-at-death data since the usual system of pro-rata allocation for specimens assigned to two or three possible stages renders the overall survivorship curves very unstable: as sample size decreases, the shape of the curve becomes increasingly dependent on the even smaller sample of specimens assigned to a single stage. An alternative system was deployed here for Belovode and Pločnik, allocating indefinite specimens based on relative stage duration without reference to the more definitively aged specimens (see Chapter 21). Unfortunately, it was impossible to apply this retrospectively to other assemblages, with the exception of Vinča-Belo Brdo, since raw ('uncorrected') data are rarely reported. Accordingly, these results should be viewed with caution, and samples <25 are plotted with dashed lines.

Starting with caprines, the problematic, if unavoidable, nature of combined sheep/goat kill-off curves was noted above, though in practice a substantial majority of specimens from Balkan sites are likely to be sheep. In this case, the data do seem to show a consistent change between the Early Neolithic and Late Vinča, with more

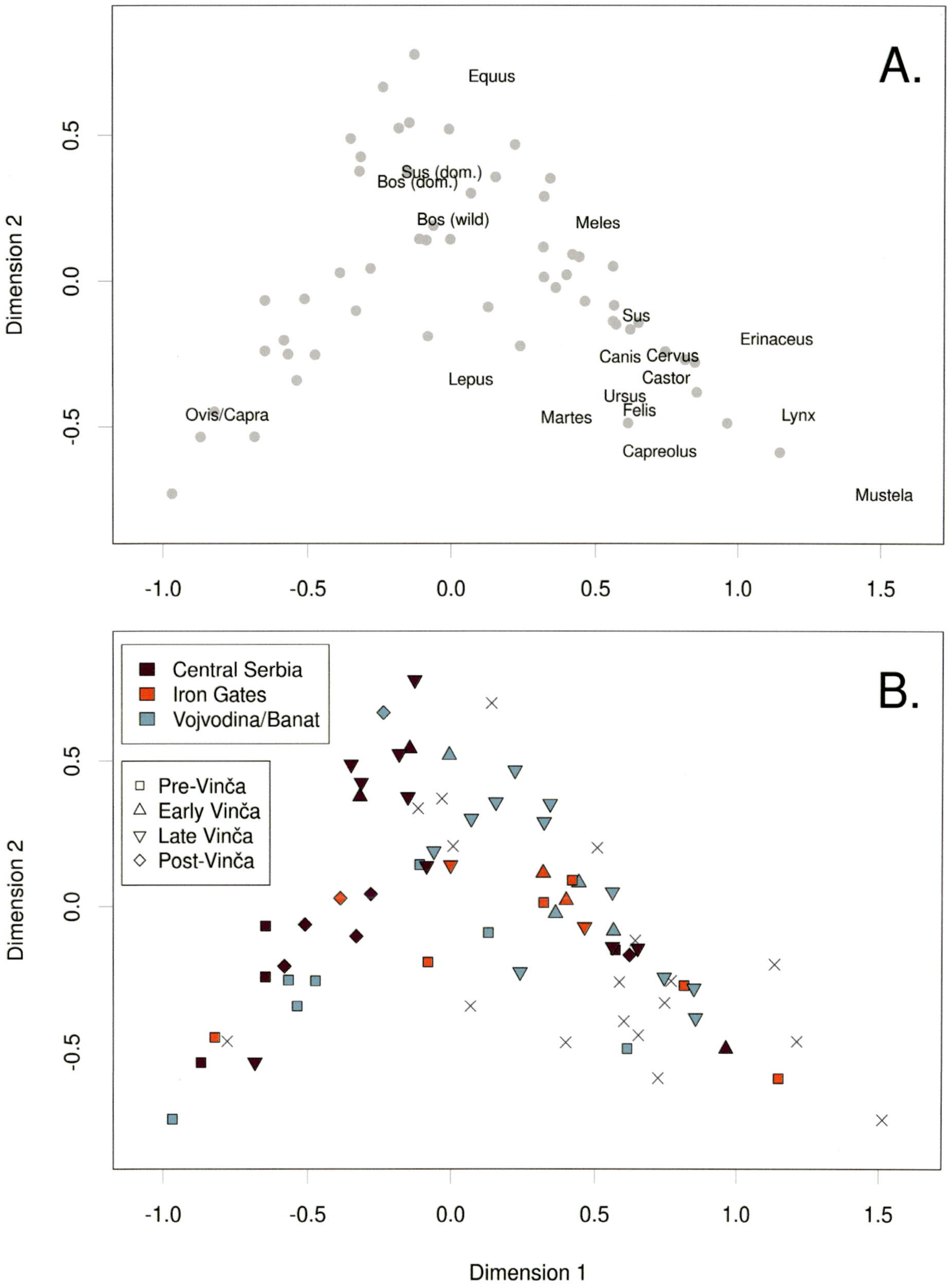

Figure 6. Correspondence analysis results: (a) factor scores for taxa (grey dots in background represent assemblages); (b) factor scores for individual assemblages plotted by period and region (grey crosses in background represent taxa).

animals surviving to early adulthood in the latter case. Unfortunately, there are only two respectable samples for the Early Vinča period – from Pločnik and Stragari (Greenfield 2005) – and even these are small, making it hard to assess at what stage the apparent shift occurred. Taken at face value, the Stragari curve would imply a similar herding strategy to that observed at Early Neolithic Blagotin and Foeni-Sălaş (both Greenfield 2005), but the sample from early Pločnik tells a very different story, sitting more comfortably with the later Vinča samples. It is worth noting that these two assemblages are also very different in terms of taxonomic composition (Figure 3).

Turning to cattle, the small number of sites with respectable samples makes it hard to draw reliable comparisons between periods. There is again some suggestion of a shift between Early Neolithic and Late Vinča, but with only one reliable sample from the former period, and only two from the intervening Early Vinča, it is impossible to speak confidently about change (or continuity) in herding practices.

Trends over time at multi-phase sites

Figure 8 shows trends over time in relative proportions of the main domesticates, large game and small game at Belovode, Pločnik, and selected other sites. Selevac is included here despite limitations in the reported data, but it should be noted that (a) game will be understated since only the major taxa were reported in full, and (b) it was necessary to estimate relative numbers of wild versus domestic cattle and pigs in each phase based on the ratios quoted for the site as a whole. Likewise, at Opovo, relative numbers of wild and domestic pigs by phase were only available in terms of Diagnostic Zones (DZ), not NISP. NISP figures were estimated based on the ratio derived from DZ, a procedure which is likely to slightly overestimate wild specimens.

The comparison emphasises the relatively low contributions of hunting at Belovode and Pločnik compared to other large settlements such as Selevac, Parţa, and Gomolava (Opovo is a smaller site, and at the upper end of wild contributions). Interestingly,

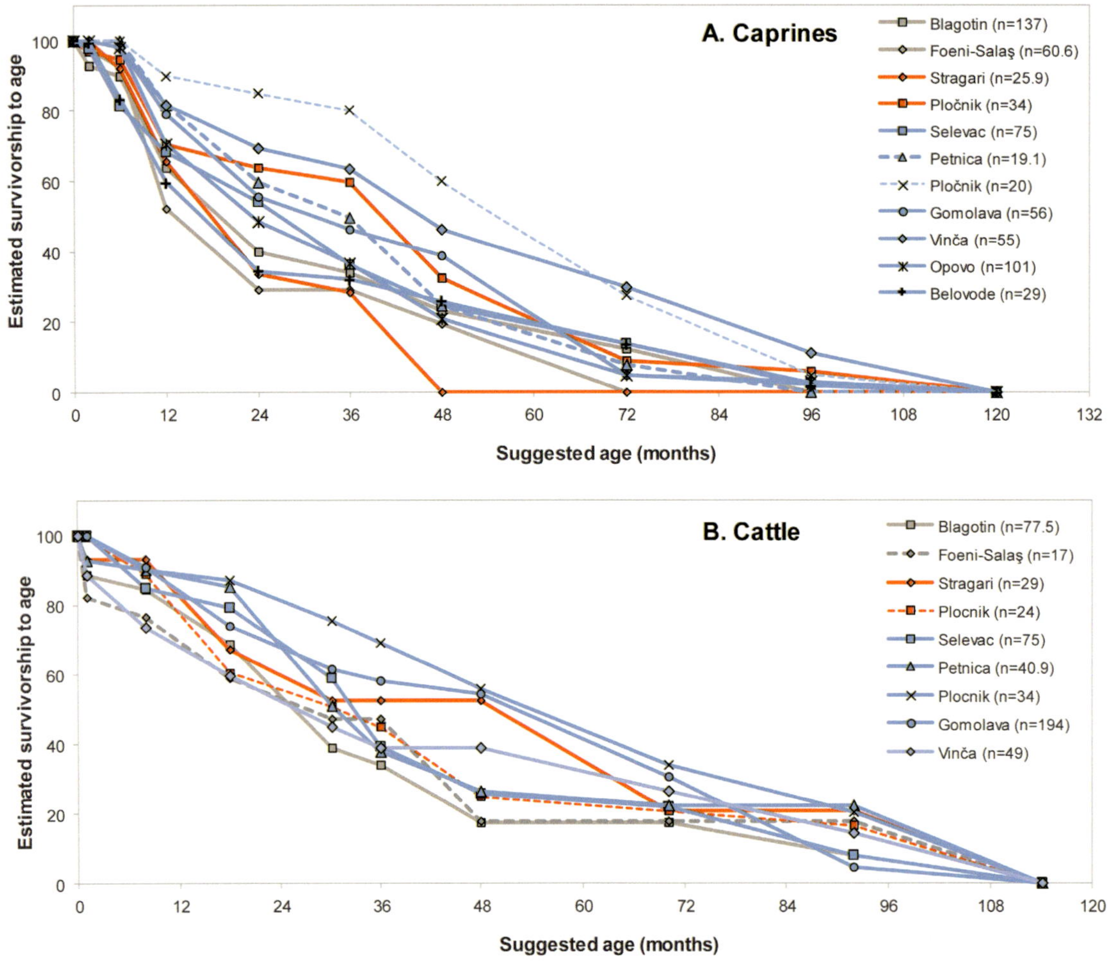

Figure 7. Survivorship curves for (a) caprines and (b) cattle from Neolithic sites, following the Payne (1973) system. Only samples with n >15 are shown, and dotted lines are used for those with n <25. Grey lines represent Early Neolithic, orange lines Early Vinča, and blue lines Late Vinča. NB data for Belovode and Pločnik are not identical to those in Stojanović and Orton (this volume), and Bulatović and Orton (this volume) since here they have been separated by slightly different periods. Other data are from Greenfield 2005, apart from Gomolava (Orton 2008), Opovo (Russell 1993), Selevac (Legge 1990), and Vinča-Belo Brdo (Bulatović 2018).

almost opposite trends are seen at the former two sites in terms of the relative numbers of the domesticates: at Belovode, both caprines and pigs increase over time at the expense of cattle; at Pločnik the contribution of cattle increases slightly over time while caprines and pigs decline. Neither site shows a clear trend in terms of the contribution of hunting, although there does seem to be a drop-off into Gradac II/III at Pločnik.

Clear trajectories are, however, seen at Selevac and Gomolava, where the contributions of wild taxa steadily decline. This might be expected if settlement populations were growing, since a practical limit on the exploitation of local wild resources would presumably be reached before the carrying capacity of the surrounding landscape in terms of livestock; there is no need to invoke actual depletion of wild resources or habitat disruption here, as Legge (1990: 221–222) does for Selevac, though this might be implied by, for example, changing red deer age profiles at Gomolava (Orton 2008: 121). These trajectories are not seen elsewhere, however. Like Belovode and Pločnik, little change is seen at Opovo, while at Parţa the relative contribution of wild taxa actually increases over the first three phases, before dropping back slightly in level 5.

It is also worth noting that where sites span the Early to Late Vinča transition, they generally exhibit considerable continuity across it. Only at Parţa does this threshold seem to coincide with an apparent switch in the direction of change. Interestingly, this is associated with a reduction in apparent density of occupation at the site, which the excavators attribute to disruption due to external factors (Draşovean 2007: 21). There are some fairly marked changes across the transition at Belovode, which must be treated with caution given its small sample size for Early Vinča, while at Pločnik the most abrupt shift takes place later, between Gradac I and Gradac II horizons.

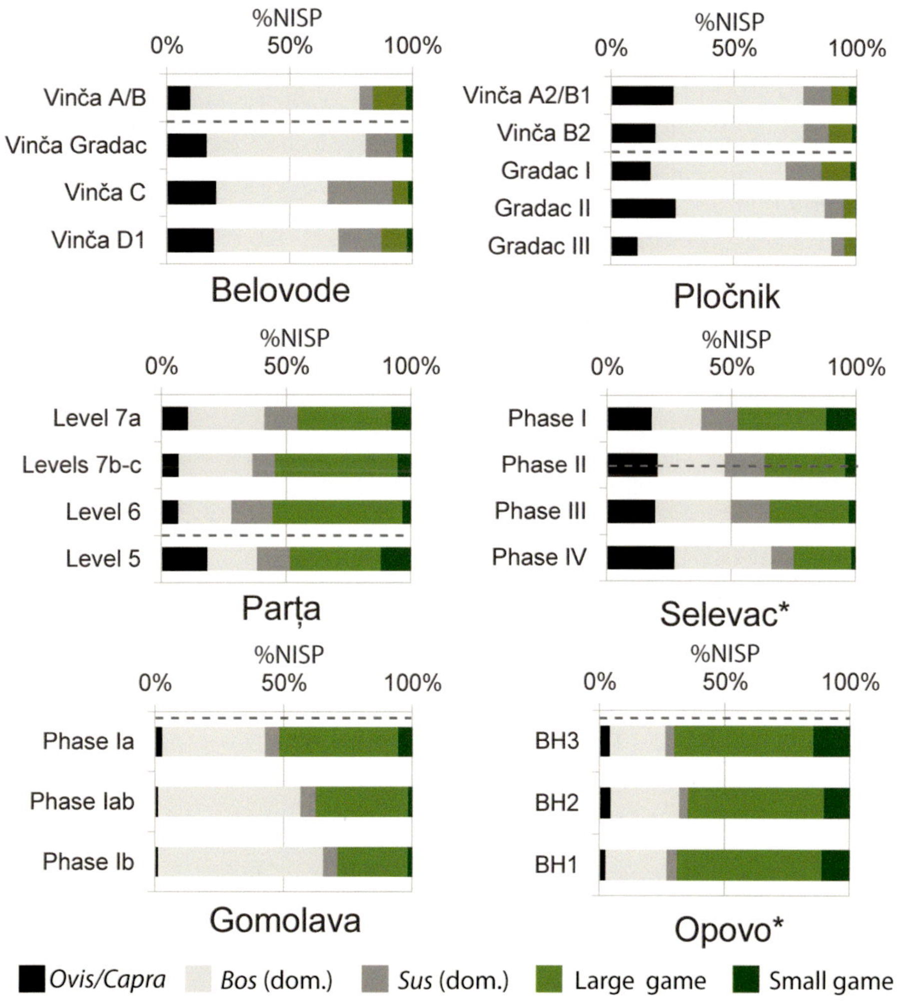

*Selevac and Opovo data involve some interpolation of frequencies due to incomplete phase-level data for wild versus domestic cattle (Selevac) and pigs (Opovo).

Figure 8. Trends over time in relative proportions (by NISP) of the main domesticates, large game, and small game at Belovode, Pločnik, and selected other multi-phase assemblages. Large game includes red deer, wild pig, aurochs, wild ass, bear, and wolf. Small game refers to roe deer, beaver, and any smaller mammals (mostly hare and small carnivores).

Conclusions

In this chapter we reviewed firstly some of the problems and challenges associated with zooarchaeology of the Neolithic in the central Balkans, and secondly the available data relating to patterns of animal use across the threshold of metallurgy in the region – the latter being informed to some extent by the former. In so doing we hope to have provided a degree of context into which the new data from Belovode and Pločnik can be placed.

To the extent that broad patterns in faunal remains from Neolithic sites might be expected to reveal wider changes in social organisation, landscape use and so-on, the overall impression within the Vinča period is one of continuity over time – certainly compared to fairly marked changes in animal use observed between the Early Neolithic and Early Vinča, especially on the southern Great Hungarian Plain. There is, however, some evidence for an increase in variability amongst assemblages in the latter part of the Vinča period, which might perhaps be linked to a diversification in the ways in which Late Vinča communities used their landscapes. Given that considerable continuity is nonetheless seen at Belovode and Pločnik – two of the sites most closely linked to metallurgical innovations – it seems unlikely that the onset of metallurgy is implicated in this apparent diversification.

Post-Vinča assemblages unfortunately remain thin on the ground for substantive comment on changing animal use into the Eneolithic, when there is a demonstrable and dramatic shift in the settlement evidence. However, recent data from Early Eneolithic Bubanj and Humska Čuka (Bulatović 2018) point to a marked post-Vinča shift back towards sheep and goat herding, at least in southern Serbia.

Acknowledgments

We would like to thank Miljana Radivojević, Ben Roberts, Thilo Rehren, and the rest of the AHRC *Rise of Metallurgy in Eurasia* project for the opportunity to study the bones from Belovode and Pločnik, and for some financial support in so doing. Miroslav Marić provided invaluable assistance with the stratigraphy and dating at each site, while we are indebted to Gordana Grabež for facilitating access to the material in the National Museum (Belgrade) after an early attempt to export it to London for analysis was foiled by the Hungarian border service. Bone recording was ultimately carried out at the Faculty of Philosophy, University of Belgrade, where we thank Vesna Dimitrijević and Sofija Stefanović for permission to use the facilities and reference collection of the Laboratory for Bioarchaeology. The work on Pločnik formed part of J. Bulatović's PhD thesis, defended at the Faculty of Philosophy in 2018.

In addition, the lead author wishes to thank Marc Vander Linden and the ERC *EUROFARM* project, under whose auspices he was able to contribute his time to the early stages of this work.

The bibliographic reference for this chapter is:

Orton, D., Bulatović, J. and Dimitrijević, I. 2021. Evidence for animal use in the central Balkan Neolithic across the early metallurgical horizon: the animal remains from Belovode and Pločnik in context, in Radivojević, M., Roberts, B. W., Marić, M., Kuzmanović Cvetković, J., and Rehren, Th. (eds) *The Rise of Metallurgy in Eurasia*: 585–598. Oxford: Archaeopress.

Part 5
The Rise of Metallurgy in Eurasia and Beyond

Photo by: Marko Djurica

Chapter 52

Balkan metallurgy in a Eurasian context

Miljana Radivojević and Benjamin W. Roberts

As outlined in Chapter 2, three key lines of enquiry shaped *The Rise of Metallurgy in Eurasia* project and underpinned the research questions. Firstly, there are competing views about whether metallurgy in Eurasia had a single origin or arose in multiple places. There are also different perspectives regarding the ways in which pre-existing technical knowledge influenced and inspired the emergence of this new technology. Further discourse relates to the manner in which this early metallurgy was organised across the *chaîne opératoire* of metal production and use and developed across a range of metals and alloys. Each of these three themes are fundamental to early metallurgy across the world (see papers in Roberts and Thornton 2014). These are areas of investigation with a deep history of scholarship and a wide range of competing explanatory models.

That these lines of enquiry can be re-evaluated in the Balkans is due to the integrated theoretical and methodological approach of *The Rise of Metallurgy in Eurasia* project, which has extended the scientific investigation beyond the 'when' and the 'where' of early metallurgy to include explorations of 'how' and 'why'. The project sits firmly within an emerging trend, highlighted by Thornton (2009) as a paradigm shift and evident in global early metallurgical research published in a special double edition of the *Journal of World Prehistory* (Thornton and Roberts 2009 and papers in volume 22) and subsequently in Roberts and Thornton (2014). There is now an identifiable convergence in the scholarship of early metallurgy towards the need to define and analyse the theories and underlying evidence surrounding concepts of invention (see papers in Roberts and Radivojević 2015; Frieman 2021) and innovation (e.g. Burmeister *et al.* 2013; Maran and Stockhammer 2017 Ottaway 2001; Rosenstock *et al.* 2016; Scharl 2016; Frieman 2021). In addition, there is a much stronger expectation that all the available evidence—from ore sources to finished objects—is analysed using a more holistic approach (cf. Ottaway 1994; Shimada 2007), and that results are then compared, contrasted and integrated with comparable analyses of contemporary craft production and consumption in other materials and in broader societal contexts (Miller 2005, 2007).

This dynamic academic environment inspired the six specific research questions that underpinned *The Rise of Metallurgy in Eurasia* project:

1. How did the mineralogical and technological basis for early metal production in the Balkans emerge and evolve during the 6th–5th millennia BC?
2. To what extent was metallurgy related to pottery technology and production, and how did pre-existing technological knowledge influence the emergence of metallurgy?
3. How were ore sources, smelting, and casting connected and organised?
4. Where did the smelted metals circulate?
5. What metal types were being made and how did these evolve?
6. Was there a close relationship between ore sources, metallurgical technology, and artefact types?

This chapter begins by considering the first two themes described above, surrounding the origins of metallurgy, and addressed through research questions 1, 2 and 5. We then turn to the final theme relating to early metallurgy and society with research questions 1, 3, 4 and 6.

Origins

Scholarship debating the origins of metallurgy in the Balkans dates to the early antiquarian period in the mid-19th to the early 20th century (Childe 1944; Grbić 1929; Von de Pulsky 1884) and consistently equated technological advancement, as expressed through evidence of pyrotechnological abilities, with temporal and societal change (see also reviews by Pearce 2019; Schier 2014a;). This model also acknowledged the unique historical circumstances that may have dictated variations in the dynamics of innovations, and in their adoption and adaptation (Childe 1951). The mid–late 20th century saw the proposal of two highly influential ideas that are still structuring the current debates: the identification of local metal-using communities representing new Balkan Copper Age, Chalcolithic or Eneolithic societies (e.g. Gimbutas 1980; Jovanović 1971; Lichardus 1991c; Renfrew 1969) and the introduction of metallurgy through an external population group, as represented through archaeological cultures migrating into the Balkans from the East (e.g. Childe 1929; Garašanin 1973).

The continued importance of Balkan metallurgical origins in archaeological scholarship is evident in the use of some of the earliest applications of radiocarbon dating in world archaeology to challenge existing '*Ex Oriente Lux*' interpretations with a model of independent invention (Jovanović and Ottaway 1976; Renfrew 1969, 1970; contra Wertime 1964, 1973). Debates on early metallurgy in the Balkans and Anatolia/Near East rapidly divided scholars into advocates of either independent invention or of migration/diffusion (see Bognár-Kutzián 1976 for a summary). Although highly controversial at the time (Makkay 1976: 263), early radiocarbon dates have now become widely accepted. The independent invention model for Balkan metallurgy was subsequently reinforced by the quantity, typology and metallurgy of cast shaft-hole axes (Charles 1969), the ability to calibrate the new radiocarbon dates using dendrochronology (Renfrew 1973) and the stratigraphical demonstration at the excavations at Sitagroi, Greece, showing that the Central Balkan Vinča culture preceded the Early Bronze Age of Anatolia (i.e., Troy I) by more than two millennia (Renfrew 1979: 139). Further supporting evidence accumulated during the 1970s with the excavation and relative dating of copper mines at Ai Bunar (Chernykh 1978a, 2008a) and the discovery of substantial numbers of the earliest gold—and copper—objects at the cemetery of Varna in Bulgaria (Ivanov 1978b; Ivanov and Avramova 2000). The '*Ex Balcanae Lux*' model (Todorova 1978) represented a new paradigm and caused a widespread re-appraisal of the origins of early technologies in the region (see Sterud *et al.* 1984 for a summary). As Thornton (2001) dryly noted, the major advocates for an independent invention of southeastern European metallurgy were also the most adamant diffusionists for Neolithic subsistence and the origins of the Indo-European languages (Renfrew 1987; Gimbutas 1973).

In addition to dominating the agenda for the 'when' and 'where' for the origins of metallurgy, Renfrew's (1969) suggestion of a direct connection between the production of graphite painted ceramics and the invention of copper metallurgy also provided a new explanatory framework for the 'how'. This pyrotechnological transfer model was subsequently also advocated by Gimbutas (1976a), however, this claim was not investigated from a comparative perspective until nearly four decades later (Amicone 2017; Amicone *et al.* 2020; see Chapter 44). A partial model for the 'why', that went beyond the hindsight of assumed ideas of value and desirability of metals as used by both advocates of independent invention and diffusion, was provided by Glumac (1991). He made the connection between the use of copper minerals for decoration and early metal use in the context of the Balkans - an approach subsequently extended across Anatolia and the Near East by Thornton (2001). The early use, across the Starčevo and subsequently early Vinča culture occupation horizons, of a specific type of copper minerals (black-and-green manganese-rich copper carbonates) that were later smelted to produce the earliest known copper metal, has most recently been shown to be a unique technological trait of Balkan copper and tin-bronze metallurgy, reinforcing the argument for independent invention (Radivojević *et al.* 2010a; Radivojević 2013, 2015).

There are consequently three definable, yet partially inter-related, models for the mineralogical and technological basis underpinning the emergence of metallurgy in the Balkans during the 6th–5th millennia BC (Research Question 1). All are primarily focussed on copper minerals/ores and the ambiguous relationship with ceramic pyrotechnology (Research Question 2). Underlying each of these models are the motivations of the people involved in identifying, selecting and smelting copper ores and their knowledge of and relationship with pre-existing pyrotechnologies. However, it is argued that the singular and uneven focus of early metallurgy scholarship has served to obscure the presence of polymetallism whereby multiple metals were being produced, circulated, used and deposited by the same communities in the same region at the same time (Radivojević *et al.* 2013). The mono-metallic perspective also serves to significantly underplay the deliberate manipulation of metal compositions whether by using complex ores or the addition of other metals, thus creating differences in colour, hardness and castability. It is therefore imperative to adapt these three models when evaluating the evidence for lead, gold, tin bronze and silver in the Balkans (Research Question 3). Each of the models will be evaluated against the pre-existing data as well as the new evidence from *The Rise of Metallurgy in Eurasia* project regarding minerals and ores, Balkan-Anatolian connections and pyrotechnology.

These models are:

1. The migration of individuals and groups from the east with the necessary knowledge relating to the selection of copper minerals/ ores and expertise in pyrotechnology.
2. The transmission, via existing socio-economic networks, of knowledge and expertise relating to copper ores, copper minerals and pyrotechnology from the east into communities in the Balkans.
3. The independent invention of metallurgy by communities in the Balkans who exploited the rich abundance of copper minerals and their knowledge of them and, through time, adapted their pyrotechnological expertise to smelt metal.

Ores, pyrotechnologies and Balkan-Anatolian connections

Evidence suggests that use of copper mineral and native copper in neighbouring Anatolia and the Near East occurred much earlier than in the Balkans as detailed in Chapter 3. This chapter also highlighted the frequency of copper minerals recorded at early Vinča culture settlements (pre-5000 BC), including the sites of Belovode and Pločnik, as demonstrated in earlier excavations (see Chapters 5 and 6). *The Rise of Metallurgy in Eurasia* project confirmed the presence of copper minerals in the pre-5000 BC phases at each site (see Chapters 11 and 26) and the careful excavation and recording techniques further demonstrated that copper minerals were extensively sourced, identified, selected, carried back to settlements and manipulated in the centuries prior to any evidence for copper smelting (cf. Radivojević 2015).

It is beyond doubt that the practice of using green minerals was transmitted to the Balkans from Anatolia, particularly given its association with early agricultural communities and attested migration and other movements (e.g. Ammerman and Cavalli-Sforza 1971; Brami 2017; Furholt 2017; Ivanova 2020; Mathieson *et al.* 2018; Racimo *et al.* 2020; Rosenstock *et al.* 2016; Shennan 2018; Silva and Vander Linden 2017). However, whilst the first model of migrating individuals and groups from the east is strongly supported with regards to identifying and exploiting copper minerals in the Balkans, the question remains as to whether the ability to undertake copper smelting derived from the same source.

As detailed in Chapter 5, the archaeometallurgical analysis of five small copper slags from Trench 3 at Belovode, together with the radiocarbon dating of the excavated horizon in which they were found, that provided evidence for copper smelting at *c.* 5000 BC (Radivojević *et al.* 2010a). The further evidence of two copper metal droplets found in a sealed refuse pit in Trench 18 at Belovode (Feature 21, see Chapter 11), excavated for *The Rise of Metallurgy in Eurasia* project and dated to the 49th century BC, provides a more contextually and chronologically secure dating for copper smelting but nonetheless confirms the earlier interpretation for *c.* 7000 year-old metallurgy. Beyond the Balkans, the closest evidence to the copper smelting to Belovode comes from Tal-i Iblis in southeastern Iran in the early centuries of the 5th millennium BC; this is only relatively dated to as it comes from the excavation of spoil heaps (Frame 2012). Until very recently, the assumed metallurgical activities at Çatalhöyük had stimulated scholarly debates due to an unusually early date of *c.* 6500 BC for a find that appeared to contain features of a metallurgical 'slag' (Cessford 2005; Mellaart 1964; Neuninger *et al.* 1964). The argument that the Neolithic Çatalhöyük communities were possibly smelting metal has since been discussed in the literature and found both support (Hauptmann 2000; Hauptmann *et al.* 1993; Strahm 1984) and open caution (Birch *et al.* 2013; Craddock 2001; Muhly 1989; Pernicka 1990; Radivojević *et al.* 2010a; Roberts *et al.* 2009; Tylecote 1976). A full re-analysis of the original metallurgical 'slag' from Çatalhöyük and revised contextualisation showed that this sample was a burnt copper mineral, probably deposited as a green pigment in a burial and subsequently baked during a destructive fire event in the dwelling in which it was discovered (Radivojević *et al.* 2017).

Further fieldwork evidence for copper smelting in Anatolia dates from late 5th / early 4th millennium BC occupation of the eastern Anatolian site of Değirmentepe (Lehner and Yener 2014; Mehofer 2014), although no analytical evidence of this find is known to the authors. In the southern Levant, the more secure evidence for copper smelting dates to the later 5th millennium BC (Golden 2010, 2014; Klimscha 2013). To the west, the earliest evidence for copper smelting in Central Europe and the Central Mediterranean also dates to the late 5th millennium BC (Dolfini 2013, 2014; Höppner *et al.* 2005; Scharl 2016; Turck 2010). The potential early 5th millennium BC date in the Western Mediterranean at Cerro Virtud, southeast Spain remains disputed and over a millennium earlier than any other copper smelting site in the region (Gauss 2013; Kunst 2013; Montero-Ruíz *et al.* 2021; Murillo-Barroso and Montero-Ruíz 2012; Rovira and Montero-Ruíz 2013; Ruíz Taboada and Montero-Ruíz 1999). For copper smelting alone, the evidence in the Balkans pre-dates neighbouring regions as well as those further afield and supports the third model, i.e., the independent invention of metallurgy.

As highlighted in Chapter 3, the smelting of copper ores was not the earliest application of pyrotechnology in either the Balkans or Anatolia. The transmission of ceramic forms and pyrotechnology from Anatolia to the Balkans occurred from *c.* 6600 BC, with ceramic production and consumption subsequently being extensively practiced and developed by early farming communities (Amicone *et al.* 2019; de Groot 2019; Spataro and Furholt 2020). Given that this process started around 1500 years before the earliest evidence for metallurgy in the Balkans or elsewhere, it leads us to the issue of the interdependence of pottery and metal pyrotechnologies (Research Question 2). The key question is whether the ability to create and manage high temperatures (exceeding *c.* 1000 °C) could have led to the transformation of copper ore to copper metal. In their research on the interdependence of pottery and metal technology at the Vinča sites of Belovode (Chapter 14) and Pločnik (Chapter 29) within *The Rise of Metallurgy in Eurasia* project, Amicone *et al.* (2020) (Chapter 43) dismissed the importance of high temperatures in pottery firing for proving this relationship in terms of the mastery of the Vinča potters' pyrotechnological skills. Instead, they highlight the more critical skill of controlling firing atmosphere conditions. The production of a functional pot required temperatures only in the range of 600–700 °C, and not in excess of 1000 °C. It is contended that previous scholarship over-estimated the role of firing temperatures in seeking the best fit with the model of interdependence between pottery and metal technologies.

The findings from *The Rise of Metallurgy in Eurasia* project (see Chapter 43) do not, however, preclude that other advances in pyrotechnology, such as mastery of fire control, could not have laid the groundwork for further technological advances such as copper extraction. The only hypothesis from previous scholarship on pottery pyrotechnology that Amicone *et al.* (2020) (and Chapter 43 of this volume) confirmed was that of Frierman (1969) and the two-step process of firing graphite-painted pottery, broadly similar to the two-step process of the earliest metal smelting (Radivojević *et al.* 2010b: 2777). The theories of graphite-rich moulds being used for metal casting used to counteract the oxidation process (Gaul 1948: 98; Ryndina and Ravich 2000: 16–17) remains to be confirmed archaeologically. However the roughly contemporary emergence of both copper smelting and the practice of graphite decoration at the dawn of the 5th millennium BC makes them equally likely to have influenced each other. This is particularly valid for settlements with strong evidence for metallurgical practice and adjacent graphite deposits such as Pločnik, on which Amicone *et al.* (2020) (and see Chapter 43) build their case. Hence, these technologies are rather seen as 'close cousins', clearly impacting each other, highlighting the need for future programs to date the emergence of graphite painted pottery in conjunction with early metallurgy in the Balkans.

It is notable that the emergence of graphite painting and copper smelting technology take place at around the same time as the appearance of a distinct phase of the Vinča culture called 'Gradac' (Vinča B2–C1). This phenomenon has been mainly characterised by the following: changes in pottery styles; house destruction horizons at several sites; an increased number of settlements erected at more dominant positions; the intensification of elaborated monumental figurine production (Garašanin 1991); and the intensification of mining activities at Rudna Glava (Pernicka *et al.* 1993; Radivojević and Kuzmanović Cvetković 2014). This pattern of a shift in some settlement and living practices has been identified across the Balkans at around the same time (Garašanin 1994/1995), prompting Jovanović (2006) to argue that the emergence of metallurgy was the driving force for the observed change. While current research, including this monograph, can certainly confirm that the emergence and intensification of copper metallurgy exhibits *correlation* with the changes in a number of material and dwelling practices, more studies to include dating and archaeomaterials analysis is required to explore the *causation* behind these phenomena across the Balkans (cf. Radivojević and Grujić 2018a).

When this evidence is evaluated against the three models outlined above it can be argued that, whilst the knowledge and practice of selecting copper minerals and the pyrotechnology to make the ceramics derived from early farming communities moving from Anatolia to the Balkans, the smelting of copper ores occurred independently in communities in the central Balkans. The nature of the pyrotechnological pathway that led to the smelting of copper metal remains unclear, as *The Rise of Metallurgy in Eurasia* project has demonstrated that the temperatures and conditions required to create contemporary and earlier ceramics are not transferrable in a straightforward manner.

Invention, innovation and polymetallism

In a special section of the *Cambridge Archaeological Journal* (Roberts and Radivojević 2015), we argued, alongside other authors analysing ceramics, gold, iron glass and glaze from across the world, that the processes of inventions in early societies could be evaluated in terms of pyrotechnologies. This contrasts with the reluctance of most archaeologists in recent decades to engage with the concept of invention. This is not only due to the partial nature of the archaeological record, which frequently reveals earlier dates for defined phenomena as investigations intensify, but also to the intellectual baggage of technological determinism. The social evolutionary schemes of Lubbock (1872), Morgan ([1877] 1985) and others, whereby 'inventions and discoveries stand in serial relations along the lines of human progress and register its successive stages' (Morgan 1985: vi), casts a long shadow. The continued influence of a classification scheme of technological stages within social evolutionary schemes (cf. Childe 1944) has contributed to this intellectual unease. The more recent exploration of the socially constructed underpinnings of modern understandings of technology and a more in-depth recognition of technology as a socially embedded phenomenon (e.g. Dobres and Hoffman 1994, 1999; Lemonnier 1993, 2012; Schiffer 2001) has reinforced the suspicion that the concept of invention is an anachronism. This suspicion is reinforced by debates over inventions within archaeology that begin with scientific criticisms of the proposed earliest dates for a phenomenon, but can rapidly become mired in academic, nationalist, and post-colonial politics. The danger of drifting into 'Originsland' as eloquently deconstructed by Gamble (2007: 61ff) is ever present.

Invention is frequently defined as the discovery of a new idea, material, or process, deliberately or by chance (cf. Renfrew 1978b). An invention may include a radically new product as much as a recombination of technological components in a novel manner (Basalla 1988; Nelson and Winter 1982; Weber *et al.* 1993). Alternatively, an invention may involve the application of an existing technology to a new purpose (Fleming and Sorenson 2004; Henrich 2010). The three stages of the process of invention (gestation, cradle, maturation) very rarely take place during the lifetime of a single inventor (Lienhard 2006: 165). It could take a few generations of inventors who contribute to accumulation of knowledge before the invention is sufficiently perfected to find a wider acceptance. These contributions are commonly mirrored in a high frequency of failure and unintended outcomes that occur over a period of uncertain length. Therefore, the biography of a past invention would have often started a couple of decades or even centuries before it became visible in the archaeological record. It is also often the case that inventions cannot find any economic use until other ideas that are yet to be discovered render practical what was once considered a 'long shot' (Wiener 1993: 145).

It has been frequently argued that when an invention affects the evolution of the system and is successfully transmitted within a population, and beyond, it is recognised as an innovation (e.g. Henrich 2001; O'Brien and Shennan 2010a; Ottaway 2001; Renfrew 1978b, 1986; van der Leeuw and Torrence 1989). A technological innovation is not usually only a monolithic entity, but an 'amalgam of units', or a 'recipe' containing ingredients required to make an object (Lyman and O'Brien 2003; Mesoudi and O'Brien 2008; Neff 1992, 1996; O'Brien and Shennan 2010b). In a technological context, this recipe is a list of what, how, and when to make something, and for how long (Krause 1985).

Michael Schiffer (2005, 2010, 2011) proposes a more nuanced definition of invention, arguing that invention is 'the creation of an idea or vision for a technology that has performance characteristics—often use-related ones—differing from those of other technologies' (Schiffer 2011: 36). The emphasis on invention as the *idea*, rather than the materialisation of the idea, is more common within industry and commerce where 'invention is the first occurrence of an idea for a new product or process, while innovation is the first attempt to carry it out into practice' (Fagerberg 2004: 3; see Schumpeter 1939). However, there remain disagreements within these disciplines, as there are within archaeology, as to whether invention should also encompass the realisation of the idea (e.g. Rogers 1962, 2003; contra Wiener 1993). Similar ambiguities and debates in the definition and evidencing of an invention also underlie the development of modern patent law (Pottage and Sherman 2010). Indeed, the intellectual history of the terms 'invention' and 'innovation' reveals ever-shifting definitions, meanings and applications relating to the societies in which they were used (Godin 2015).

The archaeological record reveals the materialisation of ideas and therefore, from this perspective, the innovations rather than the inventions (see Jones 2004; Ingold 2007; Lemonnier 2012; Malafouris 2013; Meskell 2005; Miller 2005; Renfrew *et al.* 2005). However, it is argued that the time-depth of the archaeological record can, with sufficient data resolution, provide the possibility of investigating the accumulation of knowledge or process components preceding and underlying the invention of a particular technique or material. These would be observable through experiments, re-combinations, or re-applications (to a new purpose) over a period of several decades, centuries or millennia (Basalla 1988; Henrich 2010; Lienhard 2006; Weber *et al.* 1993). Taking invention as the appearance of a new idea, process or material, the differentiation between an invention and an innovation within pyrotechnologies is potentially far less ambiguous than in other aspects of past human activities in the archaeological record. The conceptualisation of invention as a process—rather than a singular event—as has been extensively explored in archaeology by Schiffer (2005, 2010, 2011: 57–85) for mainly modern technologies. Invention and innovation in archaeological interpretative models, including those relating to the origins of metallurgy, have also been critically re-evaluated more recently by Frieman (2021). These more nuanced approaches enable the anomalies to be placed within the choices and sequences of transformative actions underlying pyrotechnology. The selection of black-and-green copper ores for smelting as opposed to the pure green minerals for decoration by communities in the Balkans represents not only a process of invention, but one in which choices and sequences can be identified.

Given the presented evidence, while the practice of sourcing copper minerals and native copper originated outside of the Balkans and was brought into the region—presumably accompanying other broadly contemporary materials that were being exploited such as obsidian—the development of copper metallurgy took a technologically distinctive and independent route in the Balkans from as early as 6200 BC (Radivojević 2015). The very moment of 'invention', though, is difficult to pinpoint but is certainly not later than 5000 BC, when we already see the developed and repetitive process of smelting under similar redox conditions and with similar 'recipes' across the Vinča culture sites (Radivojević and Rehren 2016). From this perspective, the invention of copper metallurgy could have taken place any time between 6200 BC and 5000 BC, but most likely during the second half of the 6th millennium BC.

But what was the process of invention of copper smelting?

Prior to *The Rise of Metallurgy in Eurasia* project, Radivojević and her colleagues had shown, in extensive analytical studies across the Neolithic and Chalcolithic Balkan sites in recent years (Radivojević *et al.* 2010a; Radivojević and Rehren 2016), that the use of copper minerals for decorative purposes and for copper smelting involved different selection practices and intent based on colour. The identification of dual selection (pure green copper minerals vs. black-and-green copper ores) not only indicated aesthetic differences, but also varying compositional requirements, implying that copper smiths distinguished between the material properties of differently coloured minerals. Such a technological practice and clear distinction between minerals for ornaments and those for smelting has not yet been identified in Anatolia or in the Near East. This was the foundation of Radivojević's (2012, 2015) claim that the preference for the black-and-green appearance in the selection of copper minerals by Late Mesolithic/Early Neolithic communities in the central Balkans from *c.* 6200 BC prompted early experimentation and subsequent copper smelting. This selection process is particularly evident in the abundance of manganese oxide in the post-5000 BC Vinča culture copper slags, which is known to facilitate the formation of a melt under the variable redox conditions that one would expect from hole-in-the-ground smelting installations (Huebner 1969: 463; Radivojević and Rehren, 2016: 221 ff.).

The results from *The Rise of Metallurgy in Eurasia* project, which recorded the extensive presence of copper minerals throughout the entire sequence of Belovode and Pločnik, demonstrates that the hypothesis on consistent selection of black-and-green copper minerals holds up for the newly discovered workshop (F6, see Chapter 11), while malachite beads show a mixed picture of selecting pure green and black and green minerals for decorative items. These results, however, reinforce the major tenet of the claim for the independent evolution of Balkan metallurgy: Vinča culture communities were intentionally selecting black-and-green copper minerals for copper smelting over the course of 600 years, and these were sourced in eastern Serbia, uniquely for all early copper smelting sites across Serbia and Bosnia (Chapter 41). While the exact locations of these deposits are not certain, we know that they are multiple to begin with, and contain cobalt and nickel mineralisations that have, incidentally, already been detected in the ancient mines of Rudna Glava and Majdanpek (Pernicka *et al.* 1993). Even though the pursuit for these deposits remains a task for the future, the results of *The Rise of Metallurgy in Eurasia* project reinforce previous analytical investigations (e.g. Radivojević and Grujić 2018a; Radivojević and Kuzmanović Cvetković 2014; Radivojević and Rehren 2016) by providing a higher resolution of insight into the connectedness of the early copper metal making Vinča sites and local deposits, as well as their networks of supply across the Balkans.

The selection of lead ores and the application of pyrotechnology to them, evidenced from the *c.* 5200 BC lead slag 'cake' at Belovode (Radivojević and Kuzmanović Cvetković 2014) and the use of lead ore for beads at Autoput, Selevac and Opovo in Serbia and Donja Tuzla in Bosnia—in all cases in horizons that end in 4500/4400 BC at the latest (Glumac and Todd 1987; Quitta and Kol 1969; Vogel and Waterbolk 1963), as detailed in Chapter 3—is too often overlooked. This is particularly valid when claims for the 'new earliest' are made, as with lead ore processing in Pietrele, which emerges *after* the end of the Vinča culture (Hansen *et al.* 2019). It is, however, evident that the knowledge, pyrotechnological experiments and establishment of craft and material practices surrounding vibrantly coloured minerals and later ores, whose metallurgical properties could only have been distinguished by their colours (black, green, blue and violet), were fundamental during the centuries spanning *c.* 6200–5000 BC (Radivojević 2015).

A consequence of the colour preference in (complex) copper ores is the tin bronze foil (Pločnik 63), as detailed in Chapters 3 and 6. Excavated from an undisturbed context, on the floor of a dwelling structure next to the likely copper metal workshop at the site, about 1 m from a fireplace, the foil was enclosed in several late Vinča culture pottery vessels (Radivojević *et al.* 2013: 1033, Figure 2). This securely contextualised find comes from a single, undisturbed occupation horizon at Pločnik, dated to *c.* 4650 BC. This date is, according to the field evidence, the *terminus ante quem* for the Pločnik foil at present. The composition indicated that stannite [Cu_2FeSnS_4], a copper-tin bearing mineral, was the probable ore used for making this natural alloy with *c.* 12wt% Sn and relevant traces of As, Fe, Co and Ni (see Radivojević *et al.* 2013: 1035, Table 1). As detailed in Chapter 3, there are 14 additional tin bronze artefacts known from the mid-late 5th millennium BC Serbia and Bulgaria; however, these finds only occurred together in what appears to be a short-lived tin bronze horizon in the Balkans based on geochemistry that links them with the Pločnik foil.

When the Pločnik foil date is evaluated against the earliest evidence for tin bronzes in Europe, Anatolia, and Asia (Pigott 2011; 2021; Rahmstorf 2011; Radivojević *et al.* 2013), claims from two discoveries require detailed consideration. These are the assertions about the 6th millennium BC emergence of naturally alloyed tin bronze artefacts from the sites of Tel Tsaf in Southern Levant (Garfinkel *et al.* 2014) and Aruchlo in Georgia (Hansen *et al.* 2012). The Tel Tsaf metal awl was discovered in a secondary context (burial in a silo) in what is currently claimed to be a largely Middle Chalcolithic horizon, broadly dated to between 5100 and 4600 BC. The authors ascribe the metal awl to the late 6th millennium BC (Garfinkel *et al.* 2014), despite the fact that: a) the skeletal burial had enough datable materials available between the individual and the rest of the burial offerings; and b) even if the secondary context is truly Middle Chalcolithic, its characteristics are more indicative of the end of the silo's use-life, at best, 4600 BC. The analysis of this heavily corroded awl 'with no original metal left' (Garfinkel *et al.* 2014: 3), conducted with portable ED-XRF, implied an Sn content between 3.5wt% and 7wt%. While the authors acknowledge that these figures may be overestimated given that tin is known to be relatively immobile in most burial conditions compared to copper and hence usually found enriched in corroded layers, the questions remain: how much tin was there, and was it enough for the smiths working with it to detect any difference in the performance of the artefact, or in its colour?

Comparative analysis of tin bronze artefacts using handheld XRF and EPMA (Electron Probe Micro Analyser) indicate that the former technique can differ around 20% from the true metal body values when applied to the metal surface cleaned from corrosion, or *c.* 70% or more when performed on the corroded surface of the same object (Orfanou and Rehren 2015). Although we do not know exactly the effect of the burial deposits on the enrichment of tin in the Tel Tsaf artefacts, estimates based on the reported XRF analysis in Garfinkel *et al.* (2014: 4, Table 1) indicate that the true value of tin content in the Tel Tsaf awl could potentially be between *c.*1wt% and 2wt%. While this is a speculative calculation with many unknowns, with such a composition the Tel Tsaf awl would still qualify as a tin bronze but without any indication of intentional alloying, which is a key factor for the foil from Pločnik in Serbia. The colour range of the awl would barely differ from that of common contemporary copper artefacts, even with up to *c.* 5wt% of Sn content (see Figure 6, Chapter 3 this volume), which suggests that the process of its making, if truly contextualised towards the end of the Middle Chalcolithic, probably made no difference to its appearance at the time. In addition, the performance of the awl would be dependent on the reduction in thickness by working, which is unknown due to the lack of any preserved metal body. The Aruchlo bead from the Neolithic site in Georgia is also optimistically set too early in the date range (5800–5300 BC). The handheld XRF analysis of this heavily corroded item with no metal body preserved reveals a compositional structure of what looks like predominantly malachite mineral with relevant impurities of tin, arsenic and iron, which are comparable with the polymetallic mineralisations in that area (Bastert-Lamprichs *et al.* 2012; Hansen *et al.* 2012). In sum, the claims for the early tin-bronzes in the Levant and Georgia in the 6th millennium BC require more rigorous analytical probing in order to substantiate their currently published interpretations. There is therefore no compelling evidence that either the idea or the technological expertise for the tin bronzes in the Balkans derived from communities or networks beyond this region.

In order to understand the why there is a tin bronze foil at Pločnik at *c.* 4650 BC, it is instead necessary to understand the impact on copper of major impurities such as tin, arsenic and antimony. The resulting metals not only melt at lower temperatures than pure copper objects and are easier to cast (Lechtman 1996; Northover 1989;) but also transform their colour into different shades of bright yellow, whose range has recently been experimentally demonstrated (Radivojević *et al.* 2018b). The key, however, is the use of black-and-green copper-based ores, such as stannite, which is argued as the starting point of experimentation with the new metal in the Vinča culture context (Radivojević *et al.* 2013). It is possible that this colour selection process meant the exclusion of grey-coloured copper fahlores, such as tennantite and tetrahedrite, whose successful smelting would have produced arsenical copper, a metal that is not present in the Balkans until the mid-4th millennium BC (cf. Chernykh 1978b; Pernicka *et al.* 1997; Radivojević *et al.* 2010a). Nevertheless, there are indications for the use of hydrated iron-arsenates, such as the green/blue scorodite [$FeAsO_4 \cdot 2H_2O$], which precipitates from the primary ore of arsenic, arsenopyrite [FeAsS], in the copper smelting process in Gornja Tuzla (Radivojević and Rehren 2016: 219, 225 ff, Figure 7h). This rare occurrence of utilising arsenic-rich ores has been convincingly argued to be due to the selection of an attractive mineral colour, which is admittedly the same rationale for selecting stannite for making tin-bronzes in Pločnik. Given the highly experimental nature of early metal technology in the Balkans being, we may see more arsenic present in the copper smelting process in future excavations, even if only as a colourful addition to the smelt.

As detailed in Chapter 3, the appearance of thousands of small decorative objects made of gold also dates from the mid-5th millennium BC in northeastern Bulgaria, southeastern Romania and northern Thessaly (Higham *et al.* 2007; Krauss *et al.* 2017; Makkay 1991). Although the gold from the cemetery of Varna I is claimed as the earliest known (dated most recently between 4690 and 4330 cal. BC) (Krauss *et al.* 2016), there are earlier uses of gold ornaments (although not as securely dated) in the Varna II cemetery (Todorova and Vajsov 2001: 54), as well as in the cemetery of Durankulak (Avramova 2002: 193, 202, Table 24; Dimitrov 2002: 147). The earliest gold from beyond the Balkans is in the form of eight gold and electrum rings from Nahal Qanah in the southern Levant, dating to the late 5th millennium BC (Gopher and Tsuk 1996; Klimscha 2013) with all other finds across Europe, Anatolia and Asia dating to the 4th millennium BC or later (Meller *et al.* 2014); there is no evidence for an external transmission of the idea, technology or metal into the Balkans. Colour manipulation is revealed in the carefully controlled alloying of copper with gold to create a specific colour palette from *c.* 4550 BC at Varna, Bulgaria (Leusch *et al.* 2015). As Radivojević *et al.* (2013) originally proposed, the mid 5th millennium BC sees a polymetallic horizon in the Balkans which is driven by a desire for specific colours and pyrotechnological experimentation. These metallurgical inventions and innovations were, as was long ago recognised by Cyril Stanley Smith (1981), motivated more by cultural aesthetics ('desire to beautify') than efficient functionality.

Early metals, metallurgies and societies

Despite scholarly critiques for over a century, the Four or Three Age system—depending on which scholarly tradition is followed—remains firmly embedded. The emphasis on the transition from stone to metal technologies, as identified across the majority of Europe, North Africa and Western Asia by a Copper, Eneolithic or Chalcolithic Age (cf. Pearce 2019), has often ensured that the appearance of metallurgy is connected to broader political, social and technological changes (Roberts and Frieman 2012; Schier 2014a). The influence of modern value systems mean that early metal and metallurgy was inevitably ascribed a high value by scholars relative to other contemporary artefacts and technologies in other materials. As reviewed extensively in Kienlin (2010) and in Chapter 3, the direct consequence was that, even prior to the discovery of early metal production evidence, specific interpretative narratives of metals and societies rapidly became embedded in scholarship. Drawing on the pre-existing evidence as well as new evidence from *The Rise of Metallurgy in Eurasia* project, this section addresses not only Research Questions 1, 3, 4 and 6 on the organisation of early metal production in the Balkans but subsequently re-evaluates the interpretations around early metallurgy and metal, and their relationships to the societies involved.

Sourcing ores

The extensive provenance dataset dating from the 5th millennium BC in the Balkans (cf. Pernicka *et al.* 1993, 1997; Radivojević *et al.* 2010a; Radivojević and Grujić 2018a), paired with fresh analysis of materials excavated within the remits of *The Rise of Metallurgy in Eurasia* project (see Chapter 41) largely confirmed the previous indicators for the sourcing of copper ores during this period. The higher resolution approach to exploring the early copper supply routes in the Balkans highlighted two important points: a) the significance of several east Serbian copper deposits (some yet unknown) for the early phase of metallurgy evolution in the Balkans (Early Chalcolithic, *c.* 5000–4600 BC), and b) the super-connectedness of metal producing and consuming sites in Serbia, Bosnia and Bulgaria, along important communication routes.

Both lead isotope and trace element analyses (Chapter 41) of metallurgical materials from Belovode and Pločnik reveal the complex dynamics of copper acquisition routes between *c.* 5200 BC and *c.* 4450 BC in the Balkans. Overall, the Balkan Chalcolithic communities were utilising copper from at least six (or seven) copper deposits, two of which are still unidentified. These are: Majdanpek, Ždrelo, Ai Bunar, Medni Rid, 'Group of 16', 'Cluster #8' and potentially Rudna Glava (or an associated Co/Ni rich mineralisation). Of these, Vinča culture communities were not using metal from Bulgaria's Medni Rid, while Ai Bunar's copper has only been used during the extended occupation of Pločnik, or beyond *c.* 4600 BC. While these deposits were identified based on most artefacts clustering in distinctive groups, the list is not exhaustive.

Of particularly interest were the consistencies of trace elements with metal production evidence from Rudna Glava, the earliest dated copper mine in the world thus far, with *c.* 5500 BC being the proposed date for the start of mining activities at the site. Although the mine has confirmed signs of Vinča culture mining (Jovanović 1980), no copper metal artefact has yet been confirmed to come from this mine. This is despite the partial consistency of metal production evidence (slags from Belovode, Vinča and Gornja Tuzla), with the best match regarding Co/Ni content amongst the known copper deposits in east Serbia. For future research, it would be important to analyse Ždrelo ores for trace elements and pursue similarly rich Co/Ni mineralisations in the region which, altogether, may offer clearer pointers as to which deposit (or deposits) provided the black-and-green copper ores that were consistently smelted in the Vinča culture sites (cf. Radivojević and Rehren 2016).

Regarding the interconnectedness of these early metallurgical sites, the direct ^{14}C dating of materials associated with metallurgical materials from Belovode and Pločnik provided, for the first time, high-resolution evidence of the beginning, evolution, and end of metallurgy in these settlements and, most importantly, data on the cooperation between these communities concerning access to copper ores. For instance, the close consistency of Belovode production evidence with copper implements from Pločnik, or similar correspondence of copper minerals from the earliest levels in both sites pinned to the 51st century BC, reinforces the assumption of their close connectedness and involvement in sharing metallurgical knowledge. Equally, the matching data from smelted copper in one horizon and malachite associated with another in Belovode speaks to the presence of consistent supply networks throughout the occupation of this settlement. Both of these sites exhibit more connections with Vinča and Gornja Tuzla, indirectly or directly, as well as along the lower Danube route, which was highly important in the exchange of metallurgical knowledge and/or ores and artefacts.

Making metals

What, then, did the copper smelting process look like in the first 500 years of this practice in the Balkans, based on current analytical and field evidence? In the absence of any other smelting installations, the current model of metal production inferred from the hole-in-the-ground 'furnaces' (cf. Radivojević and Rehren 2016; Rehren *et al.* 2016), including F6 in Belovode elaborated here (Chapter 11), speaks to the large quantity of extant copper metal artefacts being produced in multiple individual episodes. These smelting episodes could have been made more efficient if many were undertaken simultaneously, within each (or between many) participating household or, more precisely, within one or many of the backyards of individual dwellings, as this was an outdoor operation. The production efficiency would depend upon the smelting charge (ores + fuel) and the ability to maintain the redox conditions. While the final result could be anywhere between tens and hundreds of grams of copper metal, it is unlikely, based on the current evidence, that the heavier (1 kg plus) copper implements were produced in a single smelt. The fragmentary evidence of melting crucibles in Bulgaria and, possibly, Romania, suggests that this metal was remelted and cast (Rehren *et al.* 2020; Ryndina *et al.* 1999; Stefan 2018). We can observe evidence for the latter only from metallographic examination of as-cast objects (Kienlin 2010).

In 2013, during the excavation campaign at Pločnik, the first author ran series of smelting experiments based on these early reconstructions of metal extraction process and, with her team, managed to successfully smelt copper from local ores. One clear outcome of these experiments (the full account is currently outside the remit of this paper) is that, in order to be successful, the process demands a community of people working together in a range of co-ordinated roles (Figure 1). Six people were required to operate six blowpipes with ceramic nozzles on their tips (see Figure 4e-f, Chapter 3). These individuals were regularly replaced, as fresh blowers were needed every 15–20 minutes in a process that, on average, lasted 60 minutes. Meanwhile, a seventh member of the smelting crew ('master smelter') was engaged in maintaining the fire or regular charcoal charge. An additional, critical, member of the team was – a drummer! Well-paced and uninterrupted air blowing into the 'furnace' was crucial for the success of the smelt, ensuring that the desired temperature was reached at a rate that prevented the copper metal ending up in the slag. With a large group of people participating, it was impossible to maintain the air flow without a unifying rhythm. This phenomenon has also been observed in

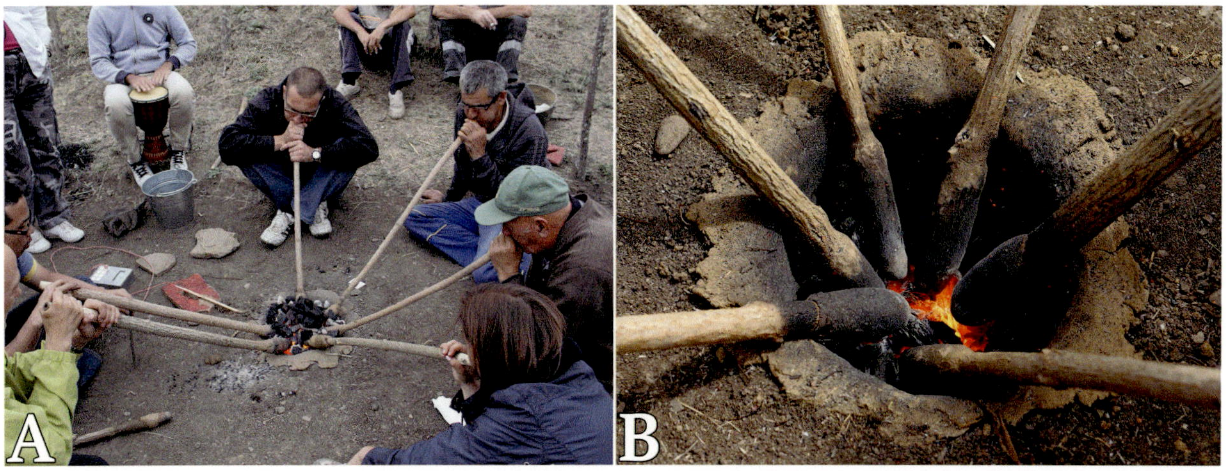

Figure 1. a) Copper smelting experiment in Serbia in 2013 aimed at reconstructing the earliest metal extraction process based on archaeological and laboratory reconstructions. Note six blowers, one drummer and one master smelter (upper left), with two people in the back waiting as a replacement for the blowers; b) Ideal reconstruction of the hole-in-the-ground smelting installation from the site of Belovode in Serbia. (Photo CC BY-NC-ND 4.0 J. Pendić and M. Djurica @Reuters).

experimental reconstructions of African iron smelting, where drums were used to regulate the use of bellows (cf. Chirikure 2010; Humphris *et al.* 2018).

These smelting experiments prompt us to consider how control of the smelting knowledge was exerted, if at all, and how the personal relationships between the many participants shaped the smelting technology. They also raise the question of how strict the replication of the 'recipe' could had been in this environment, and whether the variations that we see in the composition of colourful ores used for copper smelting (Radivojević and Rehren 2016) could be explained by human factors, such as trial and error processes. The experiments also prompt other questions: how was trust developed and what kind of ties or rituals were connecting the participants in the metal production process? In the absence of any indication showing this was a full-time occupation, were they all members of a specific group within a community which co-operated specifically for metallurgy (a 'cooperative'?), or simply a mix of family, neighbours, and friends helping each other during the metalmaking 'season'? Given the scarcity of evidence for any hierarchical structure in the Balkan Chalcolithic communities (cf. Porčić 2019a), what is the likelihood that they were organised as a collective grouping – with everyone given equal decision-making power? As Iles (2018) has convincingly demonstrated in her ethnographically informed study of the social landscapes of iron metallurgy in Africa, globally influential interpretations that invariably portray African metal smelting as a secret and exclusively male activity, do indeed stand up to detailed investigation. Future exploration of these nuances in both fieldwork and laboratory settings will allow us to gain more insight and reveal a more complete picture of how past smelters operated and interacted within the boundaries of their personal, social and environmental surroundings.

Why do we not see these hole-in-the-ground 'furnaces' in the field, other than through the indirect evidence, such as slagged pre-fragmented lining sherds (Figure 4c-d, Chapter 3)? Figure 2 shows that these structures are hardly recognisable only nine months post-smelting, although this assumes that the 'furnace' is used only once, which may not have been the case in the past. Nevertheless, these structures are ephemeral by the very rapid nature of their construction: in our experimental case it took a maximum of 30 minutes to dig a shallow hole and line it with sherds, potentially with the addition of clay as a binder. These smelting installations were clearly not intended to be durable, but rather to be ready for operation in a relatively short time. As such, they could have been built anywhere on or off the settlement site, the only evidence of their existence left for subsequent archaeologists being the ores, slags, slagged sherds and potentially metal artefacts.

This brings us to the question of how many smelting installations or 'workshop areas' we can estimate were present, based on the current evidence. In the case of Belovode or Pločnik, both of which we studied in detail and excavated, it seems very likely that that every household produced some metal in their backyard or communal area. This is corroborated by the extensive evidence for hundreds of copper ores found in every context, every feature, every dwelling, and every communal area across both sites (see Chapters 5, 6, 11, 26 and 41). These were predominantly manganese-rich, black-and-green copper ores, which we know were used for copper extraction (cf. Radivojević 2015). Slag and slagged sherd finds were notoriously rare at both sites (and beyond) prior to *The Rise of Metallurgy in Eurasia* project and its targeted methodological approach, however an excavation recovery bias must be taken into account: slag is essentially dark grey or brown, and small, crushed samples are not discernible

Figure 2. a) Installation from Figure1b post-smelting; b) Installation from Figure1b after 9 months. (Photo CC BY-NC-ND 4.0 J. Pendić and M. Radivojević).

in the soil. A salutary lesson on their potential invisibility to the excavator is that the small, less than one gram per sample, copper slags at Belovode were only identified in the field because they included some remaining copper metal with green patination (Radivojević *et al.* 2010a: 2779). These finds were held for 14 years at the National Museum in Belgrade in a box mislabelled 'copper minerals' because, at the time, nothing was known about the copper smelting in the Vinča culture save for how these early slags might have appeared. This experience highlights the need to include magnets (to detect Fe-rich slag matrix) and sieving as regular practice in future excavations targeting Chalcolithic sites in the region and beyond, otherwise evidence for early metallurgy could easily be missed.

The widespread presence of copper ores at Belovode and Pločnik calls for a different interpretation of metallurgical activities at these sites, with implications for other sites with similar evidence. We would argue that the pursuit by archaeologists of an early metallurgy specialist 'workshop' and an individual 'smith' reflect a romanticised—even mythological—ideal that may resonate in Childean narratives but is simply not reflected in the reality of the archaeological and archaeometallurgical evidence. With high-resolution fieldwork integrated with laboratory analyses and experimental reconstructions, a very different perspective emerges.

This new interpretative framework for Vinča culture metallurgy comprises:

1) multiple production episodes due to the limited scope for mass production using the hole-in-the-ground 'smelting installations' or furnaces;
2) collective and co-ordinated actions by groups of people, from the acquisition of the ore through to the production stages;
3) community-wide accessibility to the knowledge and practices of metal production; and
4) the absence of a single 'specialist smith'- due to lack of evidence.

On this last point, there is a lack of differentiation in the general material culture assemblage in the excavated dwelling features. In addition, even at the Varna 1 cemetery, Leusch *et al.* (2017) argue convincingly that there are considerable challenges to any confident identification and interpretation of a given grave as specifically belonging to a metalworker, whether through the associated artefacts or by osteological analyses. In further support of this notion, papers in Brysbaert and Gorgues (2017) demonstrate that the interpretations of a specialist craftsperson's societal identity in European prehistory through the evidence of their crafting activity are diverse and complex – showing that there can be no straightforward analogy with ethnographic case studies from other continents or myths from other societies emphasising the separation and status of the smith (cf. Budd and Taylor 1995).

The results from the excavations at Belovode and Pločnik emphasise instead the spatial, material, and technological integration of craft activities that encompasses the contemporary production of ceramics, polished stone tools, flint blades and metallurgy by individuals and groups, all within the broader settlement context at each site. The creation and/or shaping of these inorganic materials also occurred within the vicinity of cereal processing, animal butchery and other food production activities. There is no straightforward distinction between what might be classified as specialist as opposed to non-specialist craft activities either at Belovode or Pločnik or in the broader discussions of craft activities in the Central Balkans during the late 6th to 5th millennium BC (Chapters 45, 46, 47, 49). As highlighted by several scholars (Kuijpers 2018: 231–237; Molloy 2008: 174; Molloy and Mödlinger 2020), there is a difference between a specialisation (i.e. a skill at a specific activity) as opposed to a specialist (i.e. a person who is focussed exclusively on a specific craft or activity) and, as will be argued below, with the possible exception of elements of Varna goldworking (Leusch *et al.* 2015), that a community with specialisations rather than individual specialists underpinned craft production in the Balkans during the 5th millennium BC.

Using and depositing metals

The excavations at Belovode revealed 12 copper mineral ornaments, two copper metal droplets and one fragment of a finished copper metal artefact (see Chapter 11) whilst the excavations at Pločnik yielded 13 copper mineral ornaments, one copper metal droplet and fragments of five finished copper metal artefacts (see Chapter 26). The copper metal artefacts, like the minerals, are—where it is possible to identify a form—small, ornamental, and, as beads and rings or bracelets, apparently designed to adorn the human body. Given the archaeological contexts in which they were recorded, they have been analysed and interpreted within the context of copper production and circulation (see Chapter 41). Their fragmentary condition means that, whilst the copper mineral/metal objects may well have had an extensive use-life, as for instance with the loop or ring (C_P1/13) and bracelet or wire (P13/13), no current wear trace analysis approach (Dolfini and Crellin 2016) would be able to identify this.

Yet, the presence of these small ornamental copper metal and mineral artefacts raises two important considerations: 1) the recycling of copper; and 2) the types of objects being produced. Regarding the former, it is currently not possible to determine the extent of recycling as a practice during the 5th millennium BC in the Balkans. The recorded copper artefactual evidence from the region—estimated by Ryndina (2009) to be *c.* 4,300 artefacts but likely to be higher (Chapman and Gaydarska 2020)—is neither a reliable indication of the extent of copper production nor of recycling (Taylor 1999). The archaeological and archaeometallurgical evidence at Belovode, and especially at Pločnik, implies that the recycling of copper metal objects most likely occurred, even in the broader sense of mixing metals, repairing etc. When placed in the broader context of the compositional evidence and network analysis (see Chapter 41), the recycling that did occur must have been within, and not across, identified networks of supply. These supply networks have been extensively discussed in previous research (Radivojević and Grujić 2018a), where it has been convincingly shown that the complexities of copper supply (be it from a single or multiple sources), include a high degree of cultural homogeneity. Each of the connected communities, i.e., the cultures or horizons labelled on the basis of spatially and temporally coherent material culture and/or settlement practices (e.g.,Vinča, KGK VI or Bodrogkeresztúr) stuck to their own trusted sources of supply and acquisition. This kind of regularity in supply chain resulted in artefacts with similar chemical composition clustering within so-called 'modules'; these modules ultimately display either a consistent smelting of the same type of copper ores, or a homogenisation effect as a result of recycling ores from multiple resources but *within* a module (or an archaeological culture as data demonstrated). Hence, if we take this homogenisation effect to reflect a potential recycling practice, it was an activity that followed the same community-orientated connections and practices as the primary production of metal objects.

Regarding the second issue—the nature of the copper objects being made—the debates surrounding copper artefacts are dominated by the discoveries and analyses of large axes, especially during the second half of the 5th millennium BC to which the majority of those recovered are dated. As recently highlighted by Chapman and Gaydarska (2020), the depositional patterns of copper objects are dominated by large tools found in the landscape beyond settlements. This, at least partially, reflects a major recovery bias with small ornaments far less likely either to survive or to be readily identified when compared to large copper tools invariably found as 'chance discoveries' in the landscape. The near absence of an archaeologically visible funerary practice across the Vinča culture (see Chapter 4), where copper ornaments may have been worn, as evidenced at Gomolava (Stefanović 2008), means that it is primarily in excavations of settlement sites that copper ornaments are found and recorded. However, without a detailed and systematic methodological approach such as that undertaken by *The Rise of Metallurgy in Eurasia* project, it is possible that small ornaments could have been missed. For instance, the excavations at Belovode resulted in the first discovery of a copper metal object at the site despite *c.* 15 years of earlier excavations employing a different fieldwork approach. It may also be the case that copper ornaments were more widely recycled by Vinča communities and that it is fragmentary evidence of this wider practice that has been uncovered. Nonetheless, it seems very probable that early copper ornaments were far more widely made, circulated and worn than has previously been appreciated.

Interpreting metals

The interpretation of early metals and metallurgy in any region around the world must contend with narratives that were frequently well established in scholarship prior to the discovery of any clear supporting archaeological or archaeometallurgical evidence, as highlighted in Chapter 3. To re-cap from Chapter 3, these are:

1. The knowledge and expertise relating to production of metal represented a technological revolution.
2. The knowledge and expertise relating to metallurgy was restricted to specialist individuals who practiced in relative secrecy and held a distinct and elevated status.
3. The properties of metal objects—whether hardness, lustre and/or colour—ensure that they are fundamentally and consistently desirable and valuable to the predominantly farming communities.
4. The production, circulation and consumption of metals was integral to the creation and maintenance of elite status and identity in farming communities.
5. The invention and/or innovation of metallurgy impacts significantly upon the social, political and ritual lives of the farming communities.

Each of these dominant narratives will be critically evaluated against new and previously existing evidence with the underlying aim of building a new framework of interpretation from the primary data, i.e., 'bottom-up', rather than from long held theories, or 'top-down'.

1. Technological revolution?

Perhaps the most pervasive narrative is that the earliest metallurgy represents a technological revolution that is comparable to the domestication of plants and animals or the emergence of cities – influentially termed the

Neolithic and Urban revolutions respectively by V. Gordon Childe (1936). As has been highlighted in the extensive debates surrounding the conceptual and archaeological complexities of these latter phenomena, especially with respect to southeastern Europe (Chapman 2020a; Gaydarska 2017; Gaydarska *et al.* 2020; Ivanova 2020; Shennan 2018; Porčić 2019a; Whittle 2018), defining what might constitute a technological revolution is not straightforward. From a pyrotechnological perspective, it is argued to mean going beyond the 'deliberate processes utilising the control and manipulation of fire' (McDonnell 2001: 493) or more simply 'the use of fire as a tool' (Bentsen 2013) to enable the transformation of matter. This requires a shift in the community perception of the natural environment, as specific rocky outcrops, riverbanks and coasts are newly understood and exploited for the appropriation of raw materials in pyrotechnological processes (Boivin and Owoc 2004).

The concept of extractive metallurgy, just as any other *idea*, had multiple origins. In addition to other places, it found fertile ground to develop within the Vinča culture phenomenon. It evolved through *experimentation*, demonstrated by the presence of black-and-green minerals and 'slagless' extraction prior to the earliest documented smelting, but also by the selection of compositionally different yet similarly coloured ores. The Vinča culture metalworkers also developed an *understanding* of the smelting process and applied it in a consistent manner throughout the centuries of practice. The Vinča culture communities must have had social institutions in place to provide *logistics* for the distribution of metal implements to markets that desired these objects. Metallurgy *de facto* transforms the matter and is, as such, the first pyrotechnology with completely transformed products. The road to this invention is clearly demonstrated in the shift in the sourcing, collection and use of copper minerals for decoration and use of copper ores (minerals smelted to gain metal) to produce metals by communities in the Balkans. However, given that copper smelting was preceded by the pyrotechnological production of ceramics (de Groot 2019) and paralleled by the production of graphite-painted pottery (see Chapter 43), to what extent does it constitute a technological revolution of the kind that is expected to create a major and immediate impact upon societies?

Given the small-scale characteristics of metal production in evidence for the first half millennium across the Balkans, it is only with the hindsight of the metallurgical developments that would, several millennia later, impact upon world history, that this initial development could be viewed as revolutionary. In a seminal and highly influential paper, Budd and Taylor (1995) argued strongly for early metallurgy in Eurasia to be understood within an interpretational framework that drew on anthropological ideas of ritual and magic rather than the application of modern industrial and technological standards. Whilst the social evolutionary concepts of V. Gordon Childe can certainly be abandoned, it will be argued below that their replacement by a 'ritual' interpretation is contradicted by the archaeological and archaeometallurgical evidence.

2. Special smiths?

The 'metal smith' of later prehistoric Europe has been variously interpreted in scholarship as 'nomadic, a reviled outsider, elite in status, a mediator of wealth, a shaman or a proto-scientist' (Molloy and Mödlinger 2020: 169; cf. Eliade 1962), in a debate that primarily concentrates upon evidence from the European Bronze Age. Once again, the influence of V. Gordon Childe can be felt due to his interpretation of itinerant smiths whose movements between tribes were responsible for the diffusion of new ideas, technologies, and practices (Wailes 1996; Trigger 1986). Subsequent critiques of 'Childean smiths' have emphasised issues including the lack of ethnographic parallels, the over-emphasis on full-time specialisms, the lack of a ritual framework, and the need to incorporate kinship (Rowlands 1971; Budd and Taylor 1995; Kienlin 2010; Rowlands 1971).

The identification and interpretation of the 'smith' in the European Chalcolithic and Bronze Age is far from straightforward. Major surveys of the funerary evidence thought to constitute 'metalworkers graves' by Jockenhövel (2018) and Nessel (2012; 2013) – neither of which encompass the Chalcolithic in the Central or East Balkans – have consistently highlighted the complexity, ambiguity and rarity of compelling evidence. When evaluating an extensive range of evidence for smiths across Later Bronze Age Europe (1500–800 BC), Molloy and Mödlinger (2020) argue persuasively that metalworking was widely practiced and embedded within, rather than isolated from, the social lives of the communities involved. This is also the conclusion reached by (Găvan 2015) in her analysis of metal and metalworking evidence found in the Bronze Age Tell Settlements from the Carpathian Basin (c. 2500-1500 BC). The evidence excavated at Belovode (Chapter 14) and Pločnik (Chapter 29) implies that this integration of metalworking and community life occurred from the earliest metallurgy.

The only potential evidence for an identifiable 'smith' in the Balkans during the 5th millennium BC lies in the re-interpretation of the famous 'prince' from Grave 43 at Varna, who should now be more appropriately designated a 'smith', buried with his range of tools, including the mis-interpreted penis foil (Chapter 3). It is perhaps not coincidental that it is in Varna where the techniques of gold production, such as gold alloying, lost wax casting and gilding, are technologically unparalleled across the Balkan region (cf. Ivanov 1988a; Leusch *et al.* 2014, 2015). Yet these rare and complex specialisms in gold technology need to be understood, not only in the light of far more widespread metallurgical specialisms across the Balkans (see Chapters 3 and 41), but also alongside other archaeologically identifiable specialisms in other inorganic and organic materials, domestic plant and animal management, and settlement construction (Chapman 2020a). The widespread evidence for archaeologically visible *specialisms* across the communities in the Balkans, let alone those that are far more difficult to discern in the archaeological record, should be distinguished from the presence of *specialists* (Kuijpers 2018) where the evidence in any material or practice is far more limited. Within this broader perspective of specialisms throughout small-scale farming communities, the relationship between 'specialist' craft production and elites, as has been frequently proposed for early metallurgy in the Balkans (see Chapter 3) becomes difficult to sustain (Brysbaert and Gorgues 2017).

Beyond the use of metal objects for daily activities, the communal nature of metal production in settlements and the relative lack of inequalities within those settlements provides support for the notion that investment in craft production may have been prompted by demands for ceremonial communal activities. Spielmann (2002) illustrates ethnographically that economic intensification, and even craft specialisation, can evolve to meet an increased demand for food, exchangeable items and paraphernalia required to take part in collective ceremonial events. In an extensive review of social complexity and inequality in the Chalcolithic Balkans, Porčić (2019a) notes that trends in the development of copper metallurgy and other crafts, the circulation and production of items of exotic raw materials, household size, cattle husbandry and population size all increase in the 5th millennium BC in comparison to the previous period. However, these markers remain at levels too low to ascribe to them the rise of inequality. He builds on this, arguing that the presence of craftspeople is not sufficient to claim the existence of an elite that supported them, nor that the economy at the time was directed from a single centre. Rather, the incentive for craft specialisation (in our case, metallurgy) came from a socio-political arena and as such might, for example, have developed to supply the need of all participants in ceremonial events that involved metal tools.

3. *Desirable properties?*

The theoretical justification for elevated interpretative status of the metal objects in the Balkans is invariably justified by the distinctive forms produced, together with their material properties of hardness, lustre and colour. However, societies across the Balkans in the 5th millennium BC preferred brilliance, colour aesthetics, precision and geometric thinking. This preference dominates the material culture of the time (Chapman 2011) and has its roots in the Mesolithic period in the region (Chapman and Richter 2009; cf. Srejović 1972). Well-executed craftsmanship, bold colours, dramatic shapes and symmetrical design can be encountered combined in single objects in the 5th millennium BC Balkan material culture. For instance, a high degree of standardisation is seen in the production of flint blades from the Bükk culture (Vértes 1965), remarkable geometric precision in the pottery of the Cucuteni-Tripolye culture (Washburn and Crowe 2004), spectacular craftsmanship in the gold-decorated vessels in the Varna I cemetery (Ivanov 1988b), and outstanding painting techniques in the silver-sheen of graphite-painted pottery of the Karanovo-Gumelniţa-Kodžadermen (KGK) VI cultural complex, and beyond (Todorova and Vajsov 1993).

Hence, dazzling metals on the one hand and glittering black-burnished ware on the other represent only some of the spectacularly crafted objects in the wider context of the 5th millennium BC material culture in the Balkans (cf. Chapman 2011). What emerges as a pattern is not only the lustrous colour spectra, which continued to expand over the course of this period with the discovery of gold or tin-bronze, but also a specific pursuit for the ultimate expression of a completely homogenised brilliance only achieved with metals. In this light, the emergence and spread of metals—first copper, and in succession gold and tin bronzes—may be the best illustration of such a quest for the decisive material statement at the time. Radivojević and Rehren (2016) went as far to call this period the *Age of Brilliance*, to emphasise the importance of the production of highly reflective objects (e.g. metals, metal powder decorations on pottery, graphite painting) as an integral part of both cultural and technological identity.

Yet, the value of these properties may have varied during the 5th millennium BC. More generally, we see the majority of Vinča culture metal coming out of domestic contexts, while just after the mid-5th millennium BC the prevalent context for metals is burial grounds (Radivojević 2006). Examples from the domestic context emerge as heavily worked and even deformed (Šljivar *et al.* 2006), while implements from burial contexts at Varna or KGK VI burials are mostly 'as new', or with minimal traces of use wear. A future study in use-wear analysis of these implements would reveal a much needed high-resolution picture of consumer behaviour with regard to metal implements.

4. *Kings of metal?*

The evidential basis for the debates connecting metal objects to societal elites in the Balkans tends to centre upon the cemetery site of Varna, Bulgaria which rapidly came to be considered the Type-site demonstrating the relationship between early metallurgy and a high level of social differentiation (e.g. Ivanov and Avramova 2000; Renfrew 1978a, 1986; Ivanov and Avramova 2000) and Chapter 3. Whilst not diminishing neither the site of Varna nor the related research, the quantity and quality of the metal evidence has completely overwhelmed ongoing and, indeed, increasingly circular, debates on early metals, elites and social complexity in the Balkans (e.g. Hansen 2012, 2013b; Kienlin 2010; Kienlin and Zimmermann 2012).

Whilst there is no doubt that there are substantial differences in the treatment of individuals across the Varna cemetery as discussed earlier (see Krauss *et al.* 2014, 2017), there have been few cemeteries excavated in the Balkans dating to *c.* 5000–3700 BC that are of comparable size (e.g., Durankulak, Bulgaria) (Todorova 2002a), and none that are comparable in metallurgical, or indeed material, extravagance in their grave goods, especially beyond northeastern Bulgaria (Lichter 2001). In the pursuit of markers for individual wealth, the most commonly cited example is the individual in burial No. 43. Yet, this burial is one of three skeletal graves in the assemblage of the 11 richest (or Group A), the other eight being symbolic graves or cenotaphs (Leusch et *al.* 2017: 112, Table 2). An interpretation of the symbolic graves suggests that wealth may have been deposited as an expression of 'collective social identity' and, as such, did not reinforce the social order but was made by communities to strengthen their ties with the dead (Biehl and Marciniak 2000: 202). This theory is supported by the fact that none of the deposited items had been used (some were even crudely made), and there is no evidence from the settlement research to show hierarchy or any form of strong social differentiation akin to that assumed in the cemetery (Ivanova 2007; Leusch *et al.* 2017: 113). By highlighting that the Varna cemetery was the end product of a dynamic process that mobilised all available resources to define and display the community's identity, Biehl and Marciniak (2000) approach the point that we are making above about the cooperative nature of metallurgical production in the Balkans.

Beyond Varna, the quantity and size of copper tools, primarily axes, known to have circulated in the Balkans is drawn upon when relationships between metals and elites are explored (Klimscha 2020). Ryndina (2009) estimates that the amount of metal circulating in the region translates into *c.* 4,300 artefacts whilst Chernykh (1992) proposes 4.7 tonnes. The amount of extant copper metal artefacts discovered across the Balkans in the 5th millennium BC outweighs the contemporary mining and production evidence. In addition to the beads, fish-hooks and awls already known from the late 6th millennium BC, this period witnessed an outburst in the production of massive copper implements—such as hammer-axes, chisels and bracelets—from the very beginning of copper smelting practices at *c.* 5000 BC. However, the figures, artefact discoveries and distributions reported need to be considered with caution as the number of extant copper implements in the Balkans predominantly date to the second half of the 5th millennium BC and may simply represent specific depositional or recycling practices (see Chapman and Gaydarska 2020; Taylor 1999). This is certainly highly significant when contrasting the Balkan evidence with that for metal artefacts known from the European and Near Eastern Bronze Ages (Radivojević *et al.* 2019) and potentially also earlier.

In the absence of any indication of centralised decision-making or elites, or even the presence of noticeable differences in wealth, it is safest to assume that the wealth we can identify belonged to a household unit, or groups of households representing an extended family, a clan or, indeed, a cooperative community. An interesting find from the Vinča culture site of Stubline potentially sheds a novel light on this perspective. Forty-three clay figurines were recovered together with eleven miniature clay models of (copper) implements in seven or eight spatial clusters (Crnobrnja 2011; Crnobrnja *et al.* 2009). These figurines were found arranged (Figure 3) in front of a large domed oven inside a dwelling structure, surrounded by ceramic materials typologically characteristic of the Vinča D2 phase, and dated to *c.* 4650/4600 BC (Crnobrnja 2011: 132). Forty-two figurines are identical in design: crudely shaped cylindrical bodies with a bird-like head, in contrast to a single large example that was made with more technical skill. All have a hole in the right shoulder; some of these holes presumably held the tool handles. However, not all the figurines have tools associated with them, implying that possibly these suffered from post-depositional processes. Unlike the figurines, the clay models of the implements were meticulously shaped and polished, with particular attention paid to fine details. Their form even allows for the distinguishing of different types of tools, such as hammer-axes, pickaxes, long tools with a blade, mallets and a macehead or 'sceptre' (Crnobrnja 2011: 134). Interestingly, some of the miniature implement models in clay are strikingly similar to their contemporaneous counterparts in copper metal, the Pločnik hammer axes for example, while counterparts for the macehead or 'sceptre' are found at Divostin II (House 13) or in the form of a gilded hammer axe at Varna 1 (burial No. 4) (Leusch *et al.* 2017: 113, Figure 7; Porčić 2019a).

Whilst the figurines at Stubline are undoubtedly important, exactly what they represent has been a matter of debate. The tall figurine with a macehead (status marker) may be interpreted as a representation of anything from a highly ranked individual to a deity, the presentation of a (equal) community with carefully and distinctively designed miners' and metallurgists' tools may represent one of our 'cooperatives' seen

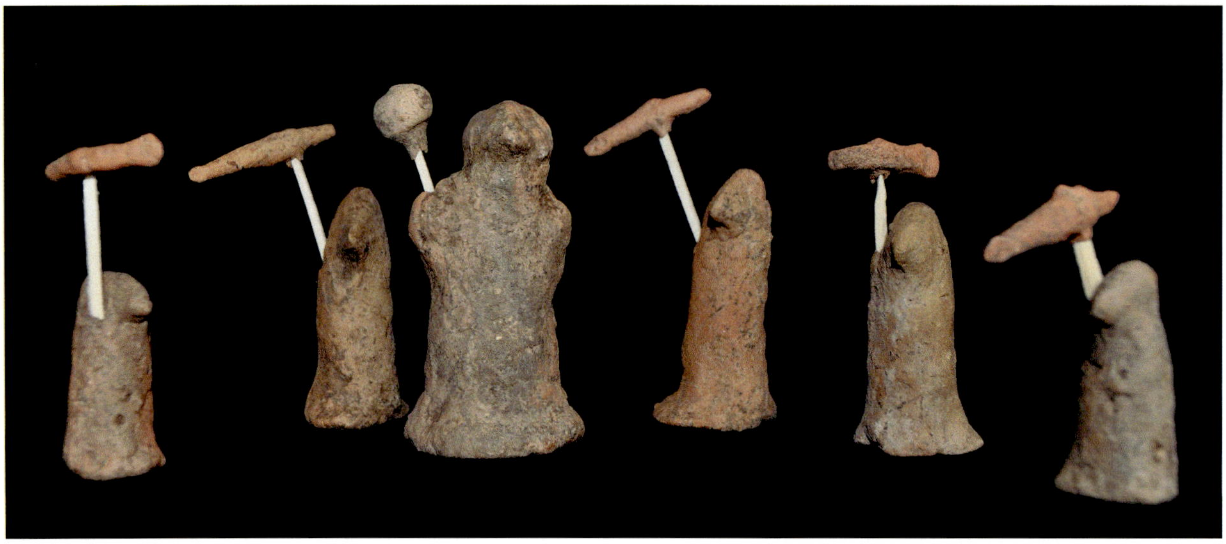

Figure 3. A selection of the Vinča culture figurines from the site of Stubline. The central figure has a clay model of a sceptre; others have clay models of hammer-axes. (After Crnobrnja 2011: 140, Figure 9; copyright by A. Crnobrnja).

through the eyes of the artisan at the time. If the possession of copper was considered an indication of prestige or wealth, then the Stubline figurines may well show us that it was equally distributed within a practicing community. Finds like this tell us who the owners of these tools were not likely to be: 'kings' or any kind of gender-exclusive community.

5. *Societal impact?*

With all the above evidence, we are increasingly witnessing a much more critical approach to long-held Childean ideas regarding early metallurgy, such as its close associations with emerging elites and major societal transformations (e.g. Bartelheim 2007; Biehl and Marciniak 2000; Chapman 1991; Kienlin 2010; Kienlin and Zimmermann 2012; Lichardus 1991c; Porčić 2012b, 2019a). The data presented here lead us to conclude that the invention of copper metallurgy at *c.* 5000 BC appears to have had very little impact on society at the time. However, the polymetallic (r)evolution with novel metals such as gold or bronze and a vast expansion of metal production and circulation from the mid-5th millennium BC played a significant role in individual and group identities. In the context of its early emergence at the sites of Belovode and Pločnik, we see a partial transformation of pottery production (diversification of forms, or recipes, see Chapter 43) and moderate changes over time in diet, subsistence and dwelling habits, which by being correlated with the presence of metals in the lives of the Belovode and Pločnik communities, do not offer enough information to argue for causation of these phenomena.

The considerably nuanced view presented in this chapter and throughout the monograph adds a particular value for future explorations of the societal impact of metallurgy. At a broader spatial and temporal scale, we can gain clearer insights into the relationships between metallurgy and metallurgists and the organisation of these communities across the Balkans – not by the vague identification of a metal-using 'elite' but by exploring how metal relates to broader demographic patterns, settlement densities and connections between communities. Belovode and Pločnik stand out for sharing the same supply networks and/or deposits in relation to copper mineral or copper ore acquisition from the very beginning (51st century BC) until the end (46th century BC) (Chapter 41). During this time, both the technology of copper making and the copper supply are consistent and unchanging. The dominant mode of production remains the hole-in-the-ground installations, implements are made from almost pure metal in a selected number of types and even finishing techniques (annealing + hot/cold working) remain the same. The continuity and consistency of copper smelting technology suggest that such knowledge in the Vinča culture was likely passed on as an 'all-in-one' package within a potentially conservative tradition. The transmission of knowledge was perhaps kept within a particular lineage of craftspeople, or a cooperative, where skills were most likely passed from a senior member or parent to the offspring or apprentice? (cf. Shennan and Steele 1999).

The size of a learning network has been shown to be very important for skills transmission in traditional and prehistoric communities (Henrich 2004; Powell *et al.* 2009; Roux 2008; Shennan 2001). As for any innovation, the spread of metallurgical skills within the Vinča culture required a sufficient number of learners. Given the consistency of selection practices of black-and-green ores and copper smelting technology throughout *c.* 600 years, this appears to have been stable during this period. The learning network of Vinča culture metalworkers ceased to exist at Belovode and Pločnik (and also Vinča and Gornja Tuzla) around the mid-5th millennium BC, with the end of the Vinča culture, so mysteriously marked by an abandonment of these and other settlements in modern day Serbia. Nonetheless, the continued production of massive copper implements across the Balkans throughout the entire 5th millennium BC, suggests that this learning network possibly continued to grow in other parts of the region.

Recent research has revealed a clear increase in the population of settlements during the Balkan Chalcolithic (Porčić 2019a) and, by extension, in population densities at settlement sites (Rosenstock *et al.* 2016). These two trends would have significantly enhanced any production activities that required communal and cooperative dynamics – as can be demonstrated for metal production. When integrated with the evidence from network analyses that communities were frequently and regularly cooperating in the production and distribution of metal artefacts (cf. Radivojević and Grujić 2018a), it is clear that the societies in the Balkans provided an institutional and technological context within which metallurgy was able to thrive. However, there is no evidence to suggest that metal played either a causal role, whether in creating a larger and more densely settled population in the region, or in the inter-connections spanning the many communities, as both trends can be seen to emerge in the 6th millennium BC (Porčić 2019a). The influence of metallurgy and metallurgists on the diversity of partially overlapping and fluctuating communities across the Balkan societies may instead have been in the development of existing areas or the creation of new areas of cooperation and connections within and between communities. It is only when (as here) metals are compared to other widely distributed materials and technologies that both pre-date and are contemporary with metallurgy—such as obsidian and graphite-painted pottery—that their role can be more thoroughly re-evaluated. Our evidence makes it increasingly problematic to argue that metal defined the organisation of these communities.

There is, however, a notable increase in wealth during the 5th millennium BC in the Balkans. Orton (2010) and Orton and colleagues (2016) indicate a general increase in the number of cattle bones in faunal assemblages, which may have been partially due to investment in the social arena, with cattle representing a form of wealth (cf. Russell 1998). Moreover, the difference in wealth can be seen in the presence of status markers such as the macehead (*sensu* Siklósi 2004; 2013) from Divostin II, found in House 13, which also differs in size and assemblage from other excavated houses at this site (Porčić 2009, 2012b). The same applies in relation to the presence of large houses and households in settlements like Divostin and Stubline, which Tripković (2009a) argues to reflect the existence of extended or multi-family households. The creation of larger basal units (such as households) and many levels of decision-making is at the core of Porčić's (2019a) argument that Vinča society was most likely organised as a sequential hierarchy (*sensu* Johnson 1982), or with decisions made by consensus within a household group, before a representative would negotiate on their behalf at a village level. This kind of organisation enabled a relatively egalitarian decision-making process. In this context, the interpretation of large buildings with house floor areas between 100 and 200 square metres is of great significance. As there is no evidence to suggest that such buildings were homes to local elite or were temples (Chapman 2010), they can perhaps be seen as communal buildings that enabled the working of sequential hierarchies, as their size fits the low level of integrative facilities (Porčić 2019a).

The Eurasian context

We now see that, across Europe and Asia in the 5th and 4th millennium BC, the introduction of different metals and technologies varies across three geographically and geologically neighbouring 'heartlands' of metallurgy: the Balkans, Anatolia and Iran. For instance, in the Balkans, the polymetallic 'revolution' occurs around the mid-5th millennium BC when, after *c.* 500 years of making only copper (and possibly lead), we see gold, tin bronze and, most likely, silver being produced before the end of the millennium. In Iran, early metal use starts as in the Balkans (with copper and some lead). however, despite evidence for the silvery alloy CuAs from the early 5th millennium BC, true polymetallism occurs in Iran towards the end of the 5th millennium BC and beginning of the 4th millennium BC with a complex range copper alloys, gold and silver; tin-bronzes do not appear before the end of the 4th millennium BC. For Anatolia, native and/or smelted copper is the primary choice until the mid-4th millennium BC, when silver-like alloys first emerge (CuAs, CuAg), followed by tin-bronze at the same time as in Iran (data from Lehner and Yener 2014; Leusch *et al.* 2015; Radivojević and Roberts 2013; Radivojević *et al.* 2013; Roberts *et al.* 2009; Thornton 2001, 2007, 2009, 2010). The different intellectual traditions, languages of scholarship and contemporary politics in each region has meant that, despite geographical proximities and interconnected underlying geologies, early metallurgical research at a broader pan-regional spatial scale is significantly under-developed. Yet, the fundamental conclusion drawn from these comparisons is that there is no single narrative for metal technology that unites these neighbouring regions , or indeed the adjacent metallurgical 'heartland' of the Caucasus (Courcier 2014) into a single entity in the 5th or 4th millennium BC.

Just as no single narrative unites the evolution of metallurgy during this period, neither does a single narrative explain and interpret the societal impact of this technology. We acknowledge that many interpretative models that include the emergence of elites and social complexity as a direct consequence of making and trading metal artefacts drew on far less data of a lower resolution than is available in the 21st century. It is on the shoulders of intellectual giants like V.G. Childe, T. Wertime, C. Renfrew, S.C. Smith, B. Jovanović, E.N. Chernykh and many others that we stand and build a new, more nuanced view of what modern archaeological perspectives on *The Rise of Metallurgy in Eurasia* meant for farming communities in the Balkans during the 5th millennium BC.

First and foremost, the emergence of metallurgy was a technological revolution. Like no other before, it transformed matter from ore to metal and demanded a skilled manipulation of fire in order to yield workable results. It built on and interacted with the exceptional crafting of shiny and colourful artefacts from clay over the course of the 5th millennium BC. Its value differed, both from earlier to later centuries and from the places of production to the places of consumption. While the produced metal objects may have played a significant role in various ceremonies, the early onset of production itself was very likely a community effort that took place in shared spaces, probably with music and accompanying rituals. This 'cooperative' practice is further underlined by the absence of evidence in the excavation record for a specialist in the Vinča culture and beyond (with Varna as the exception rather than the rule).

The increasing wealth of the metal producing and consuming societies in the Vinča culture may have been influenced by metallurgy, although probably not to the extent of causing increases in population, livestock management, expanding households or possession of other commodities, and we see no clear evidence that any of these factors worked independently. This brings us to a novel narrative of the evolution of copper metallurgy in the Balkans, the only independent feature of which lies in the invention of this technology with a particular choice of crucial ingredients, such as black-and-green copper ores, and following a unique technological trajectory within the Vinča culture community to begin with. However, detailed excavations, high resolution materials and networks analysis of archaeometallurgical materials indicate the importance of communal practices and cooperation, which also emerge as the main points from analysis of other aspects of material culture, including examples from across the 5th millennium BC Balkans – most of them previously underappreciated for their explanatory powers.

Hence, we propose that the strength of shared practices and cooperation should be brought to the core of future archaeological inquiry on the topic of early metallurgy may pursue in the future, in the Balkans, across Eurasia, and globally. As highlighted in the subsequent chapter (Chapter 53), we take inspiration from major recent developments in the understanding of the processes involved in the global domestication of animals and plants, which is now thought to have taken place independently in sixteen geographical centres (Larson *et al.* 2014). We also recognise that, following Graeber and Wengrow (2021), with access to new and better data it is imperative that we use it to critically evaluate our own ideas and established interpretations, as well as remaining open to braving new explanations.

The bibliographic reference for this chapter is:

Radivojević, M. and Roberts, B. W. 2021. Balkan metallurgy in a Eurasian context, in Radivojević, M., Roberts, B. W., Marić, M., Kuzmanović Cvetković, J., and Rehren, Th. (eds) *The Rise of Metallurgy in Eurasia*: 601–618. Oxford: Archaeopress.

Chapter 53

Where do we take global early metallurgy studies next?

Benjamin W. Roberts, Miljana Radivojević and Thilo Rehren

The results and experiences gained from the multidisciplinary and holistic approaches underlying the *Rise of Metallurgy in Eurasia* project provide an opportunity, not only to reflect on programmes of further research in the Balkans, but also on scholarship in early metallurgy across the world. This chapter outlines what might be usefully taken forward from this project, but also seeks to highlight gaps in our understandings that could be addressed. It is by no means a comprehensive agenda for global early metallurgy studies but is instead intended to stimulate further debate and discussions that lead to new programmes of research.

Ores and Metals

The analysis and interpretation of early metallurgy across the world is invariably dominated by ongoing debates of 'origins', where competing scholars representing sites or regions seek to claim the status of publishing the 'earliest'. The scholarly prestige involved and the enhanced potential for attracting funding for future projects are undeniable. Yet the scholarship of early metallurgy is frequently singly focussed, especially in perspectives beyond that of a region. Not all metals are researched or interpreted equally. Copper and tin bronze are by far the most ubiquitous early metal discoveries across Europe, North and East Africa, Asia and the Americas and it is therefore unsurprising to see their prominence in research publications. However, rarer discoveries of gold and silver will typically be given a far greater prominence in modern scholarship and past values inevitably leading to interpretations of 'metals of power' (Meller et al. 2014). In contrast, lead which remains hugely under-researched and invariably under appreciated.

Similarly, tin bronze, perhaps due to its entrenchment in the Three (or Four) Age system in Asia, Europe and North Africa, is a far more extensively researched alloy than arsenical and antimonial copper (e.g. see Wilkinson et al. 2011). This is despite the widespread and consistent use of arsenical and antimonial copper from c. 3500 BC for over a millennium prior to tin bronze adoption across an area spanning Central Europe to Central Asia (e.g. Chernykh 1992; Roberts et al. 2009). In addition, there are geologically rich regions with extensive evidence for metallurgical innovation such as Iberia and Iran where the use of arsenical copper persisted for centuries despite the widespread potential availability in the region of both copper *and* tin ores (Cuenod et al. 2015; Helwing 2013; Perucchetti et al. 2020; Rehren et al. 2012; Thornton 2009). Yet, there is neither a pan-regional synthesis or comparative analysis of arsenical and antimonial copper metal use in Europe, North Africa or Asia nor even clarity on how the alloy was consistently produced, transmitted and traded.

The potential complexities relating to the selection of certain ores or the combination of certain metals in the creation of metal alloys can be largely resolved with archaeometallurgical analyses and ideally experimental replication. However, what has less been resolvable is the broader theoretical interpretation of the appearance and disappearance of these early metal alloys. The identification of a tin bronze artefact at Pločnik dating to c. 4650 BC, which is broadly contemporary with fourteen other 5th millennium BC tin bronze artefacts scattered across the Balkans, was a huge surprise to many archaeologists who naturally associate tin bronze with the Bronze Age chronological period several millennia later and met with misplaced and easily dispelled scepticism. The Bronze Age in the central Balkans may start at c. 3200 BC but the widespread adoption of tin bronze in the Balkans occurs only at c. 1800 BC (e.g. Boyadžiev, 1995; Pare, 2000; Pernicka et al., 1997). What had not been widely understood is that the earliest known gold artefacts as excavated at the Durankulak and Varna II cemeteries were not only contemporary to the tin bronze at Pločnik but also demonstrate a consistent and deliberate manipulation of the natural gold composition with the addition of copper (Leusch et al., 2015; Radivojević et al., 2014). There are numerous comparable examples such as the small numbers of brass artefacts, an alloy of copper and zinc, being created alongside similar coloured tin bronzes from the 3rd millennium BC onwards in Europe and Southwest Asia well before the widespread adoption of brass in region during the 1st century BC (Thornton, 2007). As Killick and Fenn (2012) note in their global review of research into pre-industrial mining and metallurgy, archaeometallurgists are 'adept at discovering inventions that failed to become innovations'. The period and the metal are clearly different entities, yet the intellectual baggage of

Childe's (1944) framework, programmatically entitled 'Archaeological Ages as Technological Stages', remains deeply embedded.

To begin to address these issues, research on global early metallurgy would benefit from analysing a larger assemblage of minerals and ornaments from pre-metallurgical contexts. It could shed more light on the mineral selecting practices of these communities and improve our understanding of how the knowledge of metallurgy evolved (cf. Rapp 2010). The identification of the minerals being selected and collected by communities in the Balkans in the millennia prior to any evidence of metal smelting and exploring their visually distinguishing characteristics and their potential sources, was fundamental to discussing the processes of invention and innovation that led to copper metallurgy (Radivojević, 2015; Radivojević and Rehren, 2016). With a detailed understanding of the ores and minerals involved and what areas were being prospected, the potential metals and alloys that might have been produced can be assessed.

The processes underpinning the smelting of the selected ores to create metals remain talismanic in global early metallurgy scholarship. However, they are frequently poorly understood beyond individual assemblages. A major obstacle to further understanding is the absence of a database containing the technological data on early smelting activities that would enable comparative analysis. Such a comparative archaeometallurgy would be dedicated to exploring the questions of how these technologies evolve, which parameters shaped them and why they exhibited similarities or differences over the course of their development. When considering the smelting processes that could have occurred in the Balkans during the 5th millennium BC, it was striking how few comparative surveys providing any detail, such as in Bourgarit (2007), had been published.

Furthermore, the experimental replication of smelting processes identified in excavation and post-excavation analyses is only infrequently conducted and even less frequently fully published (e.g. Hauptmann 2020; Heeb and Ottaway 2014; Timberlake 2007). Yet it has proved to be invaluable in understanding the expertise, raw materials, logistics and organisation of activities required to create a metal object. Finally, what is frequently missing in the early metallurgical research is freely available raw data from the excavations and post-excavation programmes which would enable subsequent scholars to re-evaluate the original conclusions. As highlighted in the recent pan-European survey of Bronze Age metal production and circulation (Radivojević et al. 2019) and put into practice in this volume (see Appendices A and B), this level of data-sharing and transparency would hugely enrich the quality and standard of scholarship.

Origins and Regions

In an important global review on the current perspectives and future of domestication studies relating to plant and animal food production arising from a dedicated seminar, Larson et al. (2014) tentatively identified sixteen regions across the world where independent domestication is thought to have occurred, and identified spatial and temporal patterns of transmission. The authors then identified three challenges: filling in gaps on maps with dates and data; exploring the environmental and ecological contexts of agricultural origins; and explaining why hunters and gatherers turned to cultivation and herding. Despite the differences in a debate that is primarily biological and geographical as opposed to one that is primarily geological and metallurgical, there is much that can be learnt for global early metallurgical research.

Scholars in early metallurgy tend to focus on a particular geographical or political region, typically one that is geologically rich in metallurgical ore sources. Extensive and detailed research in one region is then published with a far more general perspective on neighbouring regions and those further afield. The comparative analysis of early metallurgy in different regions is invariably concentrated upon neighbouring regions, and/or specific metals and then really tends only to focus on potential technological connections. This undeniably has been the case for the *Rise of Metallurgy in Eurasia* project. Its authors could legitimately cite the sheer quantity of data and debate that would need to be processed in order to achieve a more global perspective as a justification for this situation. However, what is really missing is a clear global framework for early metallurgy that provides the foundations and framework for further study and debate that goes beyond edited volumes of individual regional syntheses (e.g. Roberts and Thornton, 2014).

The definition of this global framework requires careful consideration. For instance, if it is to follow Larson et al. (2014) in defining independent regions of metallurgical invention, there are different perspectives. From geographical and geological perspectives, the Balkans, Anatolia, Levant, Caucasus and Iran are inherently interconnected, yet each has its own traditions of early metallurgical scholarship and tends to be perceived as a distinct early metallurgical region. Furthermore, the *Rise of Metallurgy in Eurasia* project highlighted the independent invention of copper, tin bronze, gold and potentially lead metallurgy in the Balkans, with no chronologically comparable evidence in neighbouring western Anatolia. Then there are areas within regions whose spectacular finds can distort the interpretations and perception of the broader region. For instance, the Varna cemetery and gold in northeast Bulgaria is unthinkingly and incorrectly extrapolated to communities across the entire Balkan region when considering the making and consuming of gold artefacts. Similarly, the arsenical/antimony copper in the Nahal Mishmar hoard is taken to typify the entire Levant. Should a metallurgical invention, even one evidenced in only small quantities before disappearing again as appears to be the case with tin-bronzes in the Balkans, be given priority over a widely adopted metallurgical innovation two millennia later? Which activities relating to early metallurgy should be encompassed? Given the traditional focus on smelting as *true* metallurgy, this would exclude regions where metal artefacts were made, circulated and used without smelting, such as the Copper Culture in North America that dates from c. 6000 BC (Bebber 2021).

The *Rise of Metallurgy in Eurasia* project highlighted that the singular and uneven focus of early metallurgy scholarship has served to obscure the presence of polymetallic horizons whereby multiple metals were being produced, circulated, used and deposited by the same communities in the same region at the same time. The mono-metallic perspective also serves to significantly underplay the deliberate manipulation of individual metal compositions, whether by the use of more complex ores or the addition of other metals, thus creating differences in colour, hardness and castability. It is argued that a polymetallic together with a geological perspective would provide a strong foundation for such a truly global approach. However, these questions and others surrounding the definition of a global framework for early metallurgy are not going to be easily resolved.

Yet, when comparing Larson et al. (2014) to the current state of knowledge in global early metallurgy, there is no clear perspective in current scholarship simply when and where different metals and metal technologies were potentially independently invented and/or innovated. When reviewing their three challenges, it is evident that there is no publication outlining where the priority areas are for adding new dates and data relating to early metallurgy. The geological and archaeological contexts of metallurgical origins around the world have not yet been subjected to any systematic comparative analysis. The explanations as to why communities across the Americas, Africa, Europe and Asia used metal alongside and in place of stone, bone, wood and other materials metal are invariably founded on assumptions rather than evaluations (see Frieman 2021).

Metallurgy and Environment

The environmental impact of global early metallurgy lies not only in the carbon contributions linked to fuel procurement and burning in order to achieve the necessary temperatures required to extract and work metal but also in the changes in land use, including deforestation, which are known to be major sources of carbon release from soils, which contain more than 70% of terrestrial carbon (Liu et al., 2020b). The first major anthropogenic carbon release has been first noticed around 6000 BC, marking the impact of the advent of agriculture, in particular through the large-scale cultivation of rice (Fuller et al., 2011; Ruddiman, 2014). Yet, the findings by Brovkin et al. (2019) demonstrate that the rise of CO_2 by 20 ppm between 6000 and 2000 BC cannot be solely ascribed to either ocean and land carbon sources. Although the impact of metallurgy has been acknowledged from the Medieval period, the impact of ancient metallurgical pursuits has been discounted through a focus on the Classical World alone (Williams, 2006), while the intensification of metal production in earlier periods around the world remains to be explored. Recently, Liu et al. (2020a) identified the importance of metallurgy and environmental studies in eastern Eurasia for observing climate change adaptation and the effects of continental scale connectedness through trade, flagging up the need for research between early global metallurgy and environmental sciences.

Currently, early global metallurgy is being increasingly identified in the form of metallurgy-related air pollution by environmental scientists studying sequences from peat bogs, glaciers and lake sediments. However, it is very regionally focussed. The recent identification of lead pollution in the peat bog sequence at Crveni Potok, Serbia from c. 3600 BC is the earliest such evidence in Europe, with comparable evidence from eastern Europe only from c. 3150 BC and western Europe only from c. 2950 BC (Longman et al., 2018). Whilst this is substantially later than the 5th millennium BC direct metal production evidence studied in the *Rise of Metallurgy in Eurasia* project, it is possible to use environmental sequences to re-evaluate the earliest known evidence for metallurgy in a region. For instance, Eichler et al. (2017) controversially argued the onset of intensive copper smelting in South America from evidence trapped in an ice core from the Illimani glacier, Bolivia from c. 700 BC, substantially earlier than the currently known evidence for significant copper smelting from c. 200 BC. There is currently no global survey of existing environmental data or programme of analysis that potentially relates to early metallurgy in terms of its initial, detectable appearance or its broader environmental impact.

Mining and Trade

The scale of early metallurgical mining activities across the world, beginning with copper ores at Rudna Glava, Ai Bunar, Jarmovac and elsewhere in the Balkans in the 6th/5th millennium BC (Chernykh, 1978; Ryndina, 2009), is now known to be far more extensive than previously understood. However, the implications of the scale, complexity and distribution networks associated with these mines are yet to be fully realised. Over 50 copper mines are now dated to the Chalcolithic-Bronze Age across Europe alone with a broader trend towards the consolidation of copper mining in major centres around mid-2nd millennium BC such as Mitterberg (Austria), Great Orme (UK) or Cyprus (e.g. Pernicka et al. 2016; Radivojević et al. 2019; Williams and Le Carlier de Veslud, 2019), amongst others. However, all of these are dwarfed by the production capacity of Kargaly (south Urals, Russian Federation), a primarily Late Bronze Age (1700-1400 BC) copper mining landscape that spread across 500 km^2 (Chernykh, 1997). With c. 100,000 tonnes of copper produced only in this mid-2nd millennium BC period, Kargaly had seven times the production capacity of the Mitterberg (Austria), the largest coeval European mine (Pernicka et al., 2016). This scale of production would have required annual felling of 150 ha of woodland, up to seven times the size of Kargaly, which stands in stark contrast to the paltry 2.6% of forest coverage of Kargaly today (Díaz del Rio et al., 2006). And yet Kargaly was one of at least six Bronze Age mining centres of a similar size in the Eurasian Steppe at the time, including Askaraly in east Kazakhstan, Dzhezkazgan in central Kazakhstan, Bozshakol in northeast Kazakhstan, Kendyktas Plateau in south Kazakhstan or Zerafshan Valley mines in Uzbekistan (Alimov et al., 1998; Berdenov, 2008; Boroffka et al. 2002; Park et al., 2020; Stöllner et al., 2011).

The *how* and *why* behind the scaling up of the metal production from c. 5000 BC to c. 1000 BC across Europe, North Africa and Asia was beyond the scope of this monograph, yet, we offer a glimpse of the complexity of operations underlining the acquisition of ores, logistics of their circulation, or multiple episodes of production required to produce a large metal implement from the very beginnings of metallurgy. Large-scale production was only enabled by the evolution of furnaces, which we see in a developed form only around the early – mid-2nd millennium BC, while the 3,000 years of metal production in between are usually ascribed to small-scale or 'household' production. These three millennia of a continual innovation in metal production are yet to be addressed in an all-encompassing synthesis that addresses specific evolutionary trajectories of metal making technology across Eurasia, and globally, and draws parallels to exhibit patterns of cooperation and knowledge transmission. These networks of cooperation were most likely responsible for connecting the East and the West through the exchange of commodities, most prominently copper alloys such as tin-bronzes, which laid out the foundations of the first global economic network at the time, later identified as the Silk Roads.

These networks of exchange in commodities, most visibly in metals, but also in textiles, animals and foods, were underpinned not only by social and political relationships, but also by agreed frameworks of measurement and value. Given the physical properties of metals such as colour, lustre, malleability and their ability to be re-shaped and recycled according to the requirements of the communities involved, it is no surprise to see metals becoming central to these networks of exchange. Research has demonstrated the emergence of standardised weight systems during the 2nd millennium BC in bronze (Kujpers et al. 2020) and gold (Rahmstorf 2019) in Europe, the widespread use of silver in trade across Southwest Asia from the 4th millennium BC (Sherratt 2016) as well as the adoption of standardised weighing equipment from the Atlantic to the Indian Ocean during the 3rd-2nd millennium BC (Rahmstorf 2011). We are only just beginning to understand the dynamics of ancient trade and markets (cf. Kristiansen et al. 2018; Rahmstorf and Stratford 2019; Rahmstorf *et al.* 2021) – and how metal was valued and used within them in the past rather than how we have projected it to be.

Societies

The relationship between early metals and societies across the world has been defined primarily by 19th and early to mid-20th century perspectives from Europe and Southwest Asia where scholarship placed metal at the core of schemes of social complexity and emerging inequalities. The period of the 5th-4th millennium BC in the Balkans and Southwest Asia is especially important in this respect as successive generations of scholars have argued that in the 5th millennium BC there are no great differences between these regions, from demographic, material and environmental perspectives, in their potential to develop urbanism and civilisation (Porčić, 2019a; Tringham, 1992, pp. 133-134). However, the 4th millennium BC in the Near East yields what Graeber and Wengrow (2021) have robustly characterised as the elusive evidence for territorial attachments that lead to private ownership, and then to a surplus of food, which in turn leads to the accumulation of wealth and power beyond the immediate kin-group, and ultimately to the production of sophisticated weapons, tools, vehicles, the rise of cities and centralised governments, with bureaucrats and priests making sure that the imbalance is maintained (and women kept in harems), while inventing writing systems to record a single 'correct' version of the past, whether it is 'correct' economically, administratively, politically or religiously. This stands in complete contrast to the dispersed farming communities of the Balkans, where a negative perception is created of a peripheral region that missed its opportunity to become 'civilised'.

The data shows that the 5th millennium BC Balkan communities did not 'run headlong for their chains' (*sensu* Rousseau, 1761), or put simply, the early advancements in polymetallurgical technologies did not materialise themselves into the emergence of cities and states. Furthermore, it seems clear that one of the major reasons why the concept of an independent origin for Balkan metallurgy was for many decades considered too bold, was due to the accepted wisdom about the development of metallurgy (as much as social evolution) based on the Southwest Asian model, or indeed, deeply intertwined with the '*Ex Oriente Lux*' concept. This proposed necessary decoupling of metal, social complexity and inequality in our perspectives on the global past is given further support when viewed from the Americas (Bebber 2021; Erhardht 2014; Hosler 2014; Lechtman 2014; Zori 2019) and more global perspectives (Chernykh 2021). Taking into account the volume, depth and analytical scrutiny of archaeological and archaeometallurgical research conducted around the world over the past 50 years and briefly reviewed here, it is clear that late 19th century narratives connecting metallurgy and a single perception of social progress have no place in 21st century archaeology.

The bibliographic reference for this chapter is:

Roberts, B. W., Radivojević, M., and Rehren, Th. 2021. Where do we take global early metallurgy studies next?, in Radivojević, M., Roberts, B. W., Marić, M., Kuzmanović Cvetković, J., and Rehren, Th. (eds) *The Rise of Metallurgy in Eurasia*: 619–623. Oxford: Archaeopress.

Appendices

Appendix A – Excavation data for Belovode and Pločnik (seasons 2012 and 2013)

Avaliable online at https://doi.org/10.5522/04/14769990

Appendix B – Data relating to specific chapters

Appendix B_Ch5

Radivojević, M. 2007. Evidence for Early Copper Smelting in Belovode, a Vinča Culture Settlement in Eastern Serbia. Unpublished MSc dissertation. UCL Institute of Archaeology, London.

Avaliable online at https://doi.org/10.32028/9781803270425/AppendixB_Ch5

Appendix B_Ch11

Certified Reference Materials - basalt glass
EPMA
Optical Microscopy
SEM EDS
Technical documentation_images of studied materials_Belovode

Avaliable online at https://doi.org/10.32028/9781803270425/AppendixB_Ch11

Appendix B_Ch14

Belovode Catalogue
Belovode Petrography

Avaliable online at https://doi.org/10.32028/9781803270425/AppendixB_Ch14

Appendix B_Ch15

Belovode Figurines Database

Avaliable online at https://doi.org/10.32028/9781803270425/AppendixB_Ch15

Appendix B_Ch20

20.1. List of all taxa in the analysed samples from Belovode (prior to the taxa amalgamation and extrapolation of item counts).

20.2. Images of some of the archaeobotanical remains from Belovode: 1. einkorn grain; 2. einkorn glume base; 3., 'new type' hulled wheat glume bases; 4. 'new type' hulled wheat grains; 5. flax/linseed seeds; 6. barley rachis; 7. bitter vetch seeds; 8. Rubus idaeus/fruticosus seeds; 9. Corylus avellana nutshell fragments; 10. Prunus cf. domestica var. insititia fruit stone fragment; 11. Lapsana communis seed; 12. Chenopodium album type seeds; 13. Trifolium arvense type seed; 14. Trifolium repens type seed; 15. Cerastium seed; 16. Fallopia convolvulus seed; 17. Galium aparine seed; 18. Indeterminate plant matter (? nutmeat); 19. Indeterminate (Scrophularia type) seed.

Avaliable online at https://doi.org/10.32028/9781803270425/AppendixB_Ch20

Appendix B_Ch26

EPMA
Optical Microscopy
SEM EDS
Technical documentation_images of studied materials_Pločnik

Avaliable online at https://doi.org/10.32028/9781803270425/AppendixB_Ch26

Appendix B_Ch29

Pločnik Catalogue
Pločnik Petrography

Avaliable online at https://doi.org/10.32028/9781803270425/AppendixB_Ch29

Appendix B_Ch30

Pločnik Figurines Database

Avaliable online at https://doi.org/10.32028/9781803270425/AppendixB_Ch30

Appendix B_Ch34

34.1. List of all taxa in the analysed samples from Pločnik (prior to the taxa amalgamation and extrapolation of item counts)

34.2. Images of some of the archaeobotanical remains from Pločnik: 1. emmer grain; 2. emmer spikelet fork; 3. terminal spikelet fork (cf. emmer); 4–5. 'new type' hulled wheat spikelet forks; 6–8. tetraploid free-threshing wheat rachis segments; 9. Cornus mas fruit; 10. Rubus idaeus/fruticosus seed; 11. Corylus avellana nutshell fragments; 12. Fragaria vesca seed; 13. Solanum nigrum seed; 14a-b. possible food remains (with embedded fragment of a cereal grain visible in 14b); 15. fragment of dung pellet; 16. Hypericum seed; 17. Chenopodium album type seed.

Avaliable online at https://doi.org/10.32028/9781803270425/AppendixB_Ch34

Appendix B_Ch38

GIS data on the Late Neolithic houses of the Vinča settlements Belovode, Drenovac, Pločnik, and Stubline. The file in WTK-format contains data of all house ground plans which were reconstructed on the basis of the magnetic prospection data. The file also contains the information (size in sqm and orientation) for each house that was taken into account in the statistical evaluation. This is the size of the ground area of the houses are in square metres. The house orientation refers to the north azimuth.

Avaliable online at https://doi.org/10.32028/9781803270425/AppendixB_Ch38

Appendix B_Ch41

Tables from Chapter 41

Avaliable online at https://doi.org/10.32028/9781803270425/AppendixB_Ch41

Bibliography

A

Adams, J.L. 1988. Use-wear analyses on manos and hide-processing stones. *Journal of Field Archaeology* 15(3): 307–315.

Adams, J.L. 1989. Experimental replication of the use of ground stone tools. *Kiva* 54 (3): 261–273.

Adams, J.L. 2002. *Ground Stone Analysis: A Technological Approach*. Salt Lake City: The University of Utah Press.

Adams, J., S. Delgado, L. Dubreuil, C. Hamon, H. Plisson and R. Risch. 2009. Functional analysis of macro-lithic artefacts: a focus of working surfaces, in F. Sternke, L. Eigeland and K.-J. Costa (eds) *Non-Flint Raw Material Use in Prehistory: Old Prejudices and New Directions*: 48–50. Proceedings of the XV World Congress of the International Union for Prehistoric and Protohistoric Sciences, Vol. 11, Session C77. (British Archaeological Reports International Series S1939): 43–66. Oxford: Archaeopress.

Albarella, U. 1995. Depressions on sheep horncores. *Journal of Archaeological Science* 22: 699–704.

Alberti, G. 2013a. An R script to facilitate correspondence analysis. A guide to the use and the interpretation of results from an archaeological perspective. *Archaeologia e Calcolatori* 24: 25–53.

Alberti, G. 2013b. Making sense of contingency tables in archaeology: the aid of correspondence analysis to intra-site activity areas research. *Journal of Data Science* 11: 479–499.

Alberti, G. 2014. Modeling group size and scalar stress by logistic regression from an archaeological perspective. *PLoS ONE* 9, e91510.

Alberti, G. 2015. CAinterprTools: An R package to help interpreting Correspondence Analysis results. *Software* X(1–2): 26–31.

Altmaier, A. and C. Kany. 2002. Digital surface model generation from CORONA satellite images. *Journal of Photogrammetry and Remote Sensing* 56: 221–235.

Alimov, K., N. Boroffka, M. Bubnova, J. Burjakov, J. Cierny, J. Jakubov, J. Lutz, H. Parzinger, E. Pernicka, V. Radililovskij, V. Ruzanov, T. Sirinov, D. Starsinin, and G. Weisgerber 1998. Zinnbergbau in Mittelasien. *Eurasia Antiqua* 3: 137-199.

American Society for Metals. 1979. *Source Book on Copper and Copper Alloys: A Comprehensive Collection of Outstanding Articles from the Periodical and Reference Literature*. Metals Park (OH): American Society for Metals.

Amicone, S. 2017. Pottery Technology in the Vinča Culture at the Dawn of Metal Age. A View from Pločnik and Belovode. Unpublished PhD dissertation, University College London.

Amicone, S., P. Quinn, N. Mirković-Marić, M. Marić, and M. Radivojević (eds). 2019. *Tracing Pottery Making Recipes in the Balkans, 6th–4th Millennia BC*. Oxford: Oxbow Books.

Amicone, S., A.R. Mathur, R.D. Pavitra, N. Mirković-Marić, I. Pantović and J. Kuzmanović Cvetković. 2020a. Beneath the surface: Exploring variability in pottery paste recipes within Vinča culture. *Quaternary International* 560–561: 86-101. DOI: https://doi.org/10.1016/j.quaint.2020.04.017.

Amicone, S., M. Radivojević, P.S. Quinn, C. Berthold and Th. Rehren. 2020b. Pyrotechnological connections? Re-investigating the link between pottery firing technology and the origins of metallurgy in the Vinča culture, Serbia. *Journal of Archaeological Science* 118: 105-123. DOI: https://doi.org/10.1016/j.jas.2020.105123.

Ammerman, A.J. and L.L. Cavalli-Sforza. 1971. Measuring the rate of spread of early farming in Europe. *Man* 6: 674–688.

Amov, B.G. 1999. Lead isotope data for ore deposits from Bulgaria and the possibility for their use in archaeometry. *Berliner Beiträge zur Archäometrie* 16: 5–19.

Amov, B.G. and V.N. Văkova. 1994. A summary of lead isotope data for ore deposits in Bulgaria, in H. Todorova and P. Popov (eds) Проблеми на най-ранната металургия. Sofia: Трудове на Минно-геоложкия университе.

Andel, K. 1958. Pohrebisko z doby medenej v Tibave na Vychodnom Slovensku. *Slovenská Archeologia* 6: 39–49.

Antonijević, I. and P. Mijatović. 2014. The copper deposits of Bor, eastern Serbia: Geology and Origin of the Deposits. *Annales Géologiques de la Péninsule Balkanique* 75: 59–74.

Antonović, D. 1992. *Predmeti od glačanog kamena iz Vinče*. Beograd: Filozofski fakultet, Cenar za arheološka istraživanja.

Antonović, D. 1997. Use of light white stone in the Central Balkans Neolithic. *Starinar* 48: 33–40.

Antonović, D. 2000. Predmeti od glačanog kamena sa Belovoda. *Viminacium* 11: 23–34.

Antonović, D. 2002. Copper processing in Vinča: new contributions to the thesis about metallurgical character of the Vinča culture. *Starinar* 52: 27–45.

Antonović, D. 2003. *Neolitska industrija glačanog kamena u Srbiji*. Beograd: Arheološki institute. Posebna izdanja 37.

Antonović, D. 2011. Industrija glačanog kamena na lokalitetu Crkvine: primer sonde 5, in M. Blagojević (ed.) *Kolubara* 5: 195–212. Beograd: Republički zavod za zaštitu spomenika kulture.

Antonović, D. 2013. Predmeti od glačanog kamena sa lokaliteta Livade u Kaleniću, in M. Blagojević (ed.) *Kolubara* 6: 23–36. Beograd: Republički zavod za zaštitu spomenika kulture.

Antonović, D. 2014a. *Kupferzeitlichen Äxte und Beile in Serbien.* (Prähistorische Bronzefunde, Vol 9, Band 27). Mainz: Akademie der Wissenschaften und der Literatur.

Antonović, D. 2014b. Manufacturing of stone axes and adzes in Vinča culture, in S. Vitezović and D. Antonović (eds) *Archaeotechnology: studying technology from prehistory to the Middle Ages*: 77–88. Belgrade: Srpsko arheološko društvo.

Antonović, D. 2014c. Examination methodology for ground stone artefacts, in S. Vitezović and D. Antonović (eds) *Archaeotechnology: studying technology from prehistory to the Middle Ages*: 13–27. Belgrade: Srpsko arheološko društvo.

Antonović, D., K. Resimić-Šarić and V. Cvetković. 2005. Stone raw materials in the Vinča culture: petrographic analysis of assemblage from Vinča and Belovode. *Starinar* 55: 53–66.

Antonović, D. and J. Šarić 2011. Kameno oruđe sa lokaliteta Crkvine u Stublinama, *Starinar* 61: 7–22.

Antonović, D. and M. Vukadinović. 2012. Eneolithic mine Prljuša – Mali Šturac: archaeological and geophysical investigations. *Starinar* 62: 95–106.

Antonović, D., M. Vukadinović and A. Cicović. 2014. Praistorijski rudnik na lokalitetu Prljuša, Mali Šturac istraživanje 2012, in D. Antonović, S. Golubović and V. Bikić (eds) *Arheologija u Srbiji. Projekti Arheološkog Instituta u 2012*: 28–31. Beograd: Arheološki Institut.

Arnold, D.E. 1985. *Ceramic Theory and Cultural Process.* Cambridge: Cambridge University Press.

Arnold, D.E. 1993. *Ecology and Ceramic Production in an Andean Community.* Cambridge: Cambridge University Press.

Arnold, D.E. 2000. Does the standardization of ceramic pastes really mean specialization? *Journal of Archaeological Method and Theory* 7/4: 333–375.

Arnold, E. and H. Greenfield. 2006. *The Origins of Transhumant Pastoralism in Temperate South Eastern Europe: A Zooarchaeological Perspective from the Central Balkans* (British Archaeological Reports International Series 1538). Archaeopress: Oxford.

Arsenijević, M. and J. Živković. 1998. Methodology and the results of statistical-typological analysis of pottery from the Vinča culture settlement of Belovode, near Petrovac-upon-Mlava. *Journal of Serbian Archaeological Society* 14: 281–291.

Arsić, R., J. Miletić and V. Miletić. 2010. Zaštitna istraživanja na lokalitetu Crkvine, Mali Borak. *Kolubara* 5: 55–67.

Averbouh, A. 2000. Technologie de la Matière Osseuse Travaillée et Implications Palethnologiques. Unpublished PhD dissertation, Université de Paris I.

Avramova, M. 2002. Der Schmuck aus den Gräbern von Durankulak, in H. Todorova (ed.) *Durankulak, Band II - Die prähistorischen Gräberfelder von Durankulak, Teil 1*: 191–206. Sofia: Publishing House Anubis Ltd.

B

Babović, Lj. 1986. Zbradila – Korbovo: compte-rendu des fouilles en 1981, Đerdapske Sveske 3: 95–115. Belgrade: Archaeological Institute.

Бабовић, М. 1992. Могућност примене аероархеолошких метода, in Д. Срејовић (ed.) *Археологија и природне науке*: 181–189. Beograd: Srpska akademija nauka i umetnosti

Bachmann, H.-G. 1980. Early copper smelting techniques in Sinai and in the Negev as deduced from slag investigations, in P.T. Craddock (ed.) *Scientific Studies in Early Mining and Extractive Metallurgy*: 103–134. London: The British Museum Press.

Bachmann, H.-G. 1982. *The Identification of Slags from Archaeological Sites* (Institute of Archaeology Occasional Publication 6). London: Institute of Archaeology.

Bačkalov, A. 1979. *Predmeti od kosti i roga u predneolitu i neolitu Srbije*. Beograd: Savez arheoloških društava Jugoslavije.

Bacvarov, K. and S. Vitezović. 2014. Objects of special use from Late Neolithic/Chalcolithic site of Voden, Thrace (SE Bulgaria). Poster presented at 10th Meeting of the Worked Bone Research Group of the ICAZ, Belgrade, Serbia, 25–30 Aug 2014.

Bailey, D.W. 2000. *Balkan Prehistory: Exclusion, Incorporation and Identity*. London, New York: Routledge.

Bailey, D.W. 2005. *Prehistoric Figurines: Representation and Corporeality in the Neolithic*. London: Routledge.

Bailey, D.W. 2014. Touch and cheirotic apprehension of prehistoric figurines, in P. Dent (ed.) *Sculpture and Touch*: 17–43. London: Ashgate.

Bajnóczi, B., G. Schöll-Barna, N. Kalicz, Z. Siklósi, G.H. Hourmouziadis, F. Ifantidis, A. Kyparissi-Apostolika, M. Pappa, R. Veropoulidou and C. Ziota. 2013. Tracing the source of Late Neolithic spondylus shell ornaments by stable isotope geochemistry and cathodoluminescence microscopy. *Journal of Archaeological Science* 40: 874–882.

Baker, P. and F. Worley. 2014. *Animal Bones and Archaeology: Guidelines for Best Practice.* Swindon: English Heritage.

Balasescu, A., D. Moise and V. Radu. 2006. Une utilisation des bovins pour la traction pendant le Chalcolithique en Roumanie? In P. Pétrequin, R.-M. Arbogast, A.-M. Pétrequin, S. van Willigen and M. Bailly (eds) *Premiers chariots, premiers araires. La diffusion de la traction animale en Europe pendant les IVe et IIIe millénaires avant notre* ère. (CRA Monograph 29): 269–273. Paris: CNRS.

Balasse, M., A. Evin, C. Tornero, V. Radu, D. Fiorillo, D. Popovici, R. Andreescu, K. Dobney, T. Cucchi and A. Bălăşescu. 2016. Wild, domestic and feral? Investigating the status of suids in the Romanian Gumelniţa (5th mil. cal BC) with biogeochemistry and geometric morphometrics. *Journal of Anthropological Archaeology* 42: 27–36.

Bánffy, E. and P. Sümegi. 2011. The [environ-]mental contexts of earliest Neolithic settlement and architecture in western Hungary, in A. Hadjikoumis, E. Robinson and S. Viner (eds) *The Dynamics of Neolithisation in Europe*: 231–265. Oxford: Oxbow Books.

Bankoff, H. and F. Winter. 1990. The later Aeneolithic in southeastern Europe. *American Journal of Archaeology* 94: 175–191.

Bartosiewicz, L., V. Boronean, C. Bonsall and S. Stallibrass. 2001. New data on the prehistoric fauna of the Iron Gates: a case study from Schela Cladovei, Romania, in R. Kertesz and J. Makkay (eds) *From the Mesolithic to the Neolithic*: 15–22. Budapest: Archaeolingua.

Bar-Yosef, O. and M.P. Deborah. 2008. ASIA, WEST | Palaeolithic Cultures, in D.M. Pearsall (ed.) *Encyclopedia of Archaeology* : 865–875. New York: Academic Press.

Bar-Yosef, O., D.E. Mayer and N. Porat. 2008. Green stone beads at the dawn of agriculture. *Proceedings of the National Academy of Sciences* 105: 8548–8551.

Bar-Yosef, O. and P. Van Peer. 2009. The chaine opératoire approach in Middle Paleolithic archaeology. *Current Anthropology* 50(1): 103–131. DOI: https://doi.org/10.1086/592234.

Barge-Mahieu, H. and Y. Taborin. 1991. Fiche canines residuelles de cerf (appelées craches), in H. Camps-Fabrer (ed.) *Fiches typologiques de l'industrie osseuse préhistorique.Cahier IV. Objets de parure*. Aix-en-Provence: Université de Provence.

Bartelheim, M. 2007. *Die Rolle der Metallurgie in vorgeschichtlichen Gesellschaften*, Forschungen zur Archäometrie und Altertumswissenschaft 2. Rahden/Westf: Marie Leidorf.

Basalla, G. 1988. *The Evolution of Technology*. Cambridge: Cambridge University Press.

Bastert-Lamprichs, K., N. Benecke, S. Hansen and R. Neef. 2012. *Der Beginn der Landwirtschaftim Sudkaukasus. Die Ausgrabungen in Aruchlo in Georgien*. Berlin: DAI Eurasien Abteilung.

Baxter, M.J. and H.E.M. Cool. 2010. Correspondence analysis in R for archaeologists: an educational account. *Archeologia e Calcolatori* 21: 211–228.

Beck, A.R. and G. Philip. 2012. The archaeological exploitation of declassified satellite photography in semi-arid environments, in W.S. Hanson and I.A. Oltean (eds) *Archaeology from Historical Aerial and Satellite Archives*: 261–278. New York: Springer.

Beeston, R.F., J. Palatinus, C. Beck and E.C. Stout. 2006. Organic residue analysis of pottery sherds from Chrysokamino. *Hesperia Supplements* 36: 413–428.

Behre, K.E. 2008. Collected seeds and fruits from herbs as prehistoric food. *Vegetation History and Archaeobotany* 17: 65–73.

Beijerinck, W. 1947. *Zadenatlas der Nederlandsche flora*. Wageningen: E. Veenman and Zonen.

Beldiman, C. 2007. *Industria materiilor dure animale în preistoria României. Resurse naturale, comunităţi umane şi tehnologie din paleoliticul superior până în neoliticul timpuriu*. Bucureşti: Asociaţia Română de Arheologie, Studii de Preistorie, Supplementum 2, Editura Pro Universitaria.

Beldiman, C., D.-M. Sztancs and C. Ilie. 2012. *Artefacte din materii dure animale în colecţia Muzeului de Istorie Galaţi. Eneolitic (Osseous materials artefacts in the collection of History museum of Galaţi. Aeneolithic)*. Cluj Napoca: Editura Mega.

Benac, A. 1961. Neolitski telovi u sjeveroistočnoj Bosni i neki problemi bosanskog neolita. *Glasnik Zemaljskog Muzeja u Sarajevu* 15–16: 39–78.

Bendixen, M. 1995. Compositional perceptual mapping using Chi-squared Tree Analysis and Correspondence Analysis. *Journal of Marketing Management* 11: 571–581.

Benecke, N. 2009. Arheozoološka istraživanja. *Glasnik Zemaljskog Muzeja Bosne i Hercegovine u Sarajevu* 50–51: 124–126.

Benecke, N., S. Hansen, D. Nowacki, A. Reingruber, K. Ritchie and J. Wunderlich. 2013. Pietrele in the lower Danube region: integrating archaeological, faunal and environmental investigations. *Documenta Praehistorica* 40: 175–193.

Bentley, R.A., T.D. Price, J. Lüning, D. Gronenborn, J. Wahl and P.D. Fullagar. 2002. Human migration in early Neolithic Europe. *Current Anthropology* 43: 799–804.

Bentsen, S.E. 2013. Using pyrotechnology: Fire-related features and activities with a focus on the African Middle Stone Age. *Journal of Archaeological Research* 22: 1–35.

Bebber, M.R. 2021. The Role of Functional Efficiency in the Decline of North America's Copper Culture (8000–3000 BP): An Experimental, Ecological, and Evolutionary Approach. *Journal of Archaeological Method and Theory*. DOI: https://doi.org/10.1007/s10816-020-09497-0

Berciu, D. 1961. *Contribuţii la problemele neoliticului în Romînia în lumina noilor cercetări*. Bucureşti: Editura Academiei Republicii Populare Romine.

Berciu, D. 1967. *Romania: Before Burebista*. London: Thames and Hudson.

Berdenov, S. 2008. Kazahstanskie mestorozhdeniya medi i olova i ih razrabotka v bronzovom veke (Kazakh sources of copper and tin and their processing in the Bronze Age, in Russian). *Izvestiya Natsionalnoy Akademii nauk Respubliki Kazahstana. Seriya obschestvennyih nauk* 1: 42-55.

Bibikov, S.N. 1953. Rannetripolskoe poselenie Lukavrublevetskaya na Dnestre. *Materialy i Issledovanija po Arkheologii SSR* 38.

Biehl, P. and A. Marciniak. 2000. The construction of hierarchy: rethinking the Copper Age in Southeastern Europe, in M. Diehl (ed.) *Hierarchies in*

Action: Cui Bono? :181–209. Carbondale (IL): Center for Archaeological Investigations, Southern Illinois University.

Birch, T., Th. Rehren and E. Pernicka. 2013. The metallic finds from Çatalhöyük: a review and preliminary new work, in I. Hodder (ed.) *Substantive Technologies at Çatalhöyük*: 307–316. London, Los Angeles (CA): British Institute at Ankara, Cotsen Institute of Archaeology.

Biró, K.T. 2014. Carpathian obsidians: state of art of Central European obsidian research, in M. Yamada and A. Ono (eds) *Lithic Raw Material Exploitation and Circulation in Préhistory. A Comparative Perspective in Diverse Palaeoenvironment. ERAUL* 138: 45–67.

Bitelli G. and V.A. Girelli. 2009. Metrical use of declassified satellite imagery for an area of archaeological interest in Turkey. *Journal of Cultural Heritage* 10, Supplement 1: 35–40.

Blagojević, M. 2005. Keramičke posude iz ranoeneolitske kuće sa lokaliteta Livade, Kalenić. *Kolubara* 4: 31–78.

Blažić, S. 1985. Prilog poznavanju ostataka faune sa arheološkog lokaliteta 'Golokut'. *Rad Vojvođanskih muzeja* 29: 33–35.

Blažić, S. 1995. Ostaci životinjskih vrsta sa lokaliteta na trasi auto-puta kroz Srem, in Z. Vapa (ed.) *Arheološka istraživanja duž auto-puta kroz Srem*: 331–346. Novi Sad: Pokrajinksi zavod za zaštitu spomenika kulture.

Blažić, S. 2005. The faunal assemblage, in S. Karmanski (ed.) *Donja Branjevina: a Neolithic Settlement near Deronje in the Vojvodina (Serbia)*: 74–76. Trieste: Societa per la Preistoria e Protostoria della Regione Friuli-Venezia Giulia.

Blažić, S., and D. Radmanović. 2011. Fauna kasnovinčanskih staništa Crkvine i Belež. *Kolubara* 5: 239–250.

Bleed, P. 2008. Skill matters. *Journal of Archaeological Method and Theory* 15: 154–166.

Blondel, V.D., J.-L Guillaume, R. Lambiotte and E. Lefebvre. 2008. Fast unfolding of communities in large networks. *Journal of Statistical Mechanics: Theory and Experiment* 10: P10008.

Bloom, R., R. Crippen, C. Elachi, J. Zarins, N. Clapp and G.R. Hedges. 1997. Space technology and the discovery of lost city of Ubar. *IEEE Aerospace Conference Proceedings* 1: 19–28.

Boardman, S. and G. Jones. 1990. Experiments on the effects of charring on cereal plant components. *Journal of Archaeological Science* 17: 1–11.

Bocquet-Appel, J.-P. 2002. Paleoanthropological traces of a Neolithic demographic transition. *Current Anthropology* 43: 637–650.

Boëda, E. 1990. Identification de chaînes opératoires lithiques de paléolithique ancien et moyen. *Paléo* (2): 43–80.

Boessneck, J. 1969. Osteological differences between sheep (*Ovis aries* Linn.) and goat (*Capra hircus* Linn.), in D. Brotwell and E Higgs (eds) *Science in Archaeology*: 331–358. London: Thames and Hudson.

Boessneck, J., H.-H. Müller and M. Teichert. 1964. Osteologische Unterscheidungsmerkmale zwischen Schaf (*Ovis aries* Linné) und Ziege (*Capra hircus* Linné). *Kühn-Archiv* 78: 1–29.

Bogaard, A. 2004. *Neolithic Farming in Central Europe. An Archaeobotanical Study of Crop Husbandry Practices.* London: Routledge.

Bogaard, A., J. Bending and G. Jones. 2007. Archaeobotanical evidence for plant husbandry and use at Ecsegfalva 23, in A. Whittle (ed.) *The Early Neolithic on the Great Hungarian Plain: Investigations of the Körös Culture Site of Ecsegfalva 23, Co. Békés*: 421–445. Budapest: Institute of Archaeology, Hungarian Academy of Sciences.

Bogaard, A. and P. Halstead 2015. Subsistence practices and social routine in Neolithic southern Europe, in C. Fowler, J. Harding and D. Hofmann (eds) *The Oxford Handbook of Neolithic Europe*: 385–410. Oxford: Oxford University Press.

Bogdanov, B. 1982. Bulgaria, in F.W. Dunning, W. Mykura and D. Slater (eds) *Mineral Deposits of Europe. Volume 2: Southeast Europe*: 215–232. London: The Institute of Mining and Metallurgy and the Mineralogical Society.

Bogdanović, M. 1988. Architectural and Structural Features at Divostin, in A. McPherron and D. Srejović (eds) *Divostin and the Neolithic of Central Serbia*: 35–142. Pittsburgh: University of Pittsburgh. Kragujevac: Narodni muzej Kragujevac.

Bogdanović, M. 1990. Die Spätneolithischen Siedlüngen in Divostin, in D. Srejović and N. Tasić (eds) *Vinča and its World*: 99–106. Belgrade: Serbian Academy of Sciences and Arts, Centre for Archaeological Research, Faculty of Philosophy.

Bogdanović, M. 1992. Iskopavanja na neolitskom naselju Gruža u Grivcu. Glasnik Srpskog arheološkog društva 8: 41–45.

Bogdanović, M. 2004. Naselja Vinčanske kulture. Arhitektura, in M. Bogdanović (ed.) *Grivac. Naselje protostarčevačke i vinčanske kulture*: 157–204. Kragujevac: Centar za naučna istraživanja SANU i Univerziteta u Kragujevcu, Narodni Muzej Kragujevac.

Bognár-Kutzián, I. 1963. *The Copper Age Cemetery of Tiszapolgar-Basatanya*. Budapest: Akademiai Kiado.

Bognár-Kutzián, I. 1972. *The Early Copper Age Tiszapolgar Culture in the Carpathian Basin*. Budapest: Akademiai Kiado.

Bognár-Kutzián, I. 1976. On the origins of early copper-processing in Europe, in J.V.S. Megaw (ed.) *To Illustrate the Monuments. Essays on Archaeology Presented to Stuart Piggott*: 6–76. London: Thames and Hudson.

Bogosavljević, V. 1989. Uvod u rudarsku arheologiju na Kopaoniku (oblici starog rudarstva i metalurgije). *Journal of Serbian Archaeological Society* 9: 86–93.

Bogosavljević, V. 1995. Mining hammerstones of Prljuša-Mali Šturac site, in P. Petrović and S.

Đurđekanović (eds) *Ancient Mining and Metallurgy in Southeast Europe*: 37–44. Bor, Belgrade: Archaeological Institute Belgrade, Museum of Mining and Metallurgy Bor.

Bogosavljević, V., K. Penezić, J. Jovanović, and S. Stefanović. 2019. Human jaw bone from the Late Neolithic Pločnik site: A multidisciplinary approach. *Zbornik Narodnog Muzeja* (Belgrade) 24: 9–30.

Bogosavljević, V., D. Mrkobrad and D. Bogosavljević. 1988. Istraživanje srednjovekovnog rudarstva na zapadnom Kopaoniku (okolina Kiževaka). *Naša Prošlost (Kraljevo)* 3: 9–48.

Bogosavljević-Petrović, V. 1992. *Okresana kamena industrija sa neolitskog naselja Divlje Polje: katalog povodom izložbe 'Tehnologija obrade kamena u neolitu'*. Kraljvo: Narodoni muzej Kraljevo.

Bogosavljević-Petrović, V. 1995. Arheometalurški kompleksi na Kopaoniku (Archaeometallurgical complexes in Kopaonik, in Serbian). *Journal of Serbian Archaeological Society* 10: 58–71.

Bogosavljević-Petrović, V. 2011. Redukcija kamenih sirovina na lokalitetu Crkvine - sonda 5 sa arheološkim celinama, in M. Blagojević (ed.) *Kolubara* 5: 213–238. Beograd: Republički zavod za zaštitu spomenika kulture.

Bogosavljević-Petrović, V. and J. Marković. 2012. History and current studies of petroarcheological data from the Neolithic and Chalcolithic in Serbia *Bulgarian e-journal of Archaeology* 2(1): 51–66. http://be-ja.org/article/history-and-current-studies-of-petroarcheological-data-from-the-neolithic-and-eneolithic-in-serbia/

Bogosavljević-Petrović, V. and J. Marković. 2014. Raw material studies of West Central Serbia. *Journal of Lithic Studies* 1(1): 55–71. DOI: https://doi.org/10.2218/jls.v1i1.823.

Bogosavljević-Petrović, V., J. Marković and Lj. Rašković. 2012. Savremeni rudnik magnezita 'Lazac': izvor za arheološka proučavanja. *Naša prošlost* 13: 79–96.

Bogosavljević-Petrović, V. and A. Starović. 2013. Balkanski kremen iz zbirke za stariji neolit Narodnog muzeja u Beogradu. *Zbornik Narodnog muzeja, Serija: Arheologija* 21(1): 77–108.

Bogosavljević-Petrović, V. and M. Tomović. 1993. Kasnoantički rudarsko-metalurški kompleks na zapadnom Kopaoniku. *Journal of Serbian Archaeological Society* 9: 236–240.

Bogucki, P. 1982. *Early Neolithic Subsistence and Settlement in the Polish Lowlands.* (British Archaeological Reports International Series S150). Oxford: Archaeopress.

Boivin, N. and M.-A. Owoc. 2004. *Soil, Stones and Symbols: Cultural Perceptions of the Mineral World.* London: University College London.

Bojadžiev, J. 2002. Die absolute Datierung der neo- und äneolithischen Gräberfelder von Durankulak, in H. Todorova (ed.) *Durankulak, Band II - Die prähistorischen Gräberfelder von Durankulak, Teil 1*: 67–70. Sofia: Publishing House Anubis Ltd.

Bojňanský, V. and A. Fargašová. 2007. *Atlas of Seeds and Fruits of Central and East European Flora: the Carpathian Mountains Region*. Dordrecht: Springer.

Bökönyi, S. 1970. Animal remains from Lepenski Vir. *Science* 167: 1702–1704.

Bökönyi, S. 1974. The vertebrate fauna, in M. Gimbutas (ed.) *Obre I and II: Neolithic Sites in Bosnia*: 55–154. Wissenschaftliche Mitteilungen des Bosnisch-herzegowinischen Landesmuseums, Band 3, Heft B (Archäologie). Sarajevo: Zemaljski Muzej.

Bökönyi, S. 1976. The vertebrate fauna from Anza, in Gimbutas, M. (ed.) *Neolithic Macedonia*: 313–374. Los Angeles: University of California Press.

Bökönyi, S. 1988. The Neolithic fauna of Divostin, in A., McPherron and D. Srejović (eds) *Divostin and the Neolithic of Central Serbia*: 419–445. Pittsburgh: University of Pittsburgh:

Bökönyi, S. 1991. Prehistoric animal remains from Bubanj-Hum at Niš. *Starinar* 40–41: 89–94.

Bökönyi, S. 1992. Animal remains of Mihajlovac-Knjepište: an early Neolithic settlement of the Iron Gate Gorge. *Balcanica* 23: 77–87.

Bolomey, A. and S. Marinescu-Bîlcu. 1988. Industria osului în aşezarea cucuteniana de la Draguşeni-Ostrov. *Studii şi cercetari de istorie veche şi arheologie* 39/4: 331–353.

Bonga, L. 2013. Late Neolithic Pottery from Mainland Greece, ca., 5,300–4,300 BC. Unpublished PhD dissertation. Temple University, Philadelphia.

Bonnardin, S. 2008. From traces to function of ornaments: some Neolithic, in L. Longo and N. Skakun (eds) *Prehistoric Technology 40 Years Later: Functional Studied and the Russian Legacy*: 297–308. Oxford: Archeopress.

Bonsall, C., M. Gurova, C. Hayward, C. Nachev and N. Pearce. 2010. Characterization of 'Balkan flint' artefacts from Bulgaria and the Iron gates using LA-ICP-MS and EPMA. *Interdisciplinary Studies* 22–23: 9–18.

Borić, D., B. Hanks, D. Šljivar, M. Kočić, J. Bulatović, S. Griffiths, R. Doonan and D. Jacanović. 2018. Enclosing the Neolithic World: A Vinča Culture Enclosed and Fortified Settlement in the Balkans. *Current Anthropology* 59: 336-346.

Borojević, K. 1990. Analiza Paleobotaničkih Makroostataka sa Arheoloških Lokaliteta u Srbiji. Unpublished MA dissertation, Department of Archaeology, University of Belgrade, Serbia.

Borojević, K. 2006. *Terra and Silva in the Pannonian Plain: Opovo Agro-Gathering in the Late Neolithic* (British Archaeological Reports International Series 1563). Oxford: Archaeopress.

Borojević, K. 2013. *New plant evidence from the Middle and Late Neolithic in central Balkans.* Paper presented at the 16th Conference of the International Work

Group for Palaeoethnobotany, Thessaloniki, Greece. Abstract in *Book of Abstracts*: 99.

Borojević, K., D. Antonović, J. Vuković, V. Dimitrijević, D. Filipović, M. Marić, K. Penezić, B. Tripković, V. Bogosavljević-Petrović and N. Tasić. 2020. Use of space in a Late Neolithic/Early Eneolithic building at the site of Vinča-Belo Brdo in the Central Balkans, in N. Tasić, D. Urem-Kotsou and M. Burić (eds) *Making Spaces into Places. The North Aegean, the Balkans and Western Anatolia in the Neolithic.* (British Archaeological Reports International Series S3001): 157–168. Oxford: BAR Publishing.

Boroffka, N., Cierny, J., Lutz, J., Parzinger, H., Pernicka, E. and G. Weisgerber 2002. Bronze Age tin from Central Asia, in K. Boyle, C. Renfrew and M. Levine, M. (eds.) *Ancient Interactions: East and West in* Eurasia, 135-159. Cambridge: McDonald Institute Monographs.

Bourgarit, D. 2007. Chalcolithic copper smelting, in S. La Niece, D. Hook and P. Craddock (eds) *Metals and Mines: Studies in Archaeometallurgy*: 3–14. London: Archetype Publications.

Boyadžiev, Y. 1995. Chronology of prehistoric cultures in Bulgaria, in D.W. Bailey, I. Panayotov and S. Alexandrov (eds) *Prehistoric Bulgaria* (Vol. Monographs in World Archaeology No. 22): 149–192. Madison (WI): Prehistory Press.

Boyadžiev, Y. 2002. Die absolute Chronologie der neo- und äneolithischen Gräberfelder von Durankulak, in H. Todorova (ed.) *Durankulak- Die Prähistorischen Gräberfelder*: 67–70). Sofia: The Bulgarian Academy of Sciences.

Brami, M. 2017. *The diffusion of Neolithic practices from Anatolia to Europe. A contextual study of residential construction, 8,500-5,500 BC cal BC.* (British Archaeological Reports International Series S2838). Oxford: BAR Publishing.

Breunig, P. 1987. *14C-Chronologie des vorderasiatischen, südost- und mitteleuropäischen Neolithikums.* Wien: Fundamenta A/13, Böhlau Verlag, Köln.

Brochier, J.É. 2013. The use and abuse of culling profiles in recent zooarchaeological studies: some methodological comments on 'frequency correction' and its consequences. *Journal of Archaeological Science* 40: 1416–1420.

Bronk Ramsey, C. 2009. Bayesian analysis of radiocarbon dates. *Radiocarbon* 51: 37–60.

Brovkin, V., S. Lorenz, T. Raddatz, T. Ilyina, I. Stemmler, M. Toohey, and M. Claussen 2019. What was the source of the atmospheric CO2 increase during the Holocene? *Biogeosciences* 16: 2543-2555.

Brown, B.M. 1987. Population estimation from floor area: a restudy of 'Naroll's Constant' *Cross-Cultural Research* 21: 1–49.

Brukner, B. 1962. Praistorijsko naselje Beletinci kod Obreža. *Rad Vojvođanskih muzeja* 11: 89–122.

Brukner, B. 1977. Beitrag zur Feststellung des Beginns der Metallurgie und der Aeneolithisierung. *Archaeologia Iugoslavica* 18: 9–12.

Brukner, B. 1980. Naselje vinčanske grupe na Gomolavi (neolitski i ranoeneolitski sloj). Izveštaj sa iskopavanja 1967-1976. *Rad Vojvođanskih Muzeja* 26: 5–55.

Brukner, B. 1988. Die Siedlung der Vinča-Gruppe auf Gomolava (die Wohnschicht des spätneolithikums und frühäneolithikums – Gomolava Ia-b und Gomolava Ib) und der Wohnhorizont des äneolithischen Humus (Gomolava II), in N. Tasić and J. Petrović (eds) *Gomolava – Chronologie und Stratigraphie der vorgeschichtlichen und antiken Kulturen der Donauniederung und Südosteuropas. Internationales Symposium, Ruma, 1986*: 9–38. Novi Sad: Vojvođanski muzej.

Brukner, B. 2002. Die Vinča–Kultur in Raum und Zeit: 61–103. Sarajevo, Frankfurt A/M, Berlin, Heidelberg, Knj. 30: Godišnjak Akademije nauka i umetnosti Bosne i Hercegovine, Centar za balkanološka ispitivanja.

Brukner, B. 2003. Vinčanska kultura u prostoru i vremenu. *Rad Muzeja Vojvodine* 43–45: 7–28.

Brysbaert, A. and A. Gorgues (eds). 2017. *Artisans Versus Nobility? Multiple Identities of Elites and 'Commoners' Viewed through the Lens of Crafting from the Chalcolithic to the Iron Ages in Europe and the Mediterranean.* Leiden: Sidestone Press.

Buckley, M., S. Whitcher Kansa, S. Howard, S. Campbell, J. Thomas-Oates and M. Collins. 2010. Distinguishing between archaeological sheep and goat bones using a single collagen peptide. *Journal of Archaeological Science* 37: 13–20.

Budd, P. 1991. Eneolithic arsenical copper - heat-treatment and the metallographic interpretation of manufacturing processes, in E., Pernicka and G.A. Wagner (eds) *Archaeometry '90: International Symposium on Archaeometry*: 35–44. Heidelberg: Birkhäuser Verlag.

Budd, P. and B.S. Ottaway. 1991. The properties of arsenical copper alloys: implications for the development of eneolithic metallurgy, in P. Budd, B. Chapman, C. Jackson, R. Janaway and C.A. Ottaway (eds) *Archaeological Sciences 1989: Proceedings of a Conference on the Application of Scientific Techniques to Archaeology, Bradford, September 1989*: 132–142. Bradford: Oxbow Books.

Budd, P. and T. Taylor. 1995. The faerie-smith meets the bronze industry - magic versus science in the interpretation of prehistoric metal-making. *World Archaeology* 27: 133–143.

Budden, S. 2008. Skill amongst the sherds: understanding the role of skill in the early to late Middle Bronze Age in Hungary, in I. Berg (ed.) *Breaking the Mould: Challenging the Past through Pottery. Prehistoric Ceramics Research Group Occasional Paper 6* (British Archaeology Reports International Series S1861): 1–18. Oxford: Archaeopress.

Buder, I. 1984. Uticaj vojnih činilaca na razvoj fotogrametrije. *Zbornik radova, Vojnogeografski institut*: 89–99. Beograd.

Бугарски, И. and В. Иванишевић. 2014a. Оживљавање изгубљених крајолика. *Гласник Друштва конзерватора Србије* 38: 252–254. Београд.

Бугарски, И and В. Иванишевић. 2014b. Примена аерофотографије у српској археологији. *Саопштења* XLVI: 251–263. Београд.

Bulatović, A. 2003. *Nalazi vinčanske kulture sa Fafosa i Valača*. Vranje: Narodni Muzej.

Bulatović, A. 2007. Vranje: kulturna stratigrafija praistorijskih lokaliteta u Vranjskoj regiji. Beograd: Arheološki institut, Vranje: Narodni muzej.

Bulatović, A. 2015. The phenomenon of prehistoric ritual pits: several examples from the central Balkans (in Serbian). *Starinar* 65: 7–35. DOI: https://doi.org/10.2298/STA1565007B.

Bulatović, A. and C. Jović. 2009. Leskovac: kulturna stratigrafija praistorijskih lokaliteta u Leskovačkoj regiji. Beograd: Arheološki institut, Leskovac: Narodni muzej.

Bulatović, A. and D. Milanović. 2020. *Bubanj. The Eneolithic and the Early Bronze Age Tell in Southeastern Serbia.* (Mitteilungen der Prähistorischen Kommission, Band 90). Wien: VÖAW.

Bulatović, A., M. Vander Linden and M. Gori. 2018. New AMS dates as a contribution to the absolute chronology of the early Eneolithic in the central Balkans. *Starinar* 68: 19–32.

Bulatović, J. 2010. Ostaci faune iz zatvorenih eneolitskih celina na nalazištu Bubanj kod Niša. Unpublished MA dissertation, University of Belgrade.

Bulatović, J. 2011. Ostaci životinja iz kasnovinčanske jame na nalazištu Vitkovačko polje. *Kruševački Zbornik* 15: 237–272.

Bulatović, J. 2012. Eksploatacija domaćih životinja u kasnom neolitu i eneolitu na području centralnog Balkana. Studije slučaja: Vitkovo i Bubanj. *Glasnik Srpskog arheološkog društva* 28: 279–300.

Bulatović, J. 2018. Arheozoološki aspekti društvenih i kulturnih promena na centralnom Balkanu u petom milenijumu pre nove ere. Unpublished Ph.D. dissertation, University of Belgrade.

Bulatović, J. and S. Milošević. 2015. Životinjski ostaci iz potkapine Mokranjske Stene, in A. Kapuran and A. Bulatović (eds) *Mokranjske Stene: kulturno nasleđe Negotinske Krajine*: 41–55. Negotin: Muzej Krajine.

Bulatović, J. and M. Spasić. 2019. Životinjski ostaci sa neolitskog nalazišta Grabovac – Đurića vinogradi. *Zbornik Narodnog muzeja* 24–1 (arheologija): 63–84.

Bunardžić, R., M. Bunardžić, S. Derikonjić and P. Gavović. 2008. *Topography of Polimlje*. Belgrade: Republic Institute for Heritage Protection.

Burmeister, S., S. Hansen, M. Kunst and N. Müller-Scheeßel (eds). 2013. *Metal Matters. Innovative Technologies and Social Change in Prehistory and Antiquity* (Menschen - Kulturen - Traditionen 12, ForschungsCluster 2), Rahden/Westf: Leidorf.

Burton, J. 1984. Axe Makers of the Wahgi, Pre-Colonial Industrialists of the Papua New Guinea Highlands. Unpublished PhD dissertation, Australian National University.

C

Cameron, C.M. 1990. The effect of varying estimates of pit structure use-life on prehistoric population estimates in the American Southwest. *Kiva* 55: 155–166.

Campbell, G., L. Moffet and V. Straker. 2011. *Environmental Archaeology: A Guide to the Theory and Practice of Method, From Sampling and Recovery to Post-Excavation* (second edition). Swindon: English Heritage.

Camps-Fabrer, H. 1966. *Matière et art mobilier dans la préhistoire nord-africaine et saharienne.* Paris: Mémoires du Centre de recherches anthropologiques préhistoriques et ethnographiques.

Camps-Fabrer, H. 1979. Principes d'une classification de l'industrie osseuse néolithique et de l'âge des métaux dans le Midi méditerranéen, in H. Camps-Fabrer (ed.) *Industrie de l'os et bois de cervidé durant le néolithique et âge des métaux. Première réunion du groupe de travail no. 3 sur l'industrie de l'os préhistorique*: 17–26. Paris: Centre national de la recherche scientifique.

Cann, J.R. and C. Renfrew. 1964. The characterisation of obsidian and its application to the Mediterranean region. *Proceedings of the Prehistoric Society* 30: 111–133.

Cantacuzino, G. and S. Morintz. 1963. Die jungsteinzeitlichen Funde in Cernica (Bukureşti) *Dacia* 7.

Cantoro, G. 2012. A free set of tools for automated imagery rectification. *ISAP News* 30: 9–11.

Cantoro, G. 2015. Aerial reconnaissance in archaeology – from archives to digital photogrammetry, in A. Sarris (ed.) *Best Practices of Geoinformatic Technologies for the Mapping of Archaeolandscapes*: 103–114. Oxford: Archaeopress.

Cappers, R.T.J., R.M. Bekker and J.E.A. Jans. 2006. *Digital Seed Atlas of the Netherlands.* (Groningen Archaeological Studies 4). Eelde: Barkhuis Publishing.

Carneiro, R.L. 1986. On the relationship between size of population and complexity of social organization. *Journal of Anthropological Research* 42: 355–364.

Carneiro, R.L. 2000. The transition from quantity to quality: a neglected causal mechanism in accounting for social evolution. *Proceedings of the National Academy of Sciences of the United States of America* 97: 12926–12931.

Carter, T. 2009. L'obsidienne Égéenne: caractérisation, utilisation et culture, in M.-H. Moncel and F. Frohlich (eds) *L'Homme et Le Precieux. Matières Minérales Précieuses de La Préhistoire À Aujourd'hui*

(British Archaeological Reports International Series 1934): 199–212. Oxford: Archaeopress.

Carvajal-Ramírez, F., A.D. Navaro-Ortega and F. Agüera-Vega. 2019. Virtual reconstruction of damaged archaeological sites based on Unmanned Aerial Vehicle Photogrammetry and 3D modelling. Study case of a southeastern Iberia production area in the Bronze Age. *Measurement* 136: 225–236.

Cassana, J. and J. Cothren. 2008. Stereo analysis, DEM extraction and orthorectification of CORONA satellite imagery: archaeological applications from Near East. *Antiquity* 82(317): 732–49.

Cermanović-Kuzmanović, A. and D. Srejović. 1992. *Leksikon religija i mitova drevne Evrope.* Beograd: Savremena administracija.

Cessford, C. 2005. Absolute dating at Çatal Höyük, in I. Hodder (ed.) *Changing Materialities at Çatalhöyük: Reports from the 1995-99 Seasons*: 65–100. Cambridge: McDonald Institute Monographs, British Institute at Ankara.

Cevey, C., D. Günther, V. Hubert, K. Hunger, E. Hildbrand, M.-A. Kaeser, E. Lehmann, N. Müller-Scheessel, M. Wörle-Soares, C. Strahm and S. van Willigen. 2006. Neue archäometallurgische Untersuchungen zum Beginn der Kupferverarbeitung in der Schweiz. *Archäologie der Schweiz* 29: 24–33.

Çevik, Ö. 2018. What follows the Late Neolithic occupation in Central-Western Anatolia? A view from Ulucak, in S. Dietz, F. Mavridis, Ž. Tankosić and T. Takaoğlu (eds) *Communities in Transition: The Circum-Aegean Area in the 5th and 4th Millennia BC*: 506–512. Oxford: Oxbow.

Chapman, J. 1981. *Vinča Culture of South-East Europe: Studies in Chronology, Economy and Society* (British Archaeological Reports International Series S117). Oxford: Archaeopress.

Chapman. J. 1990. The Neolithic in the Morava-Danube confluence area: a regional assessment of settlement pattern, in D. Krstić and R. Tringham (eds) *Selevac. A Neolithic village in Yugoslavia* (Monumenta Archaeologica 15): 13–44. Los Angeles: University of California Press.

Chapman, J. 1991. The creation of social arenas in the Neolithic and Copper Age of South East Europe: the case of Varna, in P. Garwood, P. Jennings, R. Skeates and J.Toms (eds) *Sacred and Profane.* (Oxford Committee for Archaeology Monograph 32): 152–171. Oxford: Oxbow books.

Chapman, J. 1999a. The origins of warfare in the prehistory of central and eastern Europe, in J. Carman and A. Harding (eds) *Ancient Warfare: Archeological Perspectives*:101–142. Stroud: Sutton.

Chapman, J. 1999b. Deliberate house-burning in the prehistory of central and eastern Europe, in A. Gustafson and H. Karlsson (eds) *Glyfer och arkeologiska rum: En vänbok till Jarl Nordbladh*: 113–126. Göteborg: University of Göteborg Press.

Chapman, J. 2000. *Fragmentation in Archaeology. People, Places and Broken Objects in the Prehistory of South-Eastern Europe.* London and New York: Routledge.

Chapman, J. 2006. Dark burnished ware as a sign: ethnicity, aesthetics and categories in the later Neolithic of the central Balkans, in N. Tasić and C. Grozdanov (eds) *Homage to Milutin Garašanin*: 295–308. Belgrade: Serbian Academy of Sciences and Arts.

Chapman, J. 2007. The elaboration of an aesthetic of brilliance and colour in the climax Copper Age, in F. Lang, C. Reinholdt and J. Weilhartner (eds) *Stephanos Aristeios. Archäologische Forschungen zwischen Nil und Istros. Festschrift für Stefan Hiller zum 65. Geburtstag*: 64–74. Wien: Phoibos.

Chapman, J. 2010. Houses, households, villages and proto-cities in southeastern Europe, in D.W. Anthony and J.Y. Chi (eds) *The Lost World of Old Empire: The Danube Valley, 5000-3500 BC*: 74–89. Princeton (NJ): Institute for the Study of the Ancient World.

Chapman, J. 2011. Enchantment and enchainment in later Balkan prehistory: towards an aesthetic of precision and geometric order, in A. Hadjikoumis, E. Robinson and S. Viner (eds) *The Dynamics of Neolithisation in Europe. Studies in Honour of Andrew Sherratt*: 153–176. Oxford: Oxbow Books.

Chapman, J. 2013. From Varna to Brittany via Csõszhalom – Was there a «Varna Effect»? In A. Anders and V. Kulcsár (eds) *Moments in Time. Papers Presented to Pal Raczky on his 60th Birthday*: 323–335. Budapest: L'Harmattan.

Chapman, J. 2020a. *Forging Identities in Balkan Prehistory: Dividuals, Individuals and Communities, 7000-3000 BC.* Leiden: Sidestone.

Chapman, J. 2020b. The Vinča group – (Almost) 40 years on. *Quaternary International* 560-561: 5–19.

Chapman, J. and B. Gaydarska. 2007. *Parts and Wholes: Fragmentation in Prehistoric Context.* Oxford: Oxbow.

Chapman, J. and B. Gaydarska. 2020. The deposition of history in prehistory: copper objects on sites and in the landscape. *Proceedings of the Prehistoric Society* 86: 139–163.

Chapman, J. and E. Richter. 2009. Geometric order and scientific principles: a view from the Mesolithic, Neolithic and Chalcolithic of central and south east Europe, in G. Bodi (ed.) *In Medias Res Praehistorie. Miscellanea in Honorem Annos LXV Peragentis Professoris Dan Monah Oblata*: 21–58. Iaşi: Editura Universităţii 'Alexandru Ioan Cuza'.

Chapman, J. and S. Souvatzi. 2020. The Neolithic of Southeast Europe: recent trends. *Annual Review of Anthropology* 49(1): 123–140.

Chapman, J. and R.F. Tylecote. 1983. Early copper in the Balkans. *Proceedings of the Prehistoric Society* 49: 373–379.

Chapman, J., T. Higham, V. Slavchev, B. Gaydarska and N. Honch. 2006. The social context of the emergence, development and abandonment of the Varna cemetery, Bulgaria. *European Journal of Archaeology* 9: 159–183. DOI: https://doi.org/10.1177/1461957107086121.

Charles, J.A. 1969. Appendix I. A metallurgical examination of south-east European copper axes. *Proceedings of the Prehistoric Society* 35: 40–42.

Chataigner, C. 1998. Sources des artefacts Néolithiques, in M.-C. Cauvin, A. Gourgaud, B. Gratuze, N. Arnaud, P. Poupeau, J.L. Poidevin and C. Chataigner (eds) *L'Obsidienne Au Proche et Moyen Orient: Du Volcan À l'Outil* (British Archaeological Reports International Series 738: 273–324. Oxford: Archaeopress.

Chernykh, E.N. 1966. *Istorija Drevnejsej Metallurgii Vostocnoj Evropy (History of Ancient Metallurgy of Eastern Europe, in Russian)*. Moskva: Akademija Nauk SSSR.

Chernykh, E.N. 1978a. Aibunar - a Balkan copper mine of the fourth millennium BC. *Proceedings of the Prehistoric Society* 44: 203–217. DOI: https://doi.org/10.1017/S0079497X00010136.

Chernykh, E.N. 1978b. *Gornoe Delo i Metallurgiya v Dreivneishei Bolgarii*. Sofia: Bolgarskoi Akademii Nauk.

Chernykh, E.N. 1991. Frühestes Kupfer in der Steppen- und Waldsteppenkulturen Osteuropas, in J. Lichardus (ed.) *Die Kupferzeit als Historische Epoche*: 85–88. Bonn: Habelt.

Chernykh, E.N. 1992. *Ancient Metallurgy in the USSR: The Early Metal Age*. Cambridge: Cambridge University Press.

Chernykh, E.N. 1997. *Kargaly, Zabytyi mir (Kargaly, the Forgotten World, in Russian)*. Moscow: Izdatelstvo 'NOX'.

Chernykh, E.N. 2008a. Formation of the Eurasian 'steppe belt' of stockbreeding cultures: viewed through the prism of archaeometallurgy and radiocarbon dating. *Archaeology, Ethnology and Anthropology of Eurasia* 35: 36–53. DOI: http://dx.doi.org/10.1016/j.aeae.2008.11.003.

Chernykh, E.N. 2008b. The 'steppe belt' of stockbreeding cultures in Eurasia during the Early Metal Age. *Trabajos de Prehistoria* 65: 73–93.

Chernykh, E.N. 2013. *Nomadic Cultures in the Mega-Structure of Eurasian World.* Volume 1 (in Russian). Moscow: Jaziki Slavjanskoi Kulturi. Translation into English by Irina Savinetskaya and Peter N. Hommel and published also as Chernykh, E.N. 2017. Nomadic Cultures in the Mega-Structure of the Eurasian World. Boston, MA: Academic Studies Press with LRC Publishing House.

Chernykh, E.N. and S.V. Kuzminykh. 1989. *Ancient Metallurgy in the Northern Eurasia. Seyma-Turbino Phenomenon (in Russian)*. Moscow: Nauka.

Chernykh, E.N., S.V. Kuzminykh and L.B. Orlovskaya. 2004. Ancient metallurgy in northern Eurasia: from the Urals to the Sayano-Altai, in K.M. Linduff (ed.) *Metallurgy in Ancient Eastern Eurasia from the Urals to the Yellow River*: 15–36. Lewiston (NY): Edwin Mellen.

Chernykh, E. N. 2021. *The Cultures of Homo: Challenging essays about humankind's multi-million years history.* Moscow: TAUS.

Childe, V.G. 1930. *The Bronze Age*. Cambridge: Cambridge University Press.

Childe, V.G. 1929. *The Danube in Prehistory*. Oxford: Oxford University Press.

Childe, V.G. 1936. *Man Makes Himself.* London: Watts and Co.

Childe, V.G. 1936-37. Neolithic black ware in Greece and on the Danube. *The Annual of the British School at Athens* 37: 26–35.

Childe, V.G. 1944. Archaeological ages as technological stages. *Journal of the Royal Anthropological Institute of Great Britain and Ireland* 74: 7–24.

Childe, V.G. 1950. The urban revolution. *The Town Planning Review* 21: 3–17.

Childe, V.G. 1951. *Social Evolution*. London: Collins.

Chirikure, S. 2010. *Indigenous Mining and Metallurgy in Africa.* Cambridge, Cape Town: Cambridge University Press.

Chokhadzhiev, S. 2000. Development of the graphite ornamentation in the Copper Age (in the light of evidence from the Struma cultures), in L. Nikolova (ed.) *Technology, Style and Society. Contributions to the Innovations between the Alps and the Black Sea in Prehistory* (British Archaeological Reports International Series 854): 97–112. Oxford: Archaeopress.

Choyke, A. 1997. The bone tool manufacturing continuum. *Anthropozoologica* 25–26: 65–72.

Choyke, A. 2001. Late Neolithic red deer canine beads and their imitations, in A. Choyke and L. Bartosiewicz (eds) *Crafting Bone: Skeletal Technologies Through Time and Space. Proceedings of the 2nd Meeting of the (ICAZ) Worked Bone Research Group, Budapest, 31 August-5 September 1999*: 251–266. Oxford: Archaeopress.

Choyke, A. 2013. Hidden agendas: ancient raw material choice for worked osseous objects in Central Europe and beyond, in A. Choyke and S. O'Connor (eds) *From These Bare Bones: Raw Materials and the Study of Worked Osseous Objects*: 1–11. Oxford and Oakville: Oxbow Books.

Choyke, A. and J. Schibler. 2007. Prehistoric bone tools and the archaeozoological perspective: research in Central Europe, in C. Gates St-Pierre and R. Walker (eds) *Bones as Tools: Current Methods and Interpretations in Worked Bone Studies*: 51–65. Oxford: Archaeopress.

Christensen, M. 2004. Fiches caractères morphologiques, histologiques et mécaniques des matières dures d'origine animale, in D. Ramseyer (ed.) *Matières et Techniques. Fiches de la Commission de Nomenclature sur l'Industrie de l'Os Préhistorique. Cahier XI*: 17–27. Paris: Centre national de la recherche scientifique.

Christidou, R. 2001. Study of bone tools at three late/final Neolithic sites from Northern Greece, in A. Choyke and L. Bartosiewicz (eds) *Crafting Bone: Skeletal Technologies Through Time and Space – Proceedings of the 2nd Meeting of the (ICAZ) Worked Bone*

Research Group, Budapest, 31 August-5 September 1999: 41–47. Oxford: Archaeopress.
Clason, A. 1979. The farmers of Gomolava in the Vinča and La Tene period. *Palaeohistoria* 21: 42–81.
Clason, A. 1980. Padina and Starčevo: game, fish and cattle. *Palaeohistoria* 22: 142–173.
Collard, M., A. Ruttle, B. Buchananand M.J. O'Brien. 2013. Population size and cultural evolution in nonindustrial food-producing societies. *PLoS ONE* 8, e72628.
Colominas, L. 2013. Specialization or re-utilization? Study of the selection documented in a bone-working refuse. Assemblage from Roman Baetulo (Badalona, Spain), in A. Choyke and S. O'Connor (eds) *From These Bare Bones: Raw Materials and the Study of Worked Osseous Objects*: 88–95. Oxford and Oakville: Oxbow Books.
Comşa, E. 1973. Parures néolithiques en coquillages marins découvertes en territorie roumain. *Dacia* XVII: 61–76.
Comşa, E. 1990. Complexul neolitic de la Radovanu. *Cultură şi civilizaţie la Dunărea de Jos* 8: 7–126.
Comşa, E. 1991. L'utilisation du cuivre en Roumanie pendent le néolithique moyen, in J.-P. Mohen and C. Éluère (eds) *Découverte du Métal*: 77–84. Paris: Picard.
Cook, S.F. 1972. *Prehistoric Demography.* Reading (MA): Addison-Wesley Publishing Company.
Corns, A. and R. Shaw. 2013. Lidar and World Heritage Sites in Ireland: Why was such a rich data source gathered, how is it being utilized, and what lessons have been learned? In R.S. Opitz and D.C. Cowley (eds) *Interpreting Archaeological Topography*: 146–160 Oxford: Oxbow Books.
Costin, C.L. 1991. Craft specialization: issues in defining, documenting and explaining the organization of production. *Archaeological Method and Theory* 3: 1–56.
Courcier, A. 2014. Ancient metallurgy in the Caucasus from the sixth to the third millennium BCE, in B.W. Roberts and C.P. Thornton (eds.) *Archaeometallurgy in Global Perspective*: 579–664. New York: Springer.
Craddock, P.T. 1995. *Early Metal Mining and Production.* Edinburgh: Edinburgh University Press.
Craddock, P.T. 2001. From hearth to furnace: evidences for the earliest metal smelting technologies in the eastern Mediterranean. *Paléorient* 26: 151–156.
Craddock, P.T. and N.D. Meeks. 1987. Iron in ancient copper. *Archaeometry* 29: 187–204.
Crnobrnja, A. 2009. *Neolitsko naselje na Crkvina u Stublinama.* Belgrade: Muzej grada Beograda.
Crnobrnja, A. 2011. Arrangement of Vinča culture figurines: a study of social structure and organisation. *Documenta Praehistorica* 38: 131–147.
Crnobrnja, A. 2012. Group identities in the Central Balkan Late Neolithic. *Documenta Praehistorica* 39: 155–165.
Crnobrnja, A. 2014. The (E)neolithic settlement Crkvine at Stubline, Serbia, in W. Schier and F. Draşovean (eds) *The Neolithic and Eneolithic in Southeast Europe: New Approaches to Dating and Cultural Dynamics in the 6th to 4th Millennium BC*: 173–186. Rahden: Leidorf.
Crnobrnja, A., Z. Simić and M. Janković. 2009. Late Vinča culture settlement at Crkvine in Stubline: household organisation and urbanisation in the late Vinča culture period. *Starinar* 59: 9–25.
Csányi, M., P. Raczky and J. Tárnoki. 2009. Preliminary report on the cemetery of the Bodrogkeresztúr culture excavated at Rákóczifalva-Bagi-föld. *Tisicum* 38: 13–34.
Cuénod, A., Bray, P., and A. Mark Pollard 2015. The "Tin Problem" in the Prehistoric Near East: Further Insights from a Study of Chemical Datasets on Copper Alloys from Iran and Mesopotamia. *Iran* 53: 29-48.
Čuljković, B., S. Stefanović and S. Romac. 2002. Upotreba PCR-a u fizičkoj antropologiji-utvrđivanje pola (The application of PCR in physical anthropology- determination of sex, in Serbian), in *Zbornik radova sa I simpozijuma antropologa Republike Srpske*: 23–31. Banja Luka: Društvo antropologa Republike Srpske.
Cuomo di Caprio, N. 2007. *Ceramica in Archeologia.* Roma: L'Erma Bretschneider.
Custer, J.F., T. Eveleigh, V. Klemas and I. Wells. 1986. Application of LANDSAT data and synoptic remote sensing to predictive models for prehistoric archaeological sites: An example from the Delaware coastal plain. *American Antiquity* 51(3): 572–588.
Cvetković, N. 2004. Motel Slatina – Paraćin Trench: Gasovod II, Analyses of the Neolithic Fauna, in M. Vasić, P. Vučković and B. Cvetković (eds) *The Central Pomoravlje in Neolithization of South East Europe*: 75–82. Belgrade: Archaeological Institute.

Č

Čađenović G., D. Rašković and M. Bugar. 2003. *Vitkovačko polje u praistorijskom periodu. Naselje vinčanske kulture.* Kruševac: Narodni muzej.
Čović, B. 1961. Rezultati sondiranja na preistorijskom naselju u Gornjoj Tuzli. *Glasnik Zemaljskog Muzeja (Sarajevo)* 15/16: 79–139.

D

Dammers, B., A. Franz, R.G. Sobott and K. Bente. 2012. Erste Ergebnisse archäometrischer Untersuchungen zur mittel- und spätneolithischen Keramik von Uivar (Rumänien) unter besonderer Berücksichtigung der 3D-µ-Röntgen-Computertomographie, in B. Ramminger and O. Stilborg (eds) *Naturwissenschaftliche Analysen vor- und frühgeschichtlicher Keramik II*: 33–57. Bonn: Habelt.
Darvill, T. 1987 *Ancient Monuments in the Countryside: An Archaeological Management Overview.* Swindon: English Heritage.
Davenport, W.G., M. King, M. Schlesinger and A.K. Biswas. 2002. *Extractive Metallurgy of Copper (4th Edition).* Oxford: Pergamon.
Davies, O. 1937. Ancient mining in the central Balkans. *Revue Internationale des Etudes Balkaniques* 1: 405–418.

Deer, W.A., R.A. Howie and J. Zussman. 1992. *An Introduction to the Rock-Forming Minerals*. Harlow, London: Longman.

de Groot, B.G. 2019. A diachronic study of networks of ceramic assemblage similarity in Neolithic Western Anatolia, the Aegean and the Balkans (c. 6600–5500 bc). *Archaeometry* 61: 600–613. DOI: https://doi:10.1111/arcm.12450.

De Pascale, A. 2003. 'Hammerstones from early copper mines': Sintesi dei ritrovamenti nell'Europa e nel Mediterraneo orientale e prime considerazioni sui mazzuoli di Monte Loreto (IV millennio BC - Liguria). *Rivista di Studi Liguri* 69: 5–42.

Derex, M., M.-P. Beugin, B. Godelle and M. Raymond. 2013. Experimental evidence for the influence of group size on cultural complexity. *Nature* 503: 389–391.

Dergachev, V.A. 2004. Karbunskiy klad, in M. Videiko (ed.) *Encyclopedia Trypilskoi tsyvilizatsii. Volume 2.* : 213–214. Kyiv: Ukrpoligraphmedia.

Derikonjić, S. 1996. *Neolitske Zajednice Polimlja*. Proboj: Zavičajni muzej.

Derikonjić, S. 2010. *Tragom jarmovačkih rudara. Catalogue*. Priboj on Lim: Homeland Museum of Priboj.

Derikonjić, S., M. Radivojević, E. Pernicka and Th. Rehren. 2011. The Vinča culture mining complex in Jarmovac, southwest Serbia, in A. Hauptmann, D. Modarressi-Tehrani and M. Prange (eds) *Archaeometallurgy in Europe III: Proceedings of the 3rd International Conference, Deutsches Bergbau-Museum Bochum, June 29 - July 1, 2011*: 39. Bochum: Deutsches Bergbau-Museum.

Deroin, J.P., F. Téreygeol and J. Heckes. 2011. Evaluation of very high to medium resolution multispectral satellite imagery for geoarchaeology in arid regions – case study from Jabali, Yemen. *Journal of Archaeological Science* 38: 101–114.

d'Errico, F. 1993. Identification des traces de manipulation, suspension, polissage sur l'art mobilier en os, bois de cervidés, ivoire, in P. Anderson, S. Beyries, M. Otte and H. Plisson (eds) *Traces et Fonction: Les Gestes Retouvés. Actes du Colloque International de Liège*: 177–188. Liège: ERAUL.

d'Errico, F. and M. Vanhaeren. 2002. Criteria for identifying red deer (*Cervus elaphus*) age and sex from their canines. Application to the study of Upper Palaeolithic and Mesolithic Ornaments. *Journal of Archaeological Science* 29: 211–232.

Дероко, А. 1950. Кулич и Рам каструми на римском Дунавском лимесу. *Старинар* I: 169–173. Београд.

Дероко, А. 1951a. Смедеревски град. *Старинар* II: 59–98. Београд.

Дероко, А. 1951b. Град Голубац. *Старинар* II: 139–149. Београд.

Deschler-Erb, S., E. Marti-Grädel and J. Schibler. 2002. Die Knochen-, Zahn- und Geweihartefakte, in A. de Capitani, S. Deschler-Erb, U. Leuzinger, E. Marti-Grädel and J. Schibler (eds) *Die jungsteinzeitliche Seeufersiedlung Arbon / Bleiche 3: Funde*: 277–366. Archäologie im Thurgau, Band 11. Frauenfeld: Departement für Erziehung und Kultur des Kantons Thurgau.

Dewar, R.E. 1994. Contending with contemporaneity: a reply to Kintigh. *American Antiquity* 59: 149–152.

Diaconescu, D. 2014. New remarks about the typology and the chronology of the Pločnik and Čoka copper hammer-axes, in W. Schier and F. Draşovean (eds) *New Approaches to Dating and Cultural Dynamics in the 6th to 4th Millennium BC.* (Prähistorische Archäologie in Südosteuropa 28): 221–241. Rahden: Leidorf.

Díaz-Andreu, M. 2007. *A World History of Nineteenth-Century Archaeology Nationalism, Colonialism, and the Past.* Oxford: Oxford University Press.

Dimić, V. 2015. Traseological markers on ground stone cutting–implements from Lađarište site near Vrnjačka Banja. *Glasnik Srpskog arheološkog društva* 31: 39–73.

Dimić, V. 2013a Analiza glačanog i abrazivnog kamenog oruđa sa lokaliteta Lađarište kod Vrnjačke Banje. *Naša prošlost* 14: 31–56.

Dimić, V. 2013b Petrološka, trasološka i funkcionalno-tipološka studija glačanog i abrazivnog kamenog oruđa sa lokaliteta Lađarište kod Vrnjačke Banje, Odeljenje za arheologiju Filozofskog fakulteta, Univerzitet u Beogadu, Beograd.

Dimitrijević, S. 1968. *Sopotsko-Lenđelska kultura*. Zagreb: Archaeological Institute, Faculty of Philosophy.

Dimitrijević, S. 1974. Problem stupnjevanja starčevačke kulture s posebnim obzirom na doprinos južnopanonskih nalazišta rešavanju ovih problema. *Materijali* X: 59–122.

Dimitrijević, S. 1979. Sjeverna zona, in A. Benac (ed.) *Praistorija Jugoslovenskih Zemalja (Prehistory of Yugoslav Countries)*: 229–362. Sarajevo: Centar za balkanološka istraživanja Akademija nauka i umjetnosti Bosne i Hercegovine.

Dimitrijević, V. 2006. Vertebrate fauna of Vinča Belo Brdo (Excavation Campaigns 1998–2003). *Starinar* 56: 245–269.

Dimitrijević, V. and B. Tripković. 2002. New *Spondylus* findings at Vinča-Belo Brdo 1998–2001 campaigns and the regional approach to problem. *Starinar.* New series. LII: 48–62.

Dimitrijević, V. and B. Tripković. 2006. *Spondylus* and *Glycymeris* bracelets: trade reflections at Neolithic Vinča-Belo Brdo. *Documenta praehistorica* XXXIII: 237–252.

Dimitrov, K. 2002. Die Metallfunde aus den Gräberfeldern von Durankulak, in H. Todorova (ed.) *Durankulak, Band II - Die prähistorischen Gräberfelder von Durankulak, Teil 1*: 127–158. Berlin, Sofia: Publishing House Anubis.

Djurić, N. 1986. Nalazi sa neolitskog lokaliteta Kremen. *Zbornik 2*: 9–15

Dobney, K., and K. Reily. 1988. A method for recording archaeological animal bones: the use of diagnostic zones. *Cicaea* 5: 79–96.

Dobres, M.-A. and R. Hoffman. 1994. Social agency and the dynamics of prehistoric technology. *Journal of Archaeological Method and Theory* 1: 211–258.

Dobres, M.-A. and R. Hoffman (eds). 1999. *The Social Dynamics of Technology: Practice, Politics, and World Views.* Washington DC: Smithsonian Institution Press.

Dolfini, A. 2013. The emergence of metallurgy in the central Mediterranean region: a new model. *European Journal of Archaeology* 16: 21–62. DOI: https://doi.org/10.1179/1461957112Y.0000000023.

Dolfini, A. 2014. Early metallurgy in the central Mediterranean, in B.W. Roberts and C.P. Thornton (eds) *Archaeometallurgy in Global Perspective: Methods and Syntheses*: 473–506. New York: Springer.

Dolfini, A. and R.J. Crellin. 2016. Metalwork wear analysis: the loss of innocence. *Journal of Archaeological Science* 66: 78–87.

Dombay, J. 1939. *A Zengővárkonyi őskori Telep és Temető - The Prehistoric Settlement and Cemetery at Zengővárkony.* (Archaeologica Hungarica 23). Budapest: Magyar Történeti Múzeum.

Dombay, J. 1960. *Die Siedlung und das Graberfeld in Zengővárkony. Beitrage zur Kultur des Aeneolithikums in Ungarn* (Archeologica Hungarica 37). Budapest: Akadémiai Kiadó.

Draşovean, F. 2006. Nordgriechenland und der Mittlere Donauraum zum Ende des 6. und Beginn des 5. Jahrtausends v. Chr, in N. Tasić and C. Grozdanov (eds) *Homage to Milutin Garašanin*: 267–277. Belgrade: Serbian Academy of Sciences and Arts, Macedonian Academy of Sciences and Arts.

Draşovean, F. 2007. The Neolithic tells from Parţa and Uivar. Similarities and differences of the organization of the social space. *Analele Banatului* 15: 19–32.

Draşovean, F., W. Schier, A. Bayliss, B. Gaydarska and A. Whittle. 2017. The lives of houses: duration, context, and history at Neolithic Uivar, Romania. *European Journal of Archaeology* 20: 636–662. DOI: https://doi:10.1017/eaa.2017.37.

Draşovean, F., and W. Schier (eds). 2021. *Uivar "Gomila"A Prehistoric Settlement in the Romanian Banat. Vol. I: Site, Architecture, Stratigraphy and Dating.* (Prähistorische Archäologie in Südosteuropa 32). Rahden: Leidorf.

Dreibrodt, S., R. Hofmann, C. Lubos, S. Dazert, J. Zahrer, M. Fuchs and H.-R. Bork. 2013. Ergebnisse geomorphologisch-geoarchäologischer Begleituntersuchungen an den Siedlungen im Visokobecken (Bosnien-Herzegowina), in J. Müller, K. Rassmann and R . Hofmann (eds) *Okolište 1 - Untersuchungen einer spätneolithischen Siedlungskammer in Zentralbosnien*: 287–310. Bonn: Verlag Rudolf Habelt.

Drennan, R.D. 2009. *Statistics for Archaeologists. A Common Sense Approach.* New York: Springer.

Driehaus, J. 1952–55. Zur Datierung und Herkunft donauländischer Axttypen der frühen Kupferzeit. *Archaeologia Geographica* 2: 3–4: 1–8.

Dubbini, M., L.C. Curzio and A. Campedelli. 2016. Digital elevation models from unmanned aerial vehicle surveys for archaeological interpretation of terrain anomalies: case study of the Roman castrum of Burnum (Croatia). *Journal of Archaeological Science Reports* 8: 121–134.

Dubreuil, L. 2001. Functional studies of prehistoric grindingstones. A methodological research. *Bulletin du Centre de recherche français à Jérusalem* 9: 73–87.

Dumitrescu, H. 1961. Connections between the Cucuteni-Tripolie cultural complex and the neighbouring Eneolithic cultures in the light of the utilization of golden pendants. *Dacia* 5: 69–93.

Dunbar, R.I.M. 1993. Coevolution of neocortical size, group size and language in humans, *Behavioral and Brain Sciences* 16: 681–735.

E

Earle, T.K. 2018 Introduction: Artisans, technologies and consumers - a political economy approach to craft specialization, in I. Miloglav and J. Vuković (eds) *Artisans Rule: Product Standardization and Craft Specialization in Prehistoric Society*: 1–19. Newcastle upon Tyne: Cambridge Scholars Publishing.

Ecsedy, I. 1990. On the early development of prehistoric metallurgy in southern Transdanubia. *Godišnjak Akademije BiH 28. Centar za balkanološka ispitivanja (Sarajevo)* 26: 209–231.

Edwards, K.J., P. Halstead and M. Zvelebil. 1996. The Neolithic transition in the Balkans – archaeological perspectives and palaeoecological evidence: a comment on Willis and Bennet. *The Holocene* 6: 120–122.

Ehrhardt, K.L. 2014. Copper Working Technologies, Contexts of Use, and Social Complexity in the Eastern Woodlands of Native North America, in B.W. Roberts & C.P. Thornton (ed.) *Archaeometallurgy in Global Perspective: Methods and Syntheses*: 303-328. New York: Springer.

Ehrich, R.W. and H.A. Bankoff. 1992. East central and southeastern Europe, in R.W. Ehrich (ed.) *Chronologies in Old World Archaeology*: 31–375. Chicago: University of Chicago Press.

Eichler, A., G. Gramlich, T. Kellerhals, L. Tobler, Th. Rehren, and M. Schwikowski, 2017. Ice-core evidence of earliest extensive copper metallurgy in the Andes 2700 years ago. *Scientific Reports* 7: 41855.

Eliade, M. 1962. *The Forge and The Crucible.* London: Rider.

Elias, N. and N.J. Scotson. 1965. *The Established and the Outsider*. Lomdon: Frank Cass and Company Limited.

Ellenberg H., H.E. Weber, R. Düll, V. Wirth, W. Werner and D. Paulissen. 1992. Zeigerwerte von Pflanzen in Mitteleuropa. *Scripta Geobotanica* 18: 1–258.

Elliott, J.F. 1976. Phase relationships in the pyrometallurgy of copper. *Metallurgical and Materials Transactions B. Process Metallurgy and Materials Processing Science* 7: 17–33.

Elster, E. 2001. Middle Neolithic to Early Bronze Age bone tools from Sitagroi, Greece, in A. Choyke and L. Bartosiewicz (eds) *Crafting Bone: Skeletal Technologies through Time and Space - Proceedings of the 2nd Meeting of the (ICAZ) Worked Bone Research Group, Budapest, 31 August–5 September 1999*: 355–369. Oxford: Archaeopress.

El Susi, G. 2011. The management of livestock in the early Neolithic settlements (Starčevo Criş complex) from Transylvania and Banat, in S. Luca and C. Suciu (eds) *Early Neolithic (Starčevo Criş) Sites on the Territory of Romania*: 47–56. Oxford: Archaeopress.

Éluère, C. and C. J.Raub. 1991. Investigations on the gold coating technology of the great dish from Varna, in J.-P. Mohen and C. Éluère (eds) *Découverte du Métal*: 13–30. Paris: Picard.

Erkul E., D. Wilken, W. Rabbel, T. Yas, R. Hoffmann, N. Müller-Scheessel, K. Rassmann and J. Müller. 2013. Geoelektrische Untersuchugnen auf dem spätneolithischen Siedlungshügel von Okolište, in J. Müller, K. Rassmann and R. Hofmann (eds) *Okolište 1 - Untersuchungen einer spätneolithischen Siedlungskammer in Zentralbosnien. UPA 228*: 103–111. Bonn: Verlag Rudolf Habelt.

Evans, R. 1973. Craft Specialization in the Chalcolithic Period of the Eastern Portion of the Balkans Peninsula. Unpublished PhD dissertation, University of California.

Evin, A., L. Girdland Flink, A. Bălăşescu, D. Popovici, R. Andreescu, D. Bailey, P. Mirea, C. Lazăr, A. Boroneanţ, C. Bonsall, U. Strand Vidarsdottir, S. Brehard, A. Tresset, T. Cucchi, G. Larson and K. Dobney. 2015. Unravelling the complexity of domestication: a case study using morphometrics and ancient DNA analyses of archaeological pigs from Romania. *Philosophical Transactions of the Royal Society B* 370: 20130616.

F

Fagerberg, J. 2004. Innovation: a guide to the literature, in J. Fagerberg, J. Mowery and R. Nelson (eds) *The Oxford Handbook of Innovation*: 1–26. Oxford: Oxford University Press.

Feinman, G.M. 2011. Size, complexity, and organizational variation: a comparative approach. *Cross-Cultural Research* 45: 37–58.

Fewkes, V. 1936. Neolithic sites in the Morava-Danube area. *Bulletin of American School of Prehistoric Research* 12: 5–81.

Filipović, D. 2014. Southwest Asian founder and other crops at Neolithic sites in Serbia. *Bulgarian e-Journal of Archaeology* 4: 195–215.

Filipović, D., D. Challinor and M. Andrič. 2017. Vinča tell in southeast Europe: multi-proxy palaeobotanical evidence from Late Neolithic levels and the implications for the environment and economy. *Quaternary International* 429: 13–23. DOI: https://doi.org/10.1016/j.quaint.2014.09.059.

Filipović, D. and Đ. Obradović. 2013. Archaeobotany at Neolithic sites in Serbia: a critical overview of the methods and results, in N. Miladinović-Radmilović and S. Vitezović (eds) *Bioarheologija na Balkanu: bilans i perspektive*: 25–55. Beograd: Srpsko Arheološko Društvo.

Filipović, D. and N. Tasić. 2012. Vinča-Belo Brdo, a Late Neolithic site in Serbia: consideration of macro-botanical remains as indicators of dietary habits. *Balcanica* 43: 7–27.

Filipović, D., Đ. Obradović and B.Tripković. 2018. Plant storage in Neolithic southeast Europe: synthesis of the archaeological and archaeobotanical evidence from Serbia. *Vegetation History and Archaeobotany* 27: 31–44. DOI: https://doi.org/10.1007/s00334-017-0638-7.

Filipović, D., M. Marić, D. Challinor, J. Bulatović and N. Tasić. 2019. Natural environment and resources, and the long life of the Neolithic settlement at Vinča, southeast Europe. *Archaeological and Anthropological Sciences* 11: 1939-1960.

Filipovitch, L., I. Stefanova, M. Lazarova and M. Petrova. 1998. Holocene Vegetation in Stara Planina (the Balkan Range). *Phytologia Balcanica* 4/1–2: 13–25.

Fleming, L. and O. Sorenson. 2004. Science as a map in technological search. *Strategic Management Journal* 25: 909–928.

Fol, A. and J. Lichardus (eds). 1988. *Macht, Herrschaft und Gold: das Gräberfeld von Varna (Bulgarien) und die Anfänge einer neuen Europäischen Zivilisation.* Saarbrücken: Moderne Galerie des Saarland-Museums.

Food and Agriculture Organization of the UN. The Reference Soil Group of the Cambisols, viewed 05/12/2020, <http://www.fao.org/docrep/003/y1899e/y1899e08.htm>

Forenbaher, S. 1993. Radiocarbon dates and absolute chronology of the central European Early Bronze Age. *Antiquity* 67: 218–256.

Forte, V. 2019. Skilled people or specialists? Knowledge and expertise in Copper Age vessels from central Italy. *Journal of Anthropological Archaeology* 55: 1–20.

Forţiu, S. and A. Cîntar (eds). 2014. *ArheoVest, Nr. II: In Honorem Gheorghe Lazarovici. Interdisciplinaritate în Arheologie, Timișoara, 6 decembrie 2014, Volume 2: Metode Interdisciplinare, Asociația 'ArheoVest' Timișoara.* Szeged: JATEPress Kiadó.

Foster, G. 1965. The sociology of pottery: questions and hypothesis arising from contemporary Mexican work, in F. Matson (ed.) *Ceramics and Man* (Viking Fund Publications in Anthropology 4): 43–61. New York: Wenner-Gren Foundation for Anthropological Research Inc.

Fowler, M.J.F. 2012. Declassified intelligence satellite photographs, in S.W. Hanson and I.A. Oltean (eds) *Archaeology from Historical Aerial and Satellite Archives*: 47–65. New York: Springer.

Frame, L.D. 2012. Reconstructing ancient technologies: Chalcolithic crucible smelting at Tal-i Iblis, Iran, in P. Jett, B. McCarthy and J.G. Douglas (eds) *Scientific Research on Ancient Asian Metallurgy. Proceedings of the Fifth Forbes Symposium at the Freer Gallery of Art*: 183–204. Washington, DC: Archetype Publications in association with the Freer Gallery of Art, Smithsonian Institution.

Freudenberg, M. 2009. Steingerate zur Metallbearbeitung; Einige neue aspekte zum spatneolithichen und fruhbronzezeitlichen metallhanwerk von der hintergrund des Schleswig-Holsteinischen fundmaterials. Archaologisches Koresponedenzblatt 39(3): 341–359.

Frieman, C. 2021. *An archaeology of innovation: approaching social and technological change in human society*. Manchester: Manchester University Press.

Frierman, J. 1969. The Balkan graphite ware (Appendix II). *Proceedings of the Prehistoric Society* 35: 42–44.

Fuller, D.Q. and S.A. Weber. 2005. Formation processes and paleoethnobotanical interpretation in South Asia. *Journal of Interdisciplinary Studies in History and Archaeology* 2: 93–115.

Fuller, D.Q., C. Stevens and M. McClatchie. 2014. Routine activities, tertiary refuse, and labor organization. Social inferences from everyday archaeobotany, in M. Madella, C. Lancelotti and M. Savard (eds) *Ancient Plants and People: Contemporary Trends in Archaeobotany*: 174–217. Tucson: University of Arizona Press.

Fuller, D. Q., J. van Etten, K. Manning, C. Castillo, E. Kingwell-Banham, A. Weisskopf, A., L. Qin, Y-I. Sato and R.J. Hijmans, 2011. The contribution of rice agriculture and livestock pastoralism to prehistoric methane levels: An archaeological assessment. *The Holocene* 21: 743-759.

Furholt, M. 2012. Kundruci: Development of social space in a Late Neolithic tell-settlement in Central Bosnia, in R. Hofmann, F.-K. Moetz, and J. Müller (eds) *Tells: Social and Environmental Space. Proceedings of the International Workshop 'Socio-Environmental Dynamics over the Last 12,000 Years: The Creation of Landscapes II (14th-18th March 2011)' in Kiel, Vol. 3*: 203–219. Bonn: Habelt.

Furholt, M. 2013. Abseits des Weges – Prospektionen und Ausgrabungen in Kundruci, in J. Müller, K. Rassmann and R. Hofmann (eds) *Okolište I - Untersuchungen einer spätneolithischen Siedlungskammer in Zentralbosnien*: 173–206. Bonn: Habelt.

Furholt, M. 2017. *Das Ägäische Neolithikum und Chalkolithikum. Transformationen sozialer Handlungsmuster in Anatolien und Griechenland zwischen 6500 und 4000 v. Chr*, (Universitätsforschungen zur Prähistorischen Archäologie 304). Bonn: Habelt.

Furholt, M., J. Bátora, I. Cheben, H. Kroll, K. Rassmann and P. Tóth. 2014. Vráble – Veľké Lehemby: Eine Siedlungsgruppe der Linearkeramik in der Südwestslowakei. Vorbericht über die Untersuchungen der Jahre 2010 und 2012 und Deutungsansätze. *Slovenska Archaeologia* LXII(2): 227–266.

G

Gaastra, J.S., H.J. Greenfield and M.V. Linden. 2018. Gaining traction on cattle exploitation: zooarchaeological evidence from the Neolithic Western Balkans. *Antiquity* 92: 1462–1477. DOI: https://doi.org/10.15184/aqy.2018.178.

Gaastra, J.S., A. de Vareilles and M. Vander Linden. 2019. Bones and seeds: an integrated approach to understanding the spread of farming across the Western Balkans. *Environmental Archaeology*: 1–17. DOI: https://doi.org/10.1080/14614103.2019.1578016.

Gajić-Kvaščev, M., M.M.,Stojanović, Z. Šmit, V. Kantarelou, A.G. Karydas, D. Šljivar, D. Milovanović and V. Andrić. 2012. New evidence for the use of cinnabar as a colouring pigment in the Vinča culture. *Journal of Archaeological Science* 39: 1025–1033.

Gál, E. 2011. Prehistoric antler and bone tools from Kaposujlak-Vardomb (South-Western Hungary) with special regard to the Early Bronze Age implements, in J. Baron and B. Kufel-Diakowska (eds) *Written in Bones. Studies on Technological and Social Contexts of Past Faunal Skeletal Remains*: 137–164. Wrocław: Uniwersytet Wrocławski, Instytut Archeologii.

Gale, N.H., Z.A. Stos-Gale, P. Lilov, M. Dimitrov and T. Todorov. 1991. Recent studies of Eneolithic copper ores and artefacts in Bulgaria, in J.P. Mohen and C. Éluère (eds) *Découverte du Métal*: 49–76. Paris: Picard.

Gamble, C. 2007. *Origins and Revolutions: Human Identity in Earliest Prehistory*. Cambridge: Cambridge University Press.

Garašanin, D. 1961. Die Siedlung der Starčevokultur in Nosa bei Subotica und das Problem der neolitischen Lehmscheunen, in W. Dehn and G. Bersu (eds) *Bericht Uber Den V. Internationalen Kongress Fur Vor- und Fruhgeschichte Hamburg Vom 24. Bis 30. August*: 303–307. Berlin: Verlag Gebr. Mann.

Garašanin, M. 1951. Hronologija Vinčanske Grupe (Chronology of the Vinča Group, in Serbian). PhD dissertation, Univerza v Ljubljani.

Garašanin, M. 1954. Iz istorije mlađeg neolita u Srbiji i Bosni (History of the late Neolithic in Serbia and Bosnia). *Glasnik Zemaljskog Muzeja, Sarajevo* 9: 5–39.

Garašanin, M. 1973. *Praistorija na tlu Srbije (Prehistory in Serbia, in Serbian)*. Belgrade: Srpska Književna Zadruga.

Garašanin, M. 1979. Centralnobalkanska zona, in A. Benac (ed) *Praistorija jugoslavenskih zemalja* II: 144–212. Sarajevo: Svijetlost – Akademija nauka i umjetnosti Bosne i Hercegovine.

Garašanin, M. 1984. Vinča i vinčanska kultura u neolitu jugoistočne Evrope, in S. Ćelić (ed.) *Vinča u praistoriji i srednjem veku*: 42–57. Beograd: Galerija SANU.

Garašanin, M. 1990. Vinča und seine Stellung im Neolithikum Südosteuropa, in D. Srejović and N. Tasić (eds) *Vinča and Its World*: 11–16. Belgrade:

Serbian Academy of Sciences and Arts and Centre for Archaeological Research, Faculty of Philosophy.
Garašanin, M. 1991. Die Übergang vom Neolithikum zur frühen Bronzezeit auf dem Balkan und an der unteren Donau- ein Rückblick nach dreißig Jahren, in J. Lichardus (ed.) *Die Kupferzeit als Historische Epoche*: 205–216. Bonn: Dr. Rudolf Habelt Gmbh.
Garašanin, M. 1993. Zu den Problemen der Vinča-Gruppe in Rumänien. *Balcanica* 24: 7–20.
Garašanin, M. 1994/1995. Die Gradac-Stufe der Vinča-Gruppe und der Beginn des Aeneolithikums. *Dacia* 38/39: 9–17.
Garašanin, M. 1998 Vinčanska kultura i jadranski uticaji, in N. Tasić (ed.) *Arheološko blago Kosova i Metohije od neolita do ranog srednjeg veka*: 56–88. Beograd: Srpska akademija nauka i umetnosti.
Garašanin, M. and D. Garašanin. 1951. *Arheološki nalazi u Srbiji*. Beograd: Prosveta.
Garašanin, M. and D. Garašanin. 1958. Pavlovac kod Vranja – praistorijsko naselje. *Starinar* 7–8 (1956–1957).
Garašanin, M. and D. Garašanin. 1979. *Supska. 'Stublina' - praistorijsko naselje vinčanske grupe*. Beograd: Narodni muzej.
Gardner, E.J. 2003. Technical analysis of the ceramics / Appendix 7.1. Graphite-painted, in E.S. Elster and C. Renfrew (eds) *Excavations at Sitagroi, a Prehistoric Village in Northeast Greece: Vol. 2, The Final Report*: 283–298. Los Angeles: Cotsen Institute of Archaeology.
Garfinkel, Y., F. Klimscha, S. Shalev and D.Rosenberg. 2014. The beginning of metallurgy in the Southern Levant: a late 6th millennium Cal BC copper awl from Tel Tsaf, Israel. *PLoS ONE* 9: e92591. DOI: https://doi.org/10.1371/journal.pone.0092591.
Gatsov, I. 1993. *Neolithic Chipped Stone Industries in Western Bulgaria*. Krakow: Jagellonian University, Institute of Archaeology.
Gaul, J.H. 1948. The Neolithic period in Bulgaria. *Bulletin of the American Schools of Prehistoric Research* 16. Cambridge (MA).
Gauss, R. 2013. The development of metallurgy on the Iberian Peninsula: technological and social patterns of a long-term innovation process, in S. Burmeister, S. Hansen, M. Kunst and N. Müller-Scheeßel (eds) *Metal Matters. Innovative Technologies and Social Change in Prehistory and Antiquity*: 209–230. Rahden/ Westf.: Marie Leidorf.
Găvan, A., 2015. *Metal and Metalworking in the Bronze Age Tell Settlements from the Carpathian Basin*. Cluj-Napoca: Mega Publishing House.
Gaydarska, B. 2017. Introduction: European prehistory and urban studies. *Journal of World Prehistory* 30: 177–188.
Gaydarska, B., M. Nebbia and J. Chapman. 2020. Trypillia megasites in context: independent urban development in Chalcolithic Eastern Europe. *Cambridge Archaeological Journal* 30(1): 97–121.
Georgieva, P. 2012. *The Neolithic and Eneolithic in Southeast Europe*. Sofia.University Press: 'St. Kliment Ohridski'.
Gerritsen, F. 2008. Domestic times: houses and temporalities in late prehistoric Europe, in A. Jones (ed.) *Prehistoric Europe: Theory and Practice*: 143–161. Oxford: Blackwell.
Gillis, R.E., J. Bulatović, K. Penezić, M. Spasić, N.N. Tasić and C.A. Makarewicz. 2020. Herding and hunting at Vinča-Belo Brdo and Stubline during the Late Neolithic, a stable isotopic perspective, in N. Marković and J. Bulatović (eds) *Animal Husbandry and Hunting in the Central and Western Balkans Through Time*: 19–39. Oxford: Archaeopress.
Gillis, R.E., J.S. Gaastra, M. Vander Linden and J.-D. Vigne. 2019. A species specific investigation into sheep and goat husbandry during the early European Neolithic. *Environmental Archaeology* DOI: https://doi.org/10.1080/14614103.2019.1615214.
Gimbutas, M. 1973. Old Europe c. 7000–3500 BC: the earliest European civilization before the infiltration of the Indo-European peoples. *Journal of Indo-European Studies* 1: 1–21.
Gimbutas, M. 1974. *The Goddesses and Gods of Old Europe: 6500-3500 BC. Myths and Cult Images*. London: Thames and Hudson.
Gimbutas, M. (ed.) 1976a. *Neolithic Macedonia as Reflected by Excavations at Anza*. Los Angeles (CA): Insitute of Archaeology, University of California
Gimbutas, M. 1976b. Ideograms and symbolic design on ritual objects of Old Europe (Neolithic and Chalcolithic south-east Europe), in J.V.S. Megaw (ed.) *To Illustrate the Monuments. Essays on Archaeology Presented to Stuart Piggott*: 78–98. London: Thames and Hudson.
Gimbutas, M. 1977. Varna: a sensationally rich cemetery of the Karanovo civilisation, about 4500 BC. *Expedition* 19: 39–47.
Gimbutas, M. (ed.) 1980. *The Transformation of European and Anatolian Culture, 4500-2500 BC and Its Legacy. Proceedings of the Multidisciplinary International Conference in Dubrovnik, 1979* (Vol. 17, Vol. 3). Washington DC: Journal of Indo-European Studies.
Gimbutas, M. 1982. *The Goddesses and Gods of Old Europe: 6500-3500 BC. Myths and Cult Images* (Second edition). London: Thames and Hudson.
Gläser, R. 1996. Zur absoluten Datierung der Vinča-Kultur anhand von 14C-Daten, in F. Draşovean, (ed.) *The Vinča Culture, its Role and Cultural Connections*: 141–162. Timisoara: The Museum of Banat.
Glišić, J. and B. Jovanović. 1960. Eneolitsko naselje na Kormadinu kod Jakova. *Starinar* XI: 113–139.
Glumac, P. 1988. Copper mineral finds from Divostin, in D., Srejović and A. McPherron (eds) *Divostin and the Neolithic of Central Serbia*: 457–462. Pittsburgh, Kragujevac: University of Pittsburgh, Narodni muzej Kragujevac.
Glumac, P. 1991. The Advent of Metallurgy in the Vinča Culture. Unpublished PhD dissertation. University of California.

Glumac, P. and J. Todd. 1987. New evidence for the use of lead in prehistoric south-east Europe. *Archeomaterials* 2: 123–145.

Glumac, P. and J.A.Todd. 1991. Early metallurgy in southeast Europe: the evidence for production, in P. Glumac (ed.) *Recent Trends in Archaeometallurgical Research* (MASCA Research Papers in Science and Archaeology Vol. 8/1): 9–19. Philadelphia (PA): University of Pennsylvania.

Glumac, P. and R.E. Tringham. 1990. The exploitation of copper minerals, in R.E. Tringham and D. Krstić (eds) *Selevac, A Neolithic Village in Yugoslavia*: 549–563. Los Angeles (CA): University of California Press.

Godin, B. 2015. *Innovation Contested: The Idea of Innovation over the Centuries.* London: Routledge.

Golden, J.M. 2010. *Dawn of the Metal Age: Technology and Society During the Levantine Chalcolithic*: London, Oakville (CT): Equinox Publications.

Goleanu, A., A. Marian, M. Gligor, C. Florescu and S. Varvara, S. 2005. Chemical and structural features of the Neolithic ceramics from Vinţa, Lumea Noua and Petresti cultures (Roumania). *Revue Roumaine De Chimie* 50: 939–949.

Goossens, R., A. De Wulf, J. Burgeois, W. Gheyle and T. Willems. 2006. Satellite imagery and archaeology: the example of CORONA in the Altai Mountains. *Journal of Archaeological Science* 33: 745–755.

Gopher, A. and T. Tsuk. 1996. *The Nahal Qanah Cave: Earliest Gold in the Southern Levant.* Tel Aviv: Tel Aviv University.

Gorelick, L. and A.J. Gwinnett. 1990. Innovative lapidary craft techniques in Neolithic Jarmo. *Archeomaterials* 4: 25–32.

Gori, M. and M. Ivanova (eds). 2017. *Balkan Dialogues: Negotiating Identity Between Prehistory and the Present.* New York: Routledge.

Görsdorf, J. and J. Bojadžiev. 1996. Zur absoluten Chronologie der bulgarischen Urgeschichte. *Eurasia Antiqua* 2: 105–173.

Gosselain, O.P. and D. Livingstone Smith. 2005. The source: clay selection and processing practices in Sub-Saharan Africa, in A. Livingstone Smith, D. Bosquet and R. Martineau (eds) *Pottery Manufacturing Process: Reconstruction and Interpretation* (British Archaeological Reports International Series 1349): 33–47. Oxford: Archaeopress.

Govedarica, B. 2001. Zur Typologie und Chronologie der Hammeräxte vom Typ Pločnik, in R.M. Boehmer and J. Maran (eds) *Lux Orientis: Archäologie zwischen Asien und Europa*: 153–164. Rahden: Leidorf.

Govedarica, B. 2016. Das Phänomen der balkanischen Kupferzeit, in V. Nikolov and W. Schier (eds) *Der Schwarzmeerraum vom Neolithikum bis in die Früheisenzeit (6000-600 v. Chr.). Kulturelle Interferenzen in der zirkumpontischen Zone und Kontakte mit ihren Nachbargebieten.* (Prähistorische Archäologie in Südosteuropa 30): 11–22. Rahden/Westf: Verlag Marie Leidorf GmbH.

Grabundžija, A., C. Schoch and A. Ulanowska. 2016. Bones for the loom. Weaving experiment with astragali weights. *Prilozi Instituta za arheologiju u Zagrebu* 33: 287–306.

Graeber, D. and D. Wengrow. 2021. *The Dawn of Everything. A New History of Humanity.* New York: Farrar, Straus and Giroux.

Graf, M. 2013. Industrijski putevi kulture (Industrial culture pathways, in Serbian) *Energija, ekonomija, ekologija,* 15: 328–335.

Grant, A. 1982. The use of tooth wear as a guide to the age of domestic ungulates, in B. Wilson, C. Grigson and S. Payne (eds) *Ageing and Sexing Animal Bones from Archaeological Sites* (British Archaeological Reports British Series 109): 91–108. Oxford: Archaeopress.

Gratuze, B. 1999. Obsidian characterization by laser ablation ICP-MS and its application to prehistoric trade in the Mediterranean and the Near East: sources and distribution of obsidian within the Aegean and Anatolia. *Journal of Archaeological Science* 26(8): 869–881. DOI: https://doi.org/10.1006/jasc.1999.0459.

Grbić, M. 1929. *Pločnik, eine Prähistorische Ansiedlung aus der Kupferzeit.* Belgrade: National Museum Belgrade.

Grbić, M. 1934. Neolitsko naselje u Botošu kod Velikog Bečkereka. *Starinar* 8–9: 40–58.

Грбић, М. 1936. Архитектура у Басијани (Сремски Петровци) I. Реконструкција места према ситуационом плану и авионском снимку. *Гласник Историског друштва* IX/1: 19–31. Нови Сад.

Грбић, М. 1950. Градиште код Кикинде. *Старинар* I: 113–118. Београд.

Грбић, М. 1951. Градиште код Кикинде.Наставак радова у 1948 години. *Старинар* II: 133–138. Београд.

Green, F.J. 1982. Problems of interpreting differentially preserved plant remains from excavations of medieval urban sites, in A.R. Hall and H.K. Kenward (eds) *Environmental Archaeology in the Urban Context* (Council for British Archaeology Research Report 23): 40–46. London: Council for British Archaeology.

Greenacre, M. 1988. Clustering the rows and columns of a contingency table. *Journal of Classification* 5: 39–51.

Greenacre M. 2007. *Correspondence Analysis in Practice.* Boca Raton-London-New York: Chapman and Hall/CRC.

Greenfield, H.J. 1986. *The Paleoeconomy of the Central Balkans (Serbia). A Zooarchaeological Perspective on the Late Neolithic and Bronze Age (ca. 4500-1000 B.C.)* (British Archaeological Reports International Series 304 (i, ii)). Oxford: Archaeopress.

Greenfield, H.J. 1994. Faunal remains from the Early Neolithic Starčevo settlement at Bukovačka Česma. *Starinar* 43–44: 103–113.

Greenfield, H.J. 2005. A reconsideration of the Secondary Products Revolution in south eastern Europe: on the origins and use of domestic animals for milk, wool and traction in the central Balkans, in J. Mulville and A. Outram (eds) *The Zooarchaeology of Fats, Oils, Milk and Dairying*: 14–31. Oxford: Oxford Books.

Greenfield, H.J. 2014. Zooarchaeology of the Late Neolithic, Eneolithic and Middle Bronze Ages at Vinča-Belo Brdo, Serbia: the 1982 excavations, in H.J. Greenfield (ed.) *Animal Secondary Products: Domestic Animal Exploitation in Prehistoric Europe, the Near East and the Far East*: 274–334. Oxford: Oxbow.

Greenfield, H.J. 2017. Hunting and herding in the Middle Neolithic of central Serbia: a zooarchaeological analysis of Stragari-Šljivik, Serbia, in P. Rowley-Conwy, D. Serjeantson and P. Halstead (eds) *Economic Zooarchaeology: Studies in Hunting, Herding and Early Agriculture:* 183–194. Oxford: Oxbow.

Greenfield, H.J. and E. Arnold. 2014. 'Crying over spilt milk': an evaluation of recent models, methods, and techniques on the origins of milking during the Neolithic of the Old World, in H.J. Greenfield (ed.) *Animal Secondary Products: Domestic Animal Exploitation in Prehistoric Europe, the Near East and the Far East*: 130–187. Oxford: Oxbow.

Greenfield, H.J. and E. Arnold. 2015. 'Go(a)t milk?' New perspectives on the zooarchaeological evidence for the earliest intensification of dairying in south eastern Europe. *World Archaeology* 47 (5): 792–818.

Greenfield, H.J. and T. Greenfield. 2014. Subsistence and settlement in the Early Neolithic of temperate SE Europe: a view from Blagotin, Serbia. *Archaeologia Bulgarica* 18(1): 1–33.

Greenfield, H.J. and T. Jongsma. 2008. Sedentary pastoral gatherers in the early Neolithic: architectural, botanical, and zoological evidence for mobile economies from Foeni-Salaş, south-west Romania, in D. Bailey, A. Whittle and D. Hofmann (eds) *Living Well Together? Settlement and Materiality in the Neolithic of South-East and Central Europe*: 108–130. Oxford: Oxbow.

Greenshaw, S.I. 2018. Structure from motion: origins and originality. *The Photogrammetric Record* 33: 7–10.

Grigson, C. 1982. Sex and age determination of some bones and teeth of domestic cattle: a review of the literature, in B. Wilson, C. Grigson and S. Payne (eds) *Ageing and Sexing Animal Bones from Archaeological Sites* (British Archaeological Reports British Series 109): 7–23. Oxford: Archaeopress.

Guerrero, E., S. Naji and J.-P. Bocquet-Appel. 2008. The signal of the Neolithic demographic transition in the Levant, in J.-P. Bocquet-Appel and O. Bar-Yosef (eds) *The Neolithic Demographic Transition and its Consequences*: 57–80. Berlin: Springer.

Gurova, M. 2012. 'Balkan flint'—fiction and/or trajectory to Neolithization: evidence from Bulgaria. *Bulgarian e-journal of Archaeology.*

Gurova, M. 2014. Neolithic flint assemblages from Bulgaria: an overview. *Самарский научный вестник* 3(8): 94–108.

Guthrie, D.R. 1983. Osseus projectile points: biological considerations affecting raw material selection and design among Paleolithic and Paleoindian peoples, in J. Clutton-Brock and C. Grigson (eds) *Animals and Archaeology 1: Hunters and Their Prey*: 273–294. Oxford: Archaeopress.

H

Hahn, H.P. 2012. Segmentary societies as alternatives to hierarchical order: sustainable social structures or organization of predatory violence? In T. Kienlin and A. Zimmermann (eds) *Beyond Elites: Alternatives to Hierarchical Systems in Modelling Social Formations: International Conference at the Ruhr-Universität Bochum, Germany, October 22-24, 2009.* (University Research on Prehistoric Archaeology 215): 33–40. Bonn: Habelt.

Hair J, F., W.C. Black, B.J. Babin and R.E. Anderson. 2009. *Multivariate Data Analysis* (Seventh edition). London: Prentice Hall.

Halstead, P. 1996. Pastoralism or household-herding? Problems of scale and specialization in early Greek animal husbandry. *World Archaeology* 28: 20–42.

Halstead, P. 1985. A study of mandibular teeth from Romano-British contexts at Maxey, in F. Pryor and C. French (eds) *The Fenland Project No. 1: Archaeology and Environment in the Lower Welland Valley* (East Anglian Archaeology 27): 219–24. Cambridge: Cambridgeshire Archaeological Committee.

Halstead, P., P. Collins and V. Isaakidou. 2002. Sorting sheep from goats: morphological distinctions between the mandibles and mandibular teeth of adult *Ovis* and *Capra*. *Journal of Archaeological Science* 29: 545–553.

Hambleton, E. 1999. *Animal Husbandry Regimes in Iron Age Britain.* (British Archaeological Reports British Series 282). Oxford: Archaeopress.

Hampton, O.W. 1999. *Culture of Stone: Sacred and Profane Uses of Stone Among the Dani.* College Station (TX): Texas A and M University Press.

Hansen, S. 2007. Bilder vom Menschen der Steinzeit. Untersuchungen zur anthropomorphen Plastik der Jungsteinzeit und Kupferzeit in Südosteuropa. *Archäologie in Eurasien* 20. Mainz: Deutsches Archäologisches Institut.

Hansen, S. 2009. Kupfer, Gold und Silber im Schwarzmeerraum während des 5. und 4. Jahrtausends v. Chr, in J. Apakidze, B. Govedarica and B. Hänsel (eds) *Der Schwarzmeerraum vom Äneolithikum bis in die Früheisenzeit (5000-500 v.Chr.). Kommunikationsebenen zwischen Kaukasus und Karpaten. Internationale Fachtagung von Humboldtianern für Humboldtianer im Humboldt-Kolleg in Tiflis - Georgien 17.- 20. Mai 2007.* (Prähistorische Archäologie in Südosteuropa, Band 25): 11–50. Rahden: Leidorf.

Hansen, S. 2012. The archaeology of power, in T.L. Kienlin and A. Zimmermann (eds) *Beyond Elites. Alternatives to Hierarchical Systems in Modelling Social Formations. International Conference at the Ruhr-Universität Bochum, Germany, October 22-24, 2009* (Universitätsforschungen zur Prähistorischen Archäologie 215): 213–223. Bonn: Habelt.

Hansen, S. 2013a. The birth of the hero. The emergence of a social type in the 4th millennium BC, in E. Starnini (ed.) *Unconformist Archaeology. Papers in Honour of Paolo Biagi* (British Archaeological Reports International Series 2528): 101–112. Oxford: Archaeopress.

Hansen, S. 2013b. Innovative metals: copper, gold and silver in the Black Sea region and the Carpathian Basin during the 5th and 4th Millennium BC, in S. Burmeister, S. Hansen, M. Kunst, and N. Müller-Scheeßel (eds) *Metal Matters. Innovative Technologies and Social Change in Prehistory and Antiquity*: 137–170. Rahden: Leidorf.

Hansen, S. and M. Toderaş 2012. The Copper Age settlement Pietrele on the Lower Danube River (Romania), in R. Hofmann, F.-K. Moetz and J. Müller (eds) *Tells: Social and Environmental Space. Proceedings of the International Workshop 'Socio-Environmental Dynamics over the Last 12,000 Years: The Creation of Landscapes II' in Kiel, 14th -18th March 2011. Volume 3*: 127–138. Bonn: Habelt.

Hansen, S., G. Mirtskhulava, K. Bastert-Lamprichs and M. Ullrich. 2012. Aruchlo- Eine Neolithische Siedlung im Sudkaukasus. *Altertum* 57: 81–106.

Hansen, S., P. Raczky, A. Anders and A. Reingruber (eds). 2015. *Neolithic and Copper Age between the Carpathians and the Aegean Sea: Chronologies and Technologies from the 6th to the 4th Millennium BCE. International Workshop Budapest, 30 March-1 April 2012.* Bonn: Habelt.

Hansen, S., I. Montero-Ruiz, S. Rovira, D. Steiniger and M. Toderaş. 2019. The earliest lead ore processing in Europe. 5th Millennium BC finds from Pietrele on the Lower Danube. *PLoS ONE* 14: e0214218. DOI: https://doi.org/10.1371/journal.pone.0214218.

Harris, E.C. 1989. *Principles of Archaeological Stratigraphy.* London: Academic Press.

Harris, E.C., M.R. Brown and G.J. Brown (eds). 1993. *Practices of Archaeological Stratigraphy.* London: Academic Press.

Hauptmann, A. 2000. *Zur frühen Metallurgie des Kupfers in Fenan/Jordanien, Der Anschnitt, Beiheft 11.* Bochum: Deutsches Bergbau-Museum.

Hauptmann, A. 2007. *The Archaeometallurgy of Copper: Evidence from Faynan, Jordan.* Berlin: Springer.

Hauptmann, A. 2014. The investigation of archaeometallurgical slag, in B.W. Roberts and C.P. Thornton (eds) *Archaeometallurgy in Global Perspective: Methods and Syntheses*: 91–105. New York: Springer.

Hauptmann, A. 2020. *Archaeometallurgy - Materials Science Aspects.* New York: Springer.

Hauptmann, A., J. Lutz, E. Pernicka and Ü.Yalçin. 1993. Zur Technologie der frühesten Kupferverhüttung im östlichen Mittelmeerraum, in M. Frangipane, A. Hauptmann, M. Liverani, P. Matthiae and M. Mellink (eds) *Between the Rivers and over the Mountains: Archaeologica Anatolica et Mesopotamica Alba Palmieri Dedicata*: 541–563. Roma: Dipartimento di Scienze Storiche Archeologiche e Antropologiche dell'Antichità, Università di Roma 'La Sapienza'.

Hauptmann, A., C. Bendall, G. Brey, I. Japariže, I. Gambašiže, S. Klein, M. Prange and T. Stöllner. 2010. Gold in Georgien. Analytische Untersuchungen an Goldartefakten und an Naturgold aus dem Kaukasus und dem Transkaukasus, in S., Hansen, A. Hauptmann, I. Motzenbäcker and E. Pernicka (eds.) *Von Majkop bis Trialeti. Gewinnungund Verbreitungvon Metallen und Obsidian in Kaukasien im 4.-2. Jt. v. Chr. Beiträge des Internationalen Symposiums in Berlin vom 1.-3. Juni 2006. Kolloquien zur Vor- und Frühgeschichte 13*: 139–160. Bonn: Habelt.

Healey J.F. 2013. *The Essentials of Statistics. A Tool for Social Research* (Third edition). Belmont (CA): Wadsworth.

Heeb, J. 2014. *Copper Shaft-Hole Axes and Early Metallurgy in South-Eastern Europe: An Integrated Approach.* Oxford: Archaeopress.

Heeb J., and B. Ottaway 2014. Experimental Archaeometallurgy, in B.W. Roberts and C.P. Thornton (eds) *Archaeometallurgy in Global Perspective: Methods and Syntheses*: 161-192. New York: Springer.

Helmer, D. 2000. Discrimination des genres *Ovis* at *Capra* à l'aide des prémolaires inférieures 3 et 4 et interprétation des âges d'abattage: l'example de Dikili Tash (Greece). *Anthropozoologica* 31: 29–38.

Helwing, B. 2013. Early metallurgy in Iran: An innovative region as seen from the inside, in S. Burmeister, S. Hansen, M. Kunst, and N. Muller-Scheessel (eds.) *Metal Matters: Innovative Technologies and Social Change in Prehistory and Antiquity*: 105–135. Rahden.: Leidorf.

Henrich, J. 2001. Cultural transmission and the diffusion of innovations: adoption dynamics indicate that biased cultural transmission is the predominate force in behavioral change. *American Anthropologist.* 103: 992–1013.

Henrich, J. 2004. Demography and cultural evolution: how adaptive cultural processes can produce maladaptive losses: the Tasmanian case. *American Antiquity* 69: 197–214.

Henrich, J. 2010. The evolution of innovation-enhancing institutions, in M.J. O'Brien and S.J. Shennan (eds) *Innovation in Cultural Systems. Contributions from Evolutionary Anthropology*: 99–120. Cambridge, MA: The MIT Press.

Hervella, M., M. Rotea, N. Izagirre, M. Constantinescu, S. Alonso, M. Ioana, C. Lazăr, F. Ridiche, A.D. Soficaru, M.G. Netea and C. de-la-Rua. 2015. Ancient DNA from South-East Europe reveals different events during Early and Middle Neolithic influencing the European genetic heritage. *Plos ONE* 10(6): e0128810.

Herzog, I. 2012. Aus Punkten werden Flächen: Vorschläge einer Methodik zur Abgrenzung von Gebieten mit hoher Funddichte. *Bericht der Römisch-Germanischen Kommission* 91: 197–215.

Higham, T., J. Chapman, B. Gaydarska, V. Slavchev, N. Honch, Y. Yordanov and B. Dimitrova. 2007. New perspectives on the Varna cemetery (Bulgaria) - AMS dates and social implications. *Antiquity* 81: 640–654.

Higham, T., V. Slavchev, B. Gaydarska and J. Chapman. 2018. AMS dating of the Late Copper Age Varna Cemetery, Bulgaria. *Radiocarbon* 60: 493–516. DOI: https://doi.org/10.1017/RDC.2018.9.

Hillier, B. and J. Hanson. 1984. *The Social Logic of Space.* Cambridge: Cambridge University Press.

Hillman, G.C. 1981. Reconstructing crop husbandry practices from the charred remains of crops, in R. Mercer (ed) *Farming Practice in British Prehistory*: 123–162. Edinburgh: Edinburgh University Press.

Hillman, G.C. 1984a. Interpretation of archaeological plant remains: application of ethnographic models from Turkey, in W.A. Casparie and W. van Zeist (eds) *Plants and Ancient Man: Studies in Palaeoethnobotany*: 1–41. Rotterdam: Balkema.

Hillman G.C. 1984b. Traditional husbandry and processing of archaic cereals in recent times: the operations, products and equipment which might feature in Sumerian Texts. Part I: the glume-wheats. *Bulletin on Sumerian Agriculture* 1: 114–52.

Hillman, G.C. 1985. Traditional husbandry and processing of archaic cereals in modern times. Part II, the free-threshing cereals. *Bulletin on Sumerian Agriculture* 2: 1–31.

Hofmann, R. 2013a. Okolište – Spätneolithische Keramik und Siedlungsentwicklung in Zentralbosnien: Ergebnisse und Interpretationen, *Godišnjak Centra za balkanološka ispitivanja Akademije nauka i umjetnosti Bosne i Hercegovine* 42: 23–44.

Hofmann, R. 2013b. *Okolište 2 - Spätneolithische Keramik und Siedlungsentwicklung in Zentralbosnien. Neolithikum und Chalkolithikum in Zentralnosnien 2.* (Universitätsforschungen zur prähistorischen Archäologie 234). Bonn: Verlag Rudolf Habelt.

Hofmann, R. and N. Müller-Scheeßel. 2013. Der spätneolithische Tell von Okolište im Visoko-Becken in Zentralbosnien: Ergebnisse der Grabungskampagne 2006 und 2007, in J. Müller, K. Rassmann and R. Hofmann (eds) *Okolište I - Untersuchungen einer spätneolithischen Siedlungskammer in Zentralbosnien*: 69–102. Bonn: Habelt.

Hofmann, R., and N. Müller-Scheeßel. 2020. Orientation of Neolithic dwellings in Central and Southeast Europe: Common denominator between the Vinča and Linearbandkeramik worlds. *Quaternary International*: 1–12. DOI: https://doi.org/10.1016/j.quaint.2020.05.006.

Hofmann, R., Z. Kujundzic-Vejzagić, J. Müller, N. Müller-Scheeßel and K. Rassmann. 2007. Prospektionen und Ausgrabungen in Okolište (Bosnien-Herzegowina). Siedlungsarchäologische Studien zum zentralbosnischen Spätneolithikum. *Berichte der Römisch-Germanischen Kommission* 87: 43–212.

Hofmann, R., Z. Kujundžić-Vejzagić, J. Müller, K. Rassmann and N. Müller-Scheessel. 2009. Rekonstrukcija procesa naseljavanja u kasnom neolitu na prostoru centralne Bosne. *Glasnik Zemaljskog Muzeja* 50/51: 11–177.

Holmberg, E.J. 1964. The appearance of Neolithic black burnished ware in mainland Greece. *American Journal of Archaeology* 68: 343–348.

Holste, F. 1939. Zur chronologischen Stellung der Vinča-Keramik, *Wiener prahistorische Zeitschrift XXVI*. Wien.

Hopf, M. 1974. Pflanzenreste aus Siedlungen der Vinča-Kultur in Jugoslawien. *Jahrbuch des Römisch-Germanisches Zentralmuseum* 21: 1–11.

Höppner, B., M. Bartelheim, M. Huijsmans, R. Krauss, K.P. Martinek, E. Pernicka and R. Schwab. 2005. Prehistoric copper production in the Inn Valley (Austria), and the earliest copper in central Europe. *Archaeometry* 47: 293–315.

Horedt, K. 1976. Die ältesten neolithischen Kupferfunde Rumäniens. *Jahresschrift für Mitteldeutsche Vorgeschichte* 60: 175–181.

Hosler, D. 2014. Mesoamerican metallurgy: the perspective from the west, in B.W. Roberts and C.P. Thornton (eds) *Archaeometallurgy in Global Perspective: Methods and Syntheses*: 329-359. New York: Springer.

Huebner, J. 1969. Stability relations of rodochrosite in the system manganese-carbon-oxygen. *American Mineralogist* 54: 457–481.

Humphris, J., M.F. Charlton, J. Keen, L. Sauder and F.Alshishani. 2018. Iron smelting in Sudan: Experimental archaeology at the royal city of Meroe. *Journal of Field Archaeology* 43: 399–416. DOI: https://doi.org/10.1080/00934690.2018.1479085.

Hüser, A. 2005. *Die Knochen- und Geweihartefakte der lieanrbandkeramischen Siedlung Bad Nnauheim-Nieder-Mörlen in der Wetterau.* (Kleine Schriften 55). Marburg: Phillips-Universität Marburg.

I

Ifantidis, F. and M. Nikolaidou (eds). 2011. *Spondylus in Prehistory: New Data and Approaches. Contributions to the Archaeology of Shell Technologies* (British Archaeological Reports International Series 2216). Oxford: Archaeopress.

Игњатовић, М. 2008. Каталог [Katalog], in Дубравка Николић [N. Dubravka] *Винча - праисторијска метропола. Истраживања 1908-2008. [Vinča, praistorijska nekropola. Istraživanja 1908-2008]*: 203–277. Beograd: Filozofski fakultet, Narodni muzej, Muzej grada Beograda, SANU.

Iles, L. 2018. Forging networks and mixing ores: Rethinking the social landscapes of iron metallurgy. *Journal of Anthropological Archaeology* 49: 88–99. DOI: https://doi.org/10.1016/j.jaa.2017.10.001.

Ingold, T. 2007. Materials against materiality. *Archaeological Dialogues* 14: 1–16.

Inizan, M.-L., M. Reduron-Ballinger, H. Roche and J. Féblot-Augustins. 1999. *Technology and Terminology of Knapped Stone.* Nanterre: Cercle de Recherches et d'Etudes Préhistoriques.

Inizan, M.-L., M. Reduron-Ballinger, H. Roche and J. Tixier. 1995. *Technologie de la Pierre Taillée.* Paris: CNRS.

Isaakidou, V. 2006. Ploughing with cows: Knossos and the 'secondary products revolution', in D. Serjeantson and D. Field (eds) *Animals in the Neolithic of Britain and Europe*: 95–112. Oxford: Oxbow Books.

Иванишевић, В. and И. Бугарски. 2015. Дигитална фотограметријска снимања античких и средњевековних локалитета у Србији. *Саопштења* XLVII: 165–180.

Ivanišević, V. and I. Bugarski. 2015. Landscape reconstruction in the Middle Danube Roman Limes: case studies from Lederata and Smorna, in V. Ivanišević, T. Veljanovski, D. Cowely, G. Kiarszys and I. Bugarski (eds) *Recovering Lost Landscapes* (Monographs, Volume 58): 57–65. Belgrade: Institute of Archaeology.

Ivanišević V., I. Bugarski and A. Stameković. 2018. Примена историјско – археолошког географског информационог система у истраживањима Царичиног града и околине. *Саопштења* L: 259–279. Beograd.

Ivanišević, V., T. Veljanovski, D. Cowely, G. Kiarszys and I. Bugarski. 2015. Recovering lost landscapes – introduction to an aerial perspective, in V. Ivanišević, T. Veljanovski, D. Cowely, G. Kiarszys and I. Bugarski (eds) *Recovering Lost Landscapes* (Monographs, Volume 58): 7–10. Belgrade: Institute of Archaeology.

Ivanov, I. 1978a. Rannokhalkolitni grobove do grad Varna. *Izvestiya na Narodniya Muzej Varna* 14: 81–93.

Ivanov, I. 1978b. Les fouilles archéologiques de la nécropole chalcolithique à Varna. *Studia Praehistorica* 1/2: 1–26.

Ivanov, I. 1988a. Das Gräberfelds von Varna: Katalog, in A. Fol and J. Lichardus (eds) *Macht, Herrschaft und Gold: Das Gräberfeld von Varna und die Anfänge einer Neuen Europäischen Zivilisation*: 183–208. Saarbrücken: Moderne Galerie des Saarland-Museums.

Ivanov, I. 1988b. Die Ausgrabungen des Gräberfeldes von Varna, in A. Fol and J. Lichardus (eds) *Macht, Herrschaft und Gold: Das Gräberfeld von Varna und die Anfänge einer Neuen Europäischen Zivilisation*: 49–66. Saarbrücken: Moderne Galerie des Saarland-Museums.

Ivanov, I. and M. Avramova. 2000. *Varna Necropolis. The Dawn of European Civilization. Treasures of Bulgaria 1.* Sofia: Agato.

Ivanova, M. 2007. Tells, invasion theories and warfare in fifth millennium BC north-eastern Bulgaria, in T. Pollard and I. Banks (eds) *War and Sacrifice: Studies in the Archaeology of Conflict*: 33–48. Leiden: Brill.

Ivanova, M. 2012. Perilous waters: Early maritime trade along the western coast of the Black Sea (fifth millennium BC). *Oxford Journal of Archaeology* 31: 339–365. DOI: https://doi.org/10.1111/j.1468-0092.2012.00392.x.

Ivanova, M. 2013. *The Black Sea and the Early Civilizations of Europe, the Near East and Asia.* Cambridge: Cambridge University Press.

Ivanova, M. 2020. Growing societies: An ecological perspective on the spread of crop cultivation and animal herding in Europe, in K. Gron, L. Sørensen and P. Rowley-Conwy P. (eds) *Farmers at the Frontier: A Pan European Perspective on Neolithisation*: 7–44. Oxford: Oxbow.

Ivanova, M., B. Athanassov, V. Petrova, D. Takorova and P.W. Stockhammer (eds). 2018. *Social Dimensions of Food in the Prehistoric Balkans.* Oxford: Oxbow.

J

Jacanović, D. 1985. *Požarevački kraj od najstarijih zemljoradnika do dolaska Rimljana.* Požarevac: Narodni Muzej.

Jacanović, D. 1988. Neolithic sites in the Danubian region from the mouth of the Velika Morava to Golubac, in D. Srejović (ed.) *The Neolithic of Serbia*: 111–119. Belgrade: University of Belgrade.

Jacanović, D. and D. Šljivar. 2001. Nalazi astragala sa neolitskog lokaliteta Belovode. *Zbornik Narodnog muzeja* XVII-1: 31–36.

Jacanović, D. and D. Šljivar. 2003. Preliminary reports on archaeological research from 1998 to 2002, Veliko Laole, Belovode (in Serbian). *Viminacium* 13–14: 297–302.

Jackson, C.M. and J.W. Smedley. 2004. Medieval and post-medieval glass technology: melting characteristics of some glasses melted from vegetable ash and sand mixtures. *Glass Technology* 45: 36–42.

Jacomet, S. and A. Kreuz. 1999. *Archäobotanik.* Stuttgart: Eugen Ulmer.

Jakucs J. and V. Voicsek. 2015. The northernmost distribution of the early Vinča culture in the Danube valley: a preliminary study from Szederkény-Kukorica-dűlő (Baranya County, southern Hungary). *Antaeus 33*: 13–54.

Jakucs J., E. Bánffy, K. Oross, V. Voicsek, C. Bronk Ramsey, E. Dunbar B. Kromer A. Bayliss, D. Hofmann, P.

Marshall and A. Whittle. 2016. Between the Vinča and Linearband-keramik worlds: the diversity of practices and identities in the 54th–53rd centuries cal BC in south-west Hungary and beyond. *Journal of World Prehistory* 29: 267–336.

Janković, S. 1967. *Ležišta Metaličnih Mineralnih Sirovina.* Belgrade: Mining and Geology Faculty, University of Belgrade.

Janković, S. 1977. The copper deposits and geotectonic setting of the Thethyan Eurasian Metallogenic Belt. *Mineralium Deposita* 12: 37–47. DOI: https://doi.org/10.1007/BF00204503.

Janković, S. 1982. Yugoslavia, in F.W. Dunning, W. Mykura and D. Slater (eds) *Mineral Deposits of Europe. Volume 2: Southeast Europe*: 143–202. London: The Institute of Mining and Metallurgy and the Mineralogical Society.

Jelenković, R. 1999. *Ležišta metaličnih mineralnih sirovina (Metallic Mineral Deposits, in Serbian).* Belgrade: Mining and Geology Faculty, University of Belgrade.

Jevtić, M. 1986. Grobovi bakarnog doba iz Vinče. *Starinar* 37: 135–144.

Jochum, K.P., U. Weis, B. Stoll, D. Kuzmin, Q. Yang, I, Raczek, D.E. Jacob, A. Stracke, K. Birbaum, D.A. Frick, D. Günther and J. Enzweiler. 2011. Determination of reference values for NIST SRM 610-617 glasses following ISO Guidelines. *Geostandards and Geoanalytical Research* 35(4): 397–429. DOI: http://doi.wiley.com/10.1111/j.1751 908X.2011.00120.x

Jockenhövel, A. 2018. Alteuropäische Gräber der Kupferzeit, Bronzezeit und älteren Eisenzeit mit Beigaben aus dem Gießereiwesen (Gießformen, Düsen, Tiegel), in M. Overbeck (ed.) *Die Gießformen in West- und Süddeutschland (Saarland, Rheinland-Pfalz, Hessen, Baden-Württemberg, Bayern*: 213–317. Stuttgart: Franz Steiner.

Johnson, G.A. 1982. Organisational structure and scalar stress, in C. Renfrew, M.J. Rowlands and B. Segraves (eds) *Theory and Explanation in Archaeology*: 389–421. New York: Academic Press.

Jones, A. 2004. Archaeometry and materiality: materials-based analysis in theory and practice. *Archaeometry* 46: 327–338.

Jones, G. 1984. Interpretation of archaeological plant remains: ethnographic models from Greece, in W.A. Casparie and W. van Zeist (eds) *Plants and Ancient Man: Studies in Palaeoethnobotany*: 43–61. Rotterdam: Balkema.

Jones, G. 1987. A statistical approach to the archaeological identification of crop processing. *Journal of Archaeological Science* 14: 311–323.

Jones, G. 1990. The application of present-day cereal processing studies to charred archaeobotanical remains. *Circaea* 6(2): 91–96.

Jones, G. 1992. Weed phytosociology and crop husbandry: identifying a contrast between ancient and modern practice. *Review of Palaeobotany and Palynology* 73: 133–143.

Jones, G. and P. Halstead. 1995. Maslins, mixtures and monocrops: on the interpretation of archaeological crop samples of heterogenous composition. *Journal of Archaeological Science* 22: 103–114.

Jones, G., A. Bogaard, P. Halstead, M. Charles and H. Smith. 1999. Identifying the intensity of crop husbandry on the basis of weed floras. *Annual of the British School at Athens* 94: 167–89.

Jones, G., S. Valamoti and M. Charles. 2000. Early crop diversity: a 'new' glume wheat from northern Greece. *Vegetation History and Archaeobotany* 9: 133–146.

Jones, R.E. 1986. *Greek and Cypriot Pottery: A Review of Scientific Studies.* Athens: British School at Athens.

Jongsma, T. and H. Greenfield. 1996. The vertebrate fauna from middle and late Neolithic Sânandrei, in F. Draşovean (ed.) *The Vinča Culture: Its Role and Cultural Connection*: 295–308. Timişoara: Museum of Banat.

Jovanović, B. 1962. Postanak keramičkih tipova balkansko-podunavskog neolita i eneolita. *Starinar* 13–14: 9–25.

Jovanović, B. 1965. Starija vinčanska grupa u južnom Banatu. Rad vojvođanskih muzeja 14: 15–41.

Jovanović, B. 1971. *Metallurgy in the Eneolithic Period in Yugoslavia.* Belgrade: Archaeological Institute.

Jovanović, B. 1978. The oldest copper metallurgy in the Balkans. *Expedition* 21: 9–17.

Jovanović, B. 1979. Rudarstvo i metalurgija eneolitskog perioda Jugoslavije, in A. Benac (ed.) *Praistorija Jugoslavinskih Zemalja* 3: 27–85. Sarajevo: Svetlost.

Jovanović, B. 1980. The origins of copper mining in Europe. *Scientific American* 242: 152–167.

Jovanović, B. 1982. *Rudna Glava, najstarije rudarstvo bakra na centralnom Balkanu.* Belgrade: Arheološki Institut, Muzej rudarstva i metalurgije.

Jovanović, B. 1983. Mali Šturac, ein neues prähistorisches Kupferbergwerk in Zentralserbien. *Der Anschnitt* 4/5: 177–179.

Jovanović, B. 1993. Archaeological comment to E. Pernicka's et al. 'Eneolithic and Early Bronze Age copper artefacts from the Balkans and their relation to Serbian copper ores'. *Prähistorische Zeitschrift* 68: 55–57.

Jovanović, B. 1994. Gradac Phase in the relative chronology of the late Vinča culture. *Starinar (new series)* 43/44: 1–12.

Jovanović, B. (ed.) 1995. *Ancient Mining and Metallurgy in Southeast Europe.* Donji Milanovac: Archaeological Institute Belgrade and Museum of Mining and Metallurgy Bor.

Jovanović, B. 2006. Gradac Phase of the Vinča culture: origin of a typological innovation, in N. Tasić and C. Grozdanov (eds) *Homage to Milutin Garašanin*: 221–233. Belgrade: Serbian Academy of Sciences and Arts and Macedonian Academy of Sciences and Arts.

Jovanović, B. and B.S. Ottaway. 1976. Copper mining and metallurgy in the Vinča group. *Antiquity* 50: 104–113.

Jovanović, D. 2001. *Zlato i Bakar Istočne Srbije*, Belgrade: Srpsko društvo za istoriju nauke, Arheološki Institut Beograd.

Jovanović, M. 2011. *Gospodari gline i žita*. Novi Sad: Muzej Vojvodine.

Jovanović, S., M. Savić, R. Trailović, Z. Janković and D. Šljivar. 2003. Evaluations of the domestication process in Serbia: palaeozoological remnants at Neolithic settlement of Belovode. *Acta Veterinaria* 53: 427–434.

Jovanović, S., M. Savić, R. Trailović, Ž. Janković and D. Šljivar. 2004. Evaluation of the domestication process in Serbia – domestication of Neolithic cattle. *Acta Veterinaria* 54(5–6): 467–473.

Jović, V. 1996. Mineralogical study of malachite and azurite from the Belovode locality (Veliko Laole). *Préhistorie Européene* 8: 191–196.

Junghans, S., E. Sangmeister and M. Schröder. 1968. *Kupfer und Bronze in der frühen Metallzeit Europas.* (Studien zu den Anfängen der Metallurgie, 1-3). Berlin: Gebr. Mann.

Jurišić, A. 1959. Velika Gradina u Staparima, praistorijsko naselje. *Starinar* 9/10: 367–368.

K

Kaczankowska, M. and J. Kozlowski. 1990. Chipped stone industry of the Vinča culture, in D.Srejović and N. Tasić (eds) *Vinča and its World. International Symposium - The Danubian Region from 6000 to 3000 BC*: 35–47. Belgrade: Serbian Academy of Sciences and Arts and Centre for Archaeological Research.

Kaiser, T. 1984. Vinča Ceramics: Economic and Technological Aspects of Late Neolithic Pottery Production in Southeast Europe. Unpublished PhD dissertation, University of California.

Kaiser, T. 1989. Steatite-tempered pottery from Selevac, Yugoslavia: a Neolithic experiment in ceramic design. *Archeomaterials* 3/1: 1–10.

Kaiser, T. 1990. Ceramic technology, in R.E. Tringham and D. Krstiš (eds) *Selevac: a Neolithic Village in Yugoslavia*: 255–287. Los Angeles: University of California Press.

Kaiser, T. and B. Voytek. 1983. Sedentism and economic change in the Balkan Neolithic. *Journal of Anthropological Archaeology* 2: 323–353.

Kaiser, T., U.M. Franklin and V. Vitali. 1986. Pyrotechnology and pottery in the Late Neolithic of the Balkans, in J.S. Olin and M.J. Blackman (eds) *Proceedings of the 24th International Archaeometry Symposium*: 85–94. Washington DC: Smithsonian Institution.

Kapuran, A. and S. Milošević. 2013. Rockshelter Mokranjske Stene – a new late prehistoric site in eastern Serbia region. *Archaeologia Bulgarica* 17 (2): 17–37.

Kassebaum, T. 2019. Aggregation or Separation: (Re) considering Approaches Used in the Analysis of Sheep (*Ovis aries*) and Goat (*Capra hircus*) Faunal Remains. Unpublished MSc dissertation, University of York.

Kavtaradze, G.L. 1999. The importance of metallurgical data for the formation of Central Transcaucasian chronology, in A. Hauptmann, E. Pernicka, Th. Rehren and Ü. Yalçın (eds) *The Beginnings of Metallurgy* (Der Anschnitt, Beiheft 9): 67–102. Bochum: Deutsches Bergbau-Museum.

Kenoyer, J.M. and M.-L. Miller. 1999. Metal technologies of the Indus Valley tradition in Pakistan and western India, in V.C. Pigott (ed.) *The Archaeometallurgy of the Asian Old World*: 107-152. Philadelphia (PA): University of Pennsylvania Museum.

Kenoyer, J.M., M.Vidale and K.K. Bhan. 1991. Contemporary stone beadmaking in Khambhat, India: patterns of craft specialization and organization of production as reflected in the archaeological record. *World Archaeology* 23: 44–3.

Kienlin, T.L. 2010. *Traditions and Transformations: Approaches to Eneolithic (Copper Age) and Bronze Age Metalworking and Society in Eastern Central Europe and the Carphatian Basin* (British Archaeological Reports International Series 2184). Oxford: Archaeopress.

Kienlin, T.L. 2014. Aspects of metalworking and society from the Black Sea to the Baltic Sea from the fifth to the second millennium BC, in B.W. Roberts and C.P. Thornton (eds) *Archaeometallurgy in Global Perspective: Methods and Syntheses*: 447–472. New York: Springer.

Kienlin, T.L. and A. Zimmermann (eds). 2012. *Beyond Elites. Alternatives to Hierarchical Systems in Modelling Social Formations. International Conference at the Ruhr-Universität Bochum, Germany, October 22-24, 2009* (Universitätsforschungen zur Prähistorischen Archäologie 215): 15–32. Bonn: Habelt.

Kilikoglou, V., Y. Bassiakos, A.P. Grimanis, K. Souvatzis, A. Pilali-Papasteriou and A. Papanthimou-Papaefthimiou. 1996. Carpathian Obsidian in Macedonia, Greece. *Journal of Archaeological Science* 23(3): 343–349.

Killick, D. 2014. From ores to metals, in B.W. Roberts and C.P. Thornton (eds) *Archaeometallurgy in Global Perspective: Methods and Syntheses*: 11–45). New York: Springer.

Killick, D. and T.R. Fenn 2012. Archaeometallurgy: the study of preindustrial mining and metallurgy. *Annual Review of Anthropology* 41: 559-575.

Kingery, W.D. and J.D. Frierman. 1974. The firing temperature of a Karanova sherd and inferences about south-east European Chalcolithic refractory technology. *Proceedings of the Prehistoric Society* 40: 204–205.

Kintigh, K.W. 1984. Measuring archaeological diversity by comparison with simulated assemblages. *American Antiquity* 49: 44–54.

Kintigh, K.W. 1994. Contending with contemporaneity in settlement-pattern studies. *American Antiquity* 59: 143–148.

Kjeld Jensen, C. and K. Høilund Nielsen. 1997. Burial data and correspondence analysis, in C. Kjeld Jensen and K. Høilund Nielsen (eds) *Burials and Society: The Chronological and Social Analysis of Archaeological Burial Data*: 29–61. Aarhus: Aarhus University Press.

Klassen, L. 2000. *Frühes Kupfer im Norden. Untersuchungen zu Chronologie, Herkunft und Bedeutung der Kupferfunde der Nordgruppe der Trichterbecherkultur* (Jutland Archaeological Society 36). Aarhus: Aarhus University Press.

Klimscha, F. 2013. Innovations in Chalcolithic metallurgy in the Southern Levant during the 5th and 4th millennium BC. Copper-production at Tall Hujayrat al-Ghuzlan and Tall al-Magass, Aqaba Area, Jordan, in S. Burmeister, S. Hansen, M. Kunst and N. Müller-Scheeßel (eds) *Metal Matters. Innovative Technologies and Social Change in Prehistory and Antiquity*: 31–64. Rahden/West: Leidorf.

Klimscha, F. 2014. Power and prestige in the Copper Age in the lower Danube, in C.-E. Stefan, M. Florea, S.-C. Ailincăi and C. Micu (eds) *Studii Privind Preistoria Sud-estului Europei: Volum Dedicat Memoriei lui Mihai Şimon. Studies in the prehistory of southeastern Europe: volume dedicated to the memory of Mihai Şimon*. (Biblioteca Istro-Pontica 10): 131–168. Brăila: Muzeul Brăilei, Editura Istros.

Klimscha, F. 2016. *Pietrele 1: Beile und Äxte aus Stein. Distinktion und Kommunikation während der Kupferzeit im östlichen Balkangebiet* (Archäologie in Eurasien 34). Bonn: Habelt.

Klimscha, F. 2020. New research on the 'Kings of Metal': Systems of social distinction in the Copper Age of Southeastern Europe. *Journal of Archaeological Research* 28, 99–138 . DOI: https://doi.org/10.1007/s10814-019-09132-x.

Kline, M.A. and R. Boyd. 2010. Population size predicts technological complexity in Oceania. *Proceedings of the Royal Society of London B: Biological Sciences* 277: 2559–2564.

Kogălniceanu, R., A. Ilie, M. Mărgărit, A. Simalcsik and V. Dumitraşcu. 2014. A hoard of astragals discovered in the Copper Age settlement at Iepureşti, Giurgiu County, Romania. *Documenta Praehistorica* XLI: 283–304

Kohl, P.L. 2007. *The Making of Bronze Age Eurasia*. Cambridge: Cambridge University Press.

Kohler-Schneider, M. 2003. Contents of a storage pit from late Bronze Age Stillfried, Austria: another record of the 'new' glume wheat. *Vegetation History and Archaeobotany* 12: 105–111.

Kokalj, Ž. and M. Somrak. 2019. Why not a single image? Combining visualizations to facilitate fieldwork and on-screen mapping. *Remote Sensing* 11: 747.

Koryakova, L. and A. Epimakhov. 2007. *The Urals and Western Siberia in the Bronze and Iron Ages*. Cambridge: Cambridge University Press.

Kosse, K. 1990. Group size and societal complexity: thresholds in the long-term memory. *Journal of Anthropological Archaeology* 9: 275–303.

Kostov, R.I. 2013. Nephrite-yielding prehistoric cultures and nephrite occurrences in Europe: Archaeomineralogical Review. *Haemus* 2: 11–30.

Kotsakis, K., A. Papanthimou-Papaefthimiou, A. Pilali-Papasteriou, T. Savopoulou, Y. Maniatis and B. Kromer. 1989. Carbon-14 dates from Mandalo, W. Macedonia, in Y. Maniatis (ed.) *Archaeometry, International Symposium on Archaeometry*: 679–85. Amsterdam: Elsevier.

Koukouli-Chryssanthaki, C., Todorova, H. Aslanis, I. Vajsov, I. and Valla, M. 2007. Promachon – Topolnica. A Greek-Bulgarian archaeological project, in Todorova, H. Stefanovich, M. and Ivanov, G. (eds.) *The Struma / Strymon Valley in Prehistory. Proceedings of the International Symposium 'Strymon Praehistoricus'. In The Steps of James Harvey Gaul*: Vol. 2: 43-78. Sofia: Gerda Henkel Stiftung.

Knipper, C., A. Mittnik, K. Massy, C. Kociumaka, I. Kucukkalipci, M. Maus, F. Wittenborn, S.E. Metz, A. Staskiewicz, J. Krause and P.W. Stockhammer. 2017. Female exogamy and gene pool diversification at the transition from the Final Neolithic to the Early Bronze Age in central Europe. *Proceedings of the Academy of Sciences* 114(38): 10083–10088.

Krajnović, D. and S. Janković. 1995. Copper mineralisation as potential raw material source of ancient copper metallurgy in Serbia, in B. Jovanović (ed.) *Ancient Mining and Metallurgy in Southeast Europe, 1995 Donji Milanovac, Srbija*: 21–28. Belgrade, Bor: Archaeological Institute, Museum of Mining and Metallurgy.

Krause, R. 1985. *The Clay Sleeps*. Tuscaloosa: University of Alabama Press.

Krause, R. 2003. *Studien zur kupfer- und frühbronzezeitlichen Metallurgie zwischen Karpatenbecken und Ostsee* (Vorgeschichtliche Forschungen 24). Rahden: Leidorf.

Krauss, R. 2008. Karanovo und das südosteuropäische Chronologiesystem aus heutiger Sicht *Eurasia Antiqua* 14: 117–149.

Krauss, R., S. Zäuner and E. Pernicka. 2014. Statistical and anthropological analyses of the Varna necropolis, in H. Meller, R. Risch and E. Pernicka (eds) *Metals of Power - Early Gold and Silver. 6th Archaeological Conference of Central Germany, October 17-19, 2013, Halle*: 371–387. Halle: Landesmuseum für Vorgeschichte.

Krauss, R., C. Schmid, D. Ciobotaru and V. Slavchev. 2016. Varna und die Folgen. Überlegungen zu den Ockergräbern zwischen Karpatenbecken und der nördlichen Ägäis, in M. Bartelheim, B. Horejs and R. Krauss (eds) *Von Baden bis Troja, Resourcennutzung, Metallurgie und Wissenstransfer. Eine Jubiläumsschrift für Ernst Pernicka*. (Oriental and European Archaeology, Volume 3): 273–315. Rahden: Leidorf.

Krauss, R., C. Schmid, D. Kirschenheuter, J. Abele, V. Slavchev and B. Weninger. 2017. Chronology and development of the Chalcolithic necropolis of Varna I. *Documenta Praehistorica* 44: 282–305. DOI: https://doi.org/10.4312/dp.44.17.

Krauss, R., J. Bofinger and B. Weninger. 2020. Formation and Transformation of Early Neolithic Lifestyles in Europe. *Quaternary International* 560–561: 1-286

Kreiter, A., Gy. Szakmány and M. Kázmér. 2009. Ceramic technology and social process in late Neolithic Hungary, in P. Quinn (ed.) *Interpreting Silent Artefacts: Petrographic Approaches to Archaeological Ceramics*: 101–119. Oxford: Archaeopress.

Kreiter, A., I. Azbej Havancsák, P. Sipos, M. Tóth and V. Orsolya. 2011. Maroslele-Panáról származó neolitikus kerámia töredékek petrográfiai, XRF és XRD vizsgálata. (The petrographic, XRF and XRD analyses of the Neolithic pottery from Maroslele-Pana), in T. Paluch (ed.) *Maroslele-Pana. Egy Középső Neolitkus Lelőhely a Kultúrák Határvidékén*: 303–326. Szeged: Móra Ferenc Múzeum.

Kreiter, A., Á. Pető and P. Pánczél. 2013. Materializing tradition: Ceramic production in Early and Middle Neolithic Hungary, in E. Bánffy (ed.) *The Early Neolithic of the Danube-Tisza Interfluve*: 127–140. Oxford: Archaeopress.

Kreiter, A., S. Czifra, Z. Bendő, J.E. Imre, P. Pánczél and G. Váczic. 2014. Shine like metal: an experimental approach to understand prehistoric graphite-coated pottery technology. *Journal of Archaeological Science* 52: 129–142.

Kreiter, A., T. Marton, L. Gomart, K. Oross and P. Pánczél. 2017a. Looking into houses: Analysis of LBK ceramic technological change on a household level, in L. Burnez-Lanotte (ed.) *Matières à Penser. Raw Materials Acquisition and Processing in Early Neolithic Pottery Productions.* Proceedings of the Workshop of Namur (Belgium), 29–30 May 2015: 111–132. Paris: Société Préhistorique Française.

Kreiter, A., N. Kalicz, K. Kovács, Zs. Siklósi and O. Viktorik. 2017b. Entangled traditions: Tisza and Lengyel ceramic technology in a Late Neolithic settlement in northern Hungary. *Journal of Archaeological Science Reports* 16: 285–295.

Kreiter, A., T. Marton, K. Oross and P. Pánczél. 2019. Looking into pots: Understanding Neolithic ceramic technological variability from Western Hungary, in S. Amicone, P.S. Quinn, M. Marić, N. Mirković-Marić and M. Radivojević (eds) *Tracing Pottery Making Recipes in the Balkans 6th-4th Millennia BC*: 65–77. Oxford: Archaeopress.

Kreuz, A. and N. Boenke. 2002. The presence of two-grained einkorn at the time of the Bandkeramik culture. *Vegetation History and Archaeobotany* 11: 233–240.

Kreuz, A., E. Marinova, E. Schäfer and J. Wiethold. 2005. A comparison of Early Neolithic crop and weed assemblages from the Linearbandkeramik and the Bulgarian Neolithic cultures: differences and similarities. *Vegetation History and Archaeobotany* 14: 237–258.

Kroll, H. 2013. Die Wirtschaft des Umfelds von Okolište: Zagrebnice, Kundruci und Donje Moštre, in J. Müller, K. Rassman and R. Hofmann (eds) *Okolište 1 - Untersuchungen einer spätneolithischen Siedlungskammer in Zentralbosnien. Neolithikum und Chalkolithikum in Zentralbosnien 1.* (Universitätsforschungen zur prähistorischen Archäologie 228): 231–238. Bonn: Habelt.

Kristiansen, K., T. Lindkvist, and J. Myrdal. (eds) 2018. *Trade and Civilisation: Economic Networks and Cultural Ties from Prehistory to the Early Modern Era.* Cambridge: Cambridge University Press.

Kuijpers, M.H.G. 2018. *An Archaeology of Skill. Metalworking Skill and Material Specialization in Early Bronze Age Central Europe.* London: Routledge.

Kuijpers M.H.G. and C.N. Popa 2021. The origins of money: Calculation of similarity indexes demonstrates the earliest development of commodity money in prehistoric Central Europe. *PLoS ONE* 16(1): e0240462. https://doi.org/10.1371/journal.pone.0240462

Kulbicki, G. 1958. High-temperature phases in montmorillonites. *Clays and Clay Minerals*: 144–158. National Academy of Sciences – National Research Council Publication 566.

Kuleff, I. and E. Pernicka. 1995. On the instrumental neutron activation analysis of native copper: some methodological considerations. *Journal of Radioanalytical and Nuclear Chemistry* 191: 145–161.

Kuna, M. 1981. Zur Neolithischen und Äneolithischen Kupferverarbeitung im Gebiet Jugoslawiens. *Godišnjak Akademije BiH 17. Centar za balkanološka ispitivanja (Sarajevo)* 19: 13–81.

Kunst, M. 2013. The innovation of copper metallurgy on the Iberian Peninsula: its significance for the development of social complexity in the 3rd millennium BC, in S. Burmeister, S. Hansen, M. Kunst and N. Müller-Scheeßel (eds) *Metal Matters. Innovative Technologies and Social Change in Prehistory and Antiquity*: 181–208. Rahden.: Leidorf.

Kunze, R. and E. Pernicka. 2020. Pre-industrial mining in the Medni Rid, in R. Krauss, E. Pernicka, R. Kunze, K. Dimitrov and P. Leshtakov (eds) *Prehistoric Mining and Metallurgy at the Southeast Bulgarian Black Sea Coast (RessourcenKulturen Band 10)*: 398–425. Tübingen: Tübingen University Press.

Kuzmanović Cvetković, J. 1998. *Prokuplje, grad sv. Prokopija.* Prokuplje: Narodni muzej Toplice.

Kuzmanović Cvetković, J., Z. Cvetković and B. Đevori. 1988. *Arheološka nalazišta opštine Prokuplje - katalog izložbe povodom 40 godina Muzeja Toplice.* Prokuplje: Narodni Muzej Toplice.

Kuzmanović Cvetković, J. and D. Šljivar. 1998. Die Göttin aus Pločnik. *Starinar* 49: 173–178.

Kuzmina, E.E. 2008. *The Prehistory of the Silk Road.* Philadelphia (PA): University of Pennsylvania Press.

L

Lang, F. 2005. Knochen- und Geweihobjekte, in S. Hiller and N. Vassil Nikolov (eds) *Karanovo. Die*

Ausgrabungen im Nordsüd-Schnitt, 1993-1999: 263-340. Sofia, Phoibos Verlag, Wien: Archäologisches Institut der Universität Salzburg, Archäologisches Institut mit Museum der Bulgarischen Akademie der Wissenschaften.

Lankton, J.W., J. Diamanti and J.M. Kenoyer. 2003. *A Bead Timeline: A Resource for Identification, Classification and Dating*. Washington, DC: The Bead Society of Greater Washington.

Larson, G., D.R. Piperno, R.G. Allaby, M.D. Purugganan, L. Andersson, M. Arroyo-Kalin, L. Barton, C. Climer Vigueira, T. Denham, K. Dobney, A.N. Doust, P. Gepts, M.T.P. Gilbert, K.J. Gremillion, L. Lucas, L. Lukens, F.B. Marshall, K.M. Olsen, J.C. Pires, P.J. Richerson, R. Rubio de Casas, O.I. Sanjur, M.G. Thomas and D.Q. Fuller. 2014. Current perspectives and the future of domestication studies. *Proceedings of the National Academy of Sciences* 111: 6139–6146.

Lasaponara, R. and N. Masini. 2011. Satellite remote sensing in archaeology: Past, present and future perspectives. *Journal of Archaeological Science* 38(9): 1995–2002.

Lazarovici, C.-M. 2006. Absolute chronology of the late Vinča culture in Romania and its role in the development of the Early Copper Age, in N. Tasić and C. Grozdanov (eds) *Homage to Milutin Garašanin*: 277–293. Belgrade: Serbian Academy of Sciences and Arts and Macedonian Academy of Sciences and Arts.

Lazarovici, G. 1970. Cultură Vinča A in Banat. *Acta Musei Napocensis (Romania)* 7: 475–488.

Lazarovici, G. 1979. *Naoliticul Banatului. Bibliotheca Musei Napocensis IV*. Cluj-Napoca: Bibliotheca Musei Napocensis.

Lazarovici, G. 1981. Die Periodisierung der Vinča-Kultur in Rumänien. *Prähistorische Zeitschrift* 56: 169–196.

Lazarovici, G. 2014. Beginning of the Copper Age in Transylvania and some data regarding the copper and gold metallurgy, in Schier, W. and F. Draşovean (eds) *New Approaches to Dating and Cultural Dynamics in the 6th to 4th Millennium BC*. (Prähistorische Archäologie in Südosteuropa 28): 243–271. Rahden/Westf: Verlag Marie Leidorf Gmbh.

Lazarovici, Gh., Fl. Draşovean and Z. Maxim. 2001. *Parta*. Timisoara: Museum Banaticum Temesiense.

Lazić, M. 1992. Doprinos arheozoologije u proučavanju ekonomije praistorijskih društvenih zajednica, in D. Srejović (ed.) *Arheologija i prirodne nauke*: 49–60. Beograd: SANU.

Lechtman, H. 1996. Arsenic bronze: dirty copper or chosen alloy? A view from the Americas. *Journal of Field Archaeology* 23: 477–514.

Lechtman, H. 2014. Andean metallurgy in prehistory, in B.W. Roberts and C.P. Thornton (eds) *Archaeometallurgy in Global Perspective: Methods and Syntheses*: 361–422. New York: Springer.

Legge, A. 1990. Animals, economy and environment, in R. Tringham and D. Krstić (eds) *Selevac: a Neolithic village in Yugoslavia*: 215–242. Los Angeles: University of California Press.

Legrand, A. 2007. *Fabrication et utilisation de l'outillage en matières osseuses du Néolithique de Chypre: Khirokitia et Cap Andreas -Kastros*. (British Archaeological Reports International Series 1678). Oxford: Archaeopress.

Lehner, J.W. and K.A. Yener. 2014. Organization and specialization of early mining and metal technologies in Anatolia, in B.W. Roberts and C.P. Thornton (eds) *Archaeometallurgy in Global Perspective. Methods and Syntheses*: 529–558. New York: Springer.

Leković, V. 1988. Zlatara-Ruma, in D. Srejović and V. Kostić (eds) *The Neolithic of Serbia*: 108–109. Belgrade: Centre for Archaeological Research, Faculty of Philosophy, the University of Belgrade.

Leković, V. 1990. The *vinčanization* of Starčevo culture, in D. Srejović and N. Tasić (eds) *Vinča and its World, International Symposium - The Danubian Region from 6000 to 3000 B.C., October 1988*: 67–74. Belgrade: Serbian Academy of Sciences and Arts - Centre for Archaeological research - Faculty of Philosophy.

Lemonnier, P. 1993. Introduction, in P. Lemmonier (ed.) *Technological Choices: Transformation in Material Cultures since the Neolithic*: 1–35. London: Routledge.

Lemonnier, P. 2012. *Mundane Objects: Materiality and Non-verbal Communication*. Walnut Creek, CA: Left Coast Press.

Lenneis, E. and P. Stadler. 1995. Zur Absolutchronologie der Linearbandkeramik aufgrund von ^{14}C-Daten. *Archäologie Österreichs* 6: 4–12.

Leshtakov, P. 2005. The sources and distribution of graphite as a means of decoration in the Bulgarian Chalcolithic, in G. Dumitroaia, J. Chapman, O. Weller, C. Preoteasa, R. Munteanu, D. Nicola and D. Monah (eds) *Cucuteni. 120 Years of Research. Time to Sum Up*: 293–296. Piatra-Neamţ: Constantin Matasă.

Leshtakov, P. 2013. Early copper metallurgy on the west Black Sea coast: archaeological evidence on prehistoric exploitation of the Rosen ore field, in *Where are the sites? Research, Protection and Management of Cultural Heritage, Ahtopol, Bulgaria, 2013*: 35–54. Sozopol: Centre for Underwater Archaeology, Bulgaria.

Letica, Z. 1964. The Neolithic figurines from Vinča. *Archaeology* 17(1): 26–32.

Letsch, J. and W. Noll. 1978. Material und Herstellung antiker C-Schwarz-Keramik, Teil I / Teil II. *Berichte der Deutschen Keramischen Gesellschaft* 55:163–168 and 259–261.

Letsch, J. and W. Noll. 1983. Phasenbildung in einigen keramischen Teilsystemen bei 600- 1000°C in Abhängigkeit von der Sauerstofffugazität. *Ceramic Forum International* 7: 259–267

Leusch, V., E. Pernicka and B. Armbruster. 2014. Chalcolithic gold from Varna - Provenance, circulation, processing, and function, in H. Meller, R. Risch and E. Pernicka (eds) *Metals of Power - Early*

Gold and Silver. 6th Archaeological Conference of Central Germany, October 17-19, 2013, Halle: 165–182. Halle: Landesamt für Denkmalpflege und Archäologie Sachsen-Anhalt/Landesmuseum für Vorgeschichte.

Leusch, V., B. Armbruster, E. Pernicka and V. Slavčev. 2015. On the invention of gold metallurgy: The gold objects from the Varna I cemetery (Bulgaria) – technological consequence and inventive creativity. *Cambridge Archaeological Journal* 25: 353–376. DOI: https://doi.org/10.1017/S0959774314001140.

Leusch, V., M. Brauns and E. Pernicka. 2016. Precise and accurate analysis of gold alloys: Varna, the earliest gold of mankind. A case study, in L. Dussubieux, M. Golitko and B. Gratuze (eds) *Recent Advances in Laser Ablation ICP-MS for Archaeology. Natural Science in Archaeology*: 95–113. Berlin, Heidelberg: Springer.

Leusch, V., S. Zäuner, V. Slavčev, R. Krauß, B. Armbruster and E. Pernicka. 2017. Rich metallurgists' (?) graves from the Varna I cemetery - Rediscussing the social role of the earliest metalworkers, in A. Brysbaert and A. Gorgues (eds) *Artisans versus Nobility? Multiple identities of Elites and 'Commoners' Viewed Through the Lens of Crafting from the Chalcolithic to the Iron Ages in Europe and the Mediterranean*: 101–124. Leiden: Sidestone Press.

Lichardus, J. 1988. Der Westpontische Raum und die Anfänge der kupferzeitlichen Zivilisation, in A. Fol and J. Lichardus (eds) *Macht, Herrschaft und Gold : das Graberfeld von Varna (Bulgarien) und die Anfänge einer neuen europäischen Zivilisation*: 79–130. Saarbrücken: Moderne Galerie des Saarland-Museums.

Lichardus, J. (ed.) 1991a. *Die Kupferzeit als Historische Epoche. Symposium Saarbrücken und Otzenhausen 6.-13.11.1988* (Saarbrücker Beiträge zur Altertumskunde 55). Bonn: Dr. Rudolf Habelt GmbH,

Lichardus, J. 1991b. Das Gräberfeld von Varna im Rahmen des Totenrituals des Kodžadermen-Gumelniţa-Karanovo VI-Komplexes, in J. Lichardus (ed.) *Die Kupferzeit als historische Epoche: Symposium Saarbrücken und Otzenhausen 6.-13.11.1988, Teil 1*: 167–194. Bonn: Habelt.

Lichardus, J. 1991c. Die Kupferzeit als historische Epoche. Versuch einer Deutung, in J. Lichardus (ed.) *Die Kupferzeit als Historische Epoche. Symposium Saarbrücken und Otzenhausen 6.-13.11.1988* (Saarbrücker Beiträge zur Altertumskunde 55) : 763–800). Bonn: Dr. Rudolf Habelt GmbH.

Lichter, C. 1993. Untersuchungen zu den Bauten des südosteuropäischen Neolithikums und Chalkolithikums. *Internationale Archäologie* 18. Leidorf: Buch am Erlbach.

Lichter, C. 2001. *Untersuchungen zu den Bestattungssitten des südosteuropäischen Neolithikums und Chalkolithikums* (Monographien 5). Zabern, Mainz: Heidelberger Akademie der Wissenschaften, Internationale Interakademische Kommission für die Erforschung der Vorgeschichte des Balkans.

Lichter, C. 2014. Temples in the Neolithic and Copper Age in southeast Europe. *Documenta Praehistorica* 41: 119–136.

Lichter, C. 2016. Burning down the house—Fakt oder Fiktion? In K. Bacvarov and R. Gleser (eds) *Southeast Europe and Anatolia in Prehistory: Essays in Honor of Vassil Nikolov on His 65th Anniversary*: 305–317. Bonn: Habelt.

Lienhard, J.H. 2006. *How Invention Begins*. Oxford: Oxford University Press.

Liu, R., G. Dong, M. Ma, and A.M. Pollard 2020a. Introduction to the Special Issue: Correlating changes for environmental, technological and societal transformation in prehistoric eastern Asia. *The Holocene* 31: 165-168.

Liu, X., D. Chen, T. Yang, F. Huang, S. Fu, and L. Li 2020b. Changes in soil labile and recalcitrant carbon pools after land-use change in a semi-arid agro-pastoral ecotone in Central Asia. *Ecological Indicators* 110: 105925.

Liritzis, I. 2008. Assessment of Aegean obsidian sources by a portable ED-XRF analyser: grouping, provenance and accuracy, in Y. Facorellis, N. Zacharias and K. Polikreti (eds) *Proceedings of the 4th Symposium of the Hellenic Society for Archaeometry* (British Archaeological Reports International Series S1746): 399–406. Oxford: Archaeopress.

Longman, J., D. Veres, W. Finsinger, and V. Ersek 2018. Exceptionally high levels of lead pollution in the Balkans from the Early Bronze Age to the Industrial Revolution. *Proceedings of the National Academy of Sciences* 115 (25): E5661-E5668. doi:10.1073/pnas.1721546115.

Lorenzo-Seva, U. 2011. Horn's Parallel Analysis for selecting the number of dimensions in Correspondence Analysis. *Methodology* 7: 96–102.

Lubbock, J. 1872. *Pre-Historic Times, as Illustrated by Ancient Remains, and the Manners and Customs of Modern Savages*. London: D. Appleton and Company.

Luca, S.A. 1999. *Sfârşitul eneoliticului pe teritoriul intracarpatic al României - cultura Bodrogkeresztúr* (Volume 11). Alba Iulia: Bibliotheca Musei Apulensis.

Luik, H. 2007. Dazzling white. Bone artefacts in Bronze Age society – some preliminary thoughts from Estonia, in A. Merkevičius (ed.) *Colours of Archaeology. Material Culture and Society. Papers from the Second Theoretical Seminar of the Baltic Archaeologists (BASE) Held at the University of Vilnius, Lithuania, October 21–22, 2005*: 49–64. Vilnius; Helsinki; Riga; Tartu: University of Vilnius, University of Helsinki, University of Latvia, University of Tartu. (Interarchaeologia; 2).

Lunardi, A. 2008. Experimental testing with polished green stone axes and adzes: technology of use, in L. Longo and N. Skakun (eds) *'Prehistoric Technology' 40 Years Later: Functional Studies and the Russian Legacy: Proceeding of the International Congress Verona (Italy), 20–23 April 2005*. (British Archaeological Reports

International Series S1783): 369–373. Oxford: Archaeopress.

Lyman, C.E. and M.J. O'Brien. 2003. Cultural traits: units of analysis in early twentieth-century anthropology. *Journal of Anthropological Research* 59: 225–250.

Lyneis, M.M. 1988. Antler and bone artifacts from Divostin, in A. McPherron and D. Srejović (eds) *Divostin and the Neolithic of Central Serbia* (Ethnology Monographs 10): 301–319. Pittsburgh: University of Pittsburgh.

M

Madas D. 1988. Ceramic vessels from Divostin II house floors, in A. McPherron and D. Srejović (eds) *Divostin and the Neolithic of Central Serbia*: 143–157. Pittsburgh: University of Pittsburgh.

Maddin, R., J.D. Muhly and T. Stech. 1999. Early metalworking at Çayönü, in A. Hauptmann, E. Pernicka, Th. Rehren and Ü. Yalçın (eds) *The Beginnings of Metallurgy, Der Anschnitt, Beiheft 9*: 37–44. Bochum: Deutsches Bergbau-Museum.

Maggetti, M. 1982. Phase analysis and its significance for technology and origin, in J.S. Olin and A.D. Franklin (eds) *Archaeological Ceramics*: 121–133. Washington DC: Smithsonian Institute Press.

Maggetti, M. 1994. Mineralogical and petrographical methods for the study of ancient pottery, in F. Buragato, O. Grubessi and L. Lazzarini (eds) *1st European Workshop on Archaeological Ceramics*: 23–35. Rome: Università degli Studi La Sapienza.

Maigrot, Y. 2003. Etude Technologique et Fonctionnelle de l'Outillage en Matières Dures Animales. La Station 4 de Chalain (Néolithique Final, Jura, France). Unpublished PhD dissertation, Université de Paris I.

Makkay, J. 1976. Problems concerning Copper Age chronology in the Carpathian Basin. *Acta Archaeologica Hungarica* 28: 251–300.

Makkay, J. 1989. *The Tiszaszőlős Treasurel*. Budapest: Akadémiai Kiadó.

Makkay, J. 1990. The Protovinča problem as seen from the northernmost frontier, in D. Srejović and N. Tasić (eds) *Vinča and Its World*: 113–122. Belgrade: Serbian Academy of Sciences and Arts.

Makkay, J. 1991. The most ancient gold and silver in central and south-east Europe, in J.-P. Mohen and C. Éluère (eds) *Découverte du Métal*: 119–128. Paris: Picard.

Makkay, J. 1995. The rise and fall of gold metallurgy in the Copper Age of the Carpathian Basin: the background of the change, in G. Morteani and J.P. Northover (eds) *Prehistoric Gold in Europe: Mines, Metallurgy and Manufacture*: 65–76. Dordrecht: Kluwer Academic Publishers.

Malafouris, L. 2013. *How Things Shape the Mind: A Theory of Material Engagement.* Cambridge, MA: MIT Press.

Malešević M., M. Vukanović, Z. Obradinović, M. Dimitrijević, T. Brković, M. Stefanović, R. Stanisavljević, O. Jovanović, S. Trifunović, Lj. Karajičić, M. Jovanović and Z. Pavlović. 1980. *Tumač za list Kuršumlija (K 34-31)*. Osnovna geološka karta. Beograd: Savezni geološki zavod.

Mallia, K. 2015. Evaluating Methodological Bias of Identification Rates of *Ovis* and *Capra* on Age-Profiles with the Aid of Zooms, using Çatalhöyük as a Test-Case. Unpublished MSc dissertation, University of York.

Maniatis, Y. and M.S. Tite. 1975. A scanning electron microscope examination of the bloating of fired clays. *Transactions and Journal of the British Ceramic Society* 74/8: 229–232.

Maniatis, Y. and M.S. Tite 1981. Technological examination of Neolithic and Bronze Age pottery from central and southeast Europe and from the Near East. *Journal of Archaeological Science* 8: 59–76.

Mantellini, S. and A.E. Berdimuradov. 2019. Evaluating the human impact on the archaeological landscape of Samarkand (Uzbekistan): a diachronic assessment of the Taylak district by remote sensing, field survey, and local knowledge. *Archaeological Research in Asia* 20: 1–11.

Maran, J. 2000. Das ägäische Chalkolithikum und das erste Silber in Europa, in C. Işık (ed.) *Studien zur Religion und Kultur Kleinasiens und des ägäischen Bereiches. Festschrift für Baki Ögün zum 75. Geburtstag*: 179–193. Bonn: Habelt.

Maran, J. and P.W. Stockhammer (eds). 2017. *Appropriating Innovations: Entangled Knowledge in Eurasia, 5000–1500 BCE*. Oxford: Oxbow.

Mareş, I. 2002. *Metallurgiei aramei în neo-eneoliticul României. The Metallurgy of Copper in the Romanian Neo- Eneolithic*. Suceava: Editura Bucovina Istorică.

Marić, M. 2011. Arhitektura i idealna rekonstrukcija objekata od lepa. *Kolubara* 5, 69–83. Beograd: Republički zavod za zaštitu spomenika kulture.

Marić, M. 2013a. The Vinča culture climate and environment in the Danube region in the 5th and 6th millennium BC, in L.Č. Popović, M. Vidaković and Đ.S. Kostić (eds) *Resources of Danubian Region: The Possibility of Cooperation and Utilization. Proceedings of the Humboldt Kolleg Belgrade 2013* :204-232. Belgrade: Humboldt-Club Serbien.

Marić, M. 2013b. Zaštitna arheološka istraživanja na lokalitetu Jaričište 1, in V. Filipović, R. Arsić and D. Antonović (eds) *Rezultati Novih Arheoloških Istraživanja u Severozapadnoj Srbiji i Susednim Teritorijama*: 17–31. Beograd: Srpsko Arheološkog Društvo. Valjevo: Zavod za zaštitu spomenika kulture.

Marić, M. 2015. Modelling obsidian trade routes during Late Neolithic in the South-East Banat region of Vršac using GIS. *Starinar* 65: 37–52.

Marić, M. 2017. Encircled in water: Modelling watercourses of the late Neolithic – early Copper Age Vinča culture sites in the Danube region (c. 5300 – 4500 BC). *Quaternary International* 429: 54–63. DOI: https://doi.org/10.1016/j.quaint.2014.06.030.

Marić, M. and N. Mirković-Marić. 2011. Obrenovac kod Dimitrovgrada, naselje vinčanske kulture. *Glasnik Društva Konzervatora Srbije* 35: 65–68.

Marić, R. 1951. Iskopavanja na Orašju, prethodni izveštaj o radovima u 1948/49. Godini. *Starinar new series* 2: 113.

Marinković, S. 2010. Arheološki materijal sa lokaliteta Živanićeva Dolja iz zbirke Narodnog muzeja u Zrenjaninu – vinčanska kultura. *Rad Muzeja Vojvodine* 52: 21–36.

Marinova, E. 2007. Archaeobotanical data from the early Neolithic of Bulgaria, in S. Colledge and J. Conolly (eds) *The Origins and Spread of Domestic Plants in Southwest Asia and Europe*: 93–109. London: Left Coast Press.

Marinova, E. and Tz. Popova. 2008. *Cicer arietinum* (chick pea) in the Neolithic and Chalcolithic of Bulgaria: implications for cultural contacts with the neighbouring regions? *Vegetation History and Archaeobotany* 17: 73–80.

Marinova, E. and S. Thiébault. 2008. Anthracological analysis from Kovačevo, southwest Bulgaria: woodland vegetation and its use during the earliest stages of the European Neolithic. *Vegetation History and Archaeobotany* 17: 223–231.

Marinova, E., S. Tonkov, E. Božilova and I. Vajsov. 2012. Holocene anthropogenic landscapes in the Balkans. The palaeobotanical evidence from Southwestern Bulgaria. *Vegetation History and Archaeobotany* 21: 413–427.

Marinova, E., D. Filipović, Đ. Obradović and E. Allué. 2013. Wild plant resources and land use in the Mesolithic and Early Neolithic southeast Europe: archaeobotanical evidence from the Danube catchment of Bulgaria and Serbia. *Offa* 69/70: 467–478.

Maritan, L. 2004. Archaeometric study of Etruscan-Padan type pottery from Veneto region: Petrographic, mineralogical and geochemical-physical characterisation. *European Journal of Mineralogy* 16: 297–307.

Markotić, V. 1984. *The Vinča Culture.* Calgary: Western Publishers.

Marković, N., J. Bulatović, J. Jovanović and V. Katić 2018. Bioarheološka perspektiva ranoneolitskog nalazišta Bataševo, in I. Todorović, G. Jošović, I. Becić and Lj. Vasiljević (eds) *Juhorski zapis - tematski zbornik*: 29–38. Svojnovo – Kruševac: Udruženje za kulturu i umetnost Logos – Istorijski Arhiv.

Martineau, R. 2010. Brussinage, polissage et degrés de séchage. *Les nouvelles de l'archéologie* 119: 13–19.

Martinon, S. 2017. Graphite-treated pottery in the northeastern Mediterranean from the Chalcolithic to the Bronze Age. *Near Eastern Archaeology* 80: 3–13.

Martinón-Torres, M. and Th. Rehren. 2008. *Archaeology, History and Science: Integrating Approaches to Ancient Materials.* Walnut Creek (CA). Left Coast Press.

Martinón-Torres, M. and Th. Rehren. 2014. Technical ceramics, in B.W. Roberts and C.P. Thornton (eds) *Archaeometallurgy in Global Perspective: Methods and Syntheses*: 107–131. New York: Springer.

Mathieson, I., S. Alpaslan-Roodenberg, C. Posth, A. Szécsényi-Nagy, N. Rohland, S. Mallick, I. Olalde, N. Broomandkhoshbacht, F. Candilio, O. Cheronet, D. Fernandes, M. Ferry, B. Gamarra, G.G. Fortes, W. Haak, E. Harney, E. Jones, D. Keating, B. Krause-Kyora, I. Kucukkalipci, M. Michel, A. Mittnik, K. Nägele, M. Novak, J. Oppenheimer, N. Patterson, S. Pfrengle, K. Sirak, K. Stewardson, S. Vai, S. Alexandrov, K.W. Alt, R. Andreescu, D. Antonović, A. Ash, N. Atanassova, K. Bacvarov, M.B. Gusztáv, H. Bocherens, M. Bolus, A. Boroneanţ, Y. Boyadzhiev, A. Budnik, J. Burmaz, S. Chohadzhiev, N.J. Conard, R. Cottiaux, M. Čuka, C. Cupillard, D.G. Drucker, N. Elenski, M. Francken, B. Galabova, G. Ganetsovski, B. Gély, T. Hajdu, V. Handzhyiska, K. Harvati, T. Higham, S. Iliev, I. Janković, I. Karavanić, D.J. Kennett, D. Komšo, A. Kozak, D. Labuda, M. Lari, C. Lazar, M. Leppek, K. Leshtakov, D.L. Vetro, D. Los, I. Lozanov, M. Malina, F. Martini, K. McSweeney, H. Meller, M. Menđušić, P. Mirea, V. Moiseyev, V. Petrova, T.D. Price, A. Simalcsik, L. Sineo, M. Šlaus, V. Slavchev, P. Stanev, A. Starović, T. Szeniczey, S. Talamo, M. Teschler-Nicola, C. Thevenet, I. Valchev, F. Valentin, S. Vasilyev, F. Veljanovska, S. Venelinova, E. Veselovskaya, B. Viola, C. Virag, J. Zaninović, S. Zäuner, P.W. Stockhammer, G. Catalano, R. Krauß, D. Caramelli, G. Zariņa, B. Gaydarska, M. Lillie, A.G. Nikitin, I. Potekhina, A. Papathanasiou, D. Borić, C. Bonsall, J. Krause, R. Pinhasi and D. Reich. 2018. The genomic history of southeastern Europe. *Nature* 555: 197. DOI: https://doi.org/10.1038/nature25778

McCarthy, J. 2014. Multi-image photogrammetry as a practical tool for cultural heritage survey and community engagement. *Journal of Archaeological Science* 43: 175–185.

McDonnell, J.G. 2001. Pyrotechnology, in D.R. Brothwell and M. Pollard (eds) *Handbook of Archaeological Science*: 493–506. London: Wiley.

McGeehan-Liritzis, V. and N.H. Gale. 1988. Chemical and lead isotope analysis of Greek Late Neolithic and Early Bronze Age Metals. *Archaeometry* 30: 199–225.

McLaren, F.S. and R.N.L.B Hubbard. 1990. The archaeobotanical remains, in R. Tringham, D. Krstić (eds) *Selevac, A Neolithic Village in Yugoslavia*: 247–254. Los Angeles: UCLA Institute of Archaeology Press.

McPherron, A. and D. Srejović (eds). 1988. *Divostin and the Neolithic of Central Serbia.* Pittsburgh (PA): University of Pittsburgh.

McPherron A., J. Rasson and B. Galdikas. 1988. Other artefact categories, in A. McPherron and D. Srejović (eds) *Divostin and the Neolithic of Central Serbia*: 325–343. Pittsburgh (PA): University of Pittsburgh.

Medović, A., R. Hofmann, T. Stanković-Pešterac, S. Dreibrodt, I. Medović and R. Pešterac. 2014. The

Late Neolithic settlement mound Borđoš near Novi Bečej, Serbian Banat, in a multiregional context – Preliminary results of geophysical, geoarchaeological and archaeological research. *Rada Muzeja Vojvodine* 56: 53–77.

Mehofer, M. 2014. Metallurgy during the Chalcolithic and the Beginning of the Early Bronze Age in Western Anatolia, in B. Horejs and M. Mehofer (eds) *Western Anatolia before Troy - Proto-Urbanisation in the 4th millennium BC*: 463–490. Vienna: OREA 1.

Meier, J. 2013. More than fun and games? An experimental study of worked bone astragali from two Middle Bronze Age Hungarian sites, in A. Choyke and S. O'Connor (eds) *From These Bare Bones: Raw Materials and the Study of Worked Osseous Objects*: 166–173. Oxford and Oakville: Oxbow Books.

Mellaart, J. 1964. Excavations at Çatalhöyük, 1963: third preliminary report. *Anatolian Studies* 14: 39–119.

Meller, H., R. Risch and E. Pernicka (eds). 2014. *Metalle der Macht - Frühes Gold und Silber. Metals of power - Early gold and silver.* Halle: Landesamt für Denkmalpflege und Archäologie Sachsen-Anhalt - Landesmuseum für Vorgeschichte Halle.

Menghin, O. 1931. *Weltgeschichte der Steinzeit.* Wien: Schroll.

Menze, B.H. and J.A. Ur. 2012. Mapping patterns of long-term settlement in Northern Mesopotamia at a large scale. *Proceedings of the National Academy of Sciences of the United States of America* 109(14): E778–E787.

Merkyte, I. 2005. *Lîga. Copper Age strategies in Bulgaria.* (Acta Archaeologica 76(1). Acta Archaeologica Supplementa VI, Centre of World Archaeology Publications 2). København: Blackwell Munksgaard.

Merlini, M. 2005. Semiotic approach to the features of the Danube script. *Documenta Praehistorica* 32: 233–251.

Merpert, N.J. and R.M. Munchaev. 1972. The early metallurgy in Mesopotamia (Древнейшая металлургия Месопотамии, in Russian), in R.M. Munchaev and N.J. Merpert (eds) *Early Agricultural Settlements in Northern Mesopotamia. The Investigations of Societ Expeditions in Iraq (Раннеземледельческие поселения северной Месопотамии. Иследования советской экспедиции в Ираке, in Russian)*: 307–316. Miami (FL). Field Research Projects.

Merpert, N.Y. and R.M. Muncaev. 1987. The earliest levels at Yarim Tepe I and Yarim Tepe II in northern Iraq. *Iraq* 49: 1–36.

Meskell, L. (ed.) 2005. *Archaeologies of Materiality.* Oxford: Blackwell.

Mesoudi, A. and J.M. O'Brien. 2008. The learning and transmission of hierarchical cultural recipes. *Biological Theory* 3: 63–72.

Michelaki, K. 2008. Making pots and potters in the Bronze Age Maros villages of Kiszombor-Új-Élet and Klárafalva-Hajdova. *Cambridge Archaeological Journal* 18: 355–380.

Michelaki, K.V., G. Braun and G.V. Hancock. 2014. Local clay sources as histories of human-landscape interactions: a ceramic taskscape perspective. *Journal of Archaeological Method and Theory* 22: 783–827.

Miculinić, K. and M. Mihaljević. 2003. Analiza faune prapovijesnog nalazišta Slavča-Nova Gradiška. *Opvscula Archaeologica* 27: 71–80.

Mihaylov, P. 2008. New evidence for the earliest copper yielding and processing in the Upper Struma Valley, in M. Čičikova and D. Gergova (eds) *Phosphorion. Studia in Honorem Mariae Čičikova*: 35–47. Sofia: Prof. Marin Drinov Academic Publishing House.

Milić, M. 2014. PXRF characterisation of obsidian from Central Anatolia, the Aegean and Central Europe. *Journal of Archaeological Science* 41: 285–296.

Milić, M. 2015. Obsidian Exchange and Societies in the Balkans and the Aegean from the Late 7th to 5th Millennia BC. Unpublished PhD dissertation, University College London.

Milić, M., K. Brown and T. Carter. 2013. A visual characterisation of the Çatalhöyük Obsidian (Appendix 21), in I. Hodder (ed.) *Substantive Technologies at Çatalhöyük Reports from the 2000-2008 Seasons*: 417–478. Los Angeles: Monographs of the Cotsen Institute of Archaeology, University of California at Los Angeles.

Milleker, F. 1938. Vorgeschichte des Banats: Neolithikum. *Starinar* 13: 102–166.

Miller, D. (ed). 2005. *Materiality.* Durham (NC): Duke University Press.

Miller, H. 2007. *Archaeological Approaches to Technology.* London, Amsterdam: Elsevier/Academic Press.

Milojčić, V. 1949. *Chronologie der Jungeren Steinzeit Mittel- und Sudosteuropas.* Berlin: Gebr. Mann.

Mineral Deposits of Serbia, n.d., Ore Deposit Database, viewed 18 June 2020, <https://bokisingl.files.wordpress.com/2011/04/serbia-mineral-deposits.pdf>

Mioč, U.B., P. Colomban, G. Sagon, M. Stojanović and A. Rosić. 2004. Ochre decor and cinnabar residues in Neolithic pottery from Vinča, Serbia. *Journal of Raman Spectroscopy* 35: 843–846. DOI: https://doi.org/10.1002/jrs.1221.

Mittnik, A., K. Massy, C. Knipper, F. Wittenborn, R. Friedrich, S. Pfrengle, M. Burri, N. Carlichi-Witjes, H. Deeg, A. Furtwängler, M. Harbeck, K. von Heyking, C. Kociumaka, I. Kucukkalipci, S. Lindauer, S. Metz, A. Staskiewicz, A. Thiel, J. Wahl, W. Haak, E. Pernicka, S. Schiffels, P.W. Stockhammer and J. Krause. 2019. Kinship-based social inequality in Bronze Age Europe. *Science* 366: 731–734.

Montero-Ruiz, I., M. Murillo-Barroso, and M. Ruiz-Taboada 2021. The beginning of the production and use of metal in Iberia: From independent invention to technological innovation. *Eurasia Antiqua* 23: 193–208.

Molloy, B. 2008. Martial arts and materiality: a combat archaeology perspective on Aegean swords of the fifteenth and fourteenth centuries BC. *World Archaeology* 40: 116–134.

Molloy, B. and M. Mödlinger. 2020. The organisation and practice of metal smithing in Later Bronze Age Europe. *Journal of World Prehistory* 33: 169–232.

Monah, D., G. Dumitroaia, F. Monah, C. Preoeasa, R. Munteanu and D. Nicola. 2002. *Poduri-Dealul Ghindaru: o troie în Subcarpații Moldovei* (Bibliotheca Memoriae Antiquitatis 13). Piatra Neamț: Editura Constantin Matasa.

Montelius, O. 1899. *Der Orient und Europa: Einfluss der orientalischen Cultur auf Europa bis zur Mitte des letzten Jahrtausends v. Chr.* Stockholm: Königliche Akademie der schönen Wissenschaften, Geschichte und Alterthumskunde.

Monthel, J., P. Vadala, J.M. Leistel and F. Cottard. 2002. *Mineral Deposits and Mining Districts of Serbia.* Belgrade: Republic of Serbia Ministry of Mining and Energy, Geoinstitut Belgrade and Mining and Geology Faculty.

Montufo, A.M. 1997. The use of satellite imagery and digital image processing in landscape archaeology. A case study from the island of Mallorca, Spain. *Geoarchaeology: An International Journal* 12(1): 71–85.

Morgan, L.H. 1985 [1877]. *Ancient Society*. Tucson (AZ): University of Arizona Press.

Motolsci, J. 1970. Historische Erforschung der Körpergrösse des Rindes auf Grund von Ungarischem Knochenmaterial. *Zeitchr. f. Tierzüchtg. u. Züchtgsbiol.* 87 (2): 89–137.

Mrkobrad, D. 1989. Istraživanje rudarske topionice u Donjoj Zaplanini. *Glasnik Društva Konzervatora Srbije* 13: 91–92.

Mrkobrad, D., V. Bogosavljević and D. Bogosavljević. 1989. Prilog istraživanju kopaoničkog rudarstva. *Glasnik Društva Konzervatora Srbije* 13: 93–95.

Muhly, J.D. 1989. Çayönü Tepeşi and the beginnings of metallurgy in the Old World, in A. Hauptmann, E. Pernicka and G.A Wagner (eds) *Old World Archaeometallurgy* (Der Anschnitt, Beiheft 7): 1–13. Bochum: Deutsches Bergbau-Museum.

Muhly, J.D. 2002. Early metallurgy in Greece and Cyprus, in Ü. Yalçın (ed.) *Anatolian Metal II,* (Der Anschnitt, Beiheft 15): 77–82. Bochum: Deutsches Bergbau-Museum.

Muhi, B., D. Bugarin and J. Miloradić. 2018. Resursi i perspektive razvoja industrijskog i rudarskog turizma u Srbiji (Resources and perspectives of developing industrial and mining tourism in Serbia, in Serbian). *Business Economics* 23: 141–156.

Müller, J. 2006. Demographische Variablen des Bosnischen Spätneolithikums - zur Frage der Bevölkerungsrekonstruktion im Südosteuropäischen Neolithikum, in N. Tasić and C. Grozdanov (eds) *Homage to Milutin Garašanin:* 367–378. Belgrade: Serbian Academy of Sciences and Arts, Macedonian Academy of Sciences and Arts.

Müller, J. 2007. Demografski varijabiliteti u kasnom neolitu na području Bosne i pitanje gustine naseljenosti u neolitu jugoistočne Evrope, *Godišnjak Centra za balkanološka ispitivanja ANU BiH* 34: 17–33.

Müller, J. 2012. Tells, fire, and copper as social technologies, in R. Hofmann, F.-K. Moetz and J. Müller (eds) *Tells: Social and Environmental Space. Proceedings of the International Workshop 'Socio-Environmental Dynamics Over the Last 12,000 Years: The Creation of Landscapes II' in Kiel, 14th -18th March 2011. Volume 3* (Universitätsforschungen zur Prähistorischen Archäologie 207): 47–52. Bonn: Habelt.

Müller, J. 2017. Inheritance, population and social identities. Southeast Europe 5200-4300 BCE, in M. Gori and M. Ivanova (eds) *Balkan Dialogues. Negotiating Identity between Prehistory and the Present*: 156–168. London: Routledge.

Müller, R., Th. Rehren and S. Rovira. 2004. Almizaraque and the early copper metallurgy of southeast Spain: new data. *Madrider Mitteilungen* 45: 33–56.

Müller, J., K. Rassmann and R. Hofmann (eds). 2013a. *Okolište 1 - Untersuchungen einer spätneolithischen Siedlungskammer in Zentralbosnien* (Universitätsforschungen zur Prähistorischen Archäologie 228). Bonn: Verlag Rudolf Habelt

Müller, J., K., Rassmann and Z. Kujundžić-Vejzagić. 2013b. Okolište – Rekonstruktion spätneolithischer und frühchalkolithischer Siedlungsprozesse des zentralbosnischen Visokobeckens: Fragestellungen, Forschungsstrategien, Ergebnisse, in J. Müller, K. Rassmann, and R. Hofmann (eds) *Okolište 1 - Untersuchungen einer spätneolithischen Siedlungskammer in Zentralbosnien*: 11–68. Bonn: Habelt.

Müller, J., K. Rassmann and M. Videjko (eds). 2016. *Tripolye-Megasites and European Prehistory 4100-3400 BCE. Themes in Contemporary Archaeology*. London: Routledge and CRC Press.

Müller-Scheeßel, N., J. Bátora, S. Reiter, K. Radloff and P. Tóth. 2016. Joint prospections on Early Bronze Age site in the Žitava Valley by the Romano-German Commission, Frankfurt and the Archaeological Institute of the Slovak Academy of Sciences, Nitra and Comenius University, Bratislava. *Študijné Zvesti* 60: 79–96.

Müller-Scheeßel, N., J. Müller, I. Cheben, W. Mainusch, K. Rassmann, W. Rabbel, E. Corradini and M. Furholt. 2020. A new approach to the temporal significance of house orientation in European Early Neolithic settlements. *Plos One* 15(1): 1–18.

Murillo-Barroso, M. and I. Montero Ruiz. 2012. Copper ornaments in the Iberian Chalcolithic: technology versus social demand. *Journal of Mediterranean Archaeology* 25: 53–73.

N

Naumov, G., L. Fidanoski, I. Tolevski and A. Ivkovska. 2009. *Neolithic Communities in the Republic of Macedonia*. Skopje: Dante.

Neff, H. 1992. Ceramics and evolution. *Archaeological Method and Theory* 4: 141–193.

Neff, H. 1996. Ceramics and evolution, in M.J. O' Brian (ed.) *Evolutionary Archaeology: Theory and Application*: 244–269. Salt Lake City: University of Utah Press.

Nelson, R. and S. Winter. 1982. *An Evolutionary Theory of Economic Change*. Cambridge (MA): Belknap.

Nerantzis, N., Y. Bassiakos and S. Papadopoulos. 2016. Copper metallurgy of the Early Bronze Age in Thassos, north Aegean. *Journal of Archaeological Science Reports* 7: 574–580. DOI: http://dx.doi.org/10.1016/j.jasrep.2015.08.008.

Nessel, B. 2012. Metallurgen im Grab: Überlegungen zur sozialen Einstufung handwerklicher Spezialisten, in T.L. Kienlin and A. Zimmerman (eds) *Beyond Elites: Alternatives to Hierarchical Systems in Modelling Social Formations*: 423–432. Bonn: Habelt.

Nessel, B. 2013. The 'absence' of smiths and founders: Why tools are rare in Bronze Age burials, in B. Rezi, R.E. Németh and S. Berecki (eds) *Bronze Age Crafts and Craftsmen in the Carpathian Basin*: 139–148. Târgu Mureş: Editura MEGA.

Neubauer, F. and C. Heinrich. 2003. Late Cretaceous and Tertiary geodynamics and ore deposit evolution of the Alpine-Balkan-Dinaride orogen, in D. Eliopoulos (ed.) *Mineral Exploration and Sustainable Development. Proceedings of the Seventh Biennial SGA Meeting on Mineral Exploration and Sustainable Development, Athens, Greece, August 24-28, 2003*: 1133–1136. Athens: Millpress.

Neuninger, H., R. Pittioni and W. Siegl 1964. Frühkeramikzeitliche Kupfergewinnung in Anatolien. *Archaeologia Austriaca* 35: 98–110.

Nikolić, D. 2004. Keramičko posuđe, in M. Bogdanović (ed.) *Grivac. Naselja protostarčevačke i vinčanske kulture*: 203–315. Kragujevac: Narodni muzej.

Nikolić, D. (ed.). 2008. *Sto godina istraživanja arheološkog lokaliteta Vinča-Belo Brdo* (100 Years of archaeological research at Vinča- Belo Brdo, in Serbian). Belgrade: Faculty of Philosophy University of Belgrade, National Museum Belgrade, City Museum Belgrade, Serbian Academy of Arts and Sciences.

Nikolov, V. 1975. *Bogdan, Zaminec, praistorisko selishte*. Sofia.

Nikolov, V. 1984. *Krivodol-drevni kulturi*. Sofia.

Nodari, L., E. Marcuz, L. Maritan, C. Mazzoli and U. Russo. 2007. Hematite nucleation and growth in the firing of carbonate-rich clay for pottery production. *Journal of the European Ceramic Society* 27: 4665–4673.

Nørgaard, H.W., E. Pernicka and H. Vandkilde. 2019. On the trail of Scandinavia's early metallurgy: Provenance, transfer and mixing. *PLOS ONE* 14: e0219574.

Northover, J.P. 1989. Properties and use of arsenic-copper alloys, in A. Hauptmann, E. Pernicka and G.A. Wagner (eds) *Old World Archaeometallurgy*: 111–118. Bochum: Deutsches Bergbau-Museum.

Novotna, M. 1970. *Die Äxte und Beile in der Slowakei.* (Prähistorische Bronzefunde 9/3). München: Beck.

Novotný, B. 1958. *Die Slowakei in der Jungeren Steinzeit.* Bratislava: Archeologický ústav Slovenskej akadémie vied v Nitre.

O

Obelić, B., M. Krznarić-Škrivanko, B. Marijan and I. Krajcar-Bronić. 2004. Radiocarbon dating of Sopot culture sites (Late Neolithic) in eastern Croatia. *Radiocarbon* 46(1): 245–258.

Obradović, Đ. 2013. *New data on early agriculture in the Central Balkans: Archaeobotanical investigations at early and Late Neolithic sites in the Morava Valley, Serbia.* Poster presented at the 16th conference of the International Work Group for Palaeoethnobotany, Thessaloniki, Greece. Abstract in *Book of Abstracts*: 167.

Obradović, Đ. 2020. Arheobotanička istraživanja u kasnom neolitu Pomoravlja: društveno-ekonomski aspekti proizvodnje i pripreme hrane. Unpublished doctoral dissertation, University of Belgrade.

O'Brien, M.J. and S.J. Shennan (eds). 2010a. *Innovation in Cultural Systems. Contributions from Evolutionary Anthropology.* Cambridge, MA: The MIT Press.

O'Brien, M.J. and S.J. Shennan. 2010b. Issues in anthropological studies of innovation, in M.J. O'Brien and S.J. Shennan (eds) *Innovation in Cultural Systems. Contributions from Evolutionary Anthropology*: 3–17. Cambridge (MA): The MIT Press.

O'Brien, W. 2015. *Prehistoric Copper Mining in Europe: 5000-500 BC.* Oxford: Oxford University Press.

O'Connor, T. 2003. *The Analysis of Urban Animal Bone Assemblages.* York: Council for British Archaeology.

Olausson, D. 1980. Starting from scratch: the history of edge-wear research from 1838 to 1978. *Lithic Technology* 9(2): 48–60.

Olausson, D. 1983a. *Flint and groundstone axes in the Scanian Neolithic. An evaluation of raw materials based on experiment.* Lund: CWK Gleerup.

Olausson, D. 1983b. Tools and Technology: Lithic Technological Analysis of Neolithic Axe Morphology. Unpublished Phd Thesis, Lund University.

Olausson, D. 1990. Edge-wear analysis in archaeology. The current state of research. *Laborativ arkeologi* 4: 5–14.

Olofsson, A. and I. Rodushkin. 2011. Provenancing flint artefacts with ICP-MS using REE signatures and Pb isotopes as discriminants: preliminary results of a case study from Northern Sweden. *Archaeometry* 53: 1142–1170. DOI: https://doi.org/10.1111/j.1475-4754.2011.00605.x.

Orfanou, V. and Th. Rehren. 2015. A (not so) dangerous method: pXRF vs. EPMA-WDS analyses of copper-based artefacts. *Archaeological and Anthropological Sciences* 7: 387–397. DOI: https://doi.org/10.1007/s12520-014-0198-z.

Orton, C. and M. Hughes. 2013 *Pottery in Archaeology*. Cambridge: Cambridge University Press.

Orton, D.C. 2008. Beyond Hunting and Herding: Humans, Animals, and the Political Economy of the Vinča Period. Unpublished Ph.D. dissertation, University of Cambridge.

Orton, D. 2010. Both subject and object: Herding, inalienability and sentient property in prehistory. *World Archaeology* 42: 188–200. DOI: https://doi.org/10.1080/00438241003672773.

Orton, D.C. 2012. Herding, settlement, and chronology in the Balkan Neolithic. *European Journal of Archaeology* 15 (1): 5–40. DOI: https://doi.org/10.1179/1461957112Y.0000000003.

Orton, D.C. 2014. Preliminary report on Neolithic animal bones from Kočićevo, in I. Pandžić and M. Vander Linden (eds) *The Neolithic Site of Kočićevo in the Lower Vrbas Valley (Republika Srpska, Bosnia and Herzegovina): Report of the 2009-2014 Field Seasons*: 97–120. Banja Luka: Philosophy Faculty, University of Banja Luka.

Orton, D., J. Gaastra and M. Vander Linden. 2016. Between the Danube and the deep blue sea: Zooarchaeological meta-analysis reveals variability in the spread and development of Neolithic farming across the Western Balkans. *Open Quaternary* 2: 1–26. DOI: http://doi.org/10.5334/oq.28.

Ottaway, B.S. 1979. Analysis of the earliest metal finds from Gomolava. *Rad Vojvođanskih muzeja* 25: 53–59.

Ottaway, B.S. 1989. Interaction of some of the earliest copper-using cultures in central Europe, in A. Hauptmann, E. Pernicka and G.A. Wagner (eds) *Old World Archaeometallurgy* (Der Anschnitt, Beiheft 7): 19–23. Bochum: Deutsches Bergbau-Museum.

Ottaway, B.S. 1994. *Prähistorische Archäometallurgie*. Espelkamp: Marie Leidorf.

Ottaway, B.S. 2001. Innovation, production and specialisation in early prehistoric copper metallurgy. *European Journal of Archaeology* 4: 87–112.

Ottaway, B.S. and B. Roberts. 2008. The emergence of metalworking, in A. Jones (ed.) *Prehistoric Europe: Theory and Practice*: 193–225. Chichester: Wiley-Blackwell.

Özdoğan, M. 2011 Archaeological evidence on the westward expansion of farming communities from Eastern Anatolia to the Aegean and the Balkans. *Current Anthropology* 52(S4): S415–S430.

Özdoğan, M. and N.O. Dede. 1989. Excavations at Toptepe in Eastern Thrace 1989. *Arkeoloji ve Sanat* 13(46–49): 3–23.

Özdoğan, M. and A. Özdoğan. 1999. Archaeological evidence on the early metallurgy at Çayönü Tepeşi, in A. Hauptmann, E. Pernicka, Th. Rehren and Ü. Yalçın (eds) *The Beginnings of Metallurgy* (Der Anschnitt, Beiheft 9): 13–22. Bochum: Deutsches Bergbau-Museum.

P

Pakkanen, J., A. Brysbaert, D. Turner, D. and Y. Boswinkel. 2020. Efficient three-dimensional field documentation methods for labour cost studies: Case studies from archaeological and heritage contexts. *Digital Applications in Archaeology and Cultural Heritage* 17: 1–10.

Palavestra, A., I. Bogdanović and A. Starović. 1993. Bodnjik-Družetić – eneolitsko gradinsko naselje. *Glasnik Srpskog arheološkog društva* 9: 186–191.

Palavestra A., I. Bogdanović and A. Starović. 1996. Bodnjik-Družetić, kampanja 1994. *Glasnik Srpskog arheološkog društva* 11: 190–197.

Paluch T. 2011. *Maroslele-Pana: egy középső neolitikus lelő hely a kultúrák határvidékén (Maroslele-Pana: A Middle Neolithic site at the frontier of cultures).* (Monographia Archaeologica 2). Szeged: Móra Ferenc Múzeum.

Pantović, I. 2014. *Vinčanski Amulet.* Vršac: Gradski muzej.

Papadopoulos, S. 2008. Silver and copper production practices in the prehistoric settlement at Limenaria, Thasos, in I. Tzachili (ed.) *Aegean Metallurgy in the Bronze Age. Proceedings of an International Symposium held at the University of Crete, Rethymnon, Greece, on November 19-21, 2004*: 59–67. Athens: Ta Pragmata Publications.

Papathanasiou, A., W. Parkinson, D. Pullen, M. Galaty and P. Karkanas (eds). 2018. *Neolithic Alepotrypa Cave in the Mani Greece*. Oxford: Oxbow.

Parcak, S. 2009. *Satellite Remote Sensing for Archaeology.* New York: Routledge.

Pare, C.F.E. 2000. Bronze and the Bronze Age, in C.F.E. Pare (ed.) *Metals Make the World go Round. The Supply and Circulation of Metals in Bronze Age Europe*: 1–38. Oxford: Oxbow Books.

Parkinson, W.A., R.W. Yerkes and A. Gyucha. 2004. The transition from the Neolithic to the Copper Age: excavations at Vésztı-Bikeri, Hungary, 2000–2002. *Journal of Field Archaeology* 29(1/2): 101–121.

Parzinger, H. 1993. *Studien zur Chronologie und Kulturgeschichte der Jungstein-, Kupfer- und Frühbronzezeit zwischen Karpaten und Mittlerem Taurus. Volume 52.* Mainz am Rhein: von Zabern.

Parzinger, H. 2005. Die mittel- und spätneolithische Keramik aus Aşağı Pınar Grabungen 1993–1998, in H. Parzinger and H. Schwarzberg (eds) *Aşağı Pınar II. Die mittel- und spätneolithische Keramik* (Archäologie in Eurasien 18 – Studien im Thrakien-Marmara Raum 2): 1–245. Mainz: Verlag Philipp von Zabern.

Patay, P. 1974. Die hochkupferzeitliche Bodrogkeresztúr-Kultur. *Bericht der Römisch Germanischen Komission* 55: 1–71.

Patay, P. 1984. *Kupferzeitliche Meißel, Beile und Äxte in Ungarn.* (Prähistorische Bronzefunde Volume 9, Band 15). München: C.H. Beck'sche Verlagsbuchhandlung.

Patterson, C.C. 1971. Native copper, silver, and gold accessible to early metallurgists. *American Antiquity* 36: 286–321. DOI: https://doi.org/10.2307/277716.

Park, J.-S., Voyakin, D. and A. Beisenov 2020. The implication of diachronic changes reflected in LBA bronze assemblages of Central Kazakhstan. *Archaeological and Anthropological Sciences,* 12 (7). DOI: https://doi.org/10.1007/s12520-019-00989-z.

Pascual Benito, J.L. 1998. *Utillaje óseo, adornos e ídolos neolíticos valencianos.* Valencia: Disputación provincial de Valencia. Servicio de investigación prehistórica. Serie de trabajos varios 95.

Pawlik, A.F. 2007. Analysis of two polished stone adzes from Ille Cave at El Nido, Palawan Island, Philippines. *Hukay: Journal of the University of the Philippines Archaeological Studies Program* 10: 38–59.

Payne, S. 1972. Partial recovery and sample bias: the results of some sieving experiments, in E. Higgs (ed.) *Papers in Economic Prehistory*: 49–64. Cambridge: Cambridge University Press.

Payne, S. 1973. Kill-off patterns in sheep and goats: the mandibles from Aşvan Kale. *Anatolian Studies* 23: 281–303.

Payne, S. 1985. Morphological distinctions between mandibular teeth of young sheep, *Ovis*, and goats, *Capra*. *Journal of Archaeological Science* 12: 139–147.

Pearce, M. 2015. The spread of early copper mining and metallurgy in Europe: An assessment of the diffusionist model, in A. Hauptmann, D. Modarressi-Tehrani and M. Prange (eds) *Archaeometallurgy in Europe III: Proceedings of the 3rd International Conference, Deutsches Bergbau-Museum Bochum, June 29 - July 1, 2011* (Der Anschnitt 26): 45–54. Bochum: Deutsches Bergbau-Museum.

Pearce, M. 2019. The 'Copper Age'– a history of the concept. *Journal of World Prehistory* 32: 229–250. DOI: https://doi.org/10.1007/s10963-019-09134-z.

Pearsall, M.D. 2000. *Paleoethnobotany. A Handbook of Procedures*. Orlando: Academic Press.

Pecikoza, V. 2011. Deposits of copper and Vinča sites in the area of Valjevo Mountains, in Blagojević, M. (ed.) *Kolubara* 5: 39–46. Belgrade: Republički zavod za zaštitu spomenika kulture

Pelegrin, J. 1988. Débitage expérimental par pression: du plus petit au plus grand, in *Technologie Préhistorique*. (Notes et Monographies Techniques du CRA 25): 37–53. Paris: Edition du CNRS.

Pelegrin, J. 2006. Long blade technology in the Old World: an experimental approach and some archaeological results, in J. Apel and K. Knutsson (eds) *Skilled Production and Social Reproduction - Aspects of Traditional Stone-Tool Technology*: 37–68. Upsalla: Societas Archeologica Upsaliensis.

Pentedeka, A. 2011. Links of clay in Neolithic Greece: the case of Platia Magoula Zarkou, in A. Brysbaert (ed.) *Tracing Prehistoric Social Networks through Technology: A Diachronic Perspective on the Aegean*: 106–125. London/New York: Routledge.

Perić, S. 2006. The Gradac period in the Neolithic settlements in the Morava valley, in N. Tasić and C. Grozdanov (eds) *Homage to Milutin Garašanin*: 235–250. Belgrade: Serbian Academy of Sciences and Arts.

Perić, S., C. Rummel, G. Schafferer, H. Wendling and D. Winger. 2016. Geomagnetic survey of Neolithic settlements in the Middle Morava Valley – preliminary results, in S. Perić (ed.) *The Neolithic in the Morava Valley. New Insights into Settlement and Economy*: 9–27. Belgrade: Institute of Archaeology.

Perišić, S. 1984. *Predmeti od kosti, roga i kamena iz Odseka za praistoriju Muzeja grada Beograda*. Beograd: Muzej grada Beograda.

Perišić, N., M. Marić-Stojanović, V. Andrić, U.B. Mioč and L. Damjanović. 2016. Physicochemical characterisation of pottery from Vinča culture, Serbia regarding the firing temperature and decoration technique. *Journal of the Serbian Chemical Society* 81/12: 1415–1426.

Pernicheva, L. 2004. Chalcolithic anthropomorphic figurines from Tell Kirilovo, Stara Zagora District, in V. Nikolov, K. Bachvarov, and P. Kalchev (eds) *Praistoričeska Trakija*. Sofia-Stara Zagora: Arheološki institut.

Pernicka, E. 1990. Gewinnung und Verbreitung der Metalle in prähistorischer Zeit. *Jahrbuch des Römisch-Germanischen Zentralmuseums Mainz* 37: 21–129.

Pernicka, E. 1999. Trace element fingerprinting of ancient copper: a guide to technology or provenance? In S.M.M. Young, A.M. Pollard, P. Budd and R.A. Ixer (eds) *Metals in Antiquity*: 163–171. Oxford: Archaeopress.

Pernicka, E. 2020. Development of metallurgy in Eurasia. *Quaternary International* 560–561: 38–44. DOI: https://doi.org/10.1016/j.quaint.2020.03.057.

Pernicka, E., F. Begemann and S. Schmitt-Strecker. 1993. Eneolithic and Early Bronze Age copper artefacts from the Balkans and their relation to Serbian copper ores. *Prähistorische Zeitschrift* 68: 1–54.

Pernicka, E., F. Begemann, S. Schmitt-Strecker, H. Todorova and I. Kuleff. 1997. Prehistoric copper in Bulgaria. Its composition and provenance. *Eurasia Antiqua* 3: 41–180.

Pernicka, E., Lutz, J. and Stöllner, T. 2016. Bronze Age copper produced at Mitterberg, Austria, and its distribution. *Archaeologia Austriaca* 100: 19-55.

Perucchetti, L., Montero-Ruiz, I., and P. Bray 2020. Mapping archaeometallurgical data of the Iberian Copper Age: Different ways to look at a big picture. *Journal of Archaeological Science* 119: 105165. DOI: https://doi.org/10.1016/j.jas.2020.105165

Petrović, J. 1992. Архитектура куће 4 на Гомолави; насеље млађе винчанске културе, *Рад Војвођанских музеја* 34: 19–32.

Pétrequin P. and C. Jeunesse. 1995. La Hache de Pierre. *Carrières Vosgiennes et Échanges de Lames Polies pendant le Néolithique (5400-2100 avant J.-C.).* Paris: Editions Errance.

Pétrequin, P., S. Cassen, M. Errera, L. Klassen, A.M. Pétrequin and J.A. Sheridan. 2013. The value of things: the production and circulation of Alpine jade axes during the 5th-4th millennia in a European perspective, in *Economic Archaeology: From Structure to Performance in European Archaeology* (University Research on Prehistoric Archaeology 237): 65–82. Bonn: Habelt.

Petrović, J. 1999/2000. Građevinski objekti u naselju Starčevačke kulture na Golokutu. *Rad Muzeja Vojvodine* 41–42: 7–10.

Picon, M. 1995. Grises et grises: quelques réflexions sur les céramiques cuites en mode B, in Câmara Municipal de Tondela (ed.) *Actas das 1ªJornadas de Cerâmica Medieval e Pós Medieval, métodos e resultados para o seu estudo* (1992): 283–287. Tondela: Porto, Municipal de Tondela.

Pigott, V.C. 1999. The development of metal production on the Iranian Plateau: an archaeometallurgical perspective, in V.C. Pigott (ed.) *The Archaeometallurgy of the Asian World*: 73–106. Philadelphia (PA): University of Pennsylvania Museum.

Pigott, V.C. 2011. Sources of tin and the tin trade in southwest Asia: recent research and its relevance to current understanding, in P.P. Betancourt and S.C. Ferrence (eds) *Metallurgy: Understanding How, Learning Why. Studies in Honor of James D. Muhly*: 273–291. Philadelphia (PA): INSTAP Academic Press.

Pigott, V.C. 2021. The acquisition of tin in Bronze Age southwest Asia, in B. Lyonnet and N.A. Dubova (eds) *The World of the Oxus Civilization.* London: Routledge.

Pitblado, B.L., M.B. Cannon, H. Neff, C.M. Dehler and S.T. Nelson. 2013. LA-ICP-MS analysis of quartzite from the Upper Gunnison Basin, Colorado. *Journal of Archaeological Science* 40(4): 2196–2216. DOI: http://www.sciencedirect.com/science/article/pii/S0305440312005109

Plisson, H. and A. Lompré. 2008. Technician or researcher? A visual answer, in L. Longo and N. Skakun (eds) *'Prehistoric Technology' 40 Years Later: Functional Studies and the Russian Legacy: Proceedings of the International Congress Verona (Italy), 20-23 April 2005* (British Archaeological Reports International Series S1783): 497–501. Oxford: Archaeopress.

Pollard, A.M. and C. Heron. 2008. *Archaeological Chemistry.* Cambridge: The Royal Society of Chemistry.

Poplin F. 2004. Fiche éléments de nomenclature anatomique relative aux matères dures d'origine animale, in D. Ramseyer (ed.) *Matières et techniques. Fiches de la Commission de nomenclature sur l'industrie de l'os préhistorique. Cahier XI*: 11–15. Paris: CNRS.

Popova, Tz. 2014. Agriculture and subsistence economy in the territory of Bulgaria during Neo-Eneolithic, in C. Preoteasa and C.-D. Nicola (eds) *L'impact anthropique sur l'environnement durant le Néo-Énéolithique du Sud-Est de l'Europe: in honorem dr. Gheorghe Dumitroaia*: 37–47. Piatra-Neamţ: Editura 'Constantin Matasă'.

Porčić, M. 2009. Archaeology of Vinča Houses: Theoretical-Methodological Frameworks of Studying Demography and Social Structure. Unpublished PhD dissertation. Belgrade: University of Belgrade.

Porčić, M. 2011. An exercise in archaeological demography: estimating the population size of Late Neolithic settlements in the central Balkans. *Documenta Praehistorica* 38: 323–332.

Porčić, M. 2012a. Social complexity and inequality in the Late Neolithic of the Central Balkans: reviewing the evidence. *Documenta Praehistorica* 39: 167–183.

Porčić, M. 2012b. De facto refuse or structured deposition? House inventories of the Late Neolithic Vinča culture. *Starinar* 62: 19–43.

Porčić, M. 2012c. Effects of residential mobility on the ratio of average house floor area to average household size: implications for demographic reconstructions in archaeology. *Cross-Cultural Research* 46: 72–86.

Porčić, M. 2019a. Evaluating social complexity and inequality in the Balkans between 6500 and 4200 BC. *Journal of Archaeological Research*, doi:10.1007/s10814-018-9126-6.

Porčić, M. 2019b. Society and technology in the Neolithic and the Eneolithic of the Balkans, in J. Müller and S. Kadrow (eds) *Habitus? The Social Dimension of Technology and Transformation*:19–33. Leiden: Sidestone Press.

Porčić, M. 2020. Observations on the origin and demography of the Vinča culture. *Quaternary International* 560–561: 57–64. DOI: https://doi.org/10.1016/j.quaint.2020.04.012.

Porčić, M. and T. Blagojević. 2014. Fragmentation, context and spatial distribution of the Late Neolithic figurines from Divostin, Serbia, in C.E. Ursu and S. Terna (eds) *Anthropomorphism and Symbolic Behaviour in the Neolithic and Copper Age Communities of South-Eastern Europe*: 91–107. Suecedava: Editura karl A. Romstorfer.

Porčić, M., T. Blagojević and S. Stefanović. 2016. Demography of the Early Neolithic population in Central Balkans: Population dynamics reconstruction using summed radiocarbon probability distributions. *PLoS ONE* 11: e0160832. DOI: https://doi.org/10.1371/journal.pone.0160832.

Porčić, M. and M. Nikolić. 2016. The approximate Bayesian computation approach to reconstructing population dynamics and size from settlement data: demography of the Mesolithic-Neolithic transition at Lepenski Vir. *Archaeological and Anthropological Sciences* 8: 169–186.

Pottage, A. and B. Sherman. 2010. *Figures of Invention: A History of Modern Patent Law.* Oxford: Oxford University Press.

Poupeau, G., F.-X. Le Bourdonnec, T. Carter, S. Delerue, M.S. Shackley, J.-A. Barrat, S. Dubernet, P. Moretto, T. Calligaro, M. Milić and K. Kobayashi. 2010. The use of SEM-EDS, PIXE and EDXRF for obsidian provenance studies in the Near East: a case study from Neolithic Çatalhöyük (Central Anatolia). *Journal of Archaeological Science* 37(11): 2705–2720.

Powell, A., S. Shennan and M.G. Thomas. 2009. Late Pleistocene demography and the appearance of modern human behavior. *Science* 324: 1298–1301.

Pritchard-Parker, M.A. and J.A. Torres. 1998. Analysis of experimental debitage from hammerstone use and production: implications for ground stone use. *Lithic Technology* 23 (2): 139–146.

Prummel, W. 1988. Distinguishing features of postcranial skeletal elements of cattle, *Bos primigenius f. taurus*, and red deer, *Cervus elaphus. Schriftenaus der Archaologisch-Zoologishen Arbeitsgruppe Schleswig-Kiel* 12: 1–52.

Prummel, W. and H. Frisch, H. 1986. A guide for the distinction of species, sex and body size in bones of sheep and goats. *Journal of Archaeological Science* 13: 567–577.

Q

Quinn, P.S. 2013. *Ceramic Petrography: The Interpretation of Archaeological Pottery and Related Artefacts in Thin Section.* Oxford: Archaeopress.

Quinn, P.S., P. Day, V. Kilikoglou, E. Faber, S. Katsarou-Tzeveleki and A. Sampson. 2010. Keeping an eye on your pots: the provenance of Neolithic ceramics from the Cave of the Cyclops, Youra, Greece. *Journal of Archaeological Science* 37: 1042–1052.

Quitta, H. and G. Kol. 1969. Radiocarbondaten zum Neolithikum und zur frühen Bronzezeit Südosteuropas und der Sowietunion. *Zeitschrift für Archäologie* 3: 223–255.

R

Racimo, F., J. Woodbridge, R.M. Fyfe, M. Sikora, K.-G. Sjögren, K. Kristiansen and M.Vander Linden. 2020. The spatiotemporal spread of human migrations during the European Holocene. *Proceedings of the National Academy of Sciences* 117: 8989. DOI: https://doi.org/10.1073/pnas.1920051117.

Racky, P. and A. Anders. 2006. Social dimensions of the Late Neolithic settlement of Polgár-Csöszhalom (Eastern Hungary). *Acta Antiqua Academiae Scientiarum Hungaricae* 57: 17–3.

Radivojević, M. 1998. Field survey and topography of a Vinča culture settlement in Kremen- Mačina (in Serbian). *Papers from the Petnica Science Centre (Serbia)* 48: 316–340.

Radivojević, M. 2006. A contribution to the typology and distribution of hammer axes of Pločnik type in southeast Europe (in Serbian). *Journal of Serbian Archaeological Society* 22: 211–224.

Radivojević, M. 2007. Evidence for Early Copper Smelting in Belovode, a Vinča Culture Settlement in Eastern Serbia. Unpublished MSc dissertation, University College London.

Radivojević, M. 2012. On the Origins of Metallurgy in Europe: Metal Production in the Vinča Culture. Unpublished PhD dissertation, University College London.

Radivojević, M. 2013. Archaeometallurgy of the Vinča culture: a case study of the site of Belovode in eastern Serbia. *Historical Metallurgy* 47: 13–32.

Radivojević, M. 2015. Inventing metallurgy in western Eurasia: a look through the microscope lens. *Cambridge Archaeological Journal* 25: 321–338.

Radivojević, M., F. Draşovean, B. Gaydarska, J. Chapman and Th. Rehren (in preparation). Chalcolithic copper production at the Foeni-Petreşti culture site of Foeni, Romania. *Journal of Archaeological Science.*

Radivojević, M. and J. Grujić. 2017. Dataset for 'Community structure of copper supply networks in the prehistoric Balkans: An independent evaluation of the archaeological record from the 7th to the 4th millennium BC' (AHRC, Trans.). *Journal of Complex Networks* 6. DOI: https://doi.org/10.17863/CAM.9599.

Radivojević, M. and J. Grujić. 2018. Community structure of copper supply networks in the prehistoric Balkans: An independent evaluation of the archaeological record from the 7th to the 4th millennium BC. *Journal of Complex Networks* 6: 106–124. DOI https://doi.org/10.1093/comnet/cnx013.

Radivojević, M. and J. Kuzmanović Cvetković. 2014. Copper minerals and archaeometallurgical materials from the Vinča culture sites of Belovode and Pločnik: overview of the evidence and new data. *Starinar* 64: 7–30.

Radivojević, M. and Th. Rehren. 2016. Paint it black: the rise of metallurgy in the Balkans. *Journal of Archaeological Method and Theory* 23: 200–237. DOI: https://doi.org/10.1007/s10816-014-9238-3.

Radivojević, M., and Th. Rehren. 2019. *First to the finish line? Metallurgy of lead making in the 6th and 5th millennium BC Balkans.* Paper presented at the Archaeometallurgy in Europe, 5th International Conference, 19th-21st June 2019, Miskolc, Hungary.

Radivojević, M. and B.W. Roberts. 2013. *The Polychrome Beginnings of the Use of Metals and Alloys.* Paper presented at the Eighth International Conference on the Beginnings of the Use of Metals and Alloys, 10-15 September 2013, Nara, Japan.

Radivojević, M., Th. Rehren, E. Pernicka, D. Šljivar and M. Brauns and D. Borić. 2010a. On the origins of extractive metallurgy: new evidence from Europe.

Journal of Archaeological Science 37: 2775–2787: DOI: https://doi.org/10.1016/j.jas.2010.06.012.

Radivojević, M., Th. Rehren and M. Blagojević. 2010b *Arsenical Copper Smelting in Kolubara, western Serbia.* Paper presented at the 38th International Symposium on Archaeometry Abstract Handbook, Tampa, Florida.

Radivojević, M., Th. Rehren, J. Kuzmanović Cvetković, M. Jovanović and J.P. Northover. 2013. Tainted ores and the rise of tin bronze metallurgy, c. 6500 years ago. *Antiquity* 87: 1030–1045. DOI: https://doi.org/10.1017/S0003598X0004984.

Radivojević, M., Th. Rehren, J. Kuzmanović Cvetković and M. Jovanović. 2014a. Context is everything indeed: a response to Šljivar and Borić. *Antiquity* 88: 1315–1319.

Radivojević, M., Th. Rehren, J. Kuzmanović Cvetković and M. Jovanović. 2014b. Why are there tin bronzes in the 5th millennium BC Balkans? In S. Vitezović and D. Antonović (eds) *Archaeotechnology: Studying Technology from Prehistory to the Middle Ages*: 235–256). Belgrade: Srpsko Arheološko Društvo.

Radivojević, M., Th. Rehren, S. Farid, E. Pernicka and D.S. Camurcuoğlu. 2017. Repealing the Çatalhöyük extractive metallurgy: the green, the fire and the 'slag'. *Journal of Archaeological Science* 86: 101–122. DOI: https://doi.org/10.1016/j.jas.2017.07.001.

Radivojević, M., J. Pendić, A. Srejić, M. Korać, C. Davey, A. Benzonelli, M. Martinón-Torres, N. Jovanović and Ž. Kamberović. 2018. Experimental design of the Cu-As-Sn ternary colour diagram. *Journal of Archaeological Science* 90: 106–119. DOI: https://doi.org/10.1016/j.jas.2017.12.001.

Radivojević, M., B.W. Roberts, E. Pernicka, Z.A. Stos-Gale, M. Martinón-Torres, H. Vandkilde, J. Ling, D. Brandherm, P. Bray, J. Mei, K. Kristiansen, Th. Rehren, S. Shennan and C. Broodbank. 2019. The provenance, use and circulation of metals in the European Bronze Age: The state of debate. *Journal of Archaeological Research* 27: 131–185. DOI: https://doi.org/10.1007/s10814-018-9123-9.

Radimsky, W. and M. Hoernes. 1895. *Die neolithische Station von Butmir bei Sarajevo in Bosnien.* Wien: Holzhausen.

Radmanović, D., D. Kostić, J. Lujić and S. Blažić. 2014. Vertebrate fauna at the Neolithic and Eneolithic sites in Vojvodina (Serbia). *Zbornik Matice Srpske za prirodne nauke* 126: 75–86.

Radovanović, I. 1996. *The Iron Gates Mesolithic* (International Monographs in Prehistory. Archaeological Series 11). Ann Arbor (MI): Berghahn Books.

Radovanović, I., M. Kaczanowska, J.K. Kozlowski, M. Pawlikowski and B. Voytek. 1984. *The Chipped Stone Industry from Vinča.* Belgrade: Centar za arheološka istraživanja, Filozofski Fakultet.

Rahmstorf, L., 2011. Re-integrating 'diffusion': the spread of innovations among the Neolithic and Bronze Age societies of Europe and the Near East in T.C. Wilkinson, S. Sherratt and J. Bennet (eds) *Interweaving worlds: systemic interactions in Eurasia, 7th to the 1st millennia BC: 100-119. Oxford: Oxbow.*

Rahmstorf, L., 2019. Scales, weights and weight-regulated artefacts in Middle and Late Bronze Age Britain. *Antiquity* 93: 1197-1210.

Rahmstorf, L., G. Barjamović and N. Ialongo (eds) 2021. *Merchants, Measures and Money. Understanding Technologies of Early Trade in a Comparative Perspective.* Kiel/Hamburg: Wachholtz Verlag.

Rahmsdorf, L. and E.P. Stratford (eds) 2019. *Weights and Marketplaces from the Bronze Age to the Early Modern Period.* Kiel/Hamburg: Wachholtz Verlag.

Rapp, G. 2009. *Archaeomineralogy* (Second edition). Natural Science in Archaeology. Berlin, Heidelberg: Springer.

Rassmann, K., J. Ohlrau, C. Mischka, N.B. Burdo, M. Videjko and J. Müller. 2014. High precision Tripolye settlement plans, demographic estimations and settlement organization. *Journal of Neolithic Archaeology* 6: 63–95.

Rassmann, K., C. Mischka, M. Furholt, J. Ohlrau, K. Radloff, K. Winkelmann, G. Serlegi, T. Marton, A. Osztás, K. Oross and E. Bánffy. 2015. Large scale geomagnetic prospections on Neolithic sites in Hungary II. *Hungarian Archaeology Online* http://files.archaeolingua.hu/2015NY/eng_Rassmann_15S.pdf

Raub, C.J. 1995. The metallurgy of gold and silver in prehistoric times, in G. Morteani and J.P. Northover (eds) *Prehistoric Gold in Europe: Mines, Metallurgy and Manufacture*: 243–259. Dordrecht: Kluwer Academic Publishers.

Reddy, S.N. 2003. *Discerning Palates of the Past: An Ethnoarchaeological Study of Crop Cultivation and Plant Usage in India.* (International Monographs in Prehistory. Ethnoarchaeological Series 5). Ann Arbor (MI): Berghahn Books.

Reed, K. 2013. Farmers in Transition: The Archaeobotanical Analysis of the Carpathian Basin from the Late Neolithic to the Late Bronze Age (5000-900 BC). Unpublished PhD dissertation, University of Leicester.

Reed, K. 2015. From the field to the hearth: plant remains from Neolithic Croatia (ca. 6000–4000 cal BC). *Vegetation History and Archaeobotany* 24: 601–619.

Reed, W.P. 1992. Certificate of Analysis, Standard Reference Materials 612 and 613. Technical Report, National Institute of Standards and Technology.

Rehder, J.E. 1994. Blowpipes versus bellows in ancient metallurgy. *Journal of Field Archaeology* 21: 345–350.

Rehren, Th. 1997. Die Rolle des Kohlenstoffs in der Prähistorischen Metallurgie. *Stahl Und Eisen* 117: 87–92.

Rehren, Th. 2003. Crucibles as reaction vessels in ancient metallurgy, in P.T. Craddock, and J. Lang (eds) *Mining and Metal Production Through the Ages*: 207–215. London: The British Museum Press.

Rehren, Th. 2008. METALS, Primary Production Studies of, in M.P. Deborah (ed.) *Encyclopedia of Archaeology*: 1616–1620. New York: Academic Press.

Rehren, Th. 2014. Crossing boundaries, in P. Stone and Z. Hui (eds) *Sharing Archaeology: Academe, Practice and the Public*: 36–46. New York, London: Routledge.

Rehren, Th. and E. Pernicka. 2008. Coins, artefacts and isotopes-archaeometallurgy and Archaeometry. *Archaeometry* 50: 232–248. DOI: https://doi.org/10.1111/j.1475-4754.2008.00389.x.

Rehren, Th., M. Charlton, S. Chirikure, J. Humphris, A. Ige and H.A. Veldhuijzen. 2007. Decisions set in slag - the human factor in African iron smelting, in S. La Niece, D. Hook and P. Craddock (eds) *Metals and Mines: Studies in Archaeometallurgy*: 211–218. London: Archetype Publications.

Rehren, Th., Boscher, L. and E. Pernicka. 2012. Large scale smelting of speiss and arsenical copper at Early Bronze Age Arisman, Iran. *Journal of Archaeological Science* 39 (6): 717-1727. DOI:https:??doi.org/10.1016/j.jas.2012.01.009.

Rehren, Th., P. Leshtakov and P. Penkova. 2016. Reconstructing Chalcolithic copper smelting at Akladi Cheiri, Chernomorets, Bulgaria, in V. Nikolov and W. Schier (eds) *Der Schwarzmeerraum vom Neolithikum bis in die Früheisenzeit (6000-600 v.Chr.). Kulturelle Interferenzen in der zirkumpontischen Zone und Kontakte mit ihren Nachbargebieten*: 205–214. Rahden/Westf: Verlag Marie Leidorf GmbH.

Rehren, Th., P. Penkova and E. Pernicka. 2020. The Chalcolithic copper smelting at Akladi Cheiri, in R. Krauss, E. Pernicka, R. Kunze, K. Dimitrov and P. Leshtakov (eds) *Prehistoric Mining and Metallurgy at the Southeast Bulgarian Black Sea Coast.* (RessourcenKulturen Band 10): 137–156. Tübingen: Tübingen University Press.

Reimer, P.J., E. Bard, A. Bayliss, J.W. Beck, P.G. Blackwell, C. Bronk Ramsey, P.M. Grootes, T.P. Guilderson, H. Haflidason, I. Hajdas, C. Hatt, T.J. Heaton, D.L. Hoffmann, A.G. Hogg, K.A. Hughen, K.F. Kaiser, B. Kromer, S.W. Manning, M. Niu, R.W. Reimer, D.A. Richards, E.M. Scott, J.R. Southon, R.A. Staff, C.S.M. Turney and J. van der Plicht. 2013. IntCal13 and Marine13 radiocarbon age calibration curves 0-50,0000 years cal BP. *Radiocarbon* 55(4): 1869–1887.

Reingruber, A. 2014. The wealth of the tells: Complex settlement patterns and specialisations in the West Pontic between 4600 and 4250 cal BC, in B. Horejs and M. Mehofer (eds) *Western Anatolia before Troy: Proto-Urbanisation in the 4th millennium BC?*: 217–241. Vienna: Austrian Academy of Sciences Press.

Reitz, E. and E. Wing. 2008. *Zooarchaeology*. 2nd edition. Cambridge: Cambridge University Press.

Remondino, F. 2013. Worth a thousand words – Photogrammetry for archaeological 3D surveying, in R.S. Opitz and D.C. Cowley (eds) *Interpreting Archaeological Topography*: 115–122. Oxford: Oxbow Books.

Renfrew, C. 1969. The autonomy of the south-east European Copper Age. *Proceedings of the Prehistoric Society* 35: 12–47.

Renfrew, C. 1970. The place of Vinča culture in European prehistory. *Zbornik Radova Narodnog Muzeja (Belgrade)* 6: 45–57.

Renfrew, C. 1971. Sitagroi, radiocarbon and prehistory of south-east Europe. *Antiquity* 45(180): 275–282.

Renfrew, C. 1972. *The Emergence of Civilisation: the Cyclades and the Aegean in the Third Millennium B.C.* London: Methuen.

Renfrew, C. 1973. Sitagroi and the independent invention of metallurgy in Europe, in M. Garašanin, A. Benac and N. Tasić (eds) *Actes du VIIIe congrès international des sciences préhistoriques et protohistoriques: Belgrade, 9-15 septembre 1971* (Volume 2): 473–481. Belgrade: Union Internationale des Sciences Préhistoriques et Protohistoriques.

Renfrew, C. 1978b. The anatomy of innovation, in D. Green, C. Haselgrove and M. Spriggs (eds) *Social Organization and Settlement.* (British Archaeological Reports International Series 47/1): 89–117. Oxford: Archaeopress.

Renfrew, C. 1978a. Varna and the social context of early metallurgy. *Antiquity* 52: 199–203.

Renfrew, C. 1979. *Problems in European Prehistory*. Edinburgh: Edinburgh University Press.

Renfrew, C. 1986. Varna and the emergence of wealth in prehistoric Europe, in A. Appadurai (ed.) *The Social Life of Things: Commodities in Cultural Perspective*: 141–168. Cambridge: Cambridge University Press.

Renfrew, C. 1987. *Archaeology and Language: The Puzzle of Indo-European Origins*. London: Jonathan Cape.

Renfrew, C., J.E. Dixon and J.R. Cann. 1965. Obsidian in the Aegean. *Annual of the British School of Archaeology at Athens* 60: 225–247.

Renfrew, C., J.E. Dixon and J.R. Cann. 1966. Obsidian and early culture contact in the Near East. *Proceedings of the Prehistoric Society* 32: 30–72.

Renfrew, C., J.E. Dixon and J.R. Cann. 1968. Further analysis of Near Eastern obsidians. *Proceedings of the Prehistoric Society* 34: 319–331.

Renfrew, C. and E. Elster. 2003. Metal artifacts and metallurgy, in E. Elster and C. Renfrew (eds.), *Prehistoric Sitagroi: Excavations in northeast Greece, 1968-1970. Volume 2: The final report*: 301–324. Los Angeles: Cotsen Institute of Archaeology, University of California.

Renfrew, C., C. Gosden and L. DeMarrais (eds). 2005. *Rethinking Materiality.* Cambridge: McDonald Institute for Archaeology.

Renfrew, J.M. 1979. The first farmers in South East Europe, in U. Körber-Grohne (ed.) Festschrift Maria Hopf. *Archaeo-Physika* 8: 243–265.

Rice, P.M. 1981. Evolution of specialized pottery production: a trial model. *Current Anthropology* 22: 219–227.

Rice, P.M. 1991. Specialization, standardization, and diversity: a retrospective, in R.L. Bishop and F.W. Lange (eds) *The ceramic legacy of Anna O. Shepard*: 257–279. Niwot. Colorado: University Press of Colorado.

Rice, P.M. 2015. *Pottery Analysis: A Sourcebook*. Chicago: University of Chicago Press.

Ристић-Опачић, J. 2005. Топографско-хронолошке карактеристике насеља винчанске културе на територији Србије. Гласник Српског археолошког друштва 21: 71–112.

Roberts, B.W. and M. Vander Linden (eds). 2011. *Investigating Archaeological Cultures: Material Culture, Variability, and Transmission*. London: Springer.

Roberts, B.W. and C. Frieman. 2012. Drawing boundaries and drawing models: investigating the concept of the 'Chalcolithic frontier' in north-west Europe, in M.J. Allen, J. Gardiner, and A. Sheridan (eds) *Is there a British Chalcolithic? People, Place and Polity in the Late 3rd Millennium*: 27–39. Oxford: Oxbow.

Roberts, B.W. and C.P. Thornton (eds). 2014. *Archaeometallurgy in Global Perspective. Methods and Syntheses*. New York: Springer.

Roberts, B.W. and M. Radivojević. 2015. Invention as a process: pyrotechnologies in early societies. *Cambridge Archaeological Journal* 25: 299–306. DOI: https://doi.org/10.1017/S0959774314001188.

Roberts, B.W., C.P. Thornton and V.C. Pigott. 2009. Development of metallurgy in Eurasia. *Antiquity* 83: 1012–1022.

Rogers, E.M. 1962. *Diffusion of Innovations (1st edition)*. New York: Free Press of Glencoe.

Rogers, E.M. 2003. *Diffusion of Innovations (5th edition)*. New York: Simon and Schuster.

Rosenstock, E., S. Scharl and W Schier. 2016. Ex oriente lux? – Ein Diskussionsbeitrag zur Stellung der frühen Kupfermetallurgie Südosteuropas, in M. Bartelheim, B. Horejs and R. Krauss (eds) *Von Baden bis Troia: Resourcennutzung, Metallurgie und Wissenstransfer. Eine Jubiläumsschrift für Ernst Pernicka* (Oriental and European European Archaeology 3): 59–122. Rahden: Leidorf.

Rostoker, W. and J.R. Dvorak. 1990. *Interpretation of Metallographic Structures* (3rd Edition). San Diego, London: Academic Press Incorporated.

Rostoker, W., V.C. Pigott and J.R. Dvorak. 1989. Direct reduction to copper metal by oxide-sulfide mineral interaction. *Archeomaterials* 3: 69–87.

Rothermel M., K. Wenzel K., D. Fritsch D. and N. Haala. 2012. SURE: Photogrammetric surface reconstruction from imagery. *Proceedings, Low Cost 3D Workshop, Berlin, 01.12.2012.*

Rousseau, J.-J. 1761. *A discourse upon the origin and foundation of the inequality among Mankind*. London: R. & J. Dodsley.

Roux, V. 2003. Ceramic standardization and intensity of production: quantifying degrees of specialization. *American Antiquity* 68: 768–782.

Roux, V. 2008. Evolutionary trajectories of technological traits and cultural transmission, in M. Stark, B.J. Bowser and L. Horne (eds) *Cultural Transmission and Material Culture*: 82–104. Tucson: The University of Arizona Press.

Roux, V. 2010. Technological innovations and developmental trajectories: Social factors as evolutionary forces, in M. J. O'Brien and S. J. Shennan (eds) *Innovation in cultural systems. Contributions from Evolutionary Anthropology*: 217–234. Cambridge: The MIT Press.

Roux, V. 2017. *De Cramiques et des Hommes. Décoder les Assemblages Archéologiques.* Paris: Presses Universitaires de Paris Nanterre.

Roux, V. and D. Corbetta. 1989. *The Potter's Wheel: Craft Specialization and Technical Competence*. New Delhi/ Bombay/Calcutta: Oxford and IBH Publishing.

Roux, V., B. Bril and G. Dietrich. 1995. Skills and learning difficulties involved in stone knapping: the case of stone-bead knapping in Khambhat, India. *World Archaeology* 27 (1): 63-87.

Rovira, L.S. and I. Montero Ruiz. 2013. Iberia: technological development of prehistoric metallurgy, in S. Burmeister, S. Hansen, M. Kunst and N. Müller-Scheeßel (eds) *Metal Matters. Innovative Technologies and Social Change in Prehistory and Antiquity*: 231–239. Rahden/Westf.: Marie Leidorf.

Rowlands, M.J. 1971. The archaeological interpretation of prehistoric metalworking. *World Archaeology* 3: 210–224.

Rowley-Conwy, P. 2003. Early domestic animals in Europe: imported or locally domesticated? In A. Ammerman and P. Biagi (eds) *The Widening Harvest: the Neolithic Transition in Europe – Looking Back, Looking Forward*: 99–117. Boston: Archaeological Institute of America.

Rowley-Conwy, P. 2007. *From Genesis to Prehistory: The Archaeological Three Age System and its Contested Reception in Denmark, Britain and Ireland.* Oxford: Oxford University Press.

Roy, D.P., V. Kovalskyy, H.K. Zhang, E.F. Vermonte, L. Yan, S.S. Kumar and A. Egorov. 2016. Characterization of Landsat-7 to Landsat-8 reflective wavelength and normalized difference vegetation index continuity. *Remote Sensing of Environment* 185: 57–70.

Ruddiman, W. F. 2014. *Earth Transformed.* New York: W.H. Freeman & Co Ltd.

Ruíz Taboada, A. and I. Montero-Ruíz. 1999. The oldest metallurgy in Western Europe. *Antiquity* 73: 897–903.

Russell, N. 1990. The bone tools, in R. Tringham and D. Krstić (eds) *Selevac. A Neolithic village in Yugoslavia*: 521–548. Los Angeles: UCLA.

Russell, N. 1993. Hunting, Herding and Feasting: Human Use of Animals in Neolithic Southeast Europe. Unpublished PhD dissertation, University of California at Berkley.

Russell, N. 1998. Cattle as wealth in Neolithic Europe: Where's the beef? In D.W. Bailey (ed.) *The Archaeology of Value: Essays on Prestige and the Processes of Valuation* (British Archaeological Reports S730): 42–54. Oxford: Archaeopress.

Russell, N. 2001. The social life of bone: a preliminary assessment of bone tool manufacture and discard at

Çatalhöyük, in A. Choyke and L. Bartosiewicz (eds) *Crafting Bone: Skeletal Technologies Through Time and Space - Proceedings of the 2nd meeting of the (ICAZ) Worked Bone Research Group, Budapest, 31 August - 5 September 1999*: 241–250. Oxford: Archaeopress.

Russell, N., K. Twiss, D. Orton and A. Demirergi. 2013. More on the Çatalhöyük mammal remains, in I. Hodder (ed.) *Humans and Landscapes of Çatalhöyük: Reports from the 2000–2008 Seasons*: 213–258. Los Angeles: Cotsen Institute.

Ryndina, N. 2009. The potential of metallography in investigations of early objects made of copper and copper-based alloys. *Historical Metallurgy* 43: 1–18.

Ryndina, N.V. and I.J. Ravich. 2000. Eneolithic Balkan-Carpathian Metallurgical Province (BCMP) as a unique centre of the origin and development of the earliest metallurgical activity (Part I). *Bulletin of the Metals Museum* 32: 6–28.

Ryndina, N.V. and I.J. Ravich. 2001. Eneolithic Balkan-Carpathian Metallurgical Province (BCMP) as a unique centre of the origin and development of the earliest metallurgical activity (Part II). *Bulletin of the Metals Museum* 34: 1–21.

Ryndina, N.V. and I.J. Ravich. 2012. О металлопроизводстве майкопских племен Северного Кавказа (по данным химико-технологических исследований) (On Maikop metallurgy in the Northern Caucasus, in Russian). Вестник археологии, антропологии и этнографии (*Moscow*) 2: 4–20.

Ryndina, N., G. Indenbaum and V. Kolosova. 1999. Copper production from polymetallic sulphide ores in the northeastern Balkan Eneolithic culture. *Journal of Archaeological Science* 26: 1059–1068.

S

Saporta, G. 2006. *Probabilités, Analyse des Données et Statistique* (2nd Edition). Paris: Editions Technip.

Schachermeyr, F. 1955. *Die ältesten Kulturen Griechenlands.* Stuttgart: W. Kohlhammer.

Schacht, R.M. 1984. The contemporaneity problem. *American Antiquity* 49: 678–695.

Schalk, E. 1998. *Die Entwicklung der Prähistorischen Metallurgie im Nördlichen Karpatenbecken. Eine Typologische und Metallanalytische Untersuchung.* Rahden-Westfalen: Leidorf Verlag.

Scharl, S. 2016. Patterns of innovation transfer and the spread of copper metallurgy to Central Europe. *European Journal of Archaeology* 19: 215–244. DOI: https://doi.org/10.1080/14619571.2016.1147313.

Schauer, P., Shennan, S., Bevan, A., College, S., Edinborough, K., Kerig, T. and M. Parker Pearson, M. 2020. Cycles in Stone Mining and Copper Circulation in Europe 5500–2000 BC: A View from Space. *European Journal of Archaeology* 24(2), 204-225. DOI: https://doi.org/10.1017/eaa.2020.56.

Schibler, J. 1981. *Typologische Untersuchungen der cortaillodzeitlichen Knochenartefakte. Die neolithischen. Ufersiedlungen von Twann.* Bd.17. Bern: Staatlicher Lerhmittelverlag.

Schickler, H. 1981. Neolithische Zinnbronzen, in H. Lorenz (ed.) *Studien zur Bronzezeit: Festschrift für Wilhelm Albert v. Brunn*: 419–445. Mainz am Rhein: von Zabern.

Schier, W. 1991. Untersuchungen zur Keramik von Vinča-Belo Brdo. *Banatica* 10: 131–140.

Schier, W. 1995. Vinča Studien. Tradition und Innovation im Spätneolithikum des zentralen Balkanraumes am Beispiel der Gefäskeramik aus Vinča-Belo Brdo. Unpublished dissertation. Heidelberg: Heidelberg University.

Schier, W. 1996. The relative and absolute chronology of Vinča: new evidence from the type site, in F. Draşovean (ed.) *The Vinča Culture, its Role and Cultural Connections*: 141–162. Timişoara: National Museum,Timişoara.

Schier, W. 1997. Vinča studies. Tradition and innovation in the Late Neolithic of the Central Balkan region: ceramic vessels from Vinča Belo Brdo. *Archaologisches Nachrichtenblatt* 2(1): 37–46.

Schier, W. 2000. Measuring change: the Neolithic pottery sequence of Vinča-Belo Brdo. *Documenta Praehistorica* 27: 187–197.

Schier, W. 2001. Tellstratigraphien als Zeitmasstab, in Lux Orientis. Archäologie zwischen Asien und Europa, in R. Boehmer and J. Maran (eds) *Festschrift für Harald Hauptmann zum 65.* Geburtstag (Rahden): 371–379.

Schier, W. 2014a. The Copper Age in southeastern Europe - historical epoch or typo-chronological construct? In W. Schier and F. Draşovean (eds) *New Approaches to Dating and Cultural Dynamics in the 6th to 4th Millennium BC.* (Prähistorische Archäologie in Südosteuropa 28): 419–435. Rahden: Leidorf.

Schier, W. 2014b. Der spätneolithisch-kupferzeitliche Tell von Uivar (Rumänien). Prospektionen und Grabungen 1998–2009, in W. Schier and M. Meyer (eds) *Vom Nil bis an die Elbe. Forschungen aus fünf Jahrzehnten am Institut für prähistorische Archäologie der Freien Universität Berlin*: 17–36. Rahden/Westf: Leidorf.

Schier, W. and F. Draşovean. 2004. Vorbericht über die rumänisch-deutschen Prospektionen und Ausgrabungen in der befestigten Teilsiedlung von Uivar, jud. Timiş, Rumänien (1998-2002). *Praehistorische Zeitschrift* 79(2): 145–230.

Schier, W. and F. Draşovean (eds). 2014. *New Approaches to Dating and Cultural Dynamics in the 6th to 4th Millennium BC* (Prähistorische Archäologie in Südosteuropa 28). Rahden/Westf: Verlag Marie Leidorf Gmbh.

Schier, W. and V. Nikolov (eds). 2016. *Der Schwarzmeerraum vom Neolithikum bis in die Früheisenzeit (6000-600 v.Chr.). Kulturelle Interferenzen in der zirkumpontischen Zone und Kontakte mit ihren Nachbargebieten*. Rahden/ Westf: Leidorf.

Schiffer, M. (ed.) 2001. *Anthropological Perspectives on Technology*. Albuquerque: University of New Mexico Press.

Schiffer, M.B. 2005. The devil is in the details: the cascade model of invention processes. *American Antiquity* 70: 485–502.

Schiffer, M.B. 2010. Can archaeologists study processes of invention? In M.J. O'Brien and S.J. Shennan (eds) *Innovation in Cultural Systems. Contributions from Evolutionary Anthropology*: 235–250. Cambridge, MA: The MIT Press.

Schiffer, M.B. 2011. *Studying Technological Change: A Behavioral Approach*. Salt Lake City: University of Utah Press.

Schmid, E. 1972. *Atlas of Animal Bones: For Prehistorians, Archaeologists and Quaternary Geologists*. New York: Elsevier.

Schneider, G., H. Knoll, K. Gallis and J.P. Demoule. 1994. Production and circulation of Neolithic Thessalian pottery: chemical and mineralogical analyses, in J.-C. Decourt, B. Helly and K. Gallis (eds) *La Thessalie. Quinze Années de Recherches Archéologiques, 1975-1990: Bilans et Perspectives*: 61–70. Athens: Ministry of Culture.

Schoch, W.H., B. Pawlik and F.H. Schweingruber. 1988. *Botanische Makroreste: Ein Atlas Zur Bestimmung Häufig Gefundener und Ökologisch Wichtiger Pflanzensamen*. Bern: Paul Haupt.

Schroedter, T.M., R. Hofmann, M. Furholt, J. Müller and O. Nelle. 2011. Woodland and its use in central Bosnia during the late Neolithic. Results from anthracological investigations in the Visoko-basin. *Sagvntvum Extra* 11: 87–89.

Schubert, F. 1965. Zu den südosteuropäischen Kupferäxten. *Germania* 43: 274–295.

Schumpeter, J. 1939. *Business Cycles: A Theoretical, Historical, and Statistical Analysis of the Capitalist Process*. New York: McGraw-Hill.

Sériadès, M.L. 1992a. Le métal, in Treuil, R. (ed.) *Dikili Tash, Village Préhistorique de Macédoine Orientale*: 113–116. Athens: Ecole Française d'Athènes.

Séfériadès, M.L. 1992b. L'os et le bois de cervidé, in R. Treuil (ed.) *Dikili Tash. Village préhistorique de Macédoine orientale I. Fouilles de Jean Deshayes (1961-1975), Bulletin de correspondance hellénique, supplément XXIV, 1992*: 99–112.

Séfériadès, M.L. 2010. Spondylus and long-distance trade in prehistoric Europe, in D. Anthony (ed.) *The Lost World of Old Europe: The Danube Valley 5000-3500 BC*: 178–190. Princeton and Oxford: Princeton University Press.

Sekereš, L. 1967. Ludaš-Budžak, ranoneolitsko naselje. *Arheološki pregled* 9: 9–13.

Semenov, S.A. 1976. *Prehistoric Technology: An Experimental Study of the Oldest Tools and Artefacts from Traces of Manufactures and Wear*. Bradford-on-Avon. Wiltshire: Moonraker.

Sergeev, G.P. 1963. Rannetripolskii klad u s. Karbuna. *Sovetskaja Arheologija* 1: 135–151.

Shennan, S. 2000. Population, culture history and the dynamics of culture change. *Current* Anthropology 41: 811–835.

Shennan, S. 2001. Demography and cultural innovation: a model and its implications for the emergence of modern human culture. *Cambridge Archaeological Journal* 11: 5–16.

Shennan, S. 2013. Demographic continuities and discontinuities in Neolithic Europe: evidence, methods and implications. *Journal of Archaeological Method and Theory* 20: 300–311. DOI: https://doi.org/10.1007/s10816-012-9154-3.

Shennan, S. 2018. *The First Farmers of Europe: An Evolutionary Perspective*. Cambridge: Cambridge University Press.

Shennan, S. and J. Steele. 1999. Cultural learning in hominids: a behavioural ecological approach, in H.O. Box and K.R. Gibson (eds) *Mammalian Social Learning: Comparative and Ecological Perspectives*: 367–388. Cambridge: Cambridge University Press.

Shennan, S., S.S. Downey, A. Timpson, K., Edinborough, S. Colledge, T. Kerig, K. Manning and M.G. Thomas. 2013. Regional population collapse followed initial agriculture booms in mid-Holocene Europe. *Nature Communications* 4: 2486.

Shepard, A.O. 1980. *Ceramics for the Archaeologist*. Washington DC: Carnegie Institution.

Sherratt, A.G. 1981. Plough and pastoralism: aspects of the secondary products revolution, in I. Hodder, G. Isaac and N. Hammond (eds) *Pattern of the Past: Studies in Honour of David Clarke*: 261-306. Cambridge: Cambridge University Press.

Sherratt S. 2018. Why was (and is) silver sexy? Silver during the 4th-3rd millennia in the Near East and Mesopotamia, in X-L. Armada, M. Murillo-Barroso and M. Charlton (eds) *Metals, Minds and Mobility. Integrating Scientific Data with Archaeological Theory*: 97-106. Oxford: Oxbow.

Shimada, I. (ed.) 2007. *Craft Production in Complex Societies: Multi-Crafting, Sequential Production, and Producers*. Salt Lake City (UT): University of Utah Press.

Siklósi, Z. 2004. Prestige goods in the Neolithic of the Carpathian Basin. Material manifestations of social differentiation. *Acta Archaeologica Academiae Scientiarum Hungaricae* 55: 1–62. DOI: https://doi.org/10.1556/aarch.55.2004.1-2.1.

Siklósi, Z. 2013. *Traces of Social Inequality during the Late Neolithic in the Eastern Carpathian Basin. Dissertationes Pannonicae IV, 3*. Budapest: Eötvös Loránd University, Institute of Archaeological Sciences.

Siklósi Z., Prange M., Kalicz N. and P. Raczky 2015. New data on the provenance of Early Copper finds from the Great Hungarian Plain. In Hansen S., Raczky P., Anders A. and A. Reingruber (eds.) *Neolithic and*

Copper Age between the Carpathians and the Aegean Sea. Chronologies and Technologies from the 6th to the 4th Millennium BCE: 57-92. Archäologie in Eurasien 31. Bonn: Habelt.

Siklósi, Z. and M. Szilágyi 2019. New data on the provenance of copper finds from the Early-Middle Copper Age of the Great Hungarian Plain. *Archaeological and Anthropological Sciences* 11: 5275–5285. DOI: https://doi.org/10.1007/s12520-019-00867-8

Sillar, B. 2003. Technological choices and experimental archaeology, comments on M.S. Tite, V. Kilikoglou and G. Vekinis, 'Review Article: Strength, toughness and thermal shock resistance of technological choice'. *Archaeometry 43(3):* 301–324.

Sillitoe, R.H. 1983. Enargite-bearing massive sulfide deposits high in porphyry copper systems. *Economic Geology* 78: 348–352.

Silva, F. and M. Vander Linden. 2017. Amplitude of travelling front as inferred from 14C predicts levels of genetic admixture among European early farmers. *Scientific Reports* 7: 11985. DOI: https://doi.org/10.1038/s41598-017-12318-2.

Silver, I. 1969. The ageing of domestic animals, in D. Brothwell and E. Higgs (eds) *Science in Archaeology: A Survey of Progress and Research*: 283–302. London: Thames and Hudson.

Simonović, D. Č. 1967. Osnovne karakteristike jugoslovenske poljoprivredne proizvodnje, *Zbornik radova Pravno - Ekonomskog Fakulteta u Nišu* 6: 265–283.

Shott, M.J. 1996. Mortal pots: on use life and vessel size in the formation of ceramic assemblages. *American Antiquity* 61(3): 462–482.

Skakun, N. 1992. Evolution des techniques agricoles in Bulgarie chalcolithique (d'après les analyses tracéologiques), in P. Anderson (ed.) *Préhistoire de L'Agriculture: Nouvelles Approches Expérimentales et Ethnographiques* (Centre de Recherches Archéologiques Monograph 6): 289–303. Paris: Centre National de la Recherche Scientifique.

Skakun, N. 1993. Results of traseological examination of flint implements from Neolithic settlements in Western Bulgaria, in I. Gatsov (ed.) *Neolithic Chipped Stone Industries in Western Bulgaria*: 52–54. Kraków: Jagellonian University Press.

Скакун Н., Э. Ибрагимова and В.Терёхина. 2015. Некоторые результаты комплексного изучения материалов поселений Беловоде и Плочник, Неолитические культуры Восточной Европы: хронология, палеоэкология, традиции. Материалы Международной научной конференции, посвященной 75-летию Виктора Петровича Третьякова. Санкт-Петербург.

Skibo, J.M. 2013 *Understanding Pottery Function. Manuals in Archaeological Method, Theory and Technique.* New York: Springer.

Slavchev, V. (ed.) 2008. *The Varna Eneolithic Necropolis and Problems of Prehistory in Southeast Europe: Studia in Memoriam Ivan Ivanov.* (Acta Musei Varnensis 7). Varna: Regionalen Istoričeski Muzej, Varna.

Smith, C.S. 1981. On art, invention and technology, in C.S. Smith (ed.) *A Search for Structure*: 325–331. Cambridge (MA): The MIT Press.

Smith, K.Y. and F.D. Neiman. 2007. Frequency seriation, correspondence analysis, and Woodland period ceramic assemblage variation in the deep south. *Southeastern Archaeology* 26(1): 47–72.

Solecki, R.S., R.L. Solecki and A.P. Agelarakis. 2004. *The Proto-Neolithic Cemetery in Shanidar Cave.* Texas: Texas A&M University Press.

Sørensen, M.L.S. 2010. Households, in K. Kristiansen and T. Earle (eds) *Organizing Bronze Age Societies. The Mediterranean, Central Europe & Scandinavia compared:* 122–154. Cambridge: Cambridge University Press.

Spandl, K., A. Dain-Owens, C. Champness, M. Dresser, M. Hann, R. Godwin and P. Booth. 2010. Trials to identify soil cultivation practices to minimise the impact on archaeological sites. (Defra project number BD1705). *Effects of Arable Cultivation on Archaeology* (EH Project number 3874). Known collectively as: 'Trials' Appendix 3: Studying the effects of different cultivation systems on flat archaeological sites. Oxford Archaeology: 1–58.

Spasić, D. 1993. Dva neolitska lokaliteta u Pomoravlju. *Glasnik Srpskog Arheološkog Društva* 9: 180–185.

Spasić, M. 2011. Vinčanska keramika s lokaliteta Crkvine. *Kolubara* 5: 101–146.

Spataro, M. 2014. Continuity and change in pottery manufacture between the early and middle Neolithic of Romania. *Journal of Archaeological and Anthropological Sciences* 6(2): 175–197.

Spataro, M. 2017. Innovation and regionalism in the Middle/Late Neolithic of south and south-eastern Europe (ca. 5,500-4,500 cal. BC): A ceramic perspective, in L. Burnez-Lanotte (ed.) *Matières à Penser. Raw Materials Acquisition and Processing in Early Neolithic Pottery Productions.* Proceedings of the Workshop of Namur (Belgium), 29-30 May 2015: 61–72. Paris: Société Préhistorique Française.

Spataro, M. 2018. Origins of specialization: the ceramic chaîne opératoire and technological take-off at Vinča-Belo Brdo, Serbia. *Oxford Journal of Archaeology* 37: 247–265. DOI: https://doi.org/10.1111/ojoa.12140

Spataro, M. and M. Furholt (eds). 2020. *Detecting and Explaining Archaeological Innovation in Prehistory.* Leiden: Sidestone.

Spataro, M., M. Cubas, O.E. Craig, J.C. Chapman, A. Boroneanţ and C. Bonsall. 2019. Production and function of Neolithic black-painted pottery from Schela Cladovei (Iron Gates, Romania). *Archaeological and Anthropological Sciences* 11: 6287–6304. DOI: https://doi.org/10.1007/s12520-019-00918-0.

Spielmann, K.A. 2002. Feasting, craft specialization, and the ritual mode of production in small-scale societies. *American Anthropologist* 104: 195–207, DOI: https://doi.org/10.1525/aa.2002.104.1.195.

Srdoč, D., A. Sliepčević, B. Obelić and N. Horvatinčić. 1977. Rudjer-Boskovic Institute Radiocarbon Measurements IV. *Radiocarbon* 19(3): 465–475.

Srdoč, D., N, Horvatinčić, B. Obelić, I.K. Bronić and A. Sliepčević. 1987. Rudjer Bošković Institute Radiocarbon Measurements X. *Radiocarbon* 29(1): 135–147.

Srdoč, D., A. Sliepčević and J. Planinić 1975. Rudjer-Boskovic Institute Radiocarbon Measurements III. *Radiocarbon* 17(1): 149–155.

Srejović, D. 1963. Versuch einer historischen Wertung der Vinča-Gruppe. *Archaeologica Iugoslavica* 4: 5–18.

Srejović, D. 1968. Neolitska plastika centralnobalkanskog područja, in L. Trifunović (ed.) *Neolit centralnog Balkana*: 177–240. Beograd: Narodni muzej.

Srejović, D. 1969. *Lepenski Vir: Nova Praistorijska Kultura u Podunavlju*. Belgrade: Srpska književna Zadruga.

Srejović, D. 1972. *Europe's First Monumental Sculpture: New Discoveries at Lepenski Vir*. London: Thames and Hudson.

Srejović, D. (ed.). 1984a. *The Chipped Stone Industry from Vinča: Excavations 1929-1934*. Belgrade: Univerzitet u Beogradu, Centar za Arheološka Istraživanja.

Srejović, D. 1984b. Umetnost i religija, in S. Ćelić (ed.) *Vinča u praistoriji i srednjem veku*: 42–56. Beograd: Srpska akademija nauka i umetnosti.

Srejović, D. 1988. The Neolithic of Serbia: a review of research, in D. Srejović (ed.) *The Neolithic of Serbia: Archaeological Research 1948-1988*: 5–19. Belgrade: Faculty of Philosophy.

Srejović, D. 1997. *Arheološki leksikon*. Beograd: Savremena administracija.

Срејовић, Д. and Б. Јовановић. 1959. Оруђе и оружје од кости и накит из Винче [Utensils and weapons in bone and ornaments from Vinča]. *Starinar (new series)* 9-10: 181–190.

Srejović, D. and Z. Letica. 1978. *Vlasac: mezolitsko naselje u Djerdapu 2: Arheologija)*. Belgrade: Srpska Akademija Nauka i Umetnosti.

Srejović, D. and N. Tasić (eds). 1990. *Vinča and its World*. Belgrade: Serbian Academy of Sciences and Arts.

Srejović, D., B. Stojanović, N. Tasić and Dušan Krstić. 1984. *Vinča u praistoriji i srednjem veku*. Beograd: Srpska Akademija Nauka i Umetnosti.

Stalio, B. 1960. Pločnik-Prokuplje-naselje. *Arheološki pregled (Belgrade)* 2: 33–36.

Stalio, B. 1962. Pločnik, Prokuplje- naselje vinčanske grupe. *Arheološki pregled (Belgrade)* 4: 19–25.

Stalio, B. 1964. Novi metalni nalaz iz Pločnika kod Prokuplja. *Zbornik Narodnog Muzeja (Belgrade)* 4: 35–41.

Stalio, B. 1972. *Gradac. Praistorijsko naselje*. Belgrade: Narodni Muzej.

Stalio, B. 1973. Četvrti nalaz bakarnog i kamenog oruđa sa Pločnika kod Prokuplja. *Zbornik Narodnog Muzeja* (Belgrade) 7: 157–161.

Starović, A. 2004. *Signs of Civilization: Exhibition Catalogue*. Novi Sad: Serbian Academy of Sciences and Arts.

Starović, A. 2005. If the Vinča script once really existed who could have written or read it? *Documenta Praehistorica* 32: 253–260.

Stefan, C.-E. 2018. Notă asupra unor materiale preistorice de la Romula (Reşca, judeţul Olt). *Buletinul Muzeului Judeţean Teleorman* 10: 109–130.

Stefanović, S. 2008. Late Neolithic boys at the Gomolava cemetery (Serbia), in L. Oosterbeek, (ed.) *Babies Reborn: Infant/Child Burials in Pre- and Protohistory. Proceedings of the XV World Congress of the International Union for Prehistoric and Protohistoric Sciences* (British Archaeological Reports International Series 1832): 95–100. Oxford: Archaeopress.

Stefanovich, M. and C. Angelova (eds). 2007. *PRAE: In honorem Henrieta Todorova*. Sofia: Bulgarian Academy of Sciences.

Sterud, E., R.K. Evans and J.A. Rasson. 1984. Part I: Ex Balcanis Lux? Recent Developments in Neolithic and Chalcolithic Research in Southeast Europe. *American Antiquity* 49: 713–741. DOI: https://doi.org/10.2307/279738.

Stevanović, M. 1997. The age of clay: the social dynamics of house destruction. *Journal of Anthropological Archaeology* 16: 334.

Stojanović, I. and J. Bulatović. 2013. Arheozoološka istraživanja mlađe praistorije na teritoriji Srbije, in N. Miladinović-Radmilović and S. Vitezović (eds) *Bioarheologija na Balkanu. Bilans i perspektive. Radovi Bioarheološke sekcije Srpskog arheološkog društva*: 13–24. Beograd: Srpsko Arheološko Društvo.

Stojanova-Kanzurova, E. and Z. Rujak. 2016. Food procurement, storage and preparation during the Neolithic in the Republic of Macedonia, in N. Miladinović-Radmilović and S. Vitezović (eds) *Bioarheologija na Balkanu 2: Metodološke, komparativne i rekonstruktivne studije života u prošlosti*: 69–79. Beograd: Srpsko Arheološko Društvo.

Stojić, M. and D. Jacanović. 2008. *Požarevac: kulturna stratigrafija praistorijskih lokaliteta u Braničevu*. Beograd: Arheološki Institut. Požarevac: Narodni muzej.

Stojić, M. and M. Jocić. 2006. *Niš. Arheološka građa Srbije*. Beograd, Niš: Arheološki Institut, Narodni muzej.

Stöllner, T., Samaschev, Z., Berdenov, S., Cierny, J., Doll, M., Garner, J., Gontscharov, A., Gorelik, A., Hauptmann, A., Rainer, H., Kusch, G. A., Merz, V., Riese, T., Sikorski, B. and Zickgraf, B. 2011. Tin from Kazakhstan - Steppe tin for the West? *In:* Yalçın, Ü. (ed.) *Anatolian Metal* V: 231-251. Bochum: Deutsches Bergbau-Museum.

Stout, D. 2002. Skill and cognition in stone tool production: an ethnographic case study from Irian Jaya. *Current Anthropology* 45(3): 693–722.

Stout, D. 2005. The social and cultural context of stone-knapping skill acquisition, in V. Roux and B. Bril (eds) *Stone Knapping: The Necessary Conditions for a Uniquely Hominin Behaviour*: 331–340. Cambridge: McDonald Institute for Archaeological Research.

Strahm, C. 1984. Die Anfänge der Metallurgie in Mitteleuropa. *Helvetica Archaeologica* 97: 1–39.

Stratouli, G. 1998. *Knochenartefakte aus dem Neolithikum und Chalkolithikum Nordgriechenlands*. Bonn: Rudolf Habelt.

Stratton, S., Griffiths, S., Kogălniceanu, R., Simalcsik, A., Morintz, A., Ştefan, C., Dumitraşcu, V., Bronk Ramsey, C., Nehlich, O., Beavan, N., Borić, D. and A. Whittle 2019. The Emergence of Extramural Cemeteries in Neolithic Southeast Europe: A Formally Modelled Chronology for Cernica, Romania. *Radiocarbon* 61: 319-346.

Struckmeyer, K. 2011. The bone tools from the dwelling mound Feddersen Wierde, Germany, and their functions, in J. Baron and B. Kufel-Diakowska (eds) *Written in Bones. Studies on Technological and Social Contexts of Past Faunal Skeletal Remains*: 187–195. Wrocław: Uniwersytet Wrocławski, Instytut Archeologii.

Stuiver, M. and H.A. Polach. 1977. Discussion: reporting of 14C data. *Radiocarbon* 19(3): 355–363.

Sümegi, P., G. Persaits and S. Gulyás. 2012. Woodland-grassland ecotonal shifts in environmental mosaics: lessons learnt from the environmental history of the Carpathian Basin (Central Europe) during the holocene and the last ice age based on investigation of paleobotanical and mollusk remains, in R.W. Myster (ed.) *Ecotones Between Forest and Grassland*: 17–57. New York: Springer-Verlag.

Svoboda, V., J. Vuković, S. Knežević, D. Izvonar and D. Kićević. 2004/2005. Experimental archaeology: traditional production of ceramics, presentation of phase I. *Diana* 10: 123–131.

Szakmány, Gy., K. Vanicsek, Z. Bendő, A, Kreiter, A. Pető, Z. Lisztes-Szabó and F. Horváth.

2019. Petrological analysis of late Neolithic ceramics from the tell settlement of Gorzsa (south-east Hungary), in S. Amicone, P.S. Quinn, M. Marić, N. Mirković-Marić and M. Radivojević (eds) *Tracing Pottery Making Recipes in the Balkans 6th-4th Millennia BC*: 156–171. Oxford: Archaeopress.

Š

Šarić, J. 2002. Stone as material for production of chipped artifacts in Early and Middle Neolithic of Serbia. *Starinar* 52: 11–26.

Šarić, J. 2006. Typology of chipped stone artefacts in the Early and Middle Neolithic in Serbia. *Starinar* 56: 9–45. DOI: https://doi.org/10.2298/STA0656009S.

Šarić, J. 2015. *Artefakti od okresanog kamena u starijem i srednjem neolitu na tlu Srbije*. Beograd: Arheološki institut.

Šarić, K. and V. Cvetković. 2013. Mineraloško-petrografska analiza uzoraka sa eneolitskog lokaliteta Crkvine u Malom Borku (Mineralogical-petrographical analyses of samples from Eneolithic locality Crkvine (Mali Borak)), in M. Blagojević (ed.) *Kolubara* 6: 37–43. Beograd: Republički zavod za zaštitu spomenika kulture.

Šiška, S. 1964. Gräberfeld der Tiszapolgár-Kultur in Tibava. *Slovenská Archeologia* 12: 293–356.

Šljivar, D. 1993–2009. *Belovode Field Excavation Diaries*. Belgrade: Project Archives, National Museum Belgrade.

Šljivar, D. 1996. The Eastern Settlement of the Vinča culture at Pločnik: a relationship of its stratigraphy to the hoards of copper objects. *Starinar* 47: 85–97

Šljivar, D. 1999. Pločnik, naselje vinčanske kulture i problem najranije metalurgije bakra, in D. Marinković (ed.) *Prokuplje u Praistoriji, Antici i Srednjem veku*: 31–44. Prokuplje: Archaeological Institute Belgrade and Museum of Toplica.

Šljivar, D. 2006. The earliest copper metallurgy in the Balkans. *Metalurgija-Journal of Metallurgy* 12: 93–104.

Šljivar, D. and D. Antonović. 1996. Mačina – naselje vinčanske kulture kod Prokuplja. *Zbornik Narodnog Muzeja* 16(1). *Arheologija*: 61–74.

Šljivar, D. and D. Jacanović. 1995. Rezultati arheoloških istraživanja neolitskog naselja Konjušnica u Viteževu. *Glasnik Srpskog Arheološkog Društva* 10: 181–185.

Šljivar, D. and D. Jacanović. 1996a. Veliko Laole – Belovode, naselje vinčanske kulture. *Glasnik Srpskog Arheološkog Društva* 12: 55–60.

Šljivar, D. and D. Jacanović. 1996b. Veliko Laole, Belovode-Vinča culture settlement in northeastern Serbia. *Préhistorie Européene* 8: 175–188.

Šljivar, D. and D. Jacanović. 1996c. Veliko Laole, 'Belovode' naselje vinčanske grupe. *Glasnik Srpskog Arheološkog Društva* 11: 185–189.

Šljivar, D. and D. Jacanović. 1997a. Veliko Laole-Belovode, the settlement of the Vinča culture. *Journal of Serbian Archaeological Society* 13: 115–125.

Šljivar, D. and D. Jacanović. 1997b. Veliko Laole-Belovode, istraživanja u 1997. *Glasnik Srpskog Arheološkog Društva* 14: 73–78.

Šljivar, D. and D. Jacanović. 1997c. Arheometalurgija bakra na naselju vinčanske culture Belovode kod Petrovca na Mlavi, in M. Lazić (ed.) *Arheologija istočne Srbije*: 189–195. Belgrade: Faculty of Philosophy, Centre for Archaeological Research.

Šljivar, D. and D. Jacanović. 2005. Zoomorphic figurines from Belovode. *Zbornik narodnog muzeja* 18(1): 69–78.

Šljivar, D. and J. Kuzmanović Cvetković. 1996-2009. *Pločnik Field Excavation Diaries*. Belgrade: National Museum Belgrade.

Šljivar, D. and J. Kuzmanović Cvetković. 1997a. Pločnik kod Prokuplja, naselje vinčanske kulture. *Glasnik Srpskog Arheološkog Društva* 13: 103–113.

Šljivar, D. and J. Kuzmanović Cvetković. 1997b. Pločnik kod Prokuplja, naselje vinčanske kulture. *Glasnik Srpskog Arheološkog Društva* 14: 79–85.

Šljivar, D. and J. Kuzmanović Cvetković. 1998a. Najstarija metalurgija bakra na Pločniku kod Prokuplja, naselju vinčanske kulture. *Arheometalurgija* (Belgrade) 6: 1–18.

Šljivar, D. and J. Kuzmanović Cvetković. 1998b. Pločnik near Prokuplje, excavation in 1997. *Journal of the Serbian Archaeological Society* 14: 79–85.

Šljivar, D. and J. Kuzmanović Cvetković. 2009a. Pločnik, archaeology and conservation. *Diana* (Belgrade) 13: 56–61.

Šljivar, D. and J. Kuzmanović Cvetković. 2009b. Pločnik 2008. Field Journal of Trench 21. September/October. (Unpublished). Belgrade: National Museum in Belgrade.

Šljivar, D., D. Jacanović and J. Kuzmanović Cvetković. 2006. New contributions regarding the copper metallurgy in the Vinča culture, in N. Tasić and C. Grozdanov (eds) *Homage to Milutin Garašanin*: 251–266. Belgrade: Serbian Academy of Sciences and Arts and Macedonian Academy of Sciences and Arts.

Šljivar, D., J. Živković and D. Jacanović. 2011. *Belovode: naselje vinčanske kulture 5400-4600 g.p.n.e.* Petrovac na Mlavi: Zavičajni muzej.

Šljivar, D., J. Kuzmanović Cvetković and J. Živković. 2012. Belovode, Pločnik: on copper metallurgy in the Vinča culture. *Zbornik Narodnog Muzeja (Belgrade)* 20: 27–46.

Šljivar D., J.D. Živković and M.M. Svilar. 2015. Belovode, sonda XV. Prilozi stratigrafiji vinčanske kulture. *Zbornik Narodnog Muzeja* XXII-1: 9–27.

Šmilauer, P. 1992. *CanoDraw 3.0 User's Guide*. University College London: Environmental Change Research Centre.

Šumatić, N., J. Todorović, I. Komljenović and M. Marković. 1999. *Atlas korova*. Banja Luka: Glas Srpski, Šumarski fakultet, Poljoprivredni fakultet.

T

Taborin, Y. 1993. La parure en coquillage au paléolithique. *XIXe supplément à Gallia Préhistoire*. CNRS Editions.

Taborin, Y. 2004. *Langage Sans Parole. La Parure aux Temps Préhistoriques*. Paris: La Maison des roches.

Tasić, N. 1973. Neolitska Plastika. Beograd: Muzej grada Beograda.

Tasić, N. 1979. Tiszapolgar i Bodrogkeresztur kultura, in A. Benac (ed.) *Praistorija jugoslavenskih zemalja III - eneolitsko doba*: 55–85. Akademija nauka i umetnosti Bosne i Hercegovine.

Tasić, N. 1982. Naselje bakarnog doba u istočnoj Srbiji. *Zbornik radova Muzeja rudarstva i metalurgije u Boru (Serbia)* 2: 19–36.

Tasić, N. 1995. *Eneolithic Cultures of the Central and Western Balkans*. Belgrade: Institute for Balkan Studies.

Tasić, N. 2005. Vinča - the third glance, in L. Nikolova, J. Fritz and J. Higgins (eds) *Prehistoric Archaeology and Anthropological Theory and Education* (Reports of Prehistoric Research Projects 6–7): 1–8. Salt Lake City: Karlovo.

Tasić, N. 2007. Ritual pottery set from Vinča. *Glasnik Srpskog Arheološkog Društva* 23: 203–210.

Tasić, N. 2008a. Nemi svedoci jednog vremena: figuralna umetnost Vinče, in D. Nikolić (ed.) *Sto godina istraživanja arheološkog lokaliteta Vinča-Belo Brdo*: 139–163. Belgrade: Faculty of Philosophy University of Belgrade, National Museum Belgrade, City Museum Belgrade, Serbian Academy of Arts and Sciences.

Tasić, N. 2008b. *Vinča. Praistorijska metropola*. Beograd: Narodni muzej.

Tasić, N. 2015. *Metodologije i procedure na arheološkim istraživanjima u Vinči 1998-2015*. Beograd: Dosije Studio.

Tasić, N. and E. Tomić. 1969. *Crnokalačka bara – naselje starčevačke i vinčanske kulture*. Kruševac: Narodni muzej. Arheološko društvo Jugoslavije.

Tasić, N.N., M. Marić, K. Penezić, D. Filipović, K. Borojević, N. Russell, P. Reimer, A. Barclay, A. Bayliss, D. Borić, B. Gaydarska and A. Whittle. 2015. The end of the affair: formal chronological modelling for the top of the Neolithic tell of Vinča-Belo Brdo. *Antiquity* 89(347): 1064–1082. DOI: https://doi.org/10.15184/aqy.2015.101.

Tasić, N.N., M. Marić, D. Filipović, K. Penezić, E. Dunbar, P. Reimer, A. Barclay, A. Bayliss, B. Gaydarska and A. Whittle. 2016a. Interwoven strands for refining the chronology of the Neolithic tell of Vinča-Belo Brdo, Serbia. *Radiocarbon* 58: 795–831. DOI: https://doi.org/10.1017/RDC.2016.56.

Tasić, N., M. Marić, N. Bevan, C.B. Ramsey, B. Kromer, A. Barclay, A. Bayliss, B. Gaydarska and A. Whittle. 2016b. Vinča-Belo Brdo, Serbia: the times of a tell. *Germania* 93: 89–157.

Taylor, T. 1999. Envaluing metal: theorizing the Eneolithic 'hiatus', in S.M.M. Young, A.M. Pollard, P. Budd and R.A. Ixer (eds) *Metals in Antiquity* (British Archaeological Reports International Series 792): 22–32. Oxford: Archaeopress.

Teichert, M. 1969. Osteometrische Untersuchungen zur Berechnung der Wideństhöhe bei vor- und frühgeschichtlichen Schweinen. *Kühn Archiv* 83 (3): 237–292.

Teichert, M. 1975. Osteometrische Untersuchungen zur Berechnung der Widerristhöhe bei Schafen, in A. Clason (ed.) *Archaeozoological Studies*: 51–69. Amsterdam: North-Holland Publishing Company/Elsevier.

ter Braak, C.F.J. and P. Šmilauer. 1997–1999. *CANOCO for Windows version 4.02*. Wageningen: Centre for Biometry.

Ţerna, S. and B. Govedarica (eds). 2016. *Interactions, Changes and Meanings. Essays in Honour of Igor Manzura on the Occasion of His 60th Birthday*. Kishniev: High Anthropological School University, Kishniev.

Thomas, J.T. 2011. Fashioning identities, forging inequalities: Late Neolithic/Copper Age personal ornaments of the Portuguese Estremadura. *European Journal of Archaeology* 14 (1–2): 29–59.

Thomas, R. and N. Johannsen. 2011. Articular depression in domestic cattle phalanges and their archaeological relevance. *International Journal of Paleopathology* 1: 43–54.

Thomas, P., M. Radivojević, S. Derikonjić, F. Schapals, J. Pendić, A. Jablanović, K. Kosczinski, N. Heil, B.W. Roberts, M. Marić, E. Pernicka, Th. Rehren and T. Stöllner. (In preparation). Jarmovac, a new prehistoric mine in the Balkans.

Thornton, C.P. 2001. The Domestication of Metal: A Reassessment of the Early Use of Copper Minerals and Metal in Anatolia and Southeastern Europe. Unpublished MPhil dissertation, University of Cambridge.

Thornton, C.P. 2007. Of brass and bronze in prehistoric southwest Asia, in S. La Niece, D. Hook, and P. Craddock (eds.) *Metals and Mines: Studies in Archaeometallurgy*: 123–135. London: Archetype Publications.

Thornton, C.P. 2009. The Emergence of complex metallurgy on the Iranian Plateau: escaping the Levantine Paradigm. *Journal of World Prehistory* 22: 301–327. DOI: https://doi.org/10.1007/s10963-009-9019-1.

Thornton, C.P. 2010. The Rise of Arsenical Copper in Southeastern Iran. *Iranica Antiqua* 45: 31–50.

Thornton, C.P. and B.W. Roberts. 2009. Introduction: The beginnings of metallurgy in global perspective. *Journal of World Prehistory* 22: 181–184.

Thoury, M., B. Mille, T. Séverin-Fabiani, L. Robbiola, M. Réfrégiers, J.F. Jarrige and L. Bertrand. 2016. High spatial dynamics-photoluminescence imaging reveals the metallurgy of the earliest lost-wax cast object. *Nature Communications* 7: 13356. DOI: https://doi.org/10.1038/ncomms13356.

Timberlake, S., 2007. The use of experimental archaeology/archaeometallurgy for the understanding and reconstruction of Early Bronze Age mining and smelting technologies, in S. La Niece, D. Hook and P. Craddock, (eds) *Metals and Mines. Studies in Archaeometallurgy*, 27-36. London: Archetype/British Museum.

Timpson, A., S. Colledge, E. Crema, K. Edinborough, T. Kerig, K. Manning, M.G. Thomas and S. Shennan. 2014. Reconstructing regional population fluctuations in the European Neolithic using radiocarbon dates: a new case-study using an improved method. *Journal of Archaeological Science* 52: 549–557.

Todorova, H. 1978. *The Eneolithic Period in Bulgaria in the Fifth Millennium B.C.* (British Archaeological Reports International Series 49). Oxford: Archaeopress.

Todorova, H. 1981. *Die kupferzeitlichen Äxte und Beile in Bulgarien.* (Prähistorische Bronzefunde, Volume 9, Band 14). München: C.H. Beck'sche Verlagsbuchhandlung.

Todorova, H. 1982. *Kupferzeitlische Siedlungen im Nordostbulgarien.* Munchen: C.H. Beck.

Todorova, H. 1986. *Kamenno-Mednata Epokha v Bulgariya. Peto Khilyadoletie predi Novata Era.* Sofia: Nauka i Izkustvo.

Todorova, H. 1995. The Neolithic, Eneolithic and Transitional Period in Bulgarian prehistory, in D.W. Bailey, I. Panayotov and S. Alexandrov (eds) *Prehistoric Bulgaria* (Monographs in World Archaeology 22): 79–98. Madison (WI): Prehistory Press.

Todorova, H. (ed.) 2002a. *Durankulak - Die Prähistorischen Gräberfelder.* Sofia: The Bulgarian Academy of Sciences.

Todorova, H. 2002b. Chronologie, horizontale Stratigraphie und Befunde, in H. Todorova (ed), *Durankulak, Die prähistorischen Gräberfelder,* band II: 35–52. Sofia: Anubis.

Todorova, H. and I. Vajsov. 1993. *Novo-Kamennata Epokha v Bulgariya.* Sofia: Nauka i Izkustvo.

Todorova, H. and I. Vajsov. 2001. *Der kupferzeitliche Schmuck Bulgariens.* (Prähistorische Bronzefunde, Vol. 20, Band 6). Stuttgart: Steiner.

Todorova, M. 1997. *Imagining the Balkans.* Oxford: Oxford University Press.

Todorova-Simeonova, H. 1971. Spätneolithisches Gräberfeld bei der Stadt Devnja. *Bulletin du Musée National à Varna* 7: 3–26.

Tomkins, P.D. and P.M. Day. 2001. Production and exchange of the earliest ceramic vessels in the Aegean: a view from early Neolithic Knossos, Crete. *Antiquity* 75: 259–260.

Trajković-Filipović, T., D. Milanović and A. Bulatović. 2008. Reviziona arheološka iskopavanja lokaliteta Bubanj kod Niša u 2008. godini. *Zbornik Narodnog muzeja u Nišu* 16–17: 309–318.

Trbojević, N. 2005. The possibility of reconstruction of the early Eneolithic house at the locality of Livade in the village of Kalenić. *Kolubara* 4: 79–88.

Trigger, B.G. 1986. The role of technology in V. Gordon Childe's archaeology. *Norwegian Archaeological Review* 19: 1–14.

Tringham, R. 1971. *Hunters, Fishers and Farmers of Eastern Europe, 6000-3000 B.C.* London: Hutchinson and Co. Ltd.

Tringham, R. 1992. Life after Selevac: Why and how a Neolithic settlement is abandoned. *Balcanica* 23: 133–145.

Tringham, R. and D.A Krstić. 1990a. Selevac and the transformation of southeast European Prehistory, in R. Tringham and D. Krstić (eds) *Selevac: a Neolithic Village in Yugoslavia*: 567–616. Los Angeles: University of California Press.

Tringham, R. and D.A. Krstić (eds). 1990b. *Selevac: A Prehistoric Village in Yugoslavia.* Los Angeles: University of California Press.

Tringham, R. and D. Krstić. 1990c. Conclusion: Selevac in the wider context of European prehistory, in

R. Tringham and D. Krstić (eds) *Selevac. A Neolithic Village in Yugoslavia*: 567–616. Los Angeles: UCLA.

Tringham, R. and M. Stevanović. 1990. Nonceramic uses of clay, in. R. Tringham and D. Krstić (eds) *Selevac, A Neolithic Village in Yugoslavia*: 328–388. Los Angeles: University of California Institute of Archaeology.

Tringham, R.E., B. Brukner and B. Voytek. 1985. The Opovo Project: a study of socio-economic change in the Balkan Neolithic. *Journal of Field Archaeology* 12: 425–44.

Tringham, R., A. McPherron, J. Gunn and G. Odell. 1988. The flaked stone industry from Divostin and Banja, in A. McPherron and D. Srejović (eds) *Divostin and Neolithic of Central Serbia*: 203–254. Pittsburgh, Kragujevac: University of Pittsburgh, Narodni muzej Kragujevac.

Tringham, R.E., B. Brukner, T. Kaiser, K. Borojević, L. Bukvić, P. Šteli, N. Russel, M. Stafanović and B. Voytek. 1992. Excavation at Opovo, 1985–1987: Socioeconomic change in the Balkans Neolithic. *Journal of Field Archaeology* 19(3): 351–386.

Tripković, B. 2003a. Obsidian deposits in the Central Balkans? Tested against archaeological evidence. *Starinar* 53–54: 163–79.

Tripković, B. 2003b. A dialogue between the household and community: a case study of Banjica, in L. Nikolova (ed.) *Early Symbolic Systems for Communication in Southeast Europe* (British Archaeological Reports International Series 1139): 447–457. Oxford: Archaeopress.

Tripković, B. 2007. *Domaćinstvo i prostor u kasnom neolitu. Vinčansko naselje na Banjici* (Household and Space in the Late Neolithic. Vinča Settlement at Banjica). Beograd: Srpsko Arheološko Društvo.

Tripković, B. 2009a. Domaćinstvo i zajednica u kasnom neolitu Zapadnog Balkana (Household and Community in the Late Neolithic of the Western Balkans, in Serbian). Unpublished PhD dissertaion, University of Belgrade.

Tripković, B. 2009b. House(hold) continuities in the central Balkans. *Opuscula Archaeologica* 33: 7–28.

Tripković, B. 2011. Containers and grains: food storage and symbolism in the Central Balkans (Vinča period). *Documenta Praehistorica* 38: 159–172.

Tripković, B. 2013. *Domaćinstvo i zajednica: kućne i naseobinske istorije u kasnom neolitu centralnog Balkana*. Beograd: Univerzitet u Beogradu, Filozofski fakultet.

Tripković, B. and M. Milić. 2008. The origin and exchange of Obsidian from Vinča – Belo Brdo. *Starinar* 58: 71–86.

Tripković, B., and K. Penezić. 2017. On-site and off-site in western Serbia: a geoarchaeological perspective on Obrovac-type settlements. *Quaternary International* 429: 35–44.

Trubitt, M. 2003. The production and exchange of marine shell prestige goods. *Journal of Archaeological Research* 11(3): 243–277.

Tsirtsoni, Z. 2016a. The end of the Balkan Chalcolithic: A viewpoint from the south, in V. Nikolov and W. Schier (eds) *Der Schwarzmeerraum vom Neolithikum bis in die Früheisenzeit (6000-600 v. Chr.). Kulturelle Interferenzen in der Zirkumpontischen Zone und Kontakte mit ihren Nachbargebieten. Proceedings of the Humboldt-Kolleg held at Varna, May 16-20, 2012* (Prähistorische Archäologie in Südosteuropa Band 30): 339–358. Rahden/Westf: Leidorf.

Tsirtsoni, Z. (ed.) 2016b. *The Human Face of Radiocarbon. Reassessing Chronology in Prehistoric Greece and Bulgaria, 5000-3000 cal BC* (Travaux de la Maison de l'Orient 69). Lyon: Maison de l'Orient et de la Méditerranée Jean Pouilloux.

Tucakov, J. 1986. *Lečenje biljem*. Beograd: Izdavačka radna organizacija 'Rad'.

Turck, R. 2010. *Die Metalle zur Zeit des Jungneolithikums in Mitteleuropa - Eine sozialarchäologische Untersuchung*. (Universitätsforschungen zur prähistorischen Archäologie 185). Bonn: Habelt.

Tutin, T.G., H.H. Vernon, N.A. Burges, D.H. Valentine, S.M. Walters and D.A. Webb. 1964–1980. *Flora Europaea*. Cambridge: Cambridge University Press.

Tykot, R.H. 2011. Obsidian finds on the fringes of the Central Mediterranean: exotic or eccentric exchange? In A. Vianello (ed.) *Exotica in the Prehistoric Mediterranean*: 33–44. Oxford: Oxbow Books.

Tylecote, R.F. 1962. *Metallurgy in Archaeology*. London: Edward Arnold.

Tylecote, R.F. 1976. *A History of Metallurgy*. London: The Metals Society.

Tylecote, R.F., H.A. Ghaznavi and P.J. Boydell. 1977. Partitioning of trace elements between the ores, fluxes, slags and metal during the smelting of copper. *Journal of Archaeological Science* 4: 305–333.

U

Ucko, P.J. 1969. Ethnography and archaeological interpretation of funerary remains. *World Archaeology* 1: 262–280.

Ur, J. 2003. CORONA satellite photography and ancient road networks: A northern Mesopotamian case study. *Antiquity* 77(295): 102–115.

Urem-Kotsou, D. 2016. Salting the roads: connectivity in the Neolithic Balkans, in B.P.C. Molloy (ed.) *Of Odysseys and Oddities: Scales and Modes of Interaction between Prehistoric Aegean Societies and Their Neighbours* (Sheffield Studies in Aegean Archaeology 10): 123–141. Oxford: Oxbow Books.

U.S. Geological Survey, 2008, Declassified intelligence satellite photographs: U.S. Geological Survey Fact Sheet 2008-3054, viewed 21.07.2020, https://pubs.usgs.gov/fs/2008/3054/pdf/fs2008-3054.pdf

U.S. Geological Survey, 2020a, Landsat Missions, viewed 20.06.2020, https://www.usgs.gov/

U.S. Geological Survey, 2020b, Earth Explorer, viewed 20.06.2020, https://earthexplorer.usgs.gov/

V

Vajsov, I. 2007. Promachon-Topolnica. A typology of painted decorations and its use as a chronological marker, in H. Todorova, M. Stefanovich and G. Ivanov (eds) *The Struma/Strymon River Valley in Prehistory. Proceedings of the International Symposium 'Strymon Praehistoricus'*: 79–120. Düsseldorf: Gerda Henkel Stiftung.

Valamoti, S.-M. 2004. *Plants and People in Late Neolithic and Early Bronze Age Northern Greece: an Archaeobotanical Investigation* (British Archaeological Reports International Series 1258). Oxford: Archaeopress.

Valamoti, S.-M. and K. Kotsakis. 2007. Transitions to agriculture in the Aegean: the archaeobotanical evidence, in S. Colledge and J. Conolly (eds) *The Origins and Spread of Domestic Plants in Southwest Asia and Europe*: 75–92. London: Left Coast Press.

Valamoti, S-M., A. Moniaki and A. Karathanou. 2011. An investigation of processing and consumption of pulses among prehistoric societies: archaeobotanical, experimental and ethnographic evidence from Greece. *Vegetation History and Archaeobotany* 20: 381–396.

van der Leeuw, S.E. and R. Torrence (eds). 1989. *What's New? A Closer Look at the Process of Innovation*. London: Unwin Hyman.

van der Veen, M. 2005. Gardens and fields: the intensity and scale of food production. *World Archaeology* 37: 157–163.

van der Veen, M. 2007. Formation processes of desiccated and carbonised plant remains - the identification of routine practice. *Journal of Archaeological Science* 34: 968–990.

van Zeist, W. 2002. Plant Husbandry and Vegetation of Tell Gomolava, Vojvodina, Yugoslavia. *Palaeohistoria* 43/44: 87–115.

Vander Linden, M., I. Pandžić and D. Orton. 2014. New radiocarbon dates for the Neolithic period in Bosnia & Herzegovina. *Godišnjak/Jahrbuck* 43:7–34.

Varvara, S., B. Fabbri, S. Gualtieri, P. Ricciardi and M. Gligor. 2008. Archaeometric characterisation of the Neolithic pottery discovered at Alba Iulia-Lumea Noua archaeological site (Romania). *Studia Universitatis Babes-Bolyai Chemia* 53(1): 5–13.

Васић, Ч. 1987. Релативни хронолошки односи између објеката на Акропољу Царичиног града. *Саопштења* XIX: 127–138. Beograd.

Vasić, M. 1911. Gradac, preistorisko nalazište latenskog doba. *Glas Srpske Kraljevske Akademije* 86: 97–140.

Vasić, M. 1932-1936. *Preistoriska Vinča I-IV (Prehistoric Vinča, in Serbian)*. Belgrade: Državna štamparija.

Васић, М. 1932. Преисториска Винча I. Београд: Државна штампарија.

Vasić, M. 1936a. *Preistorijska Vinča II*. Beograd: Kraljevska državna štamparija.

Vasić, M. 1936b. *Preistorijska Vinča III. Plastika*. Beograd: Državna štamparija kraljevine Jugoslavije.

Vasić, M. 1936c. *Preistoriska Vinča IV. Keramika*. Belgrade: Državna Štamparija Kraljevine Jugoslavije.

Vetnić, S. 1974. Počeci rada na ispitivanju kulture prvih zemljoradnika u srednjem Pomoravlju. *Materijali* X: 123–168.

Vértes, L. 1965. The depot of silex blades from Boldogkőváralja. *Acta Archaeologica Hungarica* 17: 128–136.

Videiko, M. 2004. Absolute dating of Tripolye culture, in M. Videiko (ed.) *Encyclopedia Trypilskoi tsyvilizatsii. Volume 1*: 85–95. Kyiv: Ukrpoligraphmedia.

Vigne, J.-D. And D. Helmer. 2007. Was milk a 'secondary product' in the Old World Neolithisation process? Its role in the domestication of cattle, sheep and goats. *Anthropozoologica* 42(2): 9–40.

Virmani, S.M., K.L. Sahrawat and J.R. Burford. 1982. Physical and chemical properties of vertisols and their management, in *Proceedings of the 12th International Congress of Soil Science, 8-16 February 1982, New Delhi*. India: 80–93. New Delhi: Indian Society of Soil Science.

Vitezović, S. 2007. Koštana Industrija u Neolitu Srednjeg Pomoravlja. Unpublished MPhil dissertation, University of Belgrade.

Vitezović, S. 2011a. Koštana Industrija u Starijem i Srednjem Neolitu Centralnog Balkana. Unpublished PhD dissertation, University of Belgrade.

Vitezović, S. 2011b. The Neolithic bone industry from Drenovac, Serbia, in J. Baron and B. Kufel-Diakowska (eds) *Written in Bones. Studies on Technological and Social Contexts of Past Faunal Skeletal Remains*: 117–135. Wrocław: Uniwersytet Wrocławski, Instytut Archeologii.

Витезовић, С. 2012. Коштана индустрија са локалитета Витково-Трифуновићи (ископавања 2001. године) [Bone industry from the site Vitkovo-Trifunovići (excavation season 2001)], Крушевачки зборник *[Kruševac collection]* 15: 351–376.

Vitezović, S. 2012. The white beauty – Starčevo culture jewellery. *Documenta Praehistorica* XXXIX: 91–203

Vitezović, S. 2013a. Analiza praistorijskih koštanih predmeta – neka metodološka razmatranja, in N. Miladinović-Radmilović and S. Vitezović (eds) *Bioarheologija na Balkanu. Bilans i perspektive. Radovi bioarheološke sekcije Srpskog arheološkog društva*: 107–132. Beograd-Sremska Mitrovica: Srpsko arheološko društvo-Blago Sirmiuma.

Vitezović, S. 2013b. Osseous raw materials in the Vinča culture, in A. Choyke and S. O'Connor (eds) *From These Bare Bones: Raw Materials and the Study of Worked Osseous Objects*: 59–72. Oxford and Oakville: Oxbow Books.

Vitezović, S. 2013c. Personal ornaments in the Vinča culture: the case study of Vitkovo and Stragari, in I. Vasile Ferencz, N. Cătălin Rişcuţa and O. Tutilă Bărbat (eds) *Archaeological Small Finds and their Significance: Proceedings of the Symposium: Costume as an Identity Expression*: 9–20. Deva, Cluj-Napoca: Mega.

Vitezović, S. 2014. Pavlovac - Kovačke Njive: preliminarni izveštaj analize koštane industrije, in D. Antonović, S. Golubović and V. Bikić (eds) *Arheologija u Srbiji. Projekti Arheološkog instituta u 2012*: 16–19. Beograd: Arheološki institut.

Vitezović, S. 2016a. *Metodologija proučavanja praistorijskih koštanih industrija*. Beograd: Srpsko arheološko društvo.

Vitezović, S. 2016b. The sea within: the use of mollusc shells as ornaments in the central Balkans Neolithic, in C. Preoteasa, C.-D. Nicola (eds) *Cucuteni Culture within the European Neo-Eneolithic Context: Proceedings of the International Colloquium Cucuteni - 130. 15-17 October 2014, Piatra-Neamț, Romania: In Memoriam Dr. Dan Monah, In Memoriam Dr. Gheorghe Dumitroaia*: 237256. Piatra-Neamț: Editura 'Constantin Matasă'.

Vitezović, S. 2018. Managing raw materials in prehistory: the importance of studying osseus raw materials. *Opuscula Archaeologica* 29–40: 13–24.

Vitezović, S. and D. Antonović. 2020. Functional differentiation and possible regional specialisation of the Vinča culture settlements: viewpoint from osseous and lithic industries. *Quaternary International* 539: 39–48.

Vitezović, S., D. Antonović and V. Dimić. 2017. Life in white: symbolism and importance of the white colour in the Neolithic in the Balkans, in V.P. Sekulov (ed.) *Acta Musei Tiberiopolitani, Volume* 2: 26–37. Strumica: Institute for the Protection of Monuments of Culture and Museum Strumica.

Vitelli, K.D. 1993a. *Franchthi Neolithic Pottery, Volume 1. Classification and Ceramic Phases, 1 and 2*. Bloomington, Indiana: Indiana University Press.

Vitelli, K.D. 1993b. Power to the potters: comment on Perlés systems of exchange and organization of production in Neolithic Greece. *Journal of Mediterranean Archaeology* 6: 247–257.

Vlassa, N. 1967. Unele probleme ale neoliticuli Transilvaniei. *Acta Musei Napocensis (Romania)* 4: 403–423.

Vogel, J.C. and H.T. Waterbolk. 1963. Groningen radiocarbon dates IV. *Radiocarbon* 5: 163–202.

Vollmar, K. 2009. *Das grosse Buch der Farben*. Krummwisch: Königsfurt-Urania Verlag GmbH.

Von den Dreisch, A. 1976. *A Guide to the Measurements of Animal Bones from Archaeological Sites*. Cambridge (MA): Harvard University Peabody Museum.

Von de Pulsky, F. 1884. *Die Kupferzeit in Ungarn*. Budapest: Kessinger Publishing.

Vörös, I. 1980. Zoological and palaeoeconomical investigations of the Early Neolithic Körös culture. *Folia Archaeologica* 31: 3–64.

Voruz, J.-L. 1984. *Outillages Osseux et Dynamisme Industriel dans le Néolithique Jurassien*. Lausanne: Cahiers d'archéologie romande 29.

Voytek, B. 1985. The Exploitation of Lithic Resources in Neolithic Southeast Europe. Unpublished PhD Dissertion, University of Berkeley.

Voytek, B. 1988. Chipped stone industry of Selevac, Serbia, Yugoslavia, in T. Szelag (ed.) *Archaeologia Interregionali. Chipped Stone Industries of the Early Farming Cultures in Europe*. Warsaw: Warsaw University Press.

Vukanović, T. 1969. A Neolithic Janus from Kosovo, *Vranjski glasnik* 5: 112.

Vukmanović, M. and N. Radojčić. 1990. Quantitative analysis of the pottery, in R. Tringham and D. Krstić (eds) *Selevac: A Neolithic village in Yugoslavia*: 289–317. Los Angeles: The Institute of Archaeology, University of California.

Vuković, J. 2010. Neolithic fine pottery: properties, performance and function. *Glasnik Srpskog Arheološkog Društva 26*: 7–23.

Vuković, J. 2011. Late Neolithic pottery standardization. Application of statistical analysis. *Glasnik Srpskog archeološkog društva (Journal of the Serbian Archaeological Society)* 27: 81–101.

Vuković, J. 2013. Posude za pečenje u kasnom neolitu: đuveči iz Vinče. Zbornik Narodnog muzeja XXI/1: 129–144.

Vuković, J. 2015. Izgubljeni u tranziciji: problem prelaza ranog/srednjeg u kasni neolit centralnog Balkana u jugoslovenskoj/srpskoj arheologiji druge polovine XX veka. *Etnoantropološki problemi* 10(3): 651–673.

Vuković, J. 2018a. Erasing boundaries or changing identities? The transition from early/middle to late neolithic, new evidence from southern Serbia, in M. Gori and M. Ivanova (eds) *Balkan Dialogues: Negotiating Identity between Prehistory and the Present*: 240–253. London: Routledge.

Vuković, J. 2018b. Late Neolithic Vinča pottery firing procedure. *Opuscula Archaeologica* 39/40: 25–35.

Vuković, J. 2020. Technological innovation and social change. Early vs. Late Neolithic pottery production of the Central Balkans, in M. Spataro and M. Furholt (eds) *Detecting and Explaining Technological Innovation in Prehistory*: 135–150. Leiden: Sidestone

Vuković, J. and I. Miloglav. 2018. Part-time labor and household production: Emergence of specialized potters in the Late Neolithic Vinča (Serbia) and Late Eneolithic Vučedol (Croatia) Societies, in I. Miloglav and J. Vuković (eds) *Artisans Rule: Product Standardization and Craft Specialization in Prehistoric Society*: 120–136. Cambridge: Cambridge Scholar Publishing.

Vuković, S. and D. Marković. 2019. Ostaci životinja iz horizonta mlađe praistorije na lokalitetima Nad Klepečkom i Rit, in A. Kapuran, A. Bulatović, A. Golubović and V. Filipović (eds) *Viminacijum u praistoriji - iskopavanja 2005-2015*: 223–248. Beograd: Arheološki institut.

Vulpe, A. 1975. *Die Äxte und Beile in Rumänien*. (Prähistorische Bronzefunde Volume 9, Band 2). München: C. H. Beck'sche Verlagsbuchhandlun.

Vulpe, R. 1957. *Izvoare: Săpăturile din 1936-1948*. Bucharest: Editura Academiei Republicii Populare Romîne.

W

Wailes, B. (ed.) 1996. *Craft Specialization and Social Evolution: In Memory of V. Gordon Childe.* Philadelphia: University Museum of Archaeology and Anthropology, University of Pennsylvania.

Walker, A. and A. Bogaard. 2011. Preliminary archaeobotanical results from teleor 003/Măgura 'Buduiasca', in S. Mills and P. Mirea (eds) *The Lower Danube in Prehistory. Landscape Changes and Human-Environment Interactions*: 151–160. Bucureşti: Editura Renaissance.

Wallaert-Pêtre, H. 2001. Learning how to make the right pots: apprenticeship strategies and material culture, a case study in handmade pottery from Cameroon. *Journal of Anthropological Research* 57: 471–493.

Ward, G.K. and S.R. Wilson. 1978. Procedures for comparing and combining radiocarbon age determinations – critique. *Archaeometry* 20: 19–31.

Washburn, D. and D. Crowe (eds). 2004. *Symmetry Comes of Age.* Seattle (WA): University of Washington Press.

Watanabe, N., S. Nakamura, B. Liu and N. Wang. 2017. Utilization of structure from motion for processing CORONA satellite images: Application to mapping and interpretation of archaeological features in Liangzhu Culture, China. *Archaeological Research in Asia* 11: 38–50.

Watson, J. 1979. The estimation of relative frequencies of mammalian species: Khirokitia 1972. *Journal of Archaeological Science* 6: 127–137.

Weber, R.J., S. Dixon and A.M. Llorente. 1993. Studying invention: the hand tool as a model system. *Science, Technology, and Human Values* 18: 480–505.

Weninger, B., C. Lee, E. Rohling, O. Bar-Yosef, U. Böhner, M. Budja, M. Bundschuh, A. Feurdean, H.G. Gebe, O. Jöris, J. Linstädter, P. Mayewski, T. Mühlenbruch, A. Reingruber, G. Rollefson, D. Schyle, L. Thissen, H. Todorova and C. Zielhofer. 2009. The impact of rapid climate change on prehistoric societies during the Holocene in the Eastern Mediterranean. *Documenta Praehistorica* 36: 7–59.

Wertime, T.A. 1964. Man's first encounters with metallurgy. *Science* 146: 1257–1267.

Wertime, T.A. 1973. The beginnings of metallurgy: a new look. *Science* 182: 875–87.

White, L. 1959. *The Evolution of Culture: the Development of Civilization to the Fall of Rome.* New York: McGraw-Hill.

Whittle, A. 1996. *Europe in the Neolithic: The Creation of New Worlds.* Cambridge: Cambridge University Press.

Whittle, A. 2003. *The Archaeology of People: Dimensions of Neolithic Life.* London: Routledge.

Whittle, A. 2018. *The Times of their Lives: Hunting History in the Archaeology of Neolithic Europe*. Oxford: Oxbow.

Whittle, A., L. Bartosiewicz, D. Borić, P.B. Pettitt and M. Richards. 2002. In the beginning: new radiocarbon dates for the Early Neolithic in northern Serbia and south-east Hungary. *Antaeus (Budapest)* 25: 63–117.

Whittle, A., A. Bayliss, A. Barclay, B. Gaydarksa, E. Bánffy, D. Boric, F. Draşovean, J. Jakucs, M. Marić, D. Orton, I. Pantović, W. Schier, N. Tasić and M.Vander Linden. 2016. A Vinča potscape: formal chronological models for the use and development of Vinča ceramics in south-east Europe. *Documenta Praehistorica* 43: 1–60.

Wiener, N. 1993. *Invention. The Care and Feeding of Ideas.* Cambridge (MA): The MIT Press.

Wijnen, M. 1994. Neolithic pottery from Sesklo – technological aspects, in J.-C. Decourt, B. Helly and K. Gallis (eds) *La Thessalie. Quinze Années de Recherches Archéologiques, 1975-1990: Bilans et Perspectives*: 149–155. Athens: Ministry of Culture.

Wilkinson, T.C., Sherratt, S. and J. Bennet. (eds) 2011. *Interweaving Worlds: Systemic Interactions in Eurasia: 7th to the 1st Millennia BC.* Oxford: Oxbow.

Williams, M. 2006. *Deforesting the Earth: From Prehistory to Global Crisis, an Abridgement.* Dordrecht: Kluwer Academic Publishers.

Williams, R. A. and Le Carlier de Veslud, C. 2019. Boom and bust in Bronze Age Britain: major copper production from the Great Orme mine and European trade, c. 1600–1400 BC. *Antiquity,* 93, 1178-1196.

Williams-Thorpe, O., S.E. Warren and J.G. Nandris. 1984. The distribution and provenance of archaeological obsidian in Central and Eastern Europe. *Journal of Archaeological Science* 11: 183–212.

Willis, K.J. 1994. The vegetation history of the Balkans. *Quaternary Science Reviews* 13: 769–788.

Willis, K.J. and K.D. Bennett. 1994. The Neolithic transition – fact or fiction? Palaeoecological evidence from the Balkans. *The Holocene* 4: 326–330.

Willms, C. 1985. Neolitischer Spondylusschmuck. Hundert Jahre Forschung. *Germania* 63(2): 331–343.

Windler, A. 2018. *Der Austausch von Spondylus gaederopus in Europa zwischen 5.500 und 5.000 v. Chr. Eine ökonomische Analyse.* Rahden, Westfalen: Marie Leidorf.

Windler, A. 2019. The use of *Spondylus gaederopus* during the Neolithic of Europe. *Journal of Open Archaeology Data* 7: 7. DOI: http://doi.org/10.5334/joad.59

Winn, S. 1981. *Pre-Writing in South-eastern Europe: The Sign Systems of the Vinča Culture ca. 4000 BC.* Calgary: Western Publishers.

Wright, E. and S. Viner-Daniels. 2015. Geographical variation in the size and shape of the European aurochs (*Bos primigenius*). *Journal of Archaeological Science* 54: 8–22.

Wright, K., P., Critchley, A., Garrard, D., Baird, R., Bains and S. Groom. 2008. Stone bead technologies and early craft specialisation: insights from two Neolithic sites in eastern Jordan. *Levant* 40: 131–165.

Wright, K. and A. Garrard. 2003. Social identities and the expansion of stone bead-making in Neolithic Western Asia: new evidence from Jordan. *Antiquity* 77: 267–284.

Y

Yahalom-Mack, N., D. Langgut, O. Dvir, O. Tirosh, A. Eliyahu-Behar, Y. Erel, B. Langford, A. Frumkin, M. Ullman and U. Davidovich. 2015. The earliest lead object in the Levant. *PLoS ONE* 10(12): e0142948. DOI: https://doi.org/10.1371/journal.pone.0142948.

Yang, J., H. Shao and L.Pan. 2020. *The Metal Road of the Eastern Eurasian Steppe: The Formation of the Xiongnu Confederation and the Silk Road*. Singapore: Springer.

Yorston R.M., V.L. Gaffney and P. Reynolds. 1990. Simulation of artefact movement due to cultivation. *Journal of Archaeological Science* 17: 67–83.

Youni, P. 2000. Painted pottery from East Macedonia in North Greece: technological analysis of decorative techniques. *Documenta Praehistorica* 27: 198–214.

Yovchev, D. Native gold and platinum in stream sediments from Dvoynitsa River and right tributaries of Kamchia River, Bulgaria *'GEOSCIENCES 2014', Sofia, 2014*: 33–34. Bulgarian Geological Society.

Z

Zakšek, K., K. Oštir and Ž. Kokalj. 2011. Sky-view factor as a relief visualization technique. *Remote Sensing* 3: 398–415.

Zeder, M. 2006. Reconciling rates of long bone fusion and tooth eruption and wear in sheep (*Ovis*) and goat (*Capra*), in D. Ruscillo (ed.) *Recent Advances in Ageing and Sexing Animal Bones*: 87–118. Oxford: Oxbow.

Zeder, M. and H. Lapham. 2010. Assessing the reliability of criteria used to identify postcranial bones in sheep, *Ovis*, and goats, *Capra*. *Journal of Archaeological Science* 37: 2887–2905.

Zeder, M. and S. Pilaar. 2010. Assessing the reliability of criteria used to identify mandibles and mandibular teeth in sheep, *Ovis*, and goats, *Capra*. *International Journal of Archaeological Science* 37: 225–242.

Žeravica, Z. 1993. *Äxte und Beile aus Dalmatien und anderen Teilen Kroatiens, Montenegro, Bosnia und Herzegowina.* (Prähistorische Bronzefunde Volume 9, Band 8). Stuttgart: Steiner.

Zhang, Y., B. Guindon and J. Cihlar. 2002. An image transform to characterize and compensate for spatial variations in thin cloud contamination of Landsat images. *Remote Sensing of Environment* 82(2): 173–187.

Zhang, Y. 2002a. Problems in the fusion of commercial high-resolution satellite as well as Landsat 7 images and initial solutions, in C. Armenakis and Y.C. Lee (eds) *The International Society for Photogrammetry and Remote Sensing. Commission IV Symposium 'Geospatial Theory, Processing and Applications', 9-12 July 2002, Ottawa, Canada.* Vol. 34, Part 4.

Zhang, Y., 2002b. A new automatic approach for effectively fusing Landsat 7 as well as IKONOS images, *IEEE International Geoscience and Remote Sensing Symposium*, Vol. 4, Toronto, Ontario, Canada, 2002: 2429–2431.

Zhu, Z., M.A. Wulder, D.P. Roy, C.E. Woodcock, M.C. Hansen, V.C. Radelod, S.P. Healey, C. Schaaf, P. Hostert, P., Strobl, J.-F. Pekel, L. Lymburner, N. Pahlevan and T.A. Scambos. 2019. Benefits of the free and open Landsat data policy. *Remote Sensing of Environment* 224: 382–385.

Zidarov, P. 2005. Bone artefacts. *Acta Archaeologica* 76(1), *Acta Archaeologica Supplement* 6. *Lîga. Copper Age Strategies in Bulgaria*: 124–131, 185–187.

Živanović, Z. 2013. *Eneolitska gradina Bodnjik*. Koceljeva: Zavičajni muzej Koceljeva.

Zohary, D., M. Hopf and E. Weiss. 2012. *Domestication of plants in the Old World*. Fourth edition. Oxford: Oxford University Press.

Zori, C. 2019. Extracting Insights from Prehistoric Andean Metallurgy: Political Organization, Interregional Connections, and Ritual Meanings. *Journal of Archaeological Research* 27: 501–556. DOI: https://doi.org/10.1007/s10814-019-09128-7.